Digital Signal Processing

Thomas J. Cavicchi

Digital Signal Processing

Thomas J. Cavicchi

John Wiley & Sons, Inc.
New York / Chichester / Weinheim / Brisbane / Singapore / Toronto

ACQUISITIONS EDITOR	Bill Zobrist
MARKETING MANAGER	Katherine Hepburn
SENIOR PRODUCTION EDITOR	Robin Factor
ILLUSTRATION EDITOR	Sigmund Malinowski
SENIOR DESIGNER	Dawn Stanley
PRODUCTION MANAGEMENT	Argosy

On the Cover: The Matlab graphic on the cover is the magnitude of the transfer function $H(z)$ of the Chebyshev II digital bandpass IIR filter designed in Example 8.13. Highlighted in dark blue along the unit circle is the magnitude of the frequency response (discrete-time Fourier transform). Below is a contour plot of $|H(z)|$. Within and on the unit circle are the poles and zeros of the filter (all the zeros are on the unit circle). Notice that both the contour plot and the three-dimensional plot above it are defined only for $|z| > 0.9361$, the magnitude of the largest-magnitude pole of $H(z)$; inside, $|H(z)|$ is arbitrarily set to zero.

The book was set in TimesTen Roman by Argosy and printed and bound by Hamilton Printing. The cover was printed by Phoenix Color.

This book is printed on acid-free paper.

Library of Congress Cataloging in Publication Data:
Cavicchi, Thomas J.
    Digital signal processing / Thomas J. Cavicchi.
        p.   cm.
    ISBN 0-471-12472-9 (alk. paper)
    1. Signal processing—Digital techniques. I. Title.
TK5102.9.C37 2000
621.382'2—dc21   99-22866
                                                                                      CIP

ISBN 0-471-12472-9
ISBN 0-471-xxxxx-x (pbk)

Printed in the United States of America

10 9 8 7 6 5 4 3 2 1

Dedication

To my parents Richard H. and Mary Anne Cavicchi, my sister Elizabeth M. Cavicchi and her husband Alva L. Couch, my brother and his wife Richard E. and Clare L. Cavicchi, and their three daughters Violet K., Elise J., and Julia K. Cavicchi.

Contents

Preface

It is no longer necessary to strain one's imagination to think of practical uses for digital signal processing (DSP), because now it is everywhere. Consequently, DSP is essential material for most modern electrical engineers, somewhat in the way "power" was two generations ago. Because of its practical usefulness in the information age, people in many other scientific areas also need to know DSP.

Working proficiency in DSP may be assessed by the ability to compare, select, and properly use appropriate existing DSP algorithms for an application, to create original working algorithms or conceptual insights, to comprehend DSP research (e.g., articles in *IEEE Transactions on Signal Processing*), or to select or construct appropriate hardware implementations.

Like any highly advanced discipline, real practical applications of DSP can be developed only through a solid understanding of the fundamental principles. The main goal of this book is to convey this core knowledge and application examples that will prepare the reader for successfully developing the abilities listed above. Consequently, this book is oriented more towards fostering a deeper understanding of the basics than presenting a survey or encyclopedia of all the existing techniques.

Introductory DSP courses have traditionally been offered in the first year of graduate school. Increasingly, they are also being offered in the senior year of the undergraduate program. This book may be used in either situation, because its only prerequisite is junior-level signals and systems, yet it also includes ample materials to form the bridge from basic knowledge to advanced graduate-level studies.

Both students and teachers sometimes complain about the "dryness" of DSP textbooks. This complaint is less likely in this book because of a combination of aspects. Matlab/Simulink is used in ways that bring the formulas to life; graphical illustrations are a hallmark of this work. A conversational writing style used conveys enthusiasm about this subject. Both intuitive and analytical arguments frequently complement each other. Unexpected results and the asking and answering of questions are used occasionally to further heighten reader interest. Handwaving is avoided as much as possible within the space limitations, and often plentiful extra steps in derivations are included so the reader usually does not need pen and paper in hand to read the book. Expectations are high, to challenge the reader to advance as far as possible. Excessive tangents on material beyond the scope of a typical first course on DSP are avoided. Original commentaries offer a welcome fresh look at the subject at issue. Finally, a single central theme unifies the book, preventing it from being a compendium of specialized techniques, each of which all but a small particular interested party will find boring or useless.

This theme is, again, the essentials of digital signal processing. The reader has a right to have some basic questions fully answered in a DSP book. What are the relations between continuous-time and discrete-time signals and systems? Why the name *difference equations* when only sums, not differences, are involved? What is the purpose for having so many transforms, and how do we know when to use each one? Why the z^n for discrete time, as opposed to e^{st} for continuous time, and what really is a z-transform and why do we need it? Are there not fundamental properties that apply to all transforms, or do we really have to separately re-prove a property 12 times for each of the 12 transform relations? How can we use the computer to successfully process sampled-data signals and perform operations that traditionally were done in the analog world? Does aliasing really matter, what causes it, and what does $f = \frac{1}{2}$ mean? Is the $s - z$ aliasing relation wholly dependent on the fictitious Dirac impulse-sampled signal, or is the latter just a mathematical convenience used in the derivation? Where did the discrete Fourier transform (DFT) definition originate, and how can the DFT be applied to both aperiodic and periodic sequences? What is it that makes the fast Fourier transform (FFT) "fast"? Why such an emphasis on digital filter design and other theory when we have Matlab to do all the dirty work for us? What are some practical applications of sampling-rate conversion and the cepstrum? What exactly is a random signal, and how do we know whether to use deterministic or stochastic DSP? Can we see some examples of real-world data prediction or see a basic adaptive filter in the process of adapting?

These and many other fundamental questions are answered in this book. The following paragraphs summarize the organization. In Chapter 1, digital signal ideas are graphically introduced through weather data, musical recording signals, and biomedical brain-wave data. Chapter 2 presents an efficient review of signals and systems first for continuous time and then for discrete time. Separate coverage of continuous and discrete time will recall the usual coverage in a prerequisite signals and systems course. The reader should find the level moderately challenging and the notation and style established for the remainder of the book. Some highlights of this chapter are the discussion of convolution in Section 2.3.3, the relation between homogeneous/particular solutions and zero-state/ zero-input responses in Section 2.4.1, discretization of a practical-application analog system in Section 2.4.1, intuitive meaning of *state* in Section 2.4.6.1 (page 54ff), and the detailed Appendix 2A on the Dirac delta function—some results of which are used repeatedly later in the book.

Any substantial understanding of DSP requires confidence and competence in the area of functions of a complex variable. The review in Chapter 3 is entirely self-contained with all steps shown and builds from the simple function concept to power series, of which the z-transform is a primary tool in DSP. Reading this chapter takes the mystery out of the inverse z-transform relation and facilitates the understanding of z-transform properties and application. Only essential complex-variable information leading to the z-transform is included. An unusually complete and quantitative (Matlab-based) graphic showing z-poles and their corresponding responses is presented in Section 3.4.4.

Chapters 4 and 5 complete the job of setting the stage for DSP proper (Chapters 6 through 10). Full discussion of transforms is facilitated by the background on complex variables established in Chapter 3 and the time-domain background reviewed in Chapter 2. In Chapter 4, concepts of periodic and aperiodic functions and their transforms, the meanings of s and z, and the relation between Fourier and general power expansions are explained. Also, the defining equations of the DFT are obtained directly from the discrete-time Fourier transform (DTFT) of a periodic sequence. The concepts of system identification and transforms as defining expansions of signals over basis function sets are graphically illustrated. Table 4.1 and Section 4.2 suggest which transform is best to use in common types of application. Section 4.6 presents a pictorial tourguide of the various transforms. We analyze the simple RC circuit from many viewpoints in Section 4.7, from differential equation to discretization. Appendix 4A offers a diagram and formula summary of how to get from one transform domain to any other.

Having at hand the transform definitions and concepts from Chapter 4, the study of transform properties in Chapter 5 can be highly systematic. Rather than laboriously re-proving each property for each transform, a single property is obtained that can be applied to all or nearly all transforms by mere substitutions. Tedious algebra is dramatically reduced. With transform properties already covered in the prerequisite signals and systems course, such a new approach should be a welcome change from the usual chore that properties present. For traditionalists, several conventional property proofs are also included. The group of properties concludes with a brief review of transforms of real-valued signals and of Parseval's relation.

Digital signal processing is often used to process samples of continuous-time signals. In Section 6.2, the reason for appealing to the Fourier domain for the aliasing condition is clarified by consideration of precisely what causes aliasing. The sampling/aliasing transform formula is derived in a manner that places the Dirac-delta-sampled signal in its proper place: just a convenience for mathematically decomposing a physically existing and measurable staircase signal in a way that facilitates developing the s–z relation. Graphically oriented discussion of oversampling and an example reconstruction clarify the subtle ideas. Replacement of analog systems by equivalent digital systems is a matter of industry-wide importance that is easily understood at a basic level through the topic of simulation. Qualitative discussions of sampled-data control systems and audio compact disc systems help maintain contact with applications. Another key digital operation on analog signals is use of the DFT to estimate Fourier spectra. In Sections 6.5 and 6.6, forward and reverse operations are analyzed and illustrated with numerical examples that show the effects of aliasing and truncation. An interesting result is that the DFT rectangular-rule estimate of the continuous-time Fourier transform is exact, in the absence of either aliasing or truncation (in practice, however, one cannot simultaneously completely avoid both). Section 6.7 completes the picture, laying out graphics that show how everything in digital and continuous Fourier analysis fits together.

In Chapter 7, attention is focused on using the DFT to process sequences. Discrete-time Fourier transform estimation using the DFT (including zero-padding as an interpolation between original DFT samples), sampling rate conversion, periodic/circular convolution/correlation (including the need for zero-padding), block convolution (uniquely including a complete Matlab example), time-dependent spectra, DFT resolution, cepstral analysis (including numerical examples of echo removal), and the FFT highlight this study. As in other chapters, Matlab and other graphics are used here to great advantage (I have written or modified well over 700 Matlab routines for this work). A vivid example of time-aliasing involving the Dow Jones Industrial Average is both fun and effective. DFT time-domain interpolation is shown to compute an infinite-length sum using an N-length sum, with zero error. The presentation of the FFT features explicit identification of where the computational advantage lies and a readable explanation of bit reversal.

Digital filter design is the subject of Chapters 8 [infinite impulse response (IIR) filters] and 9 [finite impulse response (FIR) filters]. Just as continuous time is usually covered before discrete time in a signals and systems course due to students' prior circuits knowledge, it again makes sense to discuss analog-based (IIR) digital filters before purely digital-based (FIR) filters. To make the presentation self-contained (without, however, getting bogged down in nonessential math associated with Chebyshev differential equations and elliptic functions), a complete discussion of analog filter design principles with numerical Matlab examples is presented. Included is an example of intuitive pole–zero-placement design and why this is usually inadequate. With this background, the standard impulse-invariant and bilinear transform conversion procedures are clear and again illustrated with Matlab examples. The topic of prewarping and the bilinear transform (BLT) is given special attention. The chapter concludes with brief discussions on and Matlab examples of spectral transformations.

Chapter 9 includes the usual introductory materials, but with an unusually complete digital filter-oriented presentation of Gibbs's phenomenon (e.g., the maximum overshoot and its frequency depend on the filter cutoff). Following the window and frequency-sampling approaches, the McClellan–Parks method is introduced. Linear-phase filters are explained carefully, as the possibility of obtaining them by the various FIR design methods is one of the important advantages of FIR over IIR filters. Several Matlab examples of McClellan–Parks filter design are presented. Chapter 9 closes with numerical examples comparing different filter structures with respect to coefficient quantization.

The review of random vectors and processes in Section 10.2 provides an easy transition from deterministic to stochastic DSP. An interesting example on weather data illustrates the meaning of ergodicity. Next, all steps are shown to obtain the power spectral density (PSD), Wiener–Kinchin theorem, and periodogram—with particular attention to what is being averaged over. Because of its complete dominance in entry-level stochastic processing, the rationales behind and distribution functions of Gaussian random entities are discussed in detail. Clear discussions on elements of estimation theory including the Cramer–Rao bound, maximum likelihood, and maximum entropy help pave the way for easier understanding of later DSP courses. Coverage of LSI systems with stochastic inputs allows discussion of roundoff effects in digital filters (Section 10.8); the theoretical error variances based on linear analysis are verified with Simulink true nonlinear roundoff simulation. The big payoff for studying stochastic DSP comes in Sections 10.9 and 10.10, where least-squares and adaptive filters are developed and applied to data. Included are examples of deconvolution, stock market and other economic prediction, adaptive transformation of a triangle wave into a sinusoid or into a square wave, and extraction of a triangle-wave signal from overwhelming noise.

This book includes over 525 end-of-chapter problems illustrating the principles. Because the problems and their solutions were written by me, there is a high degree of correlation between the text and problems. The full range from easy calculation exercises to lab-oriented studies to difficult analytical and conceptual problems allows flexibility for the use of this book in a variety of programs and styles of teaching. Over 165 of the problems are Matlab-oriented and are so indicated with a Matlab icon. In many cases, Matlab work constitutes only part of the problem; other parts may be purely analytical. Thus, the analytical and numerical aspects of DSP problem solving support each other. In several of the Matlab problems, the student must write significant original code; in all cases, Matlab is represented as a highly useful tool, not a crutch. Other student aids include the boxing of important relations in the text, replication of cited equations at the citation to eliminate flipping of pages, text examples with all steps and relevant graphics provided, and availability to instructors of completely detailed solutions of all the problems.

A unique feature of this text is a set supplementary reference materials located at the worldwide web site for this book. Setting up this web site has allowed the book to be streamlined for readability in certain areas and the overall book length minimized, while still having the extras just a mouse click away. These web pages include both text and color graphics, and each is referenced at the appropriate place in the text. The notation and style are integrated with those of the main text, as the web pages are written exclusively as part of the total package. About 8% of the end-of-chapter problems also refer to these web pages (60% of these are the

---

Web site access information:

The author's web site is accessible from
http://www.wiley.com/college/engin/cavicchi124729

---

advanced-level problems in Chapter 5); they are so marked. The reference for the second web page for Chapter 6 is notated WP 6.2. To obtain this page, go to the author's web site, click on WEB PAGES, and than click on 6.2. In a few instances, web page equations are referenced in this book: Equation 7 on WP 6.2 would be labeled (W6.2.7).

Contents of Web Pages

Most of this book has been successfully tested in both undergraduate and graduate classes during its evolution over the past decade. Please send your comments or corrections to tjcavicchi@gcc.edu. If for any reason you have difficulty reaching me by this email address, go to the web site for this book to obtain updated information.

Finally, I would like to make the following acknowledgments. Dr. Chi-Yang Cheng and student Bob Duke were especially helpful for showing me how to convert figures to Postscript. Also, I received some helpful critical comments from the reviewers. But most of all by far, I would like to thank my family members for their continual support.

Thomas J. Cavicchi

Chapter 1

Introduction

As we enter the new millennium, digital computation—whether visible or invisible to us—becomes ever-more pervasive in our lives. Nearly every aspect of our daily personal and professional tasks involves a computer or microprocessor of some type. Magnetic cards, remote controls, software, digital entertainment systems, personal computers/networking, copiers/laser printers, new telecommunications systems (e.g., cellular phones, high-speed modems for Internet connection, video teleconferencing systems), health care apparatus, industrial control systems, automotive electronics, computerized billing/banking systems, and voice recognition/synthesis systems all affect the way we live. Each application makes use of digital signal processing (DSP) technology.

For those who wish to be part of the ongoing progress in the very wide range of areas of applications of modern electronic technology, an understanding of the processing of discrete-time data is essential. In this book, the main focus is on the conceptual and quantitative sides of DSP, which in many applications are hidden from the user by a menu-driven keypad or screen. Courses on computer architecture, design, and interfacing take up the hardware aspect, which is also hidden from the user and which is naturally essential for the real-world implementation of DSP ideas. In essence, to be involved in system design requires one to pry off the user-friendly artifices and become involved in determining what makes the system "go." Once the system is built (or fixed), the lid can be put back on and marketed.

On another level, the study of digital signal processing is intrinsically worthwhile. DSP is highly applicable to practical problems, yet it also depends on profound mathematics, sometimes hundreds or even thousands of years old. The more we study the history of DSP and math in general, the more fully we can appreciate and make use of the achievements in those fields. Thus, studying DSP helps us mentally put together the proper place of each result in history, theory, and application. For example, in many cases the techniques of DSP acquired practical significance only after the rise of computing power during the last several decades.

The study of DSP also helps us unify and modernize our approach to working in new or traditionally non-DSP areas, both applied and theoretical. For example, understanding DSP theory helps one understand and become able to contribute in other fields, such as advanced linear systems theory, Fourier analysis, state variable/control theory, approximation theory, optimization theory, and mathematical physics. Each of these disciplines comes into play at least once in this book. On the applications side, studying DSP also helps one to understand the reasons for various features in DSP algorithms and DSP chips, know how to correctly sample signals for any application, and most important, know what to do with a sampled signal once obtained from a given application. Anyone involved in predicting, influencing, or characterizing a physical system may take advantage of DSP.

Finally, a word on the DSP industry. According to Texas Instruments, the DSP worldwide market has grown at more than 30% per year since 1988 and was estimated to be $4 billion in 1998. These facts and figures give a clear indication of the growing importance of DSP in our society, as well as the potential career opportunities in this field.

Today's hardware is keeping up with the requirements of modern applications. As one example, the capabilities of Texas Instruments' multichannel multiprocessing DSP chip TMS320C6202 are astounding: 2000 million instructions per second (MIPS)/500 million multiply accumulates per second (MMAS), 250-MHz clock, eight functional processor units (automatic code parallelization), 128 Kbyte RAM on-chip (and 256 Kbyte program memory), only 1.9-W power consumption, and only $130/chip if bought in large quantities. A chip like the 6202 is useful for multichannel communications/servers, imaging, and advanced modems.

As a brief introduction to DSP, we consider later in this chapter a few practical areas where DSP comes into play. Certainly, whenever a continuous-time signal is to be sampled and fed into a computer, DSP is relevant. Industrial and consumer measurement, control, and communications systems, biomedical instrumentation, video and audio reproduction systems, meteorological analysis and prediction, and transportation systems all may make basic or extensive use of DSP. Also, inherently discrete-time systems such as financial/manufacturing forecasting and population/genetic modeling clearly make extensive use of DSP. Our review leaves out endless other applications of special interest today that make great use of DSP, including cellular and in general mobile digital communications systems, digital TV/radio and other emerging DSP-based entertainment technology, and automotive and other electronic digital control systems.

Having had such an introduction, in the remainder of the book we get into the details that make DSP work. Our approach is to build from the ground up: to understand well the most basic and important principles of modern DSP, rather than survey every important area and algorithm. We use every tool available: software visualizations, extended application examples, drill-type examples, and explanations of the rationales behind the jargon and symbology that make DSP seem so out of reach to those outside the field. It is believed that in the long run, an in-depth study of the basics is more valuable in a first course on DSP than an encyclopedia presenting all the results but lacking in depth and coherence.

In Chapters 2 and 3, we form a strong foundation for later materials with a review of signals and systems, complex variables/complex calculus, and the z-transform. The z-transform is one of many transforms in DSP. In Chapter 4, we put several of these transforms in perspective so that it is clear which transform is best to use in a given application. The reader has probably studied transform properties in a signals and systems course, but in Chapter 5 we examine these properties more closely to identify the root of each property as it applies to each transform in Chapter 4. This approach can lead to some powerful and more general approaches to further characterize and relate transforms and convolution operators.

We then get into DSP proper, by studying sampling, the DFT/FFT, and filter design. In Chapter 6, we study the effects of sampling a signal and see how a DSP system can emulate a related analog system. We also look at how the DFT/FFT can be used to estimate continuous-time Fourier transforms. Chapter 7 is devoted entirely to applications of the DFT and includes a detailed look at the famous FFT algorithm. When people think of DSP, often the first thing that comes to mind is digital filter design, the subject of Chapters 8 and 9. We take a complete view, from the requisite analog filter design theory to an important optimal FIR algorithm, to filter structures and quantization effects.

Increasingly, the limitations of traditional deterministic DSP theory are being recognized and overcome with alternative approaches. One of these, the neural network approach, is beyond the scope of this book. The other major approach is stochastic signal processing, the subject of our last chapter, Chapter 10. We assume little prior contact with stochastic theory and go all the way from basic principles to examples of actual numerical applications of stochastic signal-processing algorithms. These examples point the way toward the far-reaching potential of DSP theory to solve important and difficult real-world problems. Chapter 10 therefore serves as a handy tie between the familiar introductory DSP in this book and the rather involved mathematics of advanced graduate-level DSP courses or studies.

1.1   SIGNALS AND SYSTEMS: DEFINITIONS

A signal may be defined formally as a function of time and/or space whose values convey information about a physical or mathematical process it represents. Note the root word *sign*. A signal is called analog if the variable on which it depends (time/space) is continuous, and discrete-time if its independent time variable is discrete. The voltage across a microphone when we sing into it is analog, because that voltage is truly analogous to the original acoustical process creating the signal. As an example, Figure 1.1 shows a 3-ms segment of the great 1930s trumpeter Bunny Berigan playing "I can't get started." As can be seen in the figure, an analog signal has a definite value for every moment in time/space. The values of both the signal and the independent variable (e.g., time/space) can be quantified by any real number, even one that "never ends," such as π.

A discrete-time signal is a sequence of values defined at only discrete instants of time. (We may also discuss discrete-space signals.) An example of an inherently

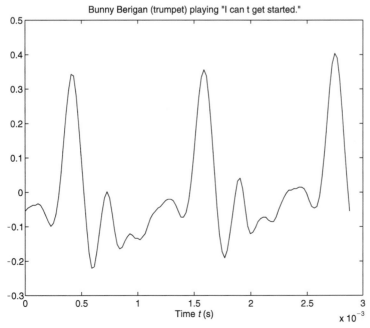

Figure 1.1

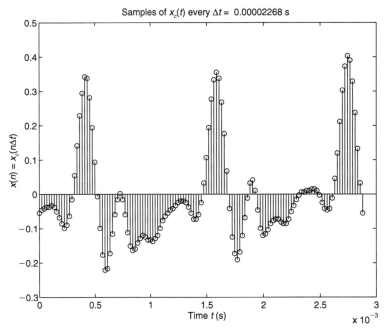

Samples of $x_c(t)$ every $\Delta t = 0.00002268$ s

Figure 1.2

discrete-time signal is the exact amount of rainfall in a city for one year (before the loss of precision incurred by measuring it). A sequence of samples of the microphone voltage also constitutes a discrete-time signal; see Figure 1.2, which shows the same trumpet signal as in Figure 1.1, but sampled[1] every $\Delta t = 22.68$ μs (i.e., the sampling frequency is $F_s = 1/\Delta t = 44.1$ kHz). If the values a discrete-time signal can take on are limited to discrete values, the signal is called a digital signal. An inherently digital signal is the price of a bushel of corn each day (discrete time, and the price can have only an integral number of cents).

For a signal to be completely representable and storable in a digital computer memory, it must be sampled in time and discretized in value. That is, it must be a practical digital signal having both finite duration (finite length) and a finite number of discrete values it can take on at any given sequence index. Very long sequences can be processed a chunk at a time. To discretize the value, a rounding or quantization procedure must be used.

A digital version of Berigan's trumpet playing is shown in Figure 1.3a. It has been quantized to six-bit accuracy using rounding. The horizontal -.-. lines show the only possible values that the 6-bit memory can store. The original samples $x(n)$ are the O stems, and the rounded samples $\hat{x}(n)$ are the * stems; we see that quantization has resulted in some error $e(n) = x(n) - \hat{x}(n)$. Although it would clutter the illustration (and anyway is already indicated by the sample stem lines), we could draw vertical grid lines in every sample; the digital signal values must lie on the intersections between the horizontal (time-sampling) and vertical (value-sampling) grid lines.

Figure 1.3b shows the quantization error sequence $e(n)$ resulting from the rounding in Figure 1.3a. For example, for the first and last samples, the nearest quantization level is below $x(n)$ versus $t(n) = n\Delta t$ in Figure 1.3a; hence $e(n)$ is positive for those values of n in Figure 1.3b. In Chapter 10, it will be important to note that $e(n)$

---

[1]We use Δt for the sampling interval and T for the period of a periodic signal. This notation avoids confusion when sampling periodic signals.

* stems = 6-bit rounded version; O = original samples of $x_c(t)$ every $\Delta t =$ 0.00002268 s

— . — . — . = possible quantization values for 6-bit rounding

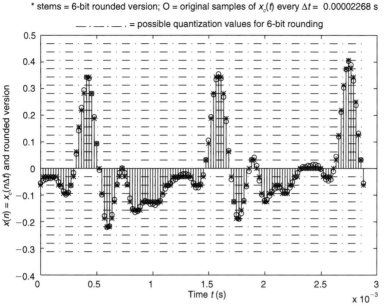

Figure 1.3*a*

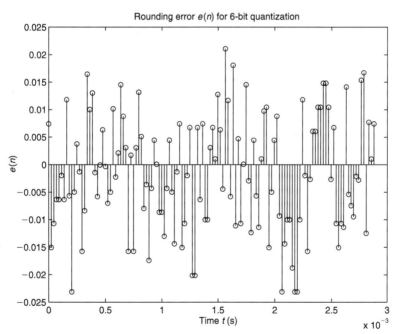

Figure 1.3*b*

is a highly unpredictable, random sequence but with definite upper and lower bounds determined by the number of significant bits retained in the quantization.

Because of the universal use of digital computers, the digital signal is now the most important signal type to be studied in signal processing. To minimize the tedious analysis of the quantization effects involved in transforming a discrete-time signal into a digital signal, however, most of the emphasis in this book is on (nonquantized) discrete-time signals (however, see Sections 9.6 and 10.8). Digital signals are very often good approximations of their usual discrete-time signal sources—which is increasingly true with the longer wordlengths of modern computers and microprocessors.

In signal-processing jargon, a system is any physical device and/or mathematical operator that transforms a signal presented to its input(s) into an output signal(s) and/or physical effect. Hopefully, the "processing" it does on the signal is in some way beneficial. In this book, we restrict our attention primarily to single-input, single-output (SISO) systems.

More crucially, a fundamental assumption behind most of the discussion in this book is the linear shift-invariance of the signal-processing system. Linear shift-invariance is important because it makes analysis of systems tractable. For example, for discrete-time systems, the input–output relation of a linear shift-invariant system is a difference equation with constant coefficients, for which comprehensive solution methods are well known. Also, the concepts of impulse response and frequency response apply in their usual forms only to linear shift-invariant systems; thus linear shift-invariance allows a simple and complete description of the system at hand. Finally, other characteristics of systems including stability and causality are straightforward to determine under the assumption of linear shift-invariance.

1.2 WHY PROCESS IN DISCRETE TIME?

Learning DSP is certainly a lot of work. Is it worth the bother? Why not just use analog processing, as had been done since the earliest days of electronic signal processing? Why be required to study the mathematical relation between a discrete-time signal and the analog signal from which it may have been obtained? The following are some advantages of digital over analog processing.

Once sampled and converted to a fixed bit-length binary form, the signal data is in an extremely convenient form: a finite array of numbers having a fixed number of binary digits. This array can be stored on hard disks or diskettes, on magnetic tape, or in semiconductor memory chips. All the advantages of digital processing are now available.

For example, analog signal-processing circuits drift with time and temperature, whereas binary (on/off) decision-making is completely immune to such distortions. Because a 1/0 pattern rather than the noise-vulnerable analog waveform is stored or transmitted, there is tremendous noise immunity possible in digital data transmission. Only a 1/0 decision must be made correctly for digital, as opposed to the need to exactly reproduce an easily distorted waveshape for analog transmission.

Analog electronics tends to be physically large and expensive compared with equivalent digital electronics. Digital processing is as flexible as your ability to program a computer. Analog processing is only as flexible as the particular hardware you have soldered into place (not very!). Also, the sky is the limit with the potential complexity available using digital signal processing, whereas practical analog processing is relatively very limited in this respect.

Analog data cannot be stored without degradation over time or during copying, as can digital data. For example, whenever a copy is made of a cassette tape of music, the fidelity is reduced in the copy. This degradation does not happen with a copy of digital data; there is essentially zero degradation. Thus the digital information can be perfectly stored "forever" (indefinitely). About the only advantage of analog processing is that it can be faster than digital processing. Even this advantage, however, is now becoming dubious with cheap, incredibly fast DSP processors.

In Chapter 2, we begin a quantitative study of signals and systems, the basis of DSP. For now, we examine a few sequences constructed from actual measurements, and conclude this chapter with a brief timeline history of technological advances that are in some way relevant to DSP.

1.3   TEMPERATURE PROFILES

One of the signals most important to people throughout the ages has been the air temperature. To introduce the power of graphical visualization of time series (discrete-time signals), let us look at some weather data. In Chapters 7 and 10, a few signal-processing procedures are applied to some weather data. For now, we will just content ourselves with the reading of information from a set of curves.

Figure 1.4 shows the following temperatures in Cleveland, Ohio, plotted versus day of the year, with January 1 labeled as "day 1" and December 31 as "day 365": record high, normal high, normal low, and record low. The 365 points for each plot have been connected with lines for legibility, as will often be done in this text. In our daily viewing of weather reports on the local television news, we see these values for only one day at a time (the current day). How much more informative the graph can be, for it shows trends!

For example, because each point on the normal high and low curves is an average of values over the same 130 years for that day, the curves are very smooth. The smoothness from day to day indicates that the process underlying the data is regular, stable, and unchanging. Of course, that is the seasonal process caused primarily by the angle of the Earth's axis and the consequent length of days. Contrarily, each data point on the record high and low temperature comes from the particular year at which that record occurred. Because of the variability in weather from year to year, and because values for adjacent days may come from different, often widely separated years, the record curves are erratic. If more years were available, these curves would be smoother if it is true that the possible variability about the normals

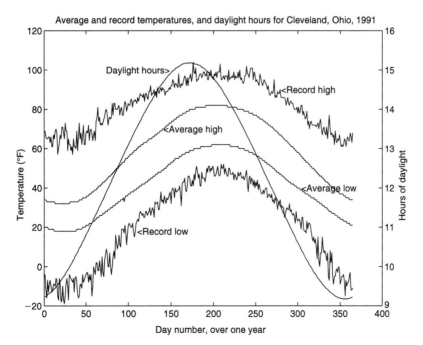

Figure 1.4

is physically limited; with more samples, the maximum or minimum would more assuredly approach the fundamental limit for each day.

Another fact we may easily see from Figure 1.4 is that the record temperature range is wider in winter than in summer. Interestingly, the opposite is true for the normal curves. Perhaps the difference in behavior is due to the possible presence of snow (for the low record) or lack of snow for many of the years, the latter of which influences the normal low relatively upwardly. The high heat-reflectivity of snow can cause extremely low temperatures in record-breaking cases. Contrarily, the presence or lack of snow is averaged out in the normal (averaged) curves. The same effect observed in the record and normal high curves cannot be explained this way, however. Weather is a complicated phenomenon, so there is probably more at issue here. Also unexplained is why the normals are farther apart in the summer than in the winter. Perhaps the presence or absence of clouds makes more of a difference in the summer heating process than in the winter.

It has been noted in writings on climate, as well as in one's memory, that often in the third or last week of January there is a thaw. This thaw is clearly visible in the record high plot but is inexplicably absent in the normal high plot, which holds steady at 32° F normal high. The study of data from many individual years may help explain this discrepancy and perhaps lead to a physical explanation.

Yet another aspect is the delay-shift of the curves with respect to the hours of sunlight per day, as seen in Figure 1.4. That is, the equinoxes (days for which the minimum and maximum of the daylight hours curve occur) do not coincide with the extrema on the temperature curves; there is always a delay. This interesting effect may be due to both the time-constant in heating/cooling the air as well as the presence or lack of snow in the winter. Perhaps the prevailing winds also come into play. We estimate this delay using a DSP technique in Chapter 7.

Finally, we clearly see in Figure 1.4 the rather monotonous situation where day after day the normal high and low temperatures remain unchanged for over a month, in January and in July. By contrast, in the other months there is a nearly linear increase or decrease of high or low temperature with the day. Specifically, the normal high goes up at about 10° F/month and down at 11° F/month, whereas the record high goes up at about 6° F/month and down at 11° F/month; similar estimates of slopes can be made for the two lower curves. The roughly ±10° F/month temperature change rule of thumb for the spring and fall months is a handy fact to remember, at least for Clevelanders.

Figure 1.4 is an example of the phrase "A picture tells a thousand words." The ability to make correct and creative inferences from inspection of a curve can be a great help when designing signal processing signals for a particular application situation. In fact, common sense plays as great a role in the analysis and design of DSP systems as it does in any scientific area.

1.4 PHONOGRAPH RECORDING RESTORATION

One area of DSP that is very interesting to lovers of vintage music is the restoration of old phonographic recordings. This book does not have audio capability, so we examine only one aspect of restoration, which can be demonstrated through plots. The author has several 78 rpm records, some of which are cracked. They still play, but there is an annoying "thump" every revolution. One solution is to sample the signal and literally remove the thumps.

In this experiment, a short sample of the recording ("Marcheta" by the Green-Arden Orchestra, circa 1922) on a cracked 78 rpm Victor record was sampled at $F_s = 44.1$ kHz (see Chapter 6 on how the sampling rate is selected). The sampled

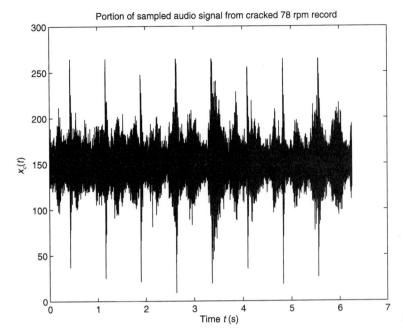

Portion of sampled audio signal from cracked 78 rpm record

Figure 1.5

sequence, with lines connecting the samples, is shown in Figure 1.5. The main feature to notice is the thumps, one for each revolution of the record. Adjacent thumps are separated by (60 s/min)/(78 rev/min) = 0.769 s/rev, the time for a revolution (the author's record player may be seen to be slightly faster than 78 rpm).

In Figure 1.6a, we zoom in on one of the thumps. In some record restoration techniques, the distorted portion (involving the large swings in Figure 1.6a) is replaced by either zero or a linear interpolation by the removed samples. Instead, in this example, a group of samples immediately preceding the removed portion equal in number to that of the defective section is replicated to replace the defective section. The material preceding rather than following the crack is used because the former signal is completely free of the "thump" transient, whereas the latter signal still has significant thump transient. Care is used to select a portion that begins and ends at the same value, so as not to introduce a sudden change in value. The result for the same time segment is in Figure 1.6b; it is reasonably consistent with the waveform away from the crack on the record.

The result for the larger time segment is shown in Figure 1.7. The periodic thump is largely, although not perfectly, removed. It is not perfectly removed because the reverberations of the thump extend sufficiently far beyond the crack that removing the thump entirely would distort the ever-changing music. Also, with more of the music replicated, an echo is audible in the reproduction. A compromise was used, but with practice a better job could probably be done. Upon playing back the 6.17-s segment of music, there is very little distortion and the thump is almost inaudible. Were this laborious process done for the entire recording, the whole broken record would sound as if it were unbroken!

It is illuminating to examine the Fourier log magnitude spectra of the "before" and "after" signals (in decibels; see Chapter 4 for more on Fourier analysis). The spectrum of the cracked record is shown in Figure 1.8, computed using the DFT (see Chapters 6 and 7). An item to notice is the steep decline with increasing frequency, except for a resonance around 8 kHz. This resonance is not part of the music, nor is it the thumping due to the crack in the record. Rather, it is the very

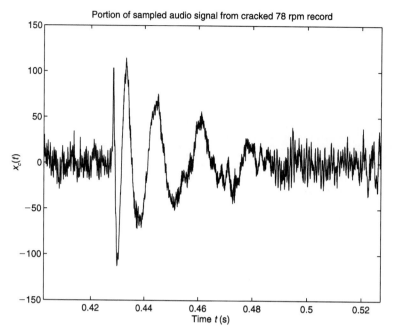

Figure 1.6a

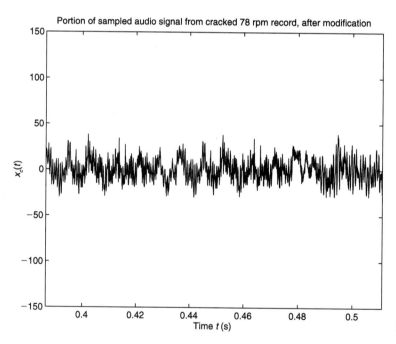

Figure 1.6b

loud record surface noise, as the author confirmed by listening to 8 kHz on a function generator and then the record (the surface noise had the same pitch). This type of noise may be largely removed using a digital filter (see Chapters 8 and 9).

The processed spectrum has an overall appearance nearly identical to that in Figure 1.8 until one zooms in on a narrow frequency range. This is done in Figures 1.9a and b, which compare "before" with "after" (here the log is not used so as to illustrate

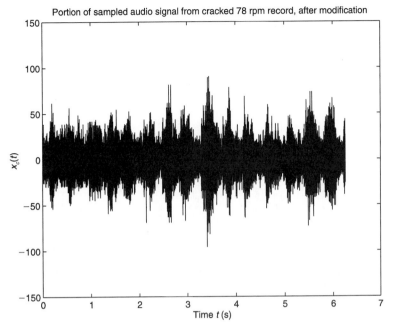

Figure 1.7

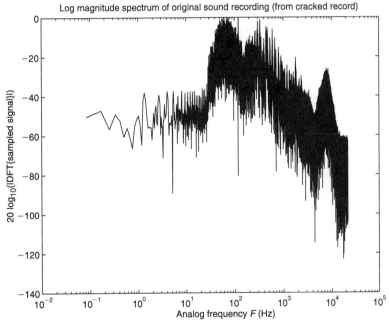

Figure 1.8

the effect more clearly). We find in Figure 1.9a that all the harmonics of the thump fundamental frequency $F_0 = 78/60 = 1.3$ Hz are represented (in different strengths). This is natural because the thumps are periodic and involve a large nearly discontinuous temporal effect as the needle jumps the crack. The discontinuity implies a very broad band of harmonics. With most of the thumps removed, in Figure 1.9b we see much less of the thump "comb" spectrum. How might we do better? That question is a matter for more advanced study in DSP. For example, we might try what is known as a comb filter, to filter out the harmonics. However, there are over 33,000 harmon-

Magnitude spectrum of original sound recording (from cracked record)

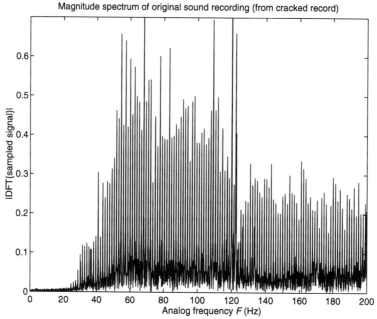

Figure 1.9a

Magnitude spectrum of modified sound recording (same segment)

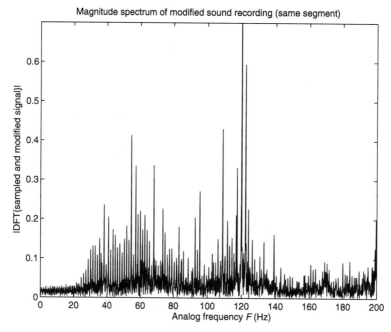

Figure 1.9b

ics in the sampled band, so traditional comb filters are impractical. Thus, the manual technique, although laborious to do, does the job reasonably well until a better technique or an automated form of the present technique is designed.

Finally, in Figure 1.10 we show the spectrum of an electrically recorded segment of rock music, recorded from a stereo FM radio. Notice how the high frequencies fall off much more slowly with frequency than in Figure 1.8 and that the annoying surface noise resonance is absent. "Marcheta" was acoustically, not electrically,

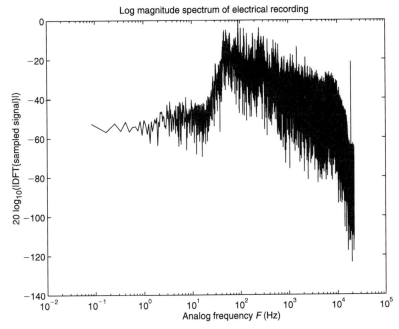

Figure 1.10

recorded; electrical recording was developed in 1925. Keep in mind that we do not expect a flat spectrum here, because this plot is a musical spectrum, not a system frequency response. A slightly flatter spectrum might be obtained were the stereo used of higher fidelity (the stereo used is the author's old compact, which has an eight-track tape recorder!).

1.5   BIOMEDICAL SIGNAL EXTRACTION

In Section 1.4, we studied an example of noise removal done by a "brute force" approach. In Chapters 8 and 9, we study the frequency-selective digital filtering approach to noise removal (and other desired ends). Another useful set of techniques exists for noise removal when a sample of the noise source by itself is available for measurement, as well as the noise-distorted signal. The traditional approach is to use subtraction techniques to cancel the noise component of the noise-corrupted signal, leaving the desired pure signal as the result. Also, adaptive techniques may be used, as investigated in Section 10.10.

In this section, we consider yet another approach presented in a recent paper on magnetoencephalography.[2] Physicians use magnetoencephalographs (magnetic brain waves, or MEG) to study which areas of the brain control which parts of the body, and, once identified, MEGs can be used for diagnosing particular regions and functions of the brain. A large number of electrodes covers the head, and each electrode generates an independent signal. The problem is that cardiac (heart) magnetic fields are much stronger than brain waves and can swamp out the brain waves signal.

Samonas et al. first segment the data, sampled at 1 kHz, into heartbeat cycles. These segments are then "resampled" into a vector of chosen, fixed length by a resampling process involving Lagrangian interpolation (the same idea is used in the

---

[2]M. Samonas, M. Petrou, and A. A. Ioannides, "Identification and Elimination of Cardiac Contribution in Single-Trial Magnetoencephalographic Signals," *IEEE Trans. on Biomedical Engineering,* 44, 5 (May 1997): #386–393.

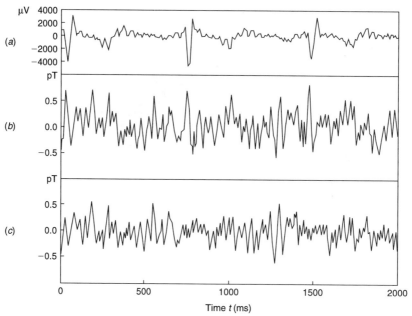

Figure 1.11 M. Samonas, M. Petrou, A.A. Ioannides, "Identification and estimation of cardiac contribution in single-trial magnetoencephalographic signals," IEEE Transactions on Biomedical Engineering (Vol. 44, No. 5). Copyright 1997 IEEE.

McClellan–Parks filter design algorithm of Section 9.5/WP 9.1 for a different application). These normalized segments are then averaged together, presumably averaging out the MEG signal content to leave a good estimate of the cardiac magnetic field waveform. The nonaveraged segments are of course also saved for signal extraction. Each of these length-N sequences is considered to be a vector of values.

The second step is to perform Gram–Schmidt orthonormalization. Simply, the portion of the noisy-signal vector that is orthogonal (in N-space) to the cardiac noise vector is the MEG signal estimate produced. That is, the component of the noisy-signal vector "along" the pure cardiac noise vector is subtracted out, leaving an MEG result that has no cardiac noise component at all (ideally).

The results of this approach are as shown in Figure 1.11. In Figure 1.11*a*, the electrocardiogram (noise) is shown as a visual reference, and in Figure 1.11*b* the actual noisy MEG signal is shown. The "cleaned" MEG signal is shown in Figure 1.11*c*; we see that the cardiac noise component has indeed been reduced or eliminated. For those interested in seeing graphics illustrating the projection idea and the reduction in noise obtained, see WP 1.1.

1.6 TIMELINE OF DISCOVERIES AND INVENTIONS RELEVANT TO DIGITAL SIGNAL PROCESSING

The names and dates below are example citations, not undisputed claims of priority. Often, more than one person contributed to the ideas, at different times. Occasionally, data are uncertain, but the name of a pioneer and an approximate date are still given. This timeline is not claimed to be comprehensive.

π (approximation): Egyptian papyrus, c. 2000 B.C.; Al-Kashi, better approximation 1436; Francois Vieta, exact expression 1579. Mnemonic (number of letters in each word equals given digit of π): "See, I have a rhyme assisting my feeble brain, its tasks ofttimes resisting." Symbol for π: William Jones, 1706; Euler, 1737.

Quadratic equation: Solved by Babylonians, c. 2000 B.C.; reduced to a formula by Brahmagupta, c. A.D. 628.

Pythagorian theorem: Pythagorus, c. 540 B.C.

Law of cosines: Euclid's *Elements*, c. 300 B.C., citing probably Pythagorus from c. 540 B.C.

Integers: Introduced as early as c. 200 B.C. by Arabs and Hindus. Leopold Kronecker (mid 1800s): "God made the integers, and all the rest is the work of man."

Sin($a + b$) addition formula: Effectively used possibly as early as c. 140 B.C., and certainly Ptolemy used it in A.D. 150. Formal introduction of sine function: India, c. A.D. 300.

$$\sum_{n=0}^{\infty} a^n = 1/(1 - a):$$ Francois Vieta, 1590 (for $a = \frac{1}{4}$, Archimedes, c. 225 B.C.).

Finite-term version calculated by Babylonians, 2000 B.C., and put in modified equation form in Euclid's *Elements*, c. 300 B.C.

$e^{j\theta} = \cos(\theta) + j\sin(\theta)$ (Euler's formula): 1743, although an equivalent form by Roger Cotes dates to 1710. Also $\cos(\theta) = \frac{1}{2}\{e^{j\theta} + e^{-j\theta}\}$.

Logarithms: Natural (base e): John Napier, 1614; common (base 10): Henry Briggs and John Napier, 1615.

Polynomials: Early developments, René Descartes, 1637; fundamental theorem of algebra (FTA) and fact that if $f(z) = 0$, then $f(z^*) = 0$, Jean le Rond d'Alembert, 1746; first satisfactory proof of FTA, Carl Gauss, 1797.

Probability: Beginnings in questions about gambling; first printed work by Christiaan Huygens in 1657; first book by Jacques Bernoulli in 1713. Also, Pierre Laplace, 1814.

Binomial theorem, numerical integration rules: Isaac Newton, 1664.

Polar coordinates: Isaac Newton, 1671.

Determinants: Gottfried Leibniz, 1693.

Taylor's series: Not originally by Brook Taylor but by James Gregory, c. 1675.

Finite differences: Henry Briggs, c. 1600; Brook Taylor's *Methodus*, 1715.

Cramer's rule: Not originally by Gabriel Cramer but by Colin Maclaurin, 1729.

Maclaurin's series: Not originally by Colin Maclaurin but by James Stirling, c. 1730.

Complex numbers: First instance of square root of negative number in Heron, c. A.D. 50; symbol i for square root of –1: Leonhard Euler, 1777; $a + bi$ notation: Caspar Wessel, 1797; theory of functions of a complex variable: Augustin-Louis Cauchy, 1825. René Descartes (1637) contributed the terms *real* and *imaginary*; Carl Gauss (1832) contributed the term *complex*.

Residue theorem: Augustin-Louis Cauchy, 1827.

Least squares: Carl Gauss, 1794 (at age 17!).

Fast Fourier transform: Carl Gauss, 1805.

Laplace transform: Pierre Laplace, 1812.

Fourier series: Daniel Bernoulli, 1738; Jean-Baptiste Joseph de Fourier, 1822. Fourier integral: Michel Plancherel, 1910.

Digital computer concept: Charles Babbage, 1833.

Telegraph: Samuel Morse, 1838.

Telephone: Alexander Bell, 1875.

Laurent series: Pierre-Alphonse Laurent, 1843.

Phonograph: Cylinder: Thomas Edison, 1877; disc: Emile Berliner, 1888; electrical recording: J. P. Maxfield and H. C. Harrison, 1925; long-play: P. C. Goldmark, 1948.

Kronecker delta function: Leopold Kronecker, c. 1880.

Vector analysis (modern): Oliver Heaviside, 1882.

Difference equations: Jules-Henri Poincaré, 1885.

Phasors: Charles Steinmetz, 1893.

Laplace transform applied to transient circuit analysis: Oliver Heaviside, 1900–1930.

AM radio: Guglielmo Marconi, 1901; public broadcast: F. Conrad, 1920; FM radio: Edwin Armstrong, 1933.

Vacuum-tube amplifier: Lee DeForest, 1907.

LC filter: George Campbell, 1911; RLC, 1931; systematic filter theory, Hendrik Bode, 1934.

Decibel unit: 1924.

Negative feedback amplifier: Harold Black, 1927.

Television, kinescope (CRT): Philo Farnsworth, 1929; first regularly scheduled broadcast, 1939; cable and color, 1947.

Harmonic analysis of stochastic signals: Norbert Wiener, 1930.

Radar: R. M. Page, 1936.

Dirac delta function: P. A. M. Dirac, c. 1940.

Transistor amplifier: Walter Brattain, 1947.

ENIAC computer: J. P. Eckert and J. W. Mauchly, 1946; IAS computer, John Von Neumann, 1952.

Integrated circuit: Robert Noyce and Jack Kilby, 1958.

Sampling theorem: Harry Nyquist, 1928; Claude Shannon, 1949.

z-transform: Generating functions, Pierre Laplace, 1766; today's z-transform: John R. Ragazzini and Lotfi A. Zadeh, 1952.

Digital filters (electronic): J. F. Kaiser, 1966.

Mouse: Doug Englebart, 1968.

Unix operating system: 1969.

Arpanet: 1969.

Microcomputer: Marcian Hoff, 1969; patented 1974.

Matlab: Late 1970s.

First word processor: Scribe, by Brian Reid, 1980.

Compact disc: Sony, 1982.

Windows: Microsoft, 1985.

TMS320 DSP chip: Texas Instruments, early 1980s.

World-wide web: 1992.

Pentium: Intel, 1993.

This timeline is intended only to represent some interesting, fundamental discoveries and inventions broadly relevant to basic DSP, not to be comprehensive or include the latest developments.

For much more on DSP history, see Frederik Nebeker, *Signal Processing: The Emergence of a Discipline, 1948–1998,* published by the IEEE History Center in 1998.

Chapter 2

Signals and Systems

2.1 INTRODUCTION

In this first substantial chapter, we review some basic concepts concerning signals and signal-processing systems. Because sampled-data signal processing involves both discrete-time and continuous-time signals and systems, it is appropriate at this stage to consider them both. The discussions in this and later chapters will assume familiarity with the basics of signals and systems analysis for both continuous- and discrete-time systems. This chapter serves as both a review and a second look at signals and systems, in order to establish notation and a smooth transition from prerequisite materials to DSP as covered in this book. We will feel free to make occasional reference to transforms and poles in this review, even though a more systematic presentation of them will be given in later chapters.

We begin by looking at several important signals used in signal-processing analysis and design. Then we focus on a few particular continuous-time systems. Next, systems are described in more general, quantitative ways. In particular, the consequences of linearity and shift invariance are shown and the conditions for linearity reviewed. Once linearity can be assumed, the characterization of the system in terms of modes, stability, and causality is possible.

Because of the general reader's familiarity with continuous-time systems, we review them first. This familiarity stems from the early study of electric circuits and mechanics. Then discrete-time systems are summarized in an analogous and concise fashion. We perform a simple discretization of one of the continuous-time systems we first studied, to show the relation between continuous-time and discrete-time signals and systems. This relation will be examined much more thoroughly later in the book.

2.2 SIGNALS OF SPECIAL IMPORTANCE

Although it is assumed that the reader has had a course in signals and systems, before reviewing system theory it is worth reviewing in one place some of the most important signals used in signal processing.

First, the unit step is useful both as a system input and as an analysis tool. It is defined by (see Figure 2.1; subscript c signifies function of continuous time)

$$u_c(t) = \begin{cases} 1, & (t \geq 0) \\ 0, & (t < 0) \end{cases} \qquad \text{Unit step function,} \atop \text{continuous time (Figure 2.1}a\text{).} \qquad (2.1a)$$

$$u(n) = \begin{cases} 1, & (n \geq 0) \\ 0, & (n < 0). \end{cases} \qquad \text{Unit step sequence,} \atop \text{discrete time (Figure 2.1}b\text{).} \qquad (2.1b)$$

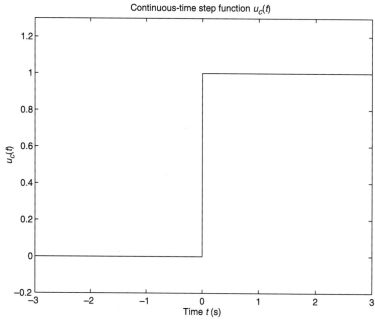

Figure 2.1a

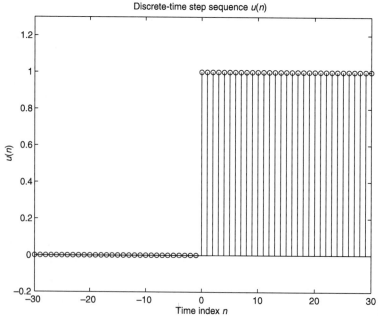

Figure 2.1b

In control systems, the response to a step input is fundamental. For example, an elevator that is requested to go up receives a step input. It is desired that the system respond quickly but smoothly. As on/off steps are the only types of input that the elevator is designed to take, the step response is everything. Alternatively, whenever we desire a function defined for all time to apply only for nonnegative time, we simply multiply it by a step function. We may also construct a wide variety of signals by adding shifted and ramped steps. The step function plays a fundamental role in signal processing.

Another signal that can be used either as an interrogating input or an analysis tool is the impulse or *delta* function (see Figure 2.2):

$$1 = \int_{-\infty}^{\infty} \delta_c(t)dt, \quad \delta_c(t \neq 0) = 0 \qquad \text{Dirac delta function, continuous time (Figure 2.2a).} \qquad (2.2a)$$

$$\delta(n) = \begin{cases} 1, & n = 0 \\ 0, & (n \neq 0). \end{cases} \qquad \text{Kronecker delta function, discrete time (Figure 2.2b).} \qquad (2.2b)$$

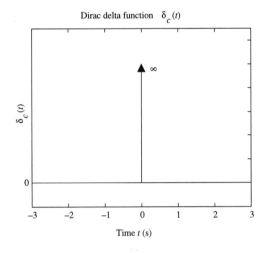

Dirac delta function $\delta_c(t)$

Figure 2.2a

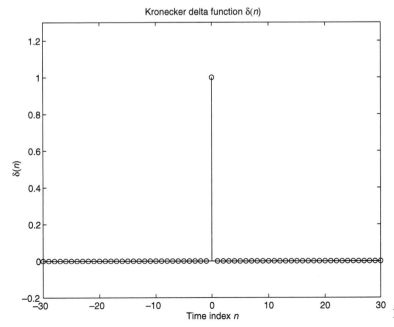

Kronecker delta function $\delta(n)$

Figure 2.2b

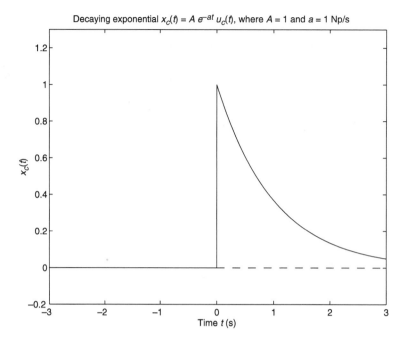

Decaying exponential $x_c(t) = A e^{-at} u_c(t)$, where $A = 1$ and $a = 1$ Np/s

Figure 2.3*a*

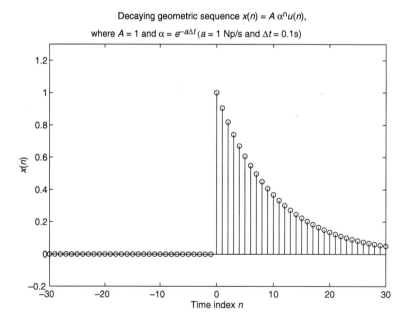

Decaying geometric sequence $x(n) = A \alpha^n u(n)$,

where $A = 1$ and $\alpha = e^{-a\Delta t}$ ($a = 1$ Np/s and $\Delta t = 0.1$s)

Figure 2.3*b*

We note that $\delta_c(t) = du_c(t)/dt$ and $\delta(n) = u(n) - u(n - 1)$—a sort of discrete-time "derivative" of $u(n)$, as we discuss in Section 2.4. The response of a system to these inputs is called the impulse response (continuous time) or the unit sample response (discrete time). The frequency response, as we will review later, is the Fourier transform of the system response to the delta function. In the time domain, the output of a system due to any input is the convolution of the input with the system response to the delta function. In the theory of sampling, $\delta_c(t)$ plays a key role, as we will see in Chapter 6. Notice that although $u(n) = u_c(n\Delta t)$ for any value of Δt (e.g., $\Delta t = 1$ s), $\delta(n)$ is *not* equal to $\delta_c(n\Delta t)$, the latter of which is infinite at $n = 0$. The Dirac delta function $\delta_c(t)$ is reviewed in more detail in Appendix 2A.

The progressive-powers function comes up repeatedly in the analysis of linear systems (see Figure 2.3):

$$x_c(t) = Ae^{-at}u_c(t), \quad \text{Exponential time function,} \atop \text{continuous time (Figure 2.3a)} \tag{2.3a}$$

$$x(n) = A\alpha^n u(n). \quad \text{Geometric time sequence,} \atop \text{discrete time (Figure 2.3b).} \tag{2.3b}$$

Note that $x(n) = x_c(n\Delta t)\big|_{a\,=\,-\ln(\alpha)/\Delta t}$ for any value of Δt. The reasons for the importance of this signal will become more apparent in the course of study, but its primary importance is that the natural modes of a stable, relaxing system are decaying-powers functions [either nonoscillating as in (2.3) or oscillating as below in (2.5)]. Typical nonoscillatory examples in electric circuits are RL and RC combinations. (For unstable systems, there are also rising exponentials.)

Finally and more generally, the sinusoid—both damped and undamped—is fundamental in signal processing (see Figures 2.4 and 2.5):

Undamped (for all t or n):

$$x_c(t) = A\cos(\Omega t + \phi), \quad \text{Undamped sinusoid, continuous time} \atop \text{(Figure 2.4a)} \tag{2.4a}$$

$$x(n) = A\cos(\omega n + \phi), \quad \text{Undamped sinusoid, discrete time} \atop \text{(Figure 2.4b).} \tag{2.4b}$$

Note that $x(n) = x_c(n\Delta t)\big|_{\Omega\,=\,\omega/\Delta t}$ for any value of Δt.

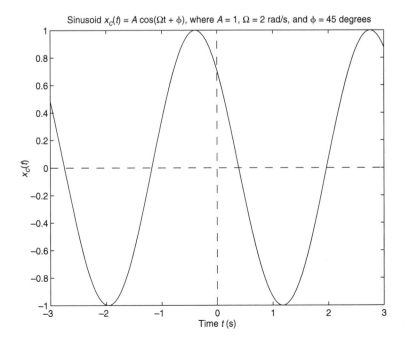

Sinusoid $x_c(t) = A\cos(\Omega t + \phi)$, where $A = 1$, $\Omega = 2$ rad/s, and $\phi = 45$ degrees

Figure 2.4a

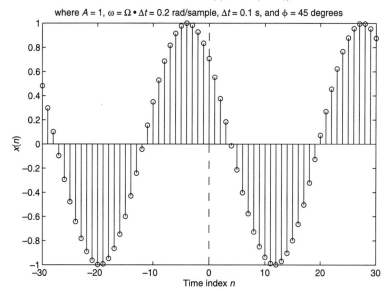

Figure 2.4b

Damped (for $t \geq 0$ and $n \geq 0$):

$$x_c(t) = Ae^{-at}\cos(\Omega t + \phi)u_c(t), \qquad \text{Damped/growing sinusoid, continuous time (Figure 2.5a)} \qquad (2.5a)$$

$$x(n) = A\alpha^n\cos(\omega n + \phi)u(n). \qquad \text{Damped/growing sinusoid, discrete time (Figure 2.5b).} \qquad (2.5b)$$

Note that $x(n) = x_c(n\Delta t)\big|_{\Omega = \omega/\Delta t,\ a = -\ln(\alpha)/\Delta t}$ for any value of Δt. The period of oscillation is $T = 1/F = 2\pi/\Omega$ s for continuous time and $N = 1/f = 2\pi/\omega$ samples

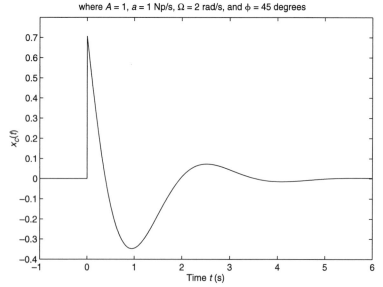

Figure 2.5a

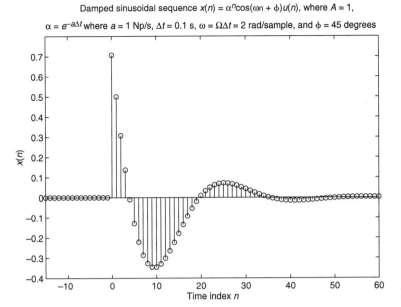

Damped sinusoidal sequence $x(n) = \alpha^n \cos(\omega n + \phi)u(n)$, where $A = 1$,

$\alpha = e^{-a\Delta t}$ where $a = 1$ Np/s, $\Delta t = 0.1$ s, $\omega = \Omega\Delta t = 2$ rad/sample, and $\phi = 45$ degrees

Figure 2.5*b*

for discrete time (and, for the sampled continuous-time function, the period is $N\Delta t$). Here, Ω is the frequency in radians per second, F is the frequency in hertz, ω is the frequency in radians per sample, and f is the frequency in cycles per sample. Notice that $x(n)$ is periodic with period $N = 1/f$ if and only if a single period ahead falls on a sample; equivalently, $x(n)$ is periodic if $1/f = 2\pi/\omega$ is an integer (the period N). Note that the nonoscillatory exponential is just a special case of (2.5) in which the frequency Ω or ω is zero.

As we will review later, the exponentially damped/growing sinusoid is the most general linear system mode, the response of a system to no input other than an initially stored amount of energy. A typical electric circuit example is the *RLC* combination. Moreover, we will find in Chapters 4 and 5 that the undamped and damped sinusoid, and the exponential are unique in that when presented to a system as the input, the output has the same form as the input (aside from scaling and phase-shifting; the form is exactly the same for bicausal versions, and approximately the same after transients for causal versions). This fact, combined with the possibility of decomposing general time functions into sums or integrals of damped and undamped sinusoids, provides a basic tool in signal and system analysis.

The basic signals just described are useful for interrogating, characterizing, and designing systems as well as characterizing, analyzing, and modeling real-world signals.

2.3 SYSTEM CHARACTERIZATION AND SOLUTION: CONTINUOUS TIME

Recall that there are essentially three categories of systems: continuous-time, discrete-time, and hybrid. The continuous-time system takes a continuous-time input and produces a continuous-time output. Examples are RLC and op-amp circuits, mass/spring/damper systems, and chemical/thermal/biological systems. The discrete-time system takes a discrete-time input and produces a discrete-time output. Examples are digital electronics/computers, the stock market, and population dynamics. Finally, the hybrid system takes a continuous-time input and produces a

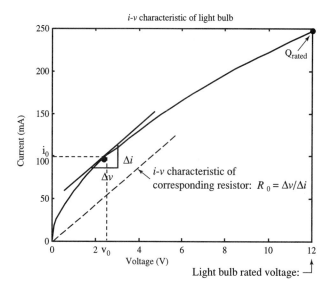

Figure 2.6

Light bulb rated voltage:

discrete-time output, or vice versa; usually these systems are called analog/digital converters and digital/analog converters. Application examples are measurement systems, digital simulations/implementations of analog systems, and compact disc and other digital audio and video systems.

2.3.1   Linearity

Let us briefly consider some examples of linear and nonlinear systems. An example of the simplest sort of linear relation is Ohm's law: $i = v/R$. Doubling the voltage v across resistor R doubles the current i in R. Suppose now that the resistor is a light bulb; the i–v relation is experimentally measured to be that in Figure 2.6. Notice that for very small v (and therefore a cold bulb), i is proportional to v, but as the voltage becomes much larger, the i–v relation becomes nonlinear. The nonlinearity occurs because R increases with temperature, and as the voltage increases, the bulb heats up due to large current and Joule's heat dissipation law $p = i^2R = v^2/R$.

The actual rated voltage of this automobile bulb is 12 V. If we consider small variations about an operating point (v_0, i_0), even in the nonlinear region of the i–v characteristic we have $\Delta i \approx \Delta v/R_0$, where $R_0 = di/dv\big|_{v_0}$. This linearization is possible because the tangent to any curve, a straight line, matches the slope of the curve at the point of tangency (operating point); again, see Figure 2.6. All linear approximations of i for this operating point lie on that tangent line.[1] Note that the bulb becomes visibly glowing around 1.2 V, well beyond the point (about 0.05 V, as data taken in that region shows) at which the i–v relation has become significantly nonlinear. The system is nonlinear because the slope R_0 is different for different operating points (values of v_0); for a linear resistor, $R_0 = R = $ constant for all operating points.

This example illustrates the fundamentally important idea of linearization about an operating point. It is so crucial, when extended to more complex systems,

---

[1] For a function f of two variables x_1 and x_2, all linear approximations of f at operating point Q fall on the plane formed by the lines passing through Q with slopes respectively equal to the partial derivatives of f with respect to x_1 and x_2, evaluated at Q.

because it allows the wealth of knowledge and techniques concerning linear system analysis and design to be applied to a nonlinear system. Naturally, if the variations about the operating point are sufficiently large, the linear approximation does not hold and other techniques of nonlinear analysis must be tried. In this book, we restrict our attention to linear systems, because intelligent design can often lead to linear systems, and nonlinear system analysis techniques assume familiarity with linear systems analysis.

Ohm's law is an example of a system without dynamics. That is, no past or future values of v affect i, and neither do any derivatives or integrals of v or i. When energy storage elements are involved, there can be exchanges of energy between the source and the element or between the energy storage elements. Such exchanges typically result in exponential time dependencies of signal variables (e.g., voltage, current, position) that are decaying, growing, oscillatory, or a combination of these. Examples of energy storage elements are inductors, capacitors, springs, and masses. Of course, arrangements of these devices can be arbitrarily complex.

As above, let the subscript c hereafter be used to denote continuous-time signals and systems [example: $x_c(t)$]. This notation is introduced to avoid confusion later on when we must deal simultaneously with discrete-time and continuous-time systems. In fact, the system in Example 2.2 will itself be discretized in Section 2.4.

EXAMPLE 2.1

Find the differential equation of the mechanical spring-mass system in Figure 2.7.

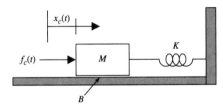

Figure 2.7

Solution

According to Newton's second law, the applied force $f_c(t)$ to the right is opposed by the inertial force,[2] the frictional force, and the spring compression. Thus

$$M\frac{d^2x_c(t)}{dt^2} + B\frac{dx_c(t)}{dt} + Kx_c(t) = f_c(t).$$

We may view $f_c(t)$ as the input variable and $x_c(t)$ as the output. The input–output relation is thus a second-order differential equation for this physical system.

We now consider a second example of a physical system whose model differential equation involves higher derivatives, a time-derivative of the input variable, and the output variable.

---

[2] If the mass is decelerating, the inertial force aids f_c. This behavior is automatically incorporated, as in that case $d^2x_c/dt^2 < 0$. The word *oppose* applies for the other terms for $dx_c/dt > 0$ and $x_c > 0$, which is the case when we analyze the motion beginning at rest.

EXAMPLE 2.2

Without using transform techniques, find the relation between $v_{RL,c}(t)$ = the output voltage across the lossy inductor and capacitor, and $v_{s,c}(t)$ = the input voltage of the circuit shown in Figure 2.8. Physically, the system represents a generator driving a resistive transmission line and an inductive, power-factor-corrected load. For simplicity, the internal impedance of the generator is neglected.

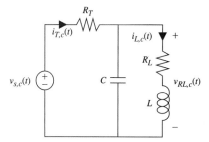

Figure 2.8

Solution

Kirchhoff's voltage law around the outer loop gives

$$v_{s,c}(t) - i_{T,c}(t)R_T = v_{RL,c}(t). \tag{2.6}$$

By the definition of $v_{RL,c}(t)$,

$$i_{L,c}(t)R_L + L\frac{di_{L,c}(t)}{dt} = v_{RL,c}(t). \tag{2.7}$$

By Kirchhoff's current law,

$$i_{T,c}(t) = i_{L,c}(t) + C\frac{dv_{RL,c}(t)}{dt}. \tag{2.8a}$$

By (2.6), we can eliminate $i_{T,c}(t)$ in (2.8a):

$$\frac{v_{s,c}(t) - v_{RL,c}(t)}{R_T} = i_{L,c}(t) + C\frac{dv_{RL,c}(t)}{dt}. \tag{2.8b}$$

To obtain a single differential equation relating $v_{RL,c}(t)$ *to* $v_{s,c}(t)$, we must eliminate $i_{L,c}(t)$. Because $d/dt\{d/dt\} = d^2/dt^2$, operational factors of the form $(d/dt + \alpha)$ can be treated algebraically. Following the rules of algebra, we "multiply" (operate on) (2.8b) by $R_L + L(d/dt)$. We choose $R_L + L(d/dt)$ rather than something else because from (2.7) we see that it will eliminate $i_{L,c}(t)$ from (2.8b):

$$\left(R_L + L\frac{d}{dt}\right)\left[\frac{v_{s,c}(t) - v_{RL,c}(t)}{R_T}\right] = \left(R_L + L\frac{d}{dt}\right)i_{L,c}(t) + \left(R_L + L\frac{d}{dt}\right)C\frac{dv_{RL,c}(t)}{dt}. \tag{2.9}$$

Thus the first of the two major terms on the right-hand side of (2.9) is, by (2.7), equal to $v_{RL,c}(t)$ so that $i_{L,c}(t)$ has been eliminated and (2.9) becomes

$$\left\{LC\frac{d^2}{dt^2} + \left(\frac{L}{R_T} + R_L C\right)\frac{d}{dt} + 1 + \frac{R_L}{R_T}\right\}v_{RL,c}(t) = \frac{R_L + L d/dt}{R_T}v_{s,c}(t), \tag{2.10}$$

or finally, division of (2.10) by LC gives

$$\left\{\frac{d^2}{dt^2} + a_1\frac{d}{dt} + a_0\right\}v_{RL,c}(t) = \left\{b_1\frac{d}{dt} + b_0\right\} v_{s,c}(t), \tag{2.11}$$

where

$$a_1 = \frac{1}{R_TC} + \frac{R_L}{L}, \quad a_0 = \frac{1 + R_L/R_T}{LC}, \quad b_1 = \frac{1}{R_TC}, \quad b_0 = \frac{R_L}{LCR_T}. \tag{2.12}$$

Thus we have seen how the differential equation arises in practice, from which the transfer function in analog signal processing theory is derived. Below, we formally verify that the differential equation (2.11) is an example of a system having a linear input–output relation. The concept of algebraic manipulation of operators is fundamental to linear systems analysis. The same basic procedure used above will later be used to identify and solve the input–output difference equation of a related discrete-time system.

It should also be noted that the variables $i_{L,c}(t)$ and $v_{RL,c}(t) = v_{C,c}(t)$ have special significance: They are the *physical variables* state variables used to easily write the state equations for this system. Typically, these "physical variables" are selected to be the result of integration of the energy-storage element relation (e.g., capacitor voltage from $i = Cdv/dt$ and inductor current from $v = Ldi/dt$). Not only are they the variables whose initial conditions are typically known, but they also quantify the amount of stored energy ($\frac{1}{2}Cv_C^2$ for capacitor, $\frac{1}{2}Li_L^2$ for inductor). In Section 2.4.6.3 (page 59ff), we examine the relation between the state–space solution and the unit sample response of a discrete-time system.

The important point in Example 2.2 is that merely by connecting elements whose constitutive relations are just simple first-order differential equations (e.g., $v = Ldi/dt$, $i = Cdv/dt$), we can end up with an overall input–output relation obeying a single, possibly high-order differential equation. Moreover, this differential equation may involve derivatives of the input as well as the output, which might not be immediately apparent.

It is from the modeling of physical systems that we obtain input–output relations such as (2.11). Although modeling is an extremely important task, it is beyond the scope of this book; see, however, Example 4.11 for how data acquisition hardware can be used for modeling an unknown network. For more details on modeling with differential equations, refer to books on circuit analysis, mechanics, chemistry, or whatever field is the basis for the phenomenon under investigation. Similarly, the detailed study of signal measurement—transducers, signal conditioning, and the like—are outside the scope of this book. A primary goal of the study of signal processing is knowing what to do with a model and signals, once obtained.

EXAMPLE 2.3

For the system in Example 2.2, suppose that the "signal" $v_{s,c}(t)$ is a 60-Hz sine wave distorted by the presence of a third-harmonic component. The third harmonic is due to one or more of the following on the generator: nonsinusoidal stator flux distribution, nonuniform reluctance of the rotor/stator flux path, and core saturation effects. Suppose $R_T = 5\ \Omega$, $R_L = 3\ \Omega$, $L = 10.9$ mH, and $C = 91/\mu$F, convenient for a scaled-down investigation. Suppose the third-harmonic distortion in $v_{s,c}(t)$ is 20% of the original sinusoidal amplitude (which in practice may be reduced using specialized pitch design techniques). What is it in $v_{RL,c}(t)$ in the steady state, that is, ignoring any start-up transients? Work only from the differential equation, and in the process, obtain an explicit expression for $v_{RL,c}(t)$.

Solution

First note that in (2.12), for the given circuit component values we have $a_1 = 2473\ s^{-1}$, $a_0 = 1.61 \cdot 10^6\ s^{-2}$, $b_1 = 2198\ s^{-1}$, and $b_0 = 6.05 \cdot 10^5\ s^{-2}$. For simplicity, assume that

$$v_{s,c}(t) = \cos(2\pi 60t) + 0.2 \cos(2\pi 180t)\ (V). \tag{2.13}$$

Because we seek only the steady-state solution, only the particular solution (sinusoids), not the homogeneous solution, is involved. Here we care only about the steady-state solution because our task is to determine how much third-harmonic distortion resides in the output voltage; we do not care about the initial conditions/start-up transient. We know that both the amplitudes and the phases of the sinusoidal components of $v_{s,c}(t)$ will be altered in $v_{RL,c}(t)$. Thus let

$$v_{RL,c}(t) = A_1\cos(2\pi 60t) + B_1\sin(2\pi 60t) + A_2\cos(2\pi 180t) + B_2\sin(2\pi 180t)\ (V). \tag{2.14}$$

Substituting $v_{RL,c}(t)$ into (2.11), we have upon taking first and second derivatives and collecting terms,

$$\{[a_0 - (2\pi 60)^2]A_1 + a_1 2\pi 60 B_1\}\cos(2\pi 60t) + \{-a_1 2\pi 60 A_1 + [a_0 - (2\pi 60)^2]B_1\}\sin(2\pi 60t)$$
$$+ \{[a_0 - (2\pi 180)^2]A_2 + a_1 2\pi 180 B_2\}\cos(2\pi 180t) + \{-a_1 2\pi 180 A_2 + [a_0 - (2\pi 180)^2]B_2\}\sin(2\pi 180t)$$
$$= b_0\cos(2\pi 60t) - b_1 2\pi 60 \sin(2\pi 60t) + b_0 0.2 \cos(2\pi 180t) - b_1 0.2(2\pi 180)\sin(2\pi 180t). \tag{2.15}$$

Inspection of (2.15) shows that the two different frequency terms are decoupled—an important characteristic of linear shift-invariant systems (superposition)—so we have for the 60-Hz sinusoid

$$\begin{bmatrix} a_0 - (2\pi 60)^2 & a_1 2\pi 60 \\ -a_1 2\pi 60 & a_0 - (2\pi 60)^2 \end{bmatrix} \begin{bmatrix} A_1 \\ B_1 \end{bmatrix} = \begin{bmatrix} b_0 \\ -b_1 2\pi 60 \end{bmatrix} \rightarrow \begin{bmatrix} A_1 \\ B_1 \end{bmatrix} = \begin{bmatrix} 0.540\ V \\ -0.214\ V \end{bmatrix}, \tag{2.16a}$$

giving an output amplitude of $\{A_1^2 + B_1^2\}^{1/2} = 0.581$ V for the 60-Hz sinusoid, whereas for the third harmonic,

$$\begin{bmatrix} a_0 - (2\pi 180)^2 & a_1 2\pi 180 \\ -a_1 2\pi 180 & a_0 - (2\pi 180)^2 \end{bmatrix} \begin{bmatrix} A_2 \\ B_2 \end{bmatrix} = \begin{bmatrix} 0.2 b_0 \\ -0.2 b_1 2\pi 180 \end{bmatrix} \rightarrow \begin{bmatrix} A_2 \\ B_2 \end{bmatrix} = \begin{bmatrix} 0.180\ V \\ 0.0217\ V \end{bmatrix}, \tag{2.16b}$$

giving an output amplitude of $\{A_2^2 + B_2^2\}^{1/2} = 0.182$ V for the third harmonic (180 Hz). Thus the third harmonic percentage of the output $v_{RL,c}(t)$ is $0.182/0.581 = 31.3\%$. We conclude that the harmonic distortion is exacerbated in the load.

When $v_{s,c}(t)$ is one or more steady-state sinusoids, the quickest method to find $v_{RL,c}(t)$ from the electrical engineer's point of view is to use Fourier transforms. This approach trivializes the problem to just a complex voltage phasor divider for each sinusoidal frequency, yet it gives the same result that formal solution of (2.11) gives. Using the bar notation of phasor analysis, we note that the $R_L LC$ load impedance is

$$\overline{Z}_{eq} = (R_L + j \cdot 2\pi FL)\ \|\ \frac{-j}{[2\pi FC]} = \begin{cases} 3.944 + j4.161\ \Omega\ \text{for}\ F = 60\ \text{Hz} \\ 17.90 - j25.30\ \Omega\ \text{for}\ F = 180\ \text{Hz}, \end{cases} \tag{2.17}$$

and thus obtain the frequency response values

$$\overline{H} = \frac{\overline{V}_{RL}}{\overline{V}_s} = \frac{\overline{Z}_{eq}}{\overline{Z}_{eq} + R_T} = \begin{cases} 0.581\ \angle\ 21.6° = 0.540 + j0.214\ \text{for}\ F = 60\ \text{Hz} \\ 0.908\ \angle\ -6.87° = 0.902 - j0.109\ \text{for}\ F = 180\ \text{Hz}. \end{cases} \tag{2.18}$$

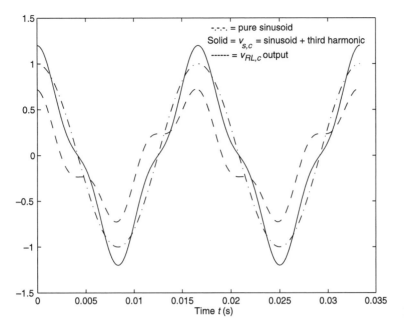

Figure 2.9

The percentage of the output that is third-harmonic is obtained as $0.2(0.908)/0.581 = 31.3\%$, which agrees with the differential equations approach. In addition, $\text{Re}\{\overline{H}\} = A_1$ and $\text{Im}\{\overline{H}\} = -B_1$ for $F = 60$ Hz as must be true (the minus sign on $-B_1$ comes out of the complex algebra); similar relations hold for $F = 180$ Hz. The original unit-amplitude pure 60-Hz sinusoid as well as $v_{s,c}(t)$ and $v_{RL,c}(t)$ are plotted in Figure 2.9; note that the phase shifts indicated in $\overline{H}$ above are also evident for each component of $v_{RL,c}(t)$.

We now examine the formal condition for linearity of a continuous-time signal-processing system. Recall from your study of circuits that a system $y_c(t) = L\{x_c(t)\}$ defined by the input–output operator L is linear if, given that $y_{1,c}(t) = L\{x_{1,c}(t)\}$ and $y_{2,c}(t) = L\{x_{2,c}(t)\}$, it follows that

$$L\{\alpha_1 x_{1,c}(t) + \alpha_2 x_{2,c}(t)\} = \alpha_1 y_{1,c}(t) + \alpha_2 y_{2,c}(t). \quad \text{Condition for a linear system.} \quad (2.19)$$

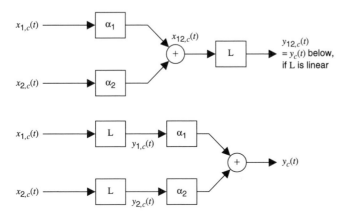

Figure 2.10

See Figure 2.10, which illustrates the meaning of (2.19): The output $y_{12,c}(t)$ of L operating on a weighted sum of inputs $x_{12,c}(t) = \alpha_1 x_{1,c}(t) + \alpha_2 x_{2,c}(t)$ is equal to the weighted sum $y_c(t)$ of the results of L operating on each original input individually.

EXAMPLE 2.4

(a) Suppose the output of a system is $y_c(t) = x_c(t-5) + (1/5)^t dx_c(t)/dt$. Is this a linear input/output relation?

(b) Repeat part (a) for $y_c(t) = x_c(t-5) + (1/5)^t dx_c(t)/dt + 7$.

Solution

(a)   Let $y_{1,c}(t)$ be the output for $x_{1,c}(t)$:

$$y_{1,c}(t) = x_{1,c}(t-5) + (1/5)^t \, dx_{1,c}(t)/dt,$$

and let $y_{2,c}(t)$ be the output for $x_{2,c}(t)$:

$$y_{2,c}(t) = x_{2,c}(t-5) + (1/5)^t \, dx_{2,c}(t)/dt.$$

Then the output $y_{12,c}(t)$ for $x_{12,c}(t) = \alpha_1 x_{1,c}(t) + \alpha_2 x_{2,c}(t)$ is

$$y_{12,c}(t) = x_{12,c}(t-5) + \left(\frac{1}{5}\right)^t \frac{dx_{12,c}(t)}{dt}$$

$$= \{\alpha_1 x_{1,c} + \alpha_2 x_{2,c}\}\big|_{t-5} + \left(\frac{1}{5}\right)^t \frac{d}{dt}\{\alpha_1 x_{1,c}(t) + \alpha_2 x_{2,c}(t)\}$$

$$= \alpha_1 x_{1,c}(t-5) + \alpha_1 \left(\frac{1}{5}\right)^t \frac{d}{dt}\{x_{1,c}(t)\} + \alpha_2 x_{2,c}(t-5) + \alpha_2 \left(\frac{1}{5}\right)^t \frac{d}{dt}\{x_{2,c}(t)\}$$

$$= \alpha_1 y_{1,c}(t) + \alpha_2 y_{2,c}(t).$$

So yes, the condition for linearity (2.19) is satisfied and the system is linear.

(b)   Defining $y_{1,c}(t), y_{2,c}(t)$ as above, with the addition of 7 in each case, we have

$$y_{12,c}(t) = \{\alpha_1 x_{1,c} + \alpha_2 x_{2,c}\}\big|_{t-5} + \left(\frac{1}{5}\right)^t \frac{d}{dt}\{\alpha_1 x_{1,c}(t) + \alpha_2 x_{2,c}(t)\} + 7$$

$$\neq \alpha_1 y_{1,c}(t) + \alpha_2 y_{2,c}(t)$$

because the "+ 7" is not multiplied by $\alpha_1 + \alpha_2$. We conclude that the system is not linear.

EXAMPLE 2.5

The generalization of the differential equation found in Example 2.2 [see (2.11)] for any maximum order N of derivatives of the output and maximum order M of derivatives of the input occurring in the model is

$$\sum_{\ell=0}^{N} a_\ell \frac{d^\ell y_c(t)}{dt^\ell} = \sum_{\ell=0}^{M} b_\ell \frac{d^\ell x_c(t)}{dt^\ell}. \tag{2.20}$$

Let us verify that this class of systems is linear.

Solution

Using the above notation for input–output pairs, if $y_{1,c}(t) = L\{x_{1,c}(t)\}$, then according to (2.20) and the linearity of the sum and derivative operators, for any constant α_1 we have (using simple distributivity)

$$\sum_{\ell=0}^{N} a_\ell \frac{d^\ell\{\alpha_1 y_{1,c}(t)\}}{dt^\ell} = \alpha_1 \sum_{\ell=0}^{N} a_\ell \frac{d^\ell\{y_{1,c}(t)\}}{dt^\ell} \qquad (2.21a)$$

$$= \alpha_1 \sum_{\ell=0}^{M} b_\ell \frac{d^\ell x_{1,c}(t)}{dt^\ell}$$

$$= \sum_{\ell=0}^{M} b_\ell \frac{d^\ell\{\alpha_1 x_{1,c}(t)\}}{dt^\ell},$$

and similarly,

$$\sum_{\ell=0}^{N} a_\ell \frac{d^\ell\{\alpha_2 y_{2,c}(t)\}}{dt^\ell} = \sum_{\ell=0}^{M} b_\ell \frac{d^\ell\{\alpha_2 x_{2,c}(t)\}}{dt^\ell}. \qquad (2.21b)$$

Adding (2.21a) and (2.21b) gives

$$\sum_{\ell=0}^{N} a_\ell \frac{d^\ell\{\alpha_1 y_{1,c}(t) + \alpha_2 y_{2,c}(t)\}}{dt^\ell} = \sum_{\ell=0}^{M} b_\ell \frac{d^\ell\{\alpha_1 x_{1,c}(t) + \alpha_2 x_{2,c}(t)\}}{dt^\ell}, \qquad (2.21c)$$

which is exactly the same form as (2.20) with $x_c(t) = x_{12,c}(t) = \alpha_1 x_{1,c}(t) + \alpha_2 x_{2,c}(t)$ and $y_c(t) = y_{3,c}(t) = \alpha_1 y_{1,c}(t) + \alpha_2 y_{2,c}(t)$, so evidently $y_{3,c}(t) = y_{12,c}(t) \equiv L\{x_{12,c}(t)\}$. Thus we conclude that indeed (2.19) is satisfied and the system is linear.

2.3.2   Time (or Shift) Invariance

Another property of (2.20), besides linearity, that is crucial for ease in signal processing is time invariance. More generally (e.g., for either time- or space-dependent systems), we use the name shift invariance. Time invariance arises when the coefficients a_ℓ and b_ℓ in (2.20) are independent of time t. The condition for time invariance is that if $y_c(t) = L\{x_c(t)\}$, then the output due to $x_c(t - \tau)$ is $y_c(t - \tau)$:

$$L\{x_c(t - \tau)\} = y_c(t - \tau). \quad \text{Condition for time invariance.} \qquad (2.22)$$

In words, it does not matter when we begin sending in a particular signal; the output response to that signal will always be the same except for a delay/advance corresponding to when the input is applied. In Figure 2.11, an example is shown of an input, its output, the same input delayed, and its output for a shift-invariant system.

Let us now determine whether the general system differential equation (2.20) is time invariant. First note that $df(t - \tau)/dt = df/dt\big|_{t - \tau}$ and $f_1(t - \tau) + f_2(t - \tau) =$

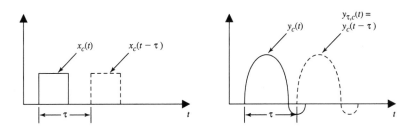

Figure 2.11

$f_1(t) + f_2(t)\big|_{t-\tau}$. With these properties in mind, we let the input in (2.20) be $x_c(t - \tau)$ and call the resulting output $y_{\tau,c}(t)$. We have

$$\sum_{\ell=0}^{N} a_\ell \frac{d^\ell y_{\tau,c}(t)}{dt^\ell} = \sum_{\ell=0}^{M} b_\ell \frac{d^\ell x_c(t - \tau)}{dt^\ell} \qquad (2.23a)$$

or, equivalently,

$$\sum_{\ell=0}^{N} a_\ell \frac{d^\ell y_{\tau,c}(t)}{dt^\ell} = \sum_{\ell=0}^{M} b_\ell \frac{d^\ell x_c(t)}{dt^\ell}\bigg|_{t-\tau} . \qquad (2.23b)$$

Because $y_c(t) = \mathrm{L}\{x_c(t)\}$ holds for any value of t, including "$t - \tau$", (2.20) asserts that

$$\sum_{\ell=0}^{N} a_\ell \frac{d^\ell y_c}{dt^\ell}\bigg|_{t-\tau} = \sum_{\ell=0}^{M} b_\ell \frac{d^\ell x_c}{dt^\ell}\bigg|_{t-\tau} . \qquad (2.23c)$$

By identifying the left-hand sides of (2.23c) and (2.23b), due to the right-hand sides being identical for all values of t, we conclude that $y_{\tau,c}(t) = y_c(t - \tau)$, and the time invariance of (2.20) is shown.

Thus continuous-time systems satisfying a linear differential equation with constant coefficients are called linear time- or shift-invariant (LTI or LSI) systems.

EXAMPLE 2.6

If the output of a system is $y_c(t) = tx_c(t)$, is the system time invariant?

Solution

No. Substituting $x_c(t - \tau)$ gives $y_{\tau,c}(t) = tx_c(t - \tau)$, which is not equal to $y_c(t - \tau) = (t - \tau)x_c(t - \tau)$. See Figure 2.12 for an illustration. Figure 2.12a shows the "unshifted"

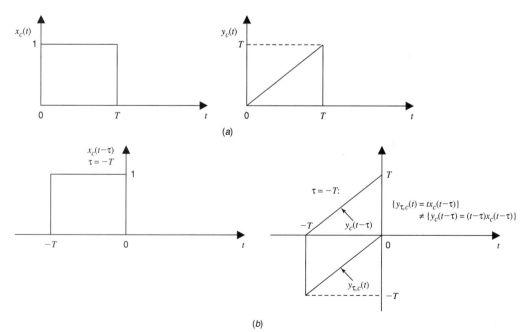

Figure 2.12

input and resulting output. In Figure 2.12b, the input is shifted to the left by T, but the output produced by $y_c(t) = tx_c(t)$ is not a shifted version of $y_c(t)$ in Figure 2.12a; thus, the system is not time invariant.

2.3.3   Solutions for General Inputs: Convolution

The model equation (2.20) is an Nth-order, nonhomogeneous linear differential equation with constant coefficients. A major parting from usual elementary differential equation theory in the solution of (2.20) for signal-processing applications is that the input, or driving function, of the system is kept general. When we study the method of undetermined coefficients or variation of parameters methods, we assume that $x_c(t)$ is a particular known function. We must re-solve the differential equation for each driving function we encounter, and such solutions can be found exactly for only a very limited set of very simple inputs. In signal processing, $x_c(t)$ could be, for example, a speech waveform—extremely complicated, time-varying in form, and unknowable beforehand or in closed form. Thus we must have methods for solving (2.20) that work for any input $x_c(t)$.

To this end, we make use of the Dirac delta function $\delta_c(t)$ (see Appendix 2A), which satisfies $\delta_c(t \neq 0) = 0$ and

$$1 = \int_{-\infty}^{\infty} \delta_c(t)dt, \qquad (2.24a)$$

or equivalently, noting that time-reversing and shifting a Dirac delta does not change its area,

$$1 = \int_{-\infty}^{\infty} \delta_c(t - \tau)d\tau. \qquad (2.24b)$$

By multiplying both sides of (2.24b) by $x_c(t)$, we have

$$x_c(t) = x_c(t)\int_{-\infty}^{\infty} \delta_c(t - \tau)d\tau. \qquad (2.25a)$$

Given that $\delta_c(t \neq 0) = 0$ implies that $\delta_c(t - \tau)\big|_{\tau \neq t} = 0$, $x_c(t)$ can be placed inside the integral without changing the result:

$$x_c(t) = \int_{-\infty}^{\infty} x_c(\tau)\delta_c(t - \tau)d\tau. \qquad (2.25b)$$

Note that we can view (2.25b) as an expansion of $x_c(t)$ over shifted Dirac delta functions weighted by x_c evaluated at the shift, times $d\tau$. We may also view (2.25b) as a sampling property of $\delta_c(t)$; see Appendix 2A. Now let $y_c(t) = \text{L}\{x_c(t)\}$, where we will be assuming that L is LTI. Then by (2.25b) we may write

$$y_c(t) = \text{L}\left\{\int_{-\infty}^{\infty} x_c(\tau)\delta_c(t - \tau)d\tau\right\}, \qquad (2.26a)$$

which, because of the assumed linearity of L, is equivalent to

$$y_c(t) = \int_{-\infty}^{\infty} x_c(\tau)\text{L}\{\delta_c(t - \tau)\}d\tau. \qquad (2.26b)$$

The system output when the input is $\delta_c(t)$ is called the impulse response. Let the impulse response, $\text{L}\{\delta_c(t)\}$, be designated $h_c(t)$: $h_c(t) = \text{L}\{\delta_c(t)\}$. Then, by the shift-invariance of L, we have $\text{L}\{\delta_c(t - \tau)\} = h_c(t - \tau)$, so that (2.26b) is equivalent to

$$y_c(t) = \int_{-\infty}^{\infty} x_c(\tau)h_c(t-\tau)d\tau \tag{2.27a}$$

$$= \int_{-\infty}^{\infty} x_c(t-\tau)h_c(\tau)d\tau \tag{2.27b}$$

$$= x_c(t) *_c h_c(t), \tag{2.27c}$$

where (2.27b) follows from (2.27a) by the change of variables $\tau_{new} \equiv t - \tau_{old}$ and where $*_c$ denotes continuous-time convolution. Equation (2.27) is the familiar convolution input–output relation. We conclude that convolution is a direct consequence of the system properties of linearity [used to obtain (2.26b)] and shift invariance [used to obtain (2.27a)].

In everyday speech, the word *convolute* means rolled or wound together one part upon another. In (2.27), we see how convolution got its name: One of the inputs is "upon the other" backwards and shifted with respect to the other and to the integration variable τ. Figure 2.13a shows an example $x_c(t), h_c(t)$, and $y_c(t) = x_c(t) *_c h_c(t)$, and

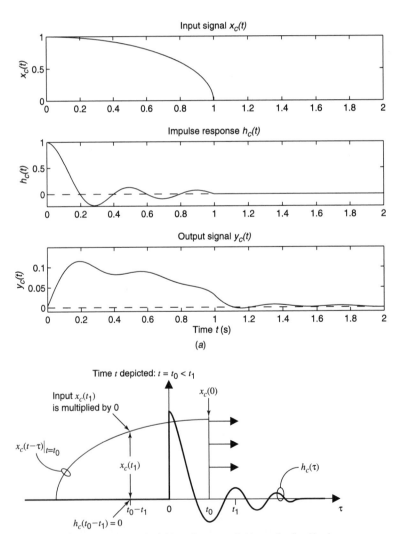

Figure 2.13

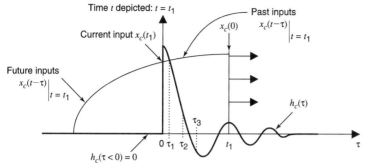

Entire waveform $x_c(t-\tau)$ moves to right as t increases; $h_c(\tau)$ remains fixed in place.

(c)

Figure 2.13 (continued)

Figure 2.13b shows the integrand factors of (2.27b) for the particular value of t, $t = t_0$. (Ignore t_1 in Figure 2.13b, and τ_1, τ_2, and τ_3 in Figure 2.13c for the moment.) The integral of the pointwise product of these factors over the entire interval of τ that the product is nonzero is $y_c(t_0)$. For each value of t, a different picture like Figure 2.13b applies; for example, Figure 2.13c shows the integrand factors for the later time $t = t_1$. Letting $t = t_\ell$ range from $-\infty$ to $+\infty$ results in the entire output waveform $y_c(t)$.

It is of interest to trace the effect of a particular input value, $x_c(t_1)$, on $y_c(t)$ in (2.27b) as t increases. Note that the argument of x_c in (2.27b), namely $t - \tau$, is equal to t_1 [thus specifying $x_c(t_1)$] for $\tau = t - t_1$. Thus for any value of t, $x_c(t_1)$ is multiplied by $h_c(\tau)\big|_{\tau = t - t_1} = h_c(t - t_1)$ to make its contribution to $y_c(t)$. In particular, in Figure 2.13b, $x_c(t_1)$ is multiplied by $h_c(t_0 - t_1)$, which is zero because the argument $t_0 - t_1$ of the depicted filter h is negative. In general, we see that $h_c(t - t_1) = 0$ for all $t < t_1$ (i.e., for $\tau = t - t_1 < 0$). Thus $x_c(t_1)$ makes no contribution at all to $y_c(t)$ for $t < t_1$; the system is causal, a property formally defined in Section 2.3.4. For example, there is no way that the output of a circuit can react to an input we have not yet presented to the circuit.

At the later time $t = t_1$(Figure 2.13c), $x_c(t_1)$ is multiplied by $h_c(0)$. For $t > t_1$—that is, beyond the time shown in Figure 2.13c—the reversed input waveform $x_c(t - \tau)$ moves farther to the right, while $h_c(\tau)$ is fixed in place. As this happens, $x_c(t_1)$ is multiplied no longer by $h_c(0)$ but by $h_c(\tau)$ for larger and larger values of $\tau = t - t_1$. In the common case that $h_c(\tau)$ becomes small (or zero) for large τ, the contribution $x_c(t_1)$ makes to $y_c(t)$ diminishes (or becomes zero) as t increases. Again, intuitively, we expect an input presented to a circuit long, long ago to have little effect on the current output. A similar argument can again also be made for (2.27a) by reversing the roles of x and h.

Alternatively, we may view the entire waveform $y_c(t)$ for all values of t in (2.27b) as an $h_c(\tau)$-weighted sum of distinct shifted-by-τ, *non*-time-reversed input waveforms $x_c(t - \tau)$ with t as the variable, as τ is varied as a shift parameter from $-\infty$ to $+\infty$. (Although $\tau < 0$ means $x_c(t - \tau)$ is advanced rather than delayed, for causal systems $x_c(t - \tau)$ is weighted by $h_c(\tau) = 0$ for $\tau < 0$.) Notice that the only place t occurs on the right-hand side of (2.27b) is in $x_c(t - \tau)$. A particular contributing, weighted waveform $x_c(t - \tau)h_c(\tau)$ determined for a particular value of τ—for example, τ_1—contributes to the total value of the waveform $y_c(t)$ for each instant (value of t) that that weighted waveform $x_c(t - \tau_1)h_c(\tau_1)$ is nonzero. See Figure 2.14, in which three example weighted and shifted input signals are shown for three values of τ. Notice how, in accordance with $\{h_c(\tau_1), h_c(\tau_2),$ and

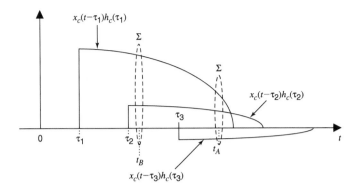

Figure 2.14

$h_c(\tau_3)\}$ shown in Figure 2.13c, the scaling of the shifted x_c waveforms in Figure 2.14 decreases and changes sign as τ increases from τ_1 to τ_2 to τ_3.

In Figure 2.14, to compute $y_c(t)$ for a particular value of t, we add up all the shifted/weighted waveforms at the desired value of t. For example, at $t = t_A$, we add all three shown contributions in the dashed oval, but at $t = t_B$, we add only $h_c(\tau_1)x_c(t_B - \tau_1)$ and $h_c(\tau_2)x_c(t_B - \tau_2)$ in the corresponding dashed oval because $h_c(\tau_3)x_c(t - \tau_3)$ equals zero at $t = t_B$. Of course, there is actually an infinite number of these waveforms to be summed within each oval, infinitesimally separated and further weighted by the infinitesimal differential $d\tau$. The weighted sum of all the shifted impulse responses at any given value of t is $y_c(t)$. Again, an analogous discussion can also be made for (2.27a) by reversing the roles of x and h.

EXAMPLE 2.7

Find $y_c(t) = x_c(t) *_c h_c(t)$ if $x_c(t) = u_c(t) - u_c(t - T)$ and $h_c(t) = e^{-\alpha t}u_c(t)$, where $\alpha > 0$. Note that $x_c(t)$ is just a boxcar starting at $t = 0$ and ending at $t = T$, whereas $h_c(t)$ is a decaying exponential for $t \geq 0$ and zero otherwise.

Solution

Sketches of $x_c(t)$ and $h_c(t)$ appear, respectively, in Figure 2.15a and Figure 2.15b. In this case, (2.27a) is more convenient than (2.27b):

$$y_c(t) = \int_0^T e^{-\alpha(t-\tau)}u_c(t - \tau)d\tau. \tag{2.28}$$

Notice that $u_c(t - \tau)$ is nonzero (and unity) only for $\tau < t$. For $t < 0$, there is no τ on $[0, T]$ that is less than t, so $y_c(t < 0) = 0$. For $0 < t < T$, the step function in the integrand changes the upper limit to t:

$$y_c(t) = \int_0^t e^{-\alpha(t-\tau)}d\tau = \frac{e^{-\alpha t}(e^{\alpha t} - 1)}{\alpha} = \frac{1 - e^{-\alpha t}}{\alpha}. \tag{2.29}$$

For all $t > T$, the upper limit remains fixed at T:

$$y_c(t) = \int_0^T e^{-\alpha(t-\tau)}d\tau = e^{-\alpha t} \cdot \frac{e^{\alpha T} - 1}{\alpha}. \tag{2.30}$$

Note that both (2.29) and (2.30) give $(1 - e^{-\alpha T})/\alpha$ at $t = T$, which is a good check on our work. A sketch of $y_c(t)$ appears in Figure 2.15c.

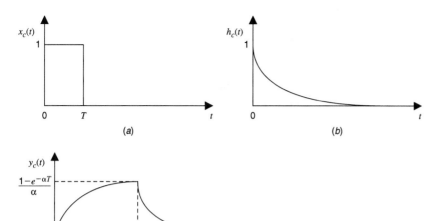

Figure 2.15

EXAMPLE 2.8

Find $y_c(t) = x_c(t) *_c h_c(t)$ if $x_c(t) = \sin(\Omega_0 t)$ and $h_c(t) = e^{-\alpha|t|}$, with $\alpha > 0$ and both functions defined for all t.

Solution

This time, let us use (2.27b). In addition, it will be convenient to use the complex sinusoid representation

$$\sin\{\Omega_0(t - \tau)\} = \frac{1}{2j}\{e^{j\Omega_0(t-\tau)} - e^{-j\Omega_0(t-\tau)}\}. \tag{2.31}$$

Thus

$$y_c(t) = \frac{1}{2j}\left\{\int_{-\infty}^{0} \left[e^{j\Omega_0 t}e^{(\alpha-j\Omega_0)\tau} - e^{-j\Omega_0 t} e^{(\alpha+j\Omega_0)\tau}\right] d\tau \right.$$
$$\left. + \int_{0}^{\infty} \left[e^{j\Omega_0 t} e^{(-\alpha-j\Omega_0)\tau} - e^{-j\Omega_0 t} e^{(-\alpha+j\Omega_0)\tau}\right] d\tau\right\}. \tag{2.32a}$$

Performing the simple integrations gives

$$y_c(t) = \frac{1}{2j}\left\{\frac{e^{j\Omega_0 t}}{\alpha - j\Omega_0} - \frac{e^{-j\Omega_0 t}}{\alpha + j\Omega_0} + \frac{-e^{j\Omega_0 t}}{-\alpha - j\Omega_0} + \frac{e^{-j\Omega_0 t}}{-\alpha + j\Omega_0}\right\}, \tag{2.32b}$$

and, upon making a common denominator,

$$y_c(t) = \frac{2\alpha}{2j}\frac{e^{j\Omega_0 t} - e^{-j\Omega_0 t}}{\alpha^2 + \Omega_0^2} \tag{2.32c}$$

$$= \frac{2\alpha}{\alpha^2 + \Omega_0^2} \sin(\Omega_0 t). \tag{2.32d}$$

Equation (2.32d) shows that this example is another special case of the result that sending a sinusoid into an LSI system modifies only its amplitude and phase by the Fourier transform of the system impulse response evaluated at the input sinusoidal frequency Ω_0. In this case, there is zero phase change from input to output, and the input magnitude is scaled by $2\alpha/(\alpha^2 + \Omega_0^2)$. Our first encounter of this magnitude-scaling of sinusoids into LSI systems was in Example 2.3, where the result was obtained in a numerical fashion via the differential equation as well as phasor analysis. The cause for zero added phase is traced to $h_c(t)$ being a real, even function of t, a property we consider in Section 5.9.

Although a systematic discussion the Laplace and Fourier transforms is given in Chapter 4, it is assumed the reader has a basic working knowledge of them from a signals and systems course. In light of the convolution relation (2.27), it is now appropriate to review the convolution theorem. Let us take the bilateral Laplace transform of both sides of (2.27b):

$$Y_c(s) = LT\{y_c(t)\} \tag{2.33a}$$

$$= \int_{-\infty}^{\infty} \left\{ \int_{-\infty}^{\infty} x_c(t - \tau) h_c(\tau) d\tau \right\} e^{-st} \, dt, \tag{2.33b}$$

which upon reversal of the order of integrations becomes

$$Y_c(s) = \int_{-\infty}^{\infty} h_c(\tau) \int_{-\infty}^{\infty} x_c(t - \tau) e^{-st} dt \, d\tau. \tag{2.33c}$$

Defining a new integration variable $v = t - \tau$,

$$Y_c(s) = \int_{-\infty}^{\infty} h_c(\tau) \int_{-\infty}^{\infty} x_c(v) e^{-s(v+\tau)} \, dv \, d\tau. \tag{2.33d}$$

Because $e^{-s\tau}$ is independent of v, it can be pulled outside the v integral:

$$Y_c(s) = \int_{-\infty}^{\infty} h_c(\tau) \, e^{-s\tau} \, d\tau \int_{-\infty}^{\infty} x_c(v) \, e^{-sv} \, dv \tag{2.33e}$$

$$= H_c(s)X_c(s), \tag{2.33f}$$

where (2.33f) follows from the definition of the Laplace transform. The analogous relation for Fourier transforms is obtained by setting $s = j\Omega$:

$$Y_c(j\Omega) = H_c(j\Omega)X_c(j\Omega). \tag{2.34}$$

The simplicity of (2.33f) and (2.34) is clear; we can use them to avoid having to perform the convolution integral. Specifically, if the Laplace/Fourier transforms of both $h_c(t)$ and $x_c(t)$ are known or easily found, just multiply them and take the inverse Laplace/Fourier transform of the result to obtain $y_c(t)$.

More important, (2.34) is the conceptual basis of linear filter analysis and design. In simplest terms, it says that to boost/reduce the energy or power of a given signal $x_c(t)$ at a desired frequency (or frequency range), just apply $x_c(t)$ to a filter whose magnitude is relatively large/small at that frequency (or range). Then (2.34) shows that the desired frequency will be boosted/reduced in the output signal $y_c(t)$ and the undesired frequencies attenuated. Equivalently, (2.34) is the basis of phasor analysis, as used in the last part of Example 2.3.

The system impulse response $h_c(t)$ can be measured experimentally by presenting an impulse-approximating signal to the input and measuring the output waveform $y_c(t) \approx h_c(t)$. Alternatively, its Fourier transform $H_c(j\Omega)$ can be measured by presenting a ramp input to a voltage-to-frequency converter whose swept-frequency sinusoidal output is in turn presented as the system input. Or, white noise can be used as the input; then, using methods presented in Chapter 10, one may estimate $|H_c(j\Omega)|^2$. As yet another alternative, if the model coefficients in the differential equation (2.20) are known, the Laplace transform of (2.20) gives

$$H_c(s) = \frac{Y_c(s)}{X_c(s)} = \frac{B_c(s)}{A_c(s)}, \qquad (2.35)$$

where

$$A_c(s) = \sum_{\ell=0}^{N} a_\ell s^\ell \text{ and } B_c(s) = \sum_{\ell=0}^{M} b_\ell s^\ell. \qquad (2.36)$$

Then the impulse response is obtained as $h_c(t) = \text{LT}^{-1}\{H_c(s)\}$.

2.3.4   Causality

Physical systems (including electronics) are reactive to, as opposed to anticipatory to, their inputs. Thus any response to an input $x_c(t_1)$ occurring at time t_1 cannot begin occurring until $t = t_1$ at the earliest. This behavior is causality. In terms of the convolution relation (2.27a), we guarantee that $x_c(t_1)$ cannot affect any $y_c(t)$ value for which $t < t_1$ by requiring that its coefficient, $h_c(t-\tau)\big|_{\tau = t_1} = h_c(t - t_1)$, be zero. But $t < t_1$ implies that $\alpha \equiv t - t_1 < 0$, so we thus require that $h_c(\alpha < 0) = 0$. Switching from α to a generic t, we have

$$h_c(t < 0) = 0 \text{ for causality.} \qquad (2.37)$$

For example, we already saw that $h_c(t)$ for the causal system shown in Figure 2.13b is zero for negative argument.

Although causality generally holds for physical temporal signal-processing systems, there is generally no such restriction for spatial processing systems. In spatial processing, $t < t_1$ translates to, for example, $x < x_1$. Clearly there is no fundamental difference between $x < x_1$ ("left") and $x > x_1$ ("right") the way there is between "past" and "future."[3] Moreover, ideal temporal filters, from which physical approximating filters are derived, are often bicausal [note that $h_c(t)$ in Example 2.8 is a bicausal system]. Thus anyone in signal processing should be comfortable working with both causal and bicausal or anticausal systems.

2.3.5   An Interpretation of Initial Conditions

Another common temporal restriction on signal processing systems is that nonzero inputs be restricted to $t > t_1$. That is, someone turns on the input signal at $t = t_1$ and before that, $x_c(t) = 0$. Often, the time variable t is defined such that $t_1 = 0$, so the inputs are nonzero only for $t > 0$.[4] Under this restriction, solution for the output in terms of the input and initial conditions is known as an initial value problem. The system output, and possibly various-order derivatives, are measured or otherwise

---

[3] In spatial systems, x and y are spatial coordinates, not input and output! We will continue to use x and y as input and output symbols in the following discussion.
[4] Defining the reference time $t_1 = 0$ is the same as defining ground (reference potential) as 0 V in a circuit. In both cases, only relative time or potential has real meaning. Absolute time $t = 0$ occurred long before any of us was born.

known at $t = 0$, and it is desired to determine the output for $t > 0$ given these initial values and the subsequent input values.

In (2.33), the bilateral Laplace transform was used. Alternatively, the unilateral Laplace transform could be used, with appropriate initial conditions on $y_c(t)$ and its derivatives. It is interesting to note that in either case, $h_c(t)$ and $H_c(s)$ are the same. The system impulse response describes the interaction between an input—not initial conditions—and the output.

For cases in which $x_c(t < 0) \neq 0$, the unilateral Laplace transform will still give correct results if the effects of $x_c(t < 0)$ on $y_c(t)$ and its derivatives are evaluated for $t = 0$, and used as "initial conditions" and if the input is then redefined to be zero for $t < 0$. Thus initial conditions may be viewed as the result of inputs occurring before $t = 0$, which physically must have occurred in order for those nonzero initial conditions to have arisen. Note that the convolution relation (2.27) is the output in the absence of initial conditions and so is called the zero-state response. We see that there is a philosophical or applications preference at play in the interpretation of "initial conditions" versus "inputs for $t < 0$."

2.3.6   Stability

The basic definition of stability in common usage is that a bounded input to a system always produces a bounded output. Specifically, a continuous-time system $h_c(t)$ is bounded-input, bounded-output (BIBO) stable if and only if for *any* and *all* $x_c(t)$ such that $|x_c(t)| < M_x < \infty$ for all t, then $|y_c(t)| < M_y < \infty$ for all t. The output magnitude is

$$|y_c(t)| = |x_c(t) *_c h_c(t)|, \tag{2.38a}$$

which by the convolution relation (2.27b) can be written

$$|y_c(t)| = \left| \int_{-\infty}^{\infty} h_c(\tau) x_c(t - \tau)\, d\tau \right|. \tag{2.38b}$$

Because $|a + b| \leq |a| + |b|$, it follows from (2.38b) that

$$|y_c(t)| \leq \int_{-\infty}^{\infty} |h_c(\tau) x_c(t - \tau)|\, d\tau \tag{2.38c}$$

$$= \int_{-\infty}^{\infty} |h_c(\tau)|\, |x_c(t - \tau)|\, d\tau. \tag{2.38d}$$

Because $|x_c(t)| < M_x$, the integral in (2.38d) cannot exceed the value it would have if M_x were substituted for all $|x_c(t - \tau)|$. Thus $|y_c(t)| \leq M_x \int_{-\infty}^{\infty} |h_c(\tau)|\, d\tau$.

> Now, if $\int_{-\infty}^{\infty} |h_c(\tau)|\, d\tau < \infty$, it follows that $|y_c(t)| < \infty \rightarrow$ system stable.

Thus we have established the "if" part of the proof; now we develop the "only if" part. Suppose we choose $x_c(t)$ such that $x_c(t - \tau) = |h_c(\tau)|/h_c(\tau)$ so that $h_c(\tau) x_c(t - \tau) = |h_c(\tau)|$ for $h_c(\tau) \neq 0$, and zero otherwise. Note that $\{x_c(t), -\infty < t < \infty\}$ will be a different waveform for each value of t selected for evaluation of $y_c(t)$, which is immaterial to the proof because we need prove violation only for any one input for any one computed time t. Clearly, $|x_c(t)|$ is bounded, because it has magnitude 1 (or 0). Then in this case, $y_c(t) = \int_{-\infty}^{\infty} |h_c(t)|\, dt$; thus if $\int_{-\infty}^{\infty} |h_c(t)|\, dt$ is unbounded, so will be $|y_c(t)|$—and if that is true for even this one input waveform $x_c(t)$, then $h_c(t)$ is not a BIBO stable system. Thus for BIBO stability, $h_c(t)$ must be an absolutely integrable waveform ("absolutely" because of the absolute value bars).

Any text on continuous-time signal processing will use either the above result or other arguments to show that equivalently a causal continuous-time system is stable if and only if all the poles of $H_c(s)$ are in the left-half plane. After having reviewed complex variables in Chapter 3, we will be in a position to use the above time-domain proof on BIBO stability to infer the equivalent pole-location criterion (left-half plane).

EXAMPLE 2.9

(a)   An LSI system has impulse response $h_c(t) = te^{-t}u_c(t)$. Is the system BIBO stable?

(b)   Repeat part (a) for $h_c(t) = e^{2t}u_c(t)$.

Solution

(a) $$\int_{-\infty}^{\infty} |h_c(\tau)|\, d\tau = \int_0^{\infty} |\tau e^{-\tau}|\, d\tau = -\tau e^{-\tau}\Big|_0^{\infty} + \int_0^{\infty} e^{-\tau}\, d\tau = 0 - 0 - e^{-\tau}\Big|_0^{\infty} = 1 < \infty,$$

where to integrate $\tau e^{-\tau}$ we used integration by parts, with $u = \tau$, $du = d\tau$, $dv = e^{-\tau}\, d\tau$, and $v = -e^{-\tau}$. Because the result of integration is finite, $h_c(t)$ is absolutely integrable and the system is BIBO stable. We note, from a transform point of view, that $H_c(s) = 1/(s + 1)^2$ has a double pole at $s = -1$. Because both of these poles are in the left-half plane, the system is stable.

(b) $$\int_{-\infty}^{\infty} |h_c(\tau)|\, d\tau = \int_0^{\infty} |e^{2\tau}|\, d\tau = \frac{1}{2}e^{2\tau}\Big|_0^{\infty} = \infty.$$

Thus $h_c(t)$ is not absolutely integrable, and the system is unstable. From a transform viewpoint, $H_c(s) = 1/(s - 2)$ has a pole at $s = 2$, in the right-half plane.

EXAMPLE 2.10

An LSI system has the system function $H_c(s) = 1/s^2$. Is the system stable?

Solution

No. Let us argue again from the time domain by noting that the inverse Laplace transform of $H_c(s)$ is $h_c(t) = tu_c(t)$. Even without integrating, we see that $h_c(t)$ blows up as $t \to \infty$, and so we suspect that $h_c(t)$ is not absolutely integrable. We check as follows:

$$\int_{-\infty}^{\infty} |h_c(\tau)|\, d\tau = \int_0^{\infty} \tau\, d\tau = \frac{t^2}{2}\Big|_0^{\infty} = \infty.$$

Thus the system is unstable due to the double pole at the origin, even though one might erroneously think that the system is *metastable* or marginally stable. This is the case for only simple (single) poles on the $j\Omega$-axis; the simple poles still fail the BIBO test, but at least their $h_c(t)$ does not grow without bound for $t \to \infty$ as does $t^\ell u_c(t)$.

2.4 SYSTEM CHARACTERIZATION AND SOLUTION: DISCRETE TIME

2.4.1 Linearity

Most of the same considerations made in the discussion of linearity for continuous-time systems also apply for discrete-time systems. For example, just as most continuous-time systems are nonlinear at an exact level, so are most discrete-time systems. Also, exactly the same condition for linearity applies as stated in (2.19) for continuous time, but it is now restated with the discrete-time variable n replacing time t. A system $y(n) = L\{x(n)\}$ is linear if, given that $y_1(n) = L\{x_1(n)\}$ and $y_2(n) = L\{x_2(n)\}$, it follows that

$$L\{\alpha_1 x_1(n) + \alpha_2 x_2(n)\} = \alpha_1 y_1(n) + \alpha_2 y_2(n). \quad \text{Condition for a linear system.} \quad (2.39)$$

EXAMPLE 2.11

   (a) If $y(n) = 30/x(n)$, is the system linear? Justify your answer.
   (b) Repeat part (a) for $y(n) = \cos(10n)x(n-1) + e^{e^n}x(n-12)$.
   (c) Repeat part (a) for $y(n) = 10 + 7x(n-4)$.

Solution

   (a) Without even formally applying the condition (2.39) for linearity, we note that doubling $x(n)$ halves $y(n)$, rather than doubling it as would happen in a linear system. We immediately conclude that the system is nonlinear. (Doubling is rigorous only for showing nonlinearity via counterexample; proving linearity rigorously requires proving satisfaction of (2.39) for two possibly *nonequal* inputs.)

   (b) This time we will apply (2.39). Let
$$y_1(n) = \cos(10n)x_1(n-1) + e^{e^n}x_1(n-12)$$
and $\quad y_2(n) = \cos(10n)x_2(n-1) + e^{e^n}x_2(n-12).$
Then the output $y_{12}(n)$ due to an input $\alpha_1 x_1(n) + \alpha_2 x_2(n)$ is
$$y_{12}(n) = \cos(10n)\{\alpha_1 x_1(n-1) + \alpha_2 x_2(n-1)\} + e^{e^n}\{\alpha_1 x_1(n-12) + \alpha_2 x_2(n-12)\}$$
$$= \alpha_1 y_1(n) + \alpha_2 y_2(n),$$
so we conclude the system is linear even though the coefficients of $x(n-1)$ and $x(n-12)$ are highly nonlinear functions of n.

   (c) Strictly, no. Doubling $x(n)$ does not double $y(n)$ due to the term 10. The proper term for this relation is *affine*.

An ordinary linear differential equation is the temporal model for a continuous-time linear system. An Nth-order linear difference equation is the temporal model for a discrete-time linear system:

$$\sum_{\ell=0}^{N} a_\ell\, y(n-\ell) = \sum_{\ell=0}^{M} b_\ell\, x(n-\ell). \quad (2.40)$$

Proof that (2.40) satisfies the condition (2.39) for linearity proceeds exactly as in the continuous-time case and is left as an exercise.

It might be asked why the linear input–output relation is a difference equation rather than some other sort of relation. Indeed, why is (2.40) called a "difference" equation when there seem to be only added terms in it? The answer to these questions becomes plain when we introduce discrete-time signals and systems in the most straightforward way: discretization of continuous-time signals and systems. In the study of simulation (see Section 6.4) and other applications, the derivative is most simply approximated by a *first difference* or *backward difference*. Because

$$\frac{d}{dt} f_c(t) = \lim_{\Delta t \to 0} \frac{f_c(t) - f_c(t - \Delta t)}{\Delta t}, \tag{2.41}$$

we can merely let Δt be small and finite, and in place of the derivative, we can use the right-hand side of (2.41), a scaled difference. Thus evaluating at $t = n \Delta t$ gives

$$\left. \frac{d}{dt} f_c(t) \right|_{t = n \Delta t} \approx \frac{f_c(n \Delta t) - f_c((n-1)\Delta t)}{\Delta t} \tag{2.42a}$$

or

$$\left. \frac{d}{dt} f_c(t) \right|_{t = n \Delta t} \approx \frac{f(n) - f(n-1)}{\Delta t}, \tag{2.42b}$$

where $f(n) = f_c(n \Delta t)$.[5]

We could similarly construct an approximation of the second derivative by just taking the first difference of the first difference:

$$\left. \frac{d^2}{dt^2} f_c(t) \right|_{t = n \Delta t} \approx \left\{ \frac{f_c(n \Delta t) - f_c((n-1)\ \Delta t)}{\Delta t} - \frac{f_c((n-1)\ \Delta t) - f_c((n-2)\ \Delta t)}{\Delta t} \right\} / \Delta t \tag{2.43a}$$

$$= \frac{f_c(n \Delta t) - 2 f_c((n-1)\Delta t) + f_c((n-2)\ \Delta t)}{\Delta t^2}, \tag{2.43b}$$

or, with $f(n) = f_c(n \Delta t)$,

$$\left. \frac{d^2}{dt^2} \{ f_c(t) \} \right|_{t = n \Delta t} \approx \frac{f(n) - 2f(n-1) + f(n-2)}{\Delta t^2}, \tag{2.43c}$$

which is known as the central difference approximation.

An important point is that the first difference approximation of the first derivative involved a weighted sum of $f(n)$ and $f(n-1)$, whereas the second derivative approximation involved a weighted sum of $f(n)$, $f(n-1)$, and $f(n-2)$. An approximation of d^m/dt^m involves $f(n), f(n-1), \ldots, f(n-m)$. Moreover, the model differential equation (2.20) involves weighted sums of d^ℓ/dt^ℓ; it may thus be seen that a discretization of (2.20) involves a weighted sum of $y(n), y(n-1), \ldots, y(n-N)$ and a weighted sum of $x(n), \ldots, x(n-M)$. Thus under the differencing, the differential equation (2.20) becomes a "difference" equation (2.40). The coefficients $\{a_\ell, b_\ell\}$ in (2.40) are obviously different from the $\{a_\ell, b_\ell\}$ of the original continuous-time system (2.20), but were given the same symbols here for notational brevity. In Example 2.12 below, however, we will prime the coefficients of the discretized system to prevent any confusion.

---

[5] It should be noted that when $f(n) = f_c(n \Delta t)$, n is also sometimes referred to as "time" instead of the correct term "sample index" or "time index," even though n is dimensionless.

It should now be clear how the term *difference equation* came about, despite the apparent lack of explicit "differences" in (2.40), and also how each of the terms in each of the difference sums in (2.40) arises under a given discretization approximation or algorithm. If we set $x(n) = x_c(n \Delta t)$, then depending on the accuracy of the discretization, we hope that the solution $y(n)$ of the difference equation (2.40) will satisfy $y(n) \approx y_c(n \Delta t)$.

EXAMPLE 2.12

Apply the backward difference discretization above to the differential equation in Example 2.2, and solve it in the steady state (no transients). Use the same component values and samples of the same input used in Example 2.3.

Solution

Recall that in Example 2.2 we obtained the form (2.11):

$$\left\{ \frac{d^2}{dt^2} + a_1 \frac{d}{dt} + a_0 \right\} v_{RL,c}(t) = \left\{ b_1 \frac{d}{dt} + b_0 \right\} v_{s,c}(t).$$

Letting $t = n \Delta t$ and substituting the differentiation approximations into (2.11) with $v_{RL}(n) \approx v_{RL,c}(n \Delta t)$ and $v_s(n) = v_{s,c}(n \Delta t)$, we obtain the difference equation

$$\frac{v_{RL}(n) - 2v_{RL}(n-1) + v_{RL}(n-2)}{\Delta t^2} + a_1 \frac{v_{RL}(n) - v_{RL}(n-1)}{\Delta t} + a_0 v_{RL}(n)$$

$$= b_1 \frac{v_s(n) - v_s(n-1)}{\Delta t} + b_0 v_s(n), \tag{2.44a}$$

where due to the discretization of derivatives, for equality in this difference equation we actually have only $v_{RL}(n) \approx v_{RL,c}(n \Delta t)$, not $v_{RL}(n) = v_{RL,c}(n \Delta t)$. Collecting terms in (2.44a) gives

$$\left\{ \frac{1}{\Delta t^2} + \frac{a_1}{\Delta t} + a_0 \right\} v_{RL}(n) + \left\{ \frac{-2}{\Delta t^2} - \frac{a_1}{\Delta t} \right\} v_{RL}(n-1) + \frac{1}{\Delta t^2} v_{RL}(n-2) \tag{2.44b}$$

$$= \left\{ \frac{b_1}{\Delta t} + b_0 \right\} v_s(n) - \frac{b_1}{\Delta t} v_s(n-1),$$

or, equivalently,

$$v_{RL}(n) + a_1' v_{RL}(n-1) + a_2' v_{RL}(n-2) = b_0' v_s(n) + b_1' v_s(n-1), \tag{2.44c}$$

where

$$a_1' = \frac{-2/\Delta t^2 - a_1/\Delta t}{\alpha}, \quad a_2' = \frac{1}{\Delta t^2 \alpha}; \quad b_0' = \frac{b_1/\Delta t + b_0}{\alpha}, \quad b_1' = \frac{-b_1}{\Delta t \alpha}, \tag{2.45a}$$

where

$$\alpha = \frac{1}{\Delta t^2} + \frac{a_1}{\Delta t} + a_0. \tag{2.45b}$$

Equations (2.44) and (2.45) are valid for $N = 2, M = 1$, and any values of $\{a_{\ell \neq 2}, b_\ell, \Delta t\}$ that one might be given. Notice how quickly the coefficients become complicated; this situation is typical of discretizations when variables are kept symbolic

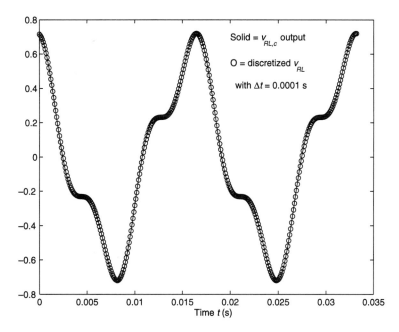

Figure 2.16*a*

rather than given numerical values. Programs like Matlab become a crucial component of verifying one's work. The important point to note in (2.45) is that the difference equation coefficients depend on the sampling interval Δt.

To solve the difference equation for $v_s(n) = v_{sc}(n\,\Delta t) = \cos(2\pi 60n\,\Delta t) + 0.2\cos(2\pi 180n\,\Delta t)$, we use the same technique that we used in Example 2.3: Let

$$v_{RL}(n) = A_1\cos(2\pi 60n\,\Delta t) + B_1\sin(2\pi 60n\,\Delta t) + A_2\cos(2\pi 180n\,\Delta t) + B_2\sin(2\pi 180n\,\Delta t). \quad (2.46)$$

When $v_{RL}(n)$ is substituted into the difference equation (2.44c), trigonometric identities are applied, and the terms collected,[6] the result is a matrix equation for each of the two frequencies, expressed here in one equation by letting the frequency be a variable F:

$$\begin{bmatrix} 1 + a_1'\cos(2\pi F\Delta t) + a_2'\cos(4\pi F\Delta t) & -a_1'\sin(2\pi F\Delta t) - a_2'\sin(4\pi F\Delta t) \\ a_1'\sin(2\pi F\Delta t) + a_2'\sin(4\pi F\Delta t) & 1 + a_1'\cos(2\pi F\Delta t) + a_2'\cos(4\pi F\Delta t) \end{bmatrix} \begin{bmatrix} A \\ B \end{bmatrix}$$
$$= \begin{bmatrix} b_0' + b_1'\cos(2\pi F\Delta t) \\ b_1'\sin(2\pi F\Delta t) \end{bmatrix}, \quad (2.47)$$

which is easily inverted for A and B, where $\{A = A_1, B = B_1\}$ for $F = 60$ Hz and $\{A = A_2, B = B_2\}$ for $F = 180$ Hz.

---

[6] The coefficients of $\cos(2\pi 60n\,\Delta t)$ in the difference equation must individually be matched, as must those of $\sin(2\pi 60n\,\Delta t)$, $\cos(2\pi 180n\,\Delta t)$, and $\sin(2\pi 180n\,\Delta t)$. The 60- and 180-Hz solutions are decoupled, so we solve the general problem $v_{s,A}(n) = V\cos(\Omega n\,\Delta t)$, where later for Ω we will substitute $\Omega = \Omega_0 = 2\pi \cdot 60$ rad/s and $\Omega = 3\Omega_0 = 2\pi 180$ rad/s. Let $v_{L,A}(n) = A\cos(\Omega n\,\Delta t) + B\sin(\Omega n\,\Delta t)$ so that $v_{L,A}(n-1)$ $= A\cos\{\Omega(n-1)\Delta t\} + B\sin\{\Omega(n-1)\Delta t\} = A[\cos(\Omega n\,\Delta t)\cos(\Omega\Delta t) + \sin(\Omega n\,\Delta t)\sin(\Omega\Delta t)]$ $+ B[\sin(\Omega n\,\Delta t)\cos(\Omega\Delta t) - \cos(\Omega n\,\Delta t)\sin(\Omega\Delta t)]$ and, similarly, $v_{L,A}(n-2) = A[\cos(\Omega n\,\Delta t)\cos(2\Omega\Delta t)$ $+ \sin(\Omega n\,\Delta t)\sin(2\Omega\Delta t)] + B[\sin(\Omega n\,\Delta t)\cos(2\Omega\Delta t) - \cos(\Omega n\,\Delta t)\sin(2\Omega\Delta t)]$. Also, $v_{s,A}(n-1)$ $= V[\cos(\Omega n\,\Delta t)\cos(\Omega\Delta t) + \sin(\Omega n\,\Delta t)\sin(\Omega\Delta t)]$. Substituting into the difference equation, we first match coefficients of $\cos(\Omega n\,\Delta t)$: $1 \cdot A + a_1'[A\cos(\Omega\Delta t) - B\sin(\Omega\Delta t)] + a_2'[A\cos(2\Omega\Delta t) - B\sin(2\Omega\Delta t)]$ $= V[b_0' + b_1'\cos(\Omega\Delta t)]$. Matching the coefficients of $\sin(\Omega\Delta t)$ gives $1 \cdot B + a_1'[A\sin(\Omega\Delta t) + B\cos(\Omega\Delta)t] + a_2'[A\sin(2\Omega\Delta t) + B\cos(2\Omega\Delta t)] = Vb_1'\cos[\Omega\Delta t]$. In the text, we use $\Omega = 2\pi F$.

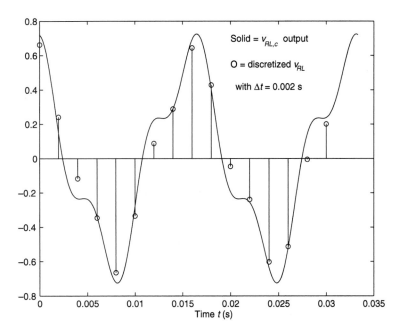

Solid = $v_{RL,c}$ output

O = discretized v_{RL}

with $\Delta t = 0.002$ s

Time t (s)

Figure 2.16b

We now substitute the component values and compute $v_{RL}(n)$ in (2.46). If a very small value for Δt is selected, such as 0.1 ms (or less), the results for (A_1, B_1) and (A_2, B_2) are about the same as in Example 2.3; therefore, $v_{RL}(n) \approx v_{RL,c}(n\,\Delta t)$. Thus in this case, the signals are indistinguishable from those in Example 2.3, as shown in Figure 2.16a. If a more moderate value of Δt is used, say 2 ms, the results are as shown in Figure 2.16b. In this latter case, there is significant error due to an insufficient sampling rate (Δt too large). We expect the results to improve for small Δt because, after all, the difference derivative approximations [(e.g., (2.42b)] used in discretizing the differential equation are exact only in $\lim_{\Delta t \to 0}$ [see, e.g., (2.41)].

EXAMPLE 2.13

Without using transforms, solve the difference equation $y(n) - (2/5)y(n-1) = x(n)$, where $x(n) = \{3(2/5)^n + 7(1/2)^n\}u(n)$ and $y(0) = 6$. Plot the results using Matlab for $n \in [0, 30]$.

Solution

We solve in three steps. **(a)** Solve the homogeneous difference equation and do not solve for the unknown constants in the homogeneous solution. **(b)** Solve the particular difference equation and do solve for the unknown constants in the particular solution. **(c)** Add the homogeneous and particular solutions and solve for the unknown homogeneous-solution constants by satisfaction of the initial conditions.

We perform these in turn.

(a) $y_h(n) - (2/5)y_h(n-1) = 0$. Try $y_h(n) = A\alpha^n$. Why? The rationale is, in general, as follows. In $\sum_m a_m y_h(n-m) = 0$, if we substitute $y_h(n) = \alpha^n$, we obtain

$$\sum_m a_m \alpha^{n-m} = \left\{ \sum_m a_m \alpha^{-m} \right\} \alpha^n = 0.\text{ Because } \alpha \neq 0, \text{ we must have } \sum_m a_m \alpha^{-m} = 0,$$

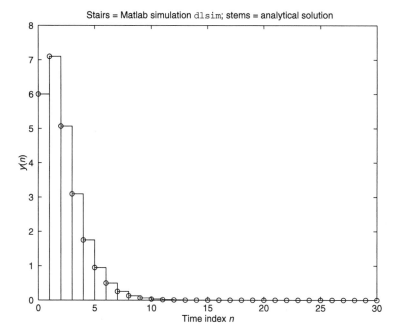

Stairs = Matlab simulation `dlsim`; stems = analytical solution

Figure 2.17

which is called the characteristic equation. It has as many solutions for α as the order of the difference equation. Thus $A\alpha^n$ is a solution of the homogeneous difference equation as long as α is a solution of the characteristic equation. In our example, substituting $y_h(n)$ gives $A\{\alpha^n - (2/5)\alpha^{n-1}\} = 0$, or $1 - (2/5)\alpha^{-1} = 0$, or $\alpha = 2/5$. So $y_h(n) = A(2/5)^n$.

(b) Normally, we *would* try $y_p(n) = c_1(2/5)^n + c_2(1/2)^n$, but we see that $(2/5)^n$ is a mode of $y_h(n)$, so we instead try $y_p(n) = c_1 n(2/5)^n + c_2(1/2)^n$. (In general, use $n^{\ell}\alpha^n$, where ℓ = multiplicity of α^n in the homogeneous solution.) Substituting gives $c_1 n(2/5)^n + c_2(1/2)^n - (2/5)[c_1(n-1)(2/5)^{n-1} + c_2(1/2)^{n-1}] = 3(2/5)^n + 7(1/2)^n$.

The left-hand side may be written as
$$[c_1 n - (2/5)(5/2)c_1 n + (2/5)(5/2)c_1](2/5)^n + c_2(1 - 4/5)(1/2)^n$$
$$= c_1(2/5)^n + (1/5)c_2(1/2)^n.$$

The terms proportional to n will always cancel because $2/5$ is a solution of the characteristic equation. Equating this result with the previous right-hand side, we determine that $c_1 = 3, c_2 = 35$. Thus $y_p(n) = 3n(2/5)^n + 35(1/2)^n$.

(c) $y(n) = [A + 3n](2/5)^n + 35(1/2)^n$. Now, $y(0) = A + 35 = 6$, so $A = -29$. Thus $y_h(n) = -29(2/5)^n$ and $y(n) = (3n-29)(2/5)^n + 35(1/2)^n$.

In Figure 2.17, we depict the solution using both our analytical result (stems) and the result of calling Matlab's `dlsim` (solid curve/stairs). The use of `dlsim` will be clearer after reading Section 2.4.6.3 (page 59ff) and (2.73). Specifically, for later reference, $H(z) = z/(z - 2/5)$, so we call $[a,b,c,d] = $ `tf2ss([1 0],[1 -2/5])` to obtain $w(n + 1) = 0.4w(n) + 1 \cdot x(n)$, $y(n) = 0.4w(n) + 1 \cdot x(n)$, so $y(0) = 0.4w(0) + 3 + 7 = 6$, giving $w(0) = -10$, from which we call `dlsim(a,b,c,d,3.*(2/5).^n+7.*(1/2).^n,-10)`. The two methods agree exactly.

In Example 2.13, we used the homogeneous and particular solutions in the classical *undetermined coefficients* approach. A different decomposition of solutions is used more commonly by electrical engineers making use of convolution, z-transform, and state–space approaches: the zero-state response (ZSR) and the zero-input response (ZIR). What are these latter solution components, and how are they related to the homogeneous and particular solutions? From their names, it would appear that the ZIR and the homogeneous solution are related and that the ZSR and the particular solution are related. The relations are not identities, however.

Consider first the homogeneous solution. We know that in solving difference/differential equations, we set the input to zero, thus it might appear that the homogenous solution is the same thing as the ZIR. Recall, however, that in the procedure, the homogeneous solution leaves undetermined constants, whose values are later determined only when the particular solution involving inputs is added. Thus once determined, the homogeneous solution contains all the system modes in the quantities dictated by both the initial conditions and the input, that is, the entire amounts of the system modes. Contrarily, the ZIR by definition includes only that portion of the total solution that would exist for the given initial conditions if zero input were applied. So the homogeneous solution and the ZIR are definitely different because the input excites the system modes.

Next consider the particular solution. Recall that in solving the difference/differential equation, the particular solution involves only the input modes, not the system modes. However, as just noted, the input still excites the system modes, so the particular solution is not a complete solution, even if the initial conditions are all zero and even after the values of its unknown constants are determined. When the initial conditions are nonzero, they affect only the system modes. Contrarily, the ZSR (which, e.g., is the result of usual convolution) is an entire solution for the case of zero initial conditions, and includes the portion of system-mode output contributed by the input. Thus the particular solution and ZSR are unequal.

In conclusion, the ZIR and ZSR are total solutions for, respectively, the cases of (a) $x(n) = 0$ for all n and (b) all initial conditions are zero. The homogeneous solution is equal to the ZIR plus the portion of the ZSR due to system modes. The particular solution is the rest of the ZSR, due only to input modes. In an end-of-chapter problem in Chapter 3, Example 2.13 is re-solved using the z-transform, and the ZSR and ZIR are identified and compared with the particular and homogeneous solutions obtained in Example 2.13.

2.4.2   Time (or Shift) Invariance

As for continuous-time systems, time invariance arises when the coefficients a_ℓ and b_ℓ in the difference equation (2.40) are independent of time index n. The condition for time invariance is that if $y(n) = L\{x(n)\}$, then the output due to $x(n - n_0)$ is $y(n - n_0)$: $L\{x(n - n_0)\} = y(n - n_0)$. Again, the proof that (2.40) is time invariant is identical to that presented for continuous time, with suitable substitutions, and is helpful to carry out as an exercise.

Thus discrete-time systems satisfying a linear difference equation with constant coefficients are called linear time- or shift-invariant (LTI or LSI) systems.

EXAMPLE 2.14

Suppose that $x(n)$ and $y(n)$ satisfy the difference equation $y(n - 3) + 2y(n - 1) - 10y(n) = 5x(n - 1) + (\frac{1}{2})^n x(n)$. Is the system shift invariant?

Solution

No; the "time"-dependent coefficient $(\frac{1}{2})^n$ violates the requirement for shift invariance. It remains $(\frac{1}{2})^n$ even when we replace $x(n)$ by $x(n - n_0)$, which will cause more than a mere shift by n_0 in $y(n)$.

2.4.3   Solutions for General Inputs: Convolution

Just as is true for continuous-time systems, it is very helpful to have at hand the zero-state solution of the system difference equation (2.40) with a general nonzero input driving sequence $x(n)$. Again use is made of a delta function, but this time it is the Kronecker delta sequence, $\delta(n)$, which is unity for $n = 0$ and zero otherwise. From this definition it follows that

$$1 = \sum_{m=-\infty}^{\infty} \delta(m), \tag{2.48a}$$

or, equivalently, for any integer n,

$$1 = \sum_{m=-\infty}^{\infty} \delta(n - m). \tag{2.48b}$$

Multiply both sides of (2.48b) by $x(n)$ to obtain

$$x(n) = x(n) \sum_{m=-\infty}^{\infty} \delta(n - m), \tag{2.49a}$$

or, given that $\delta(m \neq n) = 0, x(n)$ can be placed inside the sum without changing the result:

$$x(n) = \sum_{m=-\infty}^{\infty} x(m)\delta(n - m). \tag{2.49b}$$

As was true for (2.25b) $\left[x_c(t) = \int_{-\infty}^{\infty} x_c(\tau)\delta_c(t - \tau)\,d\tau\right]$, (2.49b) can be viewed either as an expansion of $x(n)$ over shifted Kronecker delta functions or as the sampling property of $\delta(n)$. Now if $y(n) = L\{x(n)\}$ where L is an LSI system, then by (2.49b) we may write

$$y(n) = L\left\{ \sum_{m=-\infty}^{\infty} x(m)\,\delta(n-m) \right\}, \tag{2.50a}$$

which, because of the assumed linearity of L, is equivalent to

$$y(n) = \sum_{m=-\infty}^{\infty} x(m)L\{\delta(n - m)\}. \tag{2.50b}$$

The system output when the input is $\delta(n)$ is called the unit sample response, and it is analogous to the impulse response of continuous-time systems. Let the unit sample response, $L\{\delta(n)\}$, be designated $h(n)$: $h(n) = L\{\delta(n)\}$. Then, by the shift-invariance property, we have $L\{\delta(n - m)\} = h(n - m)$, so that (2.50b) is equivalent to

$$y(n) = \sum_{m=-\infty}^{\infty} x(m)h(n - m) \tag{2.51a}$$

$$= \sum_{m=-\infty}^{\infty} x(n - m)h(m) \tag{2.51b}$$

$$= x(n) * h(n), \tag{2.51c}$$

where (2.51b) follows from (2.51a) by the change of variables $m_{\text{new}} \equiv n - m_{\text{old}}$ and * denotes discrete-time convolution. Again we see that convolution, here discrete-time, is a direct consequence of the system properties of linearity and shift invariance. Also, both views of convolution presented in Section 2.3.3 [in discrete-time notation, tracing $x(n_1)$ contributions and summing entire weighted sequences $h(m)x(n - m)$] apply to discrete-time convolution as well.

The well-known convolution theorem, which again plays a prominent role in digital signal-processing theory, will be derived in Section 5.7 in a systematic development of transform properties; the proof is very similar to that for continuous time in (2.33) but involves the z-transform rather than the Laplace transform. Also, suitable modifications of the comments at the end of Section 2.3.3 on the specialization of the convolution theorem to Fourier transforms and the importance of that relation for filter analysis and design apply for discrete-time systems.

EXAMPLE 2.15

Suppose $h(n) = 5\delta(n + 10) + 6^n u(n) + \sin(n)$. Is this system linear and shift invariant?

Solution

Yes, because $h(n)$ is the unit sample response, which satisfies linear convolution. Note, however, that it is not causal (see Section 2.4.4) due to the first and third terms, and it is unstable (see Section 2.4.5) due to the second and third terms (the sine is metastable).

EXAMPLE 2.16

Find $y(n) = x(n) * h(n)$ if $x(n)$ and $h(n)$ are as shown in Figure 2.18. Assume that both sequences are zero for all $n < 0$, that the plots continue in the steady state as shown, and that $h(n)$ is a single-term geometric sequence.

Solution

Write the given sequences in mathematical form: $x(n) = \delta(n - 1) - \delta(n - 2) + u(n - 5)$ and, noting that $h(0) = 1$ and $h(1) \approx 0.833 \approx 5/6$, we see that $h(n)$ follows the geometric pattern, $h(n) = (5/6)^n u(n)$. Recalling that $\delta(n - \ell) * f(n) = f(n - \ell)$ and using (2.51b) for the final term involving the step function, we have

$$y(n) = (5/6)^{n-1} u(n - 1) - (5/6)^{n-2} u(n - 2) + \sum_{m=0}^{\infty} (5/6)^m u(n - m - 5). \qquad (2.52)$$

A very common error is to forget to include the shifted unit step sequences. Because $u(\,\cdot\,)$ is nonzero only for nonnegative argument, the unit step sequence in the sum in (2.52) is nonzero only for $m \le n - 5$. Moreover, that condition can be satisfied for some value(s) of m on $[0, \infty]$ only when $n \ge 5$. Thus we must tack a $u(n - 5)$ onto our result for the sum:

$$y(n) = (5/6)^{n-1} u(n - 1) - (5/6)^{n-2} u(n - 2) + u(n - 5) \sum_{m=0}^{n-5} (5/6)^m. \qquad (2.53)$$

We now use an identity that is frequently convenient in the simplification of expressions involving shifted step functions [easily proved using sketches and the

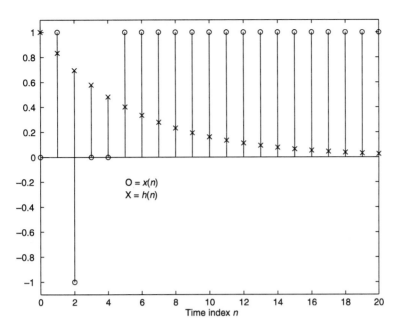

Figure 2.18

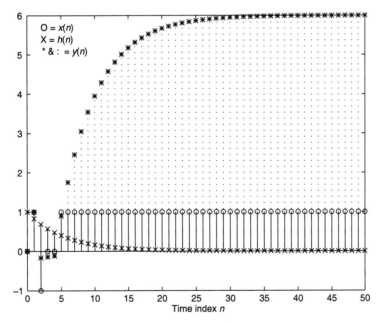

Figure 2.19

definition of the unit step sequence just by subtracting off the "first" value of $f(n)u(n - \ell)$]:

$$f(n)u(n - (\ell + 1)) = f(n)u(n - \ell) - f(\ell)\delta(n - \ell), \qquad (2.54a)$$

where for use in (2.53), we will in (2.54a) take $\ell = 1$:

$$f(n)u(n - 2) = f(n)u(n - 1) - f(1)\delta(n - 1). \qquad (2.54b)$$

Although not used in this example, we cite here for reference another special case of (2.54a) that is often useful, $\ell = 0$:

$$f(n)u(n-1) = f(n)u(n) - f(0)\delta(n). \tag{2.54c}$$

Our application uses (2.54b) "backwards": $f(n)u(n-1) = f(1)\delta(n-1) + f(n)u(n-2)$. Using this result and evaluating the geometric sum, (2.53) may be rewritten as

$$y(n) = \delta(n-1) + (5/6)^{n-2}(5/6-1)u(n-2) + \frac{1-(5/6)^{n-4}}{1-5/6}u(n-5) \tag{2.55a}$$

$$= \delta(n-1) - (1/6)(5/6)^{n-2}u(n-2) + 6 \cdot (1 - (5/6)^{n-4})u(n-5). \tag{2.55b}$$

All three sequences $x(n), h(n)$, and $y(n)$ are plotted together in Figure 2.19.

2.4.4 Causality

As is true for physical continuous-time systems, discrete-time systems tend to be causal, often because they are mere discretizations of or replacements for continuous-time physical systems. For causal discrete-time systems, any response to an input occurring at sample n_1 cannot begin occurring until $n = n_1$ at the earliest. In terms of the convolution relation (2.51b), the coefficient $h(m)$ of $x(n_1-m)$ must be zero for all $n_1 - m$ [argument of x] $> n_1$ [argument of y], or for $m < 0$. The only way this requirement is guaranteed—given that m ranges from $-\infty$ to ∞—is to have $h(m) = 0$ for all $m < 0$:

$$\boxed{h(m < 0) = 0 \text{ for causality.} \qquad (2.56)}$$

2.4.5 Stability

A discrete-time system $h(n)$ is BIBO stable if and only if for any and all $x(n)$ such that $|x(n)| < M_x < \infty$ for all n, then $|y(n)| < M_y < \infty$ for all n. Our development here is completely analogous to the analog development in Section 2.3.6. The output magnitude is

$$|y(n)| = |x(n) * h(n)|, \tag{2.57a}$$

which by (2.51b) can be written

$$|y(n)| = \left| \sum_{m=-\infty}^{\infty} h(m)x(n-m) \right|. \tag{2.57b}$$

Because $|a+b| \le |a| + |b|$, it follows from (2.57b) that

$$|y(n)| \le \sum_{m=-\infty}^{\infty} |h(m)x(n-m)| \tag{2.57c}$$

$$= \sum_{m=-\infty}^{\infty} |h(m)| \, |x(n-m)|. \tag{2.57d}$$

Because $|x(n)| < M_x$, the sum in (2.57d) will not exceed the value it would have if M_x were substituted for all $|x(n-m)|$. Thus

$$|y(n)| \le M_x \sum_{m=-\infty}^{\infty} |h(m)|. \tag{2.57e}$$

Now, if $\sum\limits_{m=-\infty}^{\infty} \left| h(m) \right| < \infty$, it follows that $\left| y(n) \right| < \infty \rightarrow$ system stable.

Thus we have established the "if" part of the proof; we now develop the "only if" part. Suppose we choose $x(n)$ such that $x(n - m) = \left| h(m) \right| / h(m)$ so that $h(m)x(n - m) = \left| h(m) \right|$ for $h(m) \neq 0$, and zero otherwise. Clearly, $\left| x(m) \right|$ is bounded, for it has magnitude 1 (or 0). Note that $\{x(n), -\infty < n < \infty\}$ will be a different sequence for each value of n selected for evaluation of $y(n)$, which is immaterial to the proof because we need prove violation only for any one input for any one computed time n. Then in this case, $y(n) = \sum\limits_{m=-\infty}^{\infty} \left| h(m) \right|$; thus if $\sum\limits_{m=-\infty}^{\infty} \left| h(m) \right|$ is unbounded, so will be $\left| y(n) \right|$ —and if that is true for even this one input sequence[7] $x(n)$, then $h(n)$ is not a BIBO stable system. Thus for BIBO stability, $h(n)$ must be an absolutely summable waveform (again, "absolutely" because of the absolute value bars).

In Section 3.4.1, we derive the equivalent z-domain criterion for stability, namely, that all the poles must be within the unit circle of the z-plane for BIBO stability.

EXAMPLE 2.17

(a)   Is the system $h(n) = (1/n)u(n - 1)$ BIBO stable?

(b)   Repeat part (a) for $h(n) = (-2)^n u(n)$.

(c)   Repeat part (a) for $h(n) = \left(\frac{1}{2}\right)^n u(n)$.

Solution

(a)   No, because the harmonic sequence $1/n$ is not absolutely summable.[8]

(b) $\sum\limits_{m=-\infty}^{\infty} \left| h(m) \right| = \sum\limits_{m=0}^{\infty} \left| (-2) \right|^m = \sum\limits_{m=0}^{\infty} 2^m,$

which diverges, because the terms always increase rather than attenuate with m.

(c) $\sum\limits_{m=-\infty}^{\infty} \left| h(m) \right| = \sum\limits_{m=0}^{\infty} \left| \left(\frac{1}{2}\right)^m \right| = \frac{1}{1 - \frac{1}{2}} = 2 < \infty,$

so this system is stable.

2.4.6   State Equations and LSI Systems

Like most of the material in this chapter, the reader has probably already encountered state–space representations in a signals and systems course. We review that topic here because of its major role in signal processing and to put it in perspective with other signal-processing representations.

As just one example of the importance of state space in practical DSP calculations, simulation algorithms in Matlab for computing the step response and general-input responses of a system use state-variable representation. Detailed knowledge of state space [in particular, (2.73), in Section 2.4.6.3] is required for using Matlab to simulate discrete-time systems having nonzero initial conditions; we already encountered this in our Matlab verification at the end of Example 2.13. Matlab also uses the state–space representation and techniques to compute the poles of a given

---

[7]Generally, if it fails for this one, it will also fail for most practical $x(n)$ sequences.

[8]See, for example, I. S. Sokolnikoff and E. S. Sokolnikoff, *Higher Mathematics for Engineers and Physicists* (New York: McGraw-Hill, 1941), p. 8.

system. An important algorithm in Matlab's Simulink for simulating continuous-time systems, the Runge–Kutta method, is based on the numerical solution of the state equations (see also Section 6.4.3). Yet another use of state space in Matlab is the transformation of normalized filters into filters having desired nonunity cutoff frequencies (see, e.g., lp2lp described in Section 8.4.1).

Also, optimal and adaptive control algorithms are often developed in state space (e.g., the Kalman filter is completely based on the state–space representation). An application of adaptive control is in automated drug-delivery systems, where the "state" of the patient is changing with time and consequently so are the parameters modeling the patient. For signal processing with nonlinear or multi-input/multi-output systems, state space is very convenient. Note that for nonlinear systems, Fourier methods and LSI theory do not apply.

Most important, state space is one of the major techniques for LSI system analysis and design. State space is a decomposition of a system that easily allows examination and control of the individual internal dynamics that combine to determine the output. Each of the individual internal processes, or state variables, satisfies a first-order difference equation in which its own next value is determined by a combination of the current values of it, the other state variables, and the input. Together, the state variables satisfy a set of coupled first-order difference equations that can easily be solved in matrix form. Sometimes these state equations are the equations most easily obtained from the physics.

Another use of state–space analysis is in the development of structures for hardware realizations of digital systems. For example, in Section 2.4.6.2 we see how a structure naturally arises from state–space techniques that in DSP jargon is called the transposed direct form II structure. We use this form in Section 2.4.6.3 to make inferences about the condition for system linearity in the presence of initial conditions. Moreover, in Section 2.4.6.4 we draw connections between the state–space solution and the unit sample response/convolution solution that help clarify the distinction between finite impulse response (FIR) and infinite impulse response (IIR) filters in reference to the difference equation coefficients.

Finally, in Section 2.4.6.5 we see how the state can be represented in a form explicitly showing the dynamics of the independent modes that make up the output response, and the explicit relation between these modes and the z-transform of the system coefficients. In particular, we find that the matrix eigenvalues are equal to the z-domain poles of the system.

Each of the major techniques for LSI analysis and design focuses on particular aspects of a system. Frequency-domain analysis takes advantage of the special properties concerning the processing of steady-state sinusoids by an LSI system. Z-domain analysis is ideal for system description when transients are involved and for innumerable cases where detailed analysis of the system is required. Both Fourier and z-domain approaches are used extensively in digital filter analysis and design. Difference-equation analysis is convenient for developing the system model from physical laws or measured data. State–space analysis can be helpful for the reasons given above and much more. One must be facile in all these approaches to be versatile and effective in the field of DSP.

2.4.6.1   The Concept of State

What, then, is the state of a system? Consider the example of a student's grades in his or her classes. The grade in the course of each discipline for semester[9] k could be considered as the value of a state variable at time (semester) k:

---

[9]For simplicity, we consider that the student takes only one course in each area per semester.

$w_1(k)$ = (numerical) grade in EE course for current semester k

$w_2(k)$ = grade in math course for current semester k

$w_3(k)$ = grade in humanities course for current semester k.

The output, or bottom line, is the grade-point average (GPA):

$$y(k) = \text{GPA} = \alpha_1 w_1(k) + \alpha_2 w_2(k) + \alpha_3 w_3(k),$$

which is a linear combination of the values of all the state variables, where α_ℓ is proportional to the credit hours for course ℓ. How well the student does in each course is itself a linear combination of his or her past grades in that area and perhaps also other areas. For example, doing well in past math courses will probably improve your grade in DSP, more so than will doing well in physical education; so the weights will vary.

Each state variable (course grade) is also a function of how much effort ("control effort") is applied to that course during the current semester:[10] $x(k)$ = amount of studying during current semester k. [If a course is too easy, its grade will depend less on $x(k)$!] Thus the new value of each state variable is a linear combination of past values of each of the state variables and the input(s).

Most employers, however, do not have time to examine the entire sequence of state vectors $\mathbf{w}(k)$ (individual grades), but only a linear combination of them: the GPA. That is, to many people, it just does not matter to them what grades you got in individual courses, as long as the average of your senior year grades—the current system output—is really high.

Another example of state is an automobile engine characterized at sampling time k. To completely describe the variables that affect $y(k)$ = gas mileage in miles/gallon, which we will consider to be an output of this system, you would have to discuss the values of:

$w_1(k)$ = air/fuel mixture ratio

$w_2(k)$ = the concentration of the exhaust gas recirculation

$w_3(k)$ = the carbon deposits on the spark plugs

$w_4(k)$ = the spark advance, which times the on/off switching of the ignition coil

$w_5(k)$ = the air pressure in each tire [actually, four state variables for $w_5(k)$]

$w_6(k)$ = the age and odometer reading of the automobile (indicating engine wear)

$w_7(k)$ = the dirtiness of the air and oil filters[11]

and so on. Relevant inputs would be

$x_1(k)$ = the driving habits of the driver (jackrabbit versus cautious driving)

$x_2(k)$ = the octane of gasoline used.

Thus $x_1(k)$ and $x_2(k)$ could be considered inputs that affect the state and thus the output (fuel efficiency). Several other outputs could also be specified:

$y_2(k)$ = pollutant concentrations (e.g., carbon monoxide)

$y_3(k)$ = speed of car

$y_4(k)$ = direction of motion

---

[10]Note: This system is multi-input only if independent effort functions exist for each class; if the inputs to all classes are linked by proportionality constants, we need only one input.

[11]It might be argued that $w_7(k)$ is an input if the oil filter is changed. I suppose I could change the oil and filter at every stoplight (example value of k)! Or, for that matter, another "input" could be that I just buy a new car!

and so on. For the last two outputs, we would need to add $w_8(k)$ = suspension condition and $x_3(k)$ = position of steering wheel.

We see that this system is not only multi-input, multi-output, but its complete description involves a large number of state variables. In general, each of the state variables affects the others as well as its own future value and the outputs. This system would not easily be modeled using a simple transfer function approach. Yet many real-world systems are multi-input, multi-output.

Nevertheless, in this book we limit our attention to single-input, single-output systems. Moreover, in DSP the state variables will usually either represent values stored in digital shift registers or, even more abstractly, they may exist only "on paper" within a state model selected for its computational advantages. However, the above examples give an intuitive idea of the meaning of the internal structure represented in a state model.

2.4.6.2 The State Equations

We have seen that the difference equation (2.40) is the input–output relation for an LSI system. The analysis leading to that conclusion did not account for initial conditions, nor did it concern itself with causality of the unit sample response $h(n)$. We shall see in Chapter 3 that there are possibly several $h(n)$ sequences associated with (2.40) but only one purely causal sequence, and only one stable unit sample response sequence—which may not be the causal one. Similarly, there is an infinite number of causal solutions of (2.40) but only one for a given set of initial conditions.

We may always scale each term in (2.40) by the same constant without changing the solution, so that the coefficient a_0 of $y(n)$ is unity. Equivalently, we may redefine the a_ℓ and b_ℓ to be $a_{\ell,\text{new}} = a_{\ell,\text{old}}/a_{0,\text{old}}$ and $b_{\ell,\text{new}} = b_{\ell,\text{old}}/a_{0,\text{old}}$ so that (2.40) may now be rewritten as

$$y(n) = \sum_{k=1}^{N} (-a_k)y(n-k) + \sum_{\ell=0}^{M} b_\ell x(n-\ell). \tag{2.58}$$

By rewriting in this way (with $a_0 = 1$), we are assuming that the system is causal; if we arranged to have $a_N = 1$, that would be convenient for analyzing an anticausal system.[12]

Suppose we subtract $b_0 x(n)$ from both sides of (2.58) and define the delay operator

$$D\{f(n)\} = f(n-1) \tag{2.59}$$

so that

$$f(n-\ell) = D^\ell f(n). \tag{2.60}$$

In practice, $D\{\cdot\}$ may either represent an actual data shift register or just a delay operator within the mathematical model of a physical system. Then, using (2.60),

$$y(n) - b_0 x(n) = \sum_{k=1}^{N} (-a_k)y(n-k) + \sum_{\ell=1}^{M} b_\ell x(n-\ell) \tag{2.61a}$$

$$= \sum_{k=1}^{N} (-a_k)D^k\{y(n)\} + \sum_{\ell=1}^{M} b_\ell D^\ell\{x(n)\}, \tag{2.61b}$$

which can be written in a nested form involving only single delay operators as

$$y(n) - b_0 x(n) = D\{-a_1 y(n) + b_1 x(n) + D\{-a_2 y(n) + b_2 x(n) + \dots$$
$$+ D\{-a_M y(n) + b_M x(n) + D\{-a_{M+1}y(n) + \dots + D\{-a_N y(n)\}\dots\}. \tag{2.61c}$$

---

[12]In control theory, $y(n+k)$ and $x(n+l)$ are conventionally used in (2.40), so in that case, for causal system analysis we set $a_N = 1$.

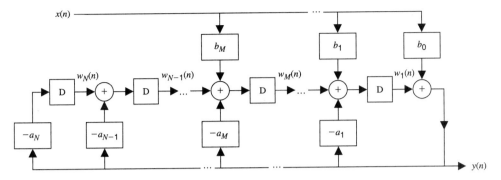

Figure 2.20

A simulation diagram of (2.61c) is shown in Figure 2.20; it is known as the transposed direct form II digital filter structure. Let us further define the inverse of D as D^{-1}; that is, $D^k D^{-m}\{f(n)\} = f(n - k + m)$. Then we can "factor out" D^N and rewrite (2.61c) as follows:

$$y(n) - b_0 x(n) = D^N\{-a_N y(n) + D^{-1}\{-a_{N-1} y(n)\} + \ldots + D^{M-N}\{-a_M y(n) + b_M x(n)\} +$$
$$D^{M-N-1}\{-a_{M-1} y(n) + b_{M-1} x(n)\} + \ldots +$$
$$D^{-(N-1)}\{-a_1 y(n) + b_1 x(n)\}\}. \tag{2.62}$$

Define the state variable $w_1(n) = y(n) - b_0 x(n)$. Often in practical system implementation, $w_1(n)$ [and $w_\ell(n)$ below] may be the contents of a data/shift register; otherwise, they may be abstract internal variables within the linear model of the physical system. Then, further decomposing (2.62) and defining another $N - 1$ state variables $w_\ell(n)$ as indicated,

$$w_1(n) = \quad D\{ \qquad \text{Braces include all covered Ds.} \tag{2.63}$$

N of these

$$\begin{aligned}
&D\{ \\
&D\{ \\
&\quad\vdots \\
&D\{ \\
&D\{-a_N[w_1(n) + \overline{b_0 x(n)}]\}\} \; w_N(n) \quad\} w_{N-1}(n) \quad\} w_M(n) \quad\} w_{M-1}(n) \quad\}[=w_1(n)], \\
&\quad -a_{N-1}[w_1(n) + b_0 x(n)]\} \\
&+\vdots \\
&+ -a_M[w_1(n) + b_0 x(n)] + b_M x(n)\} \\
&+ -a_{M-1}[w_1(n) + b_0 x(n)] + b_{M-1} x(n)\} \\
&+\vdots \\
&+ -a_1[w_1(n) + b_0 x(n)] + b_1 x(n)\}
\end{aligned}$$

we see that

$$D^{-1} w_N(n) = w_N(n + 1) = -a_N[w_1(n) + b_0 x(n)] \tag{2.64}$$
$$D^{-1} w_{N-1}(n) = w_{N-1}(n + 1) = w_N(n) - a_{N-1}[w_1(n) + b_0 x(n)]$$
$$\vdots$$
$$w_M(n + 1) = w_{M+1}(n) - a_M[w_1(n) + b_0 x(n)] + b_M x(n)$$
$$w_{M-1}(n + 1) = w_M(n) - a_{M-1}[w_1(n) + b_0 x(n)] + b_{M-1} x(n)$$
$$\vdots$$
$$w_1(n + 1) = w_2(n) - a_1[w_1(n) + b_0 x(n)] + b_1 x(n).$$

Notice that each w_ℓ is the output of a delay element in Figure 2.20. We may turn the single high-order difference equation into a set of first-order difference equations:

$$
\begin{bmatrix}
w_1(n+1) \\
w_2(n+1) \\
\vdots \\
w_{M-1}(n+1) \\
w_M(n+1) \\
\vdots \\
w_{N-1}(n+1) \\
w_N(n+1)
\end{bmatrix}
=
\begin{bmatrix}
-a_1 & 1\ 0 & \cdots & 0\ 0 & \cdots & 0\ 0 \\
-a_2 & 0\ 1 & \cdots & 0\ 0 & \cdots & 0\ 0 \\
\vdots & & & & & \\
-a_{M-1} & 0\ 0 & \cdots & 1\ 0 & \cdots & 0\ 0 \\
-a_M & 0\ 0 & \cdots & 0\ 1 & \cdots & 0\ 0 \\
\vdots & & & & & \\
-a_{N-1} & 0\ 0 & \cdots & 0\ 0 & \cdots & 0\ 1 \\
-a_N & 0\ 0 & \cdots & 0\ 0 & \cdots & 0\ 0
\end{bmatrix}
\begin{bmatrix}
w_1(n) \\
w_2(n) \\
\vdots \\
w_{M-1}(n) \\
w_M(n) \\
\vdots \\
w_{N-1}(n) \\
w_N(n)
\end{bmatrix}
+
\begin{bmatrix}
b_1 - a_1 b_0 \\
b_2 - a_2 b_0 \\
\vdots \\
b_{M-1} - a_{M-1} b_0 \\
b_M - a_M b_0 \\
\vdots \\
-a_{N-1} b_0 \\
-a_N b_0
\end{bmatrix}
x(n)
$$

$$
= \mathbf{A}\mathbf{w}(n) + \mathbf{b}x(n) \tag{2.65}
$$

for the usual case $M \leq N$, where $\mathbf{A}$ and $\mathbf{b}$ are, respectively, the matrix and column vectors shown, $\mathbf{w}(n) = [w_1(n) \ \ldots \ w_N(n)]^{\mathrm{T}}$. From the definition $w_1(n) = y(n) - b_0 x(n)$, we also have (putting into array form)

$$
\begin{aligned}
y(n) &= w_1(n) + b_0 x(n) \\
&= [1\ 0\ 0 \ldots 0]\mathbf{w}(n) + b_0 x(n) \\
&= \mathbf{c}\mathbf{w}(n) + dx(n),
\end{aligned} \tag{2.66}
$$

where $\mathbf{c} = [1\ 0\ 0 \ldots 0]$ and $d = b_0$.

Equations (2.65) and (2.66) are known in control theory as the state/output equations for observable (or dual phase-variable) form. The only difference is, again, that in control theory the difference equation (2.40) is written with $y(n+\ell)$ and $x(n+\ell)$ terms rather than $y(n-\ell)$ and $x(n-\ell)$; this changes $\{\mathbf{A}, \mathbf{b}, \mathbf{c}, d\}$ somewhat from their values given for (2.65) and (2.66). Also, in control theory, x is the symbol for state variables and u is used for representing inputs; in this book, x is the system input and u is the unit step sequence. Finally, we can conveniently obtain (2.65) and (2.66) in Matlab from the command `[A1,B1,C1,D1]=tf2ss ([b_0 ...` b_N`],[a_0 ... a_N])` (some of the last b_ℓ are zero; insert zeros for them) by writing subsequently $\mathbf{A} = $ `A1`^{T}, $\mathbf{b} = $ `C1`^{T}, $\mathbf{c} = $ `B1`^{T}, $d = $ `D1`, as Matlab provides the phase-variable form, the dual of (2.65) and (2.66).

Notice that in writing the state equations, we have done the reverse of what we did in Example 2.2. There we took several first-order differential equations and made one high-order differential equation out of it.[13] Now we are doing exactly the reverse: making several first-order difference equations out of a single high-order difference equation.

---

[13]One set of continuous-time state equations for Example 2.2 is easily read from (2.7) and (2.8b):

$$
\begin{bmatrix}
\dfrac{di_{L,c}(t)}{dt} \\[2mm]
\dfrac{dv_{RL,c}(t)}{dt}
\end{bmatrix}
=
\begin{bmatrix}
\dfrac{-R_L}{L} & \dfrac{1}{L} \\[2mm]
\dfrac{1}{C} & \dfrac{-1}{R_T C}
\end{bmatrix}
\begin{bmatrix}
i_{L,c}(t) \\[2mm]
v_{RL,c}(t)
\end{bmatrix}
+
\begin{bmatrix}
0 \\[2mm]
\dfrac{1}{R_T C}
\end{bmatrix}
v_{s,c}(t).
$$

EXAMPLE 2.18

Find the state-variable representation in (2.65) and (2.66) for the discretized system described in Example 2.3, with component values in Example 2.3 substituted into (2.12) $[a_1 = 1/(R_T C) + R_L/L,\ a_0 = (1 + R_L/R_T)/(LC),\ b_1 = 1/(R_T C),$ and $b_0 = R_L/(LCR_T)]$, and difference equation coefficients as determined in Example 2.12, in (2.45) $[a'_1 = \{-2/\Delta t^2 - a_1/\Delta t\}/\alpha,\ a'_2 = 1/(\Delta t^2 \alpha),\ b'_0 = \{b_1/\Delta t + b_0\}/\alpha,\ b'_1 = -b_1/(\Delta t \alpha),$ where $\alpha = 1/\Delta t^2 + a_1/\Delta t + a_0]$. Let $\Delta t = 2$ ms.

Solution

The $a_\ell,\ b_\ell$ coefficients in the state equations (2.65) and (2.66) are the $a'_\ell,\ b'_\ell$ of Example 2.12. Their values are, for $\Delta t = 2$ ms, $a'_1 = -0.5602,\ a'_2 = 0.0807,\ b'_0 = 0.5497,$ and $b'_1 = -0.3545$. We also need, in (2.65), $b'_1 - a'_1 b'_0 = -0.0466$ and $-a'_2 b'_0 = -0.0443$. Substituting these values into (2.65) and (2.66) gives for the state model (with $y = v_{RL}$ and $x = v_s$),

$$\begin{bmatrix} w_1(n+1) \\ w_2(n+1) \end{bmatrix} = \begin{bmatrix} 0.5602 & 1 \\ -0.0807 & 0 \end{bmatrix} \begin{bmatrix} w_1(n) \\ w_2(n) \end{bmatrix} + \begin{bmatrix} -0.0466 \\ -0.0443 \end{bmatrix} v_s(n)$$

$$v_{RL}(n) = \begin{bmatrix} 1 & 0 \end{bmatrix} \begin{bmatrix} w_1(n) \\ w_2(n) \end{bmatrix} + 0.5497 v_s(n).$$

It is left as an exercise to verify that a simulation solution of these state equations [e.g., using dlsim in Matlab and also the solution (2.68), below] gives the same steady-state solution as the analytically determined steady-state solution in Example 2.12 (stems in Figure 2.16b). Note, however, that at "start-up" there will be a difference unless the initial state-variable values [$w_1(0)$ and $w_2(0)$] are chosen exactly correctly to match [see (2.73) below]. After a few samples (or cycles), however, the agreement is very close even without such matching.

2.4.6.3 Solution of the State Equations and Linearity

Further insight can be gained by iteratively solving the state and output equations (2.65) and (2.66). Let $\mathbf{w}(0)$ be the initial value of the state vector $\mathbf{w}$ at $n = 0$. Then by the state equations (2.65),

$$\mathbf{w}(1) = \mathbf{A}\mathbf{w}(0) + \mathbf{b}x(0)$$

$$\mathbf{w}(2) = \mathbf{A}\mathbf{w}(1) + \mathbf{b}x(1) = \mathbf{A}[\mathbf{A}\mathbf{w}(0) + \mathbf{b}x(0)] + \mathbf{b}x(1)$$

$$= \mathbf{A}^2\mathbf{w}(0) + \mathbf{A}\mathbf{b}x(0) + \mathbf{b}x(1)$$

$$\mathbf{w}(3) = \mathbf{A}\mathbf{w}(2) + \mathbf{b}x(2) = \mathbf{A}[\mathbf{A}^2\mathbf{w}(0) + \mathbf{A}\mathbf{b}x(0) + \mathbf{b}x(1)] + \mathbf{b}x(2)$$

$$= \mathbf{A}^3\mathbf{w}(0) + \mathbf{A}^2\mathbf{b}x(0) + \mathbf{A}\mathbf{b}x(1) + \mathbf{b}x(2) \tag{2.67}$$

$$\vdots$$

$$\mathbf{w}(n) = \mathbf{A}^n\mathbf{w}(0) + \sum_{\ell=0}^{n-1} \mathbf{A}^{n-\ell-1}\mathbf{b}x(\ell),\ n \geq 0.$$

Let $\mathbf{A}_1^n$ be the first row of $\mathbf{A}^n$; thus $\mathbf{A}_1^{n-\ell-1}$ is the first row of $\mathbf{A}^{n-\ell-1}$. Then

$$w_1(n) = \mathbf{A}_1^n\mathbf{w}(0) + \sum_{\ell=0}^{n-1} \mathbf{A}_1^{n-\ell-1}\mathbf{b}x(\ell) \tag{2.68}$$

so that[14]

$$y(n) = w_1(n) + b_0 x(n)$$

$$= \mathbf{A}_1^n \mathbf{w}(0) + \sum_{\ell=0}^{n-1} \mathbf{A}_1^{n-\ell-1} \mathbf{b} x(\ell) + b_0 x(n). \tag{2.69}$$

Now let $m = n - \ell$ in (2.69) (and thus $\ell = n - m$) so that (2.69) becomes

$$y(n) = \mathbf{A}_1^n \mathbf{w}(0) + \sum_{m=1}^{n} \mathbf{A}_1^{m-1} \mathbf{b} x(n-m) + b_0 x(n), \quad n \geq 0. \tag{2.70}$$

Notice that if $x(n) = 0$ for all n, then $y(n) = \mathbf{A}_1^n \mathbf{w}(0)$. Thus we call $y_{ZIR}(n) = \mathbf{A}_1^n \mathbf{w}(0)$ the zero-input response for $n \geq 0$. Also, if $\mathbf{w}(0) = 0$, then $y(n)$ is equal to the last two of the three terms in (2.70). In this case, $y(n)$ is the zero-state response $y_{ZSR}(n)$ because for that solution, the initial state is zero and only the given applied input determines the output. If we further define

$$h(n) = \begin{cases} \mathbf{A}_1^{n-1} \mathbf{b}, & n > 0 \\ b_0, & n = 0 \\ 0, & n < 0, \end{cases} \tag{2.71}$$

which is composed solely of the a_ℓ and b_ℓ coefficients, then (2.70) can be written

$$y(n) = y_{ZIR}(n) + \sum_{m=0}^{n} h(m) x(n-m) = y_{ZIR}(n) + y_{ZSR}(n). \tag{2.72}$$

$y(n)$ in (2.72) is the total solution of the initial value problem. For $x(n < 0) = 0$ and $h(n < 0) = 0$, (2.72) reduces to (2.51b) $[y(n) = \sum_{m=-\infty}^{\infty} x(n-m) h(m)]$ when $y_{ZIR}(n) = 0$ for all n. Thus (2.51) is a general ZSR solution and $h(n)$ in (2.71) is the unit sample response.

Notice that if $y_{ZIR}(n)$ is nonzero, the system difference equation (2.58) is not a strictly linear system: Doubling $x(n)$ does not double $y(n)$; more formally, the linearity criterion (2.39) fails. In this case, the system would be called affine. Because $\mathbf{A}_1^n \neq 0$ if any $a_{\ell>0}$ is nonzero, it follows that $y_{ZIR}(n) = 0$—and therefore the system is strictly linear—if and only if $\mathbf{w}(0) = 0$. (This condition can be relaxed if the initial conditions are consistent with the input values; see Section 2.3.5 and Section 2.4.6.4.) What does this requirement imply with respect to initial values of $y(n)$? Assuming that $x(n < 0) = 0$ and thus from (2.66) $y(n) = w_1(n)$ for $n < 0$, we have from the state equations (2.65)

$$\mathbf{w}(0) = \mathbf{A}\mathbf{w}(-1) \tag{2.73a}$$

or, in particular, for the direct form II representation,

$$w_N(0) = -a_N w_1(-1) = -a_N y(-1). \tag{2.73b}$$

Also, first using the state equations (2.65) and then using the fact that $w_1(n) = y(n)$ and substituting a decremented version of (2.73b), we have

---

[14]Note: b_0 is not the first element of $\mathbf{b}$, which is $b_1 - a_1 b_0$; see (2.65).

$$w_{N-1}(0) \quad = -a_{N-1}w_1(-1) + w_N(-1)$$

$$= -a_{N-1}y(-1) - a_Ny(-2). \tag{2.73c}$$

Next,

$$w_{N-2}(0) \quad = -a_{N-2}w_1(-1) + w_{N-1}(-1)$$

$$= -a_{N-2}y(-1) - a_{N-1}y(-2) - a_Ny(-3)$$

$$= -\sum_{\ell=-3}^{-1} a_{N-3-\ell}y(\ell), \tag{2.73d}$$

where the last equality points out the general pattern, which follows all the way to

$$w_1(0) \quad = -a_1w_1(-1) + w_2(-1)$$

$$= -\sum_{\ell=-N}^{-1} a_{-\ell}y(\ell), \tag{2.73e}$$

where the second equality in (2.73e) makes immediate use of the generalization of the last equality in (2.73d). In general,

$$w_m(0) = -\sum_{\ell=-(N-m+1)}^{-1} a_{m-1-\ell}y(\ell). \tag{2.73f}$$

We thus conclude that for $\mathbf{w}(0)$ to be zero, and thus for the system to be strictly linear, we must have $y(-1) = y(-2) = \dots = y(-N) = 0$, if at least one of the $a_\ell > 0$ coefficients is nonzero. $\mathbf{w}(0) = \mathbf{0}$ implies a zero-state response, but again see the further discussion in Sections 2.3.5 and 2.4.6.4.

Equations (2.73) are essential for simulating discrete-time systems in Matlab whenever initial conditions are present. We use (2.73) to convert our given output initial conditions into the initial state that Matlab requires (see the problems of Chapter 3). A simple first-order example was already presented at the end of Example 2.13, and we now consider a second-order example.

EXAMPLE 2.19

Suppose $y(n) - 0.8y(n-1) + 0.15y(n-2) = 4x(n-1)$, where $x(n) = (\frac{1}{2})^n u(n)$. Find the first four values of $y(n)$ using the state equations, assuming that $y(-1) = 1$ and $y(-2) = 0$. Use the dual phase-variable canonical form in (2.65). Also find the first four values of the unit sample response of this system.

Solution

$N = 2, a_1 = -0.8, a_2 = 0.15$, and $b_1 = 4$, so (2.65) becomes

$$\begin{bmatrix} w_1(n+1) \\ w_2(n+1) \end{bmatrix} = \begin{bmatrix} 0.8 & 1 \\ -0.15 & 0 \end{bmatrix} \begin{bmatrix} w_1(n) \\ w_2(n) \end{bmatrix} + \begin{bmatrix} 4 \\ 0 \end{bmatrix} x(n) \tag{2.74}$$

and (2.66) becomes

$$y(n) = w_1(n) + 0 = [1 \quad 0]\mathbf{w}(n). \tag{2.75}$$

By (2.73), the initial value of the state vector, $\mathbf{w}(0)$, is found from

$$w_1(0) = -\{0.15y(-2) + (-0.8)y(-1)\} = -0.15 \cdot 0 + 0.8 \cdot 1 = 0.8 \tag{2.76}$$

$$w_2(0) = -0.15y(-1) = -0.15$$

so that, according to (2.67) and noting that $x(0) = 1$, the solution for the state vector is

$$\mathbf{w}(1) = \begin{bmatrix} 0.8 & 1 \\ -0.15 & 0 \end{bmatrix} \begin{bmatrix} 0.8 \\ -0.15 \end{bmatrix} + \begin{bmatrix} 4 \\ 0 \end{bmatrix} 1 = \begin{bmatrix} 4.49 \\ -0.12 \end{bmatrix} \tag{2.77}$$

$$\mathbf{w}(2) = \begin{bmatrix} 0.8 & 1 \\ -0.15 & 0 \end{bmatrix}^2 \begin{bmatrix} 0.8 \\ -0.15 \end{bmatrix} + \begin{bmatrix} 0.8 & 1 \\ -0.15 & 0 \end{bmatrix} \begin{bmatrix} 4 \\ 0 \end{bmatrix} 1 + \begin{bmatrix} 4 \\ 0 \end{bmatrix} \frac{1}{2} = \begin{bmatrix} 5.472 \\ -0.6735 \end{bmatrix}$$

$$\mathbf{w}(3) = \begin{bmatrix} 0.8 & 1 \\ -0.15 & 0 \end{bmatrix}^3 \begin{bmatrix} 0.8 \\ -0.15 \end{bmatrix} + \begin{bmatrix} 0.8 & 1 \\ -0.15 & 0 \end{bmatrix}^2 \begin{bmatrix} 4 \\ 0 \end{bmatrix} 1 + \begin{bmatrix} 0.8 & 1 \\ -0.15 & 0 \end{bmatrix} \begin{bmatrix} 4 \\ 0 \end{bmatrix} \frac{1}{2} + \begin{bmatrix} 4 \\ 0 \end{bmatrix} \frac{1}{4}$$

$$= \begin{bmatrix} 4.7041 \\ -0.8208 \end{bmatrix}$$

and the output, by $y(n) = [1 \quad 0]\mathbf{w}(n) = w_1(n)$, is

$$y(0) = w_1(0) = 0.8, \qquad y(1) = w_1(1) = 4.49, \tag{2.78}$$

$$y(2) = w_1(2) = 5.472, \quad y(3) = w_1(3) = 4.7041.$$

Because of the nonzero initial conditions, this system is not strictly linear, but rather affine. The usual linear system techniques apply, however, for the zero-state response. The zero-input response is

$$\mathbf{w}_h(1) = \mathbf{A}\mathbf{w}_h(0) = \begin{bmatrix} 0.8 & 1 \\ -0.15 & 0 \end{bmatrix} \begin{bmatrix} 0.8 \\ -0.15 \end{bmatrix} = \begin{bmatrix} 0.49 \\ -0.12 \end{bmatrix} \tag{2.79}$$

$$\mathbf{w}_h(2) = \mathbf{A}^2\mathbf{w}_h(0) = \begin{bmatrix} 0.272 \\ -0.0735 \end{bmatrix}$$

$$\mathbf{w}_h(3) = \mathbf{A}^3\mathbf{w}_h(0) = \begin{bmatrix} 0.1441 \\ -0.0408 \end{bmatrix}$$

and $y_{ZIR}(n) = w_{h,1}(n)$, so

$$y_{ZIR}(0) = 0.8, \quad y_{ZIR}(1) = 0.49, \quad y_{ZIR}(2) = 0.272, \quad y_{ZIR}(3) = 0.1441. \tag{2.80}$$

The unit sample response, by (2.71), is

$$h(n < 0) = 0, \tag{2.81}$$

$$h(0) = b_0 = 0$$

$$h(1) = \mathbf{A}_1^0 \mathbf{b} = [1 \quad 0] \begin{bmatrix} 4 \\ 0 \end{bmatrix} = 4$$

$$h(2) = \mathbf{A}_1^1 \mathbf{b} = [0.8 \quad 1] \begin{bmatrix} 4 \\ 0 \end{bmatrix} = 3.2$$

$$h(3) = \mathbf{A}_1^2 \mathbf{b} = [0.49 \quad 0.8] \begin{bmatrix} 4 \\ 0 \end{bmatrix} = 1.96.$$

Performing the convolution operation $h * x$ and adding on y_{ZIR}, it is seen numerically that the result is $y(n)$ as found in (2.78), as must be true. We may also verify (2.81) by taking the z-transform of the given difference equation (see Section 3.4.3), finding the system function, and taking its inverse z-transform. For those who are interested, the closed-form expression for $h(n)$ in this example is $h(n) = 20[(\frac{1}{2})^n - 0.3^n]u(n)$. Finally, we may obtain these results using Matlab's `dlsim` with $\mathbf{w}(0)$ as found in (2.76).

2.4.6.4   Relation between a_l and Length of Unit Sample Response

As may be assumed from the remarks at the end of Example 2.19, it is not a coincidence that the symbol $h(n)$ was used in (2.71) [$h(n) = \mathbf{A}_1^{n-1}\mathbf{b}$ for $n > 0, b_0$ for $n = 0$, 0 for $n < 0$], implying that $h(n)$ as given there is the unit sample response. For if $x(n) = \delta(n)$ and the initial conditions are all zero [so $y_{\mathrm{ZIR}}(n) = 0$], we have, in (2.72)

$$\left[y(n) = y_{\mathrm{ZIR}}(n) + \sum_{m=0}^{n} h(m)x(n-m)\right], y(n) = h(n), \text{ which is precisely the definition}$$

of a unit sample response.

It is well known and will later be verified by other means that if any of the $a_\ell > 0$ are nonzero, $h(\cdot)$ has infinite length. Such a response is commonly called an infinite impulse response (IIR; more properly, an infinite-length unit sample response). Yet (2.72) shows only a finite number of terms $n + 1$, in the convolutional sum. Is there a contradiction?

The terms for $h(m > n)$, if they existed, would all multiply $x(n < 0)$ in (2.51b)

$$\left[y(n) = \sum_{m=-\infty}^{\infty} x(n-m)h(m)\right], \text{ whereas (2.72) was derived assuming that } x(n < 0) = 0.$$

However, as noted in Section 2.3.5, initial conditions may be viewed as being caused by inputs for $n < 0$. For example, in continuous time, an initial voltage on a capacitor may be viewed as the result of applying current for $t < 0$, which in fact had to occur for there to be a nonzero voltage at $t = 0$. The form of input that brought about the initial state may (or may not) have been entirely different from that of $x(n > 0)$. Note that (2.72) does not exclude the possibility of $h(m > n) \neq 0$; it just does not use them, except if we account for their effect via $y_{\mathrm{ZIR}}(n)$. It is conventional in DSP to do this, effectively defining $y_{\mathrm{ZIR}}(n)$ to be

$$y_{\mathrm{ZIR}}(n) = \sum_{m=n+1}^{\infty} h(m)x(n-m), \tag{2.82}$$

and just write (2.51) rather than treat the input–output relation as an initial value problem as in (2.72). If the initial conditions are accounted for in this way [via (2.82)], the system is strictly linear even with nonzero "initial conditions" because they have been relegated to their source [$x(n < 0)$].

We still have not answered how it is that the $a_{\ell > 0}$ being nonzero make a difference—how they cause $h(\cdot)$ to be infinite length. If we set all the $a_{\ell > 0}$ in (2.58) to zero, we are left with the input–output relation

$$y(n) = \sum_{\ell=0}^{M} b_\ell x(n-\ell), \tag{2.83}$$

which again is of a purely convolutional form as in (2.51). We identify

$$h(n) = b_n, \tag{2.84}$$

which by (2.83) clearly has finite length and is commonly called a finite impulse response (FIR). More properly, it would be termed a finite-length unit sample response; the IIR/FIR difference is useful only for discrete-time systems anyway, which do not involve true impulses.

We may put (2.83) in the initial-value-problem form (2.72):

$$y(n) = y_{ZIR}(n) + \sum_{m=0}^{L} h(m)x(n-m),$$ (2.85)

where $L = \min(n, M)$ and

$$y_{ZIR}(n) = \begin{cases} \sum_{m=n+1}^{M} h(m)x(n-m), & n < M \\ 0, & \text{otherwise.} \end{cases}$$ (2.86)

Comparing (2.84) through (2.86) with $\{(2.71), (2.72), \text{ and } (2.82)\}$ and noting that b_m is finite length (length $M + 1$), whereas $A_1^{n-1}b$ in (2.71) continues to be nonzero for all $n > 0$ whenever at least one of the $a_{\ell>0}$ is nonzero, the distinction between the FIR and IIR cases is clear and is due to at least one of the $a_{\ell>0}$ being nonzero. Indeed, for the IIR case we may combine (2.72) and (2.82) as

$$y(n) = \sum_{m=0}^{\infty} h(m)x(n-m),$$ (2.87)

with $h(n)$ as given in (2.71). Intuitively, we may say that if future values of $y(n)$ depend on its past values, then $h(n)$ [which is $y(n)$ for an input unit sample] is self-perpetuating and therefore an IIR; such a system is called autoregressive.

It is also an interesting exercise to substitute $\{a_{\ell>0} = 0\}$ into the state equations (2.65). Now

$$\mathbf{A} = \begin{bmatrix} 0 & 1 & 0 & \dots & 0 \\ 0 & 0 & 1 & \dots & 0 \\ & & \vdots & & \\ 0 & 0 & 0 & \dots & 1 \\ 0 & 0 & 0 & \dots & 0 \end{bmatrix}.$$ (2.88)

When this new $\mathbf{A}$ is raised to the mth power, the diagonal of 1's shifts to the right $m - 1$ places. Noting that $\mathbf{A}$ is $N \times N$, it follows that $\mathbf{A}^m = 0$ for all $m \geq N$. Substituting $\mathbf{A}$ of (2.88) into (2.71) and noting that now $A_1^{n-1}b = b_n$ for $1 \leq n \leq M$ and zero otherwise, we indeed obtain (2.84) $[h(n) = b_n]$ as we must for the FIR case. It is quite interesting how everything works out to be self-consistent.

2.4.6.5 System Modes

If we express the state equations in a special diagonal form, we reveal in the time domain the independent contributing components to the output $y(n)$, known as the system modes. Let $\mathbf{w}(n) = \mathbf{M}\mathbf{q}(n)$, where $\mathbf{M} = [\mathbf{v}_1 \quad \mathbf{v}_2 \quad \dots \quad \mathbf{v}_N]$ is the matrix of eigenvectors $\mathbf{v}_\ell$: $\mathbf{A}\mathbf{v}_\ell = \lambda_\ell \mathbf{v}_\ell$, where for simplicity in this discussion, we assume distinct eigenvalues λ_ℓ. Let us examine the zero-input response $\mathbf{w}(n) = \mathbf{A}^n \mathbf{w}(0)$. Then because here $\mathbf{w}(n + 1) = \mathbf{A}\mathbf{w}(n)$, we have $\mathbf{M}\mathbf{q}(n + 1) = \mathbf{A}\mathbf{M}\mathbf{q}(n)$ or $\mathbf{q}(n + 1) = \mathbf{M}^{-1}\mathbf{A}\mathbf{M}\mathbf{q}(n)$. Now note that $\mathbf{A}\mathbf{M} = \mathbf{M}\mathbf{\Lambda}$ where $\mathbf{\Lambda} = \text{diag}(\lambda_1, \dots, \lambda_N)$, which is just a matrix form of $\mathbf{A}\mathbf{v}_\ell = \lambda_\ell \mathbf{v}_\ell$. Thus $\mathbf{M}^{-1}\mathbf{A}\mathbf{M} = \mathbf{\Lambda}$, so $\mathbf{q}(n + 1) = \mathbf{\Lambda}\mathbf{q}(n)$ or $\mathbf{q}(n) = \mathbf{\Lambda}^n \mathbf{q}(0)$, so

$$
\begin{aligned}
y_{ZIR}(n) &= \mathbf{c}\mathbf{w}(n) = \mathbf{c}\mathbf{M}\mathbf{q}(n) \\[2mm]
&= (\mathbf{c}\mathbf{M})\mathbf{\Lambda}^n \mathbf{q}(0) \\[2mm]
\hline
&= \sum_{m=1}^{N} A_m \lambda_m^n, \quad n \geq 0,
\end{aligned}
$$ (2.89)

where the last step just shows the dot product sum produced in evaluating the previous step, where $A_m = (\mathbf{cM})_m q_m(0)$. The terms λ_m^n are called the modes of the system, for they define the independent patterns that make up the total response $y_{ZIR}(n)$. The weights A_m will be different for the case of nonzero input, but the weighted modes will still be present with or without nonzero initial conditions.[15]

What are the mode constants λ_ℓ in terms of the a_ℓ and/or b_ℓ? From $\mathbf{Av}_\ell = \lambda_\ell \mathbf{v}_\ell$ we know that $|\mathbf{A} - \lambda_\ell \mathbf{I}| = 0$. With $\mathbf{A}$ as in the state equations (2.65), we have

$$
|\mathbf{A} - \lambda\mathbf{I}| =
\begin{bmatrix}
-a_1-\lambda & 1 & 0 & \cdots & 0 & 0 \\
-a_2 & -\lambda & 1 & \cdots & 0 & 0 \\
-a_3 & 0 & -\lambda & \cdots & 0 & 0 \\
& & \vdots & & & \\
-a_{N-1} & 0 & 0 & \cdots & -\lambda & 1 \\
-a_N & 0 & 0 & \cdots & 0 & -\lambda
\end{bmatrix},
\tag{2.90a}
$$

which becomes, under Laplace's expansion of a determinant,

$$
|\mathbf{A} - \lambda\mathbf{I}| = (-a_1 - \lambda)
\begin{bmatrix}
-\lambda & 1 & \cdots & 0 & 0 \\
0 & -\lambda & \cdots & 0 & 0 \\
& & \vdots & & \\
0 & 0 & \cdots & -\lambda & 1 \\
0 & 0 & \cdots & 0 & -\lambda
\end{bmatrix}
+ a_2
\begin{bmatrix}
1 & 0 & 0 & \cdots & 0 & 0 \\
0 & -\lambda & 1 & \cdots & 0 & 0 \\
& & \vdots & & & \\
0 & 0 & 0 & \cdots & -\lambda & 1 \\
0 & 0 & 0 & \cdots & 0 & -\lambda
\end{bmatrix}
$$

$$
-a_3
\begin{bmatrix}
1 & 0 & 0 & 0 & \cdots & 0 & 0 \\
-\lambda & 1 & 0 & 0 & \cdots & 0 & 0 \\
0 & 0 & -\lambda & 1 & \cdots & 0 & 0 \\
& & & \vdots & & & \\
0 & 0 & 0 & 0 & \cdots & -\lambda & 1 \\
0 & 0 & 0 & 0 & \cdots & 0 & -\lambda
\end{bmatrix}
+ \ldots \pm a_N \cdot 1 \quad \text{for } N_{\text{odd}}^{\text{even}}. \tag{2.90b}
$$

By evaluating the simple determinants, we obtain

$$
|\mathbf{A} - \lambda\mathbf{I}| = -(a_1 + \lambda)(\mp\lambda^{N-1}) + a_2(\pm\lambda^{N-2}) - a_3(\mp\lambda^{N-3}) + \ldots \pm a_N \quad \text{for } N_{\text{odd}}^{\text{even}} \tag{2.90c}
$$

$$
= \pm(\lambda^N + a_1\lambda^{N-1} + a_2\lambda^{N-2} + a_3\lambda^{N-3} + \ldots + a_N) = 0 \text{ for all N,} \tag{2.90d}
$$

or, multiplying by λ^{-N} and recalling that $a_0 = 1$,

$$
\sum_{\ell=0}^{N} a_\ell \lambda^{-\ell} = 0. \tag{2.90e}
$$

Letting $z = \lambda$, (2.90e) may be written as $A(z) = 0$, where $A(z)$ is the z-transform of the coefficients $\{a_\ell\}$, (see Chapter 3 for full details on the z-transform). Thus the modes of $y_{ZIR}(n)$ (the undriven system response) are the eigenvalues of $\mathbf{A}$, which are also the poles of the transfer function $H(z)$ of the system. (It can be shown that the eigenvalues of all state-variable representations are the same; this result is not at all limited to dual-phase-variable form.)

---

[15]Except, theoretically, for very specialized inputs that avoid excitation of the modes.

2.5 SYSTEMS AS BUILDING BLOCKS

A thorough understanding of just a simple, single LSI system opens up a world of arbitrarily complex systems by virtue of the building block nature of LSI systems. In particular, if two systems $h_1(n)$ and $h_2(n)$ are in cascade (see Figure 2.21),

$$y(n) = h_2(n) * \{h_1(n) * x(n)\} \tag{2.91}$$
$$= h_{12}(n) * x(n),$$

where

$$h_{12}(n) = h_1(n) * h_2(n). \qquad h_1(n) \text{ and } h_2(n) \text{ in cascade.} \tag{2.92}$$

Proof

$$y(n) = \sum_{m=-\infty}^{\infty} h_2(n-m) \sum_{\ell=-\infty}^{\infty} h_1(m-\ell)x(\ell). \tag{2.93a}$$

Pulling the ℓ summation outside, we have

$$y(n) = \sum_{\ell=-\infty}^{\infty} \left\{ \sum_{m=-\infty}^{\infty} h_2(n-m)h_1(m-\ell) \right\} x(\ell), \tag{2.93b}$$

which upon defining $k = m - \ell$ becomes

$$y(n) = \sum_{\ell=-\infty}^{\infty} \left\{ \sum_{k=-\infty}^{\infty} h_2(n-\ell-k)h_1(k) \right\} x(\ell) \tag{2.93c}$$

$$= \sum_{\ell=-\infty}^{\infty} h_{12}(n-\ell)x(\ell) \tag{2.93d}$$

$$= \{h_{12} * x\}(n). \tag{2.93e}$$

Notice that fundamentally we have proved the associative property for convolution: $a * \{b * c\} = \{a * b\} * c$. In terms of z-transforms, it will be clear from later chapters that $Y(z) = H_{12}(z)X(z)$, where $H_{12}(z) = H_1(z)H_2(z)$.

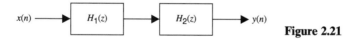

Figure 2.21

EXAMPLE 2.20

Suppose that a system $h_1(n) = a_1 \cdot z_1^n u(n)$ is cascaded with another system $h_2(n) = a_2 \cdot z_2^n u(n)$. Find the overall unit sample response $h_{12}(n)$ for general a_1, a_2, z_1, z_2 and then plot $h_1(n), h_2(n)$, and $h_{12}(n)$ for $a_1 = 2, a_2 = -3, z_1 = \frac{1}{2}$, and $z_2 = -\frac{1}{4}$.

Solution

Using (2.92), we have

$$h_{12}(n) = h_1(n) * h_2(n)$$

$$= \sum_{m=-\infty}^{\infty} h_1(m)h_2(n-m).$$

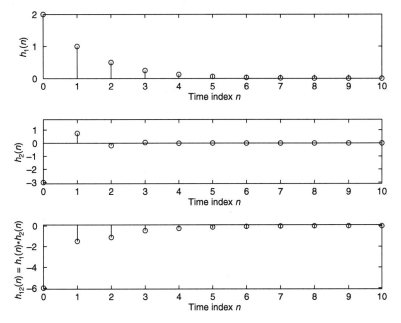

Figure 2.22

When we substitute the given sequences, $h_{12}(n)$ becomes

$$h_{12}(n) = a_1 a_2 \sum_{m=0}^{\infty} z_1^m z_2^{(n-m)} u(n-m).$$

Noting that if $n < 0$ there will not be any $m \geq 0$ satisfying $n - m \geq 0$ (i.e., $m \leq n$), then

$$h_{12}(n) = a_1 a_2 \cdot z_2^n \left\{ \sum_{m=0}^{n} \left(\frac{z_1}{z_2} \right)^m \right\} u(n),$$

which by the sum of a geometric series is equal to

$$h_{12}(n) = a_1 a_2 \cdot z_2^n \frac{1 - (z_1/z_2)^{n+1}}{1 - z_1/z_2} u(n)$$

$$= a_1 a_2 \frac{z_2^{n+1} - z_1^{n+1}}{z_2 - z_1} u(n).$$

For the given values, $h_{12}(n)$ becomes

$$h_{12}(n) = -6 \cdot \frac{(-4)^{-n-1} - 2^{-n-1}}{-\frac{3}{4}} u(n),$$

which, along with $h_1(n)$ and $h_2(n)$, is plotted in Figure 2.22. Notice that $h_2(n)$ oscillates, because $z_2 = -\frac{1}{4}$ is negative.

The other basic configuration is that of systems in parallel; see Figure 2.23. Two systems in parallel are equivalent to a single system whose unit sample response is the sum of the two unit sample responses of the two given systems:

$$y(n) = \{h_1(n) + h_2(n)\} * x(n). \qquad h_1(n) \text{ and } h_2(n) \text{ in parallel.} \qquad (2.94)$$

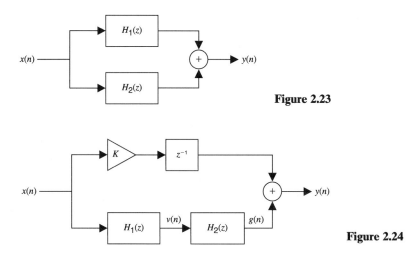

Figure 2.23

Figure 2.24

With (2.91) and (2.94) as extended to any number of interconnected systems (not just two), there are limitless possibilities for creating signal-processing schemes tailored to suit the given application. Moreover, to minimize the effect of computer finite wordlength, a single high-order system may be broken into cascade or parallel components, each of which is implemented individually.

To facilitate both analysis and design of signal-processing systems, the block diagram and signal flowgraph schematic representations are available. Both can be applied to continuous-time, discrete-time, and sampled-data (hybrid) systems.

The block diagram is the more intuitive and was already informally used in Figures 2.21 and 2.23, and even back in Figure 2.10 to illustrate the linearity property. A variable name symbolizes a signal, a line symbolizes a signal path or wire, and a block symbolizes a system.

For example, in Figure 2.24, the input divides into two branches: One consisting of a gain K followed by a delay is in parallel with another consisting of a cascade combination filter $h_{12}(n) = h_1(n) * h_2(n)$ [see (2.92)]. Parallel paths can be either introduced intentionally or can be unintentional, as in the example of multiple acoustic paths arising from reflections on walls. Usually the systems are represented in system diagrams by their z-transform rather than their unit sample responses. This representation is due to an algebraic convenience, where we may simply write $H(z) = Kz^{-1} + H_1(z)H_2(z)$, which is purely algebraic (no convolution involved in the z-domain). Of course, the z^{-1} is just the z-domain symbol for the unit delay operator D($\cdot$) in (2.59).

EXAMPLE 2.21

Using Matlab's Simulink, simulate the step response of the system in Figure 2.24 just described, for $h_1(n)$ a first-order lag network, $h_2(n)$ a first-order autoregressive system, and $K = 0.5$.

Solution

A first-order lag network is described by the difference equation $v(n) + a_1v(n-1) = x(n) + b_1x(n-1)$, where $-1 < b_1 < a_1 < 0$; the name "lag" is used because the lag network introduces a lag in the phase function of the discrete-time Fourier trans-

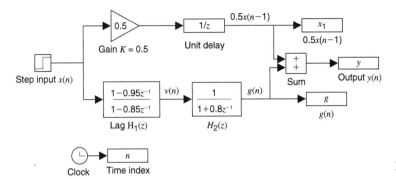

Figure 2.25

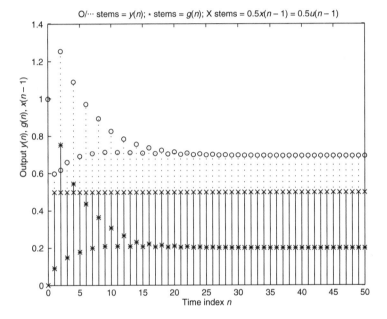

Figure 2.26

form. For $a_1 = -0.85$ and $b_1 = -0.95$, and for the first-order autoregressive system described by $g(n) + 0.8g(n-1) = v(n)$, the block diagram is as shown in Figure 2.24 and redrawn within Simulink in Figure 2.25.

It is worthwhile to examine the differences between Figure 2.24 and the Simulink block diagram shown in Figure 2.25. First, the Simulink blocks show the actual parameter values, when they fit in the box or triangle. Second, the summation symbol is a box in Simulink. Third, the intermediate results have been labeled and directed not only to the summer but also to output boxes for graphical display, as has been $y(n)$, the final output. With $x(n)$ a unit step sequence $u(n)$, which is explicitly indicated in the Simulink diagram, the intermediate outputs $0.5x(n-1)$ and $g(n)$, and the overall output $y(n)$ are as shown in Figure 2.26. To verify empirically that the system is linear, we can double the input [let $x(n) = 2u(n)$] and compare the output with that for $x(n) = u(n)$. This is done in Figure 2.27, and indeed the output doubles.

Finally, suppose a nonlinear element is added to the system: The physical system represented by $H_2(z)$ has a limited output range. This situation can be modeled in Simulink by placing a saturation block on its output; see Figure 2.28. The saturation

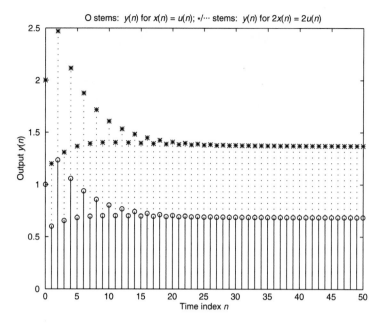

Figure 2.27

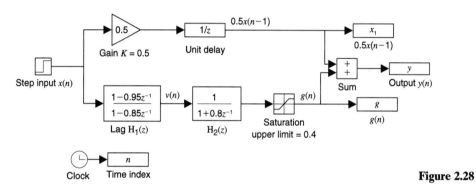

Figure 2.28

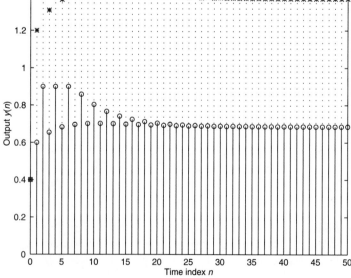

Figure 2.29

changes both $g(n)$ and $y(n)$. The results for a saturation limit of 0.4 are shown in Figure 2.29; $y(n)$ for $x(n) = 2u(n)$ is no longer twice $y(n)$ *for* $x(n) = u(n)$, and its shape is changed. This nonlinear part of this example demonstrates the versatility of Simulink for the study of a wide variety of complex systems that would otherwise be very difficult to analyze.

A signal flowgraph is merely a shorthand for the block diagram, where the system block is replaced by an arrow with the contents of the system block written above the arrow; see Figure 2.30 for the signal flowgraph corresponding to the block diagram in Figure 2.24. The main point to be aware of is that in the signal flowgraph, a summer and a branch node are both represented similarly: just an intersection of branches. The way to distinguish them is that the branch node has only one incoming branch and more than one outgoing branch, whereas the summer is the opposite. The reason they are notated the same is that they are both special cases of a multi-input node having several output branches, each of which is equal to the sum of the input signals.

There are no loops in Figure 2.30; that is, there is no feedback. In classical control systems analysis and design, the main feature is the introduction of output (and/or state) feedback. Recall that in Figure 2.20 we had signal feedback, but that was internal to the given system. Such feedback is contrasted with the high-level feedback of system outputs designed to contribute to inputs of other systems, as in controls. A helpful tool in determining the overall system function of feedback control systems is Mason's gain rule (see the problems).

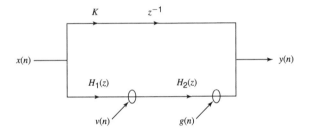

Figure 2.30

2.6   SIGNAL CHARACTERIZATION

A major goal of this book is to clarify the effects of and appropriateness for LSI systems on various signal types. Among the types of signals that have been helpfully classified are

Discrete-time signals
inherently discrete time/sampled-data
finite-length
causal/bicausal/anticausal
real-valued/complex-valued
even/odd
infinite-length
causal/bicausal/anticausal

real-valued/complex-valued

even/odd

stable/unstable

finite-energy

finite-power

periodic/aperiodic

deterministic/stochastic

Continuous-time signals

(categories mostly similar to those for discrete-time signals)

Although it is impractical to discuss each signal type in detail at this early stage, a few clarifications will now be added. Let the signal be the sequence $x(n)$.

An inherently discrete-time signal $x(n)$ is just a sequence such as the number of automobiles sold each day. It is not derived from the process of sampling a continuous-time signal, as is a sampled-data signal. An example of a sampled-data signal is a sampled capacitor voltage, giving values every Δt: $x(n) = x_c(n \Delta t)$.

A causal signal $x(n)$, like a causal unit sample response, satisfies $x(n) = 0$ for all $n < 0$, whereas an anticausal $x(n)$ satisfies $x(n) = 0$ for all $n \geq 0$. Also important are the generalizations known as right- and left-sided sequences. A right-sided $x(n)$ satisfies $x(n) = 0$ for all $n < n_0$, and a left-sided $x(n)$ satisfies $x(n) = 0$ for all $n \geq n_0$ where n_0 is any integer (positive or negative).

Although nearly all practical measured signals are real-valued, there are cases where analysis is simplified by introducing related complex-valued signals [e.g., by writing $A \cos(\Omega t + \phi)$ as $\text{Re}\{Ae^{j(\Omega t + \phi)}\}$ and then analyzing the complex quantity $Ae^{j(\Omega t + \phi)}$]. Also, the spectra of real-valued signals are generally complex, so we must allow for the analysis of complex-valued signals.

Even/odd refer to complex-conjugate symmetry of the values of $x(n)$ with respect to either side of $n = 0$. For example, an even sequence $x(n)$ satisfies $x(-n) = x^*(n)$ and an odd sequence $x(n)$ satisfies $x(-n) = -x^*(n)$. Note that a general $x(n)$ (neither odd nor even) can be decomposed into its even and odd components; for example, $x(n) = x_e(n) + x_o(n)$, where $x_e(n) = \frac{1}{2}[x(n) + x^*(-n)]$ and $x_o(n) = \frac{1}{2}[x(n) - x^*(-n)]$. Notice that upon substituting these definitions, we verify that $x_e(-n) = x_e^*(n)$ and $x_o(-n) = -x_o^*(n)$, so $x_e(n)$ and $x_o(n)$ live up to their names.

Stability/instability of a signal characterizes the (usually fictitious) system whose unit sample response is equal to $x(n)$. We discussed stability of a system in Section 2.4.5.

A finite-energy signal $x(n)$ satisfies

$$E_x = \sum_{n=-\infty}^{\infty} |x(n)|^2 < \infty. \quad \text{Finite-energy signal } x(n). \tag{2.95}$$

An infinite-duration signal that nevertheless has finite energy is a decaying exponential. A finite-(average)-power signal satisfies

$$\lim_{N \to \infty} \frac{1}{2N+1} \sum_{n=-N}^{N} |x(n)|^2 < \infty. \quad \text{Finite-power signal } x(n). \tag{2.96}$$

For example, a sinusoid or unit step violates (2.95) but satisfies (2.96).

Recall that in a resistor, power $p_c(t) = v_c(t)i_c(t) = i_c^2(t)R = v_c^2(t)/R$, both of which go as the square of the measurable signal (voltage or current). For the generic signal, we simply discard the R factor. Similar ideas apply for discrete-time signals. Energy is the integral (or running sum) of the power. So the total energy in the signal goes as the integral (or sum) of the square of the signal; hence the definition in (2.95) for "energy" E_x of a signal, which really represents a proportionality to physical energy. If the signal is periodic, then the energy, which is the integral or sum over all time of the nonnegative power, is infinite. If the signal is aperiodic, its energy may be either finite or infinite; clearly, if it has finite duration, its energy is finite.

We find it practical to treat many finite-energy signals as infinite-energy but finite-power when this is convenient. For example, in communication systems, speech signals may be so treated even though they obviously last only a finite time (although it may seem otherwise for a talkative person).

A periodic signal satisfies $x(n + N) = x(n)$ for all n (for example, a sinusoid), whereas an aperiodic signal does not (for example, a typical sampled speech waveform, excepting radio advertisements).

A deterministic signal is one that is repeatable. If we measure it 10 times, the same sequence will result every time; an example is measurements of orbit trajectories of planets (e.g., Earth to sun distance as a function of time). Equivalently, if we design, define, or otherwise know the sequence beforehand, it is considered deterministic. If it is not deterministic, it is called stochastic; an example is the noise patterns heard on a radio turned away from a station. The greater a factor noise is, the more stochastic analysis becomes preferable to deterministic analysis, for such unpredictable noise is always present in measurements. We restrict our attention to deterministic signal analysis in most of this book until Chapter 10.

APPENDIX 2A: THE DIRAC DELTA FUNCTION

In both continuous-time and sampled-data signal processing, the Dirac delta function plays an important role, coming up repeatedly in analysis. Its origin was in quantum mechanics mathematics, and its modern form was introduced in the 1940s by the physicist P. A. M. Dirac.[16] Although some mathematicians might object to the presence of "function" in its name, it can be treated like a function under an integral sign. It is the limit of a sequence of pulselike functions as they become narrower and simultaneously taller such that their area remains fixed.

An example often used to visualize the Dirac delta function $\delta_c(t)$ is the rectangle

$$\delta_c(t) = \lim_{\Delta \to 0} \begin{cases} \dfrac{1}{\Delta}, & |t| < \dfrac{\Delta}{2} \\ 0, & \text{otherwise.} \end{cases} \tag{2A.1}$$

See Figure 2A.1, in which it is clear that the area under the curve of each function in the limiting sequence is unity. Thus $\delta_c(t)$ is often defined as follows:

$$\int_{-\infty}^{\infty} \delta_c(t)dt = 1, \quad \delta_c(t \neq 0) = 0. \tag{2A.2}$$

It was shown in (2.25b) that $x_c(t) = \int_{-\infty}^{\infty} x_c(\tau) \, \delta_c(t - \tau)d\tau$. We may view this relation as a sampling property if we change τ to t and t to t_0 and use $\delta_c(-w) = \delta_c(w)$:

$$\int_{-\infty}^{\infty} x_c(t)\delta_c(t - t_0) \, dt = x_c(t_0) \tag{2A.3}$$

---

[16]Many earlier researchers, however, including Oliver Heaviside, Gustav Robert Kirchhoff, Charles Hermite, Simeon Denis Poisson, and Augustin-Louis Cauchy, studied various forms much earlier. In particular, Heaviside noted the sampling property. For a history of $\delta_c(t)$, see B. Van der Pol and H. Bremmer, *Operational Calculus Based on the Two-Sided Laplace Integral* (Cambridge: Cambridge Univ. Press, 1959), pp. 62–66.

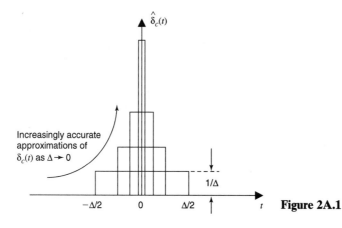

Figure 2A.1

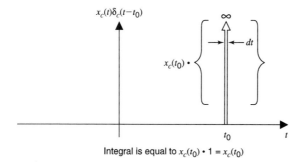

Figure 2A.2a

Figure 2A.2b

and for $t_0 = 0$ in particular,

$$\int_{-\infty}^{\infty} x_c(t)\delta_c(t)\,dt = x_c(0). \tag{2A.4}$$

The sampling property (2A.3) is illustrated in Figure 2A.2. In Figure 2A.2a, we show the factors $x_c(t)$ and $\delta_c(t - t_0)$ of the integrand in (2A.3). In Figure 2A.2b, we show the result of taking the product, namely $x_c(t)\delta_c(t - t_0)$. Recalling that the integral of a Dirac delta is unity, we obtain $x_c(t_0)$ as the integration result, as in (2A.3); a sample of $x_c(t)$, namely $x_c(t_0)$, is extracted by integrating $x_c(t)$ against a t_0-shifted Dirac delta. Further discussion of this sampling property is given at the end of this appendix.

Equation (2A.4) is yet another definition occasionally given to $\delta_c(t)$. A special case of (2A.4) is $x_c(t) = e^{-st}$, giving the Laplace transform of $\delta_c(t)$:

$$\int_{-\infty}^{\infty} \delta_c(t)e^{-st}dt = 1. \tag{2A.5}$$

We conclude that $LT\{\delta_c(t)\} = 1$ for all s and also that $FT\{\delta_c(t)\} = 1$ for all Ω. According to the uniqueness of the Fourier transform, we thus also have (citing and evaluating the inverse Fourier transform),

$$\delta_c(t) = \frac{1}{2\pi} \int_{-\infty}^{\infty} 1 \cdot e^{j\Omega t}d\Omega \tag{2A.6a}$$

$$= \lim_{\Omega_1 \to \infty} \frac{1}{2\pi} \int_{-\Omega_1}^{\Omega_1} e^{j\Omega t}d\Omega = \lim_{\Omega_1 \to \infty} \frac{\sin(\Omega_1 t)}{\pi t} = \lim_{\Omega_1 \to \infty} \frac{\Omega_1}{\pi}\text{sinc}(\Omega_1 t), \tag{2A.6b}$$

where $\text{sinc}(x) \equiv \sin(x)/x$ [which some authors denote $sa(x)$, the sampling function]. We conclude that not only is the tall, narrow rectangle a defining function for $\delta_c(t)$, but so also is the sinclike function $(\Omega_1/\pi)\text{sinc}(\Omega_1 t)$, which also becomes tall and narrow for large Ω_1. Yet again, some researchers refer to (2A.6) as the defining equation for $\delta_c(t)$.

Another fundamental property of $\delta_c(t)$ is worth showing because from it many other properties follow, and it is also good practice in this sort of analysis. Suppose the argument of $\delta_c(t)$ is $g_c(t)$, where $g_c(t)$ is a well-behaved function [i.e., $g_c(t)$ and its derivatives are continuous]. Let t_ℓ be the roots of $g_c(t)$: $g_c(t_\ell) = 0$. Then near the ℓth root t_ℓ, $g_c(t)$ can be Taylor-expanded in powers of $\epsilon = t - t_\ell$:

$$g_c(t) = g_c(t_\ell + \epsilon) = [g_c(t_\ell) = 0] + g'_c(t_\ell)\epsilon + \tfrac{1}{2}g''_c(t_\ell)\epsilon^2 + \ldots \tag{2A.7a}$$

and similarly,

$$g_c(t_\ell - \epsilon) = -g'_c(t_\ell)\epsilon + \tfrac{1}{2}g''_c(t_\ell)\epsilon^2 - \ldots \tag{2A.7b}$$

To determine an expression for $\delta_c(g_c(t))$ in terms of $\delta_c(t)$, we examine the effect it has when integrated against another function $f_c(t)$. We split the integral from $-\infty$ to $+\infty$ into long sections where $g_c(t)$ is nonzero, and over tiny sections each including only one of the t_ℓ. Over the long sections $[a, b]$ where $g_c(t) \neq 0$,

$$\int_a^b f_c(t)\delta_c(g_c(t))dt = 0 \tag{2A.8}$$

because on that range of integration the argument of δ_c is never zero. Now let us look at just the portion of the integral around one of the t_ℓ, from $t_\ell - \epsilon$ to $t_\ell + \epsilon$:

$$\int_{t_\ell - \epsilon}^{t_\ell + \epsilon} f_c(t)\delta_c(g_c(t))dt = \int_{g_c(t_\ell - \epsilon)}^{g_c(t_\ell + \epsilon)} f_c(t)\frac{\delta_c(g_c)}{g'_c(t)}dg_c, \tag{2A.9}$$

where on the right-hand side we made the change of variables from t to g_c by recalling that if $g_c = g_c(t)$, then $dg_c = g'_c(t)dt$ or $dt = dg_c/g'_c(t)$.

From (2A.7), we see that for very small ϵ for which the term linear in ϵ dominates, the signs of $g_c(t_\ell \pm \epsilon)$ are determined by the sign of $g'_c(t_\ell)$. Explicitly, if $g'_c(t_\ell) > 0$, the upper integration limit in (2A.9) is positive and the lower limit is negative; vice versa for $g'_c(t_\ell) < 0$. In the latter case, we must reverse the order of the limits by a sign change to make the lower limit to be negative and the upper limit to be positive. Then we can appeal to (2A.4), which has that polarity relation for the integration limits; no such change is required if $g'_c(t_\ell) > 0$.[17] Next bring the minus

---

[17]For essentially all functions of interest, $g_c{}'(t_l) \neq 0$.

sign, when $g'_c(t_\ell) < 0$, over to $g'_c(t)$ in the denominator of the integrand, and noting that it will be sampled at $t = t_\ell$, we use the notation $-g'_c(t_\ell) = |g'_c(t_\ell)|$, which actually holds for either polarity of $g'_c(t_\ell)$.

We then have, by the sampling property,

$$\int_{t_\ell - \epsilon}^{t_\ell + \epsilon} f_c(t)\delta_c(g_c(t))dt = \frac{f_c(t_\ell)}{|g'_c(t_\ell)|}. \tag{2A.10}$$

By doing this for each t_ℓ, we finally obtain

$$\int_{-\infty}^{\infty} f_c(t)\delta_c(g_c(t))dt = \sum_{\ell=1}^{N_r} \frac{f_c(t_\ell)}{|g'_c(t_\ell)|}, \tag{2A.11}$$

where N_r is the number of real roots t_ℓ of $g_c(t)$. From (2A.11), we conclude that

$$\delta_c(g_c(t)) = \sum_{\ell=1}^{N_r} \frac{\delta_c(t - t_\ell)}{|g'_c(t_\ell)|}. \tag{2A.12}$$

Some important special cases of (2A.12) are
1. $g_c(t) = -t$:

$$\delta_c(-t) = \delta_c(t) \quad \text{(here } t_1 = 0 \text{ is the only root } t_l\text{).} \tag{2A.13}$$

2. $g_c(t) = at$:

$$\delta_c(at) = \frac{\delta_c(t)}{|a|} \quad \text{(again, } t_1 = 0\text{).} \tag{2A.14}$$

3. $g_c(t) = t^2 - a^2$: $\delta_c(t^2 - a^2) = \dfrac{\delta_c(t-a) + \delta_c(t+a)}{2|a|}$ (here $t_1 = a, t_2 = -a$). (2A.15)

The first special case says that $\delta_c(t)$ is an even function, which we could tell just by looking at a plot of $\delta_c(t)$. The second special case is important whenever we need to scale the time variable. For example, when we scale time down ($a < 1$), the Dirac delta may yield a larger result when integrated against another function of t unless the function itself is inversely scaled [e.g., if $f_c(t) = 1/t$, then $f_c(t_1/a) = a/t_1$ is inversely scaled relative to $\delta_c(a(t - t_1))$]. The third special case is important in electromagnetic wave computations.

We can use integration by parts to prove identities involving $\delta'_c(t) = d\delta_c(t)/dt$. Left as exercises to show are $\delta'_c(-t) = -\delta'_c(t), t\delta_c(t) = 0, t\delta'_c(t) = -\delta_c(t)$,

$$\int_{-\infty}^{\infty} f_c(t)\delta'_c(t - t_0)dt = -f'_c(t_0), \tag{2A.16}$$

and

$$\int_{-\infty}^{\infty} \delta_c(a - t)\delta_c(t - b)dt = \delta_c(a - b). \tag{2A.17}$$

Consider also the following further useful results concerning the Dirac delta function. Our first concerns (2A.6) and then (2A.5). Reverse the roles of time and frequency variables in (2A.6a): $t \to \Omega$ and $\Omega \to t'$, giving the first expression in (2A.18). For the next step, define $t = -t'$ so that $dt = -dt'$ keeps the $+/-\infty$ limits from reversing under this time reversal. The remaining steps in (2A.18) are basic algebra:

$$\delta_c(\Omega) = \int_{-\infty}^{\infty} \frac{e^{j\Omega t'}dt'}{2\pi} dt = \int_{-\infty}^{\infty} \frac{e^{-j\Omega t}dt}{2\pi} dt = \lim_{t_1 \to \infty} \int_{-t_1}^{t_1} \frac{e^{-j\Omega t}dt}{2\pi} dt \tag{2A.18}$$

$$= \lim_{t_1 \to \infty} \frac{2\sin(\Omega t_1)}{2\pi\Omega} = \frac{1}{\pi} \lim_{t_1 \to \infty} t_1 \text{sinc}(\Omega t_1).$$

From the second equality in (2A.18), we have $FT\{1\} = 2\pi \cdot \delta_c(\Omega)$, a result that does not, however, extend to the Laplace transform (this will be discussed in Example 4.10). The final equality in (2A.18) explicitly shows that $\delta_c(\Omega)$ is writable as a limit of high-scaled sinc functions of Ω.

In Section 5.2, we derive the shift properties $x_c(t - \tau) \overset{\text{LT}}{\leftrightarrow} X_c(s) \cdot e^{-s\tau}$ and $X_c(s - s_0)$ $\overset{\text{LT}}{\leftrightarrow} x_c(t) \cdot e^{s_0 t}$. Therefore, (2A.5) [LT$\{\delta_c(t)\} = 1$] can be generalized to

$$LT\{\delta_c(t - \tau)\} = e^{-s\tau} \text{ and } FT\{\delta_c(t - \tau)\} = e^{-j\Omega\tau}, \tag{2A.19a}$$

and (2A.18), written as $FT\{1\} = 2\pi \cdot \delta_c(\Omega)$, can be generalized to

$$FT^{-1}\{2\pi \cdot \delta_c(\Omega - \Omega_0)\} = e^{j\Omega_0 t}. \tag{2A.19b}$$

As our next topic, we consider a more general definition of the Dirac delta function that allows the limits of an infinite variety of functions to be viewed as Dirac delta functions. Suppose we have a function $d(x)$ having unit area:

$$\int_{-\infty}^{\infty} d(x) \, dx = 1. \tag{2A.20}$$

Now consider the function $\alpha d(\alpha(x - x'))$, where we will let $\alpha \to \infty$. Define

$$I = \lim_{\alpha \to \infty} \int_{-\infty}^{\infty} \alpha d(\alpha(x - x'))f(x') \, dx' \tag{2A.21a}$$

$$= \lim_{\alpha \to \infty} \int_{-\infty}^{\infty} d(\alpha(x - x'))f(x') \, d[\alpha x']. \tag{2A.21b}$$

Let $t = \alpha(x - x')$ so that $x' = x - t/\alpha$. Then

$$I = \lim_{\alpha \to \infty} \int_{+\infty}^{-\infty} d(t) f(x - \frac{t}{\alpha}) \, [-dt]. \tag{2A.22a}$$

Swapping integration limits gives

$$I = \lim_{\alpha \to \infty} \int_{-\infty}^{\infty} d(t) f(x - \frac{t}{\alpha}) \, dt. \tag{2A.22b}$$

As $\alpha \to \infty$, $f(x - t/\alpha) \approx f(x)$ [and in practice $d(\pm\infty) = 0$ for $t = \pm\infty$, where $t/\alpha \neq 0$], so

$$I = \int_{-\infty}^{\infty} d(t) \lim_{\alpha \to \infty} \{f(x - \frac{t}{\alpha})\} \, dt \tag{2A.22c}$$

$$= f(x).$$

Comparing (2A.21a) and its result $I = f(x)$ in (2A.22c) with the Dirac delta function sampling property (2A.3), which can be taken as a definition of the Dirac delta function, we see that

$$\lim_{\alpha \to \infty} \alpha d\{\alpha(x - x')\} = \delta_c(x - x'). \tag{2A.23}$$

Thus with $d(\cdot)$ any smooth function having unit integral from $-\infty$ to $+\infty$, we can form a Dirac delta function out of it by scaling it and its argument by $\alpha \to \infty$. We have already seen in this appendix some particular examples of $d(\cdot)$. For example, in (2A.1), [$\delta_c(t) = \lim_{\Delta \to 0} 1/\Delta$, $|t| < \Delta/2$ and 0 otherwise], letting $\alpha = 1/\Delta$, we have $d(t) = 1$ for $|t| < \frac{1}{2}$ and zero otherwise; the area of $d(t)$ (i.e., its integral from $t = -\infty$ to $+\infty$) is unity and $\lim_{\alpha \to \infty} \alpha d(\alpha t) = \delta_c(t)$. We also found in (2A.6b) that $d(t) = \text{sinc}(t)/\pi$

was a Dirac delta function (where we had $\alpha = \Omega_1$). It can be verified by a number of techniques that $\text{sinc}(t)/\pi$ has unit area.

EXAMPLE 2A.1

Express $\displaystyle\sum_{n=-\infty}^{\infty} e^{j(\omega-\omega')n}$ in terms of Dirac delta functions.

Solution

By evaluating geometric sums and using trigonometric identities, we can express the given sum in terms of a sinc function whose argument is driven to infinity—which as just noted is a Dirac delta function. The result is

Complete derivation details on WP 2.1.

$$\sum_{n=-\infty}^{\infty} e^{j(\omega-\omega')n} = 2\pi\delta_c(\omega - \omega'), \quad |\omega - \omega'| < \pi. \tag{2A.24}$$

Noting that the left-hand side of (2A.24) is periodic in ω with period 2π (just adding $2\pi k$ to ω does not change any values of sum terms), the right-hand side of (2A.24) must actually be replaced by its 2π–periodic extension to be valid for all $\omega - \omega'$:

$$\sum_{n=-\infty}^{\infty} e^{j(\omega-\omega')n} = 2\pi\sum_{\ell=-\infty}^{\infty} \delta_c(\omega - \omega' + 2\pi\ell). \tag{2A.25}$$

For $\omega' = 0$, (2A.25) reduces to

$$\sum_{n=-\infty}^{\infty} e^{j\omega n} = 2\pi\sum_{\ell=-\infty}^{\infty} \delta_c(\omega + 2\pi\ell). \tag{2A.26}$$

These results are cited later in this book.

We have derived the important result (2A.25)–(2A.26) without resorting to Fourier series, by which they may also be proved. Our derivation, however, depends only on the very general definition of the Dirac delta function (2A.23), which we proved true here. Moreover, with (2A.23) as our definition, we can better visualize the Dirac delta function as a limit of functions having the sampling property when integrated against any "test" function.

This brings us at last to the concept of *distributions*. Consider the following generalization of the left-hand side of (2A.4), where we replace $\delta_c(t)$ by a generalized function $\phi(t)$ [so that the integral may no longer be equal to $f_c(0)$, and here omit the c subscripts]:

$$\int_{-\infty}^{\infty} \phi(t)f(t)\, dt. \tag{2A.27}$$

The generalized function $\phi(t)$ may be called a distribution when inserted in (2A.27). This is because it may be viewed as a "distribution" of "density" over the range $-\infty < t < \infty$, weighting the given function $f(t)$ accordingly at every value of t just as any "density" function in physics weights an integrand (in, for example, mass density, in the calculation of the center of mass of an object).

Although $\phi(t)$ may be discontinuous, if it is chosen to be the Dirac delta function $\delta_c(t)$, it is nevertheless a distribution because when it is a factor of the integrand in, for example, (2A.27), the value of the integral is well defined. Delta functions are often called generalized functions because even though they are ill-behaved outside an integral, they are well-behaved as factors of integrands of other normal functions. A shorthand way of referring to this property of an expression having meaning only in the context of integration against a well-behaved, continuous "test" function is to say that the expression has meaning only "in a distributional sense," a term we shall use in Chapter 4.

Finally, if we shift $\phi(t)$ in (2A.27), we have

$$\int_{-\infty}^{\infty} \phi(t - t')f(t') \, dt' = g(t). \tag{2A.28}$$

The integral result in (2A.28) is an operator acting on the function $f(t)$; we input the function $f(t)$ and out comes not just a value, but rather a function $g(t)$. For example, if $\phi(t) = \delta_c(t)$, we have $g(t) = f(t)$—the sampling property of the Dirac delta function, (2A.3). As another example, if $\phi(t) = h_c(t)$ (a system impulse response), then in (2A.28) $g(t)$ is the system output $y_c(t)$—the convolution of $f(t)$ and $h_c(t)$.

PROBLEMS

The icon **M** signifies that at least part of a problem makes use of Matlab.

Continuous-Time Signals and Systems

2.1. Without using transform techniques, find the differential equation of the mechanical system shown in Figure P2.1, where B_d is the damper coefficient (following the same physical law as friction B). Determine for it the values of the a_ℓ and b_ℓ coefficients in (2.20), with $a_N = 1$.

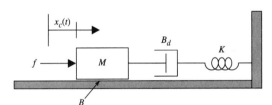

Figure P2.1

2.2. Without using transform techniques, find the differential equation of the electric circuit shown in Figure P2.2, relating the input $i_{s,c}(t)$ to the output $v_{o,c}(t)$. The input current source $i_{s,c}(t)$ models the output of an antenna, and the RLC is a tuning bank. Determine for it the values of the a_ℓ and b_ℓ coefficients in (2.20), with $a_N = 1$. Solve the differential equation, without using transform techniques, for $i_{s,c}(t) = A\cos(\Omega_0 t)$, where $\Omega_0 = 1/\{LC\}^{1/2}$ (pretend you do not know about resonance).

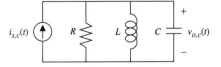

Figure P2.2

2.3. Without using the Laplace transform, find the relation between v_{RL}, the output, and v_s, the input of the circuit shown in Figure P2.3, without using transform techniques. Physically, the system represents an inductive generator driving a noninductive transmission line and an inductive, power-factor–corrected load.

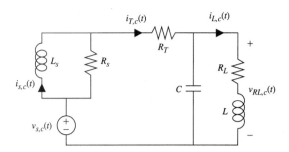

Figure P2.3

2.4. (a) Sketch force f versus displacement x for a spring, with a nonlinearity such that for x_{max}, f becomes infinite (and beyond this point the spring would break). Show two operating points for $x < x_{max}$, and indicate the nonlinearity.

(b) Using Matlab, plot the two-dimensional function $\exp\{-(x_1^2 + 2x_2^2)\}$ for $|x_1| < 2$ and $|x_2| < 2$. Show for a given operating point the region over which all linearization approximations will be for that operating point.

2.5. Consider the transistor amplifier shown in Figure P2.5, in which $R_1 = 100$ kΩ, $R_2 = 400$ kΩ, $R_C = 200$ Ω, $R_E = 500$ Ω, $R_L = 1$ kΩ, $\beta = 150$, $V_{CC} = 12$ V, $kT/q = 26$ mV. The transistor is a nonlinear device but is approximately linear near an operating point. Determine the operating point base current, and perform a linearization about the operating point to find the approximately linear signal voltage gain. Model the resulting small-signal circuit with an input impedance, a dependent source, and an output impedance (that is, with a simple linear model).

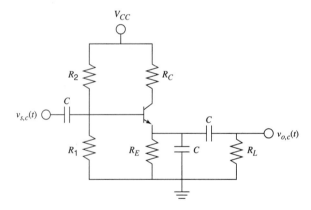

Figure P2.5

2.6. A simple equation of motion of a car is $f - F_d - F_r = Ma$, where f = thrust force (N), $F_d = \frac{1}{2}\rho C_D S v^2$ is the drag force (N), $F_r = f_r Mg$ is the rolling resistance (N) (approximately independent of speed), and Ma is the inertial force (N). In these relations, M = mass of vehicle (kg), g = acceleration of gravity = 9.81 m/s^2, a = acceleration of car (m/s^2) = dv/dt, v = speed (m/s), ρ = mass density of air = 1.23 kg/m^3, C_D = drag coefficient, S = frontal surface area of car (m^2), and f_r = rolling resistance coefficient. For a typical passenger car, $C_D = 0.4$, $S = 2$ m^2, $M = 4000$ lb = 1814 kg, and $f_r = 0.017$. Perform linearization of this nonlinear system relating f and v about an operating point of $v = V = 50$ mph. For an additional 50 N of thrust, find the increase in speed (in mph, using $v_{mph} = 0.6214 v_{kmph}$ and $v_{kmph} = 3.6 v_{m/s}$) and the time constant involved. Comment on the acceptability of your result, from a practical driving standpoint.

2.7. Perform a linearization of the function $y = \sin(2\pi x) + 0.2x^3 - 5x^{1/2} + 2 \cdot \ln(x)$ about the point $x = 1.5$. Plot both y and its linear approximation in the vicinity of $x = 1.5$, on either side. Show how near $x = 1.5$ the approximation is good and gradually falls apart on either side. Thus for small $|x - 1.5|$, a linear model for this system is valid.

2.8. For $\beta > 0$, find step response $s_c(t)$ of (a) $h_c(t) = -\beta e^{-\alpha t} u_c(t)$ (any $\alpha \neq 0$) and (b) $h_c(t) = \beta e^{-\alpha t} u_c(-t)$, $\alpha < 0$. Either plot $s_c(t)$ or describe the plot.

2.9. (a) Plot the function $x_c(t)$ defined as follows: $x_c(t)$ is a linear ramp from $t = 0$ to $t = 1$ s (maximum value 2), flat at 2 from $t = 1$ to 2 s, and a ramp back down to zero taking from 2 s to 4 s. Next, pass $x_c(t)$ through the system defined by the differential equation

$$\frac{d^3}{dt^3} y_c(t) + 5\frac{d^2}{dt^2} y_c(t) + 11\frac{d}{dt} y_c(t) + 15 y_c(t) = \frac{d}{dt} x_c(t) + 2 x_c(t).$$

Using `lsim`, simulate the output $y_c(t)$ for the ramp signal $x_c(t)$. Plot $y_c(t)$ from $t = 0$ to $t = 15$ s.

(b) Repeat part (a) for the input $x_c(t)' = x_c(t - t_0)$, where $t_0 = 5$ s. Is $y_c(t)' = y_c(t - t_0)$? That is, is the system time invariant? Demonstrate with your plots.

2.10. Suppose in Problem 2.9 that the coefficient of $dy_c(t)/dt$ is changed to -11. Carry out the simulations for $x_c(t)$ and $x_c(t - t_0)$. Conclude whether or not an unstable system is still time invariant. In this case, superimpose the simulation plots on the same axes, with $y_c(t)$ based on $x_c(t)$ shown only for $t \in [0, 10 \text{ s}]$, to facilitate your conclusions.

2.11. Provide all Matlab code for your solution of this problem.

(a) Plot the exact solution from Example 2.7 in Matlab for $T = 2$ s, $\alpha = 0.8$ Np/s, from $t = 0$ to $t = 10$ s (for about 120 equally spaced times).

(b) Estimate the results of part (a) using rectangular rule integration. Plot using "O" stems on top of the results in part (a). Examine for keeping a maximum of 100 terms and then for keeping a maximum of 10 terms. Explain any discrepancies with the results in part (a).

(c) Estimate the results of part (a) using `lsim`. Plot using "*" stems on top of the results in part (a).

2.12. (a) Determine $y_c(t) = h_c(t) *_c x_c(t)$, where $x_c(t) = t + 5$ for $t \in [1, 2]$ and zero otherwise, and $h_c(t) = e^{-t/2} u_c(t)$. Clearly, there will be different intervals to consider. In general, under what circumstances is $y_c(t)$ continuous at the interval end points? Is $y_c(t)$ continuous in this case at the interval boundaries? If so, use that as a check on your work.

(b) Plot $x_c(t), h_c(t)$, and $y_c(t)$ on the same axes using Matlab.

2.13. Find complete expressions for the output $y_c(t)$ of a system with impulse response $h_c(t) = u_c(t + 3) - u_c(t - 1)$ and input $x_c(t) = t[u_c(t + 4) - u_c(t + 1)]$. Plot $y_c(t)$ using Matlab. Is this system causal? Why?

2.14. Determine

$$y_c(t) = x_c(t) *_c h_c(t), \quad \text{where } x_c(t) = h_c(t) = \begin{cases} 1, & |t| \leq T \\ 0, & \text{otherwise.} \end{cases}$$

2.15. Find C and γ in terms of the other parameters in $c_c(t) = C \cos(\Omega t + \gamma) = a_c(t) + b_c(t)$, where $a_c(t) = A \cos(\Omega t + \theta)$ and $b_c(t) = B \cos(\Omega t + \phi)$, (a) without using phasors, and (b) with using phasors.

2.16. For the circuit in Figure P2.16, solve for B and ϕ in $v_{o,c}(t) = B \cos(\Omega t + \phi)$ via differential equation and by phasor analysis, in terms of R, L, C, and A, Ω, and θ of $v_{s,c}(t) = A \cos(\Omega t + \theta)$.

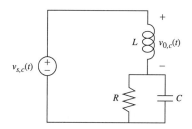

Figure P2.16

2.17. Prove that for any arbitrarily complicated network of R, L, and C (with any number and variety of R, L, and C and their connections involved) under sinusoidal excitation, the phasor relation between the total voltage across the network and current into it satisfies complex Ohm's law $\overline{V} = \overline{I}R$.

2.18. Determine under what, if any conditions that the continuous-time signal $x_c(t) = A_1\cos(\Omega_1 t + \theta_1) + A_2\cos(\Omega_2 t + \theta_2)$ is periodic in t. Under that condition, what is the period? Give examples demonstrating your results.

2.19. Suppose we wish to plot a continuous-time square wave $x_c(t)$ in Matlab. The square wave is equal to 1 for $|t| \leq \frac{1}{2}T_1$, 0 for $\frac{1}{2}T_1 < |t| < \frac{1}{2}T$, and this single period is periodically extended with period T. With $t = \texttt{linspace}(0, 5T, N)$, write a *single line* of code for $x_c(t)$ such that the correct plot results when we type $\texttt{plot}(t, x)$. No $\texttt{for}$ loops or $\texttt{if}$ blocks are allowed. Plot the square wave using this Matlab code for $T = 1$ s and $T_1 = 0.25$ s.

Discrete-Time Signals and Systems

2.20. Prove that (2.40) satisfies the condition for linearity, (2.39).

2.21. Prove that (2.40) is time invariant; use the proof for continuous time as a model.

2.22. A nonlinear discrete-time system is described by the difference equation $y(n) - 4\cos\{y(n-1)\} = 3x^2(n)$. Determine whether the system is shift invariant.

2.23 (a) Find the sum $\sum_{n=0}^{N-1} n \cdot a^{-n}$. Hint: Consider differentiation of the geometric sum without the n factor, with respect to a. Simplify as much as possible.

(b) A sequence $x(n)$ is zero for $n < 0$ and for $n \geq N$. For $n \in [0, N-1]$, $x(n) = 1 - (2/[N-1])n$. Let $h(n) = a^n u(n)$ and $y(n) = x(n) * h(n)$. Without performing any calculations, qualitatively sketch how you expect $y(n)$ to appear, being as specific as possible without calculations. Justify the characteristics of your plot verbally. Consider both a near zero and near 1; what are the limiting cases $a = 1$ and $a = 0$?

(c) Using the result in part (a), perform the convolution in part (b) (simplify as much as possible), and plot your result in Matlab for $n \in [0, 99]$, $N = 50$, and (i) $a = 0.95$ and (ii) $a = 0.4$. Does it agree with your initial assessment in part (b)? Also compare against the result of $\texttt{conv}$ using a long-truncated version of $h(n)$.

2.24. A spring/mass/friction system is as shown in Figure P2.24.

(a) Write the differential equation, and discretize it using backward differencing.

(b) Suppose the mass begins at $x = 0$ initially, and we release it at $t = 0$. Viewing this as a step input, use $\texttt{step}$ in Matlab to plot the step response for the following physical values: $M = 1$ kg, $B = 2.5$ kg/s, and $K = 160$ N/m. Then use $\texttt{dstep}$ to obtain the discrete-time step response; investigate $\Delta t = 0.0001$ s, 0.001 s, 0.01 s, and 0.1 s. Use a value of N based on the duration in seconds of the plot produced by $\texttt{step}$. Use output parameters for $\texttt{dstep}$, and make an appropriate time vector. Superimpose a plot of the $\texttt{dstep}$ output onto the $\texttt{step}$ plot (use $\texttt{plot}$ rather than $\texttt{stem}$ because of so many points), and verify agreement or disagreement, depending on the value of Δt. Comment on how large the number of samples per period of the step response is required for good agreement.

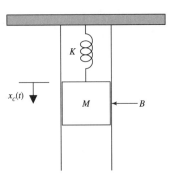

Figure P2.24

2.25. For each of the sequences in Figure P2.25, use discrete convolution to find the response to the input $x(n)$ of the linear shift-invariant system with unit sample response $h(n)$.

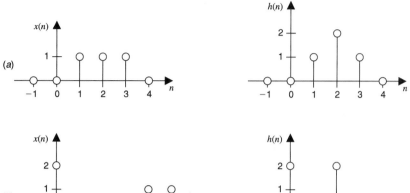

Figure P2.25

2.26. For all n, find $y(n) = x(n) * h(n)$ for
$$x(n) = \begin{cases} 1, & n \in [0, N-1] \\ 0, & \text{otherwise} \end{cases}$$
and $h(n) = a^n u(n)$.

2.27. A system has unit sample response $h(n) = (\frac{1}{2})^n u(n)$. By evaluating the convolutional sum, determine the output of this system when the input is $5u(n-3)$.

2.28. Find the unit sample response of the system $y(n) = x(n) - x(n-1) + 3x(n-6)$.

2.29. Solve for the unit sample response of the first-order difference equation $a_0 y(n) + a_1 y(n-1) = b_0 x(n)$. Assume zero initial conditions. Do not use transform methods.

2.30. A discrete-time feedback control system is as shown in Figure P2.30. Find the overall unit sample response, and from it determine the complete range of gain A for which the closed-loop system is stable. (Do not use the z-transform, introduced in Chapter 3.) Hint: Use the result from Problem 2.29, both "forward" and "in reverse"; also, gain A may be negative.

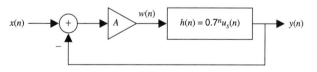

Figure P2.30

2.31. In this problem, we write an array-based routine to compute convolution of two finite-length sequences, in the time domain. Although in this case a direct sum computation would be simpler, some important ideas in array programming as well as convolution are learned in this exercise.

(a) Write a routine vrev.m that reverses the elements of an array.

(b) Write one routine that prezeropads and another that postzeropads an array out to a desired length.

(c) Write a routine conv1.m with a for loop that computes $y = x * h$ as in (2.51) with appropriate index limits, giving the same results as conv but with each $y(n)$ value computed as the result of an inner product (array multiplication) of h with a vector derived from x. Test your program for the arrays $x = [1 \quad 2 \quad 3 \quad 4]$ and $h = [-1 \quad 4 \quad -6 \quad 8$ $-10]$. In your solution, print out the array to be multiplied by h for each value of n. Do these arrays, put into a matrix form, agree with results you can obtain using the Matlab built-in function `convmtx`? Also write a simple routine giving the matrix directly, and include a general expression for the elements of the convolution matrix for a finite vector of values **x** [whose elements are $x(n)$] to be convolved with another vector **h**.

2.32. We showed in this chapter that $h(n)$ corresponds to a BIBO-stable system if and only if $h(n)$ is absolutely summable. Determine whether the following series are absolutely summable and thus whether the systems having these unit sample responses are BIBO stable.

(a) $h_1(n) = (1/n!)u(n)$

(b) $h_2(n) = [(-1)^n/n]u(n-1)$

(c) $h_3(n) = (n/3^n)u(n-1)$

Hint: Use the ratio test, developed by Augustin-Louis Cauchy in 1821.

Difference Equations

2.33. (a) Analytically solve the difference equation $a_0 y(n) + a_1 y(n-1) + a_2 y(n-2) = x(n)$ in terms of a_0, a_1, a_2, and $x(n)$ without using transform techniques; in the process, determine the system unit sample response $h(n)$. Assume that $a_1^2 > 4a_0 a_2$.

(b) Suppose that $a_0 = 1, a_1 = -0.2$, and $a_2 = -0.63$. Program your result in part (a) in Matlab, and plot $h(n)$ out to $n = 50$. Superimpose the result of `dimpulse`, as follows: `dimpulse([1 0 0],[a0 a1 a2])`. The reason for [1 0 0] rather than 1 is pursued in the problems of Chapter 3.

2.34. A linear system is expressed as $3y(n-2) - 4y(n-6) + 7y(n-3) = 2x(n-4) - 5x(n-1)$. Is the system causal? Why or why not?

2.35. For the system $y(n-5) - 1.3214y(n-4) + 0.5061y(n-3) - 0.6232y(n-2) + 0.9009y(n-1) - 0.3065y(n) = x(n) - 0.9x(n-1)$, obtain the unit sample response using the following Matlab routines:

(a) `dlsim`, (b) `filter`, and (c) `dimpulse`. Do you find discrepancies? Can you make them all agree by writing a simple routine?

Discrete-Time State Equations

2.36. Derive the phase-variable canonical form for writing of state equations for a linear shift-invariant system. To do this, write $Y(z) = [X(z)/A(z)]B(z)$.

Hint: Treat the b_0 term separately.

2.37. A discrete-time system is described by the difference equation
$20y(n) - 31y(n-1) + 12y(n-2) = x(n) - 0.4y(n-1)$.

(a) Give the state equations in dual phase-variable canonical form.

(b) Find the step response from (2.69) by programming that equation in Matlab. Compare with the result obtained from `dstep` using `dstep([b0 b1 0],[a0 a1 a2])`, where the reason for the added zero will be discussed in the problems of Chapter 3. Assume zero initial condition on the state vector.

(c) Repeat part (b) for finding the unit sample response. Use (2.69), substituting $x(n) = \delta(n)$ [of course, this calculation is effectively identical to just evaluating (2.71)].

2.38. Consider again Problem 2.37. Given that a unit sample is zero for all $n > 0$, it seems reasonable to relate the unit sample response to the homogeneous response. To do this, we merely need to obtain the equivalent initial condition produced by the input unit sample of $\delta(n)$ at $n = 0$. Doing so will allow us to use (2.89) to obtain $h(n)$ in terms of the system modes (without the z-transform theory of Chapter 3). Show what to do for this system and for a general second-order system; compare with your results in part (c) of Problem 2.37.

2.39. (a) For the system $y(n) - 1.75y(n-1) + 1.575y(n-2) - 0.6885\,y(n-3) = x(n-2) - 0.5x(n-3)$, determine the matrices $\mathbf{A}, \mathbf{B}, \mathbf{C}$, and $\mathbf{D}$ by the Matlab function `tf2ss`, and find the eigenvalues matrix $\mathbf{\Lambda}$ directly from `eig` and by using $\mathbf{\Lambda} = \mathbf{M}^{-1}\mathbf{A}\mathbf{M}$ with $\mathbf{M}$ obtained from `eig`. Your results should be the same.

(b) Let the initial state vector $\mathbf{w}(0) = \begin{bmatrix} 1 & 0 & 1 \end{bmatrix}^T$, where $\mathbf{w}(n)$ is the state vector obtained from Matlab in part (a). Determine the homogenous response both by using the routine `dlsim` and by performing the modes calculation in (2.89). What are the values of the A_m, and do you find agreement in the results?

(c) Verify numerically, by working with the originally given difference equation, that the modes of the system are indeed equal to the poles of the system.

2.40. Write a simple routine using the `min` function to determine the minimum value of a real-valued matrix. The outputs should be row, column, and minimum value.

2.41. Use Matlab to verify that the solution of the state equations in Example 2.18 gives the same steady-state solution as the analytically determined steady-state solution in Example 2.12. Note, however, that at "start-up" there will be a difference unless the initial state-variable values $[w_1(0)$ and $w_2(0)]$ are chosen exactly correctly to match. After a few samples, however, the agreement should be very close.

Multiple-Subsystem Systems

2.42. Find $y_c(t)$ in Figure P2.42 terms of real-valued $x_c(t)$, where the box marked τ signifies a delay element with delay τ seconds and $\times$ is a signal multiplier. For those with random signals background (see also Chapters 5, 7, and 10): If τ is adjustable and T is large, what practical purpose might this circuit serve?

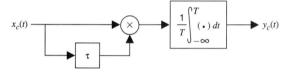

Figure P2.42

2.43. Although we will have a formal, systematic presentation of z-transform theory in Chapter 3, now is a good time to review the basic system theory results. For example, as noted in the text, the input–output relation in the z-domain for two cascaded systems $H_1(z)$ and $H_2(z)$ is $Y(z) = H_{12}(z)X(z)$, where $H_{12}(z) = H_1(z)H_2(z)$. Use this result to determine the overall input–output z-domain relation (overall transfer function) for the system whose block diagram is shown in Figure P2.43.

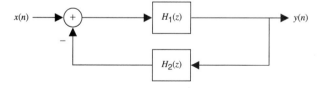

Figure P2.43

2.44. Repeat Problem 2.43 for the system shown in Figure P2.44. Manipulate the diagram into a form in which you can use the result of Problem 2.43 (e.g., use the distributive law).

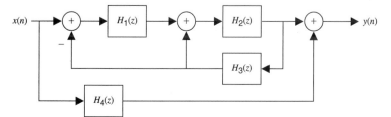

Figure P2.44

2.45. Mason's gain rule can simplify the finding of an overall transfer function when the system has many blocks and loops. The rule is that the overall transfer function T is given by (see Mason's paper and book for proof).[18]

$$T = \sum_{n=1}^{N} T_n \frac{\Delta_n}{\Delta}$$

where

N = number of forward paths

T_n = gain of nth forward path

$$\Delta = 1 - \sum_i L_i + \sum_i \sum_{\substack{j \\ j>i}} L_i L_j - \sum_i \sum_{\substack{j \\ j>i}} \sum_{\substack{k \\ k>j}} L_i L_j L_k + \dots$$

= 1 – sum of all individual loop gains

   + sum of products of nontouching loop gains taken 2 at a time

   – sum of products of non-touching loop gains taken 3 at a time + ...

in which L_i = ith loop gain (a loop originates and ends at same node) and $\Delta_n = \Delta$ for the graph in which all loops touching the nth forward path are removed. "Touch" means having at least one node in common.

Use this rule to solve Problem 2.44, working with only the original block diagram. Note that Mason's gain rule works for continuous-time systems also, because it is merely a topological graphing theorem.

Dirac Delta Function

2.46. Use (2A.1) to prove the sampling property of the Dirac delta function $\delta_c(t)$ cited in (2.25b).

2.47. Prove that

(a) $\int_{-\infty}^{\infty} f_c(t) \delta_c'(t - \tau)\, dt = -f_c'(\tau)$

(b) $t\delta_c(t) = 0$

(c) $t\delta_c'(t) = -\delta_c(t)$

(d) $\delta_c'(-t) = -\delta_c'(t)$

(e) $\int_{-\infty}^{\infty} \delta_c(a - t)\delta_c(t - b)\, dt = \delta_c(a - b)$

2.48. In advanced electromagnetic computations (e.g., Raj Mittra's spectral iterative technique), the Dirac delta function $\delta_c(k^2 - k_0^2)$ arises in solution of the wave equation for the electric field vector, where $k = k_x^2 + k_y^2 + k_z^2$ is the wave vector magnitude; k_0 is the free-space wave number of the medium in which the wave propagates; and k_x, k_y, and k_z are rectangular components of the wave vector. To facilitate evaluation of integrals over k_z for the electric field, it is desired to simplify this Dirac delta function into two Dirac deltas without squared arguments of the variable k_z. Prove (2A.15) using (2A.12), and use (2A.15) to show

that $\delta_c(k^2 - k_0^2) = \dfrac{\delta_c(k_z - \{k_0^2 - k_x^2 - k_y^2\}^{1/2}) + \delta_c(k_z + \{k_0^2 - k_x^2 - k_y^2\}^{1/2})}{2\{k_0^2 - k_x^2 - k_y^2\}^{1/2}}.$

2.49. Interpret convolution in terms of the nomenclature "operator" and "distribution."

2.50. In (2.20), let the differential operators $L_x\{\cdot\}$ and $L_y\{\cdot\}$ be defined by

$$L_y\{y_c(t)\} = \sum_{\ell=0}^{N} a_\ell \frac{d^\ell y_c(t)}{dt^\ell} \quad \text{and} \quad L_x\{x_c(t)\} = \sum_{\ell=0}^{M} b_\ell \frac{d^\ell x_c(t)}{dt^\ell}.$$ Thus we may view $x_c(t)$ as the source

and $y_c(t)$ as the response satisfying $L_y\{y_c(t)\} = L_x\{x_c(t)\}$, subject to initial conditions that must be satisfied (or, in spatial systems where t is replaced by $\{x, y, z\}$, and boundary conditions rather than initial conditions). The system (2.20) is linear and time invariant. Thus the solution for $y_c(t)$ when $x_c(t) = \delta_c(t)$, again subject to the initial conditions, is called the Green's function $g_c(t) = L_y^{-1}\{L_x\{\delta_c(t)\}\}$. Relate $g_c(t)$ to another system-related time function we discussed in this chapter.

---

[18]Mason, S. J., "Feedback theory—further properties of signal flow graphs," *Proceedings of the IRE,* 44 (July 1956): 920–926; S. J. Mason, H. J. Zimmerman, *Electronic Circuits, Signals, and Systems* (New York: Wiley, 1960).

Chapter 3

Complex Variables and the z-Transform

3.1 INTRODUCTION

Anyone intending to become involved in DSP applications and research should have a solid understanding of complex calculus and its offspring, the z-transform. Since the beginning of digital signal processing in the 1960s, complex analysis and the z-transform have played an important role. A study of journals such as the *IEEE Transactions on Signal Processing* will show just how pervasive these tools are in a wide range of DSP research and development. The reasons for going to the bother of complex analysis—even though our measurement data is usually real-valued—are similar to those for its use in the area of continuous-time signals and systems. A few of these reasons are mentioned now, although some of the terminology may be more clear when reread after completing this chapter.

The most important DSP use of complex analysis is the z-transform, the discrete-time analog of the Laplace transform of continuous-time system analysis. We study the z-transform in this chapter and throughout this book. Its purposes are many, but foremost is its ability to completely characterize signals and linear systems in the most general ways possible.

In particular, the oscillatory modes of a linear shift-invariant (LSI) discrete-time system are the *poles* of the system function: the z-transform of the system unit sample response. The locations of these poles in the complex z-plane imply the stability or instability of the system, and how resonantly the system responds to particular oscillatory components of the input.

The fact that the system function of an LSI system is a rational function of z means that the huge store of mathematical knowledge about polynomials and rational functions of a complex variable can all be applied to DSP. For example, we know that the poles can be complex-valued because for LSI systems they are the roots of a (factorable) polynomial, which in general are complex. Furthermore, if the unit sample response is real-valued, then all complex poles must occur in complex conjugate pairs. Knowledge of the properties of functions of a complex variable can facilitate linear shift-invariant system analysis.

Often it is desired to *equalize* or remove the effect that one distorting system has had on a signal. In this case, the locations of the *zeros* of the system function indicate whether this equalization can be done by straightforward pole–zero cancellation. In particular, the stability of a pole–zero cancellation system depends on the locations of the zeros of the original distorting system.

We may also obtain information about signals by examining their z-transforms. If a pole of the z-transform of the signal is near the unit circle on the z-plane, then a large concentration of energy of the signal resides in steady-state oscillations associated with that pole. Contrarily, if instead a zero of the signal z-transform is near the unit circle, we know that very little energy exists in the steady-state oscillations corresponding to that zero.

Paradoxically, complex analysis often simplifies mathematical analysis. For example, instead of having two sets of Fourier coefficients to think about, only one complex-valued frequency domain variable is necessary. Also, integrations of the discrete-time Fourier transform over frequency often most easily can be performed as contour integrals around the unit circle on the z-plane.

Yet another benefit of complex analysis is that damped or undamped complex sinusoids, and only they, are eigenfunctions of LSI systems. That is, if the input has time dependence z_1^n, then so will the output: $z_1^n * h(n) = \sum_m z_1^{n-m} h(m) = z_1^n \cdot \sum_m h(m) z_1^{-m}$ $= z_1^n \cdot H(z_1)$, where $H(z_1) = \sum_m h(m) z_1^{-m}$ is a constant with respect to n, called the eigenvalue (and in Section 3.4, the z-transform of the $h(n)$ sequence), and z_1^n is the eigenfuntion of the convolution "*" operator. This fact gives an added dimension of usefulness and convenience to Fourier and z-transform analysis. For example, it facilitates the solution of difference equations because convolution operators are transformed into simple algebraic ones.

We shall find in Section 8.6 that the z-transform can be used to obtain entire families of digital filters from one prototype design. In addition, prototype continuous-time filters can be converted into high-quality digital filters by simple mappings from the s-domain to the z-domain. In many ways, the z-transform and complex analysis play important roles in digital filter design and DSP generally.

It is assumed that the reader has had at least a course including complex numbers and arithmetic. Thus we will move directly to a brief review of functions of a complex variable, and on to complex calculus in more detail, which is the basis for the z-transform. As opposed to the sterile coverage of complex calculus in a typical course on complex analysis, this chapter is strongly directed toward the immediate *application* of the results to the z-transform for DSP. This motivation should make the math a little less painful and a lot more interesting, for there is a real reason for going through the math.

Before moving on, a Matlab note: A highly desirable aspect of Matlab is that all variables can be complex, and most available functions can be taken of a complex variable with equal ease to that for real variables. In fact, unless redefined otherwise, in Matlab j has the default value $(-1)^{1/2}$, which is of considerable convenience to electrical engineers.

3.2 FUNCTIONS OF A COMPLEX VARIABLE

First we present a graphic of a complex number. We may visualize the magnitude and phase of the complex number by plotting them three-dimensionally as the vertical heights of surfaces over a portion of the complex plane, as shown in Figures 3.1a and b, respectively. Figure 3.1 and several subsequent 3-D plots were generated

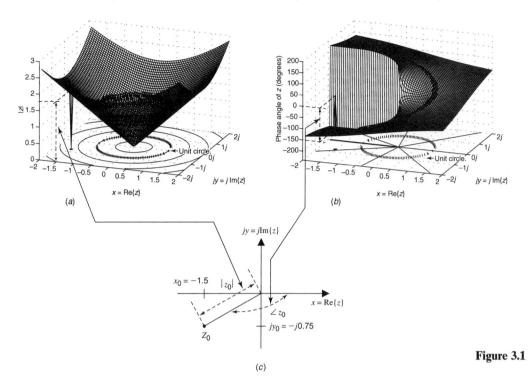

Figure 3.1

using Matlab's `meshc` command, along with some additional "tricks" to enhance the illustrations. The upper portion shows the three-dimensional plot, and below that is a contour plot of the same function on the same x- and y-axes.

In these plots, the extrema of the real and imaginary axes included are ± 2. The unit circle, roughly halfway out, is shown in the plots for reference as a raised contour on the plotted surface and as a dotted circle in the contour plot. Also, the spike on each plot is at the same (x, y) location for reference: $z = z_0 = -1.5 - j0.75$; it is also usually visible in the contour plot as a dot. Its vertical (functional) value is always artificially set to zero to help the viewer visualize the "zero level" plane on the plots.

Notice in Figure 3.1a that the height of the surface is constant on circles centered on the origin (the unit circle being one example), and it rises linearly with the distance from the origin. The surface in Figure 3.1b is a gradually sloping curve from $-\pi$ to π, rising with $\angle z$. It is constant along radial lines emanating from the origin, which is the center of the plot. In Figure 3.1c, we show the relation of the heights of these two surfaces, at the particular value z_0 of z, to the polar coordinates of z_0 on a "top view" of the z-plane.

Next, we discuss a function of a complex variable $f(z)$: a mapping from a complex plane representing the independent complex variable z to another complex plane representing the dependent variable $w = f(z)$. An example is $w = f(z) = z^2$. Depending on the position of $z = x + jy = z \angle \theta$ in the complex z-plane, $w = z^2$ will be located at a determined location in the w-plane. For instance, if $w = u + jv$, then in this example $u = x^2 - y^2$ and $v = 2xy$. Thus the position of w in the complex w-plane is at $x^2 - y^2$ along the horizontal axis and $2xy$ along the vertical axis. Equivalently, w is located at the point having polar coordinates $\left(|z|^2, 2\theta \right)$.

Figures 3.2a and b show, respectively, the magnitude and phase of w as a function of the position of z on the complex z-plane. Again on both plots the unit circle is shown for reference. Now the magnitude rises quadratically with the distance

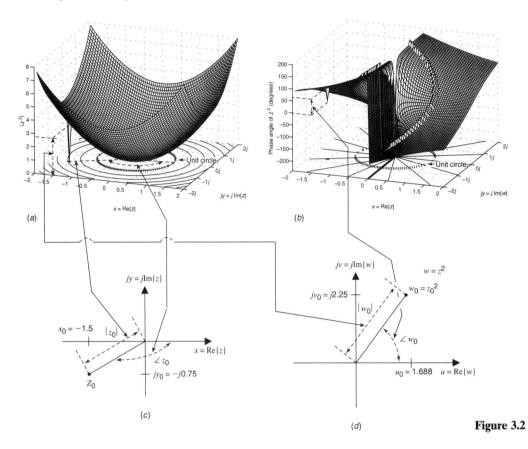

Figure 3.2

from the origin, but it again is constant on circles centered on the origin. The phase has the same form as that in Figure 3.1b except that two complete cycles occur as we traverse a path around the origin of the z-plane.

In Figure 3.2c, we show the location of the "spiked" point z_0 in the z-plane in the plots of Figures 3.2a and b. Then in Figure 3.2d we show how the heights of the surfaces in Figures 3.2a and b indicate the polar coordinates of $w_0 = z_0^2$, the corresponding value of w on the w-plane. This set of figures should clearly indicate the idea of a mapping from the z to the w-plane. The heights of the surfaces are determined by and precisely indicate the mapping function $f(z)$.

To get a feel for the consequence of raising to higher powers, Figure 3.3 shows $w = z^6$. Notice that the magnitude (Figure 3.3a) appears flatter in the middle than for Figure 3.2a because z^6 rises much more gradually than z^2 for $|z| < 1$. However, $|z^6|$ is much steeper than the plots in Figures 3.1a and 3.2a for $|z| > 1$. The phase plot in Figure 3.3b shows the six complete cycles $\angle w$ has from $-\pi$ to π as we traverse around the z-plane.

For yet another example, we consider in Figure 3.4 the function $f(z) = (z - 1)/(z + 1)$. This function is discussed further in Chapters 6 and 8 in a different form as the bilinear transform, an important function in the design of digital filters. Because the denominator of $f(z)$ is zero at $z = -1$, the magnitude (Figure 3.4a) has an infinite peak there. Note that $z = -1$ is on the unit circle. The magnitude quickly drops down to a fairly uniform value of roughly 1 for most of the rest of the z-plane.

The phase (Figure 3.4b) has a complicated shape. Note that on the real axis, $(z - 1)/(z + 1)$ is real; that is, its phase is zero or π. Also note that the phase jumps discontinuously across the real axis for $-1 < z < 1$, but varies continuously across it outside that interval. We explain this characteristic by writing the rectangular form:

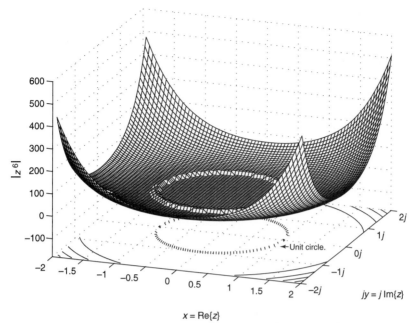

Figure 3.3a

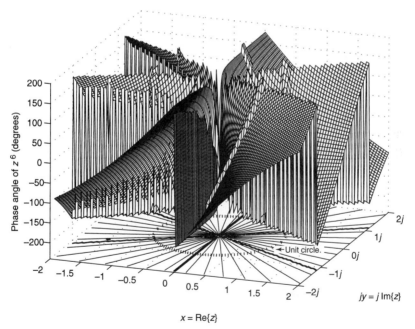

Figure 3.3b

$(z - 1)/(z + 1) = (x^2 + y^2 - 1 + j2y)/[(x + 1)^2 + y^2]$. For $|z| < 1$ and $y \approx 0$, the real part (aside from the nonnegative denominator factor) $x^2 + y^2 - 1$ is negative, so the phase $\tan^{-1}\{2y/(x^2 + y^2 - 1)\}$ will be $\pm\pi$ for y slightly positive and negative, respectively. For $|z| > 1$ and y nearly 0, however, $x^2 + y^2 - 1 > 0$, so the phase will be continuous through zero as y changes sign.

An important fact in filter design is that the negative half-plane of the w (later called s) plane maps to the interior of the unit circle in the z-plane. That is, all negative real parts of w arise from z such that $|z| < 1$. This is illustrated in Figure 3.4c, which shows the real part of w. We see that it is always negative within the unit circle and always positive outside it.[1] Contrarily, there is no such relation for the imaginary part of w (Figure 3.4d).

Transcendental functions of z are defined as well. For example, Figure 3.5 shows the function $w = \sin(z)$. Notice that due to the identities $\cos(\theta) = \frac{1}{2}(e^{j\theta} + e^{-j\theta})$ and $\sin(\theta) = [1/(2j)](e^{j\theta} - e^{-j\theta})$, we have:

$$\sin(z) = \sin(x + jy) = \sin(x)\cos(jy) + \cos(x)\sin(jy) = \tfrac{1}{2}\{\sin(x)[e^{-y} + e^{y}] - j\cos(x)[e^{-y} - e^{y}]\} = \sin(x)\cosh(y) + j\cos(x)\sinh(y).$$

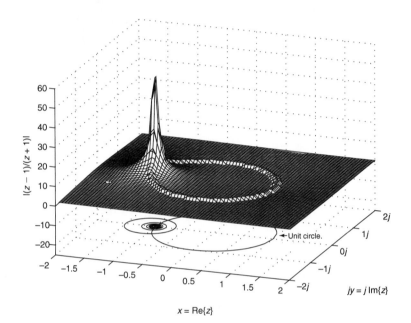

Figure 3.4a

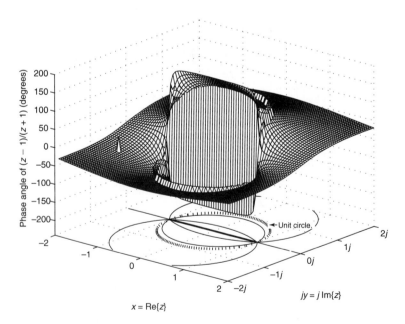

Figure 3.4b

[1]Keep in mind that although the heights of the plots are characteristics of w, the coordinates of the level plane in these plots are always x and y of $z = x + jy$.

Thus $|\sin(z)| = \{\sin^2(x)\cosh^2(y) + \cos^2(x)\sinh^2(y)\}^{1/2} = \{\sin^2(x) + \sinh^2(y)\}^{1/2}$. Therefore, $|\sin(z)|$ (see Figure 3.5a) rises exponentially with $|y|$ and oscillates with x (the oscillation being noticeable only for small $|y|$). The phase angle of $\sin(z)$ (see Figure 3.5b) is $\tan^{-1}\{\cot(x)\tanh(y)\}$, a rather complicated function to analyze. Finally, $\mathrm{Re}\{\sin(z)\}$ and $\mathrm{Im}\{\sin(z)\}$ are shown, respectively, in Figures 3.5c and d; they agree numerically with these results for their expressions $\sin(x)\cosh(y)$ and $\cos(x)\sinh(y)$, respectively.

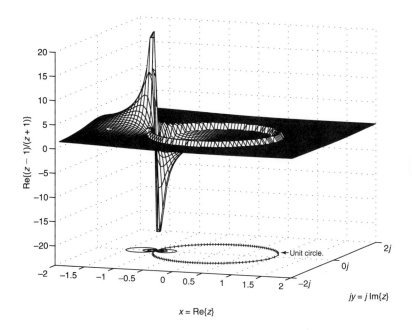

Figure 3.4c

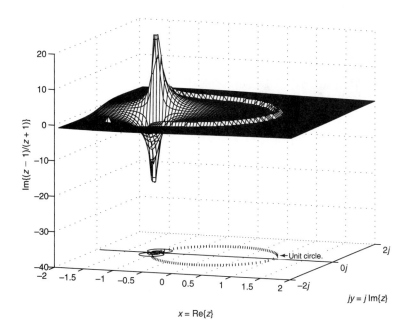

Figure 3.4d

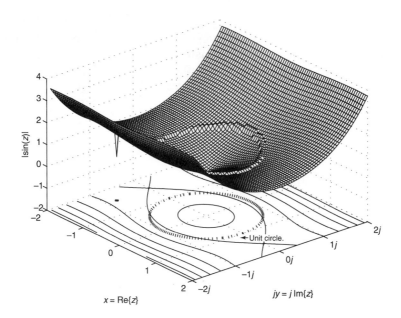

Figure 3.5*a*

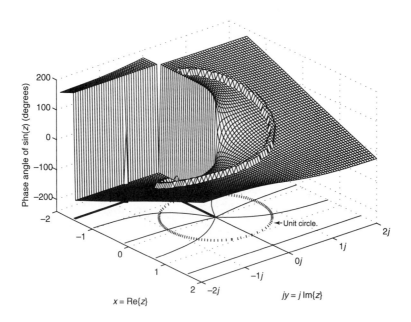

Figure 3.5*b*

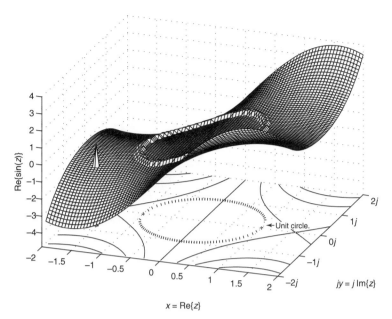

Figure 3.5c

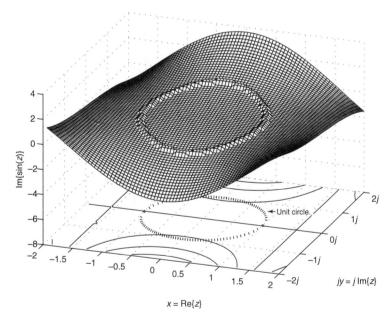

Figure 3.5d

EXAMPLE 3.1

Show the region of the z-plane satisfying $\left| \dfrac{z-4}{z-2} \right| \geq 3$.

Solution

At first, this problem seems impossible, because we would have to evaluate every point in the z-plane individually. Using some simple analysis, however, the answer is straightforward. Write the inequality as $|z-4| \geq 3|z-2|$.
Letting $z = x + jy$ and squaring both sides, we obtain

$$(x-4)^2 + y^2 \geq 9[(x-2)^2 + y^2],$$

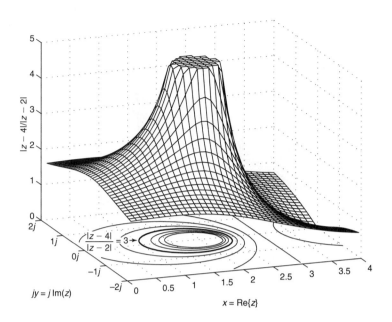

Figure 3.6

or, expanding the squares and collecting terms,

$$x^2(9-1) + x(8-36) + 36 - 16 + (9-1)y^2 \le 0,$$

or simplifying,

$$x^2 - \left(\frac{7}{2}\right)x + \frac{5}{2} + y^2 \le 0.$$

Completing the square, we have

$$x^2 - \left(\frac{7}{2}\right)x + \frac{49}{16} + \frac{40-49}{16} + y^2 \le 0,$$

or finally,

$$\left(x - \frac{7}{4}\right)^2 + y^2 \le \left(\frac{3}{4}\right)^2.$$

Thus, the region described by the original inequality is the interior of a circle of radius 3/4 centered on the point $z_0 = (7/4, 0)$; see Figure 3.6. Because the values rise to infinity at $z = 2$, all values above 5 were cut off to show the region of interest. The contour $|z-4|/|z-2| = 3$ is in bold and is notated. Also notice the contour $|z-4|/|z-2| = 1$; it is the straight contour $z = 3 + jy$. Substitution of $z = 3 + jy$ into $|z-4|/|z-2|$ indeed shows that it is unity for all y, and solution of $|z-4| = |z-2|$ using the preceding procedure yields $x = 3$ for any y.

The ideas above seem relatively straightforward. Things get more complicated, though, when the function $w = f(z)$ is a one-z-to-many-w function and we wish to invert from z back to w using $w = f(z)$. In this case, many values of w map to a single point z under the inverse of f, denoted $z = f^{-1}(w)$. Now viewing w as the "starting" plane and z as the "resulting" plane, we see that in trying to "go back" from $z = z_0$ to the w-plane via $w_0 = f(z_0)$, the result is not a single unique value of w_0.

In this situation, which arises for functions such as $w = f(z) = \pm\sqrt{z}$ and $w = f(z) = \ln(z)$, we define *branches* of the one-to-many function f. One branch is defined for each region of the w-plane over which the relation to the z-plane is one to one. Then, for a given branch, there is no ambiguity in inverting the function.

Let us consider the logarithm function. It has great importance for us in the discretization (sampling) of a continuous-time signal. In Chapter 6 we will find that the

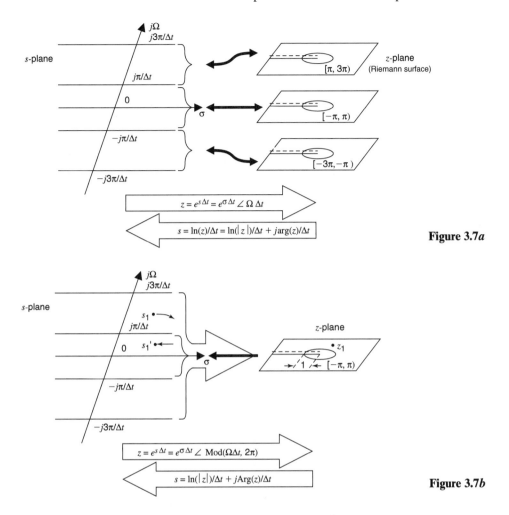

Figure 3.7a

Figure 3.7b

s-plane poles map to z-plane poles according to $z = e^{s\Delta t}$ and thus back from z to s according to $s = \ln(z)/\Delta t$, where Δt is the sampling interval. For the following discussion to be of maximum use later, in this example we change the notation from w to s for the "starting" plane and again use z for the "resulting" plane. By doing this, this discussion applies directly to the situation of sampling, which we discuss in Chapter 6.

In Figure 3.7a, we see the mapping from the s- to z-plane depicted as a one-to-one s-z mapping, where the multilayered z-plane is called a Riemann surface. The many-s-to-one-z version is shown in Figure 3.7b, but first we discuss Figure 3.7a. Because $s = \sigma + j\Omega$ and $z = e^{s\Delta t} = e^{\sigma\Delta t} \cdot e^{j\Omega\Delta t} = |z| \angle z$, we have $|z| = e^{\sigma\Delta t}$ and $\angle z = \arg(z) = \Omega\Delta t = \omega$.

However, because $e^{j[y_0 + 2\pi\ell]} = e^{jy_0}$ for any y_0, the distinction between $\Omega\Delta t$ and $\Omega\Delta t + 2\pi\ell$ is in practice lost; only $\mathrm{Arg}(z) = \mathrm{Mod}(\Omega\Delta t, 2\pi) = \Omega\Delta t - \epsilon \cdot 2\pi \cdot [(|\Omega\Delta t| + \pi)/(2\pi)]$ is retained, where $[\cdot]$ signifies the greatest integer (always rounded toward $-\infty$; floor in Matlab), $\epsilon = \mathrm{sgn}(\Omega\Delta t) = \pm 1$ for $\Omega\Delta t$ positive/negative, and $\mathrm{Mod}(\cdot)$ is the modulo of $\Omega\Delta t$ with respect to 2π as just defined so as to result in the range $(-\pi, \pi)$.

Notice the capital A on Arg signifying the principal argument, having the convenient range $(-\pi, \pi)$.[2] Given only $\mathrm{Arg}(z)$, we have no way of knowing from which $2\pi/\Delta t$-width region of the s-plane that s came. Our attention is generally focused on $(-\pi/\Delta t, \pi/\Delta t)$, the low-band signals; so usually the logarithm is taken over to Ω

---

[2]One could equally well define $\mathrm{Arg}(\cdot)$ to be on $(0, 2\pi)$, but usually we prefer $(-\pi, \pi)$.

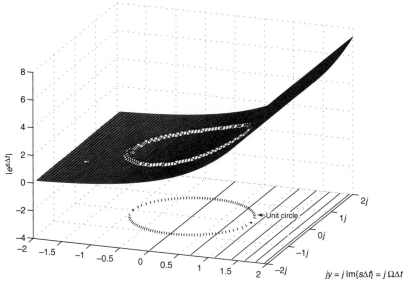

$x = \mathrm{Re}\{s\Delta t\} = \sigma\Delta t$

$jy = j\,\mathrm{Im}\{s\Delta t\} = j\,\Omega\Delta t$

Figure 3.8a

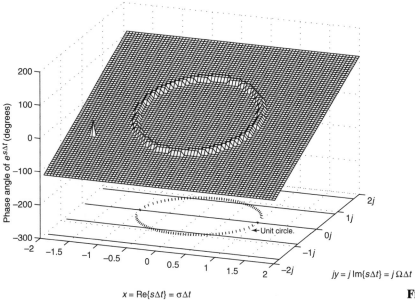

$x = \mathrm{Re}\{s\Delta t\} = \sigma\Delta t$

$jy = j\,\mathrm{Im}\{s\Delta t\} = j\,\Omega\Delta t$

Figure 3.8b

$\in(-\pi/\Delta t, \pi/\Delta t)$. In the unlikely case that we knew the absolute angle $\angle z = \arg(z)$, we could easily recover the correct principal (modulo) argument $\Omega\,\Delta t = \mathrm{Arg}(z)$ on $(-\pi, \pi)$ by the Matlab command $\Omega\Delta z = \mathtt{sign}(\angle z)*(\mathtt{rem}(\mathtt{abs}(\angle z)+ \pi, 2\pi) - \pi)$.

The ray emanating from the origin of the z-plane and cutting the plane at, for example, the angle π (as in Figure 3.7a) and delimiting the branch is called a branch cut. Thus in the case of a branch cut along the negative real axis, we define $z = e^{\sigma\Delta t}$ $\angle\mathrm{Arg}(e^{s\Delta t})$, where $\mathrm{Arg}(\cdot)$ is as defined above. For inversion, $s\Delta t = \ln(|z|) + j\,\mathrm{Arg}(e^{j\omega}) = \ln(|z|) + j\,\mathrm{Mod}(\omega, 2\pi)$; now see Figure 3.7b. This figure shows the conventional definition of the inverse of the exponential function $z = e^{s\Delta t}$. The modulo effect depicted in Figure 3.7b, in which an s-value s_1 outside $\Omega \in (-\pi/\Delta t, \pi/\Delta t)$ is transformed to the z-plane (z_1) and then back to s_1' with $\Omega' \in (-\pi/\Delta t, \pi/\Delta t)$ and not to

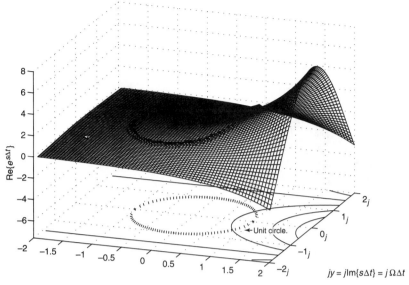

$x = \text{Re}\{s\Delta t\} = \sigma\Delta t$

$jy = j\text{Im}\{s\Delta t\} = j\,\Omega\Delta t$

Figure 3.8c

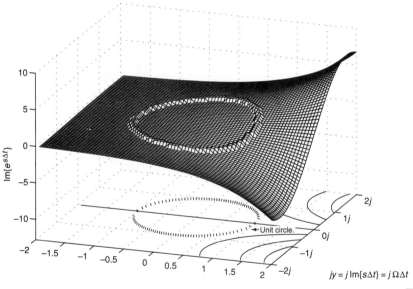

$x = \text{Re}\{s\Delta t\} = \sigma\Delta t$

$jy = j\text{Im}\{s\Delta t\} = j\,\Omega\Delta t$

Figure 3.8d

its original location, is a manifestation of the so-called aliasing effect we analyze in detail in Section 6.2. Thus aliasing is intertwined with the one-to-many multiple-branch mapping $s = \ln(z)/\Delta t$. A very important aspect of Figure 3.7b is that the entire $j\Omega$ axis ($s = j\Omega$) is mapped to $z = e^{j\Omega\Delta t} = e^{j\omega}$, where $\omega = \Omega\Delta t$. As $\left| z = e^{j\omega} \right| = 1$, we see that the entire $j\Omega$ axis maps to the unit circle in the z-plane, under the mapping $z = e^{s\Delta t}$, but with an infinite number of traversals, as in Figure 3.7a.

We show the magnitude, phase, real, and imaginary parts of $z = e^{s\Delta t}$ in Figure 3.8.[3] Noting that with $e^{s\Delta t}$ defined as $e^{\sigma\Delta t} \angle \text{Mod}(\Omega\Delta t, 2\pi)$, we see in Figure 3.8a that

---

[3]With our view of $s\Delta t$ in this case as being the "starting plane" and z the result, note that the "floor" axes in Figures 3.8 and 3.9 are thus $\sigma\Delta t$ and $j\Omega\Delta t$.

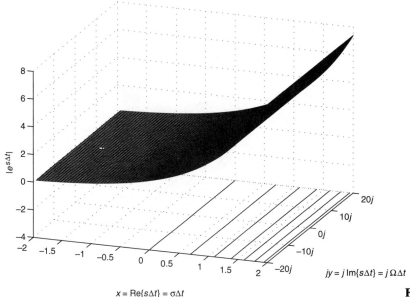

$x = \text{Re}\{s\Delta t\} = \sigma\Delta t$

$jy = j\,\text{Im}\{s\Delta t\} = j\,\Omega\Delta t$

Figure 3.9*a*

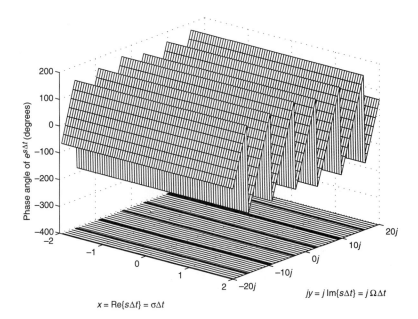

$x = \text{Re}\{s\Delta t\} = \sigma\Delta t$

$jy = j\,\text{Im}\{s\Delta t\} = j\,\Omega\Delta t$

Figure 3.9*b*

$|e^{s\Delta t}|$ rises exponentially with σ and in Figure 3.8*b* that $\angle e^{s\Delta t}$ increases linearly with Ω for $|\Omega\Delta t| < \pi \approx 3.14$. In Figures 3.8*c* and *d*, we show, respectively, the real and imaginary parts of $e^{s\Delta t}$; notice that because $\text{Re}\{e^{s\Delta t}\} = e^{\sigma\Delta t}\cos(\Omega\Delta t)$ and $\text{Im}\{e^{s\Delta t}\} = e^{\sigma\Delta t}\sin(\Omega\Delta t)$, there is a sinusoidal dependence on Ω. Because, as in the other three-dimensional plots, we have restricted $\Omega\Delta t$ to the range $|\Omega\Delta t| \leq 2$, the periodicity is not evident in Figures 3.8*b*, *c*, or *d*. Figure 3.9 shows the exponential function for the range $|\sigma\,\Delta t| \leq 2$ and $|\Omega\Delta t| \leq 20$. In this case, the 2π-periodicities in the argument, in the real part, and in the imaginary part are obvious. We shall see that $e^{s\Delta t}$ has a well-defined derivative everywhere on the *s*-plane.

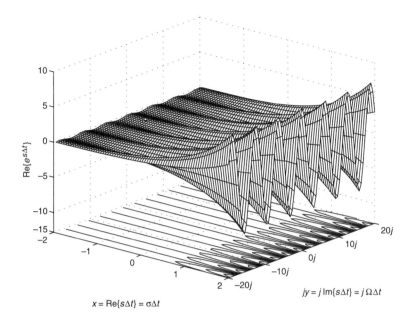

$$x = \text{Re}\{s\Delta t\} = \sigma \Delta t \qquad jy = j\,\text{Im}\{s\Delta t\} = j\,\Omega \Delta t$$

Figure 3.9c

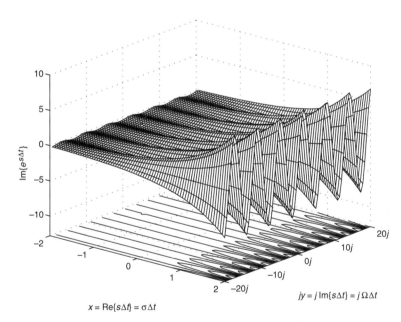

$$x = \text{Re}\{s\Delta t\} = \sigma \Delta t \qquad jy = j\,\text{Im}\{s\Delta t\} = j\,\Omega \Delta t$$

Figure 3.9d

3.3   CALCULUS OF FUNCTIONS OF A COMPLEX VARIABLE

3.3.1   Differentiation of Functions of a Complex Variable

Provided that appropriate modifications and intricacies are attended to, we may perform integral and differential calculus on functions of a complex variable. For example, a function $f(z)$ may possess a first derivative at $z = z_0$,

$$\frac{df(z)}{dz}\bigg|_{z=z_0} = f'(z_0) = \lim_{z \to z_0} \frac{f(z) - f(z_0)}{z - z_0}, \tag{3.1}$$

everywhere within a certain region of the z-plane. In such regions, the function is called analytic. If the region is the entire z-plane, the function is called entire.

EXAMPLE 3.2

Is a proper rational function entire?

Solution

No, because it has at least one pole, where the function is not analytic. Consequently, a proper rational function may not be called entire. This example is studied more quantitatively in the problems.

Unlike the case for real variables, there are innumerable directions from which to take the derivative limit $z \to z_0$. To be analytic, the derivative limit must exist and be the same from any approach direction; see Figure 3.10. However, use of the total derivative of the real and imaginary components u and v of $f(z) = u(x, y) + jv(x, y)$ with respect to x and y (where $z = x + jy$) shows that it is sufficient to show that the limit exists in the two orthogonal directions $\hat{\mathbf{x}}$ and $\hat{\mathbf{y}}$, because the derivative along any other direction can be decomposed into partial derivatives along $\hat{\mathbf{x}}$ and $\hat{\mathbf{y}}$. By equating the derivative of $f(z)$ by an approach along $\hat{\mathbf{x}}$ with that for one along $\hat{\mathbf{y}}$, we obtain the important Cauchy–Riemann equations.

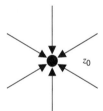

Figure 3.10

Suppose we wish to determine whether a function $f(z)$ is analytic at $z = z_0 = x_0 + jy_0$. The following two derivatives must be equal: the derivative along $\hat{\mathbf{x}}$ (see Figure 3.11a) and that along $\hat{\mathbf{y}}$ (Figure 3.11b). The derivative along $\hat{\mathbf{x}}$ is

$$f'(z_0) = \lim_{z \to z_0} \frac{u(x, y_0) + jv(x, y_0) - u(x_0, y_0) - jv(x_0, y_0)}{x - x_0}$$

$$= \lim_{z \to z_0} \frac{u(x, y_0) - u(x_0, y_0)}{x - x_0} + j \frac{v(x, y_0) - jv(x_0, y_0)}{x - x_0}. \tag{3.2a}$$

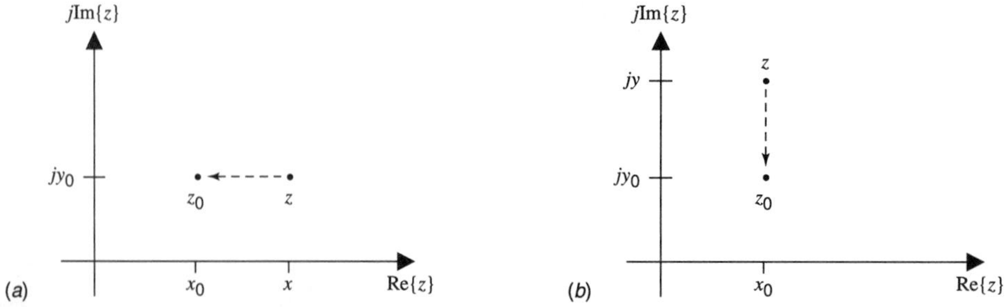

Figure 3.11

According to (3.1), the first term above is $\partial u / \partial x$ at $z = z_0$, whereas the second term is $j \partial v / \partial x$ at $z = z_0$. Consequently, (3.2a) is rewritten as

$$f'(z_0) = \left. \frac{\partial u}{\partial x} \right|_{z=z_0} + j \left. \frac{\partial v}{\partial x} \right|_{z=z_0} = \left. u_x \right|_{z=z_0} + \left. j v_x \right|_{z=z_0}. \tag{3.2b}$$

The derivative along $\hat{\mathbf{y}}$ (Figure 3.11b) is, similarly,

$$f'(z_0) = \lim_{z \to z_0} \frac{u(x_0, y) + jv(x_0, y) - u(x_0, y_0) - jv(x_0, y_0)}{j(y - y_0)}$$

$$= \lim_{z \to z_0} \frac{u(x_0, y) - u(x_0, y_0)}{j(y - y_0)} + \frac{v(x_0, y) - v(x_0, y_0)}{y - y_0}, \tag{3.2c}$$

which again by (3.1) may be written as

$$f'(z_0) = -j \left. \frac{\partial u}{\partial y} \right|_{z=z_0} + \left. \frac{\partial v}{\partial y} \right|_{z=z_0} = \left. v_y \right|_{z=z_0} - \left. j u_y \right|_{z=z_0}. \tag{3.2d}$$

When the real and imaginary parts of (3.2b) and (3.2d) are equated, as they must be for $f(z)$ to have a derivative at $z = z_0$ and be called analytic there, we obtain the Cauchy–Riemann conditions

$$u_x = v_y \text{ and } u_y = -v_x, \quad \text{Cauchy-Riemann equations.} \tag{3.3}$$

where all the derivatives in (3.3) are evaluated at $z = z_0$. These equations can be used to determine whether a function is analytic at a point or within a region.

EXAMPLE 3.3

Is $f(z) = 3x + j(x + 7y^2)$ analytic, where $z = x + jy$?

Solution

$u = 3x$, $v = x + 7y^2$, so $u_x = 3$, $v_y = 14y \neq u_x$. Also, $u_y = 0$, and $v_x = 1 \neq -u_y$. So there is no region in which $f(z)$ is analytic.

Figure 3.12 graphically illustrates the violation of the Cauchy–Riemann equations. We show $u(x,y)$ in the upper figure and $v(x,y)$ in the lower figure. Consider the

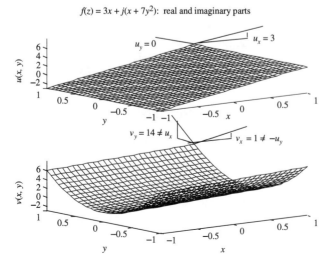

$f(z) = 3x + j(x + 7y^2)$: real and imaginary parts

Figure 3.12

particular point $z = 1 + j$; the various partial derivatives are shown, and comparison verifies the result that the Cauchy–Riemann equations are indeed violated. For example, $u_x = 3$ in the upper figure is not equal to $v_y = 14$ in the lower figure.

EXAMPLE 3.4

Is $f(z) = (1 + j)z^3 + jz + 1 - 8j$ analytic?

Solution

First, $z^3 = (x + jy)^3 = (x^2 - y^2 + j2xy)(x + jy) = x^3 - 3xy^2 + j(3x^2y - y^3)$, so $f(z) = x^3 - 3xy^2 - y + 1 - 3x^2y + y^3 + j(3x^2y - y^3 + x - 8 + x^3 - 3xy^2)$ so that $u_x = 3x^2 - 3y^2 - 6xy$, $v_y = 3x^2 - 3y^2 - 6xy = u_x$ and $u_y = -6xy - 1 - 3x^2 + 3y^2$, $v_x = 6xy + 1 + 3x^2 - 3y^2 = -u_y$. Thus the Cauchy–Riemann equations are satisfied and $f(z)$ is analytic for all z. Figure 3.13 illustrates the satisfaction of the Cauchy–Riemann equations, in particular at the point $z = 1 - j$. For example, $u_y = 5$ in the upper figure is equal to $-\{v_x = -5\}$ in the lower figure. Unlike $f(z)$ in Example 3.3, here $f(z)$ is a simple polynomial in z, so we expect it to be analytic for all z.

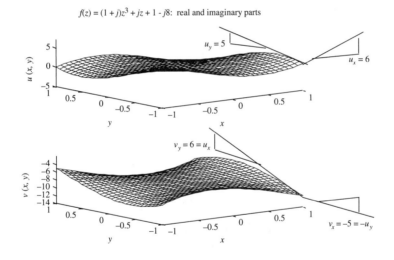

$f(z) = (1 + j)z^3 + jz + 1 - j8$: real and imaginary parts

Figure 3.13

In general, differentiation of elementary functions is straightforward: $d/dz\{g[f(z)]\} = g'[f(z)]f'(z)$ (the chain rule), $d/dz[z^n] = nz^{n-1}$, $d/dz[\sin(z)] = \cos(z)$, $d/dz[\cos(z)] = -\sin(z)$, $d/dz[e^z] = e^z$, $d/dz[\ln(z)] = 1/z$ within a particular branch, and so forth.

3.3.2 Integration of Functions of a Complex Variable

It is of considerable interest that we can view a function f of one complex variable z as a pair of real-valued functions u and v of two real variables x and y: $f(z) = u(x, y) + jv(x, y)$. The variables x and y can be thought of as orthogonal variables in two-dimensional space. The importance of this recognition is particularly manifest in the integration of a function of a complex variable. Two-dimensional integration theory is well developed, and we can call on its results without having to rederive them; they are found in any freshman calculus text.

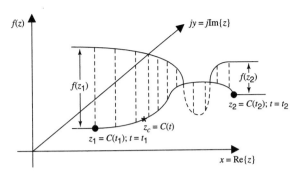

Assumed for simplicity: $f(z)$ is real-valued.   **Figure 3.14**

In analogy with two-dimensional integration, we just replace the two-dimensional position vector $\rho = (x, y)$ by the complex variable $z = x + jy$. Thus, as in two-dimensional integration theory, we can express the integration of $f(z)$ from one point z_1 in the z-plane to another point z_2 in terms of a single real-valued parameter t defining the integration contour C; see Figure 3.14. We let $z = C(t)$, where when z runs from z_1 to z_2, t runs from t_1 to t_2: $C(t_1) = z_1$, $C(t_2) = z_2$. We also have $dz = C'(t)\,dt$ because z depends on the single parameter t through $C(t)$. Think of C as a stretch of road over mountainous terrain, and we drive a car over it from $t = t_1$ to $t = t_2$, and $f(z)$ is our elevation above the flat "plain plane" [assuming that $f(z)$ is real]. We have, integrating our elevation over the stretch of road,

$$\int_C f(z)\,dz = \int_{t_1}^{t_2} f[C(t)]C'(t)\,dt. \tag{3.4}$$

Suppose that $f(z)$ is the derivative of another function of a complex variable $F(z)$, which is analytic on the contour C. Thus (3.4) can be written

$$\int_C f(z)\,dz = \int_{t_1}^{t_2} F'[C(t)]C'(t)\,dt, \tag{3.5a}$$

which by the chain rule in reverse is

$$\int_C f(z)\,dz = \int_{t_1}^{t_2} d/dt\{F(C(t)\} \,dt$$

Assumed: $f(z) = dF(z)/dz$, where $F(z)$ is analytic on contour C.

$$= F(C(t_2)) - F(C(t_1))$$

$$\tag{3.5b}$$

Fundamental theorem of calculus (complex version).

$$= F(z_2) - F(z_1).$$

Notice that if we let $z_2 = z_1$, C becomes a closed contour, and the right-hand side of (3.5) is zero. Thus

$$\oint_C f(z)\,dz = 0 \quad \text{if on } C, f(z) = d/dz\{F(z)\}, \tag{3.6}$$

Consequence of the complex fundamental theorem of calculus.

where F is a differentiable (analytic) function on C $[dF(z)/dz = f(z)]$. Note that we did not say F is analytic within the contour; that F be analytic on the contour is the only requirement because the contour itself (and an infinitesimally thick region on either side of C) is the only set of values involved in the integration and differentiability determination.

It surely seems reasonable to conclude that the closed contour "average" is zero, for we end up back where we started; we will see, however, that this is not always true. Another condition for which the closed contour integral of $f(z)$ is zero is if both on C and within C, $f(z)$ is analytic. The result is the well-known Cauchy integral theorem, which is provable using Green's theorem for two-dimensional integration and the Cauchy-Riemann equations. Notice that in (3.4), the integrand $f(z)$ times the differential dz is

$$f(z)\, dz = (u + jv)(dx + j\, dy) \tag{3.7}$$

$$= u\, dx - v\, dy + j\{u\, dy + v\, dx\}.$$

Now, by Green's theorem for two-dimensional integration,[4]

$$\oint_C u\, dx - v\, dy = \int_S \{-v_x - u_y\} dS, \tag{3.8a}$$

where S is the region enclosed by C, and

$$\oint_C u\, dy + v\, dx = \int_S \{u_x - v_y\}\, dS. \tag{3.8b}$$

By the Cauchy–Riemann equations (3.3), the right-hand sides of both (3.8a) and (3.8b) are zero if the function is analytic everywhere in S. Consequently, (3.8) and (3.7) give the Cauchy integral theorem

$$\oint_C f(z)\, dz = 0 \quad \text{if both on } C \text{ and within } C, f \text{ is analytic.} \tag{3.9}$$

$$\text{Cauchy's integral theorem.}$$

Although (3.9) might appear redundant in light of (3.6), Cauchy's integral theorem is helpful when $F(z)$ is difficult to determine and thus test for analyticity on C. It is often much easier to directly test $f(z)$ for analyticity.

It is an interesting fact that the condition for a two-dimensional vector field (e.g., the gravitational or coulomb force fields) to be conservative is identical to that for a function of a complex variable to be analytic. In fact, they are just different applications of the same statement: The integral of $f(z)\, dz$ along a path—analogously, the integral of $\mathbf{F} \cdot d\boldsymbol{\ell}$ (work) along a path— must be independent of the path for $f(z)$ to be analytic/$\mathbf{F}$ to be conservative.

Both (3.6) and (3.9) are very useful. For example, consider the integral $I = \oint_C (z - z_0)^n dz$ along a closed circular contour C enclosing z_0. We know that for $n < -1$, $f(z) = (z - z_0)^n$ is the derivative of a function analytic on C [namely, $F(z) = (z - z_0)^{n+1}/(n + 1)$], so by (3.6), $I = 0$ even though $f(z) = (z - z_0)^n$ is *not* analytic at $z = z_0$ for $n < -1$. By (3.9), we know that for $n > -1$, again $I = 0$ because $(z - z_0)^n$ is then analytic everywhere within and on C [or we could also use (3.6) here].

What about $n = -1$? Obviously, $1/(z - z_0)$ is not analytic at $z = z_0$, which is in C, so (3.9) does not hold. Also, the integral of $dz/(z - z_0)$ is $F(z) = \ln(z - z_0) = \ln(|z - z_0|) + j\angle(z - z_0)$, which is discontinuous across the branch cut on the ray emanating from z_0 parallel to the negative real axis necessarily defined for $\ln(z - z_0)$. Specifically, $\text{Im}\{\ln(z - z_0)\} = \angle(z - z_0)$ is discontinuous by 2π when that ray is crossed.

---

[4]See any calculus text for a proof of it.

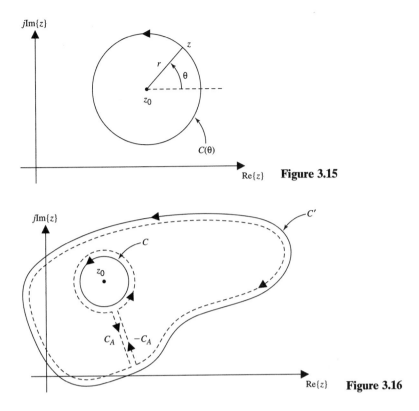

Figure 3.15

Figure 3.16

Therefore, (3.6) does not hold either because on C, $F(z)$ is discontinuous where C crosses that ray. Suppose, for simplicity, that $z_0 = 0$, as illustrated in Figure 3.7b, where $s = \ln(z)/\Delta t$ is now "$F(z)$". Then if by going counterclockwise on a contour C around the origin we cross the branch cut along the negative real axis of the z-plane, the imaginary part of $s = \ln(z)/\Delta t$ suddenly jumps from $\pi/\Delta t$ to $-\pi/\Delta t$; it is discontinuous. Thus on the branch cut (i.e., where C crosses the negative real axis for $z_0 = 0$), $\ln(z - z_0) = \ln(z)$ is not differentiable.

In fact, we will now show that $I \neq 0$ for the case $n = -1$. For $n = -1$ and for a general value of z_0, it is possible to directly evaluate I. Let the contour be defined by $z = C(\theta) = re^{j\theta} + z_0$, where our contour parameter is here called θ rather than t; see Figure 3.15. Then $dz = jre^{j\theta}\, d\theta$ and $z - z_0 = re^{j\theta}$, so that the integral of $dz/(z - z_0)$ around C is just the integral from 0 to 2π of $j\, d\theta$; the result is $I = j2\pi$.

We can obtain a more general result by making use of the deformation theorem, which we will also find useful later; see Figure 3.16. Assume that a general function $f(z)$ to be integrated is analytic everywhere within C' except at $z = z_0$. [Although $f(z) = 1/(z - z_0)$ in our discussion above, the deformation theorem applies to any qualifying $f(z)$.] With C_B defined as the dashed total closed contour avoiding z_0, C the little circular contour around z_0, and C_A as the contour in one direction from C to C', and defining all closed contours to be counterclockwise, we know that

$$\oint_{C_B} f(z)\, dz = \left\{ \oint_C + \int_{C_A} - \oint_{C'} - \int_{C_A} \right\} f(z)\, dz, \tag{3.10}$$

which by the Cauchy integral theorem is zero because the function $f(z)$ is analytic everywhere within C_B.

From this reasoning, we obtain that the integral around C' is equal to the integral around C, no matter how deformed C' is (no matter what shape it has). Thus

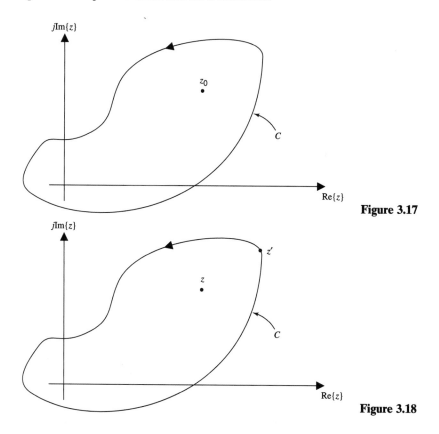

Figure 3.17

Figure 3.18

the general conclusion is the deformation theorem: The integral of $f(z)$ around any closed contour is the same, provided all such contours lie in the same region of analyticity of $f(z)$. Equivalently, the integral of $f(z)$ is independent of the contours C and C' if $f(z)$ is analytic between those contours.

With this fact, we can combine all our results for the important integral of $(z - z_0)^n$ as follows. Let C be any contour enclosing z_0; see Figure 3.17. Then

$$(I =) \oint_C (z - z_0)^n \, dz = j2\pi\delta(n + 1), \tag{3.11}$$

where $\delta(n)$ is, again, the Kronecker delta function introduced in Section 2.2. Note that if C were a contour that encircled z_0 m times rather than once as previously assumed, the result would be $j2\pi m\delta(n + 1)$.

Now suppose that C is any closed contour enclosing a particular value of z and a region in which $f(z)$ is analytic; see Figure 3.18. Then the integral of $f(z')dz'/(z' - z)$ is determined as follows:

$$\oint_C \frac{f(z')}{z' - z} \, dz' = f(z) \overbrace{\oint_C \frac{dz'}{z' - z}}^{= j2\pi} + \underbrace{\oint_C \frac{f(z') - f(z)}{z' - z} \, dz'}_{= 0}. \tag{3.12}$$

$= f'(z')$ if C is tight around z; then for other C not tight around z, just use the deformation theorem.

The first integral on the right side of (3.12) is, by (3.11), equal to $j2\pi \cdot f(z)$. The second integral is, by the deformation theorem, equal to the integral of $[f(z') - f(z)]/(z' - z)$ around a circular contour closed tightly around z. If the radius of the circular contour is made vanishingly small, by (3.1) $[df(z)/dz = \lim_{z \to z_0} \{f(z) - f(z_0)\}/\{z - z_0\}]$ we are merely integrating the derivative of $f(z)$, $f'(z)$ around C. But $f'(z) = d/dz\{f(z)\}$, and $f(z)$ is analytic on the contour C, so by (3.6) the second integral is zero.

We therefore have

$$f(z) = \frac{1}{2\pi j} \oint_C \frac{f(z')}{z' - z}\, dz'. \qquad\qquad \text{Cauchy integral formula.}$$

$$(3.13)$$

The Cauchy integral formula (3.13) is useful for, among other things, deriving Taylor's and Laurent's series for expanding functions over power series. It also shows us that we can obtain $f(z)$ for any z in a region of analyticity from its values along a finite contour in that same region, a fact we return to in Section 4.2.

EXAMPLE 3.5

Evaluate the following integrals:

(a) $I = \oint_C \frac{z'^3}{z' - j}\, dz'$,

where C is any closed contour enclosing j in the z-plane.

(b) $I = \oint_C \frac{z'^2}{z' - 1}\, dz'$,

where C is a circular contour of diameter 1 centered on the origin.

(c) $I = \oint_C \frac{\cos(z')}{z' - \pi/4}\, dz'$,

where C is any contour enclosing $\pi/4$ in the z-plane.

Solution:

(a) Because z^3 is an analytic function everywhere, we can immediately apply the Cauchy integral formula (3.13) to obtain $I = 2\pi j(j^3) = 2\pi$.

(b) We might try to apply (3.13), but we notice that C does not enclose $z = 1$. Thus the integrand is an analytic function everywhere on and within the contour, and so by the Cauchy integral theorem (3.9), $I = 0$.

(c) Again, because $\cos(z)$ is an analytic function everywhere and C encloses $\pi/4$, we can immediately apply the Cauchy integral formula (3.13) to obtain $I = 2\pi j \cos(\pi/4) = j\pi \cdot \sqrt{2}$.

By differentiating (3.13) n times, we obtain the formula for the nth derivative of an analytic function at point z in terms of an integral involving just $f(z)$, not its derivatives:

$$f^{[n]}(z) = \frac{n!}{2\pi j} \oint_C \frac{f(z')}{(z' - z)^{n+1}}\, dz', \qquad\qquad (3.14)$$

where the square brackets in an exponent will henceforth (unless otherwise noted) symbolize the derivative of order indicated within the brackets. Again, (3.14) is useful in deriving Taylor's series. It can also be used to show that if $f(z)$ is analytic, then so are its derivatives of all orders, a fact not true in general for functions of a real variable. Note, however, that if $f(z)$ is analytic, that does not imply that its integral $F(z)$ is; just recall the case $f(z) = 1/z$ (analytic for all $z \neq 0$), for which $F(z) = \ln(z)$, where $F(z)$ is not analytic across any branch cut.

EXAMPLE 3.6

Evaluate the following integral.

$$I = \oint_C \frac{z'^5 - 3\sin(z')}{(z' - \pi/6)^4}\, dz',$$

where C is the unit circle centered on the origin.

Solution

The numerator is an analytic function, and C encloses $\pi/6$. Using (3.14), we have $I = (2\pi j/3!)f^{[3]}(\pi/6)$, where $f(z) = z^5 - 3\sin(z)$ so that $f'(z) = 5z^4 - 3\cos(z), f^{[2]}(z) = 20z^3 + 3\sin(z)$, and $f^{[3]}(z) = 60z^2 + 3\cos(z)$. Substituting $z = \pi/6$ gives $f^{[3]}(\pi/6) = 5\pi^2/3 + 3\sqrt{3}/2$, so we finally obtain $I = (2\pi j/6) \cdot \left(5\pi^2/3 + 3\left(\dfrac{\sqrt{3}}{2}\right)\right) = (\pi j/18)(10\pi^2 + 9\sqrt{3})$.

3.3.3   Taylor and Laurent Series

We can use the Cauchy integral formula (3.13) to obtain the Taylor expansion of an analytic $f(z)$ about a point z_0, where z and z_0 are anywhere in a region within which $f(z)$ is analytic; see Figure 3.19. The Taylor series is an expansion of $f(z)$ over non-negative powers of $(z - z_0)$. We could define an arbitrarily shaped contour C whose entire interior lies in the region of analyticity of $f(z)$ (see Figure 3.20), but because of the deformation theorem the result of any integration will be the same as that for the circle C' centered on z_0 in Figure 3.19, on which we now focus. Note that

$$\frac{1}{z' - z} = \frac{1}{z' - z_0} \cdot \frac{1}{1 - \dfrac{z - z_0}{z' - z_0}}. \tag{3.15}$$

Note that we have $|z - z_0| < |z' - z_0|$ because z_0 is the center of C', z is within C', and z' is on C'.[5] Consequently, we can use, in "reverse," the familiar result

$$\sum_{n=0}^{\infty} a^n = \frac{1}{(1 - a)} \quad \text{for } |a| < 1 \tag{3.16}$$

with $a = (z - z_0)/(z' - z_0)$ to rewrite (3.15) as

$$\frac{1}{z' - z} = \frac{1}{z' - z_0} \cdot \sum_{n=0}^{\infty} \left\{ \frac{z - z_0}{z' - z_0} \right\}^n. \tag{3.17}$$

---

[5]For this inequality to hold all along the arbitrary contour C, z must be within the largest circle centered on z_0 that can be inscribed within C, such as r_1 in Figure 3.20. Also, for the same reason, C must enclose z_0 (for both Taylor and Laurent series, so C must enclose the origin for the inverse z-transform integral).

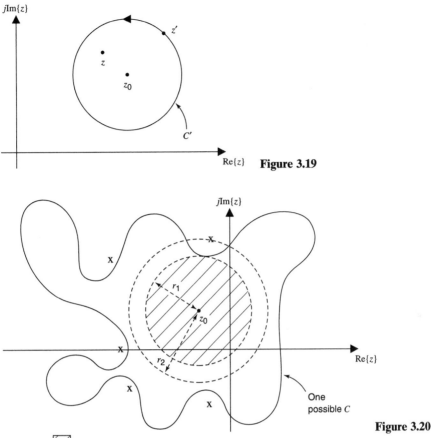

Figure 3.19

Figure 3.20

z must lie in [⧄] for C as shown; for other allowable C, z may be anywhere within circle $|z - z_0| < r_2$.
For Taylor series, C must enclose z_0 but no poles, and thus lie in the disk $|z - z_0| < r_2$.

Substituting this result into the Cauchy integral formula (3.13) gives

$$f(z) = \frac{1}{2\pi j} \oint_C \frac{f(z')}{z' - z_0} \sum_{n=0}^{\infty} \left\{ \frac{z - z_0}{z' - z_0} \right\}^n dz'. \tag{3.18a}$$

Pulling out the sum, rearranging, and using the deformation theorem to change C' to the more general C (which, however, must always be farther from z_0 than is z), we have

$$f(z) = \sum_{n=0}^{\infty} \left[\frac{1}{2\pi j} \oint_C \frac{f(z')}{(z' - z_0)^{n+1}} \, dz' \right] (z - z_0)^n \tag{3.18b}$$

or

$$f(z) = \sum_{n=0}^{\infty} a_n (z - z_0)^n, \qquad \text{Taylor series.} \tag{3.18c}$$

where by comparing the integral in (3.18b) with that in (3.14),

$$a_n = \frac{f^{[n]}(z_0)}{n!}. \qquad \text{Taylor series coefficients.} \tag{3.19}$$

Equations (3.18c) – (3.19) are the Taylor series for a function $f(z)$ about $z = z_0$, where z and z_0 lie in a region in which $f(z)$ is analytic within the contour C.[6] If $z_0 = 0$, the series is called Maclaurin.

EXAMPLE 3.7

Give the first four terms of the Taylor expansion of $f(z) = \dfrac{z^2 + 2z + 5}{(z + 2)(z^2 + 4z + 20)}$ about the point $z_0 = -2 + j2$. Also determine the region of convergence of this expansion.

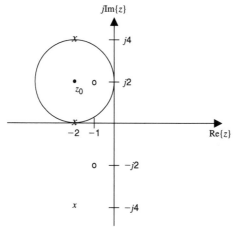

Figure 3.21

Solution

The pole–zero plot is as shown in Figure 3.21. The distance from z_0 to the poles closest to z_0 is 2. Therefore if we restrict z to the interior of the circle $|z - z_0| < 2$, then the condition for (3.16) holds for any C whose interior is in the region of analyticity of $f(z)$, namely, any simple contour in the z-plane enclosing z_0 but none of the poles of $f(z)$. Thus, the region of convergence is $|z - z_0| < 2$. Notice that even though there is a zero at a distance of only 1 from z_0, it has no bearing on the specification of the region of convergence. The coefficients are obtained from (3.19) as follows:

$$a_0 = f(z_0) = \frac{z^2 + 2z + 5}{(z + 2)(z^2 + 4z + 20)}\bigg|_{z=-2+j2} = -0.1667 - j0.041667.$$

For all our final numerical evaluations, we can conveniently use Matlab. Noting that $(z + 2)(z^2 + 4z + 20) = z^3 + 6z^2 + 28z + 40$, we obtain

---

[6]Note the distinction between the region of analyticity of $f(z)$ and the region of convergence of the series expansion. See Figure 3.20, where the region of analyticity in which the contour C in (3.18b) may be drawn is limited only to avoiding inclusion of any of the poles, whereas the region of convergence of the Taylor series is limited to $|z - z_0| < r$ where $r = \min\{r_1, r_2\}$, where r_1 = radius of largest circle centered on z_0 that can be inscribed in C and r_2 = distance from z_0 to the pole closest to z_0.

$$a_1 = f'(z_0) = \frac{(z^3 + 6z^2 + 28z + 40)(2z + 2) - (z^2 + 2z + 5)(3z^2 + 12z + 28)}{(z^3 + 6z^2 + 28z + 40)^2}\bigg|_{z=-2+j2}$$

$$= \frac{-z^4 - 4z^3 + z^2 + 20z - 60}{z^6 + 12z^5 + 92z^4 + 416z^3 + 1264z^2 + 2240z + 1600}\bigg|_{z=-2+j2} = 0.17361 + j0.05556.$$

Next,

$$a_2 = \frac{f''(z_0)}{2!}$$

$$= \frac{\begin{array}{l}(z^6 + 12z^5 + 92z^4 + 416z^3 + 1264z^2 + 2240z + 1600)(-4z^3+2z^2+2z+20) \\ - (-z^4 -4z^3 + z^2 + 20z - 60)(6z^5 + 60z^4 + 368z^3 + 1248z^2 + 2528z + 2240)\end{array}}{2(z^6 + 12z^5 + 92z^4 + 416z^3 + 1264z^2 + 2240z + 1600)^2}\bigg|_{z=-2+j2}$$

$$= 0.03241 - j0.002315.$$

The tedium became overwhelming, so the author wrote Matlab code to automate the calculation of the remaining coefficient. The result is

$$a_3 = \frac{f'''(z_0)}{3!} = -0.04128 - j0.01543.$$

We conclude that the four-term (approximate) Taylor expansion for $f(z)$ about z_0 is

$$\frac{z^2 + 2z + 5}{(z + 2)(z^2 + 4z + 20)} \approx -0.16667 - j0.041667 + (0.17361 + j0.05556)(z + 2 - j2)$$
$$+ (0.03241 - j0.002315)(z + 2 - j2)^2$$
$$- (0.04128 + j0.01543)(z + 2 - j2)^3,$$

which is accurate for z near $z_0 = -2 + j2$.

In digital signal processing, the z-transform is an essential representation for linear shift-invariant systems. We will see in Section 3.4.1 that it is merely a special Laurent series. Thus to understand the z-transform, we must first understand Laurent series. The Laurent series of $f(z)$ is an expansion of $f(z)$ about z_0 over possibly both nonnegative and negative powers of $z - z_0$. The negative powers can result from singularities between z, the point of evaluation of $f(\cdot)$, and z_0, the point about which we construct the series. One of those singularities can be z_0 itself.

There are different Laurent expansions of a single function $f(z)$ for different regions of the z-plane. The regions are distinguished by which poles of $f(z)$ lie between the point of evaluation (z) and the point about which the expansion is formed (z_0). It is most convenient to delimit these regions by concentric circles centered on z_0, of radii equal to the distances from each pole to z_0. Thus a typical situation is that shown in Figure 3.22. In this case, there are two singularities separating z_0 from the annular region including the point of evaluation, z.

We again will use the Cauchy integral formula (3.13) and the deformation theorem to determine the series expansion. Let C_z be the closed circular contour around z in Figure 3.22, and let C_1 and C_2 be, respectively, the closed circular contours around the outside and inside of the annular convergence region including z. Not only is $f(z')$ analytic for all z' within the closed dashed contour in Figure 3.22, but so also is $f(z')/(z' - z)$, because z is excluded from its interior. Therefore, the deformation theorem shows [just as it was used for (3.10)] that

$$\oint_{C_1} \frac{f(z')}{z' - z}\, dz' = \left\{ \oint_{C_2} + \oint_{C_z} \right\} \frac{f(z')}{z' - z}\, dz', \qquad (3.20)$$

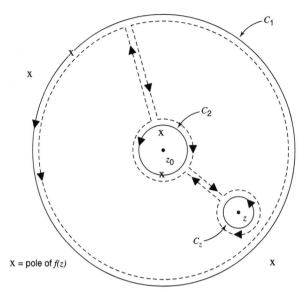

Figure 3.22

where we note that contributions along all opposing straight line segments again cancel each other.

By the Cauchy integral formula, the integral around C_z in (3.20) is equal to $2\pi j f(z)$, so that (3.20) becomes:

$$f(z) = \frac{1}{2\pi j} \left\{ \oint_{C_1} - \oint_{C_2} \right\} \frac{f(z')}{z' - z} \, dz'. \tag{3.21}$$

Consider the integral around C_1; refer to Figure 3.23. Because z is within C_1, $|z - z_0| < |z' - z_0|$, the same condition used to derive (3.18). Thus we can re-express the C_1 integral in exactly the same way that the Taylor expansion integral was reexpressed in (3.15) through (3.18). We conclude that the integral around C_1 in (3.21) is equal to the right-hand side of (3.18b) with C replaced by C_1. Notice that all we have done is modify how we express the integral factor $1/(z' - z)$. The only condition for so expressing it is $|z - z_0| < |z' - z_0|$, which does not at all require that $f(z')$ be analytic within C_1.

Now consider the integral around C_2 in (3.21); refer to Figure 3.24. Because z is outside C_2, $|z' - z_0| < |z - z_0|$. Therefore, we have

$$\frac{-1}{z' - z} = \frac{1}{z - z_0} \cdot \frac{1}{1 - \dfrac{z - z_0}{z' - z_0}}. \tag{3.22a}$$

Using the geometric sum formula in reverse, (3.22a) may be written as

$$\frac{-1}{z' - z} = \frac{1}{z - z_0} \cdot \sum_{n=0}^{\infty} \left\{ \frac{z' - z_0}{z - z_0} \right\}^n \tag{3.22b}$$

$$= \sum_{n=0}^{\infty} \frac{(z' - z_0)^n}{(z - z_0)^{n+1}}, \tag{3.22c}$$

which by letting $m = n + 1$ can be rewritten as

$$\frac{-1}{z' - z} = \sum_{m=1}^{\infty} \frac{(z' - z_0)^{m-1}}{(z - z_0)^m} \tag{3.22d}$$

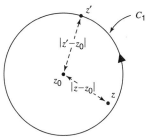

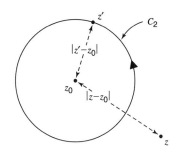

$f(z)$ is analytic within C_1.

Figure 3.23

Figure 3.24

$$= \sum_{n=-\infty}^{-1} \frac{(z - z_0)^n}{(z' - z_0)^{n+1}}, \tag{3.22e}$$

where in (3.22e) we defined $n = -m$. The last expression is identical in form to the expansion used in the integral around C_1 [see (3.18b)], except that now n takes on negative rather than nonnegative values. Again, in this reexpressing of the integral, no assumption has been made that $f(z)$ is analytic within C_2.

In (3.21), we now substitute (3.18b) into the first integral and (3.22e) into the second integral. Because $f(z')/(z' - z_0)$ is analytic between C_1 and C_2, the crucial deformation theorem allows us to bring out C_2 and bring in C_1 until they coincide,[7] in which case we call this common contour C. We therefore finally have the Laurent series:

where

$$f(z) = \sum_{n=-\infty}^{\infty} a_n(z - z_0)^n, \qquad \text{Laurent series.} \tag{3.23a}$$

$$a_n = \frac{1}{2\pi j} \oint_C \frac{f(z')}{(z' - z_0)^{n+1}} \, dz'. \qquad \text{Laurent series coefficients.} \tag{3.23b}$$

The radius r of C satisfies $r_A < r < r_B$, where r_A and r_B are, respectively, the minimum and maximum distances between z_0 and any point within the annulus of analyticity containing the point of evaluation, z.

Note that (3.19) $[a_n = f^{[n]}(z_0)/n!]$ does not apply here for any n even though (3.14) would indicate it is equal to the integral for a_n in (3.23b). This is because (3.19) and (3.13) are true only for $f(z)$ analytic everywhere within C, which does not generally hold for Laurent expansions. The fact that a_n in (3.23b) is not equal to $f^{[n]}(z_0)/n!$ is true even if $f^{[n]}(z_0)$ exists, because we are assuming that there may be poles between z and z_0. However, if $f(z)$ is analytic everywhere within C, then they are equal, and the Laurent series reduces to the Taylor series. Check for yourself that in that case, the Cauchy integral theorem will dictate that $a_{n<0} = 0$ by focusing on the integral around C_2, the integral responsible for $a_{n<0}$.

A final clarification is needed. One might argue that because in the Laurent expansion there are terms such as $a_{-3}/(z - z_0)^3$, it *must* be true that $f(z)$ is not

---

[7]For example, for C_1, draw a closed dotted C-shaped ring with infinitesimal gap whose outer radius is just inside the pole delimiting the region of convergence and whose inner radius is anywhere in the region of convergence. Then $f(z)$ is analytic within the dotted ring, so the integral around the inner part of the ring equals that around the outer portion.

analytic at $z = z_0$, because there there these terms are infinite. Keep in mind, however, that the Laurent expansion is never valid and thus never evaluated at $z = z_0$ when it includes negative powers of $(z - z_0)$, so the above argument is false. In fact, there may be negative powers of $z - z_0$ in the Laurent expansion of $f(z)$ about z_0 even if $f(z_0)$ is well defined, due to poles existing between z_0 and z.

To generate the Laurent series, we need to be able to calculate the a_n Laurent coefficients in (3.23b). For this calculation, we need the residue theorem, which we shall now present.

3.3.4   Residue Theorem

There is a special value of n in (3.23b): $n = -1$. This value is special because for $n = -1$, $(z' - z_0)^{n+1} = (z' - z_0)^0 = 1$, so we have $2\pi j \cdot a_{-1}$ equal to the integral of $f(z)$ around the closed contour C, where the second equality in (3.24) will be discussed below:

$$a_{-1} = \frac{1}{2\pi j} \oint_C f(z)\, dz = \tilde{a}_{-1}^{\{0\}} = \mathrm{Res}\{f, z_0\}, \qquad \text{if } z_0 \text{ is only pole } C \text{ encloses.} \qquad (3.24)$$

If it is assumed that there are several singularities (poles) of $f(z)$ in a region but that C is taken to enclose only the pole z_k (call it C_k), then we designate a_{-1} for that (kth) pole as $\tilde{a}_{-1}^{\{k\}}$, where the $\{\cdot\}$ refer to the pole indexed within the brackets, not a power. We call $\tilde{a}_{-1}^{\{k\}}$ the residue of $f(z)$ at z_k: $\mathrm{Res}\{f, z_k\} = \tilde{a}_{-1}^{\{k\}}$. Note that for Laurent series, z_k may be only one of several poles enclosed by C, whereas $C = C_k$ for a residue encloses only the pole z_k.

The term *residue* probably refers to "what is left over" (residue) when $f(z)$ is integrated around a closed contour C_k. This residue would be zero if $f(z)$ were analytic within C_k. Actually, $\tilde{a}_{-1}^{\{k\}}$ is the $n = -1$ coefficient of the Laurent series of $f(z)$ about z_k for z in the immediate vicinity of z_k, with no other poles intervening in the annulus between z_k and the point of evaluation of the series, z.

By the deformation theorem, we can make the conclusion that if C surrounds M poles, then the integral around C will be the sum of the residues; see Figure 3.25. Because the interior of the dashed contour interior contains no poles, the integral around the dashed contour is zero, by the Cauchy integral theorem. The integral around the dashed contour is just the difference between the integral around C and the sum of the integrals around the small circular contours. Furthermore, the inte-

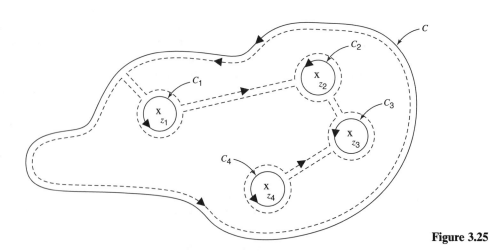

Figure 3.25

gral around the kth small contour is the residue $\tilde{a}_{-1}^{\{k\}}$. Thus we have the residue theorem

$$\oint_C f(z)\, dz = 2\pi j \sum_{k=1}^{M} \tilde{a}_{-1}^{\{k\}} = 2\pi j \sum_{k=1}^{M} \text{Res}\{f, z_k\}, \qquad \text{Residue theorem.} \qquad (3.25a)$$

where M is the number of poles enclosed by C and $\tilde{a}_{-1}^{\{k\}}$ is the residue of the kth pole enclosed by C, z_k. In particular, the Laurent coefficients a_n become, by (3.23b) and (3.25a),

$$a_n = \sum_{k=1}^{M} \text{Res}\left\{ \frac{f(z)}{(z - z_0)^{n+1}}, z_k \right\}, \qquad (3.25b)$$

where the z_k are the poles enclosed by the selected contour in the region of analyticity of the expansion of $f(z)$ in (3.23a). The residue theorem will be very useful in inverting the z-transform, as we shall see in Section 3.4.2, and in general for calculating the Laurent series coefficients.

An alternative view of the residue theorem is based on the partial fraction expansion (PFE) of $f(z)$, which will also be discussed in Section 3.4.2. In PFE, the function $f(z)$ has been expanded as a sum of weighted terms $(z - z_k)^{-n}$ over both n and k, with n ranging from 1 to the order M_k of the pole z_k, and k ranging from 1 to the number N_p of distinct poles of $f(z)$. From (3.11) $\left[\oint_C (z - z_0)^n\, dz = j2\pi\delta(n + 1) \right]$ applied individually to each PFE term, we know that if we integrate $f(z)$ around a closed contour C and substitute for $f(z)$ its partial fraction expansion, the integrals of all the terms $(z - z_k)^{-n}$ will be zero, except those for which $n = 1$, and then only if z_k is within C. Furthermore, that kth nonzero result will be the PFE coefficient $a_{-1}^{\{k\}} = \tilde{a}_{-1}^{\{k\}}$ for the PFE term $1/(z - z_k)$, where $a_{-1}^{\{k\}}$ is the notation we will use in Section 3.4.2 for the PFE coefficient of $1/(z - z_k)$ in the partial fraction expansion. Thus the integral of $f(z)$ around C is $2\pi j$ times the sum of the simple-pole PFE coefficients $a_{-1}^{\{k\}}$ for all k for which z_k is within C. Because partial fraction expansion itself is based on this complex integration theory, we have just another view of the same theory.

An important special case of $f(z)$ in the residue theorem (3.25) for DSP is $f(z) = \phi(z)/P(z)$, where $\phi(z)$ is analytic and nonzero everywhere that $P(z) = 0$, and $P(z)$ is a (factorable) polynomial having N_p distinct roots:

$$P(z) = \prod_{k=1}^{N_p} (z - z_k)^{M_k}, \qquad (3.26)$$

where M_k is the order of the kth pole of $f(z)$, z_k. The results now to be derived concerning the determination of the residue of $f(z)$ at pole z_k of order M_k apply even if $P(z)$ does not have this particular form. However, they do apply to $f(z) = \phi(z)/P(z)$ with $P(z)$ in (3.26), which includes all linear shift-invariant system functions.

Suppose that to evaluate $\oint_C f(z)dz$, we want $\tilde{a}_{-1}^{\{k\}} = \text{Res}\{f, z_k\}$, the residue of $f(z)$ at $z = z_k$ in $f(z) = \sum_{n=-\infty}^{\infty} a_n^{\{k\}}(z - z_k)^n$, where z is in the deleted neighborhood of z_k (no other poles intervening between z_k and z), so $a_{-1}^{\{k\}} = \tilde{a}_{-1}^{\{k\}}$. For $n < -M_k$, the integrand of $\tilde{a}_n^{\{k\}} = \frac{1}{2\pi j} \oint_{C_k} \frac{f(z')}{(z' - z_k)^{n+1}}\, dz'$, namely $f(z')/(z' - z_k)^{n+1}$, is an analytic function because the factor $1/(z' - z_k)^{n+1}$ cancels the M_k-order pole factor $(z' - z_k)^{M_k}$ of $f(z)$. Therefore, $\tilde{a}_n^{\{k\}}$ is zero for $n < -M_k$. Consequently, the Laurent series for $f(z)$ about $z = z_k$ in the deleted neighborhood of z_k "goes back" only to $-M_k$:

$$f(z) = \frac{\tilde{a}_{-M_k}^{\{k\}}}{(z - z_k)^{M_k}} + \frac{\tilde{a}_{-(M_k-1)}^{\{k\}}}{(z - z_k)^{M_k-1}} + \ldots + \frac{\tilde{a}_{-1}^{\{k\}}}{z - z_k} + \sum_{n=0}^{\infty} \tilde{a}_n^{\{k\}}(z - z_k)^n. \qquad (3.27)$$

Now define $\psi_k(z)$ such that

$$\psi_k(z) = (z - z_k)^{M_k} f(z). \qquad (3.28a)$$

We multiply each term of $f(z)$ in (3.27) individually by $(z - z_k)^{M_k}$ to obtain

$$\psi_k(z) = \tilde{a}_{-M_k}^{\{k\}} + \tilde{a}_{-(M_k-1)}^{\{k\}}(z - z_k) + \ldots + \tilde{a}_{-1}^{\{k\}}(z - z_k)^{M_k-1}$$

$$+ \sum_{n=0}^{\infty} \tilde{a}_n^{\{k\}}(z - z_k)^{n+M_k}. \qquad (3.28b)$$

If we take the $(M_k - 1)$st derivative of $\psi_k(z)$, the terms involving $\tilde{a}_n^{\{k\}}$ for $n < -1$ are all zero, because the derivative of a constant is zero. The term involving $\tilde{a}_{-1}^{\{k\}}$ is $(M_k - 1)! \, \tilde{a}_{-1}^{\{k\}}$, and the terms involving $\tilde{a}_n^{\{k\}}$ for $n > -1$ involve positive powers of $z - z_k$ which give zero at $z = z_k$. Therefore, $\psi_k^{[M_k-1]}(z_k) = (M_k - 1)! \, \tilde{a}_{-1}^{\{k\}}$ because all other terms are zero at $z = z_k$. That is,

$$\tilde{a}_{-1}^{\{k\}} = \frac{\psi_k^{[M_k-1]}(z_k)}{(M_k - 1)!} = \frac{\{(z - z_k)^{M_k}f(z)\}^{[M_k-1]}}{(M_k - 1)!}\bigg|_{z=z_k} = \text{Res}\{f, z_k\}. \qquad (3.29)$$

If the pole is a simple pole ($M_k = 1$), then (3.29) simplifies to

$$\tilde{a}_{-1}^{\{k\}} = \psi_k(z_k) = \{(z - z_k)f(z)\}\big|_{z=z_k} = \text{Res}\{f, z_k\}. \qquad (3.30)$$

Thus (3.29) and (3.30) are useful for calculating $\oint_C f(z)\,dz$ when their results are evaluated for all enclosed poles and substituted into the residue theorem formula (3.25a). A major application is z-transform inversion (see Section 3.4.2).[8]

Note that a special case of the residue theorem is $f(z) = 1/(z - z_0)$ for which by (3.30) $\text{Res}\{f, z_0\} = 1\big|_{z=z_0} = 1$. Thus, $\oint_C \frac{dz}{z - z_0} = 2\pi j \cdot 1 = 2\pi j$, as we explicitly found earlier in (3.11).

EXAMPLE 3.8

Compute $\oint_C \dfrac{\sin(7z)}{z^2(z + 5)(z^2 + 1)}\,dz$ for the following contours C:

(a) C = square centered on $-3.5 - j$, having sides of length 2.
(b) C = circle of radius 2 centered on $z = -1 + j$.

---

[8]It is again important not to confuse (3.29) and (3.30) with PFE formulas, which are presented later.

Solution

(a) Refer to Figure 3.26*a*. Of the five poles, C encloses only -5, so by (3.30),

$$I = 2\pi j \cdot \tilde{a}_{-1}^{\{z=-5\}} = 2\pi j \cdot \text{Res}\left\{\frac{\sin(7z)}{z^2(z+5)(z^2+1)}, -5\right\}$$

$$= \frac{2\pi j \sin(7z)}{z^2(z^2+1)}\bigg|_{z=-5} = 2\pi j \frac{\sin(-35)}{25(26)} = j0.004139.$$

(b) Refer to Figure 3.26*b*. Now C encloses the double pole at $z = 0$ and the pole at $z = j$, and neither of the other two poles. Thus

$$I = 2\pi j\{\tilde{a}_{-1}^{\{z=0\}} + \tilde{a}_{-1}^{\{z=j\}}\}$$

$$= 2\pi j \cdot \text{Res}\left\{\frac{\sin(7z)}{z^2(z+5)(z^2+1)}, 0\right\} + 2\pi j \cdot \text{Res}\left\{\frac{\sin(7z)}{z^2(z+5)(z^2+1)}, j\right\}$$

$$= 2\pi j \cdot \left\{\frac{d}{dz}\left[\frac{\sin(7z)}{(z+5)(z^2+1)}\right]\bigg|_{z=0} + \frac{\sin(7z)}{z^2(z+5)(z+j)}\bigg|_{z=j}\right\}.$$

Now note that $\sin(7j) = (e^{j7j} - e^{-j7j})/(2j) = (e^{-7} - e^7)/(2j) = j\sinh(7) = j548.3$. Thus

$$I = 2\pi j \cdot \left\{\frac{(z+5)(z^2+1)\,7\cos(7z) - \sin(7z)[2z(z+5)+z^2+1]}{(z+5)^2(z^2+1)^2}\bigg|_{z=0} + \frac{j548.3}{j^2(5+j)(2j)}\right\}$$

$$= -66.253 - j322.47.$$

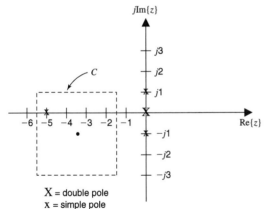

X = double pole
x = simple pole

Figure 3.26*a*

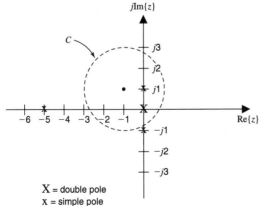

X = double pole
x = simple pole

Figure 3.26*b*

EXAMPLE 3.9

Find the Laurent expansion of $f(z) = 1/[z(z + 5)]$ about the point $z_0 = -\frac{1}{2}$ that is valid for the evaluation point $z = 4.5$ and its vicinity.

Solution

We know that $|z - z_0| = 5$. Referring to Figure 3.27, the pole–zero diagram, we see that any contour C we would choose that is drawn in the region of convergence for $z = 4.5$ (namely, $|z - z_0| > 4.5$) would enclose both poles, $z = 0$ and $z = -5$. Thus, the Laurent series coefficients formula (3.23b) becomes

$$a_n = \frac{1}{2\pi j} \oint_C \frac{dz'}{z'(z' + 5)(z' + \frac{1}{2})^{n+1}}.$$

For $n < 0$, there is no $(z' + \frac{1}{2})$ factor in the denominator, so according to (3.25b),

$$a_n = \text{Res}\left\{\frac{(z + \frac{1}{2})^{-(n+1)}}{z(z + 5)}, 0\right\} + \text{Res}\left\{\frac{(z + \frac{1}{2})^{-(n+1)}}{z(z + 5)}, -5\right\}$$

$$= \left(\frac{1}{5}\right)\left[\left(\frac{1}{2}\right)^{-(n+1)} - (4.5)^{-(n+1)}\right].$$

For $n = 0$, we find

$$a_0 = \frac{1}{(5 \cdot \frac{1}{2})} + \frac{1}{(-5) \cdot (-4.5)} + \frac{1}{(-\frac{1}{2} \cdot 4.5)}$$

$$= \frac{2}{5} + \frac{2}{45} - \frac{4}{9} = 0.$$

For $n = 1$, we have a double pole in the integrand at $z = -\frac{1}{2}$, so now

$$a_1 = \frac{1}{(5 \cdot \frac{1}{4})} + \frac{1}{(-5) \cdot (-4.5)^2} + \frac{-d/dz\{z(z + 5)\}}{\{z(z + 5)\}^2}\bigg|_{z = -\frac{1}{2}}$$

$$= \frac{4}{5} - \frac{4}{5 \cdot 81} - \frac{4}{\frac{1}{4} \cdot 81/4} = 0.$$

Remarkably, if we continued this procedure, we would find that $a_n = 0$ for all $n \geq 0$. In Section 3.6, we will see how to prove this fact without having to carry out the tedious calculations. Thus our result is

$$f(z) = \frac{1}{5} \sum_{n=1}^{\infty} \left[2^{n+1} - \left(\frac{1}{4.5}\right)^{n+1}\right]\left(z + \frac{1}{2}\right)^n.$$

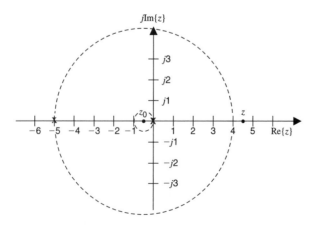

Figure 3.27

We may summarize and consolidate many of our results on the integration of a function of a complex variable as follows:

$$\oint_C f(z)\,dz = \begin{cases} 2\pi j \sum_k \text{Res}\{f, z_k\} & \text{if, within } C, f(z) \text{ has singularities at } z_k \\ & \hspace{2em} \text{[Residue theorem (3.25a)],} \hspace{2em} (3.31) \\ 0 & \begin{cases} \text{if } f(z) = F'(z) \text{ on } C \quad \text{(Fundamental theorem of calculus),} \\ \text{if } f(z) \text{ is analytic within and on } C \quad \text{(Cauchy integral theorem).} \end{cases} \end{cases}$$

Whether or not $f(z)$ is singular at $z = z_m$ or at other points between z and z_m, we can expand $f(z)$ about any point z_m over powers of $z - z_m$:

$$f(z) = \sum_{n=n_{f,m}}^{\infty} a_n^{(m)}(z - z_m)^n, \begin{cases} \text{Laurent's expansion of } f(z) \text{ about } z_m \text{ for general } n_{f,m} {}^9 \\ z\text{-transform of } a_n^{(m)} \text{ for general } n_{f,m} \text{ and } z_m = 0 \\ \hspace{2em} \text{(see Section 3.4.1)} \hspace{4em} (3.32) \\ \text{Taylor's expansion of } f(z) \text{ about } z_m \text{ for } n_{f,m} \geq 0 \\ \text{MacLaurin's expansion of } f(z) \text{ for } n_{f,m} \geq 0 \text{ and } z_m = 0, \end{cases}$$

where

$$a_n^{(m)} = \frac{1}{2\pi j} \oint_C \frac{f(z')}{(z' - z_m)^{n+1}}\,dz' \qquad \text{true for general } f(z) \tag{3.33a}$$

$$= \sum_{\substack{k, \\ z_k \text{ within } C}} \text{Res}\left\{ \frac{f(z)}{(z - z_m)^{n+1}}, z_k \right\} \qquad \text{true for general } f(z) \tag{3.33b}$$

$$= f^{[n]}(z_m)/n! \quad \text{true only if } n_{f,m} \geq 0 \quad \begin{array}{l}\text{[i.e., } f(z) \text{ is analytic within } C; \\ \text{this expression is (3.19),} \\ \text{just rewritten],}\end{array} \tag{3.33c}$$

where C is in the same region of analyticity of $f(z)$ as is z, and typically is a circle centered on z_m. It should be noted that the $\{a_n^{(m)}\}$ depend not only on f and z_m, but also on which region of analyticity z happens to be in, so there are different Laurent expansions of $f(z)$ for different regions in the z-plane.

Finally, note that for analytic $f(z)$, the only pole of $f(z)/(z - z_m)^{n+1}$ is z_m, of order $n + 1$. Substituting into (3.29), we find the completely general (3.33b) reduces to (3.33c), the Taylor coefficients for analytic functions.

3.4 THE z-TRANSFORM

Many of the uses and advantages of the z-transform were reviewed in the introduction to this chapter, and other advantages will be taken up further in Chapter 4. In this section, we introduce the z-transform (ZT), which will be an important analysis tool throughout the rest of this book.

First we define the z-transform and its regions of convergence, then we present some examples of finding the z-transform of a given sequence and finding the sequence corresponding to a given z-transform, and finally the application of the z-transform for the solution of difference equations is examined. A systematic presentation of the properties of the z-transform is presented in Chapter 5 and summarized in Appendix 5A. A table of z-transform pairs is given in Appendix 5B.

---

[9] In $n_{f,m}$, f and m refer to the fact that $n_{f,m}$ depends on both $f(z)$ and z_m about which $f(z)$ is expanded.

3.4.1 Definition of the z-Transform and Its Region of Convergence

As already hinted in Section 3.3.3 and explicitly noted in (3.32), the z-transform is merely a special case of the Laurent expansion. Specifically, if we expand $H(z)$ about $z_m = 0$ [note the notational change from $f(z)$ to $H(z)$], let $n_{H,m} = -\infty$ for generality, define $h(n) = a_{-n}^{[m]}$, and finally replace the symbol n by $-n$ so that in (3.33a) the power $-(n_{old} + 1)$ becomes $n_{new} - 1$, then (3.32) becomes

$$H(z) = \sum_{n=-\infty}^{\infty} h(n)\, z^{-n}, \qquad z \in \text{region of convergence}, \qquad (3.34)$$

and (3.33a) becomes

$$h(n) = \frac{1}{2\pi j} \oint_C H(z)\, z^{n-1}\, dz, \qquad C \text{ in region of convergence}, \qquad (3.35)$$

where C, generally a circle centered on the origin, and which may enclose any or all singularities of $H(z)$, is often taken to be the unit circle. Thus the coefficients $a_n^{[0]}$, where $z_0 = 0$, in the Laurent expansion of a function $H(z)$ about $z = 0$ are the time reversal of the sequence $h(n)$ whose z-transform is $H(z)$.

Notice that the viewpoint of Laurent's expansion has changed now that we have expressed it as the z-transform. Previously, we were interested in the power series representation of a function $f(z)$ of a complex variable z. Now we are interested in a "transform" $H(z)$ of a sequence of numbers $h(n)$. The previous view helps us understand the origin and mathematical basis for the z-transform, and the latter view is appropriate for its application. This latter view will be ours for much of the rest of this book.

The z-transform as used in DSP originally arose from attempts to apply the Laplace transform to sequences. If we mathematically construct a Dirac delta impulse train $x_{s,c}(t)$ whose impulses are Δt apart and whose strengths are equal to the given sequence $x(n)$ [often chosen to be equal to $x_c(n\,\Delta t)$, where $x_c(t)$ is an originating continuous-time signal], so that

$$x_{s,c}(t) = \sum_{n=-\infty}^{\infty} x(n)\delta_c(t - n\,\Delta t), \qquad (3.36)$$

then the Laplace transform of $x_{s,c}(t)$, $X_{s,c}(s)$, is

$$X_{s,c}(s) = \int_{-\infty}^{\infty} \sum_{n=-\infty}^{\infty} x(n)\delta_c(t - n\,\Delta t)e^{-st}\, dt. \qquad (3.37a)$$

Pulling out the sum, we have

$$X_{s,c}(s) = \sum_{n=-\infty}^{\infty} x(n) \int_{-\infty}^{\infty} \delta_c(t - n\,\Delta t)e^{-st}\, dt, \qquad (3.37b)$$

which by the sampling property of the Dirac delta function becomes

$$X_{s,c}(s) = \sum_{n=-\infty}^{\infty} x(n)e^{-s\,\Delta t \cdot n}. \qquad (3.37c)$$

By defining $z = e^{s\,\Delta t}$, we have

$$X_{s,c}(s) = \sum_{n=-\infty}^{\infty} x(n)z^{-n}\Big|_{z=e^{s\Delta t}} \qquad (3.37d)$$

or

$$X_{s,c}(s) = X(z)\Big|_{z=e^{s\Delta t}}. \tag{3.37e}$$

The series defined by $X_{s,c}(s)$ in (3.37d) is the z-transform of $x(n)$ (3.34), evaluated at $z = e^{s\Delta t}$. It must be emphasized that, in general, the z-transform is not simply related to $X_c(s)$, the Laplace transform of $x_c(t)$, but rather to $X_{s,c}(s)$, the Laplace transform of the impulse train whose coefficients are given by $x(n) = x_c(n\Delta t)$. There is nevertheless a relation between $X(z)$ and $X_c(s)$, which we will derive in Section 6.2.

It is from this signal-processing viewpoint that the z-transform as commonly used today has been defined. For specificity, in this introduction of the z-transform we will hereafter mostly consider the unit sample response $h(n)$ of a linear shift-invariant system as the sequence to be transformed [as opposed to signals such as $x(n)$ to which it applies equally]. $H(z)$ is called the system function of the LSI system and, for LSI systems, is a rational function of z. As noted, the z-transform is a special Laurent series: (3.32) with $z_m = 0$ and $h(n) = a_{-n}^{[0]}$. The only possible confusion is the "time" index reversal of the Laurent coefficients a_n of z^n compared with $h(n)$, where $h(n)$ are the coefficients of z^{-n}.

This index reversal arises from the evolution of the z-transform definition from the Laplace transform definition rather than from a Laurent series. That is, the transform function is defined as an integral (or series) over negative powers e^{-st} (or z^{-n}) for $t > 0$ (or $n > 0$), so that the time expansion functions are written in terms of positive powers e^{st} (or z^n) for $t > 0$ (or $n > 0$). The Laurent expansion, (3.32), however, expresses the function of z as a sum of powers of z with "$+n$" in the exponent. As long as this distinction is clearly kept in mind, then all the results pertaining to Laurent series can be applied to z-transforms.

Apparently, the original reason for writing the z-transform in terms of z^{-n} rather than z^{+n} even though it is "unfortunate" and relatively awkward was that the only "extensive" table of z-transforms then available was for the z^{-n} definition![10] Once it had been decided to use z^{-1} rather than z to represent a delay operation, a practical reason for writing the z-transform in terms of z^{-n} rather than z^{+n} is that z^{-1} (a delay), but not an advance element, is a physically realizable operator. Thus $a_n z^{-n}$ represents a delay of n time samples followed by scaling by a_n, both of which are readily realizable in hardware. Also, the modes of a causal system are $a_\ell z_\ell^n$ rather than a_ℓ / z_ℓ^n.

In Section 2.4.5, we showed that for a system to be stable, $h(n)$ must be an absolutely summable sequence. What is the condition on the z-transform of $h(n)$ so that the system is stable? We will find a stipulation on $H(z)$ analogous to the "all poles in the left-half plane" condition for the Laplace transform. Noting that $e^{j\omega n}$ has unit magnitude and equals z^n for $|z| = 1$ and proceeding with the assumption that $h(n)$ is absolutely summable, we have, for any finite constant M_h,

$$\infty > M_h > \sum_{n=-\infty}^{\infty} |h(n)| \tag{3.38a}$$

$$= \sum_{n=-\infty}^{\infty} |h(n)e^{-j\omega n}| \tag{3.38b}$$

$$\geq \Big| \sum_{n=-\infty}^{\infty} h(n)e^{-j\omega n} \Big|. \tag{3.38c}$$

---

[10]J. R. Ragazzini and L. A. Zadeh, "The Analysis of Sampled-Data Systems," *AIEE Proc.*, November 1952, pp. 225–232.

Noting that the argument of $|\cdot|$ in (3.38c) is the z-transform of $h(n)$ (3.34) evaluated on the unit circle, we have

$$\left| \; H(z)\Big|_{z=e^{j\omega}} \; \right| \leq \infty. \tag{3.38d}$$

We thus conclude that for the system to be stable, $H(z)$ must converge (be well defined) on the unit circle $|z| = 1$. Equivalently, we say that the region of convergence of $H(z)$ includes the unit circle. This condition must hold whether or not $h(n)$ is causal. This is one reason why the inversion contour in (3.35) is usually taken to be the unit circle: For stable systems, the region of convergence includes it.

It should also be noted that $H(z)\big|_{|z|=1} = H(e^{j\omega})$, the discrete-time Fourier transform (DTFT) of $h(n)$. Thus all stable sequences must have convergent DTFTs, just as all stable continuous-time system impulse responses must have convergent Fourier transforms. Taking C to be the unit circle makes the ZT^{-1} and $DTFT^{-1}$ formulas identical, when we consequently set $z = e^{j\omega}$ for evaluation of the ZT^{-1} integral.

Where else in the z-plane is $H(z)$ well defined? Suppose that $h(n)$ is causal. Then the sum in (3.34) effectively runs only from 0 to ∞. For finite values of n, clearly each term is finite as long as $z \neq 0$. For the sum to converge, the crucial requirement is that the terms $h(n)z^{-n}$ die away as $n \to \infty$. This requirement will be satisfied if $|z| > r_h$, where if $z_h = r_h e^{j\theta}$, then z_h^n is the largest-magnitude geometrically increasing component in $h(n)$.[11]

These geometrical components appear in the z-transform as poles: $ZT\{z_\ell^n u(n)\} = 1/(1 - z_\ell/z) = z/(z - z_\ell)$, as will be further discussed in Sections 3.4.2 and 3.4.4. Thus, the region of convergence of a causal sequence $h(n)$ is anywhere in the z-plane outside the radius of the largest-magnitude pole of $H(z)$. This is analogous to the region of convergence of the Laplace transform of a causal signal being anywhere to the right of the rightmost pole of $H_c(s)$.

Moreover, if $h(n)$ is a stable, causal sequence, we know that the unit circle must be included in the region of convergence of $H(z)$. Therefore, it must be true that $H(z)$ be defined in a region such as that shown hatched in Figure 3.28. All the poles of $H(z)$ must be within the unit circle for $h(n)$ to be a stable, causal sequence. In the time domain, this means simply that all the geometric components of $h(n)$ (*modes*) decay rather than grow with increasing n. The requirement that all the poles of $H(z)$ be within the unit circle is the analog of the Laplace transform stability requirement, in which all the poles of a stable, causal system must be in the left-half plane.

By identical arguments, if $h(n)$ is an anticausal sequence, the region of convergence of $H(z)$ will be anywhere within a circle whose radius is equal to the magnitude of the smallest-magnitude pole of $H(z)$. If $h(n)$ is also stable, this radius must be larger than the unit circle to guarantee that the unit circle is within the region of convergence. Thus a typical region of convergence for an anticausal, stable sequence is as shown hatched in Figure 3.29.

If $h(n)$ is bicausal, it can always be decomposed into the sum of a causal sequence $h_C(n)$ and an anticausal component $h_{AC}(n)$. The regions of convergence of $H_C(z)$ and $H_{AC}(z)$ may or may not overlap; they will if the magnitude of the largest-magnitude pole of $H_C(z)$ is smaller than the magnitude of the smallest-magnitude pole of $H_{AC}(z)$. Moreover, $h(n)$ is stable only if unity is between these two magnitudes

---

[11]Note that $n^m z_\ell^n$ is the most general term for impulse responses of LSI systems, to which this discussion also applies because as $n \to \infty$, z_ℓ^n behavior dominates over n^m for finite m.

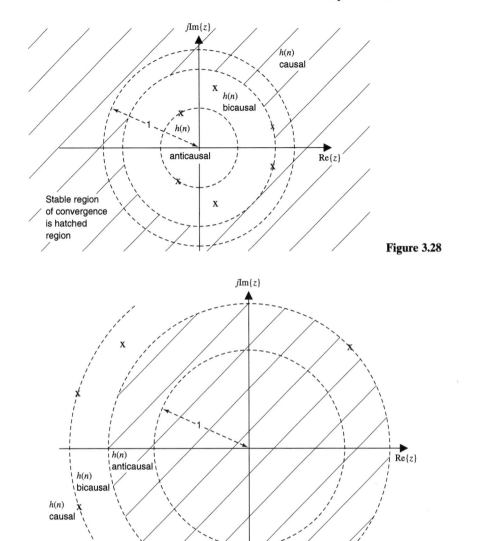

Figure 3.28

Figure 3.29

—that is, if the unit circle is within the region of convergence of $H(z)$—because only then does the ROC of $H(z)$ lie within the stable ROC of each of its poles (of all magnitudes). An example for a stable bicausal system is shown in Figure 3.30.

If we are given an $H(z)$ with several poles at different radii, there are many possible sequences having $H(z)$ as their z-transform. Determination of which of these sequences applies depends on which poles of $H(z)$ are poles of $H_C(z)$ and which are poles of $H_{AC}(z)$. There will, however, be only one stable sequence having $H(z)$ as its z-transform, because only one annulus includes the unit circle. A z-transform $H(z)$, specified together with its region of convergence, is uniquely paired with a sequence $h(n)$.

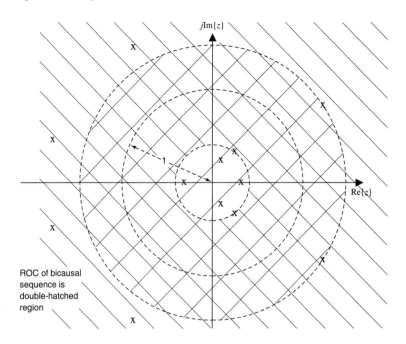

Figure 3.30

3.4.2 The Mechanics of Transforming and Inverse-Transforming

The class of sequences for which we may readily compute the z-transform in closed form is somewhat limited. The first of two main categories is the finite-length sequence. If $h(n)$ is nonzero only for $n \in [0, N-1]$, then from (3.34),

$$H(z) = \sum_{n=0}^{N-1} h(n)z^{-n}. \tag{3.39}$$

This z-transform is the simplest to compute of all: Just add the N terms. Whether or not the result can be reduced to a single closed-form expression naturally depends on $h(n)$.

Most z-transforms that appear in tables are merely variations on geometrical and exponential sequences (which may or may not be finite length), the other category for which it is easy to compute the z-transform in closed form. The basic relation used in computing such z-transform pairs is that, for any real- or complex-valued number or variable a,

$$\sum_{n=M}^{N} a^n = \begin{cases} \dfrac{a^M - a^{N+1}}{1-a}, & a \neq 1 \\ N - M + 1, & a = 1, \end{cases} \tag{3.40}$$

which has the following commonly used specializations and extensions:

$$\sum_{n=0}^{N-1} a^n = \begin{cases} \dfrac{1 - a^N}{1 - a}, & a \neq 1 \\ N, & a = 1 \end{cases} \tag{3.41a}$$

and the familiar result already used in Chapter 2,

$$\sum_{n=0}^{\infty} a^n = \frac{1}{1-a}, \quad \text{valid only for } |a| < 1. \tag{3.41b}$$

The proofs for all of them originate from the fact (obtainable by simple long division) that

$$\frac{1}{1-a} = \sum_{n=0}^{N-1} a^n + \frac{a^N}{1-a}. \tag{3.42}$$

EXAMPLE 3.10

Find the z-transform of $h(n)$ and provide the region of convergence.

(a) $h(n) = \delta(n)$.

(b) $h(n) = \begin{cases} 1, & n \in [0, N-1] \\ 0, & \text{otherwise.} \end{cases}$

(c) $h(n) = u(n)$, where $u(n) = 1$ for $n \geq 0$ and 0 otherwise.

(d) $h(n) = a^n u(n)$.

(e) $h(n) = a^n[u(n) - u(n-N)] = \begin{cases} a^n, & n \in [0, N-1] \\ 0, & \text{otherwise.} \end{cases}$

(f) $h(n) = a^{|n|}$.

(g) $h(n) = \left(\frac{1}{2}\right)^n u(n) + 2^n u(-n)$.

(h) $h(n) = \begin{cases} n, & n \in [0, N-1] \\ 0, & \text{otherwise.} \end{cases}$

Solution

(a) $H(z) = 1$, for all z.

(b) From (3.41a), letting $a = z^{-1}$, we immediately have

$$H(z) = \sum_{n=0}^{N-1} z^{-n} = \begin{cases} \dfrac{1 - z^{-N}}{1 - z^{-1}}, & z \neq 1 \text{ and } z \neq 0 \\ N, & z = 1. \end{cases}$$

(c) $H(z) = \sum_{n=0}^{\infty} z^{-n} = \dfrac{1}{1 - z^{-1}}$, valid only for $|z| > 1$. This $h(n)$ is not a stable sequence, because the region of convergence of $H(z)$ does not include the unit circle. Sometimes a system such as this one that has a simple pole on the unit circle (in this case, at $z = 1$) is called metastable.

(d) $H(z) = \sum_{n=0}^{\infty} a^n z^{-n} = \dfrac{1}{1 - a/z} = \dfrac{z}{z - a}$, valid only for $|z| > |a|$.

Recall that in Example 2.17, part (c), we had $h(n) = \left(\frac{1}{2}\right)^n u(n)$. In this case, $H(z) = z/(z - \frac{1}{2})$, $|z| > \frac{1}{2}$. Notice that even though the pole $z = +\frac{1}{2}$ is in the right-half plane, that is immaterial for the z-transform. Rather, the pole $+\frac{1}{2}$ is within the unit circle, which is the z-plane condition for stability, as we showed in Section 3.4.1. Thus, as we found in Example 2.17, part (c), by showing that $h(n)$ is absolutely summable, $h(n)$ is a stable, not an unstable system.

(e) $H(z) = \sum_{n=0}^{N-1} a^n z^{-n} = \dfrac{1 - (a/z)^N}{1 - a/z}$, valid for all $z \neq a$ and $z \neq 0$.

For $z = a$, $H(a) = \sum_{n=0}^{N-1} 1^n = N$.

(f) $H(z) = \sum\limits_{n=-\infty}^{-1} (az)^{-n} + \sum\limits_{n=0}^{\infty} (a/z)^n$

$\qquad = \sum\limits_{m=0}^{\infty} (az)^m - 1 + \sum\limits_{n=0}^{\infty} (a/z)^n.$

The first sum converges only for all $|z| < 1/a$, whereas the second sum converges only for all $|z| > a$. The two are simultaneously satisfied for all $a < |z| < 1/a$, which evidently can be met only if $|a| < 1$, in which case we see that $h(n)$ is stable because the unit circle is in the region of convergence. In this case, we have

$$H(z) = \frac{1}{1-az} - 1 + \frac{1}{1-a/z} = \frac{az}{1-az} + \frac{1}{1-a/z}$$

$$= \frac{1-a^2}{(1-az)(1-a/z)}, \quad a < |z| < 1/a \quad \text{and} \quad |a| < 1.$$

(g) Writing $h(n)$ as $\left(\tfrac{1}{2}\right)^{|n|}$, we see that this example is just a special case of part (f), with $a = \tfrac{1}{2}$. Thus we can immediately write

$$H(z) = \frac{\tfrac{3}{4}}{(1 - \tfrac{1}{2}z)(1 - 1/[2z])}, \quad \text{valid only for } \tfrac{1}{2} < |z| < 2.$$

(h) This $h(n)$ is a variation on that in part (b). Denoting the solution of part (b) by $h_b(n)$ and that of this part by $h_h(n)$ [$= $ given $h(n)$], we have $h_h(n) = n \cdot h_b(n)$. Now notice that $-z d/dz\{H_b(z)\} = \sum\limits_{n=-\infty}^{\infty} n \cdot h_b(n)z^{-n} = \mathrm{ZT}\{n \cdot h_b(n)\} = \mathrm{ZT}\{h_h(n)\} = H(z)$. Thus from the solution of part (b) we can immediately obtain the desired z-transform by negating and differentiating $H_b(z)$, and multiplying by z:

$$H(z) = -z \frac{(1-z^{-1})Nz^{-N-1} - (1-z^{-N})z^{-2}}{(1-z^{-1})^2},$$

which, upon multiplying numerator and denominator by z^2 and rearranging, gives

$$H(z) = z\frac{1 + (N-1)z^{-N} - Nz^{-(N-1)}}{(z-1)^2}, \quad \text{for all } z \neq 1 \text{ and } z \neq 0.$$

Although it may be said that this example was "tricky", it foreshadows a host of properties of the z-transform that will be systematically derived in Chapter 5. Note that for $z = 1$, we have the special case[12] $H(1) = \sum\limits_{n=0}^{N-1} n = N(N-1)/2$.

By calculating the z-transform for many basic sequences such as those in Example 3.10, a table of known z-transform/sequence pairs can be made (see Table 5B.4). Then any time either a sequence or a z-transform identical to one in the table is encountered, the z-transform or inverse z-transform can immediately be written.

---

[12]Proof that sum of n from 0 to $N-1$ is $N(N-1)/2$: for N even,

$1 + 2 + \ldots + N - 1 = (1 + N - 1) + (2 + N - 2) + \ldots + \left(\frac{N}{2} - 1 + \frac{N}{2} + 1\right) + \frac{N}{2}$

$\qquad = \left(\frac{N}{2} - 1\right)N + \frac{N}{2} = \frac{(N-1)N}{2}.$

For N odd,

$1 + 2 + \ldots + N - 1 = (1 + N - 1) + (2 + N - 2) + \ldots + \left(\frac{N-1}{2} + \frac{N+1}{2}\right) = \frac{(N-1)}{2}N.$

Also, usage of the properties of z-transforms developed in Chapter 5 can augment the table with additional z-transform/sequence pairs.

It may seem unlikely to the reader to begin with an $H(z)$ and desire to invert it to find $h(n)$. Usually for system implementation it is $H(z)$ that we want to find, not $h(n)$. There are many cases, however, where inverse z-transformation is helpful. For example, the convolution theorem for the z-transform says that convolution in the n-domain transforms to mere multiplication in the frequency domain. Therefore, when using the z-transform to find the output $y(n)$ of an LSI system, the z-transform of the signal, $X(z)$, is multiplied by the system function, $H(z)$. To obtain $y(n)$, we must then inverse-transform the result, $Y(z)$. Even if $X(z)$ and $H(z)$ are known in closed form, it is often the case that $Y(z)$ may not appear in the given z-transform table. Also, suppose that two systems $H_1(z)$ and $H_2(z)$ are cascaded. Then net system function, $H_1(z)H_2(z)$ again might not be found in the z-transform table. Finally, occasionally there are errors in the tables! Consequently, it is convenient to know alternative methods of inverse z-transforming in addition to table look-up. We now review a few of these techniques.

Contour Integration

Contour integration is the most general technique, which works for any $H(z)$ having a finite number of singularities of finite orders. Assume that the order of the kth singularity z_k of $H(z)z^{n-1}$ is M_k. Application of (3.31) and (3.29) in (3.35) gives

$$h(n) = \frac{1}{2\pi j} \oint_C H(z)\, z^{n-1}\, dz \tag{3.43a}$$

$$= \sum_k \text{Res}\{H(z)z^{n-1}, z_k\} \tag{3.43b}$$

$$= \sum_k \frac{\{(z - z_k)^{M_k} H(z)z^{n-1}\}^{[M_k-1]}}{(M_k - 1)!}\Bigg|_{z=z_k}, \tag{3.43c}$$

where as before the $[M_k - 1]$ in square brackets is the $(M_k - 1)$st derivative with respect to z, whereas the M_k superscript on $(z - z_k)$ is the M_kth power. Often this method of z-transform inversion is actually the easiest, even though it may at first seem the most difficult. Once the denominator is factored and the poles z_k enclosed by C and their orders M_k determined, we merely perform the required evaluations in (3.43c) to obtain $h(n)$.

There is, however, one difficulty: For $n < 0$, (3.43a) shows that there may be multiple-order poles at $z = 0$ [depending on $H(z)$]. The more negative n is, the higher the order of the derivative required to find the residue at $z = 0$. To get around this problem,[13] make the following change of variable in (3.43a). Let $p = 1/z$, so that $dp = -dz/z^2$ or $dz = -z^2 dp = -dp/p^2$. For simplicity, let C be the unit circle. Then note that because $z = e^{j\omega}$ on C, we have $p = e^{-j\omega}$; in effect, the direction of the new contour is reversed. We can again make it go counterclockwise by multiplying the integral by -1. Also note that $z^{n-1} = p^{-n+1}$. In general, the radius of the new contour C' in (3.44) below is the inverse of the radius of C in (3.35). However, if the unit circle

---

[13]A. V. Oppenheim and R. W. Schafer, *Discrete-Time Signal Processing*. (Englewood Cliffs, NJ: Prentice Hall, 1989)..

is used in (3.43a), it is also used in (3.44); in that case, $C' = C$. With all the substitutions, we obtain

$$h(n) = \frac{1}{2\pi j} \oint_{C'} H(1/p)p^{-n-1}\, dp. \tag{3.44}$$

Now there are no poles of p^{-n-1} for $n < 0$. It would thus appear that (3.44) is more convenient than (3.43) when $h(n)$ is a left-sided (e.g., anticausal) sequence (as in Example 3.14, below) or for finding the anticausal component of a bicausal $h(n)$. This is not, however, generally true if $H(z)$ is not a rational function, as we will show in Example 3.15.

Equation (3.44) is also often useful for determining when $h(n)$ "begins"—for what values of n is $h(n)$ guaranteed to be zero—by examination of whether any poles are enclosed by C'. In particular, all the poles z_k of $H(z)$ become poles at $p_k = 1/z_k$ in $H(1/p)$. From this fact, it is clear that because all the poles of a stable causal $H(z)$ lie within the unit circle, $h(n < 0)$ will indeed be zero because then there are no enclosed poles in (3.44) [i.e., the poles of $H(1/p)$ are all outside C' = the unit circle].

EXAMPLE 3.11

Find the stable sequence whose z-transform is $H(z) = \dfrac{1 - 2z^{-1} + 2z^{-2}}{1 - 0.3z^{-1} + 0.02z^{-2}}$.

Solution

Notice that $H(z)$ is often written in terms of z^{-1} rather than z. This is because the z-transform of a difference equation involving delayed versions of the various sequences will include polynomials in z^{-1} times the z-transform of each sequence. Thus it is z^{-1} rather than z that naturally arises in discrete-time LSI analysis. However, for inverse z-transformation, especially contour integration, it is often more convenient to express $H(z)$ in terms of a rational function of z rather than z^{-1}. Thus in this example we multiply both the numerator and denominator by z^2 and factor the denominator:

$$H(z) = \frac{z^2 - 2z + 2}{z^2 - 0.3z + 0.02}$$
$$= \frac{z^2 - 2z + 2}{(z - 0.1)(z - 0.2)}.$$

Both poles are within the unit circle, so stable $h(n)$ must be right-sided. Our contour must lie in the selected region of convergence, which for stable sequences must include the unit circle, so let C be the unit circle.

By expressing the inverse z-transform expression in the p-form (3.44), we can determine for what n we are guaranteed to have $h(n) = 0$. For this example, go back to the original expression for $H(z)$ in terms of z^{-1} and replace z^{-1} by p:

$$h(n) = \frac{1}{2\pi j} \oint_{C'=C} \frac{1 - 2p + 2p^2}{1 - 0.3p + 0.02p^2} p^{-n-1}\, dp$$

$$= \frac{1}{2\pi j} \oint_{C} \frac{1 - 2p + 2p^2}{0.02(p - 5)(p - 10)} p^{-n-1}\, dp.$$

The only possible poles are at $p = 10$ and $p = 5$ [the inverses of the poles of $H(z)$], and at $p = 0$. The first two poles are not within C (the unit circle), whereas a pole at $p = 0$ occurs *only* for $-n - 1 \leq -1$ or $n \geq 0$. Thus at the outset, we know that our results will have $u(n)$ appended to them, because only for $n \geq 0$ can there be nonzero results; $h(n < 0) = 0$.

Returning to the usual form of (3.43b) for $n > 0$ we have

$$h(n) = \text{Res}\left\{\frac{(z^2 - 2z + 2)z^{n-1}}{(z - 0.1)(z - 0.2)}, 0.1\right\} + \text{Res}\left\{\frac{(z^2 - 2z + 2)z^{n-1}}{(z - 0.1)(z - 0.2)}, 0.2\right\}$$

$$= \frac{0.1^2 - 2 \cdot 0.1 + 2}{0.1 - 0.2}(0.1)^{n-1} + \frac{0.2^2 - 2 \cdot 0.2 + 2}{0.2 - 0.1}(0.2)^{n-1}$$

$$= -181 \cdot (0.1)^n + 82 \cdot (0.2)^n.$$

At first, one might think this solution is complete. At $n = 0$, however, there is also a simple pole at $z = 0$, due to the -1 in z^{n-1} with $n = 0$. Thus we have for $n = 0$ the additional term

$$\text{Res}\left\{\frac{z^2 - 2z + 2}{z(z - 0.1)(z - 0.2)}, 0\right\} = \frac{2}{0.02} = 100$$

so that, for all n,

$$h(n) = 100\delta(n) + \{-181 \cdot (0.1)^n + 82 \cdot (0.2)^n\}u(n).$$

EXAMPLE 3.12

Find the stable sequence whose z-transform is $H(z) = \dfrac{-2z^{-1} + 2z^{-2}}{1 - 0.3z^{-1} + 0.02z^{-2}}$.

Solution

The denominator is the same as in Example 3.11. Thus

$$H(z) = \frac{-2z + 2}{(z - 0.1)(z - 0.2)}.$$

Again, both poles are within the unit circle, so $h(n)$ must be right-sided so that the region of convergence of $H(z)$ includes the unit circle. This time $h(n)$ expressed in the p-form of (3.44) gives (with $C' = C =$ unit circle)

$$h(n) = \frac{1}{2\pi j}\oint_C \frac{-2 + 2p}{0.02(p - 5)(p - 10)}p^{-n}\,dp.$$

By arguments identical to those in Example 3.11, the result will be zero due to no enclosed poles unless $-n \leq -1$, or $n \geq 1$. Thus at the outset we know that our results must be appended by $u(n - 1)$, rather than $u(n)$ as in Example 3.11.

Furthermore, this time there will be no pole at $z = 0$ for $n \geq 1$ because $n - 1$, the power of z in the integrand for $h(n)$, is ≥ 0 for $n \geq 1$. Thus

$$h(n) = \left[\frac{-2 \cdot 0.1 + 2}{0.1 - 0.2}(0.1)^{n-1} + \frac{-2 \cdot 0.2 + 2}{0.2 - 0.1}(0.2)^{n-1}\right]u(n - 1)$$

$$= [-180 \cdot (0.1)^n + 80 \cdot (0.2)^n]u(n - 1).$$

Note that if we compute $h(0)$ using the "z-form" of the inverse z-transform contour integral, there are three terms due to the simple poles at $z = 0.1, z = 0.2$, and $z = 0$. From $h(n)$ above, the first two are -180 and 80, whereas that for $z = 0$ is $2/(0.1 \cdot 0.2) = 100$. Thus the residues for $n = 0$ sum exactly to zero!

EXAMPLE 3.13

Find the stable sequence whose z-transform is $H(z) = \dfrac{1 - 0.5z^{-1}}{(1 + 0.5z^{-1})(1 - 0.3z^{-1})^2}$.

Solution

First rewrite $H(z)$ as a rational function of z and factor the numerator:

$$H(z) = \frac{z^2(z - 0.5)}{(z + 0.5)(z - 0.3)^2}.$$

Refer to the pole–zero diagram in Figure 3.31. There is a double pole at $z = 0.3$ and a single pole at $z = -0.5$; both poles are within the unit circle. The region of convergence of the stable sequence having z-transform $H(z)$—the one that includes the unit circle—is causal, or right-sided: the region of convergence is $|z| > z_h = 0.5$. Note that if $z_h > 1$, the stable sequence will *not* be causal.

The p-form of the inverse z-transform is (again, with $C' = C = $ unit circle for stable $h(n)$)

$$h(n) = \frac{1}{2\pi j} \oint_C \frac{1 - 0.5p}{(1 + 0.5p)(1 - 0.3p)^2} p^{-n-1}\, dp.$$

The poles are at $p = 2$ and $p = 3.333$, both of which are outside the unit circle. If, however, $-n - 1 \le -1$ or $n \ge 0$, there will be a pole at $p = 0$. Thus as we already saw, for stable $h(n)$ there can be nonzero results only for $n \ge 0$; consequently, there will be $u(n)$ factors in $h(n)$. There will be no pole at $z = 0$ for $n = 0$ in the z-integral for $h(n)$, due to the z^2 factor in $H(z)$. Thus

$$h(n) = \left[\text{Res}\left\{ \frac{(z - 0.5)z^{n+1}}{(z + 0.5)(z - 0.3)^2}, -0.5 \right\} + \text{Res}\left\{ \frac{(z - 0.5)z^{n+1}}{(z + 0.5)(z - 0.3)^2}, 0.3 \right\} \right] u(n).$$

The pole at -0.5 is simple and that at 0.3 is double. Evaluating the residues gives

$$h(n) = \left[\frac{-0.5 - 0.5}{(-0.5 - 0.3)^2} (-0.5)^{n+1} + \frac{d}{dz} \frac{(z - 0.5)z^{n+1}}{z + 0.5} \bigg|_{z=0.3} \right] u(n)$$

$$= \left[-1.5625(-0.5)^{n+1} + \frac{d}{dz} \frac{z^{n+2} - 0.5z^{n+1}}{z + 0.5} \bigg|_{z=0.3} \right] u(n)$$

$$= \left[0.7812(-0.5)^n + \frac{(z + 0.5)\{[n + 2]z^{n+1} - 0.5[n + 1]z^n\} - (z^{n+2} - 0.5z^{n+1})}{(z + 0.5)^2} \bigg|_{z=0.3} \right] u(n)$$

$$= \left[0.7812(-0.5)^n + \frac{0.3^n[(0.3 + 0.5)\{(n + 2)0.3 - 0.5(n + 1)\} - \{0.3^2 - 0.5 \cdot 0.3\}]}{(0.3 + 0.5)^2} \right] u(n),$$

or finally,

$$h(n) = [0.7812(-0.5)^n + (0.2187 - 0.25n)(0.3)^n]u(n).$$

Interestingly, $0.7812 + 0.2187 = 1$.

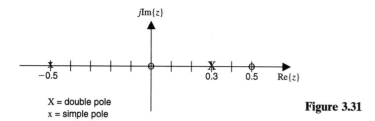

X = double pole
x = simple pole

Figure 3.31

EXAMPLE 3.14

Find the stable sequence whose z-transform is

$$H(z) = \frac{1 - 3z^{-1}}{1 - 9z^{-1} + 14z^{-2}}.$$

Solution

Factor the above as

$$H(z) = \frac{z(z - 3)}{(z - 2)(z - 7)}.$$

We see that the poles of $H(z)$ are outside the unit circle. Thus we know that the stable $h(n)$ will be an anticausal (left-sided) sequence; the region of convergence for stable $H(z)$ is $|z| < 2$. Looking at $h(n)$ in the z-form, we have

$$h(n) = \frac{1}{2\pi j} \oint_C \frac{z - 3}{(z - 2)(z - 7)} z^n dz,$$

which has a pole at $z = 0$ (the only pole within C) only if $n \leq -1$; so the terms in $h(n)$ will have the factor $u(-n - 1)$.

Using (3.44), we see immediately that at the sequence end ($n = -1$) there is no pole at $p = 0$—no special case. Thus

$$h(n) = \frac{1}{2\pi j} \oint_C \frac{1 - 3p}{1 - 9p + 14p^2} p^{-n-1} dp$$

$$= \frac{1}{2\pi j} \oint_C \frac{1 - 3p}{(1 - 2p)(1 - 7p)} p^{-n-1} dp$$

$$= \frac{1}{(14)2\pi j} \oint_C \frac{1 - 3p}{(p - 1/2)(p - 1/7)} p^{-n-1} dp.$$

Evaluating the two simple-pole residues, we obtain

$$h(n) = \frac{1}{14} \left\{ \frac{(-\frac{1}{2})}{\frac{5}{14}} \left(\frac{1}{2}\right)^{-n-1} + \frac{\frac{4}{7}}{\frac{-5}{14}} \left(\frac{1}{7}\right)^{-n-1} \right\} u(-n - 1)$$

$$= \{-0.2(2)^n - 0.8(7)^n\} u(-n - 1).$$

EXAMPLE 3.15

Find the stable sequence $h(n)$ whose z-transform is $H(z) = \cos(z)$.

Solution

We try (3.43a) with C the unit circle. Caution: Do not conclude that $\cos(z)$ on the unit circle is $\cos(\omega)$; the former is $\cos(e^{j\omega})$, a rather complicated function. Instead, write (3.43a), which for this case is

$$h(n) = \frac{1}{2\pi j} \oint_C \cos(z) z^{n-1} \, dz.$$

We see that for $n > 0$, this integral has no singularities, so that $h(n > 0) = 0$. Thus $h(n)$ must be anticausal. We might then attempt to use (3.44), which for this case is

$$h(n) = \frac{1}{2\pi j} \oint_C \cos\left(\frac{1}{p}\right) p^{-n-1} \, dp.$$

The problem is that $\cos(1/p)$ has an infinite-order pole at $p = 0$, as can be determined by writing the MacLaurin expansion of $\cos(1/p)$.

In this case, it is easiest to just use (3.43a) as written above, and for $n \le 0$ note that there is a pole of order $-(n-1)$ at $z = 0$ [whereas $\cos(z)$ is analytic for all finite z]. Thus by (3.43c), we have

$$h(n) = \{z^{-(n-1)}\cos(z)z^{n-1}\}^{[-(n-1)-1]}\Big|_{z=0} / (-n)!$$

$$= \frac{d^{(-n)}\cos(z)}{dz^{(-n)}}\Big|_{z=0} / (-n)! = \frac{\epsilon}{(-n)!},$$

where

$$\epsilon = \begin{cases} 1, & n = 0, -4, -8, \dots \\ -1, & n = -2, -6, -10, \dots \\ 0, & \text{otherwise,} \end{cases}$$

or $h(n) = (-1)^{n/2}/(-n)!$ for $\{n \le 0$ and n even$\}$; zero otherwise. This sequence $h(n)$ is nothing other than the familiar MacLaurin series for $\cos(z)$, which we should have been able to guess at the outset! That is, an anticausal $h(n)$ is really the (index-reversed) set of coefficients of the MacLaurin expansion of $H(z)$, if $H(z)$ is analytic for all z.

Partial Fraction Expansion

Partial fraction expansion is valid for proper rational functions[14] $H(z)$. The familiar partial fraction expansion of a rational function $f(z)$ is derived in the problems. Suppose that $H(z)$ has N_p poles, the kth pole z_k having multiplicity M_k. To find $h(n)$, all we need to do is (a) express $H(z)$ in its partial fraction expansion and (b) use previously derived results to determine the inverse z-transform of each term in the expansion; the resulting sum is $h(n)$.

[14]If $H(z)$ is not proper, $h(n)$ will have shifted Kronecker delta functions in addition to the geometric terms treated here; see the problems.

It is usually convenient to first divide $H(z)$ by z in the case of one or more simple poles even if multiple poles also exist, then partial-fraction-expand $H_1(z) \equiv H(z)/z$, and then finally write $H(z) = zH_1(z)$. This method allows, for simple poles, the partial fraction expansion to end up with terms such as $z/(z - z_1)$, which we immediately know has inverse z-transform $z_1^n u(n)$; similarly for double poles. Notice that this division by z plays the same role as the "z^{-1}" of "z^{n-1}" in the integrand of the inverse z-transformation contour integral (3.43a).

Thus, we write

$$H_1(z) = \frac{H(z)}{z} = \sum_{k=1}^{N_p} \sum_{i=1}^{M_k} \frac{a_{-i}^{\{k\}}}{(z - z_k)^i} \tag{3.45}$$

where

$$a_{-i}^{\{k\}} = \frac{1}{(M_k - i)!} \left\{ (z - z_k)^{M_k} \frac{H(z)}{z} \right\}^{[M_k - i]} \Bigg|_{z = z_k}, \tag{3.46}$$

in which again $[M_k - i]$ is the $(M_k - i)$th derivative with respect to z. Note in (3.46) the occurrence of $M_k - i$ rather than $M_k - 1$; only for $i = 1$ is $a_{-i}^{\{k\}}$ a residue of $H(z)$ at z_k, in which case in our earlier notation, $a_{-1}^{\{k\}} = \tilde{a}_{-1}^{\{k\}}$. Moreover, $a_{-1}^{\{k\}}$ is a residue of $H(z)$ only for a contour around z_k having no other poles inside it.

For the important practical case in which we have only simple poles ($M_k = 1$ for all k), (3.45) and (3.46) boil down to

$$H_1(z) = \frac{H(z)}{z} = \sum_{k=1}^{N_p} \frac{a_{-1}^{\{k\}}}{z - z_k}, \quad \text{where } a_{-1}^{\{k\}} = (z - z_k) \frac{H(z)}{z} \Bigg|_{z = z_k}. \tag{3.47}$$

From the linearity of the z-transform, we know that the inverse z-transform of $H(z)$ is equal to the sum of the inverse z-transforms of each of the terms in the right-hand side of (3.45). These inverse z-transforms are easy to perform using (3.35) $[h(n) = \frac{1}{2\pi j} \oint_C H(z) z^{n-1} \, dz]$.

For example, terms in $H(z)$ with $\{i = 1, M_k = 1\}$ are of the form $a_{-1}^{\{k\}}/(1 - z_k z^{-1})$ with $a_{-1}^{\{k\}} = (z - z_k)H(z)/z \big|_{z_k}$, which from Example 3.10, part (d), we immediately know has inverse z–transform $a_{-1}^{\{k\}} z_k^n u(n)$. Also, terms in $H(z)$ with $\{i = 2, M_k = 2\}$ are of the form $a_{-2}^{\{k\}} z^{-1}/(1 - z_k z^{-1})^2 = -z \, d/dz(1/[1 - z_k z^{-1}]) \cdot a_{-2}^{\{k\}}/z_k$ with $a_{-2}^{\{k\}} = (z - z_k)^2 H(z)/z \big|_{z_k}$, which from the multiplication-by-n property cited in Example 3.10, part (h), we know has inverse z-transform $a_{-2}^{\{k\}} n z_k^{n-1} u(n)$.

For poles of order three or higher, however, things get complicated in the partial fraction expansion method. For example, if $M_k = 3$ and $i = 1$, we have in (3.46) a second derivative to take in order to find $a_{-i}^{\{k\}}$, and in turn we must use the derivative property $-z \, d/dz \, X(z) \leftrightarrow n x(n)$ twice. In practice, other properties such as the shift property [see Section 5.2: $z^{-m} X(z) \leftrightarrow x(n - m)$] may also be required. Thus we must be taking derivatives not only to get $a_i^{\{k\}}$ but also to obtain the inverse z-transform of $a_{-1}^{\{k\}} z/(z - z_k)^i$ (unless we are facile with the z-transform tables, in which case we need take derivatives only once). In such cases, probably contour integration in (3.35) is more direct.

EXAMPLE 3.16

Repeat Example 3.13, this time using partial fraction expansion.

Solution

For ease in inverting the simple pole, we may define $H_1(z)$ as above, after rationalizing in terms of z rather than z^{-1}:

$$H_1(z) = \frac{z(z - 0.5)}{(z + 0.5)(z - 0.3)^2}.$$

Using (3.45) and (3.46), and defining $z_1 = -0.5$ and $z_2 = 0.3$, we have:

$$H_1(z) = \frac{H(z)}{z} = \frac{a_{-1}^{\{1\}}}{z - z_1} + \frac{a_{-1}^{\{2\}}}{z - z_2} + \frac{a_{-2}^{\{2\}}}{(z - z_2)^2}, \qquad \text{general term: } \frac{a_{-m}^{\{k\}}}{(z - z_k)^m}$$

where

$$a_{-1}^{\{1\}} = \frac{-0.5(-0.5 - 0.5)}{(-0.5 - 0.3)^2} = 0.7812,$$

$$a_{-1}^{\{2\}} = \frac{d}{dz} \frac{z(z - 0.5)}{z + 0.5} \bigg|_{z=0.3} = \frac{(z + 0.5)(2z - 0.5) - (z^2 - 0.5z) \cdot 1}{(z + 0.5)^2} \bigg|_{z=0.3}$$

$$= 0.2187,$$

and

$$a_{-2}^{\{2\}} = \frac{0.3(0.3 - 0.5)}{0.3 + 0.5} \approx -0.075.$$

Thus

$$H(z) = z H_1(z) = \frac{0.7812z}{z + 0.5} + \frac{0.2187z}{z - 0.3} - \frac{0.075z}{(z - 0.3)^2}.$$

The first two terms invert to, respectively, $0.7812(-0.5)^n u(n)$ and $0.2187(0.3)^n u(n)$. From the discussion preceding this example, we know that $\alpha z/(z - z_1)^2$ has inverse z–transform $\alpha n z_1^{n-1} u(n)$. Thus in our example the third term has the inverse z-transform $-(0.075/0.3)n(0.3)^n u(n) = -0.25n(0.3)^n u(n)$. The sum of these three terms produces the same answer as given in Example 3.13:

$$h(n) = [0.7812(-0.5)^n + (0.2187 - 0.25n)(0.3)^n]u(n).$$

Caution: In Matlab's `residuez`, the kth partial fraction expansion term is $\alpha_{i,k}/(1 - z_k z^{-1})^i$. To obtain the inverse z-transform of such a term, either use a table or write as $\alpha_{i,k} z^i/(z - z_k)^i$ and use the procedure above [possibly including the delay property $z^{-\ell}X(z) \leftrightarrow x(n - \ell)$] to obtain the inverse z-transform. Note that the coefficient $\alpha_{i,k}$ is *not* equal to $a_{-i}^{\{k\}}$![15] We can always obtain $a_{-i}^{\{k\}}$ using `residue` (as opposed to `residuez`), by giving $H_1(z)$ to `residue`. Remember to then multiply by z to have the desired partial fraction expansion of $H(z)$. In addition, all the coefficients α are called "residues" in Matlab, but as previously noted, they are not in general residues, according to the definition of a residue.

---

[15]At least initially, there appears to be no simple, general relation between $\alpha_{i,k}$ and $a_{-i}^{\{k\}}$ except for $i = 1$ for which $\alpha_{1,k} = a_{-1}^{\{k\}}/z_k$.

Long Division

Long division is valid for rational functions $H(z)$. This method will not produce a closed-form solution as will contour integration and partial fraction expansion. Also, like the partial fraction expansion method, it works only for rational $H(z)$. For example, neither of these methods could be used to solve Example 3.15. However, it is certainly the easiest method if only a few of the first numerical values of $h(n)$ are sought (the others presumably being of small magnitude), and if $H(z)$ is not in factored form and factoring routines are unavailable on the computer, which is increasingly not the case.

All that we must do is divide the numerator polynomial by the denominator polynomial. In the quotient series, the coefficient of z^{-n} is $h(n)$. Note that in continuous time, there is no counterpart of this method, whereby we effect inverse transformation by mere division. Depending on whether a causal or anticausal $h(n)$ is desired (or relevant), the arrangement of terms of the divisor and the dividend will differ. It is best demonstrated by an example.

EXAMPLE 3.17

Solve Example 3.13 again, but this time by long division.

Solution

First we expand the denominator of $H(z)$ and multiply numerator and denominator by z^3 to obtain a rational function of z:

$$H(z) = \frac{1 - 0.5z^{-1}}{(1 + 0.5z^{-1})(1 - 0.3z^{-1})^2}$$

$$= \frac{1 - 0.5z^{-1}}{1 - 0.1z^{-1} - 0.21z^{-2} + 0.045z^{-3}}$$

$$= \frac{z^3 - 0.5z^2}{z^3 - 0.1z^2 - 0.21z + 0.045}.$$

The poles are both within the unit circle and $h(n)$ is stipulated as being stable, so $h(n)$ must be right-sided. To obtain the required right-sided sequence, we require the quotient to be a series in descending powers of z. Thus we write the long division with descending powers of z from left to right. (If the required sequence were to be left-sided, we would write the divisor and dividend in ascending powers of z.) The coefficient of z^{-n} in the quotient is $h(n)$.

$$
\begin{array}{r}
1 - 0.4z^{-1} + 0.17z^{-2} - 0.112z^{-3} + 0.0425z^{-4} - 0.02692z^{-5} + \ldots \\
\hline
z^3 - 0.1z^2 - 0.21z + 0.045\,)\,z^3 - 0.5z^2 \\
z^3 - 0.1z^2 - 0.21z + 0.045 \\
\hline
-0.4z^2 + 0.21z - 0.045 \\
-0.4z^2 + 0.04z + 0.084 - 0.018z^{-1} \\
\hline
0.17z - 0.129 + 0.018z^{-1} \\
0.17z - 0.017 - 0.0357z^{-1} + 0.00765z^{-2} \\
\hline
-0.112 + 0.0537z^{-1} - 0.00765z^{-2} \\
-0.112 + 0.0112z^{-1} + 0.02352z^{-2} - 0.00504z^{-3} \\
\hline
0.0425z^{-1} - 0.03117z^{-2} + 0.00504z^{-3} \\
0.0425z^{-1} - 0.00425z^{-2} - \ldots \\
\hline
-0.02692z^{-2} + \ldots
\end{array}
$$

Thus we obtain the values $h(0) = 1, h(1) = -0.4, h(2) = 0.17, h(3) = -0.112, h(4) = 0.0425, h(5) = -0.02692$, and so on. These are exactly the values of the result in Example 3.13, when we substitute $n = 0, 1, 2, 3, 4$, and 5, respectively into $h(n) = [0.7812(-0.5)^n + (0.2187 - 0.25n)(0.3)^n]u(n)$. We see that, indeed, all we get from the long division method is values of $h(n)$ for specific sequentially increasing values of n, not a general formula as the other two methods produce.

With our experience in calculating z-transforms, we are in a better position than before to calculate any Laurent series. On WP 3.1, we present a complete example of finding all possible Laurent expansions for the $f(z)$ in Example 3.7.

3.4.3   Solution of Difference Equations Using the z-Transform

Electrical engineers are familiar with the problem of solving differential equations to determine the response to a given input of an electric circuit having energy storage elements. In that case, the unilateral Laplace transform is very convenient for circuits whose excitation begins at a particular time, defined as $t = 0$, and which may have nonzero initial stored energy. The term *unilateral* thus refers to the time integrand being run only from 0 (not $-\infty$) to ∞. Similarly, the unilateral z-transform can be used to determine the response of a discrete-time signal processing system to a given causal input, including cases with nonzero initial conditions. Again, unilateral in this case refers to the n-index running only from 0 (not $-\infty$) to ∞.

The signal-processing system can be anything from a model of a biomedical signal source to periodic payments to be made on a purchase. When considering the use of the z-transform as a means for finding the response of a system to a given input, one should first determine whether the "excitation" (input) exists for all n or, for example, just for $n \geq 0$. The bilateral transform is appropriate if we model the input as existing for all time (no practical signal does, but the model is often useful); this will be the case in our first example, Example 3.18.

EXAMPLE 3.18

An autocorrelation sequence[16] of some data we have collected, $r(k) = 0.4^{|k|}$ for all k, is to be filtered by a causal digital filter having the ARMA input–output relation $y(n) - y(n-1) + 0.21y(n-2) = x(n) - 2x(n-1)$, where here $x(n) = r(n)$. Determine $y(n)$ for all n.

Solution

Take the bilateral z-transform of both sides of the difference equation, noting that for the bilateral z-transform, $x(n-m)$ has z-transform $z^{-m}X(z)$:

$$Y(z)\{1 - z^{-1} + 0.21z^{-2}\} = R(z)(1 - 2z^{-1}),$$

or

---

[16] See Chapters 5, 7, and 10.

$$Y(z) = \frac{1 - 2z^{-1}}{1 - z^{-1} + 0.21z^{-2}} R(z)$$

$$= \frac{z^2 - 2z}{(z - 0.7)(z - 0.3)} R(z)$$

$$= H(z) \cdot R(z),$$

where from Example 3.10, part (f), we have immediately

$$R(z) = \frac{1 - 0.4^2}{(1 - 0.4z)(1 - 0.4/z)}$$

$$= \frac{-2.1z}{(z - 2.5)(z - 0.4)}, \quad 0.4 < |z| < \frac{1}{0.4} = 2.5.$$

Note that $H(z)$ is defined only for $|z| > 0.7$ because $h(n)$ is causal. Consequently, the region of convergence of $Y(z)$ is the intersection of $|z| > 0.7$ and $0.4 < |z| < 2.5$, or $0.7 < |z| < 2.5$. Because $r(n)$ is two-sided, so is $y(n)$. Also, because the region of convergence of $Y(z)$ includes the unit circle, $y(n)$ is a stable sequence. We have

$$Y(z) = \frac{-2.1(z^3 - 2z^2)}{(z - 0.7)(z - 0.3)(z - 2.5)(z - 0.4)}.$$

The solution of the original problem now reduces to one of finding the inverse z-transform of the above rational function of z. Here we use the contour integral method, with the unit circle as the contour.

The z- and p-forms can now each be used in turn to determine $y(n)$ for all n without resorting to any high-order derivative calculations. Thus, first looking at $Y(z)z^{n-1}$,

$$Y(z)z^{n-1} = \frac{-2.1(z^3 - 2z^2)z^{n-1}}{(z - 0.7)(z - 0.3)(z - 2.5)(z - 0.4)}$$

$$= \frac{-2.1(z - 2)z^{n+1}}{(z - 0.7)(z - 0.3)(z - 2.5)(z - 0.4)},$$

we see that there are poles at $z = 0$ for $n + 1 \leq -1$, or $n \leq -2$. Thus *at least* for $n \leq -2$ we will definitely use the p-form (3.44). Then looking at $Y(1/p)p^{-n-1}$,

$$Y\left(\frac{1}{p}\right)p^{-n-1} = \frac{-2.1(p^{-3} - 2p^{-2})p^{-n-1}}{(1/p - 0.7)(1/p - 0.3)(1/p - 2.5)(1/p - 0.4)}$$

$$= \frac{-2.1(1 - 2p)p^{-n}/[(-0.7)(-0.3)(-2.5)(-0.4)]}{(p - 1.429)(p - 3.333)(p - 0.4)(p - 2.5)}$$

$$= \frac{20(p - 0.5)p^{-n}}{(p - 1.429)(p - 3.333)(p - 0.4)(p - 2.5)}.$$

We see there are poles at $p = 0$ if $-n \leq -1$, or $n \geq 1$; for $n \geq 1$, use of the z-form is more convenient. For $n = -1$ and 0, either form is equally convenient; of course, both will *always* give the same answer.

First let $n \leq 0$, for which the p-form has no poles at $p = 0$. We see above that for this range of n, the only enclosed pole for a unit circle contour is the simple pole at $p = 0.4$. We obtain

$$y(n) = \frac{20(0.4 - 0.5)(0.4)^{-n}}{(0.4 - 1.429)(0.4 - 3.333)(0.4 - 2.5)}$$

$$= 0.31566(0.4)^{-n}, \quad \text{valid for all } n \le 0.$$

Now let $n \ge -1$, for which the z-form has no poles at $z = 0$. For this range of n, the enclosed poles are all simple and are located at $z = 0.7, 0.3$, and 0.4. Thus for all $n \ge -1$, we can write

$$y(n) = \left.\frac{-2.1(z - 2)z^{n+1}}{(z - 0.3)(z - 2.5)(z - 0.4)}\right|_{z=0.7} + \left.\frac{-2.1(z - 2)z^{n+1}}{(z - 0.7)(z - 2.5)(z - 0.4)}\right|_{z=0.3}$$

$$+ \left.\frac{-2.1(z - 2)z^{n+1}}{(z - 0.7)(z - 0.3)(z - 2.5)}\right|_{z=0.4}$$

$$= -8.84722(0.7)^n - 12.1705(0.3)^n + 21.333(0.4)^n, \quad \text{valid for all } n \ge -1.$$

Note that either expression for $y(n)$ gives $y(-1) \approx 0.126$ and $y(0) \approx 0.317$. The two expressions for $y(n)$ can be combined into one that is valid for all n: $y(n) = \{-8.84722(0.7)^n - 12.1705(0.3)^n + 21.333(0.4)^n\}u(n) + 0.31566(0.4)^{-n}u(-n - 1)$. Finally, note that some good practice may be obtained by taking the z-transform of the result above and verifying that $Y(z)$ as previously given is obtained. The same can be said of some of the other inverse z-transform examples presented in this section.

If the input sequence $x(n)$ begins at a definite time index, without loss of generality we may define n so that that value of n is zero. It is not necessary to do this, but it simplifies our analysis here.[17] Assuming that we have values of the output sequence $y(n)$ for a sufficient number of negative values of n (initial values), the unilateral z-transform may be applied to solve for the output due to $x(n)$ for $n \ge 0$.

As for the Laplace transform of a differentiated signal, when we take the unilateral z-transform of a delayed signal we must properly include the initial conditions. Let ZT_ul symbolize the unilateral z-transform operator. For any sequence $g(n)$, its unilateral z-transform $G_\text{ul}(z)$ is defined as

$$G_\text{ul}(z) = \text{ZT}_\text{ul}\{g(n)\} \tag{3.48}$$

$$= \sum_{n=0}^{\infty} g(n)z^{-n}.$$

Then the unilateral z-transform of $g(n - n_0)$ is

$$\text{ZT}_\text{ul}\{g(n - n_0)\} = \sum_{n=0}^{\infty} g(n - n_0)z^{-n}. \tag{3.49a}$$

Let $m = n - n_0$ to obtain

$$\text{ZT}_\text{ul}\{g(n - n_0)\} = \sum_{m=-n_0}^{\infty} g(m)z^{-(m + n_0)}. \tag{3.49b}$$

---

[17]For example, in cases where different $x(n)$ sequences, $x_i(n)$, are applied over different intervals starting at $n = n_i$, it may be more convenient to just let $x_i(n)$ begin at $n = n_i$. Each $x_i(n)$ can be considered to be acting alone if the system is linear. See also footnote 4 in Chapter 2.

We now extract $G_{ul}(z)$ in (3.48) from the sum in (3.49b) to obtain

$$ZT_{ul}\{g(n - n_0)\} = z^{-n_0} \cdot G_{ul}(z) + \sum_{m=-n_0}^{-1} g(m)z^{-(m + n_0)}. \tag{3.49c}$$

Commonly used special cases of (3.49c) are

$$ZT_{ul}\{g(n - 1)\} = z^{-1}G_{ul}(z) + g(-1) \tag{3.50a}$$

and

$$ZT_{ul}\{g(n - 2)\} = z^{-2}G_{ul}(z) + g(-2) + z^{-1}g(-1). \tag{3.50b}$$

Thus, for an ARMA(M, N) system, we need the M values of $x(n)$ $\{x(-1), \dots, x(-M)\}$, and the N values of $y(n)$ $\{y(-1), \dots, y(-N)\}$ to be able to solve for $y(n)$ for $n \geq 0$. From the above assumptions, we let $x(n < 0) = 0$, whereas $y(n < 0)$ may not be zero due to previous inputs on the system [before $x(n)$ began], the initial conditions. Recall the full discussion of initial conditions in Sections 2.3.5 and 2.4.6.4. (page 63ff).

EXAMPLE 3.19

(a) Using the z-transform, solve for $y(n)$ for $n \geq 0$ in the difference equation $y(n) + 0.1y(n - 1) - 0.72y(n - 2) = (1/4)^n u(n)$, where $y(-1) = 2, y(-2) = -1$.

(b) Verify the results using `dlsim` of Matlab; plot the two $y(n)$ sequences on the same axes to verify agreement.

Solution

(a) Taking the unilateral z-transform of the given difference equation, using (3.50a and b), gives $Y_{ul}(z)\{1 + 0.1z^{-1} - 0.72z^{-2}\} + 0.1y(-1) - 0.72[y(-2) + z^{-1}y(-1)] = 1/[1 - \frac{1}{4}z^{-1}]$. Solving for $Y_{ul}(z)$, we have

$$Y_{ul}(z) = \frac{-0.2 + 0.72(-1 + 2z^{-1}) + 1/[1 - \frac{1}{4}z^{-1}]}{1 + 0.1z^{-1} - 0.72z^{-2}}.$$

Making a common denominator gives

$$Y_{ul}(z) = \frac{(-0.92 + 1.44z^{-1})(1 - \frac{1}{4}z^{-1}) + 1}{(1 + 0.9z^{-1})(1 - 0.8z^{-1})(1 - 0.25z^{-1})}$$

$$= \frac{-0.92 + (0.23 + 1.44)z^{-1} - 0.36z^{-2} + 1}{(1 + 0.9z^{-1})(1 - 0.8z^{-1})(1 - 0.25z^{-1})}$$

$$= \frac{0.08 + 1.67z^{-1} - 0.36z^{-2}}{(1 + 0.9z^{-1})(1 - 0.8z^{-1})(1 - 0.25z^{-1})}$$

$$= \frac{0.08z\{z^2 + 20.875z - 4.5\}}{(z + 0.9)(z - 0.8)(z - 0.25)}.$$

Thus

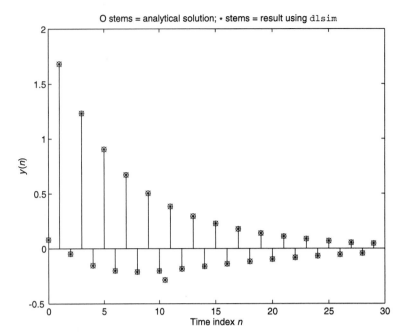

O stems = analytical solution; * stems = result using dlsim

Figure 3.32

$$\frac{Y_{\text{ul}}(z)}{z} = \frac{0.08\{z^2 + 20.875z - 4.5\}}{(z + 0.9)(z - 0.8)(z - 0.25)}$$

$$= \frac{-0.9197954}{z + 0.9} + \frac{1.0986096256}{z - 0.8} - \frac{0.09881423}{z - 0.25}, |z| > 0.9,$$

easily giving

$$y(n) = \{-0.9197954(-0.9)^n + 1.0986096256(0.8)^n - 0.09881423(0.25)^n\}u(n).$$

(b) To verify, we need to incorporate the initial conditions. Use the state–space formulation given in (2.65) and (2.66):

$$\mathbf{A} = \begin{bmatrix} -0.1 & 1 \\ 0.72 & 0 \end{bmatrix}, \quad \mathbf{B} = \begin{bmatrix} -0.1 \\ 0.72 \end{bmatrix}, \quad \mathbf{C} = [1 \ 0], \quad \mathbf{D} = 1.$$

From (2.73), the initial state vector is $w_2(0) = -(-0.72) \cdot 2 = 1.44$ and $w_1(0) = -0.1 \cdot 2 - (-0.72) \cdot (-1) = -0.92$, or $\mathbf{w}(0) = \begin{bmatrix} -0.92 \\ 1.44 \end{bmatrix}$.

With these values substituted, the results, as shown in Figure 3.32 are identical to those of part (a).

EXAMPLE 3.20

A causal sequence $x(n)$ is presented to a causal linear shift–invariant system for which the output is $y(n)$, which is describable by the difference equation $y(n) - a_1 y(n - 1) = b_0 x(n) + b_1 x(n - 1) + b_2 x(n - 2)$. Find $y(n)$ for $n \geq 0$ in terms of $x(n)$, the difference equation coefficients, and the prior initial condition $y(-1)$.

Solution

Using (3.50a and b), we have, upon taking the unilateral z-transform of the difference equation, $Y_{ul}(z)(1 - a_1z^{-1}) - a_1y(-1) = X_{ul}(z)(b_0 + b_1z^{-1} + b_2z^{-2}) + 0$, where the "$+ 0$" reflects the assumed fact that $x(n)$ is causal, so $x(-1) = x(-2) = 0$. Note that because $x(n)$ is causal, $X_{ul}(z) = X(z)$. Solving for $Y_{ul}(z)$, we have

$$Y_{ul}(z) = \frac{X(z)(b_0 + b_1z^{-1} + b_2z^{-2}) + a_1y(-1)}{1 - a_1z^{-1}}.$$

All that remains is to find the inverse unilateral z-transform of this expression.

Consider first the first term on the right-hand side, $X(z) \cdot H(z)$, where $H(z) = [b_0 + b_1z^{-1} + b_2z^{-2}]/[1 - a_1z^{-1}]$. This $H(z)$ has one simple pole at $z = a_1$. Letting $h(n)$ be the causal inverse z-transform of $H(z)$, we have $ZT_{ul}^{-1}\{X(z)H(z)\} = ZT^{-1}\{X(z)H(z)\} = x(n) * h(n)$, where by the delay property $z^{-n_0} W(z) \leftrightarrow w(n - n_0)$ (with zero "initial condition" terms because $a^nu(n)$ is zero for $n < 0$) we have for $h(n)$

$$h(n) = b_0(a_1)^nu(n) + b_1(a_1)^{n-1}u(n - 1) + b_2(a_1)^{n-2}u(n - 2)$$

$$= a_1^n\{b_0u(n) + (b_1/a_1)u(n - 1) + (b_2/a_1^2)u(n - 2)\}.$$

Furthermore, we can express $h(n)$ in terms of just $u(n)$ and two Kronecker deltas as we reduced two shifted steps to one in Example 2.16; recall (2.54c) $[f(n)u(n - 1) = f(n)u(n) - f(0)\delta(n)]$ and (2.54b) $[f(n)u(n - 2) = f(n)u(n - 1) - f(1)\delta(n - 1) = f(n)u(n) - f(0)\delta(n) - f(1)\delta(n - 1)]$:

$$h(n) = \left\{b_0 + \frac{b_1}{a_1} + \frac{b_2}{a_1^2}\right\}a_1^nu(n) - \left[\frac{b_1}{a_1} + \frac{b_2}{a_1^2}\right]\delta(n) - \frac{b_2}{a_1^2}a_1\delta(n - 1).$$

The inverse unilateral z-transform of $a_1y(-1)/(1 - a_1z^{-1})$ is just $y(-1)a_1^{n+1}u(n)$. Therefore, noting that $\delta(n - m) * x(n) = x(n - m)$ and that

$$a^nu(n) * x(n) = \sum_{m=0}^{\infty} x(m)a^{n-m}u(n - m) = a^n\sum_{m=0}^{n} x(m)a^{-m}$$

for causal $x(n)$ and $n \geq 0$, we obtain

$$y(n) = x(n) * h(n) + y(-1)a_1^{n+1}u(n)$$

$$= \left\{b_0 + \frac{b_1}{a_1} + \frac{b_2}{a_1^2}\right\}a_1^n\sum_{m=0}^{n} x(m)a_1^{-m} - \frac{1}{a_1}\left\{\left[b_1 + \frac{b_2}{a_1}\right]x(n) + b_2x(n - 1)\right\} + y(-1)a_1^{n+1},$$

where the latter expression is valid only for $n \geq 0$.

An advantage of the procedures in this example is that the input $x(n)$ can be general; we have solved the problem for all possible inputs, not just one!

Final note: It should be kept in mind that if it is desired to solve for the output in terms of initial conditions $\{y(0), y(1), \ldots; x(0), x(1), \ldots\}$ rather than $\{y(-1), y(-2), \ldots;$ $x(-1), x(-2), \ldots\}$, it may be more convenient to rewrite the equation as $\sum_{\ell=0}^{N} a_\ell y(n + \ell)$ $= \sum_{\ell=0}^{M} b_\ell x(n + \ell)$ and use the shift left (advance) formula (see Section 5.2) rather than the delay formula.

3.4.4   Meaning of a Pole of $H(z)$

In this section, we systematically examine, using pictures, the temporal characteristics of a pole of a rational transfer function $H(z)$. We limit our attention to simple poles. The goal is to enable us to "find our way around the unit disk" in the z-plane. Such understanding is crucial if one is to successfully design digital signal-processing systems.

For linear shift-invariant systems, the system transfer function $H(z)$ may be expressed as a rational function of z^{-1}. Specifically, let the output and input sequences satisfy the difference equation (2.40) $\left[\sum_{\ell=0}^{N} a_\ell y(n-\ell) = \sum_{\ell=0}^{M} b_\ell x(n-\ell) \right]$. If all initial conditions are zero and we take the z-transform of (2.40), the result may be expressed as $Y(z) = H(z) \cdot X(z)$, where

$$H(z) = \frac{B(z)}{A(z)} \tag{3.51}$$

in which

$$A(z) = \sum_{m=0}^{N-1} a_m \cdot z^{-m} \tag{3.52a}$$

$$= a_0 \prod_{m=1}^{N} (1 - z_m z^{-1}) \tag{3.52b}$$

and

$$B(z) = \sum_{m=0}^{M-1} b_m \cdot z^{-m} \tag{3.53a}$$

$$= b_0 \prod_{k=1}^{M} (1 - z_k z^{-1}), \tag{3.53b}$$

where the second equalities hold by the fundamental theorem of algebra. Because $H(z_k) = 0,$[18] z_k is a zero of $H(z)$. Moreover, because $H(z_m) = \infty$, z_m is a pole of $H(z)$. The name *pole* refers to the fact that for $z = z_m$, $H(z)$ maps to the "north" pole of the Riemann sphere in the $H(z)$ plane.[19]

Assuming uniqueness of the poles z_m, $H(z)$ may be expressed as

$$H(z) = \sum_{m=0}^{N-1} \frac{A_m}{1 - z_m z^{-1}}. \tag{3.54}$$

Assuming that $H(z)$ is defined for all z such that $|z| > |z_m|_{\max}$, where $|z_m|_{\max}$ is the largest of the magnitudes of all poles, it has the causal inverse z-transform

$$h(n) = u(n) \cdot \sum_{m=0}^{N-1} A_m z_m^{n}. \tag{3.55}$$

---

[18] We assume that none of the z_m equals z_k.
[19] The Riemann sphere is placed on the complex $H(z)$-plane so that the bottom ("south pole") is on the origin $[H(z) = 0]$. A ray can be drawn from any point in the $H(z)$-plane tangent to the ball. The intersection with the ball is unique, except that all points on the plane an infinite distance from the origin map to the top of the ball—the "north pole." [Originally, one would say "$H(z)$ has a pole at z_m" rather than the vernacular "z_m is a pole of $H(z)$."]

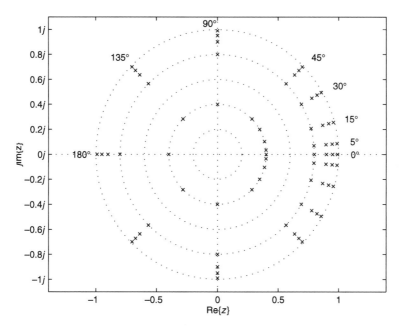

Figure 3.33

Equation (3.55) indicates that the unit sample response $h(n)$ is a linear combination of geometric sequences, each of which consists of successive powers of one of the poles z_m. Because the unit sample response $h(n)$ decays to zero as $n \to \infty$ if and only if $|z_m| < 1$ for all $m \in [0, N-1]$ and because of the previously derived requirement (Section 2.4.5) that $h(n)$ must be absolutely summable for the system to be BIBO stable, then it must be true that for any stable system $|z_m| < 1$ for all $m \in [0, N-1]$.

Therefore, in the following graphical illustrations, we consider only poles within the unit circle, because others correspond to unstable sequences/systems. Figure 3.33 shows a region of the z-plane containing the unit disk. Each "x" symbol corresponds to the location of a pole z_m and may be specified by the polar form $z_m = |z_m| \angle \theta$. To get a feel for the character of the geometric sequence corresponding to a pole having a given location in the unit disk, Figure 3.34 shows a representative group of geometric sequences that directly correspond to the pole locations in Figure 3.33. Thus for example, the second plot from the right on the bottom row in Figure 3.34 corresponds to the pole in Figure 3.33 located 135° counterclockwise from the real axis and having magnitude 0.99.

Each plot in Figure 3.34 shows only the first 100 points of each geometric sequence. Notice that poles of magnitude less than or equal to 0.4 die to nearly zero in just a few samples; such sequences are thus rapidly decaying transients often unnoticeable in the steady state (unless continually excited by the input). Contrarily, poles with magnitude 0.99 still have significant magnitude at the 100th sample and so have a more steady-state character. In the situation of convolution with an input, these more "steady-state" poles of $H(z)$ significantly involve more remotely past inputs than do the quickly decaying transient poles having small magnitudes. In the situation of signal analysis, each pole represents in the time domain a weighted decaying complex sinusoid that is one component of the total signal.

Note that the sequences in Figure 3.34 are all real; actually, each sequence represents the sum of the sequence corresponding to z_m and that corresponding to its complex conjugate z_m^*. That is, the real signal depicted in Figure 3.34, $|z_m|^n \cdot \cos(n\theta_m)$, results[20] from poles at locations z_m and z_m^*, where z_m is found at $\{|z_m|, \theta_m\}$ in Figure 3.33.

---

[20]Each complex pole term has been given magnitude coefficient $\frac{1}{2}$ for the unit-magnitude damped cosine.

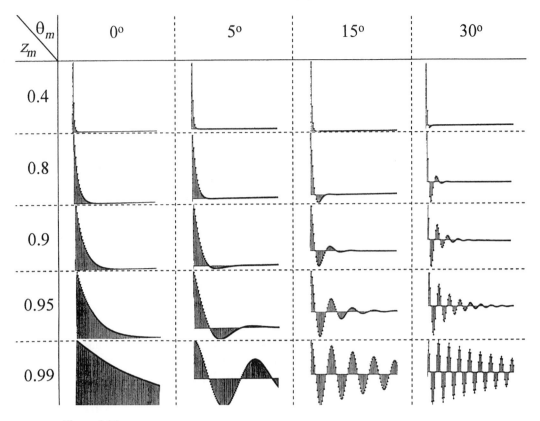

Figure 3.34

Because of the nonlinear nature of the geometric sequence, there is more visible change of behavior with $|z_m|$ for $|z_m|$ nearing unity than there is, say, for $|z_m|$ between 0 and 0.4; hence the radii chosen for the plots in Figure 3.34 were not evenly spaced. Similarly, there is much more visible change of behavior for smaller angles, so the spacing of selected angles is least for small θ_m.

The angle of z_m corresponds to the digital frequency, or cycles per sample. Thus $\theta_m = 0$ corresponds in the time domain to a nonoscillatory decaying exponential, the decay parameter being $|z_m|$. As θ_m progresses towards $180°$, the number of cycles per sample increases towards $\frac{1}{2}$. Thus for $\theta_m = 180°$ an oscillation occurs once every two samples, so in the 100-sample time window shown there are 50 decaying sinusoidal oscillations. Similarly, there are only 25 decaying sinusoidal oscillations in the window shown for $\theta_m = 90°$, for which there are four samples per cycle.

With the pictures in Figures 3.33 and 3.34 in mind, we can arrive at an understanding of how filtering works from a time-domain perspective. Suppose the filter has just one pole at $z = |z_1| \angle \omega_1$ with $|z_1|$ at or near unity. Consider what happens when it is convolved with $x(n)$.

Recall that $h(n - m)$ slides by $x(m)$ as n increases, and the inner product is taken of the vector of x values and the vector of shifted and time-reversed h values. If the input signal $x(m)$ is a sinusoid of frequency at or near ω_1, there will periodically be certain shift values n where $h(n - m)$ and $x(m)$ are "in sync," and their inner product will be large. Thus, the magnitude of the output $y(n) = (h * x)(n)$ will be large, with oscillations at the digital frequency ω_1.

The reason the output will be oscillatory is that for some shifts, the inner product will be small due to the two oscillatory sequences $h(n)$ and $x(n)$ being "out of phase," so the sample-by-sample products in the convolution, some positive and

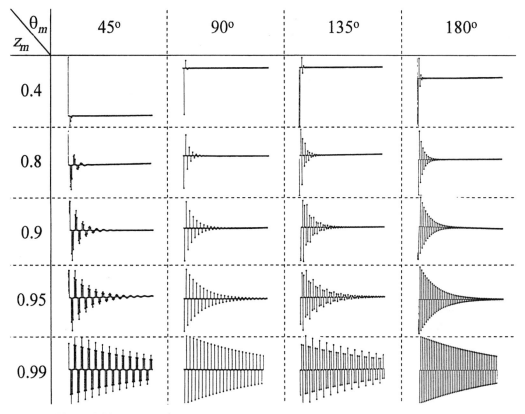

z_m \ θ_m	45°	90°	135°	180°
0.4				
0.8				
0.9				
0.95				
0.99				

Figure 3.34 (*continued*)

some negative, will mostly cancel. For other nearby shifts, they will either be in sync (large positive output) or antisync (large negative output). The period of the cycle of going in and out of phase is equal to the period of the oscillation in $x(n)$. Contrarily, if $x(n)$ is a sinusoid of frequency ω_2 far from ω_1, there will still be oscillation of frequency ω_2, but the amplitude will not be large because the point-by-point product values will tend to average to zero (they are never fully in or out of sync). Thus the output $y(n)$ will have small amplitude.

The discussion above is just the time-domain perspective of $y(n) = |H(e^{j\omega_\ell})| A \cos\{\omega_\ell n + \phi + \angle H(e^{j\omega_\ell})\}$ for $x(n) = A \cos(\omega_\ell n + \phi)$. With the single-pole filter frequency response being $H(e^{j\omega}) = \dfrac{\alpha}{1 - z_1 e^{-j\omega}}$, if for simplicity we take

$z_1 = e^{j\omega_1}$ then $|H(e^{j\omega})| = \dfrac{\alpha}{\sqrt{2[1 - \cos(\omega - \omega_1)]}}$. Thus $|H(e^{j\omega_\ell})|$ is very large for

ω_ℓ near ω_1 and small for ω_ℓ far from ω_1. this is the frequency-response view of our time-domain dicussion above. When $h(n)$ and $x(n)$ have more than one significant frequency component, it becomes nearly impossible to try to track the output in the time-domain manner given above. Herein lies the usefulness of the frequency domain, where all the required information can be determined at a glance. However, having a knowledge of the process in both domains can be very helpful in obtaining an understanding of DSP.

Finally, we note that system poles are also behavioral modes. For example, consider a home heating system. If the temperature system pole is so slow that it takes from January to March to rise and reach our desired set point of 70°F, the heating system is unsatisfactory. If it takes all winter to quit oscillating between 30°F and 90°F, the home is uncomfortable. Finally, if the system overshoot is 40°F, not many such heating systems will be sold. Analogous behaviors can be seen in the time responses in Figure 3.34 if we imagine one or more of these transients as being superimposed in various amounts on the ideal step response. Herein lies the importance of good control system design and of understanding the system modes (poles) as crucial quantifiers of performance in both analog and digital systems.

PROBLEMS

Review of Complex Numbers and Variables

3.1. Find (a) the rectangular form of $z = 2\angle 120°$ and (b) the polar form of $z = 3 + j4$.

3.2. Reduce $(1 + 3j)/(4 - j) + (2 - j)/(5j)$ to the form $z = x + jy$.

3.3. Solve for the real numbers x and y: $3x + j2y + 6y = 6 - 7j$.

3.4. Give the rectangular form for division of two complex numbers.

3.5. Prove Euler's formula via defining, as in [Franklin, G.F., Powell, J.D., Emami–Naeini, A., *Feedback Control of Dynamic Systems*, Third Edition (New York: Addison-Wesley, 1994)], $a(\theta) = \cos(\theta) + j\sin(\theta)$ and solving a differential equation.

3.6. Draw the complex number $z_1 - z_2$ as a vector in the complex plane. Does the "vector" $z_1 - z_2$ point to z_1 or to z_2?

3.7. Without explicitly converting numerator and denominator to polar form, find $|z|$ if $z = (1 + j2)/(3 + j4)$.

3.8. Find and plot in the complex plane the three cube roots of $-8 = 8 \cdot e^{j\pi}$.

3.9. Analytically determine and use Matlab to plot all the sixth roots of $z = -3 + j4$ in the z-plane.

3.10. How are the following complex numbers geometrically related to z?

(a) $-z$ (b) z^* (c) $1/z$

3.11. Solve $x^4 + 9 = 0$ for all solutions x and factor $x^4 + 9$ into two quadratic factors with real coefficients.

3.12. Determine the region in the z-plane for which $|z + 3|/|z - 1| < 2$.

3.13. What is the locus of the points z that satisfy the given conditions?

(a) $|z - 3| \le 6$

(b) $|z| = 0.5$

(c) $|z + a| = |z - a|$, a real

(d) $|z - a| = |z - b|$, any complex or real $\{a, b\}$

(e) $|z| < |z - 5|$

(f) $1 < |z - 3j| < 5$

(g) Re$\{z^2\} > 0$

(h) Re$\left\{\dfrac{1}{z}\right\} < \dfrac{1}{4}$

3.14. What geometric relation between complex numbers z_1 and z_2 must hold so that $|z_1 + z_2| = |z_1| + |z_2|$? (Hint: Think of z_1 and z_2 as vectors in the z-plane.)

3.15. For complex $w = u + jv$ and $z = x + jy$, find the maps in the w-plane of the following regions in the z-plane, where $w = z^2$; that is, find a single expression involving u, v, and constants.

(a) $-\infty < x < \infty, y = 3$

(b) $|z| < 1, -\pi \le \arg\{z\} < 0$

(c) $|z| = 5, \dfrac{\pi}{2} \le \arg\{z\} < \pi$

3.16. Consider the polynomial $A(z) = a_0 + a_1 z^{-1} + a_2 z^{-2} + \ldots + a_{N-1} z^{-(N-1)}$, where $z = r e^{j\omega}$ ($r > 0$) and the a_ℓ are all real-valued. Obtain an expression for $|A(z)|$ and $\angle A(z)$. What relevance do these results have for a special value of r?

3.17. Show the extremely important result that if $P(z_1) = 0$—that is, if z_1 is a root of $P(z)$—then $z_1{}^*$ is also a root of $P(z)$, assuming that $P(z)$ is an Nth-order polynomial in z with real coefficients a_i.

3.18. (a) Solve the quadratic equation $z^2 + bz + c = 0$, where b and c are real. Express in rectangular form for the case in which the roots are complex.

(b) Give the condition on the coefficients b and c such that z is in the left half of the z-plane (condition for a stable continuous-time pole, with z replaced by s).

(c) Express $|z|$ in terms of b and c. Using this expression, give the condition on the coefficients such that z is within the unit circle in the z-plane, the stability condition for poles of discrete-time systems.

3.19. Consider $w = (z_1 - z_2)^{1/N}$. What condition on the real and/or imaginary parts of z_1 and z_2 must hold if for one of the roots w_ℓ, $\angle w_\ell = -\pi/(2N)$?

3.20. Show that—for instance, in a power series or rational function—if all occurrences of j are replaced by $-j$, the result is the complex conjugate of the original expression. Also show that this conjugation procedure works for exponentials/sin/cos, $\cos^{-1}$, integrals, derivatives, and so on.

3.21. Another important transcendental function in DSP is the (natural) logarithm, $w = \ln(z)$. Plot in Matlab a three-dimensional plot of the magnitude, phase, real, and imaginary parts of $\ln(z)$, using the same style as in the text examples. Discuss the reasons for the appearances of magnitude, phase, real, and imaginary parts of $\ln(z)$.

Analyticity of Functions of Complex Variables

3.22. Where is the function $w = 1/z$ analytic? Argue using the Cauchy–Riemann equations.

3.23. Verify numerically, using Matlab, that $f(z)$ is analytic by computing (3.1) along lines in different directions (angles of approach to z_0) that all pass through z_0. For specificity, let $z_0 = 1 + j$, $f(z) = \cos(z)$, and $dz = 0.0004$, and let the angles from the real axis be 45°, 20°, and –70°.

3.24. Give a simple function $w(z)$ involving only very simple second-order polynomials that is analytic only along a line in the z-plane. Plot 30-contour maps of the real and imaginary parts of $w(z)$ on the same axes. Do these plots help you determine where $w(z)$ may or may not be analytic? Repeat the contour plots for $w(z) = z^2$ and compare with your plot for $w(z)$ analytic only along one line. Be quantitative with your arguments, using clabel, manual with the contours to obtain numerical values of certain desired contours.

3.25. Repeat the contours investigation of Problem 3.24 for the function $w(z) = |z|^2 + 2\angle z - j\angle z \cdot \cos(4|z|)$. Is this $w(z)$ analytic anywhere? Also plot contour plots of $|w(z)|$ and $\angle w(z)$. Can you make any inferences about analyticity from the magnitude and phase plots? Justify. (Hint: The natural log function may come in handy here.) Prove by citation of the figures in the text, which show several contour plots of magnitude and phase of various $w(z)$.

3.26. Suppose that $w = f(z) = u(z) + jv(z)$, where $z = x + jy = |z|\angle\theta_z$. Express the Cauchy–Riemann conditions (3.3) in terms of $\{|z|, \theta_z\}$ rather than $\{x, y\}$.

3.27. Show that a rational function of w is analytic everywhere except at its poles by showing that the Cauchy–Riemann equations are satisfied. Let

$$w(z) = \frac{\displaystyle\prod_{m=1}^{M} (z - z_{zm})}{\displaystyle\prod_{n=1}^{N} (z - z_{pn})} .$$

Note that this problem is a more detailed study of Example 3.2. The proof is most easily accomplished using polar form. Note: This calculation is messy, but it is manageable if intermediate quantities are defined and one has the requisite tenacity.

3.28. Let $k_x^2 + k_y^2 = k_0^2$, or equivalently, $k_y^2 = k_0^2 - k_x^2$, where k_x and k_y are components of the wave vector of a plane-wave component of a propagating electromagnetic field and k_0^2 is the constant (real-valued) wavenumber of the medium in which the wave propagates. Both k_x and k_y may be complex-valued, signifying exponentially declining/growing oscillation components of the Fourier-decomposed wave. Find two single-valued branches for k_y; where are they discontinuous (that is, where are suitable branch cuts)? Use the Cauchy–Riemann equations for $\mathrm{Re}\{k_y\}$, $\mathrm{Im}\{k_y\}$ in terms of $\{|k_x|, \theta_{kx}\}$ to prove that within the region defined for each branch, k_y is an analytic function of k_x. Hint: Write $k_y = j\{k_x^2 - k_0^2\}^{1/2} = j\{(k_x - k_0)(k_x + k_0)\}^{1/2}$, and examine the real axis for $\mathrm{Re}\{k_x\} > k_0$ and $\mathrm{Re}\{k_x\} < 0$. Express $k_x \pm k_0$ in polar form.

Complex Integration

3.29. Find the integral of $z^3/(z-j)$ around any closed contour (a) enclosing $z = j$ and (b) not enclosing j.

3.30. Evaluate exactly the value of $I = \oint_C \dfrac{dz}{(z + \frac{1}{2})^2(z + 2)}$, where C is the unit circle.

3.31. Evaluate the integral $I = \oint_C \dfrac{e^{z+a}\sin(bz)}{5z + 20}\, dz$ where a and b are positive constants and where C is a closed contour running along the $j\Omega$ axis from $-j\infty$ to $+j\infty$ and enclosing the entire (a) right half-plane and (b) left half-plane.

3.32. Find a function whose integral around a circle of radius $r > 5$ centered on the origin is equal to $e^{\cos(5)}$. The function cannot contain the constant $e^{\cos(5)}$ as a factor.

3.33. If we can represent a function $f(z)$ by the Laurent series (3.23), we might suppose that the integral around a closed contour C, where C encloses z_A among several other poles of $f(z)$, is equal to $2\pi j a_{-1}^{\{z_A\}}$. Is this correct? If so, why? If not, why not?

3.34. Use contour integration to evaluate $I = \displaystyle\int_0^\infty \dfrac{dx}{x^4 + 1}$. Hint: Show proportionality to the result of using a D-shaped contour with the straight side of D on the real axis and x replaced by the complex variable z.

3.35. Repeat Problem 3.34 for $I = \displaystyle\int_{-\infty}^\infty \dfrac{x^2\, dx}{(x^2 + 1)(x^2 + 4)}$.

3.36. Compute the following integral for $n \geq 4$: $I = \dfrac{1}{2\pi j}\oint_C \dfrac{z^{n+3}\, dz}{(z-1)^2(z + 0.4)}$, where C is (a) a circle of radius $\frac{1}{2}$ centered on $z = 0$, and (b) a circle of radius 1.5 centered on $z = 0$.

3.37. Write a Matlab program that, for a user-specified contour function, will numerically validate both the Cauchy integral theorem and the residue theorem. First let the contour be a circle, then an ellipse; you should obtain the same results for the theorems either way. First try $f(z) = 1$ (to check your program), then $\cos(z)$, and then $e^z/(z - 0.2)$, where the contour is a circle or ellipse centered on the origin containing the point 0.2 (but not necessarily centered on it). Thus we validate the seemingly nonintuitive fact that no matter what contour you use (as long as it encircles/does not encircle a pole), the results of integrating around that contour are the same and are as the theorems predict. This exercise could be extended to contours encircling numerous but not all poles of a complicated $f(z)$, and even to multiple-order poles. For example, validate the result of Example 3.6.

3.38. (a) Using (3.42), derive (3.40).

(b) Verify (3.40) numerically, using Matlab to compute the sum directly and compare against the closed form in (3.40). Let $a = 0.9$, $M = 100$, and $N = 400$. Do not use a `for` loop.

3.39. Prove the validity of (3.29) by using (3.14), by defining a new function $g(z)$ having a multiple pole at z_k, where $g(z)$ is related to the analytic $f(z)$ in (3.14).

3.40. Suppose that $f(z) = g(z)/h(z)$ is a function of z having a simple pole singularity at $z = z_m$, but the factor $z - z_m$ is not easily factored out; for example, $f(z)$ is a rational function of $e^{\alpha z}$, not of z. Find the residue of $f(z)$ at z_m in terms of g and h and/or their derivatives. Hint: The general definition of a simple pole is that if $f(z) = g(z)/h(z)$, then $g(z_m) \neq 0, h(z_m) = 0$, but $h'(z_m) \neq 0$. [This definition holds true for a simple pole term such as $1/(z + a)$, but not for a double pole term such as $1/(z + a)^2$.] The generalization of (3.30) for the definition of a residue is $\text{Res}\{f(z), z_m\} = \lim_{z \to z_m} (z - z_m)f(z)$.

Laurent Expansions

3.41. Why is the following argument incorrect? Since, by long division, $z/(2 - z) = \frac{1}{2}z + \frac{1}{4}z^2 + \frac{1}{8}z^3 + \dots$ and $z/(z - 2) = 1 + 2/z + 4/z^2 + 8/z^3 + \dots$ and since $z/(2 - z) + z/(z - 2) = 0$, it follows that $\dots + 8/z^3 + 4/z^2 + 1/z + 1 + \frac{1}{2}z + \frac{1}{4}z^2 + \frac{1}{8}z^3 + \dots = 0$.

3.42. Find the regions in which each of the following series converges absolutely

   (a) $1 - z + z^2 - z^3 + z^4 - z^5 + \dots$

   (b) $1 + (z - j - 2) + \frac{1}{4}(z - j - 2)^2 + \frac{1}{9}(z - j - 2)^3 + \frac{1}{16}(z - j - 2)^4 + \dots$

   (c) $1 + \frac{1}{2}(z + 4j) + \frac{1}{4}(z + 4j)^2 + \frac{1}{8}(z + 4j)^3 + \frac{1}{16}(z + 4j)^4 + \dots$

   Hint: Suppose that $G(z) = \sum_{n=0}^{\infty} h(n)(z - z_0)^n$. Then $G(z)$ converges absolutely if $\sum_{n=0}^{\infty} |h(n)(z - z_0)^n| < \infty$. Let $R = 1/\lim_{n \to \infty} |h(n)|^{1/n}$.

   Then the Cauchy–Hadamard theorem (merely a complex form of the ratio test) states that the circle of convergence of $G(z)$ is the circle with center at $z - z_0$ and radius R, and that $G(z)$ converges absolutely within this circle ($|z - z_0| < R$).[21]

3.43. State explicitly the significance of the results for the following special cases for (3.32) and (3.33): (a) $f(z) = 1, n_{f,m} = 0$; (b) $z = z_m, n = 0, n_{f,m} \geq 0$; (c) $n = -1, n_{f,m} \leq 0$ and $C = C_m$, where C_m is a contour in a deleted neighborhood of z_m (e.g., C_3 for z_3 in Figure 3.25), where $f(z)$ is a rational function having poles of multiplicity M_m at z_m.

Example on WP 3.1 may be helpful for Problems 3.44 and 3.45.

3.44. (a) Find all Laurent Expansions of $f(z) = (z + 2)/[z(z + 0.9)]$ about $z_0 = 1$ (that is, the Laurent expansions for all z).

   (b) Write Matlab code to check your result for any value of z by comparing a direct evaluation of $f(z)$ against the series approximation; 4000 terms should be sufficient for practically any value of z.

3.45. Repeat Problem 3.44 (both parts) for $f(z) = z/[(z + 1/2)^2(z - 0.8)]$, this time expanding about $z_0 = 1 + j$.

z-Transforms: Forward and Inverse

3.46. Find the sequence $h(n) = \text{ZT}^{-1}\{H(z)\}$ such that $\text{ZT}\{h(n)\}$ is equal to $-\frac{8}{9}$ at $z = \frac{1}{8}$, where $H(z) = z/[(z - \frac{1}{2})(z + \frac{1}{4})]$. Use contour integration.

3.47. Find the sequence $h(n)$ for which $\text{ZT}\{h(n)\} = H(z) = 1/[(z + \frac{1}{8})(z - \frac{1}{3})]$ and which corresponds to the coefficients of a power series of $H(z)$ involving only negative powers of z. Use contour integration.

3.48. Find the stable sequence $h(n)$ whose z-transform is $H(z) = z/[(z^2 - 0.9z - 2.2)(z + 1/10)]$. Use contour integration.

3.49. Find the stable sequence $h(n)$ whose z-transform is $H(z) = (z + 3)/(z^2 + 0.4z + 0.29)$. Use contour integration; no complex-valued quantities are allowed in your result.

3.50. Find the stable sequence $h(n)$ whose z-transform is $H(z) = \dfrac{z(z + 2)}{(z + 3)^2(z - 7)}$; use contour integration.

---

[21]Proved in A. I. Markushevich, *The Theory of Functions of a Complex Variable*, Vol. 1. (Englewood Cliffs, NJ: Prentice Hall, 1965), pp. 344–346.

3.51. Find the causal sequence $h(n)$ whose z-transform is $H(z) = (z + 3)/(z^2 + 0.4z + 0.29)$. Use partial fraction expansion, and compare with the results obtained using contour integration in Problem 3.49. No complex-valued quantities are allowed in your final result.

3.52. (a) Find the causal sequence $h(n)$ whose z-transform is

$H(z) = (z^2 - 0.4z + 0.68)/[(z - 1)^2(z - 0.8)]$. Use partial fraction expansion. Avoid use of the Kronecker delta in your solution. Is the system stable?

(b) Compare the first five nonzero values of $h(n)$ obtained by long division with the results in part (a).

3.53. In this problem, we bring together and compare results of Chapters 2 and 3. Let $H(z) = 3/[(z + 0.5)(z - 0.8)]$.

(a) Is this system stable? Why or why not?

(b) Write the state equations in phase-variable canonical form (see Problem 2.36), and write the output equation.

(c) Using the results of part (b), write the state equations and output equation in decoupled form.

(d) Find the step response in closed form, assuming zero initial conditions. Check using the transfer function. When will this check work? When would the state form be advantageous for finding $y(n)$?

3.54. Find the causal sequence whose z-transform is $H(z) = z^2/(z^4 + 1)$. Use partial fraction expansion. Do not use sinusoidal functions in your final result.

3.55. (a) Find $H(z) = \text{ZT}\{h(n)\}$, where $h(n) = -(\frac{1}{4})^n u(n) + 3 \cdot 2^n u(-n - 1)$; include the region of convergence.

(b) Find the purely causal $[h_1(n)]$ and purely anticausal $[h_2(n)]$ unit sample responses whose z-transforms are $H(z)$ as found in part (a). Use contour integration techniques.

(c) Verify your results using Matlab, by programming a truncated z-transform sum for z in each region of convergence and by comparing against a direct evaluation of the rational function $H(z)$ found in part (a). Your results should agree well for, for example, 4000 terms or less retained.

3.56. Suppose that $H(z) = \dfrac{1}{z(z^2 + j)}$.

(a) Expand $H(z)$ in partial fractions.

(b) Draw the pole diagram.

(c) Define the two regions in which the Laurent expansions of $H(z)$ will differ.

(d) Find the Laurent expansions about $z = 0$ in those two regions up to $z^{\pm 9}$; is a pattern for the coefficients emerging?

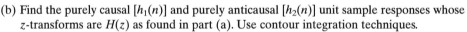

Note: $(a + b)^{-1} = \displaystyle\sum_{k=0}^{\infty} (-1)^k \, a^{-(1+k)} b^k = \dfrac{1}{a} \sum_{k=0}^{\infty} \left(\dfrac{-b}{a}\right)^k$ for $|a| > |b|$.

3.57. For $H(z) = \dfrac{z - 2}{(z^2 - 1.41z + 1.21)(z - 0.7)}$, find the BIBO stable $h(n)$, evaluating only for $n \geq 0$, using (a) partial fraction expansion and (b) contour integration.

3.58. Give the transfer function having a zero at $z = 10$, a zero at $z = 0$, a double pole at $z = 1$, a pole at -0.5, and poles at $0.9\angle\pm100°$ (assume unit magnitude of highest-power-of-z coefficients in the numerator and the denominator). Then partial fraction expand it. Compare your PFE coefficients with those obtained using `residue` (which is convenient for z-transforms expressed in terms of z rather than z^{-1}). Analytically calculate the unit sample response; plot it out to $n = 120$ using Matlab. Then graphically compare this result with that obtained using `filter` on the numerator and denominator polynomials you found.

3.59. (a) Let $x(n) = (1 - e^{-\sigma n})/n$ for all $n \geq 0$, where $\sigma > 0$. Find an analytical expression for $X(z)$, and for $\sigma = 0.8$, plot a three-dimensional plot of $|X(z)|$. Hint: In Problem 4.50, we will show that $\mathrm{ZT}\{(1/n)u(n-1)\} = \ln\{z/[z-1]\}$.

(b) Plot $|X(z)|$ versus ω, where $z = |z|e^{j\omega}$ for 600 equally spaced values of $\omega \in [0, \pi]$ for $|z| = 2$. Superimpose an estimate of $|X(z)|$ by computing the z-transform sum using $x(n)$ for $n \in [0, N-1]$, where $N = 300$. You should find agreement. Repeat for $|z| = 1$ (the DTFT). Do you still find agreement (especially near and at $\omega = 0$)? Why or why not? Also repeat for $|z| = 0.7$, and comment on this result.

z-Transform Theory

3.60. Expand out a polynomial in terms of its roots, and identify the relations between the polynomial coefficients and the roots.

3.61. The partial fraction expansion of rational functions [(P3.60.1) below] is extremely important in DSP as well as being a technique for integration. Suppose that $H(z)$ is the rational function

$$H(z) = \frac{B(z)}{A(z)} = \frac{\displaystyle\prod_{m=1}^{M}(z - z'_m)^{n_m}}{\displaystyle\prod_{\ell=1}^{L}(z - z_\ell)^{k_\ell}} \quad \text{where } k_\ell \text{ and } n_m \text{ are integers, and}$$

$$\sum_{m=1}^{M} n_m < \sum_{\ell=1}^{L} k_\ell \quad \text{so that } H(z) \text{ is a } \textit{proper} \text{ rational function.}$$

(a) Prove that

$$H(z) = \sum_{\ell=1}^{L} \sum_{i=1}^{k_\ell} \frac{a_{-i}^{\{\ell\}}}{(z - z_\ell)^i}, \tag{P3.60.1}$$

where

$$a_{-i}^{\{\ell\}} = \frac{1}{(k_\ell - i)!} \frac{d^{k_\ell - i}}{dz^{k_\ell - i}} \left[(z - z_\ell)^{k_\ell} H(z) \right]\Big|_{z = z_\ell}.$$

Thus, the first sum (ℓ sum) is over the distinct poles, and the second sum (i sum) is over all the multiplicities of each of those poles. Include a justification of the formula for $a_{-i}^{\{\ell\}}$ in (P3.60.1). Hint: Consider the Laurent expansions of $H(z)$ about, for example, z_1 (in a neighborhood of z_1) and z_2 (in a neighborhood around z_2); for example, about $z = z_1$ it is as follows [and you may assume this result without proving it; it was proved in Section 3.3.4, in the discussion leading to (3.27)]:

$$H(z) = \sum_{i=1}^{k_1} \frac{a_{-i}^{\{1\}}}{(z - z_1)^i} + \sum_{i=0}^{\infty} a_{\,i}^{\{1\}}(z - z_1)^i \tag{P3.60.2}$$

$$= P^{\{1\}}(z) + NP^{\{1\}}(z),$$

where $a_{-i}^{\{1\}}$ is as given in (P3.60.1) for $\ell = 1$ (and $a_{\,i}^{\{1\}}$ is given by the usual Laurent contour integral), $P^{\{\ell\}}(z)$, the first sum in (P3.60.2) for $\ell = 1$, is the principle part of the expansion of $H(z)$ about $z = z_\ell$, and $NP^{\{\ell\}}(z)$ is the corresponding nonprincipal part. In particular, also consider the Laurent expansion of the nonprinciple part of the Laurent expansion about z_2 of $H(z)$ about z_1 and its relation to the Laurent expansion about z_2. Once you have shown all the details for consideration of z_1 and z_2, generalize your statements to obtain (P3.60.1).

(b) For an $H(z)$ having only simple poles [$k_\ell = 1$ for all ℓ], to what does (P3.60.1) simplify? Simplify as much as possible.

3.62. In this problem, we show the easiest way by far to invert the z-transform of a system having complex poles.

(a) Find the inverse z-transform of $H_{A1}(z) = \dfrac{Az}{z - z_1} + \dfrac{A^* z}{z - z_1^*}$. Your result must have only real-valued quantities. Assume causal sequences throughout.

(b) Extend the result in part (a) to inverse z-transforming a general rational function $H(z)$ having real-valued coefficients of all powers of z.

3.63. Repeat Problem 3.62 for the Laplace transform, making all necessary changes.

3.64. (a) In (2.40), what is the condition on the coefficients for causality? Use the z-transform to argue.

(b) From your discussion in part (a), what is the condition that the order of the numerator of $H(z)$ expressed as a function of z (not z^{-1}) is equal to the order of the denominator? Does it involve M and N? What common situation in the difference equation satisfies this condition?

(c) Discuss the case $M = N$ and $b_0 \neq 0$, $a_0 \neq 0$.

3.65. In this problem, assume only simple poles.

(a) Let $H(z) = \dfrac{\displaystyle\sum_{k=0}^{M} b_k z^{-k}}{A(z)}$, where

$$A(z) = \sum_{\ell=0}^{N} a_\ell z^{-\ell} = z^{-N} \sum_{\ell=0}^{N} a_\ell z^{N-\ell} = z^{-N} \prod_{\ell=1}^{N} (z - z_\ell) = \prod_{\ell=1}^{N} (1 - z_\ell z^{-1}),$$

where $a_0 = 1$ and where in the last step we factored out z for each term, leaving $z^{-N+N} = 1$ out front. Assume for now that $N > M$. We make this assumption so that when we convert to a function of z rather than z^{-1}, we not only obtain a proper rational function, but we also have $N - M$ factors of z out front, one of which we pull out, conveniently leaving $H_1(z) = H(z)/z$. Clearly z_ℓ are the poles of the system. Show that without having to rationalize and define $H_1(z) = H(z)/z$, we can immediately obtain the correct PFE coefficients A_ℓ for the simple pole z_ℓ by just evaluating

$$A_\ell = \frac{\displaystyle\sum_{k=0}^{M} b_k z_\ell^{-k}}{\displaystyle\prod_{\substack{m=1\\m\neq\ell}}^{N} (1 - z_m z_\ell^{-1})} = H(z)(1 - z_\ell z^{-1})\Big|_{z=z_\ell}.$$

Thus we have obtained a shortcut for the PFE, when we have been given the difference equation. Just form $H(z)$ written as a function of z^{-1} and directly evaluate A_ℓ as shown. Then

$$H(z) = \sum_{\ell=1}^{N} \frac{A_\ell}{1 - z_\ell z^{-1}} = \sum_{\ell=1}^{N} A_\ell \frac{z}{z - z_\ell}.$$

(b) Extend to the case in which $M \geq N$. As an example of your results, expand

$$H(z) = \frac{2 + 3z^{-1} + 4z^{-2} + 8z^{-3}}{1 + 5z^{-1} + 6z^{-2} + 7z^{-3}}.$$

3.66. Let $x(n) = a^n u(n)$, $h_1(n) = b^n u(n)$, $h_2(n) = c^n u(n)$, and $h_3(n) = d^n u(n)$, where $a \neq b \neq c \neq d$. In Figure P3.66, find $y(n)$ in terms of a, b, c, d, and n.

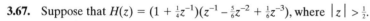

$x(n) \longrightarrow \boxed{h_1(n)} \longrightarrow \boxed{h_2(n)} \longrightarrow \boxed{h_3(n)} \longrightarrow y(n)$

Figure P3.66

3.67. Suppose that $H(z) = (1 + \frac{1}{4}z^{-1})(z^{-1} - \frac{5}{6}z^{-2} + \frac{1}{6}z^{-3})$, where $|z| > \frac{1}{2}$.

(a) Find $h(n)$ by using a direct calculation. An easy way is to express $H(z)$ as a function of z rather than z^{-1}, and then let $H_1(z) = H(z)/z^2$ in this case. Remember, the partial fraction expansion theorem is valid only for proper rational functions.

(b) What is the pole–zero excess (number of poles minus number of zeros)? Give a formula for the order of the zero at $z = 0$ in cases similar to that in part (a). The region of convergence indicates a causal sequence, yet the result in part (a) is (partly) noncausal. Comment in general about Problem 3.64 in this light. Specifically, in the text it is said that for different regions of convergence, we can make the $h(n)$ either causal, anticausal, or bicausal as we choose, independent of the a_ℓ and b_ℓ sequences. Yet in Problem 3.64 we have conditions on the a_ℓ and b_ℓ for causality. Resolve any apparent contraction.

(c) When we attempt to use `residuez` in Matlab to obtain our result numerically, it balks because the first coefficient of $A(z)$ must be nonzero, and here it is zero. Show what to do to still use Matlab in this example. Carry out your suggestion analytically, and check out the numbers on Matlab. Of course, your result should agree with that in part (a).

3.68. Let $H(z) = (1 + \frac{1}{4}z^{-1} + \frac{1}{7}z^{-2})/(1 + \frac{1}{2}z^{-1})$. Find $h(n)$ by the following approaches.

(a) Perform long division to obtain a proper rational function of z, and then use the usual approaches to obtain $h(n)$. Verify your results via Matlab using the command `residuez`. What is the order of the pole at the origin when $M > N$? Is $h(n)$ causal?

(b) Take the approach suggested in the solution of part (b) of Problem 3.67: Pull out the pole at the origin of appropriate order, inverse z-transform the result (not including that pole at the origin), and delay that sequence by the appropriate amount. Your answer should agree exactly with that in part (a).

z-Transforms for Solving Difference Equations; Miscellaneous

3.69. A unit sample response is determined to be $h(n) = \{a_1\alpha^n + a_2\beta^n\}u(n)$. Write down the difference equation, with the coefficients written in terms of those of $h(n)$.

3.70. (a) Using the z-transform, solve for $y(n)$ for $n \geq 0$ if $y(n) - 1.6y(n-1) + 0.89y(n-2) = x(n)$,

where $x(n) = u(n) - u(n-5)$ and $y(-1) = 5, y(-2) = 4$. No complex quantities are allowed in the result. In addition, use your final expression to determine the number of samples per oscillation in $y(n)$.

(b) Verify your result using `dlsim` of Matlab; plot the two $y(n)$ sequences on the same axes (out to about $n = 30$).

3.71. Repeat Problem 3.70 for $y(n) - 1.7y(n-1) + 0.72y(n-2) = x(n)$ where $x(n) = 8$

$\sin(\pi n/10)u(n)$ and $y(-1) = 6, y(-2) = 3$. No complex quantities are allowed in the result. This time, plot the first 100 points. Qualitatively, how does this solution compare with that of Problem 3.70?

3.72. Use the z-transform to solve the difference equation for $n \geq 0$

$$x(n) - 1.6x(n-1) + \tfrac{89}{100}x(n-2) = 0, x(0) = 1, x(1) = 2.$$

3.73. Use the z-transform to re-solve Example 2.13. Explicitly identify the zero-state and zero-input responses, and compare with, respectively, the particular and homogeneous solutions determined in Example 2.13. Hint: Rewrite the difference equation and use $ZT\{y(n+1)\} = z[Y(z) - y(0)]$ (proved for general shift in Problem 5.6).

3.74. For $y(n) - 0.25y(n-1) - 0.125y(n-2) = x(n)$:

(a) Let $y(-1) = 0, y(-2) = -1, x(n) = (\frac{1}{3})^n u(n)$. Find $y(n)$ for all n, using the z-transform.

(b) Find the system function $H(z)$ and the unit sample response $h(n)$, again using the z-transform.

(c) Find $y_1(n) = x(n) * h(n)$, where $x(n)$ is as in part (a) and $h(n)$ is as in part (b). Does $y_1(n) = y(n)$ from part (a)? Why or why not?

3.75. Using the z-transform, solve the difference equation for $n \geq 0, x(n) + a_1x(n-1) + a_2x(n-2) = 0$, with $x(0)$ and $x(1)$ given. Express only in terms of $a_1, a_2, x(0)$, and $x(1)$.

3.76. Use the z-transform to solve the difference equation $y(n) - \frac{1}{2}y(n-1) = x(n)$, where

$x(n) = [8(1/3)^n - 7(5/9)^{2n}]u(n)$ and $y(-1) = 6$. Hint: $a^{2n} = (a^2)^n$. Plot the result in Matlab; verify using `dlsim`.

3.77. Consider a linear shift-invariant system with input $x(n)$ and output $y(n)$ for which $y(n) - \frac{7}{12} \cdot y(n-1) + \frac{1}{12} \cdot y(n-2) = -5 \cdot x(n) + x(n-1)$. Let the system transfer function $H(z)$ be $Y(z)/X(z)$.

(a) What are the zeros of this system?

(b) What are the poles of this system?

(c) Draw the fully labeled pole–zero diagram.

(d) Find the causal unit sample response sequence $h_1(n)$. What is the corresponding region of convergence of $H(z)$?

(e) Find the bicausal unit sample response sequence $h_2(n)$. What is the corresponding region of convergence of $H(z)$?

(f) Find the anticausal unit sample response sequence $h_3(n)$. What is the corresponding region of convergence of $H(z)$?

(g) Which, if any, of $h_1(n)$, $h_2(n)$, and $h_3(n)$ are stable?

(h) Each of $h_1(n)$, $h_2(n)$, and $h_3(n)$ must satisfy the difference equation given above. For brevity, prove only that $h_1(n)$ from part (d) satisfies that difference equation by direct substitution. What should be and is the right-hand side?

3.78. Use the command dimpulse to plot $h(n) = h_1(n)$ obtained in part (d) of Problem 3.77: $H(z) = \{-5 + z^{-1}\}/\{1 - \frac{7}{12}z^{-1} + \frac{1}{12}z^{-2}\}$. For the numerator coefficients num use the format required by freqz, the frequency-response command for discrete time: num $= [-5\ 1]$. Check your result by directly plotting the sequence you obtained in Problem 3.77. If you find a discrepancy (e.g., a shift between these sequences), refer to the Control Systems Toolbox manual to determine the problem. Modify your numerator coefficients to obtain agreement, and comment on your findings. Summarize your findings with writing what the numerator and denominator arrays must be for freqz and for dimpulse, for a general transfer function.

3.79. (a) Without using the general result in (5.8), directly find the unilateral z-transform of $y(n-1)$ and of $y(n-2)$ in terms of initial conditions [e.g., $y(-1)$].

(b) Consider the difference equation $y(n) - \frac{3}{4}y(n-1) + \frac{1}{8}y(n-2) = x(n)$. Take the unilateral z-transform of this difference equation and express it in terms of the initial conditions used in part (a); let $x(n) = a^n u(n)$, $|a| < 1$, and $a \neq \frac{1}{2}$ or $\frac{1}{4}$. Solve for $y(n)$ in this difference equation by inverting the z-transform by partial fraction expansion. Determine the final value of $y(n)$ both from your expression and from the final value theorem. Do they agree?

3.80. In this problem, we wish to check the results of Problem 3.79 using Matlab. We might try the routine dlsim, except it does not allow for initial conditions when the system is in transfer function form. The only way we can integrate initial conditions is through the initial condition of the state vector, in state–space form. You will show in this problem how to use dlsim to solve any SISO difference equation, given initial values of the output.

(a) Represent the system in dual phase-variable (observable) state form, as given in (2.65) and (2.66).

(b) Convert the initial conditions on $y(n)$ to an initial condition on the state vector, using (2.73).

(c) Plot on the same axes both your result from Problem 3.79 and the result of dlsim for $n \in [0, 40]$. You should obtain perfect agreement. Choose whatever values you wish for a, $y(-1)$, and $y(-2)$; for example, $a = 0.7$, $y(-1) = 6$, and $y(-2) = -3$.

3.81. Numerically verify that the result of Example 3.18 (a) when evaluated at the times $n = 0$ and $n = -1$ gives, respectively, $y(0) = 0.317$ and $y(-1) = 0.126$ and (b) satisfies the original difference equation given in the problem statement.

3.82. The transfer function $H(z) = \sum_{n=0}^{N-1} a^n z^{-n}$, obviously FIR [$h(n)$ has length N], can also be written as $H(z) = [1 - (a/z)^N]/(1 - a/z)$. Find the corresponding difference equation. This latter expression for $H(z)$ appears to indicate an autoregressive (IIR) system, because the difference equation obtained has autoregressive terms. Is there a contradiction here? Explain. Hint: Consider long division.

3.83. In control theory, where frequently the characteristic polynomial has a variable coefficient,
we wish to determine the range of that coefficient for stability. Thus, we determine for what
range of the parameter K are all the poles within the unit circle. The computer cannot easily
provide such a range, but the Jury test can. Given $A(z) = \sum\limits_{\ell=0}^{N} a_\ell z^\ell$ (a_ℓ real and $a_N > 0$), form
the table:

a_0	a_1	a_2	$\cdots$	a_{N-1}	a_N
a_N	a_{N-1}	a_{N-2}	$\cdots$	a_1	a_0
b_0	b_1	b_2	$\cdots$	b_{N-1}	
b_{N-1}	b_{N-2}	b_{N-3}	$\cdots$	b_0	

.

.

.

p_0	p_1	p_2	p_3
p_3	p_2	p_1	p_0
q_0	q_1	q_2	$\longleftarrow$ row number $2N - 3$

where, for example,

$$b_k = \begin{vmatrix} a_0 & a_{N-k} \\ a_N & a_k \end{vmatrix}$$, and similar for other rows (beyond the second row), where the a changes to

whichever letter signifies the two upper-adjacent rows. Requirements for stability: $A(1) > 0$,
$\{A(-1) > 0, N \text{ even}; A(-1) < 0, N \text{ odd}\}$, $|a_0| < a_N$, $|b_0| > |b_{N-1}|, \dots, |q_0| > |q_2|$. Find the range
of K for stability if the characteristic equation is $0.3 - 0.2z + 0.5z^2 + (K - 2)z^3 + z^4 = 0$.
Verify using `roots` of Matlab on either side of the boundaries you determine.

Chapter 4

Why All Those Transforms?

4.1 INTRODUCTION

Many different "transforms" arise in the field of signal processing. Unless the reasons for using each of the transforms are known, one may be bewildered by the "forest of transforms." First of all, what is the purpose of a "transform"? We already looked at the z-transform and some of its uses in Chapter 3. In this chapter, the reasons for using the most basic six transforms of signal processing will be disclosed. For a given signal type and application, the best choice of transform will become clear.

In any signal-processing study, we begin with a signal to be processed. The signal may be subject to an existing signal-processing system (manufactured or natural), and the goal is to predict the form of the output for information extraction or further processing. Alternatively, a system is to be designed to implement a desired processing task. Or finally, the signal itself is the object of study. In all these cases, the ease of analysis and/or the desirability of information available in the transform (frequency) domain provides the motivation for "transforming" the signal.

One of the simplest transforms that was once useful to scientists for arithmetic was the logarithm. In the days before calculators, multiplication was a very tedious operation. However, because $\log_{10}(xy) = \log_{10}(x) + \log_{10}(y)$,[1] multiplication could be transformed into addition.

Specifically, the product xy was sought. First the logs of x and y were found using tables of the logarithm, which were then widely available. Now in the "log domain," multiplication becomes just addition, so one would add the two logs and obtain $\log(xy)$. Exponentiation of the result (inverse transformation) using the table in reverse would yield the desired product, xy.

The expansions discussed in this chapter are all over powers of a simple specified function. An operation extremely common in signal processing is convolution, discussed in Chapter 2. In the transform domains to be discussed, this operation is simplified, just as multiplication is simplified using the logarithm.

In fact, in all the transform domains we will discuss, convolution becomes multiplication. Moreover, the simplification is due to the same property of powers that makes the log transform work! That is, it is a consequence of the fact that for any complex number w, $w^{(x + y)} = w^x \cdot w^y$. This result, which transforms convolution into

---

[1] Recall that $10^u \cdot 10^v = 10^{(u + v)}$ with $u = \log_{10}(x)$, $v = \log_{10}(y)$, and $10^{\log_{10}(w)} = w$, where w is x and y so that $xy = 10^{\log_{10}(x) + \log_{10}(y)}$ and thus $\log_{10}(xy) = \log_{10}(x) + \log_{10}(y)$.

multiplication, is proved for the case of the Laplace transform in Section 2.3.3 and for the z-transform in Section 5.7; proofs for the other cases are just straightforward variations on the same method.

The application of integral transforms for the solution of boundary value problems was pioneered by a British electrical engineer, Oliver Heaviside (1850–1925). He found that these transforms could turn ordinary differential equations (ODEs) into algebraic equations. Also, partial differential equations could be turned into ODEs, which would then be solved either by standard ODE procedures or by further transformation into algebraic equations.

Thus, an understanding of "transform calculus" as discussed in this text should be of interest not only to the student of signal processing, but also to those working in mathematical physics, fluid mechanics, circuits, and so on. Naturally, the present focus will be on the use of transforms for the characterization, analysis, and design of signals and linear systems, the bearers and processors of useful information. For these tasks, expansions of general functions over ordered powers of simple functions come in handy.

As one example, consider Fourier analysis. Using Fourier spectral analysis, we can identify certain spectral bands as containing the signal of interest and others as containing primarily noise. Using the convolution theorem, one may design systems that annihilate the noise band while leaving the signal band intact. Fourier spectral analysis is also useful, among innumerable other applications, in speech synthesis/recognition and speech coding, sampling rate determination, traditional and modern radio communication techniques that focus on one frequency band per "channel," audio and video reproduction systems, vibration analysis, and a variety of applications that fill many books.

In this chapter, the various power expansions are defined,[2] related, and explored. Issues such as the meaning of frequency variables and the implications of periodicity and aperiodicity are subsequently discussed. A special section is given to help understand the discrete Fourier transform because of its preeminence in DSP. An extended pictorial "tour" of the concepts is then presented. The chapter concludes with an introduction to a general approach to "transform" representation of a time function.

A summary of many of the results of this study of transforms appears in tabular form in Appendix 4A. This appendix contains a rather complete set of relations between transforms both in the form of a diagram and the equations themselves.

4.2   TRANSFORM DEFINITIONS IN SIGNAL PROCESSING

In the following transform definitions, the label *synthesis* designates the composition of a signal from the weighted contributing basis functions chosen to represent the function. *Analysis* refers to the decomposition of the signal into its constituent components, specifying the weights of the basis functions in the expansion. For continuous-time signals and systems, the most general expansion function to be considered in this book is the *kernel* of the Laplace transform, e^{st}. The continuous-time function $x_c(t)$ is expanded as an integral over possibly complex-valued powers s of the function e^t, namely e^{st}. With $s = \sigma + j\Omega$, in practice e^{st} implies the basis functions $e^{\sigma t}\cos(\Omega t + \phi)$ (see Section 4.3 for details). The Laplace transform pair is

---

[2] On WP 5.1 and the problems, heuristic proofs of many of these transform definitions are given, with the aid of the orthogonality of the basis functions.

$$x_c(t) = \frac{1}{2\pi j} \int_C X_c(s) \cdot e^{st} \, ds \qquad \text{Laplace transform synthesis.} \qquad (4.1a)$$

where

$$X_c(s) = \int_{-\infty}^{\infty} x_c(t) \cdot e^{-st} \, dt, \qquad \text{Laplace transform analysis.} \qquad (4.1b)$$

where C is a contour in the complex s plane completely contained within the region of convergence of $X_c(s)$ and extending from $\sigma_C - j\infty$ to $\sigma_C + j\infty$, with σ_C typically fixed.[3] For any σ_C between σ_1 and σ_2, between which $X_c(s)$ is everywhere analytic, it does not matter what value σ_C takes, or even whether it is fixed over the entire contour.

The variety of C all yielding $x_c(t)$ in (4.1a) is due to a simple application of the Cauchy integral theorem (3.9), $\oint_C f(z) \, dz = 0$ if both on C and within C, f is analytic. The idea is illustrated in Figure 4.1, where C_1 and C_2 are two contours in the same region of convergence of $X_c(s)$, namely $\sigma_1 < \sigma < \sigma_2$. The left vertical side of the rectangular contour is one example of path C with $\sigma_C = \sigma_{C1}$, the right vertical side is another possible path with $\sigma_C = \sigma_{C2}$, and along the two horizontal connecting sides at infinity $X_c(s)e^{st}$ vanishes. Because the integral all around the closed contour is zero, the integral along the left contour C_1 is equal to minus that along C_2. Then we finally note that minus the integral along C_2 is equal to the inverse Laplace transform, with $\sigma = \sigma_{C2}$; the two minus signs cancel, and the inverse Laplace transform integrals for $\sigma_C = \sigma_{C1}$ and $\sigma_C = \sigma_{C2}$ are equal. Further discussion of allowable C in (4.1a) is pursued in the problems.

Finally, recall that the region of convergence for right- or left-sided $x_c(t)$ (e.g., causal or anticausal) is, respectively, to the right or left of the rightmost/leftmost pole of $X_c(s)$ (a fact that is easy to remember).

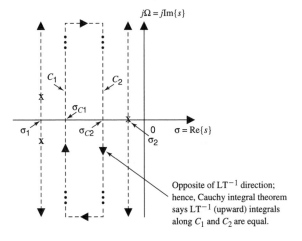

Opposite of LT^{-1} direction; hence, Cauchy integral theorem says LT^{-1} (upward) integrals along C_1 and C_2 are equal.   **Figure 4.1**

---

[3] Do not confuse the lowercase subscript c on signals denoting a continuous-time-related quantity with the uppercase subscript C on σ denoting contour.

EXAMPLE 4.1

For $x_c(t) = t \cdot w_{R,c}(t)$, where the rectangular window function $w_{R,c}(t) = 1$ for $t \in [0, T]$ and zero otherwise (see Figure 4.2a), find and plot $X_c(s)$.

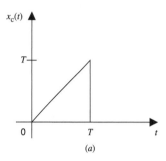

(a)

Figure 4.2a

Solution

Using integration by parts, we let $u = t$ so $du = dt$ and $dv = e^{-st} dt$, so $v = -e^{-st}/s$, giving

$$X_c(s) = \int_0^T t \cdot e^{-st} dt = -\frac{t}{s} \cdot e^{-st} \Big|_0^T + \frac{1}{s} \int_0^T e^{-st} dt$$

$$= -\frac{T}{s} e^{-sT} - \frac{1}{s^2} \{e^{-sT} - 1\} = \frac{1 - (1 + sT) \cdot e^{-sT}}{s^2},$$

which reduces to $T^2/2$ for $s = 0$ by L'Hospital's rule. For $T = 1$ s, $|X_c(s)|$ is as shown in Figure 4.2b and $\angle X_c(s)$ is in Figure 4.2c, for $|\sigma| \leq 10$ Np/s, $|\Omega| \leq 10$ rad/s.

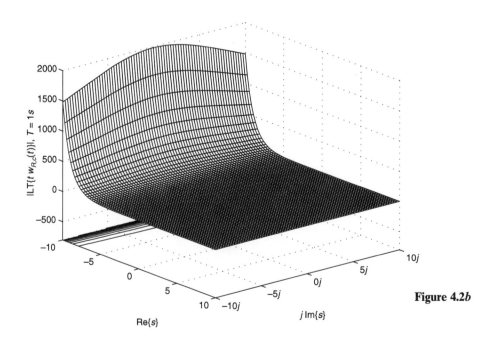

Figure 4.2b

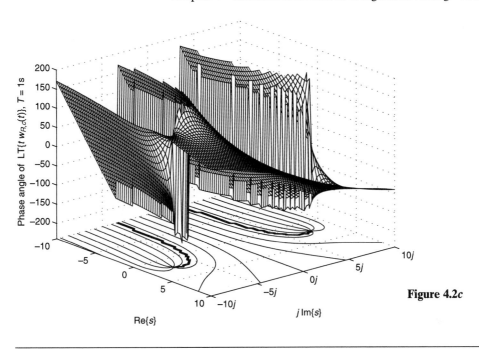

Figure 4.2c

For discrete-time signals and systems, the most general basis function for "expansion over powers" is z^n, the kernel of the z-transform (Section 3.4). In this case, the discrete-time function $x(n)$ is expanded as an integral over possibly complex-valued powers $\ln(z)$ of the function e^n [where $(e^n)^{\ln(z)} = (e^{\ln(z)})^n = z^n$]. With $z = |z| \angle \omega$, in practice z^n implies the basis functions $|z|^n \cos(\omega n + \phi)$ (again, see Section 4.3). The z-transform pair is [rewriting (3.34) and (3.35)]:

$$x(n) = \frac{1}{2\pi j} \oint_C X(z) \cdot z^n \left(\frac{dz}{z}\right) \qquad \text{z-transform synthesis.} \qquad (4.2a)$$

where

$$X(z) = \sum_{n=-\infty}^{\infty} x(n) \cdot z^{-n} \qquad \text{z-transform analysis.} \qquad (4.2b)$$

where C is a closed contour in the complex z-plane completely contained within the appropriate region of convergence of $X(z)$ and enclosing the origin $z = 0$ (usually a circle centered at the origin).

The Laplace and z expansions are guaranteed to exist if the time function being expanded is of exponential order;[4] that is, if the finite constant α exists such that

$$e^{-\alpha t} |x_c(t)| < \infty, \quad \text{all } t > T, \text{ where } T < \infty \text{ (continuous time)} \qquad (4.3a)$$

$$\alpha^{-n} |x(n)| < \infty, \quad \text{all } n > N, \text{ where } N < \infty \text{ (discrete time).} \qquad (4.3b)$$

Because the impulse responses of LSI systems are weighted sums of exponentials (possibly multiplied by t^ℓ or n^ℓ, but these are subordinate to the exponentials), they are all of exponential order and therefore have Laplace/z-transforms even when unstable.

---

[4] R. V. Churchill, *Operational Mathematics* (New York: McGraw-Hill, 1972), p. 6.

EXAMPLE 4.2

Let $x(n) = n$ for $n \in [0, N - 1]$ and zero otherwise (see Figure 4.3a), which we write as $x(n) = nw_R(n)$, where the rectangular window sequence $w_R(n) = 1$ for $n \in [0, N - 1]$ and zero otherwise. Plot $|X(z)|$ and $\angle X(z)$ for $N = 10$.

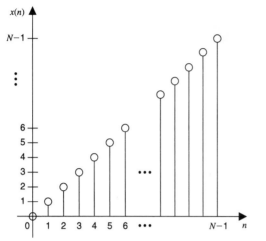

Figure 4.3a

Solution

In Example 3.10, part (h), we found that $X(z) = z \dfrac{1 + (N - 1)z^{-N} - Nz^{-(N-1)}}{z(z - 1)^2}$, for all $z \neq 1$ and $z \neq 0$, which at $z = 1$ reduces to $N(N - 1)/2$. For $N = 10$, $|X(z)|$ is as shown in Figure 4.3b and $\angle X(z)$ is in Figure 4.3c, for $|x| \leq 1.5$ and $|y| \leq 1.5$ where $z = x + jy$. The magnitude plot Figure 4.3b is cut off, however, at $|X(z)|_{\max} = 10$ to show more detail; the area shown as flat at $|X(z)| = 10$ in fact rises steeply toward infinity as $|z| \to 0$.

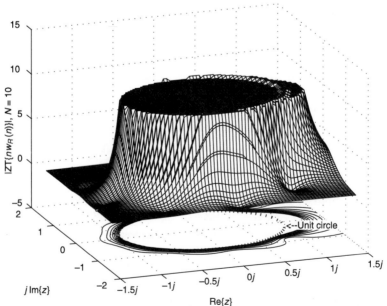

Figure 4.3b

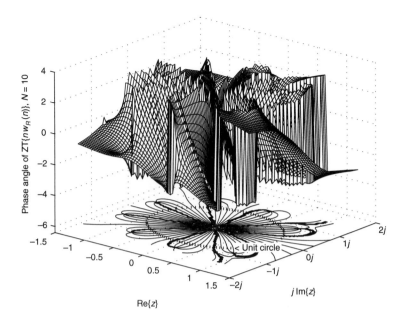

Phase angle of $ZT\{nw_R(n)\}$, $N = 10$

Re{z}   j Im{z}

< Unit circle

Figure 4.3c

It was noted above that the transform is really an expansion of the given time function over a set of *basis functions*. For example, the Laplace transform is the entire set of coefficients $X_c(s)$ (4.1b) that scale the basis functions e^{st} in the expansion of $x_c(t)$ in (4.1a). Using the computer and pictures, let us verify this claim for a particular Laplace transform pair.

EXAMPLE 4.3

Suppose $x_c(t) = e^{-2t}\cos(10t)u_c(t)$ (solid curve in Figure 4.4). Using the exponential form of cosine, it is a good exercise to prove that $X_c(s) = (s + 2)/[(s + 2)^2 + 10^2] = (s + 2)/(s^2 + 4s + 104)$. Demonstrate numerically that if we start with $X_c(s)$, we can approximately reproduce $x_c(t)$ by carrying out a numerical evaluation of (4.1a).

Solution

Matlab has a routine called `quad8` that is ideal for this purpose, but any robust numerical integration routine that can handle complex-valued functions should work. Recall from (3.4) $\left[\int_C f(z)\,dz = \int_{t_1}^{t_2} f[C(t)]C'(t)\,dt\right]$ that we can implement integration of a complex variable by evaluating an equivalent one-dimensional integral. To avoid confusion with time, rename the t variable in (3.4) v.

In agreement with (4.1a), let our contour be as shown in Figure 4.5, which also shows the poles and zero of $X_c(s)$. Obviously, the contour must be finite-length to be integrated over using a computer; we stop at $\Omega = \pm 50$ rad/s, where $|X_c(s)|$ is very small so that the contributions from $|\Omega| > 50$ will be small. The contour in Figure 4.5 can be described in terms of v as follows: $s = s_1 + j \cdot 2 \cdot 50 \cdot v$, where $s_1 = -1 - j50 = $ starting point of integral, where $0 \le v \le 1$. The integrand is $X_c(s)e^{st}/(j2\pi)$, and $ds = j \cdot 2 \cdot 50 \cdot dv$.

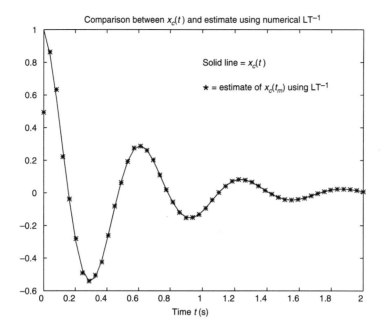

Figure 4.4

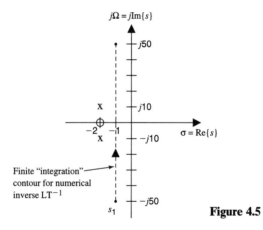

Figure 4.5

Substituting in s and defining $s_0 = -2 + j10$ as the upper-half-plane pole of $X_c(s)$, and noting that $ds/(j2\pi) = j100dv/(2\pi j) = (50/\pi)dv$, we have

$$x_c(t) \approx \frac{50}{\pi} \int_0^1 \frac{(s_1 + j100v + 2)e^{(s_1 + j100v)t}}{(s_1 + j100v - s_0)(s_1 + j100v - s_0^*)} dv.$$

Applying the routine quad8[5] to this integral for 50 equally spaced values of t from 0 to 2 s results in the asterisks ($*$) in Figure 4.4. We find that the integration was successful. The only exception is at the discontinuity around $t = 0$, where the finite approximation always produces the midpoint of the discontinuity, which in this case is $\frac{1}{2}$.

As an alternative to the numerical quadrature, we can obtain results that are nearly as accurate by just dividing up the integral (4.1a) into N_{term} terms evaluated at equal intervals along the same contour and using a rectangular rule. The advantage here is simplicity, and it allows us to look at the individual waveforms $X_c(s)e^{st}$ summing to $x_c(t)$ in (4.1a). We use the same contour as before and rewrite (4.1a) as:

---

[5]Quad8 is an adaptive Newton–Cotes quadrature rule.

$$x_c(t) \approx \frac{\Delta s}{2\pi j} \sum_{i=1}^{N_{\text{term}}} \frac{(s_1 + j100i/N_{\text{term}} + 2)e^{(s_1 + j100i/N_{\text{term}})t}}{(s_1 + j100i/N_{\text{term}} - s_0)(s_1 + j100i/N_{\text{term}} - s_0^*)},$$

where $\Delta s = j(2 \cdot 50)/N_{\text{term}}$. In Figure 4.6, we let $N_{\text{term}} = 75$ and evaluate the result at a very large number of points. That way, our estimate is drawn as a "continuous-time" function even though it is actually a very long, finely spaced sequence ranging from $t = 0$ to $t = 2$ s. As with the adaptive quadrature approach, the results are extremely good, and can be made arbitrarily accurate as N_{term} is increased and s_1 extends to nearer $-1 - j\infty$. Again notice that $\frac{1}{2}$ is produced at $t = 0$.

An array of individual terms plus (with the exception of $s = -1 + j0$) their complex conjugates in the preceding sum is presented in Figure 4.7. That is, each lower-half value of the sum index i is complemented by an upper-half value so that the sum of each pair is a real-valued time function equal to twice the real part of either of them, as must be true because $x_c(t)$ is real. For brevity, only 12 representative terms are shown out of the total number of distinct damped sinusoidal terms, $(N_{\text{term}} - 1)/2 + 1 = 38$. The vertical scales are the same for all plots.

The most interesting point to notice is that the amplitude of the terms is largest for $s = -1 + j10$ (Figure 4.7f), which is the term in the integral that is closest to the poles (see Figure 4.5). Specifically $x_c(t) = e^{-2t}\cos(10t)u_c(t)$ is a damped sinusoid of frequency $\Omega = 10$ rad/s and thus the poles of $X_c(s)$ have imaginary part 10 rad/s. Consequently, the energy of $x_c(t)$ is most concentrated—that is, the Laplace transform magnitude is largest—for the two $\sigma = 1$ Np/s complex damped sinusoidal components oscillating at ± 10 rad/s whose sum is shown in Figure 4.7f. We see this in Figure 4.8, which shows $|X_c(s)|$ for $|\sigma| < 2$ Np/s and $|\Omega| \leq 50$ rad/s. The two

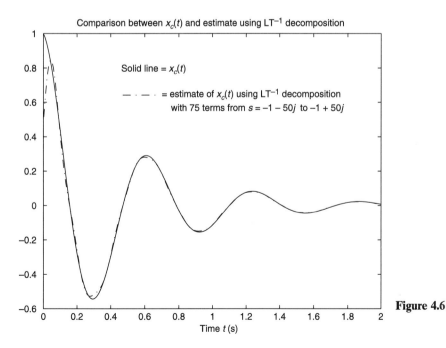

Comparison between $x_c(t)$ and estimate using LT^{-1} decomposition

Solid line = $x_c(t)$

— · — · — = estimate of $x_c(t)$ using LT^{-1} decomposition with 75 terms from $s = -1 - 50j$ to $-1 + 50j$

Time t (s)

Figure 4.6

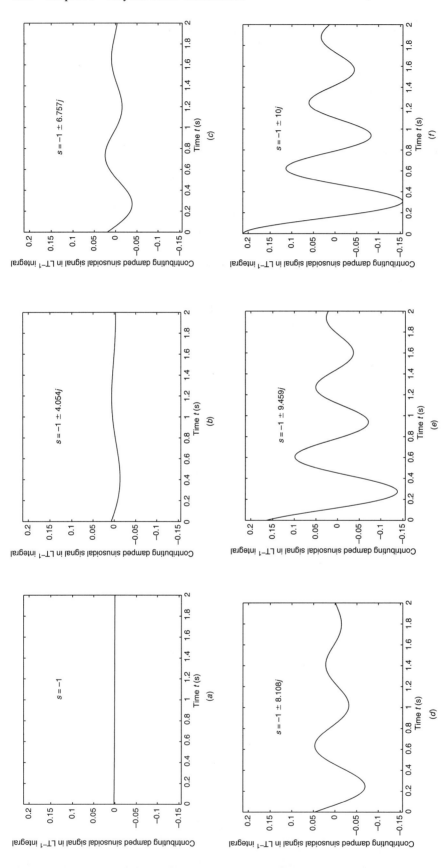

Figure 4.7a–f Some of the terms used* in the Laplace transform decomposition of $x_c(t) = e^{-2t}\cos(10t)u_c(t)$. Values of s chosen for these plots were selected for visual interest (they are nonuniformly spaced).
(a) $s = -1$, (b) $s = -1 \pm j4.054$, (c) $s = -1 \pm j6.757$, (d) $s = -1 \pm j8.108$, (e) $s = -1 \pm j9.459$, (f) $s = -1 \pm j10.00$,

*(f: $s = -1 + j10$) is plotted for illustration purposes only; $s = -1 + j10.00$ was not one of the values of s used in the sum.

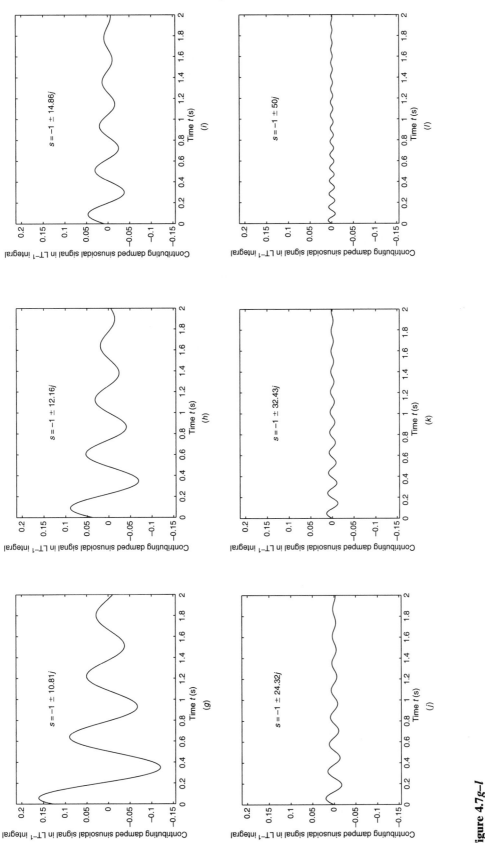

Figure 4.7g–l
(g) $s = -1 \pm j10.81$, (h) $s = -1 \pm j12.16$, (i) $s = -1 \pm j14.86$, (j) $s = -1 \pm j24.32$, (k) $s = -1 \pm j32.43$, (l) $s = -1 \pm j50.00$.

sloping ridges parallel to the real axis are $\Omega = \pm 10$ rad/s for the given range of σ depicted. The entire path of integration is shown raised in Figure 4.8 (parallel to the imaginary axis, at $\text{Re}\{s\} = -1$) and is indeed maximum for $|\Omega| = 10$ rad/s. Of course, $|X_c(s)|$ goes to ∞ at $s = s_0$ and $s = s_0^*$, but those points were not on the selected grid; thus the magnitude appears to be finite at $\{s_0, s_0^*\}$.[6]

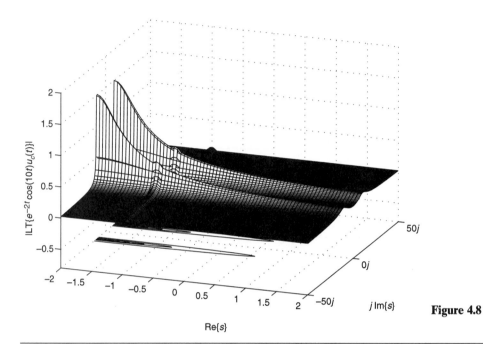

Figure 4.8

EXAMPLE 4.4

Repeat Example 4.3 for the z-transform, but this time approximate $x(n) = nw_R(n)$ (see Example 4.2) using only evenly spaced constituent components of $x(n)$ in (4.2a). Let $N = 10$, as in Example 4.2.

Solution

Let the radius of our contour in (4.2a) be $r_c = 0.9$; see Figure 4.9. The angles being uniformly spaced, we take $d\theta = 2\pi/N_{\text{term}}$, where N_{term} is the number of terms by which we approximate the integral in (4.2a). The ith value of z in the sum is $z_i = r_c \angle(i-1)2\pi/N_{\text{term}}$. Then $dz = z_{i+1} - z_i$. Using this procedure, we obtain the results shown in Figure 4.10. The original sequence $x(n)$ is shown with its values connected by lines for ease in comparing with the estimate of $x(n)$. The sequence with stems and circles is the numerical inverse z-transform of $X(z)$ calculated as just described. The results obtained using only 50 terms are essentially perfect [only a very small deviation from $x(n)$ can be seen for n near 60]. The error is substantial if N_{term} is reduced to 30 or below.

As in Example 4.3, an array of individual terms in the discretized contour plus (with the exception of $z = \pm 0.9$) their complex conjugates is presented in Figure

---

[6] Indeed, $s = s_0$ is not in the region of convergence of $X_c(s)$.

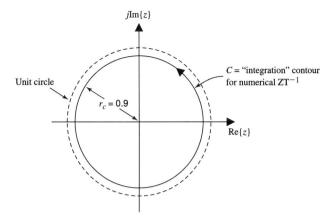

Figure 4.9

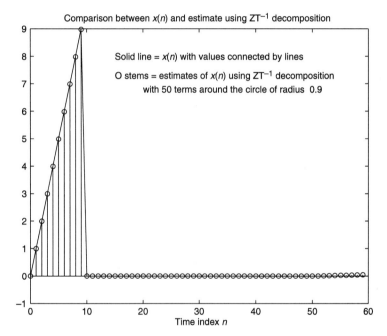

Figure 4.10

4.11. Again, $x(n)$ is real and we show twice the real part of the terms for complex values of z. For brevity, only 12 representative terms out of the total number of distinct real damped sinusoidal terms, $N_{term}/2 + 1 = 26$, are shown. From Figure 4.3b, we know that the magnitudes of individual terms will tend to be largest near the positive real axis due to the poles at $z = 1$, which is borne out in Figure 4.11. Remember that these "individual terms" are actual weighted constituent components of $x(n)$ in its expansion over the basis set z^n. Defining z_C to be a value of z of radius 0.9 on contour C, the terms shown are $2 \cdot \text{Re}\{X(z_C)z_C^n\}$, which are damped sinusoidal sequences.

An interesting feature in this case is that all the contributing terms shown in Figure 4.11 extend well beyond $n = 9$, the last nonzero value of $x(n)$; they all "miraculously" combine to zero when added up. In fact, this is also a characteristic of Fourier transforms, but there it is even more dramatic: The individual sinusoids that make up a possibly finite-length time function are themselves *infinite*-length with never-attenuating amplitudes.

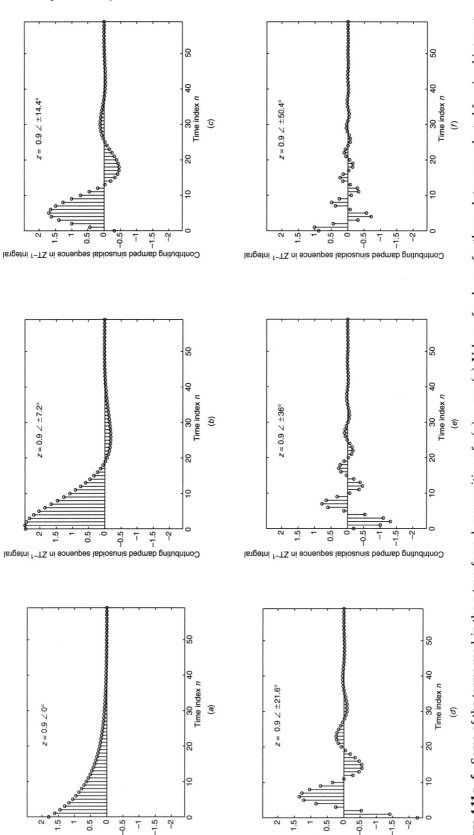

Figure 4.11a–f   Some of the terms used in the z-transform decomposition of $x(n) = nw_R(n)$. Values of z chosen for these plots were selected for visual interest (they are nonuniformly spaced).
(*a*) $z = 0.9\angle 0°$,   (*b*) $z = 0.9\angle\pm7.2°$,   (*c*) $z = 0.9\angle\pm14.4°$,   (*d*) $z = 0.9\angle\pm21.6°$,   (*e*) $z = 0.9\angle\pm36°$,   (*f*) $z = 0.9\angle\pm50.4°$,

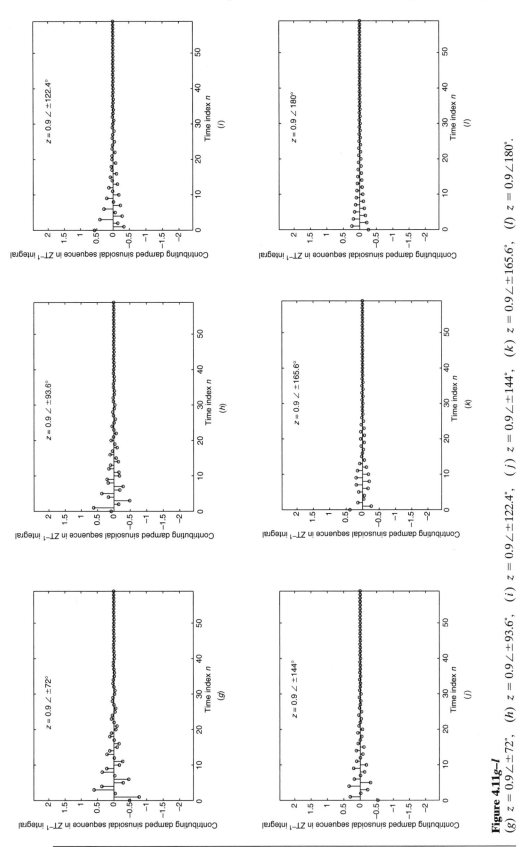

Figure 4.11g–l
(g) $z = 0.9 \angle \pm 72°$,   (h) $z = 0.9 \angle \pm 93.6°$,   (i) $z = 0.9 \angle \pm 122.4°$,   (j) $z = 0.9 \angle \pm 144°$,   (k) $z = 0.9 \angle \pm 165.6°$,   (l) $z = 0.9 \angle 180°$.

Every one of the six transforms considered in this chapter exhibits the property of a weighted sum or integral of a fundamental basis function raised to (possibly complex) powers. The Fourier expansions are particular cases of the more general Laplace and z-transforms, which are functions of a complex variable. Thus for convenience, we will call the expansions in (4.1a) and (4.2a) the general power expansions (GPEs) of the given signal or system impulse response.

Consider signals that are absolutely integrable (continuous time)

$$\int_{-\infty}^{\infty} |x_c(t)|\, dt < \infty \tag{4.4}$$

or absolutely summable (discrete time)

$$\sum_{n=-\infty}^{\infty} |x(n)| < \infty. \tag{4.5}$$

These signals are a subclass of the signals of exponential order (from Chapter 2, this means they are stable sequences; i.e., they could be impulse responses of BIBO stable systems).

Fourier's theorem states that absolutely integrable (or summable) signals may be decomposed into a superposition integral of infinite-duration sinusoids

$$x_c(t) = \frac{1}{2\pi} \int_{-\infty}^{\infty} X_c(j\Omega) \cdot e^{j\Omega t}\, d\Omega \qquad \text{(continuous time)}$$
Fourier transform synthesis. $\qquad$ (4.6a)

where

$$X_c(j\Omega) = \int_{-\infty}^{\infty} x_c(t) \cdot e^{-j\Omega t}\, dt, \qquad \text{Fourier transform analysis.} \qquad (4.6b)$$

or

$$x(n) = \frac{1}{2\pi} \int_{-\pi}^{\pi} X(e^{j\omega}) \cdot e^{j\omega n}\, d\omega \qquad \text{(discrete time)}$$
Discrete−time Fourier transform
(DTFT) synthesis. $\qquad$ (4.6c)

where

$$X(e^{j\omega}) = \sum_{n=-\infty}^{\infty} x(n) \cdot e^{-j\omega n}. \qquad \text{Discrete−time Fourier transform}$$
(DTFT) analysis. $\qquad$ (4.6d)

Fourier's theorem (4.6) holds for the vast majority of deterministic signals and systems encountered in practice. In fact, recall from Section 3.4.1 that if the Fourier transform of a unit sample response does not exist, the system is unstable.

The weighting coefficients $X_c(j\Omega)$ and $X(e^{j\omega})$ in (4.6) are the complex amplitudes of the respective expansion functions $e^{j\Omega t}$ [see (4.6a)] and $e^{j\omega n}$ [see (4.6c)]. The squared magnitudes of these coefficients represent energy densities of the respective time functions at the particular frequency (Ω or ω).

Evidently, the expansion functions of Fourier analysis are special cases of those for GPEs. As noted, for continuous-time signals they are $e^{j\Omega t}$. This Fourier transform kernel is just that of the Laplace transform, e^{st}, with $s = j\Omega$, on the imaginary axis of the s-plane. The expansion function core e^t is simply raised to imaginary powers $s = j\Omega$ to produce $e^{j\Omega t}$. Comparing (4.6b) with (4.1b), we find that if we know the Laplace transform, the Fourier transform can be found by just evaluating the Laplace transform on the imaginary axis, $s = j\Omega$:

$$X_c(j\Omega) = X_c(s)\big|_{s = j\Omega}. \tag{4.7}$$

For discrete-time signals, the expansion functions are $e^{j\omega n}$. The discrete-time Fourier transform kernel, $e^{j\omega n}$, is equal to the z-transform kernel z^n for z set equal to $e^{j\omega}$, which is located on the unit circle of the z plane. The expansion function core e^n is simply raised to imaginary powers $\ln(z) = j\omega$ to produce $e^{j\omega n}$. Comparing (4.6d) with (4.2b), we see that the discrete-time Fourier transform is just the z-transform evaluated on the unit circle, $z = e^{j\omega}$:

$$X(e^{j\omega}) = X(z)\big|_{z = e^{j\omega}}. \tag{4.8}$$

In fact, if we set $z = e^{j\omega}$ in the inverse z-transform (4.2a) and note that (a) $dz = je^{j\omega}\,d\omega$ or $dz/z = jd\omega$ and (b) as z goes around the unit circle C, ω ranges from $-\pi$ to π, we see that (4.2a) becomes the inverse discrete-time Fourier transform (4.6c).

From (4.6) through (4.8) we see the reason for the choice of notation for the Fourier transforms: The functional symbol for the relevant Fourier transform is that of the GPE (Laplace or z-transform), whereas the argument is the particular location of evaluation on the complex plane (imaginary axis: $s = j\Omega$ or unit circle: $z = e^{j\omega}$). That is, because of (4.7) and (4.8), there is no need to define distinct symbols for the Fourier transforms versus their corresponding GPEs.

EXAMPLE 4.5

Find the Fourier transform of $x_c(t) = t \cdot w_{R,c}(t)$ and the discrete-time Fourier transform (DTFT) of $x(n) = n \cdot w_R(n)$.

Solution

These transforms are now easy to find using the GPEs found in the previous examples. From Example 4.1, and noting in the denominator that $j^2 = -1$,

$$X_c(j\Omega) = X_c(s)\big|_{s = j\Omega} = \frac{(1 + j\Omega T) \cdot e^{-j\Omega T} - 1}{\Omega^2}, \Omega \neq 0 \text{ and } X_c(j0) = T^2/2,$$

which for $T = 1$ is shown in Figure 4.12a for roughly $\Omega \in [-120, 120]$ rad/s. From Example 4.2,

$$X(e^{j\omega}) = X(z)\big|_{z = e^{j\omega}} = \frac{e^{-j\omega} + (N - 1)e^{-j\omega(N+1)} - Ne^{-j\omega N}}{(e^{j\omega} - 1)^2}, \omega \neq 1 \text{ and}$$

$$X(1) = \frac{1}{2}N(N - 1),$$

which for $N = 10$ is shown in Figure 4.12b (shown for $|\omega| \leq \pi$). Both $x_c(t)$ and $x(n)$ are always nonnegative and are reasonably smooth; hence both Fourier transform magnitudes are large at and near zero frequency.

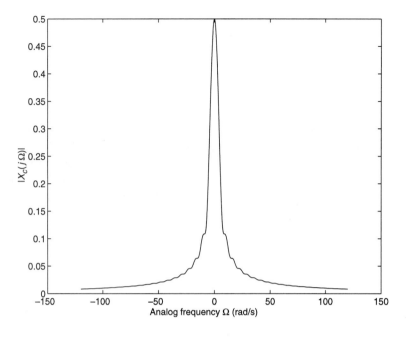

Figure 4.12*a*

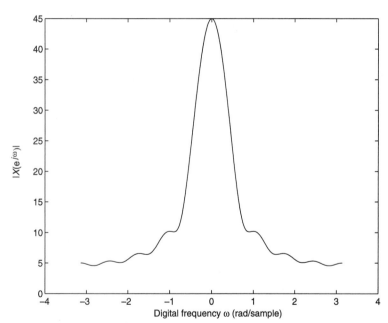

Figure 4.12*b*

If we compute the output of an LSI system when its input is one of the Fourier expansion functions [$e^{j\Omega t}$ (continuous time) and $e^{j\omega n}$ (discrete time)], we find that it is a scaled version of that expansion function.[7] That is, the expansion functions are eigenfunctions of LSI systems. The set of sinusoidal amplitudes of the general signal [$X_c(j\Omega)$ or $X(e^{j\omega})$], one for every frequency, constitutes the signal spectrum, and the system scalers [$H_c(j\Omega)$ or $H(e^{j\omega})$] are collectively called the system frequency response.

---

[7] This fact is proved on WP 5.1.

Suppose $x(n)$ has no component at $\omega = \omega_1$ until $n = n_0$, at which time a damped oscillation of frequency ω_1 is introduced into $x(n)$. It might be asked, How can the frequency component $X(e^{j\omega_1})e^{j\omega_1 n}$, defined for all n, be a component of $x(n)$ at $\omega = \omega_1$ even before $n = n_0$—a time period when we know $x(n)$ has no such component? The answer is that $X(e^{j\omega_1})e^{j\omega_1 n}$ *does* contribute with constant amplitude for *all* n including $n < n_0$, but do not forget that $x(n)$ also has an infinite number of frequency components at all other frequencies during that time. It is amazing that all of the frequency components for $n < n_0$ add up to a function that locally (i.e., for $n < n_0$) has no instantaneous content at $\omega = \omega_1$! Remember that $X(e^{j\omega_1})e^{j\omega_1 n}$ is the overall frequency component for $\omega = \omega_1$ based on the *entire* sequence $x(n)$. If we want the "instantaneous component" of $x(n)$ at $\omega = \omega_1$ for n near n_0, we can compute a *short-time Fourier transform* for that purpose: Take the DTFT of a small window of $x(n)$ centered on $n = n_0$.

Thus for discrete-time signals, the discrete-time Fourier transform $X(e^{j\omega})$ is the complex amplitude of the constituent expansion function $e^{j\omega n}$ that when summed together for all $|\omega| \leq \pi$ (actually, integrated) make up $x(n)$. For real $x(n)$, it will be shown in (5.52) that $X(e^{-j\omega}) = X^*(e^{j\omega})$; that is, $X(e^{j\omega})$ is conjugate symmetric, and similarly if x and X are replaced by, respectively, h and H. Therefore, in the inverse DTFT integral (4.6c) for real $x(n)$, the contribution to $x(n)$ due to a particular frequency $\omega = \omega_1$ is

$$X(e^{j\omega_1})e^{j\omega_1 n} + X^*(e^{j\omega_1})e^{-j\omega_1 n} = 2 \cdot \text{Re}\{X(e^{j\omega_1})e^{j\omega_1 n}\} \qquad (4.9)$$

$$= 2 \cdot |X(e^{j\omega_1})| \cdot \cos\{\omega_1 n + \angle X(e^{j\omega_1})\}.$$

Moreover, because as noted in Section 2.4.3 and proved in Section 5.7, if $x(n)$ is filtered by the digital filter $h(n)$ and the result is called $y(n)$, then $Y(e^{j\omega}) = X(e^{j\omega})H(e^{j\omega})$, which holds in particular for $\omega = \omega_1$, so that the contribution to $y(n)$ due to the frequency $\omega = \omega_1$ is, after cancellation of common factors of 2 in input and output,

$$\boxed{\begin{aligned} |Y(e^{j\omega_1})| \cdot \cos\{\omega_1 n + \angle Y(e^{j\omega_1})\} = \\ |X(e^{j\omega_1})||H(e^{j\omega_1})| \cdot \cos\{\omega_1 n + \angle X(e^{j\omega_1}) + \angle H(e^{j\omega_1})\}. \quad (4.10) \end{aligned}}$$

As $\cos\{\phi\} = \sin\{\phi + \pi/2\}$—that is, just $\sin\{\cdot\}$ with a different angle—the results in (4.9) and (4.10) can be expressed in terms of the sine function by simply adding $\pi/2$ to both phases; thus we still have the result that, e.g., $\angle H(e^{j\omega_1})$ is added to the phase of the input sine component.

Similarly, for continuous time, the input sinusoidal component $2|X_c(j\Omega_1)|\cos\{\Omega_1 t + \angle X_c(j\Omega_1)\}$ of real $x_c(t)$ results in the output sinusoidal component $2|H_c(j\Omega_1)X_c(j\Omega_1)|\cos\{\Omega_1 t + \angle X_c(j\Omega_1) + \angle H_c(j\Omega_1)\}$.

We therefore have the following result, valid for signals passing through an LSI system. The magnitude of the frequency response $|H_c(j\Omega_1)|$ or $|H(e^{j\omega_1})|$ scales (increases or decreases) the magnitude of the corresponding input sinusoidal component at frequency $\Omega = \Omega_1$ or $\omega = \omega_1$, and the phase of the frequency response $\angle H_c(j\Omega_1)$ or $\angle H(e^{j\omega_1})$ additively alters (increases or decreases) the phase of that input sinusoidal component. Because each complex sinusoidal component of the Fourier decomposition of the input is just scaled in the output, the decomposition coefficients of the output signal are the products of those of the input and the frequency response evaluated at Ω_1 or ω_1. One product applies for each component frequency Ω_1 or ω_1. Actually, this statement puts in words the convolution theorem, one of the most useful results in Fourier analysis. Analogous convolution theorems also exist for GPEs, for their expansion functions are also eigenfunctions of LSI systems.

Usually, the magnitude of the frequency response of a physical dynamic system falls to zero for infinite frequency. Why? Suppose someone tells me to "run forward" and I obediently do so. Then they tell me to "run backward" and again I do so. If, however, they tell me "run forward . . . run backward . . . run forward" too quickly, then despite my willing obedience I simply cannot keep up with the commands; *I remain motionless*, because before I get going forward, I am already told to go backward. This phenomenon is zero frequency response for high frequencies. Similarly, a motor shaft will not rotate at all if the forward and reverse commands (i.e., oscillations in the input) alternate too rapidly for the moment of inertia of the shaft to follow. As one more example, a capacitor has no time to start charging through a resistor R toward a positive voltage before the input has it start discharging toward a negative voltage, resulting in zero capacitor voltage. It is the delays in physical systems that cause the frequency response to fall off with increasing frequency [and thus $N > M$ for a continuous-time transfer function $H_c(s)$].

Another point of possible confusion should be addressed here. Note that even for a passive filter, the frequency response magnitude in a resonant circuit can be greater than 1, even though actual average power gain in a passive circuit can never exceed 1. The apparent contradiction is resolved by recalling that the "powers" in the squared frequency response ratio are *defined* to be the powers V^2/R that *would* be dissipated in a standardized resistance R by an ideal voltage source with voltage V equal to the amplitude of each signal voltage; in the power ratio, R cancels. No such R actually exists in the passive circuit, and if it were inserted, the behavior would change. The extra "oomph" to give larger unloaded output signal amplitudes than input amplitudes originates from energy stored/cyclically exchanging in the L or C and continually replenished by the source as it is dissipated in the circuit resistance. The output oscillations would continue until dissipated, even if the source were turned off!

EXAMPLE 4.6

A system has unit sample response $h(n) = 5(\frac{1}{3})^n u(n)$. Find $y(n) = x(n) * h(n)$ if $x(n) = 3\cos(\pi n/10) + 7\sin(\pi n/6 + 20°)$.

Solution

First, we write $H(e^{j\omega}) = 5\sum_{n=0}^{\infty} [e^{-j\omega}/3]^n = 5/[1 - \frac{1}{3}e^{-j\omega}]$; note that $h(n)$ and $x(n)$ are both real. Thus, (4.9) and (4.10) apply directly. Thus

$$y(n) = 3 \cdot |H(e^{j\pi/10})| \cos\left\{\frac{\pi n}{10} + 0 + \angle H(e^{j\pi/10})\right\} +$$

$$7 \cdot |H(e^{j\pi/6})| \sin\left\{\frac{\pi n}{6} + 20° + \angle H(e^{j\pi/6})\right\}$$

$$= 21.7169 \cos\left\{\frac{\pi n}{10} - 8.5766°\right\} + 47.9065 \sin\left\{\frac{\pi n}{6} + 6.8132°\right\}.$$

This approach was much more straightforward than starting off by attempting the convolution of $5\left(\frac{1}{3}\right)^n u(n)$ with $3\cos(\pi n/10) + 7\sin\{\pi n/6 + 20°\}$. Notice that there was no need to compute $X(e^{j\omega})$, which is impulsive. We see how simple Fourier decomposition makes system analysis in such simple yet common examples. In the problems, we re-solve Example 2.8 using continuous-time frequency response and observe how much easier the task is by using frequency response than by using convolution.

EXAMPLE 4.7

To visualize (4.9) and (4.10) as well as gain practice in taking inverse discrete-time Fourier transforms, we consider the following situation. Suppose that we are told that $X(e^{j\omega}) = (1 - |\omega|/\omega_c)u_c(1 - |\omega|/\omega_c)\angle -\alpha\omega$ and $H(e^{j\omega}) = (|\omega|/\omega_c)u_c(1 - |\omega|/\omega_c)\angle -\beta\omega$, where ω_c is the cutoff frequency of the triangularly shaped DTFTs and α, β are real constants. For $y(n) = x(n) * h(n)$, find $y(n)$ and graphically display the results, including showing (4.9) and (4.10) for a particular value of ω, ω_1.

Solution

The solution amounts to multiplying the two DTFTs and taking the inverse transform. The details are tedious and mainly involve integration by parts twice. Letting $\gamma = \alpha + \beta$, the result is

$$
y(n) = \frac{1}{\pi\omega_c(n - \gamma)}\left\{ \frac{-1 + \cos\{\omega_c(n - \gamma)\}}{n - \gamma} + \omega_c\sin\{\omega_c(n - \gamma)\} \right.
$$

Complete derivation details on WP 4.1.

$$
\left. -\left\{\left[\omega_c - \frac{2}{\omega_c(n - \gamma)^2}\right]\sin\{\omega_c(n - \gamma)\} + \frac{2}{n - \gamma}\cos\{\omega_c(n - \gamma)\}\right\}\right\}
$$

$$
= \frac{1}{\pi\omega_c(n - \alpha - \beta)^2}\left\{\frac{2\sin\{\omega_c(n - \alpha - \beta)\}}{\omega_c(n - \alpha - \beta)} - 1 - \cos\{\omega_c(n - \alpha - \beta)\}\right\}. \quad (4.11)
$$

If $\alpha + \beta$ is an integer, $y(\alpha + \beta) = \omega_c/6\pi$. Although tedious, it is far easier to invert the product $X(e^{j\omega})H(e^{j\omega})$ than to find $x(n)$ and $h(n)$ individually and attempt to convolve them; hence the power of the convolution theorem.

The operations we have performed are shown graphically in Figure 4.13 for $\alpha = 2$, $\beta = 3$, and $\omega_c = 2\pi \cdot f_c$ where $f_c = 0.4$ cycle/sample. On the first row of plots, we have $x(n), h(n)$, and $y(n) = x(n) * h(n)$. The n axis is restricted to $|n| \leq 15$, the range for which they are significantly nonzero. On the second row of plots are shown $|X(e^{j\omega})|$, $|H(e^{j\omega})|$, and $|Y(e^{j\omega})| = |X(e^{j\omega})| \cdot |H(e^{j\omega})|$, and the third row shows $\angle X(e^{j\omega})$, $\angle H(e^{j\omega})$, and $\angle Y(e^{j\omega}) = \angle X(e^{j\omega}) + \angle H(e^{j\omega})$.

Most interesting perhaps is the last row of plots. Here we show one component sinusoid (of an infinite number) making up $x(n), h(n)$, and $y(n)$, including the correct magnitude and phase shift. Ignore for the moment the dotted sinusoidal curves; focus on the sinusoid with stems whose envelope is shown with a dashed curve. In these plots we demonstrate (4.9) and (4.10) for a particular value of ω, $\omega = \omega_1 = 2\pi \cdot f_1$ where $f_1 = 0.15$, so $\omega_1 = 0.94$ rad/sample.

For example, consider the plot of $2|X(e^{j\omega_1})|\cos\{\omega_1 n + \angle X(e^{j\omega_1})\}$ in the lower left corner of Figure 4.13. If we go up two plots to $|X(e^{j\omega})|$ and read the magnitude at $\omega = \omega_1$, we find that that value, 0.63, is half the magnitude of the sinusoidal component of frequency ω_1 contributing to $x(n)$ in the inverse DTFT (4.6c) when we include the conjugate sinusoid as in (4.9). Quantitatively, $|X(e^{j\omega_1})| = 1 - \omega_1/\omega_c = 1 - 0.15/0.4 \approx 0.63$.

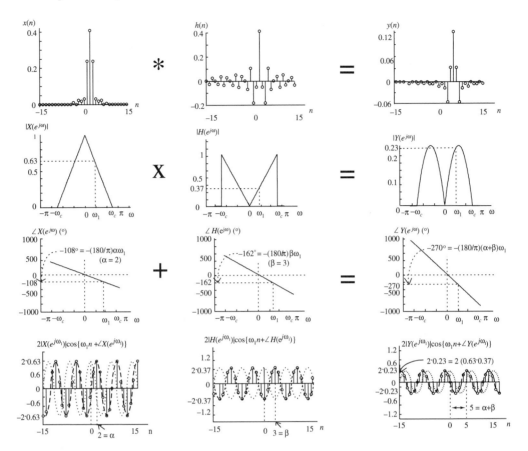

Figure 4.13

Similarly, the sinusoidal component of the system unit sample response $h(n)$ at $\omega = \omega_1$ in the bottom plot of the second column of plots is found by doubling the value $|H(e^{j\omega_1})| = 0.37$ found two plots above at ω_1 [$|H(e^{j\omega_1})| = \omega_1/\omega_c = 0.15/0.4 \approx 0.37$]. Finally, the amplitude of the component of $y(n)$ at $\omega = \omega_1$, depicted in the bottom plot of the third column is $2 \cdot 0.23$, where 0.23 is found two plots above as $|Y(e^{j\omega_1})|$. We know from (4.10), however, that we can also find it by multiplying the (half-) magnitudes of the components of $x(n)$ and $h(n)$ at ω_1: $0.23 \approx 0.63 \cdot 0.37$.

In a similar vein, we can observe the effect of the phase. It is formally shown in Section 5.2 that DTFT$\{x(n - n_0)\} = e^{-j\omega n_0} X(e^{j\omega})$. In this example, all the phases are linear. For example, the only difference between $x_0(n)$ defined as $x(n)$ for $\alpha = 0$ and $x(n)$ as depicted for $\alpha = 2$ is $x(n) = x_0(n - \alpha) = x_0(n - 2)$. For comparison, $x_0(n)$ is shown in Figure 4.14; $x(n)$ in Figure 4.13 is seen to be $x_0(n)$ shifted to the right by $\alpha = 2$. Similarly, $h(n) = h_0(n - \beta)$ and $y(n) = y_0(n - \alpha - \beta)$.

This shifting property is true not only for $x(n), h(n)$, and $y(n)$, but also for every one of their constituent sinusoidal components. This is seen in the bottom row of plots. The dotted-curve sinusoid is a zero-phase cosine, to be compared against the sinusoidal signal component given in (4.9). We indeed find for $x(n)$ that the $\omega = \omega_1$ cosinusoidal component is delayed by $\alpha = 2$ compared with the zero-phase sinusoid $2 \cdot 0.63 \cdot \cos(\omega_1 n)$. Similarly, the ω_1 component of $h(n)$ is seen to be delayed by $\beta = 3$ compared with a zero-phase sinusoid, and that of $y(n)$ is delayed by $\alpha + \beta = 5$.

It is also possible to obtain these delays from the phase plot directly above. Consider first the component of $x(n)$ at $\omega_1 = 0.94$ rad/sample. We find from the plot

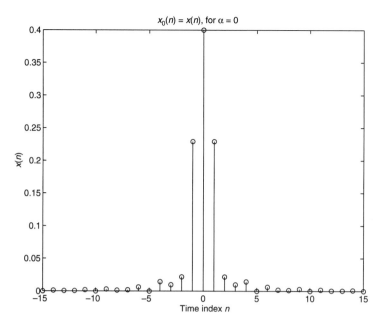

Figure 4.14

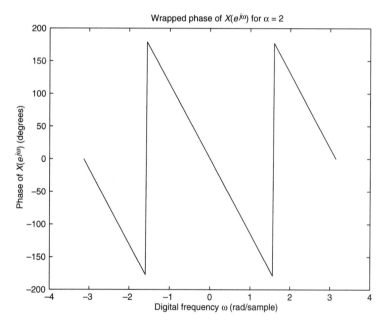

Figure 4.15

of $\angle X(e^{j\omega})$ that at $\omega = \omega_1$, the phase is $-108°$, or $-108 \cdot \pi/180 = -1.89$ rad. The slope of the linear phase curve is $-1.89/\omega_1 = -1.89/0.94 = -2 = -\alpha$. Similarly, $\angle H(e^{j\omega_1})$ $= -162°$ so that the slope of that phase curve is $-162 \cdot \pi/180/0.94 = -3 = -\beta$ and $\angle Y(e^{j\omega_1}) = -270°$, giving a slope of $-270 \cdot \pi/180/0.94 = -5 = -(\alpha + \beta)$. Notice how the slopes and delays add, as verified by (4.10) and by looking at the plots themselves.[8]

One final point. The phase curves in Figure 4.13 were *unwrapped* after being calculated as $e^{-j\omega\gamma}$. Recall that $e^{j(\theta + 2\pi k)} = e^{j\theta}$ for all integers k. Consequently, for $|\omega| > \pi/\gamma$, when the phase angle leaves the $(-\pi, \pi)$ range, evaluations of $e^{-j\omega\gamma}$ will result in $\text{Mod}(-\omega\gamma, 2\pi)$. Without "unwrapping" the phase, it appears as shown in

---

[8] The vertical axes of the phase plots were all scaled the same.

Figure 4.15. The usual graphical depiction of phase angle for digital filters shows the wrapped phase as in Figure 4.15. The phase was unwrapped in this example for clarity when discussing "linear" phase and the phase addition property. The phase in Figure 4.15 is still linear, but it has discontinuities of 2π whenever the phase hits another odd multiple of π. There can also be jumps in π where a real-valued function changes polarity; these are not due to the modulo.

The Fourier transform specifies the GPE over only an infinitesimally small region of the complex plane. One might then conclude that there is an infinite number of time functions having a given Fourier transform. That is, the Laplace (or z-) transforms for such time functions might differ anywhere in the s- (or z-) plane except along the contour $s = j\Omega$ (or along $z = e^{j\omega}$). Such is not the case, however. Throughout the entire region of analyticity of a function of a complex variable, the values of that function can be uniquely determined by its values in a neighborhood or contour within or enclosing that region.

As one example of this fact, we refer to the Cauchy integral formula (3.13) $\left[f(z) = \left(\dfrac{1}{2\pi j} \right) \oint_C \left[\dfrac{f(z')}{z' - z} \right] dz' \right]$. In that case, we showed that $f(z)$ *anywhere* within a contour C can be written purely in terms of the values of $f(z)$ on the contour, provided $f(z)$ is analytic within the contour. For example, C could be the unit circle in the z-plane so that $f(z)$ anywhere within the unit circle is expressed purely in terms of its values on the unit circle, namely, the discrete-time Fourier transform. This comment also applies for an anticausal stable sequence, whose z-transform converges within the unit circle. The idea can also be applied for the (usual) case in which $f(z)$ has all its poles within the unit circle (causal stable sequence) by writing the Cauchy integral formula for $X(1/z)$ with $|z| < 1$ because $X(z)$ is analytic for $|z| > 1$. See Appendix 4A, Formula (4) for the formula for obtaining $X(z)$ for any z from $X(e^{j\omega})$, its values on the unit circle.

In the case of continuous time, the Fourier transform specified continuously as a function of Ω fixes the Laplace transform everywhere it exists.[9] Because of this fact, it is often convenient to deal strictly with the Fourier expansion; for the specific formula to obtain $X_c(s)$ from $X_c(j\Omega)$, see Appendix 4A, Formula (16).

As we have seen in our examples, the Fourier expansion is often convenient when analyzing signals, for which purely sinusoidal, steady-state content is often of the greatest significance. It is also convenient, however, for systems analysis, where, for example, analog and digital filter design specifications are usually given on the Fourier contour (see Chapters 8 and 9). The usefulness of phasors, which are based on the Fourier transform, for sinusoidal steady-state and frequency-response analysis is universally recognized. Also, the inverse ("synthesis") transform relations are simpler for Fourier transforms, and because the transform variable is real, computations are often more convenient to perform than they are for the GPEs. Specification of a complex-valued function at every point on a complex plane (s or z) is reduced to specification of a complex-valued function at every point on an axis of a real variable (for Ω) or a subset of one: $(-\pi, \pi]$ (for ω).

The GPEs (Laplace and z-transforms) are very useful in system characterization and analysis. They are well suited to solving initial-value/transient differential equations problems defined on intervals such as $t \in [0, \infty)$ or other semibounded intervals, especially in cases where the Fourier transform integral may not converge

---

[9] This statement assumes that the Laplace transform is nonanalytic at only isolated points (poles), as is generally the case in practice.

at the upper, infinite boundary. As we saw in Section 3.4 for discrete time, they reveal how close to unstable a system may be, and they show the complete underlying (pole–zero) structure of the system. The poles are particularly illuminating, for they are the modes of the system; they explicitly dictate the transient time-domain behavior. In the frequency domain, factorization of the system function into pole and zero factors facilitates understanding the shape and bandwidth of the frequency response via Bode plot analysis. Note that the whole concept of poles and zeros has meaning and direct applicability *only* for Laplace or z-transforms. The Fourier transforms are nowhere infinite or zero as a result of poles or zeros unless the latter are located on the $j\Omega$ axis (or unit circle). Additional information involved in the issues of causality, invertibility, and so forth is also available from the GPE.

In addition, often the GPEs provide methods of implementation of a desired system behavior and filter design. In fact, some mappings used in the design of digital filters require the use of both the s- and z-planes. Finally, use of the GPEs allows the application of many results from the theory of complex variables in analysis and design of signal-processing systems. Just a few examples are results concerning the phase alterations imposed by a system on an input signal, issues in adaptive signal-processing and spectral estimation theory, and the Nyquist stability criterion.

We have just reviewed the advantages of system-function and frequency-response representations of systems. To round out this discussion, recall that in the time domain there are three more useful system representations: (a) differential/difference equations, which are convenient for modeling systems from physical laws and may be solved exactly for certain special inputs; (b) impulse/unit sample response form, which is helpful in analyzing temporal responses of systems and are the heart of convolution input–output relations, and (c) state–space form, which is convenient for computer implementation, filter structures, multi-input/multi-output systems, and so on. With experience, the most advantageous representation for the given application will usually be clear. A summary of which domain and transform may be most helpful for a variety of signal types and applications is given in Table 4.1.

EXAMPLE 4.8

Do the Fourier and Laplace transforms of $x_c(t) = e^{\beta t}u_c(t), \beta > 0$, exist?

Solution

The Fourier transform does not exist, because (4.4) $\left[\int_{-\infty}^{\infty} |x_c(t)|\, dt < \infty\right]$ is violated:

$$\int_{-\infty}^{\infty} |x_c(t)|\, dt = \int_{0}^{\infty} e^{\beta t} dt = \left(\frac{1}{\beta}\right)[\infty - 0] = \infty.$$

The Laplace transform does exist, however, for from (4.3a), $e^{-\alpha t}|x_c(t)| < \infty$ for all time if $\alpha > \beta$ (α, β real). The Laplace transform is just $(s - \beta)^{-1}$, valid for $\sigma > \beta$, where $s = \sigma + j\Omega$.

EXAMPLE 4.9

Do the discrete-time Fourier and z-transforms of $x(n) = \beta^n u(n), \beta > 1$ exist?

Table 4.1 How Do I Know Which Transform to Use? A Basic Applications Guide.

Where (CT) or (DT) appear, they apply only for, respectively, continuous or discrete time; otherwise, both CT and DT apply.

PROCEDURE/SIGNAL TYPE	MOST HELPFUL DOMAIN OR TRANSFORM TO USE
Solve initial-value differential equation (CT) or difference equation (DT)	Laplace transform (CT);
Modal signal/system analysis (e.g., in terms of $e^{\sigma t}\cos\{\Omega t + \phi\}u_c(t)$ [CT] or $a^n\cos\{\omega n + \phi\}u(n)$ [DT])	z-transform (DT)
Stability and detailed system analysis (poles/zeros, rational functions; system identification)	
Analysis/design of multiple subsystems	
Analog filter design (also Fourier transform-based) (CT)	
Digital IIR filter design (also discrete-time Fourier transform-based) (DT)	
Digital filter structures (e.g., direct forms, cascade, parallel, lattice) (DT)	
Step and impulse response determination (including steady-state accuracy)	
Impedance-to-transfer function analysis (CT)	
Analog feedback control systems (CT); digital control systems (DT)	
Analysis of unstable and noncausal systems	
Operational view of signals/systems	
Allpass/minimum phase theory	
Cepstral analysis (theory)	
Steady-state (continuously-applied) input; determine output	Fourier transform (CT);
Resonance analysis of signals; frequency response (e.g., audio systems)	discrete-time Fourier transform (DT)
Sinusoidal modulation; communications (e.g., signal riding on "carrier" wave; radar)	
Filter design specifications	
Optimal digital filter algorithms (e.g., McClellan–Parks) (DT)	
Convolutions of signals/impulse responses having simple Fourier transforms	
Phasors and impedance for sinusoidal steady-state analysis/design	
Sampling rate selection for sampled-data systems	
Speech analysis (e.g., spectrograms; speech coding)	
Stability analysis (Nyquist stability criterion) [also uses LT/ZT]	
Characteristic functions (FT of probability density functions)	
Characterization of T-periodic signal waveforms (CT); of N-periodic signal sequences (DT)	Fourier series (CT);
Convolutions involving periodic signals/systems	discrete Fourier series (DFS = DFT; FFT implementation) (DT)
Musical instruments (CT)	
Boundary-value problems (of finite domain) (CS = continuous space)	

PROCEDURE/SIGNAL TYPE	MOST HELPFUL DOMAIN OR TRANSFORM TO USE
Filtering (convolution of finite-length discrete-time signals/unit sample responses)	Discrete Fourier transform
Long correlations by block processing (e.g., overlap-add, overlap-save)	(DFT; FFT implementation)
Cross-correlation/autocorrelation of finite-length discrete-time sequences	
Frequency sampling FIR digital filter design	
Sampling rate conversion (interpolation, estimation of sequences)	
Computer estimation of continuous-time and discrete-time Fourier transforms (including zoom transform)	
Cepstral deconvolution or pitch period implementation	
Modeling of systems from physical laws (creation of differential/difference equations)	Continuous-time domain (CT);
Linearization of nonlinear systems about an operating point	discrete-time domain (DT)
Convolutions of signals/impulse responses having simple time-domain expressions	
Window method of digital filter design (DT)	
Nonlinear operations (e.g., companding)	
Time-stepping system simulation algorithms	
State-space analysis	
Statistical and model parameter estimation	
Prediction (forecasting) of time series (DT)	
Data whitening	
Adaptive filtering	

Cautionary Note: often more then one transform applies, and depending on the application, one or another may be most appropriate to use.

Solution

The discrete-time Fourier transform does not exist, because (4.5) is violated:

$$\sum_{n=-\infty}^{\infty} |x(n)| = \sum_{n=0}^{\infty} \beta^n = \lim_{N \to \infty} \frac{1 - \beta^{N+1}}{1 - \beta} = \infty.$$

The z-transform does exist, however, for from (4.3b), $\alpha^{-n}|x(n)| < \infty$ for all time if $\alpha > \beta$ (α, β real). The z-transform is just $(1 - \beta z^{-1})^{-1}$, valid for $|z| > \beta$.

EXAMPLE 4.10

Let $x_c(t) = e^{-j\Omega_0 t}$, $-\infty < t < \infty$. Find the bilateral Laplace transform of $X_c(s)$ where it exists. For what s does it exist?

Solution

Letting $s = \sigma + j\Omega$,

$$X_c(s) = \int_{-\infty}^{\infty} e^{-(j\Omega_0 + s)t} \, dt = \lim_{t \to \infty} \frac{e^{-[j(\Omega_0 + \Omega) + \sigma]t}}{-[j(\Omega_0 + \Omega) + \sigma]} - \lim_{t \to -\infty} \frac{e^{-[j(\Omega_0 + \Omega) + \sigma]t}}{-[j(\Omega_0 + \Omega) + \sigma]}.$$

For $\sigma < 0$, the upper limit is $-\infty$, and for $\sigma > 0$, the lower limit is $+\infty$. Only for $\sigma = 0$ does $X_c(s)$ exist, and then only in a distributional sense (see Appendix 2A), as a Dirac delta function. Specifically, for $\sigma = 0$,

$$X_c(s) = X_c(j\Omega) = \lim_{t \to \infty} \pi t \, \frac{2 \sin\{t(\Omega_0 + \Omega)\}}{t\pi(\Omega_0 + \Omega)} = 2\pi\delta_c(\Omega + \Omega_0),$$

where we used the fact that $\sin(x)/(\pi x)$ has unit integral from $x = -\infty$ to $+\infty$. For details, see (2A.23) and the surrounding discussion, where here we set $\alpha = t$. To verify our result, we can take the inverse Fourier transform of $2\pi\delta_c(\Omega + \Omega_0)$, which very easily yields $x_c(t) = e^{-j\Omega_0 t}$.

EXAMPLE 4.11

Draw the pole–zero plot of the system function $H_c(s) = V_{o,c}(s)/V_{i,c}(s)$ in the circuit shown in Figure 4.16. Verify your results with a data acquisition system. This circuit is a third-order passive lowpass filter. First derive $H_c(s)$ for general circuit parameter values, and then for the actual circuit use $R_1 = 200 \, \Omega$, $R_2 = 4700 \, \Omega$, $R_3 = 470 \, \Omega$, $C_1 = 100 \, \mu F$, $C_2 = 10 \, \mu F$, and $L = 0.35 \, H$ (for which the winding resistance is $R_L = 24 \, \Omega$).

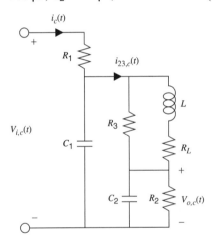

Figure 4.16

Solution

We consider a higher-order filter here to demonstrate Siglab's ability to model systems with transfer functions. Although intermediate results of the derivation are included, the reader may wish to go immediately to the transfer function $H_c(s)$ below. The parallel combination of C_2 and R_2 has impedance $Z_{pA} = R_2/(1 + R_2 C_2 s)$, whereas that of R_3 and $R_L + sL$ is $Z_{pB} = (R_3 Ls + R_3 R_L)/(Ls + R_3 + R_L)$. The series combination of Z_{pA} and Z_{pB} is

$$Z_{sA} = \frac{R_2 R_3 C_2 Ls^2 + \{(R_2 + R_3)L + R_2 R_3 R_L C_2\}s + R_L(R_2 + R_3) + R_2 R_3}{(1 + R_2 C_2 s)(Ls + R_3 + R_L)}.$$

The parallel combination of Z_{sA} and C_1 is

$$Z_{p3} = \frac{R_2 R_3 C_2 Ls^2 + \{(R_2 + R_3)L + R_2 R_3 R_L C_2\}s + R_L(R_2 + R_3) + R_2 R_3}{\begin{array}{c} R_2 R_3 C_1 C_2 Ls^3 + \{[R_2 C_2 + (R_2 + R_3)C_1]L + R_2 R_3 R_L C_1 C_2\}s^2 + \\ \{(R_3 + R_L)R_2 C_2 + [(R_2 + R_3)R_L + R_2 R_3]C_1 + L\}s + R_3 + R_L \end{array}}.$$

The total impedance $Z_T(s)$ is $R_1 + Z_{p3}$, which has the same denominator $\Delta_c(s)$ as Z_{p3} has and which has numerator

$$B_c(s) = R_1R_2R_3C_1C_2Ls^3 + \{[(R_1+R_3)R_2C_2 + (R_2 + R_3)R_1C_1]L + R_1R_2R_3R_LC_1C_2\}s^2 +$$
$$\{[R_3R_L + R_1(R_3 + R_L)]R_2C_2 + R_1[(R_2 + R_3)R_L + R_2R_3]C_1 + (R_1 + R_2 + R_3)L\}s +$$
$$R_1(R_3 + R_L) + R_L(R_2 + R_3) + R_2R_3.$$

The total current $I_c(s) = V_{i,c}(s)/Z_T(s)$. Also, $V_{o,c}(s) = R_2I_{23,c}(s)/(1 + R_2C_2s)$, where

$$I_{23,c} = I_c(s) \frac{R_2C_2Ls^2 + \{(R_3 + R_L)R_2C_2 + L\}s + R_3 + R_L}{\Delta_c(s)}.$$

Thus

$$H_c(s) = \frac{V_{o,c}(s)}{V_{i,c}(s)} = \frac{R_2(Ls + R_3 + R_L)}{B_c(s)}.$$

We measure our selected devices and obtain the following results. We measure R_1 = 199.5 Ω, but we note that the output impedance of the analog output of Siglab is 51 Ω, so we use R_1 = 250.5 Ω. Other measurements are R_2 = 4698 Ω, R_3 = 468.4 Ω, C_1 = 101.4 μF, C_2 = 10.34 μF, L = 0.34 H, and R_L = 24 Ω.

Siglab has a "virtual system identification" instrument that emits white noise up to a selected bandwidth into a connected circuit, records samples of the circuit response, performs averages (this is in doubt, because all plotted "realizations" making up the average appear identical), and discrete Fourier transforms these averages to plot the spectrum. It then performs a least-squares algorithm (see Chapter 10) to find the model difference equation coefficients that best fit the input–output discrete-time data. Finally, Siglab plots the pole–zero plot in the z-domain, or converted to the s-domain using the mapping $z = e^{s\Delta t}$ (see Chapter 6), as shown in the lower plot of Figure 4.17.

Using the analytical transfer function above with the numerical component values, we find poles at -169 Np/s $\pm$ $j553$ rad/s and at -38 Np/s, and a zero at -1447 Np/s. Compare these with the model Siglab produces: poles at $-173/6$ Np/s $\pm$ $j500.7$ rad/s and at -22.5 Np/s, and a zero at -946.2 Np/s. The zero is expected to be off by more because the transformation $z = e^{s\Delta t}$ used to obtain the s-zeros from the z-zeros is not strictly valid; see Section 6.2.

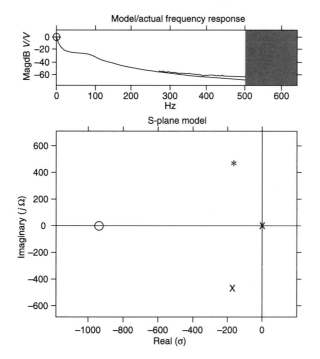

Figure 4.17

It is quite satisfying that Siglab, taking a *totally numerical* approach without any assumption of system order, produced the correct number of poles and zeros, and reasonably comparable numerical values for these. In other systems investigated, however, results are not as satisfying. The reason is that many system models of different orders and coefficient values behave similarly (e.g., frequency responses are close, as in the upper figure of Figure 4.17), so the model that theory produces will not always match that from Siglab, or from any other numerical system identifier.

It is interesting to note that sometimes an extra zero way out in the right-half plane appears in the calculated model. The difference in numerator versus denominator order might seem serious. However, the Bode-plot contribution from this extra zero "kicks in" only at that high corner frequency equal to that zero. This high frequency is way beyond the resonance band of the circuit and essentially determines only the passage (and aliasing) of high-frequency noise.

This example should help further concretely show how pole–zero models (transfer functions) are associated with real-world systems such as actual circuits. Also, with system identifiers such as in Siglab, we can come up with approximate linear models of systems whose underlying dynamics are unknown to us.

4.3   FREQUENCY VARIABLES AND THE LAPLACE AND z-TRANSFORMS

In this section, we discuss in detail continuous-time versus discrete-time frequency variables and their units. The *complex frequency* variable $s = \sigma + j\Omega$ for continuous-time signals is composed of a decay variable σ and an oscillatory variable Ω. We consider now each of these variables and their units in turn.

The real part of s, σ, is a measure of the rate of exponential decay (for $\sigma < 0$) of a damped oscillation e^{st}. The units of the attenuation parameter σ are nepers per second (Np/s), after a Latin word for John Napier, who introduced natural logarithms. Thus at $t = -1/\sigma$ (again, $\sigma < 0$, so $t = -1/\sigma > 0$), then $e^{\sigma t} = 1/e \approx 0.3679$ represents an attenuation of 1 Np (i.e., a gain of -1 Np) relative to a signal always equal to 1. More generally, we may quantify the gain exponentially in base e as the ratio of two amplitude-signals, for example, the ratio of two voltages: gain $= A = v_o/v_i = e^y$. Then the number of nepers of the gain so expressed $=$ # Np $=$ ln(signal gain) $=$ ln$(A) =$ ln$(e^y) = y$ Np.

The neper is thus analogous to the familiar decibel in that both are dimensionless logarithmic gains, positive for amplification and negative for attenuation. The difference is that the decibel quantifies the common log of the square of the signal gain [i.e., the common log of the power gain $A^2 = (v_o/v_i)^2$], not the natural log of the nonsquared signal gain A, as does the neper. Because Re$\{s\} \cdot t$ appears in the exponent of e, it is "natural" to quantify the logarithmic "gain" A or signal level of an exponential in base e, as for the neper. Contrarily, for amplifier or filter analysis, people "commonly" think in terms of orders of magnitude of power gain, which are quantified in base ten; hence decibels are then more appropriate than Np. Because our eyes respond to light and our ears to sound in a logarithmic manner (so that they are useful over extremely wide-ranging intensities), it is convenient to express signal and power gains logarithmically.

A signal gain (or level) exponentially expressed in base ten is gain $= A = v_o/v_i = 10^x$, where x is the number of orders of magnitude of the signal gain. Thus, equating the two expressions for the same gain, $A = 10^x = e^y$, we obtain $x = y \log_{10}(e) = 0.4343y$. Now, the power gain is $A^2 = 10^{2x}$. The number of bels[10] (B) is defined as

---

[10] The bel is named after Alexander Graham Bell, inventor of the telephone.

the common log of the power gain, namely # B = $\log_{10}$(power gain) = $\log_{10}(A^2)$ = $\log_{10}(10^{2x})$ = $2x$. So the number of bels of the gain is twice the exponent of 10, whereas the number of nepers of the gain is equal to the exponent of e. Thus # B = $\log_{10}(A^2)$ = $\log_{10}(e^{2y})$ = $2y \cdot \log_{10}(e)$ ≈ 0.8686 · (# Np), so nepers and bels are numerically quite similar. In particular, ±1 Np = ±0.8686 B. Now, one B of gain is associated with a gain-order-of-magnitude x_{1B} such that x_{1B} = (1 B)/2 = $\frac{1}{2}$. Thus the signal gain of 1 B is $A = 10^x\big|_{x \, = \, x_{1B}=1/2} = \sqrt{10} = 3.1623$, compared with a signal gain of $e^1 = 2.7183$ for 1 Np; again, bels and nepers are close.

More common usage than bels, however, is decibels (dB)—each decibel being one-tenth of a bel—because one B just "covers too much territory" for human purposes. For example, there are only 2 B per order of magnitude change, compared with the more convenient 20 dB per order of magnitude change [that is, the gain in dB of $10^\ell \cdot A$ is $20\log_{10}\{A\} + 20\log_{10}\{10^\ell\} = A_{dB} + 20 \cdot \ell$]. Thus, # dB = $10 \cdot$ (# B) = 8.686 · (# Np); so ±1 Np ≈ ±8.686 dB. Note that unity gain is 0 Np = 0 dB.

In conclusion, σ is a rate of exponential growth or attenuation of a mode e^{st}, and its units are nepers per second so that the units of σt are nepers. Note that it is inappropriate (however common) to refer to σ as a frequency. "Frequency" quantifies how often something recurs; an exponentially growing or decaying function does not recur (as does a sinusoid). *Complex frequency* for s *is* appropriate, because e^{st} in general does have an oscillatory component (except when $\Omega = 0$).

The frequency variable Ω is the radian frequency of the oscillatory component, $e^{j\Omega t}$, of the damped/undamped/growing oscillation e^{st}. Its units are radians per second so that if $\Omega = 1$ rad/s, a complete cycle or oscillation takes place in 2π s. The historical origin of Ω lies in the study of simple harmonic motion. If a body in rotation rotates at an angular speed Ω, it turns an angle Ω radians in 1 second. Furthermore, because one complete turn (cycle) consists of 2π rad, $F = \Omega/2\pi$ is the number of cycles or rotations per second, or in modern terminology, hertz (Hz). One cycle is completed in $T = 1/F$ seconds, where T is called the period.

Even though an oscillatory component of a signal $x_c(t)$ is not undergoing physical rotation, it can be represented by a *phasor* or vector rotating in the complex plane with angular speed Ω rad/s. Additionally, the projection of circular motion onto a line passing through its center traces out a sinusoid as the circular motion progresses: one period for each rotation. The angular frequency notation $\cos(\Omega t)$ is shorter than the equivalent $\cos(2\pi F t)$, so it is often used in analysis rather than the Hertzian frequency F. One should feel comfortable with either notation.

The situation is fundamentally different for discrete-time representations. Instead of having $s_d = \sigma_d + \Omega_d$, d standing for "discrete," the frequency domain is represented by the complex variable $z = |z|e^{j\omega} = z\angle\omega$. Why? Suppose we did try to use $s_d = \sigma_d + j\Omega_d$. Then, in analogy with the continuous-time case, the components of the signal $x(n)$ would be of the form $e^{s_d n}$. The oscillatory part would be $e^{j\Omega_d n}$, where because n is the number of samples, Ω_d must have the units radians per sample for $\Omega_d n$ to have units of radians. For continuous-time signals, Ω may take on values anywhere on $(-\infty, \infty)$. Because $e^{j(\Omega_d + 2\pi\ell)n} = e^{j\Omega_d n} \cdot e^{j2\pi n\ell} = e^{j\Omega_d n}$, however, any oscillatory components having $|\Omega_d| > \pi$ cannot be distinguished from (i.e., have the same samples as) those having the "principal radian frequency" $\omega = \text{Arg}(e^{j\Omega_d})$ = $\text{Mod}(\Omega_d, 2\pi)$. This statement is true only because n is an integer; no such uniform periodicity in Ω of $e^{j\Omega t}$ exists for all t. The units of this *principal radian frequency* or *digital radian frequency* ω are also radians per sample; the only difference is that $|\omega| \leq \pi$ because we know everything just repeats beyond π.

A *normalized* or *digital frequency* f is defined analogously to $F = \Omega/(2\pi)$ Hz as follows: $f = \omega/(2\pi)$ cycles/sample. Because $\omega \in (-\pi, \pi]$, it must be true that

$f \in (-\frac{1}{2}, \frac{1}{2}]$. The units of f (often omitted for brevity) must be (rad/sample)/(rad/cycle) = cycles/sample; evidently, any sinusoidal frequencies $F_d = \Omega_d/2\pi$ greater in magnitude than $\frac{1}{2}$ cycle/sample are indistinguishable in the sequence $e^{s_d n}$ from the "principal frequency" $f \in (-\frac{1}{2}, \frac{1}{2}]$ cycle/sample corresponding to F_d. Specifically, $f = F_d - [F_d + \frac{1}{2}]$ where again [ ] = greatest integer, always rounding toward $-\infty$. This statement holds because if $F_d = f + \ell$ with $f \in (-\frac{1}{2}, \frac{1}{2}]$, then $[F_d + \frac{1}{2}] = \ell$ because[11] $f + \frac{1}{2} \in (0, 1]$. Thus, $F_d - [F_d + \frac{1}{2}] = f + \ell - \ell = f$.

In Matlab, we can write a single line of code to implement conversion from F_d to f:

```
f = Fd - floor(Fd - sign(Fd) * 1e - 15 + 0.5)
% "floor" is the greatest integer function discussed above
   (round toward -∞)
% The -sign(Fd)·10⁻¹⁵ is a correction for thresholding on
   "floor" command at f = 1/2
```

What is the significance of the limit $f = \frac{1}{2}$? Physically, it means that samples of a sinusoidal oscillation taken fewer than twice every period are indistinguishable from those taken more frequently of a lower-frequency sinusoid. The reasoning for $\frac{1}{2}$ will become clearer in the detailed discussion of sampling and aliasing in Chapter 6. For now, let us consider examples on either side of $\frac{1}{2}$.

First suppose that $f = \frac{1}{10}$ cycle/sample. This digital frequency corresponds to a sinusoidal oscillatory component from which $x(n)$ may be derived that has been sampled 10 times each period. Now consider $f = 1$ cycle/sample. This frequency corresponds to a continuous-time sinusoidal oscillatory component from which $x(n)$ may be derived by sampling that has been sampled exactly once each cycle. Because the sinusoidal oscillation is periodic, samples taken once per cycle will all be the same. That is, the sampled component is constant and thus indistinguishable from the sampling of a constant function, corresponding to zero frequency or $f = 0$! Indeed, setting $F_d = 1$, we obtain $f = 1 - [1 + \frac{1}{2}] = 1 - 1 = 0$. Such an equivalence to the sampling of a lower-frequency sinusoid can be made for all $|f| > \frac{1}{2}$ cycle/sample.

As just noted, the implications of the idea of "principal frequency" are evident in sampling theory, which will be investigated in detail in Section 6.2. It is found there that if $x(n)$ consists of samples of a continuous-time signal $x_c(t)$ at $t = n\Delta t$, then all high frequencies in $x_c(t)$ are "aliased" into low frequencies—into just the principal frequency range $\omega \in (-\pi, \pi]$ described here! In Section 6.2, we will show that "high" means $\Omega > \pi/\Delta t = \Omega_s/2$, where $\Omega_s = 2\pi/\Delta t$ is the sampling frequency in radians per second.

Now, by examination of $e^{s_d} = e^{\sigma_d} \cdot e^{j\omega}$ and restricting $\omega \in (-\pi, \pi]$, it can be seen why z rather than s_d is used. Writing $z = e^{s_d} = |z| e^{j\arg(z)}$ with $|z| = e^{\sigma_d}$ and using $\omega = \text{Arg}(z) = \angle z$, the frequency variable ω also has the range and meaning identical to those of an angle in the complex z-plane. The units of ω are radians per sample, but as "sample" is dimensionless, we may also say the units of ω are radians. An oscillation taking exactly N samples per period has frequency $\omega = 2\pi/N$ rad/sample. Alternatively, given ω, there are $N = 2\pi/\omega$ samples per period; if $\angle z = \omega$ is expressed in degrees per sample, then $N = 360°/\angle z$. As for σ in continuous time, σ_d can range from $-\infty$ to ∞. Where $\sigma_d = 0$, $|z| = 1$; and where $\sigma_d > 0$, $|z| > 1$; and where $\sigma_d < 0$, $0 \le |z| < 1$. Thus, $|z|$ ranges from 0 to ∞, just as does the radial polar coordinate (magnitude) of a complex number in the complex z-plane. For any $|z|$, as ω ranges from $-\pi$ to π, z completely traverses a zero-centered circle of radius $|z|$ in the z-plane. Thus, instead of concentrating attention on just a "principal slice"

---

[11] The only exception is $f = \frac{1}{2}$; we artificially avoid this value in the Matlab code.

$\{-\pi < \Omega_d < -\pi, -\infty < \sigma_d < \infty\}$ of the s_d-plane, the *entire* z-plane with $z = e^{s_d}$ is equivalently covered and available for analysis.

Furthermore, now the dimensionless quantity $|z|$ quantifies attenuation of the exponential sequence $e^{s_d n} = z^n$ or the envelope of damped oscillations, whereas $\omega = \angle z$ specifies the oscillation frequency of z^n in radians per sample. We see that $|z|$ and ω are, respectively, completely analogous to the purposes σ and Ω serve in continuous time. It is interesting that for continuous time, the attenuation and frequency of a pole are given by its rectangular coordinates, whereas for discrete time they are given by the polar coordinates of the pole. Another interesting aside is that $|z| = 0$ corresponds to an infinitely damped sinusoid ($\sigma_d = -\infty$), that is, no signal component at all (beyond $n = 0$). Also, the Laplace transform expansion function e^{st} evaluated at $t = n\Delta t$ is $e^{sn\Delta t} = (e^{s\Delta t})^n$, which is equal to the z-transform expansion function z^n evaluated at $z = e^{s\Delta t}$. Finally, $|z| = 1$ and $\sigma = 0$ correspond, respectively, to the discrete- and continuous-time Fourier transforms.

Another, more pedestrian reason for using z for the discrete-time GPE variable concerns arises because the delay operation is central to discrete-time linear filtering calculations and difference equations. As noted in Section 3.4, the z-transform of $x(n - m)$ is $z^{-m}X(z)$. In signal flow diagrams, this m-delay is represented by z^{-m}, which is much more convenient to write than $e^{-s_d m}$, especially when the diagram is filled with delay operators.

An important advantage of using the z-transform for representing transfer functions from sampled inputs to sampled outputs is that the z-transform converts the transfer function, which is a rational function of $e^{s\Delta t}$—an irrational function of s—into a rational function of z. It should be noted that a rational function of $e^{s\Delta t}$ has an infinite number of s-poles and s-zeros. Contrarily, the sampled-data input–output transfer function expressed as a rational function of z allows easy determination of its finite number of z-poles and z-zeros, as well as allowing all the techniques involving rational functions to be applied to discrete-time system analysis and design.

Yet another fundamental reason for using z rather than $e^{s_d n}$ is that the modes of a discrete-time system are α^n and those of a continuous-time system are $e^{\beta t}$. It is only natural to define $z = \alpha$ and $s = \beta$, so that α_ℓ or β_ℓ are the poles of the relevant transfer function $H(z)$ or $H_c(s)$. Recall that the modes are the independent solutions of the difference/differential equation with zero input. Also, with $z_1 \equiv \alpha$, z_1^n is an eigenfunction of $H(z)$: If z_1^n is the input, then $z_1^n H(z_1)$ is the output. Although strictly true only for bicausal z_1^n [for which $H(z_1)$ can exist only for $|z_1| = 1$], the relation approximately holds for causal $z_1^n u(n)$ for any $|z_1|$ after the transient dies away, assuming that z_1^n dies away much more slowly than the system transient. This fact is easily proved using partial fraction expansion.

Finally, consider the relation between the inversion integrals in (4.1a)$[x_c(t) =$
$$\left(\frac{1}{2\pi j}\right) \int_C X_c(s) \cdot e^{st} \, ds] \text{ and (4.2a) } [x(n) = \left(\frac{1}{2\pi j}\right) \oint_C X(z) \cdot z^n (dz/z)].$$ In (4.1a), make the change of variables $z = e^{s\Delta t}$; thus $dz = \Delta t e^{s\Delta t} \, ds$, or $ds = dz/(\Delta t e^{s\Delta t}) = dz/(\Delta t \, z)$.

Evaluating (4.1a) at $t = n\Delta t$ gives $x_c(n\Delta t) = \left(\frac{1}{2\pi j}\right) \oint_{C'} X_c(\ln[z]/\Delta t) \cdot z^n \, dz/(\Delta t \, z)$.

What is C'? As noted in Section 3.2 (see Figure 3.7b), the $j\Omega$ axis maps to the unit circle under $z = e^{s\Delta t}$. More generally, contour C_1 in Figure 4.1 maps to the circle of radius $e^{\sigma_1 t}$ in the z-plane. Thus, $x(n)$ in (4.2a) is *apparently* Δt times the Laplace transform inversion integral in (4.1a) evaluated at $t = n\Delta t$, with the change of integration variable $z = e^{s\Delta t}$. However, do not be fooled! Recall from Figure 3.7b that as Ω ranges from $-\infty$ to ∞, the $e^{\sigma t}$ circle in the z-plane is traversed an infinite number of times, and for each of these traversals, a different section of $X_c(s)$ is used in the integral, so $X(z) \neq X_c(\ln[z]/\Delta t)/\Delta t$, but $X(z)$ is instead a sum involving each of these different sections of $X_c(s)$. We will show the explicit relation in Section 6.2; careful completion of the above change of variables would also prove the result.

4.4   PERIODIC AND APERIODIC FUNCTIONS

Signals can be conveniently categorized into two classes: periodic and aperiodic. Consider for the moment periodic functions for continuous time: $\tilde{x}_c(t + T) = \tilde{x}_c(t)$; that is, $\tilde{x}_c(t)$ is T-periodic in t. Note that most signals we view on the oscilloscope appear to be[12] continuous-time periodic waveforms (excepting when we use the single-trace mode). In general, for $\tilde{x}_c(t)$ to be T-periodic, it must be composed *only* of constituent functions that are also T-periodic in t. A very convenient class of constituent functions is sinusoids. For a variety of reasons—two of which are that $d^\ell/dt^\ell\{e^{j\Omega t}\} = (j\Omega)^\ell \cdot e^{j\Omega t}$ and $e^{j\Omega t} \to H_c(s) \to H_c(j\Omega)e^{j\Omega t}$, neither of which holds for real sinusoids—complex sinusoids $e^{j\Omega t}$ are more convenient to deal with mathematically than are real-valued sinusoids. A sinusoid of the fundamental frequency $\Omega_0 = 2\pi/T$ is T-periodic in t, as are all sinusoids of frequencies $\Omega_k = k\Omega_0$, which are known as the harmonics of Ω_0 (the kth harmonic is T/k-periodic, which is also T-periodic).[13]

In general, we do not know beforehand how many of the harmonics it would take to represent $\tilde{x}_c(t)$ exactly. A safe bet is to include them all; the result is the Fourier series. Fourier's theorem for both continuous- and discrete-time signals states that for periodic functions, the superposition of sinusoids composing the function is just a (possibly infinite-term) sum

$$\tilde{x}_c(t) = \sum_{k=-\infty}^{\infty} X_c(k) \cdot e^{j2\pi kt/T} \quad [\tilde{x}_c(t + T) = \tilde{x}_c(t)], \qquad \text{(continuous time)}$$
$$\text{Fourier series synthesis.} \quad (4.12a)$$

where the values of the harmonic weighting coefficients are

$$X_c(k) = \frac{1}{T} \cdot \int_0^T \tilde{x}_c(t) \cdot e^{-j2\pi kt/T} dt; \qquad \text{Fourier series analysis.} \quad (4.12b)$$

for discrete time [fundamental frequency $= \omega_0 = 2\pi/N$; $\tilde{x}(n)$ is N-periodic in n],

$$\tilde{x}(n) = \frac{1}{N} \cdot \sum_{k=0}^{N-1} \tilde{X}(k) \cdot e^{j2\pi kn/N} \quad [\tilde{x}(n + N) = \tilde{x}(n)] \qquad \begin{array}{l}\text{(discrete time)}\\ \text{Discrete Fourier}\\ \text{series synthesis.} \quad (4.12c)\end{array}$$

where

$$\tilde{X}(k) = \sum_{n=0}^{N-1} \tilde{x}(n) \cdot e^{-j2\pi kn/N}, \quad [\tilde{X}(k + N) = \tilde{X}(k)]. \qquad \begin{array}{l}\text{Discrete Fourier}\\ \text{series analysis.} \quad (4.12d)\end{array}$$

See WP 5.1 for details.   The above expressions for the values of the coefficients [$X_c(k)$ in (4.12b and $\tilde{X}(k)$ in (4.12d)] are obtained by using the orthogonality of complex exponentials. The spacing of the frequencies of the sinusoidal components (i.e., of the harmonics)

---

[12] Technically, the fact that we turn the instrument on and off excludes pure periodicity, which implies a never-beginning-or-ending signal. For all practical purposes, however, it behaves as a truly periodic signal.
[13] In the Fourier series expression, the fundamental is actually $k = 1$ (Ω_1), not $k = 0$ (dc); however, we like to reserve Ω_1 for other uses, and Ω_0 is conventional.

is the inverse of the period of the periodic time function being represented. That is, the frequency spacing equals the fundamental frequency $F_0 = 1/T$ Hz or $\Omega_0 = 2\pi/T$ rad/s, or $f_0 = 1/N$ cycles/sample or $\omega_0 = 2\pi/N$ rad/sample. This fact can be verified by comparing the frequency of $e^{j2\pi kt/T}$ (namely $\Omega_k = 2\pi k/T$) with that of $e^{j2\pi(k-1)t/T}$ [namely, $\Omega_{k-1} = 2\pi(k-1)/T = \Omega_k - \Omega_0$; also, the analogous comparison for discrete time]. The resulting sum is called a Fourier series, either continuous- or discrete-time.

The Fourier series coefficients $X_c(k)$ are all that we need to reconstruct $\tilde{x}_c(t)$ for *all* time. No fractional-harmonic frequencies (e.g., $\Omega_0 \cdot 1.3$) can be found in $\tilde{x}_c(t)$ because they are not T-periodic; if included, the expansion sum would not be T-periodic as it must be if it is to be equal to the T-periodic $\tilde{x}_c(t)$. Also, because we know that the response of an LSI system to a single steady-state sinusoid $\exp\{j\Omega_k t\}$ is just the same sinusoid scaled by $H_c(j\Omega_k)$, we can in principle calculate the total response to any periodic input by (a) Fourier-decomposing the input, (b) modifying the weight of each constituent sinusoid by the appropriate frequency response value, and (c) applying superposition to sum the individual sinusoidal responses. Specifically, if the Fourier series coefficients of $\tilde{x}_c(t)$ are $X_c(k)$, then applying the above result one component at a time, those of the output $y_c(t)$ are $Y_c(k) = H_c(j\Omega_k)X_c(k) = H_c(j2\pi k/T)X_c(k)$.

Periodic functions are not absolutely integrable/summable, because they violate the absolute integrability/summability requirements (4.4) or (4.5). Thus functions satisfying (4.4) and (4.5) are aperiodic and are representable as an integral over a continuous range of frequencies. Contrarily, Fourier's theorem for periodic functions (Fourier series) is a representation of the periodic time function as a *sum* over a discrete set of frequencies.

EXAMPLE 4.12

The E-minor chord on a guitar has the notes E_1, B_2, E_2, G_2, B_3, and E_3, where the subscripts are used to distinguish octaves (1 for lowest octave, etc.). Not only the fundamentals of these strings sound, but all the harmonics in varying strengths. Find the frequencies and periods of the various fundamentals, given that the fundamental of E_1 is 82.41 Hz. Also find the period of the total chord waveform.

Solution

The ear interprets frequency ratios, not equal-number-of-hertz intervals, as harmonically related. This is because ratios dictate the related modal patterns produced by strings, pipes, and soundboards—which form the basis for all traditional musical instruments. Specifically, the modes are determined by the constructive and destructive interferences of traveling acoustic waves in the instrument and their reflections (standing waves). Constructive interference occurs at the fundamental frequency and its harmonics; the result is a harmonically rich almost-periodic acoustic vibration. Our scale is divided up as follows:

$$A \; B\flat \; \; B \; C \; C\sharp \; D \; E\flat \; \; E \; F \; F\sharp \; G \; A\flat \; \; A_{\text{high}},$$

where $F_{A_{\text{high}}} = 2F_A = 2^{(12/12)} \cdot F_A$. Each of the intermediate notes is distributed as a percentage of this doubling of frequency. A frequency and its double are called an octave because the two notes span inclusively eight main, lettered notes in the scale.

Thus for example,

$$F_{B_2} = 2^{(7/12)} F_{E_1} = 1.4983 F_{E_1} = 1.4983(82.41) = 123.48 \text{ Hz.}$$

Notice that $F_{B_2} \approx \frac{442}{295} \cdot F_{E_1}$, or $T_{B_2} \approx \frac{295}{442} \cdot T_{E_1}$. Similarly, $F_{E_2} = 2F_{E_1} = 164.82 \text{ Hz}$ (i.e., $T_{E_2} = \frac{1}{2} T_{E_1}$), $F_{G_2} = 2^{(3/12)} F_{E_2} = 1.1892(164.82) = 196.01 \text{ Hz} \approx \frac{25}{21} \cdot F_{E_2}$ (i.e., $T_{G_2} \approx \frac{21}{2 \cdot 25} T_{E_1}$), $F_{B_3} = 2^{(7/12)} F_{E_2} = 246.95 \text{ Hz}$ (i.e., $T_{B_3} \approx \frac{295}{2 \cdot 442} \cdot T_{E_1}$), and $F_{E_3} = 2F_{E_2} = 329.64 \text{ Hz}$ (i.e., $T_{E_3} = \frac{1}{4} T_{E_1}$).

Because factors such as $2^{\frac{7}{12}}$ are irrational, the chord is actually aperiodic. However, using our approximate integer ratios above which we obtained by programming a published best-rational-number-approximation algorithm,[14] we can at least try to estimate an approximate period. Collecting our results, the total period, being the least common multiple, is

$$T \approx (295)(21)T_{E_1} \approx 21(442)T_{B_2} \approx 295(21)(2)T_{E_2} \approx 295(25)(2)T_{G_2}$$
$$\approx 21(2)(442)T_{B_3} \approx 21(295)(4)T_{E_3} \approx 75.173 \text{ s.}$$

A real guitar E-minor strum would not exhibit this periodicity, because it dies away in just a few seconds. Figure 4.18a plots the simulated undamped chord function using the exact frequency ratios (e.g., $2^{\frac{7}{12}}$ and $2^{\frac{3}{12}}$):

$$x_c(t) = \cos(2\pi F_{E_1}t) + \cos(2\pi F_{B_2}t) + \cos(2\pi F_{E_2}t) + \cos(2\pi F_{G_2}t) + \cos(2\pi F_{B_3}t) + \cos(2\pi F_{E_3}t).$$

The function $x_c(t + T)$, where $T = 75.173$ s as above is superimposed. With all our efforts to get highly accurate rational number approximations for the period ratios, the degree of disagreement is discouraging.

Instead of taking the "exact" approach, let us instead just look at the true waveform in Figure 4.18a and "eyeball" the period. The signal visually appears to be periodic with period 0.0974 s, where we estimate the period from one major maximum (e.g., at $t = 0$) to the next using ginput to obtain accurate time estimations. Notice that $0.0974/T_{E_1} = 0.0974/0.01213 \approx 8$. Moreover, $T_{B_2} \approx \frac{2}{3} T_{E_1}$, $T_{B_3} \approx \frac{1}{3} T_{E_1}$, $T_{E_2} = \frac{1}{2} T_{E_1}$, $T_{E_3} = \frac{1}{4} T_{E_1}$, so assuming that $T = 8T_{E_1}$ we look for a ratio $(8/\ell) \approx T_{G_2}/T_{E_1} = 2^{\frac{-3}{12}}/2 \approx 0.4204$, giving $\ell \approx 19$. Thus we have $19T_{G_2} \approx 24T_{B_3} \approx 12T_{B_2} \approx 32T_{E_3} = 16T_{E_2} = 8T_{E_1} \approx 0.0971 \text{ s} \approx T$.

The agreement in Figure 4.18b between $x_c(t + T)$ using $T = 0.0971$ s (dashed line) and $x_c(t)$ (solid line) is almost perfect! Yet the ratio $\frac{295}{442} = 0.667421$ is much closer than $\frac{2}{3} = 0.666667$ is to $T_{B_2}/T_{E_1} = 0.667420$, the ratio $\frac{21}{50} = 0.420000$ is much closer than $\frac{8}{19} = 0.421053$ is to $T_{G_2}/T_{E_1} = 0.420448$, and $\frac{295}{884} = 0.3337104$ is much closer than $\frac{1}{3} = 0.333333$ is to $T_{B_3}/T_{E_1} = 0.3337099$. In other words, by using poorer estimates of the period ratios, we obtained a much closer approximation of the time waveform. Why?

The answer is that what counts is not the ratio accuracy so much as the error incurred by shifting the waveform by a full period T, where T is approximated by the ratio. Consider, for example, T_{G_2}. Using the more accurate ratios, we had $T = 295(50)T_{G_2}$, and we set that equal to $295(21)T_{E_1}$. But if we substitute the exact values of T_{G_2} and T_{E_1} obtained from the irrational fractional powers of two, we find $295(50)T_{G_2} - 295(21)T_{E_1} = 15.724T_{G_2}$; that is, the error for the G_2 term over one "period" is equal to 15.724 exact periods of G_2, which is no better than a random

---

[14] O. Aberth, *Precise Numerical Analysis* (Dubuque, IA: William C. Brown, 1988).

guess for T would have been. The reason this error is so bad with the highly accurate ratio approximation is that that approximation made T be such a large number times T_{E_1} that even the small error in T_{G_2}/T_{E_1} was greatly magnified. Thus the whole idea of finding least common integer multiples is wrong because the "common multiples" are not common at all in this case. The only reason that Figure 4.18a appears at least reasonably close is that it is only T_{G_2} for which the error calculated above is bad; the others are either close (B_2, B_3) or exact (E_2, E_3).

Contrast these results with those for our "eyeball" estimate $T_{G_2}/T_{E_1} = 8/19$. Now $19T_{G_2} - 8T_{E_1} = -0.0273T_{G_2}$, which is a fairly negligible error in the G_2 waveform incurred over one "period" $T = 19T_{G_2} = 8T_{E_1}$. Consequently, Figure 4.18b is

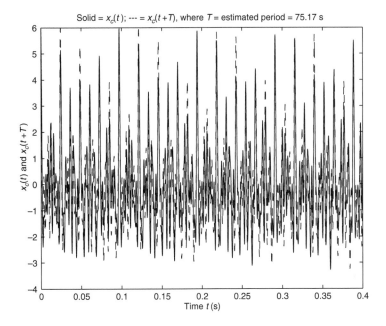

Solid = $x_c(t)$; --- = $x_c(t+T)$, where T = estimated period = 75.17 s

Figure 4.18a

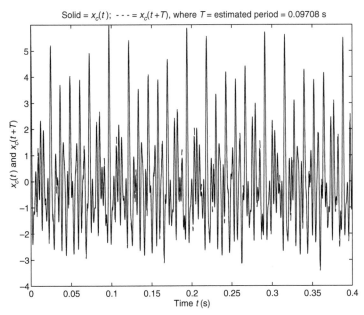

Solid = $x_c(t)$; - - - = $x_c(t+T)$, where T = estimated period = 0.09708 s

Figure 4.18b

visually nearly perfect. In this case, "common multiples" are nearly common. This example points out how crucial clear thinking about a physical problem is for obtaining meaningful results.

To conclude this example, let us note a few additional interesting facts. First, no error was introduced by omitting the harmonics in our cosine sum above. All the harmonics are multiples of the fundamental and so have an integral number of periods within each fundamental period, and the period T of the chord is—for example, above—eight fundamental periods ($8T_{E_1}$). Thus all harmonics are commensurate. Also, the magnitudes and phases of the various cosine terms are irrelevant to the periodicity of each term and to the sum of the terms, at least in the ideal case of true periodicity.

Next, notice that the cosines in Figure 4.18a all add up at $t = 0$ and that that value (six) is the maximum for all time. If instead of six cosines we added up cosines of *all* frequencies, we would find the value at $t = 0$ going to infinity $(1 + 1 + 1 + \ldots)$ and at all other times averaging out to zero. This sum is a Dirac delta,[15] and is a physically meaningful view of the Fourier transform pair $\delta_c(t) \overset{FT}{\leftrightarrow} 1$.

Finally, the same principles as investigated in this example apply for periodic and nearly periodic discrete-time signals. The only difference is the added stipulation that the prospective "period" be an integer.

The Fourier series can also be meaningfully and conveniently applied to signals of finite duration. Conventionally, we denote these finite intervals of support as [0, T] for continuous time and [0, $N - 1$] for discrete time. A finite transform such as the Fourier series (4.12b) or the discrete Fourier series (4.12d) cannot distinguish between a finite-support function and its periodic extension, for the transform integral/ sum is taken over only the finite interval of support equal to one period of the periodic extension. The Fourier series merely considers the finite-duration interval as one "period" of the periodic extension of the given function; $x_c(t)$ is one "period" of $\tilde{x}_c(t) = x_c(t_{\text{mod } T})$.

In boundary value problem terminology, we thus speak of the Fourier series as a "finite transform" defined on a finite interval with boundary conditions typically at the ends of the interval. In fact, one of the original applications Fourier himself studied was of this type. He wanted to solve a heat conduction problem on a long plate ("infinitely long"), one end of which had a constant-in-time heat applied that was a function of the distance along the edge of the end. The two long perpendicular sides were kept at constant temperature, and the object was to find the temperature at any point on the plate. To solve the partial differential equation of heat conduction for the temperature, the heat function applied to the finite-length end was expanded over an infinite sum of sinusoids.[16]

When the discrete Fourier series is used to represent the spectrum of a finite duration sequence (length N), the discrete Fourier series is called the discrete Fourier transform of length N (DFT_N or just DFT). The DFT is of tremendous importance in applications because it can be computed very efficiently by algorithms known as fast Fourier transforms (FFTs; see Sections 4.5 and 7.10). (It seems that no matter what the calculation is, researchers try to formulate it in terms of an FFT!)

---

[15] Per William Seibert, Massachusetts Institute of Technology. In $e^{j\Omega t} = \cos(\Omega t) + j \sin(\Omega t)$, the sines are all zero at $t = 0$ and for $t \neq 0$ they all average to zero as the cosines do. But the cosines are all 1 at $t = 0$, so they all add up to infinity.

[16] See R. L. Jeffrey's *Trigonometric Series: A Survey* (Toronto: University of Toronto Press, 1956), p. 4.

What is the precise relation between Fourier series and transforms? To find out, let us compare (4.12b) $\left[X_c(k) = \dfrac{1}{T} \cdot \displaystyle\int_0^T \tilde{x}_c(t) \cdot e^{-j2\pi kt/T}\, dt \right]$ with (4.6b) [here written as $X_{1,c}(j\Omega) = \displaystyle\int_{-\infty}^{\infty} x_{1,c}(t) \cdot e^{-j\Omega t}\, dt$]. Suppose that $\tilde{x}_c(t)$ is periodic in t with period T. Let $x_{1,c}(t)$ be $\{\tilde{x}_c(t), t \in [0, T]$; zero otherwise$\}$, that is, a single period of $\tilde{x}_c(t)$. Then substituting $x_c(t)$ into (4.12b) and $x_{1,c}(t)$ into (4.6b) shows that

$$X_c(k) = \frac{1}{T} \cdot X_{1,c}(j\Omega)\Big|_{\Omega = 2\pi k/T}. \tag{4.13}$$

We see that the Fourier series of a periodic function of t is equal to $1/T$ times samples of the Fourier transform of a *single period* extracted from the periodic function. The samples are spaced equally with spacing $\Omega_0 = 2\pi/T$, which is the fundamental frequency of the T-periodic function of t.

Similarly, let $\tilde{x}(n)$ be N-periodic in n and $x_1(n)$ be $\{\tilde{x}(n), n \in [0, N-1]$; zero otherwise$\}$, that is, a single period of $\tilde{x}(n)$. Then substituting $\tilde{x}(n)$ into (4.12d) $\left[X(k) = \displaystyle\sum_{n=0}^{N-1} \tilde{x}(n) \cdot e^{-j2\pi kn/N} \right]$ and $x_1(n)$ into (4.6d) $\left[X_1(e^{j\omega}) = \displaystyle\sum_{n=-\infty}^{\infty} x_1(n) \cdot e^{-j\omega n} \right]$ shows that

$$X(k) = X_1(e^{j\omega})\Big|_{\omega = 2\pi k/N}. \tag{4.14}$$

The discrete Fourier series (DFS, equivalently, the discrete Fourier transform, DFT) of a periodic function of n is thus equal to samples of the DTFT of a *single period* extracted from the periodic function. The samples are spaced equally with spacing $\omega_0 = 2\pi/N$, which is the fundamental frequency of the N-periodic sequence.

In boundary value problems, transforms with infinitely long intervals of integration are used for problems in which the desired function exists on an unbounded interval such as $(-\infty, \infty)$, often with boundary conditions at infinity. In signal processing, in this case we equivalently speak of an aperiodic function, possibly nonzero for all time.

A general aperiodic function may be thought of as a periodic function having an infinitely long period ($T \to \infty$ or $N \to \infty$). Consequently, three things happen. First, the spacing of the component harmonic frequencies [$1/T = \Omega_0/(2\pi)$ or $1/N = \omega_0/(2\pi)$] to be added in the Fourier expansion shrinks to zero, to become the incremental differentials $d\Omega/(2\pi)$ or $d\omega/(2\pi)$ we see in the inverse Fourier transform expressions. That is, the harmonics ("lines") become infinitesimally close. Second, the discrete component radian frequencies ($\Omega_k = 2\pi k/T$ and $\omega_k = 2\pi k/N$) become the continuous variables Ω and ω. Thus, the sum ("series") becomes an integral ("transform"), called the inverse Fourier transform (either continuous- or discrete-time). Finally, the integration or sum limits on the series coefficients $X_c(k)$ or $X(k)$ change from $\{[0, T]$ or $[0, N-1]\}$ both to $\{-\infty, \infty\}$.

In fact, making these substitutions in the Fourier series formulas gives precisely the Fourier transform expressions! Specifically, for continuous time, substitute (4.12b) into (4.12a) and make the above substitutions. You will end up with $x_c(t)$ expressed as the inverse Fourier transform of the forward Fourier transform as given in (4.6a and b). Similar substitutions can be made to obtain the discrete-time Fourier transform from the discrete Fourier series.

This idea was a discovery by Fourier himself. Unlike Fourier series (which had been discussed by Joseph Lagrange in 1762), the Fourier integral was originated by Fourier.[17]

The Fourier transform may also be defined for periodic functions, but only in a distributional sense (as defined in Appendix 2A). For example, because the energy of an individual sinusoidal component is concentrated at only the sinusoidal frequency, the Fourier transform must be a Dirac delta function $\delta_c(\cdot)$. We saw this result in Example 4.10; we now review it in the present context of a sinusoidal component of a more complicated function and write the corresponding result for discrete time.

Define $x_c(t; \Omega)$ as the sinusoidal component of T-periodic $\tilde{x}_c(t)$ having frequency Ω. By comparison with the inverse Fourier transform integral, we see that $\tilde{x}_c(t)$ can be decomposed over $x_c(t; \Omega)$ with $x_c(t; \Omega)$ as shown:

$$\tilde{x}_c(t) = \int_{-\infty}^{\infty} x_c(t; \Omega) \, d\Omega \text{ where } x_c(t; \Omega) = \frac{X_c(j\Omega)e^{j\Omega t}}{2\pi}. \tag{4.15a}$$

For discrete time, define $x(n; \omega)$ as the sinusoidal component of N-periodic $\tilde{x}(n)$ having frequency ω, so that by comparison with the inverse DTFT we have

$$\tilde{x}(n) = \int_{-\pi}^{\pi} x(n; \omega) \, d\omega \text{ where } x(n;\omega) = \frac{X(e^{j\omega})e^{j\omega n}}{2\pi}. \tag{4.15b}$$

Because of their periodicity, $\tilde{x}_c(t)$ and $\tilde{x}(n)$ in this discussion consist of a (possibly infinite) sum of sinusoids of respective frequencies $\Omega_k = 2\pi k/T = k\Omega_0$, where $\Omega_0 = 2\pi/T$, and $\omega_k = 2\pi k/N = k\omega_0$, where $\omega_0 = 2\pi/N$. Considering one particular frequency, Ω_k or ω_k, we have

$$\text{FT}\{x_c(t; \Omega_k) = X_c(j\Omega_k) \cdot e^{j\Omega_k t}/(2\pi)\} = X_c(j\Omega_k) \cdot \delta_c(\Omega - \Omega_k) \text{ (continuous time)} \tag{4.16a}$$

$$\text{DTFT}\{x(n; \omega_k) = X(e^{j\omega_k}) \cdot e^{j\omega_k n}/(2\pi)\} = X(e^{j\omega_k}) \cdot \delta_c(\omega - \omega_k), \tag{4.16b}$$
$$\omega, \omega_k \in [-\pi, \pi) \text{ (discrete time)}.$$

Note that here the Fourier series–Fourier transform relations (4.13) and (4.14) do not hold because we are including the entire periodic functions in the transforms $X_c(j\Omega)$ and $X(e^{j\omega})$, not just one period. It is the inclusion of all periods that is responsible for $X_c(j\Omega)$ and $X(e^{j\omega})$ blowing up at, respectively, Ω_k and ω_k whereas $X_c(k)$ and $X(k)$ are finite-interval integrals and thus finite-valued.

Recall from the end of Example 4.10 that as a check, it is easy to see that when the inverse transform of the Dirac delta function is taken, the result is exactly the corresponding term in the Fourier series:

---

[17] See Cornelius Lanczos, *Discourse on Fourier Series* (New York: Hafner, 1966), p. 2. Fourier's contribution on the series (in 1807) was the startling claim that an infinite trigonometric series could represent a *discontinuous* function. The claim is true, but was hard to accept because it meant that the (infinite) sum of analytic functions (sinusoids) could represent a nonanalytic function. An interesting comparison between Fourier series and Taylor series is made by Lanczos. The coefficients of the Taylor series are found by differentiation, whereas those of Fourier series are found by integration. Therefore, Taylor's series is "tied down" to local function behavior and of course is defined only where the function to be expanded is analytic (infinitely differentiable). Contrarily, each Fourier coefficient depends on the behavior of the function everywhere on the interval due to integration over the entire interval and is defined for all integrable (not necessarily analytic) functions.

$$\frac{X(j\Omega_k)}{2\pi}\int_{-\infty}^{\infty}\delta_c(\Omega - \Omega_k) \cdot e^{j\Omega t}\, d\Omega = X_c(j\Omega_k) \cdot e^{j\Omega_k t}/(2\pi) = x_c(t; \Omega_k)$$

(continuous time)   (4.17a)

$$\frac{X(e^{j\omega_k})}{2\pi}\int_{-\pi}^{\pi}\delta_c(\omega - \omega_k) \cdot e^{j\omega n}\, d\omega = X(e^{j\omega_k}) \cdot e^{j\omega_k n}/(2\pi) = x(n; \omega_k)$$

(discrete time).   (4.17b)

In conclusion, if $x_c(t)$ or $x(n)$ is periodic, we know from the Fourier series definitions (4.12) that it can be represented as a (possibly infinite) sum of sinusoids. Furthermore, (4.16) and (4.17) show that each of these sinusoids Fourier-transforms into a Dirac delta function. Consequently, although both aperiodic and periodic functions may be represented by Fourier transforms, the Fourier transform of a periodic function is a (possibly infinite) sum of Dirac delta functions at the harmonic frequencies $\Omega_k = 2\pi k/T$. This makes sense because only at the harmonics Ω_k can the signal have energy. An explicit expression for $X_c(j\Omega)$ for periodic $\tilde{x}_c(t)$ in terms of $X_c(k)$ is found in Appendix 4A (formula 20) and is proved in the problems. The corresponding formula for discrete time is proved in Section 4.5 and is found in (4.18f) and Appendix 4A (formula 6).

Although $\text{FT}\{x_c(t; \Omega_k)\}$ is infinite at $\Omega = \Omega_k$ and $\text{DTFT}\{x(n; \omega_k)\}$ is infinite at $\omega = \omega_k$, when these Dirac deltas are integrated in (4.15) the results are finite-valued, as they must be. To verify this, remember that the infinite-valued Dirac deltas are multiplied by, respectively, $d\Omega \to 0$ or $d\omega \to 0$. Nevertheless, the Fourier series is often more convenient in the case of periodic time functions; it is just a sum of sinusoids with finite-amplitude, constant coefficients.

We here note an interesting fact. Notice that $X(e^{j\omega})$ in the DTFT definition (4.6d) is periodic in ω with period 2π because each term in the DTFT sum is 2π-periodic in ω.[18] Notice the similarity between (4.6c) $\left[x(n) = \dfrac{1}{2\pi}\displaystyle\int_{-\pi}^{\pi} X(e^{j\omega}) \cdot e^{j\omega n}\, d\omega \right]$

and (4.12b) $\left[X_c(k) = \dfrac{1}{T}\displaystyle\int_{0}^{T} x_c(t) \cdot e^{-j2\pi kt/T}\, dt \right]$. Let $\tau = -\omega T/(2\pi)$ in (4.6c). Then

$d\omega = -2\pi d\tau/T$, the integration range becomes $\left[-\dfrac{T}{2}, \dfrac{T}{2} \right]$, and (4.6c) becomes $x(n) =$

$\dfrac{1}{T} \cdot \displaystyle\int_{-T/2}^{T/2} X(e^{-j2\pi\tau/T}) \cdot e^{-j2\pi n\tau/T}\, d\tau$. Notice that t in (4.12b) and τ in the above integral

for $x(n)$ are analogous quantities, as are $X_c(k)$ and $x(n)$, and the T-periodic functions $x_c(t)$ and $X(e^{-j2\pi\tau/T})$! The same analogies apply to the other equations of those transform pairs, (4.6d) and (4.12a). The mathematical form of the relations is identical, although the roles of "signal space" and "transform space" are reversed. We have shown that the inverse DTFT, aperiodic $x(n)$, is nothing more than the set of Fourier series coefficients of the 2π-*periodic*-in-ω DTFT, $X(e^{j\omega})$. In fact, any aperiodic sequence may be viewed as the Fourier coefficients of a periodic function of a continuous variable (in the "other" domain).

In summary, an aperiodic time function has a Fourier transform that is a continuous function of frequency. A periodic time function is most conveniently represented by its Fourier series: a sequence of coefficients or, in other words, a discrete spectrum. The same rules hold for the spectrum: If the spectrum is aperiodic, the time function must be continuous, and if it is periodic, the time function must be discrete. Concisely,

---

[18] Despite this fact, for notational brevity and agreement with existing literature, we will not place a tilde over $X(e^{j\omega})$ in this book.

Table 4.2 Periodicity, Continuity, and Transforms

Signal/Transform	Independent Variable Is	
	Continuous	Discrete
Aperiodic	FT/FT^{-1}	DTFT/FS^{-1}
Periodic	FS/DTFT^{-1}	DFT/DFT^{-1} or DFS/DFS^{-1}

Note: FT = Fourier transform, FS = Fourier series, DTFT = discrete-time Fourier transform, DFT = discrete Fourier transform, DFS = discrete Fourier series.

a function discrete in one domain, be it periodic or aperiodic, is periodic in the other, and a function continuous in one domain, be it periodic or aperiodic, is aperiodic in the other. Table 4.2 and Appendix 4A may help remembering the appropriate transforms. The time-domain/frequency-domain signal is specified as to whether it is periodic and/or discrete. The corresponding chart entry shows the appropriate transform to take one to the other domain (frequency domain/time domain).

Another way of looking at these facts concerning Fourier expansions is the following. In all the Fourier expansions [(4.6) and (4.12)], the expansion function is an exponential having an imaginary exponent of magnitude (time) · (frequency). Suppose that the function being transformed is a function of a discrete variable [e.g., $x(n)$ for discrete-time n or $X_c(k)$ for discrete-frequency Ω_k]. Then there exist equally spaced values of the "other" (resulting transform) variable (frequency ω or time t) for which the exponent increment equals $j2\pi\,\ell$, ℓ an integer, *and that spacing is the same no matter what the value of the (discrete) variable n or Ω_k being transformed, that is, that spacing is the same for all terms in the transform sum expression.* After all, for different values of, for example, n, the increment index ℓ simply takes the value $-n$ if ω is incremented by 2π. So, for example, the nth term in the DTFT sum, $x(n)e^{-j\omega n}$, is $2\pi/n$-periodic in ω, and is thus also 2π-periodic in ω; thus *all* DTFT sum terms are 2π-periodic in ω. Therefore, because each term in the transform is periodic with the same period, so is the entire result, the transform. That is, the transform [e.g., 2π-periodic $X(e^{j\omega})$ or T-periodic $\tilde{x}_c(t)$] is periodic in the resulting transform variable if the original function depends on a discrete variable [e.g., $x(n)$ or $X_c(k)$].

Conversely, if the function being transformed is a function of a continuous variable [e.g., $x_c(t)$ or $X(e^{j\omega})$], the sequential values of the resulting transform variable (e.g., Ω or n) for which the exponent equals $j2\pi k$ are equally spaced, *but that spacing is different for different values of the variable being summed over—above, t or ω, that is, that spacing is different for different terms in the transform integral.* Thus, although individual terms in the sum are periodic (above, in Ω or n), the overall transform [e.g., $X_c(j\Omega)$ or $x(n)$] is aperiodic. These ideas are intimately related to the descriptions of complex frequency given earlier.

EXAMPLE 4.13

The Fourier series coefficients $X_c(k)$ are an example of the transform of a T-periodic function of time t, $\tilde{x}_c(t)$. Discuss the reason for the periodicity of $\tilde{x}_c(t)$, given the discrete nature of this transform, $X_c(k)$.

Solution

The inverse Fourier series is

$$\tilde{x}_c(t) = \sum_{k=-\infty}^{\infty} X_c(k)\, e^{j2\pi kt/T}.$$

No matter what integer k is in the sum,

$$e^{j2\pi k(t + T)/T} = e^{j2\pi kt/T} \cdot e^{j2\pi k} = e^{j2\pi kt/T},$$

that is, the exponential is periodic *with the same period* (*here T*) *regardless of the integer value of k.* Therefore, *each* term in the k sum is periodic in t (with the same period T), so the entire sum, $\tilde{x}_c(t)$, is also T-periodic.

Contrast this example with that of the Fourier transform $X_c(j\Omega)$. The inverse Fourier transform is

$$x_c(t) = \frac{1}{2\pi} \int_{-\infty}^{\infty} X_c(j\Omega) e^{j\Omega t} d\Omega.$$

At $\Omega = \Omega_1, e^{j\Omega_1(t + T_1)} = e^{j\Omega_1 t}$ where $T_1 = 2\pi/\Omega_1$. At $\Omega = \Omega_2, e^{j\Omega_2(t + T_2)} = e^{j\Omega_2 t}$ where $T_2 = 2\pi/\Omega_2 \neq T_1$. This latter exponential is still periodic, but it has period T_2 that is different from T_1. With a different period for each "term" in the integral sum, the entire integral will be aperiodic.

4.5   MORE ON THE DFT/DFS

Because of its fundamental importance in DSP, we take here a closer look at the development of the DFT as a discrete-time Fourier series and/or a Fourier analyzer for finite-length sequences. The nomenclature "discrete Fourier transform" is somewhat unfortunate, for from above we know that the DFT is a Fourier *series* expansion—the *DTFT* is the discrete-time analog of the Fourier transform. The intention was evidently to call attention to the fact that *both* time and frequency variables in the DFT are discrete; hence the name "discrete" and not "discrete-time" or "discrete-frequency" Fourier transform. Also, they are sometimes used to estimate continuous Fourier transforms on computers (see Section 6.5), so from this viewpoint the terminology is fine. As we discuss below, sometimes we apply the name DFT instead of DFS when only the index range $[0, N - 1]$ is considered.

Until now, except for pure sinusoids, most of our time sequences have been aperiodic. In the case of sinusoids of frequencies ω_k, the DTFT has a Dirac delta at each principal-band sinusoidal frequency (i.e., at each frequency ω_k on $(-\pi, \pi]$), plus a Dirac delta at each $\omega_k + 2\pi\ell$ for all integers ℓ.[19] In short, for sinusoids, the DTFT spectrum is discrete. Just as is true in continuous time, in discrete time it is again more convenient to work with Fourier series in dealing with periodic time functions than it is to work with the impulsive Fourier transforms.

The big difference now is that the DTFT spectrum is not only discrete for the periodic time function $\tilde{x}(n)$, but it is also periodic. The latter is true because we already have shown that $X(e^{j\omega})$ is 2π-periodic in ω for *any* $x(n)$, due to $x(\text{n})$ being a function of the discrete variable n. Although our results below will apply to finite-length $x(n)$, we will begin by discussing periodic $\tilde{x}(n)$.

For a sequence $\tilde{x}(n)$ to be periodic in n, we mean that $\tilde{x}(n) = \tilde{x}(n + N\ell)$ for any integer ℓ and for N a fixed, finite integer. N is the period of $\tilde{x}(n)$. Thus, for a discrete-time periodic sequence $\tilde{x}(n)$, there is only a finite number (N) of distinct values of $\tilde{x}(n)$: $n \in [0, N - 1]$ is one such set. We then define $x_1(n) = \begin{cases} \tilde{x}(n), & n \in [0, N - 1] \\ 0, & \text{otherwise} \end{cases}$ as one period of $\tilde{x}(n)$, and which has DTFT called $X_1(e^{j\omega})$. An example $\tilde{x}(n)$ and $x_1(n)$ are shown in Figure 4.19, in which $\tilde{x}(n)$ is a 16-periodic asymmetrical triangle wave.

---

[19] In Chapter 6, these other Dirac delta functions will be seen to be the DTFTs of the entire set of corresponding aliased sampled sinusoids.

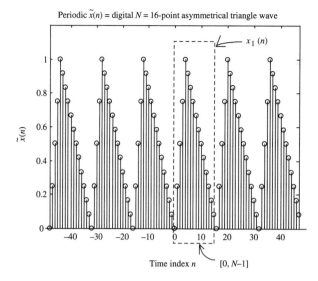

Figure 4.19

We have just said that $X(e^{j\omega}) = \text{DTFT}\{\tilde{x}(n)\}$ will in this case be both periodic and discrete. As a DTFT, it is guaranteed to be 2π-periodic in ω. Let us verify that it is discrete. Decomposing n into $n = m + \ell N$ where $m \in [0, N-1]$, we have

$$X(e^{j\omega}) = \sum_{n=-\infty}^{\infty} \tilde{x}(n)e^{-j\omega n} = \sum_{\ell=-\infty}^{\infty} \sum_{m=0}^{N-1} \tilde{x}(m + \ell N)e^{-j\omega(m + \ell N)}. \qquad (4.18a)$$

Using the periodicity of $\tilde{x}(m)$, $X(e^{j\omega})$ in (4.18a) becomes

$$X(e^{j\omega}) = \sum_{\ell=-\infty}^{\infty} \left\{ \sum_{m=0}^{N-1} \tilde{x}(m)e^{-j\omega m} \right\} e^{-j\omega\, \ell N}. \qquad (4.18b)$$

The sum in $\{\cdot\}$ is just $X_1(e^{j\omega})$, which being independent of ℓ gives

$$X(e^{j\omega}) = X_1(e^{j\omega}) \sum_{\ell=-\infty}^{\infty} e^{-j(\omega N)\ell}. \qquad (4.18c)$$

From (2A.26) $\left[\sum_{n=-\infty}^{\infty} e^{j\omega n} = 2\pi \sum_{\ell=-\infty}^{\infty} \delta_c(\omega + 2\pi\ell) \right]$, we obtain (defining $m = -\ell$)

$$X(e^{j\omega}) = X_1(e^{j\omega})\, 2\pi \sum_{m=-\infty}^{\infty} \delta_c(\omega N - 2\pi m). \qquad (4.18d)$$

Using $\delta_c(at) = \delta_c(t)/a$ for $a > 0$ and $f(\omega)\delta_c(\omega - \omega') = f(\omega')\delta_c(\omega-\omega')$, $X(e^{j\omega})$ in (4.18d) becomes

$$X(e^{j\omega}) = \frac{2\pi}{N} \sum_{m=-\infty}^{\infty} X_1(e^{j2\pi m/N})\delta_c\left(\omega - \frac{2\pi m}{N} \right). \qquad (4.18e)$$

We may decompose the sum in (4.18e) into a double sum by writing $m = k + N\ell$, where $k \in [0, N-1]$. Noting that $\exp\{j2\pi m/N\} = \exp\{j(2\pi k/N + 2\pi\ell)\} = \exp\{j2\pi k/N\}$ [i.e., using the periodicity of $X_1(e^{j\omega})$], (4.18e) becomes

$$X(e^{j\omega}) = \frac{2\pi}{N} \sum_{k=0}^{N-1} X_1(e^{j2\pi k/N}) \sum_{\ell=-\infty}^{\infty} \delta_c\left(\omega - \frac{2\pi k}{N} - 2\pi\ell \right). \qquad (4.18f)$$

For ω restricted to $[0, 2\pi)$,[20] only one δ_c for each k makes a contribution (from either $\ell = 0$ or $\ell = 1$).

---

[20] For matching usual DFT notation, we here use $[0, 2\pi)$ rather than $(-\pi, \pi]$ as the principal frequency interval. This alternative notation arises because, in a computer, indices are not allowed to be negative.

We see that $X(e^{j\omega})$ is nonzero *only* at N distinct frequencies $\omega_k = 2\pi k/N$, $k = [0, N-1]$ for $\omega_k \in [0, 2\pi)$. Substituting the right-hand side of (4.18f) into the definition of the DTFT^{-1} gives

$$\tilde{x}(n) = \frac{1}{2\pi} \int_0^{2\pi} \frac{2\pi}{N} \sum_{k=0}^{N-1} X_1(e^{j2\pi k/N}) \sum_{\ell=-\infty}^{\infty} \delta_c\left(\omega - \frac{2\pi k}{N} - 2\pi\ell\right) e^{j\omega n} \, d\omega \tag{4.19a}$$

$$= \frac{1}{N} \sum_{k=0}^{N-1} X_1(e^{j2\pi k/N}) \int_0^{2\pi} \sum_{\ell=-\infty}^{\infty} \delta_c\left(\omega - \frac{2\pi k}{N} - 2\pi\ell\right) e^{j\omega n} \, d\omega. \tag{4.19b}$$

As noted previously, only for one value of ℓ is the δ_c nonzero for the integration range $\omega \in [0, 2\pi)$: $\ell = 0$ or $\ell = 1$. For either value of ℓ, the Dirac-sampled $e^{j\omega n}$ term reduces to $e^{j2\pi nk/N}$, so (4.19b) reduces to

$$\tilde{x}(n) = \frac{1}{N} \sum_{k=0}^{N-1} X_1(e^{j2\pi k/N}) e^{j2\pi nk/N} \tag{4.19c}$$

$$= \text{DFT}^{-1}\{X_1(k)\} = \text{DFS}^{-1}\{\tilde{X}(k)\}, \tag{4.19d}$$

where $\tilde{X}(k) = X_1(e^{j2\pi k/N})$ for *all* k and $X_1(k) = X_1(e^{j2\pi k/N})$ for only $k \in [0, N-1]$ and zero otherwise. That is, $X_1(k)$ is the $k \in [0, N-1]$ period of $\tilde{X}(k)$.

Finally, by the fact that $X_1(e^{j\omega}) = \text{DTFT}\{x_1(n)\}$ and the DTFT definition,

$$X_1(k) = X_1(e^{j2\pi k/N}) = \sum_{n=0}^{N-1} x_1(n) e^{-j2\pi nk/N} = \text{DFT}\{x_1(n)\} = \text{DFS}\{\tilde{x}(n)\}. \tag{4.20}$$

We see that the DFT is used for referring to N-length aperiodic sequences, and the DFS is used for referring to their N-periodic extensions (i.e., for N-periodic sequences). *However, for n and $k \in [0, N-1]$, all values involved are identically the same.* Because we usually deal with N-length sequences in computer work, we usually use the name "DFT." In practice, for the DFT/DFS we use the symbol $x(n)$ to represent $x_1(n)$ or $\tilde{x}(n)$, and we use $X(k)$ to represent $X_1(k)$ or $\tilde{X}(k)$ for, respectively, the N-length or N-periodic sequences.

Again, exactly the same sums are involved regardless of how we interpret $\tilde{x}(n)$ for $n \in [0, N-1]$ and $\tilde{X}(k)$ for $k \in [0, N-1]$. The only difference is that for the DFS expressions to hold for a finite-length sequence, we must restrict our attention to *only* $[0, N-1]$. That is because the sinusoidal series sequence of the DFT sum is periodic outside $[0, N-1]$ (due to the function $e^{j2\pi nk/N}$), while the N-length sequence is by definition *zero* outside $[0, N-1]$. Again, because only finite sums are involved in both the DFT and DFT^{-1}, we can easily use a computer to do the transforming and inverse transforming exactly, which cannot be said of any of the other transforms we have been considering. We will discuss some properties of the DFT in Chapter 5. Its uses in estimating Fourier transforms, DTFTs, and other applications are discussed in full detail in Chapters 6 and 7.

4.6   PICTORIAL DEMONSTRATION

To help make the above mathematical ideas presented in this chapter easier to understand and remember, consider the following example. Let a continuous-time function $x_c(t)$ be given as

$$x_c(t) = \begin{cases} 2t/T, & t \in [0, T/2] \\ 2(1 - t/T), & t \in (T/2, T) \end{cases} \tag{4.21}$$

and zero otherwise, where in this example T is set to 21 s. This signal is a triangle function of duration T s and is shown in Figure 4.20a (see page 208).

Let $x(n)$ be an analogously defined discrete-time function:

$$x(n) = \begin{cases} 2n/(N-1), & n \in [0, (N-1)/2] \\ 2[1 - n/(N-1)], & n \in [(N-1)/2 + 1, N-1] \end{cases} \tag{4.22}$$

and zero otherwise, for N odd, and where in this example N is set to 21 samples. This sequence is a triangle function of duration N samples and is shown in Figure 4.21a (see page 209).

Complete step-by-step derivations of several results used in this section [e.g., (4.23), (4.24), (4.29), (4.30)] and also the discrete-time triangle and its z-transform for N even are given in Appendix 4B. The Laplace transform of $x_c(t)$ is

$$X_c(s) = \frac{(1 - e^{\frac{-sT}{2}})^2}{(T/2) \cdot s^2}, \quad s \neq 0 \text{ and } X_c(0) = \frac{1}{2}T. \tag{4.23}$$

The magnitude of this function of s rises extremely rapidly in the left-half plane $(\mathrm{Re}\{s\} = \sigma < 0)$, so in the plot of $|X_c(s)|$ in Figure 4.20b we cut off the maximum to 10.5. The center of the plot is $s = 0$, with $|\sigma| \leq 2$ Np/s, and $|\Omega| \leq 2$ rad/s. In addition, $|X_c(s)|$ on the imaginary axis, which is the Fourier transform $X_c(j\Omega)$ of $x_c(t)$, is indicated by the thick curve. Now the reason for choosing $10.5 = \frac{1}{2}T$ is evident from Figure 4.20b; it is the maximum value of $|X_c(j\Omega)|$, namely $X_c(j0)$. As can be seen, $|X_c(s)|$ decreases extremely rapidly in the right-half plane $\sigma > 0$. The dips in $|X_c(s)|$ near the imaginary axis can be seen as minima in the DTFT, and the contour plot below shows the zeros of $X_c(s)$, which by (4.23) are all on the imaginary axis at $\Omega = 4\pi \ell/T$. The large values near $s = 0$ along the imaginary axis cause the DTFT to have a "lowpass" characteristic.

The z-transform of $x(n)$ is

$$X(z) = \frac{\{1 - z^{-(N-1)/2}\}^2}{[(N-1)/2] \cdot z \cdot (1 - z^{-1})^2}, \quad z \neq 0, z \neq 1 \tag{4.24a}$$

and

$$X(1) = \frac{N-1}{2}. \tag{4.24b}$$

The magnitude of $X(z)$ rises extremely rapidly within the unit circle $(|z| < 1)$. Therefore, in the plot of $|X(z)|$ in Figure 4.21b, we cut off the maximum to $X(1e^{j0})$ $= X(1) = \frac{1}{2}(N-1) = 10$. The center of the plot is $z = 0$, and the maximum magnitude of z on any side is 1.5. In addition, $|X(z)|$ on the unit circle, which is the discrete-time Fourier transform of $x(n)$, is indicated by the thick curve. As can be seen, $|X(z)|$ decreases extremely rapidly outside the unit circle $|z| > 1$. The large values near $z = 1$ cause the DTFT to have large magnitude for low frequencies—again, a lowpass characteristic.

Substitution of $s = j\Omega$ into (4.23) gives, after some algebraic steps, the Fourier transform of $x_c(t)$:

$$X_c(j\Omega) = \frac{T}{2} \mathrm{sinc}^2\left\{\frac{\Omega T}{4}\right\} \cdot e^{-j\Omega T/2}, \tag{4.25}$$

where again, $\mathrm{sinc}(x) = \sin(x)/x$. The maximum value of $|X_c(j\Omega)|$, obtainable by L'Hospital's rule, is $T/2 = 21/2 = 10.5$, in agreement with the previous numerical results. The magnitude of $X_c(j\Omega)$ is shown in Figure 4.20c. Figure 4.20c depicts $|\Omega| \leq 2$ rad/s, to match the extent of $\mathrm{Im}\{s\}$ in the plot of $X_c(s)$. Note that the maximum value of $|X_c(j\Omega)|$ is 10.5 and that the shape of the curve agrees with that of the heavy line in Figure 4.20b.

Likewise, substitution of $z = e^{j\omega}$ into (4.24) yields the DTFT of $x(n)$:

$$X(e^{j\omega}) = \frac{[\sin\{\omega(N-1)/4\}]^2}{[(N-1)/2] \cdot \{\sin(\omega/2)\}^2} \cdot e^{-j\omega(N-1)/2}, \tag{4.26}$$

and for $\omega = 0$ we replace (4.26) by (4.24b) (recalling that $e^{j0} = 1$). The magnitude of the DTFT is shown in Figure 4.21c. Again, note the match of this curve to the heavy line in the plot of $|X(z)|$ in Figure 4.21b and the maximum value 10.0. Unlike the continuous-time case, the entire DTFT magnitude plot can be shown, because ω varies over only a finite range, $(-\pi, \pi]$.

Let the periodic extension of $x_c(t)$ with period considered to be T be denoted by $\tilde{x}_{T,c}(t)$, where the superscript ~ again indicates a periodic time function. The Fourier series of $\tilde{x}_{T,c}(t)$ is, by (4.13) $[X_c(k) = (1/T) \cdot X_{1,c}(j\Omega)|_{\Omega=2\pi k/T}]$, just $1/T$ times the Fourier transform of $x_c(t)$ (aperiodic version) evaluated at $\Omega = 2\pi k/T$:

$$X_c(k) = \frac{(T/2) \cdot \{\sin(2\pi k/4)\}^2}{T \cdot \{2\pi k/4\}^2} \cdot e^{-j(2\pi k/T)(T/2)} = \frac{2}{(\pi k)^2} \sin^2(\pi k/2) \cdot e^{-j\pi k}. \tag{4.27}$$

A plot of $|X_c(k)|$ is shown in Figure 4.20d. The maximum value of $|k|$ shown is that for which $k \cdot 2\pi/T = 2.0$ rad/s, to match the range of Ω shown in the Fourier transform plot Figure 4.20c; thus $|k|_{\max} = 13$. Just as $X_c(j\Omega)$ extends to $\Omega = \pm\infty$, there are nonzero $X_c(k)$ coefficients for all k. However, the main energy is contained within ± 2 rad/s. Note also that because $X_c(k) = X_{c,1}(j2\pi k/T)/T$, the general shape follows that of the Fourier transform; however, the shape is harder to identify because there are only $2 \cdot 13 + 1 = 27$ values of $X_c(k)$ on this range, instead of a continuous range of values.

Let the periodic extension of $x(n)$ with period considered to be N be denoted by $\tilde{x}_N(n)$. The discrete Fourier series of $\tilde{x}_N(n)$ is, by (4.14) $[X(k) = X_1(e^{j\omega})|_{\omega=2\pi k/N}]$, just the DTFT of $x(n)$ in (4.26) evaluated at $\omega = 2\pi k/N$:

$$X(k) = \frac{[\sin\{\pi k(N-1)/(2N)\}]^2}{[(N-1)/2] \cdot \{\sin(\pi k/N)\}^2} \cdot e^{-j\pi k(N-1)/N}. \tag{4.28}$$

A plot of $|X(k)|$ is shown in Figure 4.21d. There are only N distinct $X(k)$ because $e^{-j2\pi kn/N}$ is N-periodic in k. Note again that because $X(k) = X_1(e^{j2\pi k/N})$, the general shape follows that of the DTFT; however, the shape is harder to identify because there are only $N = 21$ $X(k)$ on this range, instead of a continuous range of values.

Now suppose that L successive repetitions of the triangle for $t > 0$ form $x_{L,c}(t)$ instead of only one triangle, as shown in Figure 4.20e for $L = 5$. The Laplace transform of $x_{L,c}(t)$ is now

$$X_{L,c}(s) = \frac{1 - e^{-sTL}}{1 - e^{-sT}} \cdot \frac{\{1 - e^{-sT/2}\}^2}{(T/2) \cdot s^2}, \quad s \neq 0; X_{L,c}(0) = \frac{1}{2}LT. \tag{4.29}$$

For the cases of $L = 5$ and $L = 100$, $|X_{L,c}(s)|$ is shown in Figures 4.20f and g, respectively. In this case, the Laplace transform grows so rapidly for $\sigma < 0$ that in the magnitude plots (Figures 4.20f and g), $X_{L,c}(s)$ has artificially been set to zero for $\sigma < 0$. Thus, the left edge of the nonzero portion of the plot shows the Fourier transform magnitude. Similarly, the respective Fourier transform magnitudes are shown in Figures 4.20h and i. The peaks at $\Omega = 0$ are now much sharper than before, with value $L \cdot T/2$, which is 52.5 for $L = 5$ and 1050 for $L = 100$. The case $L = 5$ is intermediate, still showing some detail in between the major peaks. On the other hand, for $L = 100$ the spectrum is extremely sharp, significantly nonzero only at *discrete* values of Ω. It would be called a "line" spectrum.

The reason for this fundamental change in behavior is that with the additional repetitions of the triangle, $x_{L,c}(t)$ is beginning to closely resemble $\tilde{x}_c(t)$, the periodic extension of $x_c(t)$. That is, $x_{L,c}(t)$ is approaching a periodic function. A quick comparison of Figure 4.20i with Figure 4.20d shows the true similarity with the Fourier series of $\tilde{x}_c(t)$. [Note: Some peaks in the plot of $|X_c(s)|$ in Figure 4.20g are missing due to quantization of s by sampling on the s-plane grid.]

Similarly, L successive repetitions of the discrete-time triangle for $n > 0$ form $x_L(n)$ shown in Figure 4.21e for $L = 5$. The z-transform of $x_L(n)$ is

$$X_L(z) = \frac{1 - z^{-NL}}{1 - z^{-N}} \cdot \frac{\{1 - z^{-(N-1)/2}\}^2}{[(N-1)/2] \cdot z \cdot (1 - z^{-1})^2}, z \neq 0, z \neq 1;$$

$$X_L(1) = \tfrac{1}{2}L(N-1). \tag{4.30}$$

For the cases of $L = 5$ and $L = 100$, $|X_L(z)|$ is shown in Figures 4.21f and g, respectively. The z-transform grows so rapidly for $|z| < 1$ that in the magnitude plots $X_L(z)$ has artificially been set to zero for $|z| < 1$. Thus, the inside edge of the plot shows the DTFT magnitude. The respective DTFT magnitudes are shown in Figures 4.21h and i. The peaks at $\omega = 0$ are now much sharper than before, with value $L \cdot (N-1)/2$ which is 50 for $L = 5$ and 1000 for $L = 100$. The case $L = 5$ is intermediate, still showing significant detail in between the major peaks. On the other hand, for $L = 100$ the spectrum is extremely sharp. It too may be roughly called a "line" spectrum. Again the aperiodic $x_L(n)$ approaches the periodic $\tilde{x}(n)$. Comparison of Figure 4.21i with Figure 4.21d shows the true similarity with the Fourier series of $\tilde{x}(n)$.

We may ask what the Fourier series of $x_c(t)$ would look like if instead of assuming the period is T, we assume MT. It will be shown in Section 5.6 that the previous coefficients based on period T remain in place as before, but now have $M - 1$ zeros placed between them (referred to as zero-interlacing). We denote this Fourier series as $X_{MT,c}(k)$. This effect is shown for $M = 5$ in Figure 4.20j, where the heavy line on the bottom is produced by the stems of all the zero coefficients in between the previous, nonzero coefficients. The maximum equivalent $|\Omega|$ shown is again 2 rad/s, but the spacing of the coefficients, including the zero coefficients, is now $2\pi/(MT)$. The amplitudes are scaled up by the factor M (compare with Figure 4.20d).

The effect is the same for the discrete-time case. A plot of $X_M(k)$ [the discrete Fourier series of $x(n)$ based on M periods] is shown in Figure 4.21j for $M = 5$. It is easy to see why the original $X(k)$ are interlaced with zeros rather than some other values: $x_M(n) = x(n)$. All that has changed is the spacing between frequency samples, $2\pi/[\text{assumed period}]$ and the amplitudes are the same as for assuming period T because of the division by the period in the FS definition (compare with Figure 4.21d). The additional samples cannot contribute, for $X(k)$ already completely represents $x(n)$ with only N coefficients. If we look at Figure 4.21j and imagine the limit as $L \to \infty$, it is clear that these intermediate values must be zero for the Fourier series and Fourier transforms to match; the DTFT is zero in between the lines of its spectrum for periodic $\tilde{x}(n)$, and the L-scaled amplitudes go to infinity just as do $\delta_c(\omega - \omega_k)$.

We have shown in Figures 4.20e through i and Figures 4.21e through i how an aperiodic function approaches the behavior of a periodic function when it is successively replicated many times. In such cases, the Fourier transform becomes identical in appearance to the Fourier series. The only difference is that the transform still remains a function of a continuous variable, and so to appear as a function of a discrete variable it has Dirac delta impulses at harmonics of the fundamental frequency $2\pi/[\text{true period}]$.

Can a function be constructed to make the Fourier series look just like the Fourier transform? Yes; all that needs to be done is to make the period of the time function be long enough. A way to do this is to leave one copy of the original triangle as in Figures 4.20*a* and 4.21*a*. Then append this function with zero for the rest of the way out to the time to be considered the new longer "period" (*MT* or *MN*). The resulting time functions are denoted $x_{zp,c}(t)$ and $x_{zp}(n)$ for "zero-padding." Examples for continuous time for $M = 2$ and $M = 10$ appear, respectively, in Figures 4.20*k* and *l*. The analogous discrete-time sequences are shown in Figs. 4.21*k* and *l*. This technique in discrete-time processing is known as zero-padding, and more will be said about it in Section 7.2.

Unlike the period extension in the previous paragraphs, the extension produced by periodic zero-padding is no longer the same function as $\tilde{x}$ because of the periodic intervals of zero. Now the integration/summation range is identical to that for the original series calculations based on period T (continuous time) or N (discrete time), for over the rest of the range we have appended zeros. However, the exponential in the Fourier series expressions (4.12b) and (4.12d) is divided by L, compared with before. The result is merely more finely spaced samples of the Fourier transform of the original aperiodic, T- or N-length time function. The effect is shown for $M = 2$ and $M = 10$ in Figures 4.20*m* and *n* for continuous time and Figures 4.21*m* and *n* for discrete time. The build-up of the Fourier transform from the Fourier series is made very clear when comparing these figures with Figures 4.20*c* and 4.21*c*.

One final pictorial example illustrates the transition from a continuous spectrum to a discrete spectrum. The continuous-time case will be described. In Figure 4.22*a* (see page 222) is shown, for $t \in [0, 10\text{ s}]$, a damped oscillatory exponential $x_c(t) = e^{\alpha t}\cos(\Omega_0 t)u_c(t)$ with $\alpha = -1.0$ Np/s and $\Omega_0 = 1.0$ rad/s. In Figure 4.22*b* is shown $|X_c(s)|$, the magnitude of the Laplace transform of $x_c(t)$, which is given by

$$X_c(s) = \frac{s - \alpha}{(s - \alpha)^2 + \Omega_0^2} \, . \tag{4.31}$$

This Laplace transform is defined only for $\sigma > \alpha$, and has been artificially set to zero for $\sigma < \alpha$ in Figure 4.22*b*. If we look on the imaginary axis we find, as noted earlier, the Fourier transform (here shown only as a white curve, for visibility). Because the peaks near $\sigma = \alpha$ dominate the plot, $X_c(s)$ appears to be zero at $\sigma = 0$. It is not zero, however, just relatively small in magnitude, and $|X_c(j\Omega)|$ is shown in Figure 4.22*c*. Note the broadness and smoothness of the peaks centered on $\pm\Omega_0$. This heavily damped sinusoid is clearly an aperiodic function of time, for as an examination of Figure 4.22*a* indicates, it is essentially zero before the finish of even one period.

If, however, the Laplace transform is just shifted to the right by considering instead $x_c(t)$ as above with $\alpha = -0.01$ Np/s, the situation is very different. Shown in Figure 4.22*d* for $t \in [0, 100\text{ s}]$, this $x_c(t)$ is still strong after 15 cycles. Its Laplace transform, shown in Figure 4.22*e*, is indeed shifted to the right so that now its peaks are very close to the imaginary axis. Therefore, the sharpness and narrowness of the peaks is manifest in the Fourier transform $X_c(j\Omega)$, shown on the $j\Omega$ axis in Figure 4.22*e* and alone in Figure 4.22*f*. The narrowness of the peaks in $X_c(j\Omega)$ compared with those in Figure 4.22*c* for larger $|\alpha|$ is the onset of a *line spectrum*, a discrete spectrum. Because of the light damping, $x_c(t)$ is similar to an undamped sinusoid of frequency Ω_0; that is, a periodic function. A similar example could be constructed for the discrete-time case.

It is hoped that with these pictures in mind, the theory as stated earlier begins to have intuitive meaning. If so, then the appropriate transform to use for any given signal will be known, as well as the relation to the less convenient transform for the case at hand.

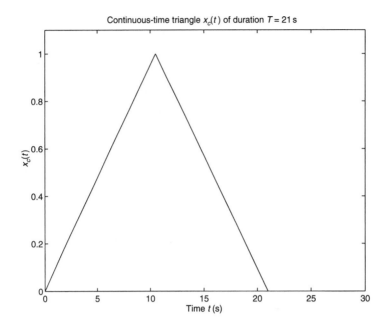

Figure 4.20*a*

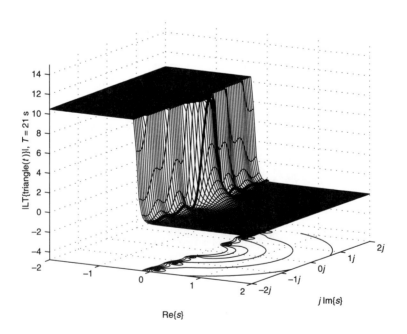

Figure 4.20*b*

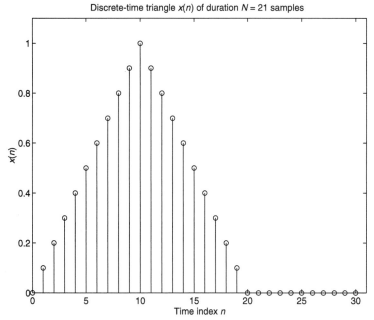

Figure 4.21*a*

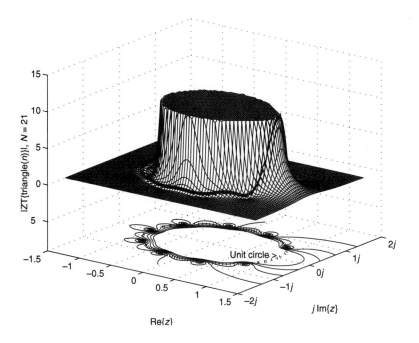

Figure 4.21*b*

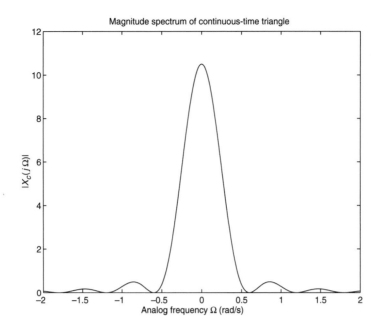

Figure 4.20c

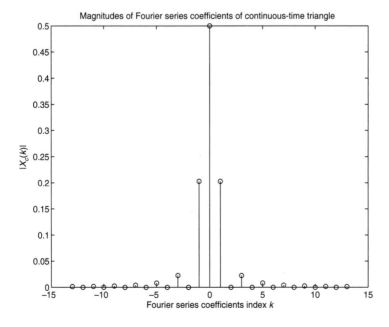

Figure 4.20d

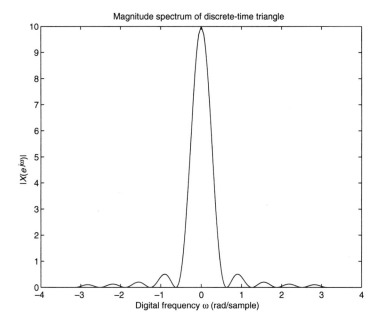

Figure 4.21c

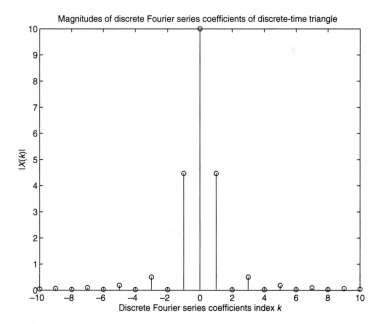

Figure 4.21d

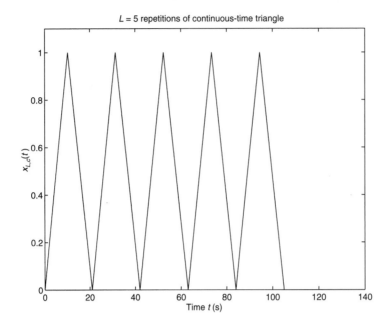

Figure 4.20*e*

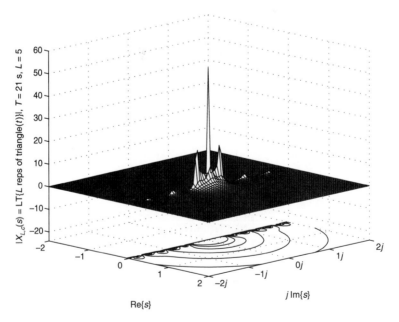

Figure 4.20*f*

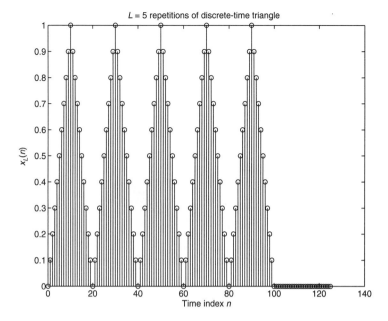

Figure 4.21e

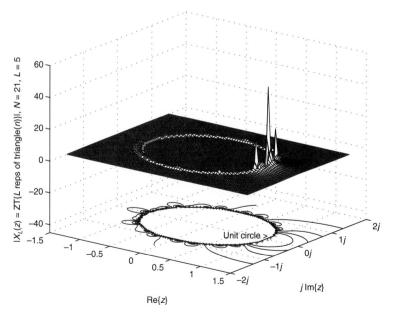

Figure 4.21f

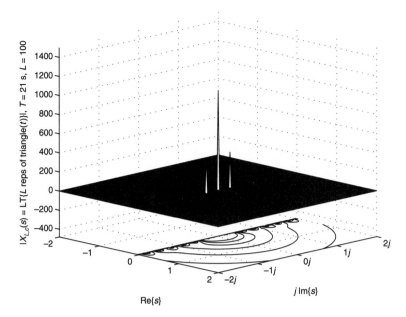

Figure 4.20*g*

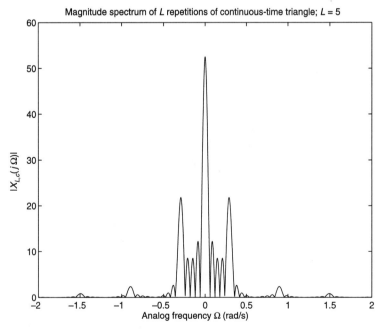

Figure 4.20*h*

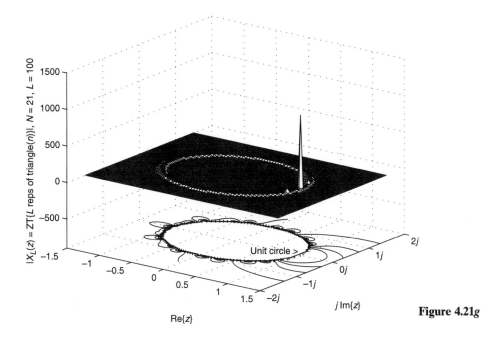

Figure 4.21g

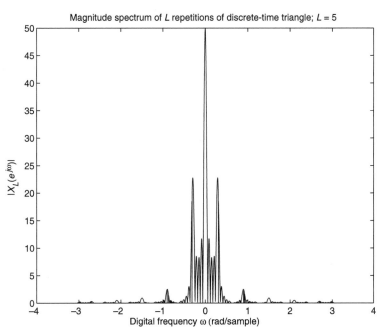

Figure 4.21h

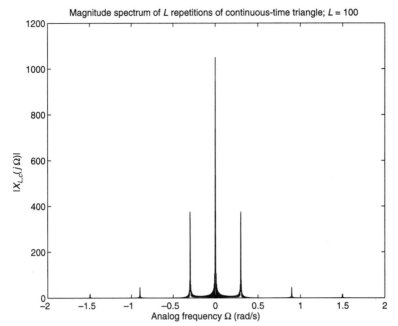

Figure 4.20*i*

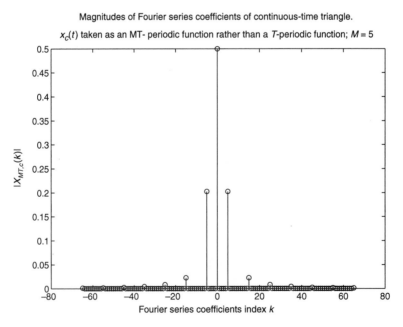

Figure 4.20*j*

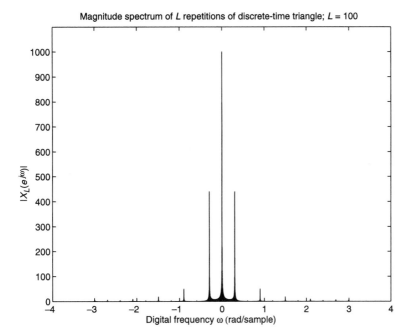

Magnitude spectrum of *L* repetitions of discrete-time triangle; *L* = 100

Figure 4.21*i*

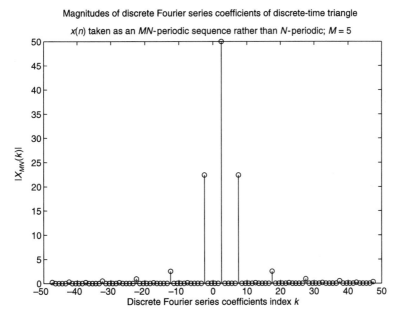

Magnitudes of discrete Fourier series coefficients of discrete-time triangle
x(*n*) taken as an *MN*-periodic sequence rather than *N*-periodic; *M* = 5

Figure 4.21*j*

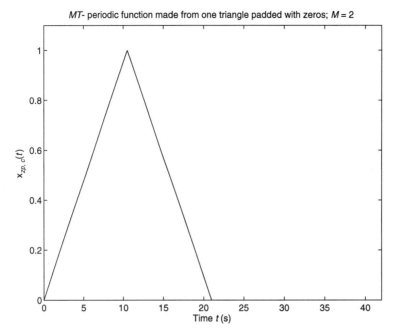

Figure 4.20*k*

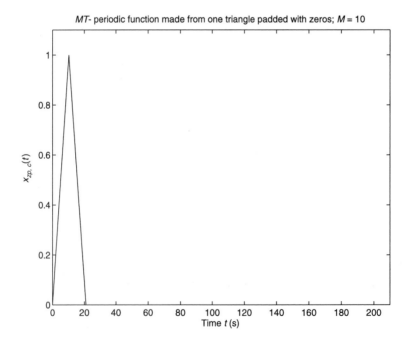

Figure 4.20*l*

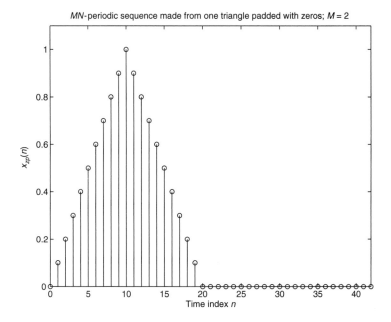

Figure 4.21*k*

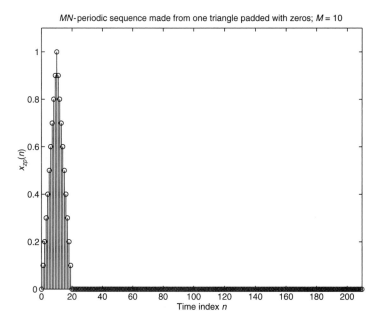

Figure 4.21*l*

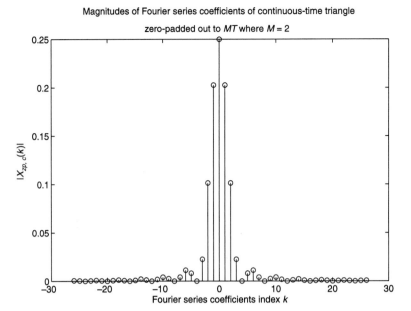

Figure 4.20*m*

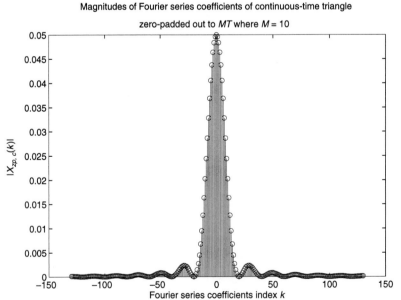

Figure 4.20*n*

Magnitudes of discrete Fourier series coefficients of discrete-time triangle
zero-padded out to $MN - 1$, where $M = 2$.

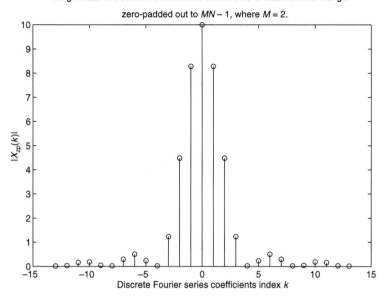

Figure 4.21*m*

Magnitudes of discrete Fourier series coefficients of discrete-time triangle
zero-padded out to $MN - 1$, where $M = 10$

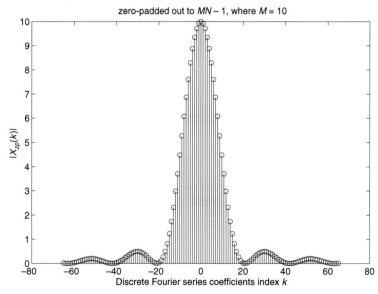

Figure 4.21*n*

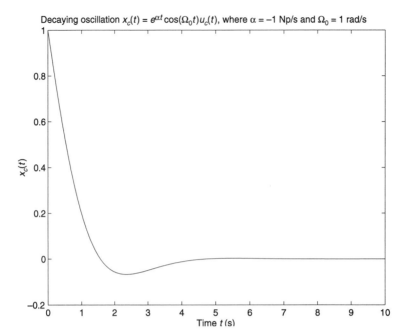

Figure 4.22*a*

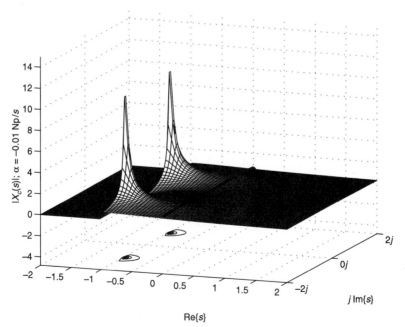

Figure 4.22*b*

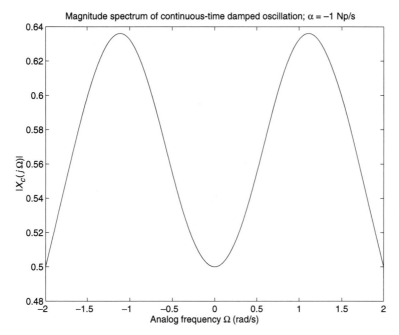

Figure 4.22*c*

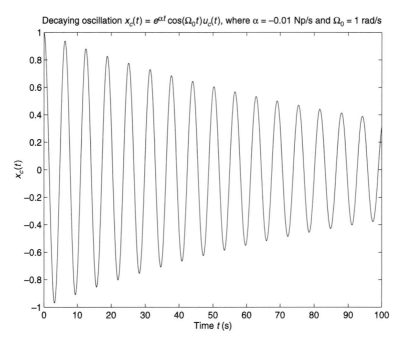

Figure 4.22*d*

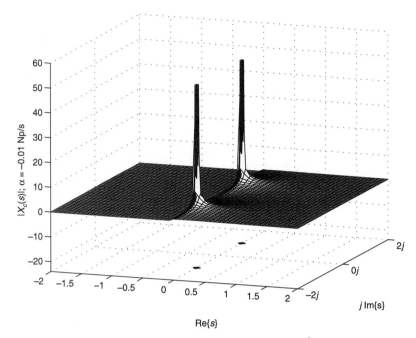

Figure 4.22e

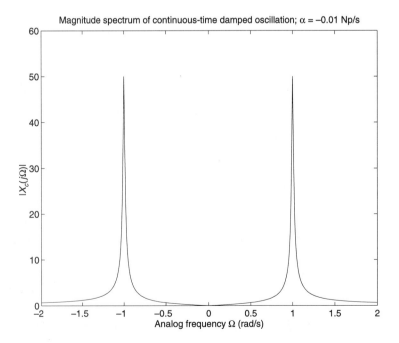

Figure 4.22f

4.7   REVIEW EXAMPLE

EXAMPLE 4.14

To further review some of the basic ideas in this book so far, we consider a very simple system, the RC circuit. Let the applied voltage be $v_{B,c}(t)$ and the voltage across the capacitor $v_{C,c}(t)$ be considered the output voltage. We will solve for $v_{C,c}(t)$ and its discretization for a variety of inputs $v_{B,c}(t)$, using a variety of methods, including conventional differential equation approaches, phasor, Laplace transform, z-transform, Fourier transform, convolution, and frequency response techniques.

Solution

(a)   Classical solution, general input

By Kirchhoff's voltage law, with polarity of $v_{C,c}(t)$ defined to match that of $v_{B,c}(t)$, we have $v_{B,c}(t) = RC dv_{C,c}(t)/dt + v_{C,c}(t)$. Define $\tau = RC$, and for now assume that $v_{B,c}(t)$ is zero for $t < 0$. We now have

$$\frac{dv_{C,c}(t)}{dt} + \frac{v_{C,c}(t)}{\tau} = \frac{v_{B,c}(t)}{\tau}. \tag{4.32}$$

We now consider solutions of this differential equation by several methods. First, for the sake of review, consider the integrating factor approach. Recall that $d/dt\left[\exp\left\{\int dt'/\tau\right\} \cdot v_{C,c}(t)\right] = \exp\left\{\int dt'/\tau\right\} \cdot [dv_{C,c}(t)/dt + v_{C,c}(t)/\tau]$. But $\exp\left\{\int dt'/\tau\right\} = \exp\{t/\tau + K\}$, where K is a constant. Substituting into (4.32),

$$\frac{d\{e^{t/\tau + K} \cdot v_{C,c}(t)\}}{dt} = \frac{e^{t/\tau + K} \cdot v_{B,c}(t)}{\tau}.$$

Dividing both sides by e^K gives

$$\frac{d\{e^{t/\tau} \cdot v_{C,c}(t)\}}{dt} = \frac{e^{t/\tau} \cdot v_{B,c}(t)}{\tau}.$$

We now integrate both sides from $t' = 0$ to $t' = t$:

$$e^{t/\tau} \cdot v_{C,c}(t) - e^0 \cdot v_{C,c}(0) = \frac{1}{\tau} \int_0^t e^{t'/\tau} v_{B,c}(t') \, dt',$$

or

$$v_{C,c}(t) = e^{-t/\tau} v_{C,c}(0) + \frac{1}{\tau} \int_0^t v_{B,c}(t') \, e^{-(t-t')/\tau} dt'. \tag{4.33}$$

We notice that the second term is $v_{B,c}(t) *_c h_c(t)$ where $h_c(t) = e^{-t/\tau} u_c(t)/\tau$. In the usual convolution situation, we let $v_{C,c}(0) = 0$. Although we have obtained convolution using the integrating factor approach, the latter approach works only for very simple differential equations, whereas convolution works for any order.

(b)   Classical solution, boxcar input

Now suppose that $v_{B,c}(t) = A\{u_c(t) - u_c(t - T)\}$, which is a boxcar of height A existing from $t = 0$ to $t = T$. Then, substituting into (4.33),

$$v_{C,c}(t) = e^{-t/\tau} \cdot v_{C,c}(0) + \frac{A}{\tau} \int_0^{\min\{t,T\}} e^{-(t-t')/\tau} dt'$$

$$= e^{-t/\tau} \cdot v_{C,c}(0) + \frac{Ae^{-t/\tau}}{\tau} \int_0^{\min\{t,T\}} e^{t'/\tau} dt'$$

$$= e^{-t/\tau} \cdot v_{C,c}(0) + \frac{Ae^{-t/\tau}}{\tau} \cdot \tau \cdot [e^{\min\{t,T\}/\tau} - 1]$$

$$= \begin{cases} e^{-t/\tau} \cdot v_{C,c}(0) + A\{1 - e^{-t/\tau}\}, & 0 \le t \le T \\ e^{-t/\tau} \cdot [v_{C,c}(0) + A\{e^{T/\tau} - 1\}], & t > T \end{cases}.$$

In the problems, the reader is encouraged to plot this result for typical values of τ, T, and $v_{C,c}(0)$ to gain an intuitive feel for the solution and to compare with approximations of it.

(c) **Laplace transform solution, general input**

Let us now solve this same problem using the unilateral Laplace transform. Taking the LT of both sides of (4.32), $sV_{C,c}(s) - v_{C,c}(0) + V_{C,c}(s)/\tau =$

$V_{B,c}(s)/\tau$, or $V_{C,c}(s) = \dfrac{v_{C,c}(0) + V_{B,c}(s)/\tau}{s + 1/\tau}$ from which we obtain

$$v_{C,c}(t) = e^{-t/\tau} \cdot v_{C,c}(0)u_c(t) + LT^{-1}\left\{\frac{V_{B,c}(s)/\tau}{s + 1/\tau}\right\}. \tag{4.34}$$

To numerically evaluate $v_{C,c}(t)$ in (4.34), we must specify $V_{B,c}(s)$.

(d) **Laplace transform solution, boxcar input**

Now let $v_{B,c}(t) = A\{u_c(t) - u_c(t - T)\}$. Then $V_{B,c}(s) = A[1/s - e^{-sT}/s]$. Then

$$LT^{-1}\left\{\frac{V_{B,c}(s)/\tau}{s + 1/\tau}\right\} = \frac{A}{\tau} LT^{-1}\left\{\frac{1}{s(s + 1/\tau)} - \frac{e^{-sT}}{s(s + 1/\tau)}\right\}.$$

But $1/[s(s + 1/\tau)] = \tau/s - \tau/(s + 1/\tau)$, so

$$LT^{-1}\left\{\frac{V_{B,c}(s)/\tau}{s + 1/\tau}\right\} = A \cdot LT^{-1}\left\{\frac{1}{s} - \frac{1}{s + 1/\tau} - e^{-sT}\left[\frac{1}{s} - \frac{1}{s + 1/\tau}\right]\right\}$$

$$= A\{(1 - e^{-t/\tau})u_c(t) - (1 - e^{-(t-T)/\tau})u_c(t - T)\}.$$

Consequently,

$$v_{C,c}(t) = e^{-t/\tau} \cdot v_{C,c}(0)u_c(t) + A\begin{cases} 1 - e^{-t/\tau}, & 0 \le t \le T \\ 1 - e^{-t/\tau} - (1 - e^{-(t-T)/\tau}) = e^{-t/\tau}(e^{T/\tau} - 1), & t > T \end{cases}$$

$$= \begin{cases} e^{-t/\tau} \cdot v_{C,c}(0) + A(1 - e^{-t/\tau}), & 0 \le t \le T \\ e^{-t/\tau}[v_{C,c}(0) + A(e^{T/\tau} - 1)], & t > T, \end{cases} \tag{4.35}$$

in agreement with the solution using the integrating factor. [We could also have used the Laplace transform convolution theorem on the product in the $LT^{-1}\{\cdot\}$ term in (4.34).]

(e)   **Classical solution, impulse response**

Now, let $v_{B,c}(t) = \delta_c(t)$, and let $v_{C,c}(0) = 0$. Using the integrating factor approach (4.33), $v_{C,c}(t) = \dfrac{1}{\tau}\displaystyle\int_0^t \delta_c(t')\, e^{-(t-t')/\tau}\, dt' = e^{-t/\tau} \cdot \dfrac{u_c(t)}{\tau} = h_c(t)$, where $h_c(t)$ is the impulse response. We may obtain the ZSR for any input (e.g., the boxcar input) by convolving that input with $h_c(t)$.

(f)   **Laplace transform solution, impulse response**

Alternatively, using the Laplace transform approach, $V_{B,c}(s) = 1$, so in (4.34) we now have

$$v_{C,c}(t) = \text{LT}^{-1}\left\{\frac{V_{B,c}(s)/\tau}{s + 1/\tau}\right\} = \text{LT}^{-1}\left\{\frac{1/\tau}{s + 1/\tau}\right\} = e^{-t/\tau} \cdot \frac{u_c(t)}{\tau} = h_c(t)$$

as above.

With an impulse in, the impulse response comes out. The solution above was a specialized result of the solution (4.34) for general inputs [for $V_{B,c}(s) = 1$], but in the general LT approach, we write for this example $sV_{C,c}(s) + (1/\tau)V_{C,c}(s) = V_{B,c}(s)/\tau$, giving the transfer function $H_c(s) \equiv V_{C,c}(s)/V_{B,c}(s) = (1/\tau)/(s + 1/\tau)$, from which again $h_c(t)$ as above results.

(g)   **Classical solution, sinusoidal response**

We next consider a sinusoidal input: $v_{B,c}(t) = A\cos\{\Omega_B t + \phi\}$ for all t. We again assume zero initial condition, for now $v_{B,c}(t)$ has been on "forever" anyway. In practical reality where sinusoids are never on "forever," the usefulness of this solution is that it is identical to that of the initial-value problem after the transients have all died away. Now the integral in (4.33) runs from $-\infty$ to t. The easiest way is to solve for $v_{C,c,+}(t)$, the output for $v_{B,c}(t) = \frac{1}{2}Ae^{j(\Omega_B t + \phi)}$ and then add on to it $v_{C,c,-}(t)$, the output for $v_{B,c}(t) = \frac{1}{2}Ae^{-j(\Omega_B t + \phi)}$. Thus we have in (4.33),

$$v_{C,c,+}(t) = \frac{Ae^{j\phi}}{2\tau}\int_{-\infty}^t e^{j\Omega_B t'} \cdot e^{-(t-t')/\tau}dt' = \frac{Ae^{(-t/\tau + j\phi)}}{2\tau}\int_{-\infty}^t e^{(1/\tau + j\Omega_B)t'}dt'$$

$$= \frac{Ae^{(-t/\tau + j\phi)}}{2\tau} \cdot \frac{e^{(1/\tau + j\Omega_B)t} - e^{-\infty}}{1/\tau + j\Omega_B} = \frac{\frac{1}{2}Ae^{j(\phi + \Omega_B t)}}{1 + j\Omega_B\tau}, \text{ for all } t.$$

We now just express the denominator $1 + j\Omega_B\tau$ in polar form and add on the solution $v_{C,c,-}(t) = v_{C,c,+}{}^*(t)$ to obtain

$$v_{C,c}(t) = 2\text{Re}\{v_{C,c,+}(t)\} = \frac{A\cos\{\Omega_B t + \phi - \tan^{-1}(\Omega_B\tau)\}}{\{1 + (\Omega_B\tau)^2\}^{1/2}}. \tag{4.36}$$

(h)   **Fourier approach, sinusoidal input**

If we now try the Laplace transform approach, we know that the LT exists only on the $j\Omega$ axis, and then only in terms of Dirac delta functions. Thus we instead just use the bilateral Fourier transform in this approach. Taking the Fourier transform of the differential equation (4.32), with $v_{B,c}(t) = A\cos\{\Omega_B t + \phi\}$, we have $j\Omega V_{C,c}(j\Omega) + (1/\tau)V_{C,c}(j\Omega) = (A\pi/\tau)[\delta_c(\Omega - \Omega_B)e^{j\phi} + \delta_c(\Omega + \Omega_B)e^{-j\phi}]$, or

$$V_{C,c}(j\Omega) = \frac{\pi A[\delta_c(\Omega - \Omega_B)e^{j\phi} + \delta_c(\Omega + \Omega_B)e^{-j\phi}]}{1 + j\Omega\tau}.$$

We find $v_{C,c}(t)$ by using FT^{-1}:

$$v_{C,c}(t) = \frac{1}{2\pi} \int_{-\infty}^{\infty} \frac{\pi A[\delta_c(\Omega - \Omega_B)e^{j\phi} + \delta_c(\Omega + \Omega_B)e^{-j\phi}]}{1 + j\Omega\tau} e^{j\Omega t} d\Omega$$

$$= \frac{1}{2}A\left[\frac{e^{j(\Omega_B t + \phi)}}{1 + j\Omega_B\tau} + \frac{e^{-j(\Omega_B t + \phi)}}{1 - j\Omega_B\tau}\right]$$

$$= 2\text{Re}\left\{\frac{1}{2}A\frac{e^{j(\Omega_B t + \phi)}}{1 + j\Omega_B\tau}\right\},$$

which again gives (4.36). Although this Fourier approach to solving a differential equation may seem harder than the phasor approach (reviewed below), remember that unlike the phasor approach it holds for general inputs—as long as we can find the Fourier transform of the input and inverse-Fourier-transform the result.

(i)   Classical undetermined coefficients approach, sinusoidal input

In the method of undetermined coefficients approach, where again here we are solving for only the particular solution, we tentatively write $v_{C,c}(t) = a\cos\{\Omega_B t\} + b\sin\{\Omega_B t\}$, so $dv_{C,c}(t)/dt = -a\Omega_B\sin\{\Omega_B t\} + b\Omega_B\cos\{\Omega_B t\}$. Substituting into (4.32) gives

$$-a\Omega_B\sin\{\Omega_B t\} + b\Omega_B\cos\{\Omega_B t\} + \frac{1}{\tau}[a\cos\{\Omega_B t\} + b\sin\{\Omega_B t\}] = \frac{1}{\tau}A\cos\{\Omega_B t + \phi\}$$

$$= \frac{A}{\tau}[\cos\{\Omega_B t\}\cos\{\phi\} - \sin\{\Omega_B t\}\sin\{\phi\}].$$

Matching coefficients of $\cos\{\Omega_B t\}$ gives $b\Omega_B + a/\tau = (A/\tau)\cos\{\phi\}$, whereas matching those of $\sin\{\Omega_B t\}$ gives $-a\Omega_B + b/\tau = -(A/\tau)\sin\{\phi\}$. Solution of these simultaneous equations gives

$$a = \frac{A}{\tau}\frac{(1/\tau)\cos\{\phi\} + \Omega_B\sin\{\phi\}}{\Omega_B^2 + (1/\tau)^2} \text{ and } b = \frac{A}{\tau}\frac{-(1/\tau)\sin\{\phi\} + \Omega_B\cos\{\phi\}}{\Omega_B^2 + (1/\tau)^2}.$$

Noting that

$$v_{C,c}(t) = a\cos\{\Omega_B t\} + b\sin\{\Omega_B t\} = \{a^2 + b^2\}^{1/2}\cos\{\Omega_B t - \tan^{-1}\{b/a\}\}, \text{ we find}$$

$$\{a^2 + b^2\}^{1/2} = \frac{A}{\tau}\frac{[\cos^2\{\phi\}/\tau^2 + \Omega_B^2\sin^2\{\phi\} + 2(\Omega_B/\tau)\cos\{\phi\}\sin\{\phi\} + (1/\tau^2)\sin^2\{\phi\} + \Omega_B^2\cos^2\{\phi\} - 2(\Omega_B/\tau)\cos\{\phi\}\sin\{\phi\}]^{1/2}}{\Omega_B^2 + (1/\tau)^2}$$

$$= \frac{A}{\tau}\frac{\{(1/\tau^2)\cdot 1 + \Omega_B^2\cdot 1\}^{1/2}}{\Omega_B^2 + (1/\tau)^2}$$

$$= \frac{A}{\tau\{\Omega_B^2 + 1/\tau^2\}^{1/2}} = \frac{A}{\{1 + (\Omega_B\tau)^2\}^{1/2}}.$$

Also,

$$\tan^{-1}\left(\frac{b}{a}\right) = \tan^{-1}\left\{\frac{-(1/\tau)\sin\{\phi\} + \Omega_B\cos\{\phi\}}{(1/\tau)\cos\{\phi\} + \Omega_B\sin\{\phi\}}\right\} = \tan^{-1}\left\{\frac{\Omega_B\tau - \tan\{\phi\}}{1 + \Omega_B\tau\tan\{\phi\}}\right\}.$$

But from trigonometry, $\tan^{-1}\{(X + Y)/(1 - XY)\} = \tan^{-1}\{X\} + \tan^{-1}\{Y\}$, so with $X = \Omega_B\tau$ and $Y = -\tan(\phi)$, $\tan^{-1}(b/a) = \tan^{-1}(\Omega_B\tau) - \phi$. With these substitutions, we again obtain (4.36).

(j)   Phasor approach, sinusoidal input

In the phasor approach, we write the complex voltage divider

$$\frac{\overline{V}_C}{\overline{V}_B} = \frac{1/(j\Omega_B C)}{1/(j\Omega_B C) + R} = \frac{1}{1 + j\Omega_B RC} = \frac{1}{1 + j\Omega_B \tau}.$$

By convention, $\overline{V}_B$ is twice the amplitude of the complex sinusoidal component $\exp\{j\Omega_B t\}$, so the input phasor voltage is $\overline{V}_B = A \exp\{j\phi\}$ [so $v_{B,c}(t) = \text{Re}\{\overline{V}_B \exp\{j\Omega_B t\}\}$], and thus $\overline{V}_C = A \exp\{j\phi\}/(1 + j\Omega_B \tau)$. Consequently, $v_{C,c}(t) = \text{Re}\{\overline{V}_C \cdot \exp\{j\Omega_B t\}\}$, again giving (4.36). The brevity and simplicity of this solution (compare with undetermined coefficients) is a testimony to its popularity.

(k)   Frequency response approach, sinusoidal input

In the frequency response approach, we compute

$$H_c(j\Omega) = \text{FT}\{h_c(t)\} = \text{FT}\left\{e^{-t/\tau} \cdot \frac{u_c(t)}{\tau}\right\} = \frac{1/\tau}{j\Omega + 1/\tau} = \frac{1}{1 + j\Omega\tau}.$$

We have shown [see paragraph following (4.10)] that with $v_{B,c}(t) = A \cos\{\Omega_B t + \phi\}$, the output is $v_{C,c}(t) = |H_c(j\Omega_B)| A \cos\{\Omega_B t + \phi + \angle H_c(j\Omega_B)\}$. Again, when we convert $H_c(j\Omega)$ evaluated at $\Omega = \Omega_B$ into polar form—that is, $H_c(j\Omega_B) = (1 + [\Omega_B \tau]^2)^{-1/2} \angle -\tan^{-1}(\Omega_B \tau)$—and substitute it into $v_{C,c}(t)$ above, we obtain (4.36). Like the phasor approach, use of the frequency response for sinusoidal inputs is quick and convenient.

(l)   Discretization, boxcar input; z-transform approach

Now we consider the basic backward-difference discretization as in Example 2.12, but unlike in Chapter 2, now we may also directly apply the z-transform and discrete-time frequency response analysis. Let $v_B(n) = v_{B,c}(n\Delta t)$ and $v_C(n) \approx v_{C,c}(n\Delta t)$ (again, we use $\approx$ because of the error in discretization; the C on $v_C(n)$ refers to "capacitor," not "continuous-time"). Replacing the derivative in (4.32) by the backward difference, we have

$$\frac{v_C(n) - v_C(n-1)}{\Delta t} + \frac{v_C(n)}{\tau} = \frac{v_B(n)}{\tau},$$

or

$$v_C(n) = \frac{\tau}{\tau + \Delta t} v_C(n-1) + \frac{\Delta t}{\tau + \Delta t} v_B(n) = a_1 v_C(n-1) + b_0 v_B(n), \quad (4.37)$$

where $a_1 = 1/(1 + \Delta t/\tau)$ and $b_0 = (\Delta t/\tau)/(1 + \Delta t/\tau)$. As earlier in this example, let $v_{B,c}(t)$ be the boxcar $v_{B,c}(t) = A[u_c(t) - u_c(t - T)]$. Let N and Δt be chosen so that $(N-1)\Delta t < T$ but $N\Delta t > T$; thus, $v_B(n) = A[u(n) - u(n - N)]$. For simplicity, let $v_C(-1) = 0$. This assumption will correspond to [assuming perfect agreement between $v_{C,c}(t)$ at $t = 0$ and $v_C(n)$ at $n = 0$, and using (4.37)] $v_{C,c}(0) = v_C(0) = b_0 v_{B,c}(0) = Ab_0$.

We now solve for $v_C(n)$ by discrete-time convolution. First we need the unit sample response $h(n)$. Take the bilateral ZT of (4.37): $V_C(z) - a_1 z^{-1} V_C(z) = b_0 V_B(z)$, so the discrete-time transfer function is $H(z) = V_C(z)/V_B(z) = b_0/(1 - a_1 z^{-1})$, which directly gives

$$h(n) = b_0 a_1^n u(n) = \frac{\Delta t}{\tau} \frac{1}{1 + \Delta t/\tau} \left\{\frac{1}{1 + \Delta t/\tau}\right\}^n u(n) = \frac{\Delta t}{\tau}\left(1 + \frac{\Delta t}{\tau}\right)^{-(n+1)} u(n).$$

Incidentally, notice that $H(0) = 0$ and $H(a_1) = \infty$, so $z = 0$ is a zero and $z = a_1 = 1/(1 + \Delta t/\tau)$ is a pole of $H(z)$. This makes sense, for in the discretization we replaced a time derivative (s) by a backward difference $(1 - z^{-1})/\Delta t$, so $s_\ell = (1 - z_\ell^{-1})/\Delta t$ or $z_\ell = 1/(1 - s_\ell \Delta t)$. Thus with $H_c(s_1 = -\infty) = (1/\tau)/(-\infty + 1/\tau) = 0$, we obtain $z_1 = 1/(1 + \infty) = 0$, and with $H_c(s_2 = -1/\tau) = \infty$, we obtain $z_2 = 1/(1 + \Delta t/\tau) = a_1$. Now,

$$v_C(n) = v_B(n)*h(n) = A\frac{\Delta t}{\tau}\sum_{m=0}^{N-1}\left(1 + \frac{\Delta t}{\tau}\right)^{-(n-m+1)}u(n - m).$$

The step function is nonzero only for $m \leq n$, so we obtain

$$v_C(n) = A\frac{\Delta t}{\tau}\left(1 + \frac{\Delta t}{\tau}\right)^{-(n+1)}\sum_{m=0}^{\min\{n,N-1\}}\left(1 + \frac{\Delta t}{\tau}\right)^{m}$$

$$= A\frac{\Delta t}{\tau}\left(1 + \frac{\Delta t}{\tau}\right)^{-(n+1)}\frac{1 - (1 + \Delta t/\tau)^{\min\{n+1,N\}}}{1 - (1 + \Delta t/\tau)}$$

$$= A\left(1 + \frac{\Delta t}{\tau}\right)^{-(n+1)}\left[\left(1 + \frac{\Delta t}{\tau}\right)^{\min\{n+1,N\}} - 1\right]$$

$$= A\begin{cases}1 - (1 + \Delta t/\tau)^{-(n+1)}, & 0 \leq n \leq N - 1 \\ (1 + \Delta t/\tau)^{-(n+1)}[(1 + \Delta t/\tau)^{N} - 1], & n \geq N\end{cases}. \qquad (4.38)$$

If we evaluate (4.35) at $t = n\,\Delta t$ and let $v_{C,c}(0) = Ab_0 = A(\Delta t/\tau)/(1 + \Delta t/\tau)$, we obtain

$$v_{C,c}(n\,\Delta t) = \begin{cases}e^{-t/\tau}\cdot A(\Delta t/\tau)/(1 + \Delta t/\tau) + A(1 - e^{-t/\tau}), & 0 \leq t \leq T \\ e^{-t/\tau}[A(\Delta t/\tau)/(1 + \Delta t/\tau) + A(e^{T/\tau} - 1)], & t > T\end{cases}\Bigg|_{t=n\,\Delta t}$$

$$= A\begin{cases}1 - e^{-t/\tau}/(1 + \Delta t/\tau), & 0 \leq t \leq T \\ e^{-t/\tau}\cdot[e^{T/\tau} - 1/(1 + \Delta t/\tau)], & t > T\end{cases}\Bigg|_{t=n\,\Delta t}$$

or

$$v_{C,c}(n\,\Delta t) = A\begin{cases}1 - e^{-n\,\Delta t/\tau}/(1 + \Delta t/\tau), & 0 \leq n\,\Delta t \leq T \\ e^{-n\,\Delta t/\tau}\cdot[e^{T/\tau} - 1/(1 + \Delta t/\tau)], & n\,\Delta t > T.\end{cases} \qquad (4.39)$$

Not only is there a formal similarity between (4.39) and (4.38), but numerically they are very close for small Δt. For example, Figure 4.23a compares $v_{C,c}(t)$ (solid) with $v_C(n)$ (the latter plotted against $n\,\Delta t$) for moderate values of T, τ, and Δt. For large Δt, the samples $v_C(n)$ will not be close to $v_{C,c}(n\,\Delta t)$.

(m)   Discretization, sinusoidal input; Fourier approach

Finally, again let $v_{B,c}(t) = A\cos\{\Omega_B t + \phi\}$ and $v_B(n) = v_{B,c}(n\,\Delta t) = A\cos(\omega_B n + \phi)$, where $\omega_B = \Omega_B\Delta t$. Using the discrete-time frequency response approach, we have $H(e^{j\omega_B}) = H(z)\big|_{z=e^{j\omega_B}} = \dfrac{b_0}{1 - a_1 e^{-j\omega_B}}$. Thus, from (4.10),

$$v_C(n) = |H(e^{j\omega_B})|A\cos\{\omega_B n + \phi + \angle H(e^{j\omega_B})\}$$

$$= \frac{Ab_0}{\{1 - a_1 e^{-j\omega_B} - a_1 e^{j\omega_B} + a_1^2\}^{1/2}}\cos\left\{\omega_B n + \phi - \tan^{-1}\left\{\frac{a_1\sin(\omega_B)}{1 - a_1\cos(\omega_B)}\right\}\right\}$$

$$= \frac{Ab_0}{\{1 + a_1^2 - 2a_1\cos(\omega_B)\}^{1/2}}\cos\left\{\omega_B n + \phi - \tan^{-1}\left\{\frac{a_1\sin(\omega_B)}{1 - a_1\cos(\omega_B)}\right\}\right\}$$

$$= \frac{A(\Delta t/\tau)/(1 + \Delta t/\tau)}{\{1 + [1/(1 + \Delta t/\tau)]^2 - [2/(1 + \Delta t/\tau)]\cos(\Omega_B \Delta t)\}^{1/2}}$$

$$\cdot \cos\left\{\Omega_B n \Delta t + \phi - \tan^{-1}\left[\frac{\sin(\Omega_B \Delta t)}{1 + \Delta t/\tau - \cos(\Omega_B \Delta t)}\right]\right\}.$$

For continuous time, we had [(4.36) with $t = n\Delta t$]:

$$v_{C,c}(n\Delta t) = \frac{A}{\{1 + (\Omega_B \Delta t)^2\}^{1/2}} \cos\{\Omega_B n \Delta t + \phi - \tan^{-1}(\Omega_B \tau)\}.$$

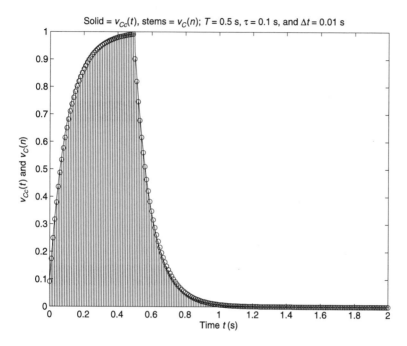

Figure 4.23a

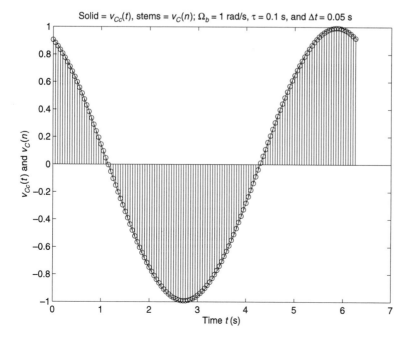

Figure 4.23b

Again, for small Δt, $v_C(n) \approx v_{C,c}(n\,\Delta t)$, as illustrated for typical parameter values in Figure 4.23b.

When one is comfortable with all the solution methods in this extended example, particularly with the transform and discretization approaches, a proper context of many signal processing algorithms and approaches may be discerned. This context as well as the techniques themselves serve as a foundation for understanding the more advanced DSP topics we discuss in later chapters.

On WP 4.2, we introduce a generalized "Fourier series" expansion of a function in terms of a sum of signal inner products and we introduce the concept of completeness. Both of these are very helpful for finding other expansions as well as the Fourier ones. Two practical non-Fourier examples given there are the Legendre polynomial and Haar wavelet expansions.

We conclude this chapter by recalling that its main purpose has been to introduce six of the most common transforms used in signal processing, both continuous and discrete time: GPEs, Fourier transforms, and Fourier series. Although the emphasis in this text will be on discrete-time signals, very frequently discrete-time signal-processing is used to simulate or replace continuous-time processing. Alternatively, the sampling of signals is performed to enable the efficient storage, transmission, or computer processing of information contained in continuous-time signals. In all cases, familiarity with the appropriate transforms for each signal type facilitates understanding the implications of performing an operation on a given signal— for both its continuous- and discrete-time formats. Finally, transforms are an indispensable tool for designing most varieties of signal-processing systems.

APPENDIX 4A: TRANSFORM RELATIONS

Short table for most appropriate uses of transforms:

Name	GPE definition/ Fourier relation to GPE	Time function continuous (C) or discrete (D)?	Time function periodic?	Transform continuous (C) or discrete (D)?	Transform periodic?
LT	$X_c(s) = \int_{-\infty}^{\infty} x_c(t) \cdot e^{-st}\, dt$	C	Y, N	C, D	N
FT	$X_c(j\Omega) = X_c(s)\big|_{s=j\Omega}$	C	Y, N	C, D	N
FS	$X_c(k) = X_c(s)\big|_{s=jk\Omega_0,\ \Omega_0 = 2\pi/T}$[a]	C	in t; per. T	D	N
ZT	$X(z) = \displaystyle\sum_{n=-\infty}^{\infty} x(n) \cdot z^{-n}$	D	Y, N	C, D	in $\angle z$; per. 2π
DTFT	$X(e^{j\omega}) = X(z)\big|_{z=e^{j\omega}}$	D	Y, N	C, D	in ω; per. 2π
DFT/DFS	$X(k) = X(z)\big|_{z=e^{j\omega_0 k},\ \omega_0=2\pi/N}$[b]	D	in n; per. N	D	in k; per. N

[a]Meaning: The Fourier series of a T-periodic function of t is equal to samples of the Laplace transform (and Fourier transform) of the *aperiodic* function of t made up of a *single period* of the given T-periodic function.

[b]Meaning: The Fourier series of an N-periodic function of n is equal to samples of the z-transform (and DTFT) of the *aperiodic* function of n made up of a *single period* of the given N-periodic function.

Extensive relationship diagram, Figure 4A.1: Relationships between all the transforms are given for general and special conditions. Some of these relations will be derived in later chapters or in the problems, and are listed here for reference. The number in Figure 4A.1 tangent to an arrow refers to the correspondingly numbered relation below. The arrow indicates the direction of operation [e.g., the equation for $x(n)$ given $X(z)$ is formula (2), but that for $X(z)$ given $x(n)$ is formula (1)].

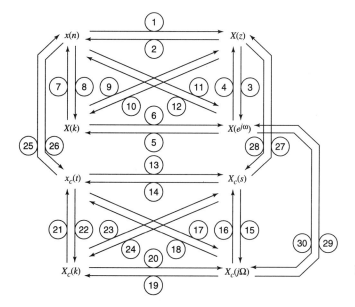

Figure 4A.1

(1) $X(z) = \sum\limits_{n=-\infty}^{\infty} x(n) \cdot z^{-n}.$ (1) and (2) are the z-transform pair.

(2) $x(n) = \dfrac{1}{2\pi j} \cdot \oint_C X(z) \cdot z^{n-1}\, dz.$

(3) $X(e^{j\omega}) = X(z)\big|_{z=e^{j\omega}}.$

(4) $X(z) = \dfrac{1}{2\pi} \cdot \displaystyle\int_{-\pi}^{\pi} \dfrac{X(e^{j\omega})}{1 - e^{-j\omega} \cdot z}\, d\omega$ (By Cauchy integral formula, valid within unit circle if $X(z)$ analytic there.)

or, when $X(z)$ has poles within the unit circle,

$X(z) = \dfrac{1}{2\pi} \cdot \displaystyle\int_{-\pi}^{\pi} \dfrac{X(e^{j\omega})}{1 - e^{j\omega}/z}\, d\omega.$ (By Cauchy integral formula, valid outside unit circle if $X(z)$ analytic there.)

We can also always write

$X(z) = \dfrac{1}{2\pi} \displaystyle\sum_{n=-\infty}^{\infty} \int_{-\pi}^{\pi} X(e^{j\omega})\, e^{j\omega n}\, d\omega \cdot z^{-n} = \mathrm{ZT}\{\mathrm{DTFT}^{-1}[X(e^{j\omega})]\}.$

Note that this formula reduces to the two preceding it for, respectively, anti-causal and causal sequences $x(n)$, upon evaluation of the geometric sum.

(5) $X(k) = X(e^{j\omega})\big|_{\omega=2\pi k/N}.$

assuming $x(n) \neq 0$ only for $n \in [0, N-1]$

(6) $X(e^{j\omega}) = \displaystyle\sum_{k=0}^{N-1} X(k) \cdot \phi\!\left(\omega - \dfrac{2\pi k}{N}\right),$ where $\phi(\omega) = \begin{cases} e^{-j\omega(N-1)/2} \cdot \dfrac{\sin(N\omega/2)}{N\sin(\omega/2)} \\[2ex] \dfrac{2\pi}{N} \cdot \displaystyle\sum_{m=-\infty}^{\infty} \delta_c(\omega - 2\pi m) \end{cases}.$

assuming $x(n)$ is N-periodic

(7) $x(n) = \dfrac{1}{N} \cdot \displaystyle\sum_{k=0}^{N-1} X(k) \cdot e^{j2\pi k n/N},$ valid only if $x(n)$ is N long or N-periodic in n (otherwise impossible).

(7) and (8a) are the DFT_N pair.

(8) $X(k) = \begin{cases} \displaystyle\sum_{n=0}^{N-1} x(n) \cdot e^{-j2\pi k n/N}, & x(n)\ N \text{ long or } N\text{-periodic} \\[2ex] \displaystyle\sum_{n=0}^{N-1} \sum_{m=-\infty}^{\infty} x(n - mN)\, e^{-j2\pi k n/N}, & \text{otherwise.} \end{cases}$

(9) $X(z) = \dfrac{1 - z^{-N}}{N} \cdot \displaystyle\sum_{k=0}^{N-1} \dfrac{X(k)}{1 - e^{j2\pi k/N}/z}.$

(10) $X(k) = X(z)\big|_{z=e^{j2\pi k/N}}.$

(11) $X(e^{j\omega}) = \displaystyle\sum_{n=-\infty}^{\infty} x(n) \cdot e^{-j\omega n}.$ (11) and (12) are the DTFT pair.

(12) $x(n) = \dfrac{1}{2\pi} \cdot \displaystyle\int_{-\pi}^{\pi} X(e^{j\omega}) \cdot e^{j\omega n}\, d\omega.$

(13) $X_c(s) = \displaystyle\int_{-\infty}^{\infty} x_c(t) \cdot e^{-st}\, dt.$ (13) and (14) are the LT pair.

(14) $x_c(t) = \dfrac{1}{2\pi j} \cdot \displaystyle\int_C X_c(s) \cdot e^{st}\, ds.$

(15) $X_c(j\Omega) = X_c(s)\big|_{s=j\Omega}$.

(16) $X_c(s) = \dfrac{1}{2\pi} \cdot \displaystyle\int_{-\infty}^{\infty} \int_{-\infty}^{\infty} X_c(j\Omega) \cdot e^{(j\Omega - s)t} \, d\Omega \, dt = \dfrac{1}{2\pi} \cdot \int_{-\infty}^{\infty} \dfrac{X_c(j\Omega)}{j\Omega - s} \cdot d\Omega$

$\qquad = \text{LT}\{\text{FT}^{-1}[X_c(j\Omega)]\}.$ By the Cauchy integral formula, valid for
$\qquad\qquad\qquad\qquad\qquad\qquad\qquad\qquad\qquad$ $\text{Re}(s) = \sigma > 0$, if $X_c(s)$ is analytic there.

(17) $X_c(j\Omega) = \displaystyle\int_{-\infty}^{\infty} x_c(t) \cdot e^{-j\Omega t} \, dt.$

$\qquad\qquad\qquad\qquad\qquad\qquad\qquad\qquad$ (17) and (18) are the FT pair.

(18) $x_c(t) = \dfrac{1}{2\pi} \displaystyle\int_{-\infty}^{\infty} X_c(j\Omega) \cdot e^{j\Omega t} \, d\Omega.$

Formulas (19) through (24) assume that unless otherwise noted, $\tilde{x}_c(t + T) = \tilde{x}_c(t)$.
Let $\Omega_0 = 2\pi/T$.

(19) $X_c(k) = X_c(j\Omega)\big|_{\Omega=k\Omega_0}$ where $X_c(j\Omega) = \text{FT}\left\{\begin{cases} \tilde{x}_c(t), & t \in [0, T] \\ 0, & \text{otherwise} \end{cases}\right\}.$

(20) $X_c(j\Omega) = \displaystyle\sum_{k=-\infty}^{\infty} X_c(k)\cdot\phi(\Omega - k\Omega_0)$ where

$\qquad\qquad\qquad \begin{cases} \phi(\Omega) = 2\pi \cdot \delta_c(\Omega), & \tilde{x}_c(t + T) = \tilde{x}_c(t) \\ \phi(\Omega) = T\left\{\dfrac{\sin(\Omega T/2)}{\Omega T/2}\right\}e^{-j\Omega T/2} \end{cases},$

$\qquad\qquad\qquad\qquad\qquad$ where this second formula for $\phi(\Omega)$ holds
$\qquad\qquad\qquad\qquad\qquad$ only if $x_c(t)$ is nonzero only for $t \in [0, T]$,
$\qquad\qquad\qquad\qquad\qquad$ and we merely use (4.12b) to define $X_c(k)$.

(21) $\tilde{x}_c(t) = \displaystyle\sum_{k=-\infty}^{\infty} X_c(k) \cdot e^{jk\Omega_0 t}.$

$\qquad\qquad\qquad\qquad\qquad\qquad$ (21) and (22) are the FS pair.

(22) $X_c(k) = \dfrac{\Omega_0}{2\pi} \cdot \displaystyle\int_{0}^{2\pi/\Omega_0} \tilde{x}_c(t) \cdot e^{-jk\Omega_0 t} \, dt.$

(23) $X_c(s) = \displaystyle\int_{-\infty}^{\infty} \sum_{k=-\infty}^{\infty} X_c(k) \cdot e^{(jk\Omega_0 - s)t} \, dt.$

(24) $X_c(k) = X_c(s)\big|_{s=jk\Omega_0}$ where $X_c(s) = \text{LT}\left\{\begin{cases} \tilde{x}_c(t), & t \in [0, T] \\ 0, & \text{otherwise} \end{cases}\right\}.$

Formulas (25) through (30) assume that $x(n) = x_c(n\Delta t)$.

(25) $x(n) = x_c(n\Delta t).$

(26) $x_c(t) = \displaystyle\sum_{n=-\infty}^{\infty} x(n) \cdot \phi(t - n\Delta t)$ where $\phi(t) = \dfrac{\sin\{\pi t/\Delta t\}}{\pi t/\Delta t}$, if $\Delta t \leq \pi/\Omega_{max}.$

(27) $X(z) = \dfrac{1}{\Delta t} \cdot \displaystyle\sum_{n=-\infty}^{\infty} X_c\left(\dfrac{\ln|z| + j[\text{Arg}(z) + 2\pi n]}{\Delta t}\right) = \dfrac{1}{\Delta t} X_c(s)\big|_{s=\ln(z)/\Delta t}$

$\qquad\qquad\qquad\qquad\qquad\qquad\qquad\qquad$ only if no aliasing for the
$\qquad\qquad\qquad\qquad\qquad\qquad\qquad\qquad$ magnitude of z specified.

(28) $X_c(s) = \Delta t \cdot X(z)\big|_{z=e^{s\Delta t}}$ for $|\Omega| = |\text{Im}(s)| < \pi/\Delta t$ only if no aliasing
$\qquad\qquad\qquad\qquad\qquad\qquad\qquad$ for the magnitude of z specified; otherwise,
$\qquad\qquad\qquad\qquad\qquad\qquad\qquad$ only very indirect relationship.

(29) $X(e^{j\omega}) = \dfrac{1}{\Delta t} \displaystyle\sum_{n=-\infty}^{\infty} X_c\left(j\dfrac{\omega + 2\pi n}{\Delta t}\right) = \dfrac{1}{\Delta t} \cdot X_c(j\omega/\Delta t), \ |\omega| \leq \pi.$

$\qquad\qquad\qquad\qquad\qquad\qquad$ only if $\Delta t \leq \pi/\Omega_{max}$

(30) $X_c(j\Omega) = \Delta t \cdot X(e^{j\Omega\,\Delta t}), |\Omega| \leq \pi/\Delta t$ only if $\Delta t \leq \pi/\Omega_{max}$;   otherwise impossible.

APPENDIX 4B: DERIVATION DETAILS FOR EXAMPLES IN SECTION 4.6

Laplace and z-Transforms of Triangles, and of Repeated Time Functions

(a) Show that the Laplace transform of

$$x_c(t) = \begin{cases} 2t/T, & t \in [0, T/2] \\ 2(1 - t/T), & t \in (T/2, T) \end{cases} \tag{4B.1}$$

as given in (4.21) equals the result in (4.23),

$$X_c(s) = \frac{\{1 - e^{-sT/2}\}^2}{(T/2) \cdot s^2}, \quad s \neq 0, \tag{4B.2}$$

which by L'Hospital's rule (twice) at $s = 0$ gives $X_c(0) = T/2$. Of use in this example is:

$$-\frac{d}{ds} X_c(s) = -\frac{d}{ds} \int_{-\infty}^{\infty} x_c(t) \cdot e^{-st} \, dt = \int_{-\infty}^{\infty} t \cdot x_c(t) \cdot e^{-st} \, dt \tag{4B.3}$$

so that

$$LT\{t \cdot x_c(t)\} = -\frac{d}{ds}\{X_c(s)\}, \tag{4B.4}$$

where $X_c(s)$ is the Laplace transform of $x_c(t)$. To avoid factors of $\frac{1}{2}$, let $a = T/2$ so that the triangle has width $2a$. For added generality, let the maximum value of the triangle (occurring at $t = a$) be A. Then $x_c(t)$ may be written either as

$$x_c(t) = A \cdot \begin{cases} 1 - |t - a|/a, & t \in [0, 2a] \\ 0, & \text{otherwise} \end{cases} \tag{4B.5a}$$

or

$$x_c(t) = A \begin{cases} t/a, & t \in [0, a] \\ 2 - t/a, & t \in [a, 2a] \\ 0, & \text{otherwise.} \end{cases} \tag{4B.5b}$$

Thus, the Laplace transform integral will have two terms. Noting that

$$LT\{\begin{cases} 1, & t \in [0, a] \\ 0, & \text{otherwise} \end{cases}\} = \int_0^a e^{-st} \, dt = \frac{1 - e^{-sa}}{s}, \tag{4B.6}$$

use of (4B.4) gives

$$LT\{\begin{cases} At/a, & t \in [0, a] \\ 0, & \text{otherwise} \end{cases}\} = \frac{A}{a} \cdot \frac{1 - (1 + sa)e^{-sa}}{s^2}. \tag{4B.7}$$

Similarly, using

$$LT\{\begin{cases} 1, & t \in [a, 2a] \\ 0, & \text{otherwise} \end{cases}\} = \int_a^{2a} e^{-st} \, dt = \frac{e^{-sa} - e^{-2sa}}{s} \tag{4B.8}$$

with (4B.4) gives

$$LT\{\begin{cases} A \cdot (2 - t/a), & t \in [a, 2a] \\ 0, & \text{otherwise} \end{cases}\} = 2A \cdot \frac{e^{-sa} - e^{-2sa}}{s} \tag{4B.9}$$

$$+ \frac{A}{a} \cdot \frac{s \cdot (2a \cdot e^{-2sa} - a \cdot e^{-sa}) + e^{-2sa} - e^{-sa}}{s^2}$$

so that, using (4B.7) and (4B.9), we obtain

$$
X_c(s) = A \cdot e^{-sa} \cdot \left\{ -\frac{(1 + sa)/a}{s^2} + \frac{2}{s} - \frac{1}{a} \cdot \frac{1+sa}{s^2} \right\} + \frac{A}{as^2}
$$

$$
+ A \cdot e^{-2sa} \left\{ -2/s + \frac{1}{a} \cdot \frac{1+2as}{s^2} \right\} \tag{4B.10a}
$$

$$
= A \cdot e^{-sa} \left\{ -\frac{1}{s} - \frac{1}{as^2} + \frac{2}{s} - \frac{1}{s} - \frac{1}{as^2} \right\} + \frac{A}{as^2} + A \cdot e^{-2sa} \left\{ -\frac{2}{s} + \frac{2}{s} + \frac{1}{as^2} \right\} \tag{4B.10b}
$$

$$
= A \cdot \frac{\{e^{-2sa} - 2e^{-sa} + 1\}}{as^2} = A \cdot \frac{(1 - e^{-sa})^2}{as^2}, \tag{4B.10c}
$$

which for $a = T/2$ and $A = 1$ gives (4B.2).

(b) Show that the z-transform of

$$
x(n) = \begin{cases} 2An/(N-1), & n \in [0, (N-1)/2] \\ 2A[1 - n/(N-1)], & n \in [(N-1)/2 + 1, N-1] \end{cases} \tag{4B.11}
$$

[as given in (4.22), in which $A = 1$] for N odd, equals the result in (4.24),

$$
X(z) = A \cdot \frac{\{1 - z^{-(N-1)/2}\}^2}{[(N-1)/2] \cdot z \cdot (1 - z^{-1})^2}, \quad z \neq 0, z \neq 1 \tag{4B.12a}
$$

and

$$
X(1) = \frac{A(N-1)}{2}. \tag{4B.12b}
$$

The formulas for N even are given at the end of this part [following (4B.19e)]. Of use in this example is the fact that

$$
-z \cdot \frac{d}{dz} \{X(z)\} = \sum_{n=-\infty}^{\infty} x(n) \cdot n \cdot z^{-n} = \text{ZT}\{n \cdot x(n)\}. \tag{4B.13}
$$

From (4B.11), the z-transform sum will have two terms. Noting that

$$
\text{ZT}\{ \begin{cases} 1, & n \in [0, (N-1)/2] \\ 0, & \text{otherwise} \end{cases} \} = \sum_{n=0}^{(N-1)/2} z^{-n} = \frac{1 - z^{-(N+1)/2}}{1 - z^{-1}}, \tag{4B.14}
$$

use of (4B.13) shows that the first ZT term is

$$
\text{ZT}\{ \begin{cases} 2An/(N-1), & n \in [0, (N-1)/2] \\ 0, & \text{otherwise} \end{cases} \}
$$

$$
= \frac{-2Az}{N-1} \cdot \frac{(1 - z^{-1})[(N+1)/2]z^{-(N+3)/2} - z^{-2}(1 - z^{-(N+1)/2})}{(1 - z^{-1})^2}. \tag{4B.15}
$$

For the second ZT term, first note that

$$
\text{ZT}\{ \begin{cases} 1, & n \in [(N-1)/2 + 1, N-1] \\ 0, & \text{otherwise} \end{cases} \} = \sum_{n=1+(N-1)/2}^{N-1} z^{-n} = \frac{z^{-(N+1)/2} - z^{-N}}{1 - z^{-1}} \tag{4B.16a}
$$

$$
= \frac{z^{-(N+1)/2} - z^{-N} - z^{-(N+3)/2} + z^{-(N+1)}}{(1 - z^{-1})^2} \tag{4B.16b}
$$

and that

$$-z \cdot \frac{d}{dz} \left\{ \frac{z^{-(N+1)/2} - z^{-N}}{1 - z^{-1}} \right\} \tag{4B.17}$$

$$= -z \cdot \frac{(1 - z^{-1}) \cdot \{-[(N + 1)/2]z^{-(N+3)/2} + Nz^{-(N+1)}\} - z^{-2}(z^{-(N+1)/2} - z^{-N})}{(1 - z^{-1})^2}.$$

Using (4B.15), (4B.16b), and (4B.17), we find that the ZT of $x(n)$ in (4B.11) is

$$X(z) = \frac{A}{(1 - z^{-1})^2} \cdot \left\{ (1 - z)\frac{(N + 1)/2}{(N - 1)/2} z^{-(N+3)/2} + \frac{z^{-1} - z^{-(N+3)/2}}{(N - 1)/2} \right.$$

$$+ 2z^{-(N+1)/2} - 2z^{-N} - 2z^{-(N+3)/2} + 2z^{-(N+1)} \tag{4B.18a}$$

$$+ (z - 1) \cdot \frac{1}{(N - 1)/2} \cdot \{-[(N + 1)/2]z^{-(N+3)/2} + Nz^{-(N+1)}\}$$

$$\left. + \frac{z^{-(N+1)} - z^{-(N+3)/2}}{(N - 1)/2} \right\}$$

$$= \frac{A}{(1 - z^{-1})^2} \cdot \left\{ \frac{N + 1}{N - 1}[z^{-(N+3)/2} - z^{-(N+1)/2}] + \frac{2z^{-1}}{N - 1} \right.$$

$$- \frac{2z^{-(N+3)/2}}{N - 1} + 2z^{-(N+1)/2} - 2z^{-N} - 2z^{-(N+3)/2} + 2z^{-(N+1)} \tag{4B.18b}$$

$$+ \frac{N + 1}{N - 1}[z^{-(N+3)/2} - z^{-(N+1)/2}] + \frac{2Nz^{-N}}{N - 1} - \frac{2Nz^{-(N+1)}}{N - 1}$$

$$\left. + \frac{2z^{-(N+1)}}{N - 1} - \frac{2z^{-(N+3)/2}}{N - 1} \right\},$$

or, collecting terms of various powers of z^{-1},

$$X(z) = \frac{A}{(1 - z^{-1})^2} \cdot \left\{ z^{-(N+3)/2}\left[\frac{N + 1 - 2 - 2N + 2 + N + 1 - 2}{N - 1} \right] \right. \tag{4B.18c}$$

$$+ z^{-(N+1)}\left[\frac{2N - 2 - 2N + 2}{N - 1} \right] + z^{-(N+1)/2}\left[\frac{-N - 1 + 2N - 2 - N - 1}{N - 1} \right]$$

$$\left. + z^{-N}\left[\frac{-2N + 2 + 2N}{N - 1} \right] + z^{-1}\frac{2}{N - 1} \right\}$$

$$= \frac{A}{(1 - z^{-1})^2} \cdot \left\{ \frac{-4}{N - 1} z^{-(N+1)/2} + \frac{2}{N - 1} [z^{-N} + z^{-1}] \right\} \tag{4B.18d}$$

$$= \frac{2z^{-1}A}{(N - 1)(1 - z^{-1})^2} \cdot \{1 + z^{-(N-1)} - 2z^{-(N-1)/2}\}, \tag{4B.18e}$$

which is equal to (4B.12a) when the $\{\cdot\}$ term is recognized as being equal to $(1 - z^{-(N-1)/2})^2$.

Finally, for the case $z = 1$ we revert to the z-transform definition:

$$X(1) = \sum_{n=0}^{N-1} x(n) \cdot 1^{-n} = \sum_{n=0}^{N-1} x(n) \tag{4B.19a}$$

$$= \frac{2A}{N - 1}\left[\frac{N - 1}{2} + 2 \cdot \sum_{n=1}^{(N-1)/2-1} n \right] \tag{4B.19b}$$

$$= \frac{2A}{N-1}\left[\frac{N-1}{2} + \frac{N-1}{2}\left(\frac{N-1}{2} - 1\right)\right] \tag{4B.19c}$$

$$= A\left[1 + \frac{N-1}{2} - 1\right] \tag{4B.19d}$$

or finally

$$X(1) = A\frac{N-1}{2}. \tag{4B.19e}$$

We now give the formulas for $x(n)$, $X(z)$, and $X(e^{j\omega})$ for the discrete-time triangle for N even. For this case,

$$x(n) = \begin{cases} \dfrac{2An}{N}, & n \in [0, N/2] \\ 2A\left(1 - \dfrac{n}{N}\right), & n \in [N/2 + 1, N - 1] \end{cases} \tag{4B.20}$$

(and zero otherwise). In this case, the z-transform of $x(n)$ is

$$X(z) = A\frac{(1 - z^{-N/2})^2}{\dfrac{N}{2}z(1 - z^{-1})^2}, \quad z \neq 0, z \neq 1 \tag{4B.21a}$$

and

$$X(1) = \frac{AN}{2}. \tag{4B.21b}$$

Finally,

$$X(e^{j\omega}) = A\frac{\sin^2(\omega N/4)}{[N/2]\sin^2(\omega/2)}e^{-j\omega N/2}. \tag{4B.22}$$

These results are proved in the problems.

(c) Equation (4.29) is due to the following property of Laplace transforms of multiple repetitions of a finite-duration function. Let the function of duration T be called $x_c(t)$, and let the function composed of L sequential repetitions of $x_c(t)$ be called $x_{L,c}(t)$. If $X_c(s)$ and $X_{L,c}(s)$ are the Laplace transforms of, respectively, $x_c(t)$ and $x_{L,c}(t)$, then

$$X_{L,c}(s) = \int_0^{LT} x_{L,c}(t) \cdot e^{-st}\, dt \tag{4B.23a}$$

$$= \sum_{m=0}^{L-1}\int_{mT}^{(m+1)T} x_{L,c}(t) \cdot e^{-st}\, dt \tag{4B.23b}$$

$$= \sum_{m=0}^{L-1}\int_0^T x_{L,c}(u + mT) \cdot e^{-s(u+mT)}\, du \tag{4B.23c}$$

$$= \sum_{m=0}^{L-1} e^{-smT} \cdot \int_0^T x_c(u) \cdot e^{-su}\, du \tag{4B.23d}$$

or finally

$$X_{L,c}(s) = \frac{1 - e^{-sLT}}{1 - e^{-sT}} \cdot X_c(s). \tag{4B.23e}$$

By L'Hospital's rule, for $s \to 0$, we have $X_{L,c}(0) = L \cdot \lim_{s \to 0} X_c(s)$. Equation (4B.23e) is the relation used to obtain (4.29).

A similar property for discrete-time signals may be derived in the same manner as in (4B.23). If $x_L(n)$ is a sequence composed of L repetitions of a sequence $x(n)$ of duration N, then the z-transform of $x_L(n)$, $X_L(z)$ in terms of that of $x(n)$, denoted $X(z)$, is

$$X_L(z) = \frac{1 - z^{-LN}}{1 - z^{-N}} \cdot X(z). \tag{4B.24}$$

Again by L'Hospital's rule, $X_L(1) = L \lim_{z \to 1} X(z)$. This equation is used in the derivation of (4.30).

PROBLEMS

Laplace transform review

4.1. Show that if $x_c(t) = e^{-2t}\cos(10t)u_c(t)$, then $X_c(s) = (s + 2)/[(s + 2)^2 + 10^2] = (s + 2)/(s^2 + 4s + 104)$, as cited in Example 4.3.

4.2. Find the LT of $r_n(t) = u_n(t) = \begin{cases} t^n, & t \geq 0 \\ 0, & t < 0 \end{cases}$. Show all steps.

4.3. Consider a constant current source $i(t) = I_0 u_c(t)$, R, and C all in parallel. Let the voltage across C be $v(t)$, and $v(0) = V_0$. Find $v(t)$ using (a) classical methods; (b) the Laplace transform.

4.4. (a) Using the partial fraction expansion theorem, apply to the continuous-time transfer function $G_c(s) = (s + 4)/[s(s + 2)^3(s^2 + 2s + 5)]$ to obtain the impulse response $g_c(t)$. (b) Verify your result by using impulse in Matlab; plot both analytical and numerical results on the same axes, out to $t = 10$ s.

4.5. Find the condition on BIBO stability for a continuous-time system, from a transform viewpoint. Find the region of convergence of the Laplace transform of a causal LSI system impulse response. Use the two results to prove that all the poles of the system function must be in the left-half-plane for a causal LSI system to be BIBO stable.

4.6. A system is described by $\dfrac{d^3 y_c(t)}{dt^3} - \dfrac{d^2 y_c(t)}{dt^2} - 20\dfrac{dy_c(t)}{dt} = 2x_c(t)$.

(a) Assuming zero ICs, find $H_c(s) = Y_c(s)/X_c(s)$.

(b) Find the causal $h_c(t)$. Is $h_c(t)$ stable? Why?

(c) Draw the pole plot and identify all ROCs for causal, anticausal, and bicausal systems. If there is more than one distinct ROC for any category, quantitatively what is the difference? Is there any ROC for stable $h_c(t)$? Why?

4.7. We find in a detailed Laplace transform table that LT$\{x_c(t)\}$, where $x_c(t) = \mathrm{sinc}(\alpha t)u_c(t)$, is $X_c(s) = (1/\alpha)\tan^{-1}\{\alpha/s\}$ for Re$\{s\} > 0$. In the text we say that LT$\{\mathrm{sinc}(\alpha t)\}$ exists only on the $j\Omega$ axis, and then only in a distributional sense. Explain why we cannot here use the property for LT$\{x_c(-t)\}$ and just add together $X_c(s)$ and LT$\{x_c(-t)\} = X_c(-s)$.

Continuous-Time Fourier Transform/Phasors/Fourier Series

4.8. If the units of $v_c(t)$ are volts, what are the units of $V_c(j\Omega)$, and of $V_c(k)$ if $v_c(t)$ is periodic? If the units of $v(n)$ are volts, what are the units of $V(z)$ and $V(e^{j\omega})$, and of $V(k)$ if $v(n)$ has length N?

4.9. (a) If the input $\tilde{x}_c(t)$ to an LSI system is T-periodic, is the output $y_c(t)$ also T-periodic?

(b) Repeat for discrete time: If $\tilde{x}(n)$ is N-periodic, is $y(n)$?

4.10. Contrast what significant quantity is replicated in the output of an LSI system when the input sequence is $\delta(n)$ versus when the input is $e^{j\omega_1 n}$.

4.11. Suppose that $x_c(t)$ and $h_c(t)$ are bandlimited functions. Explain from a frequency-domain viewpoint why $y_c(t) = x_c(t) *_c h_c(t)$ is typically a smoothed version of $x_c(t)$. Cite an exception. Extend the argument to discrete time.

4.12. Concerning continuous-time phasor analysis:

(a) Interpret phasor $\overline{A} = A\angle\theta$ in terms of the Fourier transform of $A\cos(\Omega_1 t + \theta)$. Use $FT\{e^{\pm j\Omega_1 t}\}$, a property of the Dirac delta function, and the $FT\{x_c(t - t_0)\}$ property to show that the complex Fourier coefficient is $\pi\overline{A}$ or $\pi\overline{A}^*$ for $\Omega = \pm\Omega_1$.

(b) By expressing $A\cos(\Omega_1 t + \theta)$ as a two-term complex Fourier series, interpret the phasor $\overline{A} = A\angle\theta$. Then use the Fourier series coefficients formula to explicitly show that only $k = \pm 1$ gives a nonzero coefficient—equal to $\overline{A}/2$ or $\overline{A}^*/2$ (the fundamental frequency being Ω_1).

4.13. In acoustics and imaging problems, one occasionally measures a signal composed of the sum of N sine waves reflected from the N various reflectors at different distances and reflectivities. If the originating signal was $x_c(t) = A\cos(\Omega t + \theta)$, find the received signal amplitude B and phase ϕ in $y_c(t) = \sum_{\ell=1}^{N} A_\ell \cos(\Omega t + \theta_\ell) = B\cos(\Omega t + \phi)$.

4.14. Someone adds a European ac voltage (50 Hz) to the American ac (60 Hz). Is the sum periodic? If so, find the smallest period and the frequency of the sum.

4.15. Determine, and using Matlab plot the frequency response magnitude of a system with impulse response $h_c(t) = u_c(t + 3) - u_c(t - 1)$. For what Ω is $|H_c(j\Omega)| = 0$? Also, evaluate $H_c(j0)$.

4.16. Repeat Example 2.8 using the frequency response to determine the output instead of using convolution; verify that the results are identical.

4.17. Using (4.12a), synthesize a T-periodic time function $x_c(t)$ with Fourier series coefficients $X_c(k) = \alpha^{|k|}$ for $|k| \leq 10$ and zero otherwise, for (a) $\alpha = -0.9$ and (b) $\alpha = -0.5$. Let $T = 2$ s. Evaluate on a very dense time grid, and plot your results. Is your $\tilde{x}_c(t)$ periodic with period equal to 2 s in each case? Explain the difference between $\tilde{x}_c(t)$ for the two values of α.

4.18. Find the Fourier series representation of $\tilde{a}_c(t)$, the periodic extension of the triangle

$$a_0(t) = \begin{cases} 1 - |t|/T, & |t| \leq T \\ 0, & \text{otherwise.} \end{cases}$$

Plot one period of the reconstructed triangle wave for $T = 1$ s using Matlab, keeping 2 terms, 5 terms, and 50 terms. In each case, superimpose the exact triangle for comparison.

4.19. Using the results of Problem 4.18, find the Fourier series representation of $x_c(t) = 0.4 \cdot \cos(30\pi t + 20°) + \text{perext}\{a_0(90t - 1.3)\}$. Solve by determining a net $X_c(k)$ for the entire $x_c(t)$ (rather than form two Fourier series). Display your results in Matlab by plotting the cosine, the triangle, and the cosine + triangle (truncated) Fourier series waveforms (in separate plots). Also verify that the shifted/scaled-argument triangle is correct by comparing with a direct calculation using the definition of $a_0(t)$.

4.20. Find the Fourier series representation of the T-periodic extension of the damped exponential $\{e^{-\alpha t}, t \in [0, T]; 0, \text{otherwise}\}$. Write out the Fourier series sum for the first three terms. Plot results in Matlab for $\alpha = 0.5$ and period $T = 2$ s, (a) keeping 3 terms and (b) keeping 50 terms.

4.21. Find the Fourier series coefficients of the full-wave rectified sinusoid $|\sin(\Omega_0 t/2)|$. Plot reconstruction results in Matlab for $\Omega_0 = 2\pi/T$, where $T = 1$ s, keeping 50 terms; compare against the exact $\tilde{x}_c(t)$.

4.22. Find the Fourier series coefficients of the full-wave-rectified cosinusoid $3|\cos(\Omega_0 t/2)|$. Write the three-term approximation, and plot in Matlab both the 3- and the 50-term approxima- tions against the exact $\tilde{x}_c(t)$.

4.23. Find the Fourier series coefficients of the periodic extension of a sawtooth wave, one period of which is $\tilde{x}_c(t) = t/T$ for $t \in [0, T]$. Plot reconstruction results in Matlab for $\Omega_0 = 2\pi/T$,   where $T = 1$ s, keeping 50 terms; compare against the exact $\tilde{x}_c(t)$.

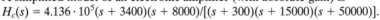

4.24. A simplified model of an electronic amplifier (with absolute gain) is
 $H_c(s) = 4.136 \cdot 10^5 (s + 3400)(s + 8000)/[(s + 300)(s + 15000)(s + 50000)]$.
Draw the asymptotic Bode plot (magnitude and phase), and verify using Matlab's bode function. From the Matlab plot, determine the dc gain of this amplifier, and check against $H_c(s = 0)$. Also, at what frequency (in hertz) does the amplifier no longer amplify (and begins to attenuate)?

4.25. An overall stereo playback system (including speakers) is modeled with the simplified quasi-normalized transfer function $H_c(s) = 6 \cdot 10^9 s(s + 1000)/[(s + 300)(s + 500)(s^2 + 16000s + 6.464 \cdot 10^9)$. Draw the magnitude and phase Bode plots using asymptotic curves (from 10 rad/s to 10^6 rad/s), and verify your results with a call to Matlab's bode. Notice the effects of poles and zeros as Ω is increased. Using the Matlab plot, find the low and high –3-dB frequencies *in hertz* (the function ginput can be helpful, as can superimposing a horizontal line at –3 dB with subplot and hold on).

Discrete-Time Fourier Transform/Frequency Response

4.26. The digital frequency variable f ranges from $-\frac{1}{2}$ to $\frac{1}{2}$. What are the units of f, and what does "$f = \frac{1}{2}$" mean?

4.27. Find $X(e^{j\omega})$ in polar form, where $x(n) = (\frac{1}{2})^n u(n)$. Only one sinusoidal function is allowed in your magnitude function.

4.28. Demonstrate the convolution theorem for the particular example of two length-4 sequences $x(n)$ and $h(n)$. That is, by multiplication of $\text{DTFT}\{x(n)\}$ and $\text{DTFT}\{h(n)\}$, show that the result is indeed equal to $\text{DTFT}\{x(n) * h(n)\}$ for this case.

4.29. The sequence $x(n)$ is an example of the transform of a 2π-periodic function of ω, $X(e^{j\omega})$. Discuss the reason for the periodicity of $X(e^{j\omega})$, given the discrete nature of this "transform" $[x(n)]$. Contrast with the aperiodicity of $x(n)$ by examining the inverse DTFT formula.

4.30. Averaging of a signal $x(n)$ intuitively smoothes the signal. Let the averaging be of the most N recent values, $x(n)$ included.

(a) Determine the unit sample response $h(n)$.

(b) Determine the transfer function $H(z)$. Does $H(z)$ have a pole at $z = 1$?

(c) Plot the magnitude and phase of the DTFT of $h(n)$ using Matlab routines, for $N = 10$ and $N = 100$. Comment on the smoothing property of the averaging operator, and comment on its value as a lowpass filter. Compare the appearances for the two values of N.

4.31. Express $H(e^{j\omega})$ in terms of real quantities only, where $h(n) = \alpha^{|n|}$ for $|n| \leq N - 1$ and zero otherwise. Simplify as much as possible.

4.32. In Example 2.21, we mention that $v(n) + a_1 v(n-1) = x(n) + b_1 x(n-1)$, $-1 < b_1 < a_1 < 0$, is a first-order lag network. Show that indeed $\angle H(e^{j\omega}) < 0$ for $\omega \in [0, \pi]$ for this system.

4.33. A digital highpass filter designed by the rectangular window method (see Chapter 9) has unit sample response $h(n) = \delta(n) - 2f_1\text{sinc}(2\pi f_1 n)$ for $n \in [-M, M]$ and zero otherwise [where $\text{sinc}(x) = \sin(x)/x$]. Using Matlab, plot the magnitude and phase frequency response plots for $f_1 = 0.3$ cycle/sample and $M = 3$; also plot $h(n)$. Repeat for $M = 100$ (same f_1); observe Gibbs's phenomenon (which is discussed in detail in Section 9.3.1). Plot $x(n) = 1.5\cos(2\pi \cdot 0.02n) + 2\cos(2\pi \cdot 0.2n + 0.2\pi) - \sin(2\pi \cdot 0.4n)$, the desired (ideal) output for $f_1 = 0.3$ cycle/sample, and the output of the truncated filter for $M = 100$ and $M = 3$ (plot ideal output on same axes for comparison). Qualitatively evaluate the results.

4.34. Suppose that we are given only the magnitude frequency response of an Nth-order FIR (i.e., length-N sequence) filter $h(n)$, where $h(n)$ are all real. How many different (independent) length-$(N+1)$ sequences have this same magnitude response, within a constant scaling factor?

4.35. (a) Find $X(z) = \text{ZT}\{x(n)\}$, where $x(n) = [\ln(a)^n/n!]u(n)$.

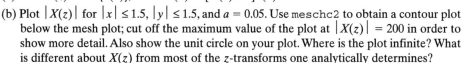

(b) Plot $|X(z)|$ for $|x| \le 1.5$, $|y| \le 1.5$, and $a = 0.05$. Use meschc2 to obtain a contour plot below the mesh plot; cut off the maximum value of the plot at $|X(z)| = 200$ in order to show more detail. Also show the unit circle on your plot. Where is the plot infinite? What is different about $X(z)$ from most of the z-transforms one analytically determines?

(c) Plot $x(n)$ versus n for $n \in [0, 20]$.

(d) Plot $|X(e^{j\omega})|$ versus ω by a direct DTFT calculation of the significantly nonzero values of $x(n)$, and also for the closed-form expression in part (a) (evaluated appropriately). Do they agree? Relate your result to that in part (b). If this sequence were a filter, what sort of filter would it be?

(e) Repeat parts (c) and (d) for $a = 10$. What changes do you notice? Can you explain them?

4.36. (a) If $X(e^{j\omega}) = 1 - |\omega|/\omega_c$ for $|\omega| < \omega_c$ and is zero otherwise for $|\omega| < \pi$, find $x(n)$ for all n.

(b) Check your result using Matlab for $\omega_c = 0.6\pi$ (i.e., $f_c = 0.3$), by calculating the DTFT directly of a truncated form of $x(n)$, and comparing with $X(e^{j\omega})$ in part (a). Verify that the phase is zero. Plot against all $|f| < \frac{1}{2}$. Also include a plot of $x(n)$ to help decide where to truncate.

(c) Repeat part (a) for the case $W(e^{j\omega}) = X(e^{j\omega})e^{-j\alpha\omega}$. It is unnecessary to repeat any detailed calculations.

4.37. (a) Find $h(n) = \text{DTFT}^{-1}\{H(e^{j\omega}) = (|\omega|/\omega_c)e^{-j\beta\omega}, |\omega| < \omega_c; 0, \text{otherwise}\}$, using whatever results you wish from Problem 4.36 to minimize effort.

(b) Using results from part (a) and from Problem 4.36, determine $y(n) = w(n) * h(n)$, using the convolution theorem.

(c) Check your result for $\alpha = \beta = 0$, using Matlab for $\omega_c = 0.6\pi$ (i.e., $f_c = 0.3$), by calculating the DTFT directly of a truncated form of $y(n)$, and comparing with $Y(e^{j\omega})$ obtained by using the convolution theorem (i.e., product of two exact spectra). Plot against all $|f| < \frac{1}{2}$. Also include a plot of $y(n)$, to help decide where to truncate. Do you prefer working in the time or frequency domain in this example?

4.38. The input to a system $H(z) = z/(z - \frac{1}{2})$ is $x(n) = \cos(2\pi \cdot 0.3n)$ for all n. We go to a z-transform table and find under the listing $\cos(\omega_1 n)$ the z-transform $[1 - \cos(\omega_1)z^{-1}]/[1 - 2\cos(\omega_1)z^{-1} + z^{-2}]$. Someone proposes that we find the system output $y(n)$ by taking the inverse z-transform of $Y(z)$ where we find $Y(z)$ by multiplying $H(z)$ by this $\text{ZT}\{\cos(\omega_1 n)\}$. Is there anything wrong with the above suggestion? Is there a simpler, correct approach? Verify your results by plotting on the same axes your analytical result against using conv of finite-length, causal versions of $x(n)$ and $h(n)$ (be sure to plot only after the transient has died out).

4.39. (a) If $x(n) = 3\cos(\pi n/10) - 4\sin(\pi n/3)$ for all n and $h(n) = [(1/4)^n + 2(1/3)^n]u(n)$, use the

frequency response to find $y(n)$; only real-valued expressions are allowed in the final result.

(b) Verify your result using `conv` in Matlab; use long sequences to minimize truncation effects resulting from using finite-length sequences. Plot results obtained via both methods on the same axes, well after steady state is reached.

4.40. For the system $y(n) - 0.85y(n-1) = 2x(n)$,

(a) Find the unit sample response $h(n)$ via the DTFT.

(b) Find the magnitude of the frequency response; simplify as much as possible. Sketch it (include maximum and minimum values and frequencies), and categorize as LPF, HPF, and so forth.

(c) Suppose $x(n) = 3\cos(\pi n/4)$. From your result in part (b), estimate the maximum value of $y(n)$. For a more exact answer, use the phase of $H(e^{j\omega})$ to find the exact maximum. Hint: Not all sinusoid values are accessed by n.

(d) Sketch the magnitude of the DTFT of $y(n) = \cos(\pi n/4) \cdot h(n)$ for $|\omega| \le \pi$. Find the most accurate maximum value of $|Y(e^{j\omega})|$ and the frequency at which it occurs, without differentiating. Justify your results.

4.41. Derive the phasor concept for digital signals. In what ways does it differ from the continuous-time phasor structure?

4.42. (a) Show that the real and delay operators $\text{Re}\{\cdot\}$ and $D\{\cdot\}$ on a sequence can be reversed.

(b) Determine $D^m\{e^{j\omega n}\}$ in terms of $e^{j\omega n}$.

(c) In terms of phasors (see Problem 4.41) and using the results in parts (a) and (b), show how to solve the difference equation (2.40) for the *steady-state* response when the input $x(n)$ is a sinusoid $X\cos(\omega n + \theta)$, where X and θ are real-valued constants. Assume that a_ℓ and b_ℓ are all real-valued.

4.43. Consider the causal system described by the difference equation

$y(n) - \frac{3}{4}y(n-1) + \frac{1}{8}y(n-2) = -\frac{1}{16}x(n) + \frac{7}{16}x(n-2)$.

(a) Find the associated transfer function, and plot the pole–zero diagram using `pzmap` (in the Controls toolbox).

(b) Plot the frequency response using `freqz`. If possible, in this and all other discrete-time frequency responses, modify `freqz` so that the digital frequency axis ranges from 0 to $\frac{1}{2}$ rather than 0 to 1. What type of filter is this?

(c) Find the unit sample response in closed form. Use PFE, and confirm your PFE coefficients using `residuez`. Then confirm your unit sample response using Matlab's `dimpulse` on the original transfer function. What advantage does "figuring it out on paper" have over just using `dimpulse`?

4.44. (a) In Problem 4.43, instead of writing $H(z)$ in terms of z, write $H(z)$ in terms of z^{-1}, and do not define $H_1(z) = H(z)/z$ (as was done in the text). Noting the improper rational function of z^{-1}, perform long division. Then perform PFE on the remainder, viewing z^{-1} as the usual z parameter. Check the unit sample response you obtain against that obtained in part (c) of Problem 4.43.

(b) We can also use the further shortcut available from Problem 3.65 that allows us to avoid long division and to avoid defining $H_1(z) = H(z)/z$ or any manipulations into standard PFE form. Show that that shortcut works here.

4.45. A system transfer function is $H(z) = \dfrac{(1 - \sqrt{2}z^{-1} + z^{-2})}{(1 - 1.5035z^{-1} + 0.64z^{-2})(1 - 0.8z^{-1})}$.

Find the output $y(n)$ if the input is $x(n) = 5\cos(\pi n/4 - 60°)$ for all n.

4.46. Let $x(n) = \sum_{\ell=0}^{\infty} a^n \delta(n - \ell N)$.

(a) Find $X(z)$ and $X(e^{j\omega})$ for real values of a ($|a| < 1$) and positive integer N. Give the region of convergence of $X(z)$.

(b) What are the poles of $X(z)$ in terms of a and N? How many poles are there?

(c) Give closed-form expressions for $|X(e^{j\omega})|$ and $\angle X(e^{j\omega})$ in terms of a and N. What are the maximum and minimum values of $|X(e^{j\omega})|$?

(d) Using Matlab, plot $x(n)$, $|X(e^{j\omega})|$, and $\angle X(e^{j\omega})$ for the case $a = 0.99$, $N = 20$. On the same axes, plot the expressions found in part (c) and plot the numerical DTFT obtained using a direct calculation of the DTFT sum of the first 1000 values of $x(n)$; your results should agree. Do the maximum and minimum values in your magnitude plot agree with those found in part (c) (give their values for the above values of a and N)?

4.47. Do the ZT and DTFT of $\left(\frac{1}{2}\right)^n u(-n)$ exist? If not, why not? If so, evaluate the transform. Is the system whose unit sample response is $\left(\frac{1}{2}\right)^n u(-n)$ stable or unstable?

4.48. Form and plot in Matlab the sequence $x(n) = \alpha^n$ for $n \in [0, N-1]$, where $N = 10$ and $\alpha = 0.9$. Then form and plot $L = 10$ repetitions of this sequence. Plot the magnitude spectrum (DTFT) for each of these sequences on the same axes [first, however, multiplying the spectrum of the nonrepeated $x(n)$ by L]. What do you find? Also plot on these axes the result given in (4B.24), with $z = e^{j\omega}$. Does it agree with your direct sum calculation of $X_{L,N}(e^{j2\pi f})$? Interpret your results in terms of periodic/aperiodic sequences. Repeat the spectral plot and comments for $L = 100$.

z-Transform Revisited

4.49. Show that the z-transform of the discrete-time triangle $x(n)$ for N even, as given in (4B.20), is as given in (4B.21a) and (4B.21b). Also show that its DTFT is as given in (4B.22).

4.50. (a) Prove that $ZT\{(1/n)u(n-1)\} = \ln\{z/[z-1]\}$. Hint: Work "backwards" and think of a familiar Taylor series. Comment on the implications of the divergence of the harmonic series in this context.

(b) Using Matlab, calculate the inverse z-transform of $\ln\{z/[z-1]\}$ using the contour $|z_c| = 1.5$; plot against $x(n)$ for $n \in [0, 29]$.

(c) Attempt to repeat part (b) with $|z_c| = 0.9$. What do you find?

4.51. In this problem, we consider the reason why transfer functions in block diagrams are labeled with the GPE rather than time-domain impulse/unit sample response. For the differential equation (2.20), define the symbol D to represent the time-differentiation operator, write the input–output relation in terms of D, and "solve" for $y_c(t)$ in terms of $x_c(t)$ in a purely operational sense. What do you find in terms of the system transfer function? Make suitable comments for the analogous statements for discrete-time systems.

4.52. To give clearer meaning to the results of Problem 4.51, in this problem we view s (rather than D) as the operator d/dt. Thus now s is a time differentiator, rather than a complex variable.

(a) Give the time-domain meaning of $1/s$ in this context.

(b) Find $\{1/s\}^\ell u_c(t)$ for $\ell \geq 0$, where $u_c(t)$ is the unit step function.

(c) Now suppose we wish to find the step response of a system described by the differential equation $dy_c(t)/dt + \alpha y_c(t) = \beta u_c(t)$, where $\alpha > 0$. First solve the equation in a symbolic, operational sense using s as viewed in part (a). Then expand the operator (a rational function of s) as a Laurent series in s, with $s_0 = 0$. We thus temporarily here *do* view s as a complex variable, but not in terms of the Laplace transform, and moreover this view may be relaxed by performing long division to obtain Laurent coefficients. Use the result of part (b) to obtain the solution, and compare with the result obtained using the Laplace transform tables.

4.53. (a) For part (c) of Problem 4.52, find the impulse response of the system described there.

(b) In this formulation, the z-transform and its inverse naturally arise. Show how they do in finding $h_c(t)$ from a stable causal system for which $M < N$. From these results, extract the formula for the inverse Laplace transform. Conclude that the operational-solution approach considered in this problem and Problems 4.51 and 4.52 is indeed equivalent to the contour-integration/Laplace transform approach. Note again that rather than contour integrals, the entire problem can be repeated for rational $H_c(s)$ by using long division. In that case, we need not consider s to be a complex variable.

Discrete Fourier series/DFT

4.54. Find the DTFT and the DFT_N of a square pulse $x(n)$ that is equal to 1 from $0 \leq n < M \, (< N)$ and is equal to zero for $M \leq n \leq N - 1$. Also find the zeros of the DTFT.

4.55. In Matlab, plot stem plots of sampled unit-amplitude, zero-phase cosinusoids having (a) 2, (b) 4, and (c) 10 samples per period. In each case, choose the total number of samples N so that there are $L = 10$ periods in the sampling frame. Plot the magnitude of the DFT_N using `fft`. Quantitatively and generally explain what you find.

4.56. Plot the real and imaginary parts of the expansion functions $W_N^{-nk} = \exp\{j2\pi nk/N\}$ versus n for $n \in [0, 4N - 1]$ for $N = 8$. Thus, there will be eight sets of plots, one for each k on $[0, N - 1]$. Superimpose on each plot a vertical dashed line passing through $n = N$. Verify that indeed all your W_N^{-nk} are N-periodic. State how many periods are contained on $[0, N - 1]$. From this you should graphically conclude that indeed all DFT^{-1}s are N-periodic. That is, all sequences that are linear combinations of the W_N^{-nk} are N-periodic. Of course, the same holds for all combinations of W_N^{+nk}, so all results of the DFT_N are N-periodic. For each k, determine how many actual periods W_N^{-nk} has for $n \in [0, N - 1]$.

Communications Applications

4.57. The Hilbert transformer is a filter $h_{H,c}(t)$ whose frequency response is $H_{H,c}(j\Omega) = -j \cdot \mathrm{sgn}(\Omega)$, where $\mathrm{sgn}(\Omega) = 1$ for $\Omega \geq 0$ and -1 for $\Omega < 0$.

(a) Determine $h_{H,c}(t)$ in closed form. To do this by a direct calculation, consider that $H_{H,c}(j\Omega) = \lim_{\alpha \to 0} -j \cdot \mathrm{sgn}(\Omega) e^{-\alpha |\Omega|}$, where $\alpha > 0$.

(b) The Hilbert transform $\hat{x}_c(t)$ of $x_c(t)$ is defined as the output of a Hilbert transformer whose input is $x_c(t)$: $\hat{x}_c(t) = x_c(t) *_c h_{H,c}(t)$. Find the Hilbert transform of $x_c(t) = \cos(\Omega_1 t)$.

(c) Let the discrete-time Hilbert transformer be defined as $H_H(e^{j\omega}) = -j \cdot \mathrm{sgn}(\omega)$, $|\omega| \leq \pi$. Determine $h_H(n)$.

(d) Repeat part (b) for the discrete-time Hilbert transformer.

(e) Analogously define the discrete Hilbert transformer for N even as

$$H_H(k) = \begin{cases} -j, & k \in [0, N/2] \\ j, & k \in [N/2+1, N - 1]. \end{cases}$$

Repeat parts (c) and (d) for this case. Simplify $h_H(n)$ as much as possible. Hint on finding the discrete Hilbert transform of $\cos(\omega_1 n)$: Choose ω_1 to give an integral number of cosinusoidal periods M for the time window $[0, N - 1]$. What happens when you do not? Verify all your results in this part with plots made using Matlab.

4.58. In communication systems, we often wish to minimize the bandwidth of the signal being transmitted to free up space for other channels. Usually the audio signal is real-valued; hence its Fourier transform is conjugate-symmetric. Consequently, one half of the spectrum ($\Omega > 0$ or $\Omega < 0$) is redundant. By forming a single-sideband version of the modulated audio signal, we can cut the required bandwidth in half. Use the example spectrum in Figure P4.58a to illustrate the stages in generating the output in Figure P4.58b [spectra of $\hat{x}_c(t)$, $v_c(t)$, $w_c(t)$, and $y_c(t)$]. Using the results of Problem 4.57, determine the output spectrum magnitude and phase plots. Does the system in Figure P4.58b reproduce in $y_c(t)$ the upper or lower sideband? For simplicity, use unwrapped phases (do not worry about taking modulo 2π). Note: The subscript c on Ω_c stands for carrier.

4.59. (a) Show that $x_{a,c}(t) = x_c(t) + j\hat{x}_c(t)$, known as the analytic signal corresponding to $x_c(t)$ [where $x_c(t)$ is either real or complex-valued], has Fourier transform $X_{a,c}(j\Omega) = 2X_c(j\Omega)u_c(\Omega)$, where $u_c(x) = 1$ for $x \geq 0$ and is zero otherwise.

(b) Let $x_c(t)$ be real-valued and be modulated on a carrier of frequency Ω_{c1}, as shown in Figure P4.59. Show the required operations on $x_c(t)$ that will change the carrier to Ω_{c2}, *without using any filtering operations* [only Hilbert transforms and modulation(s)]. You may use complex-valued signals.

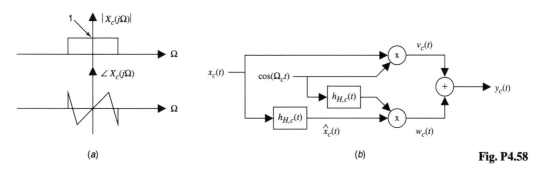

(a)

(b)

Fig. P4.58

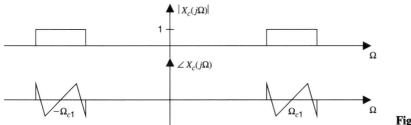

Fig. P4.59

Illustrative Verification of Results in Text

4.60. We investigate graphically the *RC* circuit and its discretization with unit-amplitude boxcar input in this problem (see Section 4.7).

(a) In Matlab, with $T = 0.5$ s, plot (4.35) from $t = 0$ to $t = 2$ s for (i) $RC = 0.1$ s, $v_{C,c}(0) = 0$ V; (ii) $RC = 0.1$ s, $v_{C,c}(0) = 2$ V; (iii) $RC = 0.01$ s, $v_{C,c}(0) = 0$ V; (iv) $RC = 2$ s, $v_{C,c}(0) = 0$ V. Interpret each result using physical arguments.

(b) Now superimpose (4.39) on top of plot (i) in part (a) for (i) $\Delta t = 5$ ms and plot out to only 1 s to avoid too many stems, (ii) $\Delta t = 0.1$ s (plot out to 2 s), and (iii) 0.2 s (again, plot out to 2 s). As in the text, choose $v_{C,c}(0)$ for exact agreement with $v_C(n)$ at $n = 0$. Again, interpret your results (and compare with the result in Figure 4.23*a*).

4.61. Show that (4.26) follows from (4.24a) by substitution of $z = e^{j\omega}$ into (4.24a).

4.62. Derive the first relation in (4) in Appendix 4A.

4.63. Derive the second relation in (4) in Appendix 4A.

4.64. Consider again the second relation in (4) in Appendix 4A, for the case in which $x(n)$ is nonzero over only the range $n \in [0, N - 1]$. Write $X(z)$ as the (finite-length) z-transform sum of $x(n)$ expressed as the DTFT^{-1} of $X(e^{j\omega})$. Your result should be

$$X(z) = \frac{1}{2\pi} \int_{-\pi}^{\pi} \frac{X(e^{j\omega})(1 - (e^{j\omega}/z)^N)}{1 - e^{j\omega}/z} \, d\omega.$$

This result appears to disagree with the (second) result in (4) in Appendix 4A, due to the second term $-(e^{j\omega}/z)^N$ in the numerator of the integrand. Resolve the apparent disagreement.

4.65. Derive the first equation in relation (20) in Appendix 4A, which holds under the assumption that $\tilde{x}_c(t + T) = \tilde{x}_c(t)$ for all t; that is, $\tilde{x}_c(t)$ is T-periodic in t. Also express $X_c(j\Omega)$ in terms of $X_{1,c}(j\Omega) = \text{FT}\{x_{1,c}(t)\}$, where $x_{1,c}(t)$ is one period of $\tilde{x}_c(t)$ [see (4.13)].

4.66. Derive the second equation in relation (20) in Appendix 4A, which holds under the assumption that the only interval on which $x_c(t)$ may be nonzero is $t \in [0, T]$.

4.67. Derive the second formula in item (6) in Appendix 4A by starting with the DFT^{-1} and DTFT formulas as given relations.

4.68. Derive item (9) in Appendix 4A.

Extension of Convergence of LT; Frequency Response for Unstable Systems, and So On

4.69. In the following series of problems, we explore some interesting issues related to the convergence of the Laplace transform. For rational functions, the Laplace transform converges only for values of s to the right of the rightmost pole $\sigma = \sigma_1$—to the right of the "abscissa of convergence." Thus, for the inversion of the Laplace transform (4.1a) the Bromwich contour (σ fixed and Ω running from $-\infty$ to $+\infty$) must be taken to the right of the abscissa of convergence. However, it is not necessary for the entire inversion integration path to lie in the region of convergence of the forward Laplace integral. Instead, we may take any path for which Im{s} runs from $-\infty$ to ∞ and merely keeps to the right of all poles where they occur; for example, the contour need not be "Bromwich" (a vertical line). Investigate this idea for the example $x_c(t) = e^t u_c(t)$. Give the Laplace transform $X_c(s)$, apply the Cauchy integral formula for inversion, and especially describe the ramifications of the above discussion for $s = j\Omega$.

4.70. It is known that a function analytic in one region but not in another may often be continued beyond the given region of analyticity. The same idea can be applied to functions of a complex variable such as the Laplace transform that exist in one region and not in another. One procedure for finding such a continuing function is by calculating Taylor expansions with shifted origins near the boundary of the original region of convergence. The circle of convergence for such origins of expansion (out to nearest pole) may include points outside the original region of convergence (Re{s} > σ_0). The result is a function equal to the original everywhere within the original region of analyticity but extended beyond. Propose how and to what extent we may by continuation extend Laplace transforms that are rational functions of s beyond the abscissa of convergence. Apply to $x_c(t) = e^t u_c(t)$.

4.71. In Problem 4.70, we saw how we may write

$$x_c(t) = e^t u_c(t) = \frac{1}{2\pi j} \int_{C_A} \frac{e^{st}}{s-1}\, ds, \tag{P4.70.1}$$

which is an expansion of $x_c(t) = e^t u_c(t)$ over the functions e^{st} (all t), *including* $e^{j\Omega t}$. For expanding $x_c(t)$ over $e^{j\Omega t}$, C_A is along the imaginary axis everywhere except near $\Omega = 0$, where it detours to the right and skirts around $s = 1$. Yet it was shown in the solution of Problem 4.69 that the expansion coefficient of $e^{j\Omega t}$ was infinity (or undefined), as given by the Laplace transform with $\sigma = 0$. Here, the expansion coefficient is, by inspection of (1), just $1/(j\Omega - 1)$. How may the apparent contradiction be resolved? Hint: We may expand $x_c(t)$ by either of two methods:

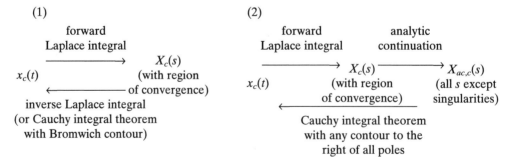

(1)

forward
Laplace integral

$x_c(t)$ $\xrightarrow{\hspace{2cm}}$ $X_c(s)$
$\xleftarrow{\hspace{2cm}}$ (with region of convergence)
inverse Laplace integral
(or Cauchy integral theorem
with Bromwich contour)

(2)

forward   analytic
Laplace integral   continuation

$x_c(t)$ $\xrightarrow{\hspace{2cm}}$ $X_c(s)$ $\xrightarrow{\hspace{2cm}}$ $X_{ac,c}(s)$
(with region   (all s except
$\xleftarrow{\hspace{2cm}}$ of convergence)   singularities)
Cauchy integral theorem
with any contour to the
right of all poles

What advantage does $X_c(s)$ have over its analytic continuation, $X_{ac,c}(s)$?

4.72. Controls practitioners routinely compute the "frequency response" of unstable open-loop systems, particularly when using the Nyquist stability criterion and performing Bode stability analysis. Is this Fourier analysis? If not, what is it? Why can the analytic continuation still provide stability information even where the Laplace transform does not exist?

Problem 4.73 involves Web materials (WP 4.2)

4.73. Prove the completeness relation (see WP 4.2) for complex sinusoids $\dfrac{1}{2\pi} \displaystyle\int_{-\infty}^{\infty} e^{j\Omega(t-t')} d\Omega =$

$\delta_c(t - t')$, which is true in a distributional sense and is of great utility in linear system theory and mathematical physics. For example, we show on WP 5.1 and in the Chapter 5 problems how the Fourier transform pair relations can be obtained using this completeness relation by writing $x_c(t)$ as an integral expansion over complex sinusoids. Under what condition on $x_c(t)$ can $x_c(t)$ be so expanded? Extend this condition to Bromwich (vertical) and other contours where the expansion function is e^{st}, and write the applicable completeness relation. Give an advantage and a disadvantage for arbitrary contours.

4.74. Assume that an LSI system $h_c(t)$ is causal, real, and stable. If it is also minimum-phase [all poles and zeros of $H_c(s)$ are in the left-half plane], then the log-magnitude and phase curves of $H_c(j\Omega)$ are tied together (see Bode's or other texts). If that is true, how then can we move the log-magnitude curve up or down without altering the phase curve? Determine the relationship, filling in as much derivation detail as you can.

Chapter 5

Transform Properties

5.1 INTRODUCTION

In Chapter 2, we reviewed the basics of signals and systems for both continuous and discrete time. The most important operation we encountered was convolution. With the background in complex analysis provided in Chapter 3, in Chapter 4 we examined the transforms that are highly useful in analyzing and designing signal-processing systems. In Chapters 3 and 4, we occasionally called on properties of these transforms.

In this chapter, we first systematically examine several transform properties. These properties arise again and again, and can be combined into compound properties when several conditions hold simultaneously. Often it is precisely because of the simplicity of the expression of a signal characteristic in another domain (via a transform property) that Fourier or GPE transforms are used. In general, without transform properties and transform pairs, the transforms would be difficult to use and interpret. Transform properties also lend insight into system behavior. For example, because $FT\{d/dt\} = j\Omega \cdot$, we know that a differentiator accentuates high frequencies; also, if we find a constant slope in the frequency response phase plot, we know that the system is providing a delay equal to minus that slope. A complete summary of properties for the six transforms appears in Appendix 5A. Short transform pair tables appear in Appendix 5B. Most readers have already seen many of these transform properties in their signals and systems course. Here we reexamine them from perhaps a more unified and interpretive viewpoint. Before deriving some commonly used transform properties, it is helpful to have close at hand the transform definitions as given in Table 5.1 (tildes on DFT sequences omitted).

In the following explanations of properties, we will form generalized properties that hold for virtually all the 12 transform relations in Table 5.1. The effort of generalizing has two benefits: (a) For nearly all cases, the analysis/algebra required for proving particular transform properties is totally eliminated, and (b) we see the underlying *reason* for the property as it applies to any transform.

Thus for generality we refer to two domains, domain 1 (variable u) and domain 2 (variable v), connected by one of the transforms in Table 5.1. For example, suppose the two domains are continuous time and discrete frequency. Then the domain 1 and domain 2 variables are, respectively, $u = t$ and $v = k$. Similarly, the domain 1 and domain 2 functions $P(u)$ and $Q(v)$ are, respectively, $P(u) = x_c(t)$ and $Q(v) = X_c(k)$.

Table 5.1   Transform Definitions

Continuous-time transforms

$$
x_c(t) = \begin{cases}
\dfrac{1}{2\pi j}\displaystyle\int_C X_c(s)\,e^{st}\,ds, & \text{where} \\
\qquad\text{LT}^{-1} & \\[2ex]
\dfrac{1}{2\pi}\displaystyle\int_{-\infty}^{\infty} X_c(j\Omega)\,e^{j\Omega t}\,d\Omega, & \text{where} \\
\qquad\text{FT}^{-1} & \\[2ex]
\displaystyle\sum_{k=-\infty}^{\infty} X_c(k)\,e^{j2\pi kt/T}, & \text{where} \\
\qquad\text{FS}^{-1} & [\tilde{x}_c(t+T)=\tilde{x}_c(t)]
\end{cases}
$$

$$X_c(s) = \int_{-\infty}^{\infty} x_c(t)\,e^{-st}\,dt$$
$$\text{LT}$$

$$X_c(j\Omega) = \int_{-\infty}^{\infty} x_c(t)\,e^{-j\Omega t}\,dt$$
$$\text{FT}$$

$$X_c(k) = \frac{1}{T}\int_0^T \tilde{x}_c(t)\,e^{-j2\pi kt/T}\,dt$$
$$\text{FS}$$

Discrete-time transforms

$$
x(n) = \begin{cases}
\dfrac{1}{2\pi j}\displaystyle\oint_C X(z)\,z^n\,\dfrac{dz}{z}, & \text{where} \\
\qquad\text{ZT}^{-1} & \\[2ex]
\dfrac{1}{2\pi}\displaystyle\int_{-\pi}^{\pi} X(e^{j\omega})\,e^{j\omega n}\,d\omega, & \text{where} \\
\qquad\text{DTFT}^{-1} & \\[2ex]
\dfrac{1}{N}\displaystyle\sum_{k=0}^{N-1} X(k)\,e^{j2\pi kn/N}, & \text{where} \\
\qquad\text{DFS}^{-1}/\text{DFT}^{-1} & [x(n+N)=x(n)] \\
& \text{and } X(k+N)=X(k)]
\end{cases}
$$

$$X(z) = \sum_{n=-\infty}^{\infty} x(n)\,z^{-n}$$
$$\text{ZT}$$

$$X(e^{j\omega}) = \sum_{n=-\infty}^{\infty} x(n)\,e^{-j\omega n}$$
$$\text{DTFT}$$

$$X(k) = \sum_{n=0}^{N-1} x(n)\,e^{-j2\pi kn/N}$$
$$\text{DFS/DFT}$$

In practice, these P and Q functions could either be signals [e.g., waveform $x_c(t)$ or its spectrum $X_c(j\Omega)$] or system-related functions [e.g., impulse response $h_c(t)$ or frequency response $H_c(j\Omega)$]. In all cases, we will start in "domain 1" [$P(u)$], so P may be either a time signal or a spectral signal. The transform relations in Table 5.1 as used from domain 1 to domain 2 generalize to

$$Q(v) = \sum_u P(u)f_{12}(u,v)\rho \tag{5.1a}$$

and as used from domain 2 to domain 1 they generalize to

$$P(u) = \sum_v Q(v)f_{21}(u,v)\gamma, \tag{5.1b}$$

where in specific cases the $\sum$ may be $\int$ and any scaling[1] is absorbed into the coefficients ρ and γ. In the above example of $\tilde{x}_c(t) \overset{\text{FS}}{\longleftrightarrow} X_c(k)$, we have $f_{12}(u,v) = e^{-j2\pi kt/T}$, $f_{21}(u,v) = e^{j2\pi kt/T}$, $\rho = dt/T$, and $\gamma = 1$. For all transforms in Table 5.1, $f_{21}(u,v) = 1/f_{12}(u,v)$. The transforms in Table 5.1 are parameterized as shown in Table 5.2. This concept of components of $P(u)$ in (5.1b) that are functions of u with v as a variable parameter was already introduced in Section 4.4, (4.15b)[2] for the DTFT^{-1} $\left[x(n) = \int_{-\pi}^{\pi} x(n;\omega)\,d\omega, \text{ where } x(n;\omega) = X(e^{j\omega})e^{j\omega n}/(2\pi)\right]$. There $x(n;\omega)$ was precisely

---

[1] Such as factors of $d\Omega/(2\pi)$, $1/N$, and $dz/(2\pi jz)$ that are independent of the variable of the function being expanded [e.g., u in (5.1b)].
[2] We remove the tilde here, because (4.15b) is valid also for aperiodic $x(n)$.

Table 5.2 Values of General Transform Parameters, for Specific Transforms

Transform	$P(u)^a$	$f_{21}(u,v)$	d	c	γ	$Q(v)^a$	$f_{12}(u,v)$	ρ
LT	$x_c(t)$	e^{st}	e	s	$\dfrac{ds}{2\pi j}$	$X_c(s)$	e^{-st}	dt
LT^{-1}	$X_c(s)$	e^{-st}	e	$-t$	dt	$x_c(t)$	e^{st}	$\dfrac{ds}{2\pi j}$
FT	$x_c(t)$	$e^{j\Omega t}$	e	$j\Omega$	$\dfrac{d\Omega}{2\pi}$	$X_c(j\Omega)$	$e^{-j\Omega t}$	dt
FT^{-1}	$X_c(j\Omega)$	$e^{-j\Omega t}$	e	$-jt$	dt	$x_c(t)$	$e^{j\Omega t}$	$\dfrac{d\Omega}{2\pi j}$
FS	$x_c(t)$	$e^{j2\pi kt/T}$	e	$\dfrac{j2\pi k}{T}$	1	$X_c(k)$	$e^{-j2\pi kt/T}$	$\dfrac{dt}{T}$
FS^{-1}	$X_c(k)$	$e^{-j2\pi kt/T}$	e	$\dfrac{-j2\pi t}{T}$	$\dfrac{dt}{T}$	$x_c(t)$	$e^{j2\pi kt/T}$	1
ZT	$x(n)$	z^n	z	1	$\dfrac{dz}{2\pi jz}$	$X(z)$	z^{-n}	1
ZT^{-1}	$X(z)$	z^{-n}	Does not hold	Does not hold	1	$x(n)$	z^n	$\dfrac{dz}{2\pi jz}$
DTFT	$x(n)$	$e^{j\omega n}$	e	$j\omega$	$\dfrac{d\omega}{2\pi}$	$X(e^{j\omega})$	$e^{-j\omega n}$	1
DTFT^{-1}	$X(e^{j\omega})$	$e^{-j\omega n}$	e	$-jn$	1	$x(n)$	$e^{j\omega n}$	$\dfrac{d\omega}{2\pi}$
DFT/DFS	$x(n)$	$e^{j2\pi kn/N}$	e	$\dfrac{j2\pi k}{N}$	$\dfrac{1}{N}$	$X(k)$	$e^{-j2\pi kn/N}$	1
DFT^{-1}/DFS^{-1}	$X(k)$	$e^{-j2\pi kn/N}$	e	$\dfrac{-j2\pi n}{N}$	1	$x(n)$	$e^{j2\pi kn/N}$	$\dfrac{1}{N}$

$^a u$ and v are the argument of P or Q as shown, except for $(a)\ j\Omega$, where u or v is just Ω and $(b)\ e^{j\omega}$, where u or v is just ω.

$Q(v)f_{21}(u,v)\gamma$ in (5.1b), divided by $d\omega$. We will continue to use the notation of (4.15) [e.g., $x(n;\omega)$] when convenient.

The basis for intuitive understanding of many of the various transform properties can now be stated as the following single principle. Below, we will be finding the transform of a "domain 1" function $P(u)$ that has been modified in some way by a linear operator $L\{\,\cdot\,\}$: $P'(u) = L\{P(u)\}$. We will seek $Q'(v)$, the transform of $P'(u) = L\{P(u)\}$ in terms of $Q(v)$, because we may already know $Q(v)$. For example, we may wish to find $Q'(v) = LT\{x_c(t - t_0)\}$ in terms of $Q(v) = X_c(s) = LT\{x_c(t)\}$, where here $L\{\,\cdot\,\}$ is the delay-by-t_0 operator. The essential principle will be that whatever the modification $L\{\,\cdot\,\}$, the transform operation transfers the modification $L\{\,\cdot\,\}$ to the expansion function $Q(v)f_{21}(u,v)\gamma$ for the given transform, and often just to $f_{21}(u,v)$. When this same modification is made to all constituent components $Q(v)f_{21}(u,v)\gamma$ of the expanded function $P(u)$, the resulting sum or integral will also have had the same operation performed on it:

$$P'(u) = L\{P(u)\} = \begin{cases} \displaystyle\sum_v L\{Q(v)f_{21}(u,v)\gamma\}, & \text{always} \quad\quad\quad\;\; (5.2a) \\[2ex] \displaystyle\sum_v Q(v)L\{f_{21}(u,v)\}\gamma, & L \text{ operates on only } u. \;\; (5.2b) \end{cases}$$

For example, if we make the same modification to all components $X_c(s)e^{st}\, ds/(2\pi j)$ in the LT^{-1} expansion of $x_c(t)$, then that operation will have been performed on $x_c(t)$ itself. If the operation is upon only t, we really just operate on e^{st}; the linear operator "passes through" $X_c(s)\, ds/(2\pi j)]$ in $L\{X_c(s)e^{st}\, ds/(2\pi j)\}$. In calculation terms, the whole game plan is always to massage $P'(u) = L\{P(u)\}$ into the form in (5.1b):

$$P'(u) = \sum_v [\,\cdot\,\cdot\,\cdot\,]f_{21}(u, v)\gamma.$$ Then whatever $[\,\cdot\,\cdot\,\cdot\,]$ ends up multiplying $f_{21}(u, v)\gamma$

we identify to be $Q'(v)$. Once we have the general property, we have all or nearly all of the 12 cases of that property for the individual transforms in Table 5.1 just by reading from Table 5.2 and substituting the values of the general parameters for the given transform.

5.2 SHIFT OF A VARIABLE

Suppose that the variable u of a function $P(u)$ in domain 1 is shifted by $-u_0$ (the minus is conventional), and it is desired to find the corresponding transform function $Q'(v)$ in domain 2 in terms of $Q(v)$, the transform of the original unshifted $P(u)$, because $Q(v)$ may already be known. Suppose that u appears in the exponent of the basis function $f_{21}(u, v)$: $f_{21}(u, v) = d^{cu}$, where c and d are constants with respect to u but may depend on v (true for all relations in Table 5.2 except ZT^{-1}). Then letting $L\{\,\cdot\,\}$ be the shift operator so $P'(u) = L\{P(u)\} = P(u - u_0)$, (5.2b) gives

$$L\{P(u)\} = \sum_v Q(v)f_{21}(u - u_0, v)\}\gamma \qquad (5.3a)$$

$$= \sum_v [Q(v)d^{-cu_0}]\, d^{cu}\gamma, \qquad (5.3b)$$

which is in the form of (5.1b) with

$$Q'(v) = Q(v)d^{-cu_0} = f_{21}(-u_0, v)Q(v). \quad \text{Shift } -u_0 \text{ of variable } u: P(u - u_0). \quad (5.4)$$

To illustrate this general result, we consider several examples. In these examples, the general shift parameter $-u_0$ will take on several different meanings. For example, the discrete-time shift is $-u_0 = -n_0$, the continuous-time shift is $-u_0 = -t_0$, the discrete-time frequency shift is $-u_0 = -\omega_0$, the Fourier series shift is $-u_0 = -k_0$, the z-shift is $-u_0 = -z_0$, and so on.

EXAMPLE 5.1

Evaluate the DTFT of $x(n - n_0)$ in terms of the DTFT of $x(n)$, namely $X(e^{j\omega})$. Investigate this property further using (4.15b) and graphics.

Solution

First we identify $P(u) = x(n)$, $Q(v) = X(e^{j\omega})$, $u_0 = n_0$, $L\{\,\cdot\,\}$ is the delay-by-n_0 operator, and $f_{21}(u, v) = e^{j\omega n}$. From (5.4) and the discussion above, we predict that $DTFT\{x(n - n_0)\} = f_{21}(-u_0, v)Q(v) = e^{-j\omega n_0}X(e^{j\omega})$. Let us verify this prediction by direct calculation. By the definition of the DTFT, we write

$$DTFT\{x(n - n_0)\} = \sum_{n=-\infty}^{\infty} x(n - n_0)e^{-j\omega n}. \qquad (5.5a)$$

Let $n' = n - n_0$. Then noting that $e^{-j\omega n} = e^{-j\omega(n'+n_0)}$ and that $e^{-j\omega n_0}$ is independent of the new sum index n', we have

$$\text{DTFT}\{x(n - n_0)\} = e^{-j\omega n_0} \sum_{n'=-\infty}^{\infty} x(n')e^{-j\omega n'} \qquad (5.5b)$$

$$= e^{-j\omega n_0}X(e^{j\omega}). \qquad (5.5c)$$

In agreement with (5.4), the DTFT of $\{x(n)$ with a shift of $n_0\}$ is the transform of the unshifted $x(n)$multiplied by the expansion function of the DTFT^{-1}, $e^{j\omega n}$, evaluated at the shift argument $n = -u_0 = -n_0$. Note that all the above work in this example is obviated by application of our general result (5.4).

Let us now arrive at an intuitive understanding of (5.5c) using (4.15b) $\left[x(n) = \int_{-\pi}^{\pi} x(n; \omega)\ d\omega, \text{ where } x(n; \omega) = X(e^{j\omega})e^{j\omega n}/(2\pi) \right]$. Let $g(n) = x(n - n_0)$ so that, by (5.5c), $G(e^{j\omega}) = \exp\{-j\omega n_0\}X(e^{j\omega})$ and by (4.15b),

$$g(n; \omega) = \frac{G(e^{j\omega})e^{j\omega n}}{2\pi} \qquad (5.6a)$$

$$= \frac{e^{j\omega(n - n_0)}X(e^{j\omega})}{2\pi}. \qquad (5.6b)$$

Now we notice that $g(n; \omega)$ is exactly equal to $x(n; \omega)|_{n - n_0}$, by evaluating the expression for $x(n; \omega)$ in (4.15b) at $n - n_0$ instead of n. As implied in the introduction to this chapter, the DTFT has transferred the shift from $x(n)$ to its constitutive components $x(n; \omega)$. With every $x(n; \omega)$ shifted by n_0, the entire integral in the expression for $x(n)$ in (4.15b) is also shifted by n_0, yielding $x(n - n_0)$. Thus, the transform property (5.5c) has the form shown because that form guarantees that each Fourier component of $x(n)$ is shifted by $-n_0$, which must be true if the resulting integral of components is to sum to $x(n - n_0)$.

That a time shift results in phase augmentation that is linear in ω [namely, $-\omega n_0$ for the DTFT; see (5.5c)] is fundamental and very important in digital filter design. We already observed this effect in Example 4.7 and its Figure 4.13. In that example, $H(e^{j\omega})$ had the linear phase $-\beta\omega$, where in Figure 4.13 we set $\beta = 3$. In the bottom row of plots in Figure 4.13, we saw that this linear phase resulted in the Fourier component ω_1 of the output $y(n)$ being delayed by $\beta = 3$ relative to that of the input $x(n)$. Because the magnitude function $|H(e^{j\omega})|$ was not the same for all ω, it follows that $y(n) \neq x(n - \beta)$.

However, we see that we could delay a total general signal by n_0 by passing it through a system with a frequency response having unit magnitude for all ω and phase equal to $-\omega n_0$. In practice, sequences are easily delayed by integer amounts n_0 using shift registers in computer systems. Thus, knowledge of the effect on a transform due to passing a signal through a shift register is clearly of practical significance in digital signal-processing system design.

To get a feeling for the linear phase augmentation produced by delay of a sequence, we consider the example $x(n) = (\frac{1}{2})^n u(n)$, whose DTFT is $X(e^{j\omega}) = 1/(1 - e^{-j\omega}/2)$. It is a good exercise to verify that in polar form, $X(e^{j\omega}) = 2/\{5 - 4\cos(\omega)\}^{1/2}$ $\angle -\tan^{-1}\{\sin(\omega)/[2 - \cos(\omega)]\}$. The magnitude $|X(e^{j\omega})|$ is plotted in Figure 5.1; it falls from a maximum of 2 at $\omega = 0$ to a minimum of $\frac{2}{3}$ for $\omega = \pm\pi$. When $x(n)$ is shifted to form $x(n - n_0)$, the magnitude $|X(e^{j\omega})|$ does not change, as examination of (5.5c) affirms. Instead, the phase $-n_0\omega$ is added to the phase $\angle X(e^{j\omega}) =$

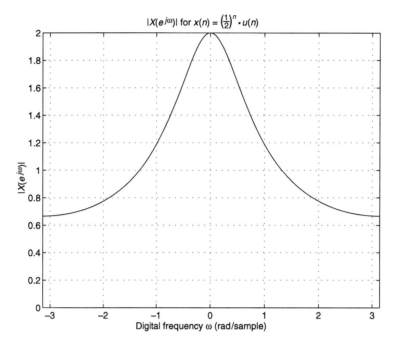

Figure 5.1

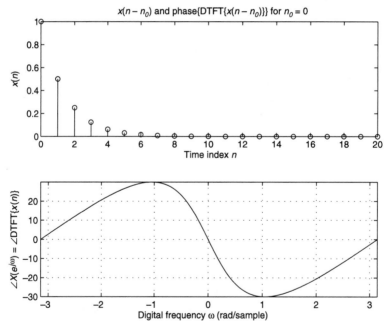

Figure 5.2a

$-\tan^{-1}\{\sin(\omega)/[2 - \cos(\omega)]\}$. In Figures 5.2$a$ through d, we show $x(n - n_0)$ and the phase $\angle \mathrm{DTFT}\{x(n - n_0)\}$ in degrees versus ω for $n_0 = 0, 1, 3,$ and 15.

To facilitate viewing the linear phase, the phase was unwrapped; from Figure 4.15, we know that in the usual phase plots there will be discontinuities of 2π whenever the angle magnitude reaches π. By looking at the unwrapped phase, we see that

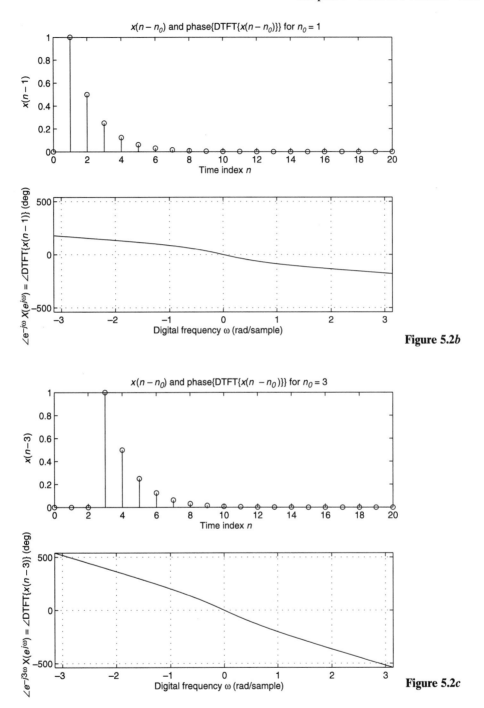

Figure 5.2b

Figure 5.2c

the more $x(n)$ is delayed, the greater the slope $-n_0$ of added phase; for $n_0 > 1$ or 2, the original shape of $\angle X(e^{j\omega})$ for $n_0 = 0$ is essentially lost, and the phase appears to be dominated by $-n_0\omega$. However, it is actually the modulo phase that affects the distribution between the real and imaginary parts of DTFT$\{x(n - n_0)\}$, so in fact the modulo counts. Under modulo, even the small phase $\angle X(e^{j\omega})$, which ranges over only $\pm 30°$ can be noticed in the sawtooth peak heights when the phase is wrapped—even for $n_0 = 15$, as shown in Figure 5.2d.

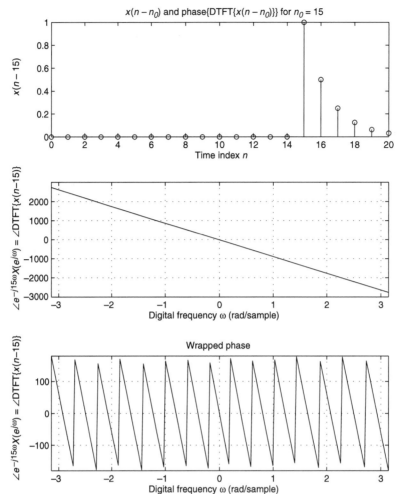

Figure 5.2d

EXAMPLE 5.2

Find the Fourier series of $\tilde{x}_c(t - t_0)$, in terms of $X_c(k)$, the Fourier series of $\tilde{x}_c(t)$, where $\tilde{x}_c(t + T) = \tilde{x}_c(t)$ [$\tilde{x}_c(t)$ is T-periodic in t].

Solution

From Table 5.2, $P(u) = \tilde{x}_c(t)$, $Q(v) = X_c(k)$, $f_{21}(u, v) = e^{j2\pi kt/T}$, and $-u_0 = -t_0$. Thus from (5.4) we immediately may conclude that $FS\{\tilde{x}_c(t - t_0)\} = e^{-j2\pi kt_0/T} \cdot X_c(k)$. However, for those wishing a traditional proof, we now provide it. By the definition of the Fourier series,

$$FS\{\tilde{x}_c(t - t_0)\} = \frac{1}{T} \int_0^T \tilde{x}_c(t - t_0)e^{-j2\pi kt/T}dt. \tag{5.7a}$$

Now let $t' = t - t_0$ so that in the exponent of (5.7a), $-t = -t_0 - t'$. Then

$$FS\{\tilde{x}_c(t - t_0)\} = \frac{e^{-j2\pi kt_0/T}}{T} \int_{-t_0}^{T-t_0} \tilde{x}_c(t')e^{-j2\pi kt'/T}dt' \tag{5.7b}$$

$$= e^{-j2\pi kt_0/T} X_c(k), \tag{5.7c}$$

where although in (5.7b) the integration limits are changed, the integral is still over exactly one period of $\tilde{x}_c(t)$, so the result is indeed $X_c(k)$ and the result is again proved. In later examples, we will often leave traditional proofs as exercises, and take advantage of our general results to eliminate the algebra. Traditional proofs are given in the basic signals and systems course.

As in Example 5.1, the transform of the shifted $\tilde{x}_c(t)$ is equal to that of the unshifted $\tilde{x}_c(t)$, namely $X_c(k)$, multiplied by the expansion function of $\tilde{x}_c(t)$ over discrete frequency, $e^{j2\pi kt/T}$, evaluated at the shift value $t = -u_0 = -t_0$. Exactly the same intuitive argument can be made as in Example 5.1 for the need for the multiplicative factor $\exp\{-j2\pi kt_0/T\}$ for delaying $\tilde{x}_c(t)$ by t_0. That factor guarantees that each Fourier series component of $\tilde{x}_c(t)$ is delayed by the same amount, t_0.

On a practical note, the main usage of oscilloscopes concerns the measurement and comparison of periodic continuous-time signals. Each screen trace shows one or several periods of the periodic signal being measured, and is triggered at a chosen instant of each period. Often the oscilloscope is used to measure t_0, the time delay between two otherwise identical or similar signals—the situation considered in this example.

EXAMPLE 5.3

If the z-transform of $x(n)$ is $X(z)$, what is the z-transform of $x(n - n_0)$, where n_0 is a constant integer?

Solution

For direct substitution into (5.4) we identify $f_{21}(-u_0, v) = z^{-n_0}$ and $Q(v) = X(z)$, giving

$$ZT\{x(n - n_0)\} = z^{-n_0}X(z). \tag{5.8a}$$

Note that the region of convergence (ROC) of $ZT\{x(n - n_0)\}$ is the same as that of $X(z)$, because z^{-n_0} is analytic everywhere. The only exception is the origin, which is not part of the ROC of $X(z)$ for causal $x(n)$.

The time-shift property must be modified for the unilateral z-transform. Moreover, it differs for n_0 positive and negative. For $n_0 > 0$, we determined the property in Chapter 3; see (3.49c) and (3.50), the former of which we reproduce here:

$$ZT_{ul}\{x(n - n_0)\} = z^{-n_0}X_{ul}(z) + \sum_{n=-n_0}^{-1} x(n)z^{-(n + n_0)}. \tag{5.8b}$$

If $x(n)$ is causal, the sum in (5.8b) disappears and we have (5.8a) also for ZT_{ul}.

Rather than write the delay theorem for $n_0 < 0$, we will write it as an advance theorem for $n_0 > 0$. It is left as an exercise to show that if $X_{ul}(z)$ is the unilateral z-transform of $x(n)$, then the unilateral z-transform of the advanced-by-n_0 version of $x(n)$ is

$$ZT_{ul}\{x(n + n_0)\} = z^{n_0}X_{ul}(z) - \sum_{n=0}^{n_0-1} x(n)\, z^{n_0 - n}, \tag{5.9}$$

by using the same techniques used to prove (5.8b). This time there is no further simplification when $x(n)$ is causal because it is generally assumed that $x(n \geq 0) \neq 0$; hence the advantage of working with delays rather than advances when unilateral ZTs are involved and sequences are causal. Note that this time the "extra" terms are subtracted, not added as in (5.8b). However, for the bilateral transform we have $\text{ZT}\{x(n + n_0)\} = z^{n_0}X(z)$, similar to the case of delay, (5.8a).

When applying this general shift property, it does not matter in which direction the transform is taken; the same rule holds either way as long as the shifted variable appears in an exponent. We simply change what we call "domain 1," as illustrated in Example 5.4 for the DTFT^{-1}, FS^{-1}, and ZT^{-1}.

EXAMPLE 5.4

Determine the appropriate inverse transform of the following: **(a)** $X(e^{j(\omega-\omega_0)})$ in terms of $x(n)$, the DTFT^{-1} of $X(e^{j\omega})$; **(b)** $X_c(k - k_0)$ in terms of $\tilde{x}_c(t)$, the FS^{-1} of $X_c(k)$ [where $\tilde{x}_c(t + T) = \tilde{x}_c(t)$]; and **(c)** $X(z - z_0)$ in terms of $x(n)$, the ZT^{-1} of $X(z)$.

Solution

(a)   Again using Table 5.2 and (5.4), we identify $f_{21}(-u_0, v) = e^{-j(-\omega_0)n}$ and $Q(v) = x(n)$, giving

$$\text{DTFT}^{-1}\{X(e^{j(\omega-\omega_0)})\} = e^{j\omega_0 n} \cdot x(n). \tag{5.10}$$

We again obtain the "transform" $x(n)$ of the given DTFT function $X(e^{j\omega})$, multiplied by the expansion function of the original function $X(e^{j\omega})$ (namely $e^{-j\omega n}$) evaluated at the shift $\omega = -\omega_0$. In the case where the time-domain function is multiplied by a (complex) sinusoid, property (5.10) is usually referred to not as the shift property but rather as the modulation property. In communications, multiplication or modulation of a signal by a sinusoid *carrier* is fundamental to the broadcasting process.

(b)   From Table 5.2 we have $f_{21}(-u_0, v) = e^{-j2\pi(-k_0)t/T}$ and $Q(v) = x_c(t)$, so from (5.4),

$$\text{FS}^{-1}\{X_c(k - k_0)\} = e^{j2\pi k_0 t/T} \cdot x_c(t). \tag{5.11}$$

(c)   This example is the only exception to the multiplication-by-shifted-expansion-function rule, because the shifted variable, $u = z$, does not itself appear in an exponent as we noted that (5.4) assumed; that is, $f_{21}(u, v) = z^{-n}$ for ZT^{-1} is not of the form d^{cu}, where $u = z$. In fact, there is no simple, general rule for this property valid for all $x(n) \leftrightarrow X(z)$ pairs; it must be derived anew for each desired $x(n) \leftrightarrow X(z)$. However, the case of the following specific example exhibits a similarity to the stated property. From Chapter 3 (e.g., Example 3.10d), we know that the inverse z-transform of $X_1(z) = z/(z - z_A)$, where $|z| > |z_A|$ and $|z_A| < 1$, is

$$x_1(n) = \text{Res}\left\{\frac{z}{z - z_A} \cdot \frac{z^n}{z}\bigg|_{z=z_A}\right\} = z_A^n u(n). \tag{5.12a}$$

The inverse z-transform of $X_1(z - z_0) = (z - z_0)/(z - z_0 - z_A)$, where $|z| > |z_0 + z_A|$ and $|z_0 + z_A| < 1$, is

$$\text{ZT}^{-1}\{X_1(z - z_0)\} = \sum \text{Res}\left\{\frac{(z - z_0)}{z - z_0 - z_A} \cdot \frac{z^n}{z}\bigg|_{z=z_0 + z_A, z=0}\right\}. \tag{5.12b}$$

In (5.12b) there is a residue at $z = 0$ only for $n = 0$. Because the form of the inverse z-transform differs for $n > 0$, we consider $n = 0$ and $n > 0$ separately. For $n = 0$, we have by partial fraction expansion

$$\sum \text{Res}\left\{\frac{z - z_0}{(z - z_0 - z_A)z}\right\} = \text{Res}\left\{\frac{z_A/(z_0 + z_A)}{z - z_0 - z_A}\bigg|_{z = z_0 + z_A} + \frac{-z_0/(-z_0 - z_A)}{z}\bigg|_{z=0}\right\} \quad (5.12c)$$

$$= \frac{z_A}{z_0 + z_A} + \frac{z_0}{z_0 + z_A} = 1.$$

For $n \geq 1$ we obtain $z_A(z_0 + z_A)^{n-1}$, so that the inverse transform is

$$\text{ZT}^{-1}\{X_1(z - z_0)\} = \delta(n) + z_A(z_0 + z_A)^{n-1}u(n - 1). \quad (5.12d)$$

As usual, we use $f(n)u(n - 1) = f(n)u(n) - f(0)\delta(n)$ [i.e., (2.54c)]:

$$\text{ZT}^{-1}\{X_1(z - z_0)\} = \left(1 - \frac{z_A}{z_0 + z_A}\right)\delta(n) + z_A(z_0 + z_A)^{n-1}u(n) \quad (5.12e)$$

$$= \frac{z_0\delta(n) + z_A(z_0 + z_A)^n u(n)}{z_0 + z_A}. \quad (5.12f)$$

For $n > 0$, the combination of terms in (5.12f), $[z_A/(z_0 + z_A)] \cdot (z_0 + z_A)^n$ bears some similarity to the expected result, $z_0^n \cdot x_1(n) = (z_0 z_A)^n u(n)$ [see (5.12a)], but is definitely distinct. Aside from different scaling, the sum of z_0 and z_A, rather than their product, is raised to the nth power.

For comparison, plots of $z_0^n \cdot x_1(n)$ and $\text{ZT}^{-1}\{X_1(z - z_0)\}$ for two example real-valued z_0 and z_A pairs are shown in Figure 5.3. To help distinguish the two sequences while still using "stem" plots, the values of $z_0^n x_1(n)$ are connected by lines. In Figure 5.3a, $z_A = 0.8$, $z_0 = 0.1$; the sequence $z_0^n x_1(n)$ decays much faster than $\text{ZT}^{-1}\{X_1(z - z_0)\}$ because $z_0 z_A \ll z_0 + z_A$ so that $(z_0 z_A)^n \ll (z_0 + z_A)^n$. In Figure 5.3b, $z_A = 0.9$, $z_0 = -0.5$; now $|z_0 z_A| \approx z_0 + z_A = 0.4$, so that now $z_0^n x_1(n)$ and $\text{ZT}^{-1}\{X_1(z - z_0)\}$ decay at about the same rate. Note, however, the (damped) oscillation of $z_0^n x_1(n)$ due to z_0 being negative real.

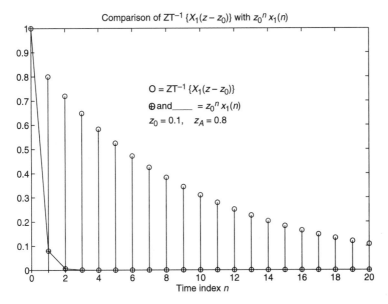

Figure 5.3a

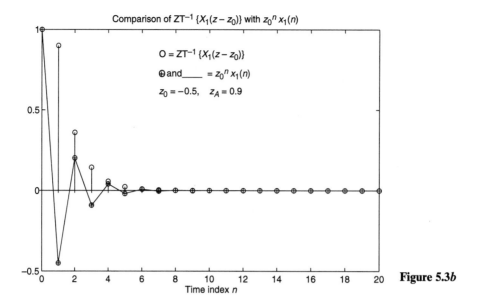

Figure 5.3b

The natural question to ask is, If $z_0^n x_1(n)$ is not $ZT^{-1}\{X_1(z - z_0)\}$, what is $ZT\{z_0^n x_1(n)\}$? In other words, if we do not have a z-transform "z-delay" property, do we have a time-domain modulation property? The answer, by the definition of the ZT, is affirmative, for this example:

$$ZT\{z_0^n \cdot z_A^n u(n)\} = \sum_{n=0}^{\infty} (z_0 z_A)^n z^{-n} \tag{5.13a}$$

$$= \frac{1}{1 - z_0 z_A/z} = \frac{z/z_0}{z/z_0 - z_A} = X_1(z/z_0). \tag{5.13b}$$

This is a special case of the general result for z-transforms that if the ZT of $x(n)$ is $X(z)$, then the ZT of $z_0^n x(n)$—that is, $x(n)$ multiplied by a geometric sequence z_0^n—is

$$ZT\{z_0^n x(n)\} = \sum_{n=-\infty}^{\infty} z_0^n x(n) z^{-n} = \sum_{n=-\infty}^{\infty} x(n)\{z/z_0\}^{-n} = X(z/z_0). \tag{5.14}$$

If the region of convergence of $x(n)$ is $r_1 < |z| < r_2$, then the region of convergence of $ZT\{z_0^n x(n)\}$ is $|z_0|r_1 < |z| < |z_0|r_2$.

We thus see that rather than an arithmetic shift in z by z_0, the analogous property for the z-transform holds for division of z by z_0. This makes sense when it is noted that $e^{x-y} = e^x/e^y$, where e^x is z and e^y is z_0; the shift is of logarithms of z, namely $x = j\omega$ and $y = j\omega_0$, for the DTFT. Moreover, an arithmetic shift in the z-plane is not really that common because the z-plane is primarily polar-oriented, so our study of it is more educational than applicational.

The above six examples constitute only half of the results for all six transform pairs. All results derived above as well as the others appear in the table of transform properties provided in Appendix 5A.

5.3   DIFFERENTIATION/DIFFERENCING OF A FUNCTION

One of the primary reasons that transforms originally attained such popularity was the ability to transform differential/difference equations into algebraic equations. This tranformation is possible because of the general rule that differentiation/differencing with respect to the original variable transforms to multiplication of the transform by a simple factor. There are related useful properties for integration of a continuous-time function or summation of a discrete-time function, a few of which are covered in the problems.

To obtain the general rule, we apply the differentiator $L\{\cdot\} = d/du$ or first differencer $L\{\cdot\} = 1 - D$, where $1\{f(n)\} = f(n)$ and $D\{f(n)\} = f(n-1)$. For all cases except ZT^{-1}, we have $f_{21}(u, v) = d^{cu}$. From (5.2b) $[P'(u) = L\{P(u)\} = \sum_{v} Q(v)L\{f_{21}(u, v)\}\gamma]$, we note that for continuous variables, $d/du\{f_{21}(u, v)\} = c \cdot f_{21}(u, v)$ so that

$$Q'(v) = cQ(v) \quad \text{Differentiation: } \frac{dP(u)}{du} \text{ (u is a continuous variable).} \tag{5.15}$$

and for discrete variables, $(1 - D)\{f_{21}(u, v)\} = (1 - d^{-c})f_{21}(u, v)$ so that

$$Q'(v) = (1 - d^{-c})Q(v). \quad \text{Differencing: } P(u) - P(u-1) \text{ (u is a discrete variable).} \tag{5.16}$$

Thus we see that if the transform of the derivative/difference of a function in domain 1 is desired, the result will be the transform of the undifferentiated/undifferenced function multiplied by a simple factor. Because of this, repeated differentiation/differencing operations merely multiply the original transform by higher powers of this factor. This property is the means by which high-order differential/difference equations may be transformed into algebraic equations.

Again some examples serve to illustrate these general comments. Results for all the transforms in Table 5.1 again appear in Appendix 5A.

EXAMPLE 5.5

Find the ZT of $\{x(n) - x(n-1)\}$ in terms of $X(z)$, the z-transform of $x(n)$.

Solution

The expression $x(n) - x(n-1)$ is the discrete-time analog of differentiation. Using (5.16) and from Table 5.2, $Q(v) = X(z)$, $d = z$, and $c = 1$, we have

$$ZT\{x(n) - x(n-1)\} = (1 - z^{-1}) \cdot X(z). \tag{5.17}$$

Equivalently, we could have used the shift property (5.8a), with $n_0 = 1$, to obtain (5.17). Recall that $x(n) - x(n-1)$ divided by sampling interval Δt is just (2.42), a simple discretization of $dx_c(t)/dt$ evaluated at $t = n\,\Delta t$, if $x(n) = x_c(n\,\Delta t)$. We conclude that for sequences, first-backward-differencing becomes just multiplication of $X(z)$ by the factor $1 - z^{-1}$. As in Example 5.3, the ROC of $ZT\{x(n) - x(n-1)\}$ is the same as that of $X(z)$.

EXAMPLE 5.6

Find the Fourier transform of $d/dt\{x_c(t)\}$ in terms of $X_c(j\Omega)$, the Fourier transform of $x_c(t)$.

Solution

Here is a case where our general result really reduces our work. We simply identify $Q(v) = X_c(j\Omega)$ and $c = j\Omega$, which put into (5.15) gives *immediately*

$$\text{FT}\left\{\frac{d}{dt}\,x_c(t)\right\} = j\Omega \cdot X_c(j\Omega). \tag{5.18}$$

Naturally, (5.18) generalizes to $\text{FT}\{d^\ell x_c(t)/dt^\ell\} = (j\Omega)^\ell X_c(j\Omega)$ for the ℓth derivative of $x_c(t)$. Deriving (5.18) the traditional way involves integration by parts and is left as an exercise.

An intuitive reason for the factor $j\Omega$ in (5.18) is that with each Fourier component $x_c(t; \Omega) = X_c(j\Omega)e^{j\Omega t}/(2\pi)$ multiplied by $j\Omega$, we are merely forming/obtaining $dx_c(t; \Omega)/dt = j\Omega X_c(j\Omega)e^{j\Omega t}/(2\pi)$ as each Fourier component of $dx_c(t)/dt$. With each component differentiated with respect to time t, we are guaranteed that the entire integral sum is differentiated, yielding $dx_c(t)/dt$.

EXAMPLE 5.7

Consider the time function $x_c(t) = \text{sinc}(\Omega_1 t) = \sin(\Omega_1 t)/(\Omega_1 t)$. We readily calculate $dx_c(t)/dt = [\Omega_1 t\cos(\Omega_1 t) - \sin(\Omega_1 t)]/(\Omega_1 t^2)$. Find $\text{FT}\{dx_c(t)/dt\}$ and plot for $\Omega_1 = 5$ rad/s.

Solution

The reader may already be aware that $X_c(j\Omega) = \text{rect}_{\Omega 1,c}(\Omega)$, where $\text{rect}_{x1,c}(x) = u_c(x + x_1) - u_c(x - x_1)$, usually proved by taking $\text{FT}^{-1}\{\text{rect}_{\Omega 1,c}(\Omega)\}$. However, it is unusual to see a proof in the "forward" direction, so we will present one here and along the way develop yet another property. We will then apply (5.18) to our result.

We notice that $\text{sinc}(\Omega_1 t) = \dfrac{1}{\Omega_1} \cdot \left\{\dfrac{\sin(\Omega_1 t)}{t}\right\}$. Noting that $\sin(\Omega_1 t) = \left(\dfrac{1}{2j}\right)\{e^{j\Omega_1 t} - e^{-j\Omega_1 t}\}$ and, from Appendix 2A (2A.19b), $\text{FT}\{e^{j\Omega_1 t}\} = 2\pi\delta_c(\Omega - \Omega_1)$, we have $\text{FT}\{\sin(\Omega_1 t)\} = j\pi\{\delta_c(\Omega + \Omega_1) - \delta_c(\Omega - \Omega_1)\}$. All we now need is $\text{FT}\{x_c(t)/t\}$. Write

$$\text{LT}\left\{\frac{x_c(t)}{t}\right\} = \int_{-\infty}^{\infty} x_c(t)\left\{\frac{e^{-st}}{t}\right\}dt, \tag{5.19}$$

and note that the term in brackets can be written

$$\frac{e^{-st}}{t} = \frac{e^{-s't}}{t}\Bigg|_s^{\infty} = \int_s^{\infty} e^{-s't} ds', \tag{5.20}$$

where the contour of this integral is any one that starts on s and goes to ∞ in the right half-plane (RHP) so that the "∞" boundary term goes to zero. Thus (5.19) can be written, upon switching the order of the integrations, as

$$LT\left\{\frac{x_c(t)}{t}\right\} = \int_s^{\infty} \int_{-\infty}^{\infty} x_c(t) e^{-s't} dt\, ds' = \int_s^{\infty} X_c(s')\, ds'. \tag{5.21}$$

(This "divide-by-t" property is no easier to obtain using the $P(u)$, $Q(v)$ approach.) Thus, letting $s' = j\Omega'$, we obtain[3]

$$FT\{\mathrm{sinc}(\Omega_1 t)\} = \frac{j\pi}{\Omega_1} \int_{j\Omega}^{\infty} \{\delta_c(\Omega' + \Omega_1) - \delta_c(\Omega' - \Omega_1)\}\, j d\Omega'. \tag{5.22a}$$

There are three cases: (a) If $\Omega < -\Omega_1$, both delta functions are included on the integration range, yielding zero; (b) if $\Omega > \Omega_1$, neither is included, again yielding zero; and (c) if $|\Omega| < \Omega_1$, only the $-\delta_c(\Omega' - \Omega_1)$ is included, yielding

$$FT\{\mathrm{sinc}(\Omega_1 t)\} = \begin{cases} \dfrac{\pi}{\Omega_1}, & |\Omega| < \Omega_1 \\[2mm] 0, & \text{otherwise.} \end{cases} \tag{5.22b}$$

It is much easier to prove (5.22b) by going "backwards," but suppose that we did not already know the result? However, we could also use duality.

We can now immediately obtain the FT of $d\{\mathrm{sinc}(\Omega_1 t)\}/dt$. By using (5.18), we have

$$FT\left\{\frac{d[\mathrm{sinc}(\Omega_1 t)]}{dt}\right\} = j\Omega \cdot FT\{\mathrm{sinc}(\Omega_1 t)\}$$
$$= \begin{cases} \dfrac{j\pi\Omega}{\Omega_1}, & |\Omega| < \Omega_1 \\[2mm] 0, & \text{otherwise.} \end{cases} \tag{5.23}$$

Notice that the expression in the problem statement, $d[\mathrm{sinc}(\Omega_1 t)]/dt = [\Omega_1 t \cos(\Omega_1 t) - \sin(\Omega_1 t)]/(\Omega_1 t^2)$ was not helpful. Although presumably it could be used to directly calculate $FT\{d[\mathrm{sinc}(\Omega_1 t)]/dt\}$, the calculation would be difficult and is unnecessary with the help of the transform property (5.18). In Figure 5.4a is shown $\mathrm{sinc}(\Omega_1 t)$ and the magnitude and phase of $FT\{\mathrm{sinc}(\Omega_1 t)\}$, whereas Figure 5.4b shows $d\{\mathrm{sinc}(\Omega_1 t)\}/dt$ and the magnitude and phase of $FT\{d[\mathrm{sinc}(\Omega_1 t)]/dt\}$, for $\Omega_1 = 5$ rad/s.

---

[3] All we need for convergence is for s' to hug the right edge of the $j\Omega'$ axis for $\Omega' \gg \Omega_1$.

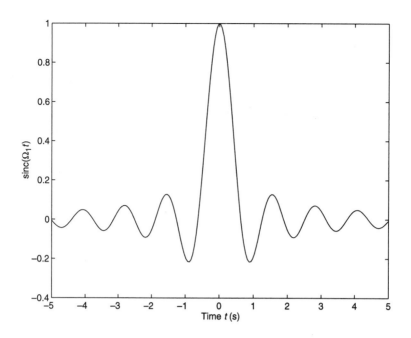

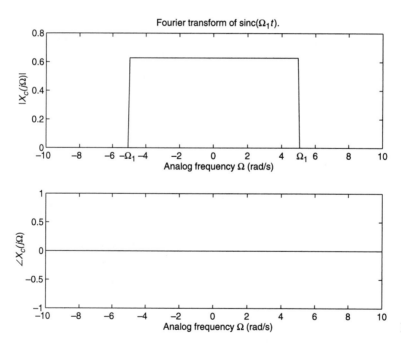

Figure 5.4a

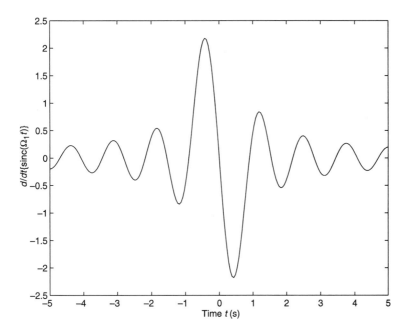

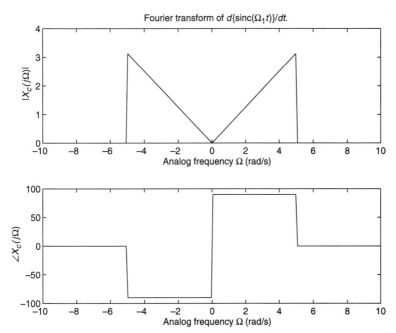

Figure 5.4*b*

EXAMPLE 5.8

Find the inverse Fourier transform of $d/d\Omega\{X_c(j\Omega)\}$ in terms of $x_c(t)$, the inverse Fourier transform of $X_c(j\Omega)$.

Solution

Again, using our general result avoids integration by parts. We simply identify $Q(v) = x_c(t)$ and $c = -jt$, which put into (5.15) gives

$$\text{FT}^{-1}\left\{\frac{d}{d\Omega} X_c(j\Omega)\right\} = -jt \cdot x_c(t). \tag{5.24a}$$

We may obtain a neater-looking form of this result from the point of view of the time domain by writing (5.24a) as

$$\text{FT}^{-1}\left\{j\frac{d}{d\Omega}X_c(j\Omega)\right\} = t \cdot x_c(t), \tag{5.24b}$$

which may be generalized to

$$\text{FT}^{-1}\left\{j^m\frac{d^m}{d\Omega^m}X_c(j\Omega)\right\} = t^m \cdot x_c(t), \quad m \geq 0. \tag{5.24c}$$

EXAMPLE 5.9

Determine $\text{DTFT}^{-1}\{dX(e^{j\omega})/d\omega\}$ in terms of $x(n)$, the DTFT^{-1} of $X(e^{j\omega})$.

Solution

We identify $Q(v) = x(n)$ and $c = -jn$, which put into (5.15) gives

$$\text{DTFT}^{-1}\left\{\frac{d}{d\omega} X(e^{j\omega})\right\} = -jn \cdot x(n). \tag{5.25}$$

Again, (5.25) may be rearranged and generalized to a form analogous to (5.24b and c). Note that in Example 3.10, part (h), we already showed the related z-transform property, $\text{ZT}^{-1}\left\{-z\frac{d}{dz}\{X(z)\}\right\} = nx(n)$.

EXAMPLE 5.10

Determine $\text{DFS}^{-1}\{[\tilde{X}(k) - \tilde{X}(k-1)]/(2\pi/N)\}$ in terms of $\tilde{x}(n)$, the inverse DFS of $\tilde{X}(k)$, where $\tilde{x}(n+N) = \tilde{x}(n)$ and $\tilde{X}(k+N) = \tilde{X}(k)$.

Solution

This property is the difference form analogous to differentiation for the discrete Fourier series, where the $2\pi/N$ is the ω-frequency spacing between samples. We identify $Q(v) = \tilde{x}(n)$, $d = e$, and $c = -j2\pi n/N$, which put into (5.16) gives

$$\text{DFS}^{-1}\left\{\frac{\tilde{X}(k) - \tilde{X}(k-1)}{2\pi/N}\right\} = \frac{N}{2\pi}(1 - e^{j2\pi n/N}) \cdot \tilde{x}(n). \tag{5.26}$$

It is interesting to consider the case $n \ll N$ because (5.26) then reduces to approximately $N/(2\pi) \cdot [1 - (1 + j2\pi n/N)] \cdot \tilde{x}(n) = -jn \cdot \tilde{x}(n)$, which matches (5.25). Physically, for small n and large N, the N-periodic sequence $\tilde{x}(n)$ behaves essentially like an aperiodic sequence.

Naturally, (5.26) holds also for the DFT because the DFT and DFS are equivalent. The only difference is that in practice, the DFT is used to represent finite-length sequences (length N). Hence all arguments of both $X(k)$ and $x(n)$ must be taken modulo N (see Section 4.5). Recall that "modulo N" means to add or subtract a sufficient integer multiple of N to the index argument to make the result lie on $[0, N-1]$. For example, if $k = 0$, we write on the left-hand side of (5.26), $X(0) - X(-1, \text{modulo } N) = X(0) - X(N-1)$, where in the second term we added N to -1. Both $X(0)$ and $X(N-1)$ are available from a computer-generated DFT array as, respectively, the first and last array elements.

5.4   SIGN REVERSAL OF A VARIABLE

One of the simplest yet most commonly used operations on a signal is sign reversal of the argument variable. One example of time reversal we shall use below occurs in the derivation of the deterministic correlation property from the convolution property. As is (even more) true for the other properties, knowing time reversal immediately doubles the number of transforms we know.

Assuming $f_{21}(u, v) = d^{cu}$, we have $L\{f_{21}(u, v)\} = d^{c(-u)} = d^{(-c)u}$. Except for the cases ZT and ZT^{-1}, reversal of the domain 1 variable merely reverses the domain 2 variable v, so $L\{f_{21}(u, v)\} = f_{21}(u, -v)$. When substituted into (5.2b) $[P'(u) = L\{P(u)\} = \sum_v Q(v)L\{f_{21}(u, v)\}\gamma]$ and again reversing v, we obtain $P'(u) = \sum_v Q(-v)f_{21}(u, v)$, from which we identify

$$Q'(v) = Q(-v). \qquad \text{Sign reversal of variable } u \text{: } P(-u). \tag{5.27}$$

EXAMPLE 5.11

Find the DTFT of $x(-n)$ in terms of $X(e^{j\omega})$, the DTFT of $x(n)$.

Solution

We have $Q(v) = X(e^{j\omega})$, so we immediately obtain from (5.27)

$$\text{DTFT}\{x(-n)\} = X(e^{-j\omega}). \tag{5.28}$$

This example also tells us immediately the sign-reversal-on-ω property: $\text{DTFT}^{-1}\{X(e^{-j\omega})\} = x(-n)$.

EXAMPLE 5.12

Find the z-transform of $x(-n)$.

Solution

We have $Q(v) = X(z)$ and $f_{21}(u, v) = z^n$. We can apply the general approach, but (5.27) will not apply because in $f_{21}(u, v) = d^{cu} = z^n$, sign reversal of z is not equivalent to sign reversal of n. Instead, note that $f_{21}(-u, v) = z^{-n} = f_{21}(u, 1/z)$, which when substituted into (5.2b) will mean we must change integration variables to $w = 1/z$. Although doing this gives the correct result, a direct ZT calculation avoids the calculus. Defining $n' = -n$, we obtain

$$ZT\{x(-n)\} = \sum_{n=-\infty}^{\infty} x(-n)\, z^{-n} = \sum_{n'=-\infty}^{\infty} x(n')z^{n'} = \sum_{n'=-\infty}^{\infty} x(n')\left[\frac{1}{z}\right]^{-n'} = X\left(\frac{1}{z}\right) \quad (5.29)$$

Note that substitution of $z = e^{j\omega}$ into (5.29) yields (5.28) directly; thus often in proving properties it is desirable to just find the z-transform property from the beginning to avoid having to do a second derivation later. If the ROC of $X(z)$ lies for $|z|$ between r_1 and $r_2 > r_1$, the ROC of $ZT\{x(-n)\} = X(1/z)$ is $1/r_2 < |z| < 1/r_1$.

EXAMPLE 5.13

Find the inverse z-transform of $X(-z)$.

Solution

Again, (5.27) does not apply. We have $f_{21}(-u, v) = (-z)^{-n} = (-1)^n z^{-n}$. Therefore, substitution into (5.2b) reveals

$$ZT^{-1}\{X(-z)\} = (-1)^n x(n). \quad (5.30)$$

Note that because $X(-z) = X(z/(-1))$, (5.30) is just a special case of (5.14) $[ZT\{z_0^n x(n)\} = X(z/z_0)]$ with $z_0 = -1$. The result in (5.30) differs significantly from $\text{DTFT}^{-1}\{X(e^{-j\omega})\} = x(-n)$ because of the differing placement of the reversed variable.

5.5 CONJUGATION OF A FUNCTION

Another common operation in DSP analytical manipulations is the complex conjugation of a function $P(u)$. This time, L operates on the entire expression in (5.1b) $[P(u) = \sum_v Q(v)f_{21}(u, v)\gamma]$, not just on $f_{21}(u, v)$ as previously [so (5.2a) applies, not (5.2b)]. Thus

$$P'(u) = L\{P(u)\} = P^*(u) \tag{5.31a}$$

$$= \sum_v Q^*(v) f_{21}^*(u, v) \gamma^*. \tag{5.31b}$$

For 11 of the 12 transform relations in Tables 5.1 and 5.2, $f_{21}(u, v) = d^{cu}$, giving

$$P^*(u) = \sum_v Q^*(v) d^{*c^*u^*} \gamma^*. \tag{5.31c}$$

In practice, (5.31c) translates to

$$Q'(v) = \begin{cases} Q^*(-v) & \text{for Fourier relations} \\ Q^*(v^*) & \text{for GPE relations (ZT and LT).} \end{cases} \qquad \text{Conjugation: } P^*(u). \tag{5.31d}$$

EXAMPLE 5.14

Find the inverse Fourier series of $X_c^*(k)$ in terms of $\tilde{x}_c(t)$, the inverse Fourier series of $X_c(k)$, where $\tilde{x}_c(t + T) = \tilde{x}_c(t)$.

Solution

We have $\gamma = \gamma^*$ and $d^{*c^*u^*} = e^{j2\pi kt/T} = e^{-j2\pi k(-t)/T}$. When these values are substituted into (5.31c) and compared with (5.1b), we conclude in agreement with (5.31d) that

$$\text{FS}^{-1}\{X_c^*(k)\} = x_c^*(-t). \tag{5.32}$$

EXAMPLE 5.15

Find the z-transform of $x^*(n)$ in terms of $X(z)$, the z-transform of $x(n)$.

Solution

Here (5.31c) becomes

$$P^*(u) = x^*(n) = \oint_C X^*(z)(z^*)^n \left[\frac{-dz^*}{2\pi j z^*} \right],$$

which upon the substitution $w^* = z$ and noting that the negative sign in $[\cdot]$ above is canceled by the reversal in direction of the resulting contour gives

$$x^*(n) = \oint_C X^*(w^*) \frac{w^n dw}{2\pi j w} = \text{ZT}^{-1}\{X^*(z^*)\},$$

or as in (5.31d),

$$\text{ZT}\{x^*(n)\} = X^*(z^*) \qquad [\text{ROC same as for } X(z)]. \tag{5.33}$$

Note that if $z = e^{j\omega}$ is put into (5.33), we obtain directly [as in (5.31d)] that $\text{DTFT}\{x^*(n)\} = X^*(e^{-j\omega})$.

EXAMPLE 5.16

Find the inverse DFS of $\tilde{X}^*(k)$ in terms of $\tilde{x}(n)$, the inverse DFS of $X(k)$, where $\tilde{x}(n + N) = \tilde{x}(n)$ and $\tilde{X}(k + N) = \tilde{X}(k)$.

Solution

With the same steps as in Example 5.14, we obtain immediately

$$\text{DFS}^{-1}\{\tilde{X}^*(k)\} = \tilde{x}^*(-n). \qquad (5.34)$$

Again, for the DFT we would use $x^*(-n,$ modulo $N)$, for example, $x^*(N - n)$ for $n \in [1, N - 1]$.

5.6 SCALING OF A VARIABLE

When a variable is stretched in domain 1, it is squeezed in domain 2. If it is squeezed in domain 1, it is stretched in domain 2. That is the basic principle behind the effect on a transform of scaling the variable of the original function. In addition, the transform may be divided by the scale factor.

Specifically, for u scaled by α, (5.2b) becomes

$$P'(u) = \sum_v Q(v)f_{21}(\alpha u, v)\gamma \qquad (5.35a)$$

$$= \sum_v Q(v)d^{(c\alpha)u}\gamma. \qquad \text{Scaling of } u \text{ by } \alpha: P(\alpha u). \qquad (5.35b)$$

$[Q'(v) \text{ varies; often } Q'(v) = Q(v/\alpha)/|\alpha|, \text{ or similar.}]$

The particular form of this property varies widely among the transforms in Table 5.1. For example, we do not expect scaling of the DTFT variable ω and the ZT variable z to yield similar time-domain results because scaling ω in $z = e^{j\omega}$ is very different from scaling z itself. Also, when periodic functions are involved [FS: $\tilde{x}_c(t)$, DTFT: $X(e^{j\omega})$, and DFS: $\tilde{x}(n)$ and $\tilde{X}(k)$], special conditions must be imposed on the scaling factor for a simple property to result. Such scaling properties for the DTFT and DFT are derived in Chapter 7. Here we shall consider some other interesting examples. Typically, in making (5.35b) look like (5.1b) we use $v_{\text{new}} = \alpha v_{\text{old}}$ and end up with $Q(v_{\text{old}})$ replaced by $Q(v_{\text{new}}/\alpha)$, as stated in words above.

EXAMPLE 5.17

Find the Laplace transform of $x_c(\alpha t)$, where α is a real-valued nonzero constant, in terms of $X_c(s)$, the Laplace transform of $x_c(t)$.

Solution

First assume that $\alpha > 0$. With $c = s$, in (5.35b) we define $w = \alpha s$, so $ds = w/\alpha$ so $\sum_v Q(v)d^{(c\alpha)u}\gamma = \int [X_c(w/\alpha)/\alpha] e^{st} ds/(2\pi j)$ and we identify $\text{LT}\{x_c(\alpha t)\} = X_c(s/\alpha)/\alpha$. For

$\alpha < 0$, the only change is that the integration limits reverse sign, which when restored produce negation, giving $LT\{x_c(\alpha t)\} = -X_c(s/\alpha)/\alpha$. Thus for all α,

$$LT\{x_c(\alpha t)\} = \frac{X_c(s/\alpha)}{|\alpha|}. \tag{5.36}$$

If the ROC of $X_c(s)$ is $\sigma_1 < \sigma < \sigma_2$, then because now s/α must be between σ_1 and σ_2, it follows that the ROC of $LT\{x_c(\alpha t)\}$ is $\alpha\sigma_1 < \sigma < \alpha\sigma_2$ for $\alpha > 0$, and it is $\alpha\sigma_2 < \sigma < \alpha\sigma_1$ for $\alpha < 0$.

Let us once again make an intuitive argument for our transform property (5.36), time scaling. In analogy with (4.15), for the LT case we consider the complex sinusoidal component $x_c(t; s)ds = X_c(s)e^{st} ds/(2\pi j)$, where we include the differential ds this time, because it will be needed to make our intuitive argument for the scaling property (5.36). Replacing t by αt gives $x_c(\alpha t; s)ds = X_c(s)e^{s\alpha t}ds/(2\pi j) = X_c(w/\alpha)e^{wt}dw/(2\pi j\alpha) = [X_c(w/\alpha)/\alpha]e^{wt}dw/(2\pi j)$, where again $w = \alpha s$. Thus, yet again, the property in (5.36) merely guarantees that the time-scaling operation is performed on each component of $x_c(t)$ in its LT^{-1} expansion.

EXAMPLE 5.18

Find the Fourier series of $\tilde{x}_c(Mt)$ in terms of $X_c(k)$, the Fourier series of $\tilde{x}_c(t)$, where $\tilde{x}_c(t + T) = \tilde{x}_c(t)$. We constrain M to be an integer so as to obtain a simple result.

Solution

In this case, the solution is *much* simpler using our general approach than conventional methods. We have $d^{\alpha cu} = e^{j2\pi kMt/T}$. For convenience, rename k to be ℓ. Thus (5.2b) becomes

$$\tilde{x}_c(Mt) = \sum_{\ell=-\infty}^{\infty} X_c(\ell)\, e^{j2\pi(\ell M)t/T}.$$

To make the above look like the usual Fourier series (i.e., matching the crucial exponential), define $k = \ell M$, so $\ell = k/M$ if and only if $k = \ell M$; otherwise, we simply substitute zero for the Fourier series coefficient $X_c'(k) \equiv FS\{\tilde{x}_c(Mt)\}$ in the FS^{-1} expression $\tilde{x}_c(Mt) = \sum_{k=-\infty}^{\infty} X'_c(k)\, e^{j2\pi kt/T}$. We thus immediately obtain

$$FS\{\tilde{x}_c(Mt)\} = \begin{cases} X_c\left(\dfrac{k}{M}\right), & k = \ell M \\ 0, & \text{otherwise.} \end{cases} \tag{5.37}$$

Equation (5.37) is the *zero-interlace* formula discussed in Section 4.6; see Figure 4.20j and, for discrete time, Figure 4.21j. Notice that defining $\tilde{x}_c(Mt)$ and finding its Fourier series, assuming it is T-periodic, is equivalent to determining the Fourier series of T-periodic $\tilde{x}_c(t)$ under the assumption that it is MT-periodic, as done in Section 4.6. This derivation was promised in that discussion.

EXAMPLE 5.19

Find the z-transform of $x(Mn)$ in terms of $X(z)$, the z-transform of $x(n)$.[4]

Solution

For this application, (5.2b) becomes $P'(u) = x(Mn) = \oint_C X(z) z^{Mn} dz/(2\pi j z)$. Let $w = z^M$, so $dw = Mdz \cdot z^{M-1}$ or $dz/z = dw/(Mz^M) = dw/(Mw)$. With C the unit circle, for z traversing the unit circle once, w traverses it M times. We can write this traversal of w as the sum of w going M times around the primary unit circle provided we append $m \cdot 2\pi$ to the phase of w on the mth time around. That seems superfluous until we recall that $z = w^{1/M}$, so on the mth time around the argument of $X(\cdot)$ is $z = \{w \cdot e^{j2\pi m}\}^{1/M}$. Finally, replacing the symbol w by the generic (new) $z \equiv w$, we have

$$x(Mn) = \oint_C \left[\frac{1}{M} \sum_{m=0}^{M-1} X(z^{1/M} \cdot e^{j2\pi m/M}) \right] z^n \frac{dz}{2\pi j z}$$

and thus

$$\mathrm{ZT}\{x(Mn)\} = \frac{1}{M} \sum_{m=0}^{M-1} X(z^{1/M} \cdot e^{j2\pi m/M}). \qquad (5.38)$$

Equation (5.38) is fundamental in the *downsampling* procedure we discuss in Section 7.5.2.

As a particular example of this scaling of n by the factor M, let $x(n) = 0.95^n \cos(15° \cdot [\pi/180°]n)u(n)$, a damped sinusoidal sequence; see Figure 5.5a for the first 100 points. The time-scaled ("downsampled") version for $M = 3, x(3n)$, is shown in Figure 5.5b; we merely take every third value of $x(n)$ in Figure 5.5a. It is left as an exercise to show that the z-transform of $x(n)$ is $X(z) = \{1 - 0.95 \cos(15 \cdot \pi/180)/z\}/ \{1 - 2 \cdot 0.95 \cos(15 \cdot \pi/180)/z + (0.95/z)^2\}$, which converges only for $|z| > 0.95$.

The magnitude of $X(z)$ is shown in Figure 5.5c, and the magnitude of $\mathrm{ZT}\{x(3n)\}$ as given in (5.38) is shown in Figure 5.5d. Because $X(z)$ does not converge for $|z| < 0.95$, it is artificially set to zero there. In Figure 5.5c, we see the poles at 0.95 $\angle \pm 15°$, whereas the poles in Figure 5.5d are evidently at angles $\pm 45°$. It makes sense that the poles of $x(3n)$ are at angles three times those of the poles of $x(n)$ because $\cos(\omega[3n]) = \cos([3\omega]n)$; by taking every third point, we "get to" the next period "faster."

Similarly, we expect the magnitudes of the poles in Figure 5.5d to be $0.95^3 = 0.857$ due to the $z^{1/M}$ term in (5.38). That is, if $X(z_1) = \infty$, then $X(z_A^{1/M}e^{j2\pi m/M}) = \infty$ and thus $\mathrm{ZT}\{x(Mn)\} = \infty$ for $z_A = e^{-j2\pi M/M} z_1^M = z_1^M$, which explains both the magnitude and phase alterations of the poles. This relation is again evident in Figure 5.5d in that the poles are seen to be farther within the unit circle than those of Figure 5.5c. Specifically, $x(3n) = 0.857^n \cos(45° \cdot [\pi/180]n)u(n)$. Evidently, $\mathrm{ZT}\{x(3n)\}$ has a larger region of convergence than $\mathrm{ZT}\{x(n)\}$: $|z| > 0.857$. Therefore in Figure 5.5d, the magnitude is set to zero only for $|z| < 0.857$; the contour plots below which show the unit circle are especially helpful here.

It would be an interesting although perhaps tedious exercise to verify that the expression for $X(z)$ above—with 0.95 replaced by 0.857 and 15° replaced by 45°, namely $\mathrm{ZT}\{x(3n)\} = \{1 - 0.857 \cos(45 \cdot \pi/180)/z\}/\{1 - 2 \cdot 0.857 \cos(45 \cdot \pi/180)/z +$

---

[4]For an alternative derivation that does not involve complex integration, see the problems of Chapter 7.

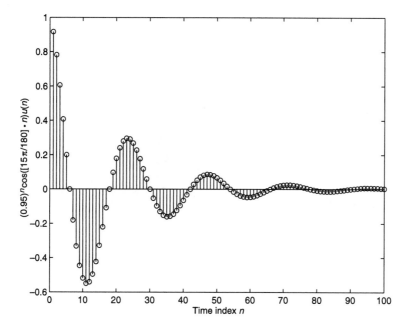

Figure 5.5*a*

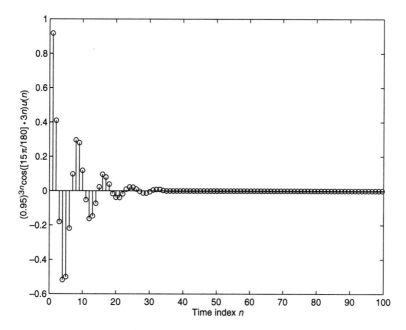

Figure 5.5*b*

$(0.857/z)^2\}$—is in fact the result obtained by substituting the expression for $X(z)$, namely $X(z) = \{1 - 0.95 \cos(15 \cdot \pi/180)/z\}/\{1 - 2 \cdot 0.95 \cos(15 \cdot \pi/180)/z + (0.95/z)^2\}$, into (5.38) with $M = 3$. We indeed find that the magnitude and phase plots for those two expressions are identical.

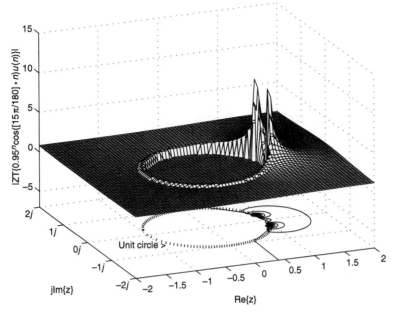

Figure 5.5c

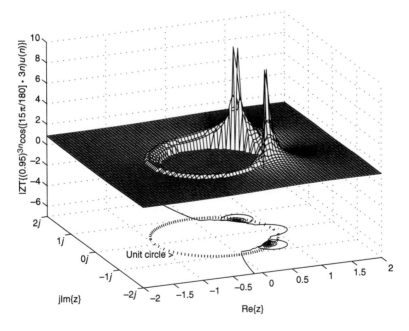

Figure 5.5d

5.7   MULTIPLICATION↔CONVOLUTION OF FUNCTIONS

Of all the transform properties, the most well-known and commonly used is the convolution ↔ multiplication property. In this section, we derive a few example forms of this property. Our general approach in previous sections has served us well. Although it can be extended to convolution,[5] in this case it is perhaps harder because of adaptational complexities. So we now return to traditional derivations.

Although we like to "start off with" convolution in the time domain due to its physical meaning for linear shift-invariant systems, the common form for all transforms is the product of functions. That is, the particular form of "convolution" varies; in the case of $ZT\{x(n)y(n)\}$ (see Appendix 5A and the problems), it differs to the point where the word "convolution" may no longer seem very descriptive. Therefore, in this section and in Appendix 5A, we always state the theorem by taking the transform of a product of functions, which is always the same.

EXAMPLE 5.20

Find the DTFT of $x(n)y(n)$ in terms of $X(e^{j\omega})$ and $Y(e^{j\omega})$, the respective DTFTs of $x(n)$ and $y(n)$.

Solution

$$\text{DTFT}\{x(n)y(n)\} = \sum_{n=-\infty}^{\infty} x(n)y(n)e^{-j\omega n}, \tag{5.39a}$$

which if we express $x(n)$ as the DTFT^{-1} of $X(e^{j\omega})$ (see, for example, Table 5.1) becomes

$$\text{DTFT}\{x(n)y(n)\} = \sum_{n=-\infty}^{\infty} \frac{1}{2\pi} \int_{-\pi}^{\pi} X(e^{j\omega'})e^{j\omega'n}d\omega' y(n)e^{-j\omega n}. \tag{5.39b}$$

By reversing the sum and integral, we obtain

$$\text{DTFT}\{x(n)y(n)\} = \frac{1}{2\pi} \int_{-\pi}^{\pi} X(e^{j\omega'}) \sum_{n=-\infty}^{\infty} y(n)\, e^{-j(\omega-\omega')n}d\omega' \tag{5.39c}$$

$$= \frac{1}{2\pi} \int_{-\pi}^{\pi} X(e^{j\omega'})\, Y(e^{j(\omega-\omega')})\, d\omega', \tag{5.39d}$$

where in (5.39d) we recognized the DTFT of $y(n)$ appearing in (5.39c). The DTFT of the product of two sequences is the convolution of their DTFTs. Again, this property is often called a modulation property because, for example, $y(n)$ modulates $x(n)$. An illuminating short exercise is to substitute $\text{DTFT}\{y(n) = e^{j\omega_0 n}\}$ into (5.39d)

---

[5]We give the extension here for $f_{21}(u,v) = d^{cu} = d^{avu}$ where $a = c/v$ is a constant for each transform, true for all but the ZT and ZT^{-1} relations. Then (5.2b) for $P(u)$ multiplied by $g(u)$ and the transform of $g(u)$ denoted $G(v)$, $P'(u) = L\{P(u)\} = P(u)g(u) = \sum_v Q(v)f_{21}(u,v)g(u)\gamma = \sum_v Q(v)f_{21}(u,v)\left[\sum_{v'} G(v')f_{21}(u,v')\gamma\right]\gamma$.
Now, $f_{21}(u,v)f_{21}(u,v') = d^{a(v+v')u}$ so $P'(u) = \sum_v Q(v)\sum_{v'} G(v')d^{a(v+v')u}\gamma^2$, which when we define $v'' = v + v'$ becomes $P'(u) = \sum_{v''}\left[\sum_v Q(v)\, G(v''-v)\gamma\right]d^{av''u}\gamma$, giving $Q'(v) = \sum_v Q(v)\, G(v''-v)\gamma = Q*G$, which is the modulation/convolution relation.

and arrive at the conclusion that the DTFT shift-in-ω property (5.10) $[\text{DTFT}^{-1}\{X(e^{j(\omega-\omega_0)})\} = e^{j\omega_0 n} \cdot x(n)]$ is just a special case of (5.39d).

<hr>

EXAMPLE 5.21

Find the ZT^{-1} of $X(z)Y(z)$ in terms of $x(n)$ and $y(n)$, the inverse ZTs of $X(z)$ and $Y(z)$, respectively.

Solution

By the inverse z-transform integral (see, for example, Table 5.1),

$$\text{ZT}^{-1}\{X(z)Y(z)\} = \frac{1}{2\pi j} \oint_C X(z)Y(z)z^n \frac{dz}{z}, \tag{5.40a}$$

which when we express $X(z)$ and $Y(z)$ as z-transforms of, respectively, $x(n)$ and $y(n)$ (see, for example, Table 5.1) becomes

$$\text{ZT}^{-1}\{X(z)Y(z)\} = \frac{1}{2\pi j} \oint_C \sum_{n'=-\infty}^{\infty} x(n')\, z^{-n'} \sum_{m=-\infty}^{\infty} y(m)\, z^{-m} z^n \frac{dz}{z}, \tag{5.40b}$$

or, rearranging,

$$\text{ZT}^{-1}\{X(z)Y(z)\} = \frac{1}{2\pi j} \oint_C \sum_{n'=-\infty}^{\infty} x(n') \sum_{m=-\infty}^{\infty} y(m)\, z^{-(n'+m)} z^n \frac{dz}{z}. \tag{5.40c}$$

Upon making the substitution $n_A = n' + m$ (and thus $m = n_A - n'$), we obtain

$$\text{ZT}^{-1}\{X(z)Y(z)\} = \frac{1}{2\pi j} \oint_C \left\{ \sum_{n_A=-\infty}^{\infty} \left[\sum_{n'=-\infty}^{\infty} x(n')y(n_A - n') \right] z^{-n_A} \right\} z^n \frac{dz}{z}. \tag{5.40d}$$

Lo and behold, we have the ZT^{-1} of the ZT of the quantity in $[\,\cdot\,]$, giving

$$\boxed{\text{ZT}^{-1}\{X(z)Y(z)\} = \sum_{n'=-\infty}^{\infty} x(n')y(n - n') = x(n) * y(n). \tag{5.40e}}$$

If we instead take the z-transform of the convolution in (5.40e), we obtain $X(z)Y(z)$ with slightly less algebra (as is shown in most signals and systems courses). Here, however, we began with the simpler product $X(z)Y(z)$ and obtained convolution as our result.

If now $z = e^{j\omega}$ is substituted into (5.40e), the familiar result cited in Chapter 4 is that $\text{DTFT}^{-1}\{X(e^{j\omega})Y(e^{j\omega})\} = x(n) * y(n)$. Recall that we had a complete graphical demonstration of the DTFT convolution theorem in Example 4.7, with all the details illustrated in Figure 4.13.

<hr>

EXAMPLE 5.22

If $y(n) = x(n)h(n)$, what is the unilateral z-transform $Y(z)$ in terms of $X(z)$ and $H(z)$? Assume that $x(n < 0) = h(n < 0) = 0$, although the result holds generally, with suitable modifications.

Solution

First note that if the radius of the largest-radius pole of $x(n)$ is r_x [so the radius of convergence of $X(z)$ is r_x] and that of $h(n)$ is r_h, then $x(n)$ has a geometric term r_x^n

and $h(n)$ has a term r_h^n. Thus in $y(n) = x(n)h(n)$, there will be the geometric terms $r_x^n, r_h^n,$ and $(r_x r_h)^n$, where it might be true that $r_x r_h > \max(r_x, r_h)$ if both r_x and r_h are > 1 [in which case the system $h(n)$ and the signal $x(n)$ are unstable]. Consequently, we can well guess that the radius of convergence of $Y(z)$ will be $\max(r_x, r_h, r_x r_h)$.

Now write $Y(z)$ as the z-transform of $\{y(n) = x(n)h(n)\}$, and then write $x(n)$ as the ZT^{-1} of $X(z)$ (with z replaced by w):

$$Y(z) = \sum_{n=0}^{\infty} x(n)h(n)z^{-n} = \frac{1}{2\pi j} \oint_C X(w) \sum_{n=0}^{\infty} h(n)w^{n-1}z^{-n}dw. \qquad (5.41a)$$

Recognizing the sum as $1/w$ times the z-transform of $h(n)$ evaluated at z/w gives

$$Y(z) = \text{ZT}\{x(n)h(n)\} = \frac{1}{2\pi j} \oint_C X(w) H\left(\frac{z}{w}\right)\frac{dw}{w}, \quad \text{valid for } |z| > \max(r_x, r_h, r_x r_h). \quad (5.41b)$$

Where is the w-contour C run? It must simultaneously reside in the ROCs of $X(w)$ ($|w| > r_x$) and $H(z/w)$ ($|z/w| > r_h$ or $|w| < |z|/r_h$); that is, $r_x < |w| < |z|/r_h$, where from above $|z| > \max\{r_x, r_h, r_x r_h\}$. It is seen that in this case the modulation property does not result in familiar convolution (in the z-domain).

5.8 DETERMINISTIC CORRELATION OF FUNCTIONS (CROSS- AND AUTO-)

A pervasive operation in statistical signal processing is the correlation between two signals. In a deterministic, discrete-time setting, the complex cross-correlation between $x(n)$ and $y(n)$ is defined as

$$r_{xy}(k) = \sum_{m=-\infty}^{\infty} x(m + k)y^*(m). \qquad (5.42)$$

It provides a measure of the similarity of sequences $x(n)$ and $y(n)$ as one is shifted with respect to the other; if $r_{xy}(k)$ has large magnitude, $x(n + k)$ and $y^*(n)$ are similar. Often, to normalize $r_{xy}(k)$ and/or to estimate the stochastic cross-correlation (see Section 10.2), a factor $1/N$ is placed out front in (5.42) and the sum run over only N samples. Alternatively, we may divide the sum in (5.42) by the norms of x and y for the same purpose. For this chapter, we will just use (5.42) as shown.

The familiar complex inner product $<x, y> = \sum_{m=-\infty}^{\infty} x(m)y^*(m)$ is just a special case: $r_{xy}(0)$. Furthermore, if in (5.42) we replace $y(m)$ by $x(m)$, the deterministic autocorrelation function (or sequence) is defined:

$$r_{xx}(k) = \sum_{m=-\infty}^{\infty} x(m + k)x^*(m). \qquad (5.43)$$

The special case $r_{xx}(0) = \sum_{m=-\infty}^{\infty} |x(m)|^2$ is the total energy in the signal.

We have much more to say about correlation functions in Chapters 7 and 10. They have application in finding pitch period in speech waveforms, modeling signals with simple ARMA models, prediction of time series, digital communications, and so on. Here we present one example of a deterministic autocorrelation function and its transform. Suppose that $x(n) = 0.85^{|n|} \cos([10° \cdot \pi/180°]n)$ for all n. In the smaller

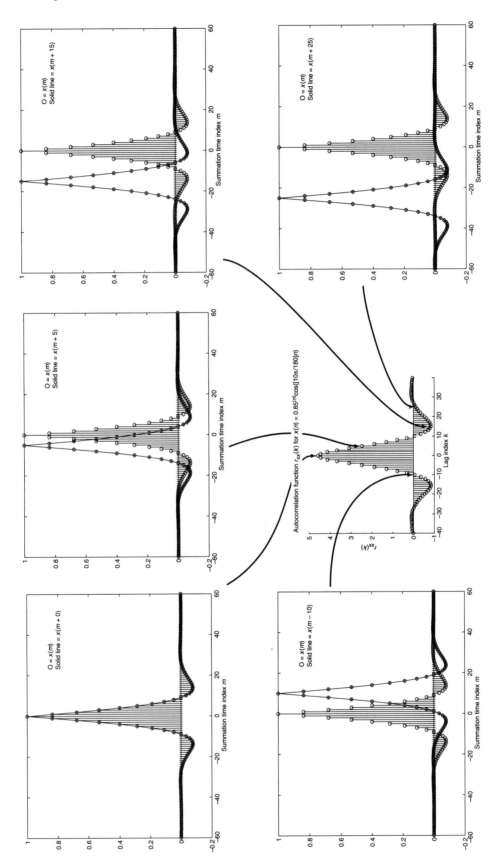

Figure 5.6

plots in Figure 5.6, we show the sum term factors for (5.43) $x(m)$ and $x(m + k)$ on the same axes, with the samples $x(m + k)$ connected with a line for $k = -10, 0, 5, 15$, and 25 (clockwise from lower left). The central plot in Figure 5.6 is $r_{xx}(k)$, for convenience calculated using a truncated version of (5.43), rather than performing the tedious task of generating the closed-form expression of $r_{xx}(k)$ for $x(n)$ as given.

First notice in (5.43) that $r_{xx}(-k) = r_{xx}^*(k)$, which for real $x(n)$ becomes $r_{xx}(-k) = r_{xx}(k)$. We see that when the functions are lined up ($k = 0$), the inner product of the sequences $x(m)$ and $x(m + k)$ calculated in (5.43) is maximized. Also, for $k = \pm 20$, notice that the overlap of $x(m)$ and $x(m \pm 20)$ occurs for $x(m)$ and $x(m \pm 20)$ having opposite signs; hence (5.43) will give a negative value for $r_{xx}(\pm 20)$. Moreover, because the values of $x(m)$ and $x(m \pm 20)$ within the overlap are minimal, $|r_{xx}(\pm 20)|$ is small.

Just as convolution was transformed into simple multiplication of transforms, the same is true for correlation. The only difference is that the transform of the domain 1 signal conjugated in the correlation expression [$y(n)$ in (5.42)] is itself conjugated (in domain 2). Again there are some strange-looking "correlations" for the GPEs, so it is preferred to state each correlation theorem in terms of the transform of the product of one domain-1 function by the conjugate of another domain-1 function. The rule for determining the appropriate property shall now be stated.

We need not rederive the correlation property for each case if the corresponding convolution theorem has already been derived. Notice that (5.42) can be written

$$r_{xy}(k) = \sum_{m=-\infty}^{\infty} x(-(n-m))y^*(m)\Big|_{n=-k}, \tag{5.44}$$

which says that correlation is just the time-reverse of the convolution of two functions: one conjugated and the other time-reversed. Each of these three modifications can be made to the convolution theorem result using previously derived properties concerning sign reversal of a variable and conjugation of a function. These rules are used in all cases, including GPEs.

EXAMPLE 5.23

Find the ZT correlation property analogous to (5.40e) for convolution.

Solution

We may rewrite (5.40e) as $\text{ZT}^{-1}\{X(z)Y(z)\} = \sum_{m=-\infty}^{\infty} x(n-m)y(m)$. We conjugate $y(m)$, time-reverse $x(n-m)$, and evaluate the convolution at $n = -k$ to obtain $r_{xy}(k)$. From the conjugation property, conjugating $y(n)$ amounts to replacing $Y(z)$ by $Y^*(z^*)$. Time-reversing $x(n)$ amounts to replacing $X(z)$ by $X(1/z)$, and time-reversing the entire expression amounts to replacing all instances of z in the product by $1/z$. Consequently, we have $X(1/[1/z])Y^*(1/z^*) = X(z)Y^*(1/z^*)$, and thus

$$\text{ZT}\{r_{xy}(k)\} = R_{xy}(z) = X(z)Y^*\left(\frac{1}{z^*}\right). \tag{5.45}$$

By evaluating (5.45) at $z = e^{j\omega}$, we immediately obtain

$$\text{DTFT}\{r_{xy}(k)\} = R_{xy}(e^{j\omega}) = X(e^{j\omega})Y^*(e^{j\omega}). \tag{5.46}$$

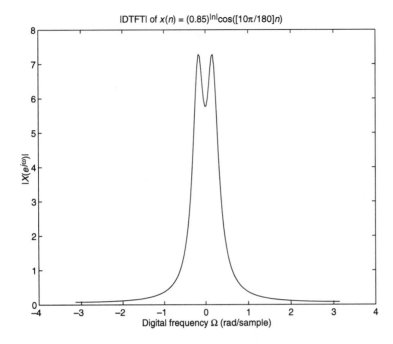

$|\text{DTFT}|$ of $x(n) = (0.85)^{|n|}\cos([10\pi/180]n)$

Figure 5.7a

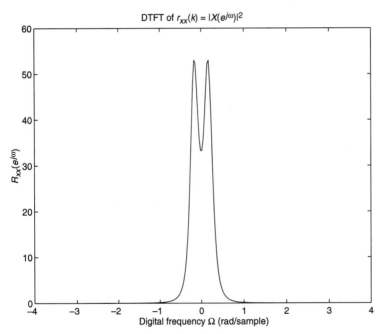

DTFT of $r_{xx}(k) = |X(e^{j\omega})|^2$

Figure 5.7b

As noted above, if we set $y(n) = x(n)$, we obtain autocorrelation theorems. For example, (5.45) becomes

$$\text{ZT}\{r_{xx}(k)\} = R_{xx}(z) = X(z)X^*\left(\frac{1}{z^*}\right), \tag{5.47}$$

and (5.46) becomes

$$\text{DTFT}\{r_{xx}(k)\} = R_{xx}(e^{j\omega}) = |X(e^{j\omega})|^2. \tag{5.48}$$

We now briefly return to our example of a two-sided damped oscillation $x(n)$ $= 0.85^{|n|}\cos([10° \cdot \pi/180°]n)$. Calculation of the transform of a more general $x(n) = r^{|n|}\cos(\omega_1 n)$ gives (see the problems)

$$X(e^{j\omega}) = \frac{r\{\cos(\omega - \omega_1) - r\}}{1 + r^2 - 2r\cos(\omega - \omega_1)} + \frac{1 - r\cos(\omega + \omega_1)}{1 + r^2 - 2r\cos(\omega + \omega_1)} \quad (\text{for } |r| < 1),$$

where in our example $r = 0.85$ and $\omega_1 = 10\pi/180$ rad. From this expression for $X(e^{j\omega})$, we conclude that $X(e^{j\omega})$ is real; in fact, it is always positive. From (5.48), we conclude that we just take the magnitude squared of $X(e^{j\omega})$, or in this case, just square $X(e^{j\omega})$ to obtain $R_{xx}(e^{j\omega})$. Both $X(e^{j\omega})$ and $R_{xx}(e^{j\omega})$ for this example are shown for $|\omega| < \pi$ in, respectively, Figures 5.7a and b. As could already be guessed from examination of Figure 5.6, $r_{xx}(k)$ has damped oscillations at the same digital frequency as does $x(n)$. This is clearly seen in the even more accentuated peaks at $\omega_1 = 10\pi/180 \approx 0.175$ rad/sample in Figure 5.7b than in 5.7a.

5.9 TRANSFORMS OF REAL-VALUED FUNCTIONS; PARSEVAL'S RELATION

Again, three essential elements of signal processing are signals, system operators, and transforms. On WP 5.1, we consider the properties of system operators. Such concepts as Hermiticity or self-adjointness, completeness relations, unicity, and definiteness are defined and discussed there for relevant signal processing operators. This latter background is helpful to have when these issues arise in your readings or in the evaluation of a new proposed operation or transform. It also ties DSP theory in with controls and communications analysis techniques. Furthermore, issues of orthogonality and completeness arise in signal coding, a commonly used technique in speech and image processing.

On WP 5.1, we extend the operator L to include not only the previous operations discussed but also the transforms that take us from domain 1 to domain 2. This allows us to study in a unified way both Fourier and convolution operators, which are the two most important in DSP. Use of vector and inner product notation facilitates unity not only for the various operators but for discrete time versus continuous time. We thereby obtain for real-valued functions the corresponding condition on their transform, as well as Parseval's relations. Furthermore, we find that the forward and reverse definitions of the Laplace, z-, and Fourier transforms "drop right out" of the so-called Hermitian condition. Also, we find that eigenanalysis is the connection between convolution and Fourier analysis: Fourier analysis is the eigendecomposition for LSI systems. The concept of positive and negative definiteness is defined and then explored by means of four specific examples; it tells us when a transform is not only real, but also always positive or always negative.

For brevity, here we will cover a bit of the same ground by a simpler and briefer, but much less illuminating viewpoint. Those readers who are already familiar with the transform properties of real-valued functions and with Parseval's theorem are invited to explore WP 5.1 to advance their understanding of signal-processing operators.

If $h(n)$ is real-valued—that is, $h^*(n) = h(n)$—what properties do $H(z)$, $H(e^{j\omega})$, and $H(k)$ have? To answer this question, set the inverse z-transform equal to its conjugate:

$$\left[\frac{1}{2\pi j} \oint_C H(z)\, z^{n-1} dz\right]^* = \frac{1}{2\pi j} \oint_C H(z)\, z^{n-1} dz. \tag{5.49}$$

The left-hand side of (5.49) is equal to $\dfrac{-1}{2\pi j} \oint_C H^*(z)(z^*)^{n-1}\, dz^*$. Let $z_{\text{new}} = z_{\text{old}}{}^*$ and thus $z_{\text{old}} = z_{\text{new}}{}^*$ and $z_{\text{old}}{}^* = z_{\text{new}}$. This substitution will change the direction of C, and thus the sign of the result, so that (5.49) becomes

$$\frac{1}{2\pi j} \oint_C H^*(z^*) z^{n-1} dz = \frac{1}{2\pi j} \oint_C H(z)\, z^{n-1} dz. \tag{5.50}$$

If (5.50) is to be true for all n, then we must have

$$H^*(z^*) = H(z) \qquad \text{for real } h(n). \tag{5.51}$$

Setting $z = e^{j\omega}$, we have for the DTFT

$$H^*(e^{-j\omega}) = H(e^{j\omega}) \qquad \text{for real } h(n), \tag{5.52}$$

and thus using the polar form of (5.52) we conclude that $|H(e^{j\omega})|$ is an even function of ω and $\angle H(e^{j\omega})$ is an odd function of ω. Finally, assuming $h(n)$ has length N, set $\omega = 2\pi k/N$ in (5.52) and use $H(-k) = H(N - k)$ to obtain

$$H^*(-k) = H^*(N - k) = H(k) \qquad \text{for real } h(n). \tag{5.53}$$

*See
WP 5.1
for
details.* These conjugate-even conditions on the transforms are called Hermitian conditions on the associated convolution operator. Similarly, if $H(e^{j\omega})$ is real-valued, then $h(n)$ is called a Hermitian sequence [i.e., conjugate-even: $h^*(-n) = h(n)$].

The results above have the following practical importance. We know that the values $k \in [0, N - 1]$ correspond to $\omega_k = 2\pi k/N$ (or $f_k = k/N$); that is, $H(k) = H_1(e^{j2\pi k/N})$ where $H_1(e^{j\omega})$ is the DTFT of $h_1(n)$, the period of the N-periodic $h(n)$ on $n \in [0, N-1]$ (see Section 4.5). Consider N even, which is the usual case because that is a requirement for the efficient power-of-two FFT (see Section 7.10). For N even, the positive-frequency values of k are $k \in [0, N/2]$, corresponding to $f_k = \{0, \ldots, \frac{1}{2}\}$, where $f = \frac{1}{2}$ we know to be the highest nonaliased digital frequency: two samples per sinusoidal period. Thus in DFT magnitude and phase plots, it is often convenient to show $k \in [0, N/2]$ or $f \in [0, \frac{1}{2}]$ on the abscissa. The remaining DFT values merely document the negative frequencies for which we already know $H(k)$ due to conjugate symmetry, assuming $h(n)$ is real.

All the comments above apply not only to a filter $h(n)$ but to any signal sequence $x(n)$. For example, consider the asymmetrical triangle wave $\tilde{x}(n)$ presented in Figure 4.19. The N-periodicity relationship $\tilde{X}(-k) = \tilde{X}(N - k)$ [which of course also holds for $\tilde{x}(n)$] is shown in Figure 5.8 ($N = 16$). With $\tilde{x}(n)$ real-valued, its DFT $X(k)$ (we now omit the tilde) additionally satisfies $X(-k) = X(N - k) = X^*(k)$; this relationship is shown in Figure 5.9. As noted above for the DTFT, the DFT magnitude function is an even function of k [$|X(N-k)| = |X(k)|$], whereas the phase is odd [$\angle X(N - k) = -\angle X(k)$]. The arrows in Figure 5.9 indicate the matching pairs. Notice the unmatched values $k = 0$ and $k = N/2$; these are not replicated in "redundant" k values in the second half of the sequence.

We now know that a (conjugate-) even time function $h(n)$ has a real-valued transform, and a real-valued time function has a (conjugate-) even transform. It follows, then, that if $h(n)$ is real *and* even, then so is its transform (and vice versa).

Magnitude of periodic DFS coefficients $\tilde{X}(k)$ of periodic $\tilde{x}(n)$ (asymmetrical triangle wave). Periodicity relationship for DFS coefficients holding for any $\tilde{x}(n)$.

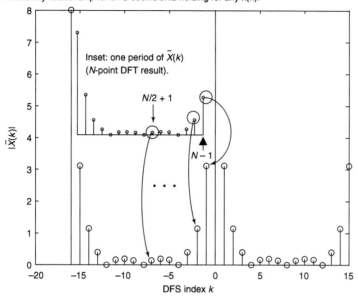

Figure 5.8

Magnitude of DFT coefficients $X(k) = fft(x)$ where $x(n) =$ asymmetrical triangle wave. Even symmetry relationship holding only when $x(n)$ is real-valued.

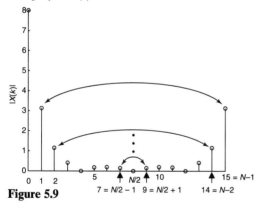

Figure 5.9

Phase-angle of DFT coefficients $X(k) = fft(x)$ where $x(n) =$ asymmetrical triangle wave. Odd symmetry relationship holding only when $x(n)$ is real-valued.

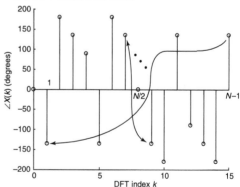

Although this property may not usually hold for a signal $x(n)$, it often does hold for ideal filters $h(n)$. Whenever we specify a filter having a desired even real-valued frequency response—that is, a filter with phase always equal to zero or π—the corresponding $h(n)$ is real and even. We return briefly to these issues in Chapters 9 and 10.

Finally, we close with Parseval's relation. We write the "energy" of $x(n)$ in terms of its DTFT (and note that $|x|^2 = xx^*$):

$$\sum_{n=-\infty}^{\infty} |x(n)|^2 = \sum_{n=-\infty}^{\infty} \frac{1}{2\pi} \int_{-\pi}^{\pi} X(e^{j\omega})e^{j\omega n}d\omega \, \frac{1}{2\pi} \int_{-\pi}^{\pi} X^*(e^{j\omega'}) \, e^{-j\omega' n}d\omega'$$

$$= \frac{1}{4\pi^2} \int_{-\pi}^{\pi} \int_{-\pi}^{\pi} X(e^{j\omega})X^*(e^{j\omega'}) \sum_{n=-\infty}^{\infty} e^{j(\omega-\omega')n}d\omega \, d\omega'.$$

The sum, by (2A.24), is $2\pi \cdot \delta_c(\omega - \omega')$ for $|\omega - \omega'| < \pi$. Using the Dirac delta sampling property, we finally obtain

$$\sum_{n=-\infty}^{\infty} |x(n)|^2 = \frac{1}{2\pi} \int_{-\pi}^{\pi} |X(e^{j\omega})|^2 d\omega. \quad \text{Parseval's relation for the DTFT.} \quad (5.54)$$

By the identical approach, we may also show Parseval for the DFT:

$$\sum_{n=0}^{N-1} |x(n)|^2 = \frac{1}{N} \sum_{k=0}^{N-1} |X(k)|^2. \quad \text{Parseval's relation for the DFT.} \quad (5.55)$$

We conclude that energy in one domain is preserved in the other for Fourier operators; such operators are called unitary.

Parseval's relation (5.54) can be used to help further interpret the DTFT in physical terms. If real $x(n)$ is presented to a very narrow-band digital filter passing $\omega \in [\pm\omega_1 - \Delta\omega/2, \pm\omega_1 + \Delta\omega/2]$ with unity gain, then the spectrum $Y(e^{j\omega})$ of the output $y(n)$ will be, approximately, $Y(e^{j\omega}) = X(e^{j\omega_1})$ for $\omega \in [\pm\omega_1 - \Delta\omega/2, \pm\omega_1 + \Delta\omega/2]$ and zero elsewhere. Now, the right-hand side of Parseval (5.54) applied to $y(n)$ is $2 \cdot |X(e^{j\omega_1})|^2 \Delta\omega/(2\pi)$, but by the left-hand side we know this is equal to the energy in $y(n)$—the energy in $x(n)$ near ω_1. We conclude that $|X(e^{j\omega})|^2/\pi$ is the energy density in $x(n)$ at frequency ω. Moreover, $|X(e^{j\omega})|^2$ is the DTFT of the temporal autocorrelation function of $x(n)$, as shown in (5.48).

APPENDIX 5A: BASIC TRANSFORM PROPERTIES

Balthazar van der Pol and H. Bremmer amusingly refer to their table of properties as the rules of "grammar" and their table of transform pairs (see Appendix 5B) as the "dictionary."

The following properties are listed for the various transforms:

(1)   Shift of independent variable

(2)   Differentiation/differencing

(3)   Integration/summing (if any)

(4)   Sign reversal of independent variable

(5)   Conjugation

(6)   Scaling of independent variable

(7)   Hermitian condition

(8) L repetitions of time function (if applicable)

(9)   Multiplication $\leftrightarrow$ convolution of functions

(10)   Deterministic correlation (cross- and auto-)

(11)   Parseval's relation (auto- and cross- forms).

(12)   Initial and final value theorems (if applicable)

This appendix is organized as follows:

Table	Transform Properties
5A.1	Laplace transform[6]
5A.2	Fourier transform
5A.3	Fourier series
5A.4	z-transform[6]
5A.5	Discrete-time Fourier transform
5A.6	Discrete Fourier series/transform

---

[6]See the text for example determinations of the effect of an operation on the region of convergence.

Table 5A.1 Laplace Transform $\qquad$ Assumed: $x_c(t) \overset{\text{LT}}{\longleftrightarrow} X_c(s)$.

$\text{LT}\{x_c(t - t_0)\} = e^{-st_0}X_c(s)$	$\text{LT}^{-1}\{X_c(s - s_0)\} = e^{s_0 t}x_c(t)$	(5A.1.1)

$$\text{LT}\left\{\frac{d}{dt}x_c(t)\right\} = sX_c(s) - x_c(0) \qquad \text{LT}^{-1}\left\{\frac{d}{ds}X_c(s)\right\} = -tx_c(t) \qquad (5A.1.2)$$
[unilateral LT; bilateral LT
involves no initial condition, but requires
exponential decay of $x_c(t)$ at $|t| \to \infty$]

$$\text{LT}\left\{\int_{-\infty}^{t} x_c(t')dt'\right\} = \frac{X_c(s)}{s} \qquad \text{LT}^{-1}\left\{\int_{s}^{\infty} X_c(s')ds'\right\} = \frac{x_c(t)}{t} \qquad (5A.1.3)$$

$\text{LT}\{x_c(-t)\} = X_c(-s)$	$\text{LT}^{-1}\{X_c(-s)\} = x_c(-t)$	(5A.1.4)
$\text{LT}\{x_c{}^*(t)\} = X_c^*(s^*)$	$\text{LT}^{-1}\{X_c^*(s)\}$ — No simple form.	(5A.1.5)

$$\text{LT}\{x_c(\alpha t)\} = \frac{X_c(s/\alpha)}{|\alpha|} \qquad \text{LT}^{-1}\{X_c(\alpha s)\} = \frac{x_c(t/\alpha)}{|\alpha|} \qquad (5A.1.6)$$

No simple condition on $x_c(t)$ to make $X_c(s)$ real.	$X_c^*(s^*) = X_c(s) \to x_c(t)$ real	(5A.1.7)

$\text{LT}\{L$ repetitions of $x_c(t)$,
where $x(n)$ has duration $T\}$ $\qquad$ — $\qquad$ (5A.1.8)
$= X_c(s) \cdot \dfrac{1 - e^{-sLT}}{1 - e^{-sT}}$

$\text{LT}\{x_c(t)y_c(t)\}$ $\qquad\qquad$ $\text{LT}^{-1}\{X_c(s)Y_c(s)\}$ $\qquad$ (5A.1.9)
$\quad = \dfrac{1}{2\pi j}\displaystyle\int_C X_c(s')Y_c(s - s')ds' \qquad = \displaystyle\int_{-\infty}^{\infty} x_c(t')y_c(t - t')dt'$
$\quad = \dfrac{X_c(s)*_c Y_c(s)}{2\pi j} \qquad\qquad = x_c(t)*_c y_c(t)$

$\text{LT}\{x_c(t)y_c{}^*(t)\}$ $\qquad\qquad$ $\text{LT}^{-1}\{X_c(s)Y_c^*(-s^*)\}$
$\quad = \dfrac{1}{2\pi j}\displaystyle\int_C X_c(s')Y_c^*((s - s')^*)ds' \qquad = \displaystyle\int_{-\infty}^{\infty} x_c(t + t')y_c^*(t')dt' \qquad (5A.1.10a)$
$\qquad\qquad\qquad\qquad\qquad\qquad = r_{xy,c}(t)$

$\text{LT}\{|x_c(t)|^2\}$ $\qquad\qquad$ $\text{LT}^{-1}\{X_c(s)X_c^*(-s^*)\}$
$\quad = \dfrac{1}{2\pi j}\displaystyle\int_C X_c(s')X_c^*((s - s')^*)ds' \qquad = \displaystyle\int_{-\infty}^{\infty} x_c(t + t')x_c^*(t')dt' \qquad (5A.1.10b)$
$\qquad\qquad\qquad\qquad\qquad\qquad = r_{xx,c}(t)$

$$\int_{-\infty}^{\infty} |x_c(t)|^2\, dt = \frac{1}{2\pi j}\int_C X_c(s)\, X_c^*(-s^*)ds \qquad \text{Parseval's theorem} \qquad (5A.1.11a)$$

$$\int_{-\infty}^{\infty} x_c(t)\, y_c^*(t)\, dt = \frac{1}{2\pi j}\int_C X_c(s)Y_c^*(-s^*)ds \qquad \text{Generalized Parseval} \qquad (5A.1.11b)$$

$$x_c(0) = \lim_{s\to\infty} sX_c(s);\ x_c(\infty) = \lim_{s\to 0} sX_c(s) \qquad \text{Initial/final value theorems} \qquad (5A.1.12)$$

Table 5A.2   Fourier Transform   Assumed: $x_c(t) \overset{\text{FT}}{\leftrightarrow} X_c(j\Omega)$.

$\text{FT}\{x_c(t - t_0)\} = e^{-j\Omega t_0}X_c(j\Omega)$	$\text{FT}^{-1}\{X_c(j[\Omega - \Omega_0])\} = e^{j\Omega_0 t}x_c(t)$	(5A.2.1)
$\text{FT}\left\{\dfrac{d}{dt}x_c(t)\right\} = j\Omega X_c(j\Omega)$	$\text{FT}^{-1}\left\{\dfrac{d}{d\Omega}X_c(j\Omega)\right\} = -jtx_c(t)$	(5A.2.2a)
$\text{FT}\left\{\dfrac{d^m}{dt^m}x_c(t)\right\} = (j\Omega)^m X_c(j\Omega) \quad (m \geq 0)$	$\text{FT}^{-1}\{j^m\dfrac{d^m}{d\Omega^m}X_c(j\Omega)\} = t^m x_c(t)$	(5A.2.2b)
$\text{FT}\left\{\displaystyle\int_{-\infty}^{t} x_c(t')dt'\right\} = \dfrac{X_c(j\Omega)}{j\Omega} + \pi X_c(0)\delta_c(\Omega)$	$\text{FT}^{-1}\left\{\displaystyle\int_{-\infty}^{\Omega} X_c(j\Omega')d\Omega'\right\} = \dfrac{jx_c(t)}{t}$	(5A.2.3)
$\text{FT}\{x_c(-t)\} = X_c(-j\Omega)$	$\text{FT}^{-1}\{X_c(-j\Omega)\} = x_c(-t)$	(5A.2.4)
$\text{FT}\{x_c^*(t)\} = X_c^*(-j\Omega)$	$\text{FT}^{-1}\{X_c^*(j\Omega)\} = x_c^*(-t)$	(5A.2.5)
$\text{FT}\{x_c(\alpha t)\} = \dfrac{X_c(j\Omega/\alpha)}{|\alpha|}$	$\text{FT}^{-1}\{X_c(j\alpha\Omega)\} = \dfrac{x_c(t/\alpha)}{|\alpha|}$	(5A.2.6)
$x_c^*(-t) = x_c(t) \to X_c(j\Omega)$ real	$X_c^*(-j\Omega) = X_c(j\Omega) \to x_c(t)$ real	(5A.2.7)
$\text{FT}\{L$ repetitions of $x_c(t)$, where $x(n)$ has duration $T\}$ $= X_c(j\Omega) \cdot \dfrac{1 - e^{-j\Omega LT}}{1 - e^{-j\Omega T}}$	—	(5A.2.8)
$\text{FT}\{x_c(t)y_c(t)\}$ $= \dfrac{1}{2\pi}\displaystyle\int_{-\infty}^{\infty} X_c(j\Omega')Y_c(j(\Omega - \Omega'))d\Omega'$ $= \dfrac{X_c(j\Omega) *_c Y_c(j\Omega)}{2\pi}$	$\text{FT}^{-1}\{X_c(j\Omega)Y_c(j\Omega)\}$ $= \displaystyle\int_{-\infty}^{\infty} x_c(t')y_c(t - t')dt'$ $= x_c(t) *_c y_c(t)$	(5A.2.9)
$\text{FT}\{x_c(t)y_c^*(t)\}$ $= \dfrac{1}{2\pi}\displaystyle\int_{-\infty}^{\infty} X_c(j\Omega')Y_c^*(j(\Omega' - \Omega))d\Omega'$	$\text{FT}^{-1}\{X_c(j\Omega)Y_c^*(j\Omega)\}$ $= \displaystyle\int_{-\infty}^{\infty} x_c(t + t')y_c^*(t')dt'$ $= r_{xy,c}(t)$	(5A.2.10a)
$\text{FT}\{|x_c(t)|^2\}$ $= \dfrac{1}{2\pi}\displaystyle\int_{-\infty}^{\infty} X_c(j\Omega')X_c^*(j(\Omega' - \Omega))d\Omega'$	$\text{FT}^{-1}\{|X_c(j\Omega)|^2\}$ $= \displaystyle\int_{-\infty}^{\infty} x_c(t + t')x_c^*(t')dt'$ $= r_{xx,c}(t)$	(5A.2.10b)
$\displaystyle\int_{-\infty}^{\infty} |x_c(t)|^2\,dt = \dfrac{1}{2\pi}\displaystyle\int_{-\infty}^{\infty} |X_c(j\Omega)|^2\,d\Omega$	Parseval's theorem	(5A.2.11a)
$\displaystyle\int_{-\infty}^{\infty} x_c(t)y_c^*(t)\,dt = \dfrac{1}{2\pi}\displaystyle\int_{-\infty}^{\infty} X_c(j\Omega)Y_c^*(j\Omega)\,d\Omega$	Generalized Parseval	(5A.2.11b)

Table 5A.3 Fourier Series   Assumed: $\tilde{x}_c(t) \overset{\text{FS}}{\longleftrightarrow} X_c(k)$ and $\tilde{x}_c(t + T) = \tilde{x}_c(t)$.

$\text{FS}\{\tilde{x}_c(t - t_0)\} = e^{-j2\pi k t_0/T} X_c(k)$

$\text{FS}^{-1}\{X_c(k - k_0)\} = e^{j2\pi k_0 t/T} \tilde{x}_c(t)$ (5A.3.1)

$\text{FS}\left\{\dfrac{d}{dt} \tilde{x}_c(t)\right\} = j\dfrac{2\pi k}{T} X_c(k)$

$\text{FS}^{-1}\left\{\dfrac{X_c(k) - X_c(k - 1)}{2\pi/T}\right\} = \dfrac{T}{2\pi}(1 - e^{j2\pi t/T}) \tilde{x}_c(t)$ (5A.3.2)
$\approx -jt\tilde{x}_c(t) \quad$ for $|t| \ll T$.

$\text{FS}\{\tilde{x}_c(-t)\} = X_c(-k)$

$\text{FS}^{-1}\{X_c(-k)\} = \tilde{x}_c(-t)$ (5A.3.3)

$\text{FS}\{\tilde{x}_c^*(t)\} = X_c^*(-k)$

$\text{FS}^{-1}\{X_c^*(k)\} = \tilde{x}_c^*(-t)$ (5A.3.4)

$\text{FS}\{\tilde{x}_c(Mt)\} = \begin{cases} X_c\left(\dfrac{k}{M}\right), & k = mM \\ & (M, m \text{ integers} \\ & \text{only}) \\ 0, & \text{otherwise} \end{cases}$

$\text{FS}^{-1}\{X_c(Mk)\} = \dfrac{1}{M} \displaystyle\sum_{m=0}^{M-1} \tilde{x}_c\left(\dfrac{t + mT}{M}\right)$ (5A.3.5)

$\tilde{x}_c^*(-t) = \tilde{x}_c(t) \to X_c(k)$ real

$X_c^*(-k) = X_c(k) \to \tilde{x}_c(t)$ real (5A.3.6)

$\text{FS}\{\tilde{x}_c(t)\tilde{y}_c(t)\} = \displaystyle\sum_{k'=-\infty}^{\infty} X_c(k')Y_c(k - k')$
$= X_c(k) * Y_c(k)$

$\text{FS}^{-1}\{X_c(k)Y_c(k)\}$ (5A.3.7)
$= \dfrac{1}{T} \displaystyle\int_0^T \tilde{x}_c(t')\tilde{y}_c(t - t')dt'$
$= \dfrac{\tilde{x}_c(t) *_c \tilde{y}_c(t)}{T} \quad$ (over 1 period)

$\text{FS}\{\tilde{x}_c^*(t)\tilde{y}_c(t)\}$
$= \displaystyle\sum_{k'=-\infty}^{\infty} X_c^*(k')Y_c(k + k')$

$\text{FS}^{-1}\{X_c(k)Y_c^*(k)\}$
$= \dfrac{1}{T} \displaystyle\int_0^T \tilde{x}_c(t + t')\tilde{y}_c^*(t')dt'$ (5A.3.8a)
$= r_{\tilde{x}\tilde{y},c}(t) \quad$ (over 1 period)

$\text{FS}\{|\tilde{x}_c(t)|^2\}$
$= \displaystyle\sum_{k'=-\infty}^{\infty} X_c^*(k')X_c(k + k')$

$\text{FS}^{-1}\{|X_c(k)|^2\}$
$= \dfrac{1}{T} \displaystyle\int_0^T \tilde{x}_c(t + t')\tilde{x}_c^*(t')dt'$ (5A.3.8b)
$= r_{\tilde{x}\tilde{x},c}(t) \quad$ (over 1 period)

$\dfrac{1}{T} \displaystyle\int_0^T |\tilde{x}_c(t)|^2 \, dt = \displaystyle\sum_{k=-\infty}^{\infty} |X_c(k)|^2$

Parseval's theorem (5A.3.9a)

$\dfrac{1}{T} \displaystyle\int_0^T \tilde{x}_c(t)\tilde{y}_c^*(t) \, dt = \displaystyle\sum_{k=-\infty}^{\infty} X_c(k)Y_c^*(k)$

Generalized Parseval (5A.3.9b)

Table 5A.4 z-Transform   Assumed: $x(n) \overset{\text{ZT}}{\leftrightarrow} X(z)$.

$\text{ZT}\{x(n - n_0)\} = z^{-n_0}X(z)$	$\text{ZT}^{-1}\{X(z - z_0)\}$ No simple form; (5A.4.1) see right column of (5A.4.6) for analogous property.
$\text{ZT}\{x(n) - x(n - 1)\} = (1 - z^{-1})X(z)$	$\text{ZT}^{-1}\left\{\dfrac{d}{dz}X(z)\right\} = (1 - n)x(n - 1)$ (5A.4.2a) or $\text{ZT}^{-1}\left\{\left[-z\dfrac{d}{dz}\right]^m X(z)\right\} = n^m x(n)$ $(m \geq 0)$ (5A.4.2b)
$\text{ZT}\left\{\displaystyle\sum_{\ell=-\infty}^{n} x(\ell)\right\} = \dfrac{X(z)}{1 - z^{-1}}$	— (5A.4.3)
$\text{ZT}\{x(-n)\} = X\left(\dfrac{1}{z}\right)$	$\text{ZT}^{-1}\{X(-z)\} = (-1)^n x(n)$ (5A.4.4)
$\text{ZT}\{x^*(n)\} = X^*(z^*)$	$\text{ZT}^{-1}\{X^*(z)\}$ No simple form. (5A.4.5)
$\text{ZT}\{x(nM)\} = \dfrac{1}{M}\displaystyle\sum_{m=0}^{M-1} X([ze^{j2\pi m}]^{1/M})$ [M an integer only]	$\text{ZT}^{-1}\left\{X\left(\dfrac{z}{\alpha}\right)\right\} = \alpha^n x(n)$ (5A.4.6)
No simple condition on $x(n)$ to make $X(z)$ real.	$X^*(z^*) = X(z) \rightarrow x(n)$ real (5A.4.7)
$\text{ZT}\{L$ repetitions of $x(n)$, where $x(n)$ has duration $N\}$ $= X(z) \cdot \dfrac{1 - z^{-LN}}{1 - z^{-N}}$	— (5A.4.8)
$\text{ZT}\{x(n)y(n)\} = \dfrac{1}{2\pi j}\displaystyle\oint_C X(z')Y\left(\dfrac{z}{z'}\right)\dfrac{dz'}{z'}$	$\text{ZT}^{-1}\{X(z)Y(z)\}$ (5A.4.9) $= \displaystyle\sum_{n'=-\infty}^{\infty} x(n')y(n - n') = x(n) * y(n)$
$\text{ZT}\{x(n)y^*(n)\}$ $= \dfrac{1}{2\pi j}\displaystyle\oint_C X(z')Y^*\left(\left(\dfrac{z}{z'}\right)^*\right)\dfrac{dz'}{z'}$	$\text{ZT}^{-1}\left\{X(z)Y^*\left(\dfrac{1}{z^*}\right)\right\}$ (5A.4.10a) $= \displaystyle\sum_{n'=-\infty}^{\infty} x(n + n')y^*(n')$ $= r_{xy}(n)$
$\text{ZT}\{|x(n)|^2\}$ $= \dfrac{1}{2\pi j}\displaystyle\oint_C X(z')X^*\left(\left(\dfrac{z}{z'}\right)^*\right)\dfrac{dz'}{z'}$	$\text{ZT}^{-1}\left\{X(z)X^*\left(\dfrac{1}{z^*}\right)\right\}$ (5A.4.10b) $= \displaystyle\sum_{n'=-\infty}^{\infty} x(n + n')x^*(n')$ $= r_{xx}(n)$
$\displaystyle\sum_{n=-\infty}^{\infty} |x(n)|^2 = \dfrac{1}{2\pi j}\displaystyle\oint_C X(z)X^*\left(\dfrac{1}{z^*}\right)\dfrac{dz}{z}$	Parseval's theorem (5A.4.11a)
$\displaystyle\sum_{n=-\infty}^{\infty} x(n)y^*(n) = \dfrac{1}{2\pi j}\displaystyle\oint_C X(z)Y^*\left(\dfrac{1}{z^*}\right)\dfrac{dz}{z}$	Generalized Parseval (5A.4.11b)
$x(0) = \lim_{z\to\infty} X(z)$; $x(\infty) = \lim_{z\to 1}(1 - z^{-1})X(z)$	Initial/final value theorems (5A.4.12)

Table 5A.5
Discrete-Time Fourier Transform $\qquad$ Assumed: $x(n) \overset{\text{DTFT}}{\longleftrightarrow} X(e^{j\omega})$.

$\text{DTFT}\{x(n - n_0)\} = e^{-j\omega n_0}X(e^{j\omega})$	$\text{DTFT}^{-1}\{X(e^{j(\omega - \omega_0)})\} = e^{j\omega_0 n}x(n)$	(5A.5.1)
$\text{DTFT}\{x(n) - x(n - 1)\} = (1 - e^{-j\omega})X(e^{j\omega})$	$\text{DTFT}^{-1}\left\{\dfrac{d}{d\omega}X(e^{j\omega})\right\} = -jnx(n)$	(5A.5.2a)
	or	
	$\text{DTFT}^{-1}\left\{\left[j\dfrac{d}{d\omega}\right]^m X(e^{j\omega})\right\} = n^m x(n)$	(5A.5.2b)
$\text{DTFT}\{\displaystyle\sum_{\ell=-\infty}^{n} x(\ell)\} = \dfrac{X(e^{j\omega})}{1 - e^{-j\omega}}$	—	(5A.5.3)
$\text{DTFT}\{x(-n)\} = X(e^{-j\omega})$	$\text{DTFT}^{-1}\{X(e^{-j\omega})\} = x(-n)$	(5A.5.4)
$\text{DTFT}\{x^*(n)\} = X^*(e^{-j\omega})$	$\text{DTFT}^{-1}\{X^*(e^{j\omega})\} = x^*(-n)$	(5A.5.5)
$\text{DTFT}\{x(nM)\} = \dfrac{1}{M}\displaystyle\sum_{m=0}^{M-1} X(e^{j(\omega + 2\pi m)/M})$ (M an integer only)	$\text{DTFT}^{-1}(e^{j\omega M}) = \begin{cases} X\left(\dfrac{n}{M}\right), & n = mM \\ 0, & \text{otherwise} \end{cases}$	(5A.5.6)
$x^*(-n) = x(n) \rightarrow X(e^{j\omega})$ real	$X^*(e^{-j\omega}) = X(e^{j\omega}) \rightarrow x(n)$ real	(5A.5.7)
$\text{DTFT}\{L$ repetitions of $x(n)$ where $x(n)$ has duration $N\}$ $= X(e^{j\omega}) \cdot \dfrac{1 - e^{-j\omega LN}}{1 - e^{-j\omega N}}$	—	(5A.5.8)
$\text{DTFT}\{x(n)y(n)\}$ $= \dfrac{1}{2\pi}\displaystyle\int_{-\pi}^{\pi} X(e^{j\omega'})Y(e^{j(\omega - \omega')})d\omega'$ $= X(e^{j\omega}) *_c Y(e^{j\omega})/(2\pi)$	$\text{DTFT}^{-1}\{X(e^{j\omega})Y(e^{j\omega})\}$ $= \displaystyle\sum_{n'=-\infty}^{\infty} x(n')y(n - n') = x(n) * y(n)$	(5A.5.9)
$\text{DTFT}\{x(n)y^*(n)\}$ $= \dfrac{1}{2\pi}\displaystyle\int_{-\pi}^{\pi} X(e^{j\omega'})Y^*(e^{j(\omega' - \omega)})d\omega'$	$\text{DTFT}^{-1}\{X(e^{j\omega})Y^*(e^{j\omega})\}$ $= \displaystyle\sum_{n'=-\infty}^{\infty} x(n + n')y^*(n')$ $= r_{xy}(n)$	(5A.5.10a)
$\text{DTFT}\{|x(n)|^2\}$ $= \dfrac{1}{2\pi}\displaystyle\int_{-\pi}^{\pi} X(e^{j\omega'})X^*(e^{j(\omega' - \omega)})d\omega'$	$\text{DTFT}^{-1}\{|X(e^{j\omega})|^2\}$ $= \displaystyle\sum_{n'=-\infty}^{\infty} x(n + n')x^*(n')$ $= r_{xx}(n)$	(5A.5.10b)
$\displaystyle\sum_{n=-\infty}^{\infty} |x(n)|^2 = \dfrac{1}{2\pi}\displaystyle\int_{-\pi}^{\pi} |X(e^{j\omega})|^2\, d\omega$	Parseval's theorem	(5A.5.11a)
$\displaystyle\sum_{n=-\infty}^{\infty} x(n)y^*(n) = \dfrac{1}{2\pi}\displaystyle\int_{-\pi}^{\pi} X(e^{j\omega})Y^*(e^{j\omega})d\omega$	Generalized Parseval	(5A.5.11b)

Table 5A.6   Assumed: $\tilde{x}(n) \overset{\text{DFS}}{\longleftrightarrow} \tilde{X}(k)$ and

Discrete Fourier Series/Transform $\tilde{x}(n + N) = \tilde{x}(n), \tilde{X}(k + N) = \tilde{X}(k)$.

Note: For DFT representation for finite-length sequences, all arguments of functions are replaced by their value modulo N.

$\text{DFS}\{\tilde{x}(n - n_0)\} = e^{-j2\pi kn_0/N}\tilde{X}(k)$ $\text{DFS}^{-1}\{\tilde{X}(k - k_0)\} = e^{j2\pi k_0 n/N}\tilde{x}(n)$ (5A.6.1)

$\text{DFS}\{\tilde{x}(n) - \tilde{x}(n - 1)\}$ $\text{DFS}^{-1}\{\tilde{X}(k) - \tilde{X}(k - 1)\}$ (5A.6.2)

$\quad = (1 - e^{-j2\pi k/N})\tilde{X}(k)$ $\quad = (1 - e^{j2\pi n/N})\tilde{x}(n)$

$\quad \approx j\dfrac{2\pi k}{N}\tilde{X}(k) \quad \text{for } k << N.$ $\quad \approx -j\dfrac{2\pi n}{N}\tilde{x}(n) \quad \text{for } n << N$

$\text{DFS}\{\tilde{x}(-n)\} = \tilde{X}(-k)$ $\text{DFS}^{-1}\{\tilde{X}(-k)\} = \tilde{x}(-n)$ (5A.6.3)

$\text{DFS}\{\tilde{x}^{*}(n)\} = \tilde{X}^{*}(-k)$ $\text{DFS}^{-1}\{\tilde{X}^{*}(k)\} = \tilde{x}^{*}(-n)$ (5A.6.4)

$\text{DFS}\{\tilde{x}(nM)\} = \dfrac{1}{M}\displaystyle\sum_{m=0}^{M-1} \tilde{X}(k + mL)$ $\text{DFS}^{-1}\{\tilde{X}(kM)\} = \displaystyle\sum_{m=0}^{M-1} \tilde{x}(n + mL)$ (5A.6.5)

$\quad (ML = N = \text{period of } \tilde{x}(n); M, L \text{ integers})$ $(ML = N = \text{period of } \tilde{X}(k); M, L \text{ integers})$

$\tilde{x}^{*}(-n) = \tilde{x}(n) \rightarrow \tilde{X}(k) \text{ real}$ $\tilde{X}^{*}(-k) = \tilde{X}(k) \rightarrow \tilde{x}(n) \text{ real}$ (5A.6.6)

$\text{DFS}\{\tilde{x}(n)\tilde{y}(n)\}$ $\text{DFS}^{-1}\{\tilde{X}(k)\tilde{Y}(k)\}$ (5A.6.7)

$\quad = \dfrac{1}{N}\displaystyle\sum_{k'=0}^{N-1} \tilde{X}(k')\tilde{Y}(k - k')$ $\quad = \displaystyle\sum_{n'=0}^{N-1} \tilde{x}(n')\tilde{y}(n - n') = \tilde{x}(n) * \tilde{y}(n)$

$\quad = \dfrac{\tilde{X}(k) * \tilde{Y}(k)}{N} \quad \text{over 1 period}$ over 1 period

$\text{DFS}\{\tilde{x}(n)\tilde{y}^{*}(n)\}$ $\text{DFS}^{-1}\{\tilde{X}(k)\tilde{Y}^{*}(k)\}$

$\quad = \displaystyle\sum_{k'=0}^{N-1} \tilde{X}(k')\tilde{Y}^{*}(k' - k)$ $\quad = \displaystyle\sum_{n'=0}^{N-1} \tilde{x}(n + n')\tilde{y}^{*}(n')$ (5A.6.8a)

                                         $\quad = r_{\tilde{x}\tilde{x}}(n) \quad [\text{over 1 period}]$

$\text{DFS}\{|\tilde{x}(n)|^2\}$ $\text{DFS}^{-1}\{|\tilde{X}(k)|^2\}$

$\quad = \displaystyle\sum_{k'=0}^{N-1} \tilde{X}(k')\tilde{X}^{*}(k' - k)$ $\quad = \displaystyle\sum_{n'=0}^{N-1} \tilde{x}(n + n')\tilde{x}^{*}(n')$ (5A.6.8b)

                                         $\quad = r_{\tilde{x}\tilde{x}}(n) \quad [\text{over 1 period}]$

$\displaystyle\sum_{n=0}^{N-1} |\tilde{x}(n)|^2 = \dfrac{1}{N}\displaystyle\sum_{k=0}^{N-1} |\tilde{X}(k)|^2$ Parseval's theorem   (5A.6.9a)

$\displaystyle\sum_{n=0}^{N-1} \tilde{x}(n)\tilde{y}^{*}(n) = \dfrac{1}{N}\displaystyle\sum_{k=0}^{N-1} \tilde{X}(k)\tilde{Y}^{*}(k)$ Generalized Parseval   (5A.6.9b)

APPENDIX 5B: BASIC TRANSFORM PAIRS

Note: The discrete Fourier series is usually calculated numerically, so a table is not particularly helpful and for brevity it is omitted. For N-length sequences, it is directly obtainable as samples of the DTFT at $\omega = 2\pi k/N$.

Table 5B.1 Laplace Transform

$x_c(t)$	$X_c(s) = \text{LT}\{x_c(t)\}$
$\delta_c(t)$	$1, \quad \text{all } s$
$u_c(t)$	$\dfrac{1}{s}, \quad \text{Re}\{s\} > 0$
$t u_c(t)$	$\dfrac{1}{s^2}, \quad \text{Re}\{s\} > 0$
$t^m e^{s_1 t} u_c(t)$	$\dfrac{m!}{(s - s_1)^{m+1}}, \quad \text{Re}\{s\} > \text{Re}\{s_1\}$
$e^{s_1 t} u_c(t)$	$\dfrac{1}{s - s_1}, \quad \text{Re}\{s\} > \text{Re}\{s_1\}$
$e^{\sigma_1 t} \cos(\Omega_1 t) u_c(t)$	$\dfrac{s - \sigma_1}{(s - \sigma_1)^2 + \Omega_1^2}, \quad \text{Re}\{s\} > \sigma_1$
$e^{\sigma_1 t} \sin(\Omega_1 t) u_c(t)$	$\dfrac{\Omega_1}{(s - \sigma_1)^2 + \Omega_1^2}, \quad \text{Re}\{s\} > \sigma_1$
$1, t \in [0, \Delta t]; 0, \text{otherwise}$	$\begin{cases} \dfrac{1 - e^{-s\Delta t}}{s} & s \neq 0; \\ \Delta t, & s = 0 \end{cases}$
$t, t \in [0, T]; 0, \text{otherwise}$	$\begin{cases} \dfrac{1 - (1 + sT) \cdot e^{-sT}}{s^2}, & s \neq 0 \\ \dfrac{T^2}{2}, & s = 0 \end{cases}$
Triangle: $\begin{cases} \dfrac{2t}{T}, & t \in \left[0, \dfrac{T}{2}\right]; \\ 2\left(1 - \dfrac{t}{T}\right) & t \in \left(\dfrac{T}{2}, T\right) \end{cases}$	$\begin{cases} \dfrac{(1 - e^{-sT/2})^2}{\dfrac{T}{2} \cdot s^2}, & s \neq 0; \\ \dfrac{T}{2}, & s = 0 \end{cases}$

Table 5B.2   Fourier Transform

$$\left[\text{Note: Below, } \operatorname{sinc}(x) = \frac{\sin(x)}{x}. \right]$$

$x_c(t)$	$X_c(j\Omega) = \text{FT}\{x_c(t)\}$
1	$2\pi\delta_c(\Omega)$
$\delta_c(t)$	1
$u_c(t)$	$\dfrac{1}{j\Omega} + \pi\delta_c(\Omega)$
$tu_c(t)$	$\dfrac{-1}{\Omega^2} + j\pi\delta_c'(\Omega)$
$t^m e^{s_1 t} u_c(t), \quad \text{Re}\{s_1\} < 0$	$\dfrac{m!}{(j\Omega - s_1)^{m+1}}$
$e^{s_1 t} u_c(t), \quad \text{Re}\{s_1\} < 0$	$\dfrac{1}{j\Omega - s_1}$
$e^{\sigma_1 t}\cos(\Omega_1 t)u_c(t), \quad \sigma_1 < 0$	$\dfrac{j\Omega - \sigma_1}{(j\Omega - \sigma_1)^2 + \Omega_1{}^2}$
$e^{\sigma_1 t}\sin(\Omega_1 t)u_c(t), \quad \sigma_1 < 0$	$\dfrac{\Omega_1}{(j\Omega - \sigma_1)^2 + \Omega_1{}^2}$
$\exp\{j\Omega_1 t\}, \quad \text{all } t$	$2\pi\delta_c(\Omega - \Omega_1)$
$\cos(\Omega_1 t), \quad \text{all } t$	$\pi\{\delta_c(\Omega - \Omega_1) + \delta_c(\Omega + \Omega_1)\}$
$\sin(\Omega_1 t), \quad \text{all } t$	$\left(\dfrac{\pi}{j}\right)\{\delta_c(\Omega - \Omega_1) - \delta_c(\Omega + \Omega_1)\}$
$\operatorname{sinc}(\Omega_1 t), \quad \text{all } t$	$\dfrac{\pi}{\Omega_1}$ for $|\Omega| \le \Omega_1; 0$, otherwise
$1, t \in [0, T]; 0$, otherwise	$Te^{-j\frac{1}{2}\Omega T}\operatorname{sinc}(\tfrac{1}{2}\Omega T)$
$1, t \in [-\tfrac{1}{2}T, \tfrac{1}{2}T]; 0$, otherwise	$T\operatorname{sinc}(\tfrac{1}{2}\Omega T)$
$t, t \in [0, T]; 0$, otherwise	$\begin{cases} \dfrac{(1 + j\Omega T)\cdot e^{-j\Omega T} - 1}{\Omega^2}, & \Omega \ne 0 \\[2mm] \dfrac{T^2}{2}, & \Omega = 0 \end{cases}$
Triangle: $\begin{cases} \dfrac{2t}{T}, & t \in \left[0, \dfrac{T}{2}\right]; \\[2mm] 2\left(1 - \dfrac{t}{T}\right), & t \in \left(\dfrac{T}{2}, T\right) \end{cases}$	$\tfrac{1}{2}T \cdot \operatorname{sinc}^2\left\{\dfrac{\Omega T}{4}\right\} \cdot e^{-j\Omega T/2}$
$\displaystyle\sum_{\ell=-\infty}^{\infty} \delta_c(t - \ell T)$	$\dfrac{2\pi}{T}\displaystyle\sum_{\ell=-\infty}^{\infty} \delta_c\left(\Omega + \dfrac{2\pi\ell}{T}\right)$
$e^{-\frac{1}{2}at^2}, \quad \text{all } t$	$\left\{\dfrac{2\pi}{a}\right\}^{1/2} e^{-\frac{1}{2}\Omega^2/a}, \quad \text{all } \Omega$

Note: If the LT is known, just set $s = j\Omega$ to obtain the FT.

Table 5B.3   Fourier Series

$$\left[\text{Note: Below, sinc}(x) = \frac{\sin(x)}{x}.\right] \qquad\qquad \text{Assumed:}\quad \tilde{x}_c(t + T) = \tilde{x}_c(t).$$

$\tilde{x}_c(t)$ for $t \in [0, T]$ (T-periodic extension of function shown)	$X_c(k) = \text{FS}\{\tilde{x}_c(t)\}$
$\delta_c(t)$	$\dfrac{1}{T}$ for all k
$1, t \in [0, T_1]; 0$, otherwise where $T_1 < T$	$\left(\dfrac{T_1}{T}\right)e^{-j\pi k T_1/T} \cdot \text{sinc}\left\{\dfrac{\pi k T_1}{T}\right\}$, general value of T_1. $= \frac{1}{2}e^{-j\pi k/2} \cdot \text{sinc}\{\frac{1}{2}\pi k\}, \; T_1 = \frac{1}{2}T.$
$1, t \in [T_1 - \frac{1}{2}T_2, T_1 + \frac{1}{2}T_2]; 0$, otherwise with $T = 2T_1$.	$\begin{cases} T_2/(2T_1), & k = 0 \\ \left(\dfrac{T_2}{T_1}\right)(-1)^k \text{sinc}\left(\dfrac{\pi k T_2}{2T_1}\right), & k \neq 0 \end{cases}$
$1 - \dfrac{\|t\|}{T}, \quad \|t\| \leq T; \quad 0$, otherwise $= \text{tri}\left(\dfrac{t}{T}\right)$	$\frac{1}{2}\delta(k) + \begin{cases} \dfrac{2}{(\pi k)^2}, & k \text{ odd} \\ 0, & k \text{ even} \end{cases}$
$0.4\cos(30\pi t + 20°) + \text{tri}(90t - 1.3)$	$\begin{cases} 0.5(0.4)e^{\pm j20°}, & k = \pm 1 \\ X_c(3k) = e^{-j1.3\pi k} \cdot \left[X_c(k) \text{ for tri}\left(\dfrac{t}{T}\right)\right] \end{cases}$
$e^{-\alpha t}$	$\dfrac{1 - e^{-\alpha T}}{\alpha T + j2\pi k}$
$\left\|\sin\left(\dfrac{\pi t}{T}\right)\right\|$	$\dfrac{2}{\pi(1 - 4k^2)}$
$\dfrac{t}{T}$	$\begin{cases} \frac{1}{2}, & k = 0 \\ \dfrac{j}{2\pi k}, & k \neq 0 \end{cases}$
$\begin{cases} 1, & t \in \left[0, \dfrac{T}{6}\right] \\ -2, & t \in \left[\dfrac{T}{6}, \dfrac{T}{3}\right] \\ 0, & \text{otherwise} \end{cases}$	$\begin{cases} 0, & k = 0 \\ \dfrac{3\cos\left(\dfrac{\pi k}{3}\right) - 2\cos\left(\dfrac{2\pi k}{3}\right) - 1}{-j\pi k}, & k \neq 0 \end{cases}$

Table 5B.4: *z*-Transform

$x(n)$	$X(z) = \mathrm{ZT}\{x(n)\}$
$\delta(n)$	1, for all z
$u(n)$	$\dfrac{z}{z-1}$, $\quad |z| > 1$
$a^n u(n)$	$\dfrac{z}{z-a}$, $\quad |z| > |a|$
$a^n u(-n-1)$	$\dfrac{-z}{z-a}$, $\quad |z| < |a|$
$na^n u(n)$ [set $a = 1$ for $nu(n)$]	$\dfrac{az}{(z-a)^2}$, $\quad |z| > |a|$
$a^n \cos(\omega_1 n)u(n)$ [set $a = 1$ for $\cos(\omega_1 n)u(n)$]	$\dfrac{z[z - a\cos(\omega_1)]}{z^2 - 2az\cos(\omega_1) + a^2}$, $\quad |z| > |a|$
$a^n \sin(\omega_1 n)u(n)$ [set $a = 1$ for $\sin(\omega_1 n)u(n)$]	$\dfrac{az\sin(\omega_1)}{z^2 - 2az\cos(\omega_1) + a^2}$, $\quad |z| > |a|$
a^n, $\quad n \in [0, N-1]$; 0, otherwise [set $a = 1$ for $1, n \in [0, N-1]$]	$\begin{cases} \dfrac{1 - (a/z)^N}{1 - a/z}, z \neq a \text{ and } z \neq 0 \\ N, \qquad\qquad z = 1 \end{cases}$
$a^{|n|}$, all n, $|a| < 1$	$\dfrac{1 - a^2}{(1 - az)(1 - a/z)}$, $\quad a < |z| < \dfrac{1}{a}$
n, $\quad n \in [0, N-1]$; 0, otherwise	$\begin{cases} \dfrac{z[1 + (N-1)z^{-N} - Nz^{-(N-1)}]}{(z-1)^2}, z \neq 1 \text{ and } z \neq 0 \\ \frac{1}{2}N(N-1), \quad z = 1. \end{cases}$

Triangle, N odd:

$$\begin{cases} \dfrac{2n}{N-1}, & n \in \left[0, \dfrac{N-1}{2}\right] \\ 2\left[1 - \dfrac{n}{N-1}\right], & n \in \left[\dfrac{N-1}{2}+1, N-1\right] \end{cases}$$

$$\begin{cases} \dfrac{\{1 - z^{-(N-1)/2}\}^2}{\dfrac{(N-1)}{2} \cdot z \cdot (1 - z^{-1})^2}, & z \neq 0, z \neq 1 \\ \frac{1}{2}(N-1), & z = 1 \end{cases}$$

Triangle, N even:

$$\begin{cases} \dfrac{2n}{N}, & n \in \left[0, \dfrac{N}{2}\right] \\ 2\left[1 - \dfrac{n}{N-1}\right], & n \in \left[\dfrac{N}{2}+1, N-1\right] \end{cases}$$

$$\begin{cases} \dfrac{(1 - z^{-N/2})^2}{\frac{1}{2}Nz(1 - z^{-1})^2}, & z \neq 0, z \neq 1 \\ \frac{1}{2}N, & z = 1 \end{cases}$$

Table 5B.5: Discrete-Time Fourier Transform

$$\left[\text{Note: Below, sinc}(x) = \frac{\sin(x)}{x}. \right]$$

$x(n)$	$X(e^{j\omega}) = \text{DTFT}\{x(n)\}, \|\omega\| \le \pi.$ Periodically extend for $\|\omega\| > \pi.$
1	$2\pi \, \delta_c(\omega)$
$\delta(k)$	1
$u(n)$	$\dfrac{1}{1 - e^{-j\omega}} + \pi\delta_c(\omega)$
$a^n u(n)$	$\dfrac{1}{1 - ae^{-j\omega}}$
$a^n u(-n - 1)$	$\dfrac{-1}{1 - ae^{-j\omega}}$
$na^n u(n)$ [set $a = 1$ for $nu(n)$]	$\dfrac{ae^{j\omega}}{(e^{j\omega} - a)^2}$
$a^n \cos(\omega_1 n)u(n)$ [set $a = 1$ for $\cos(\omega_1 n)u(n)$]	$\dfrac{e^{j\omega}[e^{j\omega} - a\cos(\omega_1)]}{e^{j2\omega} - 2ae^{j\omega}\cos(\omega_1) + a^2}$
$a^n \sin(\omega_1 n)u(n)$ [set $a = 1$ for $\sin(\omega_1 n)u(n)$]	$\dfrac{ae^{j\omega}\sin(\omega_1)}{e^{j2\omega} - 2ae^{j\omega}\cos(\omega_1) + a^2}$
$e^{j\omega_1 n}$, all n	$2\pi\delta_c(\omega - \omega_1)$
$\cos(\omega_1 n)$, all n	$\pi\{\delta_c(\omega + \omega_1) + \delta_c(\omega - \omega_1)\}$
$\sin(\omega_1 n)$, all n	$\pi j\{\delta_c(\omega + \omega_1) - \delta_c(\omega - \omega_1)\}$
$\text{sinc}(\omega_1 n)$, all n	(π/ω_1), $\omega \in [-\omega_1, \omega_1]$; 0, otherwise
1, $n \in [0, N-1]$; 0, otherwise	$e^{-j\omega(N-1)/2}\dfrac{\sin(N\omega/2)}{\sin(\omega/2)}$

Triangle, N odd:

$$\begin{cases} \dfrac{2n}{N-1}, & n \in \left[0, \dfrac{N-1}{2}\right] \\ 2\left[1 - \dfrac{n}{N-1}\right], & n \in \left[\dfrac{N-1}{2} + 1, N-1\right] \end{cases} \qquad \dfrac{[\sin\{\omega(N-1)/4\}]^2}{[(N-1)/2] \cdot \{\sin(\omega/2)\}^2} \cdot e^{-j\omega(N-1)/2}$$

Triangle, N even:

$$\begin{cases} \dfrac{2n}{N}, & n \in \left[0, \dfrac{N}{2}\right] \\ 2\left(1 - \dfrac{n}{N}\right), & n \in \left[\dfrac{N}{2} + 1, N-1\right]. \end{cases} \qquad \dfrac{\sin^2(\omega N/4)}{[N/2]\sin^2(\omega/2)} \cdot e^{-j\omega N/2}$$

$$\sum_{\ell=-\infty}^{\infty} \delta(n - \ell N) \qquad\qquad \dfrac{2\pi}{N}\sum_{\ell=-\infty}^{\infty} \delta_c\left(\omega + \dfrac{2\pi\ell}{N}\right)$$

Note: If the ZT is known, just set $z = e^{j\omega}$ to obtain the DTFT.

PROBLEMS

Time Shift

5.1. Plot $x(n - n_0)$ and $\angle\text{DTFT}\{x(n - n_0)\}$ for $n_0 = 2$ and 10, for $x(n) = (\frac{1}{2})^n u(n)$ (as in Example 5.1).

5.2. (a) Show that the property analogous to $d^\ell/d\Omega^\ell\{e^{j\Omega t}\} = (j\Omega)^\ell e^{j\Omega t}$ used in continuous-time phasor analysis *is* $D^\ell\{e^{j\omega n})\} = e^{-j\omega\ell} \cdot e^{j\omega n}$ for discrete-time analysis, where $D\{x(n)\} = x(n - 1)$.

(b) Given that $\text{Re}\{\,\cdot\,\}$ and $D\{\,\cdot\,\}$ operators are reversible, demonstrate use of the phasor to solve the difference equation $a_1 y(n - 1) + a_0 y(n) = A_1\cos(\omega_1 n + \theta_1)$ in the sinusoidal steady state.

5.3. (a) Find the unilateral z-transform of $u(n + n_0)$, where $u(n)$ is the unit step sequence and $n_0 > 0$. Express your result in terms of $U(z)$, the z-transform of $u(n)$. Are your results consistent with the advance theorem, (5.9)?

(b) Find the unilateral z-transform of $g(n + n_0)$, where $g(n) = \alpha^n u(n)$. Again, base your results on (5.9).

5.4. Prove (5.8a) using direct substitution of $x(n - n_0)$ into the forward z-transform expression.

5.5. Prove (a) (5.10) and (b) (5.11) using direct substitutions into the inverse DTFT and FS expressions.

5.6. Show that if $X_{\text{ul}}(z)$ is the unilateral z-transform of $x(n)$, then the unilateral z-transform of the advanced-by-n_0 version of $x(n)$ is as given by (5.9).

5.7. Let $y(n) = e^{j\omega_0 n}$ and verify using (5.39d) [via convolution of $X(e^{j\omega})$ with a shifted Dirac delta function] that the DTFT shift-in-ω property (5.10) is just a special case of (5.39d).

5.8. (a) Let $U(z) = \text{ZT}_{\text{ul}}\{u(n)\}$, where $u(n)$ is the unit step sequence. Find $\text{ZT}_{\text{ul}}\{u(n + 1)\}$.

(b) Generalize your result in part (a) to find $\text{ZT}_{\text{ul}}\{g(n + n_0)\}$ in terms of the unilateral $G(z)$, where $g(n) = a^n u(n)$.

5.9. Using the unilateral z-transform, find $y(n)$: $y(n + 2) + 5y(n + 1) + 6y(n) = cx(n + 1) + dx(n)$, where $x(n) = a^n u(n)$, $y(0)$ and $y(1)$ are left as general, and a is not equal to either -3 or -2. You may use the results found in Problem 5.8. Perform a check to see that you reproduce the initial conditions $y(0)$ and $y(1)$ in your analytical solution.

Differentiation/Integration

5.10. Prove (5.18) using a direct substitution of $dx_c(t)/dt$ into the forward Fourier transform expression.

5.11. Prove (5.24a) using a direct substitution of $dX_c(j\Omega)/d\Omega$ into the inverse Fourier transform expression.

5.12. If the unit sample response of a system is $h(n)$, find the unit step response $s(n)$ in terms of $h(n)$. Also find $S(z)$ in terms of $H(z)$.

5.13. Using Matlab, illustrate and verify the property $\text{DTFT}\{nx(n)\} = j\dfrac{\partial}{\partial\omega} X(e^{j\omega})$ [(5.25)] for $x(n)$

$= u(n) - u(n - N)$, where $N = 30$. Plot smooth (e.g., 256-point) plots of the magnitude and phase of the DTFT of $\{y(n) = nx(n)\}$ using either a direct DTFT formula calculation or, better, using the `fft` command. Against this, plot the exact values obtained from the property formula, applied to this particular $x(n)$. Include a calculation of the exact value of $Y(e^{j0})$.

5.14. It was shown in (5.19) through (5.21) that $LT\{x_c(t)/t\} = \int_s^{\infty} X_c(s')\, ds'$, where the contour of the integral is any one that starts on s and goes to ∞ in the right half-plane. Find the comparable property for z-transforms, namely $ZT\{x(n)/n\}$. Be careful to specify any integration contour used.

5.15. Assuming bicausal signals and the bilateral transforms,

(a) Using the result (5.21), find the LT of $x_c(t)/(t - \tau)$, where τ is a constant time parameter.

(b) Using the result for the ZT of $x(n)/n$ developed in Problem 5.14, find the ZT of $x(n)/(n - n_0)$.

5.16. In Problem 4.57, we investigated the Hilbert transformer $h_{H,c}(t)$, whose frequency response is $H_{H,c}(j\Omega) = -j \cdot \text{sgn}(\Omega)$, where $\text{sgn}(\Omega) = 1$ for $\Omega \geq 0$ and -1 for $\Omega < 0$. Determine $h_{H,c}(t)$ in closed form, this time using a transform property rather than by a direct calculation.

5.17. Derive the right-hand side of (5.22b) by first differentiating with respect to Ω the Fourier integral $X_c(j\Omega)$ with $\text{sinc}(\Omega_1 t)$ substituted for $x_c(t)$.

5.18. If $LT_{u\ell}\{x_c(t)\} = X_c(s)$, find $LT_{u\ell}\{dx_c(t)/dt\}$. Extend your result to $LT_{u\ell}\{d^n x_c(t)/dt^n\}$. Why is the unilateral Fourier transform not very helpful for solving initial-value problems?

5.19. Using the results of Problem 5.18, solve

$$\frac{d^3 y_c(t)}{dt^3} + 4\frac{d^2 y_c(t)}{dt^2} + 21\frac{dy_c(t)}{dt} + 34 y_c(t) = \frac{dx_c(t)}{dt} + 30 x_c(t),$$

where $y_c(0) = 2$, $\left.\dfrac{dy_c(t)}{dt}\right|_{t=0} = -4$, $\left.\dfrac{d^2 y_c(t)}{dt^2}\right|_{t=0} = 0$, and $x_c(0^-) = 0$.

Verify your results using dlsim in Matlab, and plot against a direct calculation using the $y_c(t)$ you analytically determine. Verify initial and final values.

5.20 If $LT\{x_c(t)\} = X_c(s)$ and $FT\{x_c(t)\} = X_c(j\Omega)$, find the (a) LT and (b) FT of $\int_{-\infty}^{t} x_c(\tau)d\tau$.

Time Reversal

5.21. Prove (5.28) using a direct substitution of $x(-n)$ into the forward DTFT expression.

5.22. Prove (5.30) using a direct substitution of $X(-z)$ into the inverse ZT expression.

5.23. (a) Using Matlab, plot stem plots of the real and imaginary parts of $x(n)$, which is composed

of the sum of a unit-amplitude triangle of length 20 [see (4B.20)], padded out to $N = 100$ with zeros, and a boxcar for the last 20 samples that is modulated by the function j^n. This gives an interesting time function whose spectrum is nonsymmetric. Plot $X(e^{j2\pi f})$ versus f for $f \in [-\frac{1}{2}, \frac{1}{2}]$. Notice that the negative as well as positive values of f are requested here.

(b) Repeat part (a) for $x(-n)$. Form $x(-n)$ by literally reversing the order of $x(n)$. Note that, however, the default Matlab time index will still be positive. In your time plot, use the correct negative time indices. Also, be sure to use the negative time indices in your DTFT calculation, and again show $f \in [-\frac{1}{2}, \frac{1}{2}]$. Verify that the time-reversal leads to a frequency-reversal, as proved in (5.28).

5.24. If $G(z) = ZT\{g(n)\}$ and $g(n)$ is real, causal, and stable, what do we know about the poles and zeros of the z-transforms of the sequences

(a) $g_1(n) = (-1)^n g(n)$ and   (b) $g_2(n) = g(-n)$? Are $g_1(n)$ and $g_2(n)$ stable? Causal?

5.25. Prove that any function of time can be decomposed as the sum of a conjugate-even function of time and a conjugate-odd function of time. Assume discrete time.

5.26. Find the sequence related to $x(n)$ whose z-transform has poles at conjugate-reciprocal locations relative to those of $X(z)$.

Conjugation of a Function

5.27. (a) Prove (5.32) using a direct substitution of $X_c^*(k)$ into the inverse FS expression. (b) Prove (5.33) using a direct substitution. (c) Prove (5.34) by a direct substitution.

Variable Scaling

5.28. Prove (5.36) using a direct substitution of $x_c(\alpha t)$ into the forward LT expression.

5.29. Prove (5.37) using a direct substitution of $\tilde{x}(Mt)$ into the forward FS expression.

5.30. Prove (5.38) using a direct substitution and complex integration techniques.

5.31. Find the z-transform of $h_\alpha(n) = h(n) \cdot \alpha^n u(n)$ in terms of $H(z)$ and α. Suggest under what circumstance this identity might arise in filter design.

5.32. Using Matlab, verify (5A.6.5) for $x(n)$ a 400-point triangle [see (4B.20)]. Let $x_M(n) = x(4n)$ (thus $M = 4$). Compare DFT$\{x_M(n)\}$ with the aliasing sum in (5A.6.5); plot magnitudes and phases. Explain any discrepancies.

Modulation

5.33. Form the sequences $x(n)$ = triangle of length 32 [see (4B.20)], $y(n) = 0.95^n u(n)$ again of length 32, and $w(n) = x(n)y(n)$. In Matlab, plot the stem plots of $x(n)$, $y(n)$, and $w(n)$. Next, taking the fft of x, y, and w, verify the property (5A.6.7) by comparing fft($w(n)$) with the formula in (5A.6.7). Remember to take modulo N (i.e., add N for negative indices) for all expressions. Plot the magnitudes and phases on the same axes (using stem plots) to verify agreement.

5.34. Find the inverse FT of $X_c(j\Omega)$, a triangle from $-\Omega_0$ to $+\Omega_0$ and of height A. Use the time-multiplication theorem and the result from Problem 2.14.

5.35. Find the Fourier transform of $\tilde{x}_c(t)y_c(t)$, where $\tilde{x}_c(t)$ has period T and $y_c(t)$ is aperiodic, in terms of the Fourier series coefficients of $\tilde{x}_c(t)$ and the Fourier transform of $y_c(t)$. If in a previous course you have already seen the frequency-domain sampling formula (6.17) and the line following it, comment on the relation between it and your result here for a specific choice of $\tilde{x}_c(t)$.

5.36. Derive the unilateral Laplace transform for the product $y_c(t)$ of two causal time functions $x_c(t)$ and $h_c(t)$. [Equivalently, derive the bilateral LT for the product of causal signals $x_c(t)$ and $h_c(t)$.] In particular, what is the region of convergence of $Y_c(s)$?

Periodic Extension

5.37. Suppose that $X_c(s) = \text{LT}\{x_c(t)\}$ and that $x_c(t)$ has interval of support $t \in [0, T]$. Find (a) the bilateral LT and (b) the unilateral LT of $\tilde{x}_{T,c}(t)$, wherever in the s-plane they exist, if $\tilde{x}_{T,c}(t)$ is the T-periodic extension of $x_c(t)$. Where relevant, use the symbol Ω_0 for the fundamental frequency $2\pi/T$.

5.38. Suppose that $X(z) = \text{ZT}\{x(n)\}$ and that $x(n)$ has interval of support $n \in [0, N-1]$. Find (a) the bilateral ZT and (b) the unilateral ZT of $\tilde{x}_N(n)$, wherever in the z-plane they exist, if $\tilde{x}_N(n)$ is the N-periodic extension of $x(n)$. Where relevant, use the symbol ω_0 for the fundamental frequency $2\pi/N$.

5.39. Repeat the ZT of the periodic extension as in Problem 5.38 (both parts), but this time let $\tilde{x}_N(n + N) = -\tilde{x}_N(n)$ (antiperiodic extension).

Meaning of Frequency Response

5.40. In this problem, we consider the practical meaning of the frequency response for continuous-time systems. It is straightforward to show that $e^{j\Omega t}$ is an eigenfunction of an LSI system $H_c(s)$ and that the frequency response is the associated eigenvalue: $x_c(t) = \exp(j\Omega_1 t) \to H_c(s) \to y_c(t) = H_c(j\Omega_1)\exp(j\Omega_1 t) = |H_c(j\Omega_1)| \cdot \exp\{j[\Omega_1 t + \angle H_c(j\Omega_1)]\}$. Now suppose, for example, that the input to the system $x_c(t)$ is the more practical cosine function—which, unlike $\exp(j\Omega_1 t)$, we may easily view on an oscilloscope: $x_c(t) = A\cos(\Omega_1 t + \theta)$. Find the (real-valued) steady-state output $y_c(t)$ (also observable on an oscilloscope), and interpret frequency response in the context of a general real-valued signal. Show why the polar form of frequency response is so helpful. Is $A\cos(\Omega_1 t + \theta)$ an eigenfunction of the LSI system?

5.41. Repeat Problem 5.40 for discrete time. It is easily shown that $e^{j\omega n}$ is an eigenfunction of an LSI system $H(z)$ [this proof is found in (W5.1.8)] and that the frequency response is the associated eigenvalue: $x(n) = e^{j\omega_1 n} \to H(z) \to y(n) = H(e^{j\omega_1})e^{j\omega_1 n} = |H(e^{j\omega_1})| \cdot \exp(j[\omega_1 n + \angle H(e^{j\omega_1})])$. Now suppose, for example, that the input to the system $x(n)$ is the more practical cosine function—which, unlike $\exp(j\omega_1 n)$, we may easily view on a stored-sampled-data plot: $x(n) = A\cos(\omega_1 n + \theta)$. Find the (real-valued) steady-state output $y(n)$ (also observable as a real discrete-time signal), and interpret frequency response in the context of a general real-valued signal. Show why the polar form of frequency response is so helpful. Is $A\cos(\omega_1 n + \theta)$ an eigenfunction of the LSI system?

5.42. (a) Using the result of Problem 5.41, find the output $y(n)$ of a system whose unit sample response is $h(n) = (\frac{1}{4})^n u(n)$ if the input is $x(n) = 3\cos\{2\pi \cdot (\frac{1}{8})n + \pi/3\}$. Simplify your expression as much as possible before using numerical approximations.

(b) Check your result with Matlab by performing a direct convolution calculation, (using 1000 points) using `conv` and plotting that result against your result in part (a). What, if any, discrepancies do you find (for certain values of n), and how do you explain them?

(c) Repeat parts (a) and (b) for $h(n) = (\frac{7}{8})^n u(n)$. Comment on how $\frac{7}{8}$ as opposed to $\frac{1}{4}$ alters any discrepancy found in part (b).

5.43. Define system bandwidth.

Miscellaneous or Multiple Properties/Transforms

5.44. (a) Usually, we assume that the order N of the denominator of $H_c(s)$ is larger than the order of the numerator M, for a physical system whose transfer function is $H_c(s)$. Explain why this is so.

(b) Contrast the above fact with impedance. For example, the impedance of an inductor is $R + sL$, for which $(M = 1) > (N = 0)$. Is there any contradiction?

5.45. Derive (a) initial value and (b) final value theorems for causal discrete-time signals.

5.46. Derive (a) initial value and (b) final value theorems for causal continuous-time signals.

5.47. (a) Show that for $\tilde{x}_c(t)$ real and T-periodic (with $\Omega_0 = 2\pi/T$), we may write $\tilde{x}_c(t) = X_c(0) + 2\sum_{k=1}^{\infty} |X_c(k)| \cos\{\Omega_0 kt + \angle X_c(k)\}$ where $X_c(k)$ are the Fourier series coefficients for $\tilde{x}_c(t)$.

(b) Under what conditions on $\tilde{x}_c(t)$ does the factor $\cos\{\Omega_0 kt + \angle X_c(k)\}$ reduce to (i) $\pm\cos(\Omega_0 kt)$ and (ii) $\pm\sin(\Omega_0 kt)$? Argue from both time-domain and frequency-domain viewpoints. If $\tilde{x}_c(-t) = -\tilde{x}_c(t)$, what must be true about $X_c(0)$?

5.48. If $x(n) = 0.95^n\cos(15° \cdot [\pi/180]n)u(n)$, show that $X(z) = \{1 - 0.95\cos(15 \cdot \pi/180)/z\}/\{1 - 2 \cdot 0.95 \cos(15 \cdot \pi/180)/z + [0.95/z]^2\}$, which converges only for $|z| > 0.95$. This result was used in the solution of Example 5.19.

5.49. Prove the transform pair cited in the text just after (5.48), namely that if $x(n) = r^{|n|}\cos(\omega_1 n)$, then

$$X(e^{j\omega}) = \frac{r\{\cos(\omega - \omega_1) - r\}}{1 + r^2 - 2r\cos(\omega - \omega_1)} + \frac{1 - r\cos(\omega + \omega_1)}{1 + r^2 - 2r\cos(\omega + \omega_1)} \qquad \text{(for } |r| < 1\text{)}.$$

5.50. Determine whether the following statements are true or false. For each statement, if it is true, derive it; if it is false, cite an example showing it to be false.

(a) Conjugate-even, complex sequences lead to purely real DTFTs.

(b) The DTFT of a real causal signal is always real.

(c) The phase spectrum of any sequence is real and is an odd function of ω.

(d) Im$\{H(e^{j\omega})\}$ is an odd function of ω when $h(n)$ is real.

(e) The magnitude spectrum of a sequence is always an even function of ω.

5.51. Using various properties of the z-transform, determine the z-transform of $y^2(-n) \cdot x^*(-n-5)$ in terms of $X(z)$ and $Y(z)$. Your result may involve contour integrals.

5.52. 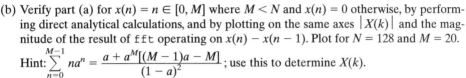 (a) Prove that DFT$_N\{x(n) - x(n-1)\} = (1 - e^{-j2\pi k/N})X(k)$ using (5.16) (all arguments are modulo N).

(b) Verify part (a) for $x(n) = n \in [0, M]$ where $M < N$ and $x(n) = 0$ otherwise, by performing direct analytical calculations, and by plotting on the same axes $|X(k)|$ and the magnitude of the result of fft operating on $x(n) - x(n-1)$. Plot for $N = 128$ and $M = 20$.

Hint: $\displaystyle\sum_{n=0}^{M-1} na^n = \frac{a + a^M[(M-1)a - M]}{(1-a)^2}$; use this to determine $X(k)$.

Problems 5.53 through 5.77 involve Web materials (WP 5.1)

Parseval's Relation; Angle Between Signals

5.53. (a) Without assuming the isometry relation (W5.1.30), prove the generalized Parseval relationship for the DFT [generalization of (W5.1.34)]:

$$\sum_{n=0}^{N-1} x_1(n)x_2^*(n) = \frac{1}{N}\sum_{k=0}^{N-1} X_1(k)X_2^*(k),$$

where $X_1(k)$ and $X_2(k)$ are, respectively, the N-point DFTs of the N-long sequences $x_1(n)$ and $x_2(n)$.

(b) If $x_1(n) = a^n$ and $x_2(n) = b^n$ ($n \in [0, N-1]$ and a, b real), express $\displaystyle\sum_{n=0}^{N-1} x_1(n)x_2^*(n)$ in the simplest terms possible. Then, using Parseval's relation, determine an interesting identity involving a, b, N, and a sum over k.

(c) Part (b) is an example where the Parseval sum is much easier to compute in the n domain than in the k domain. Can you think of a case where the reverse would be true?

5.54. Numerically verify (W5.1.34), Parseval's relation for finite-length sequences. Use $x(n) = $ a 100-point triangle [see (4B.20)]. Give your numerical value for each side of (W5.1.34). Letting $y(n) = 0.9^n u(n)$, verify the generalized DFT Parseval in (5A.6.9b). Verify by computing both left-hand and right-hand sides of the expressions; they should be the same.

5.55. (a) Using Parseval's relation for the DFT, express what the DFT does to an input vector **x** whose nth component is $x(n)$ in terms of stretching and/or rotating. Determine an explicit expression for the cosine of the angle between **x** and **X** = DFT{**x**} that involves only sums of products of direction cosines and angles.

5.56. In the context of Problem 5.55, in Matlab find the angle between $x(n) = $ a 64-length triangle sequence [see (4B.20)] and $h(n) = $ a 40-length boxcar (zero-padded out to 64). Then compare this angle with the angle between the respective DFT$_{64}$s. You will find that they are the same. Why? Prove your result. Repeat for other sequences, such as $x(n)$ with $h_1(n) = $ an FIR filter obtained by the routine fir1 and $h(n)$ with $h_1(n)$. Note: The transpose operator (prime) in Matlab already includes complex conjugation. Also, find the angle between $x(n)$ and its DFT$_{64}$, between $h(n)$ and its DFT, and between $h_1(n)$ and its DFT; comment on whether they are complex and why.

5.57. Show that the inverse Fourier transform is the adjoint of the Fourier transform, and therefore it is unitary and so Parseval's relation holds for it. Interpret Parseval's relation in intuitive terms for the FT, as was done for the DTFT in the text.

5.58 Give Parseval's relation for the Laplace transform and for the z-transform. Hint: The convolution theorem is in this case easier than trying to figure out the correct generalization of "norm" in the GPE domains.

Transforms as Expansions/Completeness

5.59. Derive the discrete Fourier series transform pair using eigenvalue/eigenfunction techniques as on WP 5.1.

5.60 In this problem, we relate a result in vector spaces to Fourier transforms, by means of an analogy between vectors and functions. A vector $\mathbf{u}$ can be represented by its coordinates a_ℓ in a nonstandard coordinate system having basis vectors $\{\mathbf{v}_\ell\}$ as $\mathbf{u} = \sum_{\ell=1}^{N} a_\ell \mathbf{v}_\ell$; that is, $\mathbf{u}$ is representable as a linear combination of the $\{\mathbf{v}_\ell\}$. If the $\mathbf{v}_\ell$ happen to be orthogonal, then the inner product of $\mathbf{v}_k$ with $\mathbf{u}$ is, from the equation above, $<\mathbf{v}_k, \mathbf{u}> = \sum_{\ell=1}^{N} a_\ell <\mathbf{v}_k, \mathbf{v}_\ell> = a_k |\mathbf{v}_k|^2$. Thus, we can easily obtain the coordinates of $\mathbf{u}$ with respect to the $\{\mathbf{v}_\ell\}$ as $a_k = <\mathbf{v}_k, \mathbf{u}>/|\mathbf{v}_k|^2$. Notice that no matrix inversion is required, as is the case if the $\{\mathbf{v}_\ell\}$ are not orthogonal. Using an infinite-interval integration for inner product, develop an interpretation of Fourier transforms in terms of the above discussion. You should obtain forward and reverse FT formulas in the process.

5.61. Repeat Problem 5.60 for the DFT.

5.62. In this problem, we further study the implications of completeness. For simplicity, we first consider the vector case. Suppose we have an orthonormal set of L vectors $\{\mathbf{v}_\ell\}$ each having N components: $<\mathbf{v}_\ell, \mathbf{v}_m> = \delta(\ell - m)$ where ℓ and m range from 1 to L. For clarity, we will use L, but for simplicity, we will assume that $L = N$. A complete set of vectors also satisfies the orthonormality constraint $\sum_{\ell=1}^{L} \mathbf{v}_{\ell,p} \mathbf{v}_{\ell,q}^* = \delta(p - q)$, where, for example, $\mathbf{v}_{\ell,p}$ is the pth component of $\mathbf{v}_\ell$ (p, q are here integers ranging from 1 to N). Define the "outer product" $\mathbf{v}_\ell >< \mathbf{v}_m$ as $\mathbf{v}_\ell >< \mathbf{v}_m = \mathbf{v}_\ell \mathbf{v}_m^{*T}$. Then, the $N \times N$ matrix $\mathbf{W} = \sum_{\ell=1}^{L} \mathbf{v}_\ell >< \mathbf{v}_\ell = \sum_{\ell=1}^{L} \mathbf{v}_\ell \mathbf{v}_\ell^{*T}$.

(a) What is the value of the pqth matrix element w_{pq}, when $\{\mathbf{v}_\ell\}$ satisfy the above completeness constraint?

(b) Using the result in part (a)—that is, under the assumption that $\{\mathbf{v}_\ell\}$ are complete— prove the generalized Fourier series expression (W4.2.2), namely (in our current notation):

$$\mathbf{x} = \sum_{\ell=1}^{L} <\mathbf{v}_\ell, \mathbf{x}> \mathbf{v}_\ell.$$

5.63 Extend the result of Problem 5.62 to the case of functions rather than vectors. Take, for example, the function "$\mathbf{x}$" $= X(e^{j\omega})$, the DTFT of a sequence $x(n)$ and "$v_{\ell,n}$" $= e^{-j\omega n}/\{2\pi\}^{1/2}$. To show that the "$v_{\ell,n}$" are complete in this case, the outer product should be a (sum of) Dirac, not Kronecker, deltas. Your "matrix" $\mathbf{W}$ will this time be infinite-dimensional, and the sum over ℓ will become an infinite-index-range sum; what is the "matrix element" w_{pq}? You will derive the DTFT pair in a manner that is the dual of the method presented in Example W5.1.1. What does "$<v_\ell, \mathbf{x}>$" represent in this example?

5.64. Concerning the completeness relations $\int_{C_{i.a.}} e^{s(t-t')} ds = \delta_c(t - t')$ for the functions e^{st} and $\oint_{C_{u.c.}} z^{n-\ell} (dz/z) = \delta(n - \ell)$ for the functions z^n ($C_{i.a.}$ is the imaginary axis and $C_{u.c.}$ is the unit circle), what sort of generalization of inner product seems to be indicated for functions of a complex variable? Recall that in all our discussions the inner product has been over complex- (or real-) valued functions of a real variable.

5.65. Under what condition is continuous-time convolution a Hermitian operator? What "transform of a real function" property follows from your work? Prove the continuous-time Fourier transform pair using orthogonality of $e^{j\Omega t}$.

5.66. Derive the discrete-time Fourier transform pair by analyzing $L_{*,\omega}$ (convolution in the ω domain) rather than L_* (convolution in the n domain, as in the text).

5.67. Heuristically derive the z-transform pair, using/generalizing the following result used in deriving the DTFT pair:

$$\frac{1}{2\pi} \cdot \int_{-\pi}^{\pi} e^{j\omega(n-\ell)} d\omega = \delta(n - \ell).$$

5.68. Heuristically derive the Laplace transform pair, using/generalizing the following result used in the solution of Problem 5.65 in deriving the FT pair:

$$\delta_c(t - t') = \frac{1}{2\pi} \int_{-\infty}^{\infty} e^{j\Omega(t-t')} d\Omega.$$

5.69. Is the Fourier transform Hermitian? Note that in this case Ω and t are both real variables ranging from $-\infty$ to ∞ and $x_c(t)$ and $X_c(j\Omega)$ are both complex-valued, so the immediate dismissal given for the DTFT on WP5.1 following Example W5.1.2 does not apply here.

Matrix Problems

5.70. (a) Write a Matlab routine that produces a 2×2 matrix **A** whose eigenvalues are equal to the values specified in the given input. Hint: There is not one unique answer; you have flexibility. Use the characteristic equation to see how to obtain **A**.

(b) Select an eigenvalue pair for which one of the eigenvalues is negative. Then try out several **x** vectors in the quadratic form $Q(\mathbf{x})$ in (W5.1.36) until you find a Q value that is negative. Thus you will have demonstrated that with a negative eigenvalue, **A** is not positive definite. Repeat for both eigenvalues positive; this time you should not be able to find an **x** value for which $Q(\mathbf{x}) < 0$.

5.71. No study of MATrix LABoratory would be complete without an example of numerical properties of matrices, which as seen on WP 5.1 are important in signal processing. Let **A** be defined by its elements $a_{mn} = (-1)^n(1 - jm) - \delta(m - n)$, where $j = \sqrt{-1}$ and $0 \leq m, n \leq N$. Note that the matrix size should be $(N + 1) \times (N + 1)$.

(a) In Matlab, form the matrix **A** for $N = 3$, without using any `for` statements (in 1 to 3 statements) and without manually filling in the numbers.

(b) Determine the inverse, determinant, rank, and condition number (`cond`) of **A**, and also its eigenvalues (display *only* the eigenvalues using the `diag` command, and do not display eigenvectors). How many distinct eigenvalues are there? Repeat for $N = 9$; now how many distinct eigenvalues are there?

(c) In this part, you will examine at least simplistically the meaning of "ill-conditioning" of a matrix. The basic idea is that when a matrix is ill-conditioned, two or more columns are nearly linearly dependent. As a result, if we have modeled a physical system by the linear equations $\mathbf{Ax} = \mathbf{b}$, where **x** is an unknown vector, **b** is a vector of measurements, and **A** models the relations between the unknown variables in **x** and the measurements in **b**, then small errors in the measurements b_m will result in very large errors in the solution **x** (obviously, a bad situation).

Now let $a_{mn} = (-1)^n - jm - \alpha\delta(m - n)$ and $N = 3$, where values of α are given below. Let $\mathbf{b} = [5.9998 \quad 6.0001 - 6j \quad 5.9997 - 12j \quad 6 - 18j]^T$.

Note: To demonstrate the ill-conditioning properly, you must enter the value of **b** exactly as shown above; if you do not, the ill-conditioning will still be there of course, but the effect will not be so clearly displayed by the techniques below. Let $\mathbf{\Delta b} = \{\delta \cdot \text{norm(b)}/(N + 1)\} \cdot \text{rand}(N + 1, 1)$, where we will set $\delta = 10^{-3}$. $\mathbf{\Delta b}$ will represent an error in measurements, as follows. Inclusion of the factor "$\delta \cdot \text{norm(b)}/(N + 1)$" makes the elements of $\mathbf{\Delta b}$ on the order of δ times those of **b**, in a mean-square average sense. The use of `rand` gives a random error vector, simulating errors in the measurements.

norm(b) $= |\mathbf{b}| = \{\sum b_\ell^2\}^{1/2}$, the 2-norm. Now define $\mathbf{x}$ satisfying $\mathbf{Ax} = \mathbf{b}$. Then $\mathbf{x}'$ in $\mathbf{Ax}' = \mathbf{b}'$ is the solution of the system for the perturbed measurements, where $\mathbf{b}' = \mathbf{b} + \Delta\mathbf{b}$. [In Matlab, calculate $\mathbf{x}'$ using inv($\mathbf{A}$); list both $\mathbf{x}$ and $\mathbf{x}'$. Specific values of the latter will differ from run to run because random numbers are involved.] If $\mathbf{A}$ is a well-conditioned matrix, then the percent difference between the solution when $\mathbf{b}$ is perturbed ($\mathbf{x}'$) and when it is not ($\mathbf{x}$) will not be many orders of magnitude greater than percent difference between the norms of $\mathbf{b}$ and of $\mathbf{b}'$. But when $\mathbf{A}$ is ill-conditioned, the fractional error in the solution when $\mathbf{b}$ is perturbed can be many orders of magnitude greater than that of $\mathbf{b}$. Perform the calculations to verify this fact, using norm of Matlab, as follows. With $\alpha = 10^{-4}$, first calculate $dx = 100|\Delta\mathbf{x}|/|\mathbf{x}|$, where $\Delta\mathbf{x} = \mathbf{x} - \mathbf{x}'$, and compare with db $100|\Delta\mathbf{b}|/|\mathbf{b}|$ [$\approx 100\delta/(N+1)$], where the "100" gives percent changes. You should find that dx is many orders of magnitude greater than db. Then change α to 1, which will make $\mathbf{A}$ be very well conditioned [$= \mathbf{A}$ in part (a)]. Repeat the above fractional error calculations, and you will find that dx is only about an order of magnitude greater than db. In fact, you will find that in this case, $C(\mathbf{A}) \cdot db$ is an upper bound for dx, where $C(\mathbf{A}) =$ cond($\mathbf{A}$) is the condition number of the matrix $\mathbf{A}$. That is, when $\mathbf{A}$ is reasonably well-conditioned, we have[6] $dx < C(\mathbf{A}) \cdot db$. For both values of α (ill-conditioned and well-conditioned $\mathbf{A}$), list $\mathbf{A}$, $C(\mathbf{A})$, db, dx, and $C(\mathbf{A}) \cdot db$ and verify the inequality $dx < C(\mathbf{A}) \cdot db$ (how close an upper bound is $C(\mathbf{A}) \cdot db$ on dx?). Also display $\mathbf{b}'$ and $\hat{\mathbf{b}}' = \mathbf{Ax}'$. Even in the case of severe ill-conditioning (experiment with smaller and smaller α!), $\hat{\mathbf{b}}'$ should equal $\mathbf{b}'$, because Matlab is a highly accurate program, capable of solving highly ill-conditioned problems. The smaller $C(\mathbf{A})$ is, the better conditioned $\mathbf{A}$ is, and the larger $C(\mathbf{A})$ is, the more ill-conditioned $\mathbf{A}$ is. Thus, the condition number $C(\mathbf{A})$ gives us a handle on how ill-conditioned a matrix $\mathbf{A}$ is—how sensitive the solution is to errors in the measurement vector—and in other cases will even indicate how accurate the numerical solution itself is (e.g., in cases where numerical calculations will yield $\mathbf{A}[\text{inv}(\mathbf{A})\mathbf{b}'] \neq \mathbf{b}'$, occurring when $\mathbf{A}$ is ill-conditioned and roundoff errors are significant).

Other Transforms

5.72. Consider the two-dimensional spatial Fourier transform

$$F(k_x, k_y) = \int_{-\infty}^{\infty} \int_{-\infty}^{\infty} f(x, y) \, e^{-j(k_x x + k_y y)} dx dy$$

with inverse two-dimensional Fourier transform

$$f(x, y) = \frac{1}{(2\pi)^2} \int_{-\infty}^{\infty} \int_{-\infty}^{\infty} f(x, y) \, e^{j(k_x x + k_y y)} dk_x dk_y.$$

The variables k_x, k_y are spatial frequencies in the x- and y-directions. Often they become wave-vector components in the propagation of waves. Write the expansion functions and completeness relation, and use them to write the expansion relation for this transform.

5.73. The "generating function" $\gamma(z)$ is important in many areas of mathematical physics and probability. The generating function "generates" the sequence of coefficients $g(n)$ in its power series expansion: $\gamma(z) = \sum_{n=-\infty}^{\infty} g(n)z^n$. We see that $\gamma(z)$ is the z-transform of $g(-n)$ or, equivalently, $\gamma(1/z)$ is the z-transform of $g(n)$; thus $\gamma(z)$ is really nothing other than a Laurent series. Show that the z-transform of the sequence of nth-order Bessel functions of the first kind $J_n(k\rho)$ evaluated at $1/z$ is the generating function $\exp\left\{\frac{k\rho}{2}(z - 1/z)\right\}$. Here $k\rho$ is a dimensionless quantity, whereas ρ has the dimension of distance and k has the dimensions of radians/distance. In practical problems, k is a wavenumber and ρ is the distance from the origin of the x-y–plane.

---

[6] S. Barnett, *Matrices: Methods and Applications* (Oxford: Clarendon Press, 1990), p. 399.

Hints: (i) View $k\rho$ as a constant parameter. (ii) Expand $\gamma(z)$ in a Maclaurin series over powers of the exponent $(k\rho/2)(z - 1/z)$. (iii) use Newton's binomial theorem for expansion of powers of $1 + a$:

$$(1 + a)^N = \sum_{k=0}^{N} \binom{N}{k} a^k, \text{ where } \binom{N}{k} = \frac{N(N-1)(N-2)\cdots(N-k+1)}{k!}.$$

(iv) Recognize the series definition for $J_n(k\rho)$: $\quad J_n(k\rho) = \sum_{m=0}^{\infty} \frac{(-1)^m (k\rho/2)^{n+2m}}{m!(n+m)!}$.

The z-transform/generating function will be bilateral ($-\infty < n < \infty$).

5.74. In Problem 5.73, we showed that $e^{\frac{k\rho}{2}(z-1/z)} = \sum\limits_{n=-\infty}^{\infty} J_n(k\rho)z^n$, where $J_n(k\rho)$ is the nth-order Bessel function of the first kind, with dimensionless argument $k\rho$.

(a) By differentiating this generating function with respect to z, show that $J_{n+1}(k\rho) + J_{n-1}(k\rho) = (2n/[k\rho])J_n(k\rho)$. This recursion formula may be used to numerically compute higher-order Bessel functions from low-order Bessel functions.

(b) By differentiating the generating function with respect to $k\rho$, show that $\partial J_n(k\rho)/\partial\rho = (k/2) \cdot [J_{n-1}(k\rho) - J_{n+1}(k\rho)]$. This formula is helpful in, for example, electromagnetic field calculations involving cylindrical wave expansions, where boundary conditions across cylindrical boundaries involve derivatives of the field with respect to ρ.

(c) Prove the "addition theorem" for Bessel functions, again using the generating function identity, to obtain $J_n(k[\rho_1 + \rho_2]) = \sum\limits_{m=-\infty}^{\infty} J_m(k\rho_1) \cdot J_{n-m}(k\rho_2)$.

5.75. The generating function identity for Bessel functions was shown in Problem 5.73:

$$e^{\frac{k\rho}{2}(z-1/z)} = \sum_{n=-\infty}^{\infty} J_n(k\rho)z^n,$$ where $J_n(k\rho)$ is the nth-order Bessel function of the first kind, with dimensionless argument $k\rho$.

(a) Let $z = je^{j\theta}$ in the generating function, where θ represents the counterclockwise angle from the $+\hat{\mathbf{x}}$ direction in the x-y-plane. Now, $\boldsymbol{\rho} = x\hat{\mathbf{x}} + y\hat{\mathbf{y}}$ and $x = \rho\cos(\theta)$, and $y = \rho\sin(\theta)$. Find the expansion of a plane wave in the $\hat{\mathbf{x}}$ direction $[e^{jk\rho\,\cos(\theta)}]$ in terms of $J_n(k\rho)e^{jn\theta}$, which are called cylindrical harmonics (or cylindrical waves). This expansion again is helpful in field problems in which a plane wave must be expanded in terms of cylindrical waves in order to be able to satisfy boundary conditions at a cylindrical boundary.

(b) Repeat part (a) for $h = e^{j\theta}$ to find the expansion of a plane wave in the y-direction $[e^{jk\rho\sin(\theta)}]$ in terms of $J_n(k\rho)e^{jn\theta}$.

(c) Repeat part (a) for a general direction $\mathbf{k} = k_x\hat{\mathbf{x}} + k_y\hat{\mathbf{y}}$ of the plane wave in two dimensions: $\boldsymbol{\rho} = x\hat{\mathbf{x}} + y\hat{\mathbf{y}}$. First express in terms of $k, \rho, \theta_\rho - \theta_\mathbf{k}$, where $\theta_\rho = \tan^{-1}(y/x)$ (previously denoted just θ) and $\theta_\mathbf{k} = \tan^{-1}(k_y/k_x)$, and of course in terms of Bessel functions. Then express your final answer in terms of $x, y, k_x, k_y, k, \rho,$ and θ.

5.76. Using the inverse z-transform relation (3.43) on the generating function identity,

$$e^{-\frac{k\rho}{2}(z-1/z)} = \sum_{n=-\infty}^{\infty} J_n(k\rho)z^{-n}$$ with z taken over the unit circle ($z = e^{j\theta}$), express $J_n(k\rho)$ as an integral over θ from $-\pi$ to π.

5.77. In this problem, we explore the cylindrical harmonics completeness relation.

(a) Write the expansion functions, two-dimensional completeness relation, and expansion relation for cylindrical harmonics. Hint: Consider two polar form vectors $\boldsymbol{\rho} = (\rho, \theta_\rho)$ and $\boldsymbol{\rho}' = (\rho', \theta_{\rho'})$. First express $\delta_c(x - x')\delta_c(y - y')$ in terms of $\delta_c(\rho - \rho')$ and $\delta_c(\theta_\rho - \theta_{\rho'})$. Substitute this expression into the completeness relation for complex exponentials (in rectangular form) derived in Problem 5.72. Then use the result for expansion of a plane wave in terms of cylindrical harmonics found in Problem 5.75 (polar form version) to express the completeness relation in terms of Bessel functions. Finally, carry out the integration over angle to obtain the intermediate result

$$\frac{\delta_c(\rho - \rho')\delta_c(\theta_\rho - \theta_{\rho'})}{\rho} = \frac{1}{2\pi} \sum_{n=-\infty}^{\infty} \int_0^\infty J_n(k\rho)J_n(k\rho')e^{j(\theta_\rho - \theta_{\rho'})n}k\,dk.$$

Then use this result to obtain the expansion relation for expansion of a general function of space $f(\rho, \theta_\rho)$ in terms of the Bessel transform coefficients. Comment on the relation between these Bessel transform coefficients and the true polar-coordinates Fourier transform coefficients.

(b) Multiply the completeness relation in part (a) by $\exp\{-j(\theta_\rho - \theta_{\rho'})m\}$, and integrate the result with respect to θ_ρ to obtain an expression for $\delta_c(\rho - \rho')/\rho$ in terms of an integral over k of the product of two Bessel functions. Using this result in the result of part (a), show by this procedure that (2A.24) is true.

(c) Specialize the results in parts (a) and (b) for the case in which $f(\rho, \theta_\rho)$ is independent of θ_ρ.

(d) Using the result in part (c), determine the Bessel and Fourier transforms of the function $f(\rho, \theta_\rho)$ for $\rho \leq a$ and 0 otherwise where a is a constant. Hint: By direct differentiation of the Bessel series, it is easily shown that $\partial\{xJ_1(x)\}/\partial x = xJ_0(x)$.

Chapter 6

The Continuous–Time/
Discrete–Time Interface

6.1 INTRODUCTION

If digital computers are to be of service in real–world signal processing, the signal must be converted from its original format, typically a continuous–time signal, into a sequence that may be stored in computer memory. The hardware for performing this conversion, the analog–to–digital converter, is discussed from design to performance in great detail elsewhere. Although we will examine some basic conversion hardware, our main focus here is on effects of this conversion on the signal and how to best accomplish desired tasks using the computer.

In Chapter 4, the appropriate expansions for the main classes of signals encountered in signal processing are given. These signal classes are continuous–time signals, both periodic and aperiodic, and discrete–time signals, both periodic and aperiodic. For each class, one Fourier–type transform and one general power expansion (GPE) are appropriate. For periodic signals, the Fourier series and discrete Fourier series are appropriate, whereas for aperiodic signals, the Fourier transform and discrete–time Fourier transform are suitable. For continuous–time signals, the Laplace transform is used, whereas for discrete–time signals, the z–transform is used. But what happens when signal content in a continuous–time signal is transferred to another format, namely a discrete–time signal? How are the two signals related? Will there be distortions? Is the conversion process reversible? These are a few of the questions that will be investigated in this chapter.

The effects of time sampling in both the time and frequency domains will first be investigated. We will find that provided the appropriate sampling criterion is satisfied, a continuous–time signal can in principle be exactly reconstructed from its samples without error. We arrive at this criterion by determining the relation between the GPEs of $x_c(t)$ and $x(n) = x_c(n\Delta t)$. An example of reconstruction with a zero–order–hold will give a sense of the relation between practical and ideal reconstruction.

Next, we briefly examine analog–to–digital and digital–to–analog circuitry, to get a feel for how the continuous–time/discrete–time interface is practically implemented. We will look at the sample–and–hold circuit, which serves in both practical sampling and reconstruction systems.

In the remainder of this chapter, we will focus on two signal–processing tasks that involve signals of both continuous–time and discrete–time formats. The first is sampled–data processing, such as the simulation or replacement of a continuous–time LSI system by a discrete–time LSI system. We look at impulse invariance, numerical integration of state equations, and operational (substitution) approaches to approximate integration as means of discretizing an analog system.

The second task is the estimation of continuous–time Fourier integrals using samples of the continuous–time signal and the discrete Fourier transform. Specifically, we investigate use of the DFT (via FFT) to estimate samples of $X_c(j\Omega)$ given a finite number of samples of $x_c(t)$ and also the reverse process: to estimate samples of $x_c(t)$ given samples of $X_c(j\Omega)$. To help assess the accuracy of these estimations and appreciate a fundamental property of Fourier transforms, the concept of the time–bandwidth product is discussed and on WP 6.2 is related to its equivalent in quantum mechanics: the uncertainty principle. Also, we discuss the concept of resolution. An application example in ultrasonic field calculations illustrates the procedure of estimating a time function given frequency samples.

In the final section, many of the relationships developed between time– and frequency–domain functions are illustrated by means of a detailed graphic. The process of using the DFT to estimate either time– or frequency–samples given samples of the other domain is considered.

6.2 SAMPLING AND ALIASING

We begin our study of the continuous–time/discrete–time interface by examining the process of sampling a continuous–time signal. In this process, samples of a continuous–time signal are acquired using circuits we will consider in Section 6.3.3. For now, we restrict our attention to the principles of sampling and aliasing. We assume the samples are uniformly spaced in time by Δt (s).

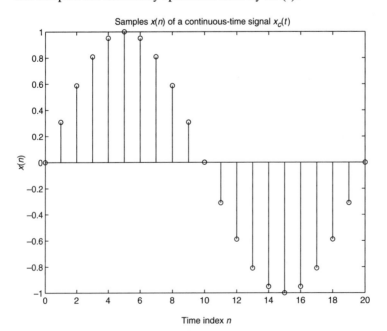

Figure 6.1

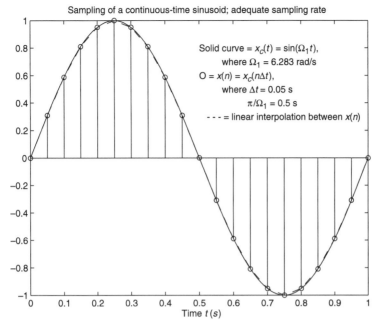

Figure 6.2

Let us start with an intuitive example: sampling of a sinusoid. Suppose that we are given a sequence of $N = 20$ samples $x(n)$ as shown in Figure 6.1, and we are not told from what continuous–time signal $x_c(t)$ it was derived. We see that $N = 20$ is the number of samples per cycle of the sinusoidal sampled sequence. If $x_c(t)$ is slowly varying compared with the sampling rate, the samples $x(n)$ are accurately indicative of the $x_c(t)$ that was used to generate them.

For example, suppose that $x_c(t) = \sin(\Omega_1 t)$, where $\Omega_1 = 2\pi/T = 2\pi$ rad/s (i.e., $F_1 = 1\,\text{Hz}$) where the period of the sinusoid is $T = 1$ s. The waveform $x_c(t)$ is the solid curve in Figure 6.2. With $\Delta t = T/N = (1\,\text{s})/20 = 0.05$ s, we obtain the samples in Figure 6.1 from $x_c(t)$ in Figure 6.2 as $x(n) = x_c(n\,\Delta t) = \sin(\Omega_1 n\,\Delta t) = \sin(2\pi n\,\Delta t) = \sin(2\pi n T/N) = \sin(2\pi n/N)$ for $T = 1$ s. In Figure 6.2, these samples $x(n)$ are super-imposed on $x_c(t)$ at $t = n\,\Delta t$. It is graphically evident that if we were given only the sampled values in Figure 6.1, linear interpolation between them would provide a reasonably accurate picture of the original continuous–time waveform $x_c(t)$ in Figure 6.2. This is illustrated by the dashed curve in Figure 6.2, which shows the linear interpolation between the samples $x(n)$ in Figure 6.1; the agreement with $x_c(t)$ is quite close.

If, however, the signal varies too rapidly during the sampling interval, a high frequency in the original signal looks like a low frequency in the samples. Suppose now that the period of $x_c(t)$ is no longer $T = 1$ s, but rather $T' = (1\,\text{s})/(N + 1) = 0.0476$ s, or equivalently, $\Omega_1' = 2\pi/T' = 131.9$ rad/s. Let the sampling interval Δt remain at $\Delta t = (1\,\text{s})/N = 0.05$ s; there is still exactly one cycle of the sampled sequence every N samples. This time a linear interpolation between the samples no longer resembles $x_c(t)$ at all, as seen in Figure 6.3. The problem in Figure 6.3 is evident: Each sample $x(n) = x_c(n\,\Delta t)$ is taken at a slightly later time on a successive period of the continuous–time signal $x_c(t)$.

Under inadequately rapid sampling, the high–frequency sinusoid $x_c(t)$ appears as a low–frequency sinusoid when its samples are linearly interpolated. This effect is called aliasing, because the actual high–frequency component is aliased to a low frequency ("alias" meaning "an assumed name" or an apparently reasonable

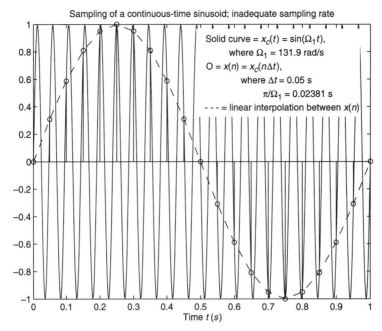

Figure 6.3

representation that is deceptive). In fact, information necessary to tell us that the samples represent the solid–curve sinusoid in Figure 6.3, as opposed to that in Figure 6.2, has been lost. Knowing only the sampled values, we thus have no hope of reconstructing the original waveform by simple interpolation.

In this case of a sinusoid, the explanation for the problem is very simple:

$$\sin(\Omega'_1 n\,\Delta t) = \sin\left[\left[2\pi \cdot (N+1)\right] \cdot \frac{n}{N}\right] \tag{6.1}$$

$$= \sin(2\pi n + 2\pi n/N)$$

$$= \sin\left(\frac{2\pi n}{N}\right)$$

$$= \sin(\Omega_1 n\,\Delta t).$$

The aliasing problem is a *direct result* of the 2π–periodicity of the sinusoid. We may also conclude that any sinusoid having frequency $\Omega_{1,\ell} = 2\pi(N\ell + 1)$ rad/s, ℓ any integer [including our original $\Omega_1 = 2\pi$ rad/s $= \Omega_{1,0}$ and our modified $\Omega'_1 = 2\pi(N+1)$ rad/s $= \Omega_{1,1}$], will result in the same sequence $x(n)$ for $\Delta t = 1/N$ s.

Notice that the sampling frequency in this example is $\Omega_s = 2\pi/\Delta t = 2\pi N$ rad/s (i.e., exactly N samples per sinusoidal period if the original continuous–time sinusoid were 1Hz or $\Omega_1 = 2\pi$ rad/s). More generally, a continuous–time sinusoid of frequency $\Omega_{1,\ell} = \Omega_1 + \Omega_s \cdot \ell$ will have the same samples at $t = n\,\Delta t$ as those of the sinusoid having *principal frequency* Ω_1 because $\sin\{(\Omega_1 + \Omega_s\ell)n\,\Delta t\} = \sin\{\Omega_1 n\,\Delta t + (2\pi/\Delta t)\ell n\,\Delta t\} = \sin\{\Omega_1 n\,\Delta t + 2\pi\ell n\} = \sin(\Omega_1 n\,\Delta t)$. Thus under sampling, any sinusoid $\Omega_{1,\ell}$ is aliased down to Ω_1, where $|\Omega_1| < \Omega_s/2$. Although this result may seem to apply only to sinusoidal $x_c(t)$, we may well venture that it must apply to every Fourier component of a general signal (the "Ω_1" being different for each component), as we clarify and demonstrate below.

By looking at the temporal waveforms, it is not so apparent what is going on when the signal is more complicated, such as speech or music waveforms. A simpler example illustrating the problem is presented in Figure 6.4. In Figure 6.4 is shown

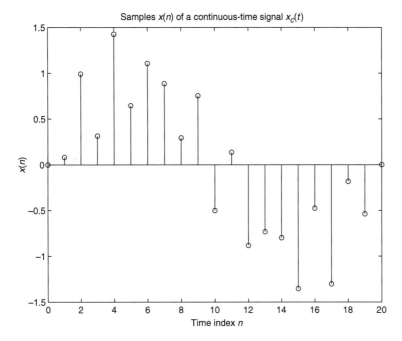

Figure 6.4

the result of sampling the sum of two sinusoids, one being low frequency and the other high frequency with respect to the sampling frequency.

Sampling of the low–frequency sinusoid follows the behavior of Figure 6.2, whereas that of the other follows Figure 6.3. It is more difficult to discern and predict what is happening overall. In Figure 6.5, we clearly see that the linear interpolation does not match $x_c(t)$, but there is no simple pattern telling us why, as there was in Figure 6.3.

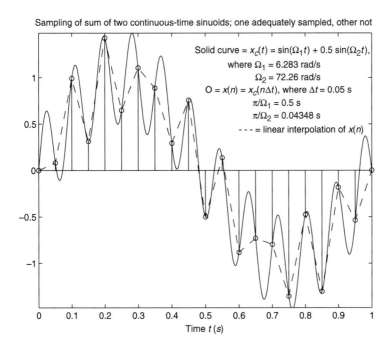

Figure 6.5

However, the only feature distinguishing the behaviors for the two sinusoids is the difference in their frequencies, because the whole aliasing problem stems from the periodicity of sinusoids. Thus it is more natural to consider the effects of sampling using the decomposition of the total signal into its constituent frequency components. Then for each sinusoidal component, we can identify whether Figure 6.2 or Figure 6.3 pertains. Thus the frequency domain is more illuminating in addressing the effects of sampling than is the time domain. Furthermore, analysis in the frequency domain will provide a quantitative criterion for identification of whether aliasing will occur for any general (non–sinuosoidal) signal sampled at any rate $\Omega_s = 2\pi/\Delta t$. Finally, it will provide in theory, under the condition for no aliasing, a means of reconstructing the original signal $x_c(t)$ *exactly* (not the approximate linear interpolation) from only its periodic samples!

We now investigate in more detail what will happen in the frequency domain when a continuous–time signal $x_c(t)$ is periodically sampled. For simplicity, assume that $x_c(t)$ is aperiodic and finite–energy, and thus possesses a Fourier transform (FT) $X_c(j\Omega)$ and Laplace transform $X_c(s)$. Suppose that $x_c(t)$ is sampled every Δt s. If the samples are called $x(n)$, then again we let $x(n) = x_c(n\,\Delta t)$.

Specifically, let us determine the relation between the z–transform $X(z)$ of the sampled sequence $x(n)$ and the Laplace transform $X_c(s)$ of the original continuous–time signal $x_c(t)$. The relation between the GPEs is the most general and therefore most desirable. Once it is known, setting $z = e^{j\omega}$ and $s = j\Omega$ will then provide the relation between the DTFT of $x(n)$, $X(e^{j\omega})$, and the FT of $x_c(t)$, $X_c(j\Omega)$.

To begin our discussion from a practical viewpoint, consider the output of the sampler in an analog–to–digital converter (before encoding into binary). The most common hardware used for sampling is the sample–and–hold circuit (see Section 6.3.3 for details). It samples $x_c(t)$ every Δt and holds that value until the next sampling instant, Δt later. Let this sampled–and–held signal be called $x_{\Delta t,c}(t)$. Pictorially, $x_{\Delta t,c}(t)$ has the appearance of a staircase that goes up and down, along with samples of $x_c(t)$ (stair heights may vary from one step to the next). In equation form,

$$x_{\Delta t,c}(t) = x_c(n\,\Delta t) \text{ for all } n\,\Delta t \le t < (n+1)\Delta t, \text{ for all } n. \tag{6.2}$$

Figure 6.6 shows a typical continuous–time signal $x_c(t)$, the sampled sequence $x(n)$, and the staircase sampled–and–held signal $x_{\Delta t,c}(t)$.

We can more conveniently express $x_{\Delta t,c}(t)$ as a sum of boxcars, each weighted by $x(n) = x_c(n\,\Delta t)$, by introducing the boxcar pulse function $p_{n,\Delta t}(t)$ as follows:

$$x_{\Delta t,c}(t) = \sum_{n=-\infty}^{\infty} x_c(n\,\Delta t)p_{n,\Delta t}(t), \tag{6.3}$$

where the shifted boxcar function or rectangular pulse is defined as[1]

$$p_{n,\Delta t}(t) = \begin{cases} 1, & n\,\Delta t \le t < (n+1)\Delta t \\ 0, & \text{otherwise.} \end{cases} \tag{6.4}$$

Equations (6.3) and (6.4) are illustrated in Figure 6.7a. In Figure 6.7a, only one of the infinite number of terms summed in (6.3) (the ℓth) is pictured: The shifted

---

[1]For brevity, we omit the subscript c for "continuous time" on $p_{n,\Delta t}(t)$ and below on $\delta_{\Delta t}(t)$, because there is no danger of confusion with discrete time. For continuity, the subscript c will be retained on $\delta_c(t)$ and, of course, on all continuous-time signals.

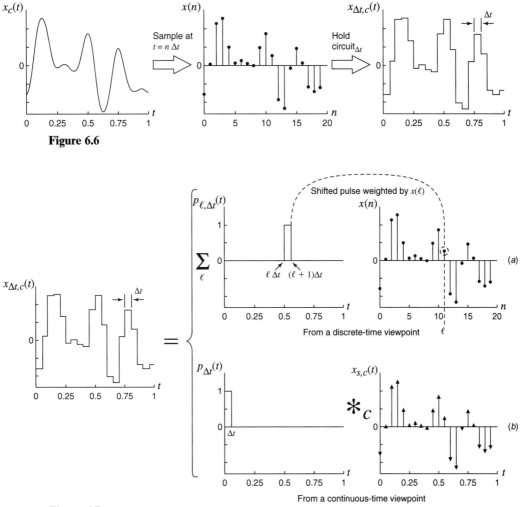

Figure 6.6

Figure 6.7

boxcar $p_{\ell,\Delta t}(t)$ is scaled by $x(\ell) = x_c(\ell \Delta t)$; the sum of such shifted and scaled "stairs" over all ℓ is $x_{\Delta t,c}(t)$.

A still more convenient representation for $x_{\Delta t,c}(t)$ is possible, which will lead us closer to the relation between $X(z)$ and $X_c(s)$. In this representation, $x_{\Delta t,c}(t)$ will be seen to be equal to the convolution of a single unshifted boxcar with a series of impulses weighted by $x(n)$, as illustrated in Figure 6.7b. To obtain this representation, we begin by showing that we can represent a shifted boxcar $p_{n,\Delta t}(t)$ in terms of an unshifted boxcar $p_{\Delta t}(t) \equiv p_{0,\Delta t}(t)$, which is unity for $0 \le t < \Delta t$ and zero otherwise. The unshifted boxcar $p_{\Delta t}(t)$ is just the impulse response of a zero–order hold (ZOH) circuit.

Recall that convolving a shifted Dirac delta with any function just shifts the latter over to the shift in the Dirac delta. In particular, by the sampling property of the shifted Dirac delta function $\delta_c(t - n\Delta t)$, we can write

$$p_{n,\Delta t}(t) = p_{\Delta t}(t - n\Delta t) \tag{6.5a}$$

$$= \delta_c(t - n\Delta t) *_c p_{0,\Delta t}(t) \tag{6.5b}$$

$$= \delta_c(t - n\Delta t) *_c p_{\Delta t}(t). \tag{6.5c}$$

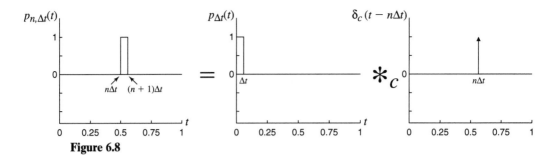

Figure 6.8

Proof: Using (6.4) with $n = 0$ for $p_{\Delta t}(t)$, we have

$$\delta_c(t - n\Delta t) *_c p_{\Delta t}(t) = \int_{-\infty}^{\infty} \delta_c(\tau - n\Delta t) \left\{ \begin{array}{ll} 1, & 0 \cdot \Delta t \leq t - \tau < 1 \cdot \Delta t \\ 0, & \text{otherwise} \end{array} \right\} d\tau, \qquad (6.6)$$

which by the Dirac delta sampling property (2A.3) $\left[\int_{-\infty}^{\infty} \delta_c(\tau - t_0)f_c(\tau)\, d\tau = f_c(t_0) \right]$ is just equal to the bracketed term evaluated at the shift parameter $\tau = n\Delta t$, or

$$\delta_c(t - n\Delta t) *_c p_{\Delta t}(t) = \left\{ \begin{array}{ll} 1, & 0 \leq t - n\Delta t < \Delta t \\ 0, & \text{otherwise} \end{array} \right. = \left\{ \begin{array}{ll} 1, & n\Delta t \leq t < (n + 1)\Delta t \\ 0, & \text{otherwise} \end{array} \right. = p_{n,\Delta t}(t).$$
$$(6.7)$$

The relation (6.7) is illustrated in Figure 6.8. Thus (6.3) can be rewritten as

$$x_{\Delta t,c}(t) = \sum_{n=-\infty}^{\infty} x_c(n\Delta t)\delta_c(t - n\Delta t) *_c p_{\Delta t}(t). \qquad (6.8a)$$

Because convolution is linear, the sum of convolutions in (6.8a) is equal to the convolution of the sums. Thus,

$$x_{\Delta t,c}(t) = \left\{ \sum_{n=-\infty}^{\infty} x_c(n\Delta t)\delta_c(t - n\Delta t) \right\} *_c p_{\Delta t}(t). \qquad (6.8b)$$

Noting that because $\delta_c(x \neq 0) = 0$, it follows that $x_c(n\Delta t)\delta_c(t - \tau) = x_c(t)\delta_c(t - n\Delta t)$, (6.8b) becomes[2]

$$x_{\Delta t,c}(t) = \left\{ x_c(t) \cdot \sum_{n=-\infty}^{\infty} \delta_c(t - n\Delta t) \right\} *_c p_{\Delta t}(t). \qquad (6.8c)$$

Now, (6.8c) may be rewritten as

$$x_{\Delta t,c}(t) = \{x_c(t) \cdot \delta_{\Delta t}(t)\} *_c p_{\Delta t}(t), \qquad (6.8d)$$

where the train of Dirac delta impulses, each separated by Δt, is

$$\delta_{\Delta t}(t) = \sum_{n=-\infty}^{\infty} \delta_c(t - n\Delta t). \qquad (6.9)$$

---

[2] Note that $x_{c,\Delta t}(t) \neq x_c(t) \cdot \left\{ \sum_{n=-\infty}^{\infty} \delta_c(t - n\Delta t) *_c p_{\Delta t}(t) \right\}$ because in (6.8c) $x_c(t)$ is under the convolutional integral sign and cannot be pulled out.

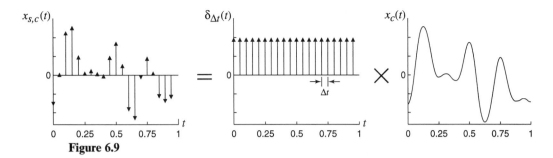

Figure 6.9

The quantity in $\{\cdot\}$ in (6.8d) is called the sampled representation of $x_c(t)$ and is denoted

$$x_{s,c}(t) = x_c(t) \cdot \delta_{\Delta t}(t) \qquad\qquad (6.10a)$$

$$= \sum_{n=-\infty}^{\infty} x_c(n\Delta t)\delta_c(t - n\Delta t), \qquad\qquad (6.10b)$$

where (6.10b) recalls the same quantity in $\{\cdot\}$ in (6.8b). The relation (6.10a) is illustrated in Figure 6.9. Thus by (6.8d) and (6.10a),

$$x_{\Delta t,c}(t) = x_{s,c}(t) *_c p_{\Delta t}(t), \qquad\qquad (6.11)$$

where again $x_{\Delta t,c}(t)$ is the staircase version of $x_c(t)$, and $x_{s,c}(t)$ is a string of impulses at $n\Delta t$ of weights $x_c(n\Delta t)$.

Note that $x_{s,c}(t)$, a signal that exists only mathematically (not physically), arose naturally from the mathematical decomposition of the practical, measurable staircase signal $x_{\Delta t,c}(t)$—the output of a ZOH circuit. Also note that $x_{s,c}(t)$ contains all the information in samples $x_c(n\Delta t)$ of $x_c(t)$ every Δt, and no more information, unlike $x_c(t)$ itself, which contains all information including between the samples. As noted above, the relation (6.11) is illustrated in Figure 6.7b.

The Laplace transform of $x_{s,c}(t)$ is obtained by first substituting (6.10b) into the Laplace transform definition

$$X_{s,c}(s) = \sum_{n=-\infty}^{\infty} x_c(n\Delta t) \int_{-\infty}^{\infty} \delta_c(t - n\Delta t)\, e^{-st} dt \qquad\qquad (6.12a)$$

and then using the sampling property of the shifted Dirac delta function to obtain

$$X_{s,c}(s) = \sum_{n=-\infty}^{\infty} x_c(n\Delta t)\, e^{-sn\Delta t} \qquad\qquad (6.12b)$$

$$= X(z)\Big|_{z\,=\,e^{s\Delta t}}, \qquad\qquad (6.12c)$$

where $X(z) = \text{ZT}\{x(n)\}$ and $x(n) = x_c(n\Delta t)$. Thus with $X_{s,c}(s) = X(e^{s\Delta t})$, we see that $X_{s,c}(s)$ and $X(z)$ are really the same function, except for an exponential difference/scaling in arguments. So, as is no surprise, $X(z)$ and $X_{s,c}(s)$ each contain all the information in the samples $x_c(n\Delta t)$, and no more. Note also that we may write (6.12c) on the unit circle (by setting $s = j\Omega = j\omega/\Delta t$) as $X(e^{j\omega}) = X_{s,c}(j\omega/\Delta t)$.

The relation in (6.12b and c) is useful, but we still do not have $X(z)$ in terms of $X_c(s)$; we have $X(z)$ only in terms of $X_{s,c}(s)$. A common error is to assume incorrectly that (6.12) holds with $X_{s,c}(s)$ on the left–hand side replaced by $X_c(s)$. We obtain our desired relation by using the periodicity of $\delta_{\Delta t}(t)$.[3]

Because $\delta_{\Delta t}(t)$ is just a Dirac delta every Δt, it is periodic in time with period Δt: $\delta_{\Delta t}(t + \Delta t) = \delta_{\Delta t}(t)$. Therefore, it can be written as a Fourier series:

$$\delta_{\Delta t}(t) = \sum_{k=-\infty}^{\infty} a_k e^{j(2\pi k/\Delta t)t}, \tag{6.13a}$$

where

$$a_k = \frac{1}{\Delta t} \int_{-\Delta t/2}^{\Delta t/2} \delta_{\Delta t}(t)\, e^{-j(2\pi k/\Delta t)t} dt = \frac{1}{\Delta t}. \tag{6.13b}$$

It is not surprising that a_k do not diminish with k, given the infinite sharpness of $\delta_{\Delta t}$. Thus,

$$\delta_{\Delta t}(t) = \frac{1}{\Delta t} \sum_{k=-\infty}^{\infty} e^{j(2\pi k/\Delta t)t}. \tag{6.14}$$

Note that (6.14) is just (2A.26) $\left[\sum_{n=-\infty}^{\infty} e^{j\omega n} = 2\pi \sum_{\ell=-\infty}^{\infty} \delta_c(\omega + 2\pi\ell) \right]$ rewritten with ω replaced by $2\pi t/\Delta t$, and using (2A.14) $[\delta_c(at) = \delta_c(t)/|a|]$ with $a = 2\pi/\Delta t$; we have merely reproved it here using Fourier series. Thus (6.10a) becomes

$$x_{s,c}(t) = x_c(t) \cdot \delta_{\Delta t}(t) \tag{6.15}$$

$$= \frac{1}{\Delta t} \sum_{k=-\infty}^{\infty} x_c(t)\, e^{j(2\pi k/\Delta t)t}.$$

Using the multiply–by–sinusoid theorem for the LT, $x_c(t)e^{j\Omega_1 t} \leftrightarrow X_c(s + j\Omega_1)$, and taking the LT of (6.15) term by term give

$$X_{s,c}(s) = \frac{1}{\Delta t} \sum_{k=-\infty}^{\infty} X_c(s + jk\Omega_s) \tag{6.16a}$$

$$= X(z)\Big|_{z=e^{s\Delta t}}, \tag{6.16b}$$

where again $\Omega_s = 2\pi/\Delta t$ and where in (6.16b) we simply recall (6.12c).

Noting that $X_{s,c}(s) = X(e^{s\Delta t})$, it is therefore often cited that the mapping from the s– to the z–plane is $z = e^{s\Delta t}$. The Fourier specialization of $z = e^{s\Delta t}$ is obtained by setting $z = e^{j\omega}$ and $s = j\Omega$, giving $\omega = \Omega \cdot \Delta t$. This relation is very useful for frequency responses in digital filter design and other applications. Thus the assumptions $z = e^{s\Delta t}$ and $\omega = \Omega \cdot \Delta t$ go together. The mapping $z = e^{s\Delta t}$ is valid only for relating $X_{s,c}(s)$ to $X(z)$, not $X_c(s)$ to $X(z)$, which are related by equating the right–hand sides of (6.16a) and (6.16b). In particular, if s_p is a pole of $X_c(s)$, then it will be a pole of $X_{s,c}(s)$ because any one term in (6.16a) going to infinity causes the sum to also. Thus by (6.16b), $z_p = e^{s_p \Delta t}$ is a pole of $X(z)$. However, if s_z is a zero of $X_c(s)$, that does *not* imply $X_{s,c}(s_z) = 0$, because one term in the sum being zero does not imply that the entire sum is zero, so $z = e^{s_z \Delta t}$ is *not* a zero of $X(z)$.

---

[3]One reference on this approach is S. M. Shinners, *Modern Control System Theory and Design* (NY: Wiley, 1992), pp. 682-683. An earlier, more original reference for this approach is C. L. Phillips, J. L. Lowry, and R. K. Cavin, "On the starred transform," *IEEE Trans. on Automatic Control,* Vol. AC-11, 4 (Oct. 1966) 760. It includes a proof of the result that is valid when $x_c(t)$ is discontinuous at sampling instants. The added term in (6.16a) is $\frac{1}{2} \sum_{k=-\infty}^{\infty} \{x_c(k\,\Delta t^+) - x_c(k\,\Delta t^-)\}\exp(-k\,\Delta ts)$.

Does this mean that the mappings $z = e^{s\Delta t}$ and $\omega = \Omega \cdot \Delta t$ are useless? Not at all. For the modes (poles) do transform this way, so if the poles s_p and Δt are known, then under sampling, the discrete–time system modes are $z_p^n = e^{s_p n \Delta t}$. Thus, although the apportionments are controlled by the zeros and so are not transferred to $X(z)$ via an easy mapping, the overall modal behavior including stability is, through $e^{s\Delta t}$. Furthermore, when we wish to design a discrete–time subsystem in a sampled–data system to behave in a specified transient manner, we can use $e^{s\Delta t}$ as a ballpark guide. Finally, in situations of high oversampling (i.e., little or no aliasing), in the primary frequency strip only the $k = 0$ term in (6.16a) is significant—which implies by the above reasoning that the zeros also map approximately according to $z = e^{s\Delta t}$. That is, if the $k = 0$ term is zero, then approximately so will be the entire sum on the primary strip, because all the aliasing terms are nearly zero there (even though not equal to zero there). Consequently, we conclude that for little or no aliasing, $z = e^{s\Delta t}$ and $\omega = \Omega \Delta t$ are valid for the entire transfer function/frequency response transformation.

The ramifications of the relation $z = e^{s\Delta t}$ are enormous. In the problems, we develop the concept of a pulse transfer function between input and output samples of the continuous–time functions $y_c(t)$ and $x_c(t)$ for a sampled–data system. Because of the sampler, there are no LSI relations between $y_c(t)$ and $x_c(t)$. We may express the relation between output and input samples using an LSI relation—and we obtain a rational function of $e^{s\Delta t}$ relating $Y_{s,c}(s)$ to $X_{s,c}(s)$—which has an infinite number of poles and zeros, because of the periodicity of the exponential. However, by defining $z = e^{s\Delta t}$, we obtain a simple rational function of z, $H(z)$, as the transfer function between output and input transforms $Y(z)$ and $X(z)$ which has a finite number of poles and zeros.

Always remember that the impulsive signal $x_{s,c}(t) \leftrightarrow X_{s,c}(s)$ does not even physically exist. It is merely a mathematical tool used in decomposing $x_{\Delta t,c}(t)$, which *does* physically exist for (a) relating the spectrum of a sampled signal to that of the continuous–time signal from which it was derived, and (b) extracting linear shift–invariant relations between sampled input and output signals via the z–transform in sampled–data systems. Like negative frequencies and complex sinusoids, Dirac delta trains do not physically exist, but they are likewise very convenient to use in analysis!

We may alternatively express the relation between $X_c(s)$ and $X(z)$ in (6.16 a and b) as

$$X(z) = \frac{1}{\Delta t} \sum_{k=-\infty}^{\infty} X_c(s + jk\Omega_s)\Big|_{s=\ln(z)/\Delta t}. \tag{6.16c}$$

Note that in (6.16c) we have eliminated the nonphysical $X_{s,c}(s)$ from the fundamental transform sampling relation. The only purpose of $X_{s,c}(s)$ was in decomposing a physical signal $x_{\Delta t,c}(t)$ [see (6.11)] in a way convenient for deriving (6.16c). Our next step is to set $s = j\Omega = j\omega/\Delta t$ in (6.16a and b) and note that in that case $z = e^{s\Delta t} = e^{(j\omega/\Delta t)\Delta t} = e^{j\omega}$. The relation $z = e^{s\Delta t}$ will yield the corresponding relation between analog and digital Fourier spectra. More on the exact meaning and origin of $\Omega = \omega/\Delta t$ in addition to that following (6.16b) will be given below. With $s = j\Omega = j\omega/\Delta t$, (6.16c) becomes

$$X(e^{j\omega}) = \frac{1}{\Delta t} \sum_{k=-\infty}^{\infty} X_c\left(j\left\{\frac{\omega + 2\pi k}{\Delta t}\right\}\right) \tag{6.17}$$

Note that by (6.16a), we also have $X(e^{j\omega}) = X_{s,c}(j\omega/\Delta t) = \text{FT}\{x_{s,c}(t)\}\big|_{\Omega=\omega/\Delta t}$.

Equation (6.17) indicates that the DTFT of $x(n)$, $X(e^{j\omega})$, is a sum of shifted replicas of $X_c(j\Omega)$ that are scaled both in argument and amplitude. This result is hardly surprising in light of our earlier discussions of the aliasing effect sampling can have.[4] But why was it that Figure 6.2 involved no aliasing, whereas (6.17) evidently implies aliasing for all cases? The reason is that the replicas in (6.17) may not always overlap; when they do not, no aliasing occurs.

Specifically, if the spectrum of $x_c(t)$, $X_c(j\Omega)$, is zero outside $\Omega \in (-\Omega_0, \Omega_0]$, then it may be possible that the shifted replicas of $X_c(j\Omega)$ in (6.17) do not overlap (alias). Such a signal is called bandlimited, with an absolute, not just -3–dB bandwidth equal to $2\Omega_0$. Each replica of $X_c(j\Omega)$ is shifted from the adjacent replicas by $\Omega_s = 2\pi/\Delta t$ (equivalently, on the ω axis, by $\omega = 2\pi$). Whether or not interference occurs is determined by Δt, which we see in (6.17) scales the spectral width: The larger Δt is, the wider each replicated/shifted spectrum extends. This makes sense when we consider that the larger Δt is, the more cycles occur per sample for each constituent frequency Ω in $X_c(j\Omega)$.

In fact, the very presence of the spectral replications in (6.17) makes sense because every nonaliased sampled sinusoid (within $|\omega| < \pi$) is indistinguishable from the samples of an infinite number of periodically spaced high–frequency sinusoids [recall (6.1)] for which $|f| > \frac{1}{2}$ cycle/sample ($|\omega| > \pi$). Thus all those higher–frequency sinusoids must be represented as nonzero components of the sequence, as guaranteed by (6.17); the sequence has no "preference" for "baseband" versus aliasable sinusoidal components, all of which are equally compatible with and contribute to the measured samples. Furthermore, although each replicated spectrum is itself aperiodic, the periodic replication in (6.17) results in $X(e^{j\omega})$ being 2π–periodic, which it must be, as the transform of a discrete–time signal (see Section 4.4).

Figure 6.10a shows the spectrum of samples of a typical bandlimited but undersampled $x_c(t)$. If, as in Figure 6.10a, the point on the ω axis that the $k = 0$ replica extends farthest to the right ($\omega_0 = \Omega_0 \Delta t$) exceeds the point that the adjacent $k = -1$ replica extends farthest to the left ($2\pi - \Omega_0 \Delta t$), aliasing will occur. That is, aliasing will occur if $\Omega_0 \Delta t > 2\pi - \Omega_0 \Delta t$, or $\Delta t > \pi/\Omega_0$. For example, the largest negative frequency $-\Omega_0 \Delta t$ from the $k = -1$ replica of $X_c(j\omega)$ is aliased such that within $\omega \in (-\pi, \pi]$ it shows up in Figure 6.10a as a contribution to the positive–frequency component of the spectrum, $X(e^{j\omega})$, at $\omega = 2\pi - \Omega_0 \Delta t$. The individual aliasing contributions in Figure 6.10a must be added to produce the total $X(e^{j\omega})$, as indicated.

Equivalently, no aliasing will occur if, as in Figure 6.10b, $2\pi - \Omega_0 \Delta t > \Omega_0 \Delta t$, or

$$\Delta t \leq \pi/\Omega_0 \quad \text{or} \quad \pi/\Delta t \geq \Omega_0. \quad \text{Nyquist criterion expressed as requirement on } \Delta t. \qquad (6.18a)$$

---

[4]Indeed, we can intuitively and quickly obtain (6.17) by the following argument. Recall from

$x(n) = \dfrac{1}{2\pi} \displaystyle\int_{-\pi}^{\pi} X(e^{j\omega})e^{j\omega n}d\omega$ that $X(e^{j\omega_A})d\omega/(2\pi)$ is the coefficient of $e^{j\omega_A n}$ in the sinusoidal decomposition of

$x(n)$. Similarly, from $x_c(t) = \dfrac{1}{2\pi} \displaystyle\int_{-\infty}^{\infty} X_c(j\Omega)e^{j\Omega t}d\Omega$, we know that $X_c(j\Omega_A)d\Omega/(2\pi)$ is the coefficient of $e^{j\Omega_A t}$

in the sinusoidal decomposition of $x_c(t)$. From Section 6.1, we know what happens to sinusoids under sampling: $x_A(n) = e^{j\Omega_A n \Delta t} = e^{j\omega_A n}$, where $\omega_A = \Omega_A \cdot \Delta t$, but also $e^{j(\Omega_A + k\Omega_s)n \Delta t} = x_A(n)$. Thus the Ω_A and $\Omega_A + \Omega_s \cdot k$ ($-\infty < k < \infty$) components in $x_c(t)$ all contribute to the ω_A component in $x(n) = x_c(n \Delta t)$. Moreover, in sampling, just as $\omega = \Omega \cdot \Delta t$, also an increment $d\Omega$ in Ω corresponds to an increment $d\omega = d\Omega \cdot \Delta t$ in ω, or $d\Omega = d\omega/\Delta t$. Thus the coefficient $X(e^{j\omega_A})d\omega/(2\pi)$ of $e^{j\omega_A n}$ in $x(n)$ is the sum of all con-

tributing coefficients $X_c(j[\Omega_A + \Omega_s \cdot k]) \dfrac{d\Omega}{2\pi} = \left[X_c\left(j\dfrac{\omega_A + 2\pi k}{\Delta t}\right)\dfrac{1}{\Delta t}\right] \cdot \dfrac{d\omega}{2\pi}$, over all k. Consequently, divid-

ing by $d\omega/(2\pi)$ and replacing ω_A by generic ω, we obtain $X(e^{j\omega}) = \dfrac{1}{\Delta t} \displaystyle\sum_{k=-\infty}^{\infty} X_c\left(j\left\{\dfrac{\omega + 2\pi k}{\Delta t}\right\}\right)$, which is (6.17).

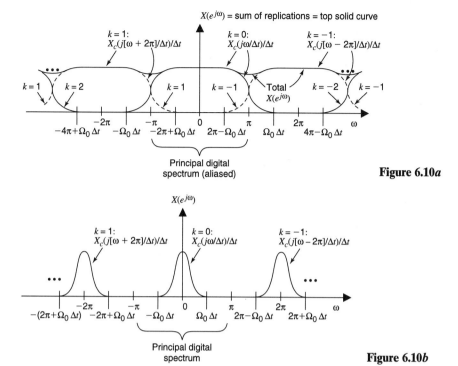

Figure 6.10*a*

Figure 6.10*b*

As noted, periodic sampling implies that a sample is measured every $\Delta t = 1/F_s$ s, where F_s is the sampling frequency in hertz, and in rad/s, $\Omega_s = 2\pi \cdot F_s = 2\pi/\Delta t$. Therefore, the condition for no aliasing (6.18a) can also be written

$$\Omega_s \geq 2\Omega_0 \quad \text{or} \quad \Omega_s \geq \text{BW}, \quad \text{Nyquist criterion expressed as} \atop \text{requirement on } \Omega_s. \qquad (6.18b)$$

where $\text{BW} = 2\Omega_0$ is the (absolute) bandwidth of $x_c(t)$. The inequality (6.18a and b) is often called the Nyquist sampling criterion after Harry Nyquist, a pioneer in signal processing and controls. For equality in (6.18b), $\Omega_s = 2\Omega_0 = \text{BW}$ is called the Nyquist sampling rate. Therefore, if the bandwidth of $x_c(t)$, $2\Omega_0$, is known, aliasing can be avoided in the sampling process simply by choosing $\Delta t \leq \pi/\Omega_0$ or, equivalently, $\Omega_s \geq \text{BW}$.[5]

The principal *baseband* spectrum, centered on dc, is usually the only replication of interest in basic signal processing. As is evident from Figure 6.10*a*, in cases of moderate aliasing, the worst aliasing typically occurs at the high frequencies within the principal spectrum. This is because for most signals the magnitude of the spectrum gradually decreases with increasing Ω, and the $k = \pm 1$ terms contribute most; if $\Omega_0 \Delta t$ just exceeds π, then the lowest aliased frequency, $2\pi - \Omega_0 \Delta t$ for $k = -1$, is just below $\omega = \pi$, the highest digital frequency in the principal spectrum.

Before some further comments on the sampling theorem, it is worthwhile to pause and review the detailed discussion of the differences between Ω and ω in

---

[5]We assume that $x_c(t)$ is real, so that $X_c(j\Omega)$ is centered at $\Omega = 0$. Otherwise, one may arrange for no aliasing if (6.18a) is satisfied with $\Omega_0 = \frac{1}{2}\{\Omega^+ - \Omega^-\}$, where $X_c(j\Omega) = 0$ for $\Omega < \Omega^-$ and for $\Omega > \Omega^+$. Equivalently, (6.18b) is satisfied with $\text{BW} = \Omega^+ - \Omega^-$. If $X_c(j\Omega)$ is double-side-banded, one can demodulate and lowpass filter before sampling, and during reconstruction (see Section 6.3) one can use modulation to shift the spectrum back up to match that of $x_c(t)$; equivalently, one can sample the original, use digital bandpass filtering, and remodulate during reconstruction.

Section 4.3. With our understanding of aliasing, we can also make the following new comments on this topic, which help clarify the relation $\Omega = \omega/\Delta t$.

Recall that $\Omega = 2\pi F$, where F is the usual Hertzian frequency (cycles/s). On the other hand, $\omega = 2\pi f$, where f is the normalized "digital" frequency (cycles/sample). The maximum frequency Ω_{max} that can be sampled at a given sampling frequency $\Omega_s = 2\pi/\Delta t = 2\pi F_s$ (where $F_s = 1/\Delta t$ samples/s) without being aliased into a lower frequency due to the 2π–periodicity of the sinusoid is $\Omega_{max} = \Omega_s/2$ rad/s; equivalently, $F_{max} = F_s/2$ cycles/s. The corresponding value of f is $f_{max} = \{F_{max} \text{ cycles/s}\} \cdot \{\Delta t \text{ s/sample}\} = \{F_s/2\} \cdot \{1/F_s\} = \frac{1}{2}$ cycle/sample (or $\omega_{max} = 2\pi f_{max} = \pi$ rad/sample). For all lower frequencies $F < F_{max}$, the relation between f and F is similarly linear: $f = F \cdot \Delta t = F/F_s$.

We conclude that if we are to associate a sampled sinusoid of digital frequency f cycles/sample with the unique continuous–time sinusoid that is not aliased under the given sampling frequency F_s samples/s, the frequency of that continuous–time sinusoid must be $F = fF_s = f/\Delta t$ (equivalently, $\Omega = \omega/\Delta t$), where F ranges from 0 to $F_s/2$ hertz as f ranges from 0 to $\frac{1}{2}$ cycle/sample. The meaning of the maximum value of f, $\frac{1}{2}$ cycle/sample, should be clearer now; again, for any continuous–time sinusoid of frequency F to be sampled without aliasing, it must be sampled at least twice per cycle: $F_s \geq 2F$. Also, the basis of the relation $\Omega = \omega/\Delta t$ should now be clear: It has meaning only for nonaliased frequencies as the relation between a digital frequency and the only analog frequency that under Δt sampling could have produced it without being aliased.

An equivalent expression for the sampling theorem can be obtained by noting that $\Omega_0 = 2\pi/T_0$, where T_0 is the period of the highest–frequency component in $x_c(t)$. Here (6.18) reads $\Delta t \leq T_0/2$, which again says that there must be at least two samples for every cycle of the highest–frequency component of $x_c(t)$. This is equivalent to the restriction of unique f being on $(-\frac{1}{2}, \frac{1}{2}]$.

The spatial analog of this interpretation may be familiar to those in electromagnetics: An electromagnetic field having wavelength λ can be reconstructed if the spatial samples separation h satisfies $h \leq \lambda/2$.

In optics, the Abbe principle determines the resolution limit of coherent imaging of an aperture object to again be approximately $\lambda/2$, where λ is the wavelength of the point source.[6] Also, the Rayleigh criterion applied to the optical microscope again yields a resolution limit of about $\lambda/2$, where λ is the wavelength of the self–luminous object/incoherent source.[7] Turning the idea around, suppose that we were to use a sinusoid of frequency $F_0 = 1/T_0$ to investigate a system whose impulse response is known to be composed of spikes at intervals in time. Then we could hope to discriminate between two of those spikes by looking at the output maxima or minima only if those spikes were spaced at most $T_0/2$ apart. Thus the significance of the $\leq$ half–sample concept comes up in many applications other than just time–sampling.

EXAMPLE 6.1

The magnitude spectrum of a continuous–time boxcar of duration $T = 1$ s is sought. With the sampling interval chosen as $\Delta t = 0.51$ s, the magnitude of the DTFT of the samples, multiplied by Δt, appears as the dashed curve in Figure 6.11 (the solid curve will be discussed below). What may be concluded? Investigate the spectral estimate for other values of Δt.

---

[6] E. G. Steward, *Fourier Optics: An Introduction* (New York: Wiley, 1983), pp. 92–93.

[7] Steward, pp. 35–36.

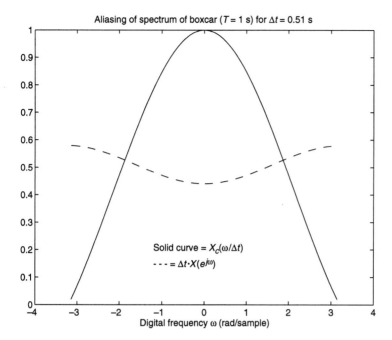

Aliasing of spectrum of boxcar ($T = 1$ s) for $\Delta t = 0.51$ s

Solid curve = $X_c(\omega/\Delta t)$

- - - = $\Delta t \cdot X(e^{j\omega})$

Digital frequency ω (rad/sample)

Figure 6.11

Solution

A continuous–time pulse $x_c(t)$ of duration T centered on $t = 0$ has FT magnitude $|X_c(j\Omega)| = T \cdot \sin(\Omega T/2)/(\Omega T/2)$, which bears little resemblance to $\Delta t \cdot X(e^{j\omega})$, the dashed curve in Figure 6.11. For reference, $\mathrm{sinc}(x) = \sin(x)/x$ is plotted in Figure 6.12. $X_c(j\omega/\Delta t)$, the desired quantity obtained from $\Delta t \cdot X(e^{j\omega})$ in the absence of aliasing, is the solid curve in Figure 6.11. Due to the large sampling interval, the range of analog frequencies depicted in Figure 6.11 ($|\Omega| < \omega_{max}/\Delta t = \pi/\Delta t$) is small; the solid curve is just part of the central peak in Figure 6.12.

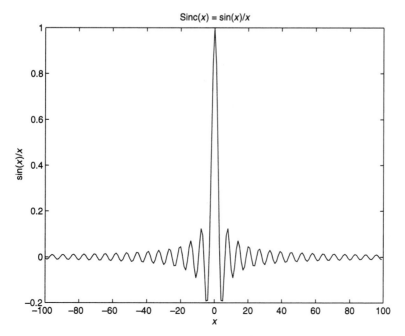

Sinc(x) = sin(x)/x

$\sin(x)/x$

x

Figure 6.12

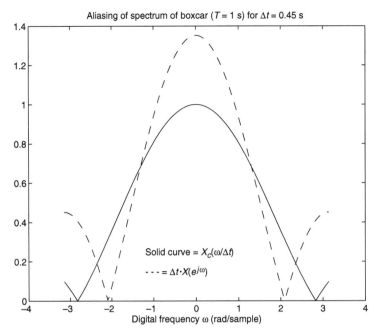

Figure 6.13

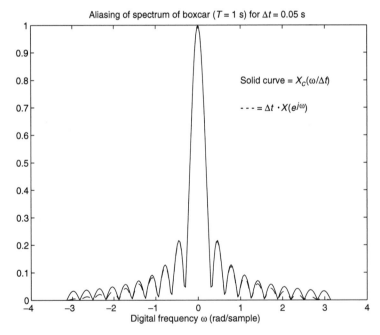

Figure 6.14

We find that our sampled spectrum is aliased, and we should decrease Δt and try again until the magnitude of our spectrum is very small for $|\omega|$ nearly π. For the same duration pulse, if Δt is just slightly decreased to $\Delta t = 0.45$ s, the DTFT of the sampled pulse (times Δt) is the dashed curve in Figure 6.13, whereas again the solid curve is $X_c(j\omega/\Delta t)$. Aliasing is still significant, and, as expected, worst for high frequencies but now there is at least some resemblance to $X_c(j\omega/\Delta t)$.

For accurate results, Δt is further reduced to 0.05 s, yielding the close agreement shown in Figure 6.14. Notice that in cases of moderate or low aliasing (such as Figure 6.14), the aliasing is again seen to be worst at high frequencies, as expected.

For the purposes of illustration of aliasing effects, we can use the first zero of $T \cdot \sin(\Omega T/2)/(\Omega T/2)$ as a very rough indicator of bandwidth of $X_c(j\Omega)$ for this example: $\Omega_{max} \approx 2\pi/T$. For $T = 1$ s, we thus should choose $\Delta t \leq \pi/\Omega_{max} = 0.5$ s for little aliasing. $\Omega_{max} = 2\pi/T$ is not a very satisfactory estimate of an effective "maximum frequency," as there are numerous sidelobes beyond the main lobe of $T \cdot \sin(\Omega T/2)/(\Omega T/2)$. However, comparison of the case $\Delta t = 0.51$ s in Figure 6.11, which is just above 0.5 s, with the case $\Delta t = 0.45$ s in Figure 6.13, which is just below 0.5 s, does show a "turning point" difference.

For cases where the detail of the true spectrum is unknown beyond the fact that it *is* known to be monotonically decreasing for high frequencies (the majority of practical and useful situations), we may look at $\omega \approx \pi$ to assess aliasing. If the magnitude of $X(e^{j\omega})$ is large there, it is almost certain that aliasing is acute and Δt must be reduced. Contrarily, if it is nearly zero for all ω near $\omega \approx \pi$, then it is likely that aliasing is slight and the sampling rate sufficient. This rule of thumb is evident in each figure in this example.

Now that both the mechanism for aliasing and a means for avoiding it have been described, the next question to be answered is whether the original signal $x_c(t)$ is recoverable from $x(n)$. Intuition would lead one to think that the signal in Figure 6.2 could be, whereas that in Figure 6.3 could not, because in Figure 6.3 vital information regarding the true frequency of $x_c(t)$ had been lost in the too–infrequent sampling process.

Exactly the same conclusion can be reached in the frequency domain, where in Figure 6.2 there is no aliasing (see Figure 6.15a), whereas in Figure 6.3 there is (see Figure 6.15b). In Figure 6.15a is shown a plot of the magnitude of the DTFT of $x(n)$ $= x_c(n\Delta t)$ [(6.17)] for the case of Figure 6.2 [$x_c(t) = \sin(\Omega_1 t)$]. The replication centered on 2π does not interfere with the zero–centered replication; the lines at low frequencies are indeed due to the original low frequency of $x_c(t)$—the $k = 0$ replication in the aliasing sum.

The appearance of Figure 6.15b is identical to that of Figure 6.15a, for they are the magnitudes of the DTFTs of identical sequences. Yet the original $x_c(t)$ of Figure 6.3 producing Figure 6.15b, denoted $x_c'(t)$, is vastly different from that of Figure 6.2 producing Figure 6.15a: $x_c'(t) = \sin(\Omega_1' t)$ where $\Omega_1' = \Omega_1 + \Omega_s$. Now the spectral lines at low frequencies are actually due to aliasing of the $k = \pm 1$ replications of the (high–frequency) $x_c'(t)$. Notice that the aliasing in Figure 6.15b is much worse than that in Figure 6.10a, by noting the positions of the replications. Given only Figure 6.15b, we have no way of knowing that the $x_c(t)$ from which $x(n)$ was obtained is not that of Figure 6.2, even though it is really from that of Figure 6.3. The aliasing in Figure 6.15b, which occurs when (6.18) is violated, is irreversible and constitutes a loss of important information about $x_c(t)$.

An advantage of performing the analysis in the frequency domain is that now the signal $x_c(t)$ can be arbitrarily complicated and the conclusion about whether perfect reconstruction is possible is just as easy to determine as for the case of an individual sinusoid. All that must be known about $x_c(t)$ is its bandwidth. Contrastingly, we could never have guessed the requirement for all signals given by (6.18), using time–domain analysis.

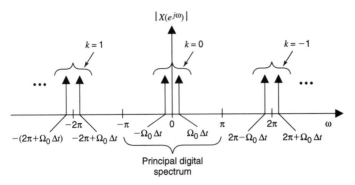

Figure 6.15a

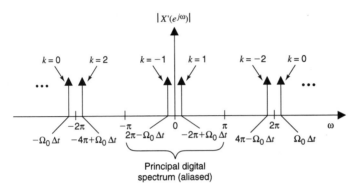

Figure 6.15b

In practice, the band of frequencies for which $x_c(t)$ has nonzero energy may be extremely wide or infinite. However, nearly all the signal energy is usually contained within a bandwidth more reasonable with respect to sampling frequencies readily obtainable with available hardware. As a safeguard, usually *oversampling* is performed, whereby Δt is chosen to be substantially smaller than the estimated "Nyquist" sampling interval. Oversampling minimizes the effect of this aliasing of tails of $X_c(j\Omega)$ and will be discussed in Section 6.3.2. In addition, to preclude any possibility that for a fixed Δt aliasing might occur for a given signal $x_c(t)$, $x_c(t)$ is prefiltered before sampling with a continuous–time lowpass filter having cutoff bandwidth below the sampling frequency. Such a filter is appropriately called an antialiasing filter.

From (6.16), even if (6.17) indicates little or no aliasing, there could theoretically be aliasing at other values of z off the unit circle. We are therefore tempted to formulate a "global" nonoverlap condition. That is, if the bandwidth of $X_c(s)$ for a given $\sigma = \ln(|z|)/\Delta t$ were $2\Omega_0(\sigma) = 2\Omega_0(\ln(|z|)/\Delta t)$, then aliasing would occur if $\Omega_0(\ln(|z|)/\Delta t) \cdot \Delta t > 2\pi - \Omega_0(\ln(|z|)/\Delta t) \cdot \Delta t$, or $\Delta t > \pi/\Omega_0(\ln(|z|)/\Delta t)$. To avoid any aliasing, this inequality would have to be satisfied for all $|z|$, not just $|z| = 1$, which is the usual Nyquist sampling criterion. Equivalently, no aliasing would occur anywhere in the region of convergence if

$$\Delta t \le \min_{|z|}\left\{\frac{\pi}{\Omega_0\left(\frac{\ln(|z|)}{\Delta t}\right)}\right\} \quad \text{or} \quad \Omega_s \ge \max_{|z|}\left\{2\Omega_0\left(\frac{\ln(|z|)}{\Delta t}\right)\right\}, \tag{6.19}$$

where $e^{\sigma_A \Delta t} < |z| < e^{\sigma_B \Delta t}$ and $X_c(s)$ converges for $\sigma_A < \sigma < \sigma_B$.

A question is whether or not this generalized condition is meaningful in signal filtering and reconstruction. The answer with respect to perfectly bandlimited signals is no, because such signals do not possess a Laplace transform for any $\sigma \neq 0$ (see Section 7.4). The answer with respect to the usual non–perfectly–bandlimited signals encountered in practice depends on whether high–frequency, moderately damped sinusoids might be "hidden" by a Fourier–dominating lightly damped low-frequency sinusoid. In practice, oversampling will generally guarantee against any such "hidden modes." However, in the problems we study an example where (6.19) can be used to prevent drastically inaccurate reconstructions due to hidden modes caused by using a fixed dB–down–from–maximum "Nyquist" criterion in the case of high–frequency, highly damped sinusoids.

The fundamental reason for such masking of modes is the spreading (and attenuating) of the Laplace transform of a heavily damped sinusoid $x_c(t)$ for σ increasingly far from the σ_0 of the damped sinusoid. When we try to represent $x_c(t)$ using basis functions $\exp\{(\sigma + j\Omega)t\}$ in the LT^{-1} integral, where $\sigma >> \sigma_0$, the form of these basis functions is increasingly dissimilar from the $\exp\{(\sigma_0 + j\Omega)t\}$ of $x_c(t)$ (e.g., for $\sigma = 0$ they are sinusoids), so it takes "a lot more of them" to produce the heavily damped shape of $x_c(t)$ (hence the spread of the spectrum as σ increases), and not any single one of them is "close" to $x_c(t)$ so there will be no strong peaks in the spectrum when $\sigma >> \sigma_0$. When $x_c(t)$ is combined with another sinusoid of lower frequency but closer to the $j\Omega$ axis than σ_0, the original $x_c(t)$ can be masked, as shown in the problems.

A practical example of the use of the Nyquist theorem in the reconstruction of $x_c(t)$ from $x(n)$ (without significant masking) will be given after consideration of the relevant theory, which will now be presented.

6.3   RECONSTRUCTION OF THE CONTINUOUS–TIME SIGNAL

6.3.1   Reconstruction Filter and the Sampling Theorem

Effects of sampling in the frequency domains, both Fourier and GPE, have been explored in detail. It remains to be shown precisely under what conditions and how the continuous–time signal that has been sampled, $x_c(t)$, can be reconstructed from its samples $x(n) = x_c(n\Delta t)$. First, we examine ideal reconstruction for perfectly bandlimited signals. Then we consider a more practical though imperfect reconstruction procedure in common usage.

Suppose that (6.18) has been satisfied—that is, $\Delta t \leq \pi/\Omega_0$—so that sampling has introduced no aliasing of high frequencies into low frequencies. The replications of $X_c(j\Omega)$ in $X(e^{j\omega})$ do not overlap, but they still exist.

We are immediately faced with what would appear to be an insurmountable problem, as far as reconstruction of $x_c(t)$ from its samples $x_c(n\Delta t)$ is concerned: We know that the DTFT of $x(n)$, $X(e^{j\omega})$, is periodic, yet $X_c(j\Omega)$ is aperiodic! If $x_c(t)$ is to be reconstructed using $x(n)$, then $x(n)$ will have to be filtered with a filter having an aperiodic Fourier spectrum (frequency response) to restore the aperiodic character of $X_c(j\Omega)$. This filter is called the reconstruction filter, and by virtue of its aperiodicity it must be continuous–time (if it were discrete–time, its Fourier transform would be a periodic function of frequency).

If we were to remove all replications except the $k = 0$ replication from $X(e^{j\omega})$ in (6.17), all that would be left is $(1/\Delta t) \cdot X_c(j\omega/\Delta t)$. Because (6.18b) has been satisfied, as ω ranges from $-\pi$ to π, $\omega/\Delta t = \omega\Omega_s/(2\pi)$ with $\Omega_s > 2\Omega_0$ ranges over at least the full bandwidth of the original $x_c(t)$, $|\Omega| \leq \Omega_0$.

The method of removing all the $k \neq 0$ replications is to multiply $X(e^{j\omega})$ by zero at all frequencies for which the shifted replications are nonzero. The $k = 0$ replication is alone retained simply by multiplying $X(e^{j\omega})$ by a nonzero *constant* at all frequencies for which that replication is nonzero. Because there is no overlap in the case of no aliasing we are assuming, these simultaneous requirements are not contradictory. These requirements can in principle be met by an ideal continuous–time lowpass filter

$$R_c(j\Omega) = \begin{cases} \Delta t, & |\Omega| \leq \Omega_s/2 = \pi/\Delta t \\ 0, & |\Omega| > \Omega_s/2, \end{cases} \qquad (6.20)$$

where the value Δt of $R_c(j\Omega)$ is the nonzero constant that cancels the $1/\Delta t$ factor introduced in the sampling process. For filtering $x(n)$, $R_c(j\Omega)$ is evaluated at $\Omega = \omega/\Delta t$ so as to express its behavior in terms of the variable ω of $X(e^{j\omega})$, and to perform the desired annihilation operations. Upon filtering $x(n)$ with $R_c(j\Omega)$, ideally all that is left is $X_c(j\omega/\Delta t)$—the Fourier transform of $x_c(t)$. That is, the result of the filtering is, in the time domain, an exact reconstruction of the original continuous–time signal $x_c(t)$.

It is of interest to express this filtering operation in the time domain. The continuous–time filter $R_c(j\Omega)$ is represented in the time domain by its inverse FT, which is calculated using basic integration and complex algebra:

$$r_c(t) = \frac{\Delta t}{2\pi} \cdot \int_{-\pi/\Delta t}^{\pi/\Delta t} e^{j\Omega t} d\Omega = \frac{\Delta t}{2\pi j t} \cdot \{e^{j\pi t/\Delta t} - e^{-j\pi t/\Delta t}\}$$

$$= \frac{2j \sin(\pi t/\Delta t)}{(1/\Delta t)2\pi j t} = \frac{\sin(\pi t/\Delta t)}{\pi t/\Delta t} = \mathrm{sinc}\left(\frac{\pi t}{\Delta t}\right). \qquad (6.21)$$

Actually, we already proved this result in Example 5.7 by a direct calculation of $\mathrm{FT}\{\mathrm{sinc}(\Omega_1 t)\}$, where in (5.22b) $[\mathrm{FT}\{\mathrm{sinc}(\Omega_1 t)\} = \pi/\Omega_1, \ |\Omega| < \Omega_1; 0, \text{ otherwise}] \ \Omega_1 = \pi/\Delta t$ so that $\pi/\Omega_1 = \Delta t$ is the passband gain. In (6.21), we have verified the result by an inverse Fourier transform calculation.

To what does the multiplication by $R_c(j\Omega)$ in the frequency domain correspond in the time domain? Remember that we are multiplying a DTFT by an FT—which are unmatched transform types—so the form of the convolution theorem that applies may not be immediately obvious. Assuming that $X_c(j\Omega) = 0$ for $|\Omega| > \pi/\Delta t$, we note that multiplying X_c by $R_c/\Delta t$ does not change the value of X_c for any ω:

$$X_c\left(\frac{j\omega}{\Delta t}\right) = \frac{1}{\Delta t}X_c\left(\frac{j\omega}{\Delta t}\right) \cdot R_c\left(\frac{j\omega}{\Delta t}\right), \quad \text{all } \omega. \qquad (6.22a)$$

Recall that $X(e^{j\omega})$ is 2π–periodic in ω and is formed by an infinite number of shifted replications of $X_c(j\omega/\Delta t)/\Delta t$. Also recall that $R_c(j\omega/\Delta t)$ in (6.20) annihilates all replications in $X(e^{j\omega})$ except the primary one ($k = 0$) centered on $\omega = 0$. Again assuming no aliasing, (6.22a) becomes

$$X_c\left(\frac{j\omega}{\Delta t}\right) = X(e^{j\omega}) \cdot R_c\left(\frac{j\omega}{\Delta t}\right), \quad \text{all } \omega, \qquad (6.22b)$$

which because as previously noted, $X(e^{j\omega}) = X_{s,c}(j\omega/\Delta t)$ [from (6.16a and b)] may also be written as

$$X_c\left(\frac{j\omega}{\Delta t}\right) = X_{s,c}\left(\frac{j\omega}{\Delta t}\right) \cdot R_c\left(\frac{j\omega}{\Delta t}\right), \quad \text{all } \omega. \tag{6.22c}$$

If we restrict ω to $|\omega| < \pi$, then because $R_c(j\omega/\Delta t) = \Delta t$ on this frequency range, (6.22b) may be rewritten as

$$X_c\left(\frac{j\omega}{\Delta t}\right) = X(e^{j\omega}) \cdot \Delta t, \quad |\omega| < \pi \text{ only and no aliasing.} \tag{6.22d}$$

Although (6.22d) is valid only for $|\omega| < \pi$, we normally confine our attention to $|\omega| < \pi$; thus, (6.22d) is quite useful in its proper context (no aliasing) as a relation between the DTFT and the FT.

Now, to obtain the time–domain expression for $x_c(t)$ in terms of $r_c(t)$ and $x(n)$ $= x_c(n\,\Delta t)$, rewrite (6.22b) with $\Omega = \omega/\Delta t$ as

$$X_c(j\Omega) = X(e^{j\Omega\,\Delta t})R_c(j\Omega), \tag{6.22e}$$

which by the definition of the DTFT may be rewritten as

$$X_c(j\Omega) = \sum_{n=-\infty}^{\infty} x_c(n\,\Delta t)e^{-j\Omega(n\,\Delta t)}R_c(j\Omega). \tag{6.22f}$$

Notice that by the FT delay theorem, $e^{-j\Omega(n\,\Delta t)} \cdot R_c(j\Omega) = \text{FT}\{r_c(t - n\,\Delta t)\}$. Using this fact to take the FT^{-1} of both sides of (6.22f) term by term, we finally obtain

$$x_c(t) = \sum_{n=-\infty}^{\infty} x_c(n\,\Delta t) \cdot r_c(t - n\,\Delta t). \tag{6.23}$$

By the inverse Fourier transform of (6.22c), the mixed–domain convolution in (6.23) is mathematically equivalent to the purely continuous–time convolution $x_c(t) = x_{s,c}(t) *_c r_c(t)$, for substitution of $x_{s,c}(t)$ in (6.10b) into this latter convolution indeed yields the right–hand side of (6.23).

It is worthwhile to state now a general result we have proved. We have shown above that the convolution of a sequence $x(n)$ and a continuous–time function $r_c(t)$,

$$y_c(t) = \sum_{n=-\infty}^{\infty} x(n) \cdot r_c(t - n\,\Delta t) \tag{6.24}$$

is [see (6.22e)] the inverse Fourier transform of

$$Y_c(j\Omega) = X(e^{j\Omega\,\Delta t}) \cdot R_c(j\Omega). \tag{6.25}$$

Although (6.24) and (6.25) hold for any $x(n)$ and $r_c(t)$, we have shown that for $x(n)$ $= x_c(n\,\Delta t)$ and $r_c(t) = \text{sinc}(\pi t/\Delta t)$ and no aliasing, we have $y_c(t) = x_c(t)$. The question of how such a convolution could be approximately implemented in practice is answered below and in the problems.

Substituting $r_c(t) = \text{sinc}(\pi t/\Delta t)$, (6.23) becomes

$$x_c(t) = \sum_{n=-\infty}^{\infty} x_c(n\,\Delta t) \cdot \text{sinc}\left\{\pi\left(\frac{t}{\Delta t} - n\right)\right\}. \tag{6.26}$$

In summary, the *Shannon sampling theorem* states that $x_c(t)$ bandlimited to frequency Ω_0 can in principle be perfectly reconstructed from its samples using (6.26), provided that

1. (6.18) has been satisfied: $\Delta t \le \pi/\Omega_0$.
2. $R_c(j\Omega)$ is a lowpass filter with cutoff $\pi/\Delta t$ and equal to Δt in the passband.

(6.27)

Incidentally, as $\Delta t \to 0$, the sinc function approaches a Dirac delta function times Δt, the samples become infinitesimally spaced, and the sum in (6.26) becomes an integral with Δt becoming dt. The mixed–type convolution sum in (6.26) becomes continuous–time convolution integral of what are now $x_c(t)$ and $\delta_c(t)$, to again yield $x_c(t)$. This limiting case thus correctly corresponds to no sampling at all. In the frequency domain, the replications become infinitely far apart as $\Delta t \to 0$, and only the zero–centered replication, $X_c(j\Omega)$, is left.[8]

In practice, (6.26) poses four problems. First, what input is physically given to the analog filter to implement (6.26)? We noted after (6.23) that the answer would be $x_{s,c}(t)$, which does not exist! In practice, finite–height–and–width sample–weighted pulses could be used to approximate $x_{s,c}(t)$. Of course, in a computer simulation, we would instead just perform a direct calculation of (6.26) using the programmed sinc function and the known $x_c(n\Delta t)$.

Second, $r_c(t)$ is infinite–length, so if implemented on a computer it must be truncated. Letting $[t/\Delta t]$ denote the greatest integer of $t/\Delta t$—the integer that when multiplied by Δt is closest to but less than t—the truncated estimate is

$$\hat{x}_c(t) = \sum_{n=[t/\Delta t]-N}^{[t/\Delta t]+N} x_c(n\Delta t) \cdot \text{sinc}\left\{\pi\left(\frac{t}{\Delta t} - n\right)\right\}. \tag{6.28a}$$

Now define $m = [t/\Delta t] - n$, so that the summation index runs over the more convenient range from $-N$ to N:

$$\hat{x}_c(t) = \sum_{m=-N}^{N} x_c\left(\left(\left[\frac{t}{\Delta t}\right] - m\right)\Delta t\right) \cdot \text{sinc}\left\{\pi\left(\frac{t}{\Delta t} - \left[\frac{t}{\Delta t}\right] + m\right)\right\}. \tag{6.28b}$$

How do we choose N? An extension of Figure 6.12 shows, for example, that the peak values of oscillations (envelope) of the sinc function $\text{sinc}(\pi t/\Delta t)$ die away to less than 1% of the maximum of the sinc function at approximately $t = 32.5\Delta t$ ($x = 102.1$ in Figure 6.12), so a reasonable choice of N for this 1% criterion would be 32. Therefore, $2 \cdot 32 + 1 = 65$ samples are required per evaluation of $\hat{x}_c(t)$. If (6.28) were carried out on a computer for simulation purposes, 65 multiply/adds per evaluation would suffice.

Third, the filter $\text{sinc}(\cdot)$ is bicausal. Thus to compute $x_c(t)$, values of $x_c(\tau > t)$ are needed but would not be available if the filtering were to be performed in "real time" (they generally would be for "off–line" processing). This problem is overcome by shifting the filter so that all of its values within the truncation window occur for $m \ge 0$. This shift is achieved by replacing $\text{sinc}\{\pi(t/\Delta t - [t/\Delta t] + m)\}$ in (6.28b) by its $N\Delta t$–delayed version $\text{sinc}\{\pi(t/\Delta t - [t/\Delta t] + m - N)\}$, which merely delays the result because convolution is time invariant:

$$\hat{x}_{\text{delayed},c}(t) = \sum_{m=0}^{2N} x_c\left(\left(\left[\frac{t}{\Delta t}\right] - m\right)\Delta t\right) \cdot \text{sinc}\left\{\pi\left(\frac{t}{\Delta t} - \left[\frac{t}{\Delta t}\right] + m - N\right)\right\}. \tag{6.29}$$

Notice that we must wait a time $2N \cdot \Delta t$ s beyond $t = 0$ until we can begin to compute $\hat{x}_{\text{delayed},c}(t)$ if only $x_c(n\Delta t)$ for $n \ge 0$ are available [rather than waiting only

---

[8]However, $X_{s,c}(s)$ does *not* approach $X_c(s)$, because the replications in $X_{s,c}(s)$ have infinite height as $\Delta t \to 0$; rather, $\Delta t \cdot X_{s,c}(s) \to X_c(s)$ for $\Delta t \to 0$ [see (6.16a)].

$N \Delta t$ as in (6.28b)], because only thereafter do we have all of the samples necessary for its calculation. Notice that the samples of $x_c(t)$ in (6.29) range from $([t/\Delta t] - 2N) \Delta t$ to $[t/\Delta t] \Delta t$, which are all in the present or past with respect to the current time, t.

Let us reexamine the reasoning leading to (6.29). We know that by reconstructing $x_c(t)$ delayed by a time $N \Delta t$, we suspect that we can use the symmetrically placed filter weights on only past, not future data. In (6.29), which is just (6.28b) with t replaced by $t - N \Delta t$ in the sinc argument with the corresponding shift in the samples required, let us verify the resulting limits on the summation index m. Let the lower limit be m_1 and the upper limit be m_2. Assume for simplicity that $t/\Delta t = [t/\Delta t]$. For the lowest filter weight to be sinc$\{-\pi N\}$ in (6.29), we must have $m_1 - N = -N$, or $m_1 = 0$. Similarly, at the upper limit m_2, we must have $m_2 - N = N$, or $m_2 = 2N$. These are the limits shown on the sum in (6.29). If $t/\Delta t \neq [t/\Delta t]$, then the sinc function is shifted *properly* from the above calculations by the amount $t/\Delta t - [t/\Delta t]$; by ignoring $t/\Delta t - [t/\Delta t]$, we have merely stipulated that the terms included in the finite sum in the case $t/\Delta t \neq [t/\Delta t]$ be, within one sample, symmetrically arranged. The *only* error in the results is due to N being finite.

The progression from (6.26) to (6.28) and then to (6.29) (truncation and shift for causality) is similar to part of a technique often used in the *window method* of digital filter design (Section 9.3). The only difference is that for digital filters, $t = n \Delta t$ so that $t/\Delta t = [t/\Delta t]$ always.

The fourth problem is that an actual circuit closely approximating even a truncated sinc function filter is too complex to be cost–efficient; there are no ideal analog lowpass filters, truncated or otherwise! Thus, the above calculations are very meaningful and useful for numerical sinc interpolation between a given set of samples using a computer, but they are not exactly implementable in analog hardware. Instead, as discussed below, sample–and–hold circuits followed by finite–order lowpass approximating filters are used in practice. We examine some typical systems in Section 6.3.3.

6.3.2 Oversampling, Antialiasing, and Digital Filtering

The accuracy of reconstruction using a zero–order–hold can be improved by substantially oversampling the original continuous–time signal—even as much as 20 to 30 times the conventional –3–dB bandwidth. This is because when the zero–order hold is used, the cutoff is not sharp and erroneous harmonics in the replicated spectra will still be significantly present in the reconstructed signal, unless the replications are very widely separated.

These ideas are illustrated in Figure 6.16. In Figure 6.16a, exact Nyquist sampling has been used; that is, no oversampling and no aliasing, but no intervals of zero between replications. Application of the ideal lowpass filter $R_c(j\Omega)$ will yield $x_c(t)$ with zero error. If instead the zero–order hold is used, as illustrated in Figure 6.16b, we see that adjacent replications will be passed through the filter and the reconstruction result will not be a good approximation of $x_c(t)$. In the time domain, we would see coarse stairsteps, possibly with large discontinuities and poor representation of the higher-frequency content of $x_c(t)$.

We now verify that the first zero of the zero–order hold is at $\omega = 2\pi$, as depicted in Figure 6.16b. The impulse response of the unshifted zero–order hold is obtained by substituting $n = 0$ into the shifted boxcar definition (6.4):

$$p_{\Delta t}(t) = \begin{cases} 1, & 0 \leq t < \Delta t \\ 0, & \text{otherwise.} \end{cases} \qquad (6.30)$$

Exact Nyquist sampling; ideal lowpass filter reconstruction

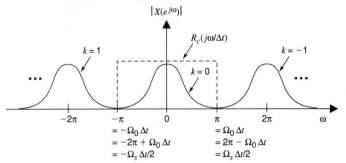

Figure 6.16a

Exact Nyquist sampling; nonideal ZOH reconstruction. Undesired replications passed

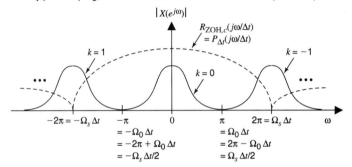

Figure 6.16b

Oversampling; nonideal ZOH reconstruction. Portions/levels of undesired replications passed are minimal

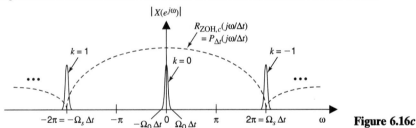

Figure 6.16c

Its Laplace transform is

$$P_{\Delta t}(s) = \int_0^{\Delta t} 1 \cdot e^{-st} dt = \frac{1 - e^{-s\,\Delta t}}{s}. \tag{6.31a}$$

Its Fourier transform is obtained by setting $s = j\Omega$:

$$P_{\Delta t}(j\Omega) = \frac{1 - e^{-j\Omega\,\Delta t}}{j\Omega} = \frac{\Delta t}{2} e^{-j\Omega\,\Delta t/2} \frac{e^{j\Omega\,\Delta t/2} - e^{-j\Omega\,\Delta t/2}}{j\Omega\Delta t/2} \tag{6.31b}$$

$$= \Delta t\, e^{-j\Omega\,\Delta t/2} \mathrm{sinc}\left(\Omega\frac{\Delta t}{2}\right),$$

or, equivalently,

$$P_{\Delta t}\left(\frac{j\omega}{\Delta t}\right) = \Delta t\, e^{-j\omega/2}\, \mathrm{sinc}\left(\frac{\omega}{2}\right). \tag{6.31c}$$

Thus, on the ω axis, the first zero of $P_{\Delta t}$ occurs for $\omega/2 = \pi$ or $\omega = 2\pi$. Note also from (6.31c) that the maximum value of $P_{\Delta t}(j\omega/\Delta t)$ is Δt, just as is true of the ideal $R_c(j\omega/\Delta t)$; the dc gains of the zero–order hold and ideal $R_c(j\omega/\Delta t)$ match.

Figure 6.16c shows use of the zero–order hold in the case of oversampling. This time, the values of the zero–order hold filter multiplying the replications are substantially reduced, and the resulting reconstruction result is a much better approximation of $x_c(t)$. Specifically, the area of the magnitude squared of the FT is an energy, so the proportion of total energy in, for example, the portion of the shown $k \neq 0$ spectrum replication that is passed is strongly reduced in Figure 6.16c compared with the corresponding proportion in Figure 6.16b. This improvement may be "eyeballed" by noting that in both cases essentially the entire principal spectrum is passed, but the portion of the "area" of the nearest $k \neq 0$ spectrum replication passed is much reduced. Keep in mind that the $1/\Delta t$ factor in $X_{s,c}$ from sampling is in both cases canceled by $P_{\Delta t}$ at $\omega = 0$. Intuitively, oversampling clearly helps because $x_c(t)$ changes very little from sample to sample, so holding steady from sample to sample introduces little error; $\lim_{\Delta t \to 0} x_{\Delta t,c}(t) = x_c(t)$.

As an added bonus, with oversampling the sinc function filter is much more nearly constant over the entire nonzero portion of the principal ($k = 0$) spectrum, which further increases the accuracy of the reconstruction. In fact, it is even possible to digitally prefilter the sequence before reconstruction to equalize the subsequent ZOH reconstruction filter magnitude distortion, which will yield an even better reconstruction. Such considerations may be useful in demanding reconstruction situations such as compact disc players.

The accuracy of reconstruction can also be further enhanced by filtering $x(n)$ with a high–quality *digital lowpass* filter before applying the reconstruction filter—another approach used in compact disc systems. For simplicity, in the following discussion we will ignore the zero–order–hold distortions. Because sharp analog filters can be sensitive to parameter variations due to thermal effects, component aging and imperfections, and so forth, and because they are expensive to build, it would be desirable to require only simple low–grade analog filters. The following procedure allows the use of low–grade analog filters by performing the required sharp filtering with a digital filter.

Suppose that $x_c(t)$ is *not* bandlimited to $\Omega_s/2$ (e.g., due to high–frequency noise); that is, $\Omega_s < 2\Omega_0$. Equivalently, Δt has been chosen to match some other signal truly bandlimited to $\Omega_s/2$, or Δt is constrained in value by the existing hardware. Figure 6.17 shows the spectrum of such a sampled signal, having aliasing. Although the Ω_0 in Figure 6.17 is shown for convenience to be not very much higher than $\Omega_s/2$, in practice the value of Ω_0 for a large number of decibels down may be unreasonably high (essentially infinite) to match with cost–effective sampling hardware. However, the high–frequency content may be noise,which we do not want anyway. Under sampling, it would alias into the signal low-frequency range and distort our signal.

To reduce the aliasing, let $x_c(t)$ be prefiltered with an antialiasing filter $A_c(j\Omega)$ having rounded magnitude cutoff at $\Omega_1 < \Omega_s/2$, very small magnitude for $\Omega \geq \Omega_s - \Omega_1$, and magnitude approximately unity in the passband $\Omega < \Omega_1$. We assume that all the *desirable* signal content in $x_c(t)$ resides in the range $|\Omega| < \Omega_1$ even though

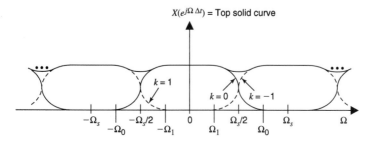

Figure 6.17

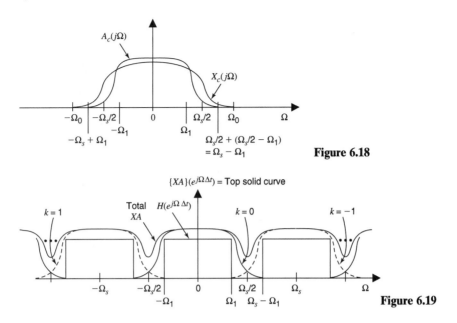

Figure 6.18

Figure 6.19

$|X_c(j\Omega)|$ extends to about Ω_0; hence we *want* to eliminate all higher–than–Ω_1 frequency content. Figure 6.18 shows the $|X_c(j\Omega)|$ of Figure 6.17 (before sampling) and $|A_c(j\Omega)|$. In between Ω_1 and $\Omega_s - \Omega_1$, we will allow $A_c(j\Omega)$ to moderately pass and distort signals. The result of this analog antialiasing filtering upon the subsequent result of sampling at Ω_s is shown in Figure 6.19, where for brevity, we write sampled $X_c \cdot A_c$ as XA. Notice that we allow aliasing from Ω_1 to $\Omega_s/2$, which we will annihilate with $H(e^{j\omega})$; this allowance permits $A_c(j\Omega)$ to have maximum transition width and therefore lowest cost and complexity. Keep in mind that the aliasing terms shown are actually to be added to obtain $\{XA\}(e^{j\Omega \Delta t})$, as shown.

Let the reconstruction filter $R_c(j\Omega)$ be identical to $A_c(j\Omega)$ except that it has magnitude Δt in the passband. Before the introduction of digital filtering, $R_c(j\Omega)$ would have had to be flat in the passband and have very sharp cutoffs (i.e., be nearly ideal and thus high–order) to retain all of the passband and still eliminate the other replicas in the aliasing sum, as well as the undesirable noisy and distorted components above Ω_1. Also, there may be nonlinear phase distortion in the high–order analog antialiasing filter just beyond the cutoff; we do not want the reconstructed signal to include this distortion.

However, we assign this requirement for sharp, high–quality analog filters instead to $H(e^{j\omega})$. The digital filter can be, for example, a fast, cheap, and high–quality digital filter with sharp cutoff at $\omega = \Omega_1\Delta t$ with purely linear phase distortion (pure delay). The idealized frequency response of such an $H(e^{j\omega})$ is also shown in Figure 6.19. The result of digital filtering is shown in Figure 6.20; the unwanted, magnitude– and phase–distorted and aliased band of XA has been annihilated by H.

With sharp $H(e^{j\Omega \Delta t})$, there again is "plenty of room for error" (transition) between Ω_1 and $\Omega_s - \Omega_1$ to allow for the transition band of the low–grade reconstruction filter $R_c(j\Omega)$. That is, the transition band of $R_c(j\Omega)$ multiplies XAH where the latter is essentially zero. This is seen in Figure 6.20, which includes a plot of the reconstruction filter $R_c(j\Omega)$. [Recall that the purpose of $R_c(j\Omega)$ is to remove the $k \neq 0$ replications without distorting the $k = 0$ spectrum.]

The final output, whose spectrum is shown in Figure 6.21, will be rather precisely that part of $x_c(t)$ having frequencies less than or equal to Ω_1. If the value of

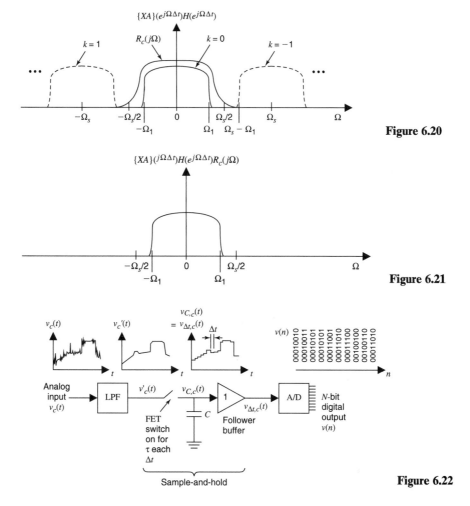

Figure 6.20

Figure 6.21

Figure 6.22

Ω_1 required to have a reasonably cheap analog filter is below the maximum frequency content we wish to preserve, then Δt will have to be decreased.

Naturally, any other digital filtering or other discrete–time signal processing can be performed on $x(n)$ before restoration to continuous time (see Section 6.4). Alternatively, the digital–to–analog (D/A) conversion (with reconstruction) may not be necessary in a given application, particularly when the end result is computer processing and display of the sampled data.

6.3.3 Basic Practical Sampled–Data System Configurations

From our discussions of the principles of sampling and reconstruction, we may now consider briefly the sort of circuitry necessary for their implementation. An example block diagram for sampling of a continuous–time signal is shown in Figure 6.22, along with typical signal waveforms. An antialiasing filter nearly removes the frequencies in the input voltage $v_c(t)$ beyond half the sampling frequency, to minimize aliasing. The sample–and–hold circuit performs the temporal discretization, and the quantizing circuit (A/D converter) performs the value discretization and binary encoding. The result is an N–bit digital signal, discretized in both time (resolution Δt) and value (resolution $V_{\mathrm{max}}/2^N$, where V_{max} is the set maximum input analog voltage to be applied to the A/D). "Sampling" A/D converter integrated circuits include an on–chip sample–and–hold circuit, thus allowing $v_c'(t)$ in Figure 6.22 as their input.

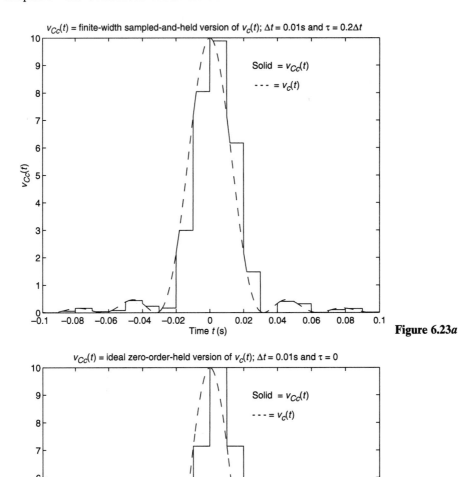

$v_{Cc}(t)$ = finite-width sampled-and-held version of $v_c(t)$; $\Delta t = 0.01$s and $\tau = 0.2\Delta t$

Solid = $v_{Cc}(t)$

- - - = $v_c(t)$

Figure 6.23a

$v_{Cc}(t)$ = ideal zero-order-held version of $v_c(t)$; $\Delta t = 0.01$s and $\tau = 0$

Solid = $v_{Cc}(t)$

- - - = $v_c(t)$

Figure 6.23b

   Fundamental to the temporal discretization process is the sample–and–hold circuit. Its purpose is to hold a stable, constant value representing the input signal
while the quantization (value discretization) is performed. A simplified version is
shown in Figure 6.22. Digital control signals control the MOSFET
(high–input–impedance) switch to either connect the filtered analog input $v_c'(t)$ to
the capacitor (during the sample mode) or disconnect it (during the hold mode).
   During sample mode, of duration $\tau << \Delta t$, the capacitor charges through a
very small RC time constant (essentially, R equals the output impedance of the
source) to $v_c'(t)$. During the hold mode, of duration $\Delta t - \tau$, the capacitor has nowhere
to discharge, and so $v_{C,c}(t)$ remains at the held value $v_c'(n\Delta t)$ for the remainder of

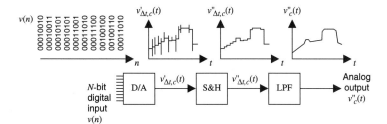

Figure 6.24

the sample–and–hold cycle. The sample–and–hold cycle thus has duration Δt. Note that the sample–and–hold operation is linear, but not time–invariant.

A typical $v_{C,c}(t)$ would appear as shown in Figure 6.23a (ignoring droop during hold and a host of other imperfections). As the sampling time τ approaches zero, $v_{C,c}(t)$ appears as in Figure 6.23b; this is the normal appearance for small sampling time. In practice, $v_{C,c}(t)$, a staircase signal, is applied to the input of an impedance buffer (follower circuit) so that later circuitry will not discharge the capacitor during the hold mode. The resulting staircase analog voltage is $v_{\Delta t,c}(t) = v_{C,c}(t)$. As can be seen from this discussion, the "sample" and (zero–order) "hold" operations of the actual sample–and–hold circuit occur on the same device: the capacitor.

A variety of circuits performing the value discretization (quantization) exists, a few of which we now mention. One, called the *flash* A/D converter, is a set of parallel–operating comparator circuits whose output patterns classify the input as being between two adjacent quantization levels. Its output patterns are not in binary code, and must be so encoded by combinational logic circuits. It is called "flash" because only one operational amplifier delay is involved; hence it is relatively very fast.

A slower but cheaper quantizer is the successive approximation circuit, which continually halves the possible range of an N–bit approximation of the sampled–and–held analog input voltage $v_{\Delta t,c}(t)$ until the N–bit resolution is achieved. On each range–halving, the current most significant bit (MSB) in the approximation register is tentatively set to 1. Then $v_{\Delta t,c}(t)$ is compared against the tentative N–bit approximation, obtained by D/A conversion (D/A is described below). The result of that comparison determines whether to keep the current MSB set to 1, thereby determining the remaining search range. All bits from MSB to least significant bit are processed in sequence; the result is stored in the output register.

Yet another quantizer is the dual–slope ramp converter. A capacitor is charged for a fixed time by integrating $v_{\Delta t,c}(t)$ and thus reaches a value proportional to $v_{\Delta t,c}(t)$. Then the capacitor is discharged at a fixed rate, and the time to reach zero is measured by a digital counter. The time to reach zero is proportional to $v_{\Delta t,c}(t)$, and thus the count is a quantization of $v_{\Delta t,c}(t)$.

For all these quantizers, the input $v_{\Delta t,c}(t)$ must be held steady for the duration of the quantization—hence the need for the "hold" function in the sample–and–hold circuit. The final digital signal $v(n)$ is then in the proper format for computer processing: an N–bit–wide parallel binary word. Although in this book we usually ignore value quantization effects and work in base 10, $v(n) \approx v_c'(n\Delta t)$ is in practice a sequence of N–bit binary words (1–0 patterns).

The reverse process is shown in Figure 6.24. A digital input $v(n)$ is converted to an analog voltage $v_c''(t)$ that approximates $v_c'(t)$. The given binary number in N–bit form is presented to the input of a D/A converter. An example is the $R/2R$ ladder D/A converter shown in Figure 6.25a, which appears in the AD7524 integrated circuit ($N = 8$). The op–amp in Figure 6.25a, used as an inverting summer, is external to the D/A chip; the latter includes all the R and $2R$ resistors.

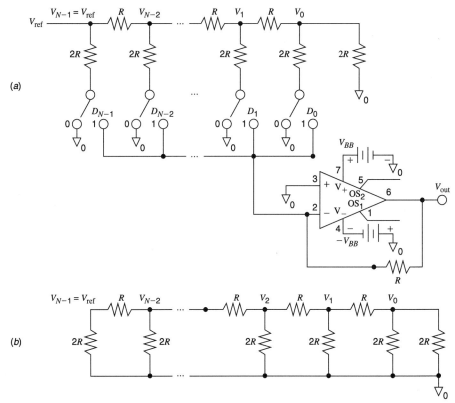

Figure 6.25

We will now describe the basic D/A operation, ignoring the host of control signals for the chip that are the domain of a course on microprocessor interfacing. Let $D_\ell = 1$ (i.e., $v_{D_\ell} = 5$ V) imply that the lower end of the ℓth $2R$ is connected to the op–amp inverting input terminal, and $D_\ell = 0$ (i.e., $v_{D_\ell} = 0$ V) mean it is connected to ground (via the ℓth MOSFET switch). Whether $D_\ell = 0$ or 1, the lower side of the ℓth $2R$ is at 0 V, either via ground or virtual ground. Thus, for determination of the numbered node voltages, the circuit in Figure 6.25b is equivalent.

Starting from the right in Figure 6.25b, $2R \parallel 2R = R$, which in series with the rightmost horizontal R gives $V_0 = V_1/2$. This series combination, $2R$, is in parallel with the next $2R$ to the left, which in turn is in series with the next rightmost horizontal R, giving $V_1 = V_2/2$, and so on. Substitution gives $V_0 = V_1/2 = \ldots = V_{\text{ref}}/2^{N-1}$, and each successive V_ℓ is twice $V_{\ell-1}$. If $D_\ell = 0$, the ℓth branch current does not appear in Kirchhoff's current law; otherwise it does. Putting these ideas together and noting that for each term the inverting summer has a gain of $-\frac{1}{2}$ gives $v_{\text{out}} = -V_{\text{ref}}\{D_{N-1}/2 + \ldots + D_0/2^N\}$. If V_{ref} is set to -10 V, then v_{out} varies over the standard range from 0 to 10 V $\cdot (1 - \frac{1}{2^N}) \approx 10$ V for, for example, $N = 12$ or 16.

The result $v'_{\Delta t, c}(t) = v_{\text{out}}$ is a staircase signal, except that it may have "glitches" (see Figure 6.24). When the input binary number changes value, not all the bits change simultaneously, or there may be timing asymmetries in the digital switching front end of the D/A converter circuit. Consequently, the binary number actually presented to the $R/2R$ ladder momentarily takes on unpredictable values. The result is wild fluctuations in the output (glitches), during the changes in the binary input (most wild from the MSBs) possibly occurring every Δt.

These glitches can be eliminated with another sample–and–hold circuit, timed to be "holding" while all the glitching takes place, so that the glitches are ignored and thus removed. The result, $v''_{\Delta t,c}(t)$ (see Figure 6.24) still includes the quantization and timing errors; $v_{\Delta t,c}(t)$ can never be exactly recovered. Finally, if desired, $v''_{\Delta t,c}(t)$ can be lowpass–filtered to smooth out the unnatural "stair" corners in $v''_{\Delta t,c}(t)$. The result, $v''_c(t)$ (again see Figure 6.24) is an approximate version of $v'_c(t)$, the lowpass version of the original analog signal $v_c(t)$.

From this simplified discussion of reconstruction circuitry, we see that the ideal sinc–interpolation filter is not used in practice, although in some cases its effect may be well approximated by high–quality analog approximating lowpass reconstruction filters. The main value in learning about the sinc interpolator is in understanding the ideal or best–possible result. The zero–order–hold is only an approximation, but it has some of the same characteristics that an ideal reconstructor has—in particular, greatly attenuating the replicated spectra introduced by sampling. As noted, the output of the A/D circuit is always staircase, so we always "start out" with effectively a zero–order–hold, which may be smoothed by a subsequent reconstruction filter.

It is interesting to note that the zero–order–hold impulse response has the same funtional form as the Fourier transform of the sinc interpolator impulse response, and vice versa. Thus, the ideal reconstruction impulse response and its spectrum have been swapped in the interest of practical expedience.

6.3.4   Example of a Practical Reconstruction

EXAMPLE 6.2

In Figure 6.26 is shown a 4096–sample section of approximately 3/4 s of a speech signal (the word "OK"), sampled at $F_s = 5.51$ kHz.[9] The magnitude of the DFT_{4096} of this section appears as in Figure 6.27. Discuss the presence or lack of presence of aliasing, and explore zero–order–hold reconstruction of the samples.

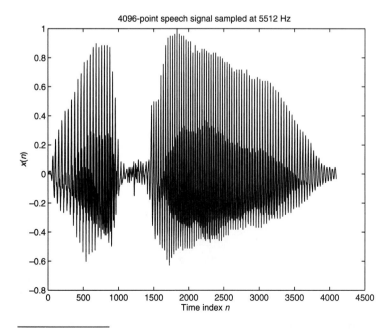

4096-point speech signal sampled at 5512 Hz

Figure 6.26

---

[9] Extracted from an original 11.025-kHz-sampled signal. Only lines joining samples are shown.

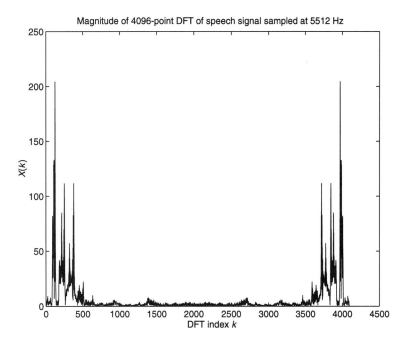

Figure 6.27

Solution

From earlier reasoning, we may be confident that there is little aliasing for $F_s = 5.51$ kHz because $\left| X(e^{j\omega}) \right|$ is very small for $\omega \approx \pi$ in Figure 6.27. If instead we sample at 0.689 kHz (5.51 kHz/8), the 512 samples appear as in Figure 6.28, and the magnitude of the DFT_{512} of these samples is as in Figure 6.29. Because there is high energy density at $\omega \approx \pi$, and because an unaliased spectrum is already known from Figure 6.27, we may conclude that for this much lower sampling rate, there is significant aliasing.

Consider now the reconstruction of a subsection of the speech signal approximately 22 ms long. The object will be to compare reconstructions based on one set of

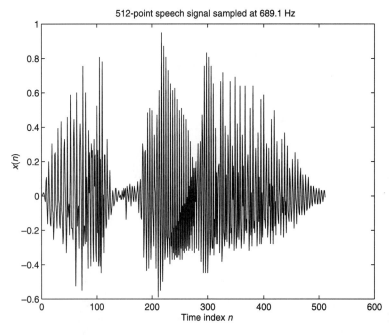

Figure 6.28

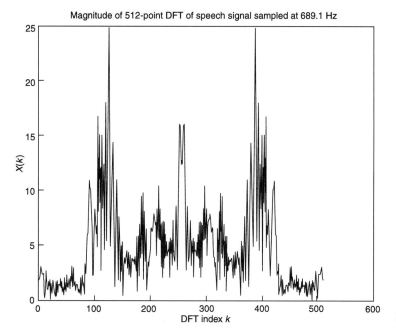

Figure 6.29

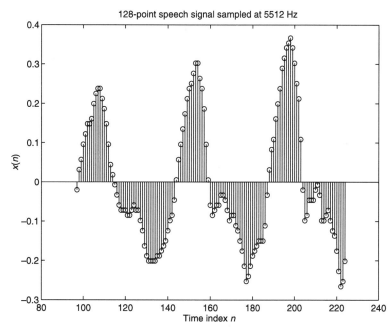

Figure 6.30

samples having essentially no aliasing with another set having serious aliasing. In this speech subsection, there are 128 samples at the 5.51–kHz rate; these are shown in Figure 6.30. At the 0.689–kHz rate, there are only 16 samples, shown in Figure 6.31.

Figure 6.32 shows (solid curve) the sinc basis reconstruction of the 5.51–kHz samples in Figure 6.30 using (6.29)

$$[\hat{x}_{\text{delayed},c}(t) = \sum_{m=0}^{2N} x_c(([t/\Delta t] - m)\Delta t) \cdot \text{sinc}\{\pi(t/\Delta t - [t/\Delta t] + m - N)\}],$$

with the exception that the sum limits are extended to always use all 128 available samples. This waveform appears much as would the continuous–time speech signal, as reconstructed from its samples using an ideal analog filter with cutoff at 5.51 kHz.

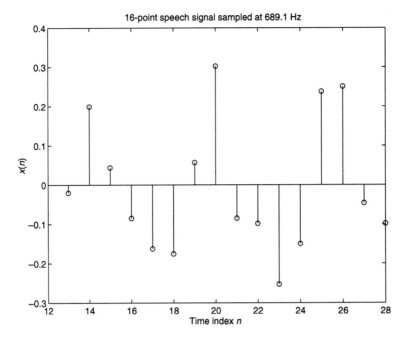

Figure 6.31

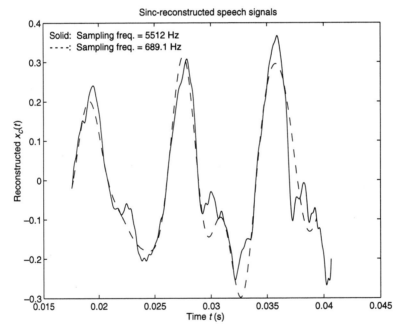

Figure 6.32

More practically, it resembles the output of a zero–order–hold equalized (digitally inverse filtered) that is then zero–order–held and lowpass filtered by a high–quality analog lowpass filter. For comparison, Figure 6.32 also shows (dashed curve) the sinc basis reconstruction based on the subsampled 0.689-kHz sequence. As we might expect, the high frequencies present in the "exact" waveform have been transferred to lower frequencies.

The result is similar to a lowpass filtering operation in this case. This lowpass effect may be explained as follows. The maximum analog Hertzian frequency representable in the 128–point sequence is 5.51 kHz/2 = 2.76 kHz, whereas that of the

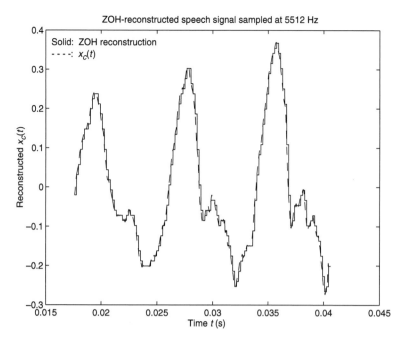

Figure 6.33*a*

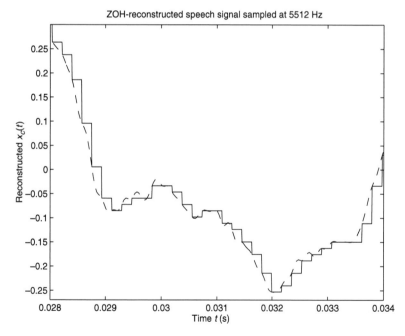

Figure 6.33*b*

subsampled sequence is only 0.689 kHz/2 = 344.6 Hz, eight times smaller. Smooth interpolation between samples using the corresponding low–frequency sinc functions cannot increase this 344.6-Hz maximum frequency of the subsampled sequence, but can only fill in more values of these lower frequency oscillations. Based on visual comparison, the reconstruction may appear adequate. However, when this continuous–time signal is converted to sound, it may seem distorted to the human ear because of the aliasing distortion of low frequencies and lost high frequencies.

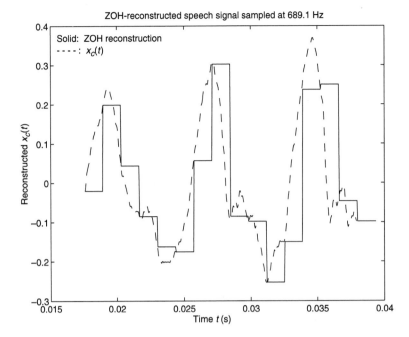

Figure 6.34

It was noted that sinc basis reconstructions are usually not available in practice. Instead, in Figure 6.33a is shown a zero–order–hold reconstruction based on the 5.51–kHz sampling rate and superimposed on the $x_c(t)$ waveform.[10] Comparing (a) (6.3) with (6.5a) substituted [i.e., $x_{\Delta t,c}(t) = \sum_{n=-\infty}^{\infty} x_c(n\Delta t)p_{\Delta t}(t-n\Delta t)$] with (b) (6.26) $[x_c(t) = \sum_{n=-\infty}^{\infty} x_c(n\Delta t)\text{sinc}\{(\pi/t)(t-n\Delta t)\}]$, we see that the zero-order-hold interpolation is equivalent to convolving the sequence with a boxcar instead of with a sinc function. Figure 6.33b shows a detail of Figure 6.33a; the two ideal and ZOH reconstructions always agree at $t = n\Delta t$, but the ZOH remains constant during the hold phase. In Figure 6.34, the zero–order–hold reconstruction of the 0.689–kHz sequence is superimposed on $x_c(t)$. The agreement is generally much worse, as expected, but there is still always exact agreement for $t = n\Delta t$.

Notice in Figures 6.33 and 6.34 that there appears to be an average delay of about $\Delta t/2$ with respect to $x_c(t)$. This half–sample delay is a general characteristic of the zero–order hold. Let us see how this arises mathematically. From an overall "average" or low–frequency viewpoint, that is, small $|s|$—we may expand $P_{\Delta t}(s)$ in (6.31a) over just the first few terms of its Taylor series:

$$\text{LT}\{p_{\Delta t}(t)\} = P_{\Delta t}(s) = \frac{1 - e^{-s\Delta t}}{s} \tag{6.32}$$

$$\approx \frac{1 - [1 - s\Delta t + \frac{1}{2}s^2\Delta t^2]}{s}$$

$$= \Delta t\left(1 - \frac{1}{2}s\Delta t\right)$$

$$\approx \Delta t \cdot e^{-s\Delta t/2}$$

$$= \Delta t \cdot \text{LT}\left\{\delta_c\left(t - \frac{1}{2}\Delta t\right)\right\},$$

---

[10]$x_c(t)$ was derived from the original 11.025-kHz-sampled sequence by sinc interpolation.

where $\delta_c(t - \Delta t/2)$ is the impulse response of a pure delay of $\Delta t/2$. Thus, to a very rude approximation $p_{\Delta t}(t) \approx \Delta t \cdot \delta_c(t - \Delta t/2)$, so the overall appearance of the output of a zero–order hold reconstructor after smoothing corresponds fairly closely to the original $x_c(t)$, smoothed and delayed by half a sample.

6.4   SAMPLED–DATA SIGNAL PROCESSING

6.4.1   Introduction

A major application of discrete–time analysis that involves the continuous–time interface is the discrete–time implementation or digital simulation of continuous–time systems. By implementation we mean that we replace an analog filter or system by a digital filter—either in hardware or software. In today's industrial and consumer electronics, whenever possible an analog system is replaced by a digital equivalent. Simulating an analog system contrastingly usually refers to a software emulation in which improvements in an analog design can be created and tested without having to physically build the system. Implementation and simulation are very closely related, and both result in tremendous cost savings. Both may be included in the term *sampled–data processing*, which implies digital signal processing of sampled analog data.

In Section 6.3.3, we examined typical hardware for converting analog signals into digital signals, as well as the reverse process. We had little to say then about what sort of processing takes place on the digitized signal. Although the remainder of this book is devoted to this subject, in this section we provide some basic approaches and ideas. Digital simulation or replacement of analog systems is a good and intuitive place to start. We thus consider what sort of computer processing will mimic the effects of a physical analog system—either an analog filter or another type of physical system.

Long ago, analog computers were the prevailing method for simulating continuous–time systems. The solution of what would typically be a differential equation model could be expressed in the form of an integral or integrals of known or other computed quantities. Inputs to the analog computer would be initial conditions and the input voltage waveform; the procedure was relatively straightforward. The benefits of simulation for system design mentioned above were obtained.

There were, however, several disadvantages. Some of these include (a) low accuracy—two or three significant figures at best, due to inherent approximation of desired operations using analog circuits, voltage drift and noise, and so forth; (b) high cost and physical size (a separate operational amplifier is devoted to each operation, such as multiplication, addition, integration, and nonlinear functions); (c) relative difficulty of system reconfiguration or change of model because electric connections must be physically broken, rearranged, and reconnected; and (d) general limitations in complexity, reproducibility, and versatility.

Each one of these limitations and problems is overcome by using a digital computer rather than an analog computer for system simulation. Accuracy is limited only by the number of bits used in the number representation, resulting in many reliable significant figures, and by the accuracy of the original analog model and of discrete approximation of continuous–time operators. Because the variables are binary rather than continuous voltage levels, noise and drift have greatly diminished impact. The digital computer has become exceptionally cheap to buy and operate,

(a) Sampled-data system:

(b) Equivalent analog system:

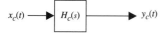

Figure 6.35

and is computationally very powerful yet extremely small. All that is usually necessary for reconfiguration/model alteration is program modification. Previous advantages of analog computers such as speed and parallelism have largely disappeared because of improved digital computer hardware/architectures.

6.4.2 Basic Sampled–Data Processing and the Impulse–Invariant Method

We know from the Shannon sampling theorem in (6.26) and (6.27) that the reconstruction of a perfectly bandlimited signal from its samples using the ideal reconstruction filter is perfect if the sampling frequency is Nyquist or better. That is, the equivalent overall *analog* filter acting on the input $x_c(t)$ has impulse response $\delta_c(t)$. In Section 6.3, we showed how inserting a sharp digital lowpass filter could maximize the effective approximation of $\delta_c(t)$ in the reconstruction process in the practical case where $R_c(j\Omega)$ is not ideal and $x_c(t)$ is not perfectly bandlimited.

There is, however, no reason why $H(e^{j\omega})$ need be restricted to only a sharp lowpass filter. By inserting a general digital filter $h(n)$ before $r_c(t)$, the net analog filter acting on the input $x_c(t)$ can be whatever we wish. In particular, we may digitally implement or simulate an existing analog filter (continuous–time system) having impulse response $h_c(t)$, rather than just $\delta_c(t)$. This idea is illustrated in Figure 6.35. When the analog system $h_c(t)$ is a physical system that cannot be replaced, this process is called simulation; when $h_c(t)$ is a filter or other analog system to be replaced or implemented by a digital system, we call it filter discretization.

To summarize the process, there are now three steps to digital implementation of an analog system, the first of which is data sampling. We have shown in (6.17) [repeated in (6.33a) below] that sampling produces a periodic spectrum of possibly overlapping replications of the value–and–argument–scaled continuous–time Fourier transform of $x_c(t)$. Furthermore, by (6.22d) [repeated in (6.33b) below], we know that if the Nyquist sampling criterion (6.18) is satisfied, the principal digital spectrum matches the original analog spectrum:

$$(1)\ \text{Sample: } x(n) = x_c(n\,\Delta t) \xleftrightarrow{\text{DTFT}} X(e^{j\omega}) = \frac{1}{\Delta t}\sum_{k=-\infty}^{\infty} X_c\left(j\left\{\frac{\omega + 2\pi k}{\Delta t}\right\}\right) \qquad (6.33a)$$

$$= \underbrace{\frac{1}{\Delta t}\cdot X_c\left(\frac{j\omega}{\Delta t}\right), \quad |\omega| < \pi}_{\text{if } \Delta t < \dfrac{\pi}{\Omega_0}}. \qquad (6.33b)$$

The second step is to filter $x(n)$ with a digital filter $h(n)$, which may be chosen to implement an existing or desired analog filter $h_c(t)$. (In practice, we may augment the simulating digital filter with the sharp lowpass filter discussed previously.) In the time and frequency domains, we have

(2) Filter with digital filter

$$h(n): y(n) = x(n) * h(n) \xrightarrow{\text{DTFT}} Y(e^{j\omega}) = X(e^{j\omega}) \cdot H(e^{j\omega}). \qquad (6.34)$$

The final step is to reconstruct the output sequence $y(n)$, if desired. Ideally, the reconstruction will provide a continuous–time output $y_c(t)$ that at each instant of time (not only at $t = n\Delta t$) is equal to the output of the desired analog filter $h_c(t)$ having input $x_c(t)$. In accordance with (6.25) [written $\hat{Y}_c(j\Omega) = Y(e^{j\Omega \Delta t}) \cdot R_c(j\Omega)$] and substituting (6.34), we have

(3) Filter with reconstruction filter $r_c(t)$:

$$\hat{y}_c(t) = \sum_{n=-\infty}^{\infty} y(n) r_c(t - n\Delta t) \xrightarrow{\text{FT}} \hat{Y}_c(j\Omega) = Y(e^{j\Omega \Delta t}) \cdot R_c(j\Omega)$$

$$= X(e^{j\Omega \Delta t}) \cdot H(e^{j\Omega\Delta t}) \cdot R_c(j\Omega). \qquad (6.35)$$

Again assuming the Nyquist sampling condition (6.18) is satisfied, writing $y(n) = x(n) * h(n)$ explicitly as a sum, and recalling (6.33b), the overall effect of the three operations is, in the time and frequency domains,

$$\hat{y}_c(t) = \sum_{n=-\infty}^{\infty} \sum_{m=-\infty}^{\infty} x_c(m\Delta t) h(n - m) \cdot r_c(t - n\Delta t) \xrightarrow{\text{FT}}$$

$$\hat{Y}_c(j\Omega) = \frac{X_c(j\Omega) H(e^{j\Omega \Delta t}) R_c(j\Omega)}{\Delta t}, \quad |\Omega| < \frac{\pi}{\Delta t} = \frac{\Omega_s}{2}, \qquad (6.36)$$

where in (6.36) all functions except the digital filter are written as or in terms of continuous–time functions.

We may characterize the overall process in the frequency domain by writing (6.36) as

$$\boxed{\hat{Y}_c(j\Omega) = H_c(j\Omega) \cdot X_c(j\Omega), \qquad (6.37)}$$

where $H_c(j\Omega)$ is the overall (net) analog filter acting on $X_c(j\Omega)$. Next, note that by the convolution theorem, if $x_c(t)$ is Nyquist–bandlimited, $y_c(t)$ is also guaranteed to be, regardless of $h_c(t)$: $Y_c(j\Omega) = X_c(j\Omega) \cdot H_c(j\Omega)$. Assuming no aliasing, we may identify the overall analog filter in terms of the digital filter as

$$\boxed{\begin{aligned} H_c(j\Omega) &= \frac{H(e^{j\Omega \Delta t}) R_c(j\Omega)}{\Delta t}, \quad |\Omega| < \frac{\pi}{\Delta t} = \frac{\Omega_s}{2} \qquad (6.38a) \\ &= H(e^{j\Omega \Delta t}), \qquad\quad |\Omega| < \frac{\pi}{\Delta t} = \frac{\Omega_s}{2}, \qquad (6.38b) \end{aligned}}$$

where the latter expression holds only when $R_c(j\Omega)$ is an ideal lowpass filter with zero phase and gain Δt in the passband $|\Omega| < \Omega_s/2$.

In the practical case where $R_c(j\Omega)$ is an approximate lowpass filter, (6.38b) is only approximately true because aliasing terms are passed and $R_c(j\Omega)$ is not flat in the passband, whereas (6.38a) still holds for the given $R_c(j\Omega)$ when there is no aliasing. The design of good analog filters is a highly developed subject; several popular approximations are given in the review of analog lowpass filter design in Section 8.4.

For simplicity, let us for now assume that $R_c(j\Omega)$ is an ideal lowpass filter and that there is no aliasing. If it is desired to implement an analog filter $H_c(j\Omega)$ via sampling, digital filtering, and reconstruction, the required digital filter is, from (6.38b),

$$\boxed{H(e^{j\omega}) = H_c\left(\frac{j\omega}{\Delta t}\right), \quad |\omega| < \pi. \qquad (6.39)}$$

Under the assumptions above, we will have $\hat{y}_c(t) = y_c(t)$, where $y_c(t) = x_c(t) *_c h_c(t)$.

The design of $H(e^{j\omega})$ is a major part of the subject of Chapters 8 and 9. One simple approach, however, is to assume that $H_c(j\Omega)$ is determined and known. Then we use (6.39) and view it in the same spirit as (6.33b): Interpret $H(e^{j\omega})$ on $|\omega| < \pi$ as the lone replication on $|\omega| < \pi$ of $H_c(j\Omega)$ due to sampling $h_c(t)$, analogous to (6.33a) but without the $1/\Delta t$ out front. That is, not only is $x(n)$ a sampled version of $x_c(t)$, but we now view $h(n)$ as a sampled version of $h_c(t)$. Again noting the absence of $1/\Delta t$ in (6.39), this interpretation implies [by the left side of (6.33a)] that we need

$$h(n) = \Delta t \cdot h_c(n\Delta t). \tag{6.40}$$

This approach is known as the *impulse invariance* method of filter design because we require that $h(n)$ be tied exactly to samples of $h_c(n\Delta t)$, the impulse response of the continuous–time system. Actually, $h_c(t)$ need not necessarily exist as a physical waveform, but may be derived from an existing analog approximating function "on paper." In that case, $h(n)$ is mathematically determined, not measured. We will discuss impulse invariance much further in Section 8.5.1.

It is worthwhile to ask how the discrete–time convolution *before* reconstruction (conversion back to continuous time) is related to samples of the continuous–time filtered output, assuming (6.40) to be satisfied (i.e., impulse invariance). We define $\hat{y}(n) = x(n) * h(n)$, as in Figure 6.35a. Intuitively, we might guess that because both $x_c(t)$ and $h_c(t)$ have been sampled, the result of discrete–time convolution $\hat{y}(n)$ would be samples of $y_c(t)$, the continuous–time convolution of $x_c(t)$ and $h_c(t)$. That is, it would seem that in Figures 6.35a and b, the sampled–data and analog system outputs agree at the sampling instants when $h(n) = h_c(n\Delta t) \cdot \Delta t$: $\hat{y}(n) = y_c(n\Delta t)$, where again $y_c(t) = x_c(t) *_c h_c(t)$. We already argued after (6.39) that for no aliasing and ideal reconstruction $\hat{y}_c(t) = y_c(t)$, which implies that $\hat{y}(n) = y_c(n\Delta t)$. Let us further justify and qualify this result.

It is assumed that Δt is chosen small enough that the sampling criterion (6.18) is satisfied for both $x_c(t)$ and $h_c(t)$: If BW is the larger of the bandwidths of $x_c(t)$ and $h_c(t)$, then BW $\leq 2\pi/\Delta t$. In practice, this inequality can never be true when $H_c(s)$ is a finite–order rational function, because all such functions have infinite bandwidths. However, in practice, $H_c(s)$ can certainly be approximately bandlimited, and we will assume that to be true in the following. Now, $y_c(t)$ is also bandlimited to $\leq 2\pi/\Delta t$, because its Fourier transform $Y_c(j\Omega)$ is the product of the transforms of the bandlimited $x_c(t)$ and $h_c(t)$. Thus, $y_c(t)$ can in principle be perfectly reconstructed from periodic samples of it Δt apart. We may still be in doubt as to whether these samples are obtainable by merely carrying out discrete–time convolution of $x(n)$ with $h(n)$.

To prove the affirmative under the above ideal assumptions, all we need to do is recall the convolution theorem for discrete–time signals: $\hat{y}(n) = x(n) * h(n) \overset{\text{DTFT}}{\longleftrightarrow} \hat{Y}(e^{j\omega}) = X(e^{j\omega})H(e^{j\omega})$. If $x(n) = x_c(n\Delta t)$ is Nyquist–sampled, then by (6.33b) $X(e^{j\omega}) = X_c(j\omega/\Delta t)/\Delta t$, for $|\omega| < \pi$. Similarly, if $h(n) = \Delta t \cdot h_c(n\Delta t)$ is Nyquist–sampled, then as in (6.39), $H(e^{j\omega}) = H_c(j\omega/\Delta t)$ for $|\omega| < \pi$. Therefore,

$$\hat{Y}(e^{j\omega}) = X(e^{j\omega})H(e^{j\omega}) = \left\{ \frac{X_c\left(\dfrac{j\omega}{\Delta t}\right)}{\Delta t} \right\} \cdot H_c\left(\frac{j\omega}{\Delta t}\right). \tag{6.41}$$

By the convolution theorem, the right–hand side of (6.41) is $Y_c(j\Omega)/\Delta t \big|_{\Omega = \omega/\Delta t}$ so

$$\hat{Y}(e^{j\omega}) = \frac{Y_c\left(\dfrac{j\omega}{\Delta t}\right)}{\Delta t} \tag{6.42a}$$

$$= \text{DTFT}\{x(n) * h(n)\}, \tag{6.42b}$$

where (6.42b) follows by the definition of $\hat{y}(n)$. Given that $Y_c(j\Omega)$ is bandlimited to at most $\pi/\Delta t$, by (6.33b) we also know that for $|\Omega| < \pi/\Delta t$, $Y(e^{j\omega}) = Y_c(j\omega/\Delta t)/\Delta t$, where by definition $y(n) = \mathrm{DTFT}^{-1}\{Y(e^{j\omega})\} = y_c(n\Delta t)$. Consequently, by (6.42b) and the uniqueness of Fourier expansions, $\hat{Y}(e^{j\omega}) = Y(e^{j\omega})$ implies $\hat{y}(n) = y(n) = y_c(n\Delta t)$, and the matter is proved. See the problems for another interesting proof of this fact.

In summary, under impulse invariance, whether or not we choose to reconstruct, under the ideal assumption of no aliasing, the discrete–time convolution of the sampled sequences indeed provides samples of the corresponding continuous–time convolution. With $x(n) = x_c(n\Delta t)$ and $h(n) = \Delta t \cdot h_c(n\Delta t)$, we have

$$
\left. x(n) * h(n) \right|_{n=n_0} = \left. x_c(t) *_c h_c(t) \right|_{t=n_0\Delta t} \quad \begin{array}{l} \text{assuming no aliasing of either} \\ x_c(t) \text{ or } h_c(t) \text{ under sampling.} \end{array} \quad (6.43a)
$$

Equivalently, replacing n_0 by generic n,

$$
\hat{y}(n) = y_c(n\Delta t). \tag{6.43b}
$$

The relation (6.43) is a conclusive reason for multiplying $h_c(n\Delta t)$ by Δt in $h(n) = \Delta t \cdot h_c(n\Delta t)$. Again, (6.43) holds only approximately in practice, because $H_c(j\Omega)$ is never perfectly bandlimited [nor for that matter is $X_c(j\Omega)$].

EXAMPLE 6.3

To illustrate the basic impulse–invariant method, suppose that $H_c(s)$ is a sixth–order Butterworth lowpass filter with cutoff frequency 10 Hz. We will compare the output of the analog filter with the reconstructed output of the digital filter designed using impulse invariance. Let the input signal be $x_c(t) = \{e^{-2t}\cos(2\pi \cdot 3t - 30°) + 1.4e^{-0.5t}\cos(2\pi \cdot 20t)\}u_c(t)$, and $\Delta t = 0.01$ s.

Solution

In this example, we will not get into all the details of impulse–invariant design, which are covered fully in Section 8.5.1. Rather, we will use Matlab to do those details for us, and focus on the results as they relate to our discussion of performing analog filtering by sampling, digitally filtering, and reconstructing the result. With $\Delta t = 0.01$ s—that is, the sampling frequency is $F_s = 1/\Delta t = 100$ Hz—the maximum nonaliased frequency in any sampled signal (or impulse response) is $\frac{1}{2} \cdot F_s = 1/(2\Delta t) = 50$ Hz. Thus the main energy of both damped sinusoids is within the sampling bandwidth, but keep in mind that both damped causal sinusoids actually have infinite absolute bandwidth. The digital frequencies of the spectral peaks of the two damped sinusoids are, respectively, $f_1 = F_1/F_s = 3/100 = 0.03$ cycle/sample and $f_2 = 20/100 = 0.2$ cycle/sample. Moreover, the cutoff frequency of the digital filter obtained by impulse invariance should end up at about $f_0 = F_0/F_s = 10/100 = 0.1$ cycle/sample, which is in between f_1 and f_2.

In Figure 6.36, we see the analog Butterworth frequency response plotted against F (Hz). The cutoff at 10 Hz is clear. Figure 6.37 shows the digital Butterworth filter obtained by using the Matlab command `impinvar` that does the impulse–invariant transformation derived above (but with shortcuts, as discussed in Section 8.5.1). As predicted, the digital filter cutoff is at $f = 0.1$ cycle/sample. Comparing the two plots, there is strong agreement in both magnitude and phase, as indicated by (6.39). The frequency axis in Figure 6.36 was purposely continued beyond

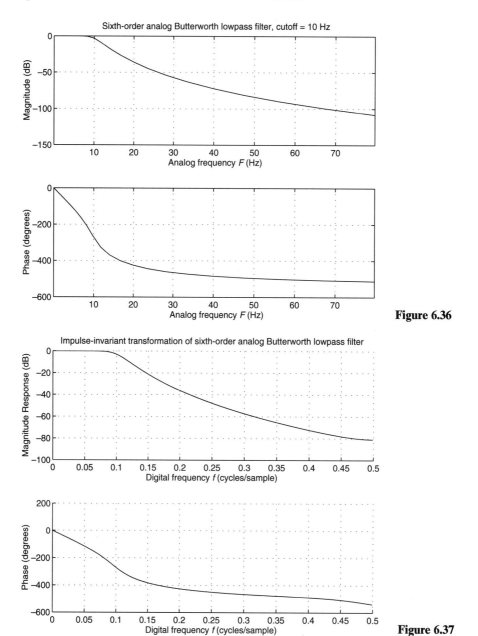

Figure 6.36

Figure 6.37

50 Hz (corresponding to digital frequency $f = \frac{1}{2}$), to emphasize that that portion is aliased into lower frequencies $(f < \frac{1}{2})$ in the digital frequency response in Figure 6.37.

In Figure 6.38 (main plots), we compare the various formats of input signal, for $0 < t < 0.1$ s. The solid curve is $x_c(t)$, the exact continuous–time signal defined in the problem statement. Its samples, $x(n) = x_c(n\Delta t)$ are shown as a stem plot. A truncated numerical sinc interpolation of the samples $x(n)$ is shown as "–.–.–.", whereas the zero–order–hold (boxcar) reconstruction is shown as the staircase signal. For comparison, the linear interpolation between the samples is shown as "– – – –". It comes as no surprise that all representations agree exactly at $t = n\Delta t$. The inset of Figure 6.38 shows a plot of only $x_c(t)$, over a larger time duration (1 s). In it, the overall high– and low–frequency damped sinusoids are recognizable.

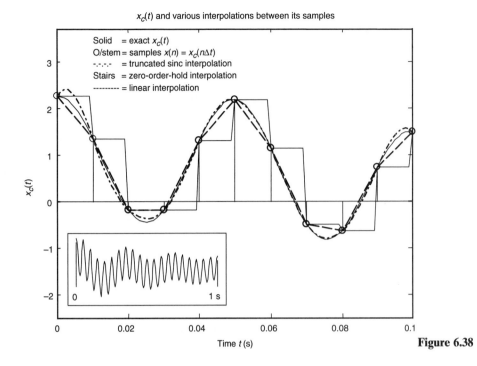

Figure 6.38

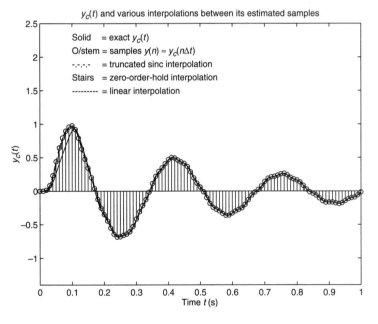

Figure 6.39

We expect the lowpass filter to filter out the higher–frequency sinusoid, because F_2 = 20 Hz > F_0 = 10 Hz is attenuated, whereas F_1 = 3 Hz < F_0 is passed. This is indeed the case in Figure 6.39, which shows all the various representations of the output $y_c(t)$. That is, aside from a delay for causality, $y_c(t)$ in Figure 6.39 is the lower–frequency sinusoidal component of $x_c(t)$ in the inset of Figure 6.38; the higher– frequency component is gone. Moreover, in Figure 6.39 there is modest disagreement between the analog–filtered and digitally–filtered/reconstructed signal, only during the first half–period. All interpolations and the stems in Figure 6.39 are made from $y(n)$, the output of the digital filter; only the solid curve is exact analog-filtered $y_c(t)$.

It may be asked how the exact solution $y_c(t)$ was obtained. Recall that $y_c(t)$ does not include zero–order–hold effects, but only the pure analog filtering. Let the analog Butterworth filter be $H_c(s)$. Then $Y_c(s) = H_c(s)X_c(s)$. Because $x_c(t)$ is of the form $x_c(t) = [M_1 e^{-\alpha_1 t}\cos(\Omega_1 t + \theta_1) + M_2 e^{-\alpha_2 t}\cos(\Omega_2 t + \theta_2)]u_c(t)$, we know its Laplace transform. Recalling that $\cos(\Omega t + \theta) = \cos(\Omega t)\cos(\theta) - \sin(\Omega t)\sin(\theta)$, and knowing the LT of $e^{-\alpha t}\cos(\Omega t)u_c(t)$ and $e^{-\alpha t}\sin(\Omega t)u_c(t)$, we obtain

$$X_c(s) = M_1 \frac{\cos(\theta_1)s + \cos(\theta_1)\alpha_1 - \sin(\theta_1)\Omega_1}{s^2 + 2\alpha_1 s + \alpha_1^2 + \Omega_1^2} + M_2 \frac{\cos(\theta_2)s + \cos(\theta_2)\alpha_2 - \sin(\theta_2)\Omega_2}{s^2 + 2\alpha_2 s + \alpha_2^2 + \Omega_2^2}.$$

In Matlab, the lowpass Butterworth filter transfer function $H_c(s)$ is easily available, so multiplication of the two rational functions gives $Y_c(s)$. Finally, to obtain the inverse Laplace transform, we merely invoke the `impulse` command on $Y_c(s)$. This series of operations was done to obtain the "exact" $y_c(t)$ plot. Use of the routine `filter` gave $y(n)$ computed by digitally filtering $x(n)$ with the digital filter transfer function $H(z)$ obtained by impulse invariance.

Why is this not the end of the filter design problem? After all, now samples of continuous–time convolution may be obtained using discrete convolution, under the proper assumptions. In turn (see Section 7.6), this convolution calculation may in general be performed with an FFT if the effective lengths of $x(n)$ and $h(n)$ are finite and reasonable. In fact, the digital filter is already "designed"—it is just samples of the analog filter $h_c(t)$. Of course, this begs the question of how $h_c(t)$ is designed, a topic reviewed in Section 8.4.

Also, there are constraints on the use of impulse invariance in practice. For example, we will show in Section 8.5.1 that impulse invariance can be applied to only those rational system functions whose numerator order is less than the denominator order. This requirement guarantees that $h_c(t)$ is at least approximately bandlimited, so that approximate versions of (6.33a) and (6.33b) hold, with all symbols "X" replaced by "H." Consequently, only lowpass and bandpass filters may be designed by this procedure; highpass and bandstop filters will be sure to alias so that (6.33a) and (6.33b) are not equivalent.

Impulse invariance is limited to linear shift–invariant systems with a single input and a single output. Also, in practice, $h_c(t)$ may sometimes have very long duration, slowing down what might otherwise be "real–time" computation. This latter limitation exists only for the cases in which an IIR impulse–invariant filter cannot be calculated in rational function form, which involves only a finite number of coefficients and thus can be efficiently implemented—for example, the case in which $h_c(t)$ is measured experimentally. Alternatively, the impulse response may somehow not be easily available or conveniently transformable. Another approach based on the bilinear transform avoids some of the problems of impulse invariance (see Section 8.5.2).

Finally, design of digital filters using approximations of existing analog filter designs is not the only design technique—for example, FIR filter design is very popular (see Chapter 9).

6.4.3   Simulation, Integration, and the Runge–Kutta Method

The above limitations on the applicability of the impulse–invariant method lead us in such cases to seek other approaches. Depending on the form of the model and the information available about it, myriad solution techniques exist sufficient to fill many volumes.[11] In this section and Section 6.4.4, we review a few of the popular alternative

---

[11]A good reference for this subsection is J. S. Rosko, *Digital Simulation of Physical Systems* (Reading, MA: Addison-Wesley, 1972).

approaches. Then an example using each of the methods on the same analog system is presented in Section 6.4.5. The purpose is to both inform and inspire, not to be in any sense comprehensive. Nor is a rigorous comparison made between the methods. For example, for simplicity we will not detail adaptive step–size techniques.

Although the approaches that follow naturally also work for analog filter discretization, we begin by reviewing the general steps in carrying out a digital simulation of a continuous–time system. The first step is isolation of the physical system to be modeled and specification of the inputs and outputs. Next, a model of that continuous–time system is postulated, which may involve differential, integral, algebraic, and nonlinear operations. The numerical constant parameters of the model tie the abstract operational model to the physical system. These parameters may be determined by either theoretical considerations or by measurements made on the physical system. Examples 2.1, 2.2, and 4.11 illustrate this process.

Once the model is completely specified, a discrete equivalent must be derived or constructed from the continuous model. This step involves approximation—for example, of integrations by summations. Then an algorithm (often recursive) for obtaining estimates of the output due to specified inputs is extracted from the discrete model. This algorithm is merely an arrangement of calculations that efficiently implements the discrete model and is translatable into programming code. The code is executed on a microprocessor, DSP chip, PC, or other digital computer. Finally, if the resulting sequence is to be used as a continuous–time controlling input, ZOH reconstruction via D/A converters and power amplifiers will be used.

As a means of ascertaining the accuracy and reliability of the discrete model and computational system, identical test inputs may be given to the physical continuous system and its discrete approximation. Based on the error between the output of the physical continuous system and the discretized model of the continuous system, modifications are made on the model parameters and/or the form of the continuous–time model itself, the sampling interval, discretization order, or discretization algorithm to maximize agreement between the discrete system output and the original physical continuous system output.

Suppose that the continuous system is modeled by a linear shift–invariant system, expressible as an ordinary differential equation with independent variable time t, input $x_c(t)$ and output $y_c(t)$, as in (2.20) $\left[\sum_{\ell=0}^{N} a_\ell \frac{d^\ell y_c(t)}{dt^\ell} = \sum_{\ell=0}^{M} b_\ell \frac{d^\ell x_c(t)}{dt^\ell} \right]$, with $a_N = 1$ and $N \geq M$.

With nonzero initial conditions, the system is also representable in the Laplace domain. The Laplace transform of (2.20) is

$$Y_c(s) = H_c(s) \cdot X_c(s) + Y_{\mathrm{IC},c}(s), \qquad (6.44a)$$

where

$$H_c(s) = \frac{B_c(s)}{A_c(s)} = \frac{\displaystyle\sum_{m=0}^{M} b_m s^m}{\displaystyle\sum_{m=0}^{N} a_m s^m} \quad \text{and} \quad Y_{\mathrm{IC},c}(s) = \frac{\mathrm{IC}_x(s) - \mathrm{IC}_y(s)}{A_c(s)},$$

in which

$$\mathrm{IC}_x(s) = -\sum_{\ell=0}^{M} b_\ell \sum_{k=1}^{\ell} s^{k-1} x_c^{[\ell-k]}(0) \qquad (6.44b)$$

and

$$\mathrm{IC}_y(s) = -\sum_{\ell=0}^{N} a_\ell \sum_{k=1}^{\ell} s^{k-1} y_c^{[\ell-k]}(0), \qquad (6.44c)$$

where as previously $[\ell - k]$ refers to the $(\ell - k)$th derivative.

By the convolution theorem, we also have the third representation, time–domain convolution:

$$y_c(t) = \int_{-\infty}^{\infty} h_c(t-\tau)x_c(\tau)d\tau + y_{\text{ic},c}(t), \tag{6.45}$$

where $h_c(t) = \text{LT}^{-1}\{H_c(s)\}$ and $y_{\text{ic},c}(t) = \text{LT}^{-1}\{Y_{\text{IC},c}(s)\}$.

A fourth representation of the system is the state variable form, similar to the discrete–time state variable form derived in Section 2.4.6. The inclusion of initial conditions is inherently convenient in state–space. Derived in much the same manner as for the discrete–time case, the results are

$$\frac{\partial}{\partial t}\mathbf{w}_c(t) = \mathbf{A}\mathbf{w}_c(t) + \mathbf{b}x_c(t), \quad y_c(t) = \mathbf{c}\mathbf{w}_c(t), \tag{6.46a}$$

where for phase–variable canonical form,[12]

$$\mathbf{A} = \begin{bmatrix} 0 & 1 & 0 & \dots & 0 \\ 0 & 0 & 1 & \dots & 0 \\ \cdot & \cdot & \cdot & & \cdot \\ \cdot & \cdot & \cdot & & \cdot \\ \cdot & \cdot & \cdot & & \cdot \\ 0 & 0 & 0 & \dots & 1 \\ -a_0 & -a_1 & -a_2 & \dots & -a_{N-1} \end{bmatrix}, \quad \mathbf{b} = \begin{bmatrix} 0 \\ 0 \\ \cdot \\ \cdot \\ \cdot \\ 0 \\ 1 \end{bmatrix}, \quad \mathbf{c} = [b_0 \quad b_1 \, . . b_M \quad 0 \dots 0]. \tag{6.46b}$$

We may solve (6.46a) for $\mathbf{w}_c(t)$ [and thus $y_c(t)$] by viewing it as a set of coupled first–order ordinary differential equations, or we may write the formal convolutional solution

$$\mathbf{w}_c(t) = e^{\mathbf{A}t} \cdot \mathbf{w}_c(0) + \int_0^t e^{\mathbf{A}(t-\tau)}\mathbf{b}x_c(\tau)d\tau \tag{6.47}$$

and then use numerical methods to evaluate the integral. (There are many ways to evaluate $e^{\mathbf{A}t}$; see the problems.) This latter approach is used in Matlab for simulation (e.g., `lsim`).

Thus there are at least four possible representations of analog systems. All are mathematically equivalent, and each has analogous continuous–time and discrete–time versions: (a) differential/difference equation, (b) Laplace/z–transform system function representation, (c) continuous–time/discrete–time convolution, and (d) continuous–time/discrete–time state equation. In this section, our simulation technique will be based on the state–space representation.

If we begin with the state equation [e.g., (6.46a)], we see that only a first derivative appears. To integrate and find the solution vector $\mathbf{w}_c(t)$ and hence $y_c(t)$, the building block for all work is the numerical solution of

$$\frac{\partial}{\partial t}w_{i,c}(t) = g_{i,c}(w_{1,c}(t), w_{2,c}(t), \dots, w_{N,c}(t), x_c(t)) \tag{6.48a}$$

$$= g_{i,c}\big|_t, \tag{6.48b}$$

where $w_{i,c}(t)$ is the ith element of the state vector $\mathbf{w}_c(t)$ and where (6.48b) is an abbreviation of (6.48a) that will be convenient to use in the following. The function $g_{i,c}$ may in general be nonlinear, and in the linear case is obtainable from the matrix representation in (6.46). The problem of simulation of continuous–time systems has been reduced to solution of a system of simultaneous first–order ordinary differential equations. With a solution of (6.48) for each $w_{i,c}(t)$, the entire solution can be constructed. It merely remains to specify the method of integration, the initial conditions, and the discretization interval Δt.

---

[12] A. D. Poularikas and S. Seely, *Signals and Systems,* 2nd ed. (Malabar, FL: Krieger, 1994) p. 856.

Some basic integration formulas are presented on WP 6.1: Euler, predictor–corrector, and a fully detailed derivation of nonadaptive Runge–Kutta. Here we will merely cite the Runge–Kutta formula resulting from that derivation, which we will use in Example 6.4:

Complete derivation details on WP 6.1.

$$
\text{Second–order Runge–Kutta estimate.} \qquad (6.49)
$$

$$
w_{i,c}((n+1)\Delta t) = w_{i,c}(n\Delta t) + \frac{1}{2}\Delta t \cdot \Bigg[g_{i,c}\Big|_{n\Delta t} + g_{i,c}\Big\{ w_{1,c}(n\Delta t) + g_{1,c}\Big|_{n\Delta t} \cdot \Delta t,
$$

$$
w_{2,c}(n\Delta t) + g_{2,c}\Big|_{n\Delta t} \cdot \Delta t, \dots, w_{N,c}(n\Delta t) + g_{N,c}\Big|_{n\Delta t} \cdot \Delta t,
$$

$$
x_c(n\Delta t) + \frac{\partial x_c}{\partial t}\Big|_{n\Delta t} \cdot \Delta t\Big\}\Bigg].
$$

We can obtain Euler, trapezoidal, and Simpson's rule integration approximations as special cases of Runge–Kutta formulas.

The final equations of both the predictor–corrector (see the problems) and Runge–Kutta methods are easy to program, because they are simple difference equations. They and a host of other numerical integration methods may be used for linear or nonlinear systems given in state-equation format.

6.4.4   Operational Approach

Another effective approach to analog system discretization is the operational method. In the basic method, it is assumed that $H_c(s)$ is a proper rational function of s.[13] If the numerator and denominator of $H_c(s)$ in $H_c(s) = B_c(s)/A_c(s)$ are divided by s^N, then $H_c(s)$ is expressed as a rational function of $1/s$. Because $1/s$ is the Laplace transform of the integration operator, a discrete–time approximation of the overall continuous–time operator $H_c(s)$ may be found by individually replacing each $1/s^m$ factor in $H_c(s)$ by a discrete–time approximating operator $F(z)$ for the mth–order integration.

As one example, consider the Bilinear transform for $m = 1$:

$$
F(z) = \frac{\Delta t}{2} \cdot \frac{z+1}{z-1}. \qquad \text{Bilinear transform approximator for } \frac{1}{s}. \qquad (6.50)
$$

The bilinear transform can be shown to be the transform representation of trapezoidal rule integration, as follows. If

$$
y_c(t) = \int_0^t x_c(\tau)\,d\tau, \qquad (6.51)
$$

then the trapezoidal rule estimates $\hat{y}(n)$ for $y(n) = y_c(n\Delta t)$ and $y(n+1)$ in terms of $x(n) = x_c(n\Delta t)$ are

$$
\hat{y}(n) = \left(\frac{\Delta t}{2}\right) \cdot \{x(0) + 2x(1) + \cdots + 2x(n-1) + x(n)\} \qquad (6.52a)
$$

$$
\hat{y}(n+1) = \left(\frac{\Delta t}{2}\right) \cdot \{x(0) + 2x(1) + \cdots + 2x(n-1) + 2x(n) + x(n+1)\} \qquad (6.52b)
$$

$$
= \hat{y}(n) + \left(\frac{\Delta t}{2}\right) \cdot \{x(n) + x(n+1)\}. \qquad (6.52c)
$$

---

[13]We assume a linear shift-invariant system; however, the general operational approach described here may be extended to both linear shift-variant systems and nonlinear systems.

If we were to plot $x_c(t)$, the area of the trapezoid formed by the t axis, $x_c(n\,\Delta t)$, $x_c((n+1)\,\Delta t)$ and the line joining the latter two is $(\Delta t/2) \cdot \{x(n) + x(n+1)\}$. Thus indeed (6.52c) is properly called trapezoidal–rule integration.

Taking the transform of the equality in (6.52c) gives

$$\frac{\hat{Y}(z)}{X(z)} = \frac{\Delta t}{2} \frac{z+1}{z-1} = F(z). \tag{6.53}$$

If we let $1/s$, which is pure integration in the s–domain, be replaced by the trapezoidal rule integration operator $F(z) = (\Delta t/2)(z+1)/(z-1)$, we can operationally discretize integration under the trapezoidal rule using only sample values to estimate the integral. Thus, if we substitute $[F(z)]^m$ into $H_c(s)$ everywhere $1/s^m$ appears, then within the validity of the trapezoidal rule, the resulting function of z, $H(z)$, is a discrete–time transfer function operationally approximately equivalent to the continuous–time transfer function $H_c(s)$. Moreover, the bilinear transform maps stable $H_c(s)$ into stable $H(z)$ (see Section 8.5.2), a fact not true of all discrete integration approximations.

From the overall z–transform $H(z)$ obtained by substituting $F(z)$ in for $1/s$ everywhere in $H_c(s)$, we may obtain a difference equation relating the current output to past output values and past and present input values. Although $F(z)$ is known as the bilinear transformation, as the z–transform of the trapezoidal rule it is also called the first–order Tustin integration.

Just as we noted a variety of techniques for performing the numerical integration of the state equations, there are many options other than the trapezoidal rule for estimation of $1/s$. For example, there exist better approximations of $1/s^m$ than operating with $F(z)$ m times—there are specialized approximations of $1/s^m$ for various values of m. If, however, the same first–order approximation (whatever it is— Tustin, first difference, etc.) *is* always repeated m times to approximate $1/s^m$, then that approximator constitutes a simple mapping from the s–plane to the z–plane. Note that in this case one always has the option of decomposing $H_c(s)$ into the product $H_{1,c}(s) \cdot H_{2,c}(s)$. By so doing, each of $H_{1,c}(s)$ and $H_{2,c}(s)$ may be low–order systems so that m in $1/s^m$ that must be approximated [e.g., by $[F(z)]^m$ in (6.50)] remains small. The result of this factorization of $H_c(s)$ may be superior numerical performance.

The selection of Δt is another parameter that in all digital simulation methods must be chosen by the user. Other than by initially following an approximate Nyquist criterion, its choice is largely empirical. Usually the tradeoff is between high accuracy (for small Δt) and high computation speed (for large Δt). Additionally, one must be mindful that too small a Δt may cause rounding problems or ill–conditioning in any algorithms involving matrix inversion. For small Δt the user must be more of an expert on quantization effects, whereas for large Δt the user must be more of an expert on such things as aliasing and bilinear transform distortions (see Chapter 8).

Because large Δt is associated with cheaper hardware, generally we try to use the largest Δt that will not cause aliasing or other distortions resulting from large Δt. Smaller Δt requires a faster processor and often longer wordlength due to quantization errors. Note that if poles are mapped according to $e^{s\Delta t}$, then for small Δt all the poles are mapped to $z \approx 1$, which again is undesirable as it nullifies the filter. Often a good compromise is to sample at 10 to 30 times the half–power (−3–dB) bandwidth.

6.4.5   Illustrative Simulation Example

The following numerical example illustrates the procedures given in Sections 6.4.2 through 6.4.4, with each technique applied to the same continuous–time system.

EXAMPLE 6.4

To illustrate the above methods of numerically solving initial value problems—that is, simulating continuous–time systems using sampled data sequences—we shall now solve a continuous–time problem using several discrete–time methods. First the exact solution will be presented, then each of the approximate formulas, and finally a graphical presentation of the results of the estimates versus the exact solution, for a few representative values of Δt.

Consider the following second–order system:

$$\frac{d^2}{dt^2}y_c(t) + 2 \cdot \frac{d}{dt}y_c(t) + y_c(t) = x_c(t), \quad \text{where } x_c(t) = [e^{-t/4} + V \cdot \sin(\Omega t)]u_c(t) \quad (6.54)$$

with initial conditions

$$y_c(0) = y_0, \quad \frac{d}{dt}y_c(t)\Big|_{t=0} = y_0' \quad (6.55)$$

and where V and Ω are fixed, chosen parameters.

Solution

This initial value problem may be solved exactly by the method of undetermined coefficients. The result is then

$$y_c(t) = (A + Bt) \cdot e^{-t} + \frac{16}{9} \cdot e^{-t/4} + \frac{V}{(1 + \Omega^2)^2} \cdot \{(1 - \Omega^2) \cdot \sin(\Omega t) - 2\Omega \cdot \cos(\Omega t)\}, \quad (6.56)$$

where A and B are selected to satisfy the initial conditions, as follows. First, equating $y_c(0)$ with y_0, we have $A + \frac{16}{9} - 2\Omega V/(1 + \Omega^2)^2 = y_0$, giving

$$A = \frac{2\Omega V}{(1 + \Omega^2)^2} - \frac{16}{9} + y_0. \quad (6.57)$$

To satisfy the derivative initial condition, we first note that

$$\frac{d}{dt}y_c(t) = \{(1 - t)B - A\} \cdot e^{-t} - \frac{4}{9} \cdot e^{-t/4} +$$

$$\frac{\Omega V}{(1 + \Omega^2)^2} \cdot \{(1 - \Omega^2) \cdot \cos(\Omega t) + 2\Omega \cdot \sin(\Omega t)\}. \quad (6.58)$$

Thus at $t = 0$, we have $B - A - \frac{4}{9} + \Omega V(1 - \Omega^2)/(1 + \Omega^2)^2 = y_0'$, giving

$$B = \frac{\Omega V}{1 + \Omega^2} - \frac{4}{3} + y_0 + y_0'. \quad (6.59)$$

Having the exact solution available for this example facilitates evaluation of the numerical solution.

Because the zero–input (initial conditions) solution is the same for any general input $x_c(t)$, we will here compute it in closed form. Then in deriving the transfer function–based estimates, we will assume for convenience that the initial values are zero; we merely need add the zero–input response to our zero–state response. Noting that we have $IC_x(s) = 0$ because no derivatives of $x_c(t)$ appear in (6.54), and also in this case $IC_y(s) = -2y_0 - 1 \cdot (y_0' + y_0 \cdot s)$, we can obtain $y_{ic,c}(t)$ directly:

$$y_{ic,c}(t) = LT^{-1}\{Y_{IC,c}(s)\} \quad (6.60a)$$

$$= LT^{-1}\left\{\frac{y_0 s + 2y_0 + y_0'}{s^2 + 2s + 1}\right\}. \quad (6.60b)$$

By basic inverse Laplace transform techniques, we have

$$y_{ic,c}(t) = (2y_0 + y_0')te^{-t} + y_0 \cdot \frac{d}{dt}\{te^{-t}\} \tag{6.60c}$$

$$= (2y_0 + y_0')te^{-t} + y_0 \cdot (1 - t)e^{-t} \tag{6.60d}$$

$$= [y_0 + (y_0 + y_0')t]e^{-t}, \quad t \geq 0. \tag{6.60e}$$

Thus, onto a zero–state response output sequence generated by *any* of the approximate methods, we may add

$$y_{ic}(n) = y_{ic,c}(n\,\Delta t) = [y_0 + (y_0 + y_0')n\,\Delta t]e^{-n\,\Delta t}, \quad n \geq 0. \tag{6.61}$$

First, the impulse–invariant solution is presented. Noting that $H_c(s) = 1/(s + 1)^2$, we have $h_c(t) = te^{-t}u_c(t)$. According to the impulse–invariant method, $h(n) = \Delta t \cdot h_c(t) = \Delta t(n\,\Delta t)e^{-n\,\Delta t}u(n)$. The z–transform of $h(n)$ is

$$H(z) = \Delta t^2\left[-z\frac{d}{dz}\{z/(z - e^{-\Delta t})\}\right] \tag{6.62}$$

$$= \Delta t^2 \cdot \frac{ze^{-\Delta t}}{(z - e^{-\Delta t})^2}$$

$$= \frac{\Delta t^2 z^{-1}e^{-\Delta t}}{1 - 2z^{-1}e^{-\Delta t} + e^{-2\Delta t}z^{-2}}$$

$$= \frac{Y(z)}{X(z)}.$$

The above leads directly, upon cross–multiplication and inverse z–transformation, to the recursion relation

$$y(n) = 2e^{-\Delta t}y(n - 1) - e^{-2\Delta t}y(n - 2) + \Delta t^2 e^{-\Delta t}x(n - 1). \tag{6.63}$$

As noted above, to this solution we can add $y_{ic}(n)$, in the case of nonzero initial conditions.

We may alternatively use Runge–Kutta method described in Section 6.4.3 if we first convert the differential equation (6.54) to state–variable form. By direct application of (6.46a) $\left[\frac{\partial}{\partial t}\mathbf{w}_c(t) = \mathbf{A}\mathbf{w}_c(t) + \mathbf{b}x_c(t), \quad y_c(t) = \mathbf{c}\mathbf{w}_c(t)\right]$ in which

$\mathbf{A} = \begin{bmatrix} 0 & 1 \\ -a_0 & -a_1 \end{bmatrix}$, $\mathbf{b} = \begin{bmatrix} 0 \\ 1 \end{bmatrix}$, and $\mathbf{c} = [b_0 \ b_1]$, the phase–variables state variable

form of (6.54) is

$$\begin{bmatrix} dw_{1,c}(t)/dt \\ dw_{2,c}(t)/dt \end{bmatrix} = \begin{bmatrix} 0 & 1 \\ -1 & -2 \end{bmatrix}\begin{bmatrix} w_{1,c}(t) \\ w_{2,c}(t) \end{bmatrix} + \begin{bmatrix} 0 \\ 1 \end{bmatrix}x_c(t) \tag{6.64a}$$

$$y_c(t) = [1 \ 0] \cdot \begin{bmatrix} w_{1,c}(t) \\ w_{2,c}(t) \end{bmatrix}, \tag{6.64b}$$

and thus $w_{1,c}(t) = y_c(t)$ and $dw_{1,c}(t)/dt = y_c'(t) = w_{2,c}(t)$. From these we may write

$$g_{1,c}\big|_t = w_{2,c}(t) \tag{6.65a}$$

$$g_{2,c}\big|_t = -w_{1,c}(t) - 2 \cdot w_{2,c}(t) + x_c(t). \tag{6.65b}$$

Each equation of the integration method is written for $w_{1,c}(t)$ and $w_{2,c}(t)$. For each equation in the iteration, the forms for $w_{1,c}(t)$ and $w_{2,c}(t)$ are both computed before proceeding to the next equation.

Application of the second–order Runge–Kutta to solution of the (first–order) state equations can be made as follows. This method is "self–starting," that is, it needs only the initial values $w_{1,c}(0) = y_0$ and $w_{2,c}(0) = y_0'$, and not Taylor series extrapolations of $w_{i,c}(t)$ out to Δt (as are required in the predictor–corrector approach). The initialization for $n = 0$ is

$$w_{i,c}(n\,\Delta t)\big|_{n=0} = w_{i,c}(0) \tag{6.66}$$

for $i = 1, 2$. Now the recursion (6.49) may be used along with (6.65), where the subscript rk stands for Runge–Kutta and all $w_{i,c}$ variables are not exact values but rather the current best estimate of them:

$$w_{1,rk,c}((n + 1)\Delta t) = w_{1,c}(n\,\Delta t) + \frac{1}{2}\Delta t[g_{1,c}\big|_{n\,\Delta t} + \tag{6.67}$$

$$g_{1,c}\{w_{1,c}(n\,\Delta t) + g_{1,c}\big|_{n\,\Delta t} \cdot \Delta t, \quad w_{2,c}(n\,\Delta t) + g_{2,c}\big|_{n\,\Delta t} \cdot \Delta t, \quad x_c(n\,\Delta t) + \partial x_c/\partial t\big|_{n\,\Delta t} \cdot \Delta t\}]$$

$$= w_{1,c}(n\,\Delta t) + \frac{1}{2}\Delta t[w_{2,c}(n\,\Delta t) + w_{2,c}(n\,\Delta t) + \{-w_{1,c}(n\,\Delta t) - 2w_{2,c}(n\,\Delta t) + x_c(n\,\Delta t)\} \cdot \Delta t]$$

$$= w_{1,c}(n\,\Delta t) + \Delta t \cdot w_{2,c}(n\,\Delta t) + \frac{1}{2}\Delta t^2 \cdot \{-w_{1,c}(n\,\Delta t) - 2w_{2,c}(n\,\Delta t) + x_c(n\,\Delta t)\}$$

and[14]

$$w_{2,rk,c}((n + 1)\Delta t) = w_{2,c}(n\,\Delta t) + \frac{1}{2}\Delta t[g_{2,c}\big|_{n\,\Delta t} + \tag{6.68}$$

$$g_{2,c}\{w_{1,c}(n\,\Delta t) + g_{1,c}\big|_{n\,\Delta t} \cdot \Delta t, \quad w_{2,c}(n\,\Delta t) + g_{2,c}\big|_{n\,\Delta t} \cdot \Delta t, \quad x_c(n\,\Delta t) + \partial x_c/\partial t\big|_{n\,\Delta t} \cdot \Delta t\}]$$

$$= w_{2,c}(n\,\Delta t) + \frac{1}{2}\Delta t\Big[-w_{1,c}(n\,\Delta t) - 2w_{2,c}(n\,\Delta t) + x_c(n\,\Delta t) -$$
$$\{w_{1,c}(n\,\Delta t) + w_{2,c}(n\,\Delta t) \cdot \Delta t\} - 2\{w_{2,c}(n\,\Delta t) + [-w_{1,c}(n\,\Delta t) -$$
$$2w_{2,c}(n\,\Delta t) + x_c(n\,\Delta t)] \cdot \Delta t\} + x_c(n\,\Delta t) + \partial x_c/\partial t\big|_{t=n\,\Delta t} \cdot \Delta t\Big]$$

$$= w_{2,c}(n\,\Delta t) + \Delta t \cdot [-w_{1,c}(n\,\Delta t) - 2w_{2,c}(n\,\Delta t) + x_c(n\,\Delta t)] +$$
$$\frac{1}{2}\Delta t^2 \cdot [-w_{2,c}(n\,\Delta t) - 2\{-w_{1,c}(n\,\Delta t) - 2w_{2,c}(n\,\Delta t) + x_c(n\,\Delta t)\} -$$
$$\frac{1}{4} \cdot e^{-n\,\Delta t/4} + \Omega V\cos(\Omega n\,\Delta t)].$$

The iteration proceeds to the new value $n_{\text{new}} = n_{\text{old}} + 1$ after setting

$$w_{i,\text{new},c}(n_{\text{new}}\,\Delta t) = w_{i,rk,c}((n_{\text{old}} + 1)\,\Delta t), \quad i = 1, 2. \tag{6.69}$$

Finally, if we express the differential equation (6.54) in terms of the system function $H_c(s) = B_c(s)/A_c(s)$, we find that

$$H_c(s) = \frac{1}{1 + 2 \cdot s + s^2} = \frac{1/s^2}{1/s^2 + 2/s + 1}. \tag{6.70}$$

---

[14]When only numerical values of $x_c(t)$ are available, a numerical estimate of $\partial x_c/\partial t$ must be computed.

Substituting the Tustin integrator $(\Delta t/2) \cdot (z + 1)/(z - 1)$ for $1/s$ and $[F(z)]^2 = \{(\Delta t/2) \cdot (z + 1)/(z - 1)\}^2$ for $1/s^2$, the discrete–time transfer function $H(z)$ is obtained after algebra as

$$H(z) = \Delta t^2 \cdot \frac{1 + 2 \cdot z^{-1} + z^{-2}}{\Delta t^2 + 4 \cdot \Delta t + 4 + (2 \cdot \Delta t^2 - 8) \cdot z^{-1} + (\Delta t^2 - 4 \cdot \Delta t + 4) \cdot z^{-2}} \quad (6.71)$$

$$= \frac{Y(z)}{X(z)}$$

with the initial conditions assumed to be zero. Thus the difference equation approximating the differential equation is

$$y_c((n + 1)\Delta t) \approx \left(\frac{1}{\Delta t^2 + 4 \cdot \Delta t + 4}\right) \cdot \{(8 - 2 \cdot \Delta t^2) \cdot y_c(n\Delta t) +$$

$$(-\Delta t^2 + 4 \cdot \Delta t - 4) \cdot y_c((n - 1)\Delta t) + \quad (6.72)$$

$$\Delta t^2 \cdot [x_c((n + 1)\Delta t) + 2 \cdot x_c(n\Delta t) + x_c((n - 1)\Delta t)]\}.$$

To start this algorithm, assume zero for output and input for $t < 0$. As with the other algorithms, we add on the zero–input response afterwards (valid for linear systems). Incidentally, numerical results identical to those using (6.72) can be obtained using c2dm in Matlab with the tustin option, followed by dlsim to simulate the resulting discrete–time system for the specified input.

Alternatively, we may substitute the Boxer–Thaler integral approximators:[15] $(\Delta t/2) \cdot (z + 1)/(z - 1)$ for $1/s$ and $(\Delta t^2/12) \cdot (z^2 + 10z + 1)/(z - 1)^2$ for $1/s^2$. In this case, the discrete–time transfer function is

$$(6.73)$$

$$H(z) = \Delta t^2 \cdot \frac{1 + 10 \cdot z^{-1} + z^{-2}}{\Delta t^2 + 12 \cdot \Delta t + 12 + (10 \cdot \Delta t^2 - 24) \cdot z^{-1} + (\Delta t^2 - 12 \cdot \Delta t + 12) \cdot z^{-2}}$$

so that the Boxer–Thaler estimate is

$$y_c((n + 1)\Delta t) \approx \frac{(24 - 10 \cdot \Delta t^2)y_c(n\Delta t) + (-\Delta t^2 + 12\Delta t - 12)y_c((n - 1)\Delta t)}{\Delta t^2 + 12 \cdot \Delta t + 12}$$
$$\frac{+ \Delta t^2[x_c((n + 1)\Delta t) + 10x_c(n\Delta t) + x_c((n - 1)\Delta t)]}{\Delta t^2 + 12 \cdot \Delta t + 12}. \quad (6.74)$$

Now that all the approximate methods have been outlined, a few sample numerical results will be shown for representative values of Δt. As a balance between the decaying exponential in the input, $e^{-t/4}$, and the sinusoid $\sin(\Omega t)$, the following parameter values were chosen: $V = 20, \Omega = 2\pi \cdot (1 \text{ Hz})$. The initial values are (arbitrarily) selected as follows: $y_0 = 2, y'_0 = -5$. Three values of Δt were selected: 0.1 s in Figure 6.40a, 0.2 s in Figure 6.40b, and 0.3 s in Figures 6.40c and d. Overall, all methods have fairly comparable results for each value of Δt, so for brevity only one method is shown for each value of Δt (except two are shown for $\Delta t = 0.3$ s). Impulse invariance is shown in Figure 6.40a, Runge–Kutta in Figure 6.40b, Tustin $H(z)$ in Figure 6.40c, and Boxer–Thaler $H(z)$ in Figure 6.40d. In all cases, the estimates are shown out to $t = 10$ s (thus, for larger Δt, there are fewer total estimated samples plotted). The exact solution is always shown as the continuous solid curve, whereas the estimated values are shown as "*" connected by straight dashed lines for clarity.

The overall results are predictable: As Δt increases, the accuracy decreases for all methods. For any Δt less than or equal to 0.1 s, we find very high accuracy for all

---

[15]In the past, these approximations for $1/s^\ell$ were called "z-forms"; in fact, it was Boxer and Thaler who proposed the general method of replacing integrators by discrete approximators.

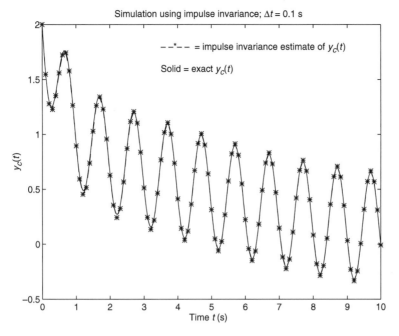

Figure 6.40*a*

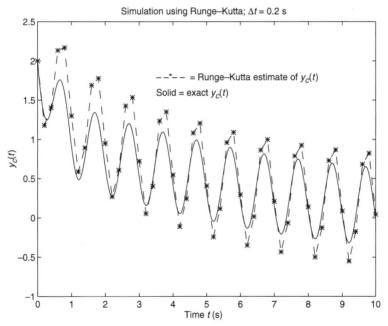

Figure 6.40*b*

methods. With $\Delta t = 0.2$ s, there is quite significant error, although for some methods still a reasonable following of the exact solution. For Δt greater than or equal to 0.3 s, we find high error and ultimately no agreement at all between estimates and the exact solution. However, notice that the Boxer–Thaler approach is superior to the Tustin result for $\Delta t = 0.3$ s.

It is interesting to attempt to determine a condition on Δt from a Nyquist viewpoint. The simplest approach is to find $\Omega_{0,a} = $ "Ω_0" for which the Fourier transform

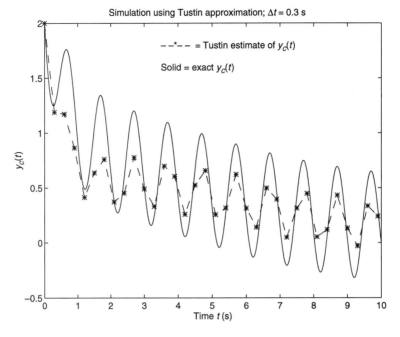

Figure 6.40c

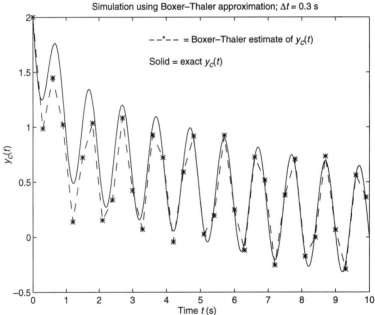

Figure 6.40d

of $x_c(t)$ is "small." For simplicity, assume that the sinusoid is the spectrally dominating term. By setting the unilateral Fourier transform $FT_{ul}\{\sin(\Omega_1 t)\} = 1/N$ at a frequency $\Omega = \Omega_{0,a}$, it is easily determined that $\Omega_{0,a} = \{\Omega_1^2 + N\Omega_1\}^{1/2}$ (for $\Omega_{0,a} > \Omega_1$). Thus, at $\Omega_{0,a}$, the FT of the sinusoid has dropped from ∞ to $1/N$. Suppose we say that the "maximum frequency" goes out to where the FT falls to just below 0.1 (i.e., take $N = 11$), which for $\Omega_1 = 2\pi \cdot 1$ (as in this example, where for brevity we used the symbol Ω for Ω_1) gives $\Omega_{0,a} = 10.42$ rad/s or $\Delta t_{Nyq} = \pi/\Omega_{0,a} = 0.30$ s. It is clear by looking at $\Delta t = 0.3$ s (Figure 6.40c), all approximations are inaccurate, whereas all are excellent for $\Delta t = 0.1$ s (Figure 6.40a), which would correspond to $N = 151$. For $\Delta t = 0.2$ s (corresponding to $N = 33$), the Boxer–Thaler estimate and the predictor–corrector estimate (see the problems) are closer to the exact solution than are the others.

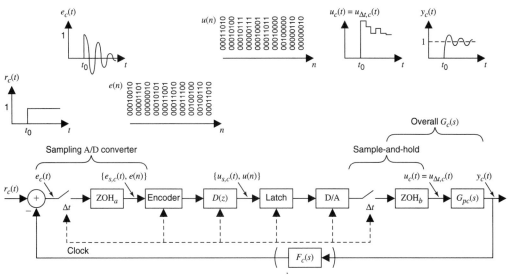

Figure 6.41

We would expect the predictor–corrector method (see the problems) to perform better than other methods in this example because its simplest form is already a Δt^3 approximation; the Runge–Kutta and Tustin estimates we derived above are Δt^2 approximations. Boxer–Thaler is a mixture of Δt^2 and a higher–order approximation of $1/s^2$. The limitation of impulse invariance is of a different sort, being dictated by the degree of aliasing, which we know is also governed by the value of Δt. The issue of the relation between the threshold of Δt for convergence and approximate Δt_{Nyq} is an interesting research topic.

Finally, we note that several additional simulations using different algorithms and values of Δt are found at the end of WP 6.1.

The simple discretization/solution methods discussed in this section are only an entry point into a highly developed subject. Again, such important details as adaptive step size have not been addressed, and no rigorous efforts at strictly fair comparison of the methods were attempted. With adaptive step size, the accuracy for a given value of maximum step size will be far higher than indicated here. Fair comparison of methods involves a formal protocol of inputs, systems, and approximation orders inappropriate for this introductory discussion. Nevertheless, the reader should now be in a position to comprehend and use more advanced approaches and topics in this field.

6.4.6   Application Example: Sampled–Data Control System

A typical linear SISO sampled–data control system has the appearance and signals shown in Figure 6.41:[16] In the canonical feedback system, it is desired that the output of a continuous–time plant $G_{pc}(s)$ follow as closely as possible a command input $r_c(t)$. [$G_{pc}(s)$ may also include the transfer function of a power amplifier and/or actuator preceding the physical plant.] For example, $G_{pc}(s)$ may be a power amplifier/motor

---

[16]See K. Ogata, *Discrete-time Control Systems,* 2nd ed. (Englewood Cliffs, NJ: Prentice Hall 1995), p. 6. Note that the subscript c here on $G_c(s)$ does *not* mean "compensator"; it signifies, as usual, a continuous-time system.

and $r_c(t)$ be a desired angular position or speed of the motor shaft. The digital controller $D(z)$—where the computer and DSP come into play—is used to compensate the plant to yield a stable, fast–responding closed–loop system. A/D and D/A converters and associated circuitry serve as the continuous–time/discrete–time interface.

The command input, $r_c(t)$, may in some cases instead be digital—for example, derived from a computer keyboard sequence. In that case, the A/D converter would be moved to follow $F_c(s)$ and we would replace $r_c(t)$ by $r(n)$. As shown, the difference between $r_c(t)$ and $y_c(t)$, sometimes called the "error," is $e_c(t)$. The sampling A/D converter samples, holds, and encodes its input signal $e_c(t)$ into a sequence $e(n)$. The digital filter $D(z)$ produces from $e(n)$ the controller output sequence $u(n)$.

The latch ensures that the number that the computer has just calculated [$u(n)$] is held constant during the D/A conversion (D/A converters often come with front–end latches). All the digital elements—A/D, computer, and D/A—usually function in synchronization with the clock signal.

The second zero–order hold (or other interpolator) transforms the sequence $u(n)$ into the continuous–time input $u_c(t) = u_{\Delta t,c}(t)$ to the plant. [Because $u_c(t)$ is staircase, in the application of a motor $u_c(t)$ may be termed a winding staircase.] No special holding circuit is theoretically necessary if zero–order holding is desired; the D/A does that automatically. However, as noted previously, it may be added to smooth out overshoots and other spikes that might originate in the D/A converter. Alternatively, a reconstruction filter other than the zero–order hold may be used.

For system analysis, ZOH_b is mathematically cascaded with $G_{pc}(s)$ to form the net analog forward transfer function $G_c(s)$. The closed–loop system output is $y_c(t)$. The transducer (sensor) and/or feedback compensating filter $F_c(s)$ is also often an analog system. Alternatively, if the output sensor has a digital output, it can function as an A/D converter of $y_c(t)$ or some modification of $y_c(t)$ so that no A/D converter is required in the entire system (as r would then logically also be discrete–time).

A fundamental point is that because of the sampler there are no linear–shift–invariant relations between $y_c(t)$ and $r_c(t)$. However, there are LSI relations between $y_{s,c}(t)$ and $r_{s,c}(t)$. These are equivalently expressed as relations between $y(n) = y_c(n \Delta t)$ and $r(n) = r_c(n \Delta t)$ or in the z–domain as LSI relations between $Y(z)$ and $R(z)$.

As indicated in Figure 6.41, in sampled–data control systems, the reconstructed signal is usually presented as the input to a dynamic "plant"—the object to be controlled. Consequently, the effects of sampling and reconstruction on cascaded systems in closed loops must be analyzed in full detail. Many interesting results, such as the step invariance of the sample–and–hold–plant combination, require systematic study of mixed–data–type signals and systems. Conversely, in introductory DSP, for brevity we mainly limit our attention to open–loop systems.

6.4.7   Application Example: Compact Disc Digital Audio Systems

Another application example of the continuous–time/discrete–time interface known and enjoyed by many people is the compact disc audio system. A full study of the compact disc system would require a whole book or more, with the assumption of this entire book as a prerequisite. We consider only the most basic elements here. Consider the digital stereo recorder/playback system shown in Figure 6.42. As in all audio systems, there is high–frequency noise whose effects we wish to minimize. Let us discuss the function of each block in both record and playback systems. We first discuss the record system (Figure 6.42a).

The initial analog filter has a wide transition band, to be inexpensive. For example, its cutoff frequency would be around $F_c = 22$ kHz (maximum audio frequency) and its

(*a*) Digital stereo recording system

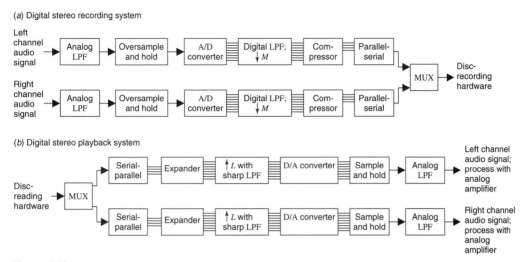

Figure 6.42

rejection frequency at maybe $F_r = 66$ kHz. Thus, for $F > 66$ kHz, there is negligible signal or noise passed. Between 22 kHz and 66 kHz, there is attenuated but nonzero noise.

We then oversample at $F_s = 176$ kHz $= 4F_c$—twice the Nyquist rate that applies to a maximum frequency of 22 kHz (oversampling is not a requirement, but can help). Such an A/D system is called an oversampling A/D converter and is often used for recording. Note that the sample and hold holds the signal steady so the A/D can convert properly. With oversampling, between 66 kHz and 88 kHz $\left(\frac{1}{2}F_s\right)$ there is negligible signal (or noise), giving a buffer for guaranteed no aliasing. For digital frequencies, $f = F/F_s$, we have zero "signal" between $f = \frac{66}{176} = \frac{3}{8}$ and $f = \frac{1}{2}$. Another advantage of the oversampling scheme is that the quantization errors in sampling are spread over the full band—most of which will be chopped off—resulting in a higher signal–to–noise ratio (proof of this intuitively understandable property is left to another course). Yet another advantage of the oversampling is that the nonlinear phase distortion in analog filters is much lower for low–order analog filters with $\Omega < \Omega_c$ than for the high–order filters that would be required without oversampling.

For the digital filter, we can use linear–phase filters that introduce no such undesirable phase distortion. In our example, the sharp, high–quality digital filter cuts off anything above 22 kHz—that is, anything above the digital frequency $f = \frac{1}{8}$. So, for $0 \leq f \leq \frac{1}{8}$ we have all the audio signal, and no noise for $f > \frac{1}{8}$ is kept even though an inexpensive analog filter was used. (We always neglect noise aliasing, because the noise floor is low–level and has very wide bandwidth anyway.) Next we downsample (see Section 7.5.2) by the factor $M = 4$, leaving us with audio up to $f = \frac{4}{8} = \frac{1}{2}$; we thus have "exact Nyquist"—a full–band signal. The new effective sampling rate of $\frac{176}{4} = 44$ kHz (for which the maximum audio signal frequency of 22 kHz corresponds to $f = \frac{1}{2}$) is the minimal sampling rate for no aliasing and thus is most efficient for minimizing the amount of disc space for recording. Next, we logarithmically compress the signal in order to minimize the bits/sample for storage on the CD. (In analog recording, companding serves to lift the low–level audio over any noise that is introduced by the recording medium, as in Dolby.) Compression/expansion or *companding* is investigated in the problems.[17]

---

[17]It is not the initial quantization during sampling of the continuous-time signal that requires companding because we can have plenty of bits/sample in our main digital processing; moreover, (nonlinear) companding there would preclude digital filtering. It is for *storage* (or, in other applications, transmission) where we need to minimize bits/sample and thus need companding. This bit reduction is represented in Figure 6.42 by a smaller number of parallel lines for the compressed signal relative to the expanded signal.

We can now convert the parallel data to serial, send through a multiplexer to handle both left and right channels, and send the result to make the pits on the compact disc. Omitted are the steps of error–correction coding or other added control bits (e.g., CD song number) and modulation. The corresponding required decoding and demodulation are omitted in the discussion below of playback. Effective coding is one of the most sophisticated aspects of the compact disc system.

For the playback system (Figure 6.42b), we first read the bits from the pits in the CD. These are demultiplexed to left and right channel data and converted to parallel data for digital processing. They are then expanded (the inverse of compression). Again to allow for an inexpensive analog filter, we first upsample by—for example, $L = 4$ (see Section 7.5.3). We must have a very sharp digital lowpass filter to remove all the replications resulting from upsampling (and we have left ourselves very little or no room for error by downsampling by $M = 4$ during recording). We then have a baseband spectrum up to 22 kHz, and the next image is at 176 kHz, which allows a very low order analog reconstruction filter. The result is maximally efficient storage of the data, minimum hardware cost, and maximum noise reduction and dynamic range, which make for a very pleasing playback.

6.5 APPROXIMATION OF $X_c(j\Omega)$ BASED ON N SAMPLES $x_c(n\Delta t)$

6.5.1 Derivation and Example

In Section 6.4, we discussed the implementation of analog filtering by digital means. Another operation involving the continuous–time/discrete–time interface is approximation of the continuous–time Fourier transform $X_c(j\Omega)$ based on samples of $x_c(t)$.

Let us begin by examining the DTFT $X(e^{j\omega})$ of the samples $x(n) = x_c(n\Delta t)$. Recall (6.17) $\left[X(e^{j\omega}) = \dfrac{1}{\Delta t} \cdot \displaystyle\sum_{\ell=-\infty}^{\infty} X_c\left(j\left\{\dfrac{\omega + 2\pi\ell}{\Delta t}\right\}\right) \right]$. Assuming no aliasing [i.e., $x_c(t)$ has finite maximum frequency Ω_0 and $\Omega_s > 2\Omega_0$], the $\ell = 0$ term in (6.17) is the only term contributing to $X(e^{j\omega})$ on $|\omega| < \pi$:

$$X(e^{j\omega})\Big|_{|\omega|<\pi} = \frac{1}{\Delta t} X_c\left(\frac{j\omega}{\Delta t}\right) \tag{6.75a}$$

$$= \frac{1}{\Delta t} X_c(j\Omega)\Big|_{\Omega=\omega/\Delta t} \tag{6.75b}$$

which may be equivalently rewritten

$$X(e^{j\Omega\,\Delta t})\Big|_{|\Omega|<\frac{1}{2}\Omega_s} = \frac{1}{\Delta t} X_c(j\Omega). \tag{6.75c}$$

Using the DTFT definition we have

$$X_c(j\Omega) = \Delta t \cdot X(e^{j\Omega\,\Delta t}) = \Delta t \sum_{n=-\infty}^{\infty} x_c(n\Delta t)\, e^{-j(\Omega\,\Delta t)n}, \tag{6.76}$$

$$\text{where } |\Omega| < \frac{1}{2}\Omega_s = \frac{\pi}{\Delta t} \text{ and } \Omega_s > 2\Omega_0.$$

Now observe that the right–hand side (sum) of (6.76) is the rectangular rule approximation of the Fourier integral, $X_c(j\Omega) = \displaystyle\int_{-\infty}^{\infty} x_c(t)e^{-j\Omega t}dt$. Noting the equality with the left–hand side of (6.76), we thus find the amazing result that under no aliasing and *only* for $|\Omega| < \frac{1}{2}\Omega_s$, rectangular–rule approximation of the Fourier integral is exact (zero error)! Moreover, the DTFT scaled by Δt provides this exact estimation by its very definition.

Another interesting fact about (6.76) is that if we take the inverse FT of both sides (on the right side using zero for Ω outside $|\Omega| \leq \Omega_s/2$), we obtain *exactly* the Shannon sinc–reconstruction formula (6.26). This is proved in the problems.

Recall that for an N–length sequence, the DFT provides exact values of the DTFT at $\omega_k = 2\pi k/N$. Therefore, for a general infinite–duration $x_c(t)$, the DFT can provide a truncated version of the above exact estimate of $X_c(j\Omega)$ at $\Omega = \Omega_k = \omega_k/\Delta t = 2\pi k/(N\Delta t)$, where $k \in [0, N-1]$. If $x_c(t)$ has finite duration $t \in [0, T]$ and we choose N and Δt so that $N\Delta t > T$, we might think that the DFT, which in that case gives exact DTFT values (no truncation), would provide a set of exact samples of $X_c(j\Omega)$ at Ω_k. However, there is a theorem stating that whenever $x_c(t)$ is time–limited, it cannot be bandlimited (see both Sections 6.5.2 and 7.4). Consequently, if $x_c(t)$ is time–limited, then there must be aliasing, because $\Omega_0 = \infty$ but $\Omega_s = \pi/\Delta t$ is finite. In practice, however, we can choose Δt small enough so that aliasing is small, and therefore we must choose N very large so that $N\Delta t > T$ for no truncation.

More specifically, if only N samples of the estimate of $X_c(j\Omega)$ are needed at the equally spaced frequencies $\Omega_k = 2\pi k/(N\Delta t), k \in [0, N-1]$, then the truncated version of (6.76) is

$$\hat{X}_{N,\Delta t,c}\left(j\frac{2\pi k}{N\Delta t}\right) = \Delta t \cdot \sum_{n=0}^{N-1} x_c(n\Delta t)e^{-j2\pi nk/N} \tag{6.77a}$$

$$= \Delta t \cdot \mathrm{DFT}_N\{x_c(n\Delta t)\}. \tag{6.77b}$$

Because the DFT can be computed extremely efficiently via the FFT [see Section 7.10], there is good reason to try to use it whenever possible. The result (6.77) is thus a windfall: An intuitive approach to approximation of the Fourier transform yields a formula that can be computed by a very efficient algorithm, the FFT.

For reference in practice, it is helpful to remember that the frequencies in the DFT spectrum correspond to analog frequencies $\Omega_k = 2\pi k/(N\Delta t) = k\Omega_s/N$, so their spacing in hertz is $1/(N\Delta t) = F_s/N$. Recall that (6.76) is valid only for $|\Omega| < \Omega_s/2$, which, using $\Omega_k = k\Omega_s/N$, translates to (6.77) being valid only for $\frac{N}{2} - 1 \leq k \leq \frac{N}{2}$ (N even) or $-\left(\frac{N-1}{2}\right) \leq k \leq \frac{N-1}{2}$ (N odd). However, we said above that $k \in [0, N-1]$. Nevertheless, recall that these conditions are actually equivalent because the DFT_N is N-periodic, so the second half of samples corresponds to negative frequencies. We return to this fact from a time-domain perspective in Section 6.5.3.

In this approach, two operations on the original signal $x_c(t)$ are performed before taking the DFT in (6.77):

(1) Sampling: $x(n) = x_c(n\Delta t) \overset{\text{DTFT}}{\longleftrightarrow} X(e^{j\omega})$,   (6.78a)

   where $X(e^{j\omega}) = \dfrac{1}{\Delta t} \cdot \displaystyle\sum_{\ell=-\infty}^{\infty} X_c\left(j\left\{\dfrac{\omega + 2\pi\ell}{\Delta t}\right\}\right)$ (6.78b)

and

(2) Windowing: $x_N(n) = x(n) \cdot w_N(n)$, where in particular for the rectangular window,

$$w_N(n) = w_{N,R}(n) = 1, \quad n \in [0, N-1]; 0, \text{ otherwise} \tag{6.79a}$$

   and[18]

---

[18]In principle, $x(n), X(e^{j\omega}), x_N(n)$, and $X_N(e^{j\omega})$ should all be subscripted with Δt, but this would complicate the notation.

$$x_N(n) = x(n) \cdot w_N(n) \overset{\text{DTFT}}{\longleftrightarrow}$$

$$X_N(e^{j\omega}) = \frac{1}{2\pi} \cdot \int_{-\pi}^{\pi} X(e^{j\omega'}) \cdot W_N(e^{j(\omega-\omega')}) d\omega' \tag{6.79b}$$

$$= \frac{X(e^{j\omega}) *_{c\omega} W_N(e^{j\omega})}{2\pi}, \tag{6.79c}$$

where $*_{c\omega}$ denotes continuous convolution over the (continuous) variable ω and where for the rectangular window which unless stated otherwise we will use henceforth,

$$W_{N,R}(e^{j\omega}) = \sum_{n=0}^{N-1} e^{-j\omega n} = \frac{1 - e^{-j\omega N}}{1 - e^{-j\omega}} \tag{6.80}$$

$$= \frac{\sin(N\omega/2)}{\sin(\omega/2)} \cdot e^{-j\omega(N-1)/2}.$$

We can obtain a single convolutional expression for $\hat{X}_{N,\Delta t,c}(j\Omega)$ in terms of $X_c(j\Omega)$ as follows. Define for convenience the integration variable $\Omega' = \omega'/\Delta t$ in (6.79b), and thus $\omega' = \pm\pi \to \Omega' = \pm\pi/\Delta t = \pm\Omega_s/2$, and $d\omega' = \Delta t \cdot d\Omega'$. Also substitute (6.78b) into (6.79b). Then before sampling at the N frequencies $\Omega_k = 2\pi k/(N\Delta t)$ (again see note above on range of k), $\hat{X}_{N,\Delta t,c}(j\Omega)$ in (6.77) may be written

$$\hat{X}_{N,\Delta t,c}(j\Omega) = \Delta t \cdot X_N(e^{j\Omega\Delta t})$$

$$= \frac{\Delta t}{2\pi} \cdot \int_{-\Omega_s/2}^{\Omega_s/2} \frac{1}{\Delta t} \sum_{\ell=-\infty}^{\infty} X_c\left(j\left\{\Omega' + \frac{2\pi\ell}{\Delta t}\right\}\right) \cdot W_{N,R}(e^{j(\Omega-\Omega')\Delta t}) \, d\Omega' \Delta t. \tag{6.81}$$

Now the definition of a new variable $\Omega'_{\text{new}} = \Omega'_{\text{old}} + 2\pi\ell/\Delta t = \Omega'_{\text{old}} + \ell\Omega_s$ allows the finite integral of an infinite sum to be converted into a single infinite–length integral. Noting that $W_{N,R}(e^{j(\omega+2\pi\ell)}) = W_{N,R}(e^{j\omega})$, or, equivalently, $W_{N,R}(e^{j(\Omega+\ell\Omega_s)\Delta t}) = W_{N,R}(e^{j\Omega\Delta t})$, $\hat{X}_{N,\Delta t,c}(j\Omega)$ in (6.81) may be rewritten as

$$\hat{X}_{N,\Delta t,c}(j\Omega) = \frac{\Delta t}{2\pi} \cdot \int_{-\infty}^{\infty} X_c(j\Omega') \cdot W_{N,R}(e^{j(\Omega-\Omega')\Delta t}) \, d\Omega' \tag{6.82a}$$

$$= \sum_{n=0}^{N-1} x_c(n\Delta t) \cdot e^{-j\Omega n \Delta t} \cdot \Delta t, \tag{6.82b}$$

where (6.82b) is just (6.77a) evaluated for general values of Ω. The DFT_N of $x_N(n) = x_c(n\Delta t)$ for $n \in [0, N-1]$ will provide the N samples $\hat{X}_{N,\Delta t,c}(j2\pi k/(N\Delta t))$ as in (6.77b) and can now be written in the following convolutional form:

$$\Delta t \cdot \text{DFT}_N\{x_N(n)\} = \hat{X}_{N,\Delta t,c}\left(j\frac{2\pi k}{N\Delta t}\right) \tag{6.83}$$

$$= \frac{\Delta t}{2\pi} \cdot X_c(j\Omega) *_{c\Omega} W_{N,R}(e^{j\Omega\Delta t})\Big|_{\Omega=2\pi k/(N\Delta t)}.$$

We see that $\hat{X}_{N,\Delta t,c}(j2\pi k/(N\Delta t))$ is the convolution of $X_c(j\Omega)$ with the window function $W_{N,R}(e^{j\Omega\Delta t})$. Again, the frequencies of evaluation are $2\pi k/(N\Delta t) = \Omega_s \cdot k/N$. The second half of the DFT_N spectral samples will be the negative–frequency components of the replication of X_c centered on Ω_s.

The transformation of the infinite sum and finite integral in (6.81) into an infinite–length integral in (6.83) has *not* eliminated any aliasing effects. It has merely produced a mathematically equivalent expression. In fact, $\hat{X}_{N,\Delta t,c}(j\Omega)$ is a *periodic* function of frequency Ω, with period $\Omega_s = 2\pi/\Delta t$. This is because $W_{N,R}$ is also $2\pi/\Delta t$–periodic in Ω—and is the only function containing Ω on the right–hand side of (6.82a). The aliasing effects now "reside" in $W_{N,R}$, which over the infinite range of Ω slides along in the convolution picking up all the aliasing terms originally found in the finite–integration aliasing sum.

Although $\hat{X}_{N,\Delta t,c}(j\Omega)$ is $2\pi/\Delta t$–periodic in Ω, the function it is estimating, $X_c(j\Omega)$, is aperiodic. Again, the approximation can possibly be good only on the band $|\Omega| < \Omega_s/2$ (or $|\omega| < \pi$ or $k \in [0, N-1]$ with second half of samples interpreted as negative frequencies); that is, on the zero–centered replication in (6.17). If $N \to \infty$ (no windowing), $W_{N,R}$ becomes a Dirac delta train, and the original aliasing sum results:

$$\hat{X}_{N,\Delta t,c}(j\Omega) \xrightarrow[N \to \infty]{} \sum_{m=-\infty}^{\infty} X_c\left(j\left\{ \Omega + \frac{2\pi m}{\Delta t} \right\} \right) \tag{6.84}$$

Otherwise, $W_{N,R}$ has the effect of smoothing each replication.

Thus we may conclude two ways to improve estimation of the Fourier transform based on taking the DFT of Δt–spaced time samples: (a) To minimize aliasing, we decrease Δt so that the spectral replication interval widens, and (b) to decrease erroneous smoothing (truncation effects), we increase N so that $W_{N,R}$ becomes more like $\delta_c(\Omega)$. Be sure not to miss the full presentation of the DFT used for estimating $X_c(j\Omega)$ in Section 6.7 [in particular, (6.108)].

To obtain a qualitative feel for the accuracy of the DFT approximation of the Fourier transform, consider the following example.

EXAMPLE 6.5

Consider L repetitions of the basic triangle function

$$x_c(t) = \begin{cases} 2t/T, & t \in [0, T/2] \\ 2(1 - t/T), & t \in (T/2, T) \end{cases} . \tag{6.85}$$

Denote L repetitions of $x_c(t)$ by $x_{L,c}(t)$. Now suppose that we define $x(n) = x_{L,c}(n\Delta t)$, $n \in [0, N-1]$ and take the DFT of $x(n)$ multiplied by Δt as in (6.77b). Compare DFT estimations of $X_{L,c}(j\Omega)$ with the exact values. Vary L, and study aliasing without truncation, and truncation without aliasing. Include plots of individual aliasing terms. Let $T = 1$ s.

Solution

In Appendix 4B, the Laplace transform of L repetitions of $x_{L,c}(t)$ was found to be as follows [and was used in Section 4.6; see (4.29)]:

$$X_{L,c}(s) = \frac{1 - e^{-sTL}}{1 - e^{-sT}} \cdot \frac{\{1 - e^{-sT/2}\}^2}{(T/2) \cdot s^2}, \quad s \neq 0; X_{L,c}(s)\big|_{s=0} = \tfrac{1}{2}LT. \tag{6.86}$$

Substitution of $s = j\Omega$ and a small amount of algebra yields for the Fourier transform of $x_{L,c}(t)$

$$X_c(j\Omega) = \frac{(T/2) \cdot \{\sin(\Omega T/4)\}^2}{\{\Omega T/4\}^2} \cdot \frac{\sin(\Omega TL/2)}{\sin(\Omega T/2)} e^{-j\Omega LT/2}. \tag{6.87}$$

Let us begin our investigation by considering aliasing without truncation. To do this, we vary Δt while leaving L fixed at 1 and N large enough to cover the entire triangle for all Δt values to be investigated. Figures 6.43 through 6.45 show estimates for $L = 1$, $N = 64$ using the following values of $\Delta t/T$: $\Delta t/T = 0.1$ in Figure 6.43, $\Delta t/T = 0.3$ in Figure 6.44, and $\Delta t/T = 0.5$ in Figure 6.45. In part (a) of each figure is shown the sequence $x(n) = x_c(n\Delta t)$ (stems) and $x_c(t)$ (–.–.–.–.). In part (b) is shown $|\hat{X}_{N,\Delta t,c}(j2\pi k/[N\Delta t])|$ (stems) and $|X_c(j\Omega)|$ (solid curve). For easy viewing, the spectral estimate sequences were rearranged to appear "zero–centered." Part (c) of each figure shows $|X_c(j\Omega)|$ (solid curve), the $m = 1$ aliasing term in (6.84),

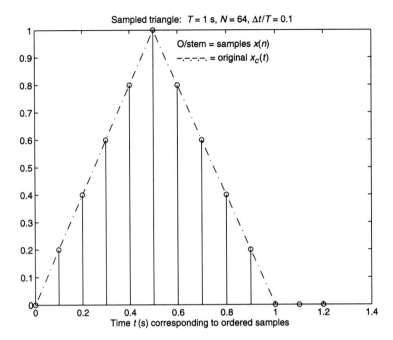

Figure 6.43a

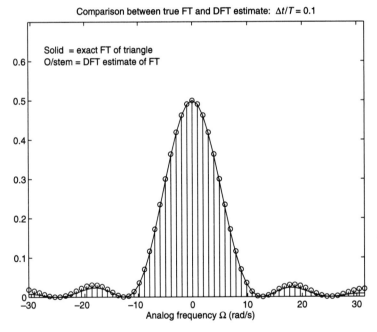

Figure 6.43b

$|X_c(j(\Omega + 2\pi/\Delta t))|$ (– – – –) and the $m = -1$ aliasing term in (6.84), $|X_c(j(\Omega - 2\pi/\Delta t))|$ (–.–.–.–.). In the absence of truncation, which is the case in Figures 6.43 through 6.45, these aliasing terms make up $X(e^{j\omega})$, of which the DFT values in part (b) of those figures are samples.

If we were to use the first zero crossing to approximate the bandwidth for finding Δt_{Nyq}, we would obtain $\Delta t_{Nyq} = T/4$. However, the $sinc^2$ function drops off much

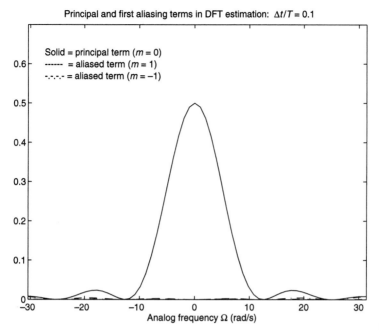

Figure **6.43c**

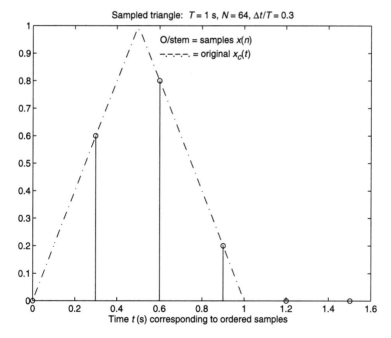

Figure **6.44a**

more sharply than sinc, so we may use a value of Δt larger than $T/4 = 0.25T$ and still obtain acceptable estimations, as in Figure 6.44 ($\Delta t = 0.3T$). As we would expect, though, as Δt continues to increase (Figure 6.45), the estimate degrades to "garbage." The display of the most significant individual aliasing terms in part (c) of Figures 6.43 through 6.45 makes clear the reason for the shape of the DFT values in part (b) of the figures. Just add the $m = -1, 0$, and 1 terms together at each frequency

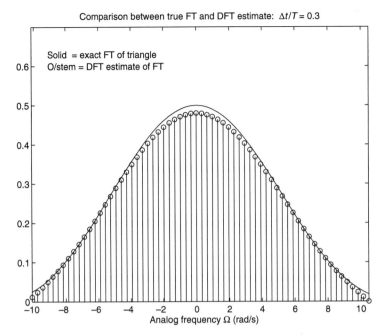

Figure 6.44b

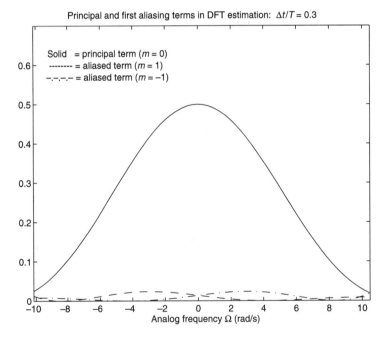

Figure 6.44c

and the curve in part (*b*) approximately results because all other *m* terms are nearly zero in this range of Ω.

Figures 6.44*c* and 6.45*c* clarify why $\pi/\Delta t = \Omega_s/2$ is sometimes called the *foldover frequency*. Notice that, for example, the $m = 1$ term is an *unbroken continuation* of the true spectrum $X_c(j\Omega)$, but *folded over backwards* with respect to the $m = 0$ term. Thus, the "folding" frequencies in Figures 6.43 through 6.45 are, respectively, 31.42, 10.47, and 6.28 rad/s.

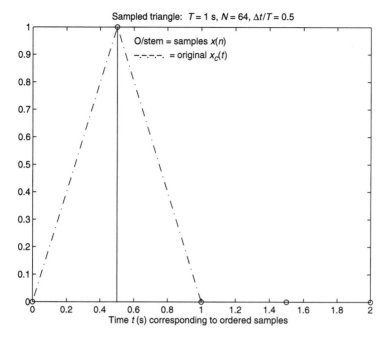

Figure 6.45a

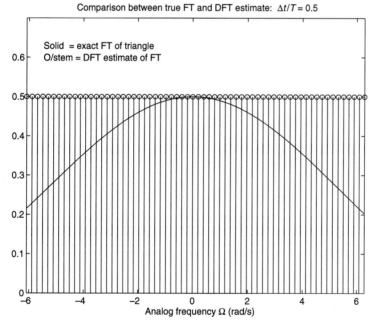

Figure 6.45b

Notice also how for the most severe case of aliasing, Figure 6.45b, that the aliasing terms shown add to a constant, independent of Ω. This could be predicted, because for $\Delta t = T/2, x(n) = x_c(nT/2)$ is just $\delta(n-1)$, whose magnitude spectrum is constant. It is nevertheless quite interesting to see in Figure 6.45c just how the aliasing terms add up to precisely $\frac{1}{2}T = \frac{1}{2}$ for all Ω.[19]

---

[19]Precisely $\frac{1}{2}\text{T} = \frac{1}{2}$ is attained for all Ω only when all aliasing terms are added together, not just the depicted terms $m = -1, 0$, and 1. Notice that each of the aliasing terms in Figure 6.46c is precisely zero at $\Omega = 0$, so for this one frequency there is no aliasing even though the signal is extremely undersampled.

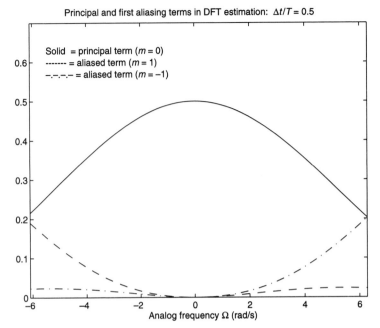

Principal and first aliasing terms in DFT estimation: $\Delta t/T = 0.5$

Figure 6.45c

Figures 6.43 through 6.45 have illustrated the aliasing resulting from improper choice of Δt, with no window effect. Now we consider the effect of truncation without significant aliasing. Let $\Delta t/T = 0.1$ for insignificant aliasing, the number of triangle repetitions in $x_{L,c}(t)$ be $L = 10$, and the number of repetitions retained in $x(n)$ after truncation be $L_{\text{keep}} = 10, 2,$ and 1 in, respectively, Figures 6.46 through 6.48. Thus in Figure 6.46, the entire waveform is included in $x(n)$, whereas portions of it are windowed out by truncation in Figures 6.47 and 6.48. Effectively,

$$x(n) = x_{L,c}(n\Delta t) \cdot w_{N,R}(n), \tag{6.88}$$

where the length of the window is N, the number of samples included in the 10, 2, and 1 replications in Figures 6.46 through 6.48. Before using the DFT to compute samples of the DTFT and estimates of $X_c(j\Omega)$, all N–length sequences were zero–padded out to $M = 1024$ points; this allows easier comparison of the complicated waveforms. Zero–padding simply gives finer–spaced values of the DTFT of $x(n)$, a fact thoroughly investigated in Section 7.2.

Parts (a) of Figures 6.46 through 6.48 show the first 100 points of $x(n)$, the sequence whose 1024–point DFT is taken after zero–padding. Also shown in these figures is the original $x_c(t)$ that has been sampled/truncated (–.–.–.). In parts (b), the exact spectrum (–.–.–.) and estimated DFT spectrum (solid) are plotted together. Naturally, as there is almost no aliasing for $\Delta t/T = 0.1$, the exact and estimated spectra are nearly identical in Figure 6.46 because the entire waveform is included, as there is no windowing for $L_{\text{keep}} = L$. Contrarily, for the other two cases, the estimate is progressively worse as the windowing is more severe. The important behavior to notice is the *smoothing* effect described analytically above by the convolution with the window function in (6.83).

With many repetitions (ten) in $x_{10,c}(t)$, the true Fourier transform has become much sharper (approaching a series of spikes) than with only one repetition as well

Sampled, repeated triangle: $T = 1$ s, retained $L = 10$, $\Delta t = 0.1$ s

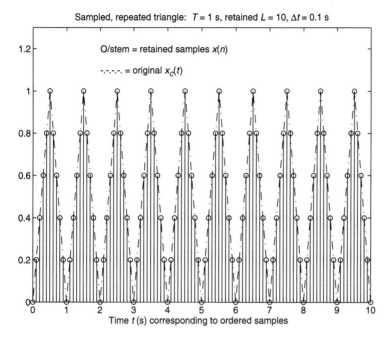

O/stem = retained samples $x(n)$

-.-.-. = original $x_C(t)$

Time t (s) corresponding to ordered samples

Figure 6.46a

Comparison between true magnitude of FT and DFT estimate: $\Delta t/T = 0.1$

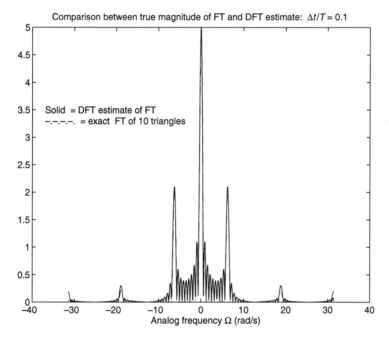

Solid = DFT estimate of FT

−.−.−. = exact FT of 10 triangles

Analog frequency Ω (rad/s)

Figure 6.46b

as $L = 10$ times taller, as described in Section 4.6. Under the smoothing effect of windowing, each spectral spike serves to replicate the smooth $\sin(N\omega/2)/\sin(\omega/2)$ window function in the DFT result [see (6.80)]. The result is an estimate distorted by the smoothing effect of the convolution of the true spectrum with the window spectrum. When only one repetition is kept (Figure 6.48), we are merely redisplaying the single–triangle spectrum for $\Delta t = 0.1$ s initially shown in Figure 6.43b.

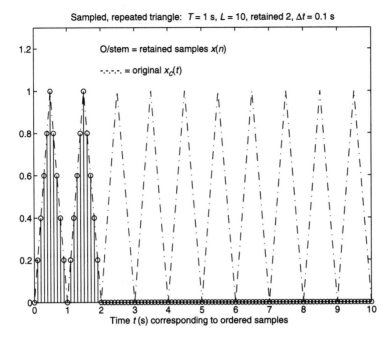

Figure 6.47*a*

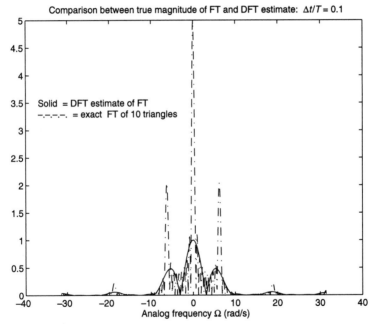

Figure 6.47*b*

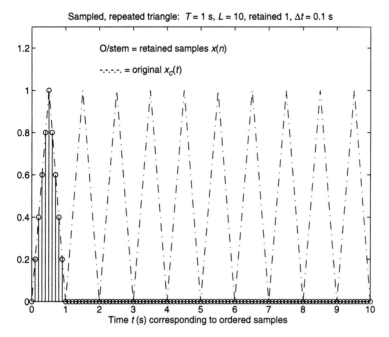

Sampled, repeated triangle: $T = 1$ s, $L = 10$, retained 1, $\Delta t = 0.1$ s

O/stem = retained samples $x(n)$

-.-.-.- = original $x_C(t)$

Time t (s) corresponding to ordered samples

Figure 6.48*a*

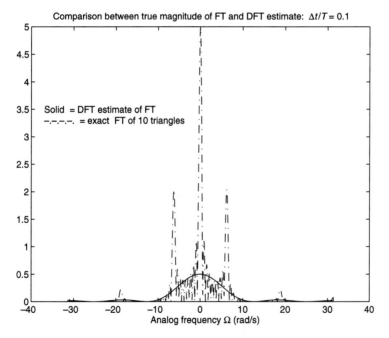

Comparison between true magnitude of FT and DFT estimate: $\Delta t / T = 0.1$

Solid = DFT estimate of FT
-.-.-.-. = exact FT of 10 triangles

Analog frequency Ω (rad/s)

Figure 6.48*b*

6.5.2   Durations of $x_c(t)$ and $X_c(j\Omega)$:   The Time–Bandwidth Product

It now may be asked from a more general viewpoint, how good an approximation of the Fourier transform does the DFT approach yield? To understand the answer, it helps to consider the signal time–bandwidth product. This concept is covered on WP 6.2 all the way from first principles and its origins in quantum mechanics to the final formulas and a detailed example of calculation of time and frequency durations for a typical signal: $x_c(t) = e^{-at}\sin(bt)u_c(t)$. The time–bandwidth product inequality actually holds for any functions that are a Fourier transform pair (e.g., position and momentum wave functions in quantum mechanics). The time or frequency durations essentially amount to deterministic normalized standard deviations of time or frequency, for the given time or frequency waveforms that form a Fourier pair. Here we merely repeat the formulas for the time–bandwidth product inequality and signal durations:

$$D_t \cdot D_\Omega \geq \frac{1}{2}, \qquad\qquad \text{Time-band-width inequality.} \qquad (6.89)$$

Complete derivation details on WP 6.2.

where

$$D_t = \left\{ \frac{\int_{-\infty}^{\infty} |tx_c(t)|^2 dt}{\int_{-\infty}^{\infty} |x_c(t)|^2 dt} \right\}^{1/2} \qquad\qquad (6.90)$$

and

$$D_\Omega = \left\{ \frac{\int_{-\infty}^{\infty} |\Omega X_c(j\Omega)|^2 d\Omega}{\int_{-\infty}^{\infty} |X_c(j\Omega)|^2 d\Omega} \right\}^{1/2}. \qquad\qquad (6.91)$$

In words, (6.89) says that the narrower the time function, the wider its spectrum, and vice versa; a signal and its spectrum cannot both be too narrow. This concept is the same as the time–scaling property $\text{FT}\{x_c(\alpha t)\} = X_c(j\Omega/\alpha)/|\alpha|$, where stretching t squeezes the spectrum and vice versa. It is further shown on WP 6.2 that the minimum value of time–bandwidth product occurs for the Gaussian waveform.

6.5.3   Approximation of $X_c(j\Omega)$ Based on N Samples $x_c(n\,\Delta t)$:   Accuracy

Now that the time–bandwidth product has been defined, the accuracy of the approximation of $X_c(j\Omega)$ based on $x_c(n\,\Delta t)$ can be discussed. Let the time–bandwidth product for $x_c(t)$ be called $Q = D_t \cdot D_\Omega$. We know that the sampling criterion must be (in practice, approximately) satisfied if there is any hope of obtaining a good estimate of $X_c(j\Omega)$. That is, we need $\Delta t \leq \pi/\Omega_0$, where Ω_0 is in practice an estimated maximum frequency for which $x_c(t)$ has significant energy.

Notice that $mD_\Omega/2$ (where in Example W6.2.1, $m = 5$) and Ω_0 are both maximum frequency measures. The measure $mD_\Omega/2$ is based on a normalized root–mean–square bandwidth D_Ω (standard deviation radian frequency), multiplied by $m > 1$ to be useful as a "maximum" frequency. On the other hand, Ω_0 is typically determined as a frequency yielding a specified decay in $|X_c(j\Omega)|$, typically in dB. In cases where the relation between time and frequency durations is important, D_Ω

may be useful; in such cases, the sampling criterion would read $\Delta t \leq 2\pi/(mD_\Omega)$ or $\Omega_s \geq mD_\Omega$, analogous to $\Omega_s \geq \text{BW}$ [(6.18b)], where BW is the absolute bandwidth for the case of an ideal bandlimited signal or otherwise an approximate bandwidth. In any case, the time duration of $x_c(t)$ is $\ell D_t = \ell Q/D_\Omega$. If Q is large, then a given band-width translates to a relatively large time duration.[20]

Suppose that Δt is fixed at a value satisfying the sampling criterion $\Delta t \leq 2\pi/(mD_\Omega)$ (or $\Delta t \leq \pi/\Omega_0$). Now also suppose N is chosen such that $N\Delta t > \ell D_t$, or $N\Delta t > T$, where T is an otherwise estimated duration of $x_c(t)$. In this case, the esti-mate $\hat{X}_{N,\Delta t,c}(j2\pi k/[N\Delta t])$ in (6.77b) [namely, $\Delta t \cdot \text{DFT}_N\{x_c(n\Delta t)\}$] should be quite good, for effects of both windowing and aliasing are minimal. Specifically, (a) an estimate based on an infinite–length window, $\hat{X}_{\infty,\Delta t,c}(j\Omega)$ would approximately equal our estimate based on N samples, $\hat{X}_{N,\Delta t,c}(j\Omega)$ (little windowing effect) and (b) by supposition there is little aliasing. Thus neither of the degrading effects of the (truncated) rectangular–rule method is significant.

The choice of N for the case $\{\Delta t < 2\pi/(mD_\Omega), N\Delta t > \ell D_t\}$ is $N > \ell m D_t \cdot D_\Omega/(2\pi)$ $= \ell m Q/(2\pi)$ (or $N > T\Omega_0/\pi$ if $\ell D_t = T$ and $mD_\Omega = 2\Omega_0$). This requirement on N is only a ballpark approximation and, in general, an underestimate. For example, if $x_c(t)$ is a Gaussian pulse, $Q = \frac{1}{2}$, and so with $\ell = m = 5$ we require only $N > m\ell/(4\pi)$ $= 25/(4\pi) \approx 2$. This may seem incredible, but in fact computation shows that even with $N = 4$ and $\Delta t = 2\pi/(5D_\Omega) \approx 2$ s for $k = 0.8$ in $x_c(t) = e^{-kt^2/2}$, the DFT approx-imation of the Fourier transform is reasonable, and for only $N = 12$ and $\Delta t = 0.4$ s, the results are excellent. In practice, of course, care must be used in applying the above requirement on N as an absolute lower limit on N; for example, always use graphical checks.

Beyond aliasing and truncation, another parameter characterizing a discrete spec-tral estimation is resolution. Let us consider the effect in the best–case scenario defined above: $\Delta t < 2\pi/(mD_\Omega), N\Delta t > \ell D_t$—that is, little aliasing or truncation effects. In so doing, it is helpful to know the minimum spacing of oscillations (variations) in $X_c(j\Omega)$, which we would desire to resolve in $\hat{X}_{N,\Delta t,c}(j\Omega)$. That spacing is determined by the temporal duration D_t of the signal $x_c(t)$, because $x_c(t)$ may be thought of as a "spectrum" of $X_c(j\Omega)$.

The width of the main lobe width of $W_{N,R}$ determines the resolving capability (resolution) of $\hat{X}_{N,\Delta t,c}(j\Omega)$. This is because even if $x_c(t)$ contained a pure (sinusoidal) oscillation manifest in $X_c(j\Omega)$ as an infinitesimal–width (Dirac delta) pulse, the result of convolving that with $W_{N,R}$ in (6.83) would be a smooth pulse having width equal to the width of the main lobe of $W_{N,R}$. However, if the conditions of the pre-ceding paragraph hold, it is impossible for $X_c(j\Omega)$ to have such a spike, for the spike has $D_t = \infty$ (because a pure oscillation has infinite duration). Rather, N and Δt have been chosen such that the main lobe of $W_{N,R}$ *approximately* matches the width of the highest-frequency ripples in $X_c(j\Omega)$. The main lobe width $\Delta\Omega$ of the rectangular window $W_{N,R}$ is determined as $\sin(N\Delta\Omega\Delta t/2) = 0 \rightarrow N\Delta\Omega\Delta t/2 = \pi \rightarrow \Delta\Omega = 2\pi/(N\Delta t)$, which is precisely the spacing of analog frequencies representing adjacent samples of the DFT spectrum estimate. The resolution limit, $\Delta\Omega = 2\pi/(N\Delta t)$ [or $\Delta F = 1/(N\Delta t)$], is approximately equal to the spacing of detail in $X_c(j\Omega)$ that can be resolved; we want $\Delta\Omega = 2\pi/(N\Delta t)$ to be small for "high" resolution.

In practice, one may not always be in a position to satisfy $N\Delta t \geq$ the true dura-tion of $x_c(t)$. The true duration may be hard to determine or may require excessively large N. However, the resolution will remain $2\pi/(N\Delta t)$ even as $X_c(j\Omega)$ becomes more ripply. Thus the window will have a smoothing distortion on the estimate of

---

[20]Because the Gaussian pulse has the minimum value of Q of all signals ($\frac{1}{2}$), its Fourier transform is best approximated by the DFT of samples method.

$X_c(j\Omega)$. The larger $N\Delta t$, the better the resolution. In fact, for a smaller Δt with N fixed, the smoothing effect becomes *worse* due to severe temporal truncation. The width of the main lobe of $W_{N,R}$ in (6.80) $[W_{N,R}(e^{j\omega}) = e^{-j\omega(N-1)/2}\sin(N\omega/2)/\sin(\omega/2)]$ gets larger, and in the limiting case $\Delta t \to 0$, with N fixed, $W_{N,R}(e^{j\Omega\,\Delta t}) \to W_{N,R}(1) = N$— no resolution at all. With N practically restricted, over–reducing Δt is wasteful, because presumably for some reasonably sized Δt, aliasing already became minimal. To obtain the highest–resolution estimate of $X_c(j\Omega)$ for a given value of N, we should choose the largest Δt for which aliasing is acceptably small. To achieve a resolution of at least $\Delta\Omega$, and with $\Delta t < \pi/\Omega_0$, we require $2\pi/(N\Delta t) < \Delta\Omega$, or $N > 2\pi/(\Delta\Omega\,\Delta t) = \Omega_s/\Delta\Omega > 2\Omega_0/\Delta\Omega$.

One further point of great importance in the interpretation of Fourier estimations via the DFT is the fact that the DFT_N is actually a Fourier series of a periodic sequence having period N. Figure 6.49 shows the finite–length sequence $x(n)$ and the N–periodic extension of it arising in the DFT_N^{-1} definition, denoted $\tilde{x}(n)$. Notice that the second half of samples on $n \in [0, N-1]$ is equivalent to samples for negative n in the periodic extension. Specific ranges for example even N and odd N are shown in Table 6.1. Analogous statements pertain to the DFT frequency domain: The members of the second half of samples are equivalent to amplitudes of negative frequency components; simply replace n by k in Table 6.1.

If one is not aware that (a) the DFT_N sequences are N–periodic and (b) the second half of samples corresponds to negative index values, then the estimates of $X_c(j\Omega)$ obtained from the DFT_N of the N samples may be grossly inaccurate. For example, Figure 6.50a shows a sequence obtained by sampling an aperiodic, bicausal signal

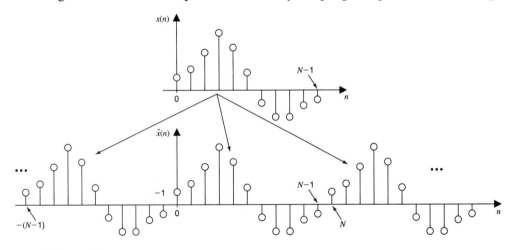

Figure 6.49

Table 6.1   Equivalent Temporal Index Ranges for the DFT

N even:		
$N = 8$: Original index n:	0 1 2 3 4	5 6 7
DFT_N equivalent index n or $n - N$:	<u>0 1 2 3 4</u>	<u>–3 –2 –1</u>
	n for $n \in [0, N/2]$	$n - N$ for $n \in [N/2 + 1, N - 1]$
N odd:		
$N = 7$: Original index n:	0 1 2 3	4 5 6
DFT_N equivalent index n or $n - N$:	<u>0 1 2 3</u>	<u>–3 –2 –1</u>
	n for $n \in [0, (N-1)/2]$	$n - N$ for $n \in [(N-1)/2 + 1, N - 1]$

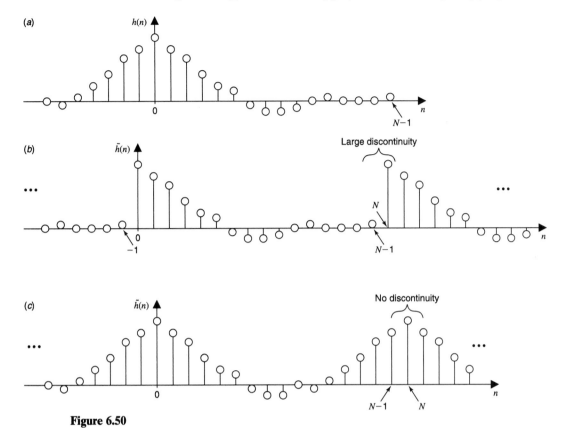

Figure 6.50

$x_c(t)$. If we were to blindly use samples $[0, N-1]$, the periodic extension would appear as shown in Figure 6.50b. Notice that sample $N-1$ will be identical to that for $n = -1$ in the N–periodic extension, whether N is even or odd. If the sequence in Figure 6.50b is used for computing the DFT_N, the DFT_N will "see" a huge discontinuity in the periodic extension between points $N-1$ and N, as indicated in Figure 6.50b. This discontinuity would show up as erroneous high–frequency components in $X(k)$.

A spectrum more faithful to the original $X_c(j\Omega)$ would be obtained by replacing the second half of samples $x_N(n)$ (which were small in magnitude anyway in Figure 6.50a) with samples of $x(n)$ for negative n, the periodic extension of which is shown in Figure 6.50c. This periodic extension has no unnatural discontinuity in it, and thus its DFT_N should give a better approximation of samples of $X_c(j\Omega)$. Alternatively, just shift the original sequence to the right (i.e., delay it) and take the group of N containing most of the energy of $x_c(n\Delta t)$ as $x_N(n)$; be sure to account for the phase shift that this delay implies in the spectrum.

6.6 APPROXIMATION OF $x_c(t)$ BASED ON N SAMPLES $X_c(jk\,\Delta\Omega)$

In this section, we discuss the approximation problem that is the dual of that discussed in Section 6.5: Given N samples of the continuous–time Fourier transform $X_c(j\Omega)$, approximate N samples of the continuous–time signal $x_c(t)$ using the DFT. Again, a simple approach is to apply the rectangular rule of numerical integration.

We approximate the continuous–time inverse Fourier integral

$$x_c(t) = \frac{1}{2\pi} \int_{-\infty}^{\infty} X_c(j\Omega)e^{j\Omega t}\, d\Omega \tag{6.92}$$

by the truncated rectangular rule estimate

$$\hat{x}_{N,\Delta\Omega,c}(t) = \frac{\Delta\Omega}{2\pi} \sum_{k=0}^{N-1} X_c(jk\,\Delta\Omega)\, e^{jtk\,\Delta\Omega}. \tag{6.93}$$

There is again the possibility of aliasing, from the sampling of the spectrum. This time the aliasing is in the time domain. We assume that $x_c(t)$ is nonzero only for $t \in [0, T]$. For $-\infty < k < \infty$ in (6.93), [i.e., for the non–truncated version of (6.93); if $x_c(t)$ is practically nearly bandlimited, not a large difference], we can form a relation analogous to (6.17) $[X(e^{j\omega}) = (1/\Delta t) \sum_{k=-\infty}^{\infty} X_c(j\{(\omega + 2\pi k)/\Delta t\})]$ in the time domain (see the problems). That is, defining $X_{s1,c}(j\Omega) = X_c(j\Omega) \cdot \delta_{\Delta\Omega}(\Omega)$ as a Dirac–train–sampled spectrum,[21] we find $\tilde{x}_{s1,c}(t) = FT^{-1}\{X_{s1,c}(j\Omega)\}$ is a sum of replications of $x_c(t)$ (possibly aliasing), with adjacent replications separated by the sampling "frequency" $t_s = 2\pi/\Delta\Omega$ (analogous to $\Omega_s = 2\pi/\Delta t$ for time-sampling). Specifically,

$$\tilde{x}_{s1,c}(t) = \frac{1}{\Delta\Omega} \sum_{k=-\infty}^{\infty} x_c(t + kt_s), \tag{6.94}$$

where if desired the $1/\Delta\Omega$ factor may be replaced by $t_s/(2\pi)$. Being t_s–periodic, $\Delta\Omega \cdot \tilde{x}_{s1,c}(t)$ has a Fourier series. Using (4.12a) as a template $[\tilde{x}_c(t) = \sum_{k=-\infty}^{\infty} X_c(k) \cdot e^{j2\pi kt/T}]$, in (4.12a) replace T by $t_s = 2\pi/\Delta\Omega$:

$$\Delta\Omega \cdot \tilde{x}_{s1,c}(t) = \sum_{\ell=-\infty}^{\infty} \beta_\ell\, e^{j\ell t\,\Delta\Omega}, \tag{6.95}$$

where

$$\beta_\ell = \frac{1}{t_s} \int_0^{t_s} \{\Delta\Omega \cdot \tilde{x}_{s1,c}(t)\} e^{-j\ell t\,\Delta\Omega} dt. \tag{6.96}$$

If $x_c(t)$ is zero for t outside the interval $[0, T]$, where $T < t_s$, then according to (6.94) we have $\tilde{x}_{s1,c}(t) = (1/\Delta\Omega) \cdot x_c(t)$ for the interval $[0, t_s]$. Thus for $t \in [0, t_s]$ and no aliasing (which results if $t_s > T$, or equivalently and more usefully in terms of $\Delta\Omega$, $\Delta\Omega < 2\pi/T$), we have $\Delta\Omega \cdot \tilde{x}_{s1,c}(t) = x_c(t)$. Consequently, (6.96) becomes (where again $t_s = 2\pi/\Delta\Omega$)

$$\beta_\ell = \frac{1}{t_s} \int_0^{t_s} x_c(t)\, e^{-j\ell t\,\Delta\Omega} dt. \tag{6.97}$$

Because $x_c(t)$ is time–limited to $T < t_s$, by the definition of the Fourier transform, we may identify β_ℓ in (6.97) as $\beta_\ell = (1/t_s)X_c(j\ell\,\Delta\Omega) = \{\Delta\Omega/(2\pi)\} \cdot X_c(j\ell\,\Delta\Omega)$. Thus with no time–aliasing, we replace $x_{s1,c}(t)$ in (6.95) by $x_c(t)/\Delta\Omega$ for $t \in [0, t_s]$, so (6.95) becomes [compare with (6.76)]

$$x_c(t) = \frac{\Delta\Omega}{2\pi} \cdot \sum_{\ell=-\infty}^{\infty} X_c(j\ell\Delta\Omega)e^{j\ell t\,\Delta\Omega}, \quad t \in [0, t_s] \text{ and } \left\{t_s > T; i.e., \Delta\Omega < \frac{2\pi}{T}\right\}. \tag{6.98}$$

Thus in (6.93), $\hat{x}_{N,\Delta\Omega,c}(t)$ is seen to be a truncated estimate of $x_c(t)$.

If only N samples of the estimate $\hat{x}_{N,\Delta\Omega,c}(t)$ are needed at the equally spaced times $n\,\Delta t = 2\pi n/(N\,\Delta\Omega)$, $n \in [0, N-1]$, then the truncated estimate in (6.93) becomes [compare with (6.77)]

$$\hat{x}_{N,\Delta\Omega,c}\left(\frac{2\pi n}{N\Delta\Omega}\right) = \frac{\Delta\Omega}{2\pi} \cdot \sum_{k=0}^{N-1} X_c(jk\,\Delta\Omega)e^{j2\pi nk/N} \tag{6.99a}$$

$$= \frac{N\Delta\Omega}{2\pi} \cdot DFT_N^{-1}\{X_c(jk\,\Delta\Omega)\}, \tag{6.99b}$$

---

[21]The subscript 1 is used to avoid confusion with $X_{s,c}(j\Omega)$ and $x_{s,c}(t)$ associated with temporal sampling.

or, in terms of the temporal estimation interval $\Delta t = 2\pi/(N\,\Delta\Omega)$,

$$\hat{x}_{N,\Delta\Omega,c}(n\,\Delta t) = \frac{1}{\Delta t} \cdot \mathrm{DFT}_N^{-1}\left\{X_c\left(\frac{jk2\pi}{N\Delta t}\right)\right\}. \tag{6.99c}$$

Note, for use below, that $t_s = 2\pi/\Delta\Omega = N\,\Delta t$.

Again, two operations on the original signal $X_c(j\Omega)$ are performed before taking the DFT_N^{-1} in (6.99b), where here we will use the more familiar DTFT, $X(e^{j\theta})$, rather than the equivalent $\tilde{x}_{s1,c}(t)$ used above. Note also that we use θ rather than ω to avoid confusion with previous results. Specifically, if $X_1(z) = \mathrm{ZT}\{X_c(jk\,\Delta\Omega)\}$ and $X_1(e^{j\theta}) = X_1(z)$ evaluated at $z = e^{j\theta}$, we can show (see the problems) that $X_1(e^{-jt\Delta\Omega}) = 2\pi \cdot \tilde{x}_{s1,c}(t)$. Thus from (6.94) and from $2\pi/\Delta\Omega = t_s = N\,\Delta t$ [compare with (6.78) and (6.79)],

1. Sampling: $X(k) = X_c(jk\Delta\Omega) \overset{\mathrm{DTFT}}{\longleftrightarrow} X_1(e^{j\theta}),$ $\qquad$ (6.100a)

   where

$$X_1(e^{-jt\,\Delta\Omega}) = t_s \cdot \sum_{\ell=-\infty}^{\infty} x_c(t + t_s \cdot \ell) \tag{6.100b}$$

$$= N\,\Delta t \cdot \sum_{\ell=-\infty}^{\infty} x_c(t + [N\,\Delta t]\cdot\ell). \tag{6.100c}$$

2. Windowing: $X_N(k) = X(k) \cdot w_N(k)$, where in particular for the rectangular window, $w_N(k) = w_{N,R}(k) = 1, \quad k \in [0, N-1]; 0,$ otherwise.

$$\qquad\qquad\qquad\qquad\qquad\qquad\qquad\qquad\qquad\qquad\qquad\qquad (6.101\mathrm{a})$$

$$X_N(k) = X(k) \cdot w_N(k) \overset{\mathrm{DTFT}}{\longleftrightarrow} X_{N1}(e^{-jt\,\Delta\Omega}) = \frac{1}{2\pi} \cdot \int_{-\pi}^{\pi} X_1(e^{j\theta}) \cdot W_N(e^{j(-t\,\Delta\Omega-\theta)})d\theta$$

$$= \frac{1}{2\pi}\int_{-\pi}^{\pi} X_1(e^{-j\theta'})W_N(e^{-j(t\Delta\Omega-\theta')})d\theta' \tag{6.101b}$$

$$= X_1(e^{-j\theta}) *_{c\theta} W_N(e^{-j\theta})/(2\pi)\Big|_{\theta=t\,\Delta\Omega}, \tag{6.101c}$$

where in (6.101b) we introduced $\theta' = -\theta$ to make use of (6.100b), below.

Now note from (6.101a) that $X_{N1}(e^{-jt\,\Delta\Omega}) = \mathrm{DTFT}\{X_N(k)\} = \mathrm{DTFT}\{X_c(jk\,\Delta\Omega) \cdot w_N(k)\}$. Consequently, noting that the *sum* in (6.93) is the DTFT of $X_c(jk\,\Delta\Omega) \cdot w_N(k) = X_N(k)$, we have from the left–hand side of (6.93) that $X_{N1}(e^{-jt\,\Delta\Omega}) = (2\pi/\Delta\Omega) \cdot \hat{x}_{N,\Delta\Omega,c}(t)$. Using this fact, we can obtain a single convolutional expression for $\hat{x}_{N,\Delta\Omega,c}(t) = (\Delta\Omega/2\pi) \cdot X_{N1}(e^{-jt\,\Delta\Omega})$ in terms of $x_c(t)$ (that holds with or without aliasing), as follows. Define for convenience $t' = \theta'/\Delta\Omega$ in (6.101b), and thus $\theta' = \pm\pi \rightarrow t' = \pm t_s/2$, and $d\theta' = \Delta\Omega \cdot dt'$. Then before sampling at the N frequencies $\Omega_k = 2\pi k/(N\,\Delta t)$, $\hat{x}_{N,\Delta\Omega,c}(t)$ may be written [see (6.101b) and henceforth assume the rectangular window]

$$\hat{x}_{N,\Delta\Omega,c}(t) = \frac{\Delta\Omega}{2\pi} X_{N1}(e^{-jt\,\Delta\Omega})$$

$$= \frac{\Delta\Omega}{2\pi} \cdot \frac{1}{2\pi} \cdot \int_{-t_s/2}^{t_s/2} X_1(e^{-jt'\Delta\Omega})W_{N,R}(e^{-j(t-t')\Delta\Omega}) \, (\Delta\Omega \, dt'). \tag{6.102}$$

Now using (6.100b) and $t_s = 2\pi/\Delta\Omega$, we obtain [compare with (6.81)]

$$\hat{x}_{N,\Delta\Omega,c}(t) = \left\{\frac{\Delta\Omega}{2\pi}\right\}^2 \int_{-t_s/2}^{t_s/2} \frac{2\pi}{\Delta\Omega} \sum_{\ell=-\infty}^{\infty} x_c(t' + t_s \cdot \ell) \cdot W_{N,R}(e^{-j(t-t')\Delta\Omega}) \, dt'. \tag{6.103}$$

Now the definition of a new variable $t'_{\text{new}} = t'_{\text{old}} + t_s\ell$ allows the finite integral of an infinite sum to be converted into a single infinite–length integral. As before, upon noting that $t'_{\text{old}} \cdot \Delta\Omega = t'_{\text{new}} \cdot \Delta\Omega - \ell t_s \Delta\Omega = t'_{\text{new}} \cdot \Delta\Omega - \ell 2\pi$, the argument of $W_{N,R}$ is unchanged after the substitution. Thus,

$$\hat{x}_{N,\Delta\Omega,c}(t) = \frac{\Delta\Omega}{2\pi} \cdot \int_{-\infty}^{\infty} x_c(t') \cdot W_{N,R}(e^{-j(t-t')\Delta\Omega}) \, dt' \qquad (6.104a)$$

$$= \frac{\Delta\Omega}{2\pi} x_c(t) \overset{*}{_c} W_{N,R}(e^{-jt\,\Delta\Omega}) \qquad (6.104b)$$

$$= \frac{\Delta\Omega}{2\pi} \sum_{k=0}^{N-1} X_c(jk\,\Delta\Omega) \cdot e^{jtk\,\Delta\Omega}, \qquad (6.104c)$$

where (6.104c) simply recalls/reproduces (6.93).

The DFT_N^{-1} of $X_N(k) = X_c(k\,\Delta\Omega)$ for $k \in [0, N-1]$ will provide the N samples $\hat{x}_{N,\Delta\Omega,c}(2\pi n/(N\Delta\Omega)) = \hat{x}_{N,\Delta\Omega,c}(n\Delta t)$ as in (6.99c). With or without aliasing, using (6.104a) the estimate can be written in terms of $x_c(t)$, in the following convolutional form:

$$\frac{1}{\Delta t} \cdot \text{DFT}_N^{-1}\{X_N(k)\} = \hat{x}_{N,\Delta\Omega,c}\left(\frac{2\pi n}{N\Delta\Omega}\right) = \hat{x}_{N,\Delta\Omega,c}(n\Delta t)$$

$$= \frac{\Delta\Omega}{2\pi} \cdot x_c(t) \overset{*}{_c} W_{N,R}(e^{-jt\,\Delta\Omega})\Big|_{t=2\pi n/(N\,\Delta\Omega)}, \qquad (6.105)$$

so we see that $\hat{x}_{N,\Delta\Omega,c}$ is the convolution of $x_c(t)$ with the window function $W_{N,R}(e^{-jt\,\Delta\Omega})$. Remember, the times of evaluation are $2\pi n/(N\,\Delta\Omega) = t_s \cdot n/N$. The second half of the DFT_N^{-1} temporal samples will be the negative–time components of the replication of x_c centered on t_s.

The remarks at the end of Section 6.5.3 that the second half of the samples given to the DFT is equivalent to samples at negative index values is particularly important for estimation of the time function using samples of $X_c(j\Omega)$. Often $x_c(t)$ is real–valued, so $X_c(j\Omega)$ will be a conjugate–symmetric function of Ω. For $\hat{x}(t)$ to also be real–valued, the conjugate–symmetric property must be guaranteed by proper choice of $X(k) = X_c(jk\,\Delta\Omega)$. Specifically, just choosing $X_c(jk\,\Delta\Omega)$ for $k\,\Delta\Omega > 0$ will not result in a real–valued time function upon DFT^{-1}. Rather, instead we replace the second half of samples by their conjugate counterparts (see Table 6.1 with n replaced by k). Specifically, set $[X(-k) = X(N-k)] = X^*(k)$ for the second half of samples. Be sure not to miss the full presentation of the DFT used for estimating $x_c(t)$ in Section 6.7 [in particular, (6.109)].

EXAMPLE 6.6

Consider again $x_{L,c}(t)$, the triangle function repeated L times (see Example 6.5). Suppose now that rather than being given $x(n) = x_{L,c}(n\Delta t)$ we are given $X(k) = X_{L,c}(jk\,\Delta\Omega) = X_{L,c}(j2\pi k/(N\Delta t))$. Investigate the estimation of $x_c(t)$ using these frequency samples.

Solution

The time function $x_{L,c}(n\Delta t)$ is again estimated by taking the inverse N–point DFT of $X(k)$ and dividing by $\Delta t = 2\pi/(N\Delta\Omega)$. By analogy with the Nyquist theory for the sampling of time functions, (6.94) $[\tilde{x}_{s1,c}(t) = \frac{1}{\Delta\Omega} \sum_{k=-\infty}^{\infty} x_c(t + kt_s)]$ indicates that no aliasing will take place if [replication interval $= t_s = 2\pi/\Delta\Omega = N\Delta t$] > [duration of $x_{L,c}(t) = L \cdot T$].

Figures 6.51 through 6.54 show estimations for various L, $\Delta\Omega$, and truncation of spectrum. In each figure, part (a) shows the samples of the spectrum (and in the truncated cases below, zero–padding out to N), whereas part (b) shows the estimate of $x_{L,c}(n\Delta t)$ (stems) overlaid on the exact solution $x_{L,c}(n\Delta t)$ (solid). The spectra are shown zero–centered for ease in viewing. Note that now L affects the amount of aliasing, not truncation [as it did for estimating $X_c(j\Omega)$], for a given value of $\Delta\Omega$.

For $\{N = 64, T = 1 \text{ s, and } L = 1\}$, $\Delta\Omega = 2\pi/(N\Delta t)$ is chosen in Figure 6.51 as $\Delta\Omega = 2\pi \cdot (0.5 \text{ Hz})$ so that $N\Delta t = 2 \text{ s} = 2 \cdot T$—that is, *spectral* oversampling by the factor two. As expected, in Figure 6.51b there is perfect agreement with $x_{1,c}(t)$. If we instead choose $\Delta\Omega = 2\pi \cdot (1.25 \text{ Hz})$ as in Figure 6.52, then $N\Delta t = 0.8 \text{ s} = 0.8 \cdot T < T$ and there is time–aliasing. In this case, there is only a single contributing aliasing

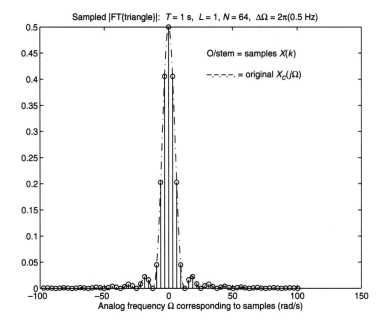

Figure 6.51a

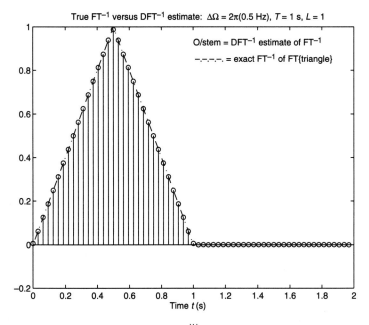

(b)

Figure 6.51b

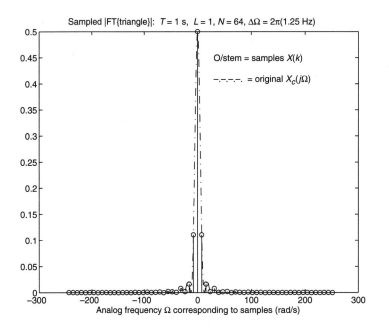

Figure 6.52*a*

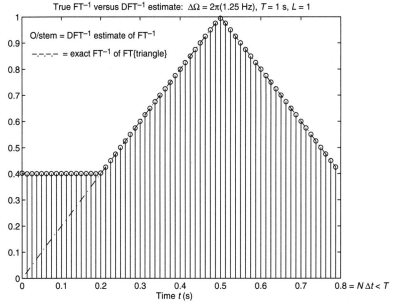

Figure 6.52*b*

term ($m = 1$). Furthermore, this term is linear and of opposite slope, causing the aliased sum to be perfectly constant in time over the aliased interval. Finally, we expect the aliasing to be worst for small t because unlike spectral replications which are centered on zero, the temporal replications *begin* at $t = 0$, as in this example $x_c(t) \neq 0$ for $t \in [0, T]$. Thus, when the adjacent left replication extends too far to the right, it distorts the primary waveform for small $t > 0$. Corresponding estimates for the case $L = 4$ are briefly studied in the problems.

Having considered the aliasing effect due to sampling in inverse Fourier transform estimation, now the smoothing effect due to windowing will be illustrated. Let

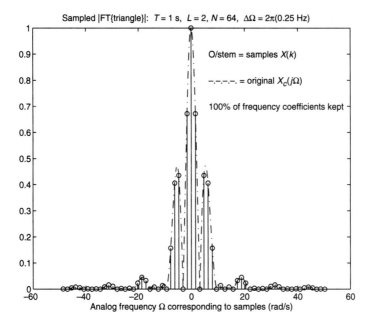

Figure 6.53a

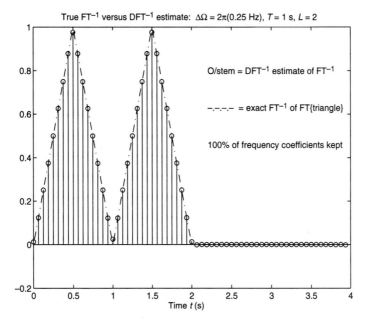

Figure 6.53b

$L = 2, N = 64$, and $\Delta\Omega = 2\pi \cdot (0.25 \text{ Hz})$ so that $N\Delta t = 2 \cdot (2T)$ so that there is no aliasing. Figures 6.53a and 6.54a keep, respectively 100% and 10% of the N DFT samples obtained by sampling $X_{2,c}(j\Omega)$. The resulting reconstruction is perfect in Figure 6.53b, whereas there is extreme smoothing in Figure 6.54b due to the severe truncation (effectively, a form of lowpass filtering). These pictorial examples should give the reader a better feel for all the mathematical results previously derived concerning the effects of sampling and windowing. Another example of this DFT method of estimation of the inverse Fourier transform is shown in Example 6.7, in a practical application.

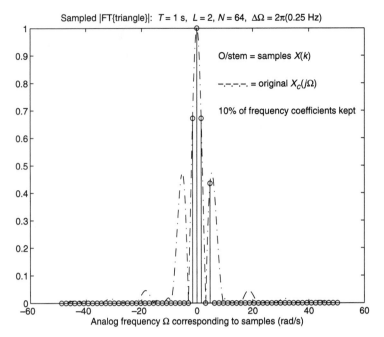

Figure 6.54a

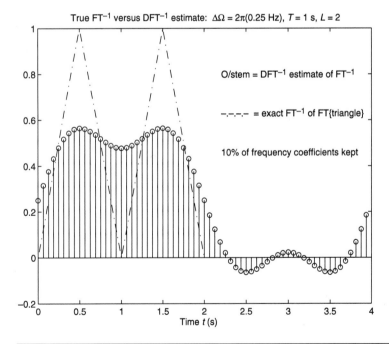

Figure 6.54b

In summary, in approximating either the forward or inverse Fourier integrals by means of a DFT_N, sampling $\leftrightarrow$ aliasing, and windowing $\leftrightarrow$ convolutional smoothing are involved and unavoidable. In either case, the resolution may be improved by increasing N. For reduction of aliasing, however, the conditions on Δt are opposite: For the forward Fourier transform estimation, Δt should be minimized; for the inverse Fourier transform estimation, Δt should be maximized ($\Delta\Omega$ minimized). We must always be aware of the implications that the N–periodicities assumed by the DFT_N have on both the choice of sequence to supply to the DFT_N and the resulting transformed sequence. Finally, for a given $\Delta\Omega$ in estimating the FT^{-1}, the temporal resolution is $\Delta t = 2\pi/(N\Delta\Omega)$, and all the comments in Sections 6.5.2 and 6.5.3 concerning the time–bandwidth product and resolution apply also to this case of inverse Fourier transform estimation. We close this section with an application example of FT^{-1} estimation using the DFT.

EXAMPLE 6.7

In this example, we refer briefly to a practical application of the results of this section, in the author's work on ultrasonic medical imaging.[22] For brevity here, the complete details of this study are found on WP 6.3. We will show here only the graphical results, which give an idea about the possibilities of using the DFT to estimate continuous–time waveforms from spectral samples.

In medical ultrasonic inverse scattering, we need to test our algorithms against known solutions, such as a right circular cylindrical scatterer. The solution is known in closed form only in the frequency domain, for plane–wave incident fields. We therefore take the inverse DFT of these known samples at each point on an imaging grid to arrive at a "known" solution against which to compare estimates via inverse scattering algorithms that use scattering data. The total ultrasonic field in Figure 6.55 calculated by this DFT^{-1} procedure is shown at times t_0 (Figure 6.55a) and $t_0 + 2\mu s$ (Figure 6.55b), where the latter time is 19 time samples away in the final array. The pictures agree with our analogous experience with water waves hitting a cylindrical obstacle.

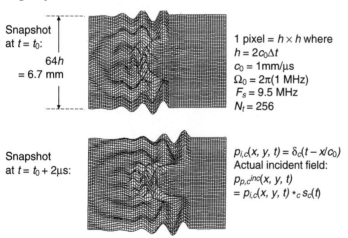

Snapshot at $t = t_0$:

64h = 6.7 mm

1 pixel = $h \times h$ where
$h = 2c_0\Delta t$
$c_0 = 1\,mm/\mu s$
$\Omega_0 = 2\pi(1\ MHz)$
$F_s = 9.5\ MHz$
$N_t = 256$

Snapshot at $t = t_0 + 2\mu s$:

$p_{i,c}(x, y, t) = \delta_c(t - x/c_0)$
Actual incident field:
$p_{p,c}^{inc}(x, y, t)$
$= p_{i,c}(x, y, t) *_c s_c(t)$

Incident wave pulse shape: $s_c(t) = Se^{-(c_0 t/\sigma)^2}\sin(\Omega_0 t)u_c(t)$

Figure 6.55

6.7   PICTORIAL SUMMARY OF DETERMINISTIC DFT SPECTRUM ESTIMATION

Figure 6.56 depicts many of the relationships we have been investigating in this chapter respecting sampling and FT or FT^{-1} estimation. Each plot in Figure 6.56 is labeled with a letter $[(a)$ through $(k)]$. Beginning at the top of the figure, we see an example $x_c(t)$ [left: plot (a)] and its FT, $X_c(j\Omega)$ [right: plot (b)]. We wish either to estimate $X_c(jk\Delta\Omega)$ using $DFT\{x_c(n\Delta t)\}$ or to estimate $x_c(n\Delta t)$ using $DFT^{-1}\{X_c(jk\Delta\Omega)\}$.

[22] T. J. Cavicchi, "Transient High-Order Ultrasonic Scattering," *J. Acoust. Soc. Amer.,* 88, 2 (Aug. 1990): pp. 1132–1141; T. J. Cavicchi, "Matrix Solution of Transient High-Order Ultrasonic Scattering," *J. Acoust. Soc. Am.,* 90, 2 (Aug. 1991): pp. 1085–1092.

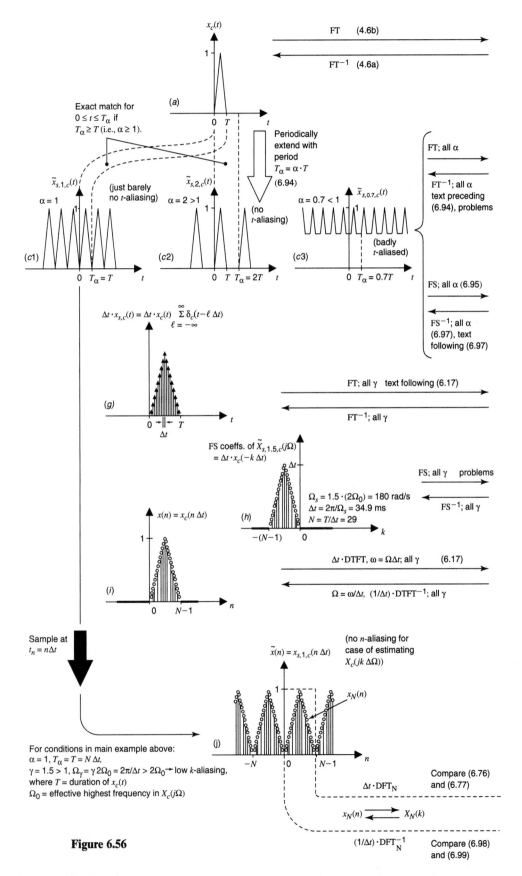

Figure 6.56

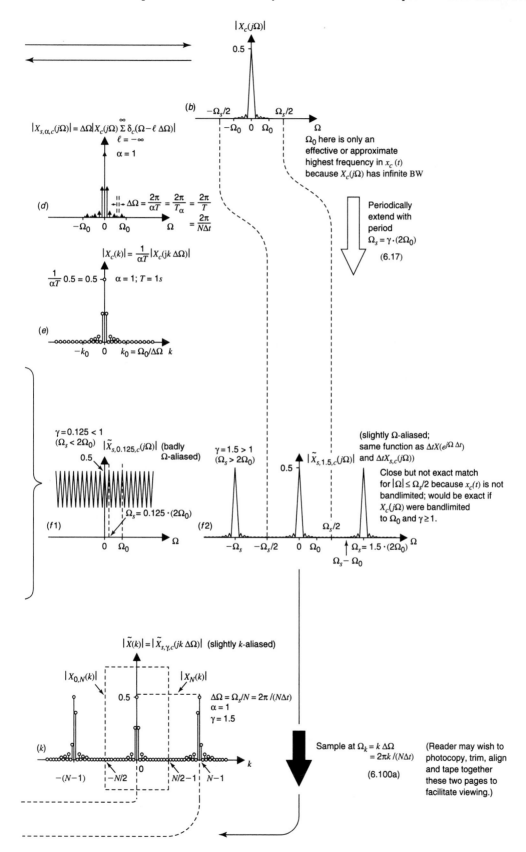

In Figure 6.56, we assume that the temporal duration of $x_c(t)$ is limited to $t \in [0, T]$. Notice that for practical convenience, we also assume that $x_c(t)$ is approximately bandlimited to $|\Omega| < \Omega_0$ [even though we know that because $x_c(t)$ is time–limited, it cannot be perfectly bandlimited]. This assumption will incur minimal aliasing errors if Δt is correctly chosen.

The first operation depicted below the top row [on the left: plots $(c1)$, $(c2)$ and $(c3)$] is the periodic replication of $x_c(t)$ every $T_\alpha = \alpha T = t_s = N\Delta t$, where we have defined the parameter $\alpha \equiv t_s/T$. This periodic replication is in anticipation of the frequency sampling we ultimately have with our N spectrum samples $X_c(k\,\Delta\Omega)$ obtained for the purpose of estimating $x_c(t)$ [see (6.100): $X(k) = X_c(jk\,\Delta\Omega) \xrightleftharpoons{\text{DTFT}} X_1(e^{j\theta})$,

$$X_1(e^{-jt\,\Delta\Omega}) = N\Delta t \cdot \sum_{\ell=-\infty}^{\infty} x_c(t + [N\Delta t] \cdot \ell)].$$ For estimation of $X_c(jk\,\Delta\Omega)$ by DFT$\{x_c(n\Delta t)\}$, α is controlled by $N\Delta t$: $N\Delta t = \alpha T$ and thus $T_\alpha = N\Delta t$, or $\alpha = N\Delta t/T$. If $\alpha < 1$, "t–aliasing" occurs in the T_α–periodic extension: That is, values of $x_c(t)$ at higher–magnitude times alias into smaller times. No time–aliasing will occur when sampling $X_c(j\Omega)$ for the estimation of $x_c(n\Delta t)$ if $T_\alpha = t_s = 2\pi/\Delta\Omega > T$; that is, if $\Delta\Omega < 2\pi/T$;

see (6.94) $[\tilde{x}_{s1,c}(t) = \dfrac{1}{\Delta\Omega} \sum_{k=-\infty}^{\infty} x_c(t + kt_s)]$ and plot (d).

However, assuming we choose $N\Delta t > T$ so that $\alpha > 1$, there is no such thing as time–aliasing ("t–aliasing") when the signal we are sampling is $x_c(t)$ [for the estimation of $X_c(jk\,\Delta\Omega)$], so in this case, $\alpha \geq 1$. Plot $(c3)$ $(\alpha < 1)$ is considered below, when we discuss sampling $X_c(j\Omega)$ to estimate $x_c(t)$.

Concerning windowing, again consider the case of estimation of $X_c(jk\,\Delta\Omega)$ via DFT$\{x_c(n\Delta t)\}$. Plot (a) in Figure 6.56 is drawn for the simpler case of no truncation: $N\Delta t \geq T$. However, it can be made to indicate truncation by merely renaming $x_c(t)$ in plot (a) as the windowed signal $x_{R,c}(t) = \{x_c(t)$ for $t \in [0, N\Delta t]$ and 0 otherwise$\}$ $= x_c(t) \cdot w_{R,c}(t)$, where $x_c(t)$ would be some longer signal (not pictured) that is nonzero outside $t \in [0, N\Delta t]$. More specifically, the window $w_{R,c}(t)$ is selected to be large enough to include the samples $x_c(n\Delta t)$ for $n \in [0, N-1]$; to avoid edge–of–window problems, it is chosen to extend $-\Delta t/2$ below $t = 0$ and $\Delta t/2$ above $t = (N-1)\Delta t$. Then the label of the continuous–time FT in plot (b), $X_c(j\Omega)$, would be replaced by $X_{R,c}(j\Omega) = W_{R,c}(j\Omega) *_{c\Omega} X_c(j\Omega)/(2\pi)$, where

$$W_{R,c}(j\Omega) = N\Delta t\, \frac{\sin\{\Omega N\,\Delta t/2\}}{\Omega N\,\Delta t/2} \cdot e^{-j\Omega(N-1)\,\Delta t/2} \tag{6.106}$$

$$= N\Delta t\, \text{sinc}\{\Omega N\,\Delta t/2\} \cdot e^{-j\Omega(N-1)\,\Delta t/2}.$$

The *order* of operations of sampling and truncation is exactly reversible, as proved in the problems using $W_{R,c}(j\Omega)$. This agrees with common sense, because in either case exactly the same set of samples results.

When we (a) set $X_{R,c}(j\Omega) = W_{R,c}(j\Omega) *_{c\Omega} X_c(j\Omega)/(2\pi)$, (b) sample/alias $X_{R,c}(j\Omega)$ using (6.17), and (c) equate the result with $\hat{X}_{N,\Delta t,c}(j\Omega)$ in (6.82a) for all Ω, we obtain the requirement that the following identity holds, where we define $\theta = \Omega\Delta t/2$ (again, see the problems for details of the derivation):

$$\sin(\theta) \cdot \sum_{m=-\infty}^{\infty} \frac{(-1)^m}{\theta + \pi m} = 1, \tag{6.107}$$

which may be verified to hold for all θ. That is, the truncation function $W_{N,R}(e^{j\omega}) = \{\sin(N\omega/2)/\sin(\omega/2)\} \cdot e^{-j\omega(N-1)/2}$ [from (6.80)] *after* sampling produces exactly the same overall effect as does $W_{R,c}(j\Omega)$ *before* sampling.

For estimation of $x_c(n\Delta t)$ by $\text{DFT}^{-1}\{X_c(jk\,\Delta\Omega)\}$, α is controlled by $\Delta\Omega$: $2\pi/\Delta\Omega = N\Delta t = \alpha T$, or $\alpha = 2\pi/(T\Delta\Omega)$. Now, we may have any value of α, depending on the value chosen for $\Delta\Omega$. In this case, it is Δt that is not independently controllable, being fixed by the points of time evaluation generated by the DFT_N^{-1}: $\Delta t = 2\pi/(N\Delta\Omega)$ [or $\Delta t = 1/(N\Delta F)$].

Three example periodic extensions of $x_c(t)$ for different values of α are shown in Figure 6.56: $(c1)$: $\alpha = 1$, $(c2)$: $\alpha = 2$, and $(c3)$: $\alpha = 0.7$. For $\alpha = 1$, the adjacent replications just barely touch and thus there is no "t–aliasing." For $\alpha = 2$, there is plenty of spread between replications and thus no t–aliasing, whereas for $\alpha = 0.7$, there is severe overlapping (t–aliasing). Again, $\alpha = 0.7 < 1$ can occur only for estimation of $x_c(n\Delta t)$ from samples of $X_c(j\Omega)$.

The periodic extension of $x_c(t)$ is labeled as $\tilde{x}_{s,\alpha,c}(t)$; in the notation of (6.94) it is $\Delta\Omega \cdot \tilde{x}_{s1,c}(t)$. The Fourier transform of $x_{s,\alpha,c}(t)$ is denoted $X_{s,\alpha,c}(j\Omega)$ in Figure 6.56, which by the paragraph preceding (6.94) is equal to $\Delta\Omega \cdot X_{s1,c}(j\Omega) = \Delta\Omega \cdot X_c(j\Omega) \cdot \delta_{\Delta\Omega}(\Omega)$ [plot (d)], a series of Dirac delta functions weighted by the samples of $\Delta\Omega \cdot X_c(j\Omega)$ every $\Delta\Omega = 2\pi/(\alpha T)$. Alternatively, because $\tilde{x}_{s,\alpha,c}(t)$ is periodic, it has a Fourier series [plot (e)]. The FS formula is given in (6.97) $\left[\beta_\ell = \frac{1}{t_s} \int_0^{t_s} x_c(t) e^{-j\ell t\,\Delta\Omega} dt \right]$ and the FS^{-1} is in (6.95) $\left[\tilde{x}_{s,\alpha,c}(t) = \Delta\Omega \cdot \tilde{x}_{s1,c}(t) = \sum_{\ell=-\infty}^{\infty} \beta_\ell e^{j\ell t\,\Delta\Omega} \right]$. See also the text following (6.97), by which we know that the FS coefficients β_k are $\{\Delta\Omega/(2\pi)\} \cdot X_c(jk\,\Delta\Omega)$, where again $\Delta\Omega = 2\pi/(\alpha T)$ so that the Fourier series coefficients β_k may be written as $X_c(jk\,\Delta\Omega)/(\alpha T)$ as indicated at the top of plot (e) in which β_k is renamed $X_c(k)$.

In anticipation of the effects of time–sampling, we next consider the periodic extension of $X_c(j\Omega)$ [Figure 6.56, plots $(f1)$ and $(f2)$], denoted $\tilde{X}_{s,\gamma,c}(j\Omega)$, where γ will be defined momentarily. We know that there will be some Ω–aliasing in our example of a triangle because $x_c(t)$ is time–limited so its bandwidth $2\Omega_0 = \infty$, but we can arrange for it to be small by choosing Δt small [and thus also choosing N large to cover $x_c(t)$ completely]. We also know that under Ω–aliasing, the replications are separated by $\Omega_s = 2\pi/\Delta t$. We let γ be defined as $\gamma = \Omega_s/(2\Omega_0)$ so that at or above Nyquist sampling, $\gamma \geq 1$ (no spectral aliasing) and for aliasing, $\gamma < 1$. [In analogy with $T_\alpha = \alpha \cdot T$ in plots (c), we could have defined $\Omega_\gamma = \gamma \cdot (2\Omega_0)$, but from the standpoint of estimating $X_c(j\Omega)$ from samples of $x_c(t)$, we know Ω_γ has the value $\Omega_\gamma = \gamma \cdot 2\Omega_0 = (\Omega_s/[2\Omega_0]) \cdot 2\Omega_0 = \Omega_s = 2\pi/\Delta t$.] For estimation of $X_c(jk\,\Delta\Omega)$ by $\text{DFT}\{x_c(n\Delta t)\}$, γ is controlled by Δt: $2\pi/\Delta t = \Omega_s = \gamma \cdot (2\Omega_0)$, or $\gamma = \pi/(\Omega_0\Delta t)$. For estimation of $x_c(n\Delta t)$ by $\text{DFT}^{-1}\{X_c(jk\,\Delta\Omega)\}$, γ is controlled by $N\Delta\Omega$: $N\Delta\Omega = \gamma(2\Omega_0)$, or $\gamma = N\Delta\Omega/(2\Omega_0)$. Again, there is no Ω–aliasing in the case of estimating $x_c(n\Delta t)$ from $\text{DFT}^{-1}\{X_c(jk\,\Delta\Omega)\}$, as long as $N\Delta\Omega > 2\Omega_0$.

In this example, we chose Ω_0 to mean not absolute bandwidth but practical bandwidth: $\Omega_0 = 60$ rad/s, beyond which $|X_c(j\Omega)|$ is quite small. Thus $\gamma = 1$ corresponds to $\Omega_s = 120$ rad/s. Two examples of replicated spectra for different values of γ are shown in Figure 6.56: $\gamma = 0.125$ [severe Ω–aliasing: plot $(f1)$] and $\gamma = 1.5$ [minimal Ω–aliasing: plot $(f2)$]. Were there no aliasing at all, the periodic extension of $X_c(j\Omega)$, denoted $\tilde{X}_{s,\gamma,c}(j\Omega)$, would be equal to $X_c(j\Omega)$ for $|\Omega| < \Omega_s/2$. The relevant mathematical relationship is given in (6.17), where here $\tilde{X}_{s,\gamma,c}(j\Omega) = \Delta t \cdot X(e^{j\Omega\Delta t}) = \Delta t \cdot X_{s,c}(j\Omega)$. This is illustrated in plot $(f2)$.

By its definition, we know that the inverse Fourier transform of $X_{s,c}(j\Omega)$, $x_{s,c}(t)$, is the "sampled signal," which from (6.10a) is a series of Dirac delta functions weighted by samples of $x_c(t)$ every Δt [see plot (g)]. Because $\tilde{X}_{s,\gamma,c}(j\Omega) = \Delta t X(e^{j\Omega\,\Delta t})$, we know that $\tilde{X}_{s,\gamma,c}(j\Omega)$ is periodic in Ω, so it has a Fourier series. We show in a homework problem that its Fourier series coefficients are $\Delta t \cdot x_c(-k\,\Delta t)$—that is, samples of the time–reversed version of $x_c(t)$—as depicted in plot (h). More important for our purposes is the (equivalent) DTFT relationship in (6.17) between the *samples* $x(n)$ of $x_c(t)$ and $\tilde{X}_{s,\gamma,c}(j\Omega)$, shown in plot (i).

Finally, putting together the N time samples and N frequency samples, we have

$$(6.77b) \quad \left[\hat{X}_{N,\Delta t,c}\!\left(j\frac{2\pi k}{N\Delta t} \right) = \Delta t \cdot \text{DFT}_N\{x_c(n\,\Delta t)\} \right] \text{ for use of the forward DFT}_N \text{ to}$$

estimate samples of $X_c(j\Omega)$ from samples of $x_c(t)$ and (6.99b) $\left[\hat{x}_{N,\Delta\Omega,c}\!\left(\dfrac{2\pi n}{N\,\Delta\Omega} \right) = \right.$

$\left. \dfrac{N\Delta\Omega}{2\pi} \cdot \text{DFT}_N^{-1}\{X_c(jk\,\Delta\Omega)\} \right]$ for use of the DFT_N^{-1} to estimate samples of $x_c(t)$ from

samples of $X_c(j\Omega)$. The following operations summarize the relationship between the N time samples [plot (j)] and their DFT_N [plot (k)], for the case of estimating $X_c(jk\,\Delta\Omega)$ from the DFT_N of the time samples $x_c(n\,\Delta t)$, where in (6.108b) we cite (6.83):

FT estimation using DFT.

$$\tilde{x}_N(n) = N\text{–PERIODIC EXT. } \{\text{SAMPLE}_{t_n=n\,\Delta t;\, n\in[0,\,N-1]}\, \{x_c(t) \cdot w_{R,c;[0,\,N\Delta t]}(t)\}\} \quad (6.108a)$$

$$x_N(n) = x_c(n\,\Delta t) = \tilde{x}_N(n), \quad n \in [0, N-1]$$

$$\to \Delta t \cdot \text{DFT}_N/\text{DFS}_N \to$$

$$\tilde{X}_N(k) = \text{SAMPLE}_{\Omega_k=2\pi k/(N\Delta t)=k\Omega_s/N;\, k\in[0,\,N-1]} \left\{ \frac{\Delta t}{2\pi} X_c(j\Omega) *_{c\Omega} W_{N,R}(e^{j\Omega\,\Delta t}) \right\} \Bigg|_{\substack{\Omega_s=2\pi/\Delta t\text{–periodic in }\Omega, \\ N\text{–periodic in }k}} \quad (6.108b)$$

$X_N(k) = \tilde{X}(k), \quad k \in [0, N-1];$
note that $\Delta\Omega = 2\pi/(N\Delta t)$ and $\Omega_s = 2\pi/\Delta t = N\,\Delta\Omega$.

For the case of estimating $x_c(n\,\Delta t)$ from the DFT_N^{-1} of the frequency samples $X_c(jk\,\Delta\Omega)$, in which we, for example, define $w'_{R,c}(\Omega)$ to be 1 for $|\Omega| \le \Omega_0$ and zero otherwise, and where in (6.109b) we cite (6.104b) $[\hat{x}_{N,\Delta\Omega,c}(t) = \left(\dfrac{\Delta\Omega}{2\pi} \right) x_c(t) *_c W_{N,R}(e^{-jt\,\Delta\Omega})]$,

---

FT^{-1} estimation using DFT^{-1}.

$$\tilde{X}_N(k) = N\text{–PERIODIC EXT. } \{\text{SAMPLE}_{\Omega_k=k\Delta\Omega;\, k\in[0,\,N-1]}\, \{X_c(j\Omega) \cdot w'_{R,c;[-N\Delta\Omega/2,\,N\Delta\Omega/2]}(\Omega)\}\} \quad (6.109a)$$

$$X_N(k) = X_c(k\,\Delta\Omega) = \tilde{X}_N(k), \quad k \in [0, N-1]$$

$$\to (N\,\Delta\Omega/[2\pi]) \cdot \text{DFT}_N^{-1}/\text{DFS}_N^{-1} \to$$

$$\tilde{x}_N(n) = \text{SAMPLE}_{t_n=2\pi n/(N\Delta\Omega)=nt_s/N;\, n\in[0,\,N-1]} \left\{ \frac{\Delta\Omega}{2\pi} x_c(t) *_c W_{N,R}(e^{-jt\,\Delta\Omega}) \right\} \Bigg|_{\substack{t_s=2\pi/\Delta\Omega\text{–periodic in }t \\ N\text{–periodic in }n}} \quad (6.109b)$$

$x_N(n) = \tilde{x}(n), \quad n \in [0, N-1];$
note that $\Delta t = 2\pi/(N\,\Delta\Omega)$ and $t_s = 2\pi/\Delta\Omega = N\Delta t$.

These are the complete relations relevant for, respectively, estimating the FT or FT^{-1}. The DFT relation for continuous–time spectrum estimation in which truncation is replaced by periodic extension is presented in Section 7.3 [see (7.18)]. These basic mathematical relations, along with the graphical ones in Figure 6.56, should give a clear idea about what is going on in DFT deterministic spectrum estimation.

PROBLEMS

Sampling and Aliasing

6.1. Define the sampling operators $S\{\cdot\}$ and $s\{\cdot\}$ defined by $X_{s,c}(s) = S\{X_c(s)\}$ and $x_{s,c}(t) = s\{x_c(t)\}$. Explain in equations and words what these operators do to their inputs.

6.2. Although the reconstruction filter in (6.20) is most convenient to analyze, it would be better to reduce the bandwidth of the filter to accommodate only the signal, when the signal has been oversampled [$\Delta t < \pi/\Omega_0$, where Ω_0 is the maximum frequency in the original signal $x_c(t)$ whose samples we wish to reconstruct]. So, in (6.20) let $R_c(j\Omega) = \Delta t$ for only $|\Omega| \leq \Omega_0$. Find the equation corresponding to (6.23) (with $r_c(t)$ determined) for this case.

6.3. (a) In the discussion following (6.1), the generalization of (6.1) for an arbitrary continuous–time sinusoid was given. As in that discussion, let the frequency of the sinusoid be Ω_1. Let the sampling interval be Δt, any value for which there is no aliasing; $\Omega_s \equiv 2\pi/\Delta t > 2\Omega_1$. Write an expression for all the frequencies $\Omega_{1,\ell>1}$ producing the same sequence of values under sampling every Δt (that is, prove the stated generalization to be true).

    (b) Using this result, suppose that an $F_1 = 5$–kHz sinusoid $\sin(2\pi F_1 t)$ is sampled every $\Delta t = 50$ μs. Give the next three frequencies (in hertz) higher than 5 kHz resulting in the same samples as those of the 5–kHz sinusoid. What repeating pattern do these sample values form? Would you consider this sampling rate and set of obtained samples qualitatively sufficient for representing the 5–kHz sinusoid? What does the reconstruction formula (6.26) tell us about this set of samples in relation to the original CT sinusoid?

    (c) Using (6.26), prove by direct evaluation of the sinc sum that the single sinusoid $\sin(2\pi \cdot 5000t)$ can be obtained at any time t by using the samples found in part (b). Hints: (i) It is found in a table of sums of trigonometric series that

$$\sum_{n=-\infty}^{\infty} \frac{\sin\{n\theta + \alpha\}}{n\theta + \alpha} = \pi \frac{\sin\{[2k + 1]\pi\alpha/\theta\}}{\theta\sin\{\pi\alpha/\theta\}}, \text{ where } 2k\pi < \theta \leq (2k + 2)\pi.$$

(ii) $\sin(3x) = 3\sin(x) - 4\sin^3(x)$ and $\sin^2(x) = \frac{1}{2} - \frac{1}{2}\cos(2x)$.

6.4. (a) In this chapter, we showed that frequencies larger in magnitude than $\Omega_s/2$ are aliased into the principal frequency band $|\Omega| < \Omega_s/2$, yet we showed in the text and Problem 6.3 that the aliased versions of a sinusoid are separated by $\ell\Omega_s$. Plot the magnitude spectrum (versus Ω, using $\omega = \Omega\Delta t$) of $\exp\{j\Omega_{1,1}n\,\Delta t\}$, where $\Delta t = 2\pi/\Omega_s$ and $\Omega_{1,1} = \Omega_1 + 1 \cdot \Omega_s$ is just slightly above $\Omega_s/2$. Also show the aliased version in the principal frequency band. Is Ω_1 positive or negative, and what frequency is it near?

    (b) Repeat part (a) but this time for $\sin\{\Omega_{1,1}n\,\Delta t\}$, and plot the imaginary part of the spectrum. Also, now let $\Omega_{1,1}$ be just slightly above Ω_s, not $\Omega_s/2$. Again, what frequency is Ω_1 near? What happens if $\Omega_{1,1}$ is just above $\Omega_s/2$? Interpret your results.

6.5. If we focus near the rim of a 27-in. diameter bicycle wheel, how fast must it be going (minimum speed) for its spokes (about 1.5 in. apart at the rim) to appear to be still when in fact in motion? Assume that the brain receives a snapshot 30 times every second. Use reasonable approximations, and give your answer in miles per hour. Does it matter whether we look at the outer versus inner part of the wheel? Note that the effect is in practice somewhat difficult to observe because of the thinness of the spokes and the tendency to focus our eyes beyond them. The effect is very easy to observe in old movie projector take–up reels during rewinding.

6.6. Show/describe an example of how a spoked wheel may appear to be slowly rotating backwards under aliasing (when spinning rapidly).

6.7. Find LT{sampled–impulsive step function} $= U_{s,c}(s)$, where $u_c(t)$ is the unit step function [and so $U_c(s) = 1/s$]. Using (6.12b) and (6.16a) to obtain two expressions for $U_{s,c}(s)$, obtain an interesting identity.

6.8. Suppose $X_c(j\Omega)$ is a triangle of unity height, centered on $\Omega = 0$, and extending to $\pm\Omega_1$. Give the range of Δt for no aliasing of $x_c(t)$, of $dx_c(t)/dt$, and of $\int_0^t x_c(t')dt'$. Comment on the results for practical sampling for general $x_c(t)$.

6.9. Suppose the spectrum of $x_c(t)$ is as described in Problem 6.8, and let $H_c(j\Omega)$ be an ideal unity–passband–gain bandpass filter with lower cutoff $\Omega_1/2$ and upper cutoff $1.5\Omega_1$ (and symmetrical for $\Omega < 0$). Draw $Y_c(j\Omega)$ for (a) $y_c(t) = x_c(t) *_c h_c(t)$ and (b) $y_c(t) = x_c(t)h_c(t)$. In each case, give the condition on Δt for no aliasing in the sampling of $y_c(t)$.

6.10. Develop the sampling relation (6.17) without reference to $x_{s,c}(t)$. This will give only (6.17), not the s–z relation (6.16). Hint: Use (2A.25) and (2A.14).

6.11. Find $\Delta_{\Delta t}(s)$, the unilateral Laplace transform of $\delta_{\Delta t}(t)$.

6.12. Consider sampling of a sinusoid of frequency Ω_0 using a sampling frequency $\Omega_s = 2\Omega_0$.

(a) What digital frequencies f and ω are associated with this sinusoid under the given sampling frequency?

(b) Suppose that the sinusoid is $x_c(t) = \cos(\Omega_0 t)$. List the pattern of sample values for this situation of just barely Nyquist–rate sampling. Write a simple formula for the pattern, not involving the cosine function.

(c) Specialize the Shannon reconstruction formula to this case. What identity do you obtain?

(d) Attempt to repeat parts (b) and (c) for $x_c(t) = \sin(\Omega_0 t)$.

6.13. A triangle wave has period $T = 0.1$ s and is sampled at the rate $\Omega_s = 2\pi/\Delta t$ where $\Delta t = 0.01$ s. Is there aliasing? Why or why not?

6.14. Suppose real–valued, Ω_0–bandlimited $x_c(t)$ has the spectrum $X_c(t)$ whose magnitude $|X_c(j\Omega)|$ is as shown in Figure P6.14a.

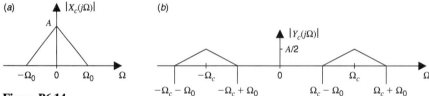

Figure P6.14

(a) What sort of processing of $x_c(t)$ would result in $y_c(t)$ whose magnitude spectrum $|Y_c(j\Omega)|$ is as shown in Figure P6.14b? Your answer should involve only real–valued time functions.

(b) Suppose that we wish to obtain from $y_c(t)$ sample values $x_c(n\Delta t)$ [assuming that $x_c(t)$ is unavailable]. Explain how this could be done using the largest Δt that will avoid aliasing; specify Δt_{max}. Be specific, and include the spectrum of any intermediate signal you need. (The appropriate ideal bicausal lowpass filter impulse response is permitted in your answer.) You should in the process be able to recover $x_c(t)$ exactly for all t.

6.15. For $X_c(j\Omega)$ equal to 1 for $|\Omega| \le \Omega_1$ and zero otherwise, and $x(n) = x_c(n\Delta t)$,

(a) Plot $X(e^{j\omega})$ for (i) $\Omega_s = \frac{4}{3}\Omega_1$ and (ii) $\Omega_s = \frac{7}{3}\Omega_1$.

(b) Suppose that for $\Omega_s = \frac{7}{3}\Omega_1$, an unwanted noise signal $0.2\cos(16.6\Omega_1 t)$ is added onto $x_c(t)$. Either modify your plot of $X(e^{j\omega})$ quantitatively for part (a, ii) and say how to eliminate the problem or explain why no modification is necessary.

6.16. A communication receiver picks up a Gaussian pulse $x_c(t) = \left(\dfrac{1}{\sigma\{2\pi\}^{1/2}}\right)e^{-t^2/[2\sigma^2]}$. In Chapter

10, we will show that $X_c(j\Omega) = e^{-(\Omega\sigma)^2/2}$. That is, the Fourier transform of a Gaussian is another Gaussian. Let $x(n) = x_c(n\Delta t)$ with $\sigma = 5$ ms. How small must Δt be to avoid aliasing components larger than 1% of the maximum value of $X_c(j\Omega)$? Repeat for 10%. Using Matlab, plot $x_c(t)$ for -0.02 s $\le t \le 0.02$ s, $X_c(j\Omega)$ for -1000 rad/s $\le \Omega \le 1000$ rad/s, and for both cases, $x(n)$ for $-10 \le n \le 10$ and $X(e^{j\omega})$ for -4 rad/cycle $\le \omega \le 4$ rad/cycle. To see the effect of negligible versus noticeable aliasing, plot the sinc reconstruction of $x(n)$ for each case, and superimpose $x_c(t)$. Finally, suppose the circuit generating $x_c(t)$ also produces a modulated version of $x_c(t)$; the modulating frequency is 6 kHz. How would that noise appear in $X(e^{j\omega})$ for the 1% case, where would it be centered, and could it be eliminated? If it could be eliminated, how? This problem considers the real–world situation in which total avoidance of aliasing is impossible, but by choosing Δt small enough, it can be negligible.

6.17. Given $H_c(s) = 5/[(s + 1)(s + 2)]$,

(a) Find the causal $h_c(t)$.

(b) Let $h(n) = \Delta t \cdot h_c(n\Delta t)$; find $H(z)$.

(c) Using Matlab, plot the magnitudes of the Fourier transform of $h_c(t)$ and the discrete–time Fourier transform of $h(n)$ for $\Delta t = 0.1$ s and $\Delta t = 0.5$ s. Compare the results. If you find aliasing, explain why it is most marked for certain frequency ranges.

6.18. The Fourier transform $X_c(j\Omega)$ of a signal $x_c(t)$ is two symmetrical audio bands, each of total width $2\Omega_A$; one is centered on $\Omega = \Omega_1$ and the other is centered on $-\Omega_1$; $\Omega_1 \gg \Omega_A$.

(a) Find the minimum value of $\Omega_s = 2\pi/\Delta t$ to avoid aliasing when sampling $x_c(t)$ [without any preprocessing of $x_c(t)$]. Show a typical DTFT spectrum for this example where moderate aliasing occurs; identify critical frequencies. Choose a simple–looking audio spectrum.

(b) We want to sample the audio information in $X_c(j\Omega)$, but we cannot sample any faster than $\Omega_s = 3\Omega_A \ll$ minimum Ω_s in part (a). Show *explicitly* how we could get these samples without aliasing. Draw a block diagram, and plot the spectrum of the signal to be actually sampled and the DTFT of the resulting samples $y(n)$. Show numerical values of all spectrum maxima. Ω_A should not appear anywhere on the ω axis of your plot of $Y(e^{j\omega})$. All critical frequencies in your plots must be labeled, and only real–valued time signals are allowed. Ideal filters are allowed. Assume $\Omega_s = 3\Omega_A$, but also give the minimum possible value of Ω_s.

6.19. A causal damped oscillation is said to be "Nyquist sampled" if there are two samples per oscillation. Comment on the deficiency of such sampling and what related damped oscillating signal (having the same damping and oscillation frequency) would be more closely bandlimited to the oscillation frequency. Justify your comments using Matlab.

6.20. An aircraft chassis is being tested for unwanted resonances that could eventually weaken it. Sensors are attached and the sensor output is measured every $\Delta t = 0.05$ s. A resonance is observed that has exactly 4 oscillations for every 9 samples.

(a) Give the resonant frequency in Hz, assuming no aliasing.

(b) In reality, the sampling rate was insufficient. In the notation following (6.1), the actual frequency is $\Omega_{1, \ell}$ where $\ell = 2$. Evaluate this frequency in Hz and give the required sampling frequency in Hz for no aliasing.

(c) One worker argues that if there is aliasing, the true frequency is just a harmonic of the one observed in part (a). Compare relevant harmonics with the actual frequency in part (b). Are they close? If not, when would they be, and under what conditions would the worker be exactly correct?

6.21. Suppose that $x_c(t) = [A_1e^{\sigma_1 t} \cos(\Omega_1 t) + A_2 t e^{\sigma_2 t} \cos(\Omega_2 t)]u_c(t)$.

(a) Find $X_c(s)$, the Laplace transform of $x_c(t)$. (The t in the second term provides double poles, which enhance the effect to be investigated.)

(b) Show that the peak of the Fourier transform of the first damped sinusoid in $x_c(t)$ can be forced to be approximately unity by choosing $A_1 = |\sigma_1| \cdot \{(1 + [2\Omega_1/\sigma_1]^2)/(1 + [\Omega_1/\sigma_1]^2)\}^{1/2}$. This value of A_1 is easily found by assuming the peak of $|X_{1,c}(j\Omega)| =$ FT $\{A_1e^{\sigma_1 t}\cos(\Omega_1 t)u_c(t)\}$ occurs at $\Omega_{max} = \Omega_1$. Also find the exact value of Ω_{max}, and from it the exact value of A_1 giving $|X_{1,c}(j\Omega_{max})| = 1$.

(c) Let $\sigma_1 = -0.01$ Np/s, $\Omega_1 = 0.1$ rad/s, A_1 as chosen in part (b), $\sigma_2 = -0.8$ Np/s, $\Omega_2 = 5$ rad/s, and $A_2 = 0.5A_1$. Using Matlab, plot $20\log_{10}\{|X_c(j\Omega_0)|\}$. Add a dashed horizontal line 30 dB down from the maximum value of $|X_c(j\Omega)|$ (in this case, -30 dB because the maximum magnitude is approximately 0 dB); determine the frequency Ω_0 for which $20\log_{10}\{|X_c(j\Omega_0)|\}$ $= -30$ dB. Use Ω_0 as a reasonable "bandwidth" of $x_c(t)$, and select Δt from it according to Nyquist sampling. Then plot on the same axes $x_c(t)$, $x_c(n\Delta t)$, and the (truncated) sinc reconstruction of $x_c(n\Delta t)$, where $t \in [0, 10$ s$]$. Is your reconstruction accurate?

(d) Now evaluate and plot (versus Ω) the magnitude of $X_c(s)$ at $s = \sigma_a + j\Omega$, where $\sigma_a = -0.9$ Np/s. Again draw a horizontal line 30 dB down from the maximum, and base Δt on that, and show the time plot as made in part (c). Is the reconstruction accurate now? Explain all your results.

6.22. For Example 6.1, provide the same plots as in Figures 6.11, 6.13, and 6.14 but for $\Delta t = 0.1$ s (an intermediate value).

6.23. In Example 6.6, we noted that for FT^{-1} approximation, L controls the amount of aliasing [not truncation, as in estimation of $X_c(j\Omega)$] for a given $\Delta\Omega$. Here we vary $\Delta\Omega$ to examine aliasing, with L held at 4. Provide the same types of plots as in, for example, Figures 6.51a and b, but with $L = 4$, $N = 64$, and (a) $\Delta\Omega = 2\pi \cdot (0.125$ Hz$)$ and (b) $\Delta\Omega = 2\pi \cdot (0.3125$ Hz$)$. These values of $\Delta\Omega$ give the same choices for $N\Delta t/(LT)$ as in the text, respectively, but for $L = 4$: $N\Delta t/(LT) = 2$ and 0.8, or equivalently, $\Delta\Omega = 2\pi/(N\Delta t) = 2\pi/(2LT) = 2\pi(0.125$ Hz$)$ and $2\pi/(0.8LT) = 2\pi(0.3125$ Hz$)$. Comment on the aliasing or lack thereof for each value of $\Delta\Omega$.

Reconstruction; Pulse Transfer Function

6.24. We may represent zero–order–hold reconstruction using the same analysis as was used in (6.2) through (6.16) to represent zero–order–hold sampling. In both cases, the signal $x_{\Delta t,c}(t)$ in (6.2) is formed from samples $x_c(n\,\Delta t)$ of the original continuous–time signal $x_c(t)$. Express $X_{\Delta t,c}(s)$ in terms of $X_{s,c}(s)$ and $P_{\Delta t}(s)$, where $P_{\Delta t}(s)$ is as given in (6.31a). Then verbally describe the reconstructive filtering operation inherent in this equation, including the impulse response of the zero–order–hold filter.

6.25. The first–order hold is a more accurate reconstructor than the zero–order hold. The first–order hold uses the line passing through $x_c((n - 1)\Delta t)$ and $x_c(n\Delta t)$ to extrapolate. Specifically, the output $x_{\Delta f1,c}(t)$ of the first–order hold is an extrapolation of this line for $n\,\Delta t < t < (n + 1)\Delta t$.

(a) Sketch a typical sequence and the first–order–hold–reconstructed signal $x_{\Delta f1,c}(t)$. For comparison, also show the zero–order–hold reconstruction on the same axes (e.g., using a dotted line).

(b) Find the Laplace transform of this filter, $P_{\Delta f1}(s)$, whose impulse response $p_{\Delta f1}(t)$ satisfies the equation analogous to (6.11): $x_{\Delta f1,c}(t) = x_{s,c}(t) *_c p_{\Delta f1}(t)$. To do this, follow the development in (6.2) through (6.8) to get started, making changes wherever appropriate for this case.

Hint: The factor $t - n\,\Delta t$ multiplying $\delta_c(t - n\,\Delta t) *_c p_{\Delta t}(t)$ in the equation analogous to (6.8a) can be moved inside, replacing $(t - n\,\Delta t)\delta_c(t - n\,\Delta t) *_c p_{\Delta t}(t)$ by $\delta_c(t - n\,\Delta t) *_c [tp_{\Delta t}(t)]$; why?

(c) Find and plot the impulse response of this filter, $p_{\Delta f1}(t)$. Then verbally describe the filtering operation in $x_{\Delta f1,c}(t) = x_{s,c}(t) *_c p_{\Delta f1}(t)$, including the impulse response of the filter.

6.26. Suppose that a signal $a_c(t)$ is sampled and passed through a zero–order–hold reconstructor; call the output $x_c(t)$. We know that $x_c(t)$ will be staircase. Suppose now that $x_c(t)$ is sampled–and–held in exact synchrony with the sample–and–holding of $a_c(t)$. Then the output $x_{\Delta t,c}(t)$ is again staircase and is equal to $x_c(t)$ because $x_c(t)$ is staircase: $x_{\Delta t,c}(t) = x_c(t)$. We know by (6.11) that $x_{\Delta t,c}(t)$ can be decomposed into $x_{\Delta t,c}(t) = x_{s,c}(t) *_c p_{\Delta t}(t)$. Now let $x_{\Delta t,c}(t)$ be processed by an analog filter $G_c(s)$, and let the output be $y_c(t) = g_c(t) *_c x_{\Delta t,c}(t) = g_c(t) *_c \{x_{s,c}(t) *_c p_{\Delta t}(t)\} = \{g_c(t) *_c p_{\Delta t}(t)\} *_c x_{s,c}(t) = h_c(t) *_c x_{s,c}(t)$, where $H_c(s) = G_c(s)P_{\Delta t}(s)$. For example, $G_c(s)$ could represent an analog amplifier, the final subsystem in a digital audio system.

(a) If $y_c(t)$ is then sampled, giving $y_{s,c}(t)$ (before ZOH reconstruction), express $Y_{s,c}(s)$ in terms of $X_{s,c}(s)$ and $H_c(s)$ and/or $H_{s,c}(s)$.

(b) Draw general frequency–domain conclusions about the "sampling" operator S defined by $X_{s,c}(s) = S\{X_c(s)\}$. In particular, does $S\{X_{s,c}(s)H_c(s)\} = S\{X_c(s)\}S\{H_c(s)\}$? If so, what was the requirement for separability? If not, what can be said?

6.27. Using unilateral Laplace transforms and the property of the "sampling operator" S{·} concerning S{$X_{s,c}(s)H_c(s)$} proved in Problem 6.26 [part (b)], prove that although $\{P_{\Delta t}(s)\}^n \neq P_{\Delta t}(s)$, it is true that $\{S\&H\}^n = S\&H$, where S&H is the sample–and–hold operator. That is, if $x_{\Delta t,c}(t)$ is passed through a sample–and–hold circuit, the output will be $x_{\Delta t,c}(t)$; the sample–and–hold operator does nothing to a sampled–and–held signal (assuming the various sample–and–holds are synchronized). Note that in the process you must show that $P_{\Delta t,s}(s) = 1$ (Re{s} > 0). From your results, conclude on for what class of signals a S&H does nothing.

6.28. Using the results of Problem 6.26, introduce the concept of the "pulse transfer function" $H_{s,c}(s) = S\{H_c(s)\}$ as the ratio $Y_{s,c}(s)/X_{s,c}(s)$, where $H_c(s) = G_c(s)P_{\Delta t}(s)$ and $G_c(s)$ is an analog transfer function. Using this pulse transfer function—or rather, its z–domain equivalent $H(z)$—explicitly show the z–domain relation between the sequence of samples of the continuous–time output and that of the input, with a simple rational function of z, namely $H(z)$. Given that $H_c(s) = G_c(s)P_{\Delta t}(s)$, express $H(z)$ in terms of $G_c(s)$ in a single, maximally–convenient expression.

6.29. Suppose we have a signal $x_c(t)$ that is sampled and digitally filtered by a digital filter $D(z)$, or equivalently, $D_{s,c}(s)$ [see (6.12c) for this equivalence]. The output of the digital filter is reconstructed by a ZOH and processed by an analog system $G_c(s)$. Find the relation between samples of the output of $G_c(s)$ and the samples $x(n) = x_c(n\Delta t)$. Refer to Figure P6.29.

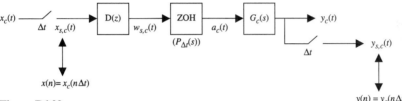

Figure P6.29

6.30. Prove, using Fourier transforms of sinc functions, that samples of continuous–time convolution are obtainable by discrete–time convolution, under the assumption that both functions being convolved are perfectly bandlimited. In practice, this result will only approximately hold.

6.31. Graphically depict examples where the spectrum of sampling of the convolution of two continuous–time signals is equal to the product of the spectra of the sampled versions of the individual signals (case of Nyquist sampling), and a case where equality does not hold (non–Nyquist sampling). Show by sketches of sampled–signal spectra versus Ω.

6.32. (a) Show the step–invariance property of the ZOH–analog system combination. That is, show that $s(n) = s_c(n\Delta t)$, or more specifically, $u(n) * h(n) = u_c(t) *_c g_c(t)|_{t=n\Delta t}$, where

$$H(z) = (1 - z^{-1})\text{ZT}\{\text{LT}^{-1}\{G_c(s)/s\}|_{t=n\Delta t}\} \text{ and } G_c(s) \text{ is the analog system.}$$

Draw block diagrams showing and explain in words the significance of this result.

(b) Generalize the result in part (a), if possible. In other words, for what class of inputs are sample values of the output of the analog system output the same whether or not the analog system is preceded by a S&H?

6.33. Suppose a continuous–time system has transfer function $G_{1,c}(s) = 1/[s(s + a)]$, where a is a positive real number.

(a) Find $g_{1,c}(t)$.

(b) Sketch the pole–zero diagram for $G_{1,c}(s)$.

(c) Sketch the pole diagram for $G_{s1,c}(s)$ for a typical sampling interval Δt. Can you determine the zeros of $G_{s1,c}(s)$? Give an explicit expression for the (possibly infinite–order) polynomial whose roots are the zeros.

6.34. Suppose $G_{2,c}(s) = A/\{(s^2 + \Omega_s^2)(s^2 + 2as + \Omega_s^2 + a^2)\}$, where, as in Problem 6.33, a is a positive real number and $\Omega_s = 2\pi/\Delta t$.

(a) Sketch the pole–zero diagram for $G_{2,c}(s)$.

(b) Sketch the pole diagram for $G_{s2,c}(s)$ for a typical sampling interval Δt. Compare with $G_{s1,c}(s)$ in Problem 6.33. What would you conclude, within a scaling factor? Find the scalar A to make your conclusion precisely true.

(c) Using Matlab, plot $g_{1,c}(t)$ and $g_{2,c}(t)$ for typical values of a and Ω_s and using A in part (b) to verify your conclusion in part (b). What must be true about the poles of $G_{2,c}(s)$ to avoid aliasing, that has here been violated?

6.35. Find $X_{\text{perext}}(e^{j\omega})$, the DTFT of the periodic extension $\tilde{x}(n)$ of a length–N sequence $x_N(n)$ in terms of $X_N(e^{j\omega}) = \text{DTFT}\{x_N(n)\}$.

6.36. In (6.24), we show the convolution of a sequence $x(n)$ with a continuous–time signal $r_c(t)$. For $x(n) = x_c(n\Delta t)$, express (6.24) in terms of $x_{s,c}(t)$. Although $x_{s,c}(t)$ is not available physically, suggest how at least in principle (6.24) might be implemented approximately with a physically obtainable signal.

Simulation

6.37. Give the time–domain relation associated with (6.53); that is, express trapezoidal rule integration as a convolution.

6.38. To approximate continuous–time second derivatives $y_c(t) = d^2 x_c(t)/dt^2$, the central difference formula is often used: $\hat{y}_c(t) = (1/\Delta t^2)[x_c(t + \Delta t) - 2x_c(t) + x_c(t - \Delta t)]$, where hopefully $\hat{y}_c(t) \approx y_c(t)$. If $y(n) = \hat{y}_c(n\Delta t) = (1/\Delta t^2)[x_c(\{n + 1\}\Delta t) - 2x_c(n\Delta t) + x_c(\{n - 1\}\Delta t)]$ then

(a) Find $y(n)$ if $x_c(t) = t^2$.

(b) Find $y(n)$ if $x_c(t) = \sin(\Omega_o t)$; express as $y(n) = A \cdot x_c(n\Delta t)$, where A is *a single constant term* (not the sum of two terms).

(c) How close is $y(n)$ to $y_c(n\Delta t)$ in parts (a) and (b)? Can you give a reason for the difference in quality? Under what condition is the central difference formula a good approximation of $y_c(t)$ in part (b)? In general?

(d) Given that in part (b) $x_c(t)$ is a sinusoid, can you give a meaning to the constant A? Be precise, and prove your assertion! [Hint: Find $H(e^{j\omega})$, where $h(n)$ is the unit sample response of the central difference filter.] State the relevant eigenvalue/eigenfunction pair.

Hints: (i) $\cos^2(a/2) = [1 + \cos(a)]/2$; (ii) $\sin^2(a/2) = [1 - \cos(a)]/2$; (iii) $\cos(2a) = \cos^2(a) - \sin^2(a)$.

6.39. We considered mapping $1/s$ to the z–domain by the bilinear transformation: $1/s = F(z) = (\Delta t/2)(z + 1)/(z - 1) = (\Delta t/2)(1 + z^{-1})/(1 - z^{-1})$. Wherever $1/s$ appears in $H_c(s)$ we substitute $F(z)$; the bilinear transform was seen to be equivalent to trapezoidal rule integration. In this problem, we consider the backward difference mapping $1/s = F(z) = \Delta t/(1 - z^{-1})$ or, equivalently, $s = (1 - z^{-1})/\Delta t$.

(a) To where in the z–plane does the left half of the s–plane map? Do stable $H_c(s)$ always transform into stable $H(z)$? Does this mapping preserve the frequency response characteristics in $H_c(j\Omega)$?

(b) Consider $H_c(s) = a/(s + a)$. By finding the difference equation and integrating to find $y(n) = y_c(n\Delta t)$ in terms of $x(n) = x_c(n\Delta t)$, show how a "backwards rectangular rule" results in substitution of $F(z) = \Delta t/(1 - z^{-1})$ for $1/s$.

(c) Determine the region of the s–plane that maps to within the unit circle in the z–plane for $F(z) = \Delta t/(1 - z^{-1})$. Where must the poles of $H_c(s)$ *not* be for $H(z)$ to be stable? Is it possible to start with an unstable $H_c(s)$ and obtain a stable $H(z)$ under this transformation?

6.40. (a) Repeat part (a) of Problem 6.39 for the forward difference $1/s = F(z) = \Delta t/(z-1)$. What sort of integration is associated with this $F(z)$? Does a stable $H_c(s)$ translate to a stable $H(z)$? Is this a desirable discretization mapping?

(b) What region in the s–plane maps to $|z| < 1$? Can you use this result to qualify your answer to part (a)?

6.41. Given $dy_c(t)/dt + ay_c(t) = bx_c(t)$ (where a and b are real and positive), use trapezoidal rule integration directly to obtain the bilinear–transform discrete–time approximation of this continuous–time system.

6.42. In Section 6.4.3, we note that there are many ways to compute functions of matrices such as $\exp\{\mathbf{A}t\}$. Suppose that matrix $\mathbf{A}$ has distinct eigenvalues $\{\lambda_\ell\}$ and a full set of eigenvectors $\{\mathbf{v}_\ell\}$, and that $f(\cdot)$ is any entire function. Show that $f(\mathbf{A}) = \mathbf{P}\mathrm{diag}[\,f(\lambda_1), \ldots, f(\lambda_N)]\mathbf{P}^{-1}$, where the modal matrix $\mathbf{P} = [\mathbf{v}_1 \ldots \mathbf{v}_N]$.

Problems 6.43 and 6.44 Involve Materials from WP 6.1

6.43. In (W6.1.4) and (W6.1.5), the prediction and correction estimates were respectively given for the predictor–corrector method of simulation. In the following sequence of problems, we develop this approach and apply it to Example 6.4. In (W6.1.5), recall that (for brevity, we will drop all c subscripts here):

$$g_i\big|_{(n+1)\Delta t} = g_i(w_1((n + 1)\Delta t), w_2((n + 1)\Delta t), \ldots, w_N((n + 1)\Delta t), x((n + 1)\Delta t)).$$

For $w_i((n + 1)\Delta t)$, we can use the "prediction" estimate obtained in (W6.1.4). This combination of "prediction" and "correction" forms the predictor–corrector method. When the predictor and corrector have the same order, useful estimates of the approximation error are available. Rosko (see Footnote 11) shows for this approximation that the error, $e_i((n + 1)\Delta t) = w_{i,\text{true}}((n + 1)\Delta t) - w_{i,\text{corr}}((n + 1)\Delta t)$, is approximately $\hat{e}_i((n + 1)\Delta t) = \frac{1}{5} \cdot \{w_{i,\text{pred}}((n + 1)\Delta t) - w_{i,\text{corr}}((n + 1)\Delta t)\}$, where again, $w_{i,\text{pred}}((n + 1)\,\Delta t)$ and $w_{i,\text{corr}}((n + 1)\,\Delta t)$ are the left–hand sides of (W6.1.4) and (W6.1.5), respectively. Using these facts, determine an improved estimate of $w_{i,\text{true}}((n + 1)\,\Delta t)$.

6.44. We now apply the predictor–corrector approach to Example 6.4 (two state variables). To get the predictor–corrector method started, we need two initial values for $w_1(t)$ and $w_2(t)$, but we are given only $w_1(0) = y_c(0)\big|_{t=0}$ and $w_2(0) = dy(t)/dt\big|_{t=0}$; we also need estimates of $w_1(\Delta t)$ and $w_2(\Delta t)$. To get around this problem, MacLaurin expansions for $w_1(t)$ and $w_2(t)$ may be used to obtain estimates of $w_1(\Delta t)$ and $w_2(\Delta t)$ as follows. To expand $w_1(t)$ and $w_2(t)$ about $t = 0$, we need successively higher time derivatives evaluated at $t = 0$. Recall that the time derivative d/dt of a function of several variables may be written as a total derivative. For this case, $d/dt = \partial/\partial w_1(t) \cdot dw_1(t)/dt + \partial/\partial w_2(t) \cdot dw_2(t)/dt + \partial/\partial x(t) \cdot dx(t)/dt = \partial/\partial w_1(t) \cdot g_1\big|_t + \partial/\partial w_2(t) \cdot g_2\big|_t + \partial/\partial x(t) \cdot dx(t)/dt$.

(a) Specialize this formula to Example 6.4.

(b) Use the result of part (a) to find the MacLaurin expansions of $w_1(\Delta t)$ and $w_2(\Delta t)$ up to order Δt^4 in Δt. Solve first using general initial conditions, and then specialize to zero initial conditions (for finding the ZSR).

(c) Using the method of first prediction and then correction as outlined above, now write the explicit recursion equations for $n = 1$ and for $n > 1$.

(d) Code your algorithm and numerically solve Example 6.4 using the predictor–corrector method with $\Delta t = 0.1$ s, 0.2 s, and 0.3 s as in the text. Plot the exact solution on the same axes. Compare against results of other techniques presented in Example 6.4.

Material on WP 6.2 Is Required for Problem 6.45

Time–Bandwidth Product

6.45. Check numerically that $x_c(t)$ is small for $t > 5D_t$ and $|X_c(j\Omega)|$ is small for $\Omega > 5D_\Omega$ using Matlab for typical numerical examples of the signal analyzed in Example W6.2.1. Examine three examples with widely different a and b values to see whether the formulas are fairly general.

Practical/Computer Applications

6.46. Suppose that $x_c(t) = e^{\sigma t}\cos(\Omega_0 t)u_c(t)$, where $\sigma = -0.5$ Np/s, and $\Omega_0 = 2\pi \cdot 0.4$ rad/s. Let $x(n) = x_c(n\Delta t)$, where $\Delta t = 0.1$ s. First plot $x(n)$ using `stem` for $N = 124$ samples, and superimpose a dense plot (e.g., $\Delta t_{\text{dense}} = \Delta t \cdot 0.05$) of $x_c(t)$ using `plot`. Your $x(n)$ should be stems passing through the solid curve $x_c(t)$. Next, estimate $X_c(j\Omega)$ using the `fft` command in Matlab. Plot both the estimate and the exact function $X_c(j\Omega) = (-\sigma + j\Omega)/([-\sigma + j\Omega]^2 + \Omega_0^2)$. You may find the routine `fftshift` helpful. How closely do they agree, and where is the error greatest?

6.47. (a) Estimate Δt required to sample a square wave (50% duty cycle) and get good spectral

estimates using the DFT. Using Matlab plots, compare the DFT results with Fourier series coefficients and with the Fourier transform of the square wave. If possible, in this problem use actual measured data samples. If you find normalization necessary, explain fully the normalization and why it is needed.

(b) Suppose that the square wave drives a motor angle position control system; plot the response and revise Δt if necessary in order to represent without significant aliasing the output angle overshoot oscillations riding on the square wave. Again, if possible, use actual data.

6.48. In sampling, storing, and processing signals, often the compander is used. A compander is the

combination of a value–compressor, followed later by a value expander, which restores the original value before compression. A typical compressor has input–output relation

$$y = \frac{\log(1 + Ax)}{\log(1 + A)}.$$

(a) Plot y versus x using Matlab for $A = 100$ and $0 \le x \le 1$. In what sense does this function "compress"?

(b) Find the equation for the expander, and plot the output using as your input the y values in part (a).

(c) Given that quantization noise in, for example, audio processing is much more noticeable at low signal levels, how does the compander reduce quantization noise? Why do you think quantization noise is more noticeable at low levels?

(d) Give a typical scenario of use of the compander. What must be avoided between the compressor and the compander?

6.49. To investigate aliasing with a real–world signal, it is convenient to consider the AM radio audio signal. Recall that AM radio audio is bandlimited before broadcasting to only 5 kHz by FCC regulation, so as to allow many radio stations on the AM band. Using an A/D converter and D/A converter, with variable Δt, select values of $F_s = 1/\Delta t$ above and below the Nyquist rate of 10 kHz. Plot and examine the DFT spectrum and listen to the reconstructed version of music or other AM radio station material.

6.50. Simulink provides a simulation of continuous–time signals and systems, with samples chosen

(in general) at adaptively selected times. The spacing between these times ("step size") falls between the user–specified limits set in "Parameters" and varies during the course of the simulation. In this problem, we wish to sample a "continuous–time" sinusoid produced by Simulink. We want one sample at (approximately) every Δt (the user specifies Δt). The problem with using the ZOH function is that it provides a number greater than one of these samples, whatever "integration times" happen to be within that Δt. (In general dynamical problems, this number will vary from one Δt–sample to the next one.) Write a Matlab routine that extracts one sample every Δt, taking the value nearest $n\Delta t$ (or $n\Delta t$ + some fraction of Δt) from the "continuous–time" sequence produced by Simulink. Of course, choose the maximum Simulink step size to be much smaller than Δt. Test your routine out on a sine wave (generated within Simulink), and plot sampled sequence you obtain against the "result of simulation" (a "continuous–time" sinusoid directly from Simulink). Use of canned linear interpolation routines or the "To Workspace/maximum number of rows" third argument option are allowed in this problem only for checking the results of your programming of the sampling routine.

6.51. (a) What is the maximum value of Δt for which the continuous–time function $x_c(t) =$

$\sin(2\pi F_{\text{sinc}}t)/[2\pi F_{\text{sinc}}t]$ may be sampled without aliasing? [In this problem, we will consider this sinc function $x_c(t)$ to be the "signal." Also note that in Matlab, $\texttt{sinc}(x) = \sin(\pi x)/(\pi x)$ as opposed to the definition in this text, $\text{sinc}(x) = \sin(x)/x$.] Give this value of Δt for $F_{\text{sinc}} = 1$ Hz. (In all Nyquist calculations, ignore truncation effects.)

(b) Using Matlab, plot $x(n) = x_c((n - N/2)\Delta t)$ for $\Delta t = 0.05$ s, and plot the frequency response magnitude, after smoothing with a Hamming filter (use the $\texttt{hamming}$ Matlab command). Have the abscissa be f in cycles/sample, for $f \in [0, \frac{1}{2}]$. To obtain sharp corners, use a very large value of N, for example, 8192. For the time plot, then, show only the range of n for which $x(n)$ is significantly nonzero. Use $\texttt{plot}$ rather than $\texttt{stem}$, due to so many samples.

(c) Repeat part (b) for Δt set to the maximum possible value. Explain the appearance of the plots. Give the reconstruction formula for this case.

6.52. (a) Suppose that $x_{\mathrm{mod},c}(t) = \cos(2\pi F_c t)x_c(t)$, where $x_c(t)$ (considered to be the "signal") is as in Problem 6.51. Let the carrier frequency be $F_c = 8$ Hz. Plot $x_{\mathrm{mod}}(n) = x_{\mathrm{mod},c}((n - N/2)\Delta t)$ for $\Delta t = 0.05$ s and its DFT magnitude [again, showing $x_{\mathrm{mod}}(n)$ only on the range where it is significantly nonzero, smoothing with Hamming window before DFT, and again use $N = 8192$].

(b) Under straightforward sampling, give the maximum value of Δt to avoid aliasing.

(c) We now begin discussion of the use of modulation with a known carrier sinusoid in an attempt to relax the constraint on the sampling interval Δt. Modulate $x_{\mathrm{mod},c}(t)$ with $\cos(2\pi F_c t)$ [call the result $x_{\mathrm{mod2},c}(t)$] before sampling, in an attempt to remodulate down to a zero–centered band. Plot $x_{\mathrm{mod2}}(n)$ and its DFT magnitude. What problem do you find, and what is its cause? Be completely quantitative in describing the major artefact. This artefact will be removed in Problem 6.53.

6.53. (a) To remove the large artefact from the result of Problem 6.52, we need one more operation. Tell what this operation is, and perform it. For this operation, you may need a package such as Simulink to proceed, and the results of Problem 6.50. Show all results with plots. Note: This problem is quite computationally intensive and required a fair amount of time on a Pentium. Hint: Although we study filters in detail in Chapters 8 and 9, the analog Butterworth filter obtained by the Matlab code

```
[Nfilt,wn]=buttord(2*pi*1.0,2*pi*2.8,0.5,30,'s')

[bfilt,afilt]=butter(Nfilt,wn,'s')
```

may be helpful. Plot its frequency response and argue why each of the parameters in the above lines is correct, and give the numerical value of the filter order. Describe the procedure to exactly recover $x_{\mathrm{mod},c}(t)$ from these new samples. Carry out this procedure, and show plots verifying its correctness.

(b) What is the new maximum allowable value for Δt to avoid aliasing, with this procedure? Verify by simulation.

6.54. (a) Use Simulink to obtain a "continuous–time" triangle wave. To do this, use the "Signal Generator" block. Note, however, that it has only sine, square, and sawtooth waveforms. Using one of these three waveforms (with frequency = 1 rad/s and peak = 1) and a second block from "Linear," obtain the triangle wave. You will find that it is always negative. Within Simulink, add the appropriate constant to make the average value of the triangle wave equal to zero. Plot the "continuous–time" waveforms of your result for $t \in [0, 100$ s$]$, and let the step size range from 0.0003 s to 0.003 s. What is the amplitude of your triangle wave, and why? What is the period in seconds? Finally, modify the simulation parameters (e.g., start at a later time) so that the triangle wave begins at zero value and is increasing (note that no phase shift can be added in "Signal generator"). Print out the final block diagram, and replot—with the value of the added constant changed to give zero average value.

(b) Sample the triangle wave using the technique developed in Problem 6.50, using $\Delta t = 0.04$ s. Extract from $x(n)$ one half–period (the positive–going half) from one of the later periods (later to avoid any startup transients in the simulation when dynamics are later added and this procedure is repeated). Scale this sequence so that its maximum value is approximately unity; by what factor should it be scaled? If necessary, readjust any necessary programmed parameters so that the first sample is zero or nearly zero. Plot the (zero–padded out to 1024) DFT magnitude versus digital frequency $f \in [0, \frac{1}{2}]$. Superimpose the exact magnitude spectrum given in (4.25) evaluated at $\Omega = \omega/\Delta t$; you should have good agreement.

6.55. In this problem, we perform antialiasing filtering using Simulink to simulate the actual continuous–time filter action.

(a) In Problem 6.54, let $\Delta t = T/3$, where T is as appears in (4.25), and repeat part (b) of Problem 6.54. Note the aliasing in the sampled spectrum (even though we are sampling greater than twice one "period"—because the signal is not a sinusoid but a triangle, having high–frequency content).

(b) Using `butter`, create an analog antialiasing filter, and filter the triangle with it before sampling. You may wish to have the stopband begin slightly greater than $F_s/2$ so that the filter order is low (and thus lower phase distortion). Implement the analog filter in Simulink, and print out the block diagram and filter magnitude and phase frequency response. Then sample the result at $\Delta t = T/3$, plot the output sequence, and again plot the DFT magnitude against (4.25). Have you nearly eliminated the aliasing? Finally, plot the magnitude spectra of the "continuous–time" input and output of the continuous–time filter, along with the filter frequency response verses F (hertz) all on one set of axes. Plot a vertical line at $F_s/2 = 1/(2\,\Delta t)$, and verify that the behavior is nearly as desired. (This plot can also help you see how far beyond $F_s/2$ your stopband may begin and will help explain any remaining aliasing.) Is it permissible to plot DFTs of the "continuous–time" sequences produced in Simulink? Why or why not?

6.56. As a practical application of sinusoidal sampling and processing, consider the determination of the direction θ of an incoming propagating wave beam. Suppose that a linear array of $N_r = 100$ receivers is aligned with the x axis at y–coordinate $y = 0$. The spacing of the receivers is $\Delta x_r = 0.05\lambda$, where λ is the wavelength of he incoming wave. Each receiver samples the incoming plane wave every $\Delta t = 1$ s, for $N_t = 200$ samples. The temporal frequency of the plane wave is $\Omega = 2\pi F$, where $F = 0.1$ Hz. The plane wave at position x,y is $\cos\{(2\pi/\lambda)[x\cos(\theta) + y\sin(\theta)] - \Omega t\}$. If each receiver is delayed by a special, distinct number of samples, then within the discretization of the sampling, each receiver will output nearly the same sinusoid. So, when added together, the normalized–to–length norm of the sum of all measurements will be maximum. The amount each receiver output is to be delayed depends on the estimate of the angle θ. If we iterate through all possible angles from 90° to 180° (in 1° steps), the maximum norm for all possible angles will be our best estimate of θ. Write a Matlab routine that implements this algorithm, and plot the error of the output versus the true angle θ. You should obtain a maximum error of only a few degrees except for $174° \le \theta \le 180°$ and mostly an error of 0°. (Note: This processing should be done on a Pentium or better, to minimize waiting for computation.)

6.57. (a) Make a Simulink block diagram that produces the step response of the analog system $H_c(s) = 25(s + 2)(s + 4)/[(s + 8)(s^2 + 0.4s + 25.04)]$. Also plot the output of a zero–order hold with sampling time 0.5 s [fed the output of $H_c(s)$]. Let the maximum time step be 0.001 s, and plot out to $t = 40$ s. Begin the step at 1 s so the first sample is zero.

(b) By using two simultaneously plotted DFT spectra (one of which is the DFT of a sequence obtained from the ZOH output), show how you can determine whether there is appreciable aliasing taking place.

6.58. Suppose that we wish to accurately calculate the period of a sinusoid of unknown frequency, from its samples every Δt—a fairly common requirement in practical signal processing applications. Using a property of the sinusoid, suggest a procedure that will do this task without using zero–crossing counting. Test your procedure in Matlab, and state your results.

6.59. In this and the following problem, we measure the frequency responses of various speakers you may have around the lab or at home. We require a microphone and either a data acquisition board (e.g., the [12–bit] Metrabyte DAS–16) or a user–friendly replacement such as the [16–bit] Siglab. We also require a frequency sweep generator. In Siglab, there is an internal repetitive chirp pulse which serves the purpose. Let the sweep be over the entire audio range (e.g., 0.1 Hz to 20 kHz). Play the chirp pulse over the speaker, and record the output of the microphone held close to the speaker. Plot the recorded waveform. Digitize the recording at, for example, 51.3 kHz ($\Delta t = 19.5$ μs). Doing so will guarantee that there is no aliasing of audio frequencies. Collect a fairly long pulse (e.g., 0.13 s, yielding a zero–buffered data record of 8192 samples). Plot the magnitude spectrum of the input and the output, estimated using the `fft` routine of Matlab. Try for different speakers and compare the spectra. Plot the non–dB magnitude for this first effort; see Problem 6.60 for log–magnitude plots.

6.60. Repeat Problem 6.59, but this time calculate the spectrum on a logarithmic frequency scale. (Do this calculation by direct evaluation of the DTFT sum.) Reduce the number of frequency values to 800–1000, to minimize computation but still provide a very smooth spectrum appearance. This time plot the magnitude in decibels, and normalize to the magnitude of the input spectrum. You will have a true and professional–looking frequency response of the microphone–speaker combination. Show the input chirp spectrum in decibels (relative to its maximum value or to its plateau). Comparing your results in this problem with those in Problem 6.59, why do we use a logarithmic frequency scale?

6.61. In Problem 6.60, we used the Matlab command `logspace`. Write the exact expression giving the values that the command x=`logspace(a,b,N)` stores in x. Also, give the slope and intercept of the straight line $\log_{10}\{x(k)\}$ versus k under this command.

Illustrative Analysis of Sampling

6.62. Because $X_{s,c}(j\Omega)$ is periodic in Ω, it has a Fourier series. Show that its Fourier series coefficients are $\Delta t \cdot x_c(-k\,\Delta t)$. This result is illustrated in Figure 6.56, parts h and f. Also show that the rectangular rule approximation of $X_c(j\Omega)$ is equivalent to the Fourier series representation of the periodic extension of $\{X_c(j\Omega),\ |\Omega| \le \Omega_s/2\}$, under no aliasing and Nyquist–or–better sampling. Thus, again verify that the non–truncated rectangular rule approximation of $X_c(j\Omega)$ is exact for $|\Omega| \le \Omega_s/2$ under Nyquist sampling [(6.76)].

6.63. Prove that the inverse Fourier transform of both sides of (6.76) with $|\Omega| < \frac{1}{2}\Omega_s$ gives the Shannon reconstruction formula (6.26).

6.64. Starting with the definition $X_{s1,c}(j\Omega) = X_c(j\Omega) \cdot \delta_{\Delta\Omega}(\Omega)$ [that is, $X_{s1,c}(j\Omega)$ is a series of impulses weighted by samples of $X_c(j\Omega)$], prove (6.94), using the same method given in the text to derive (6.17).

6.65. Starting with $X_{s1,c}(j\Omega) = X_c(j\Omega) \cdot \delta_{\Delta\Omega}(\Omega)$, show that $X_1(e^{-jt\,\Delta\Omega}) = 2\pi \cdot x_{s1,c}(t)$ [cited just before (6.100a)], where $X_1(z) = ZT\{X_c(jk\,\Delta\Omega)\}$. Thus we have the interesting result that $x_{s1,c}(t)$, the periodically extended version of $x_c(t)$ (with possible aliasing) is the DTFT of samples of the Fourier transform of $x_c(t)$, a property we saw in reverse for temporal sampling in (6.17).

6.66. Prove that the order of operations of sampling and truncation is reversible, by showing that equating the two leads to (6.107), an identity numerically verifiable in, for example, Matlab.

6.67. Does $P_{\Delta t}(s) = (1 - e^{-s\,\Delta t})/s$ have a pole at $s = 0$? If not, prove that it is analytic everywhere by using the Cauchy–Riemann equations (3.3). Note: For "x" and "y," use, respectively, "σ" and "Ω."

6.68. (a) We know from (6.12c) that $X_{s,c}(s) = X(z)\big|_{z\,=\,e^{s\,\Delta t}}$. Using the Laplace transform property for multiplication of two functions on $x_{s,c}(t) = x_c(t)\delta_{\Delta t}(t)$ and assuming that $X_c(s)$ is a factorable rational function, show that we may also write [for causal $x_c(t)$]

$$X_{s,c}(s) = \sum_{\substack{\text{left–half–}\\ \text{plane poles } w_\ell \text{ of } X_c(w)}} \operatorname{Res}\left\{\frac{X_c(w)}{1 - e^{-(s-w)\Delta t}}, w_\ell\right\} \quad (w \text{ is a dummy integration variable}).$$

[Note that usually signals are not of the form of a rational function, nor even available as the arbitrarily high–order rational function that could model them arbitrarily accurately; however, in the processing of sampled data using a digital filter, we sometimes need $G_{s,c}(s)$, where $G_c(s)$ is a continuous–time system. Thus the above result could be applied to G_c rather than X_c.] The formula above is a shortcut, for it allows us to find the z–transform $X(z)$ (by substituting z for $e^{s\,\Delta t}$ in the result of the formula above) without having to (i) inverse Laplace–transform $X_c(s)$, (ii) sample, and (iii) z–transform the resulting sequence. Hint: Close the contour in the left–half plane, and use the fact that the integrand in the portion of the contour C_R at $|w| = \infty$ in the left–half plane is zero, so the integral along the $j\Omega$ axis portion of the closed contour is equal to the entire closed–contour integral [which may be evaluated using the residue theorem (3.25a)].

      (b) Close the contour in the right–half plane, and use the generalized formula for residues found in Problem 3.40. What is your result?

6.69. Derive the mathematical relations between Figures 6.56c and d. Also comment on the relation between Figures 6.56d and e. Also, where there are differences in notation, connect the simpler notation used in earlier sections of this chapter with the more complicated (but more precise) notation used in Figure 6.56.

6.70. Comment quantitatively on Figure 6.56j as being samples of $\tilde{x}_{s,\alpha,c}(t)$ (see, e.g., Figure 6.56c1, in which $\alpha = 1$) in terms of the relation between the continuous–time FS coefficients of $\tilde{x}_{s,\alpha,c}(t)$, and the DFT_N coefficients of one period of $\tilde{x}(n)$ where $\tilde{x}(n) = \tilde{x}_{s,\alpha,c}(n\,\Delta t)$. Review quantitatively how both DFT_N and DFT_N^{-1} relations are developed, in this context. Thus, derive the DFT relations by means of appropriate continuous–variable Fourier–series relations.

6.71. Show by a direct forward Fourier transform calculation that Figure 6.56d is indeed the Fourier transform of Figure 6.56c, without appeal to previously derived results. Note that by identical arguments, we can show that Figure 6.56g is the inverse Fourier transform of Figure 6.56f.

6.72. Replacing $\Delta\Omega$ by $2\pi/(N\,\Delta t)$, obtain (6.105) expressed exclusively in terms of Δt rather than $\Delta\Omega$ by performing a direct evaluation of the $(1/\Delta t) \cdot \text{DFT}_N^{-1}$ of the $1/\Delta t$-scaled spectral samples $X_c(j2\pi k/[N\,\Delta t])/\Delta t$. Your result should be an expression for $\hat{x}_{N,\Delta t,c}(n\,\Delta t)$. Express both as a finite–interval and an infinite–interval convolution.

Problem 6.73 Refers to WP 6.2

6.73. Derive the values of the integrals I_A and I_B cited in the solution of Example W6.2.1.

Chapter 7

Discrete-Time Processing with the DFT

7.1 INTRODUCTION

In this chapter, we consider implementations of various operations on sequences in which the discrete Fourier transform is applied. The range of applications of the DFT is far too broad to cover comprehensively, but some of the most common applications will be discussed here. In particular, digital frequency response or spectrum estimation from time-domain samples, time-domain sequence estimation from frequency response samples, interpolation/sampling rate conversion, convolution, correlation, and cepstral processing are some of the more common operations that can make effective use of the DFT.

The reason for the great popularity of the DFT is the existence of fast algorithms for its computation. Procedures normally requiring number of operations of the order N^2 may be performed exactly using only on the order of $N \log_2(N)$ operations. This class of algorithms, called the fast Fourier transform (FFT), is described in detail in Section 7.10.

7.2 FREQUENCY RESPONSE/DTFT ESTIMATION BY THE DFT

The frequency response $H(e^{j\omega})$ of an LSI discrete-time system is the DTFT of the unit sample response $h(n)$. Similarly, the DTFT of a signal sequence $x(n)$ is referred to as its spectrum $X(e^{j\omega})$. As noted in previous chapters, the digital frequency response/spectrum is a function of the continuous, real variable ω and is periodic in ω with period 2π. Usually, the frequency response or spectrum is displayed graphically on two plots: the magnitude expressed as either $|H(e^{j\omega})|$ or $20 \cdot \log_{10}|H(e^{j\omega})|$, and the phase $\angle H(e^{j\omega})$.

For filtering applications, the magnitude $|H(e^{j\omega})|$ indicates to what extent various frequency components present in the input $x(n)$ are passed to the output $y(n)$ = $x(n) * h(n)$. The phase $\angle H(e^{j\omega})$ indicates to what extent the various frequency components are delayed by the filter. For signal analysis applications, the magnitude $|X(e^{j\omega})|$ indicates to what extent various frequency components are present in the signal $x(n)$, and the phase $\angle X(e^{j\omega})$ describes to what extent the various frequency components of $x(n)$ are time-delayed relative to each other.

When we compare the DTFT of a sequence $x_N(n)$ of length N,

$$X_N(e^{j\omega}) = \sum_{n=0}^{N-1} x_N(n) \cdot e^{-j\omega n}, \tag{7.1}$$

with the DFT$_N$ of the same length-N sequence,

$$X_N(k) = \sum_{n=0}^{N-1} x_N(n) \cdot e^{-j2\pi kn/N}, \tag{7.2}$$

it is evident that the DFT$_N$ is a set of N samples of the DTFT: $X_N(k) = X_N(e^{j2\pi k/N})$, uniformly spaced for $\omega \in (-\pi, \pi)$. The spacing of these DTFT samples is $2\pi/N$. If N is small, then the DFT$_N$ does not yield many values of $X_N(e^{j\omega})$. To obtain a more complete picture of $X_N(e^{j\omega})$, we may *zero-pad* and obtain $M > N$ samples of $X(e^{j\omega})$, as follows. Append $M - N$ zeros to the sequence $x_N(n)$ to obtain the longer sequence $x_{zp;M}(n)$ and take the DFT$_M$ of $x_{zp;M}(n)$. The result is

$$X_{zp;M}(k) = \sum_{n=0}^{N-1} x_N(n) \cdot e^{-2\pi kn/M} = X_N(e^{j2\pi k/M}), \tag{7.3}$$

where the sum runs over only $[0, N - 1]$ because zero-padding has not increased the set of nonzero values of $x_{zp;M}(n)$. The form of the sum is identical to that of the DTFT of $x_N(n)$, and so the sum is $X_N(e^{j\omega})$ evaluated at $\omega = 2\pi k/M$. Because by assumption $M > N$, the DFT$_M$ of $x_{zp;M}(n)$ has yielded additional, and more closely spaced values of $X_N(e^{j\omega})$ than did the DFT$_N$ of $x_N(n)$.

EXAMPLE 7.1

As an extreme case of the above results, consider $x_N(n) = 1$ for all $n \in [0, N - 1]$ (and zero otherwise). Examine the DFT$_N\{x_N(n)\}$ and its zero-padded version.

Solution

For $N = 10$, $x_N(n)$ is plotted in Figure 7.1a. By Example 3.10, part (b), we know that $X_N(z) = N$ for $z = 1$ and otherwise $X_N(z) = (1 - z^{-N})/(1 - z^{-1})$ (for $z \neq 0, 1$). We also know that $X_N(e^{j\omega}) = X_N(z)$ evaluated at $z = e^{j\omega}$, so $X_N(e^{j\omega}) = N$ for $\omega = 0$, and otherwise $X_N(e^{j\omega}) = (1 - e^{-j\omega N})/(1 - e^{-j\omega})$. Furthermore, $X_N(k) = X_N(e^{j\omega})$ evaluated at $\omega = 2\pi k/N$. We thus obtain $X_N(k) = N$ for $k = 0$, and otherwise $X_N(k) = (1 - e^{-j2\pi k})/(1 - e^{-j2\pi k/N}) = 0$, or $X_N(k) = N\delta(k)$.

With the view that $X_N(k)$ is a set of N samples of $X_N(e^{j\omega})$, we plot $X_N(k)$ in Figure 7.1b versus $f = \omega/(2\pi)$ rather than versus k. Thus, the frequencies plotted are $f_k = \omega_k/(2\pi) = k/N$. In addition, we show the spectrum only for $f \geq 0$, for we know the magnitude of the DTFT of any real sequence [such as $x_N(n)$] is an even function of f; thus showing $f < 0$ would be redundant and waste space in our plot. Indeed, in Figure 7.1b, we have $N/2 + 1$ samples of the DTFT of $x_N(n)$, but because all f_k but one happen to fall where $X_N(e^{j2\pi f}) = 0$, we do not get much of a feel for the shape of $X_N(e^{j\omega})$ in Figure 7.1b, to put it lightly.

To get a better picture, we zero-pad $x_N(n)$ out to $M - 1$, where $M = 128$.

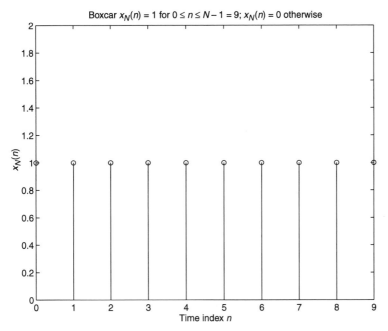

Figure 7.1*a*

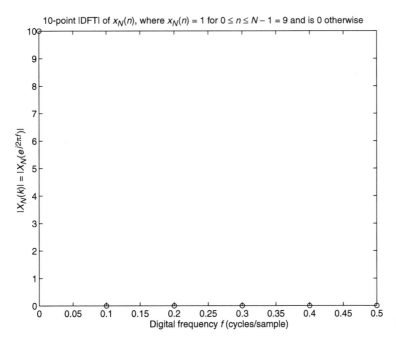

Figure 7.1*b*

The zero-padded sequence $x_{zp;M}(n)$ is shown in Figure 7.2*a*. This is probably more what we had in mind anyway for $x_N(n)$—a boxcar function of length 10 and zero otherwise. This time, the DFT_M gives $X_{zp;M}(k) = N$ for $k = 0$, and otherwise

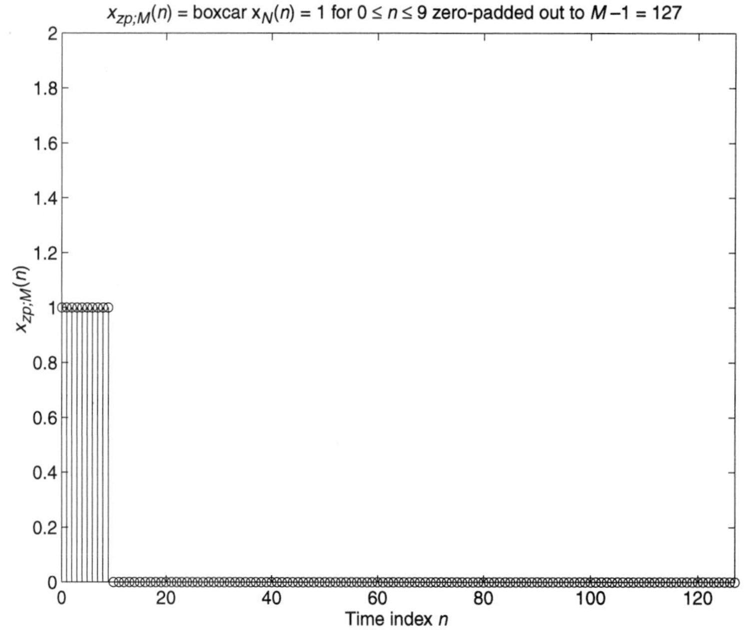

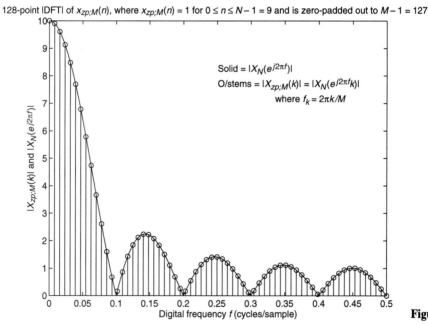

Figure 7.2a

Figure 7.2b

$X_{zp;M}(k) = (1 - e^{-j2\pi kN/M})/(1 - e^{-j2\pi k/M}) = e^{-j\pi(N-1)k/M}\sin(\pi Nk/M)/\sin(\pi k/M)$.
Figure 7.2b shows $\left|X_{zp;M}(k)\right| = \left|X_N(e^{j\omega})\right|\Big|_{\omega=\omega_k=2\pi k/M}$, again versus the corresponding values of f (now $f_k = k/M$). Also shown, as the solid curve, is a finely spaced

set of values of $|X_N(e^{j\omega})|$, obtained by zero-padding $x_N(n)$ out to 1024. We now see the $\sin(\pi Nk/M)/\sin(\pi k/M)$ shape, not only its zero values, which are all that is displayed in Figure 7.1*b*.

As a final note on this example, recall from Section 4.5 that the DFT_N is actually a discrete Fourier series of the N-periodic extension of $x_N(n)$ [see also the inverse and forward DFS definitions (4.12c) and (4.12d)]. The N-periodic extension of $x_N(n)$ in this example is 1 for all N. This is just dc, a zero-frequency sinusoid. Note that the DFT coefficients $X_N(k)$ are actually N times what would otherwise be the discrete-time analog of the Fourier series coefficients of this periodic extension of $x_N(n)$ (in which the factor of 1/period = $1/N$ would normally appear[1]). It is therefore no surprise to find that $X_N(k) = N\delta(k)$ for the non-zero-padded $x_N(n)$. Once we zero-pad out to M and calculate the DFT_M, we see that the M-periodic extension of $x_{zp;M}(n)$ is *not* dc (Figure 7.2*a*), so we no longer expect $N\delta(k)$ for $X_{zp;M}(k)$.

The effect of zero-padding is nothing more than an exact interpolation of $X_N(e^{j\omega})$, the DTFT of the N values of $x_N(n)$. Because these N values $x_N(n)$ are the inverse DFT_N of the N samples $X_N(k)$ of $X_N(e^{j\omega})$, effectively the inverse DFT_N followed by zero-padding and then by the DFT_M is the interpolation (with zero error) of the function $X_N(e^{j\omega})$ based on N values of $X_N(e^{j\omega})$. Because $X_N(e^{j\omega})$ is the DTFT of $x_N(n)$ which has only N values, there are only N "degrees of freedom" (independent basis functions) making up $X_N(e^{j\omega})$.

Specifically, $X_N(e^{j\omega})$ may be interpolated exactly and explicitly from its DFT_N samples $X_N(k) = X_N(e^{j2\pi k/N})$ as follows. First, substitute $x_N(n)$ written as the DFT_N^{-1} of $X_N(k)$ into the forward definition of the DTFT:

$$X_N(e^{j\omega}) = \sum_{n=0}^{N-1} \frac{1}{N} \sum_{k=0}^{N-1} X_N(k) \cdot e^{j2\pi kn/N} \cdot e^{-j\omega n} . \tag{7.4a}$$

Next, reverse the order of summations:

$$X_N(e^{j\omega}) = \frac{1}{N} \sum_{k=0}^{N-1} X_N(k) \sum_{n=0}^{N-1} e^{-j(\omega - 2\pi k/N)n} . \tag{7.4b}$$

Evaluating the resulting geometric sum gives

$$X_N(e^{j\omega}) = \frac{1}{N} \sum_{k=0}^{N-1} X_N(k) \frac{1 - e^{-j(\omega - 2\pi k/N)N}}{1 - e^{-j(\omega - 2\pi k/N)}} , \tag{7.4c}$$

or, factoring out half of each exponent in the numerator and denominator,

$$X_N(e^{j\omega}) = \frac{1}{N} \sum_{k=0}^{N-1} X_N(k) \frac{e^{-j(\omega - 2\pi k/N)N/2}}{e^{-j(\omega - 2\pi k/N)/2}} \cdot \frac{\sin\{(\omega - 2\pi k/N)N/2\}}{\sin\{(\omega - 2\pi k/N)/2\}} \tag{7.4d}$$

$$= \sum_{k=0}^{N-1} X_N(k) \cdot \phi\left(\omega - \frac{2\pi k}{N}\right) , \tag{7.4e}$$

[1]Unlike the continuous-time FS, the DFT places the $1/N$ normalization in the inverse rather the forward expression. This is done so that the DFT and DFT^{-1} directly provide rectangular-rule approximations of the FT and DTFT.

where

$$\phi(\theta) = e^{-j\theta(N-1)/2} \cdot \frac{\sin(N\theta/2)}{N\sin(\theta/2)}, \tag{7.5}$$

which is a proof of the second formula of item 6 in Appendix 4A.

In an interesting "coincidence," in reference to (6.79b), (6.79c), and (6.80),

$$x_N(n) = x(n) \cdot w_{N,R}(n) \overset{\text{DTFT}}{\longleftrightarrow} X_N(e^{j\omega}) = \frac{1}{2\pi} \cdot \int_{-\pi}^{\pi} X(e^{j\omega'}) \cdot W_{N,R}(e^{j(\omega-\omega')}) d\omega' \tag{6.79b}$$

$$= \frac{X(e^{j\omega}) *_{c\omega} W_{N,R}(e^{j\omega})}{2\pi}, \tag{6.79c}$$

where

$$W_{N,R}(e^{j\omega}) = \frac{\sin(N\omega/2)}{\sin(\omega/2)} \cdot e^{-j\omega(N-1)/2}, \tag{6.80}$$

we see that $\phi(\omega) = W_{N,R}(e^{j\omega})/N$, the latter being the rectangular window function that distorts the DTFT estimate when $x(n)$ is N-truncated. Note, however, that in (6.79b,) the convolution is an integral over the continuous frequency variable ω', whereas in (7.4), the "convolution" is a mixed-variable sum over the discrete frequency variable k. More important, the operation in (6.79b) is a smearing distortion, whereas that in (7.4) is an exact interpolation.

Equations (7.4) and (7.5) form the interpolation (reconstruction) of $X_N(e^{j\omega})$ from its N samples $X_N(k) = X_N(e^{j2\pi k/N})$. The situation is similar to that of polynomial interpolation. For example, an Nth-order polynomial can be interpolated without error from $N + 1$ values of the polynomial [but not necessarily specified at equally spaced abscissa values as was done in reconstructing $X_N(e^{j\omega})$]. The interpolation in (7.4) and (7.5) is also reminiscent of the reconstruction of $x_c(t)$ from its samples $x_c(n\,\Delta t)$ in (6.26) $[x_c(t) = \sum\limits_{n=-\infty}^{\infty} x_c(n\,\Delta t) \cdot \text{sinc}\{\pi(t/\Delta t - n)\}]$. Two differences are that the reconstruction of $x_c(t)$ in (6.26) involves an infinite sum rather than the finite sum in (7.4), and it has $\sin(\pi t/\Delta t)/[\pi t/\Delta t]$ as its interpolation function instead of the $\sin(N\theta/2)/[N\sin(\theta/2)]$ and the phase factor $e^{-j\theta(N-1)/2}$ in (7.5). The phase factor arises as follows. Recall that $x_N(n)$ is nonzero for $n \in [0, N-1]$, and is centered on $n = (N-1)/2$ rather than on $n = 0$. It is this shift of $(N-1)/2$ that accounts for the presence of the phase factor $e^{-j\theta(N-1)/2}$.

M equally spaced samples of $X_N(e^{j\omega})$ are available from $X_{zp;M}(k)$. Expressing $X_{zp;M}(k)$ in the interpolation form (7.4) and (7.5), we simply obtain [note renaming of k to m in (7.4e)]

$$X_{zp;M}(k) = X_N(e^{j\omega}) \Big|_{\omega = 2\pi k/M} \tag{7.6a}$$

$$= \sum_{m=0}^{N-1} X_N(m) \cdot \phi\left(2\pi\left[\frac{k}{M} - \frac{m}{N}\right]\right) \tag{7.6b}$$

where (7.6a) is just a restatement of (7.3). We see that by choosing M large enough, we can in principle obtain $X_N(e^{j\omega})$ at ω specified as precisely as required. For a length-N sequence $x_N(n)$, zero-padding combined with use of the FFT to compute the DFT_M is the efficient and standard method of calculating spectra $X_N(e^{j\omega})$ for plotting purposes or other detailed analysis.

Notice that for the type of interpolation in (7.6b), the desired abscissa values (desired ω) are all 2π times the rational number k/M. That is, ω is limited to a discrete grid of abscissa values for a given value of M. ω may be even more restricted due to limited values of M for which the DFT_M may be available in FFT form, for the given hardware or computer program. Interpolation from a coarse grid onto a finer grid by zero-padding is a common operation in discrete-time signal processing, in other contexts such as the discrete-time increase of effective sampling rate and Chebyshev approximation digital filter design.

Now suppose that the sequence $x(n)$ from which $X(e^{j\omega})$ was generated is no longer only N points long, but is much longer or even of infinite length. What is the result of still using only N values of $x(n)$ [denoted by $x_N(n)$] to estimate $X(e^{j\omega})$ by the DFT_N? In this case, the number of "degrees of freedom" in $X(e^{j\omega})$ exceeds the number of independent basis functions with which the DFT_N attempts to estimate $X(e^{j\omega})$. Moreover, the basis functions $e^{-j\omega n}$ of $X(e^{j\omega})$ for $n \in [N, M - 1]$ that are missing from $X_N(e^{j\omega})$ are the ones most rapidly oscillating with ω. Therefore, the result $X_N(e^{j\omega})$ will be a smoothed, imprecise estimation: $X_N(e^{j\omega}) \neq X(e^{j\omega})$. Actually, the same type of analysis applies here as did for using the DFT_N to approximate $X_c(j\Omega)$ as discussed in Section 6.5, except that no time-domain "sampling" has taken place, only windowing. Thus, (6.79b), (6.79c), and (6.80) (cited above) again apply.

EXAMPLE 7.2

Consider estimation of the spectrum of the long (8192-length) section of speech data shown in Figure 7.3a. The single spoken word "show" is displayed. For the purposes of this example, consider this sequence (in Figure 7.3a) to be the "total"

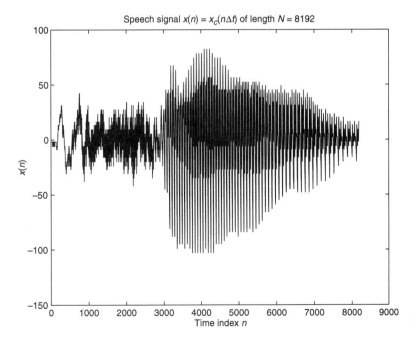

Figure 7.3a

sequence $x(n)$. The sampling frequency is $F_s = 11.025$ kHz (so $\Delta t = 90.7\mu s$); thus the section seen is about 3/4 s. Investigate trying to use different portions of $x(n)$ to estimate the spectrum, and explain the results.

Solution

The first portion of the sequence is unvoiced speech, identifiable by its "white noise" appearance; it is created by sounds such as "ff" or "sh" (in Figure 7.3a, the "sh" in "show"). Contrarily, the second half is "voiced" speech—for example, a vowel—and is identifiable by its quasi-periodic character (in Figure 7.3a, the "ow" in "show").

The $|DFT_{8192}\{x(n)\}|$ is shown in Figure 7.3b. In all figures for this example, the points of sequences have been connected together with lines for clarity. Because the frequency spacing ($\Delta\omega = 2\pi/8192$) is so small, we may consider the curve shown in Figure 7.3b to represent the magnitude of the DTFT of $x(n)$ for all ω on $[0, \pi]$. Recall that the maximum frequency displayed, $f = \frac{1}{2}$, corresponds to $F = F_s/2 \approx 5.51$ kHz.

Now suppose that we wish to estimate this DTFT magnitude spectrum using only a small number of values from $x(n)$, say 512. The 512 values are chosen to begin at sample number 1800, and this sequence, $x_{512}(n)$, is contructed from the portion of $x(n)$ shown in Figure 7.4a. The magnitude of the DTFT of $x_{512}(n)$, obtained by zero-padding $x_{512}(n)$ out to $n = 8192$ and taking the DFT_{8192}, is shown in Figure 7.4b and denoted $|X_1(e^{j2\pi f})|$. Again, because the resulting spacing is so small, we simply label Figure 7.4b as the DTFT of $x_{512}(n)$ (8192/2 + 1 values of it are displayed).

We are first struck by the grossly different appearance between Figures 7.3b and 7.4b; Figure 7.4b has a far larger fraction of its energy at high frequencies than does Figure 7.3b and a far lower proportion of energy at the low frequencies where Figure 7.3b has most of its energy. The reason for this difference is revealed by noting the starting time sample, 1800. This section of data happens to lie within the unvoiced portion of $x(n)$; that is the cause for the high frequency content in Figure 7.4b. The

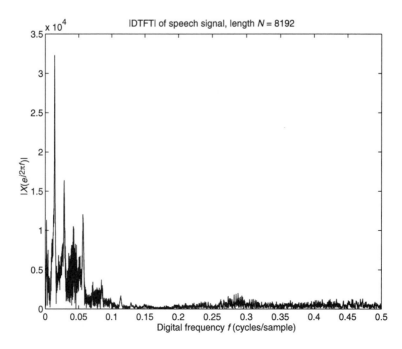

Figure 7.3b

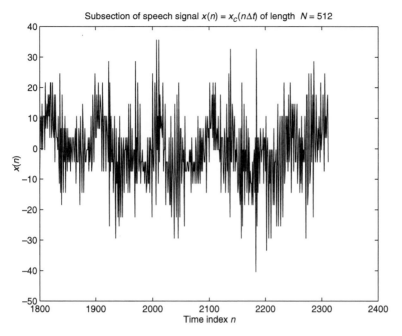

Figure 7.4*a*

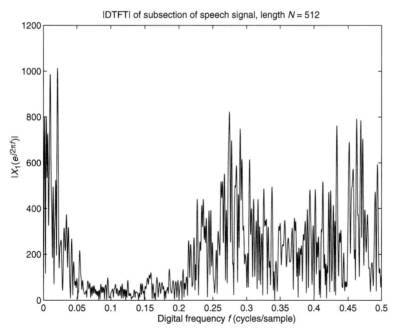

Figure 7.4*b*

discrepancy between estimated spectra for different subsections is inherent in speech: The instantaneous "spectrum" changes with time!

Similarly, Figure 7.5*a* shows a 512-point voiced subsection, this time beginning at $n = 3000$-part of the "ow" in "show." Its DTFT (i.e., $\text{DFT}_{8192}\{\text{zero-padded subsection}\}$) is shown in Figure 7.5*b* and denoted $\left| X_2(e^{j2\pi f}) \right|$. Notice the extreme dissimilarity relative to the DTFT of the unvoiced segment in Figure 7.4*b*, due to the

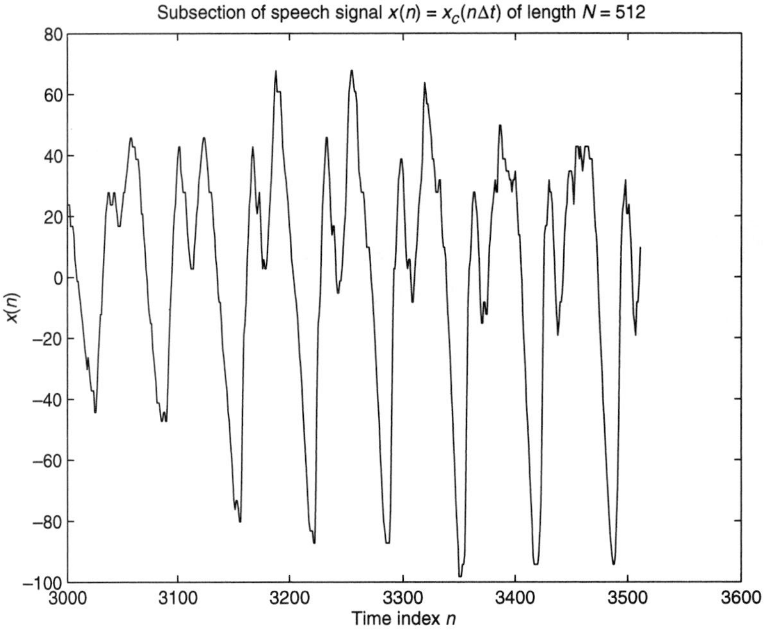

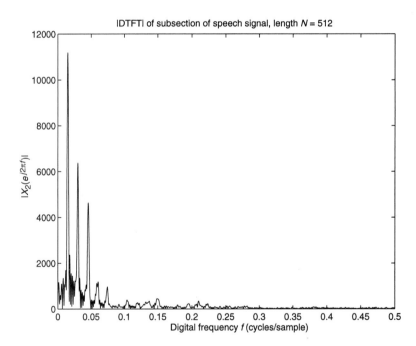

Figure 7.5a

Figure 7.5b

dissimilarity in the sequences (in Figures 7.4a and 7.5a). Also, the low-frequency region in Figure 7.5b is at least similar to that in Figure 7.3b. The proportion of high-frequency content in Figure 7.5b is minimal, because "ow" is voiced. The reason the proportion of high-frequency content is also low in the full word, Figure 7.3b, is that

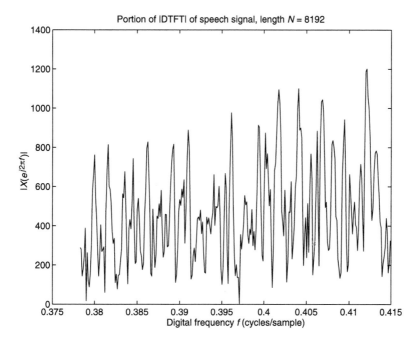

Figure 7.6*a*

"sh" is quieter than "ow," the latter of which is boosted by our vocal cords. Note that Figure 7.3b is very roughly the sum of the spectra in Figures 7.4b and 7.5b.

In Figure 7.6a, we "zoom in" on $3100 \leq k \leq 3400$, which is 300 samples of $X(e^{j2\pi f})$, the DTFT of the "total" signal $x(n)$. This range of k corresponds to $3100/8192 = 0.378 < f < 0.415$ cycle/sample, or $0.378 \cdot 11025 = 4.17$ kHz $< F < 4.58$ kHz. Figure 7.6b shows the same frequency range for the DTFT of the unvoiced 512-subsequence. Comparing these two plots, we see that the unvoiced DTFT is, on this relatively high frequency range, a modified lowpass version of the original DTFT. If we had included all the unvoiced portion, it would be a fairly accurate lowpass version of $X(e^{j\omega})$ on this frequency range. The "lowpass" (i.e., smoothing) effect on the spectrum is due to the time-domain truncation in (6.79b), (6.79c), and (6.80). Notice that the peaks in Figure 7.6b are no closer than $\Delta f = 1/512 = 0.002$ cycle/sample corresponding to $\Delta F = 1/512/\Delta t = 21.5$ Hz, whereas those in Figure 7.6a are no closer than $\Delta f = 1/8192 = 0.00012$ cycle/sample corresponding to $\Delta F = 1/8192/\Delta t = 1.35$Hz.

Finally, Figures 7.7a and b show the same information as Figures 7.6a and b for the true DTFT and the voiced subsection, on the range $100 \leq k \leq 400$ (or 134 Hz $\leq F \leq 538$ Hz). Again we see the DTFT of the 512-point subsection as a sort of lowpass version of the true DTFT, on this frequency range.

Recall that windowing in the time domain corresponds in the frequency domain to merely convolution of $X(e^{j\omega})$ with $W_{N,R}(e^{j\omega})$ in (6.79b), (6.79c), and (6.80). Then how can the spectra of different subsections of equal lengths differ as they must (in Figures 7.4b and 7.5b) if the same spectrum (namely Figure 7.3b) is in both cases convolved with the same, albeit shifted window [as in (6.79b)]? The answer is that the shifting of the time-window from one subsection to the other corresponds to introduction of a phase factor in the frequency domain [onto $W_{N,R}(e^{j\omega})$]. This phase factor will cause terms in the convolution integral to accumulate or cancel in such a way as to produce the spectrum of the subsection. It seems incredible, but that is precisely what happens!

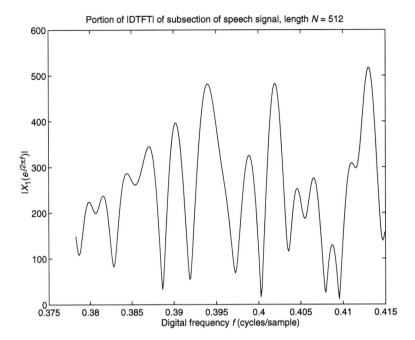

Figure 7.6*b*

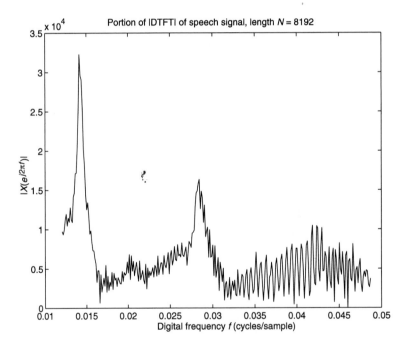

Figure 7.7*a*

The above is an example of a *time-dependent instantaneous "spectrum."* Except for smoothing, the spectra of Figures 7.4*b* and 7.5*b* are approximate representations of the spectral content at different times. As we pronounce different syllables, change the pitch of our voice, sing, and so forth, the instantaneous spectral content changes.

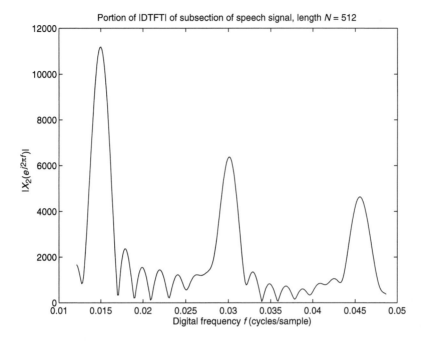

Figure 7.7*b*

This is a fact over which the signal-processing practitioner has no control; for speech, the spectral content may change significantly over intervals 10 ms or shorter.

For reasonable values of N and Δt, if we try to take a very long segment to obtain an overall representative signal sample by letting Δt be large, we risk aliasing. If we take a short sample with small Δt to be able to say that the spectral content has changed little over our sample, the truncation effect is very large and we will end up with a very distorted, smoothed spectrum. The only real solution for the most basic DFT calculation is to make Δt as small as possible and N large enough so that our transformed subsection is reasonably representative of the spectrum of the "entire" signal under consideration (unlike the case in this example). In many cases, however, we *want* this "instantaneous spectrum" and refer to it as the discrete-time short-time Fourier transform.

It is appropriate now to discuss the resolution of the DFTN with respect to digital frequencies. The question here is whether spectral details are resolvable, which could be crucial in practical situations such as mechanical vibration analysis or biomedical diagnostics. We will see below (as hinted in the discussion of Figure 7.6 in Example 7.2) that the resolution of digital frequencies is inversely proportional to N. Thus, to be able to resolve (i.e., be able to distinguish correctly) two separate spectral components of a given sequence, N must be chosen sufficiently large. To make our study of resolution concrete, we will discuss the ideas of resolution by means of the following simple example: the DFT estimation of the spectrum of a sequence having only two purely sinusoidal components.

EXAMPLE 7.3

Consider the sum of two discrete-time sinusoids, for example, $x(n) = \cos(\omega_1 n) + \cos(\omega_2 n)$, where ω_1 and ω_2 are chosen to be close together and, for simplicity, chosen such that an integral number of periods of each sinusoid fits in the N-time-window. Determine a formula for resolution, based on a detailed spectral analysis of $x(n)$. Plot the estimated spectra to support the resolution formula.

Solution

First, recall that $\cos(\omega_1 n) = \frac{1}{2}(e^{j\omega_1 n} + e^{-j\omega_1 n})$. Now recall that $\delta_c(-x) = \delta_c(x)$ in (2A.24) to obtain

$$\text{DTFT}\{e^{j\omega_A n}\} = \sum_{n=-\infty}^{\infty} e^{j(\omega_A - \omega)n} = 2\pi\delta_c(\omega - \omega_A), \qquad |\omega - \omega_A| < \pi. \tag{7.7}$$

Noting that the DTFT of $\cos(\omega_1 n)$ contains sums of the form in (7.7) with $\omega_A = \pm\omega_1$, we have

$$\cos(\omega_1 n) \xleftrightarrow{\text{DTFT}} \pi[\delta_c(\omega - \omega_1) + \delta_c(\omega + \omega_1)] \tag{7.8}$$

and a similar expression for $\text{DTFT}\{\cos(\omega_2 n)\}$. Thus, an infinite-length DFT would have one infinite-height sample at $k = k_1$ associated with ω_1 and another with $k = k_2$ associated with ω_2, and zero elsewhere. (Note, however, that for N truly infinite, k_1 and k_2 would also both be infinite.)

Suppose that we let $N = 4096$; in most cases we prefer selecting a power of 2 for sequence lengths because the FFT algorithm for implementing the DFT is far more

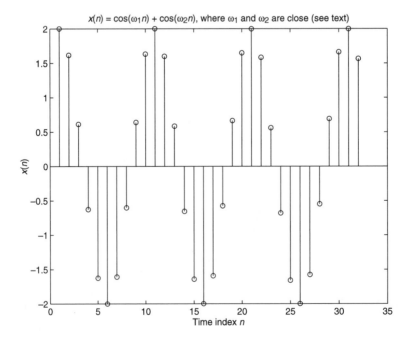

Figure 7.8a

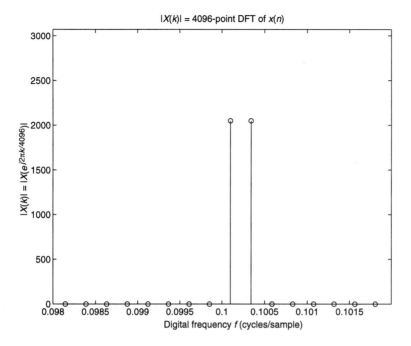

Figure 7.8*b*

efficient for those lengths than for others; see Section 7.10. Further, let us select ω_1 such that in the N points there are exactly ℓ cosinusoidal periods: $\omega_1 N = 2\pi \cdot \ell$, or $\omega_1 = 2\pi\ell/N$, or equivalently, $f_1 = \ell/N$. For $\ell = 410$, we obtain $f_1 = 0.10009765625$ cycle/ sample. Let ω_2 be chosen to include exactly one more period in the same N samples: $\omega_2 = 2\pi(\ell + 1)/N$, or for $\ell = 410$, $f_2 = 0.100341796875$ cycle/sample. Note that both of these sinusoids thus have approximately $1/0.1 = 10$ samples per oscillation period so aliasing will not be a problem. The first 32 samples of $x(n)$ are shown in Figure 7.8*a*, and the N-point $|\text{DFT}_{4096}\{x(n)\}|$ in the vicinity of $f = 0.1$ is shown in Figure 7.8*b*.

In Figure 7.8*b*, we see one sample at each of the two consecutive k values associated with ω_1 and ω_2; call them k_1 and k_2 $(= k_1 + 1)$. The height of each sample is $N/2 = 2048$ [see discussion on discrete Fourier series near the end of Example 7.1, and note the $\frac{1}{2}$ factor in $\cos(x) = \frac{1}{2}\{e^{jx} + e^{-jx}\}$]. Because of our choices of ω_1 and ω_2, we have, for example, $k_1 = \omega_1 N/(2\pi) = \{2\pi\ell/N\}N/(2\pi) = \ell$, and similarly, $k_2 = \ell + 1$. The frequency spacing between adjacent "bins" of the DFT_N is $\Delta\omega = 2\pi/N$; no sinusoidal components between these may be distinguished using a set of N samples.

Although we *can* interpolate in between ω_1 and ω_2 by zero-padding $x_N(n)$, as we saw before, doing so will not increase the resolution. We immediately know from (7.3) that all this procedure will yield is more finely spaced samples of $X_N(e^{j\omega})$ in (6.79b)—*not* samples of $X(e^{j\omega})$. If the new length after zero-padding is $M > N$, one may wish to view $x_{zp;M}(n)$ as $x_M(n)$ [the M-truncated version of $x(n)$] further truncated to $x_N(n)$ with zeros appended. Thus, the "second" windowing (to N) reduces what might have been $2\pi/M$ resolution back down to $2\pi/N$. The frequency domain effect obtained by zero-padding can be viewed only as interpolation of $X_N(e^{j\omega})$ (note the subscript N!). If the original sequence $x(n)$ is of only length N—that is, $x(n) = x_N(n)$—then zero-padding does decrease the spacing of frequency samples of $X(e^{j\omega})$ below $2\pi/N$, but in that case there can be no peaks in $X(e^{j\omega})$ to resolve that are spaced closer than $2\pi/N$.

In our example of the sum of two infinite-duration cosinusoids, $x_{zp;M}(n)$ will not have the same DTFT as $x(n)$ has (namely, the two Dirac delta functions). Instead, by (6.79b), $X_{zp;M}(e^{j\omega})$ will be the convolution of those Dirac deltas with $W_{N,R}(e^{j\omega})$. By the sampling property of the Dirac delta function (2A.3), we can easily obtain

$$X_{zp;M}(e^{j\omega}) = \tfrac{1}{2}\{W_{N,R}(e^{j(\omega-\omega_1)}) + W_{N,R}(e^{j(\omega-\omega_2)}) + W_{N,R}(e^{j(\omega+\omega_1)}) + W_{N,R}(e^{j(\omega+\omega_2)})\}, \tag{7.9}$$

and thus the M-point DFT of $x_{zp;M}(n)$ is

$$X_{zp;M}(k) = X_{zp;M}(e^{j\omega})\Big|_{\omega=2\pi k/M}, \tag{7.10}$$

and we just substitute $2\pi k/M$ for ω the four places it appears in (7.9).

For example, with $N = 4096$ and $M = 16N = 65536$, we obtain the plot in Figure 7.9. In Figure 7.9, the $|\text{DFT}_{4096}|$ of the non-zero-padded $x_{4096}(n)$ is left in for reference. From this figure and from (7.9) and (7.10), we conclude that by zero-padding, every nonzero sample of $X_N(k)$ hosts a shifted replication of $W_{N,R}(e^{j\omega})$, and the sum of all these replications is $X_{zp;M}(e^{j\omega})$. In turn, samples of $X_{zp;M}(e^{j\omega})$ at $\omega = 2\pi k/M$ are given by $X_{zp;M}(k)$. If instead we had $x(n) = \cos(\omega_1 n)$, there would be only one replication, at ω_1; in that case, it would be perfectly centered on ω_1. The peak lobes are not centered on ω_1 and ω_2 in Figure 7.9 because they comprise the addition of two replications of $W_{N,R}(e^{j\omega})$ at $\omega = \omega_1$ and $\omega = \omega_2$. These two replications are also present in Figure 7.8b, but each is zero for all k not equal to, respectively, ℓ and $\ell + 1$.

From our discussion above, we might argue that because the *distinguishing* samples are $2\pi/N$ apart, any value of N we would choose below 4096 will be unable to resolve the sinusoids ω_1 and ω_2 (whose separation is fixed at $2\pi/4096$ rad/sample). This definition would give $2\pi/N$ as the resolution. It is not quite that cut-and-dried. In Figure 7.10a is shown, in the neighborhood of ω_1 and ω_2, the $|\text{DFT}_{4080}|$ of a 4080-point segment of $x(n)$ in normal stems ("O" with solid-line stems). Also shown in Figure 7.10a in modified stems ("*" with -.-.- stems) is the $|\text{DFT}_{4096}|$ that was shown in Figures 7.8b and 7.9. Finally, Figure 7.10a also shows the $|\text{DFT}_{65536}|$ of the zero-padded $x_{4080}(n)$ sequence (solid curve). We see that the DFT_{4080} values (again, marked

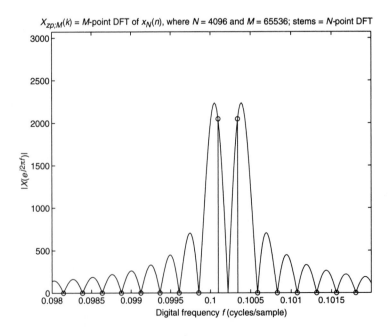

Figure 7.9

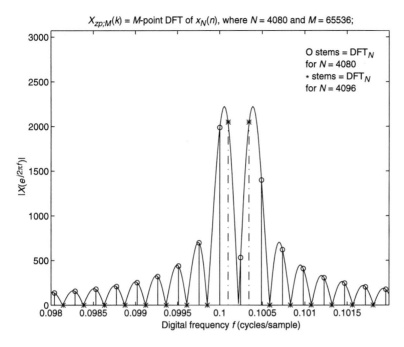

Figure 7.10*a*

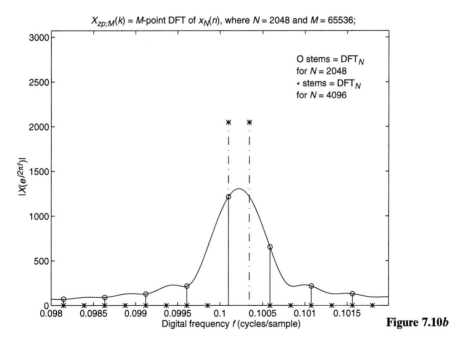

Figure 7.10*b*

with "O") miss ω_1 and ω_2, but the interpolation by zero-padding still mostly identifies the presence of two nearby sinusoids. [It also has a lot of unhelpful sidelobes, as does Figure 7.9, due to the N-truncation beyond which zeros were padded for $X_{zp;M}(k)$].

We might alternatively say that with $N \leq 4096/2 = 2048$ we may not distinguish sinusoids. This is certainly true, as Figure 7.10*b* shows. Figure 7.10*b* shows the $|DFT_{2048}|$ of a 2048-segment of $x(n)$ ("O" stems), the $|DFT_{4096}|$ ("*") shown in previous figures, and the DFT_{65536} of the zero-padded $x_{2048}(n)$ sequence (solid

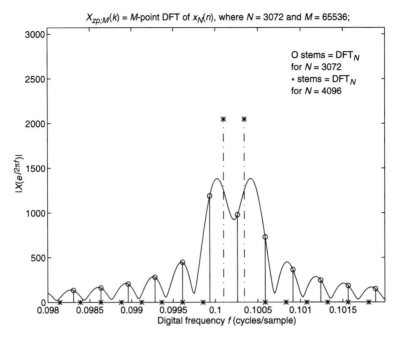

$X_{zp;M}(k) = M$-point DFT of $x_N(n)$, where $N = 3072$ and $M = 65536$;

O stems = DFT_N
for $N = 3072$
* stems = DFT_N
for $N = 4096$

$|X(e^{j2\pi f})|$

Digital frequency f (cycles/sample)

Figure 7.10c

curve). This time, ω_1 coincides with one of the DFT_{2048} samples, but ω_2 is exactly halfway to the next DFT_{2048} sample. In the interpolation, the distinction between the two original sinusoids is entirely smoothed out; they are no longer resolvable.

Finally, half-way between 2048 and 4096 is 3072; this value of N is shown in Figure 7.10c. Here we see that while the peaks of the two sinusoids are discernable, the valley between the peaks is shallow, and also the peak values of the zero-padded DFT of $x_{3072}(n)$ are both roughly $\frac{1}{2}$ a sampling interval outside the true peaks (again, located at the asterisks). Depending on one's requirements, we see there is some variety in what one may define as resolution-somewhere between $2\pi/N$ and π/N as the smallest frequency difference between distinguishable sinusoids using an N-point DFT.

When the sequence is windowed, we may express the resolution in terms of the window function $W_{N,R}$. The first zero-crossing of $W_{N,R}$ occurs at $N\omega/2 = \pi$, or $\omega = 2\pi/N$. From above, then, the available resolution using the rectangular window is roughly equal to half the main lobe width of $W_{N,R}$. Let us denote the length of the true $x(n)$ by P. If $P \gg N$, then the minimum spacing between peaks in $X(e^{j\omega})$ is $2\pi/P \ll 2\pi/N$. Moreover, if $P \to \infty$, $2\pi/P \to 0$. Consequently, the smoothing by $W_{N,R}$ prevents the distinguishing of individual sinusoidal components in $X(e^{j\omega})$.

Suppose that $x_N(n)$ is replicated L times, rather than zero-padded, from N out to $M = LN$. What would be the effect, and would there be an improvement in estimation of the DTFT? The result was given in Section 4.6 and will now be proved. With the L-repetitions extension of $x_N(n)$ designated $x_{N,L}(n)$, the DFT_M of this sequence, denoted $X_{M;N,L}(k)$ and where $M = NL$, is

$$X_{M;N,L}(k) = \sum_{n=0}^{M-1} x_{N,L}(n) \cdot e^{-j2\pi nk/M}. \tag{7.11a}$$

Noting the N-periodicity of $x_{N,L}(n)$ $[x_{N,L}(\ell N + m) = x_{N,L}(m)]$ on only the range $m \in [0, M-1]$, we can break up the M-sum in (7.11a) into L sums each of length N:

$$X_{M;N,L}(k) = \sum_{\ell=0}^{L-1} \sum_{m=0}^{N-1} x_{N,L}(m) \cdot e^{-j2\pi(\ell N + m)k/M}, \tag{7.11b}$$

or, reversing the summation symbols,

$$X_{M;N,L}(k) = \sum_{m=0}^{N-1} x_{N,L}(m) \cdot e^{-j2\pi mk/M} \cdot \sum_{\ell=0}^{L-1} e^{-j2\pi N\ell k/M} \tag{7.11c}$$

and, evaluating the geometric sum,

$$X_{M;N,L}(k) = \sum_{m=0}^{N-1} x_{N,L}(m) \cdot e^{-j2\pi mk/M} \cdot \begin{cases} L, & k = nL, 0 \le n \le N-1 \\ \dfrac{1 - e^{-j2\pi k}}{1 - e^{-j2\pi k/L}} = 0, & \text{otherwise,} \end{cases} \tag{7.11d}$$

which incidentally recalls the orthogonality of the complex exponentials. Finally,

$$X_{M;N,L}(k) = \begin{cases} L \cdot \sum_{m=0}^{N-1} x_{N,L}(m) \cdot e^{-j(2\pi/N) \cdot (k/L)m}, & k = nL, 0 \le n \le N-1 \\ 0, & \text{otherwise} \end{cases} \tag{7.11e}$$

$$= \begin{cases} L \cdot X_N(k/L) & k = nL, \text{ where } 0 \le n \le N-1 \\ 0, & \text{otherwise.} \end{cases} \tag{7.11f}$$

The result (7.11f) is that $X_{M;N,L}(k)$ is a scaled version of the original DFT_N values interlaced with zeros, as was claimed in Section 4.6. Clearly, this procedure is of no benefit for estimating the DTFT $X(e^{j\omega})$ from N samples of $x(n)$. It does not even provide noise averaging, for the same data is used repeatedly. If different data were used on each L-length segment and $x(n)$ was an L-periodic signal plus noise, then we would be doing the usual M-point DFT and the noise-averaging improvement would be attained. The derivation in (7.11) "in reverse" will be recalled in our discussion in Section 7.5.3 of an algorithm for time-domain interpolation.

Before moving on to DTFT^{-1} estimation (Section 7.3), let us look at a different but related forward DTFT estimation problem. Suppose that we had N correct values $X_{N,a}(k)$ of $X(e^{j\omega})$ and wished to interpolate these using the DFT by zero-padding in the time domain. We introduce the N,a subscript to avoid confusion with $X_N(k)$, the DFT_N of $x_N(n)$, the N-truncated version of $x(n)$. They are not equal when [length of $x(n) = P] > N$, for we showed above that $X_N(k) \ne X(e^{j2\pi k/N})$ in that case, whereas we define $X_{N,a}(k) \equiv X(e^{j2\pi k/N})$ regardless of P. Let us assume that $X(e^{j\omega})$ derives from a sequence of length $P \gg N$ points, so we expect that the interpolation using the DFT and zero-padding will not be exact.

First note, however, that if $P = N$, then $x_N(n) = x(n)$ and $X_N(k) = X(e^{j2\pi k/N})$. Thus, recognition that the DFT_N^{-1} in (7.4a), $x_N(n)$, is equal to $x(n)$ immediately shows (by definition of the DTFT) that the interpolation of the $X_N(k)$ as written in (7.4e) and (7.5) is exact. When $P \gg N$, it will be shown in Section 7.3[2] that the relation between $x_{N,a}(n) = \mathrm{DFT}_N^{-1}\{X(e^{j2\pi k/N})\} = \mathrm{DFT}_N^{-1}\{X_{N,a}(k)\}$ and $x(n)$ is $x_{N,a}(n) = \sum_{m=-\infty}^{\infty} x(n + mN)$, which is time-aliasing. Using this fact, we may determine the relation between the interpolation in between the given samples $X_{N,a}(k) = X(e^{j2\pi k/N})$, and the true $X(e^{j\omega})$ obtained by the operations (i) DFT_N^{-1}, (ii) zero-padding to M, and (iii) DFT_M. The DTFT of $x_{N,a}(n)$ is

$$\hat{X}(e^{j\omega}) = \sum_{n=0}^{N-1} \sum_{m=-\infty}^{\infty} x(n + mN) \cdot e^{-j\omega n}. \tag{7.12a}$$

---

[2]The result is fundamental to Section 7.3, so it is proved there.

Define $\ell = n + mN$. Then, using $[\cdot]$ to indicate the greatest integer, we may rewrite (7.12a) as

$$\hat{X}(e^{j\omega}) = \sum_{\ell=-\infty}^{\infty} x(\ell) \cdot e^{-j\omega\{\ell - [\ell/N]N\}}. \tag{7.12b}$$

Suppose that the forward DFT is $M = NL$ points long. This DFT_M will produce M samples of $\hat{X}(e^{j\omega})$ evaluated at $\omega_k = 2\pi k/M$ for $0 \le k \le M - 1$. Thus, (7.12b) at ω_k becomes the DFT_M of the DFT_N^{-1} of the given $X_{N,a}(k)$, which we denote as $X_{N,a,zp;M}(k)$:

$$X_{N,a,zp;M}(k) = \hat{X}(e^{j2\pi k/M})$$

$$= \sum_{\ell=-\infty}^{\infty} x(\ell) \cdot e^{-j2\pi k\{\ell - [\ell/N]N\}/M} \tag{7.13a}$$

$$= \sum_{\ell=-\infty}^{\infty} x(\ell) \cdot e^{-j2\pi k\ell/M} e^{j2\pi k[\ell/N]/L}. \tag{7.13b}$$

Notice that if $k = mL$, then the second exponent in (7.13b) is $j2\pi m[\ell/N]$, an integer multiple of $j2\pi$ because $[\ell/N]$ is an integer, by the definition of greatest integer $[\cdot]$. Consequently, that exponential is unity for $k = mL$. Also for $k = mL$, the first exponent in (7.13b) is $-j2\pi m\ell/N$, identical to the exponent in the true DTFT expression for $X(e^{j\omega})$ for $\omega = 2\pi m/N$; thus, the result for this case is that the interpolation is exact. This means that the interpolation, although being incorrect in between the original samples (i.e., for $k \ne mL$), does reproduce the original given samples $X(e^{j2\pi m/N})$ exactly. For $k \ne mL$, the second exponential factor in (7.13b) distorts what otherwise would be a perfect interpolation.

When $X(e^{j\omega})$ is due to a sequence longer than N and the DFT is used to interpolate N exact samples of $X(e^{j\omega})$ at $\omega = 2\pi k/N$ where $0 \le k \le N - 1$, the interpolation relation in (7.4e) and (7.5) $[X_N(e^{j\omega}) = \sum_{k=0}^{N-1} X_N(k) \cdot \phi(\omega - 2\pi k/N)$ where $\phi(\theta) = e^{-j\theta(N-1)/2} \sin(N\theta/2)/\{N\sin(\theta/2)\}]$ still applies. However, the result of this interpolation is no longer equal to $X(e^{j\omega})$ except at the originally given frequencies $\omega = 2\pi k/N$.

EXAMPLE 7.4

Estimate (interpolate) the $|\text{DTFT}|$ of $x(n) = a^n u(n)$ using $N = 64$ exact samples of $X(e^{j\omega})$ by zero padding in the time domain, and compare against the exact $|\text{DTFT}|$ for interesting values of a. In the process, plot and comment on the DFT_N^{-1} of the N samples of $X(e^{j\omega})$.

Solution

We know that $X(e^{j\omega}) = 1/\{1 - ae^{-j\omega}\}$, so $X_{N,a}(k) = 1/\{1 - ae^{-j2\pi k/N}\}$. It is expected that if $x(N) = a^N << x(0) = 1$, then $x_{N,a}(n) = \text{DFT}_N^{-1}\{X_{N,a}(k)\} \approx x(n)$ because the time-aliasing referred to above will be minimal. Defining $\alpha = a^N$ so that $x(N) = \alpha \cdot x(0)$, we have $a = e^{\ln(\alpha)/N}$, and thus we can find the value of a yielding a desired value of $\alpha = x(N)/x(0)$. For $\alpha = 0.01$ and thus $a = 0.9306$, Figure 7.11a shows $x(n)$ ("O" stems) and $x_{N,a}(n)$ ("*" stems). Indeed, we find close agreement between

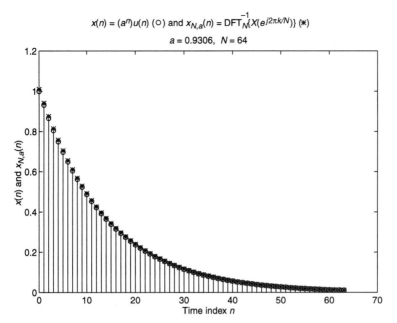

$x(n) = (a^n)u(n)$ (○) and $x_{N,a}(n) = \text{DFT}_M^{-1}\{X(e^{j2\pi k/N})\}$ (✳)

$a = 0.9306,\ N = 64$

Figure 7.11*a*

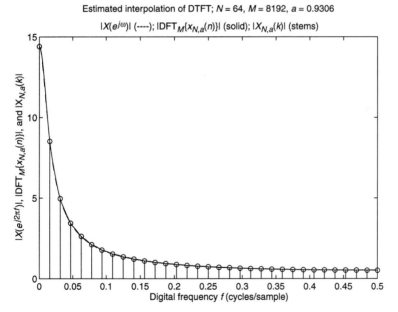

Estimated interpolation of DTFT; $N = 64$, $M = 8192$, $a = 0.9306$

$|X(e^{j\omega})|$ (----); $|\text{IDFT}_M\{x_{N,a}(n)\}|$ (solid); $|X_{N,a}(k)|$ (stems)

Figure 7.11*b*

$x_{N,a}(n)$ and $x(n)$; that is, the time-domain aliasing of the infinite-length $x(n)$ is negligible for this combination of α and N. In Figure 7.11*b* are shown, for $M = 8192$, $|X(e^{j\omega})|$ (- - - - -), $|X_{N,a,zp;M}(k)|$ (solid curve—samples joined with straight lines),

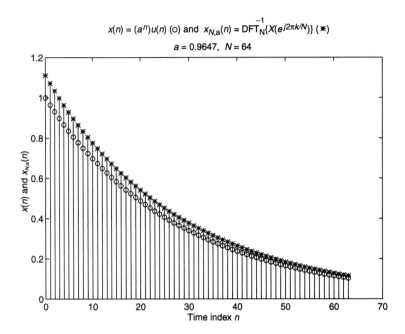

$x(n) = (a^n)u(n)$ (○) and $x_{N,a}(n) = \text{DFT}_N^{-1}\{X(e^{j2\pi k/N})\}$ (✻)

$a = 0.9647,\ N = 64$

Figure **7.12a**

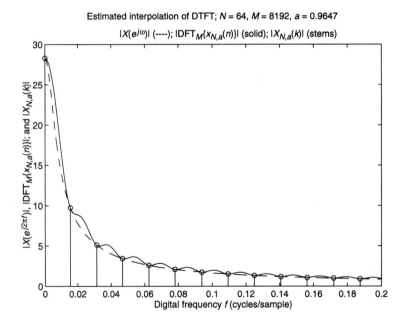

Estimated interpolation of DTFT; $N = 64$, $M = 8192$, $a = 0.9647$

$|X(e^{j\omega})|$ (----); $|\text{IDFT}_M\{x_{N,a}(n)\}|$ (solid); $|X_{N,a}(k)|$ (stems)

Figure **7.12b**

and the original $|X_{N,a}(k)| = |X(e^{j2\pi k/N})|$ (stems). The accuracy of the spectrum interpolation is excellent for this $\{a, N\}$ pair.

Figures 7.12a and b show, respectively, the time- and frequency-domain plots for a chosen to give $\alpha = 0.1$ so that $x(N) = 0.1 = x(0)/10$. Now, the time-aliasing is quite noticeable in Figure 7.12a, and the $|\text{DTFT}|$ estimated interpolation in Figure 7.12b

$x(n) = (a^n)u(n)$ (○) and $x_{N,a}(n) = \text{DFT}_N^{-1}\{X(e^{j2\pi k/N})\}$ (✳)

$a = 0.9892, \quad N = 64$

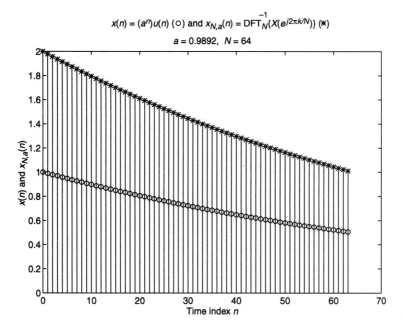

Figure 7.13a

Estimated interpolation of DTFT; $N = 64$, $M = 8192$, $a = 0.9892$

$|X(e^{j\omega})|$(----); $|\text{IDFT}_M\{x_{N,a}(n)\}|$ (solid); $|X_{N,a}(k)|$ (stems)

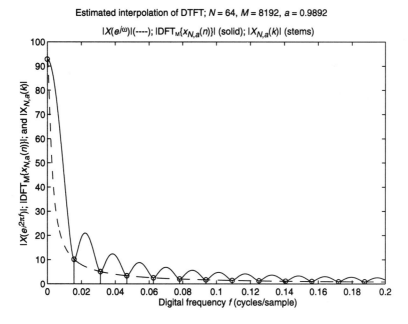

Figure 7.13b

is also noticeably inaccurate (only the more inaccurate interval $0 \leq f \leq 0.2$ is shown). Finally, if we set $\alpha = 0.5$ so that $x(N) = \frac{1}{2}x(0)$, the aliasing is horrible in Figure 7.13a, as is the estimated interpolation in Figure 7.13b. However, in all cases, including $\alpha = \frac{1}{2}$, notice that the interpolation reproduces the original $X_{N,a}(k)$ exactly.

7.3 INVERSE DTFT ESTIMATION BY THE DFT

In the previous section, we investigated the tasks of estimating $X(e^{j\omega})$ from N samples of either $x(n)$, denoted $x_N(n)$ or interpolating between a given set of N samples of $X(e^{j\omega})$, denoted $X_{N,a}(k)$ $k \in [0, N-1]$.[3] If the original sequence $x(n)$ is nonzero only for $n \in [0, N-1]$ [$x_N(n) = x(n)$], then (7.5) and (7.6) [$X_{zp;M}(k) = \sum_{m=0}^{N-1} X_N(m) \cdot \phi(2\pi[(k/M) - (m/N)])$, where $\phi(\theta) = e^{-j\theta(N-1)/2}\sin(N\theta/2)/\{N\sin(\theta/2)\}$] show how zero-padding provides an exact interpolation of $X(e^{j\omega})$ between the N values $X_N(k) = \text{DFT}_N\{x(n)\}$. When the length of $x(n)$, P, exceeds N, then the DTFT of $x_N(n)$ no longer is equal to $X(e^{j\omega})$; the distortion is quantified in (6.79b) [$X_N(e^{j\omega}) = X(e^{j\omega})*_{c\omega}W_{N,R}(e^{j\omega})/(2\pi)$]. If we happen to be given N samples $X_{N,a}(k)$ of $X(e^{j\omega})$ and $P > N$, we described in the discussion following (7.13) how distortion results when we attempt to use time-domain zero-padding of $\text{DFT}_N^{-1}\{X_{N,a}(k)\}$ and the DFT_M to interpolate between $X_{N,a}(k)$.

Now suppose that again we select N equally spaced values of $X(e^{j\omega})$, $X_{N,a}(k)$, but now our purpose is expressly to estimate $x(n) = \text{DTFT}^{-1}\{X(e^{j\omega})\}$. The DFT_N^{-1} of $X_{N,a}(k)$ is certainly equal to $x(n)$ possibly with zeros padded at the end if the true $x(n)$ has length $P \le N$. But what if $P > N$? We cited the time-aliasing result in Section 7.2 when we evaluated the interpolation of $X(e^{j\omega})$ from $X_{N,a}(k)$. This result shall now be derived and discussed. By the definition of the DFT_N^{-1},

$$\text{DFT}_N^{-1}\{X_{N,a}(k)\} = \frac{1}{N} \sum_{k=0}^{N-1} X(e^{j2\pi k/N}) \, e^{j2\pi nk/N}. \tag{7.14a}$$

Substitution of the definition of the DTFT for $X(\cdot)$ in (7.14a) gives

$$\text{DFT}_N^{-1}\{X_{N,a}(k)\} = \frac{1}{N} \sum_{k=0}^{N-1} \sum_{\ell=-\infty}^{\infty} x(\ell)e^{j2\pi k(n-\ell)/N}, \tag{7.14b}$$

or, reversing the order of summations,

$$\text{DFT}_N^{-1}\{X_{N,a}(k)\} = \sum_{\ell=-\infty}^{\infty} x(\ell) \frac{1}{N} \sum_{k=0}^{N-1} e^{j2\pi k(n-\ell)/N}. \tag{7.14c}$$

The second sum in (7.14c) is easily evaluated by geometric series and embodies the orthogonality property of complex exponentials. The sum is equal to N if $n - \ell = -mN, -m$ any integer. That is, the sum is equal to N if $\ell = n + mN$, and it is zero otherwise. In general, for all integers i and m,

$$\{\delta(i + mN), \text{all } m\} = \delta(i) + \delta(i + N) + \delta(i - N) + \delta(i + 2N) + \delta(i - 2N) + \ldots \tag{7.15a}$$

$$= \sum_{m=-\infty}^{\infty} \delta(i + mN) = \frac{1}{N} \sum_{k=0}^{N-1} e^{j2\pi ki/N}. \tag{7.15b}$$

Replacing i by $n - \ell$ in (7.15b) and substituting (7.15b) times N into (7.14c), (7.14c) becomes

$$\text{DFT}_N^{-1}\{X_{N,a}(k)\} = \sum_{\ell=-\infty}^{\infty} x(\ell) \sum_{m=-\infty}^{\infty} \delta(n - \ell + mN). \tag{7.16a}$$

For given values of n and m [move the m sum in (7.16a) to the left of the ℓ sum], only one value of ℓ satisfies the condition for the Kronecker delta to be nonzero: $\ell = n + mN$. Thus, we obtain

---

[3]Recall that the additional a subscript on $X_{N,a}(k)$ is added to prevent confusion between it and the DFT_N of $x_N(n)$.

$$\text{DFT}_N^{-1}\{X_{N,a}(k)\} = \sum_{m=-\infty}^{\infty} x(n + mN). \qquad (7.16b)$$

The sum in (7.16b) is a superposition of shifted versions of $x(n)$, with each replica separated from the adjacent replicas by N time-index values. It is thus clear that aliasing errors will occur whenever P exceeds N. How serious the degradation is depends on how large the interfering "end" values are and by how much P exceeds N (recall Figures 7.11a, 7.12a and 7.13a in Example 7.4). This result (7.16b), used in Section 7.2, is also of use in explaining DFT convolution, as discussed in Section 7.6.

It is interesting to note that $\text{DFT}_N^{-1}\{X_{N,a}(k)\}$ in (7.14) and (7.16b) is precisely the N-point rectangular rule approximation of the DTFT^{-1} integral, with $\Delta\omega = 2\pi/N$ [where the 2π in $\Delta\omega$ cancels the $1/(2\pi)$ of the DTFT^{-1} definition, leaving $1/N$; see footnote 1]. Evidently, the condition $P \leq N$ for no aliasing translates to $2\pi/N \leq 2\pi/P$; that is, $\Delta\omega \leq 2\pi/P$. Notice the clear analogy with Nyquist's sampling criterion ($\Delta t \leq 2\pi/\text{BW}$). Under the condition $\Delta\omega \leq 2\pi/P$ ($N \geq P$), rectangular rule estimation of the DTFT^{-1} integral is exact just as it is for FT estimation (over the appropriate ranges of, respectively, time index n and frequency Ω). Also, it will yield precisely zeros for $n \in [P, N-1]$, and an N-periodic extension for n outside $[0, N-1]$. The estimation is exact because only the $m = 0$ term in (7.16b) contributes for $n \in [0, N-1]$. Mathematically, we may write the above statements in terms of $X(e^{j\omega})$ as

$$x(n), n \in [0, N-1] = \frac{1}{2\pi} \int_{-\pi}^{\pi} X(e^{j\omega})e^{j\omega n} \, d\omega \qquad (7.17a)$$

$$= \frac{1}{N} \sum_{k=0}^{N-1} X(e^{j2\pi k/N})e^{j2\pi kn/N} \quad \text{if } N \geq P. \qquad (7.17b)$$

We can also now make an interesting identification relevant to discussion of the discrete-time–continuous-time interface in Chapter 6. Letting $x(n) = x_c(n\,\Delta t)$, rewrite (7.16b) and recall that $X(e^{j\omega})$ is an aliased sum of replications of $X_c(j\Omega)$ to obtain

$$\text{DFT}_N \left\{ \sum_{m=-\infty}^{\infty} x_c(\{n + mN\}\Delta t) \right\} = X_{N,a}(k) = X(e^{j2\pi k/N})$$

$$= \frac{1}{\Delta t} \sum_{m=-\infty}^{\infty} X_c\left(j\frac{2\pi}{N\Delta t}\{k + mN\} \right) \qquad (7.18)$$

which is a precise statement of the relation between the discrete Fourier transform and the Fourier transform. A version of the DFT relation in which periodic replication in the "initial" domain was replaced by truncation (the more usual situation, valid when we have only N samples of the signal to be transformed) was presented in Section 6.7 and Figure 6.56 [(6.108) and (6.109)].

Equation (7.18) says that the replicated sum of a sampled version of $x_c(t)$ and the replicated sum of a sampled version of $X_c(j\Omega)/\Delta t$ are a DFT_N pair. Either or both could be aliased. The replicas of $x_c(n\,\Delta t)$ are $t_s = N\Delta t$ apart, whereas those of $X_c(j2\pi k/\{N\Delta t\})/\Delta t$ are $\Omega_s = 2\pi/\Delta t$ apart. Equation (7.18) suggests one way of getting around the windowing problem (caused by $P > N$, where N is the available DFT length), provided we have available the entire sequence desired to be transformed and we are capable of computing the aliasing sum on the left-hand side of (7.18). However, because there is also aliasing in the other domain, Δt and N must be negotiated in the manner discussed in Section 6.5 to minimize the resulting aliasing and obtain the desired resolution.

Applying this same suggestion to the DTFT, (7.16b) provides a way of estimating the DTFT using only a single N-point DFT: First calculate the aliasing sum in

(7.16b), and then take the DFT_N of the result. That DFT_N will yield N exact samples of the DTFT of $x(n)$, at $\omega = 2\pi k/N, k \in [0, N-1]$. This method could be of use when all data are known and the data size is moderately larger than the available DFT size so that calculation of the aliasing sum is feasible.

As noted above, we have already seen an illustration of the aliasing in (7.16b) in Example 7.4. In that example, $P = \infty$ so that there is time-aliasing for any finite value of N. Practically, however, for N suitably chosen, [so that $x(N) \ll x(0)$] the aliasing was not noticeable (see Figure 7.11a), whereas it was severe for large $x(N)$ (Figure 7.13a). In Example 7.5, we clarify the time-aliasing in (7.16b) by estimating the DTFT for a finite-length sequence $x(n)$.

EXAMPLE 7.5

Suppose that $x(n)$ is the triangle sequence

$$x(n) = \begin{cases} 2n/P, & n \in [0, P/2] \\ 2(1 - n/P), & n \in [P/2 + 1, P - 1], \end{cases} \tag{7.19}$$

for P even. This triangle sequence has exactly one zero per period when P-periodically extended; it thus differs a little from the odd-length triangle we analyzed in in Appendix 4B. Let the triangle be of duration $P = 64$ samples. Suppose that **(a)** $N = 64$, **(b)** $N = 32$, or **(c)** $N = 50$ samples of the DTFT are taken to form the sequence $X_{64,a}(k)$. Estimate the DTFT^{-1} in each case using these samples of the DTFT, and compare against the exact DTFT^{-1}.

Solution

The true $x(n)$ with $P = 64$ is shown in Figure 7.14. It was proved in the problems of Chapter 4 [see also (4B.21) and (4B.22)] that the z-transform of $x(n)$ is

$$X(z) = \frac{2}{P} z^{-1} \cdot \left(\frac{1 - z^{-P/2}}{1 - z^{-1}} \right)^2 \tag{7.20}$$

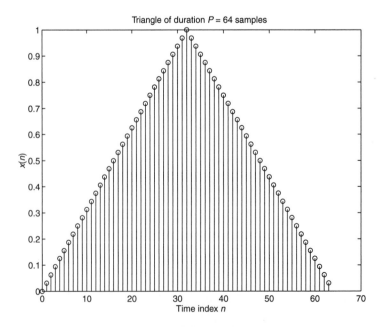

Triangle of duration $P = 64$ samples

Figure 7.14

and that the DTFT of $x(n)$ is $X(z)$ with $z = e^{j\omega}$:

$$X(e^{j\omega}) = \frac{2}{P} \cdot \left(\frac{\sin(\omega P/4)}{\sin(\omega/2)} \right)^2 \cdot e^{-j\omega P/2} \tag{7.21}$$

for $\omega \neq 0$ and $P/2$ for $\omega = 0$.

(a)   The magnitude of $X_{64,a}(k)$ is shown in stems in Figure 7.15a, with $X(e^{j\omega})$ superimposed. The inverse DFT$_{64}$ of $X_{64,a}(k)$, $x_{64,a}(n)$, is shown in Figure 7.15b. As we would expect, $x_{64,a}(n) = x(n)$, so $x(n)$ is exactly reproduced. This is expected because the length of $x(n)$, $P = 64$, is less than or equal to (in this case, equal to) N, the number of spaces between replications of $x(n)$ in (7.16b).

(b)   Now, if only 32 equally spaced DTFT samples are used (Figure 7.16a), then the result is just the constant (unity) sequence shown in Figure 7.16b. It so happens in this case that only the $k = 0$ frequency sample $X_{32,a}(0)$ is nonzero and equal to $P/2 = N = 32$; the other samples $\omega_k = 2\pi k/N$ are all zeros of $\sin(P\omega/4)$, a factor in the numerator of $X(e^{j\omega})$. Consequently, $x_{32,a}(n)$ is the inverse DFT$_N$ of $N \cdot \delta(k)$, which is unity for all $n \in [0, N-1]$.

(c)   Finally, in Figure 7.17a we choose the intermediate value $N = 50$, to represent the case of moderate time-aliasing. The aliasing appears for small n, as indicated in the plot of $x_{50,a}(n)$ in Figure 7.17b. To help explain this, in Figure 7.17c are shown the replications, in reference to (7.16b), $m = 0$ ("O" with stems), $m = -1$ ("*") and $m = 1$ ("+"). The sum of the replications shown in Figure 7.17c is equal to $x_{50,a}(n)$ in Figure 7.17b. In this case of moderate aliasing, we see that on the range $[0, N-1]$ that the DFT$_N^{-1}$ yields, there is no overlap on the high end because the nearest replication ($m = -1$) begins at $n = N$, which is outside $[0, N-1]$. The $m = 1$ replication, however, begins at $-N$ and ends at $-N + P - 1 = -50 + 63 = 13$. Thus, for $0 \le n \le 13$, the $m = 1$ produces aliasing distortion. Notice in Figure 7.17c that the only two terms contributing to $x_{50,a}(n)$, namely $m = 0$ and $m = 1$, add to a constant on $[0, 13]$ [as in part (b)]. Therefore $x_{50,a}(n)$ is constant on that range,

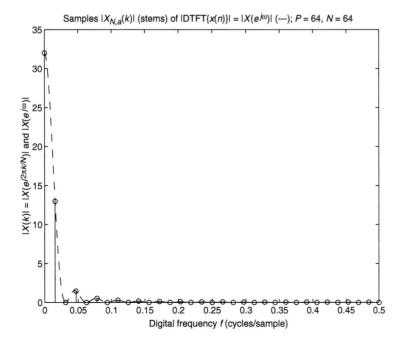

Samples $|X_{N,a}(k)|$ (stems) of $|\text{DTFT}\{x(n)\}| = |X(e^{j\omega})|$ (---); $P = 64$, $N = 64$

Figure 7.15a

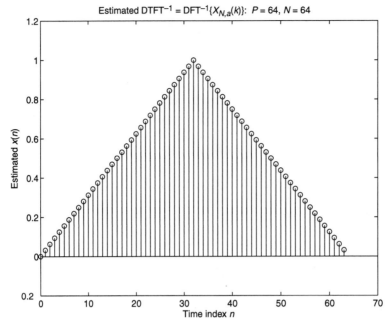

Figure 7.15b

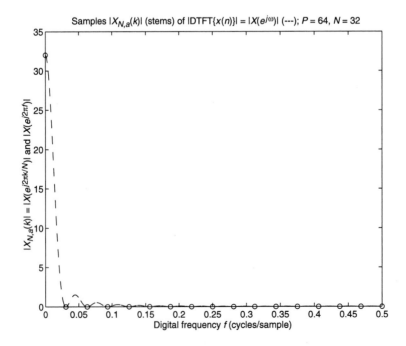

Figure 7.16a

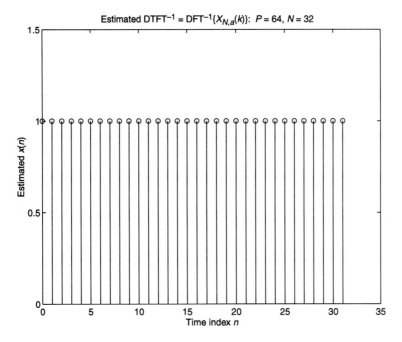

Figure 7.16*b*

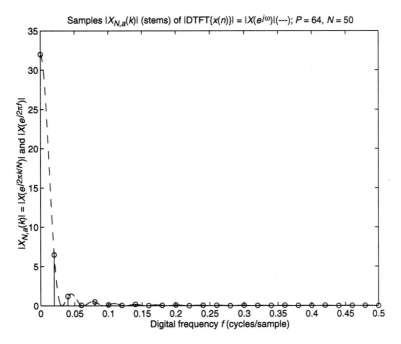

Figure 7.17*a*

as seen in Figure 7.17*b*. Similar plots could be used to explain the appearance of Figure 7.16*b*; contrast this explanation with the one offered in part (b), namely $\mathrm{DFT}_{32}^{-1}\{32 \cdot \delta(k)\} = 1$.

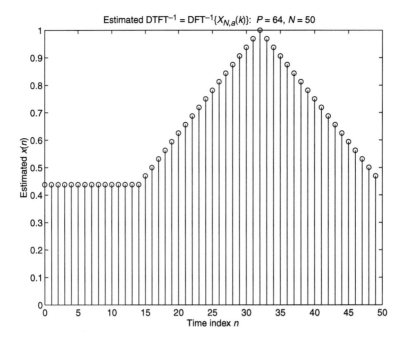

Figure 7.17*b*

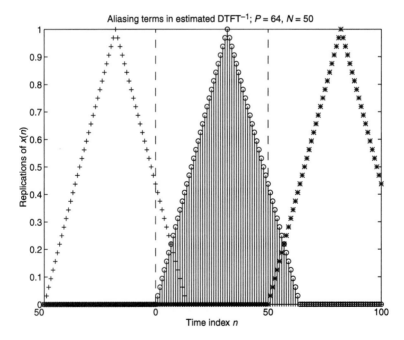

Figure 7.17*c*

EXAMPLE 7.6

As a dramatic example of the same concept, consider the Dow Jones Industrial Average. Investigate time-aliasing by insufficient spectral sampling.

Solution

A plot of the closing value every day from 1920 to the middle of 1947 (8192 days) is shown in Figure 7.18a. Note the Roaring Twenties incline, the crash in 1929 and continual decline until well into 1932, and finally the beginnings of the World War II/post-World War II boom. Each point in the plot actually corresponds to about 1.23 days, when one accounts for all the holidays and weekends the stock market is closed. Because 8192 stems are impractical to depict, the sequence points are connected with lines.

Figure 7.18b shows the DFT_{8192} of the Dow Jones averages. This plot is $20 \cdot \log_{10}\{|X(k)|/|X(0)|\}$. The log scale is used because all values but the lowest frequencies are very small. The normalization to $|X(0)|$ is used to facilitate judging how far down the high-frequency components are relative to "dc": for the most part, 60–100 dB down. We might take some comfort in the large proportion of energy in low frequencies—that would seem to reflect stability in the market—until we recall that this segment includes the 1929 crash! That the energy at higher frequencies is so uniformly distributed (no outstanding peaks) is indicative of the difficulty of predicting the stock market. Regular patterns such as marked periodicities are not evident over even as long a time as 27 years.

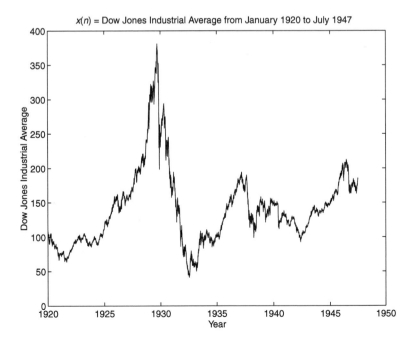

Figure 7.18a

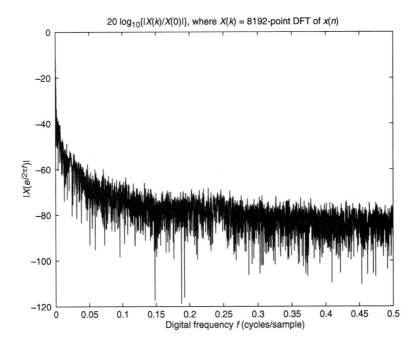

Figure 7.18*b*

The *k*th "bin" in Figure 7.18*b* corresponds to a sinusoidal component having period $(P\Delta t)/k = 27.5$ years/k. Thus, the lowest nonzero frequency ($k = 1$) is one having period equal to the entire length of the sequence. This is true in general for the DFT of any sequence; remember, the DFT is a discrete Fourier series. The longest period being $P\Delta t$ is the equivalent of the period of the "fundamental frequency" of the continuous-time Fourier series, which is equal to the entire duration T of the period of the (periodic) function whose Fourier series is being computed. Also, the highest frequency represented is $(P/2) \cdot (1/[P\Delta t]) = 1/(2\,\Delta t)$, having period $2\,\Delta t \approx 2.46$ days (the calendar time of two market days).

Let us consider this DFT_{8192} to be the DTFT of the "entire" sequence $x(n)$ shown in Figure 7.18*a*. Of course, it is not entire, for it does not extend back to 1885 when the Dow Jones averages began nor does it extend out to $t = \infty$. Nevertheless, 27 years is an interval sufficiently large that this sequence has general characteristics typical of the entire Dow Jones sequence, including the crash itself.

Now suppose that we blindly apply a 1024-point DFT^{-1} to 1024 equally spaced samples of this DTFT as shown in Figure 7.19*a*. Specifically, we take one sample from every eight of the original DFT_{8192}, $X(k)$. It certainly seems a reasonable thing to do if we had only a 1024-point DFT available, given the huge 8192 sequence. The resulting sequence $x_{1024,a}(n)$ is shown in Figure 7.19*b*. It is garbage! The Roaring Twenties boom has turned into a steep decline; the crash of 1929, a 90% drop, is replaced by a period of "inactivity"; and the "Dow" values (in the 1200 range) are more like those of the early 1980s than the 1930s (which were in the 100 range). Thus, the error incurred is not just a matter of decreased resolution (the spacing between time samples is now 9.8 days), but is severe time-aliasing.

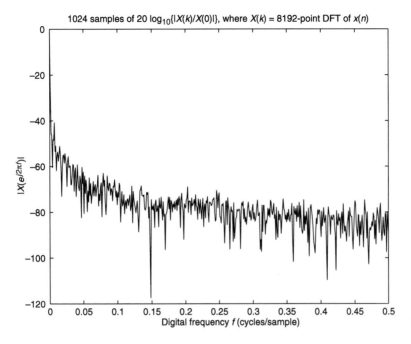

Figure 7.19*a*

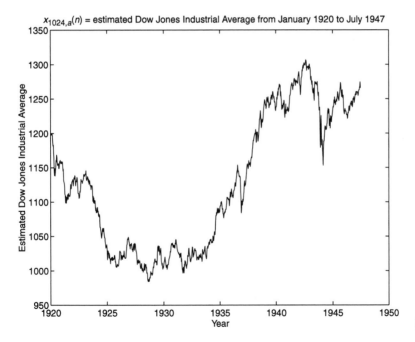

Figure 7.19*b*

To clarify this aliasing, Figure 7.19*c* shows the contributing aliasing terms in (7.16b) for the Dow sequence, just as displayed for a different sequence (the triangle) in Figure 7.17*c* of Example 7.5. As in that case, these aliasing terms add up to $x_{1024,a}(n)$ in Figure 7.19*b*, but in this case there are many more (8) contributing

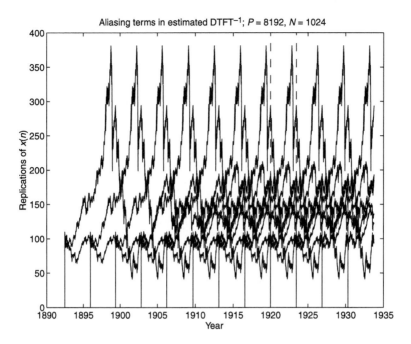

Figure 7.19c

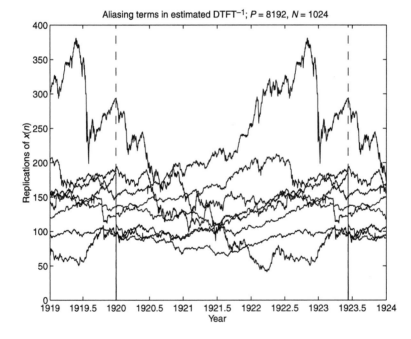

Figure 7.19d

terms because $P/N = 8$. Again, the range "$[0, N - 1]$" is demarcated by vertical dashed lines at its boundaries. If we focus in on that range, we obtain Figure 7.19d, which clearly shows the eight contributing terms. Seeing these, which at each time must be added up to produce Figure 7.19b, explains why Figure 7.19b is physically meaningless.

7.4 DFT TIME-DOMAIN INTERPOLATION[4]

We have just seen that to exactly recover $x(n)$ from N samples of its DTFT, the sequence $x(n)$ must have length $P \le N$. If $x(n)$ is longer $(P > N)$, then some or all of the time- $(n$-)domain DFT_N^{-1} values obtained will be aliased. In this section, we consider the situation where $N = P$, and we modify $X_N(k)$ to look between the original samples $x(n)$. This is the time-domain interpolation problem.

For example, suppose that we are given only a set of time samples, and the continuous-time $x_c(t)$ from which they were taken is no longer available. Can we use the DFT to smooth out the coarse data and have the smooth result be an accurate approximation of samples of the original $x_c(t)$ in between (as well as on) the original Δt samples? Thus, for every one sample, we hope to obtain L samples in the time domain. That is, assuming that the given $x_N(n)$ were obtained as $x_N(n) = x_c(n\Delta t)$, $n \in [0, N-1]$, we wish to estimate $x_c(n\Delta t/L)$.

Exactly the same idea can be used for interpolation of an N-length discrete-time sequence $x(n)$ as was used for interpolation of its DTFT using the DFT. Instead of zero-padding (real-valued) $x_N(n)$, we zero-pad the given N frequency-samples $X_N(k)$ to make an array $X_{ZP;M}(k)$ of length $M = LN$. (For this method, the now-capitalized ZP subscript refers to zero-padding in the frequency domain, not in the time domain.)

Before proceeding, however, let us review a few relevant ideas concerning Fourier series (FS). A continuous-time FS may be formed for a T-periodic $\tilde{x}_c(t)$ $[\tilde{x}_c(t + \ell T) = \tilde{x}_c(t)]$. A Fourier series may also be defined for an $x_c(t)$ that is nonzero only for $t \in [0, T]$ by viewing $x_c(t)$ as one period of its T-periodic extension $\tilde{x}_c(t) = x_c(t_{\text{modulo } T})$. Similarly, we may define a discrete Fourier series (DFS) for an N-length sequence $x(n)$ by viewing $x(n)$ as one period of its N-periodic extension $\tilde{x}_N(n) = x(n_{\text{modulo } N})$, for which $\tilde{x}_N(n + \ell N) = \tilde{x}_N(n)$. As noted previously, this DFS is the DFT_N.

Further recall that a function periodic in one domain is discrete in its transformed domain, and so if the time function is both periodic and discrete, so will be its transform (the DFT): $X(k) = X(k + N\ell)$. Thus, both $x(n)$ and $X(k)$ may be viewed as discrete, N-periodic functions.

Notice that the formula $DFT_N\{x(n)\} = X(k) = X(e^{j2\pi k/N}) = DTFT\{x(n)\}\big|_{\omega=2\pi k/N}$ holds only for $x_N(n)$ an N-length sequence, not for its N-periodic extension $\tilde{x}_N(n)$. Specifically, the DTFT of $\tilde{x}_N(n)$ is a series of weighted and shifted Dirac delta functions [proved in the problems; see also Appendix 4A, formula (6)]:

$$DTFT\{\tilde{x}_N(n)\} = (2\pi/N) \cdot \sum_{k=0}^{N-1} X(k) \sum_{\ell=-\infty}^{\infty} \delta_c(\omega - 2\pi k/N - 2\pi\ell), \text{ where } X(k) \text{ is the}$$

DFT_N of $x_N(n)$.

For simplicity, let us assume for now that N is odd. Otherwise, construction of the correct real-valued time function from use of all the original N samples in the frequency domain requires a special $\frac{1}{2}$ coefficient for the term $k = N/2$ and its conjugate pair.[5] The interpolation formula for N even will also be given, in Example 7.7.

It is next helpful to recall from Section 5.9 that if $x^*(n) = x(n)$ [i.e., $x(n)$ is real-valued], then $X^*(-k) = X(k)$, or, noting that $X(k)$ is N-periodic,

$$X^*(N - k) = X(k). \tag{7.22}$$

---

[4] This section in combination with WP 7.1 and WP 7.2, is based on and is an expansion of T. J. Cavicchi, "DFT Time-Domain Interpolation," *IEE Proceedings*, Part F, 139, 3 (June 1992): 207–211. This paper is referred to as [1] in the text.
[5] D. Fraser, "Interpolation by the FFT Revisited—An Experimental Investigation," *IEEE Trans. Acoust., Speech, and Signal Processing*, ASSP–37 (May 1989): 665–675.

Now construct the sequence

> **DFT temporal interpolation, N odd.**
>
> $$X_{ZP;M}(k) = \begin{cases} L \cdot X_N(k), & k \in [0, (N-1)/2] & (7.23a) \\ 0, & k \in [(N-1)/2 + 1, M - (N-1)/2 - 1] & (7.23b) \\ L \cdot X_N(k - M + N), & k \in [M - (N-1)/2, M-1] & (7.23c) \end{cases}$$

and take the DFT_M^{-1}, which we denote as $x_{ZP;M}(k)$ (again, note that ZP is capitalized for zero-padding in the frequency domain), where again $M = NL$. No conjugation sign belongs in (7.23c) because the latter, already conjugate-symmetric half of the Hermitian (conjugate-even) $X_N(k)$ is used.

Note in (7.23b) that we zero-pad the *middle* of $X_N(k)$ rather than at the end. By doing so we preserve the conjugate symmetry in $X_N(k)$ which applies for the usual case in which $x_N(n)$ is real-valued.

Let us put in perspective this zero-padding of $X_N(k)$ at its center from our knowledge of zero-padding an N-length discrete-time sequence $x_N(n)$. For the case of zero-padding a time sequence $x_N(n)$ there is usually no constraint or desire that the DFT sequence be real-valued, so we usually zero-pad $x_N(n)$ at the end, not the center. Instead, recalling that the second half of $x_N(n)$ is equivalent to negative time in the periodic extension that the DFT assumes, zero-padding the end of $x_N(n)$ out to $M - 1 > N - 1$ is equivalent to merely requiring that $\tilde{x}(n)$ be zero for the smallest (closest-to-zero) $M - N$ negative values of n (probably true already). This end-padding would be inappropriate for constructing $X_N(k)$, which, rather than being real or "causal," must be conjugate-even for real $x_N(n)$.

We show step by step on WP 7.1 that by evaluating geometric sums, our DFT_M^{-1} estimate $x_{ZP;M}(n) = \hat{x}_c(n \Delta t / L)$ of $x_c(n \Delta t / L)$ is

<table>
<tr><td>Complete
derivation
details on
WP 7.1</td><td>

$$\hat{x}_c\left(\frac{n\Delta t}{L}\right) = \text{DFT}_M^{-1}\{X_{ZP;M}(k)\} = x_{ZP;M}(n) \qquad (7.24)$$

$$= \sum_{m=0}^{N-1} x_N(m) \cdot \phi_N\left(\pi \cdot \left[\frac{n}{M} - \frac{m}{N}\right]\right),$$

where

$$\phi_N(\theta) = \frac{\sin(N\theta)}{N \cdot \sin(\theta)}. \qquad (7.25)$$

</td></tr>
</table>

Note that (7.24) does *not* have a convolutional form. For N even, slightly more algebra than used above shows that $\phi_N(\theta)$ is instead $\phi_N(\theta) = \cot(\theta)\sin(N\theta)/N$ (see the problems).

How good is the interpolation defined in (7.24)? In preparation for explicitly answering this question, we need to review the correct interpretation of DFT results in general. Because the DFT values in both the time and frequency domains are N-periodic and discrete, each may individually be considered a sequence of samples of a periodic function of a continuous variable.

In particular, the periodic extension of $x_N(n), \tilde{x}_N(n)$, may be considered to be samples of a 2π-periodic function of continuous time, $\tilde{x}_{N,c}(t)$, evaluated at $t = n \Delta t$ where $\Delta t = 2\pi/N$. [The subscript N signifies that there are exactly N samples spaced every Δt on one period of $\tilde{x}_{N,c}(t)$, and $x_N(n)$ is obtained from it.] With $\Delta t = 2\pi/N$, we see that equivalently, $\tilde{x}_{N,c}(t)$ is $N\Delta t$-periodic. [More generally, $x_N(n)$ could be samples of a T-periodic $\tilde{x}_c(t)$ with $\Delta t = T/N$, where T is determined by the time calibration of the application. It still, however, is $N\Delta t$-periodic. We assume $T = 2\pi$ here to help see an analogy with the DTFT pointed out below.] So, assuming that $\tilde{x}_{N,c}(t)$ is 2π-periodic in t, it can be exactly represented by a Fourier series with fundamental frequency $2\pi/T = 2\pi/2\pi = 1$ rad/s. Therefore, the spacings $\Delta\Omega$ between frequencies of nonzero energy of its continuous-time Fourier transform are unity: $\Omega_k = k$ rad/s. Thus whenever $N\Delta t = T$, the frequencies of possibly nonzero energy of the DTFT of $\tilde{x}_N(n)$, with spacing $\Delta\omega = \Delta\Omega \cdot \Delta t = 1 \cdot 2\pi/N$, are the ω_k of the DFT. A periodic time function has a line spectrum, and if $N\Delta t = T$, the lines fall on the DFT frequencies (whether or not $T = 2\pi$).

Equivalently, we know that for $\Delta t = T/N = 2\pi/N$, the DFT frequencies $\omega_k = 2\pi k/N$ correspond to analog frequencies $\Omega_k = \omega_k/\Delta t = (2\pi k/N)/(2\pi/N) = k$ rad/s, which precisely agrees with the continuous-time FS and FT frequencies of nonzero energy. Furthermore, because the expansion functions of the FS for $t = n\Delta t$ and the DFT^{-1} both are the same functions of n, namely $e^{j2\pi kn/N}$, and both produce the same values $\tilde{x}_{N,c}(n\Delta t)$, the uniqueness of Fourier expansions allows us to conclude that $X(k) = X_c(k)$. That is, the DFT coefficients are equal to the continuous-time FS coefficients. Once it is established that $X(k) = X_c(k)$, it likewise follows that the FS and DFT^{-1} reconstructions are equal not only at $t = n\Delta t$, but also at $t = n\Delta t/L$ when (7.24) is used for the DFT^{-1} reconstruction.

[All the above discussion is completely analogous to recognition that the DFT samples $X_N(k)$ produced by the forward DFT_N are samples of the 2π-periodic function of ω, the $\text{DTFT}\{x_N(n)\} = X_N(e^{j\omega})$, evaluated at $k\Delta\omega$, where $\Delta\omega = 2\pi/N$.]

Assuming that $\tilde{x}_{N,c}(t)$ is bandlimited, its maximum frequency Ω_{max} corresponds to k_{max} in its Fourier series expansion. It is assumed in this entire discussion that there is no aliasing, so the maximum frequency Ω_{max} in $\tilde{x}_{N,c}(t)$ satisfies $\Omega_{max} = k_{max} \le [\Omega_s/2] = [\pi/\Delta t] = [N/2] = (N-1)/2$ for N odd, where $[\cdot]$ designates the greatest integer. Thus, with $\Omega_{max} < [\pi/\Delta t]$, we are guaranteed that the maximum frequency and all of the finite number of frequencies of nonzero energy are covered by the DFT_N. (More generally, $\Omega_{max} < [\pi N/T]$.) Alternatively, for *any* value of N satisfying this requirement, there will be no aliasing and reconstruction will be exact. Thus, (7.23) to (7.25) have the potential for exact reconstruction of $\tilde{x}_{N,c}(t)$ on a fine grid. Below, we more explicitly prove that in (7.24), $\hat{x}_c(n\Delta t/L) = \tilde{x}_{N,c}(n\Delta t/L)$.

To use the continuous-time Fourier series in either "direction" (i.e., Fourier series of a periodic function of t or Fourier series of a periodic function of ω), we assume that the samples in the discrete domain are actually samples of an *aperiodic* discrete-variable function (in this case, of length N). Otherwise the other domain, which is continuous-variable, would have to be discrete-variable. This interpretation opposes the periodic-extension point of view of the DFT.

Thus, for interpolation in the frequency domain [i.e., of $X_N(e^{j\omega})$ in Section 7.2], $x_N(n)$ is considered to be an aperiodic finite-length (N long) Fourier series sequence, whereas for interpolation in the time domain [i.e., of $\tilde{x}_{N,c}(t)$ of this section], $X_N(k)$ is considered to be an aperiodic finite-length (N long) Fourier series sequence. As noted in Section 4.4, any aperiodic discrete-variable function is just a Fourier series representation of a related periodic continuous-variable function. We shall see that the interpretation of periodic against aperiodic is crucial in assessing the quality of reconstruction in (7.24).

It was shown in Section 7.2 that if $x(n) = x_N(n)$ [that is, $x(n)$ is in fact only length N], then the interpolation of $X_N(e^{j\omega})$ formed by zero-padding of $x_N(n)$ yields exact values of $X(e^{j\omega})$. Similarly, the interpolation of $\tilde{x}_{N,c}(t)$ formed by zero-padding of $X_N(k)$ is exact if $X(k) = X_N(k)$—that is, if we have included in $X_N(k)$ all the nonzero $X(k)$. Recall from Section 7.2 that both $X(e^{j\omega})$ and $X_N(e^{j\omega})$ are 2π-periodic functions of ω. Similarly, for the interpolation to be exact we must take $\tilde{x}_{N,c}(t)$, and $x_c(t)$ to be

2π-periodic (i.e., $N\Delta t$-periodic or, more generally, T-periodic but with $N\Delta t = T$) functions of t, just as $X(e^{j\omega})$ is 2π-periodic in ω. All these statements reduce to the following: A periodic function has a line spectrum, and if those lines are all within a bandlimit and we include them all in the DFT, then the DFT^{-1} reconstruction must be exact because the entire signal has been represented in $X(k)$.

With $x_c(t)$ assumed to be $N\Delta t$-periodic in t, we write $x_c(t)$ in the following as $\tilde{x}_c(t)$. Assuming that no aliasing has occurred, we know from (6.26) that the periodic function $\tilde{x}_c(t)$ may be exactly reconstructed from its samples as

$$\tilde{x}_c(t) = \sum_{m=-\infty}^{\infty} \tilde{x}_c(m\Delta t) \cdot \text{sinc}\left\{\pi\left(\frac{t}{\Delta t} - m\right)\right\}, \tag{7.26}$$

which when evaluated at $t = n\Delta t$ gives

$$x(n) = \tilde{x}_c(n\Delta t) = \sum_{m=-\infty}^{\infty} \tilde{x}_c(m\Delta t) \cdot \text{sinc}\{\pi(n-m)\}. \tag{7.27a}$$

It is here our goal to evaluate $\tilde{x}_c(t)$ at $t = n\Delta t/L$, which under exact interpolation is

$$\tilde{x}_c\left(\frac{n\Delta t}{L}\right) = \sum_{m=-\infty}^{\infty} \tilde{x}_c(m\Delta t) \cdot \text{sinc}\left\{\frac{\pi}{L}(n-mL)\right\}. \tag{7.27b}$$

This is the equation computationally equivalent to both the Shannon sampling theorem and the zero-interlace method, the latter of which we discuss in Section 7.5.3. How does (7.27b) compare with (7.24), which takes a different point of view? Consider the relation [(6.17), reproduced here as (7.28)] between the DTFT of $x(n)$ and $X_c(j\Omega)$, the FT of $x_c(t)$, where we here assume that $x_c(t) = \tilde{x}_c(t)$:

$$X(e^{j\omega}) = \frac{1}{\Delta t} \sum_{k=-\infty}^{\infty} X_c\left(j\left[\frac{\omega + 2\pi k}{\Delta t}\right]\right). \tag{7.28}$$

Recall that (7.26) resulted by noting that in (7.28) (a) the magnitude of $X_c(j\Omega)$ is scaled by $1/\Delta t$, (b) $X_c(j\Omega)$ is replicated every $\Omega_s = 2\pi/\Delta t$, and (c) the argument Ω is scaled by $1/\Delta t$. So, to exactly recover $x_c(t)$ from $X(e^{j\omega})$ in (7.26) under the condition of no aliasing ($\Omega_{max} < \pi/\Delta t$), the magnitude of $X(e^{j\omega})$ is scaled by Δt, and $x(n)$ is lowpass filtered by a continuous-time filter with cutoff at $\pi/\Delta t$ (to eliminate all replication terms in (7.28) except the one centered on $\Omega = 0$). If Δt is scaled by $1/L$, then from (7.28), the replications of $X_c(j\Omega)$ in $X(e^{j\omega})$ will be compressed by the factor L, with intervals of zero value from the cutoff [e.g., at $\omega = \Omega_{max} \cdot (\Delta t/L)$] out to the next replication [e.g., at $\omega = 2\pi - \Omega_{max} \cdot (\Delta t/L)$]. They will also be scaled larger by the factor L due to the larger factor $1/\{\Delta t/L\}$ out front in (7.28). Such is the appearance of $X(e^{j\omega})$ when the sampling rate is increased by the factor L, regardless of whether the time-domain sequence is periodic or finite-length.

If we had an $X(e^{j\omega})$ with this appearance, the $M = NL$ samples of it would be the DFT_M of $x_c(n\Delta t/L)$. But the above is an exact description of the zero-padding construction in (7.23)! Therefore, the interpolation given by (7.24) is exact *provided* $x_c(t)$ is $N\Delta t$-periodic—that is, if $x_c(t) = \tilde{x}_c(t)$. Again, the $N\Delta t$-periodicity is required so that all the signal energy resides in the N values of the DFT. We shall further argue below that the bandlimited periodic function passing through the given N time samples is unique, so that $\tilde{x}_c(t) = \tilde{x}_{N,c}(t)$.

It is very interesting to note that the interpolation in (7.24) is a finite sum, whereas that in (7.27b) involves an infinite number of terms, yet the two sums have the same value. Computationally, this is an advantage of (7.24) over the zero-interlace or sinc basis interpolation methods. The advantage stems from tacit use of an identity mentioned in Section 6.7 [(6.107)], which we will now derive for this application. Noting that $\tilde{x}_c(m\Delta t) = \tilde{x}_N(m)$, rewrite (7.27b) as

$$\tilde{x}_c\left(\frac{n\,\Delta t}{L}\right) = \sum_{m'=-\infty}^{\infty} \tilde{x}_N(m') \cdot \text{sinc}\left\{\frac{\pi}{L}(n - m'L)\right\}. \tag{7.29a}$$

Next, break up the m' sum into chunks N long: $m' = mN + i$, where i ranges over $[0, N-1]$ and m ranges from $-\infty$ to ∞. Also, write the explicit expression for sinc:

$$\tilde{x}_c\left(\frac{n\,\Delta t}{L}\right) = \sum_{m=-\infty}^{\infty} \sum_{i=0}^{N-1} \tilde{x}_N(mN + i) \cdot \frac{\sin\{(\pi/L) \cdot [n - (mN + i)L]\}}{(\pi/L) \cdot [n - (mN + i)L]}. \tag{7.29b}$$

Now reverse the order of summation, and use the periodicity of $\tilde{x}_N(n)$ [i.e., $\tilde{x}_N(n + mN) = \tilde{x}_N(n)$]:

$$\tilde{x}_c\left(\frac{n\,\Delta t}{L}\right) = \sum_{i=0}^{N-1} x_N(i) \cdot \sum_{m=-\infty}^{\infty} \frac{\sin\{(\pi/L) \cdot [n - (mN + i)L]\}}{(\pi/L) \cdot [n - (mN + i)L]}. \tag{7.29c}$$

Finally, define $j = n - iL$:

$$\tilde{x}_c\left(\frac{n\,\Delta t}{L}\right) = \sum_{i=0}^{N-1} x_N(i) \cdot \sum_{m=-\infty}^{\infty} \frac{\sin\{(\pi/L) \cdot (j - mNL)\}}{(\pi/L) \cdot (j - mNL)}. \tag{7.29d}$$

Now rewrite (7.24) as

$$\tilde{x}_c\left(\frac{n\,\Delta t}{L}\right) = \sum_{i=0}^{N-1} x_N(i) \cdot \frac{\sin\{(\pi/L) \cdot (n - iL)\}}{N \sin\{(\pi/[NL]) \cdot (n - iL)\}} \tag{7.30a}$$

$$= \sum_{i=0}^{N-1} x_N(i) \cdot \frac{\sin\{\pi j/L\}}{N \sin\{\pi j/(NL)\}}, \tag{7.30b}$$

where again $j = n - iL$. Evidently, if (7.29d) and (7.30b) are to be equal for all $\tilde{x}_c(t)$ satisfying the Nyquist sampling criterion for Δt and the $N\,\Delta t$-periodicity requirement, then the coefficients of $x_N(i)$ must be the same in those two equations:

$$\frac{\sin\{\pi j/L\}}{N \sin\{\pi j/(NL)\}} = \sum_{m=-\infty}^{\infty} \frac{\sin\{(\pi/L) \cdot (j - mNL)\}}{(\pi/L) \cdot (j - mNL)}. \tag{7.31a}$$

Now, in the numerator of the right side of (7.31a), use the identity for the sine of a sum [and then $\sin(\pi M) = 0$]. Also factor out πN in the denominator and multiply through by the $1/L$ to obtain

$$\frac{\sin\{\pi j/L\}}{N \sin\{\pi j/(NL)\}} = \sin(\pi j/L) \cdot \sum_{m=-\infty}^{\infty} \frac{\cos(\pi m N)}{\pi N \cdot (j/[NL] - m)}. \tag{7.31b}$$

For N odd, (7.31b) is equivalent to the following identity:

$$M \sin\left\{\frac{\pi j}{M}\right\} \sum_{m=-\infty}^{\infty} \frac{(-1)^m}{j - mM} = \pi, \tag{7.32}$$

where for genericness NL was replaced by $M = NL$. Equation 7.32 holds for any integers j and M for which $j \neq nM$. Indeed, (7.32) is really just the same as (6.107) which we required for sampling and truncation to be reversible (Section 6.7). We obtain (7.32) from (6.107) by replacing m in (6.107) by $-m$ and θ in (6.107) by $\pi j/M$.

In conclusion, although both the interpolation method of (7.24) and methods equivalent to (7.27b) such as the zero-interlace method presented in Section 7.5.3 produce the same results, the FFT method accomplishes them with zero error yet by way of a finite (length N) sum, because of (7.32). Note finally that if $x_c(t)$ were not periodic as required but nonzero only on $[0, N-1]$,[6] the interpolation would no

---

[6] This assumption is made in K.P. Prasad and P. Satyanarayana, "Fast interpolation algorithm using FFT," *Electronics Letters*, 22, 4 (1986), 185–187.

longer be exact (see below), and (7.27b) would not be equivalent to (7.24). The interpolation function for (7.26) could now only be written as [see (7.29a)]

$$\frac{\sin\{(\pi/L) \cdot (n - iL)\}}{(\pi/L) \cdot (n - iL)}, \tag{7.33}$$

whereas that in (7.30a) is

$$\frac{\sin\{(\pi/L) \cdot (n - iL)\}}{N \sin\{(\pi/[NL]) \cdot (n - iL)\}}, \tag{7.34}$$

where (7.33) has no N and (7.34) has a trigonometric function in the denominator yet in both cases i runs only from 0 to $N - 1$.

We know from Section 4.4 that if $x_c(t)$ and therefore $x(n)$ are aperiodic, their spectra are continuous, not discrete. Therefore, the discrete spectrum formed in (7.23) is only an approximation of the true spectrum. That is, the DFT is a Fourier series and can represent exactly only periodic functions of n. In any case, if $x_c(t)$ [and thus $x(n)$ if, as defined above, $x(n) = x_c(n\Delta t)$ for all $\infty < n < \infty$] is of finite duration, it is impossible to satisfy the Nyquist sampling criterion. We will now examine the reason for this.

As a preliminary, we need to recall a few results from the theory of functions of a complex variable. Consider a function of a complex variable that is analytic in some region within the complex plane. If that function is zero everywhere in some neighborhood or along some contour within the region of analyticity, then it is zero everywhere within the whole region of analyticity. This is because knowledge of the function on the curve or neighborhood is sufficient to evaluate the function anywhere else in the region of analyticity in terms of those known values alone [see the Cauchy integral formula (3.13) and the discussion in Section 4.2].

If $x_c(t)$ is time-limited, then its Laplace transform must exist and be analytic for all finite s, for $X_c(s)$ is just the result of a finite integration.[7] Thus, if such an analytic function is precisely zero on any continuous interval, no matter how small, it must be zero over the entire complex plane; that is, $X_c(s) = 0$ for all s and thus $x_c(t) = 0$ for all t. A special case of this fact is that $X_c(j\Omega)$ cannot be precisely zero over any continuous interval of Ω if $x_c(t)$ is nonzero.

In particular, that interval cannot be $|\Omega| > \Omega_0$, where Ω_0 is the maximum frequency for which $|X_c(j\Omega)| \neq 0$. That is, if $x_c(t)$ is time-limited, then it cannot be bandlimited—a fact argued from a different perspective in Sections 6.5.2 and 6.6. Note the interesting corollary: If $x_c(t) \neq 0$ is bandlimited [$X_c(j\Omega) = 0$ for $|\Omega| > \Omega_0$], then $X_c(s)$ cannot be an analytic function! Equivalently, no analytic signal transforms $X_c(s)$ are perfectly bandlimited. In fact, the Laplace transforms of perfectly bandlimited signals exist only on the $j\Omega$-axis because their perfectly sharp cutoffs require either (a) periodic time functions (which are discrete sums of bicausal sinusoids, each of which whose LT exists only on the $j\Omega$ axis) or (b) a product or convolution involving the frequency-rectangular window, which itself exists only on the $j\Omega$ axis.[8]

Because both the DFT method (7.23) to (7.25) and the zero-interlace method (7.27b) assume bandlimitedness, both may result in errors in the reconstruction of

---

[7] J. Arsac, *Fourier Transforms and the Theory of Distributions* (Englewood Cliffs, NJ: Prentice Hall, 1966), p. 227.

[8] That is, LT{sinc($\Omega_1 t$)} exists only on the $j\Omega$ axis; see Balthasar Van der Pol and H. Breener, *Operational Calculus Based on the Two-Sided Laplace Integral* (Cambridge, U.K.: Cambridge University Press, 1959), pp. 113–114.

time-limited functions (as opposed to 2π- or T-periodic functions). To be sure that we are satisfying the Nyquist criterion within desired precision, we may initially greatly oversample a representative $x_c(t)$, determine its time-bandwidth product, and choose N and Δt in the manner discussed in Section 6.5.3. Another approach is to first greatly oversample and then increase Δt until noticeable interpolation degradation appears.

Another relevant point is closely related to the analyticity argument just made: There is only one bandlimited function that is determined at every point on a finite interval. That is, if $f_c(t)$ is given on $t \in [0, T]$, there is only one extension of $f_c(t)$ outside $[0, T]$ that is bandlimited. This is because, except for that one function, there is at least one discontinuity in at least one order derivative at the boundary $t = 0$ and/or $t = T$. From Arsac (p. 214), we know that every bandlimited function must be infinitely differentiable in the time domain. The uniqueness of $f_c(t)$ allows us to conclude that $\tilde{x}_{N,c}(t) = \tilde{x}_c(t)$.

If $x_c(t)$ is aperiodic and infinite-length, it is possible that $x_c(t)$ is bandlimited (one example being the sinc function). It will not, however, then match the periodic function that (7.24) yields for $0 \le t \le N\Delta t$. In all cases, (7.24) produces equispaced samples of one period of the unique, $N\Delta t$-periodic, Nyquist-bandlimited (i.e., highest frequency $\le \pi/\Delta t$) function $\tilde{x}_{N,c}(t)$ and thus the unique unaliased periodic function that passes through the original samples. Inherent in this statement is the unique identification of the "original samples" sequence with the bandlimited function via Shannon's expansion.

Therefore, any discussion of error should actually refer to the difference between $\tilde{x}_{N,c}(t)$ and $x_c(t)$. For instance, in evaluating the accuracy of interpolation formulas using test cases, it is advisable to specify an example periodic function known to be bandlimited so that the "error" is defined in reference to the exact solution rather than to the value of a truncated sum, as some researchers have done.

One may argue that whether the actual $x_c(t)$ is periodic or is finite-length is irrelevant to (7.24), which refers in either case only to the time samples on $[0, N-1]$. Consequently, interpolations made within the interval $[0, N-1]$, the only ones made in (7.24), are the same in either case. However, it will reproduce correct interpolated values of the desired continuous-time function $x_c(t)$ on the range $t \in [0, N\Delta t]$ only if on $t \in [0, N\Delta t]$ $x_c(t)$ is equal to one period of the Nyquist-bandlimited function $\tilde{x}_{N,c}(t)$ that (7.24) interpolates.

Note that both (7.33) and (7.34) are largest for the interval for which n is near iL. In that region, the argument of the sine in the denominator of (7.34) is small. Because for small θ, $\sin(\theta) \approx \theta$, (7.33) and (7.34) are thus close where the maximum contributions to the two interpolation sums occur. Only in this sense are (7.24) and (7.27b) approximately equivalent. In particular, for the special cases $n = mL$, all terms except $m' = m$ in (7.29a) and $i = m$ in (7.30a) are zero, and the nonzero coefficient is unity. Thus, the original samples are reproduced exactly in the interpolated sequence. For L small, however, there may be no small arguments for values of n other than mL. Nevertheless, for $x_c(t)$ Nyquist-bandlimited and $N\Delta t$-periodic in t, the interpolation given by (7.24) is exact and is a practical way of computing the infinite sum in (7.29a).

EXAMPLE 7.7

Reconsider the same section of $M = 128$ samples of speech used in Example 6.2. It is shown again in Figure 7.20a and denoted $x(n)$, along with its DFT$_{128}$ in Figure 7.20b.[9] Suppose we subsample this sequence: One sample is taken from every two,

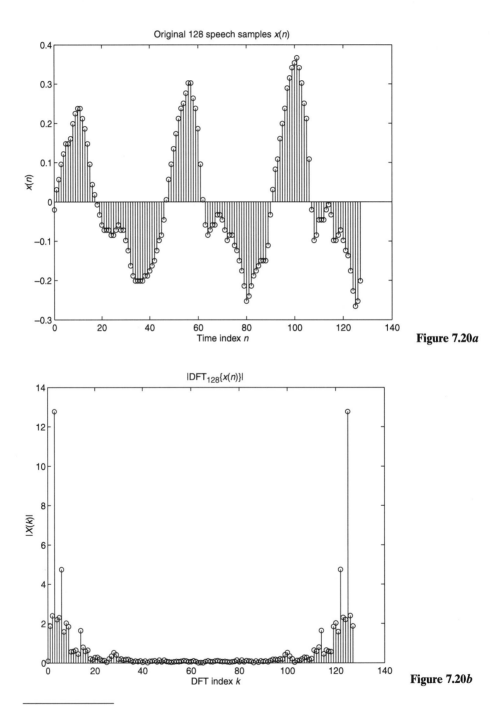

Figure 7.20a

Figure 7.20b

[9]Because in this example we are especially interested in the formation of DFT sequences, the horizontal axis of all DFT plots will be all k, not just the first half or nonnegative values of ω or f. Also, we choose a small number of points, 128, to facilitate graphical assessment of interpolation quality.

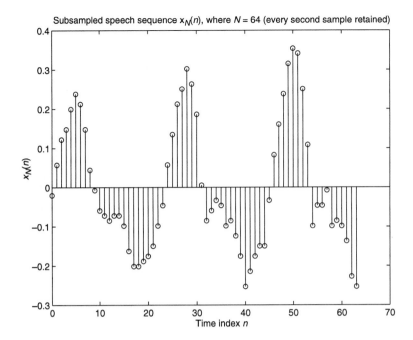

Figure 7.21*a*

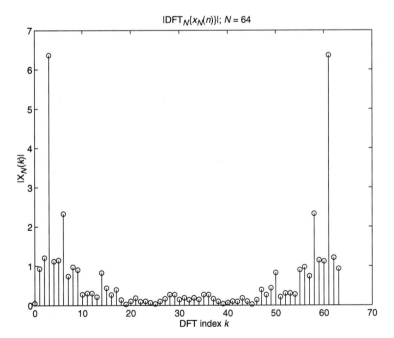

Figure 7.21*b*

resulting in the $N = 64$-point sequence $x_N(n)$ in Figure 7.21*a* with DFT_{64} in Figure 7.21*b*. Interpolate this sequence back to its original size using the DFT method, so that for every sample of the subsampled sequence, two samples are produced in the resulting sequence.

Solution

This example is the case $M = 128$, $L = 2$, and $N = M/L = 64$. The construction in (7.23) is used, but the sample at $k = N/2$ is halved for this example in which $N = 64$ is even. Specifically, in place of (7.23), we thus have for N even

---

DFT temporal interpolation, N even.

$$X_{ZP;M}(k) = \begin{cases} L \cdot X_N(k), & k \in [0, N/2 - 1] \\ \frac{1}{2}L \cdot X_N(N/2), & k = N/2 \\ 0, & k \in [N/2 + 1, M - N/2 - 1] \\ \frac{1}{2}L \cdot X_N(N/2), & k = M - N/2 \\ L \cdot X_N(k - M + N), & k \in [M - N/2 + 1, M - 1]. \end{cases}$$

(7.35)

The resulting 128-point sequence $X_{ZP;128}(k)$ obtained from (7.35) is shown in Figure 7.22a. We now take the $M = NL$-point DFT^{-1} of $X_{ZP;M}(k)$, with $M = 128$; the result is shown in Figure 7.22b (O stems), in comparison with the original 128-point sequence $x(n)$ in Figure 7.20a (* stems). The reproduction quality appears reasonably good. Recalling (7.24) and the text following (7.25)

$$[x_{ZP;M}(n) = \sum_{m=0}^{N-1} x_N(m) \cdot \phi_N(\pi \cdot [n/M - m/N]), \text{ where } \phi_N(\theta) = \cot(\theta) \cdot \sin(N\theta)/N],$$

it should be no surprise that for every Lth (in this case, every second) point in Figure 7.22b there is exact agreement between $x_{ZP;M}(n)$ and the original $x(n/L)$. This is because for every $L = (M/N)$th point $n = iL$, ϕ_N has argument $\pi(i - m)/N$ so that ϕ_N becomes $\cot\left\{\dfrac{\pi(i-m)}{N}\right\} \sin\{\pi(i-m)\}/N = \delta(i-m)$; thus there is only one term in the m sum of (7.24), and this term is $x_N(n/L)$.

Why is there disagreement for the interpolated points? The reason will become clearer after reading Section 7.5, but may be stated as follows. Examination of Figure 7.21b shows that there is significant high-frequency content in $X_{64}(k)$ of the subsampled sequence $x(n)$; $|X_N(k)|$ has very significant magnitudes for k near $N/2$; we noted in Section 6.2 the likelihood of aliasing under this condition. It will be shown in Section 7.5 that after subsampling the original sequence by taking every Lth sample, the DTFT of the small sequence is an aliased version of the true spectrum. The high frequency content in Figure 7.20b is aliased into low frequencies in Figure 7.21b in the same way that continuous Fourier transforms are aliased under Δt-sampling. Similarly to the sampling of continuous-time signals, the digital spectral aliasing is irreversible. If the original 128-point sequence $x(n)$ were very smooth (bandlimited to $|\omega| < \pi/L$), then there would have been no aliasing, and the estimate based on (7.35) would have *exactly* reproduced the original sequence.

The problem can be exacerbated by poor choice of the end points of the original subsequence $x(n)$. Examination of $x(n)$ in Figure 7.20a reveals that there is a moderately large discontinuity between the last sample $x(127)$ and the first sample of the next period, $x(128) = x(0)$. Consequently, there is an artificial high frequency content in the periodic extension of $x(n)$. In this example this end points problem is not extreme, but is noticeable in Figure 7.20b in nonnegligible $|X(k)|$ for k near $128/2 = 64$ that will alias under subsampling.

Zero-padded IDFTI sequence for interpolation; $N = 64$, $L = 2$, $M = NL = 128$

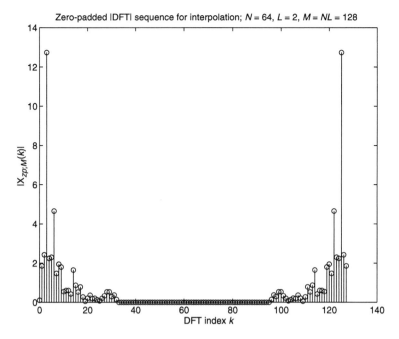

Figure 7.22a

$x(n)$ (O) and estimation of $x(n)$ by DFT interpolation of $x_N(n)$ (*); $M = 128$, $N = 64$

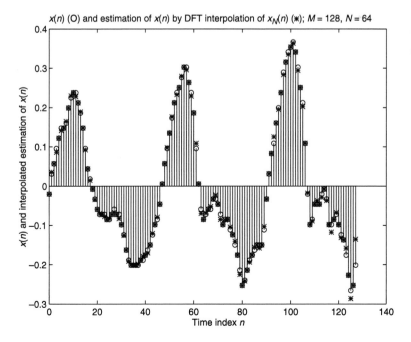

Figure 7.22b

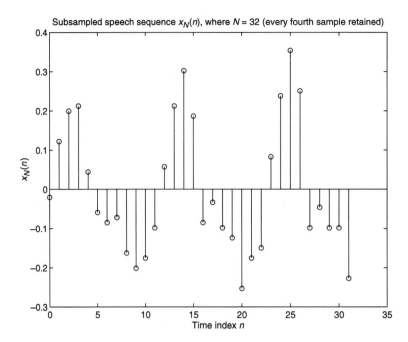

Figure 7.23a

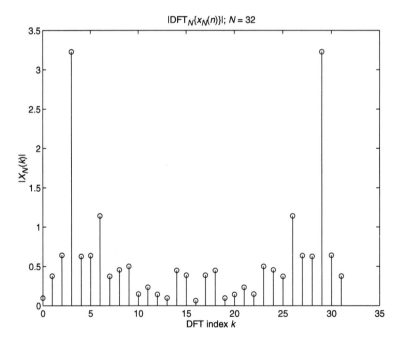

Figure 7.23b

A somewhat more extreme example is obtained by setting $L = 4$. Figure 7.23a shows the subsampled sequence $x_N(n)$ obtained by taking every fourth point of $x(n)$ in Figure 7.20a; now $N = 32$. Figure 7.23b shows the DFT magnitude; we predict much worse behavior with such large $|X(k)|$ near $k = N/2$. Figure 7.24a shows the

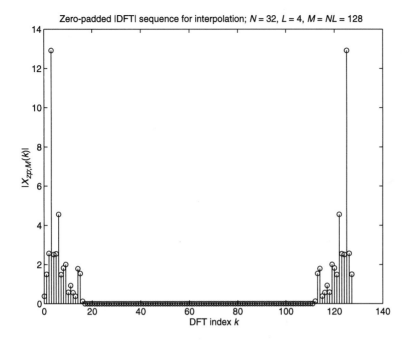

Figure 7.24*a*

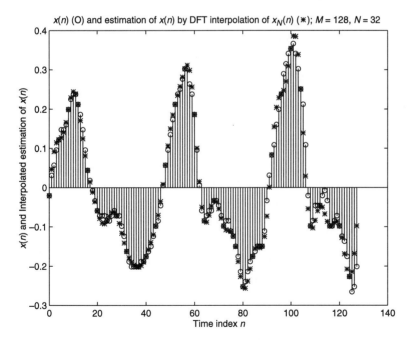

Figure 7.24*b*

zero-padded sequence obtained from (7.35), and Figure 7.24*b* shows its 128-point DFT^{-1} (O stems), the interpolated sequence, against the original $x(n)$ (* stems). Although there is still exact agreement at every $L = 4$th point, the disagreement at other values is now quite significant.

This important aspect of "downsampling" or decimation used for obtaining our sparse (subsampled) sequence to be interpolated, in which steps must be taken to avoid aliasing of high digital frequencies, is discussed in more detail in Section 7.5. Also, another detailed example of the DFT interpolation method of this section is presented on WP 7.2. In that example, this method will be compared with the zero-interlace interpolation technique presented in Section 7.5.3.

7.5   SAMPLING RATE CONVERSION

7.5.1   Introduction

Sampling rate conversion is a general term for the process of changing the time interval between the adjacent elements in a sequence consisting of samples of a continuous-time function. It is a natural generalization of interpolation that also includes "subsampling," where the resulting samples become farther apart. We have already discussed one interpolation procedure in Section 7.4, and we briefly considered downsampling in Example 7.7. In this section, we examine sampling rate conversion in more detail and present another popular interpolation method.

Why is digital sampling rate conversion needed? In many practical applications, digital filtering may be effectively and economically performed at a lower sampling rate than that of the input sequence. The process of lowering the sampling rate is called decimation. Subsequently, samples may be needed at the original or some other higher sampling rate (interpolation). It is usually beneficial to be able to achieve decimation ("downsampling") and interpolation ("upsampling") or some combination of the two entirely within the discrete-time domain. Doing so eliminates the need for a D/A–A/D pair with unequal reconstruction/sampling rates and the resulting increased cost and distortion.

Examples of *multirate processing* are found in digital audio (e.g., compact disc systems), digital communications, long-distance telephone transmission, speech coding,[10] and zoom transforms (described below). In digital audio, the recording company may have versions of the same recording at different sampling rates: one for broadcast, one for storage, one for the consumer product, and so forth. Also, for special effects it may be desirable to vary the speed of the music. These can be done conveniently all in the discrete-time domain without the degrading effects of D/A–A/D conversions.

In digital communications, A/D conversion may be economically performed using delta modulation. Delta modulation is two-level coding at a very high, over-sampling rate.[11] Contrarily, the subsequent signal processing (filtering) must be done with pulse code modulation (2^N-level coding at lower rates). Conversion between these formats requires sampling rate conversion.

Long-distance telephone systems handling several signals simultaneously use frequency division multiplexing (FDM, in which the signals occur in different frequency bands on one line) for transmission and time division multiplexing (TDM, in which

---

[10] J. S. Lim and A. V. Oppenheim, *Advanced Topics in Signal Processing* (Englewood Cliffs, NJ: Prentice Hall, 1988).

[11] See, for example, L. R. Rabiner and R. W. Schafer, *Digital Processing of Speech Signals* (Englewood Cliffs, NJ: Prentice Hall, 1978).

the signals occur at different time instants on separate lines) for switching. The sampling rate of the single-line FDM must far exceed that of the TDM signals, yet they must be interfaced. This interfacing is achieved with digital sampling rate conversion.

A similar principle is involved in the *subband coding* technique in speech processing/transmission. In this case, one begins with a single speech waveform that is modeled by a number of subbands (an FDM model). One then performs FDM/TDM conversion, codes each line for transmission and multiplexes, and reconstructs at the receiving end with TDM/FDM operations.

In zoom transform analysis, a wide-band signal is bandpass filtered, leaving only the desired frequency interval. The remaining band is modulated down to low frequencies and then downsampled to obtain a focused view of only the desired band of the original signal, which now occupies the full band because of the stretching effect of downsampling. In the problems, we will see that a simple change in the intuitive order of operations just described for the zoom transform can result in a significant computational savings.

7.5.2   Decimation or Downsampling

For the case of downsampling, we desire $x_d(n) = x_c(n\{M\,\Delta t\})$ given $x(n) = x_c(n\Delta t)$, where the subscript d stands for "decimated" and M is an integer. Thus, $x_d(n) = x(nM)$. The z-transform of $x_d(n)$ in terms of $X(z) = ZT\{x(n)\}$ was derived in Example 5.19 [we reproduce (5.38) here]

$$X_d(z) = \frac{1}{M} \sum_{m=0}^{M-1} X(z^{1/M} \cdot e^{j2\pi m/M}). \tag{7.36}$$

Setting $z = e^{j\omega}$, (7.36) becomes

$$X_d(e^{j\omega}) = \frac{1}{M} \cdot \sum_{m=0}^{M-1} X(e^{j(\omega + 2\pi m)/M}). \tag{7.37}$$

Not surprisingly, (7.36) and (7.37) are sums of shifted replications, just as was the case for sampling of a continuous-time function. For no aliasing in $X(e^{j\omega})$, we found in Section 6.2 that the requirement was $\Delta t \le \pi/\Omega_{max}$. Now suppose that the maximum freqency in $X(e^{j\omega})$ is $\omega_{max} = \Omega_{max}\Delta t$, from sampling $x_c(t)$. Then the $m = 1$ replication in (7.37) will not overlap the $m = 0$ replication if $-2\pi + M\omega_{max} < -M\omega_{max}$, that is, if $\omega_{max} = \Omega_{max}\Delta t < \pi/M$. Thus for no aliasing to occur in $X_d(e^{j\omega})$, we see that the requirement is $\Delta t \le \pi/(M\Omega_{max})$. Thus for no aliasing to occur in the decimated (downsampled) signal, the original sequence $x(n)$ must have been sampled M times more finely than is necessary for avoiding aliasing in $x(n)$. We may arrive at this requirement more simply by noting that $x_d(n) = x_c(n\{M\,\Delta t\})$, which will not be aliased if $\{M\,\Delta t\} < \pi/\Omega_{max}$, or $\Delta t < \pi/(M\Omega_{max})$. The method based on analyzing (7.37), however, giving $\omega_{max} < \pi/M$ is useful when $x(n)$ is all we have, and we know nothing about the originating $x_c(t)$ (if any).

Often this requirement $\omega_{max} < \pi/M$ cannot be assumed to have been exactly met, in which case aliasing distortion results. Thus, it is essentially always the case that it is better to have an unaliased lowpass version of the original signal than a full-band aliased version. In this common case, the "downsampler" is preceded by an antialiasing digital lowpass filter having cutoff $\omega_c = \pi/M$. In practice, the lowpass filter-downsampler combination is commonly called a "decimator."

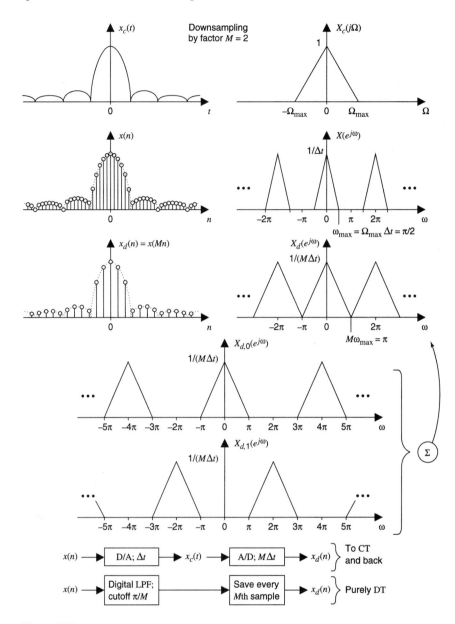

Figure 7.25

Notice that each term in the aliasing sum (7.37) is periodic with period $2\pi M$, not 2π! While the time-domain signal has been compressed, the frequency domain terms have been "stretched" or expanded [just as is true for Δt-sampling of $x_c(t)$]. This is particularly important to remember when using DTFT or DFT sketches in analyzing the behavior of a system involving sampling rate conversion. Remember to draw several complete periods of each aliasing term, or you may miss some contribution on the region of interest $\omega \in [0, 2\pi]$ (equivalently, $\omega \in [-\pi, \pi]$). An example of the time–domain and frequency-domain relations between $x_c(t)$, $x(n)$, $x_d(n)$, $X_c(j\Omega)$, $X(e^{j\omega})$, and $X_d(e^{j\omega})$ is given in Figure 7.25 for the case $M = 2$ and $\Delta t = \pi/(2\Omega_{max})$

(equivalently, for $\omega_{max} = \Delta t \cdot \Omega_{max} = \pi/2$). Notice the individual replications in the lower part of the figure, which sum to $X_d(e^{j\omega})$. Also, at the bottom of Figure 7.25 we contrast the awkward reconstruction/resampling system with the equivalent and much more efficient decimation system. Finally, note that we can obtain $X_d(e^{j\omega})$ without using (7.37) by directly taking the DTFT of $x_c(n \cdot [M \Delta t])$. However, (7.37) allows us to obtain $X_d(e^{j\omega})$ from $X(e^{j\omega})$ when, as is common in practice, only $x(n)$ is available and neither $x_c(t)$ nor $X_c(j\Omega)$ is available.

EXAMPLE 7.8

To construct an approximately bandlimited signal, let $x_r(n)$ be a truncated version of the sequence $\sin(\omega_{cr}n)/(\pi n)$. The subscript r on $x_r(n)$ denotes the rectangular shape of $X_r(e^{j\omega})$, the DTFT of $x_r(n)$, as opposed to the triangular shape of $X(e^{j\omega})$ introduced below. Also, the subscript cr on ω_{cr} denotes the "cutoff" frequency of $X_r(e^{j\omega})$; without truncation, $X_r(e^{j\omega}) = \{1, |\omega| < \omega_{cr}; 0, \text{otherwise}\}$ [equivalently, define $f_{cr} = \omega_{cr}/(2\pi)$]. Investigate the spectral effects of downsampling ($M = 2$) of $x(n) = x_r^2(n)$, which has a maximum ("cutoff") frequency of $f_1 = \omega_{cr}/\pi$ (see below), for the cases **(a)** $\omega_{cr} = 0.25\pi$, or equivalently, $f_1 = 0.25$, and **(b)** $\omega_{cr} = 0.4\pi$ or equivalently, $f_1 = 0.4$. For brevity, we omit the units of f (cycles/sample) and of ω (rad/sample).

Solution

(a) The number of points kept in $x_r(n)$ is $N = 128$. If the argument of $x_r(n)$ is shifted by $N/2$ so that $x_r(0)$ and $x_r(N-1)$ are both nearly zero, then the DFT_N of a zero-padded version of this new $x_r(n)$ will give a good representation of the complete DTFT of the original non-shifted, non-truncated $x_r(n)$. Thus, ignoring the minimal truncation effect, we use zero-padding out to 4096 points (although only a few hundred would be sufficient to obtain a smooth graph). Figure 7.26a shows $x_r(n)$, and Figure 7.26b shows $|X_r(e^{j\omega})|$.

The peak of $x_r(n)$ is $\sin(\omega_{cr}n)/(\pi n)\big|_{n=0} = \omega_{cr}/\pi = 0.25$, which agrees with the time plot (Figure 7.26a). Now set $x(n) = [x_r(n)]^2 = [\sin(\omega_{cr}n)/(\pi n)]^2$. As implied in the problem statement, the spectrum of $x(n)$ should be twice as wide as that of $x_r(n)$, because the product $x_r(n) \cdot x_r(n)$ transforms to convolution of $X_r(e^{j\omega})$ with itself: $X(e^{j\omega}) = X_r(e^{j\omega}) *_{c\omega} X_r(e^{j\omega})/(2\pi)$. Call f_1 the cutoff of $x(n)$, so $f_1 = 2 \cdot f_{cr} = \omega_{cr}/\pi = 0.25$. For general conclusions to be drawn from this example, it is f_1 (not f_{cr}) that is relevant because it is $x(n)$ that we will be downsampling.

Figure 7.27a shows $x(n)$ and Figure 7.27b shows $|X(e^{j\omega})|$ [again, by zero-padding $x(n)$]. The peak of $x(n)$ in Figure 7.27a is just the square of that of $x_r(n)$: $(\omega_{cr}/\pi)^2 = f_1^2 = 0.0625$. By ignoring the truncation, the peak of the plot of $|X(e^{j\omega})|$, $|X(e^{j0})|$, may be found from the convolution theorem as [using $X_r(e^{j\omega}) = e^{-j\omega N/2}$ for $|\omega| < \omega_{cr}$ and noting that $X_r(e^{-j\theta}) = X_r^*(e^{j\theta})$]

$$X(e^{j0}) = \frac{1}{2\pi} \int_{-\pi}^{\pi} X_r(e^{j\theta})X_r(e^{j(\omega-\theta)}) \, d\theta \bigg|_{\omega=0} = \frac{1}{2\pi} \cdot \int_{-\pi}^{\pi} |X_r(e^{j\omega})|^2 \, d\omega$$

$$= \frac{2\omega_{cr}}{2\pi} = \frac{\omega_{cr}}{\pi} = f_1 = 0.25,$$

which is verified numerically in Figure 7.27b.

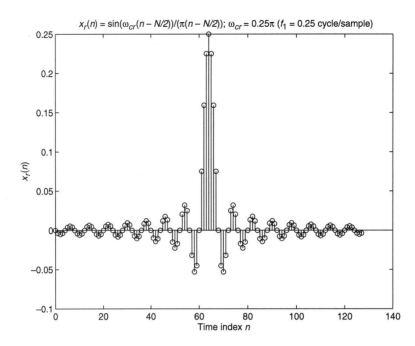

Figure 7.26a

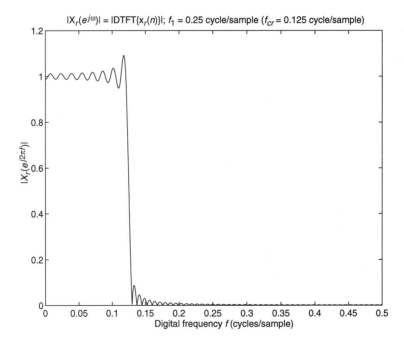

Figure 7.26b

Also note that Figure 7.27b is much more nearly ideal (as a "triangle") than Figure 7.26b is (as a "boxcar"). This is because $x(n) = \{\sin(\omega_{cr}n)/(\pi n)\}^2$ approaches zero much more quickly than does $x_r(n) = \sin(\omega_{cr}n)/(\pi n)$ as n is

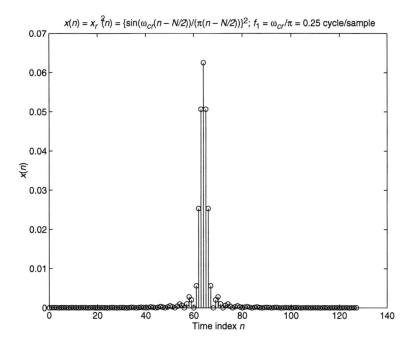

$x(n) = x_r^{\,2}(n) = \{\sin(\omega_{cr}(n - N/2))/(\pi(n - N/2))\}^2$; $f_1 = \omega_{cr}/\pi = 0.25$ cycle/sample

Figure 7.27a

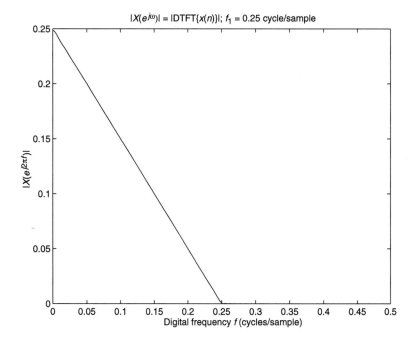

$|X(e^{j\omega})| = |\text{DTFT}\{x(n)\}|$; $f_1 = 0.25$ cycle/sample

Figure 7.27b

increased. Thus, for N fixed, the truncation error is much smaller in Figure 7.27b than in Figure 7.26b. However, there is still truncation error, as we find when we "zoom in" on the transition region $0.245 < f < 0.3$; see Figure 7.26c.

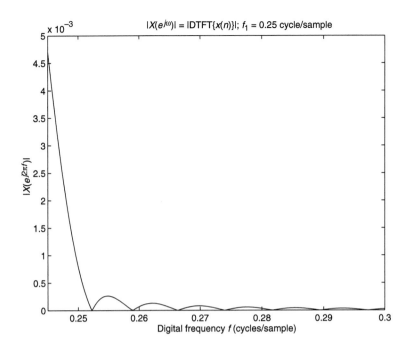

Figure 7.27c

Next, we downsample $x(n)$ by factor $M = 2$: $x_d(n) = x(2n)$; $x_d(n)$ is shown in Figure 7.28a. The magnitude of the DTFT of its 4096-zero-padded version is plotted in Figure 7.28b. In downsampling $x(n) = [\sin(\omega_{cr}n)/(\pi n)]^2$ by the factor M, we obtain $x_d(n) = (1/M^2) \cdot [\sin\{(M\omega_{cr})n\}/(\pi n)]^2$, which by the previous reasoning has a maximum value of $(1/M^2) \cdot [M\omega_{cr}/\pi]^2 = f_1^2 = 0.0625$, regardless of the value of M. Of course, downsampling (taking every Mth point) will not affect the maximum of the time sequence, as long as the maximum sample of the original sequence is kept. However, by (7.37), the magnitude of the spectrum will be reduced by the factor $1/M$ due to downsampling of $x(n)$; hence the maximum of the magnitude spectrum plot (Figure 7.28b) is $\omega_{cr}/(M\pi) = f_1/M = 0.125$. We see in Figure 7.28b that there is just barely no aliasing after downsampling; this case is the one illustrated in Figure 7.25.

(b)   The above downsampling procedure ($M = 2$) is now repeated for $\omega_{cr} = 0.4\pi$, or $f_1 = 0.4$. Now the maximum of the time sequence $x(n)$ as well as that of $x_d(n) = x(2n)$ is $f_1^2 = 0.16$, whereas that of the magnitude spectrum is $f_1 = 0.4$ for $X(e^{j\omega})$ and $f_1/2 = 0.2$ for $X_d(e^{j\omega})$. Figure 7.29a shows $X(e^{j\omega})$, Figure 7.29b shows $x_d(n)$ and Figure 7.29c shows $X_d(e^{j\omega})$.

Notice the spectral aliasing in $X_d(e^{j\omega})$ (Figure 7.29c). As we have seen before (e.g., the time-aliasing in Figure 7.17b), because linear functions of ω having opposite slopes are being added, by (7.37) the effect is a constant-in-ω bias for $\omega > 2\pi - \omega_{max} M = 2\pi(1 - f_1M) = 2\pi(1 - 0.4 \cdot 2) = 2\pi \cdot 0.2$, equal in value to the unaliased spectrum at $\omega = 2\pi \cdot 0.2$. We may explain this aliasing with reference to Figure 7.25, where the individual replications are shown. Figure 7.25 was drawn for $\omega_1 (= 2\pi f_1) = 0.5\pi$ [same as part (a) of this

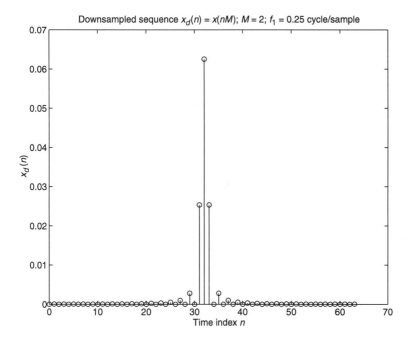

Figure 7.28*a*

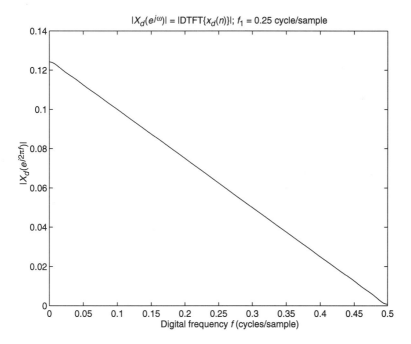

Figure 7.28*b*

example], which under downsampling with $M = 2$ gives just barely no aliasing; in particular, the replication centered on $\omega = 2\pi$ extends back to π. In
Figure 7.29, $\omega_1 = 2\pi f_1 = 0.8\pi$. Under the "stretching" of downsampling, the

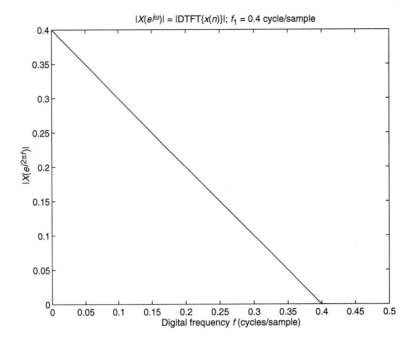

Figure 7.29*a*

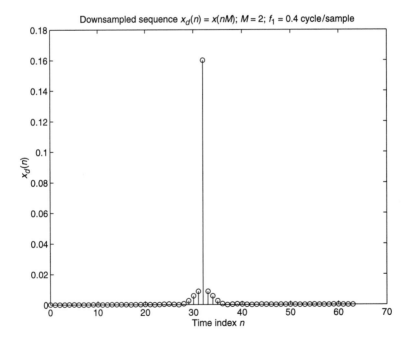

Figure 7.29*b*

0.8π half-width of a replication becomes $M \cdot 0.8\pi = 1.6\pi$. Thus, the replication centered on 2π now extends back to $\omega = 2\pi - 1.6\pi = 0.4\pi$, or to $f = 0.2$. Consequently, we begin to see aliasing in Figure 7.29*c* at $f = 0.2$. Thus we see

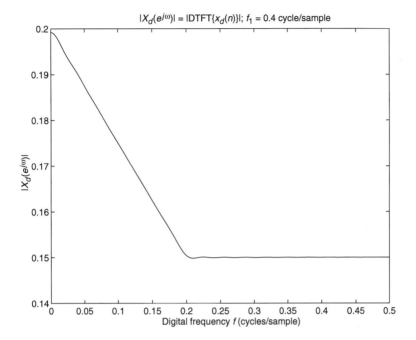

$|X_d(e^{j\omega})| = |\text{DTFT}\{x_d(n)\}|;\ f_1 = 0.4$ cycle/sample

Figure 7.29c

that for moderate downsampling aliasing, the aliasing occurs at high frequencies. We must lowpass pre-filter $x(n)$ to have a maximum digital frequency of $\omega_1 = \pi/M$ or $f_1 = 1/(2M)$ to avoid aliasing under M-downsampling.

We can now see how downsampling (decimation) can be used to reduce the rate of computations (and thus reduce costs). If a discrete-time signal is narrow-band, why wastefully process all the zero-content frequencies? Just modulate the signal (if necessary) to baseband, downsample by a factor so that the signal becomes full-band, and do subsequent processing at this low sampling rate. Then later, if required, the remodulation and interpolation to original sampling rate may be performed, which leads us to our next topic.

7.5.3   Interpolation or Upsampling

The method of interpolation of a sequence discussed in Section 7.4, in which we zero-pad the DFT, is appropriate for (a) exact interpolation of periodic sequences and (b) approximate interpolation of samples of finite/moderate-duration continuous-time signals. However, for infinite-length aperiodic sequences or aperiodic sequences of length greater than the available DFT size, another method is commonly used. In this second method, the fundmental modification to the sequence is made in the time domain rather than the frequency domain, as was the case in Section 7.4. Consequently, the processing may proceed in a continuous stream, which is important in the applications mentioned in the introduction to this section, where the data length is indefinite.

As previously, it is desired to obtain L output samples for every one input sample, yielding $L - 1$ interpolations between the given input values. Thus we desire $x_i(n) = x_c(n\{\Delta t/L\})$, given $x(n) = x_c(n\,\Delta t)$, where the subscript i stands for "interpolated." In all discussions here we assume that $X(e^{j\omega})$ has no aliasing; if $x(n)$ is finite-length, we assume that the inevitable aliasing is negligible. "At any rate," we certainly at least require $x_i(n) = x(n/L)$ for $n = mL, m$ any integer. To accomplish this requirement as well as exact interpolation for $n \neq mL$, we proceed as follows.

Recall from Section 7.4 that if Δt is scaled by $1/L$, then from the aliasing formula reproduced in (7.28) and the subsequent discussion, the replications of $X_c(j\Omega)$ in $X(e^{j\omega})$ will be compressed by the factor L, with intervals of zero value from the cutoff [e.g., at $\omega = \Omega_{max}(\Delta t/L)$] out to the next replication [e.g., at $\omega = 2\pi - \Omega_{max}(\Delta t/L)$]. They will also be scaled larger by the factor L. Such is the appearance of $X(e^{j\omega})$ when the sampling rate is increased by the factor L, regardless of the periodicity/aperiodicity of $x(n)$.

Consider for the moment interpolation of a sequence N points long. Recall from Section 7.2 that if we replicate an N-long sequence L times, then by (7.11f) $[X_{M;N,L}(k) = L \cdot X_N(k/L)$ for $k = nL$, where $0 \leq n \leq N - 1$; 0, otherwise], the $M = LN$-point DFT of that sequence consists of the N values of the DFT_N of the original N-point sequence interlaced with $L - 1$ zeros in between each value. The DFT satisfies duality, so if in the time domain we interlaced the original N time values with $L - 1$ zeros, the DFT_M should consist of L repetitions of the N values of the DFT_N of the original sequence. That is almost what we would like: The DFT is squashed to $1/L$ of its original width and replicated L times. However, to match the required form for exact interpolation, the other non-baseband replications should not be there.

These replications can be removed by lowpass filtering this M-point sequence with a digital filter having cutoff π/L. We now see that for an N-periodic sequence, the end result of the process of interleaving zeros and lowpass filtering is precisely equivalent to that of the DFT method in Section 7.4 of zero-padding the DFT in its center (high-frequency) region, except that the DFT method did not require any explicit filtering because the unwanted replications were never formed.

What if the sequence is not periodic and has indefinite length? Recall that the appearance of the DTFT of the sequence obtained by sampling the continuous-time signal at $\Delta t/L$ is fundamentally independent of the length or periodicity of $x_c(t)$. We might then infer that the same technique as described above would work for indefinite-length sequences: Insert L zeros between each time-sample and lowpass filter the resulting sequence [called $x_e(n)$ with DTFT $X_e(e^{j\omega})$] with a lowpass filter having cutoff π/L. The analysis is as follows:

$$x_e(n) = \begin{cases} x(n/L), & n = mL, \text{ where } m = 0, \pm 1, \pm 2, \dots \\ 0, & \text{otherwise} \end{cases} \tag{7.38a}$$

$$= \sum_{m=-\infty}^{\infty} x(m)\delta(n - mL). \tag{7.38b}$$

We take the DTFT:

$$X_e(e^{j\omega}) = \sum_{n=-\infty}^{\infty} \sum_{m=-\infty}^{\infty} x(m)\,\delta(n - mL)\,e^{-j\omega n} \tag{7.39a}$$

$$= \sum_{m=-\infty}^{\infty} x(m) \sum_{n=-\infty}^{\infty} \delta(n - mL)\,e^{-j\omega n}. \tag{7.39b}$$

Now note that the only nonzero term in the n sum is $n = mL$, therefore yielding for the sum the single exponential term $e^{-j(\omega L)m}$. The resulting m sum we recognize to be the DTFT evaluated at ωL. We therefore obtain

$$X_e(e^{j\omega}) = X(e^{j\omega L}). \tag{7.39c}$$

Thus, while the time-domain signal is expanded with zeros (and ultimately will be interpolated), the spectrum is compressed and thus replicated, for $\omega \in [-\pi\ \pi]$.

All that now needs to be done is to scale $X_e(e^{j\omega})$ by the factor L and remove the undesired replications, to match the resulting DTFT to that of $x_i(n) = x_c(n\Delta t/L)$, as discussed above. Therefore,

$$X_i(e^{j\omega}) = H(e^{j\omega}) \cdot X_e(e^{j\omega}), \tag{7.40}$$

where

$$H(e^{j\omega}) = \begin{cases} L & \omega \in [0, \pi/L], [2\pi - \pi/L, 2\pi] \\ 0, & \omega \in (\pi/L, 2\pi - \pi/L), \end{cases} \tag{7.41}$$

or, equivalently,

$$H(e^{j\omega}) = \begin{cases} L & \omega \in [-\pi/L, \pi/L] \\ 0, & \text{otherwise } (\omega \in [-\pi, \pi]), \end{cases} \tag{7.42}$$

due to the 2π-periodicity of the DTFT.

This development makes no assumption about whether $x_c(t)$ is periodic. Only the assumptions of true or effective bandlimitedness and Nyquist sampling are made. Provided that all nonzero $x(n)$ are used from $n = -\infty$ to ∞ and the filter is ideal, it is always exact, for it produces exactly the same modification on the DTFT that replacing Δt by $\Delta t/L$ does in the original aliasing formula (7.28). The inverse DTFT of $H(e^{j\omega})$ is

$$h(n) = \frac{L}{2\pi} \int_{-\pi/L}^{\pi/L} e^{j\omega n}\, d\omega = L \cdot \frac{\sin(\pi n/L)}{\pi n} = \text{sinc}\left(\frac{\pi n}{L}\right). \tag{7.43}$$

As multiplication of the DTFTs corresponds to convolution of the discrete-time functions,

$$x_i(n) = x_e(n) * h(n) = \sum_{m=-\infty}^{\infty} x(m) \cdot \text{sinc}\left\{\frac{\pi}{L}(n - mL)\right\}, \tag{7.44}$$

where in the second equality in (7.44), use was made of (7.38b). But (7.44) is identical to the exact $\Delta t/L$-sampling-based interpolation in (7.27b)![12] Thus, if and only if $H(e^{j\omega})$ is an *ideal* lowpass filter, the interpolation is indeed exact and requires no D/A–A/D hardware.

Figure 7.30 summarizes the process in both time and frequency domains for $L = 2$ and $\Delta t = \pi/\Omega_{\max}$ ($\omega_{\max} = \pi$), and indicates the equivalent processes in continuous and discrete time. Zero-interlacing of time samples followed by ideal lowpass filtering is equivalent to zero-padding the center of the DFT of the noninterlaced time sequence if $x_c(t)$ is $N\Delta t$ – periodic.

Note that both procedures may require the DFT—the zero-interlacing method may use the DFT for the required lowpass filtering. However, although that filtering

---

[12]Where here we ignore the periodic designations in (7.27b).

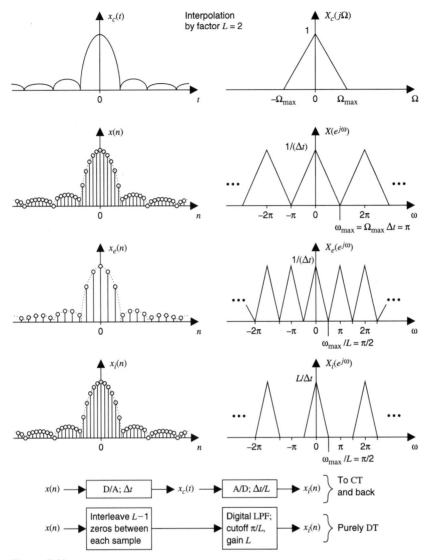

Figure 7.30

can be done in sections or blocks (see Section 7.7), it appears that the DFT method requires us to take the DFT of the entire sequence at once. This is the price to be paid for no required lowpass filtering.[13]

EXAMPLE 7.9

Plot the DTFT of the zero-interlaced signal $x_e(n)$ for $L = 4$ obtained from the sequence $x(n)$ in Example 7.8. Use $\omega_{cr} = \frac{1}{2}\pi$ (equivalently, $f_1 = \frac{1}{2}$). Then lowpass-filter $x_e(n)$ to obtain $x_i(n)$, the estimated interpolation between the original values $x(n)$. Compare with the exact interpolation, using the continuous-time function used to form $x(n)$.

---

[13] In the DFT method, we could interpolate chunks at a time, but there will be end-effect errors at the chunk ends.

Solution

Recall that $x_e(n)$ is the zero-interlaced signal before lowpass filtering at π/L; $x_e(n)$ is shown in Figure 7.31a; to help see the interlaced zeros, only a small portion of $x_e(n)$ near its maximum is shown. In addition, Figure 7.31a shows the continuous-time sinc^2 function (dashed curve) whose samples are $x(n)$, from which $x_e(n)$ was formed.

What is the precise expression for that continuous-time sinc^2 function? Let us first determine the sinc function whose samples are $x_r(n)$ and then square the result. We know from Example 7.8 that $x_r(n) = \sin(\omega_{cr}n)/(\pi n)$ (truncated). Define $\Omega_{cr} = \omega_{cr}/\Delta t$. Arbitrarily set $\Omega_{cr} = 1$ rad/s, so $\Delta t = \omega_{cr}$ (s). Then $x_r(n) = x_{r,c}(n\Delta t)$, where $x_{r,c}(t) = \sin(\omega_{cr}[t/\Delta t])/\{\pi[t/\Delta t]\} = (\omega_{cr}/\pi) \cdot \sin(t)/t = (\omega_{cr}/\pi) \cdot \text{sinc}(t)$, which in this example with $\omega_{cr} = \frac{1}{2}\pi$ is $x_{r,c}(t) = 0.5 \cdot \text{sinc}(t)$.

Notice that $x_e(n) = 0$ for some of the n values where we expect to see $x_e(n) \neq 0$ (e.g., $n = 240$ and 248). These values are zero due not to zero-interlacing but to $\{\sin(\omega_{cr}n)/(\pi n)\}^2$ being zero for such n. The DTFT of $x_e(n)$, $X_e(e^{j\omega}) = X(e^{j\omega L})$, is shown in Figure 7.31b. In Figure 7.31b, we see $L/2 = 2$ repetitions of the spectral triangle for $\omega \in [0, \pi]$ ($f \in [0, 0.5]$). The maximum in the magnitude spectrum plot is unaffected by the zero-interlacing [see (7.39c)], so $\omega_{cr}/\pi = 0.5$, as we would calculate for the maximum of $|X(e^{j\omega})|$ in Example 7.8.

Now consider the lowpass filtering of $x_e(n)$ with a filter having ω cutoff $\pi/L = \pi/4$, or equivalently, an f cutoff of $0.5/L = 0.125$ required to annihilate the undesired replications that zero-interlacing produces. A very simple approach is just to multiply the DFT values by L in the passband and zero in the stopband. As will be discussed in Section 9.4, this simple-minded approach results in large ripple in between the exactly fixed frequency samples. The result would be distortion of all the passband frequency components other than the DFT frequency components $\omega_k = 2\pi k/N$ or $f_k = k/N$. The intersample ripple can be reduced by making a few transition samples between the passband and stopband.

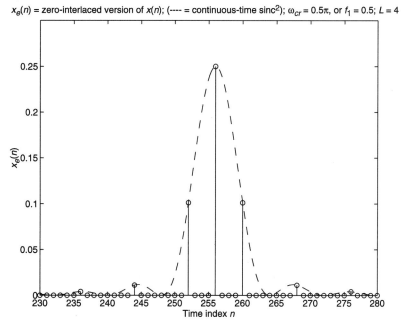

$x_e(n)$ = zero-interlaced version of $x(n)$; (---- = continuous-time sinc^2); ω_{cr} = 0.5π, or f_1 = 0.5; L = 4

Figure 7.31a

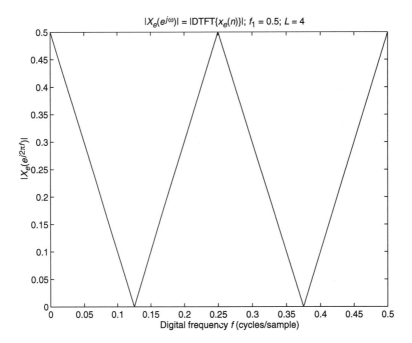

Figure 7.31*b*

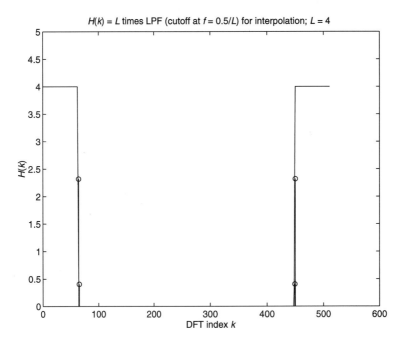

Figure 7.32

In our example, quasi-optimal values are 0.58 for the first transition sample and 0.1 for the second (see Section 9.4 for details). The resulting lowpass filter is shown in Figure 7.32. Because there are too many samples (512) to clearly show all stems, the samples have been connected by lines instead—except for the transition samples, which are displayed as stems.

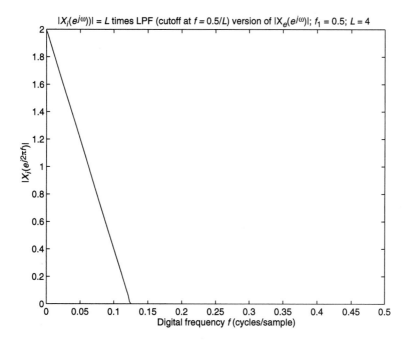

Figure 7.33*a*

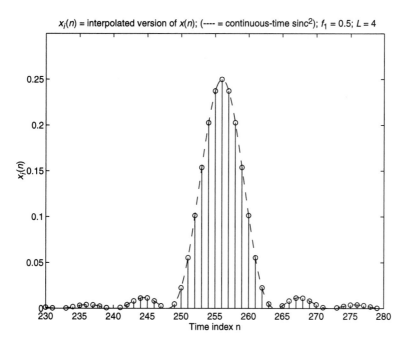

Figure 7.33*b*

We now combine the zero-interlacing with the lowpass filtering to obtain the interpolation, $x_i(n)$. The DTFT of $x_i(n)$, the result of the above lowpass filtering, is shown in Figure 7.33*a*. The maximum magnitude $X_{i,\max}$ is increased by the factor L by the lowpass filter relative to the maximum of $\left|X_e(e^{j\omega})\right|$ and so is $X_{i,\max} = L \cdot f_1 = 2.0$. The interpolated sequence $x_i(n)$ is shown in Figure 7.33*b* for the same range of n as used in Figure 7.31a [where $x_i(n)$ is significantly nonzero]. Also, as in Figure 7.31a, Figure 7.33b includes (- - - -) the exact sinc^2 function whose samples in Figure 7.31a we have interpolated to obtain $x_i(n)$ in Figure 7.33b. The accuracy of the interpolations is excellent.

As a further example of basic sampling rate conversion, we consider in detail on WP 7.2 the process of downsampling ($M = 4$) followed by upsampling ($L = 4$) the sum of five sinusoids. The methods of upsampling used are (a) the DFT method (7.24) described in Section 7.4, and (b) the zero-interlace/lowpass filtering method of this section. This example points out a significant advantage of the DFT method over the zero-interlace method for finite-length sequences (one replication of a periodic signal): The DFT method requires no digital filter to implement. The example also very clearly illustrates the different effects arising for having versus not having an integral number of signal periods in the time window of samples used for the DFT. Many intuitive insights in this example are made concrete by plentiful graphics.

Although several applications were mentioned at the beginning of this section, the focus has been on basic principles. A full discussion of hardware considerations in specific examples that mandate sampling-rate conversion is beyond the scope of this book, but it should now be understandable to the interested reader.

Furthermore, in practice, many implementation issues come up that we have ignored. For example, in practical applications where the sampling rate is to be changed only slightly—by the factor $\alpha = L/M$ near unity—L and M will have to be large integers. In such cases, the interpolation followed by decimation operations are done far more efficiently in stages than in one fell swoop as implied by our previous discussion. The reason for this is that when done in stages, the lowpass filters act when the sample rate is high (and thus the digital spectra are narrow); thus, they can be allowed to have relatively large transition widths. The resulting aliasing is designed to occur only in nonsignal bands that will later be chopped off anyway. Wide-transition-width filters can be very simple and require only minimal computation.

Minimizing computation at the higher sampling rate is advantageous because that is where filtering computations have to be done most frequently and rapidly. The sharper filtering is done at the low sampling rate. Overall, the reduction in computation achievable by multistage sampling rate conversion can be quite dramatic. The price to be paid is a more complicated processing and control structure.

Although decimation and interpolation do not necessarily require the DFT [excepting (7.23) and (7.35), which are based on the DFT], the convolution of the lowpass filter with $x_e(n)$ is often performed using the DFT because of the efficiency of the FFT. As we shall see in Section 7.6, care must be used when implementing convolution with the DFT, but it is possible to use the DFT and thus the FFT even for input sequences of indefinite length (Section 7.7). Thus, the techniques of convolution using the DFT are of general value, and are the subjects of the following two sections.

7.6 APERIODIC CONVOLUTION ESTIMATION AND WRAPAROUND

It was shown in Section 2.4.3 that the output of a linear shift-invariant system is the convolution of the input with the system unit sample response. It was further shown in Section 5.7 that for such systems the convolution theorem holds: In the frequency domain, the convolution is transformed into multiplication. Because of the efficiency of the FFT, it is desirable to use the DFT to implement discrete-time convolution. In this section, we consider the case for which $x(n)$ and $h(n)$ are of respective lengths N_x and N_h, and have respective intervals of support (IOS) $n \in [0, N_x - 1]$ and $n \in [0, N_h - 1]$. The case of infinite- or indefinite-length $x(n)$ is considered in Section 7.7. Thus for this discussion we wish to compute

$$y(n) = x(n) * h(n) = \sum_{\ell=0}^{N_x-1} x(\ell)h(n - \ell), \qquad (7.45)$$

where for $n - \ell$ outside the range $[0, N_h - 1]$ we substitute zero in (7.45).

The obvious approach to computing (7.45) would be to appeal to the convolution theorem in its DFT form. It is proposed that (7.45) be estimated by (a) taking the DFT$_M$s of $x(n)$ and $h(n)$ with both zero-padded out to $M - 1$, where we assume at least that $M > \max\{N_x, N_h\}$, but the exact requirement on M is found below, (b) multiplying the DFT$_M$s, and (c) taking the DFT$_M^{-1}$ of the result. What is the inverse DFT$_M$ of $X(k) \cdot H(k)$?

$$\mathrm{DFT}_M^{-1}\{X(k) \cdot H(k)\} = \frac{1}{M} \sum_{k=0}^{M-1} X(k) \cdot H(k)\, e^{j2\pi nk/M}. \qquad (7.46a)$$

Writing $X(k)$ and $H(k)$ as forward DFT$_M$s of, respectively, $x(n)$ and $h(n)$,

$$\mathrm{DFT}_M^{-1}\{X(k) \cdot H(k)\} = \frac{1}{M} \sum_{k=0}^{M-1} \sum_{\ell=0}^{N_x-1} x(\ell)e^{-j2\pi \ell k/M} \sum_{i=0}^{N_h-1} h(i)\, e^{-j2\pi ik/M} e^{j2\pi nk/M}. \qquad (7.46b)$$

Collecting the exponentials together and reordering the sums,

$$\mathrm{DFT}_M^{-1}\{X(k) \cdot H(k)\} = \sum_{\ell=0}^{N_x-1} x(\ell) \sum_{i=0}^{N_h-1} h(i) \cdot \frac{1}{M} \sum_{k=0}^{M-1} e^{j2\pi k(n-\ell-i)/M}. \qquad (7.46c)$$

The third sum in (7.46c) is equal to M if $n - \ell - i = -mM$, $-m$ any integer—that is, if $i = n - \ell + mM$; it is zero otherwise. By (7.15b) $\left[\frac{1}{M} \sum_{k=0}^{M-1} e^{j2\pi ki/M} = \sum_{m=-\infty}^{\infty} \delta(i + mM)\right.$ with i here replaced by $n - \ell - i\Big]$, it thus follows that $(1/M)\sum_{k=0}^{M-1} e^{j2\pi k(n-\ell-i)/M} = \sum_{m=-\infty}^{\infty} \delta(n - \ell - i + mM)$. Upon substitution, (7.46c) becomes

$$\mathrm{DFT}_M^{-1}\{X(k) \cdot H(k)\} = \sum_{\ell=0}^{N_x-1} x(\ell) \sum_{i=0}^{N_h-1} h(i) \sum_{m=-\infty}^{\infty} \delta(n - \ell - i + mM) \qquad (7.46d)$$

$$= \sum_{m=-\infty}^{\infty} \sum_{\ell=0}^{N_x-1} x(\ell) \sum_{i=0}^{N_h-1} h(i)\, \delta(\{n + mM\} - \ell - i). \qquad (7.46e)$$

For each value of m, only one value of $i \in [0, N_h - 1]$, namely $i = \{n + mM\} - \ell$, makes the δ nonzero, because we assumed that $M > N_h$. Thus,

$$\mathrm{DFT}_M^{-1}\{X(k) \cdot H(k)\} = \sum_{m=-\infty}^{\infty} \sum_{\ell=0}^{N_x-1} x(\ell)h(\{n + mM\} - \ell), \quad n \in [0, M-1]. \qquad (7.46f)$$

As in (7.45), we substitute zero whenever $\{n - mM\} - \ell$ is not on $[0, N_h - 1]$. Comparing with (7.45), the mth term in the m sum in (7.46f) is $y(n + mM)$; thus,

$$\mathrm{DFT}_M^{-1}\{X(k) \cdot H(k)\} = \sum_{m=-\infty}^{\infty} y(n + mM), \qquad n \in [0, M-1]. \qquad (7.46g)$$

We now recall that the DFT_M produces sequences in both domains that are M-periodic, due to the M-periodicity of the exponentials $e^{\pm j2\pi kn/M}$. So far we have not made use of this fact. Let us call the M-periodic extensions of $x(n)$ and $h(n)$, respectively, $\tilde{x}_M(n)$ and $\tilde{h}_M(n)$. If we replace $h(\cdot)$ by $\tilde{h}_M(\cdot)$ in (7.46f),[14] we can use the periodicity of $\tilde{h}_M(n)$ to note that $\tilde{h}_M(n + mM - \ell) = \tilde{h}_M(n - \ell)$. Thus, we are able to dispense with mM and the m sum and write (7.46f) as

$$DFT_M^{-1}\{X(k) \cdot H(k)\} = \sum_{\ell=0}^{N_x-1} \tilde{x}_M(\ell) \cdot \tilde{h}_M(n - \ell), \quad \text{all } n. \tag{7.46h}$$

The reason we can eliminate the m sum is that the various values of m in (7.46f), via mM, select out from $n - \ell + mM$ a value on the range $[0, N_h - 1]$, which gives the same h value selected from the relevant period of $\tilde{h}$ in (7.46h). We can also use the form in (7.46h) for the finite-length versions $x(n)$ and $h(n)$, as long as the argument of h is taken to be $(n - \ell)_{\text{modulo } M}$. The convolution in (7.46h) is called periodic convolution. We see that the inverse DFT_M of the product of their DFT_Ms is just the usual convolutional sum of $\tilde{x}_M(n)$ and $\tilde{h}_M(n)$ except that the convolution sum index ℓ *runs over only one period*.

Let us again examine (7.46g). This infinite sum of replications of the thing we want [here $y(n)$] is reminiscent of the frequency-aliasing sum that arises in sampling a continuous-time signal $x_c(t)$ [where we want $X_c(j\Omega)$]. We now derive the condition for no time-aliasing of $y(n)$ in (7.46g).

The IOS of $y(n)$ is by definition that range of n for which $y(n)$ is possibly nonzero. If $n < 0$, then the argument $n - \ell$ of h in (7.45) must be negative, which is outside the range $[0, N_h - 1]$ of nonzero $h(n)$. Thus, $y(n < 0) = 0$. If $n > N_x + N_h - 2$, then the argument $n - \ell$ of $h(n)$ in (7.45) must be greater than $N_h - 1$ for which again $h(n)$ is zero. Therefore, the IOS of $y(n)$ is $[0, N_x + N_h - 2]$.

Now note that the separation of repetitions in (7.46g) is equal to the DFT size M—yet the length of a single y sequence is $N_x + N_h - 1$, which might exceed M—so there may be time-aliasing! We might call this convolutional aliasing, as opposed to sampling aliasing, the latter of which we previously considered exclusively. In applications, either or both types of aliasing may be present if precautions are not taken. We will, however, see below that convolution aliasing is in fact a form of sampling aliasing.

The replications of $y(n)$ begin at $n = mM, m = 0, \pm 1, \pm 2, \ldots$ For example, the $m = 1$ replication begins at $n = -M$ and ends at $n = -M + N_x + N_h - 2$. The $m = 0$ replication begins at $n = 0$ and ends at $n = N_x + N_h - 2$, and the $m = -1$ replication begins at $n = M$ and ends at $M + N_x + N_h - 2$ (all ranges are inclusive). To make sure that the $m = 1$ replication does not overlap onto the $m = 0$ replication, the ending point of the $m = 1$ replication must be negative: $-M + N_x + N_h - 2 < 0$, or $M \geq N_x + N_h - 1$. This inequality is the requirement to avoid time-aliasing.

Suppose for simplicity that $N_x = N_h = N$, and we naively choose $M = N$. Then the three IOSs for $m = 1, 0$, and -1 become $[-N, N - 2]$, $[0, 2N - 2]$, and $[N, 3N - 2]$. We see in this case that there is no overlap for only one point on any M-interval (e.g., neither $m = -1$ nor $m = 1$ includes $N - 1$). Thus, for the DFT range $n \in [0, N - 1]$, the only nonaliased point is $n = N - 1$, the last DFT point. Therefore, for this example in which the lengths of the DFT and the two sequences $x(n)$ and $h(n)$ are all N, we find that there is only one value of n, namely $N - 1$, for which $y(n)$ in (7.45) and $DFT_M^{-1}\{X(k) \cdot H(k)\}$ agree. This is hardly acceptable!

---

[14]We also replace $x(\ell)$ by $\tilde{x}_M(\ell)$, but this has no effect because ℓ ranges only over $[0, N_x - 1]$.

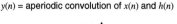

$y(n)$ = aperiodic convolution of $x(n)$ and $h(n)$

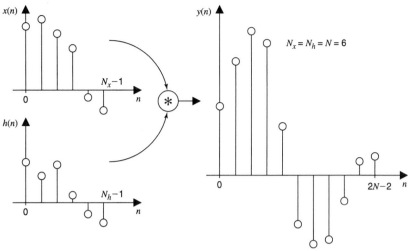

Figure 7.34a

If instead we choose for the DFT length $M \geq N_x + N_h - 1$ (in the above example, $M \geq 2N - 1$) and zero-pad both $x(n)$ and $h(n)$ out to $n = M - 1$ before taking their DFT_Ms, then there will be no overlaps, and we may write $y(n) = \text{DFT}_M^{-1}\{X(k) \cdot H(k)\}$ for $n \in [0, M - 1]$, which includes the entire range $[0, N_x + N_h - 1]$ for which $y(n) \neq 0$.

We now illustrate these ideas graphically. Figure 7.34a shows a typical aperiodic convolution $y(n)$ (7.45) for $N_x = N_h = N = 6$, for typical (though very short) $x(n)$ and $h(n)$. In Figure 7.34b are shown the contributions to the periodic convolution sum (7.46g) for $M = N$. The $m = 1$ term is shown in black dots, $m = 0$ is in white dots, and $m = -1$ is in asterisks. We see that (a) $n = N - 1$ is the only point on $[0, N - 1]$ that is not aliased (has only one contributing term), and (b) the sum of contributions repeats at $n = \ell M = \ell N$.

For $M = N$, we see that because $y(n)$ is nonzero on only $[0, 2N - 2]$, for any given n only at most *two* values of m will bring the argument of y in (7.46g) into its IOS $[0, 2N - 2]$. For example, for $n \in [0, N - 2]$ the two contributing replications are $m = 0$ (its first half) and $m = 1$ (its second half), whereas for $n \in [N, 2N - 2]$, the two contributing replications are $m = 0$ (its second half) and $m = -1$ (its first half). (In cases of more severe aliasing, many more replications may contribute at a given point.) Because the end values of $y(n)$ (for $n \geq M - 1$) are replicated back into the beginning of the aliased result, the phenomenon is called "wraparound." Upon summation of the replications shown in Figure 7.34b, the aliased result of periodic convolution for $M = N$ is shown in Figure 7.34c. Figure 7.34d shows the periodic convolution for $M = 13 > 2N - 1$, in which case there is no aliasing.[15]

Finally, note that from (7.46h), the estimate of aperiodic convolution is M-periodic in n, as it must be, because it is an inverse DFT_M. In (7.46h), shifting n by M is equivalent to shifting m by 1 in (7.46g). The aliasing sum thus "accomplishes" this M-periodicity out of the aperiodic $y(n)$. Symbolically, we may write

---

[15] The values of M chosen in Figure 7.34, namely 6 and 13, would of course in practice always be chosen to be powers of 2; they were chosen here for expedience in graphically showing the periodic convolution effects.

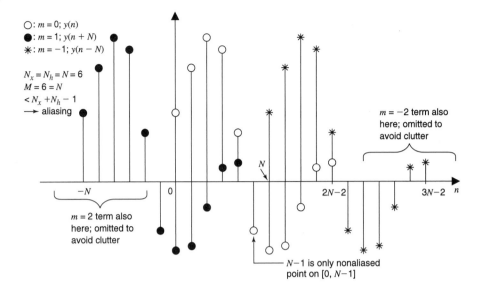

Figure 7.34b

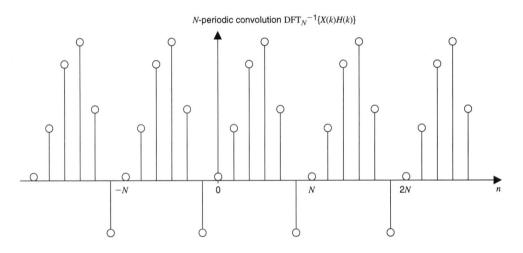

Figure 7.34c

$$x(n) *_M h(n) \Big|_n = \sum_{m=-\infty}^{\infty} \{x(n) * h(n)\} \Big|_{n+mM}, \qquad (7.47)$$

where $*_M$ indicates M-periodic convolution obtained from the M-point DFT^{-1} of the DFT$_M$s of x and h, whereas $*$ indicates normal aperiodic convolution. Note that (7.47) is just an alternative way of writing (7.46g).

The M-periodic convolution is also abbreviated "periodic convolution" or called circular convolution, whereas the usual convolution in (7.45) is called aperiodic convolution. The name circular convolution stems from the periodic convolution result being the same as the one obtained if we take a single period of each of $\tilde{x}(n)$ and $\tilde{h}(-n)$, zero-pad them out to $M - 1$, wrap them in concentric circles, and

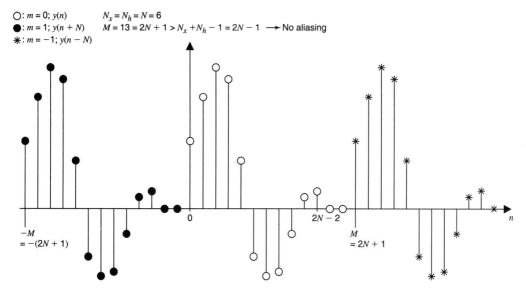

Figure 7.34d

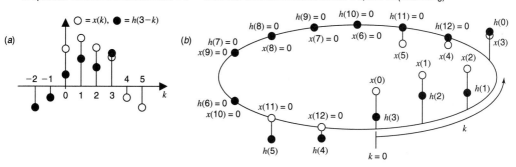

Figures 7.35a and b

sum the product of the matching pairs. The next value in the periodic convolution can be found merely by shifting one of the ringed sequences (circles) one point with respect to the other and again forming the sum of products.

This idea of circular convolution is illustrated in Figure 7.35. In Figure 7.35a are shown, for the same $x(n)$ and $h(n)$ as in Figure 7.34a, the terms involved in the aperiodic convolution sum (7.45) for $n = 3$. The corresponding terms in circular convolution (7.47) as implemented using the DFT in (7.46h) are shown in Figure 7.35b for $M = 13$ and in Figure 7.35c for $M = N = 6$. For $M = 6$, the erroneous aliasing terms in circular convolution not appearing in the aperiodic convolution are identified. All such terms are zero when $M \geq N_x + N_h - 1$ as in Figure 7.35b.

Upon reflection, the circular convolution/wraparound phenomenon (7.46g) is seen to be merely a particular case of (7.16b) $[\text{DFT}_N^{-1}\{X_{N,a}(k)\} = \sum_{m=-\infty}^{\infty} x(n + mN)]$, where x in (7.16b) plays the role of y in (7.46g). Recall that (7.16b) quantifies the aliasing that can occur when estimating the DTFT^{-1} based on N samples of the DTFT of that sequence, $X_{N,a}(k) = X(e^{j2\pi k/N})$. In (7.16b), time-aliasing occurs if the length P of $x(n)$ exceeds M, the DFT length.

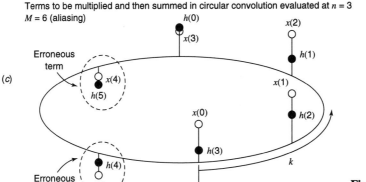

Terms to be multiplied and then summed in circular convolution evaluated at $n = 3$

Figure 7.35c

We know that if $y(n) = x(n) * h(n)$, then $Y(e^{j\omega}) = X(e^{j\omega}) \cdot H(e^{j\omega})$. By taking the M-point DFT of $x(n)$, we obtain M values of $X(e^{j\omega})$, at $\omega = 2\pi k/M$, and similarly for $H(e^{j\omega})$. Thus, *regardless of the value of M* [as long $M \geq \max(N_x, N_h)$], $X(k)H(k)$ gives M exact samples of $Y(e^{j\omega})$. The length of $y(n)$ is $N_x + N_h - 1$. We see that we already could have predicted from (7.16b) that there will not be aliasing if $M \geq N_x + N_h - 1$ and that otherwise there will be aliasing ("wraparound"). We conclude that convolutional aliasing is nothing more than a different view of time-aliasing, where the length of our time sequence, $y(n)$, may exceed the number of its DTFT samples we have [produced by $X(k)H(k)$].

EXAMPLE 7.10

Consider the simplest possible example: $x(n)$ and $h(n)$ are unity for $0 \leq n \leq N_x - 1$ and $0 \leq n \leq N_h - 1$, respectively, and zero otherwise. They are respectively shown zero-padded out to $N = 128$ in Figures 7.36a and b, where $N_x = 100$ and $N_h = 20$. Plot the aperiodic convolution $y(n) = x(n) * h(n)$, and investigate use of the DFT to reproduce it, pointing out convolution wraparound (aliasing) for values of M for which it occurs.

Solution

The aperiodic convolution is shown zero-padded out to $N = 128$ in Figure 7.37 and may be verified to be expressible as

$$y(n) = \begin{cases} n+1, & 0 \leq n \leq N_h - 1 \\ N_h, & N_h \leq n \leq N_x - 1 \\ N_x + N_h - 1 - n, & N_x \leq n \leq N_x + N_h - 2 \\ 0, & \text{otherwise.} \end{cases}$$

If we take the M-point DFT^{-1} of the product of the M-point DFTs of $x(n)$ and $h(n)$, the result is shown in Figure 7.38a ($M = 128$) and Figure 7.38b ($M = 110$). In the first case, Figure 7.38a, $M = 128 > N_x + N_h - 1 = 119$, whereas the condition is violated in Figure 7.38b (even though $M = 110$ is greater than both N_x and N_h). Therefore, Figure 7.38a is identical to Figure 7.37.

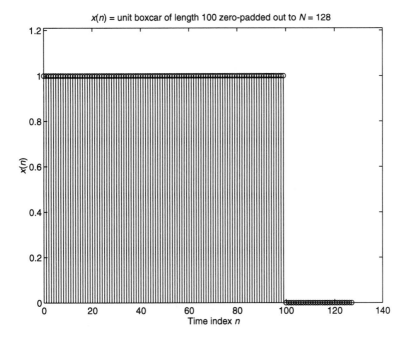

Figure 7.36a

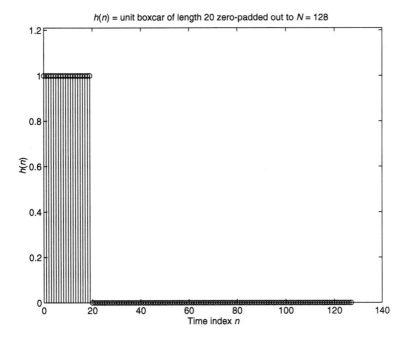

Figure 7.36b

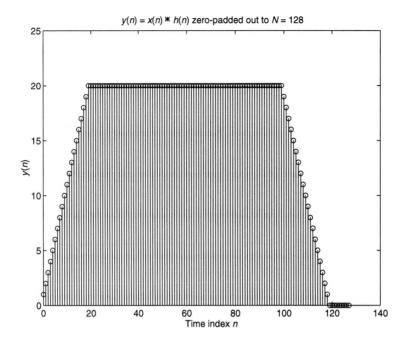

Figure 7.37

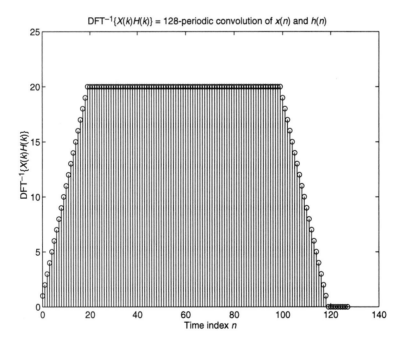

Figure 7.38*a*

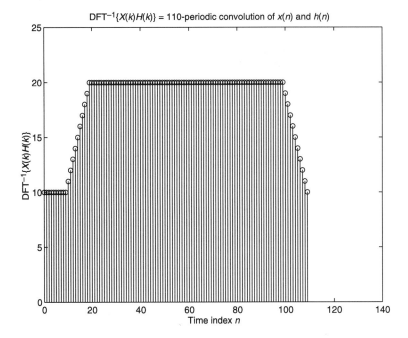

Figure 7.38b

However, Figure 7.38b is quite different for the first $N_x + N_h - 1 - M = 9$ points; they are considered erroneous for purposes of implementing aperiodic convolution with the DFT. All the rest of the points are in perfect agreement with aperiodic convolution; however, because $M < N_x + N_h - 1$, the last nine values of $y(n)$ are missing in Figure 7.38b. Those "missing" $y(n)$ values are precisely the wrapped-around values that corrupted the first 9 DFT convolution values! We may mentally add replications of $y(n)$ separated by $M = 110$ and add them together to verify Figure 7.38b. Note that the minimum value of Figure 7.38b, unlike the other figures, is not zero. On the left, it results from adding portions of two overlapping replications ($m = 0$ and $m = 1$) having opposite slopes to form a constant with respect to n. At the end of the sequence, merely note that it agrees with the aperiodic convolution value $y(110)$. Thus, because the two sequences being convolved are causal, the aliasing occurs at the beginning of the sequence.

7.7   BLOCK CONVOLUTION WITH THE DFT

We have just seen that if we choose the length of the DFT sufficiently large, periodic and aperiodic convolutions will be the same in the region of support of the aperiodic convolution. However, the required size may in practice be much too large to be handled by the given hardware. An obvious example would be data coming in continuously for an indefinite period. It is our goal in this section to decompose the aperiodic convolution of an indefinitely long (data) sequence with a finite length (filter) sequence in such a way that the DFT may be used.

The focus shall be on one of two popular methods: the *overlap-save* method (the other is called *overlap add* and is investigated in the problems). Using the results of the previous section, we will obtain this method of block convolution, which yields a continuous stream of exactly correct output values yet uses only N-point DFTs even though $N << N_x$, where N is selected by the user (we still require $N > N_h$). All that is required are that the data segments overlap and that we save the correct points— hence the name. It is easiest to understand this method by example.

We will consider a very small DFT size to more easily graphically depict the required processing for block convolution. Suppose that the available DFT length is $N = 32$. Define the infinite-length sequence $\tilde{x}(n)$ as the weighted sum of four sinusoids from which we will extract one period:

$$\tilde{x}(n) = \sum_{\ell=1}^{4} a_\ell \sin(2\pi f_\ell n + \theta_\ell),$$

where $a_1 = 2$, $a_2 = -1.5$, $a_3 = 0.7$, $a_4 = 1$, $f_1 = 1/N_x$ where $N_x = 5N = 160$ (thus indeed, $N << N_x$), $f_2 = 4/N_x$, $f_3 = 1/8 = 0.125$, $f_4 = 1/4 = 0.25$, $\theta_1 = 0$, $\theta_2 = \pi$, $\theta_3 = 0$, and $\theta_4 = \pi$. Suppose that $x(n)$ is obtained as one period of $\tilde{x}(n)$, that is, $x(n) = \tilde{x}(n)$ for $0 \leq n \leq N_x - 1$ and $x(n) = 0$ otherwise. Figure 7.39 shows a plot of $x(n)$ (notice for reference below that $x(0) = 0$). It is desired to filter out the two higher-frequency sinusoids $\ell = 3$ and $\ell = 4$ using a filter of length $N_h = 8$ ($< N$). First investigate the filtering process using the methods of Section 7.6 (all in one block), and then perform aperiodic convolution in sections.

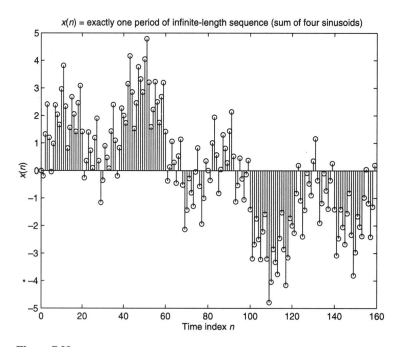

Figure 7.39

Solution

We begin by first considering the aperiodic convolution of $x(n)$ with $h(n)$. According to the discussion in Section 7.6, our DFT length must be at least $N_x + N_h - 1$ to avoid time-aliasing in the convolution result. However, forming the required DFTs of such length involves zero-padding, which implies a rectangular truncation of $\tilde{x}(n)$, from which $x(n)$ is extracted. The truncation smoothes the spectrum, and the resulting DTFT is shown in Figure 7.40a. For the DTFT of four pure sinusoids $\tilde{x}(n)$, we would expect four positive-frequency Dirac delta functions and for the DFT we would expect four Kronecker deltas; instead we have four smooth peaks in the DTFT of $x(n)$. This effect is sometimes called spectral leakage, because the energy exclusively in the original sinusoids has "leaked" into nearby frequencies. Spectral leakage also occurs without zero-padding if the array $x(n)$ given to the DFT does not include an integral number of periods of $\tilde{x}(n)$.

If instead of zero-padding we take the N_x-point DFT of $x(n)$, we obtain Figure 7.40b. Because exactly one period of $\tilde{x}(n)$ is contained within the sequence given to the DFT, we obtain pure lines (Kronecker deltas). Specifically, for the finite-length DFT sum, we obtain the discrete Fourier series coefficients, which are represented by Kronecker delta functions rather than Dirac deltas. In Figure 7.40b, we label the ordinate by $|X(k)|$ where the digital frequency f corresponding to k is $f_k = k/N$, or $k = Nf_k$ where for now we let $N = N_x = 160$ (below, we will set $N = 32$ for the overlap save method). We wish to filter out the lines at $f = 0.125$ and $f = 0.25$.

The digital filter $h(n)$ is designed to pass the two lower-frequency sinusoids and attenuate the higher-frequency sinusoids. In the case of a periodic sequence having an integral number of periods in the time window, we again could get away with just multiplying the stopband samples by zero and the passband samples by unity, and obtain perfect results. That is because all the energy is confined to the values of k corresponding to the sampled sinusoids, when an integral number of periods is included. If $X(e^{j\omega})$ had energy for any $\omega \neq \omega_k$, then the multiplying-samples idea would incur errors because of the ripples in $H(e^{j\omega})$ in between ω_k (see Section 9.4).

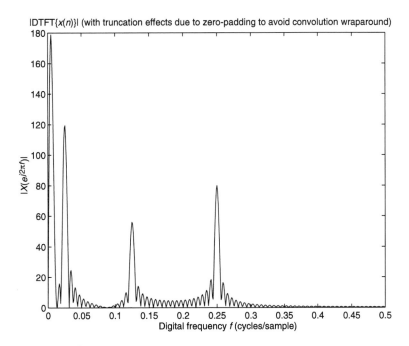

IDTFT{x(n)}I (with truncation effects due to zero-padding to avoid convolution wraparound)

Digital frequency f (cycles/sample)

Figure 7.40a

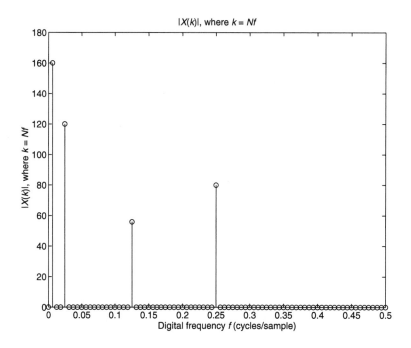

Figure 7.40*b*

For the sake of variety, we will now instead use the McClellan-Parks algorithm, which is discussed in Section 9.5. The details of that design procedure need not concern us here, except to note that it can be very efficient in cases other than this ideal situation of a periodic function with an integral number of periods in the time window. A compromise on passband and stopband is necessary for the very low order $N_h = 8$. The passband and stopband edges the McClellan-Parks algorithm requires were adjusted so that two of the nulls of the DTFT $H(e^{j\omega})$ occur fairly near f_3 and f_4. In Figure 7.41 we show a scaled version of the DTFT of Figure 7.40*b* with $|H(e^{j2\pi f})|$ superimposed. We do see that at least approximately, the lower-frequency sinusoids will be passed and the higher-frequency sinusoids will be attenuated.

If we zero-pad $x(n)$ and $h(n)$ out to $N = N_x + N_h - 1 = 167$ and multiply their N-point DFTs, the result $Y(k)$ is as shown in Figure 7.42*a*. From Figure 7.40*a* we expect the smeared spectrum rather than the pure lines of 7.40*b* or 7.41, due to the truncation resulting from zero-padding. Nevertheless, we see in Figure 7.42*a* that the high-frequency sinusoids are quite attenuated, whereas the lower-frequency sinusoids are mostly passed.

If we take the 160-point DFT of the true aperiodic convolution of $x(n)$ [i.e., one period of $\tilde{x}(n)$] with $h(n)$, we obtain the result shown in Figure 7.42*b*. It might at first be surprising that we do not have pure sinusoidal lines (though Figure 7.42*b* is much closer to pure lines than is Figure 7.42*a*), because the DFT length is exactly equal to one period of $\tilde{x}(n)$ and therefore one period of $\tilde{y}(n) = \tilde{x}(n) * h(n)$. That is, regardless of the length of $h(n)$, if $\tilde{x}(n)$ is N_x-periodic, then so will be $\tilde{y}(n) = \tilde{x}(n) * h(n)$. However, these inferences leave out the fact that the aperiodic convolution $y(n) = x(n) * h(n)$ is *not* 160-periodic; $y(n)$ is aperiodic and has length $N_x + N_h - 1 = 167$. Only the convolution of the infinite-length periodic $\tilde{x}(n)$ is periodic and with $h(n)$ is 160-periodic. Because of the truncation to only one period in $x(n)$, we are missing some terms in the $\tilde{y}(n)$ convolution sum near the ends of the "period" $[0, 159]$ due to the required $\tilde{x}(n)$ values being outside those retained in our finite-length single-period $x(n)$.

These end effects can be eliminated by defining $x'(n)$ to be several periods of $\tilde{x}(n)$, carrying out the aperiodic convolution $y'(n) = x'(n) * h(n)$, and taking the

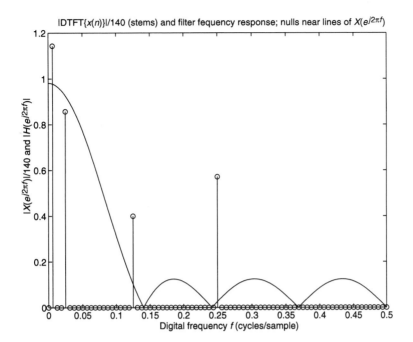

Figure 7.41

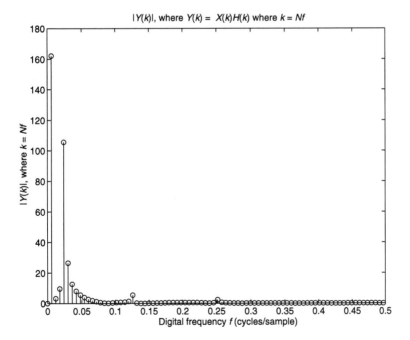

Figure 7.42*a*

160-point DFT of one centrally located "period" in $y'(n)$. The result is the pure lines spectrum in Figure 7.42*c*, which again shows the effects of filtering. [Given that when $x(n)$ is periodic, the result of normal "aperiodic" convolution is periodic, perhaps we need a new name, such as "convolution without DFT-imposed periodicity"!]

The 167-point DFT^{-1} of $Y(k) = X(k)H(k)$, $y(n)$, is shown in Figure 7.43 (stems and "O"). Also plotted on the same range of n (* stems) is $\tilde{y}(n)$, which is computed using the extension $x'(n)$ to ensure that no convolution sum terms are omitted on

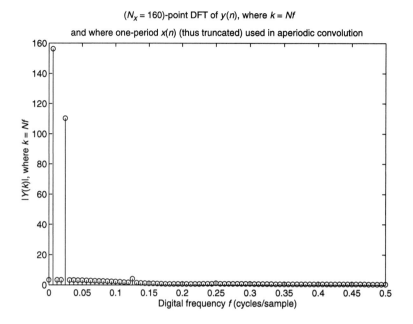

$(N_x = 160)$-point DFT of $y(n)$, where $k = Nf$
and where one-period $x(n)$ (thus truncated) used in aperiodic convolution

Figure 7.42*b*

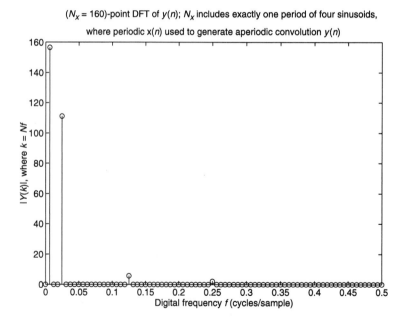

$(N_x = 160)$-point DFT of $y(n)$; N_x includes exactly one period of four sinusoids,
where periodic $x(n)$ used to generate aperiodic convolution $y(n)$

Figure 7.42*c*

the ends. We see perfect agreement except for the first and last $N_h - 1$ points. Disagreement is expected, due to the difference between Figures 7.42a and c.

It must be recognized that because we selected the DFT length to be $N = N_x + N_h - 1$, there is no convolutional (time-)aliasing; aliasing is not the cause of the discrepancy in Figure 7.43. Rather, there is disagreement for the first $N_h - 1$ terms because $x(m) = 0$ for $m < 0$, whereas this is not true of $\tilde{x}(m)$ or $x'(m)$; nonzero terms in the convolutional sum for $\tilde{y}(n)$ are zero in the sum for $y(n)$, and thus $y(n) \neq \tilde{y}(n)$ at the low end of n. The disagreement at the high end of n occurs because we limited $x(n)$ to be

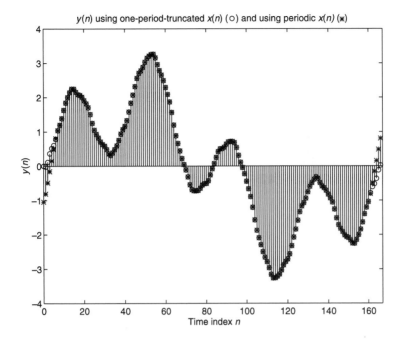

$y(n)$ using one-period-truncated $x(n)$ (○) and using periodic $x(n)$ (∗)

Figure 7.43

nonzero only out to $N_x - 1$, which is less than $N = N_x + N_h - 1$ (to avoid time-aliasing). Consequently, for the last $N_h - 1$ points, again $\tilde{x}(n)$ is nonzero but $x(n) = 0$. These missing values are called in $\tilde{y}(n)$ for $N_x \le n \le N_x + N_h - 1$ [they are drawn from the next period of $\tilde{x}(n)$, and help form part of the next period of the N_x-periodic $\tilde{y}(n)$].

Nevertheless, henceforth in our discussions, we will designate the aperiodic convolution $y(n)$ as the "true" solution, not $\tilde{y}(n)$. We do so because we are interested in checking aperiodic convolution using the block approach versus using either the convolutional sum or an all-in-one large DFT approach (the latter two are equal if $N \ge N_x + N_h - 1$). The main reason we chose the original signal $\tilde{x}(n)$ as being periodic is because the lines-spectrum makes the filtering action easy to see in both the time and frequency domains.[16] It is clear that $y(n)$ in Figure 7.43 is a smoothed version of $x(n)$ in Figure 7.39; indeed, in Figure 7.43 we can see four complete "periods" of f_2 and one complete "period" of f_1 that have escaped annihilation.

Let us now consider the block convolution approach to obtaining $y(n)$. In the overlap-save method, we break $x(n)$ into sections of length $N = 32$, sections much smaller than the total length of $x(n)$, $N_x = 160$. We somewhat arbitrarily choose the nonzero data to begin at $n = N_h - 1$ rather than at $n = 0$ so that the valid output points will begin at the same value of n for which $x(n)$ "begins" (see below).

We now break up the data sequence $x(n)$ into sections indexed by the integer k, denoted $x_k(n_k)$. Figure 7.44 shows all the segments $k = 0$ through $k = 6$; in practice there may be an indefinite number of segments. Notice that the first $N_h - 1$ data points of segment $k + 1$, $[0, N_h - 2]$, are chosen to overlap the last $N_h - 1$ data points of segment k, $[N - N_h + 1, N - 1]$. Thus, the data sections and indexes are subscripted by k: $x_0(n_0) = x(n_0)$, $x_1(n_1) = x(n_1 + N - N_h + 1)$, $x_k(n_k) = x(n_k + k[N - N_h + 1])$, where $0 \le n_k \le N - 1$ for all k. Thus the range of actual time index n corresponding to $x_k(n_k)$ is $n \in [k(N - N_h + 1), k(N - N_h + 1) + N - 1]$.

---

[16]Another reason is the insights the previous discussion may afford.

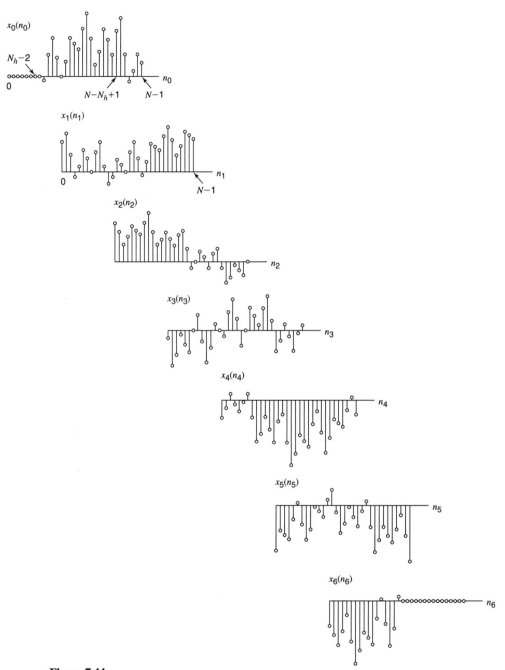

Figure 7.44

Figure 7.45a shows the $k = 2$ aperiodic convolution $(x_2 * h)(n_2)$, and Figure 7.45b shows the corresponding N-periodic convolution result $(x_2 *_N h)(n_2)$. Because the length of h is N_h and the length of x_k is N, their aperiodic convolution extends from $n_k = 0$ to $n_k = N + N_h - 1$. Contrarily, the interval of repetition of aperiodic convolution results in the N-periodic convolution is N, which is less than $N + N_h - 1$.

If we periodically extend the aperiodic section convolution, with replications beginning at $n_k = -N, n_k = 0, n_k = N$, and so forth, we see that the first $N_h - 1$ points in any replication are aliased by the last $N_h - 1$ points of the left-adjacent replication. Equivalently, the last $N_h - 1$ points in the aperiodic convolution in Figure 7.45a

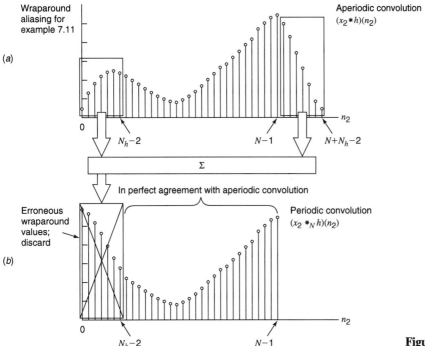

Figure 7.45

are "wrapped around" to add to the first $N_h - 1$ points. (We saw this effect before in Figure 7.38b of Example 7.10.) Therefore, we will discard the first $N_h - 1$ points in the aperiodic convolution, $[0, N_h - 2]$. The remaining points $n_k \in [N_h - 1, N - 1]$ are not aliased and are equal to the desired aperiodic convolution values. Therefore, there are $(N - 1) - (N_h - 1) + 1 = N - N_h + 1$ points of agreement per segment.

What are the corresponding values of n? Does there result an uninterrupted flow of valid aperiodic convolution values? Following the pattern in the definition of segments 0, 1, and 2 above we see that the set of points of agreement $n_k \in [N_h - 1, N - 1]$ corresponds to the following set of original time index values:

$$n \in [N_h - 1 + k(N - N_h + 1), N - 1 + k(N - N_h + 1)]$$
$$= [kN - (k - 1)(N_h - 1), (k + 1)N - k(N_h - 1) - 1]. \tag{7.48}$$

By substituting $k - 1$ for k in (7.48) we find that the points of agreement for segment $k - 1$ are

$$n \in [(k - 1)N - (k - 2)(N_h - 1), kN - (k - 1)(N_h - 1) - 1]. \tag{7.49}$$

Notice that the last valid point in segment $k - 1$ (7.49) is one less than the first valid point in segment k (7.48)! The points of agreement for segment $k + 1$ are

$$n \in [(k + 1)N - k(N_h - 1), (k + 2)N - (k + 1)(N_h - 1) - 1]. \tag{7.50}$$

The first valid point in segment $k + 1$ (7.50) is one higher than the last valid point in segment k (7.48). Thus, if the valid points from each segment are abutted, the result will be a continuous stream of valid aperiodic convolution values. One may also explicitly verify that the values of $x(m)$ contributing to $y(n)$ in the aperiodic sum kth segment are precisely those $x(m)$ values in its kth segment.

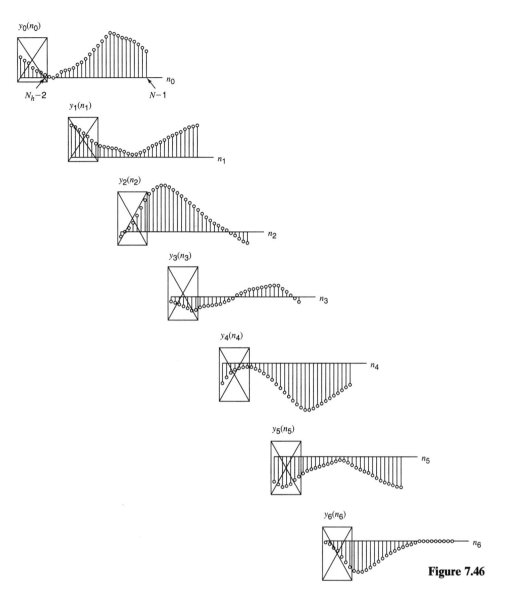

Figure 7.46

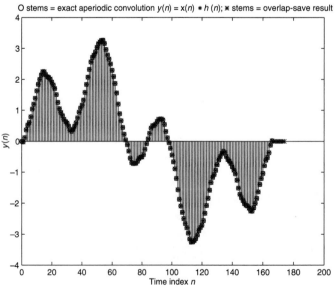

O stems = exact aperiodic convolution $y(n) = x(n) * h(n)$; ✳ stems = overlap-save result

Figure 7.47

488

The N-periodic convolutions for each of the seven segments are shown in Figure 7.46. For each segment, the discarded values are boxed with an "X" through them. The remaining, correct values are abutted together, resulting in the overlap-save-calculated true aperiodic convolution shown in Figure 7.47 in * stems. Also shown in Figure 7.47 (O stems) is the result obtained by normal aperiodic convolution; the results are identical (except for the zero-padding at the end of the overlap-save result).

As an example of an application of block convolution in a practical computational situation, the problem of two-dimensional convolutions for forward and inverse ultrasonic scattering is explored on WP 7.3. The computation of ultrasonic fields within an object to be imaged is a necessary task for quantitative imaging algorithms. The computational reductions of the FFT have been realized by adapting DFT convolution to the aperiodic convolutions appearing in the scattering equations[17]. On WP 7.3, some interesting and perhaps surprising details of this application of DFT convolution are presented and illustrated with graphics. We show here in Figure 7.48 the final results: the real and imaginary parts of the actual convolutions carried out in the spatial (above) and DFT (below) domains for an 11 $\times$ 11 grid and DFT size of 32. The aperiodic (spatial domain) and periodic (DFT domain) results agree exactly in the region of interest (object region, whose boundary is shown in bold). All garbage values from the DFT method outside the object region may be discarded. The computational complexity of the inverse scattering algorithm has thereby been reduced from N^5 to $N^3\log_2(N)$, making larger and more practically sized problems within reach of present-day computers.

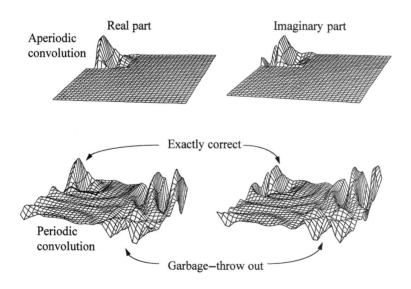

Figure 7.48

---

[17] T. J. Cavicchi and W. D. O'Brien, Jr., "Numerical Study of Higher-Order Diffraction Tomography via the Sinc Basis Moment Method," *Ultrasonic Imaging*, 11 (1989): 42–74; T. J. Cavicchi, S. A. Johnson, and W. D. O'Brien Jr., "Application of the Sinc Basis Moment Method to the Reconstruction of Infinite Circular Cylinders," *IEEE Trans. on Ultrasonics, Ferroelectrics, and Frequency Control*, 35, 1 (1987): 22–33. See these publications for additional references.

7.8 APERIODIC CORRELATION ESTIMATION

Discrete (cross-) correlation $r_{xh}(n)$ of x and h is defined by (as in Section 5.8, we here omit $1/N$ or other scaling):[18]

$$r_{xh}(k) = \sum_{m=-\infty}^{\infty} x(m+k) \cdot h^*(m). \qquad (7.51)$$

The correlation of x and h plays an important role in many data analysis situations, particularly in statistical analysis and the study of the effect of LSI systems on random sequences. It will come up repeatedly in our studies of random processes, signal modeling, and random signal processing in Chapter 10. In stochastic analysis, (7.51) represents an approximation of the true cross-correlation sequence to the degree of the validity of the assumption of ergodicity, for which statistical averages may be replaced by temporal averages (see Section 10.2 for details and definitions).

What is the DTFT of $r_{xh}(k)$ in terms of $X(e^{j\omega})$ and $H(e^{j\omega})$? We solved this problem in Example 5.23. The result from (5.46) is repeated here:

$$R_{xh}(e^{j\omega}) = X(e^{j\omega})H^*(e^{j\omega}). \qquad (7.52)$$

Thus, the only difference in the frequency domain from the convolution theorem is that one of the transforms is conjugated.

An important special case of (7.51) is where $h = x$, giving the autocorrelation formula:

$$r_{xx}(k) = \sum_{m=-\infty}^{\infty} x(m+k) \cdot x^*(m), \qquad (7.53a)$$

for which the DTFT is

$$\text{DTFT}\{r_{xx}(k)\} = R_{xx}(e^{j\omega}) = |X(e^{j\omega})|^2. \qquad (7.53b)$$

In the time domain, the differences between convolution and correlation are (a) the conjugation of h in correlation [(7.51)] and (b) more important, the argument of x, which in correlation is $k + m$ rather than $n - m$ for convolution. Note that $r_{hx}(k) = r_{xh}^*(-k)$, whereas with convolution it is immaterial which function gets the shifted $(n - m)$ argument. Notice that in correlation, unlike in convolution, $x(n)$ is not time-reversed when shifted and placed against $h(n)$ for multiplication and summation.

This last fact leads to a slightly different method when using the DFT_N to compute correlation. If x is of length N_x and h is of length N_h, as with convolution the length of $r_{xh}(k)$ is $N_x + N_h - 1$. As for convolution, N (the DFT size, which for convenience in clarifying wraparound we called M in Section 7.6) must be chosen greater than or equal to $N_x + N_h - 1$ to avoid aliasing, because $r_{xh}(k)$ is replicated every N points as discussed for convolution earlier.

However, if the IOSs of x and h are, respectively, $[0, N_x - 1]$ and $[0, N_h - 1]$, the IOS of $r_{xh}(k)$ is $[-(N_h-1), N_x - 1]$ instead of $[0, N_x + N_h - 2]$ as for convolution. Thus, using the DFT_N with x zero-padded from N_x to $N - 1$ and h zero-padded from N_h to $N - 1$ will result in the negative index portion of r_{xh} $[-(N_h - 1), -1]$ being placed at the end of the DFT^{-1} result, namely the interval $[N - (N_h - 1), N - 1]$, due to the N-periodicity of the DFT_N.

---

[18]Do not confuse the time-lag index k in $r_{xh}(k)$ with the DFT index k; if referencing the index of $\text{DFT}\{r_{xh}(k)\}$, use a different DFT index (e.g., ℓ).

This reordering is no problem if we are willing to mentally shift the last portion of the sequence to negative k, but we can instead shift x to obtain a contiguous result, which is easier to view. If we zero-pad x at its *beginning*, $[0, N - N_x - 1]$, and append the original x, then r_{xh} will have IOS $[N - \{N_x + N_h - 1\}, N - 1]$ and will be prepadded with zeros from 0 to $N - (N_x + N_h)$. Here we assume the usual post-zero-padding of $h(n)$. Let us compare the actual interval of support $k \in [-(N_h - 1), N_x - 1]$ with the range obtained by the DFT method. Because Matlab does not allow index 0, r_{xh} from the DFT will have Matlab array index IOS $k_M \in [N + 1 - \{N_x + N_h - 1\}, N]$. Thus, $k_M = N$ from the Matlab r_{xh} array obtained using the DFT method maps to the actual $k = N_x - 1$, and $k_M = N + 2 - N_x - N_h$ maps to actual $k = -(N_h - 1)$. For these and all values of $k_M \in [1, N]$, the actual k is obtained from k_M as $k = k_M + N_x - N - 1$.

EXAMPLE 7.12

Using the same $x(n)$ and $h(n)$ considered in the aperiodic/periodic convolution example, Example 7.10, compute the aperiodic correlation, and investigate reproduction of it by the DFT.

Solution

The aperiodic correlation is shown in Figure 7.49 and has the form

$$r_{xh}(k) = \begin{cases} N_h + k, & -(N_h - 1) \leq k \leq -1 \\ N_h, & 0 \leq k \leq N_x - N_h \\ N_x - k, & N_x - N_h + 1 \leq k \leq N_x - 1 \\ 0, & \text{otherwise.} \end{cases}$$

Again, $N_x = 100$ and $N_h = 20$. The DFT correlation with $x(n)$ post-zero-padded is shown in Figure 7.50. It agrees with the aperiodic Figure 7.49, only when we recall that the negative values of k in Figure 7.49 appear in the high values of k in the DFT

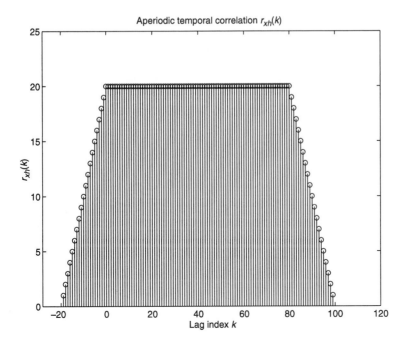

Aperiodic temporal correlation $r_{xh}(k)$

Figure 7.49

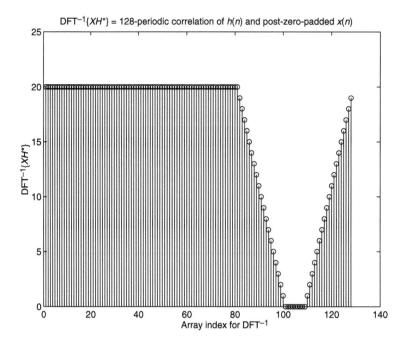

Figure 7.50

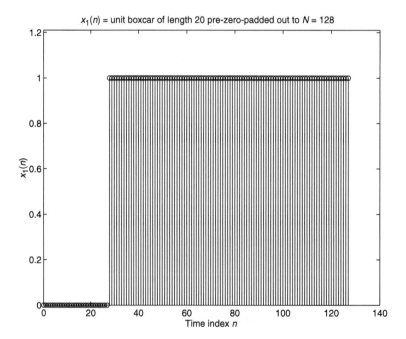

Figure 7.51

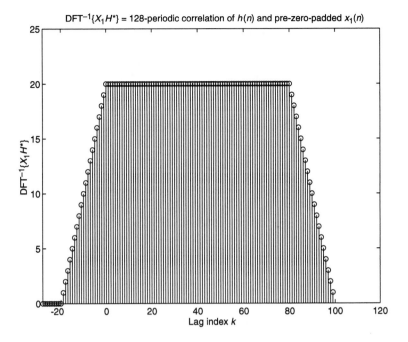

Figure 7.52

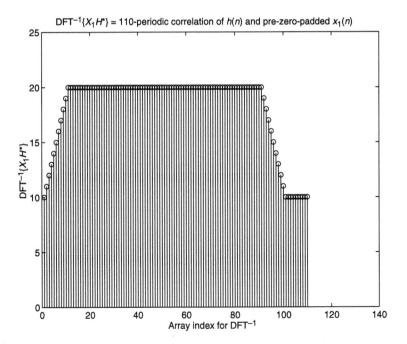

Figure 7.53

result of Figure 7.50. To make the DFT result visually similar to the true aperiodic correlation in Figure 7.49, we pre-zero-pad $x(n)$ as discussed above. With the prepadding with zeros, $x(n)$ is as shown in Figure 7.51 and is designated $x_1(n)$. The input $h(n)$ is postpadded with zeros in the usual manner.

The periodic correlation obtained from $\mathrm{DFT}_N^{-1}\{\mathrm{DFT}_N\{x_1(n)\} \cdot [\mathrm{DFT}_N\{h(n)\}]^*\}$ is shown in Figure 7.52 for the case $N = 128$. Notice as discussed above that the aperiodic correlation may be found shifted such that the first nonzero value occurs at $k_M = N - N_x - N_h + 2 = 128 - 100 - 20 + 2 = 10$, as opposed to the case of DFT convolution, where the first nonzero value is located at $k = 0$ (i.e., $k_M = 1$). For Figure 7.52, we use $k = k_M + N_x - N - 1$ to generate the abscissa values; the exact match with k in Figure 7.49 is gratifying. Finally, the case $N = 110$ is shown in Figure 7.53. Notice the aliasing, which takes place at the *end* of the sequence (as opposed to wraparound for periodic convolution, where it occurs at the beginning; see Figure 7.38*b*). Also, the leftmost values of the aperiodic correlation are missing just as the rightmost values were for DFT convolution with wraparound in Figure 7.38*b*.

EXAMPLE 7.13

Using the data in Figure 1.4, estimate the time lag from daylight hours to resulting average high temperature. Use DFT correlation.

Solution

We use the same procedure as in Example 7.12, with $x(n)$ = average high temperature and $h(n)$ = daylight hours. However, to remove the artifact of having only one year of data and the artificial cutoff to zero, we subtract off the minimum from each of $x(n)$ and $h(n)$ before correlating. The resulting correlation sequence is shown in Figure 7.54 (with so many sequence values, we show only the lines connecting points). We use a numerical maximum finder and then $k_{\max} = k_{M,\max} + N_x - N - 1$ to obtain $k_{\max} = 31$ days, precisely one month of delay from daylight hours to result-

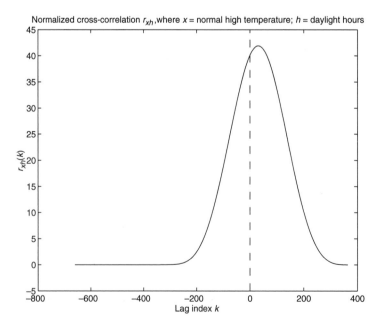

Figure 7.54

ing temperature. Note that when $k_{max} > 0$, $x(n)$ is delayed relative to $h(n)$, because $k_{max} > 0$ means that $x(n)$ must be advanced to most closely match $h(n)$. Also, we used $N = 1024$; notice the zero-padding at the beginning (most negative k).

7.9 CEPSTRUM ("HOMOMORPHIC") PROCESSING

Essentially all our work so far has been based on linear shift-invariant processing. The linear superposition of different signals has made input-output and filtering operations straightforward. In particular, if a signal has been corrupted with additive noise and the noise is on a different frequency band from that of the signal, we can filter out the noise by multiplying the spectrum of the distorted signal by zero on the noise band and unity on the signal band. However, what if the corruption of the signal is not additive, but convolutional?

Recall that if $x(n)$ is convolved with $h(n)$, the output $y(n) = x(n) * h(n)$ has z-transform

$$Y(z) = X(z)H(z). \tag{7.54}$$

Thus, in the z-domain, the "corrupting" signal $H(z)$ multiplies rather than adds to $X(z)$ as would be the case for additive noise.

If we were to take the logarithm of $Y(z)$, we would have

$$\ln\{Y(z)\} = \ln\{X(z)H(z)\} \tag{7.55a}$$

$$= \ln\{X(z)\} + \ln\{H(z)\}. \tag{7.55b}$$

As a result of these operations, we are now back in the "additive corruption" situation! Thus, if the goal is to retrieve $x(n)$ from $y(n)$ and if the spectra of $\ln\{X(z)\}$ and $\ln\{H(z)\}$ were on different bands, we could perform deconvolution by using linear filtering.

Specifically, the cepstrum of $x(n)$, denoted $x_{cep}(m)$, is the inverse z-transform of $\ln\{X(z)\}$, and $h_{cep}(m)$ is the cepstrum of $h(n)$. We call $x_{cep}(m)$ the cepstrum because it is the spectrum of $\ln\{X(z)\}$, which itself is a log-spectrum of $x(n)$. Thus, $x_{cep}(m)$ is a sort of "reverse" spectrum, implied by the reversal of the letters *spec* of spectrum in the word *cepstrum*. Because $x_{cep}(m)$ is an inverse z-transform [of $\ln\{X(z)\}$], we might call m the index of a sort of "time domain." However, $x_{cep}(m)$ is not a physical time signal that could be observed or measured; hence we use the letter m rather than n for the cepstral index.

If $x_{cep}(m)$ and $h_{cep}(m)$ were nonzero on different cepstral index (m) ranges, we could multiply $y_{cep}(m) = x_{cep}(m) + h_{cep}(m)$ by zero on the range where the distortion term $h_{cep}(m)$ is nonzero and by unity elsewhere. The result, denoted $x_{rec,cep}(m)$, could then be z-transformed, exponentiated, and inverse z-transformed to obtain the recovered sequence $x_{rec}(n) \approx x(n)$. Thus, the signal $x(n)$ distorted by convolution with $h(n)$ could be recovered by this cepstral technique.

This would be a great idea, except that unfortunately there appears to be no useful or simple correspondence between $x(n)$ and the IOS of $x_{cep}(m)$ [or between $h(n)$ and the IOS of $h_{cep}(m)$]. We will examine this fact by considering some particular examples.

Before so doing, let us review the procedures for forward and inverse cepstral operations. All z-transforms below may in practice be implemented by the DFT, with all of the previously discussed practical issues kept in mind.

Forward cepstrum:

Given $x(n)$,

$$X(z) = ZT\{x(n)\} \tag{7.56a}$$

$$X_{cep}(z) = \ln\{X(z)\} \quad \text{with phase unwrapping (see below)} \tag{7.56b}$$

$$x_{cep}(m) = ZT^{-1}\{X_{cep}(z)\}. \tag{7.56c}$$

Inverse cepstrum:

Given $x_{cep}(m)$,

$$X_{cep}(z) = ZT\{x_{cep}(m)\} \tag{7.57a}$$

$$X(z) = \exp\{X_{cep}(z)\} \tag{7.57b}$$

$$x(n) = ZT^{-1}\{X(z)\}. \tag{7.57c}$$

The reason the phase must be unwrapped is that we restrict all cepstral sequences to be stable.[19] Equivalently, the z-transform of $x_{cep}(m)$ must be analytic on the unit circle, and therefore the real and imaginary parts of $X_{cep}(z)$ must be continuous there. Now,

$$X_{cep}(z) = \ln\{X(z)\} \tag{7.58a}$$

$$= \ln\{|X(z)|\} + j\angle X(z). \tag{7.58b}$$

Thus, for $X_{cep}(z)$ to be analytic, $\ln\{|X(z)|\}$ and $\angle X(z)$ must be continuous (single-valued) on the unit circle.

For $\ln\{|X(z)|\}$ to be analytic on the unit circle, $X(e^{j\omega}) \neq 0$ for any ω, because the log of zero is undefined. Thus, cepstra are not defined for ideal highpass, ideal lowpass, or any sequences having true stopbands. For $\angle X(z)$ to be analytic on the unit circle, none of the 2π jumps normally encountered in phase computation when $|\angle X(z)|$ exceeds π are allowed. Instead, integral multiples of 2π are added to make the phase continuous for all ω. Various phase-unwrapping algorithms are available, and in Matlab the routine is called `unwrap`. It should be further noted that not only must the phase be unwrapped, but it also must be zero at $\omega = 0$ and $\omega = \pi$; otherwise, it would be impossible for the phase, which is odd for real-valued $x_{cep}(n)$, to be continuous at $\omega = 0$. We see that cepstral work is rather restrictive.

Because frequency-selective filtering, the subject of Chapters 8 and 9, is the most common type of filtering, we will now examine the inverse cepstrum of some cepstral-index-limited sequences. According to our previous discussion, if the distorting sequence $h(n)$ had a cepstrum nonzero at, say, only high cepstral indices (called "quefrencies"), then it could be filtered ("liftered") out of the distorted cepstrum $y_{cep}(m)$. Assuming the desired signal $x(n)$ was "low-band" in the cepstral

---

[19] J. M. Tribolet, *Seismic Applications of Homomorphic Signal Processing* (Englewood Cliffs, NJ: Prentice Hall, 1979), p. 19.

domain, it would be all that is left after "liftering." The inverse cepstrum of the "liftered" sequence would then be the recovered (deconvolved) signal $x(n)$.

It is thus natural to see what sort of time-domain sequences $x(n)$ are "low-band" in the cepstral domain. We may begin by taking the inverse cepstrum of $x_{cep}(m) = \{1, m \in [0, M - 1]; 0, \text{ otherwise up to } N - 1\}$. This $x_{cep}(m)$ is shown in Figure 7.55a for $M = 10$ and $N = 128$; its inverse cepstrum $x(n)$ is shown in Figure 7.55b. We conclude that $x_{cep}(m)$ being low-band does not help much in identifying or narrowing down the associated type or character of $x(n)$.

If we choose $x_{cep}(m) = \{0, m \in [0, M - 1]; 1, \text{ otherwise up to } N - 1\}$ (see Figure 7.56a) for a "high-band" cepstrum, the inverse cepstrum $x(n)$ is as shown in Figure 7.56b. From these two plots we might only conclude that the energy in $x(n)$ for the cepstral low-/high-band sequence seems to be at the low/high end of the time window. Thus, we might expect that short-duration sequences (short with respect to sequence length N) tend to have low-band cepstra. However, there is no theorem guaranteeing this. Finally, note that if we make the nonzero portion of the high-band $x_{cep}(m)$ one half-period of a sinusoid instead of unity (Figure 7.57a), then the inverse cepstrum $x(n)$ (Figure 7.57b) is almost identical in appearance to (except much lower in amplitude than) the constant high-band $x(n)$ in Figure 7.56b.

It would appear that the cepstral band-filtering idea is not highly useful for general $x(n)$ and $h(n)$ likely to be encountered in practice. However, one application for which the cepstrum has been useful is in the recovery of an echo-distorted signal. We will analyze the situation for a single echo.[20]

We give here a discrete-time version of the derivation of the cepstrum for an echo-distorted signal, which is derived for continuous time in the reference in footnote 20. Let $y(n)$ be the distorted signal, where n_0 is the delay index and α is the relative amplitude of the echo:

$$y(n) = x(n) + \alpha x(n - n_0) \tag{7.59a}$$

$$= x(n) * h(n), \tag{7.59b}$$

where

$$h(n) = \delta(n) + \alpha\delta(n - n_0). \tag{7.60}$$

We need $y_{cep}(m)$ to see how to remove the echo. We have

$$Y(e^{j\omega}) = X(e^{j\omega})H(e^{j\omega}), \tag{7.61}$$

where

$$H(e^{j\omega}) = 1 + \alpha e^{-jn_0} \tag{7.62}$$

so that

$$Y_{cep}(e^{j\omega}) = \ln\{X(e^{j\omega})\} + \ln\{1 + \alpha e^{-j\omega n_0}\}. \tag{7.63}$$

Using the power series $\ln\{1 + x\} = x - x^2/2 + x^3/3 - \ldots$ for $|x| < 1$, applicable to our application for $|\alpha| < 1$, we have

$$\ln\{1 + \alpha e^{-j\omega n_0}\} = \alpha e^{-j\omega n_0} - \tfrac{1}{2}\alpha^2 e^{-j\omega 2n_0} + \tfrac{1}{3}\alpha^3 e^{-j\omega 3n_0} - \ldots \tag{7.64}$$

[20] See R. C. Kemerait and D. G. Childers, "Signal Detection and Extraction by Cepstrum Techniques," *IEEE Trans. on Information Theory*, IT-18, 6 (Nov. 1972), 745–759 for generalizations to multiecho.

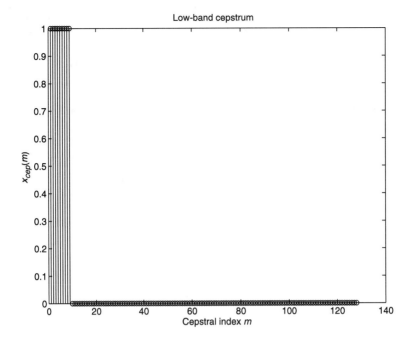

Figure 7.55*a*

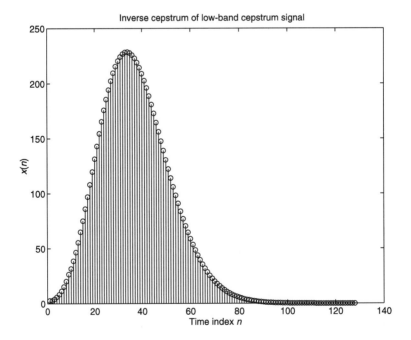

Figure 7.55*b*

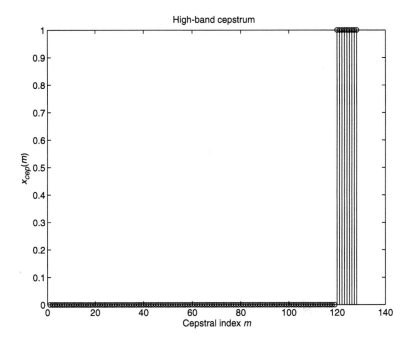

Figure 7.56*a*

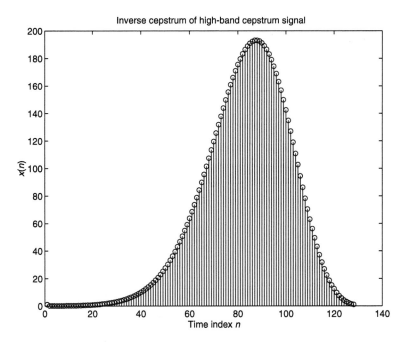

Figure 7.56*b*

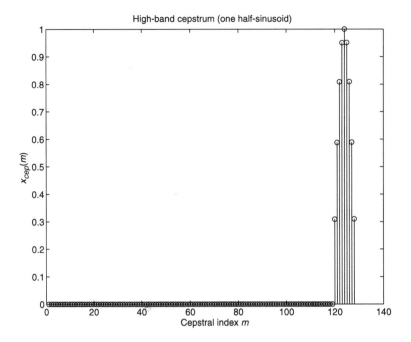

Figure 7.57*a*

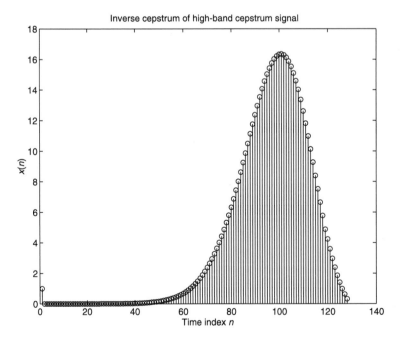

Figure 7.57*b*

so that

$$y_{cep}(m) = DTFT^{-1}\{Y_{cep}(e^{j\omega})\}$$

$$= x_{cep}(m) + \alpha\delta(n - n_0) - \tfrac{1}{2}\alpha^2\delta(n - 2n_0) + \tfrac{1}{3}\alpha^3\delta(n - 3n_0) - \dots. \quad (7.65)$$

We see that the distorted cepstrum is the desired one plus a sequence of shifted Kronecker delta functions of declining magnitude and alternating sign. Noting that there is an infinite number of delta functions, from Section 7.3 we know that all those beyond N will be cepstrally aliased. However, because $|\alpha| < 1$ these aliased deltas will diminish in significance.

For echo removal, we replace every n_0th sample in the cepstrum by a simple linear interpolation of its nearest neighbors. The result will be close to $x_{cep}(m)$, for those samples are the only ones affected by the echo. Naturally, this statement holds only if echoing is the only distortion; if the echo does not have the same shape as $x(n)$ and thus $h(n)$ is not simply a few Kronecker deltas, this procedure will not work. We now illustrate these ideas further through some numerical examples.

EXAMPLE 7.14

(a) Suppose that $x(n)$ is a geometric sequence $x(n) = 0.7^n u(n)$, and $h(n) = \delta(n)$ $+ 0.9\delta(n - n_0)$ (so $\alpha = 0.9$), where (i) $n_0 = 10$ and (ii) $n_0 = 20$. Remove the echo in $y(n) = x(n) * h(n)$ as completely as possible, using cepstral deconvolution and DFT size $N = 128$.

(b) Repeat part (a) for a segment of speech, again with $n_0 = 20$.

Solution

(a) Figure 7.58a shows $x(n)$ (circle stems) and $0.9x(n - 10)$ (* stems), whereas Figure 7.58b shows $y(n)$, the given (measured) sequence. The operations in (7.56) are carried out on $y(n)$. Figure 7.59a shows $\ln\{|Y(e^{j\omega})|\}$ at the DFT frequencies $\omega_k = 2\pi k/N$, while Figure 7.59b shows $\angle Y(e^{j\omega})$, both for k out to $N/2$. Notice that in Figure 7.59b, the unwrapped and wrapped phases are identical; this will not be the case in part (b) [the discontinuities in Figure 7.59b are due to polarity reversals and not phase-wrapping—at these phase discontinuities, the magnitude is zero; moreover, the DTFT phase is *not* discontinuous, as filling in for $\omega \neq \omega_k$ would show]. Figures 7.59a and b, respectively, are the real and imaginary parts of $Y_{cep}(e^{j\omega})$.

The cepstrum $y_{cep}(m)$ is shown in Figure 7.60a (O stems), along with $x_{cep}(m)$ (* stems); notice the Kronecker delta functions that are part of only $y_{cep}(m)$, not $x_{cep}(m)$. They appear to be every five samples, but the ones "in between" are due to aliasing. That is, we have exact values of the DTFT $Y_{cep}(e^{j\omega})$ of $y_{cep}(m)$, where $y_{cep}(m)$ is an infinite-length sequence, but we are using the finite-length DFT $\frac{-1}{128}$ to estimate $y_{cep}(m)$. By (7.16b) [written for this usage as $DFT_N^{-1}\{Y_{cep}(e^{j\omega_k})\} = \sum_{\ell=-\infty}^{\infty} y_{cep}(m + \ell N)$], there will be aliasing as described in the discussion of (7.16b).

Figure 7.60b shows the modified $y_{cep}(m)$ [denoted $x_{rec,cep}(m)$] (O stems), obtained by replacing every n_0th (10th) sample by the linear interpolation of

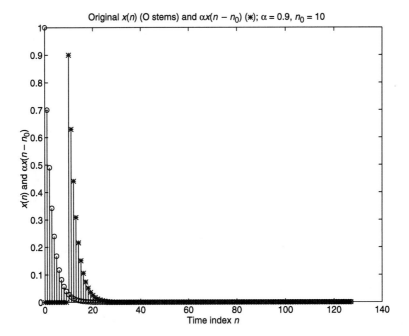

Figure 7.58*a*

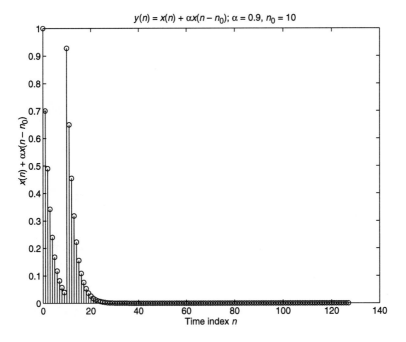

Figure 7.58*b*

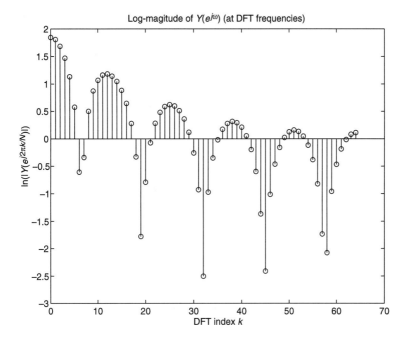

Figure 7.59*a*

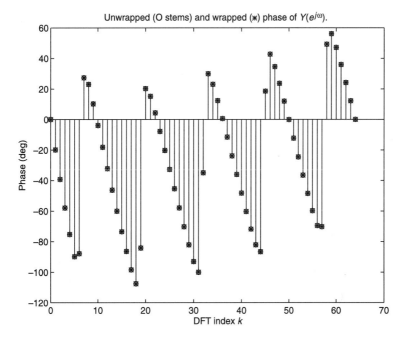

Figure 7.59*b*

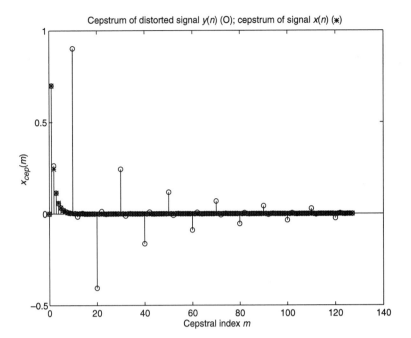

Figure 7.60*a*

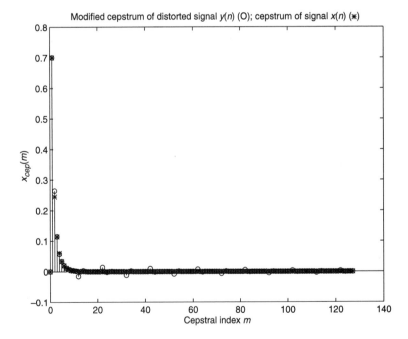

Figure 7.60*b*

its nearest neighbors, along with $x_{cep}(m)$ (* stems). The routine `interp1` in Matlab was used to perform the linear interpolation (`spline` may also be used, for cubic spline interpolation). For interpolating at index m, the values $y_{cep}(m-3), y_{cep}(m-1), y_{cep}(m+1)$, and $y_{cep}(m+3)$ were used (spacing = 2; note that $y_{cep}(m)$, the distorted value, is intentionally omitted from the input values). Matlab then produces estimates of $y_{cep}([m-3:m+3])$ (spacing = 1), so the desired replacement value $y_{cep}(m) = y_{cep}(m-0)$ is the fourth output value. For further accuracy, the indices of the significantly contributing aliased deltas in Figure 7.60a could also have been calculated and those y_{cep} values also replaced by interpolated estimates. This was not done, so the aliased deltas still appear in Figure 7.60b.

Finally, using (7.57), the estimate $x_{rec}(n)$ of $x(n)$ is obtained, yielding Figure 7.61 (O stems). Also shown in Figure 7.61 are $x(n)$ (* stems) and for comparison, the originally given $y(n)$ (dashed curve). Note that the dashed curve is just lines between points of the sequence $y(n)$ in Figure 7.58b. We see that a substantial improvement (echo reduction) has been achieved for recovery of $x(n)$.

For brevity here, the second case, $n_0 = 20$, is covered pictorially on WP 7.4. This time, there is significant error, due to the nonannihilated aliased deltas. We show here the final result in Figure 7.62.

It should be noted that, just given $y(n)$, we cannot assume to know n_0 beforehand. If n_0 is small, it may not at all be evident from examination of a plot of $y(n)$. However, n_0 will be obvious by examination of the deltas (spikes) in $y_{cep}(m)$. This identification and ability to remove the delay spikes is the whole reason for using the cepstrum in this application. In the above example with $n_0 = 20$, again we could also annihilate the obviously aliased deltas for more accurate results.

(b)   Again for brevity, we present the details of this part on WP 7.4. We find that this task is somewhat more challenging than that in part (a), due to the

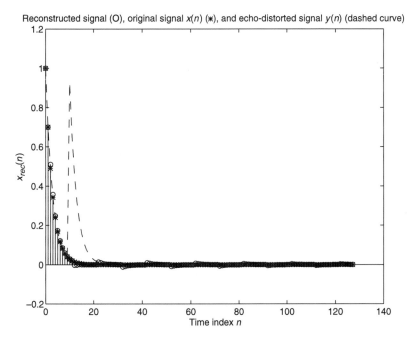

Reconstructed signal (O), original signal $x(n)$ (∗), and echo-distorted signal $y(n)$ (dashed curve)

Time index n

Figure 7.61

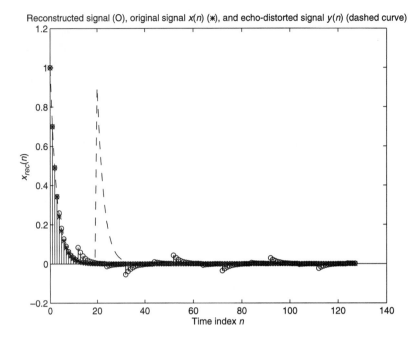

Reconstructed signal (O), original signal $x(n)$ (∗), and echo-distorted signal $y(n)$ (dashed curve)

Figure 7.62

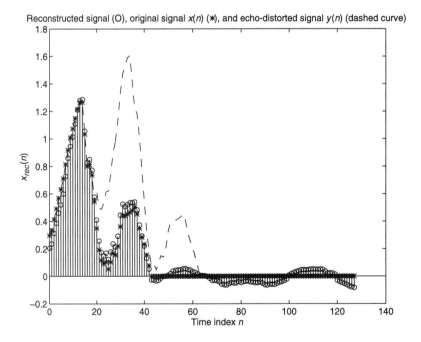

Reconstructed signal (O), original signal $x(n)$ (∗), and echo-distorted signal $y(n)$ (dashed curve)

Figure 7.63

nature of the phase of $Y(e^{j\omega})$. For its interest, the final results are shown in Figure 7.63. The reconstruction $x_{\text{rec}}(n)$ (O stems) is plotted against both $x(n)$ (∗ stems) which it is estimating and $y(n)$ (dashed curve) from which it was made. Despite the obvious errors, the agreement is fairly good and is a dramatic improvement over the given signal $y(n)$, as far as approximating the desired signal $x(n)$. Again, further improvement is possible by removing the aliased deltas in $y_{cep}(n)$.

7.10   THE FAST FOURIER TRANSFORM (FFT)

7.10.1   Introduction

In this chapter, we have focused on applications of the DFT for the processing of discrete-time signals. In previous chapters, we have also considered its use in the approximation of integral transforms and continuous-time systems. It has already been claimed that the order of computations for a special fast algorithm that computes the DFT is $N \log_2(N)$. More precisely, we will see that for the basic FFT algorithm, the order is about $\frac{1}{2}N \log_2(N)$.

A quick glance at the DFT definition will indicate the required number of multiplications if all N DFT coefficients are to be computed directly from the formula. Because each coefficient requires on the order of N multiplications, it follows that such direct computation requires on the order of N^2 multiplications. Consequently, a savings of roughly the factor $2N/\log_2(N)$ can be obtained by using the efficient algorithm (FFT). This ratio grows rapidly with N; for $N = 256$, it is 64; for $N = 1024$, it is 205; and for $N = 8192$ it is 1260—over a thousand times faster than direct calculation!

In 1965, a paper by J. W. Cooley and J. W. Tukey set off a storm of activity by describing the algorithm in terms comprehensible to the modern scientific community. However, Gauss had discovered the idea in 1805, 160 years earlier. Furthermore, that paper was written two years before Fourier had presented his trigonometric series representation of arbitrary functions on a finite interval! One author has therefore been tempted to rename the FFT the DGT, discrete Gauss transform. In any case, the algorithm is here, and the purposes of this section are to explain the decomposition behind the FFT and derive from that the basic "butterfly" flowgraph and algorithm code on which special hardware and software implementations are based.

7.10.2   Decimation-in-Time Decomposition

The only requirement on N to gain great efficiency over direct DFT calculation is that N be highly composite: $N = N_1 \cdot N_2 \cdot \ldots \cdot N_p$, N_i all integers. However, for the basic FFT algorithm, we will require that N be a power of two. For now, let $N = LM$, L and M integers, and define $W_N = e^{-j2\pi/N}$. Then the DFT_N sum may be decomposed into a double sum as follows:

$$X(k) = \sum_{n=0}^{LM-1} x(n) \cdot W_{LM}^{nk} \tag{7.66a}$$

$$= \sum_{\ell=0}^{L-1} \sum_{m=0}^{M-1} x(\ell + Lm) \cdot W_{LM}^{(\ell + Lm)k}. \tag{7.66b}$$

For the purpose of keeping track of terms in this decomposition, it is helpful to define the two-dimensional array $\mathbf{x}(\ell, m) = x(\ell + Lm)$, where the boldface indicates two-dimensional array representations. Equivalently, we may form a table from

Table 7.1 Decimation-in-Time Temporal Index Decomposition (First Level) (Table entries are values of $n = \ell + Lm$)

$\ell \backslash m$	0	1	2	...	$M - 1$
0	0	L	$2L$	...	$(M - 1)L$
1	1	$L + 1$	$2L + 1$	...	$(M - 1)L + 1$
2	2	$L + 2$	$2L + 2$	...	$(M - 1)L + 2$
.	.	.	.	...	.
.	.	.	.	...	.
.	.	.	.	...	.
$L - 1$	$L - 1$	$2L - 1$	$3L - 1$	...	$ML - 1$

Table 7.2 Decimation-in-Time Frequency Index Decomposition (First Level) (Table entries are values of $k = Mp + q$)

$p \backslash q$	0	1	2	...	$M - 1$
0	0	1	2	...	$M - 1$
1	M	$M + 1$	$M + 2$	...	$2M - 1$
2	$2M$	$2M + 1$	$2M + 2$	...	$3M - 1$
.	.	.	.	...	.
.	.	.	.	...	.
.	.	.	.	...	.
$L - 1$	$(L - 1)M$	$(L - 1)M + 1$	$(L - 1)M + 2$	...	$ML - 1$

which we may obtain $n = \ell + Lm$, as shown in Table 7.1. Similarly, k may be decomposed as $k = Mp + q$, and we may write $\mathbf{X}(p, q) = X(Mp + q)$, as shown in Table 7.2.

Notice that there are L distinct values of ℓ and p, whereas there are M distinct values of m and q. Also, notice how the decomposition of k is "transposed" compared with that for n. This transposition of M and L *must* occur to obtain the subsequent simplifications of the complex exponentials found below. Furthermore, it is this transposition that is directly responsible for bit reversal of order when $N = 2^m$. (The symbol $\underline{m}$ will be used in this section to signify which power of two N is if N is a power of two.)

Forming the decompositions of $x(n)$ and $X(k)$ as above leads to *decimation in time*, as we will see below. We could just as easily let $x(\ell, m) = x(M\ell + m)$ and $\mathbf{X}(p, q) = X(p + Lq)$; this alternative decomposition would lead to *decimation in frequency* (see the problems). For brevity, we will restrict our attention here to the decimation-in-time decomposition.

Using the decimation-in-time decompositions above, (7.66b) may be rewritten as

$$\mathbf{X}(p, q) = X(Mp + q) = \sum_{\ell=0}^{L-1} \sum_{m=0}^{M-1} \mathbf{x}(\ell, m) \cdot W_{LM}^{(\ell + Lm)(Mp + q)}. \tag{7.67}$$

But

$$e^{-j2\pi(\ell + Lm)(Mp + q)/(LM)} = e^{-j2\pi(M\ell p + \ell q + MLmp + Lmq)/(LM)} \tag{7.68a}$$

$$= W_L^{\ell p} \cdot W_{LM}^{\ell q} \cdot W_M^{mq}, \tag{7.68b}$$

where (7.68b) follows because the third exponential term on the right side of (7.68a) is unity. Consequently, (7.67) becomes

$$\mathbf{X}(p, q) = \sum_{\ell=0}^{L-1} \left\{ W_{LM}^{\ell q} \left[\sum_{m=0}^{M-1} \mathbf{x}(\ell, m) \cdot W_M^{mq} \right] \right\} \cdot W_L^{\ell p}. \tag{7.69}$$

Equation (7.69) may be considered a three-stage process:

1. The DFT_M of $\mathbf{x}(\ell, m)$ over m, in the square brackets. This step involves M multiplications for each DFT_M value. There are M different values of q and L different values of ℓ, so this stage requires $M^2 L$ multiplications. That is, there are L complete DFT_M sets to do, each of which requires M^2 multiplications (think of ℓ as a parameter that can take on L different values). This step produces LM distinct results (one for each ℓ, q pair).

2. Complex scaling by $W_{LM}^{\ell q}$. These scalars are known as *twiddle factors*. There is one multiplication for each ℓ and q; therefore, there are $ML = N$ multiplications in this step.

3. The outer DFT_L of the "innards." Here there are L multiplications for each DFT_L sum computed, and [see the left-hand side of (7.69)] there are distinct DFT_L sums for each p and q. Therefore, in the third step there are $L \cdot L \cdot M = L^2 M$ multiplications (i.e., there are N DFT_L sums to do).

What have we gained? As mentioned before, for direct computation of the DFT_N we require $N \cdot N = (ML)^2$ multiplications (i.e., there are N distinct values of k and for each k, there are N terms in the DFT_N sum). Contrarily, with the above decomposition we need $M^2 L + ML + L^2 M = N(M + 1 + L)$ multiplications, which is less than N^2 if $M + L + 1 < ML$. For example, if $N = 1024$, $L = 2$ (we will keep $L = 2$, below), and thus $M = 512$, then there are about $1.05 \cdot 10^6$ multiplications for direct computation and only $0.527 \cdot 10^6$ multiplications for the above decomposition. This amounts to a savings of about a factor of two. [Below, we shall use the above formula for the number of multiplications, written in the form $L\{M^2 + (1 + L)M\}$.]

Why is there a savings? Referring to (7.69), we see that for different values of p, the expression in $\{\cdot\}$,

$$W_{LM}^{\ell q} \cdot \sum_{m=0}^{M-1} \mathbf{x}(\ell, m) \cdot W_M^{mq}, \tag{7.70}$$

is used over and over again. By calculating it once, saving it, and reusing it whenever it is needed instead of recalculating it—which is implied in our three-step process outlined above—we avoid redundant computations. Without the decomposition, these basic subcalculations could not be identified and saved for efficient computation of the DFT_N; we implicitly recalculate them every time in calculating the original DFT_N formula $\sum_{n=0}^{N-1} x(n) \cdot W_N^{nk}$.

There is nothing to prevent us from decomposing further; for example, we may decompose the DFT_M in "[ ]" in (7.69), analyzed in Step 1 above. Let $N' = M, M' = N'/L' = N/4$, where we set $L' = 2$ and recall that $M = N/L = N/2$. Then the number of multiplications, which from above was $L\{M^2 + (1+L)M\}$, becomes under the second decomposition $L\{L'\{M'^2 + (1+L')M'\} + (1+L)M\}$

$$\underbrace{\text{reduction of what was } M^2}_{\text{in first decomposition}}$$

$$= 2 \{ [2 \cdot ((N/4)^2 + 3 \cdot (N/4))] + 3 \cdot (N/2) \},$$

$$\begin{array}{ccccccc} \uparrow & \uparrow & \uparrow & \uparrow & \uparrow & \uparrow & \uparrow \\ L & L' & M' & (1+L') & M' & (1+L) & M \end{array}$$

which for $N = 1024$ equals about $0.265 \cdot 10^6$, a savings of a factor of 3.91 over direct computation.

If we continue further decompositions, each time using $L' = 2, L'' = 2, L''' = 2, \ldots,$ so that the length of the final DFT is 2 (which can be decomposed no further), the required number of multiplications is, for $N = 1024 = 2^{10}$ ($m = 10$),

$$2[2\{2[2\{2[2\{2[2\{2[(N/512)^2 + 3(N/512)] + 3(N/256)\} + 3(N/128)] + 3(N/64)\} +$$
$$3(N/32)] + 3(N/16)\} + 3(N/8)] + 3(N/4)\} + 3(N/2)]$$
$$= 29{,}696 \approx 1/35 \text{ the number for direct computation.} \qquad (7.71)$$

The above was the nine-stage breakdown for $N = 1024 = 2^{10}$; we may break down into $m - 1$ stages for $N = 2^m$. Note that we do not reduce the number of multiplications by a factor of two for each subcomputation, particularly as the M involved becomes small.

We have not yet fully exploited this method for "radix 2" FFTs, where N must be a power m of 2: The promised $\frac{1}{2}N \log_2(N)$ for 1024 is only 5120, just over $\frac{1}{6}$ of 29,696. Consider that $W_2^\ell = e^{-j\pi\ell} = (-1)^\ell$, which is equal to 1 for ℓ even and -1 for ℓ odd. Thus, the fundamental two-point DFTs in the full decomposition of the $N = 2^m$ DFT require zero multiplications. Therefore, the term "$(N/512)^2$" in (7.71) may be removed. Similarly, the $W_L^{\ell p}$ factors in the third step of (7.69) may also be removed from the multiplication count, as $L = 2$, so $W_L^{\ell p} = \pm 1$. Because this is true also for $L', L'', L''', \ldots,$ we may change all coefficients "$L + 1$" [or 3 in (7.71)] to 1. Consequently, the only multiplications necessary are those by the "twiddle factors" $W_{LM}^{\ell q}$. Thus the number of multiplications becomes, for $N = 2^m$,

$$2^{m-1} \cdot 2 + 2^{m-2} \cdot \frac{N}{2^{m-2}} + \ldots + 2 \cdot \frac{N}{2} = N(m-1) = N \cdot [\log_2(N) - 1], \qquad (7.72)$$

which for $N = 1024$ is equal to 9216.

Even more gains can be made. Recognizing that half of the twiddle factors are unity (which will be shown in the case study $N = 8$ below), the total reduces to $(N/2) \cdot [\log_2(N) - 1]$ or $\approx (N/2) \cdot \log(N)$ for N large. The reason half the twiddle factors $W_N^{\ell q}$ are unity is that throughout the decomposition, ℓ (or ℓ', or ℓ'', etc., or also $m' = 0$) is equal to zero for half the twiddle factors (because $L = L' = L'' = \cdots = 2$), and $W_K^0 = 1$ for any integer K.

Let us now examine a case study of this type of decomposition for $N = 8 = 2^m$, where $m = 3$. We will obtain the standard "butterfly" configuration. For this case, the first decomposition (7.69) becomes, with $L = 2$ and $M = N/L = 8/2 = 4$,

$$\mathbf{X}(p,q) = \sum_{\ell=0}^{1} \left\{ W_8^{\ell q} \left[\sum_{m=0}^{3} \mathbf{x}(\ell, m) \cdot W_4^{mq} \right] \right\} \cdot W_2^{\ell p}, \qquad (7.73)$$

where $\mathbf{x}(\ell, m) = x(\ell + Lm) = x(\ell + 2m)$ and $\mathbf{X}(p, q) = X(Mp + q) = X(4p + q)$. Thus, the sequence $n = 0, 1, 2, 3, 4, 5, 6, 7$ maps to ℓ and m in Table 7.3, and the sequence $k = 0, 1, 2, 3, 4, 5, 6, 7$ maps to p and q in Table 7.4.

We now decompose again. Specifically, we decompose m in the manner that n was decomposed previously: $m = \ell' + L'm' = \ell' + 2m'$. Also, we decompose q in

Table 7.3   Decimation-in-Time Temporal Index Decomposition
(First Level, $N = 8$) (Table entries are values of n)

$\ell \backslash m$	0	1	2	3
0	0	2	4	6
1	1	3	5	7

Table 7.4   Decimation-in-Time Frequency Index Decomposition
(First Level, $N = 8$) (Table entries are values of k)

$p \backslash q$	0	1	2	3
0	0	1	2	3
1	4	5	6	7

the manner that k was decomposed previously: $q = M'p' + q' = (N/4)p' + q' = 2p' + q'$.

$$\mathbf{X}(p, q) = \sum_{\ell=0}^{1} \left\{ W_8^{\ell q} \left[\sum_{\ell'=0}^{1} \left\{ W_4^{\ell'q'} \left[\sum_{m'=0}^{1} \mathbf{x}(\ell; \ell', m') \cdot W_2^{m'q'} \right] \right\} W_2^{\ell'p'} \right] W_2^{\ell p} \right\}, \quad (7.74)$$

where $\mathbf{x}(\ell; \ell', m') = \mathbf{x}(\ell, \ell' + 2m') = x(2^2 m' + 2^1 \ell' + \ell)$. Notice that we effectively have broken up the input array into two groups: $\ell = 0$ and $\ell = 1$. The $\ell = 0$ group is the even-indexed set of $x(n)$, and the $\ell = 1$ group is the odd-indexed set of $x(n)$. Writing the outer ℓ sum as two separate terms, the first term involving only the even-indexed $x(n)$ and the second term involving only the odd-indexed $x(n)$ is commonly referred to as "decimation in time." This decimation is also done at all levels of the decomposition.

Our time decomposition new map is $m = \ell' + 2m'$, as shown in Table 7.5, so that n is found in Table 7.6 for $\ell = 0$ and in Table 7.7 for $\ell = 1$.

Table 7.5   Decimation-in-Time Temporal Index Decomposition
(Second Level, $N = 8$) (Table entries are values of m)

$\ell' \backslash m'$	0	1
0	0	2
1	1	3

Table 7.6   Decimation-in-Time Temporal Index Decomposition
(Second Level, $N = 8$, Relation to n: $\ell = 0$)
(Table entries are values of n)

$\ell' \backslash m'$	0	1
0	0	4
1	2	6

Table 7.7   Decimation-in-Time Temporal Index Decomposition
(Second Level, $N = 8$, Relation to n: $\ell = 1$)
(Table entries are values of n)

$\ell' \backslash m'$	0	1
0	1	5
1	3	7

As noted above, the new frequency index decomposition can similarly be represented with $q = M'p' + q' = (N/4)p' + q' = 2p' + q'$, and we may write

$$\mathbf{X}(p; p', q') = \mathbf{X}(p, 2p' + q')$$
$$= X(2^2 p + 2^1 p' + 2^0 q')$$

resulting in p' and q' in Table 7.8. From p' and q', we can obtain the value of k from Tables 7.9 and 7.10 for, respectively, $p = 0$ and $p = 1$. All the above decomposition results may be put into the illuminating form shown in Table 7.11. The binary forms of n and k are given in Table 7.11, and we see that the binary digits for n are from left to right, m', ℓ', ℓ whereas those of k are p, p', q'. All of these values are in agreement with the formulas $n = 2^2 m' + 2^1 \ell' + 2^0 \ell$ and $k = 2^2 p + 2^1 p' + 2^0 q'$ shown above.

The origin of the evident bit reversal between input and output orders is the transpositional nature of the decomposition that was necessary to exploit the periodicity of complex exponentials and thus make the computational savings. In this case, $n = 2(2m' + \ell') + \ell = 2^2 m' + 2^1 \ell' + 2^0 \ell$, and $k = 4p + (2p' + q') = 2^2 p + 2^1 p' + 2^0 q'$; we see that the indices most deep in the decomposition are highest significant bits for n and lowest for k. Similarly, indices least deep in the decomposition are the lowest significant bits for n and highest for k.

Finally, from the above collected results, based on (7.74), the signal flowgraph in Figure 7.64 should now be readily understood. Recall from Section 2.5 that arrow directions indicate the direction of signal flow within the given branch of the signal-processing system. Two lines coming into a junction indicate a summer, whereas two lines coming out of a junction merely indicate a branch point.

In the second column of "butterflies," the two branches coming from either of the branch points on the input side of the butterfly contribute for values of k having

Table 7.8   Decimation-in-Time Frequency Index Decomposition (Second Level, $N = 8$) (Table entries are values of q)

$p' \backslash q'$	0	1
0	0	1
1	2	3

Table 7.9   Decimation-in-Time Frequency Index Decomposition (Second Level, $N = 8$, Relation to k: $p = 0$) (Table entries are values of k)

$p' \backslash q'$	0	1
0	0	1
1	2	3

Table 7.10   Decimation-in-Time Frequency Index Decomposition (Second Level, $N = 8$, Relation to k: $p = 1$) (Table entries are values of k)

$p' \backslash q'$	0	1
0	4	5
1	6	7

Table 7.11   Summary of Decimation-in-Time Index Decompositions, $N = 8$

ℓ	ℓ'	m'	n	p	p'	q'	k
0	0	0	0 = 000	0	0	0	0 = 000
0	0	1	4 = 100	0	0	1	1 = 001
0	1	0	2 = 010	0	1	0	2 = 010
0	1	1	6 = 110	0	1	1	3 = 011
1	0	0	1 = 001	1	0	0	4 = 100
1	0	1	5 = 101	1	0	1	5 = 101
1	1	0	3 = 011	1	1	0	6 = 110
1	1	1	7 = 111	1	1	1	7 = 111

the same value of q' but differing values of p'. [The values of p' and q' are traceable from the right—from the values of $X(k)$ to which the branches under consideration contribute.] Therefore, the multiplicative factor $W_4^{\ell'q'}$ may be placed *preceding* the branch point (and consequently written only once), whereas the $W_2^{\ell'p'}$ factor must be written separately for each of the two branches. The same comment applies for

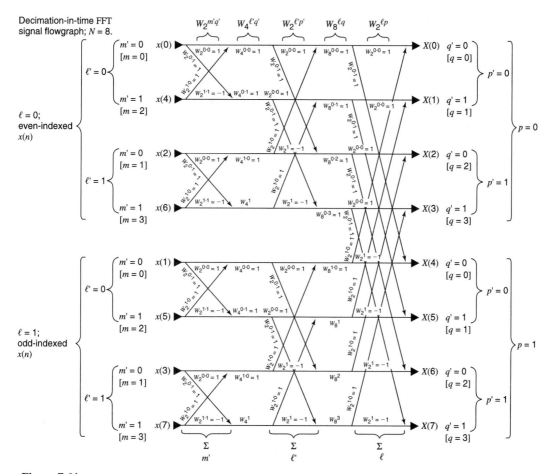

Figure 7.64

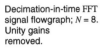

Decimation-in-time FFT signal flowgraph; $N = 8$. Unity gains removed.

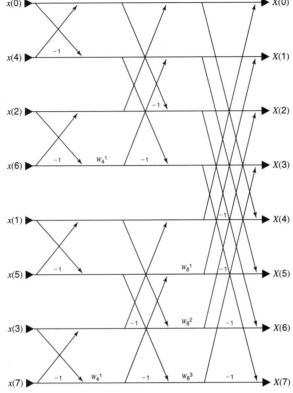

Figure 7.65

column 3, where $W_8^{\ell q}$ may be "pulled out" (the q are common), whereas $W_2^{\ell p}$ cannot (the p are different). Also keep in mind that $W_2^0 = 1$, $W_2^1 = -1$, and $W_4^1 = W_8^2$.

As promised earlier, we find that indeed over half of all the twiddle factors are equal to 1; in fact, only five of the coefficients are other than ±1. (For larger N, more nearly exactly half will be nonunity.) The flowgraph with all unity coefficients removed for simplicity and "−1" written as "−1" rather than, for example, W_2^1 is shown in Figure 7.65. So-called in-place variations of this flowgraph can be made so that $x(n)$ is in normal counting order from the top down.

7.10.3 Bit Reversal

From the butterfly flowgraph it is clear that the input data array must be presented in bit-reversed form. A computer program implementation will thus need to automatically perform bit-reversal in the input array before executing the flowgraph computations. Probably the part of the FFT Cooley-Tukey code that is most difficult to decipher is the 12 or so lines comprising the bit-reversal counter. It is made further cryptic by the common version in Fortran (until recently, the most universal computing language, but soon may be obsolete), where the index zero for arrays is avoided, so unity is always added to the index value (as noted previously Matlab also prohibits the zero array index).

Consider the following bit-reversal problem. We wish to have two counters with associated input and output vectors: The input vector is in normal counting order, and the output vector is to be put into bit-reversed order. The goal is to swap correct

Table 7.12   Bit-Reversal Process

i	I	j	J	Action to be taken (in downward order stepwise, from $i = 0$ to $i = 15$)
0 = 0000	1	0000 = 0	1	No swap; there is no reason to: $i = j$
1 = 0001	2	1000 = 8	9	Swap values of $A(2)$ and $A(9)$
2 = 0010	3	0100 = 4	5	Swap values of $A(3)$ and $A(5)$
3 = 0011	4	1100 = 12	13	Swap values of $A(4)$ and $A(13)$
4 = 0100	5	0010 = 2	3	Don't reswap $A(5)$, $A(3)$—already done
5 = 0101	6	1010 = 10	11	Swap values of $A(6)$ and $A(11)$
6 = 0110	7	0110 = 6	7	No swap; there is no reason to: $i = j$
7 = 0111	8	1110 = 14	15	Swap values of $A(8)$ and $A(15)$
8 = 1000	9	0001 = 1	2	Don't reswap $A(9)$, $A(2)$—already done
9 = 1001	10	1001 = 9	10	No swap; there is no reason to: $i = j$
10 = 1010	11	0101 = 5	6	Don't reswap $A(11)$, $A(6)$—already done
11 = 1011	12	1101 = 13	14	Swap values of $A(12)$ and $A(14)$
12 = 1100	13	0011 = 3	4	Don't reswap $A(13)$, $A(4)$—already done
13 = 1101	14	1011 = 11	12	Don't reswap $A(14)$, $A(12)$—already done
14 = 1110	15	0111 = 7	8	Don't reswap $A(15)$, $A(8)$—already done
15 = 1111	16	1111 = 15	16	No swap; there is no reason to: $i = j$

elements of the input array so that the final output array order is bit-reversed with respect to the original order. Let $N = 2^4 = 16$, the non-bit-reversed counter index be i, the non-bit-reversed array index be $I = i + 1$, the bit-reversed counter index be j, and the bit-reversed array index be $J = j + 1$. Notate by $A(\cdot)$ the input data array, which is to be modified into bit-reversed order. Table 7.12 shows the indices and corresponding bit patterns as we *proceed downward* step by step (forward counting of I and i). In the rightmost column is shown the sequential actions to be taken for the current value of I to implement bit reversal.

Everyone knows how to count forward in binary. Add 1 to the least significant bit (LSB). If the LSB was originally 0, change it to 1 and we are done. Otherwise, leave the LSB at 0 and carry-over begins, and we look at next higher LSB. If it is 0, change it to 1 and we are done. Otherwise, carry-over proceeds to the next higher LSB.

Because the bit-reversed counter is just a backwards version of normal binary counting, the same process as described in the preceding paragraph may be used to describe bit-reversed counting, but with the words "least significant bit (LSB)" replaced by "most significant bit (MSB)" and "next higher" by "next lower." The trick is how to implement this process, working with decimal numbers. Consider again $N = 16$. As shown in Table 7.12, 0000 remains unchanged under bit reversal. Now add 1 to i: 0001. For the bit-reversed counter, the MSB of 0000 is 0, so we change the MSB to 1 to obtain 1000 and are done. We change the MSB to 1 by adding $N/2 = 1000$. For the next increment in i, 0001 becomes 0010 in the non-bit-reversed counter. In the bit-reversed counter, we need to change 1000 to 0100. We can do this by subtracting $N/2$, yielding 0000, and adding $N/4 = 0100$. Suppose now that we count from 1011 to 1100 in the non-bit-reversed counter. In the bit-reversed counter, we thus change from 1101 to 0011. This is done by subtracting $N/2 = 1000$ from 1101 (yielding 0101), subtracting from that $N/4 = 0100$ (yielding 0001), and finally adding $N/8 = 0010$ (yielding 0011 as required).

A pattern emerges: We keep subtracting {N divided by successively higher powers m of 2} until the $(m + 1)$st MSB of the bit-reversed number is zero. When it is, we add on $2/N^{m+1}$ and are done. Thus the test is to compare the current value of the bit-reversed number (modified by the subtractions just referred to, if any) with $K =$ {N divided by successively higher powers m of 2}. We are now in a position to understand the bit-reversal code[21] (the entire FFT code is translated into Matlab code in the problems).

```
1               J=1
2               DO 7 I=1, N-1
3                   IF (I.GE.J) GO TO 5
4                      T=A(J)
5                      A(J)=A(I)
6                      A(I)=T
7       5           K=N/2
8       6           IF (K.GE.J) GO TO 7
9                      J=J-K
10                     K=K/2
11                     GO TO 6
12      7           J=J+K
```

The variable $J = j + 1$ is the current value of the bit-reversed counter, and $I = i + 1$ is the current value of the non-bit-reversed counter. $A(\cdot)$ is the array, which has index argument from 1 to N. The range of I and J is also from 1 to N, a range shifted higher by 1 than that of i and j. Line 1 initializes J to 1, or equivalently, j to 0. Line 2 is the beginning of the main loop for the non-bit-reversed index I, starting at 1 (i starts at 0). Notice that the I counter stops at $I = N - 1$ rather than N, because no bit reversal of $i = N - 1 = 1111 \ldots 111$ is necessary. Line 3 determines whether to swap. If $I > J$, then the swapping has already been done, so do not re-reverse and revert to the original sequence (see Table 7.12). If $I = J$, there is no need to swap because i and j are identical; "swapping" would do nothing. Lines 4, 5, and 6 are the standard method for swapping: Put one variable in a temporary, put the other variable value into the first variable, and finally put the temporary value into the second variable.

From this point down to label 7, we are done with the consideration of the value of I. The rest of the code involves the calculation of the bit-reversed index for the next value of I. Line 7 initializes K to $N/2$ to compare j with the most significant bit. In line 8, K.GE.J translates to $K > j$. If $K > j$, then as previously discussed, the current MSB of j as it now stands must be zero and there is no carry over. If there is no carry over, we add the current value of K to j (and of course to J) in order to change this current MSB from 0 to 1 (see line 12). The result is the actual next bit-reversed count. From here, we proceed on to the next value of I. This step is done in line 12 after replacing J by $J + K$.

If a carry over does occur, then K is subtracted from j (and from J) in line 9 to change that bit to a zero. Notice the word "current" before "MSB of j." The current MSB of j refers to the position of the unity bit in the binary form of K. As K is successively divided by 2 in line 10, the "MSB" examined becomes successively less significant. Line 11 begins a new "carry over" to the next lower MSB of j, to now be considered starting at line 8. This completes the discussion of the bit-reversal lines in the FFT code.

---

[21]L. R. Rabiner and B. Gold, *Theory and Application of Digital Signal Processing* (Englewood Cliffs, NJ: Prentice Hall, 1975), p. 367.

7.10.4   The Remainder of the Code

The entire code of the FFT subroutine is as follows;[22] again, conversion into Matlab is done in the problems.

```
1          SUBROUTINE FFT(A,M)
2          COMPLEX A(1024),U,W,T
3          N=2**M
4          NV2=N/2
5          NM1=N-1
6          J=1
7          DO 7 I=1,NM1
8             IF (I.GE.J.) GO TO 5
9                T=A(J)
10               A(J)=A(I)
11               A(I)=T
12   5        K=NV2
13   6        IF (K.GE.J) GO TO 7
14               J=J-K
15               K=K/2
16               GO TO 6
17   7        J=J+K
18          PI=4*ATAN(1)
19          DO 20 L=1,M
20             LE=2**L
21             LE1=LE/2
22             U=(1.0,0.0)
23             W=CMPLX(COS(PI/FLOAT(LE1)),SIN(-PI/FLOAT(LE1)))
24             DO 20 J=1,LE1
25                DO 10 I=J,N,LE
26                   IP=I+LE1
27                   T=A(IP)*U
28                   A(IP)=A(I)-T
29   10             A(I)=A(I)+T
30   20          U=U*W
31          RETURN
32          END
```

The part the FFT code other than the bit-reversal section (now lines 6 through 17) can be readily understood by consideration of the butterfly flowgraph. The array A to be transformed has length $N = 2^M$ (line 3), where M (which we denoted $\underline{m}$ above) is not to be confused with our variable M in previous discussions. So that $N/2$ and $N - 1$ do not have to be repetitively recomputed, they are respectively stored in the variables $NV2$ and $NM1$ in lines 4 and 5.

With bit-reversal accomplished in lines 6 through 17, we may now access $A(\cdot)$ in "normal order" and obtain the correct bit-reversed value. For example, for our $N = 8$ example after bit reversal $A(1)$ contains $x(0)$, $A(2)$ contains $x(4)$, $A(3)$ contains $x(2)$,

---

[22] Rabiner and Gold, p. 367.

$A(4)$ contains $x(6)$, and so on. Line 18 puts π into the variable *PI*. In the flowgraph we naturally work our way from left to right. The butterfly "column" is indexed with variable L in line 19 (do not confuse with L in our decomposition discussion) and runs from 1 to M. There are three such columns in Figures 7.64 and 7.65; in general, there will be $M = \log_2(N)$. In lines 20, 21, and 23 we define the constants for this column $LE = 2^L$, $LE1 = 2^{L-1}$, and $W = e^{-j2\pi/\{2L\}} = e^{-j\pi/LE1}$.

A fundamental idea in this code is that, for example, in Figure 7.64, array variable $A(1)$ is always (for the whole DFT computation) located on the top row, $A(2)$ on the second row, and so on. (This characteristic is known as *in-place* computation.) We work our way from the left [where $A(\cdot)$ contain the bit-reversed input] to the right [where $A(\cdot)$ contain the non-bit-reversed DFT values]. At each butterfly column, we update all the $A(\cdot)$ according to the butterfly signal flowgraph for that column, and then go on to the next column to the right.

Examination of the butterfly flowgraph (for rigor, use induction) shows that in column L there are 2^{L-1} distinct W_{LE} ("twiddle factor") values preceding the butterfly of that column. These twiddle factors are computed in the variable U, which in line 22 is initialized to 1. Because the sine and cosine functions are most time consuming to evaluate, the loop for their computation, indexed by J, is placed as far out as possible—lines 24 through 30.

To further minimize twiddle factor computation, direct computation of W_{LE}^{J-1}, $J \in [1, 2^{L-1}]$ is replaced by the repetitive multiplication of U by W—the new value of $U = W_{LE}^J = W \cdot W_{LE}^{J-1}$, which is done at the end of the J loop (line 30). Thus, each twiddle factor is used in a number of butterflies within the given column (value of L). Examination of the flowgraph[23] shows that the index of the upper legs of these butterflies, variable I in the loop from lines 25 through 29, increases from J to N in jumps of $2^L = LE$; hence line 25.

The index of the lower leg of the butterfly, IP, is 2^{L-1} larger than the upper leg. It is the input to the bottom leg of the butterfly that is multiplied by the (nonunity) twiddle factor; this product is computed in line 27 and stored in the temporary variable T to avoid doing this multiplication twice. The two outputs of the butterfly are (lower leg) $A(IP) = $ old $A(I) - T$ and (upper leg) $A(I) = $ old $A(I) + T$. Note that the lower leg had to be done first because it needs the old value of $A(I)$. Upon cycling through all the butterflies for this twiddle factor, all the twiddle factors, and all the columns (left to right), the FFT is complete and the DFT coefficients in non-bit-reversed order appear in $A(\cdot)$. It is recommended that the reader mentally "work through" the signal flowgraph in Figure 7.64, with the computer code in hand, and verify that indeed, the code is merely implementing the calculations dictated by the signal flowgraph.

Finally, it should be noted that there is a great variety of FFT algorithms in existence. The particular algorithm just discussed is known as the decimation-in-time algorithm for radix two. Other variations include decimation-in-frequency, other radix number-based algorithms, so-called prime factor algorithms, the Winograd algorithm, the chirp z-transform algorithm [particularly efficient if only a subset of the $X(k)$ are required], and myriad special-purpose algorithms optimal in particular situations.

Several chapters could be developed to all the details on all available algorithms. However, the Cooley-Tukey algorithm described in detail in this section, or slight variations on it, is the workhorse for most computational situations. Moreover, most engineers are more interested in using the FFT than in investigating all the details of a particular implementation. The purpose of this section is to expose the main principles behind the computational gains achievable by implementing the DFT using the FFT algorithm.

---

[23] To be certain, consider a flowgraph for a larger value of N or use induction.

PROBLEMS

Nonintegral Number of Periods

7.1. Find the exact expression for the DFT$_N$ of a sampled sinusoid $x(n) = \cos(\omega_1 n), n \in [0, N-1]$, when digital frequency ω_1 and (positive integer) number of data samples N are such that (a) $\omega_1 N = 2\pi L$, L a positive integer; (b) $\omega_1 N = 2\pi L + \theta$, where $|\theta| < \pi$. For both cases, numerically compare the maximum values of $|X(k)|$ as predicted in your equations with the actual DFT maximum magnitudes for $N = 64$, $L = 10$, and ω_1 as given in cases (a) and (b) with $\theta = \pi/4$ in case (b). Also plot the DFT magnitudes in each case. Can you infer the meaning of the term *spectral leakage* from this example?

7.2. In this problem, we consider in detail the very practical problem of sampling and estimating the spectrum of a periodic (or nearly periodic) signal. See Problem 7.3 for an example application. In practice, we sample a periodic time function $x_c(t) = x_c(t + T)$ and compute its DFT to estimate its spectrum $X_c(j\Omega)$. For this problem, however, we will consider only continuous-time transforms to simplify the discussion, knowing that the DFT will simply be a rectangular-rule truncated version of the final results (assuming no aliasing), within $|\Omega| < \Omega_s/2$. Recall that the FS coefficients are $X_c(k) = \dfrac{1}{T}\displaystyle\int_0^T x_c(t)\, e^{-j2\pi kt/T}\, dt$, and that if we define $X_{1,c}(j\Omega) = \displaystyle\int_0^T x_c(t)\, e^{-j\Omega t}\, dt$, then $X_c(k) = X_{1,c}(j2\pi k/T)/T$ [see (4.13)]. Next, suppose that we replace $X_{1,c}(j\Omega)$ by $X_{2,c}(j\Omega) = \displaystyle\int_0^{T_f} x_c(t)e^{-j\Omega t}\, dt$, where T_f is the length of our time window of samples. (Thus in the DFT spectral estimating process, $T_f = N\Delta t$.) Suppose that $T_f \geq LT$, where L is a positive integer. In other words, we have in our time window at least L periods of the periodic $x_c(t)$. Then $X_{2,c}(j2\pi k/[LT]) = \displaystyle\int_0^{T_f} x_c(t)e^{-j2\pi kt/(LT)}\, dt$. If $T_f = LT$, then $X_{2,c}(j2\pi k/[LT]) = LT \cdot X_{L,c}(k)$ where $X_{L,c}(k)$ are the FS coefficients of $x_c(t)$ where we assume $x_c(t)$ is just LT-periodic rather than T-periodic.

(a) Find $X_{L,c}(k)$ in terms of $X_c(k)$, as defined above.

(b) Find $X_{2,c}(j2\pi k/[LT])$ in terms of $X_c(k)$ when $T_f \neq LT$—that is, when $LT < T_f < (L+1)T$. Interpret in terms of what happens to the shape of the spectrum $X_{2,c}(j\Omega)$ as L increases.

Practical Problem with Oversampling

7.3. A guitar string ("A") is plucked and recorded by a microphone. The DFT of an N-point sample is as shown in Figure P7.3a for a sampling frequency of 51.28 kHz and $N = 8192$. The

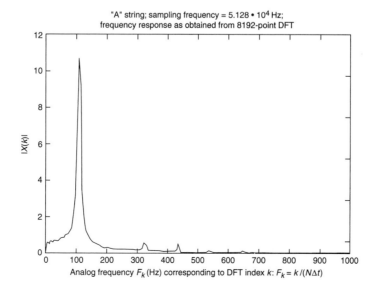

Figure P7.3a

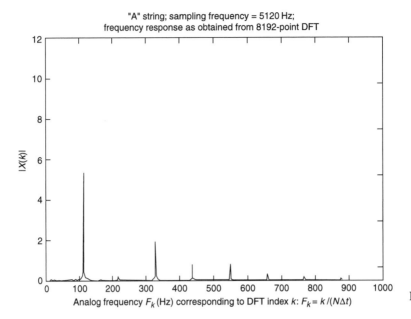

"A" string; sampling frequency = 5120 Hz;
frequency response as obtained from 8192-point DFT

Analog frequency F_k (Hz) corresponding to DFT index k: $F_k = k/(N\Delta t)$

Figure P7.3b

procedure is repeated at 5.12 kHz using the same number of samples, and the result is as in Figure P7.3b. In the context of the results in Problem 7.2 and any other relevant facts, explain these results. Is there a danger in oversampling too much? Is each of the harmonics an octave away from the adjacent harmonies?

Use of the DFT to Compute the DTFT

7.4. Let $h(n) = u(n) - u(n - 4)$.

(a) Find the DTFT of $h(n)$; express in terms of digital frequency f.

(b) Plot the magnitude of the DTFT for $0 \le f \le \frac{1}{2}$ by writing a Matlab m-file to evaluate the magnitude of your closed-form expression in part (a) for, for example, 300 values of f.

(c) Repeat part (b) by using Matlab to calculate first the exponential sum term by term [i.e., using the DTFT definition, not the closed-form result of part (b)], and then take the magnitude. Plot on the same axes, and verify that the two are the same.

(d) Repeat part (b) by computing the 32-point DFT of $h(n)$ (zero-padded) using `fft` and by plotting the magnitude of only the values of f on the range $0 \le f \le \frac{1}{2}$. Show these values by using asterisks, again on the same axes. Do you expect exact agreement?

7.5. Suppose that $x(n) = n$ for $n \in [0, N_1 - 1]$ and $x(n) = 0$ otherwise.

(a) Analytically find $X(k) = \text{DFT}_N\{x(n)\}$, where $N > N_1$. Also analytically find $X(e^{j\omega})$. Then for $N_1 = 32$ and $N = 64$, compare your result for $X(k)$ against use of `fft` operating on $x(n)$ in Matlab by plotting magnitudes and phases against each other.

(b) Verify (7.4e) and (7.5) by performing the operations there on your DFT result in part (a) and comparing with your DTFT result in part (a), and against a direct calculation of the DTFT sum using $x(n)$. Show the results graphically by plotting magnitude and phase; also include $X(k)$ in your plots (using correct frequencies corresponding to k).

7.6. Let $x(n)$ be a discrete-time triangle of length $N = 16$ [see (4B.20)]. First, plot $|X(e^{j2\pi f})|$ versus f using (i) the analytical form in (4B.22) and (ii) computing the DTFT-definition sum directly, using $x(n)$. Your results should agree. Next, zero-pad $x(n)$ out to $M = 1024$. Using `fft`, calculate $X_{zp;M}(k)$ [see (7.3)], and overlay its magnitude onto the above plots (again, there should be perfect agreement). Next, compute $|X(k)| = |\text{DFT}_N\{x(n)\}|$ using `fft`, and correctly superimpose these points onto the plot. Finally, use the interpolation formula in (7.6b) on $X(k)$, and superimpose the magnitudes of these values onto the plot. Again there should be agreement, thereby verifying (7.6b).

7.7. (a) A 128-point sample of speech signal $x(n)$ ($\Delta t = 91$ μs) is zero-padded and the 4096-point DFT is taken. What is the resolution of the DFT spectrum in this case (in cycles/sample and in hertz)?

(b) What is the frequency in hertz corresponding to DFT sample $k = 1334$?

(c) To improve the resolution to 10 Hz, what must be done for fixed Δt? Be quantitative.

7.8. Let $h(n)$ be equal to the FIR filter obtained from the Matlab statement

`h=fir1(N-1,[0.6 0.9])`, where $N = 32$.

(a) Using Matlab help, determine what type of this filter $h(n)$ is and what its cutoff frequencies are in cycles/sample.

(b) Form the sequence $h_L(n)$ composed of $L = 5$ repetitions of $h(n)$. Plot $1/L$ times the magnitude of its $M = NL$-point DFT using stems, and plot in the background the DTFT of $h(n)$, using a dashed line. Compare the two and comment.

(c) Using (4B.24), determine $H_L(e^{j\omega})$ for all ω, writing the modification to $H(e^{j\omega})$ in polar form. Plot $H_L(e^{j\omega})$ [using $H(e^{j\omega})$ obtained as the numerical DTFT of $h(n)$] on top of the others, and comment on the agreement or disagreement you find. Also, check your result by performing a numerical DTFT on $h_L(n)$ (it should agree with the other curve obtained using your general result). Is the DTFT zero for all ω between $f_k = k/N$?

Truncation

7.9. By actually carrying out the convolution, verify (6.79b) in the context of this chapter (with X replaced by a purely digital filter H) by computing the DTFT two ways. Specifically, let $H(e^{j\omega}) = 1$ for $|\omega| \leq \omega_A$ and zero otherwise. Let the FIR digital filter $h_N(n)$ be zero-phase of odd length $N = 2M + 1$, nonzero for $|n| \leq M$, and equal to $h(n) = \text{DTFT}^{-1}\{H(e^{j\omega})\}$ for this range of n (and zero otherwise). Perform the frequency-domain convolution in (6.79b), and compare with a direct evaluation (i.e., slight simplification of the finite sum) of the DTFT$\{h(n)\}$ sum using an analytical evaluation of $h_N(n)$.

7.10. Let $x(n) = a^n u(n)$ and $x_N(n) = x(n) \cdot w_{N,R}(n)$, where $w_{N,R}(n) = u(n - N) - u(n)$ is the rectangular window of length N.

(a) Find closed-form expressions for $X(e^{j\omega})$ and $X_N(e^{j\omega})$.

(b) Show numerically that $X(e^{j\omega})$ and $X_N(e^{j\omega})$ are related by (6.79b) and (6.80). Let $a = 0.9$ and $N = 10$, and write a Matlab routine to compute the right-hand side of (6.79b) using any other Matlab routines you wish. Plot the numerical-integration estimate versus the closed-form version of $|X_N(e^{j\omega})|$ on the same axes. Also plot the nontruncated $|X(e^{j\omega})|$ on the same axes to see the effect of truncation.

7.11. Suppose that $x(n) = a^n u(n)$. For $a = 0.94$ and $N = 16$, plot $X_N(k)$ and $X_N(e^{j2\pi f})$ against the analytically determined $X(e^{j2\pi f})$, where $X_N(k) = \text{DFT}_N\{x_N(n)\} = \text{DFT}$ of first N values of $x(n)[= x_N(n)]$ using `fft` and $X_N(e^{j2\pi f}) = $ analytically determined DTFT$\{x_N(n)\}$. Take care to correctly plot the $X(k)$ versus the correct digital frequencies f_k. Use $256/2 + 1$ values for which $f \geq 0$ for $X(e^{j2\pi f})$ and $X_N(e^{j2\pi f})$; plot only magnitudes. What is the relation between $X(k)$ and $X_N(e^{j2\pi f})$? Onto this plot, superimpose the result of (6.79b), (6.79c), and (6.80), using `quad8` in Matlab. Repeat for ($a = 0.94, N = 128$) and for ($a = 0.5, N = 16$). Make conclusions about truncation error versus N and a. Note that the case $N = 128$ will take a fair amount of computation time, so debug your code for $N = 16$ first.

DFT Convolution; Need for Zero-Padding

7.12. (a) Efficiently program $x(n)$ to have the values shown in Figure P7.12, using repeated calls to the `boxcar` function.

(b) Obtain a length-30 lowpass filter from `fir1` whose cutoff frequency is $f_c = 0.25$. Compare filtering with `conv` against taking the DFT^{-1} of the product of the two 30-point DFTs of, respectively, $x(n)$ and $h(n)$. Explain any discrepancies, and repeat the DFT procedure so that it gives exact agreement for all n.

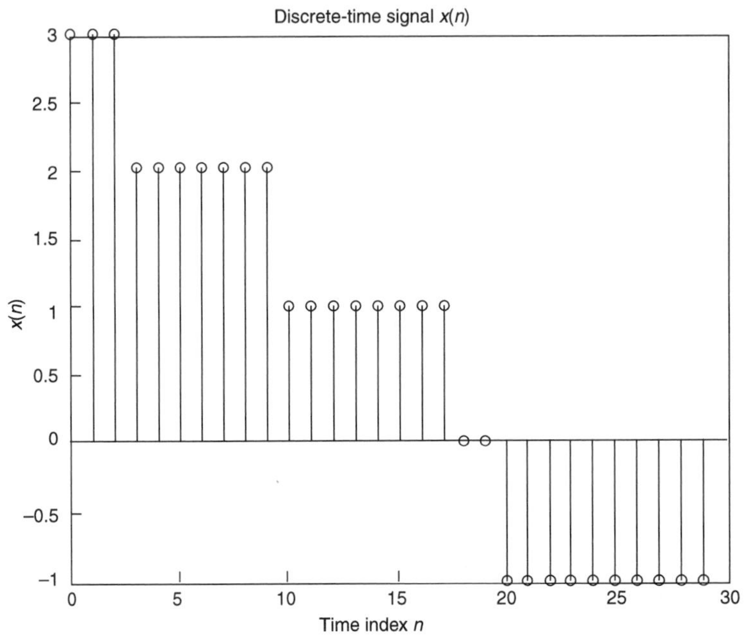

Discrete-time signal $x(n)$

Figure P7.12

7.13. Let $x(n) = \cos(5\pi n/[3N]) - 2\sin(5\pi n/[2N] + \pi/4)$, $n \in [0, N - 1]$, where $N = 64$, and let $h(n)$ be obtained from the Matlab `remez` algorithm as follows: `h=50.*remez(N-1,[0 0.44 0.52 0.9 0.97 1],[0 0 1 1 0 0])`. Plot $x(n)$ and $h(n)$ (on separate plots). To see what you can expect from this filter, plot its frequency response magnitude. Predict how the convolution of $x(n)$ and $h(n)$ should appear, and then reconcile with what you find. Then, using N-point DFTs, attempt to compute $y(n) = x(n) * h(n)$ via the `fft` routine; plot the convolution result $y_N(n)$. Next, zero-pad $x(n)$ and $h(n)$ by the appropriate number of zeros, and plot the magnitude DFTs of these sequences versus the corresponding f values. Argue why $|X_{zp}(k)|$ is for certain values of k much larger than for others. Also plot the magnitude of the product of $X_{zp}(k)$ and $H_{zp}(k)$ versus f. Then plot the DFT^{-1} of this product against $y(n)$ obtained from $x(n)$ and $h(n)$ using the `conv` command; these results should agree. Finally, use (7.46g) with the true $y(n)$ to reproduce the N-periodic $y_N(n)$ originally obtained with DFT length = N; you should find agreement.

7.14. Derive the requirement on DFT length M for no wraparound in DFT convolution, by determining explicitly which (if any) values of m in (7.46e) can possibly bring the argument of the Kronecker delta function to zero. Argue that only one value of m is permitted, and that is $m = 0$. Show how this requirement forces one to conclude that $M \geq N_x + N_h - 1$. If necessary, make tables of the Kronecker delta index for the possible values of n, ℓ, and m—identifying which such index values are acceptable (on $[0, N_h - 1]$). For a more analytical approach, it may be helpful to investigate various index range limits, and define such variables as $\alpha = n/M$ and $\beta = \ell/M$.

Sampling of the DTFT/Time-Domain Aliasing

7.15. Let $x(n) = a^n u(n)$. Let $X_N(k) = X(e^{j2\pi k/N})$ for $k \in [0, N - 1]$. Plot $\text{DFT}_N^{-1}\{X_N(k)\}$ versus n and $x(n)$ versus $n \in [0, N - 1]$ on the same axes for $a = 0.9$ and (i) $N = 10$, (ii) $N = 30$, and (iii) $N = 100$. Compare these results against those you calculate using (7.16b); do they agree for all values of N? Do you expect the aliasing to be where you see it (and where is this)?

7.16. Suppose that $x(n) = a^n$ for $n \in [0, N-1]$ and is zero otherwise, with $a = 0.98$ and $N = 64$.

(a) Write the analytical expression for $X(e^{j\omega})$.

(b) Numerically evaluate the expression in part (a) at $f_k = k/M$, where $M = 128$, and compute x_1, the inverse DFT_M of those samples. Plot and compare with $x(n)$. What do you expect to find, and what do you see in the plot?

(c) Repeat part (b) for $M = 32$. Explain quantitatively any discrepancies between the two sequences.

Sampling Rate Conversion

7.17 In Section 7.5, we discussed the "zoom transform" as a way of focusing in on a small portion of the spectrum. In this problem, we first implement it according to the directions given there. We will see that there is an easier, more efficient way.

(a) Acquire a 512-point discrete-time signal $x(n)$ (e.g., a sample of speech). First plot the entire sequence and entire DFT magnitude spectrum. Now, we wish to show the spectrum centered on $f_c = 0.4$ and $(\Delta f = 0.02)$ on either side; that is, for $f \in [0.38, 0.42]$. Use a huge-order FIR (finite-length) bandpass filter $h(n)$ designed from `fir1` using the default Hamming window (just follow the help directions; we discuss digital FIR filters in detail in Chap. 9) that passes only this range of frequencies.

(b) Let $y(n) = x(n) * h(n)$. Define the new sequence $y_1(n) = y(n) \cdot e^{-j2\pi f_{\mathrm{shift}} n}$. What is the appropriate value of f_{shift} so that the resulting spectrum is located appropriately for the next step (downsampling)? Remember that downsampling stretches the spectrum by the downsampling factor M. (Write your own downsampling routine rather than use `decimate`, because the latter has a built-in lowpass filter and we wish to investigate pure downsampling.) What difficulty will you find that requires further filtering? What kind of filtering? Does this difficulty give a hint as to how the calculations could more efficiently be done?

(c) Take the `fft` of the M-downsampled, frequency-shifted sequence $y_d(n)$, and plot the magnitude. Explain what your plot represents.

(d) Compare with the appropriate range of magnitude plot of the `fft` of the appropriately zero-padded sequence. Do they agree? Can you prove that they must be equal [within the accuracy of the filters needed in parts (a) and (b)]? What are the relative advantages and disadvantages of the two approaches?

Example on WP 7.2 may be helpful for Problems 7.18 and 7.20

7.18. Let $x_c(t) = 8\,\mathrm{sinc}(t \cdot 0.05\pi) - 4\,\mathrm{sinc}(t \cdot 0.4\pi) + \frac{1}{2}\sin(t \cdot 0.3\pi)$.

(a) Plot $x(n) = x_c(n\Delta t)$ for $\Delta t = 1$ s, $n \in [0, N-1]$, where $N = 100$. Also plot its N-point DFT magnitude. Does there appear to be any aliasing?

(b) Plot $x_c(t)$ [i.e., $x_c(n\Delta t_1)$ for $\Delta t_1 \ll \Delta t$ (e.g., $\Delta t_1 = \Delta t/10$)], and on the same axes, plot the sinc reconstruction of $x_c(t)$ based on the samples $x(n)$ in part (a), the zero-padded-DFT time-domain interpolation of Section 7.4 from these samples with $L = 10$, and also (using stems) the samples themselves. In this plot, observe the entire time in part (a) but print the plot only for $t \in [0, 30$ s$]$.

7.19. (a) Perform the same interpolation as in Problem 7.18, but this time use the zero-interlace method in (7.38) et seq. Plot the zero-interlaced sequence, its DFT, its lowpass version, and your interpolation results against those of Problem 7.18 to verify correctness. Use a simple frequency-sample approach to obtain the required lowpass filter.

(b) Downsample the result obtained in part (a) by the factor $M = L = 10$. Plot against $x(n) = x_c(n\Delta t)$ as in Problem 7.18. Your results should agree essentially exactly.

7.20. A length-N discrete-time real-valued signal $x(n)$ is exactly one period of samples of a periodic continuous-time function $x_c(t)$, where $N = 32$. The "nonnegative k" portion of the DFT_N of $x(n)$ is as follows:

```
X₁=[1 2-j*2 7+j*5 -8+j*6 -j*4 10 5 3 -15-j*4 -8 -11 -8 2 . . .
    9-j*10 1-j*2 0 2].
```

(a) Form $X(k)$, the complete DFT$_{32}$ of $x(n)$. Then use the inverse FFT Matlab routine called `ifft` to obtain $x(n)$. In terms of the polar form of $X(k)$, express $x(n)$ in closed form in terms of only real quantities. To check your result, program it in Matlab and compare plots of $x(n)$ obtained using `ifft` against the closed-form expression you obtain for $x(n)$.

(b) Determine the unique continuous-time signal $x_c(t)$ that when sampled every Δt is equal to $x(n)$, without any aliasing. Write an expression for $x_c(t)$ involving only real-valued quantities [base it on your result in part (a)]. Let $\Delta t = 0.05$ s. Plot the samples $x(n)$ (correctly against continuous time), and superimpose $x_c(t)$ evaluated on a dense grid $\Delta t_{dense} = \Delta t/10$. In your plot, the samples $x(n)$ should fall on the curve $x_c(t)$.

7.21. Let $x_c(t)$ and $x(n)$ be as determined in Problem 7.20. In this problem we will change the sampling rate working only with $x(n)$, but comparing the results against the original $x_c(t)$.

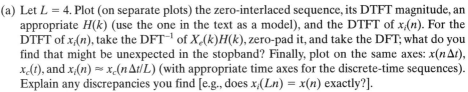

(a) Let $L = 4$. Plot (on separate plots) the zero-interlaced sequence, its DTFT magnitude, an appropriate $H(k)$ (use the one in the text as a model), and the DTFT of $x_i(n)$. For the DTFT of $x_i(n)$, take the DFT^{-1} of $X_e(k)H(k)$, zero-pad it, and take the DFT; what do you find that might be unexpected in the stopband? Finally, plot on the same axes: $x(n\Delta t)$, $x_c(t)$, and $x_i(n) \approx x_c(n\Delta t/L)$ (with appropriate time axes for the discrete-time sequences). Explain any discrepancies you find [e.g., does $x_i(Ln) = x(n)$ exactly?].

(b) Downsample $x_i(n)$ by factor $M = 3$ to obtain $x_d(n)$. Plot $x_c(t)$ and $x_d(n)$ on the same axes [with appropriate time values for $x_d(n)$]. You should find that $x_d(n)$ fall almost exactly on the curve. What approximate relation holds between $x_d(n)$ and $x_c(t)$, and what overall has been accomplished relative to the original samples $x(n) = x_c(n\Delta t)$?

7.22. Develop the downsampling relation (7.36) without using complex integration.

Attempt at Zero-Phase Filter

7.23. Someone suggests that a simple way to obtain a zero-phase FIR filter from one with nonzero phase (and having the same magnitude response as the original filter) is to use the following approach. If $h(n)$ is the FIR filter and $H(k)$ its DFT, define $H_L(k) = \log\{H(k)\}$. Then define $H_L'(k) = \text{real}\{H_L(k)\}$, $H'(k) = \exp\{H_L'(k)\}$, and finally $h'(n) = \text{DFT}^{-1}\{H'(k)\}$. What is the intent, and why does it fail? Demonstrate failure by choosing $h(n)$ to be a 39-point causal triangle and using the technique on it. Plot the resulting magnitude and phase of DTFT$\{h'(n)\}$ and a tell-tale sequence on the same axes. Also plot $h(n)$ and $h'(n)$ versus n on the same axes.

Correlation Using the DFT; Signal-to-Noise Ratio

7.24. The systems in Figures P7.24a and b are reputed to produce the same output. In particular, $y_1(n)$ in Figure P7.24a is claimed to be the nonaliased, downsampled (by factor 2) version of the autocorrelation of $x(n)$. (The autocorrelation function operator is symbolized by $\circledast$.) This downsampled version may be, if necessary, a filtered autocorrelation. In the following, draw detailed sketches of the magnitude spectra of *all* signals in each system, paying attention to amplitude scaling, for the spectrum $X(e^{j\omega})$ of $x(n)$ as given in Figure P7.24c. Express the spectrum of each signal in terms of $X(e^{j\omega})$.

(a) Is system 1 (in Figure P7.24a) linear? Is it shift-invariant? Why or why not? If $H_1(e^{j\omega})$ is required to avoid aliasing, specify it. If $H_1(e^{j\omega})$ is not required, state why not. With $X(e^{j\omega})$ as shown, how would you describe $y_1(n)$ in terms of $r_{xx}(n)$?

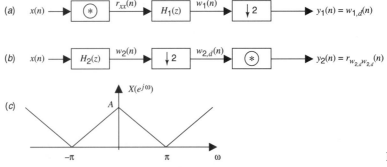

Figure P7.24

(b) Next consider system 2, shown in Figure P7.24b. Can $H_2(e^{j\omega})$ be chosen so that $y_2(n) = y_1(n)$? If not, why not? If so, specify $H_2(e^{j\omega})$.

(c) Is there an advantage of either system 1 or system 2? If so, which is better, and why?

7.25. Create (for an interesting finite-length sequence) a 50-length lowpass filter $h(n)$ with cutoff f_c

= 0.25 as follows: h = fir1(49,0.25*2). Then make a delayed version of $h(n)$: $h_1(n) = h(n - n_0)$, filling in missing values with zeros. (Choose, for example, $n_0 = 8$.) Plot the two sequences against n, using the same axes. Form the cross-correlation sequence $r_{hh1}(k)$. Use the DFT method as described in the text, using adequate zero-padding. Numerically determine the maximum value r_{max} of $r_{hh1}(k)$, and determine the corresponding Matlab array index $k_{max,M}$ and the actual lag value k_{max}. Can you quantitatively explain or predict the values of r_{max} and actual k_{max} in terms of n_0? Consider both $N_{DFT} = 2N_h - 1$ and $N_{DFT} > 2N_h - 1$. Also plot the autocorrelation sequence $r_{hh}(k)$ (also obtained using the DFT method) superimposed on $r_{hh1}(k)$. Check whether these sequences are also separated by n_0 in your plot.

7.26. Load into Matlab a long array of, for example, sampled speech data or other sampled data. Somewhere in the midst of this array, define a small array $x(n)$ that is N samples of the big array. Also define $x_1(n)$ to be a version of $x(n)$ delayed by $n_0 = 10$ samples; unlike Problem 7.25, do not pad with zeros but use data further back in the long data array for needed data that is not in $x(n)$. Subtract out the mean, if necessary, and normalize by dividing each element of the arrays by the norm of the original array (use mean and norm of undelayed sequence for both arrays, for comparability). First, plot $x(n)$ and $x_1(n)$ versus n on the same axes. Next, compute $r_{xx1}(k)$ via (7.51). In your code, use only one for loop and no if blocks (compare with using two for loops and an if block, and you will see a noticeable difference in computation time). Then compare with the result of the DFT method for calculating $r_{xx1}(k)$ (remember to zero-pad in the manner described in the text). Pretend that you do not know the value of n_0, and use your results from the DFT version of $r_{xx1}(k)$ to determine the value of n_0. Make any necessary modifications. Although n_0 in this case could easily be determined from the time plots directly both here and in Problem 7.25, why in practice would the correlation technique be useful?

7.27. In practice, the delayed signal data $x_1(n)$ will often be corrupted by noise and by distorting filtering of the propagation path (assuming the delay is due to signal propagation). Thus, the

received signal $x_1'(n)$ will be a distorted version of $x_1(n) = x(n - n_0)$. To $x_1(n)$ in Problem 7.26, we will here add noise and lowpass filter it as follows. The signal-to-noise ratio $SNR_{dB} = 20 \log_{10}\{|v_{sig}|/|v_{noise}|\}$, where $|v_{sig}|$ is the amplitude of the signal and $|v_{noise}|$ is the (root mean-square) amplitude of the additive noise. The distorting filter will be specified below.

(a) First, give a formula for the required $|v_{noise}|$ to obtain a specified signal-to-noise ratio, SNR_{dB}, (in terms of SNR_{dB} and $|v_{sig}|$). Let there be only additive noise (no filtering). Add sufficient noise to bring the SNR_{dB} to the given values specified below; for simplicity, use the randn with unit variance to produce a sequence of normally distributed random numbers typically having on the order of unit amplitude. Plot for $n_0 = 0$ the noise-corrupted $x_1(n)$ with SNR_{dB} = (i) 20 dB, (ii) 10 dB, and (iii) 0 dB against the original, uncorrupted $x_1(n)$ to get a feel for what SNR_{dB} means. Also list $|v_{sig}|/|v_{noise}|$ for these and other common values of SNR_{dB}. Give a reasonable estimate to use for $|v_{sig}|$ (an RMS value calculated from the data).

(b) For $SNR_{dB} = 20$ dB and below, use the DFT method to compute $r_{xx1'}(k)$ for $n_0 = 10$. Is the estimate of n_0 produced by the calculation in Problems 7.25 and 7.26 accurate here? For brevity, plot $r_{xx1'}(k)$ only for a noise-level case in which the n_0 estimate disagrees with the true value. Why might the $r_{xx1'}(k)$ calculation of n_0 be somewhat robust?

(c) Now let $SNR_{dB} = 10$ dB, and filter $x_1'(n)$ with the highpass filter obtained from the command filtfilt(h,1,x1) for h = fir1(31,2*f_c,'high'), where (i) $f_c = 0.02$ and (ii) $f_c = 0.1$. Plot $x_1(n)$ and $x_1''(n)$ = the noise/convolution-corrupted version of $x_1(n)$ on the same axes for comparison. Then plot the cross-correlation function of $x(n)$ and $x_1''(n)$, and attempt to estimate n_0 as before. Comment on your results. The filtfilt command filters forwards and then backwards, to give zero phase distortion (zero additional delays). What would happen if we just used conv(h,x1)?

Block Convolution

7.28. As noted in the text, there is a second approach to block convolution known as the overlap-add method. In this case, successive sections $x_k(n)$ of data of length $N - N_h + 1$ are each zero-padded out to $N = $ DFT length. Then each zero-padded section is convolved with the zero-padded $h(n)$ via the FFT algorithm. The product is a section $y_k(n)$. These output sections are overlapped as follows: The last $N_h - 1$ points of section k are overlapped and added with the first $N_h - 1$ points of section $k + 1$ to obtain valid output values; the remaining points are valid as is. Write code for the overlap-add method, and apply to either an example of your own or $x(n)$ as specified in Example 7.11. Make plots analogous to Figures 7.44 and 7.46, and as in Figure 7.47, compare the resulting convolution result with an all-in-one aperiodic convolution to verify correctness.

7.29. Extend the overlap-save technique of block convolution described in Section 7.7 to the case for which the data block size M is less than the available FFT size (N), a situation that might occur due to hardware constraints. Consider the case where $N - M < N_h$, where N_h is the length of the FIR filter $h(n)$. Does increasing N increase the number of valid output points? For continuous valid output, where should a new section begin?

7.30. In Problem 7.29, the overlap-save technique of block convolution was extended to include the case for which the block size M is less than the available FFT size N. There we considered only the case $N - M < N_h$, where N_h is the length of the filter. In this problem, let $N - M > N_h$. Draw pictures representing $x_\ell(n_\ell) * h(n_\ell)$ and the circular convolution $x_\ell(n_\ell) *_N h(n_\ell)$, identifying the boundary points of where the convolutions are zero and where the N-periodic convolution gives valid results. Show all relevant detail. For example, are there regions of wraparound, and if so, where do they occur? Has the number of valid points increased by increasing N?

7.31. Consider again the overlap-save method of sectioned convolution. Recalling how under the DFT the aperiodic convolutions are aliased (replicated), modify the algorithm so that the output values to be discarded appear at the *end* of the output sequence rather than at the beginning, as previously. Let the DFT size equal the section length $= N$, and the filter length be N_h. Show in full detail all your reasoning, using fully labeled diagrams, equations, or both. Fully specify the elements of the arrays whose DFTs are to be computed, where the valid output points lie, and which output points are to be discarded.

7.32. Show that the section $x_k(n_k)$ as defined in the text for the overlap-save method for doing block convolution is exactly the set of input values needed to compute the kth output section. That is, using the aperiodic convolution expression for $y(n)$ and the values of n in (7.48) for which $y(n)$ is obtained for section k, determine the range of m for which $x(m)$ are required. Show that this range is the same range that defines x_k in the text.

Fast Algorithms/Hartley Transform

7.33. Translate the FFT code in the text into working Matlab code. Note that there is no "GO TO" in Matlab. Use Matlab's complex variables to slightly simplify the code. Test by comparing numerical results for a test sequence with those using Matlab's internal `fft` routine.

7.34. Derive the signal flowgraph for the $N = 8$ decimation in frequency FFT algorithm, using the notation in Section 7.10. Hint: Use the suggestion for decomposition given in the text. Recognition of smaller-sized DFTs is more subtle here than for decimation in time, but with care it is possible. Specifically, examine different groups of output rather than different groups of input.

7.35. (a) Because all the imaginary parts of a real array are zero, it would seem that we should be able to implement the FFT using a complex array of length equal to half of that of the real sequence. Let the real array $x_R(n)$ have length $2N$. Using (7.66b) with $L = 2$ and $M = 2N/2 = N$, show what FFT should be taken and what further processing is needed to obtain $X_R(k)$, the $2N$-point DFT of $x_R(n)$. Hint: Put the even-indexed values of $x_R(n)$ in the real part of the N-length complex array $x_A(n)$ and the odd-indexed values of $x_R(n)$ in its imaginary part, and use the properties of DFT{Re{$x(n)$}} and DFT{Im{$x(n)$}}.

(b) Implement your algorithm in part (a) using Matlab code; verify with a typical real data array by comparing with the result of `fft`. Avoid using any `for` loops. In particular, explicitly indicate how the values of $X_R(k)$ for $k \in [N, 2N - 1]$ are obtained.

7.36. (a) How many multiplications are predicted for the eight-point FFT algorithm, using the formula that does not assume that N is large? How does this compare with the number of non-± 1 coefficients in the decimation-in-time algorithm as seen in Figure 7.65?

(b) Draw the signal flowgraph for the decimation-in-time FFT algorithm for $N = 4$. Follow the derivation procedure in the text. Draw both the fully labeled form as in Figure 7.64 and the simplified form as in Figure 7.65. How many non-± 1 coefficients are there, and how many are predicted by the formula developed in the text?

7.37. Use a Matlab version of the bit-reversal code in the text to obtain the bit-reversed index sequence for $N = 32$.

7.38 The chirp z-transform is an efficient way of computing a small number of the DFT coefficients without having to compute them all. As the name implies, it can also be used to compute the z-transform off the unit circle.

(a) Expand $(n - k)^2$, and solve for nk as a function of all the other terms, including $(n - k)^2$.

(b) Substitute this result into the DFT definition. Express the result as the convolution of a modified $x(n)$ with a "chirp" signal. Why is the signal given the name chirp?

(c) Implement the algorithm in part (b) in Matlab. Test on a sequence $x(n)$ of your choice of length 2^{14}, finding $X(k)$ for only $k \in [0, 9]$. Compare with the results of `fft`, noting also the comparison of how long it takes to obtain each result. What results do you find, and how might the situation for the chirp z-transform be improved?

(d) Modify the algorithm to "start" computing samples at $\omega_0 = 2\pi m/N$ where m is any user-selected integer between 0 and N. Which indices of the usual $X(k)$ will be obtained from the chirp procedure? Comment on the practical use of starting at ω_0.

7.39. A faster way of computing convolutions that involves no complex arithmetic is by using the fast Hartley transform (FHT). The discrete Hartley transform (DHT)

$$X_H(k) = \sum_{n=0}^{N-1} x(n)\{\cos(2\pi nk/N) + \sin(2\pi nk/N)\}, \tag{P7.39.1}$$

which the FHT computes, has the same relationship to the continuous Hartley transform that the DFT has with the Fourier transform. Specifically, the continuous Hartley transform presented by R. V. L. Hartley in 1942 (same Hartley as the Hartley oscillator) may be defined as

$$X_{H,c}(\Omega) = \int_{-\infty}^{\infty} x_c(t)\{\cos(\Omega t) + \sin(\Omega t)\} \, dt.$$

It is easily verified that $X_H(k)$ is N-periodic, by just replacing k by $k + N$ in (P7.39.1). Therefore, it is understood in the following that all indexes are to be evaluated modulo N (e.g., "$-k$" below means "$N - k$" except for $k = 0$). In this entire problem, we assume that $x(n)$ is real, which is the overwhelmingly usual case.

(a) Prove that $H_N(nk) = \cos(2\pi nk/N) + \sin(2\pi nk/N)$ and $H_N(mk)$ are orthogonal. Using this property, determine the correct formula for the inverse DHT.

(b) Show that for real $x(n)$ we can obtain $\text{DFT}_N\{x(n)\} = X(k) = X_r(k) + jX_i(k)$ from $X_r(k)$ $= X_{He}(k)$ and $X_i(k) = -X_{Ho}(k)$, where $X_{He}(k) = \frac{1}{2}[X_H(k) + X_H(-k)]$ is the even part of $X_H(k)$ and $X_{Ho}(k) = \frac{1}{2}[X_H(k) - X_H(-k)]$ is the odd part of $X_H(k)$ and thus $X_H(k) =$ $X_{He}(k) + X_{Ho}(k)$.

(c) Show that $\text{DHT}_N\{x_1(n) *_N x_2(n)\} = X_{H1}(k)X_{He2}(k) + X_{H1}(-k)X_{Ho2}(k)$, where $*_N$ denotes periodic convolution. Two approaches are possible: The easy approach is to use the results in part (b) along with the DFT convolution theorem; the brute-force approach, although harder, does not assume validity of the DFT convolution theorem.

(d) Explain why only half as many multiplications are required to compute the DHT_N as are required to compute the DFT_N, for real $x(n)$.

7.40. In this problem, we derive the fast Hartley transform algorithm. Let N be a power of 2. Also, all arguments of indexes are assumed to be taken modulo N.

(a) How many multiplications are required for the DHT_N, without using symmetries?

(b) What is the DHT of a length-2 sequence? Specifically, if $x_2(n)$ has DHT $X_{H,2}(k)$ for n and k taking values 0 and 1, give separate expressions for $X_{H,2}(0)$ and $X_{H,2}(1)$.

(c) Prove the addition theorem for the Hartley kernel $H_N(m + n) = H_N(m)\cos(2\pi n/N) + H_N(-m)\sin(2\pi n/N)$. Of course, m and n are obviously interchangeable.

(d) Break $X_H(k)$ into two sums, one involving $x(n)$ for n even and one involving $x(n)$ for n odd. For the rest of this problem, consider the case $N = 8$ when convenient, but wherever feasible retain the variable N. How many real multiplies are required at this first stage of decomposition (in terms of N)? For what range of N is a savings made?

(e) Perform the next stage of decomposition, following the procedure in part (d). From this secondary flowgraph, put all your results together into a flowgraph for the 8-point DHT.

Cepstrum

7.41 In speech processing, the determination of the pitch frequency (or period) is important for speech analysis and synthesis. We investigate here the use of the cepstrum for its estimation. Obtain a sample of speech data, and select a segment of 1024 samples that is voiced (very regular, periodic pattern). We may model such a sequence as $x(n) = s(n)w(n)$ where

$$w(n) = \sum_{\ell=0}^{L} \alpha_\ell \delta(n - \ell M),$$ where $T_{\text{pitch}} = M\Delta t$.   The low-quefrency signal $s(n)$ modulates the pulse train (basic vocal excitation) to make it appear as it does in $x(n)$; it includes vocal tract effects and other speech detail factors.

(a) Show that $\log\{W(z)\} = \sum_{\ell=1}^{L} \log\{1 + \beta_\ell z^{-M}\} + \text{constant}$; you need not determine the β_ℓ.

From this result, argue that a significant portion of the cepstrum $x_{\text{cep}}(m)$ will be roughly a periodic pulse train. Moreover, you will find that it is unmodulated, so it is much easier to analyze in the cepstral domain than it is in the normal time (n) domain.

(b) Use your method to estimate the pitch period and frequency of the actual data segment you collected. To improve the results, first apply a Hamming window to $x(n)$. Use only the "real" cepstrum: $x_{\text{cep}}(m) = DFT^{-1}\{\log\{|X(k)|\}\}$, so that you need not worry about the difficult matter of phase unwrapping. Describe a procedure for semiautomatically determining M and therefore T_{pitch}.

Illustrative Proof of Result Cited in Text

7.42. Derive the interpolation formula for DFT interpolation when N is even; in particular, what is $\phi_N(\theta)$ in this case?

Chapter 8

Digital Filter Design: Introduction and IIR Filters

8.1 INTRODUCTION

We began our study of DSP with a review of signals and systems, both continuous-time and discrete-time. Because of its great importance in z-transform theory and signal processing in general, we summarized some of the main results of complex variable theory. In Chapters 4 and 5, we compared several signal transforms with respect to both their applicability and their properties. In particular, some transforms are appropriate for discrete-time signals and systems, whereas others apply only for continuous time. This comparison naturally leads to the all-important continuous-time, discrete-time interface: sampling and reconstruction (Chapter 6). Applications discussed in Chapter 6 included digital simulation and replacement of continuous-time systems, and approximations of forward and inverse Fourier transforms using samples in the other domain. In Chapter 7, we focused on discrete-time processing, often without regard to whether the origin of the signal was a continuous-time or a discrete-time signal. We were able to bring to light some fast algorithms based on the DFT/FFT as well as discuss the FFT algorithm itself.

With this background, which includes a few examples of filtering in previous chapters, in this chapter and Chapter 9 we consider the filtering problem systematically and in more detail. In other words, previously we saw how to efficiently convolve one signal with, say, the unit sample response of a discrete-time system. We now focus more on the question, "*What* discrete-time system?" Recall that a filter—be it a coffee filter, an oil filter, a photographer's optical filter, or an electronic filter—is intended to in some way "purify" or modify the substance it filters. A coffee filter filters out the large coffee grounds, an oil filter filters out particles and sludge from the oil, an optical filter emphasizes/deemphasizes certain colors, and an electronic/numerical filter filters out unwanted signal components.

Because of its great importance in communication, control, and practically all electronic systems, electronic filter design has been developed to a very advanced state. First, analog filters, both passive and active, were perfected. Then their discrete-time equivalents were derived. Now design of digital filters is often carried out completely independently of analog systems. In addition, all the same principles used to design and apply one-dimensional filters can be applied to multidimensional signals (e.g., in image processing); we saw an example of two-dimensional filtering in Section 7.7.

A complete study of digital filters would easily fill several books and for the average DSP user would seem bewildering and overwhelming. In this and the following chapter, we confine our attention to only some of the more frequently used filter types and principles, to get a flavor of what is involved in digital filter design. We will, for example, look at some of the filter types implemented in Matlab. Although it is true that digital filter design is becoming much easier as a result of software packages, the better engineer will want to know how they work so that application, modification, debugging, and interpretation of results can be successfully achieved.

8.2   INTUITIVE APPROACHES

It may be asked what we are trying to remove from the signal in the first place. There are many possible answers to this question; a few of the more common ones follow. We may wish to eliminate additive noise in the signal contributed by power equipment, other instrumentation, communications or computer systems, and so forth. Such a noise source may be additive because it induces its contamination by electromagnetic induction within the signal circuit. Alternatively, the noise may be internally generated within the measurement equipment used to detect the signal. Additive noise could easily be eliminated simply by subtracting it, except for the fact that we generally do not know the noise signal. However, sometimes it can be estimated or predicted; such estimates can be the basis of a noise-cancellation system. With the exception of Example 8.1, these systems are outside the scope of chapters 8 and 9, but will be considered again in Chapter 10.

EXAMPLE 8.1

A constant-magnitude 60-Hz power noise component is superimposed on the desired signal $1000te^{-100t} u_c(t)$ in Figure 8.1a as shown. Devise a plan to recover the lost signal.

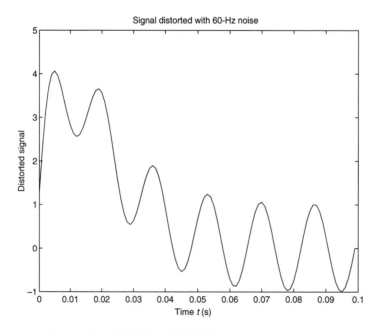

Signal distorted with 60-Hz noise

Time t (s)

Figure 8.1a

Solution

In the simple case that the phase of the superimposed sinusoid is equal to that of the power supply voltage, simple subtraction may work. We could use analog electronics to add a voltage or current of variable amplitude and opposite polarity to that in the given corrupted signal. Under the above assumption, this sample of the power supply voltage could be taken directly from the power supply ac outlet. By adjusting the amplitude, the 60-Hz noise may be nulled; see Figure 8.1b. In practice, the ideal results shown here will not be possible; some remnant of the noise will generally still exist.

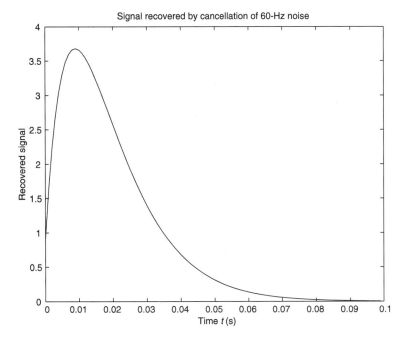

Figure 8.1b

Another major source of signal corruption is convolutional distortion. This type of distortion arises because either the original signal exists "behind" an LSI system or because our measuring equipment performs LSI distortion caused by its imperfect response characteristics. In either case, the signal we are given is equal to the original signal convolved with the unit sample response of the distorting system. Now to recover the original signal, some sort of deconvolution, or "inverse filtering" process, must be attempted. In some cases, this can be easy; in other cases, it is difficult or impossible to do accurately. We looked at cepstral processing as one approach in Section 7.9. A more simple-minded approach is taken in the following example, where it is assumed that we know the unit sample response of the distorting system.

EXAMPLE 8.2

A signal $x(n)$ is distorted by a system whose unit sample response is $h(n) = a^n u(n)$. Find the unit sample response of the filter $h_i(n)$ whose output is $x(n)$—or close to $x(n)$—if its input is $y(n) = x(n) * h(n)$.

Solution

A simple approach to deconvolution, that is not always possible, is to convolve $y(n)$ with a system whose z-transfer function is the multiplicative inverse of $H(z)$ for all z: $H_i(z) = 1/H(z)$. In this case, $H(z) = 1/(1 - az^{-1})$, so $H_i(z) = 1 - az^{-1}$. Taking the inverse z-transform gives $h_i(n) = \delta(n) - a\delta(n - 1)$. Check:

$$h_i(n) * y(n) = \sum_{m=-\infty}^{\infty} [\delta(m) - a\delta(m - 1)] \sum_{\ell=0}^{\infty} a^\ell x(n - m - \ell) \qquad (8.1a)$$

$$= \sum_{\ell=0}^{\infty} a^\ell x(n - \ell) - a^{\ell+1} x(n - [\ell + 1]). \qquad (8.1b)$$

Now notice that in (8.1b), the two terms always cancel for adjacent values of ℓ, except for the first term with $\ell = 0$—which is $x(n)$! Thus the deconvolution is exact in this example. This approach cannot be applied when $H(z)$ has zeros outside the unit circle, because then $H_i(z)$ has poles outside the unit circle and is therefore unstable.

The results of this example are presented in Figure 8.2 for $a = 0.93$ and $x(n)$ a segment of sampled speech signal having of length $N = 128$. Figure 8.2a shows $x(n)$, and Figure 8.2b shows the distorted signal $y(n) = x(n) * h(n)$. We see in Figure 8.2b that this convolutional distortion by $a^n u(n)$ is a sort of lowpass filtering, which may be predicted by examining $|H(e^{j\omega})|$. The convolution was performed in Matlab using `filter`, which simply solves the time-domain difference equation using the data as input and the filter coefficients as the difference equation coefficients (and

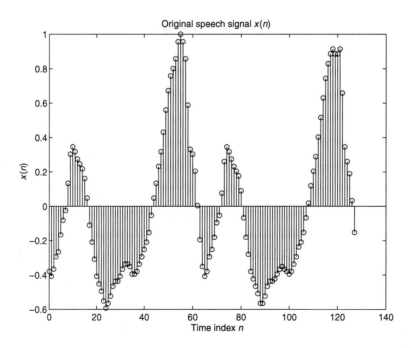

Original speech signal $x(n)$

Figure 8.2a

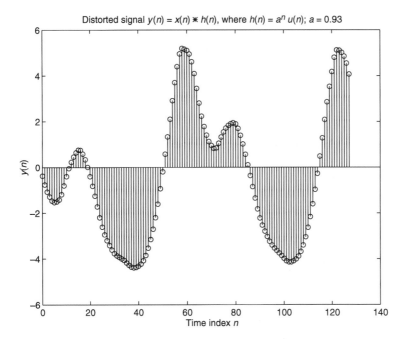

Figure 8.2*b*

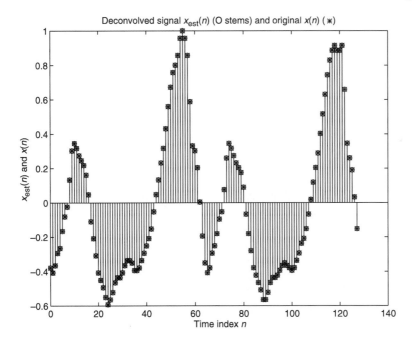

Figure 8.2*c*

assumes zero initial conditions). Finally, Figure 8.2*c* shows (O stems) $x_{\text{est}}(n) = h_i(n) * y(n) = h_i(n) * h(n) * x(n) = x(n)$; for verification, $x(n)$ is shown in * stems.

In practice, $h(n)$ could represent a single-time-constant continuous-time distorting system that has been discretized using impulse invariance for the purpose of determining the digital deconvolution filter $h_i(n)$. In this event, from Section 6.4.2 we know that the deconvolution will be imperfect because discrete-time convolution does not exactly equal samples of continuous-time convolution when one or more of the signals is nonbandlimited [e.g., $e^{s_1 t} u_c(t)$ has infinite bandwidth, where above, we used $e^{s_1 \Delta t} = a = 0.9$].

In the more common case where neither the distorting $h(n)$ nor the additive distorting signal is available to us, we turn to more general filtering techniques. The most highly developed and easily used design procedures give frequency-selective filters. These are filters that select only a certain band or bands of frequencies that they allow to pass on to their output. In many cases of practical interest, the frequency band of the signal does not overlap with that of the (additive) noise. In such cases, we simply chop off any and all energy in the "noisy" band and pass everything else. This general procedure is so commonly used that frequency-selective filters have come to dominate the topic of digital filter design.

An ideal frequency-selective filter has perfectly sharp transitions from passband to stopband and perfectly flat passbands and stopbands with zero signal passage in the stopbands. In the usual case when the desired signal has at least some energy in the band to be chopped off, distortion results even in the case of such an ideal filter. Alternatively, when the noise signal has energy in the desired signal band, the notion of a "cutoff" frequency is blurred. In such cases, an overemphasis on sharp, highly attenuating filters may be unwarranted, and approaches other than that of frequency-selective filter might be appropriate. However, these considerations do not diminish the importance of striving for the ideal filter because of its many uses where passing *any* stopband energy causes distortion (e.g, lowpass antialiasing filters used in multirate systems).

To the practical user of DSP, covering every type of frequency-selective digital filter would be both wearisome and irrelevant. In applications, we desire a simple-to-use, effective digital filter. At the same time, the techniques used to design frequency-selective filters are often prerequisite to or at least facilitate learning about the designing of other types of filters. When digital filter terminology is thrown about in books, articles, or conversation, it is helpful to know what it means and how it originates. Finally, anyone entering the field of DSP must have a solid understanding of frequency-selective filter design.

The most intuitive and qualitative technique for designing frequency-selective filters is pole- and zero-placement. In this technique, we recall that at a pole, $H(z)$ is infinite, and near it $|H(z)|$ is large, whereas at a zero, $H(z) = 0$, and near it $|H(z)|$ is small. We also know that $H(e^{j\omega}) = H(z)\big|_{z=e^{j\omega}}$, so for frequencies we want emphasized, we put a pole nearby, and for frequencies we want attenuated, we put a zero nearby. Unfortunately, this method does not provide any systematic procedure by which to choose the precise locations of the zeros, poles, or the filter orders, nor does it provide a means to quantitatively assess or improve the quality of the results. We might rightly conclude that the only sense in which this technique is optimum is that the thought involved is minimized!

Actually, for designs such as notch filters, this technique *is* useful. In Example 8.3, we consider using this technique for designing a bandpass filter. To assess the quality of our results, we will simply plot the frequency response. It may be asked why subsequently we usually plot only the filter frequency response and not the actual output sequence of the filter $y(n) = h(n) * x(n)$. The reason is that it would be harder to evaluate the quality of the filter for general input signals that way. By looking at the frequency response and comparing it with that of other filters, we can easily and fairly compare the quality of the designed filters.

EXAMPLE 8.3

Suppose that we want to pass all components of the input having frequencies between $f = \frac{1}{4}$ and $f = \frac{3}{8} = 0.375$, and highly attenuate all others. Design a filter having two zeros and four poles that approximately provides this response.

Solution

The most obvious approach is to choose (a) poles at (or near) the angles corresponding to the two cutoff frequencies and their complex conjugates, (b) and force the response to zero at the extreme low and high frequencies $f = 0$ and $f = \frac{1}{2}$ by putting (real) zeros there. Let the radius of the poles be, somewhat arbitrarily, 0.9. In general, placing them close to the unit circle makes strong peaks and valleys and narrow "transition bands," whereas placing them near the origin makes for a smooth frequency response but with wide transition bands and less marked distinction between "passbands" and "stopbands." With many more poles and zeros, a more nearly ideal response would be possible. With the radius of the poles equal to 0.9,

$$H(z) = \frac{(z-1)(z+1)}{(z - 0.9\angle\pi/2)(z - 0.9\angle -\pi/2)(z - 0.9\angle 3\pi/4)(z - 0.9\angle -3\pi/4)}. \tag{8.2}$$

The results are in Figure 8.3, which are seen to be at least approximately what is sought. However, the passband is far from flat, it has magnitude greater than unity at the cutoffs, and its phase is nowhere nearly linear (linear phase is a desirable property that will be fully explained in Section 9.5.1). The plots in Figure 8.3 were made using Matlab's `freqz` command, except that the frequency axis was divided by two to match the notation in this book (and of most other DSP practitioners).

Let us consider why we divide the `freqz` axis by 2. The so-called "Nyquist frequency" Ω_N is the highest frequency Ω_0 present in a signal, when Ω_0 is sampled exactly twice per cycle (and thus $\Omega_N = \Omega_0$ corresponds to $f = \frac{1}{2}$ cycle/sample or $\omega = \pi$ rad/sample). (The "Nyquist rate" is the value of sampling frequency Ω_s that accomplishes this, namely $\Omega_s = 2\Omega_0$.) Matlab's "normalized frequency (Nyquist = 1)" in

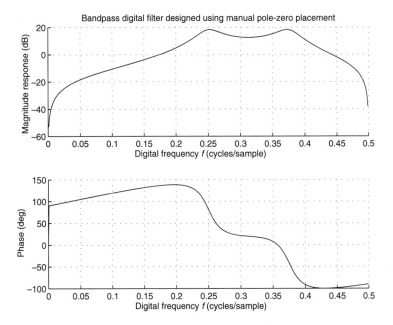

Figure 8.3

`freqz` is actually the ratio of the given frequency to the Nyquist frequency: Normalized frequency $= f/(\frac{1}{2}) = \omega/\pi = \Omega/\Omega_N$. Thus the normalized frequency is neither f [which has the intuitive meaning of number of cycles/sample of a frequency component in $x_c(t)$] nor ω (which has the intuitive meaning $\omega = \angle z = 2\pi f$). Thus in this book we use either f or ω to indicate digital frequency.

To conclude this example, in practice we probably would want the maximum log-magnitude gain to be 0 dB (i.e., unity gain), not nearly 20 dB as in Figure 8.3. This characteristic is possible by simply dividing $H(z)$ by $\max\{|H(e^{j\pi/2})|, |H(e^{j3\pi/4})|\}$.

It may be noted that the heuristic technique used in Example 8.3 would work well for Example 8.1, where we knew the exact value of the corrupting noise (60 Hz), for then we could place a zero on the unit circle at exactly the digital frequency corresponding to 60 Hz. Such a filter is called a first-order notch filter. Naturally, the 60-Hz component of the original signal would be annihilated, too, as would other frequency components near 60 Hz.

A typical application of a notch filter is in the elimination of the 60 Hz from biomedical signals such as the electrocardiogram (ECG). The opposite of a notch filter is a comb filter, which passes only a number of quite narrow bands. An example use of the comb filter is in the elimination of background EEG and other extraneous noise from visual evoked brain potentials. These brain signals, on the 5-to-20-Hz range, have their energy mainly in narrow, spaced bands. By cutting out all the rest, we can increase the signal-to-noise ratio (SNR).

Although such "fancy" multiband filters find their way into many applications, including also speech coding, and although adaptive filters are increasingly taking over (see Section 10.10), our focus in this chapter will be on the design and evaluation of basic filters such as the lowpass filter. We will see that once we have a good lowpass filter, spectral transformations can convert it to an equally high-quality highpass filter, bandpass filter, or other more complex filter. In fact, the lowpass filter has many applications in its own right. For example, the background muscle noise in an ECG signal can be removed by lowpass filtering. We have already seen that high-quality lowpass filters are necessary for the avoidance of aliasing in sampling and sampling-rate conversion, and for the elimination of other distortion.

As we saw in Section 2.4.6.4 (page 63ff), there are two fundamental categories of digital filter: FIR and IIR. In this chapter and in Chapter 9, we will look at techniques for designing filters of each type. As noted, in this chapter, we limit our attention to LSI filters. In Section 10.10, we broaden our discussion to filters whose coefficients vary with time (adaptive filters), which also are distinguishable as being FIR or IIR.

8.3   EVALUATION OF FREQUENCY-SELECTIVE FILTERS

As noted, the best way to compare filters is not to look at their outputs, which could be complicated, nonreproducible sequences, but instead at their frequency responses. In the frequency-selective filter design problem, we begin after already having decided what type of frequency response would be best to have. Some examples of this ideal filter determination process have been mentioned in Section 8.2. Once the ideal filter has been identified, the task reduces to an approximation problem. With so many design methods available, how can we choose which one to use? That question is sometimes partly answered by determining how much energy we would be willing to expend on the design. The design effort consideration was more limiting in the past than it is now. For example, with software such as Matlab, an optimal FIR filter is essentially just as easy to design as a simple truncation filter.

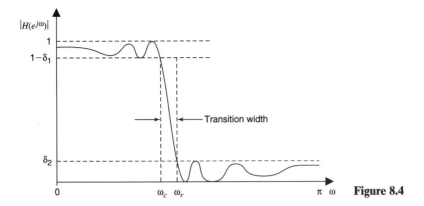

$|H(e^{j\omega})|$

1

$1-\delta_1$

Transition width

δ_2

0   ω_c ω_r π ω **Figure 8.4**

Another issue is the quality of a given filter versus that of one of the same order produced by a different design method. For frequency-selective filters, the main parameters used to assess the quality are (see Figure 8.4) maximum passband ripple δ_1 or maximum passband attenuation, maximum stopband ripple δ_2 or minimum stopband attenuation, and transition width from passband to stopband. Usually ripple and transition parameters conflict: Making one small tends to make the other larger. The transition width is quantifiable for digital filters as $\omega_c - \omega_r$, where the cutoff frequency ω_c is the maximum frequency for which $|H(e^{j\omega})|$ is above the minimum value $1 - \delta_1$ allowed in the passband and the rejection frequency ω_r is the maximum frequency for which $|H(e^{j\omega})|$ is above the maximum value δ_2 allowed in the stopband.

The number of coefficients in an FIR filter is M and in an IIR filter is $N + M$, where as before, M and N are, respectively, the numerator and denominator orders of $H(z)$. It is desirable to minimize the total number of filter coefficients, as doing so will minimize the computation time required to filter the data as well as the memory storage required. However, reducing filter order degrades the figures of merit just considered. As in any engineering design problem, all the factors must be weighed and compromised. On top of these considerations is the expense of time lost in design, which as noted above is increasingly being alleviated by user-friendly software.

8.4   ANALOG LOWPASS FILTER DESIGN

IIR digital filters are an efficient type of digital filter that can often be obtained from already designed analog filters. Historically, a large body of hard-won filter design knowledge was acquired for these analog filters, in the days before the growth of DSP. Today, this information is converted for use in the design of digital IIR filters by using various discretizations. For piece-wise constant-magnitude responses, three common analog filters are the Butterworth, Chebyshev, and elliptic filters. These analog filters are themselves useful for implementing the antialiasing and reconstruction filters in our earlier discussions, not just for designing digital filters. Because the analog filters are the sources of these digital IIR filters, we must first understand the analog approximations in order to understand their digital descendants.

As previously noted, the focus of filter approximation theory is on the lowpass filter because once designed, it can be transformed into an equivalent-quality highpass, bandpass, bandstop, or other filter by frequency transformation techniques. The lowpass filter is an even function of Ω. In the Butterworth, Chebyshev, and elliptic filters, only the magnitude function of Ω is approximated, not the phase. Thus the ideal lowpass filter magnitude function we desire to approximate is as shown in Figure 8.5.

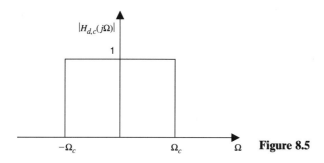

Figure 8.5

Notice that for this function, $|H_{d,c}(j\Omega)|^2 = |H_{d,c}(j\Omega)|$ (where the d,c subscript stands for "desired, continuous" filter response); consequently, it is possible to work with the more convenient $|H_{d,c}(j\Omega)|^2$. Thus we seek $|H_c(j\Omega)|^2 \approx |H_{d,c}(j\Omega)|^2$, where $H_c(j\Omega)$ is our approximating filter. The reason the comparison of the square of the magnitude is more convenient than just the magnitude is that the coefficients of $H_c(j\Omega)$ are what must be determined, and $|H_c(j\Omega)|$ is the square root of the sum of products and squares of them—nonlinear and complicated. Design is easier without the square root.

We will limit our detailed discussion to the Butterworth and Chebyshev analog filters and leave the interested reader to seek out details on the more difficult elliptic approximation problem (however, we will evaluate its performance graphically and note that Butterworth and Chebyshev filters are actually special cases of the elliptic filter). For the former two filter types, we use the following idea for lowpass filters. Because a constant divided by an Nth-order polynomial in Ω falls off as Ω^N, it will be an approximate lowpass function for large Ω. Therefore, an all-pole analog filter $H_c(s) = 1/A_c(s)$ is a good and simple choice for a lowpass filter form and is used in both the Butterworth and type I Chebyshev filters. Moreover, for a given denominator order, having the numerator constant (order zero) gives the greatest attenuation for large Ω for a given number of filter coefficients; with the numerator a constant, the denominator order maximally exceeds the numerator order.

The question thus becomes how to choose $A_c(s)$. We now embark on a systematic approach that, although mathematical, will allow us to quantitatively and efficiently meet specifications—which would be impossible using an ad-hoc approach. First, note that because the coefficients of $h_c(t)$ are real-valued, $H_c(-j\Omega) = H_c^*(j\Omega)$ so that $|H_c(j\Omega)|^2 = H_c(j\Omega)H_c(-j\Omega)$. If we replace $j\Omega$ by s [i.e., analytically continue the function $H_c(j\Omega)H_c(-j\Omega)$ throughout the s-plane except at the poles], the quantity $H_c(j\Omega)H_c(-j\Omega)$ becomes $H_c(s)H_c(-s)$. However, even for real-valued $h_c(t)$ [for which $H_c^*(s^*) = H_c(s)$], the function $H_c(s)H_c(-s) \neq |H_c(s)|^2$ except for $s = j\Omega$. Nevertheless, $H_c(s)H_c(-s)$ is a convenient even rational function of s that has the quadrantal symmetry property: Its pole configuration is symmetrical with respect to both the real and imaginary axes. Thus if s_1 is a pole of $H_c(s)H_c(-s)$, then so are s_1^*, $-s_1$, and $-s_1^*$. We know s_1^* is a pole because $h_c(t)$ is real, and we know $-s_1$ and $-s_1^*$ are poles because $H_c(s)H_c(-s)$ is an even function of s. We will find that $H_c(s)H_c(-s)$ is a convenient vehicle for translating conditions on $|H_c(j\Omega)|^2 \big[= H_c(s)H_c(-s)\big|_{s=j\Omega}\big]$ into numerator and denominator polynomials and thus pole- and zero-factors from which we may construct $H_c(s)$. We have

$$H_c(s)H_c(-s) = \frac{1}{A_c(s)A_c(-s)}. \tag{8.3}$$

Note that our design procedure will not directly result in a formula for $H_c(s)$—only in one for $|H_c(j\Omega)|^2$ and in turn an expression for $H_c(s)H_c(-s)$. Thus in implementing the filter $H_c(s)$, all the left-half-plane poles and zeros of $H_c(s)H_c(-s)$ are assigned to $H_c(s)$ so that $H_c(s)$ is a stable, causal, minimum-phase filter. In addition, we may scale the all-pole filter by a constant such as the inverse of the product of these poles so that the lowpass filter has unity dc gain. That is how we can construct an actual filter transfer function $H_c(s)$, given the poles and zeros of $H_c(s)H_c(-s)$. In terms of the ideal $A_c(j\Omega)$, denoted $A_{d,c}(j\Omega)$, Figure 8.5 is redrawn to become Figure 8.6; $A_c(j\Omega)$ must be infinity in the stopband so that $H_c(j\Omega)$ is zero, and $A_c(j\Omega)$ must be unity in the passband.

Because we prefer to find roots of a polynomial over finding where it is equal to unity, define the *characteristic function*[1] or attenuation function $K_c(s)$ for which

$$K_c(s)K_c(-s) = A_c(s)A_c(-s) - 1, \quad \text{and thus } |K_c(j\Omega)|^2 = |A_c(j\Omega)|^2 - 1. \quad (8.4)$$

Equation (8.4) is known as the Feldtkeller equation [note that we assume in (8.4) that $K_c(j\Omega)$ is conjugate-symmetric]. Now Figure 8.6 is redrawn as Figure 8.7 for the ideal $|K_c(j\Omega)|$ for a lowpass filter, denoted $|K_{d,c}(j\Omega)|$. Rational functions are smooth functions of Ω and can only approximate perfectly sharp curves such as those in Figures 8.5 through 8.7. Thus the practical $|K_c(j\Omega)|$ will be similar to the typical example shown in Figure 8.8a, where ϵ is the maximum value of $|K_c(j\Omega)|$ in the passband $(|\Omega| < \Omega_c)$, K_{min} is the minimum value of $|K_c(j\Omega)|$ in the stopband, and Ω_r is the rejection frequency defined earlier to allow a finite-width transition band from passband to stopband (where the stopband is $|\Omega| > \Omega_r$). The shaded areas show "forbidden" regions for $|K_c(j\Omega)|$.

Note that we can relate ϵ and δ_1 as follows. Because in the passband $K_{max} = \epsilon$, we have $\epsilon^2 = |A_c(j\Omega)|^2_{max} - 1$, or $|A_c(j\Omega)|^2_{max} = \epsilon^2 + 1$, or $|H_c(j\Omega)|^2_{min} =$

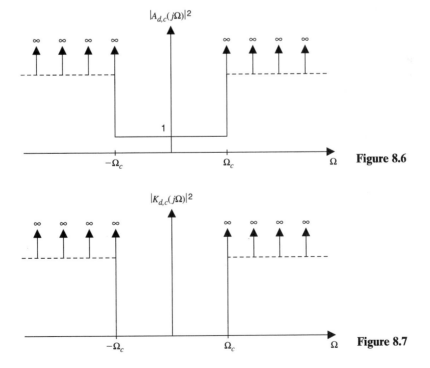

Figure 8.6

Figure 8.7

<hr />

[1]Do not confuse this with the characteristic function of probability theory nor with the characteristic polynomial of a linear system.

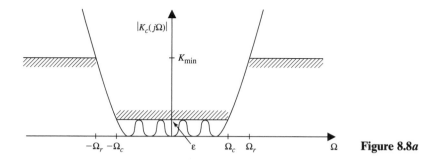

Figure 8.8a

$1/|A_c(j\Omega)|^2_{max} = 1/(1 + \epsilon^2)$, or $|H_c(j\Omega)|_{min} = 1/|A_c(j\Omega)|_{max} = 1/\{1 + \epsilon^2\}^{1/2} = 1 - \delta_1$, or, solving for ϵ, we have $\epsilon = \{1/(1 - \delta_1)^2 - 1\}^{1/2}$. By the same kind of analysis, we have in the stopband $1/\{K^2_{min} + 1\}^{1/2} = \delta_2$, or $K_{min} = \{1/\delta_2^2 - 1\}^{1/2}$. Thus we see that the new parameters ϵ and K_{min} are direct descendants of, respectively, δ_1 and δ_2.

8.4.1 Butterworth Filter

The Butterworth filter is defined as being maximally flat in the passband. For many applications, this characteristic is exactly what is needed, for the passband behavior may be all that we care about as long as the transition width is not excessive. The way the Butterworth filter achieves maximal flatness is by enforcing that $H_c(j0) = 1$ and $(|H_c(j\Omega)|^2)^{[m]}|_{\Omega=0} = 0$ for $0 < m \leq N - 1$, where as previously the superscript $[m]$ denotes the mth derivative. Noting that $|H_c(j\Omega)|^2 = 1/[|K_c(j\Omega)|^2 + 1]$, we see that the conditions on $|H_c(j\Omega)|^2$ are satisfied if $(|K_c(j\Omega)|)^{[m]}|_{\Omega=0}$ for $0 \leq m \leq N - 1$ [note that this set of requirements includes $K_c(j0) = 0$]. Requiring all these derivatives to be zero at $\Omega = 0$ uses up at the one frequency $\Omega = 0$ all but one of our degrees of freedom of our Nth-order approximating polynomial.

We also must choose for $|K_c(j\Omega)|$ a polynomial that grows monotonically for $|\Omega| > \Omega_c$ so that the stopband requirement in Figure 8.7 is met. We further want it to exceed ϵ *only* for $|\Omega| > \Omega_c$; thus, we will allow equality at $\Omega = \Omega_c$ [i.e., $|K_c(j\Omega_c)| = \epsilon$] and let it increase out from there. We will set $|K_c(j\Omega)|$ equal to a scaled version of the Butterworth polynomial, $B_N(\Omega)$, now to be determined. We first write $B_N(\Omega)$ as a general Nth-order polynomial in Ω:

$$B_N(\Omega) = \sum_{\ell=0}^{N} \alpha_\ell \Omega^\ell. \tag{8.5}$$

For determining the coefficients α_m of $B_N(\Omega)$, let $|K_c(j\Omega)| = B_N(\Omega)$ for the case $\epsilon = 1$, $\Omega_c = 1$; see Figure 8.8b. The requirement $K_c(0) = 0$ gives $\alpha_0 = 0$. The requirements on the derivatives $(|K_c(j\Omega)|)^{[m]}|_{\Omega=0}$ for $0 < m < N$ give $\alpha_m = 0$, as taking successive derivatives of (8.5) and setting $\Omega = 0$ in the results show. Thus $B_N(\Omega) = \alpha_N \Omega^N$. The requirement[2]

$$|K_c(j\Omega_c)| = |K_c(j1)| = B_N(1) = \alpha_N = 1 \; (= \epsilon) \tag{8.6}$$

gives $\alpha_N = 1$, yielding

$$B_N(\Omega) = \Omega^N. \tag{8.7}$$

For general values of Ω_c and ϵ, we merely scale (8.7):

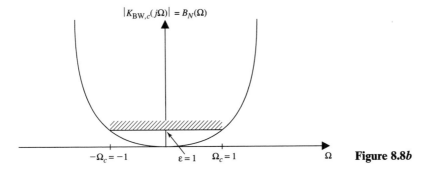

Figure 8.8*b*

$$|K_c(j\Omega)| = \epsilon B_N\left(\frac{\Omega}{\Omega_c}\right) = \epsilon\left(\frac{\Omega}{\Omega_c}\right)^N. \tag{8.8}$$

Thus

$$|H_c(j\Omega)|^2 = \frac{1}{|A_c(j\Omega)|^2}, \tag{8.9a}$$

which from (8.4) may be rewritten

$$|H_c(j\Omega)|^2 = \frac{1}{1 + |K_c(j\Omega)|^2}, \tag{8.9b}$$

and (8.8) gives

$$|H_c(j\Omega)|^2 = \frac{1}{1 + \epsilon^2(\Omega/\Omega_c)^{2N}}, \tag{8.9c}$$

or, equivalently,

$$\boxed{|H_c(j\Omega)| = \frac{1}{[1 + \epsilon^2(\Omega/\Omega_c)^{2N}]^{1/2}},} \quad \text{Butterworth magnitude response.} \tag{8.9d}$$

which is the Butterworth filter magnitude response. As predicted, for large Ω, $|H_c(j\Omega)|$ falls as $1/\Omega^N$—a good lowpass filter for large Ω, and $|H_c(j0)| = 1$ and maximally level at $\Omega = 0$—a good lowpass filter for small Ω. Unfortunately, there is no minimization of transition width narrowness, which is a main limitation of the Butterworth filter.

Note that at $\Omega = \Omega_c$, $|H_c(j\Omega_c)| = 1/\{1 + \epsilon^2\}^{1/2}$. For the case $\epsilon = 1$, Ω_c is the –3 dB frequency because in that case $|H_c(j\Omega_c)| = 1/\sqrt{2}$. Matlab assumes that $\epsilon = 1$, so if you want to design a filter with $\epsilon \neq 1$—that is, a filter whose passband cutoff frequency is defined as where the filter log magnitude is other than –3 dB—you will have to design it yourself. The only way we have control over Ω_r is by our choice of N. The higher the value of N we choose, the closer Ω_r will be to Ω_c, so the narrower will be the transition band.

In practice, specifications are not usually in the $\{\epsilon, N\}$ form. For example, suppose that we are to design a Butterworth filter having maximum attenuation A_c [a positive decibel attenuation value, not to be confused with $A_c(s)$ discussed previously] in

---

[2] Note in (8.6) that only $|K_c(j\Omega)| = \Omega^N$ holds, and *not* the formula $K_c(j\Omega) = \Omega^N$ found in some textbooks (e.g., Daniels in Footnote 7). The latter does not satisfy $K_c(-j\Omega) = K_c^*(j\Omega)$, as required in (8.4). In fact, for $K_c(j\Omega) = \Omega^N$, for N odd, $1 + K_c(j\Omega) K_c(-j\Omega) = 1 - \Omega^{2N}$, in which the minus sign is wrong. Because $K_c(s)$ and thus $K_c(j\Omega)$ may be difficult (and unneccessary) to determine, in this book we specify only $|K_c(j\Omega)|$ and not $K_c(s)$ or $K_c(j\Omega)$.

the passband and A_r (also a positive decibel value) minimum attenuation in the stopband. Thus for example, the magnitude of the frequency response at the cutoff frequency (nondecibel) is $10^{-A_c/20}$. Let us find the values of ϵ and N in terms of A_c, A_r, Ω_r, and Ω_c. First, notice that

$$A_c = -20 \log_{10}\{|H_c(j\Omega_c)|\} \quad [A_r = -20 \log_{10}(1 - \delta_1)] \qquad (8.10)$$
$$= -10 \log_{10}\{|H_c(j\Omega_c)|^2\}$$
$$= 10 \log_{10}\{1 + \epsilon^2\},$$

which when solved for ϵ gives

$$\epsilon = \{10^{A_c/10} - 1\}^{1/2}. \qquad (8.11)$$

Also, using the inequality form because N must be the nearest higher integer,

$$A_r \leq -10 \log_{10}\{|H_c(j\Omega_r)|^2\} \quad [A_r = -20 \log_{10}(\delta_2)] \qquad (8.12)$$
$$= 10 \log_{10}\left\{1 + \epsilon^2\left(\frac{\Omega_r}{\Omega_c}\right)^{2N}\right\},$$

or

$$\left(\frac{\Omega_r}{\Omega_c}\right)^{2N} \geq \frac{10^{A_r/10} - 1}{\epsilon^2}, \qquad (8.13)$$

or, taking logs,

$$2N \log_{10}\left(\frac{\Omega_r}{\Omega_c}\right) \geq \log_{10}\left\{\frac{10^{A_r/10} - 1}{\epsilon^2}\right\} \qquad (8.14)$$

or, finally,

$$N \geq \frac{\log_{10}\{[10^{A_r/10} - 1]/\epsilon^2\}}{2 \log_{10}\{\Omega_r/\Omega_c\}} \qquad (8.15a)$$
$$= \frac{\log_{10}\{(10^{A_r/10} - 1)/(10^{A_c/10} - 1)\}}{2 \log_{10}\{\Omega_r/\Omega_c\}}. \qquad (8.15b)$$

Notice that for zero transition width ($\Omega_r = \Omega_c$) and thus $\log_{10}\{\Omega_r/\Omega_c\} = 0$, $N = \infty$ (impossible in practice). We have now converted practical specifications into $\{\epsilon, N\}$ values that will achieve them.

In Matlab, the Butterworth approximation is calculated only for $\Omega_c = 1$ rad/s. Notice that $B_N(\Omega; \Omega_c) = B_N(\Omega/\Omega_c; 1 \text{ rad/s})$. Also notice that, expressing $H_c(s)$ in state–space form, $H_c(s/\Omega_c) = \mathbf{c}([s/\Omega_c]\mathbf{I} - \mathbf{A})^{-1}\mathbf{b} + d = \mathbf{c}(s\mathbf{I} - [\Omega_c\mathbf{A}])^{-1}[\Omega_c\mathbf{b}] + d$. Thus to achieve the frequency scaling, in 1p21p, Matlab converts the $\Omega_c = 1$ rad/s solution to state-space form, scales $\mathbf{A}$ and $\mathbf{b}$ by Ω_c, and then converts back to transfer-function form. This procedure is used for all filter approximations in Matlab (e.g., Chebyshev and elliptic designs).

8.4.1.1 Pole Locations

Determination of the pole locations is vital, because this information is required to construct an actual filter transfer function $H_c(s)$, as we will demonstrate in Example 8.4.

To determine the Butterworth filter pole locations, we can again analytically continue the function $|H_c(j\Omega)|^2 = H_c(j\Omega)H_c(-j\Omega)$ to $H_c(s)H_c(-s)$ by replacing Ω by s/j in (8.9c):

$$H_c(s)H_c(-s) = \frac{1}{1 + \epsilon^2 (s/[j\Omega_c])^{2N}}, \tag{8.16}$$

so that the overall function has poles at $\epsilon^2 (s_k/[j\Omega_c])^{2N} = -1$:

$$s_k = \epsilon^{-1/N} j\Omega_c (-1)^{1/(2N)} \tag{8.17a}$$

or, recalling that the L roots of -1 may be written as $\exp\{j(\pi + 2\pi k)/L\}$ for $0 \leq k \leq L - 1$,

$$s_k = \epsilon^{-1/N} j\Omega_c \exp\left\{\frac{j(\pi + 2\pi k)}{2N}\right\}, \qquad 0 \leq k \leq 2N - 1 \tag{8.17b}$$

$$= \epsilon^{-1/N} \Omega_c \exp\left\{j\frac{\pi}{2}\left[1 + \frac{2k+1}{N}\right]\right\}, \qquad 0 \leq k \leq 2N - 1, \tag{8.17c}$$

which is $2N$ poles, uniformly spaced on the $\epsilon^{-1/N}\Omega_c$-radius circle, with spacing of π/N radians, and beginning for $k = 0$ at the angle $(\pi/2)(1 + 1/N)$. Because in (8.16) $(-s)^{2N} = s^{2N}$, if s_k is a pole, so is $-s_k$, as required.

For a Butterworth pole s_k to be real, the argument of the exponential in (8.17c) must be $\pi\ell$, ℓ an integer so that $\angle s_k$ is either 0 or π:

$$\frac{\pi}{2}\left(1 + \frac{2k+1}{N}\right) = \pi\ell, \tag{8.18a}$$

or, solving for k,

$$k = N\ell - \frac{N+1}{2}, \quad \text{where } k \text{ must be an integer.} \tag{8.18b}$$

For (8.18b) to be satisfied with k and ℓ integers, N must be odd so that $(N + 1)/2$ is an integer. Thus s_k can be real only if the filter order N is odd.

Can s_k be imaginary? We certainly do not want imaginary poles, for that would mean a metastable filter! Now the requirement is, analogously,

$$\frac{\pi}{2}\left(1 + \frac{2k+1}{N}\right) = \pi\ell/2, \quad \ell \text{ odd,} \tag{8.19a}$$

or, again solving for k,

$$k = \frac{N(\ell - 1) - 1}{2}, \quad \ell \text{ an odd integer and } k \text{ must be an integer.} \tag{8.19b}$$

Because $\ell - 1$ is even for ℓ odd, so is $N(\ell - 1)$ whether or not N is even—and thus $N(\ell - 1) - 1$ is odd. Consequently, (8.19b) is not satisfied for any integer N. Thus Butterworth poles are never on the imaginary axis.

Examples of the Butterworth filter for various orders are shown in Figure 8.9. Figure 8.9 shows, for three values of N (5, 8, 17), the magnitude $|H_c(j2\pi F)|$, the log magnitude $20 \log_{10}\{|H_c(j2\pi F)|\}$, and the phase angle $\angle H_c(j2\pi F)$ in degrees versus F in hertz. The cutoff frequency Ω_c was arbitrarily selected to be $2\pi \cdot F_c$ where $F_c = 1$ Hz. Also, $\epsilon = 1$ (as always for Matlab's `butter` command), so that at F_c the log magnitude is -3 dB ($= 1/\sqrt{2} = 0.707$ for the top, nondecibel plot). As N increases, the sharpness of the filter increases, and with it the attenuation within a reasonable stopband. There is clearly no ripple in either the passband or the stopband. Notice that the phase is nonlinear, particularly near F_c for large N. However, in the passband it is quite nearly linear. We will examine the significance of linear phase in Section 9.5.1.

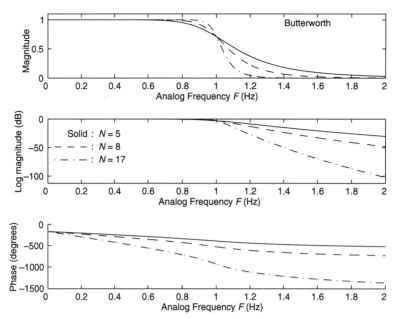

Figure 8.9

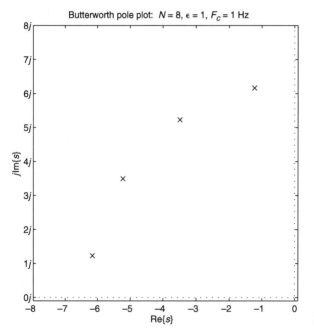

Figure 8.10

For the middle value of N, $N = 8$, the poles of the Butterworth filter are plotted in the second quadrant of the s-plane in Figure 8.10 using a slightly modified version of Matlab's `pzmap` [specifically, the axis labels were changed and `axis('image')` was used to show that the poles fall on circle]. It is clear in Figure 8.10 that all the poles are on a circular ring of radius $\Omega_c = 2\pi \cdot 1 \approx 6.28$ rad/s. Also, because $N = 8$ is even, there are no poles on the real axis; for no values of N are there poles on the imaginary axis. Only the second quadrant is shown, because the third is its mirror image, and $H_c(s)$ has no poles in the first and fourth.

EXAMPLE 8.4

Design an analog Butterworth filter whose passband cutoff is defined at the 0.5-dB attenuation level (rather than the usual 3-dB attenuation) and occurs at $F_c = 250$ Hz, and that has a minimum attenuation of 50 dB at $F_r = 350$ Hz. Plot all the poles, and write the explicit expression for $H_c(s)$, specifying all numerical values of the poles. Plot the magnitude and phase of the frequency response using $H_c(j\Omega)$ as obtained from the poles and, on the same axes, the magnitude as calculated from (8.9d) $[\, |H_c(j\Omega)| \, = 1/[1 + \epsilon^2(\Omega/\Omega_c)^{2N}]^{1/2}]$. Show explicitly that the required decibel attenuations at F_c and F_r are obtained.

Solution

We must use the design equations in the text rather than the Matlab design routine `butter`, because the passband cutoff attenuation level is other than –3 dB. We have from (8.11) $\epsilon = \{10^{0.5/10} - 1\}^{1/2} = 0.3493$. From (8.15a), $N \geq \log_{10}\{[10^{50/10} - 1]/0.3493^2\}/[2 \log_{10}(350/250)] = 20.2342$, so use at least $N = 21$. Such a high-order filter would be difficult to implement with analog electronics. However, when discretized, it merely means a bit more computation than for a less sharp (lower-order) filter. We will discretize this filter in Examples 8.8 and 8.9 (Section 8.5.2) using the bilinear transform. With $N = 21$, we have from (8.17c) $[s_k = \epsilon^{-1/N}\Omega_c \exp\{j(\pi/2)(1 + [2k + 1]/N)\}]$ that the Butterworth poles have the locations

$$s_k = 0.3493^{-1/21}(2\pi \cdot 250) \cdot \exp\{j(\pi/2)(1 + [2k + 1]/21)\}, \qquad k \in [0, 41]$$
$$= \pm 123.4 \pm j1646.9, \pm 367.5 \pm j1610.1, \pm 603.4 \pm j1537.3,$$
$$\pm 825.7 \pm j1430.2, \pm 1029.7 \pm j1291.2, \pm 1210.6 \pm j1123.3,$$
$$\pm 1364.5 \pm j930.3, \pm 1487.9 \pm j716.5, \pm 1578.1 \pm j486.8,$$
$$\pm 1633.0 \pm j246.1, \text{ and } \pm 1651.5 \text{ (rad/s)}.$$

Thus the poles are on a circle of radius $\epsilon^{-1/N}\Omega_c = 1651.5$ rad/s, and the left-half-plane poles which we select for $H_c(s)$ are as shown in Figure 8.11. We have (where here $2N = 42$)

$$H_c(s) = \cfrac{1}{\displaystyle\prod_{\substack{k=0 \\ k \text{ for left-half-plane } s_k \text{ only}}}^{2N-1} (1 - s/s_k)} = \cfrac{\displaystyle\prod_{\substack{k=0 \\ s_k \text{ in left-half-plane}}}^{2N-1} (-s_k)}{\displaystyle\prod_{\substack{k=0 \\ s_k \text{ in left-half-plane}}}^{2N-1} (s - s_k)}.$$

Construction of *any* all-pole Nth-order analog filter (not just Butterworth), from its left-half-plane poles, extracted from quadrantally symmetrical poles s_k. This filter has unity dc gain. [For Chebyshev type I, divide by $1/\{1 + \epsilon^2\}^{1/2}$ for unity maximum gain.]

Finally, the frequency response is plotted. The first plot, Figure 8.12a, shows the interval including F_c, and the second, Figure 8.12b, shows the rejection frequency F_r on appropriate magnitude axis scales. The required agreement between the two methods of calculation is exact using (a) the pole-factor expression above and (b) the Butterworth frequency response expression in (8.9d) (the solid and dashed curves are indistinguishable). We note that we exactly met the –0.5-dB specification at F_c.

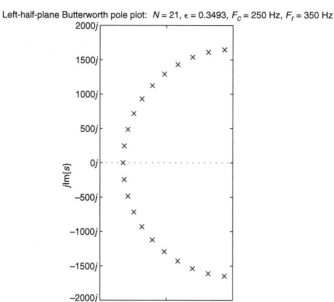

Left-half-plane Butterworth pole plot: $N = 21$, $\epsilon = 0.3493$, $F_c = 250$ Hz, $F_r = 350$ Hz

Figure 8.11

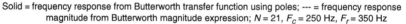

Solid = frequency response from Butterworth transfer function using poles; --- = frequency response magnitude from Butterworth magnitude expression; $N = 21$, $F_c = 250$ Hz, $F_r = 350$ Hz

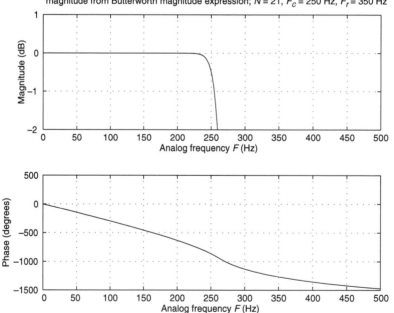

Figure 8.12a

Also, we slightly exceeded the –50-dB specification at F_r because $N = 21$ is greater than 20.2342. Keep in mind that without the preceding theoretical development of analog filters—for example, if we used the intuitive pole-placement technique—we could never have had any hope of reliably getting what we want: meeting quantitative specifications. Herein lies the payoff for the hard analytical work.

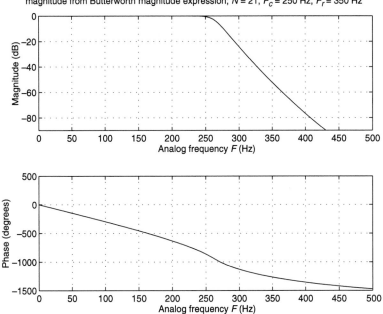

Solid = frequency response from Butterworth transfer function using poles; --- = frequency response magnitude from Butterworth magnitude expression; $N = 21$, $F_c = 250$ Hz, $F_r = 350$ Hz

Figure 8.12b

EXAMPLE 8.5

Design a third-order analog Butterworth filter using op-amps, resistors, and capacitors that has -1 dB at 500 Hz. Use PSpice for all final figures.

Solution

The cutoff frequency is $\Omega_c = 2\pi \cdot 500 = 3141.6$ rad/s. From (8.11), $\epsilon = \{10^{1/10} - 1\}^{1/2} \approx 0.51$. From (8.17c), the poles are at $s_k = \epsilon^{-(1/N)} \Omega_c \, e^{j(\pi/2)\{1+[2k+1]/N\}}$, $k \in [0, 2N-1]$. We are given $N = 3$. Define $\Omega_0 = \epsilon^{-(1/3)}\Omega_c = 0.51^{-(1/3)}3141.6 = 3935$ rad/s. The poles are then $s_0 = \Omega_0 e^{j(\pi/2)(1+1/3)} = \Omega_0 e^{j2\pi/3}$, $s_1 = \Omega_0 e^{j(\pi/2)(1+3/3)} = \Omega_0 e^{j\pi} = -\Omega_0$, $s_2 = \Omega_0 e^{j4\pi/3}$, $s_3 = \Omega_0 e^{j5\pi/3}$, $s_4 = \Omega_0 e^{j2\pi} = \Omega_0$, and $s_5 = \Omega_0 e^{j\pi/3}$. For $H_c(s)$, we select the left-half-plane poles, which are $k = 0, 1, 2$. For use below, $s_0 s_1 s_2 = \Omega_0^3 e^{j(2\pi/3+\pi+4\pi/3)} = -\Omega_0^3$. Also, $s - s_1 = s + \Omega_0$ and $(s - s_0)(s - s_2) = (s - s_0)(s - s_0^*) = s^2 + 2\text{Re}\{s_0\}s + |s_0|^2 = s^2 + \Omega_0 s + \Omega_0^2$. We now write

$$H_c(s) = \frac{1}{\displaystyle\prod_{k=0}^{2}(1 - s/s_k)} = \frac{\displaystyle\prod_{k=0}^{2}(-s_k)}{\displaystyle\prod_{k=0}^{2}(s - s_k)} = \frac{\Omega_0^3}{(s - s_1)(s - s_0)(s - s_2)}$$

$$= \left\{\frac{\Omega_0}{s + s_0}\right\}\left\{\frac{\Omega_0^2}{s^2 + \Omega_0 s + \Omega_0^2}\right\}.$$

Usually, it is convenient to build a higher-order filter in first- and second-order stages. A first-order circuit is shown in Figure 8.13a. Using a complex voltage divider and $v_{a,c} = v_- = v_+$, the input–output phasor relation[3] is $\dfrac{\overline{V}_a}{\overline{V}_s} = \dfrac{1/(RC)}{s + 1/(RC)}$. Comparing with the first-order factor of $H_c(s)$ above, we set $1/(RC) = \Omega_0 = 3935$ rad/s. Let $C = 0.01\ \mu F$; then $R = 25.4\ k\Omega$.

For the second-order filter, we use the circuit shown in Figure 8.13b. We use Kirchhoff's current law to determine the input–output relation. Omitting the circuit analysis algebra for brevity, we obtain $\dfrac{\overline{V}_o}{\overline{V}_a} = \dfrac{1/[R_1 R_2 C_1 C_2]}{s^2 + \{(R_1+R_2)/(R_1 R_2 C_1)\}s + 1/[R_1 R_2 C_1 C_2]}$.

Equating coefficients with the second-order factor in $H_c(s)$ and letting $R_2 = R_1$, we have $1/[R_1^2 C_1 C_2] = \Omega_0^2, 2R_1/(R_1^2 C_1) = 2/(R_1 C_1) = \Omega_0 = 1/[R_1\{C_1 C_2\}^{1/2}]$. Let $C_1 = 4C_2$ and $C_2 = 0.01\ \mu F$, so $C_1 = 0.04\ \mu F$. Then $R_1 = 2/(\Omega_0 C_1) = 12.7\ k\Omega\ (= R_2)$. The PSpice schematic for this filter is shown in Figure 8.14, and the overall lowpass magnitude frequency response generated by PSpice is shown in Figure 8.15a. In Figure 8.15b, we zoom in near F_c and indeed find that the magnitude response passes through -1 dB at $F_c = 500$ Hz.

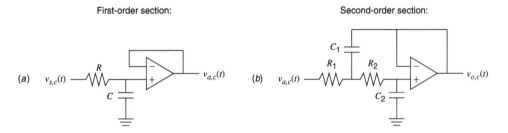

Figure 8.13

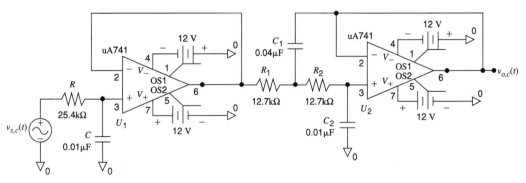

Figure 8.14

---

[3] We omit the subscript c on phasors in this example.

PSpice-generated third-order Butterworth frequency response

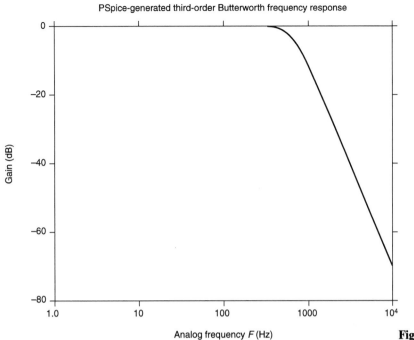

Analog frequency *F* (Hz)

Figure 8.15*a*

PSpice-generated third-order Buterworth Frequency response

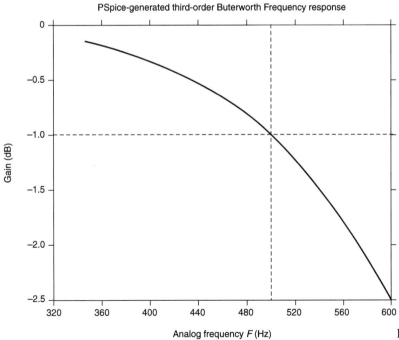

Analog frequency *F* (Hz)

Figure 8.15*b*

8.4.2   Chebyshev Filter

8.4.2.1   Type I

When the transition band of the Butterworth filter is found to be too wide for the allowable order N, the Chebyshev filter may be used. For analog filters, N may be limited by circuit expense and complexity, whereas for digital implementations, N may be limited by computation time and economically acceptable resources. For the Chebyshev filter, the derivative requirements at $\Omega = 0$ are replaced by requirements on the error in the approximation. The term *Chebyshev approximation* refers to Chebyshev's approximation techniques that minimize the maximum error in the given approximation interval (hence sometimes called minimax approximation). The minimax error occurs when the error is spread uniformly over the interval. Thus the Chebyshev or minimax approximation is obtained by an equiripple design. It so happens that for all-pole lowpass filter approximation, the equiripple chracteristic is achieved using Chebyshev rather than Butterworth polynomials.

As for the Butterworth design, we begin by setting $\epsilon = 1$ and $\Omega_c = 1$, and generalize later. We again seek a family of polynomials for $|K_c(j\Omega)|$, which monotonically increases for $|\Omega| > \Omega_c = 1$, but as noted, the Chebyshev filter has equiripple (that is, uniform-magnitude ripple) in the passband. Specifically, we seek a polynomial for $|K_c(j\Omega)|$ that oscillates between -1 and 1 for $|\Omega| < 1$ [and thus $0 \leq |K_c(j\Omega)|^2 \leq 1$ for $|\Omega| < 1$] and whose zeros all lie within $|\Omega| < 1$. As shown on WP 8.1, these requirements lead to the following differential equation for the polynomials $C_N(\Omega)$: $dC_N(\Omega)/\{1 - C_N^2(\Omega)\}^{1/2} = Nd\Omega/\{1 - \Omega^2\}^{1/2}$. Simple integration gives an inverse cosine on each side: $\cos^{-1}\{C_N(\Omega)\} = N\cos^{-1}(\Omega)$. Solving for $C_N(\Omega)$ then gives the so-called Chebyshev polynomial:

$$C_N(\Omega) = \cos(N\phi), \tag{8.20a}$$

where

$$\phi = \phi(\Omega) = \cos^{-1}(\Omega). \tag{8.20b}$$

We immediately notice that for $|\Omega| < 1$, $C_N(\Omega)$ does in fact oscillate. In particular, $\cos^{-1}(-1) = \pi$ and $\cos^{-1}(1) = 0$; thus, $C_N(-1) = \cos(N\pi) = (-1)^N$ and $C_N(1) = \cos(0) = 1$. For $|\Omega| < 1$, $C_N(\Omega)$ oscillates[4] with Ω between the extreme values -1 and 1. There is a total of $N + 1$ maxima and minima in $|\Omega| < 1$ because $C_N(\Omega)$ is an Nth-order polynomial in Ω with all its zeros in that range. Specifically, $C_N(\Omega_\ell) = 0$ for $N\phi = N\cos^{-1}(\Omega_\ell) = \frac{1}{2}\pi(2\ell + 1)$ or $\Omega_\ell = \cos\{\pi(\ell + \frac{1}{2})/N\}$, which are all on $[-1, 1]$. Note that $s_\ell = j\Omega_\ell$ are not poles of $H_c(s)$, but rather are the values of s for which $H_c(s) = 1$.

It is easy to obtain a simple recursion relation to express $C_N(\Omega)$ in terms of powers of Ω (i.e., as an Nth-order polynomial in Ω) by appealing to the addition formula for the cosine function:

$$C_{N+1}(\Omega) = \cos\{(N + 1)\phi\} \tag{8.21}$$
$$= \cos(N\phi)\cos(\phi) - \sin(N\phi)\sin(\phi),$$

whereas

$$C_{N-1}(\Omega) = \cos\{(N - 1)\phi\} \tag{8.22}$$
$$= \cos(N\phi)\cos(\phi) + \sin(N\phi)\sin(\phi),$$

so that by using (8.22) as a replacement for $\sin(N\phi)\sin(\phi)$ in (8.21),

$$C_{N+1}(\Omega) = 2\cos(N\phi)\cos(\phi) - C_{N-1}(\Omega). \tag{8.23a}$$

---

[4]It oscillates sinusoidally with respect to $N\phi$, but not sinusoidally with respect to Ω.

We now obtain the desired recursion relation from (8.23a) by use of $\cos(N\phi) = C_N(\Omega)$ and $\cos(\phi) = \cos\{\cos^{-1}(\Omega)\} = \Omega$. Substituting gives

$$C_{N+1}(\Omega) = 2\Omega C_N(\Omega) - C_{N-1}(\Omega). \tag{8.23b}$$

Equation (8.23b) is a recursion relationship to get higher-order Chebyshev polynomials from the lower ones. To get started, use (8.20) to obtain

$$C_0(\Omega) = \cos(0) = 1 \tag{8.24a}$$

$$C_1(\Omega) = \cos\{\cos^{-1}(\Omega)\} = \Omega \tag{8.24b}$$

and use (8.23b) to obtain

$$C_2(\Omega) = 2\Omega \cdot \Omega - 1 = 2\Omega^2 - 1 \tag{8.24c}$$

$$C_3(\Omega) = 2\Omega \cdot (2\Omega^2 - 1) - \Omega = 4\Omega^3 - 3\Omega, \tag{8.24d}$$

and so forth. It is quite clear from (8.24) that despite the appearance of the definition of $C_N(\Omega)$ in (8.20), in fact $C_N(\Omega)$ is an Nth-order polynomial in Ω.

Now in complete analogy with the Butterworth filter, we generalize and let $|K_c(j\Omega)| = \epsilon C_N(\Omega/\Omega_c)$, so that

$$|H_c(j\Omega)|^2 = \frac{1}{1 + |K_c(j\Omega)|^2} \tag{8.25a}$$

$$= \frac{1}{1 + \epsilon^2 C_N^2(\Omega/\Omega_c)},$$

or, equivalently,

$$|H_c(j\Omega)| = \frac{1}{\{1 + \epsilon^2 C_N^2(\Omega/\Omega_c)\}^{1/2}}, \tag{8.25b}$$

which is the Nth-order type I Chebyshev filter magnitude response.

First, note that at $\Omega = \Omega_c$, $|H_c(j\Omega_c)| = 1/\{1 + \epsilon^2\}^{1/2}$ because as already noted, $C_N(1) = 1$. Again, for the case $\epsilon = 1$, Ω_c is the –3-dB frequency because in that case $|H_c(j\Omega_c)| = 1/\sqrt{2}$. Because C_N is *squared* in (8.25), the Chebyshev term oscillates between 0 and ϵ^2 so that in the passband, $|H_c(j\Omega)|$ oscillates between 1 and $1/\{1 + \epsilon^2\}^{1/2}$, with approximately $N/2$ complete oscillations from $\Omega = 0$ to $\Omega = \Omega_c$. Numerically, we find that the "period" of the oscillations $|\Omega_\ell - \Omega_{\ell-1}| = |\cos\{\pi(\ell + \frac{1}{2})/N\} - \cos\{\pi(\ell - \frac{1}{2})/N\}| = 2\sin(\pi\ell/N)\sin\{\pi/(2N)\}$ decreases as ℓ decreases from $N/2 - 1$ or $(N - 1)/2$ for N even or odd [corresponding to $\Omega = 0$ or the first zero beyond it] to 0 corresponding to the zero nearest $\Omega = \Omega_c$; the oscillation periods[5] are not uniform (they depend on ℓ). The ripple amplitude, however, is uniform and equal to $\delta_1 = 1 - 1/\{1 + \epsilon^2\}^{1/2}$ and in decibels is A_c (the same as the maximum passband attenuation). A_c is equal to the ripple because the decibel of ripple is the decibel maximum value (0 dB) minus the decibel minimum value in the passband, $-A_c$.

By identical analysis, we again end up with $\epsilon = \{10^{A_c/10} - 1\}^{1/2}$ [see (8.11)] for ϵ and the same expression $A_c = 10\log_{10}\{1 + \epsilon^2\}$ [see (8.10)]. Note that for $\epsilon = 1$, again $A_c = 10\log_{10}(2) \approx 3.01$ dB.

Again, the only way we have control over Ω_r is by our choice of Ω_c and N. The higher value of N we choose, the closer Ω_r will be to Ω_c, so the narrower will be the

---

[5] Multiply the expressions for $|\Omega_\ell - \Omega_{\ell-1}|$ by Ω_c for $\Omega_c \neq 1$ rad/s. Also note that for N even or odd, $N/2$ or $(N + 1)/2 \leq \ell \leq N - 1$ occur on the range $-\Omega_c < \Omega < 0$.

transition band. Naturally, the expression for N will differ from that for the Butterworth filter because $C_N(\Omega)$ rather than $B_N(\Omega)$ is involved.

To begin, we note that the "cosh" form of $C_N(\Omega)$ is appropriate for $|\Omega| > 1$ in the practical evaluation of $C_N(\Omega)$ because the $\cos^{-1}$ is imaginary. [Note that for $C_N(\Omega)$ we end up taking cos of N times that imaginary number, which gives the same real number as the cosh expression gives; the cosh expression is just an equivalent that avoids complex numbers.] The condition $|\Omega| > 1$ translates in $H_c(\cdot)$ to $|\Omega| > \Omega_c$; that is, Ω in the stopband. To obtain the "cosh" form of $C_N(\Omega)$, note that if $\cos^{-1}(\Omega) = \phi$, then $\cosh^{-1}(\Omega) = j\phi$ because $\Omega = \cos(\phi) = \cosh(j\phi)$. Thus

$$C_N(\Omega) = \cos(N\phi) \tag{8.26a}$$

$$= \cosh\{jN\phi\} \tag{8.26b}$$

$$= \cosh\{N(j\phi)\}. \tag{8.26c}$$

Noting from above that $j\cos^{-1}(\Omega) = j\phi = \cosh^{-1}(\Omega)$, we obtain

$$C_N(\Omega) = \cosh\{N\cosh^{-1}(\Omega)\}. \tag{8.26d}$$

As in (8.12), we have

$$A_r \leq -10\log_{10}\{|H_c(j\Omega_r)|^2\} \tag{8.27}$$

$$= 10\log_{10}\{1 + \epsilon^2 C_N^2(\Omega_r/\Omega_c)\},$$

or, analogous to (8.13),

$$C_N^2\left(\frac{\Omega_r}{\Omega_c}\right) = \cosh^2\left\{N\cosh^{-1}\left(\frac{\Omega_r}{\Omega_c}\right)\right\}$$

$$\geq [10^{A_r/10} - 1]/\epsilon^2, \tag{8.28}$$

or, taking the square root and then the inverse cosh of both sides, and finally dividing by $\cosh^{-1}(\Omega_r/\Omega_c)$,

$$N \geq \frac{\cosh^{-1}\{([10^{A_r/10} - 1]/\epsilon^2)^{1/2}\}}{\cosh^{-1}(\Omega_r/\Omega_c)} \tag{8.29a}$$

$$= \frac{\cosh^{-1}\{[(10^{A_r/10} - 1)/(10^{A_c/10} - 1)]^{1/2}\}}{\cosh^{-1}(\Omega_r/\Omega_c)}. \tag{8.29b}$$

We find that the only differences from the Butterworth filter order (8.15b) are a square root instead of factor of two and the replacement of $\log_{10}$ by $\cosh^{-1}$.

The Chebyshev filters thus far considered, called type I, have ripple in the passband and monotonic decrease in magnitude in the stopband. Figure 8.16 displays the magnitude, log magnitude, and phase-angle plots of the frequency response of Chebyshev I filters for the same orders examined for the Butterworth filter (in Figure 8.9). As in Figure 8.9, the cutoff was set to 1 Hz, with 3 dB down at the cutoff. Notice the ripple in the passband; contrarily, the Butterworth (see Figure 8.9) has no ripple anywhere. Because on a log scale the passband ripple is very small, it is hardly noticeable in the log-magnitude plot in Figure 8.16. Especially for large N and f near f_c, the phase is quite nonlinear. Figure 8.17 shows the pole locations on the second quadrant of the s-plane for $N = 8$, $\epsilon = 1$, and $F_c = 1$ Hz. This time it is evident that the poles lie on an ellipse, as will be systematically discussed in Section 8.4.2.3.

In Matlab, two commands are used to design Chebyshev I analog filters (be sure to specify s for s-plane design, as opposed to a digital filter design). To obtain N, we

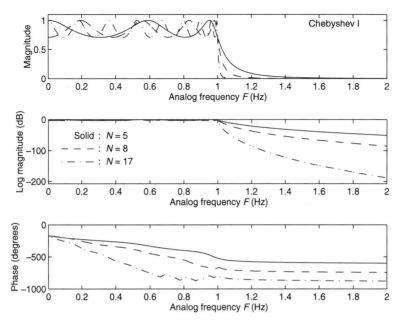

Figure 8.16

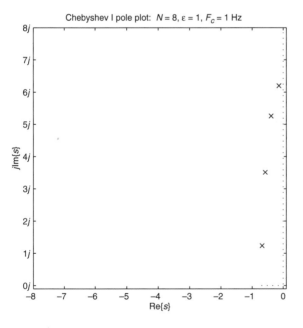

Figure 8.17

give to the routine cheb1ord the parameters wp = Ω_c, ws = Ω_r, rp = A_c, rs = A_r, opt = 's'. Unlike butter, the Matlab Chebyshev I design does not assume that A_c = 3 dB, probably because 3-dB passband ripple is unacceptably large (whereas for Butterworth, $A_c = 3$ dB just means a gradual decline in the passband to -3 dB). The resulting order [which Matlab obtains via (8.29b)] is put in the output variable order and Ω_c is stored in the output variable wn. We then give to routine cheby1 the values n = order, r = A_c, and Wn = wn = Ω_c, and again use 's' to signify an analog filter design. cheby1 calls another routine cheb1ap, and it produces the required filter via the poles, as we derive in Section 8.4.2.3. The Matlab Chebyshev I filter is thus exactly what we have designed in this discussion.

8.4.2.2 Type II (Inverse Chebyshev)

There is another filter closely related to the Chebyshev filter just described. By inverting the argument to $C_N(\cdot)$, we obtain the so-called type II or inverse Chebyshev filter. It has monotonic decline in the passband and equiripple in the stopband. The form of type II Chebyshev is conventionally defined as

$$|H_c(j\Omega)| = \frac{1}{\{1 + \epsilon^2\, C_N^2(\Omega_r/\Omega_c)/C_N^2(\Omega_r/\Omega)\}^{1/2}}. \qquad (8.30)$$

As can be seen by multiplying top and bottom of (8.30) by $C_N^2(\Omega_r/\Omega)$, this filter has not only poles (which are all that the Butterworth and Chebyshev type I filters have) but also zeros. First, note that in (8.30), $|H_c(j0)| = 1/\{1 + 0\}^{1/2} = 1$, because $C_N(x)$ goes as x^N for large x.

Next, we find that $|H_c(j\Omega_c)| = 1/\{1 + \epsilon^2\}^{1/2}$, also as before, only now the decline in $|H_c(j\Omega)|$ from $\Omega = 0$ to Ω_c is monotonic. This can be an advantage because usually, smoothness in the passband (which contains the desired signal) is much more important than in the stopband. Again, ϵ directly controls the attenuation at Ω_c. Notice, however, that the rippling occurs for $|\Omega| > \Omega_r$, not $|\Omega| < \Omega_c$ as in type I; this is really how we would want the stopband (Ω_r) to be defined in this case. Note that for Chebyshev I, ϵ and Ω_c determined the attenuation and frequency at which rippling ends and smooth decline begins, whereas Ω_r is left unspecified except in determining N. That is, Ω_r in type I is not a juncture of different behaviors (ripple/smooth). Now for type II the reverse is true: Ω_r determines where rippling begins, and now Ω_c is not a juncture of different behaviors.

The situation in Matlab is somewhat different. In Matlab, the type II Chebyshev filter is determined by replacing $C_N^2(\Omega/\Omega_c)$ in the type I filter relation in (8.25) by $C_N^2(\Omega_r/\Omega)$ and subtracting that modified right-hand side of (8.25) from unity. The result of these operations is that Ω_r (called Wn in Matlab's cheby2) is again the juncture between monotonic and oscillatory behavior, but its attenuation is directly controlled via A_r the way that passband attenuation in cheby1 is controlled by A_c. A cutoff frequency Ω_c is not specified in cheby2. Instead, cheb2ord modifies the user-supplied Ω_r, which is subsequently passed on to cheby2 to guarantee that the passband specification A_c at Ω_c is met exactly, in a manner that guarantees that the user-required A_r attenuation is achieved for $\Omega = \Omega_{r,\text{modif.}} \le \Omega_{r,\text{user}}$. (Again, specifications are exceeded when the design equations give noninteger N, where the selected N is the next larger integer.) Specifically, we give A_c, Ω_c, A_r, and Ω_r to cheb2ord exactly as we give them to cheb1ord, and the result is N and Wn —but this time Wn is the modified Ω_r value to be given to cheby2.

Matlab uses this procedure because the passband edge behavior naturally takes precedence in practice over that of the stopband edge and because it is a natural modification to the Chebyshev I code. In the end, the user designs a Chebyshev II filter in Matlab in exactly the same way he or she designs a Chebyshev I filter, except that A_r is given as the attenuation input to cheby2, whereas A_c is given to cheby1. Although it is a little harder to see what is going on in Matlab than with the conventional formula (8.30) (indeed, Matlab documentation regrettably gives no formulas at all!), essentially the same results as in our discussion can be produced if one knows exactly what one is doing. In Example 8.7, we will directly contrast the method in the text with Matlab's approach.

For (8.30), the requirement on N is the same as it is for type I, which is verified as follows. The requirement analogous to (8.27) is

$$A_r \le -10 \log_{10}\{ |H_c(j\Omega_r)|^2 \} \tag{8.31a}$$

$$= 10 \log_{10}\left\{ 1 + \epsilon^2 \frac{C_N^2(\Omega_r/\Omega_c)}{C_N^2(\Omega_r/\Omega_r)} \right\}. \tag{8.31b}$$

Noting that $C_N^2(\Omega_r/\Omega_r) = C_N^2(1) = 1$, (8.31b) reduces to

$$A_r \le 10 \log_{10}\left\{ 1 + \epsilon^2 C_N^2\left(\frac{\Omega_r}{\Omega_c}\right) \right\}, \tag{8.31c}$$

which is exactly the same requirement as (8.27) and therefore the required value of N in (8.29) also applies here. The scaling factor $C_N^2(\Omega_r/\Omega_c)$ in (8.30) makes this agreement with the corresponding Chebyshev I design process possible.

8.4.2.3 Pole and Zero Locations

We will first determine the pole locations for the type I filter. We can again analytically continue the function $|H_c(j\Omega)|^2 = H_c(j\Omega)H_c(-j\Omega)$ to $H_c(s)H_c(-s)$ by replacing Ω with s/j in (8.25a):

$$H_c(s)H_c(-s) = \frac{1}{1 + \epsilon^2 C_N^2(s/[j\Omega_c])}. \tag{8.32}$$

From (8.32), we see that the requirement for the overall function $H_c(s)H_c(-s)$ to have denominator equal to zero and thus have a pole at $s = s_k$ is :

$$C_N\left(\frac{s_k}{j\Omega_c}\right) = \cos\left\{ N \cos^{-1}\left(\frac{s_k}{j\Omega_c}\right) \right\} \tag{8.33a}$$

$$= \cos(x_k + jy_k) \tag{8.33b}$$

$$= \pm\frac{j}{\epsilon}, \tag{8.33c}$$

where

$$x_k + jy_k = N \cos^{-1}\left(\frac{s_k}{j\Omega_c}\right). \tag{8.34}$$

Expanding out the cosine in (8.33b) [recall that $\sin(jy_k) = j \sinh(y_k)$],

$$\cos(x_k)\cosh(y_k) - j \sin(x_k)\sinh(y_k) = 0 \pm \frac{j}{\epsilon}. \tag{8.35}$$

Equating real parts, and noting that $\cosh(y_k) \ne 0$ for any y_k, we obtain

$$x_k = \frac{(2k + 1)\pi}{2}, \qquad \text{For Chebyshev I poles.} \tag{8.36}$$

whereas equating imaginary parts and using $\sin(x_k) = (-1)^k$ for x_k in (8.36) and noting that either sign is allowed $[\pm(-1)^k = \pm 1$ for all $k]$ gives

$$y_k = \pm\sinh^{-1}\left(\frac{1}{\epsilon}\right). \qquad \text{For Chebyshev I poles.} \tag{8.37}$$

Finally, solving (8.33a) and (8.33b) for s_k gives

$$s_k = j\Omega_c \cos\left(\frac{x_k + jy_k}{N}\right), \qquad \text{Chebyshev I poles.} \tag{8.38}$$

with x_k, y_k as in (8.36) and (8.37), as the pole locations of the Chebyshev I filter. It is left to the problems to show that they fall on an ellipse with minor axis

$\Omega_c\sinh(y_k/N)$ and major axis $\Omega_c\cosh(y_k/N)$. By allowing k to range from 0 to $N - 1$ and using "+" in (8.37), we obtain both left- and right-half-plane poles and all the unique real and imaginary parts. By next using "−" in (8.37), we obtain the same poles as before, but with opposite-signed real parts.[6] For designing the filter, we select from all these the left-half-plane poles.

Now consider the pole locations for type II filters. We again analytically continue the function $|H_c(j\Omega)|^2 = H_c(j\Omega)H_c(-j\Omega)$ to $H_c(s)H_c(-s)$:

$$H_c(s)H_c(-s) = \frac{1}{1 + \epsilon^2[C_N^2(\Omega_r/\Omega_c)]/[C_N^2(j\Omega_r/s)]}. \tag{8.39}$$

Again, we set the denominator of $H_c(s)H_c(-s)$ equal to zero [now in (8.39)] to find the poles, s_k; now the requirement is

$$C_N\left(\frac{j\Omega_r}{s_k}\right) = \cos\left\{N\cos^{-1}\left(\frac{j\Omega_r}{s_k}\right)\right\} \tag{8.40a}$$

$$= \cos(x_k + jy_k) \tag{8.40b}$$

$$= \pm j\epsilon C_N\left(\frac{\Omega_r}{\Omega_c}\right), \tag{8.40c}$$

where $x_k + jy_k$ now take on different values from before, satisfying

$$x_k + jy_k = N\cos^{-1}\left(\frac{j\Omega_r}{s_k}\right). \tag{8.41}$$

Expanding out the cosine in (8.40b) gives

$$\cos(x_k)\cosh(y_k) - j\sin(x_k)\sinh(y_k) = 0 \pm j\epsilon C_N\left(\frac{\Omega_r}{\Omega_c}\right). \tag{8.42}$$

Equating real parts, and noting that $\cosh(y_k) \neq 0$ for any y_k, we again obtain

$$x_k = \frac{(2k + 1)\pi}{2}, \qquad \text{For Chebyshev II poles.} \tag{8.43}$$

whereas equating imaginary parts and using $\sin(x_k) = (-1)^k$ for x_k as determined in (8.43) gives

$$y_k = \pm\sinh^{-1}\left\{\epsilon C_N\left(\frac{\Omega_r}{\Omega_c}\right)\right\}. \qquad \text{For Chebyshev II poles.} \tag{8.44}$$

Notice that there are only two distinct values of y_k, and one is −1 times the other. Finally, solving the right sides of (8.40a and b) for s_k gives

$$s_k = \frac{j\Omega_r}{\cos([x_k + jy_k]/N)}, \qquad \text{Chebyshev II poles.} \tag{8.45}$$

with x_k, y_k as in (8.43) and (8.44) as the pole locations s_k of the Chebyshev type II filter. They no longer fall on any simple-shaped contour, because of the main quotient bar in (8.45).

The zeros for the type II filters can be obtained from (8.39) by multiplying top and bottom of (8.39) by $C_N^2(j\Omega_r/s)$, which shows that the zeros occur for

---

[6] It is left to the problems to show that $\text{Re}\{s_k\} = \pm\Omega_c\sin\{(2k + 1)\pi/[2N]\}\sinh\{\sinh^{-1}(1/\epsilon)/N\}$.

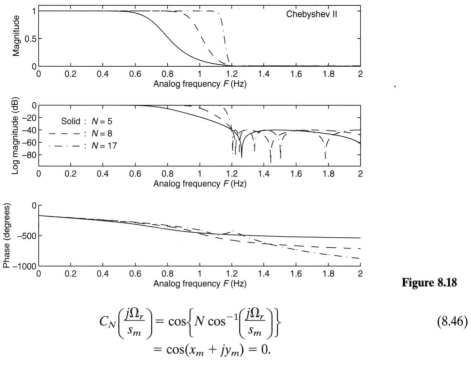

Figure 8.18

$$C_N\left(\frac{j\Omega_r}{s_m}\right) = \cos\left\{N\cos^{-1}\left(\frac{j\Omega_r}{s_m}\right)\right\} \qquad (8.46)$$
$$= \cos(x_m + jy_m) = 0.$$

Again expanding gives $x_m = (2m + 1)\pi/2$ and this time $y_m = 0$, which gives

$$s_m = \frac{j\Omega_r}{\cos\{(2m + 1)\pi/(2N)\}}, \qquad \text{Chebyshev II zeros.} \qquad (8.47)$$

and so the zeros are all on the imaginary axis.

The characteristics of the Chebyshev II filter for nearly the same conditions examined in Figures 8.9 and 8.16 are plotted in Figure 8.18. To at least approximately match the specifications with those of the previous filters, the rejection frequency was chosen to be 1.2 Hz (slightly above the desired cutoff of 1.0 Hz). First, notice that for Chebyshev II there is no ripple at all in the passband, but there is ripple in the stopband. Because the attenuation is so high in the stopband, ripple here is noticeable only in the decibel (not the nondecibel) plot. The minimum attenuation in the stopband was set to 40 dB for all values of N. Again there is phase nonlinearity for large N near the cutoff frequency.

The second-quadrant poles and zeros are plotted in Figure 8.19. Note that, as analyzed above, all the zeros are on the $j\Omega$ axis. The zeros occur at $\Omega = [38.6\ 13.6\ 9.1\ 7.7]$ rad/s. The effects of the zeros at 7.7 and 9.1 rad/s are seen in the $N = 8$ middle (log) plot Figure 8.18 at $f = \Omega/(2\pi) = [1.22\ 1.44]$ Hz: There are spikes going toward $-\infty$. If the frequency axis were extended to $F = 38.6/(2\pi) = 6.15$ Hz, all four dips would be seen, within the plot resolution.

Again we select the left-half-plane poles for $H_c(s)$. Notice that because in the rationalized form of (8.39) the numerator term $C_N^2(\Omega_r/s)$ is squared, we use all the zeros for $H_c(s)$ as simple zeros; the repeated zeros go to $H_c(-s)$. We will carry out an example of an digital Chebyshev II filter design in Section 8.5.

8.4.3   Elliptic Filter

The Butterworth filter is maximally flat in the passband. The Chebyshev filters are maximally flat in one band and equiripple in the other. All three filters are special

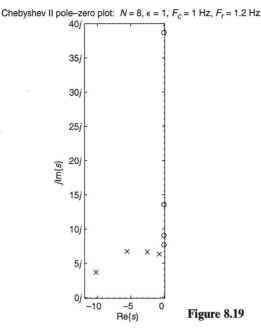

Chebyshev II pole–zero plot: $N = 8$, $\epsilon = 1$, $F_c = 1$ Hz, $F_r = 1.2$ Hz

Figure 8.19

cases of the elliptic filter (also known as the Cauer filter), which is equiripple in both bands. If for the elliptic filter we let the edge for the upper rippling band go to infinity and the ripple go to zero, we get Chebyshev I; if we let the edge of the lower ripple band and the ripple go to zero we have Chebyshev II; if we let both band edges go to these limits with zero ripple everywhere, we have Butterworth—that is, the monotonic "transition band" becomes the entire range of Ω from 0 to ∞.

The equiripple requirements for the elliptic filter are similar to those made for the Chebyshev filter. However, the requirement of two equiripple bands leads to the requirement for a pole–zero rational function for $|K_c(j\Omega)|$, not a polynomial (Chebyshev I) or an all-pole rational function (Chebyshev II). Again, a differential equation results. Although there are some similarities in form to that for the Chebyshev filter, it is much more complicated and will not be quoted.[7] The solution is called a Chebyshev rational function, which involves elliptic integrals and Jacobian elliptic functions. To adequately present these functions would go well beyond the scope of a book on DSP. Suffice it to say that mathematicians and numerical specialists have derived and implemented these functions so that the user can obtain them by a simple call to the Matlab function `ellip`.

By distributing the error over the widest frequency interval possible—both the passband and the stopband—the overall approximation is "best" (minimax) and the transition band is minimized. The "Chebyshev approximation idea" of equiripple error will be seen again when we study optimal digital FIR filters in Section 9.5/WP 9.1.

Again, the characteristics of the elliptic filter are represented in Figure 8.20, showing the same orders as before. The cutoff was set to 1 Hz with 3-dB attenuation, and the stopband was defined as starting where the attenuation reaches –80 dB. Except for the presence of passband ripple and sharper characteristics for lower N, the plot looks similar to the Chebyshev II. The passband ripple is evident in only the nondecibel magnitude plot, whereas the stopband ripple is noticeable in only the log-magnitude plot. The second-quadrant pole–zero plot is given in Figure

---

[7] See R.W. Daniels, *Approximation Methods of Electronic Filter Design* (New York: McGraw Hill, 1974), pp. 58–59.

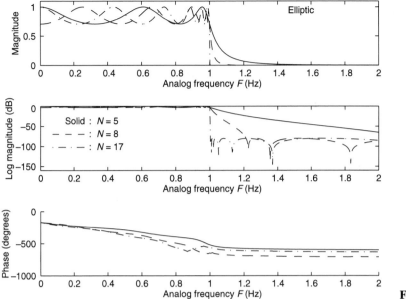

Figure 8.20

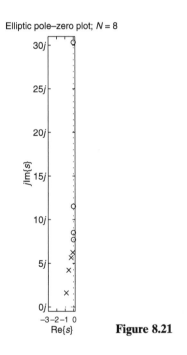

Figure 8.21

8.21. Again, all the zeros are on the $j\Omega$ axis. The elliptic filter poles fall on circles or ellipses only for the special cases of Butterworth and Chebyshev I. In general, they are functions of Jacobian elliptic sine, cosine, and derivative functions.

8.4.4 Comparison of Analog Filters

As just implied, the elliptic filter is optimal in having the minimum transition band for a given order N. Thus if a sharp filter is required, the elliptic filter is best. If the passband must be smooth, naturally Chebyshev II is preferred over the elliptic. Presumably, we care more about the smoothness of the passband, because that is the band containing our signal. A smooth passband means no magnitude distortion—no frequencies being "favored" over others in the output. Ripple in the stopband just means distortion of what is actually being annihilated anyway. If we additionally require the stopband to be smooth (which in practice would be less likely), the Butterworth is best. Sometimes in an analog implementation the Butterworth is preferred because of its simplicity.

Such issues are largely irrelevant for digital implementations, which we discuss in Section 8.5. With programs like Matlab at the designer's hand, it is no more difficult to carry out a Chebyshev or even elliptic design than to carry out a Butterworth design. However, we must always be wary of numerical problems, which are identified by examining the resulting frequency response. If the specifications have not been met, we may have to carry out the design "by hand" using, for example, improved root-finders or special function charts and routines.

To summarize the above comparison of the analog filters, Figure 8.22 compares the four analog filter types we have discussed. Each plot set shows a different order; both odd and even orders have been selected to show the differences. In particular, for N even, the magnitude response at $\Omega = 0$, if nonunity, is at its passband minimum, whereas for N odd, it is at its passband maximum at $\Omega = 0$. The specific values of N selected are $N = 3$ for Figure 8.22a, $N = 6$ for Figure 8.22b; and $N = 20$ for Figure 8.22c. An intermediate (odd) case, $N = 13$, is found on WP 8.2.

The Matlab programming to obtain the plots in Figure 8.22 was not entirely trivial. Among the programming modifications to alter what Matlab normally produces are

1. The nonlogarithmic gain (top) plot was added.

2. The logarithmic plot was modified to be expressed in decibels.

3. The frequency scale was changed from logarithmic to linear and from radian frequency to Hertzian frequency.

4. Multiple plots were called, while previously plotted curves were held in place.

5. Curves were coded and labed with which type of filter was used.

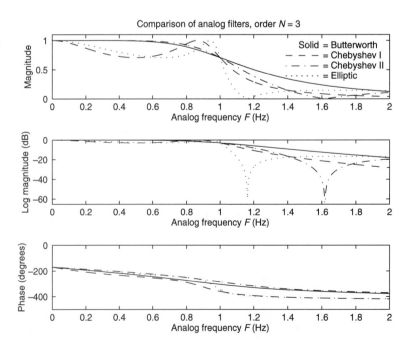

Figure 8.22a

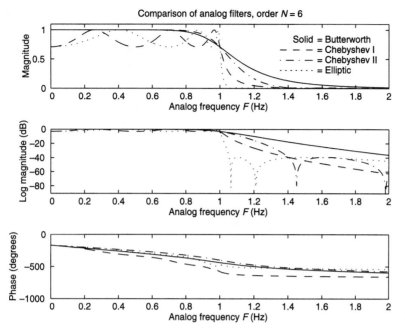

Figure 8.22*b*

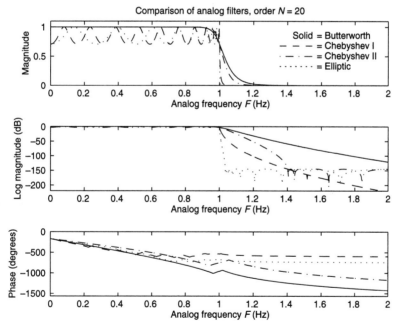

Figure 8.22*c*

6. The phase was unwrapped to avoid jagged phase curves and to facilitate assessment of phase linearity or lack thereof.

7. The number of frequency points was increased to 1000 for high resolution.

8. The axes limits were overridden to provide maximum usable viewing area, based on curve extrema.

9. The grid lines were removed for added legibility, and other minor necessary changes were made.

The following rules were used to make the filters as comparable as possible, given the differences in specifications for the different filter types as outlined in our previous discussions. For all plots, $F_c = 1$ Hz ($\Omega_c = 2\pi$ rad/s) and N was held to the same value. These two parameters totally specify the Butterworth filter. For Chebyshev I, the decibel down at Ω_c was set to 3 dB, which is identical to the value tacitly set for the Butterworth (recall that $\epsilon = 1$ in Matlab's `butter` command). For Chebyshev II, the stopband portion of the curve was selected to agree with that of the Chebyshev I at the (arbitrarily chosen) value of 1.4 Hz. Doing this made the two Chebyshev responses quite comparable, except that the equiripple and monotonic bands are reversed.

The elliptic filter was again designed to be 3 dB down at the end of its passband, $F_c = 1$, and the decibels down in the stopband was set equal to the minimum decibels down in the Chebyshev II stopband (which also has equiripple in the stopband). Thus the elliptic filter has the same sidelobe heights as the Chebyshev II. The resulting narrower transition width obtainable with the elliptic filter for the same order and all other things being equal is striking, especially for higher-order filters.

8.5   DIGITAL IIR FILTER DESIGN

Having reviewed analog filters in quite some detail, we now consider how these filters can be transformed into equivalent IIR digital filters. It makes sense that they should be transformable, because the analog and the digital IIR filter transfer functions are both ratios of polynomials—just of different variables (s or z). We will find that a good lowpass analog filter can be transformed into a good digital IIR filter. Actually, we already discussed this topic in Section 6.4 in the context of simulation of continuous-time systems. Here, however, we will revisit a few of the issues from a filter design viewpoint. We may consider transformations of analog filters into digital filters as falling into two classes: time-domain and frequency-domain approaches.

In the time-based approach, we insist that either the analog impulse response or the analog step response is exactly retained in the corresponding response of the digital filter. In the frequency-based approach, we replace either s or $1/s$ (differentiation or integration) by a discrete-time approximation of it, wherever it appears in the filter transfer function $H_c(s)$.

Recall that the "selectivity" idea of frequency-selective filters applies only to the frequency domain (not the time domain), because only in the frequency domain are the signal characteristics simply *multiplied* by those of the filter; time-domain convolution is more complicated. In the time domain, the filter impulse response affects the output signal in the manner described in Sections 2.3.3 and 2.4.3. Although digital filter design from the DSP perspective is usually done with frequency-domain behavior in mind, the control systems engineer is interested in the temporal response. Compensators are designed for control systems so that the overall closed-loop system has the desired time-domain specifications such as peak overshoot, settling time, steady-state error, and other figures of merit. In such applications, step invariance plays a very prominent role because of the practical importance of the step response as a desired step change in output value and because of the step-invariant characteristic of the zero-order hold.

The simplest digital filter, the digital lowpass filter, can be used to produce other filters. For example, once we have a digital IIR lowpass filter, we can use special frequency-domain transformations to obtain highpass, bandpass, bandstop, or multiband filters having performance characteristics similar to those of the original lowpass filter. This capability is again why in Section 8.4 we were able to focus our attention exclusively on lowpass analog filter design.

8.5.1   Impulse Invariance

If we are especially interested in preserving the temporal characteristics of the analog filter in the digital filter, a time-based approach is appropriate. The impulse-invariant method of IIR digital filter design from a given analog filter is useful both in our present discussion of filter design and in discrete-time simulation of continuous-time systems (Section 6.4). It avoids the frequency warping of the bilinear transform and preserves both magnitude and phase characteristics. Thus it does give good frequency-selective filters, despite its "time orientation."

As in Section 6.4.2, to preserve the impulse response aside from the constant scalar Δt, we set [see (6.40): $h(n) = \Delta t \cdot h_c(n\,\Delta t)$]

$$h(n) = \Delta t \cdot h_c(n\,\Delta t) \qquad\qquad (8.48)$$
$$= \Delta t \cdot \mathrm{LT}^{-1}\{H_c(s)\}\big|_{t=n\,\Delta t}.$$

Recall that in the sampling process, a spectral factor of $1/\Delta t$ is introduced; the Δt in (8.48) is intended to cancel it, giving

$$H(e^{j\omega}) = \sum_{k=-\infty}^{\infty} H_c\left(j\left\{\frac{\omega + 2\pi k}{\Delta t}\right\}\right) \qquad\qquad (8.49a)$$

In the absence of aliasing, which can never be achieved for a finite-order transfer function $H_c(s)$ but can be approximately true for lowpass-type $H_c(j\Omega)$ and over-sampling, (8.49a) reduces to

$$H(e^{j\omega}) \approx H_c(j\omega/\Delta t), \quad |\omega| < \pi \ \text{(little or no aliasing).} \qquad\qquad (8.49b)$$

Under the assumption of no aliasing of either $x_c(t)$ or $h_c(t)$, we showed in Section 6.4.2 that $y(n) = x_c(n\,\Delta t) * h_c(n\,\Delta t)$ is equal to $y_c(t)\big|_{t=n\,\Delta t} = x_c(t) \overset{*}{_c} h_c(t)\big|_{t=n\,\Delta t}$, where $\overset{*}{_c}$ represents continuous-time convolution—that is, exact simulation. However, even in the (unachievable) case of no aliasing, the frequency axis is linearly scaled by $\Omega = \omega/\Delta t$, and the matching of values occurs only for $|\omega| < \pi$; beyond π, the digital filter is periodic, whereas the analog filter is not. Although (8.49b) would seem to indicate perfect frequency matching, impulse invariance is more "temporally oriented" in the usual case of aliasing, where $h(n)$ still equals $\Delta t \cdot h_c(n\,\Delta t)$, whereas (8.49b) does not exactly hold.

As noted in Section 6.4.2, the requirement of no or little aliasing for (8.49b) to apply implies that analog highpass filters cannot be discretized using impulse invariance, because the analog filter is not even approximately bandlimited to any maximum frequency. Thus no value of Δt even approximately satisfies the Nyquist sampling criterion. Another problem with impulse invariance in some cases is that the analog impulse response tends to have infinite duration. Thus unless an analytical closed form can be found for $H(z)$ from the infinite-length $\{h(n) = \Delta t \cdot h_c(n\,\Delta t)\}$, some sort of truncation/windowing of $\{h_c(n\,\Delta t)\}$ will have to be done. Fortunately, for approximations of basic piecewise constant filters, $h_c(t)$ tends to involve exponentials and thus $h(n)$ involves geometric sequences, the z-transforms of which are very easy to find.

We saw in Section 6.2 that the general s-z relation between $H_c(s)$ and $H(z)$ [with $h(n) = \Delta t \cdot h_c(n\,\Delta t)$] is

$$H(z) = \sum_{k=-\infty}^{\infty} H_c(s + jk\Omega_s)\big|_{s=\ln(z)/\Delta t}, \qquad\qquad (8.50)$$

where $\Omega_s = 2\pi/\Delta t$. Consequently, if $H_c(s)$ has a pole at $s = s_m$, then it follows from (8.50) that $H(z)$ has a pole at $z_m = e^{s_m \Delta t}$; just note that the $k = 0$ term in the sum in (8.50) is $H_c(s_m) = \infty$ for $z = z_m$. Regardless of the values of the other terms, if this one term is infinite, so will be the sum.

There is an important consequence of this result: A stable $H_c(s)$ transforms to a stable $H(z)$ with impulse invariance. This is because under the pole transformation $z_m = e^{s_m \Delta t}$, the left half of the s-plane transforms into the interior of the unit circle in the z-plane (and the right half of the s-plane into the exterior of the unit circle in the z-plane).

However, the zeros s_ℓ of $H_c(s)$ do *not* transform to zeros at $z_\ell = e^{s_\ell \Delta t}$. Just because the $k = 0$ term in (8.50) is zero at $z = z_\ell$, any or all of the other terms in the sum can be nonzero, making the entire sum $H(z_\ell)$ nonzero.

An obvious approach to performing a direct impulse-invariant transformation is to simply transform all the poles s_m to $z_m = e^{s_m \Delta t}$. This was the approach taken by Matlab's impinvar command to find $H(z)$ from $H_c(s)$ in releases through 12/31/1996 (Student Edition 5); it has been corrected in version 5.3 (7/10/1998). It performed a partial fraction expansion of $H_c(s)$ and used the same partial fraction expansion coefficients for $H(z)$ but with the transformed poles. This procedure, however, will not work for multiple-order poles.[8] Multiple-order poles can arise either by "coincidence" or more commonly in systems having two or more integrators. Complete details for impulse invariance for multiple-order poles are provided on WP 8.3, which also includes a numerical filter discretization example illustrating by graphics the failure of old versions of Matlab and the correct solution.

EXAMPLE 8.6

Suppose that we have a section of speech data $x(n)$. See Figure 8.23, which shows $x(n)$, a portion of the word "processing" (part of "ing") of length $N = 8192$ samples selected from a recording of the author's voice saying "digital signal processing." The sampling rate is 11.025 kHz (so $\Delta t = 90.703$ µs). We therefore know that the highest unaliased frequency in the speech signal is $0.5 \cdot 11.025$ kHz $= 5.513$ kHz; any higher-frequency components of the signal would be aliased for this sampling rate. The magnitude of the DFT of $x(n)$ is shown in Figure 8.24 (only positive frequency is included).

Although we should properly denote the DFT of $x(n)$ by $X(k)$, in this example we will write the symbol $X(f)$ (with $f = f_k = k/N$) to mean $X(k)/N = X(z)\big|_{z=e^{j2\pi k/N}}/N$ to emphasize the dependence on digital frequency f. The DFT is divided by N in this example so that the sum of the squares of the scaled DFT represents the total energy in the time sequence. Remember, $X(f)$ is really a *sequence* of spectral values, $X(e^{j2\pi f_k})$, even though we show only the curve passing through f_k and label the frequency axis with f (8192 plotted stems would be very messy).

Recall that in the absence of aliasing, $f = F/F_s$. By some means, a high-frequency-noise damped sinusoid $w(n) = Ae^{\sigma_a t} \sin(\Omega_1 t)u_c(t)$ has been introduced additively, for which $f_1 = F_1/F_s = 0.45$ and thus $F_1 = 0.45F_s = 4.96$ kHz and $\Omega_1 = 2\pi F_1 = 31.17$ krad/s; $\sigma_a = -33.075$ Np/s. The amplitude A was chosen so that its spectral peak is on the same order of magnitude as the largest spectral peak of $x(n)$. Figure 8.25 shows the noise-corrupted spectrum $|V(f)| = |V(k)/N|$, where

---

[8] T. J. Cavicchi, "Impulse Invariance and Multiple-Order Poles," *IEEE Trans. on Signal Processing*, 44, 9 (September 1996): 2344–2347.

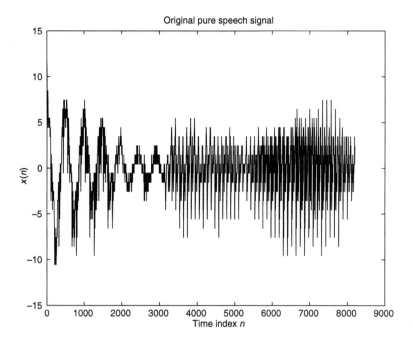

Figure 8.23

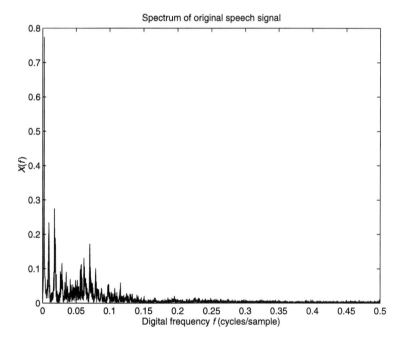

Figure 8.24

$v(n) = x(n) + w(n)$. Note that a small amount of aliasing has been introduced by $w(n)$—compare $V(0.5)$ with $X(0.5)$; it is due to the only gradual decline of $|W(f)|$.[9] Using the impulse-invariant method, design an effective lowpass filter to remove this noise. The stopband is to have an attenuation of at least 40 dB, and the passband is to have no more than 1 dB of attenuation.

---

[9] However, for the most part this aliasing merely distorts the noise itself, because it is confined to f near 0.5 cycle/sample.

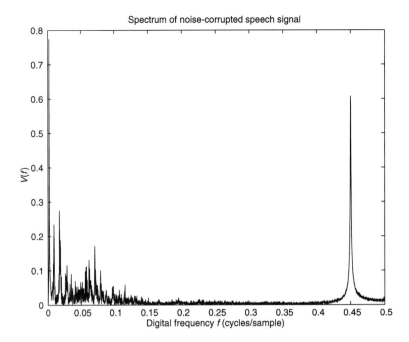

Figure 8.25

Solution

We will first attempt a Chebyshev II design, because of the desirable features that its passband is quite flat (no ripples) yet its transition band is narrow relative to that of a Butterworth filter. We inspect the discrete Fourier transform of the section (Figure 8.25) and find there is noise above the digital rejection frequency we select to be $f_r = 0.4$ cycle/sample, whereas our speech signal is essentially confined to f below our chosen cutoff frequency $f_c = 0.3$ (again for brevity, we will often omit the units of f, cycles/sample). In this application, we desire the cutoff frequency as high as possible so that we keep as much as possible of the weak spectral components of $x(n)$, which actually extend in Figure 8.24 all the way to $f = 0.5$. We will lose all these components that lie in the stopband.

Keep in mind that these ranges of f are determined by the original sampling interval Δt. Do not confuse the original Δt value with the Δt value chosen for the impulse-invariant transformation (denote as Δt_{impinv}). We are at complete liberty to choose the latter value; for simplicity, we usually choose $\Delta t_{\text{impinv}} = 1$ s. From (8.49b), assuming no aliasing we have the relation $\Omega = \omega/\Delta t_{\text{impinv}} = 2\pi f/\Delta t_{\text{impinv}} = 2\pi f$. Thus the cutoff frequency of the required (fictitious) analog filter is $\Omega_c' = 2\pi \cdot 0.3 \approx 1.89$ rad/s and the rejection frequency is $\Omega_r' = 2\pi \cdot 0.4 \approx 2.51$ rad/s. The analog cutoff/rejection frequencies for actual processing of $\Delta t = 90.703$-μs sampled-data signals are $\Omega_c = 2\pi \cdot 0.3/\Delta t = 20782$ rad/s ($F_c = 3.308$ kHz) and $\Omega_r = 2\pi \cdot 0.4/\Delta t = 27709$ rad/s ($F_r = 4.41$ kHz).

We are now in a position to immediately determine the poles and zeros, or equivalently, the filter coefficients of the analog Chebyshev II filter satisfying the given analog filter specifications. First, we use either (8.29b) $[N \geq \cosh^{-1}\{([10^{A_r/10} - 1]/[10^{A_c/10} - 1])^{1/2}\}/\cosh^{-1}(\Omega_r/\Omega_c)]$ or Matlab's `cheb2ord` (which uses the same formula) to obtain the required order $N = 8$, where N is rounded to the next higher integer than (8.29) produces (the Matlab command `ceil` does this rounding). Also, given that $A_c = 1$ dB, $\epsilon = \{10^{A_c/10} - 1\}^{1/2} = 0.5088$.

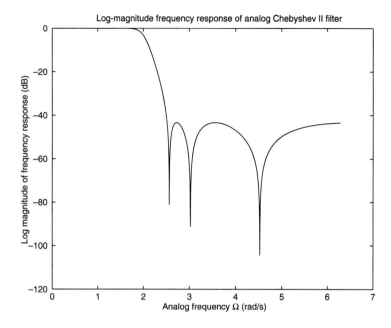

Figure 8.26

Thus from (8.30) [in our current notation,
$|H_c(j\Omega)| = 1/\{1 + \epsilon^2 C_N^2(\Omega_r'/\Omega_c')/C_N^2(\Omega_r'/\Omega)\}^{1/2}]$, we have
$|H_c(j\Omega)| = 1/\{1 + [0.5088 C_8(0.4/0.3)/C_8(0.4 \cdot 2\pi/\Omega)]^2\}^{1/2}\}$.

The curve produced by this expression and that produced by the $[b, a]$ coefficients from Matlab's cheby2 are effectively equivalent, though not identical due to the different parameterizations. The poles and zeros as determined by (8.40) through (8.47) also nearly agree with those of Matlab's [b, a]. In this example, the results will be presented for use of the equations developed in this chapter. Figure 8.26 shows the log-magnitude frequency response (in decibels) of the analog Chebyshev II filter as just designed. We see in Figure 8.26 that both passband ripple and stopband rejection specifications are achieved.

It now remains to inverse-Laplace-transform $H_c(s)$, sample every Δt_{impinv}, and z-transform the result. Because the Chebyshev II filter has no repeated poles, all we really need to do is partial-fraction-expand $H_c(s)$ into the form

$$H_c(s) = \sum_{\ell=0}^{7} \frac{A_\ell}{s - s_\ell}, \tag{8.51}$$

where because the poles are all simple,

$$A_\ell = H_c(s)(s - s_\ell)\big|_{s=s_\ell}. \tag{8.52}$$

Noting that $ZT\{LT^{-1}\{A_\ell/(s - s_\ell)\}\big|_{t=n\Delta t}\} = A_\ell/(1 - e^{s_\ell \Delta t} z^{-1})$, where $\Delta t = \Delta t_{\text{impinv}}$,

we see that we will end up with eight terms of this same form:

$$H(z) = \sum_{\ell=0}^{7} \frac{A_\ell}{1 - e^{s_\ell \Delta t} z^{-1}}. \tag{8.53}$$

To obtain $H(z)$ in rational function form, we simply combine all terms to a common denominator:

$$H(z) = \frac{\sum_{\ell=0}^{7} A_\ell \prod_{m=0, m \neq \ell}^{7} (1 - e^{s_m \Delta t} z^{-1})}{\prod_{\ell=0}^{7} (1 - e^{s_\ell \Delta t} z^{-1})}. \tag{8.54}$$

This exact procedure is automated with the Matlab command `impinvar`. First, we present the numerical PFE results:[10]

$$A_0 = 689.5 - j1221.5 \qquad\qquad s_0 = 3.1550 \angle 161.97° \qquad (8.55)$$

$$A_1 = -699.3 - j136.1 \qquad\qquad s_1 = 2.6417 \angle 132.45°$$

$$A_2 = -45.1 + j238.7 \qquad\qquad s_2 = 2.2147 \angle 112.22°$$

$$A_3 = 49.0 + j18.6 \qquad\qquad s_3 = 2.0132 \angle 96.93°$$

$$A_4 = A_0^*, A_5 = A_1^*, A_6 = A_2^*, A_7 = A_3^*, \qquad s_4 = s_0^*, s_5 = s_1^*, s_6 = s_2^*, s_7 = s_3^*.$$

These poles in the s'-plane are shown in Figure 8.27; remember, we choose only the left-half-plane poles for a stable filter $H_c(s')$. By looking near $\sigma' = 0$, as previously noted, we see that the poles of the Chebyshev II do not lie on an ellipse, as do the Chebyshev I poles. Also, as indicated earlier, the zeros of the Chebyshev II analog filter lie on the imaginary axis.

The poles in the z-plane are just $z_{p_\ell} = e^{s_\ell \Delta t}$:

$$z_{p0} = e^{s_0 \cdot 1} = 0.0498 \angle 55.9°, \quad z_{p1} = 0.1681 \angle 111.67°, \qquad (8.56)$$
$$z_{p2} = 0.4328 \angle 117.47°, \quad z_{p3} = 0.7843 \angle 114.50°,$$

and the other four poles are the complex conjugates of these. The zeros must be determined from the recombined numerator of the rational function $H(z)$ in (8.54). Numerically, we find them to be

$$z_{z0} = -4.45 \cdot 10^{-5}, \ z_{z1} = 5.724, \ z_{z2} = -0.181, \ z_{z3} = -0.0425, \qquad (8.57)$$
$$z_{z4} = 1.044 \angle 130.5°, \ z_{z5} = 0.595 \angle 155.3°,$$

and the remaining two zeros are complex conjugates of z_{z4} and z_{z5}. These poles and zeros of the digital Chebyshev II filter are as shown in Figure 8.28, which also shows

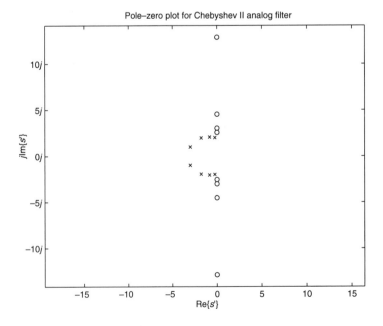

Pole–zero plot for Chebyshev II analog filter

Figure 8.27

---

[10] The poles, which can also be obtained identically from (8.43) through (8.45), are given in polar form for readability.

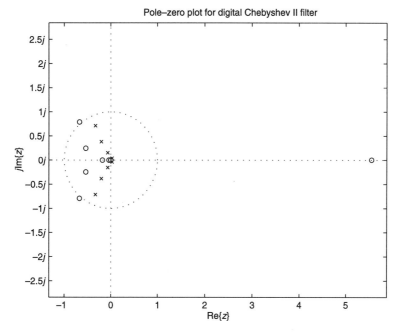

Figure 8.28

(dashed) the unit circle. They agree closely with those obtained by using `cheby2` and then `impinvar`. Again, the only difference is due to the minor difference in initial filter parameterization specification format.

Now that the poles and zeros are known, all that is left is to determine the scaling constant. To obtain unity magnitude at dc for the digital filter, we note the following. Each pole and zero term is of the form $(z - z_{z/p\ell})$, so at dc $(z = 1)$ it becomes $(1 - z_{z/p\ell})$. Therefore, we should divide the numerator by the product of all the numerator factors of this form, and multiply it by the product of all the denominator factors of this form. We initially obtain

$$H(z) = \frac{\displaystyle\prod_{\ell=0}^{7}(z - z_{z\ell})(1 - z_{p\ell})}{\displaystyle\prod_{\ell=0}^{7}(z - z_{p\ell})(1 - z_{z\ell})}, \tag{8.58}$$

where $z_{z\ell}$ and $z_{p\ell}$ are numerically as given in (8.56) and (8.57). The frequency response is as shown in Figure 8.29. The results are disappointing; there is nowhere nearly 40 dB of attenuation in the stopband. Also, the magnitude of the frequency response is not exactly unity at zero frequency, despite our efforts. Consequently, the result of filtering $v(n)$ with $h(n)$ therefore does not achieve a very good approximation of $x(n)$. This deficiency is seen most easily in the frequency domain, where the spectrum of $y(n) = v(n) * h(n)$ (Figure 8.30) shows that $w(n)$ has not been fully eliminated. The appearance of $y(n)$ is similar to that of $x(n)$, but when "played back," the annoying sound of $w(n)$ will still be heard.

No, there were no errors in calculation or in following the procedure. For example, by our very description, the digital filter "sampling frequency" $1/\Delta t_{\text{impinv}}$ should be large enough, because supposedly beyond $f = 0.4$, we are 40 dB down—near zero at $f = 0.5$. Where is the problem? The answer is, another error in `impinvar`! (This problem persists in the latest version of Matlab.)

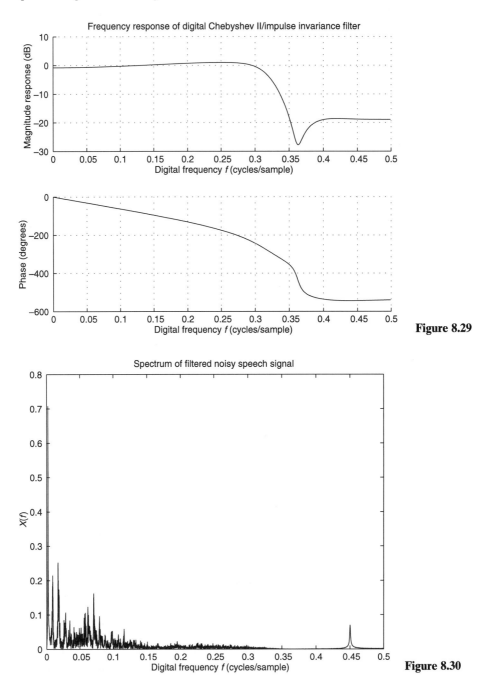

Figure 8.29

Figure 8.30

Notice that the order of the numerator is equal to the order of the denominator for a Chebyshev II filter. Consequently, in a partial fraction expansion, there is a constant term K,[11] whose inverse Laplace transform is $K\delta_c(t)$. Matlab replaces the Dirac delta function $\delta_c(t)$ by the Kronecker delta function $\delta(n)$, which is wrong. In fact, impulse invariance cannot be used for rational functions having order of numerator $M \geq N$, where N = order of denominator. Notice that this limitation

---

[11] We neglected this term in deriving the scaling in (8.58); that is the cause for nonunity magnitude at dc.

excludes both Chebyshev II and elliptic filters from impluse-invariant transformation, and also the lag and lead filters of control system design from impulse-invariant transformation.

In general, for $M \geq N$, the impulse response has "direct terms" $a_\ell d^\ell/dt^\ell \{\delta_c(t)\}$. Impulse invariance cannot be used in such cases, because one would have to sample the Dirac delta function $\delta_c(t)$ and/or its derivatives in $h_c(t)$ at $t = 0$ to obtain $h(n = 0)$, which is impossible. The terms are referred to as "direct" in Matlab's residue routine because they are terms direct from input to output, not involving the output feedback inherent in a transfer function denominator.

Another way of observing this fact is to examine frequency responses. A Chebyshev II or elliptic lowpass filter has equiripple in the stopband. Low as the value is made, *it does not decline with* Ω. Hence in the aliasing sum (8.49a), an infinite number of such tails is added, yielding the infinity that is produced by sampling $\delta_c(t)$ at $t = 0$. No value of Δt can correct this. Impulse invariance requires the frequency response to monotonically decline with Ω for large Ω if any hope of an approximate Nyquist rate can be proposed. The only reason we obtained a finite-valued frequency response for the digital filter is that Matlab makes the error of replacing $\delta_c(t)$ with $\delta(n)$, which of course gives finite (but incorrect) values for the resulting frequency response.

As noted above, strictly the impulse-invariant method cannot be applied without aliasing problems to *any* realizable filter, because all such filters have infinite bandwidth. That would seem to say that the impulse-invariant method is useless. Practically, an indefinitely declining magnitude function has an effective bandwidth, beyond which its magnitude is below a specified value. In all such cases, the impulse-invariant method can be applied successfully.

In view of the above problems, we will now redesign the filter using a Chebyshev I filter which, being all-pole *does* approach zero magnitude for large Ω. We will be able to make use of the analog Chebyshev II filter we just designed above when in Section 8.5.2 we study the bilinear transform method, which is immune to aliasing problems. The only change there will be our values of Ω_c and Ω_r.

Let us, then, begin the redesign using Chebyshev I. Although we will now have ripple in the passband (1 dB), we will still meet the specifications as originally stated. The analog specifications should be retained in the resulting digital filter. First of all, exactly the same expressions for N and ϵ apply as before, so again we have $N = 8$, $\epsilon = 0.5088$. Thus from (8.25b) [here, $|H_c(j\Omega)| = 1/\{1 + \epsilon^2 C_N^2(\Omega/\Omega_c')\}^{1/2}]$ we have $|H_c(j\Omega)| = 1/\{1 + [0.5088 C_8(\Omega/[0.3 \cdot 2\pi])]^2\}^{1/2}$ for our filter. There are no zeros, and the poles are as given in (8.33) through (8.38). Numerically, we have

$$s_0 = 1.879 \angle 92.01°, \ s_1 = 1.603 \angle 96.73°, \\ s_2 = 1.101 \angle 104.81°, \ s_3 = 0.500 \angle 131.60°, \tag{8.59}$$

and the other four poles are the complex conjugates of s_0 through s_3. The left-half-plane poles are shown in Figure 8.31. Under partial fraction expansion, we have

$$A_0 = -0.025 + j0.014, \ A_1 = 0.0586 - j0.061, \tag{8.60} \\ A_2 = -0.0559 + j0.121, \ A_3 = 0.0223 - j0.159,$$

and again the other A_ℓ are the complex conjugates of A_0 through A_3.

The log-magnitude response of this filter is shown in Figure 8.32; again it cleanly meets the specifications of –1 dB at 1.89 rad/s and −40 dB at 2.51 rad/s. Although it seems as though there is minimal ripple distortion, we may change our mind if we plot the magnitude itself, rather than the log-magnitude. This is done in Figure 8.33,

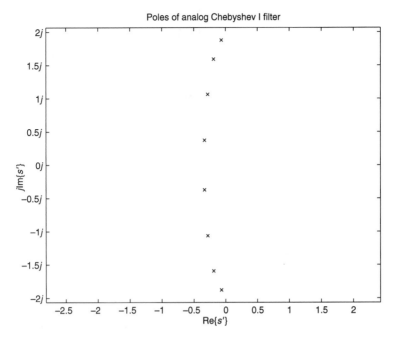

Figure 8.31

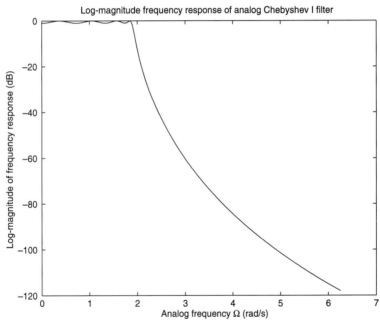

Figure 8.32

and we indeed see what appears to be an 11% ripple (magnitude of frequency response = 0.89) in the passband—which is about 1 dB ripple, as specified for the design. If this ripple is subsequently deemed unacceptable, we of course may redesign using a different Butterworth or Chebyshev I filter. We will now complete this design and obtain the digital filter that impulse invariance produces for this Chebyshev I filter.

The poles of the digital filter, again using $z_{p\ell} = e^{s_\ell \Delta t}$, are

$$z_{p0} = 0.9361 \angle 107.6°, \quad z_{p1} = 0.8287 \angle 91.23°,$$
$$z_{p2} = 0.7548 \angle 60.96°, \quad z_{p3} = 0.7177 \angle 21.41°, \tag{8.61}$$

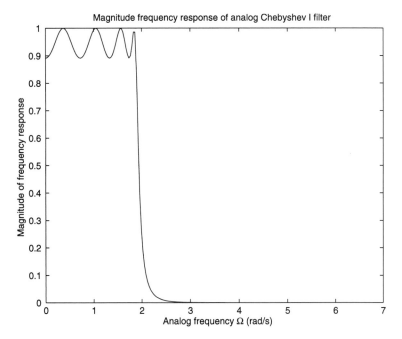

Magnitude frequency response of analog Chebyshev I filter

Figure 8.33

and again the other $z_{p\ell}$ are the complex conjugates of z_{p0} through z_{p3}. When we combine all terms for a common denominator, we obtain the numerator and subsequently the z-domain zeros numerically. However, we find that one of the zeros is out at $j10^{11}$, with no conjugate pair. Suspicious about its validity, we find that the sum of the A_ℓ is in this case zero (or nearly so in the computer, about 10^{-16}). Consequently, by (8.54), the leading coefficient of the numerator of the digital filter is ideally zero.

Because of the fortuitous effects of forming the common denominator, there is no way of beforehand analytically determining the number of zeros. To prevent the numerical artifact, we simply strip off the almost-but-not-exactly-zero leading coefficient of the numerator coefficients array in Matlab. When this is done, the zeros obtained are

$$z_{z0} = -66.23, \ z_{z1} = -4.910, \ z_{z2} = -1.348,$$
$$z_{z3} = -0.494, \ z_{z4} = -0.131, \ z_{z5} = -0.0096. \tag{8.62}$$

Thus although the analog Chebyshev I is an all-pole filter, the digital Chebyshev I has zeros as well as poles. Notice that all zeros are real. The pole–zero plot, excluding the zero at -66.23 for facilitating viewing, is in Figure 8.34. There being no simple formula for the leading nonzero coefficient, we just divide the numerator by the numerically determined leading nonzero coefficient, along with performing the scaling in (8.58).

This time the results are excellent (Figure 8.35): The filter has met the desired specifications of less than 1 dB of attenuation in the passband (up to $f = 0.3$) and at least 40 dB of attenuation in the stopband (starting at $f = 0.4$). Note that the frequency response magnitude peak value is slightly above unity. This is because the Chebyshev I before scaling has for dc not unity but the –1-dB value. If having the magnitude go over unity is undesired, just divide the numerator by the maximum value of the frequency response and then the maximum will be unity. The spectrum of the filtered output, shown in Figure 8.36, indicates that the noise $w(n)$ has been totally eliminated. The only price paid is that *all* the frequency

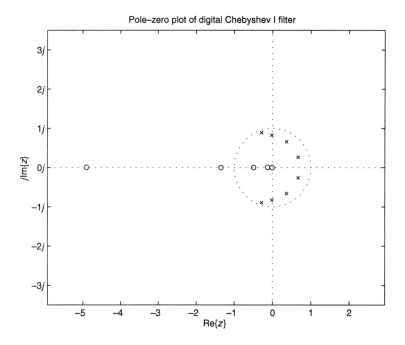

Figure 8.34

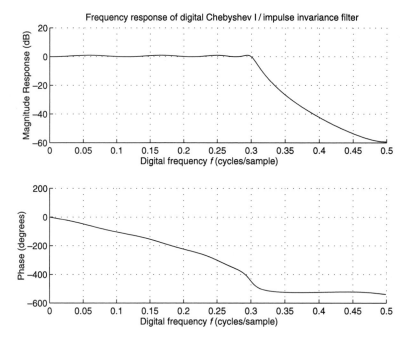

Figure 8.35

content (including that of the signal) above $f = 0.3$ has been annihilated. If this set of specifications is found unsatisfactory, the filter can be sharpened with a higher-order filter upon redesign.

Careful study of this example or, better, personal use of Matlab to repeat it brings up a large number of practical issues in the design of filters using modern software. A combination of facility with the software tool and knowledge of theory is required to bring about a successful design. Blind "plugging in" is highly prone to error.

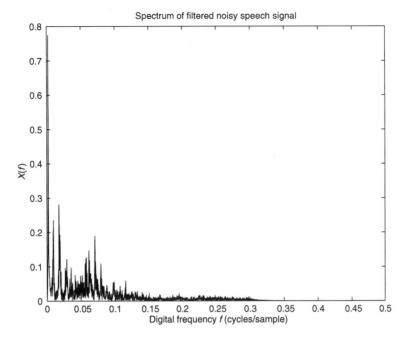

Figure 8.36

8.5.2   Bilinear Transform

As pointed out in Example 8.6, the aliasing problem can be a barrier to carrying out the otherwise straightforward impulse-invariant design. An alternative transformation for converting a well-designed analog filter into a digital filter having nearly the same characteristics is the bilinear transform. It is the most popular method for doing digital IIR design using analog filters. In fact, the Matlab programs tacitly use the Matlab routine `bilinear` to perform all digital Butterworth, Chebyshev, and elliptic filter designs.

Before beginning, it is worth pointing out that we cannot perform Butterworth, Chebyshev, and elliptic designs in the z-plane, with $\omega = \Omega\,\Delta t$. The reason is that those analog filters are based on polynomials in frequency. However, for LSI systems, $H(z)$ is a rational function of z^{-1}, which on the unit circle is a rational function of $e^{j\omega}$, *not* a rational function of ω. Incidentally, it is for this reason that the usual Bode straight-line asymptotes never occur in digital system frequency responses. We conclude that we need an "analog" domain where we have rational functions of a frequency variable in order to use analog filter designs as a basis for designing digital filters.

We have already considered the bilinear transform in Section 6.4.4 in the context of analog system simulation. There the procedure for bilinear transformation was given: Replace all instances of $1/s$ in $H_c(s)$ by $1/s' = F(z) = (\Delta t_{\mathrm{blt}}/2)(z + 1)/(z - 1)$, where to avoid confusion later (see discussion below), we introduce the prime on s and the subscript blt on Δt. This substitution amounts to global use of the trapezoidal rule approximation of integration wherever integration appears in $H_c(s)$, as proved in Section 6.4.4. A new issue arises when we apply the idea to piecewise-constant filter design. There is a warping (a nonlinear transformation) of the frequency axis under the bilinear transformation that must be accounted for to produce predictable designs. Let us discuss this problem now.

If we substitute $z = e^{j\omega}$ into the bilinear transform, we obtain

$$F(e^{j\omega}) = \frac{\Delta t_{\text{blt}}}{2} \frac{e^{j\omega} + 1}{e^{j\omega} - 1} \tag{8.63a}$$

$$= \frac{\Delta t_{\text{blt}}}{2} \frac{e^{j\omega/2}}{e^{j\omega/2}} \frac{e^{j\omega/2} + e^{-j\omega/2}}{e^{j\omega/2} - e^{-j\omega/2}} \tag{8.63b}$$

$$= \frac{\Delta t_{\text{blt}}}{2j \tan(\omega/2)} = 1/s', \tag{8.63c}$$

so that $s' = j(2/\Delta t_{\text{blt}})\tan(\omega/2) = j\Omega'$, in which

or, equivalently,

$$\Omega' = \frac{2}{\Delta t_{\text{blt}}}\tan\left(\frac{\omega}{2}\right), \tag{8.64a}$$

$$\omega = 2 \cdot \tan^{-1}\left(\frac{\Omega'\Delta t_{\text{blt}}}{2}\right). \tag{8.64b}$$

Thus the imaginary axis of the s'-plane maps to the unit circle in the z-plane under the bilinear transformation. It is left to the problems to show that the left half of the s'-plane maps to the interior of the unit circle in the z-plane and the right half of the s'-plane maps to the exterior of the unit circle in the z-plane. Therefore, a stable analog system (filter) transforms into a stable digital filter. Unlike sampling, the simple mapping in (8.64) $\left[\text{or more generally, } z = \dfrac{1 + s'\Delta t_{\text{blt}}/2}{1 - s'\Delta t_{\text{blt}}/2}\right]$ holds for zero factors as well as pole factors; there is no complicating aliasing relation involved in the bilinear transform.

Note that Δt_{blt} is not associated with any physical or even conceptual sampling, except as it exists implicitly in the trapezoidal rule of integration. Also recall the impulse invariance case Δt_{impinv}, where we set $h(n) = \Delta t_{\text{impinv}} \cdot h_c(n\,\Delta t_{\text{impinv}})$. Actually, both the bilinear transform "sampling interval" Δt_{blt} and the impulse-invariance "sampling interval" Δt_{impinv} are in practice for filter transformation purposes only. The analog filter is actually fictitious—it exists for digital filter design purposes only, a fact that we already noted is also true for impulse invariance.

Contrast this situation with that of the actual sampling of signals at time intervals Δt. We designate by the s-plane the Laplace domain of analog signals and physically existing analog filters used to perform analog filtering on them. For analog filters that exist only "on paper" solely for the purpose of designing digital filters, we define the s'-plane, and we subscript all Δt parameters by the name of the discretization transform used.

We came across this idea in Example 8.6 in which we set $\Delta t_{\text{impinv}} = 1$s, whereas $\Delta t = 90.7\mu$s for signal sampling. We defined Ω'_c and Ω'_r for the s' plane; the relation between s' and s for impulse invariance is $s = s'\,\Delta t_{\text{impinv}}/\Delta t$.

What is the relation between s' and s for the bilinear transform (BLT)? Substitute $z = e^{s\Delta t}$ (the relation for pole transformation under sampling) into $1/s' = \frac{1}{2}\Delta t_{\text{blt}}(z + 1)/(z - 1)$ to obtain

$$s' = \frac{2}{\Delta t_{\text{blt}}} \frac{e^{s\Delta t_{\text{blt}}} - 1}{e^{s\Delta t_{\text{blt}}} + 1} = \frac{2}{\Delta t_{\text{blt}}} \frac{e^{s\Delta t_{\text{blt}}/2} - e^{-s\Delta t_{\text{blt}}/2}}{e^{s\Delta t_{\text{blt}}/2} + e^{-s\Delta t_{\text{blt}}/2}} \tag{8.65a}$$

or

$$s' = \frac{2}{\Delta t_{\text{blt}}} \tanh\left(\frac{s\Delta t}{2}\right) \quad \text{(poles only).}[12] \tag{8.65b}$$

---

[12] Although this relation strictly holds for poles only, as discussed in Section 6.2 it approximately holds for the entire transformation when there is little aliasing (and holds exactly for no aliasing) for $|\Omega| < \Omega_s/2 = \pi/\Delta t$.

Thus $s' \neq s$ for the BLT. Relation (8.65) is reversible $[s = (2/\Delta t)\tanh^{-1}\{\frac{1}{2}s'\Delta t_{\text{blt}}\}]$only for $|\text{Im}\{s\}| < \pi/\Delta t$.

Now note that varying Δt_{blt} will just squash or stretch the $j\Omega'$ axis when digital filter specifications are transformed to analog ones via (8.64a). Upon application of the bilinear transform to the resulting analog filter, the same Δt_{blt} will stretch or squash in an exactly inverse manner, so nothing is gained or lost by varying Δt_{blt}. Thus as is true for impulse invariance, Δt_{blt} is often set to unity.

Again, the analog system function is not intended to represent a physical analog system. Rather, the Laplace s'-plane is a "transform domain" where filter design is relatively easy—just as the frequency domain is more convenient than the time domain for convolution.

By inverting $1/s' = F(z)$, we find

$$z = \frac{1 + \frac{1}{2}s'\Delta t_{\text{blt}}}{1 - \frac{1}{2}s'\Delta t_{\text{blt}}}. \tag{8.66}$$

The example uniformly spaced imaginary values of s' in Figure 8.37 are shown mapped into the z-plane by the bilinear transform in Figure 8.38 for $\Delta t_{\text{blt}} = 1$ s. All frequencies higher than the largest one in Figure 8.37, 10 rad/s (all the way to ∞), are on the indicated portion of the unit circle in Figure 8.38. Again, it must be kept in mind that although Δt_{blt} will control the location of the mapping of a given imaginary s' value onto the unit circle, there is no aliasing at all under the bilinear transform between the s'- and z-planes because there is no sampling operation taking place. Note the nonlinear distribution of the frequencies on the z-plane unit circle in Figure 8.38.

The nonlinear relation $\Omega' = (2/\Delta t_{\text{blt}})\tan(\omega/2)$ [(8.64a)] contrasts with the linear relation $\Omega = \omega/\Delta t$, which is at least approximately valid under an impulse-invariant transformation. Consider the mapping of the s'-values in Figure 8.37 to the z-plane under the impulse-invariant transformation when Δt_{impinv} is chosen to be $\Delta t_{\text{impinv}} = 1/F_{\text{"Nyq"}}$ (see below) $= 1/\{2 \cdot (10 \text{ rad/s})/(2\pi)\} = \pi/10 = 0.3142$ s. The mapped frequencies are now uniformly spaced around the unit circle; see Figure 8.39. Note,

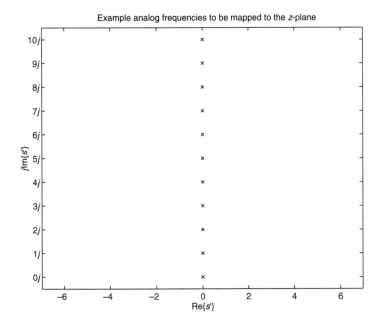

Example analog frequencies to be mapped to the z-plane

Figure 8.37

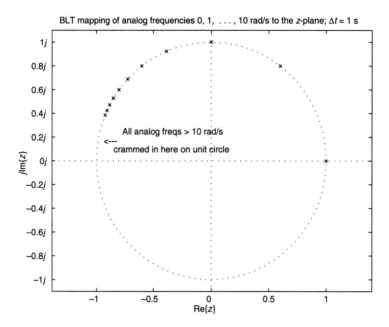

Figure 8.38

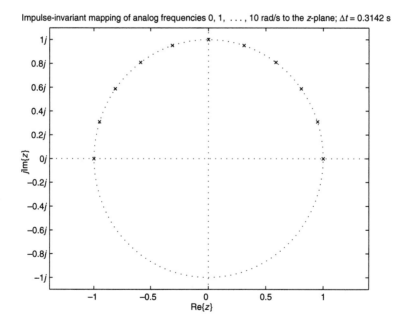

Figure 8.39

however, that unlike the bilinear transformation, strictly only the pole terms and not the zero terms undergo this impulse-invariant mapping (again see footnote 12).

Also note that although we call this "sampling interval" the "Nyquist rate" sampling interval $1/F_{\text{``Nyq''}}$, it really is not Nyquist sampling if the "x" values in Figure 8.37 are the frequencies of corresponding left-half-plane s' poles to be mapped. Specifically, the magnitude of a rational function with a pole having $\Omega = 10$ rad/s will be nonzero out to $\Omega = \infty$ and significantly nonzero well beyond 10 rad/s. Thus under these assumptions, the plot in Figure 8.39 actually represents a case of aliasing. Impulse-invariant transformation of any finite-order rational function involves at least some aliasing.

With impulse invariance, we use $\Omega = \omega/\Delta t$ to map desired digital cutoff frequencies to the required analog counterpart frequencies. The question is how to handle the nonlinearity evident in (8.64) when designing a digital filter using the bilinear transform. It is helpful to view the situation in either of two application situations. First, suppose that we wish to design a digital filter having digital cutoff frequency $\omega_c = 2\pi f_c$. Instead of designing the original analog filter to have a cutoff at $\Omega_c = \omega_c/\Delta t_{\text{blt}}$ as we would for impulse invariance, we use (8.64a): $\Omega'_c = (2/\Delta t_{\text{blt}}) \cdot \tan(\omega_c/2)$. Then, using $1/s' = F(z)$ to obtain the digital filter from the designed analog one, (8.64b) applies and we obtain:

$$\omega_{c,\text{designed}} = 2 \cdot \tan^{-1}\left\{ \frac{\Omega'_c \, \Delta t_{\text{blt}}}{2} \right\} \tag{8.67}$$

$$= 2 \cdot \tan^{-1}\left\{ \frac{[(2/\Delta t_{\text{blt}}) \cdot \tan(\omega_c/2)] \, \Delta t_{\text{blt}}}{2} \right\}$$

$$= \omega_c,$$

as desired. Selection of $\Omega'_c = (2/\Delta t_{\text{blt}}) \cdot \tan(\omega_c/2)$ is an example of a technique known as *prewarping* so that under bilinear transformation the warping is canceled.

The degree of warping depends on how large ω_c is (on the range $[-\pi, \pi]$). Suppose that $\Delta t_{\text{blt}} = 0.1$ s. Then if $\omega_c = 2\pi \cdot 0.05$ rad/sample, then $\Omega'_c = (2/0.1) \cdot \tan(\pi \cdot 0.05)$ = 3.1677 rad/s, whereas under the linear assumption $\Omega_{c,\text{lin}} = \omega_c/\Delta t_{\text{blt}} = 2\pi \cdot 0.05/0.1$ = π = 3.14159 rad/s; not much different. For $\omega_c = 2\pi \cdot 0.45$ rad/sample, the two corresponding analog frequencies are $\Omega'_c = 126.28$ rad/s and $\Omega_{c,\text{lin}} = 28.27$ rad/s— extremely different. The similarity in the first case stems from the fact that for $\omega_c/2$ small, $2 \cdot \tan(\omega_c/2) \approx \omega_c$ so that $\Omega'_c \approx \Omega_{c,\text{lin}}$. A plot of $\Omega'\Delta t$ versus ω and $\Omega_{\text{lin}}\Delta t$ versus ω is shown in Figure 8.40; it also has the digital frequency scale f for convenience. The linear plot $\Omega_{\text{lin}}\Delta t$ versus ω has a slope of unity. Notice that ω only up to 0.92π is shown for the bilinear transform curve because for $\omega \to \pi, \Omega' \to \infty$ under the BLT. That is, the unit circle in the z-plane maps to the entire $j\Omega'$ axis under the BLT.

Another view of the warping situation is the sampled-data system in which we design a digital filter to carry out the work of a desired ideal analog filter. That is, the analog signal is to be sampled every Δt, digitally filtered, and reconstructed.

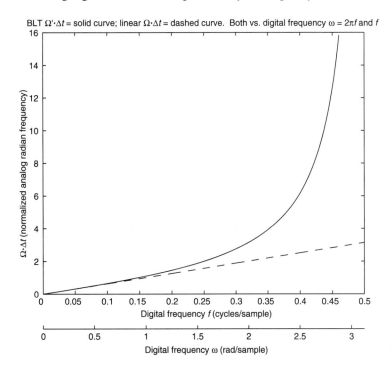

BLT $\Omega'\cdot\Delta t$ = solid curve; linear $\Omega\cdot\Delta t$ = dashed curve. Both vs. digital frequency $\omega = 2\pi f$ and f

Figure 8.40

Suppose that the desired analog filter cutoff frequency is Ω_c. We now design in a fictitious s'-plane an analog filter and use the bilinear transform to obtain from it the required digital filter $H(z)$. Let the transformation "sampling interval" to the fictitious s'-plane be $\Delta t_{blt} = \Delta t_{sampling} = \Delta t$ for convenience; as noted above, its value is arbitrary. If we wish to design the digital filter using the bilinear transformation, we need to design an analog filter with a warped cutoff,

$$\Omega_c' = \left(\frac{2}{\Delta t}\right) \tan\left(\frac{\Omega_c \Delta t}{2}\right); \tag{8.68}$$

this is a second example of prewarping. Then, under the bilinear transformation with this prewarping, we obtain the digital cutoff frequency

$$\omega_c = 2 \cdot \tan^{-1}\left(\frac{\Omega_c' \Delta t}{2}\right) \tag{8.69a}$$

$$= 2 \cdot \tan^{-1}\left\{\left[\left(\frac{2}{\Delta t}\right)\tan\left(\frac{\Omega_c \Delta t}{2}\right)\right]\frac{\Delta t}{2}\right\} = \Omega_c \Delta t. \tag{8.69b}$$

Now, we saw in Section 6.4.2 that under no aliasing, the A/D–digital filter–D/A combination acts like an analog filter with the relation $\Omega = \omega/\Delta t$. Thus the equivalent cutoff will be, under prewarping, $\Omega_{obtained} = \omega_c/\Delta t = \Omega_c \Delta t/\Delta t = \Omega_c$, as desired. We have thus shown how and why prewarping is used in the two most common application situations.

The various relationships regarding sampling and the bilinear transform are illustrated in Figure 8.41. On the left is the usual analog s-plane. From it, under sampling, we obtain the z-plane. For LSI systems relating output to input, the mapping of the system function for a sampled-data system involving sample-and-hold circuitry is $H(z) = (1 - z^{-1})ZT\{LT^{-1}\{H_c(s)/s)\}|_{t\,=\,n\Delta t}\}.$[13] Thus just as $H_c(s)$ is used to relate $y_c(t)$ to $x_c(t)$, $H(z)$ is used to relate $y(n) \cong y_c(n\Delta t)$ to $x(n) = x_c(n\Delta t)$. Or, if we

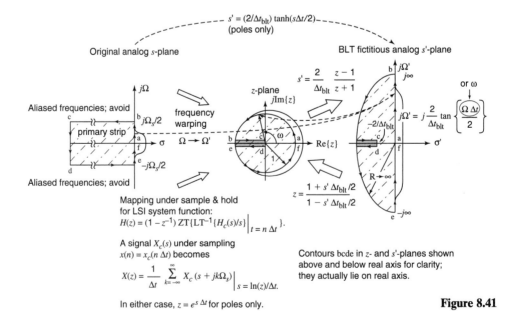

Figure 8.41

---

[13] Notice the differencing operator $(1 - z^{-1})/\Delta t$ out front, where the $1/\Delta t$ factor is implicit, arising naturally from sampling. It approximately cancels the integration of $1/s$ in this relationship. The cancellation is perfect for a step or, more generally, for a staircase input (which is the form of any output of a zero-order hold).

are analyzing the signal transforms, then under sampling $x(n) = x_c(n \Delta t)$, $X_c(s)$ transforms to $X(z) = \frac{1}{\Delta t} \sum_{k=-\infty}^{\infty} X_c(s + jk\Omega_s)\big|_{s=\ln(z)/\Delta t}$. In either case, we know that if $X_c(s)$ is rational, its poles transform according to $z = e^{s\Delta t}$, whereas its zeros do not. Also, the arrows in Figure 8.41 go only one way from s to z because of the possibility of aliasing; the process is in general not reversible. Notice that a typical lowpass filter frequency response is shown on the right side of the $j\Omega$ axis in the s-plane, and the corresponding frequency response is shown outside the unit circle in the z-plane.

Figure 8.41 also shows the primary strip with its outer edge traversed counterclockwise, along the path abcdef (in the direction of the arrows). We know from our previous studies that this contour maps as shown to abcdef in the z-plane. Next, the mapping from the z-plane to the s'-plane for the bilinear transform is shown, with the different forward and reverse variable transformations as shown; this process is entirely reversible. Notice how the abcdef path transforms into the s'-plane—notably, that the origin $z = 0$ (points c and d) transforms into $s' = -2/\Delta t_{\text{blt}}$. As proved in the problems, the entire quarter-circle in the s'-plane occurs around the single point b in the z-plane.

Again, in the s'-plane we see the corresponding lowpass filter frequency response shown drawn to the right of the $j\Omega'$ axis. In fact, most important for our current discussion is the frequency warping illustrated in Figure 8.41 for $(\Omega \to) \omega \to \Omega'$. A typical frequency Ω in the s-plane is transformed into the frequency $\Omega' = (2/\Delta t_{\text{blt}})\tan(\Omega \Delta t/2)$ in the s'-plane; the two are joined by the dotted curve across the center of the figure. Also, $\Omega = \Omega_s/2$ transforms to $\Omega' = \infty$. The digital frequency $\omega = \Omega \Delta t$ (for a pole) is shown as the angle in the z-plane of the corresponding location on the unit circle, and the warping of it into Ω' is also shown with a second dotted curve. Understanding all these relationships helps give a clear understanding of the bilinear transform and how it may be used for approximately implementing analog filtering with a digital filter, as discussed above.

An interesting fact is that the pole mappings for impulse invariance and for the bilinear transform are nearly the same for small $|s|$ (low frequencies). Here let $\Delta t_{\text{impinv}} = \Delta t_{\text{blt}} = \Delta t$. Under sampling, the mapping of an analog pole s_m is $z_{m,\text{samp}} = e^{s_m \Delta t}$, whereas the bilinear transform mapping is $z_{m,\text{blt}} = f_{\text{blt}}(s_m) = (1 + [\Delta t/2]s_m)/(1 - [\Delta t/2]s_m)$. The Taylor expansion of $z_{m,\text{blt}} = f_{\text{blt}}(s_m)$ about $s_m = 0$ is (with the algebraic details left as an exercise)

$$z_{m,\text{blt}} = \frac{1 + \Delta t \cdot s_m/2}{1 - \Delta t \cdot s_m/2} \tag{8.70a}$$

$$= f_{\text{blt}}(0) + f'_{\text{blt}}(0)s_m + \tfrac{1}{2}f''_{\text{blt}}(0)s_m^2 + \tfrac{1}{6}f'''_{\text{blt}}(0)s_m^3 + \tfrac{1}{24}f^{\text{iv}}_{\text{blt}}(0)s_m^4 + \ldots \tag{8.70b}$$

$$= 1 + s_m \Delta t + \tfrac{1}{2}(s_m \Delta t)^2 + \tfrac{1}{4}(s_m \Delta t)^3 + \tfrac{1}{8}(s_m \Delta t)^4 + \ldots , \tag{8.70c}$$

whereas

$$z_{m,\text{samp}} = e^{s_m \Delta t} \tag{8.71a}$$

$$= 1 + s_m \Delta t + \tfrac{1}{2}(s_m \Delta t)^2 + \tfrac{1}{6}(s_m \Delta t)^3 + \tfrac{1}{24}(s_m \Delta t)^4 + \ldots . \tag{8.71b}$$

We see that the first three terms in both Taylor series [(8.70c) and (8.71b)] are identical, and only for $|s|$ sufficiently large that the differences between the differing cubic and higher terms are significant do the two pole mappings significantly differ. Remember, though, that (8.71b) assumes no aliasing and does not address the mapping of zeros, for which there is no simple rule. It is the poles, however, that dictate basic modal behavior; the zeros merely determine proportions between the modes. Thus only for high oversampling where $|s_m \Delta t| < 1$ for all poles (and where the zeros *do* approximately map according to $z = e^{s\Delta t}$) is the bilinear transform "like sampling $h_c(t)$ at $t = n \Delta t$."

EXAMPLE 8.7

Using the bilinear transform, carry out the digital Chebyshev II design example that failed under the impulse-invariant technique in Example 8.6. Use the same specifications [at most 1-dB attenuation at $\Omega_c = 2\pi \cdot 0.3/\Delta t$ and at least 40-dB attenuation at $\Omega_r = 2\pi \cdot (0.4)/\Delta t$,where $\Delta t = 90.703$ μs].

Solution

Instead of mapping $f_c = 0.3$ to $\Omega_c' = \omega_c/\Delta t_{\text{impin}} = 2\pi f_c/\Delta t_{\text{impin}}$ as in impulse invariance, we use (8.64a) $[\Omega' = (2/\Delta t_{\text{blt}})\tan(\omega/2)]$. Using $\Delta t_{\text{blt}} = 1$ s, we obtain $\Omega_c' = (2/1)\tan(2\pi \cdot 0.3/2) = 2.7528$ rad/s and $\Omega_r' = (2/1)\tan(2\pi \cdot 0.4/2) = 6.1554$ rad/s. When we use (8.29) $[N \geq \cosh^{-1}\{([10^{4_r/10} - 1]/[10^{A_c/10} - 1])^{1/2}\}/\cosh^{-1}(\Omega_r/\Omega_c)]$, we get $N_{\text{blt}} = 5$ (next integer above 4.138), compared with $N_{\text{impinv}} = 8$ in Example 8.6 for impulse invariance. We see that, typically, $N_{\text{blt}} < N_{\text{impinv}}$ because $\Omega_{r,\text{blt}}'/\Omega_{c,\text{blt}}' > \Omega_{r,\text{impinv}}'/\Omega_{c,\text{impinv}}'$ as is evident from Figure 8.40; for the same digital filter specifications, we can use a "wider" transition band for our (fictitious) analog filter when we use the bilinear transform. The inverse BLT warping relatively "sharpens" (compresses) the analog filter transition band!

This difference in band width is due to the nonlinearity of (8.64) relative to $\Omega = \omega/\Delta t$ for impulse invariance. The difference in width increases for digital cutoff and rejection frequencies that are near the limit $f = 0.5$ and diminishes to no differences for very low digital cutoff frequencies; again, see Figure 8.40. Naturally, the lower value of N is one reason for the popularity of the bilinear transform.

Using the same analog design procedure as in Example 8.6, the designed analog filter is

$$H_c(s') = \frac{\sum_{\ell=0}^{4} b_\ell s'^\ell}{\sum_{\ell=0}^{5} a_\ell s'^\ell}, \tag{8.72}$$

where, with the numerator scaled so that $H_c(j0) = 1$, the coefficients are

$$a_0 = 407.4, \ a_1 = 367.4, \ a_2 = 179.1, \ a_3 = 54.93, \ a_4 = 10.48, \ a_5 = 1; \tag{8.73}$$
$$b_0 = 407.4, \ b_1 = 0, \ b_2 = 13.44, \ b_3 = 0, \ b_4 = 0.089.$$

The frequency response of this Chebyshev II analog filter is shown in Figure 8.42a. It differs from the plot in Figure 8.26 from Example 8.6 because of prewarping. In Figures 8.42b and c, we compare the results of our design with those obtained by Matlab's cheby2. Because the passband parameter ε need not be an integer, the –1-dB passband specification at $\Omega_c' = 2.7528$ rad/s is met exactly using either method; see the zoomed-in plot in Figure 8.42b.

Concerning the rejection frequency, $H_c(j\Omega')$ as we designed it has nearly 50-dB attenuation at $\Omega_r' = 6.1554$ rad/s, because in (8.29), $N = 4.138$ so that choosing $N = 4$ would have been almost good enough and $N = 5$ results in an overdesign. Matlab also takes the next integer over 4.138, but notice the difference in Figure 8.42c. Matlab guarantees exactly 40 dB of attenuation for all sidelobes; its overdesign is manifested in that the first occurrence of –40 dB is lower than that for our design. Matlab's overdesign is exclusively in the transition width. For our design, both narrowness and sidelobes were overdesigned. However, for either method the rejection specification has been exceeded. The pole–zero plot for $H_c(s')$ is as shown in Figure 8.43.

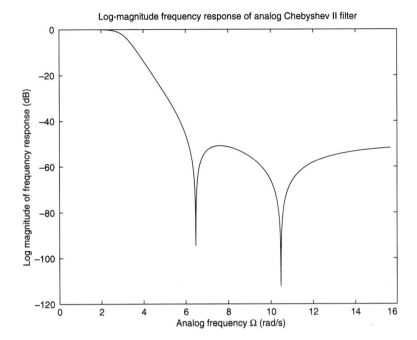

Figure 8.42*a*

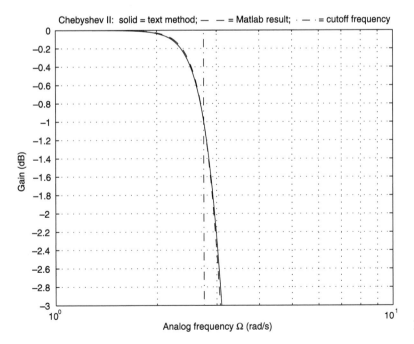

Figure 8.42*b*

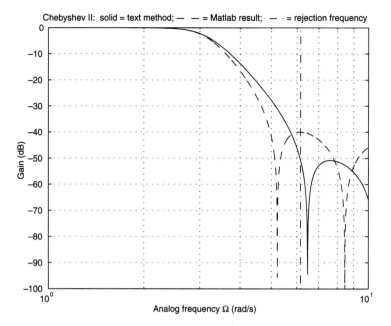

Figure 8.42c

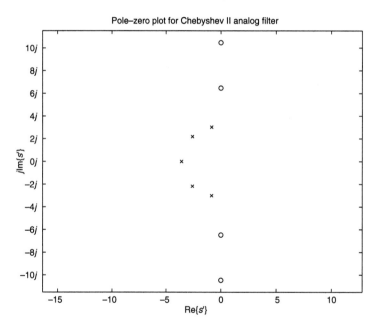

Figure 8.43

Under the bilinear transform, we have

$$H(z) = H_c(s')\big|_{s'=1/F(z)} \tag{8.74a}$$

$$= H_c(s')\big|_{s'=(2/\Delta t_{\text{blt}})(z-1)/(z+1)} \tag{8.74b}$$

$$= \frac{\displaystyle\sum_{\ell=0}^{4} b_\ell \{(2/\Delta t_{\text{blt}}) \cdot (z-1)/(z+1)\}^\ell}{\displaystyle\sum_{\ell=0}^{5} a_\ell \{(2/\Delta t_{\text{blt}}) \cdot (z-1)/(z+1)\}^\ell} \tag{8.74c}$$

$$= \frac{\displaystyle\sum_{\ell=0}^{4} b_{d,\ell} z^{\ell}}{\displaystyle\sum_{\ell=0}^{5} a_{d,\ell} z^{\ell}}, \tag{8.74d}$$

where

$$a_{d,0} = 0.03414, \ a_{d,1} = 0.2584, \ a_{d,2} = 0.8273, \ a_{d,3} = 1.5565, \ a_{d,4} = 1.5435, \ a_{d,5} = 1;$$
$$b_{d,0} = 0.1852, \ b_{d,1} = 0.8354, \ b_{d,2} = 1.5893, \ b_{d,3} = 1.5893, \ b_{d,4} = 0.8354, \ b_{d,5} = 0.1852. \tag{8.75}$$

Note that in (8.74), with the coefficients in (8.75), indeed the dc gain is $H(z)\big|_{z=1} = 1$; that is, the sum of the $b_{d,\ell}$ is equal to the sum of the $a_{d,\ell}$. As may be surmised from (8.74c), the computation of the $a_{d,\ell}$ and $b_{d,\ell}$ is messy and tedious. It is best done on a computer, after trying out a sufficient number of low-order examples to gain confidence in the procedure. Keep in mind that all the coefficients in (8.75) are real, and so all the complex poles and complex zeros occur in complex conjugate pairs.

The pole-zero plot of $H(z)$ is shown in Figure 8.44. All the zeros are on the unit circle because the imaginary axis of the s'-plane maps to the unit circle under the bilinear transform, and we showed before that a Chebyshev II analog filter has all its zeros on the imaginary axis (see also Figure 8.43).

The frequency response of the resulting digital filter is shown in Figure 8.45, and it clearly meets and even exceeds the digital specifications (-1 dB at $f_c = 0.3, < -40$ dB at $f_r = 0.4$), with an order three lower than was possible with the Chebyshev I impulse invariance design. Now with $f_c = 0.3$ and $f_r = 0.4$ and recalling the actual sampling $\Delta t = 90.703$ µs, we meet the original analog cutoff/rejection specifications at $\Omega_c = 20782$ rad/s ($F_c = 3.308$ kHz) and $\Omega_r = 27709$ rad/s ($F_r = 4.41$ kHz). The spectrum of $y(n)$ using this filter appears identical to that shown in Figure 8.36, because the differences due to the passband ripple of the Chebyshev I impulse invariant filter are not easily discerned graphically. If the end use of $y(n)$ is an audio reconstruction, then hearing the result will verify whether the ripple using Chebyshev I (with impulse invariance) is discernable. With our BLT Chebyshev II design, there is no passband ripple at all. In general, one should experiment with the

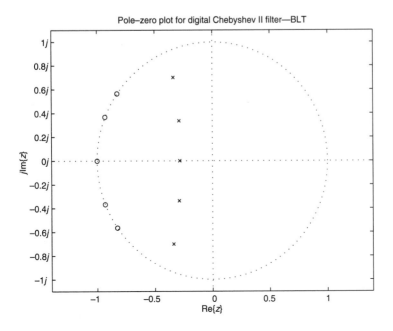

Pole–zero plot for digital Chebyshev II filter—BLT

Figure 8.44

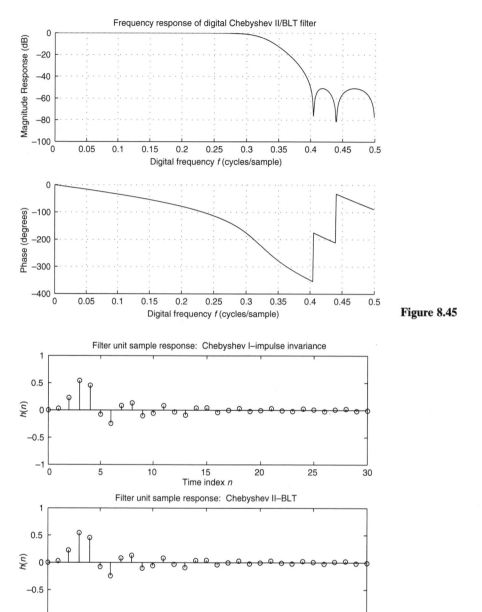

Figure 8.45

Figure 8.46

end usage of the filtered data to determine what values of specifications are appropriate and where the law of diminishing returns comes into play.

It is also interesting to examine the filter unit sample response. In Figure 8.46, we compare the unit sample response for the Chebyshev I/impulse invariance filter with that of the Chebyshev II/bilinear transform filter. The Chebyshev I/impulse invariance unit sample response appears to be much longer than that of the Chebyshev II/bilinear transform filter. However in all cases, the IIR filter unit sample response is *effectively* finite-length even though it strictly has infinite length. This is because its values decline strongly with n. Note that we could, if desired, carry out an "FIR version" (truncated unit sample response) of an IIR filter, a fact that raises several interesting implementation issues and invites comparisons with designs obtained using the window method of FIR design in Section 9.3.

A shortcut for obtaining the bilinear transform by hand calculation is possible if the analog $H_c(s)$ is put in factored form. For brevity, we write Δt_{blt} as Δt. Write

$$H_c(s) = G_c \frac{\displaystyle\prod_{\ell=1}^{M} (s - s_{z\ell})}{\displaystyle\prod_{m=1}^{N} (s - s_{pm})}, \tag{8.76}$$

where for multiple-order poles or zeros some of the $s_{z\ell}$ and s_{pm} may be repeated.[14] Then, under the bilinear transform,

$$H(z) = H_c(s)\Big|_{s = \frac{[(2/\Delta t)(z - 1)]}{[z + 1]}}, \tag{8.77a}$$

which when applied to (8.76) gives

$$H(z) = G_c \frac{\displaystyle\prod_{\ell=1}^{M}\left\{\frac{[(2/\Delta t)(z - 1)]}{z + 1} - s_{z\ell}\right\}}{\displaystyle\prod_{m=1}^{N}\left\{\frac{[(2/\Delta t)(z - 1)]}{z + 1} - s_{pm}\right\}}. \tag{8.77b}$$

Before proceeding, it is illuminating to evaluate (8.77b) at $z = 1$: We obtain $H(1) = G_c \Pi\{-s_{z\ell}\}/\Pi\{-s_{pm}\}$, which is exactly equal to $H_c(s)$ in (8.76) evaluated at $s = 0$. Thus dc gains are preserved under the bilinear transformation.

Now, multiplication of numerator and denominator in (8.77b) by $(z + 1)^N$ gives

$$H(z) = G_c \frac{(z + 1)^{N-M} \cdot \displaystyle\prod_{\ell=1}^{M} \{(2/\Delta t)(z - 1) - s_{z\ell}(z + 1)\}}{\displaystyle\prod_{m=1}^{N} \{(2/\Delta t)(z - 1) - s_{pm}(z + 1)\}}, \tag{8.77c}$$

where if $M = N$, then there is no $(z + 1)^{N-M}$ factor (and we assume $N \geq M$). Rearranging terms gives

$$H(z) = G_c \frac{(z + 1)^{N-M} \cdot \displaystyle\prod_{\ell=1}^{M} \{z(2/\Delta t - s_{z\ell}) - (2/\Delta t + s_{z\ell})\}}{\displaystyle\prod_{m=1}^{N} \{z(2/\Delta t - s_{pm}) - (2/\Delta t + s_{pm})\}}, \tag{8.77d}$$

or, finally,

$$H(z) = G \frac{(z + 1)^{N-M} \cdot \displaystyle\prod_{\ell=1}^{M} (z - z_{z\ell})}{\displaystyle\prod_{m=1}^{N} (z - z_{pm})}, \tag{8.77e}$$

where

$$G = G_c \frac{\displaystyle\prod_{\ell=1}^{M} (2/\Delta t - s_{z\ell})}{\displaystyle\prod_{m=1}^{N} (2/\Delta t - s_{pm})}, \tag{8.78a}$$

---

[14] The subscript on constant G_c here means "continuous time," not "cutoff."

$$z_{z\ell} = \frac{2/\Delta t + s_{z\ell}}{2/\Delta t - s_{z\ell}}, \tag{8.78b}$$

and

$$z_{pm} = \frac{2/\Delta t + s_{pm}}{2/\Delta t - s_{pm}}. \tag{8.78c}$$

These formulas give the same $H(z)$ as do the "brute force" calculations done above in Example 8.7, but may be easily calculated without a lot of algebraic work, or root finding, if $H_c(s)$ is already factored.

The bilinear transform is equally easily applied to any rational-function analog filter $H_c(s)$. Were there a need, one could transform digital filters into analog ones; however, given the advantages of digital systems, that is not generally done. One exception is in digital controller design, where a discretized (z-domain) plant is "analogized" by the "inverse" bilinear transform into the so-called w-plane (another name for the s'-plane). In the analog w-plane, analog controller design may be carried out in a straightforward manner; then a "forward" bilinear transform converts the designed analog controller into a digital controller.

EXAMPLE 8.8

Using the bilinear transform and $\Delta t_{\text{blt}} = 0.001$ s, obtain the corresponding digital filter $H(z)$ from the Butterworth filter $H_c(s)$ designed in Example 8.4. Use the short-cut formulas (8.77) and (8.78), and compare $H(z)$ and its frequency response plots with those obtained using the Matlab routine c2dm. Show the pole–zero plot of $H(z)$.

Solution

We have $G = 5.6617 \cdot 10^{-7}$ and

$$z_{p\ell} = 0.9291 \angle\pm 79.07°, \ 0.8008 \angle\pm 78.82°, \ 0.6870 \angle\pm 78.31°,$$

$$0.5843 \angle\pm 77.46°, \ 0.4904 \angle\pm 76.16°, \ 0.4036 \angle\pm 74.19°,$$

$$0.3227 \angle\pm 71.12°, \ 0.2473 \angle\pm 66.06°, \ 0.1784 \angle\pm 56.83°,$$

$$0.1213 \angle\pm 37.73°, \ 0.0954.$$

Note that there are $N = 21$ zeros at $z = -1$, due to the factor $(z + 1)^{N-M}$ in (8.77e), where here $M = 0$, and no other zeros in $H(z)$.

The frequency response plots in Figure 8.47 obtained using our analytical approach and using Matlab's c2dm command with tustin method are identical (except near $f = \frac{1}{2}$, where the magnitude is down several hundred decibels and numerical roundoff causes discrepancies). Also, the $H(z)$ numerator and denominator polynomials calculated by the two methods are identical, within computer roundoff of order 10^{-13}. In Figure 8.47 we limit the vertical axis in the plot to -90 dB; if shown all the way to $f = \frac{1}{2}$, it goes down to nearly -400 dB, which causes detail in the area of interest to be lost. The pole–zero plot for $H(z)$ is shown in Figure 8.48. Again, in Figure 8.48, the only zero is the (multiple-order) zero at $z = -1$.

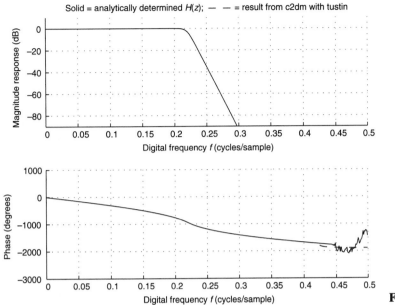

Solid = analytically determined $H(z)$; — — = result from c2dm with tustin

Figure 8.47

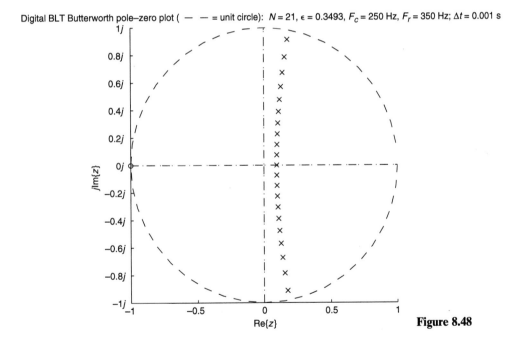

Digital BLT Butterworth pole–zero plot (— — = unit circle): $N = 21$, $\epsilon = 0.3493$, $F_c = 250$ Hz, $F_r = 350$ Hz; $\Delta t = 0.001$ s

Figure 8.48

EXAMPLE 8.9

(a)   Suppose that in Example 8.8, the digital filter is used in a sampled-data system. In that case, neglecting aliasing effects, the digital frequencies f correspond to actual analog frequencies $F = f/\Delta t$. Find the actual analog cutoff and rejection frequencies (in hertz) that were designed using the bilinear transform of the analog Butterworth filter designed in Example 8.8 (use $\Delta t_{\text{blt}} = \Delta t$). As a check, in the process compare the digital critical frequencies with those observed in the final frequency response plot of Example 8.8.

(b)   You should find a significant discrepancy with the desired analog filter critical frequencies. Redesign the digital filter using prewarping so that the actual analog frequencies in the sampled-data system are $F_c = 250$ Hz and $F_r = 350$ Hz, and use $\Delta t = 0.001$ s. Again assume that $\Delta t_{\text{blt}} = \Delta t$.

(c)   What happens if in the prewarping procedure in part (b) we instead use (i) $\Delta t = 0.0005$ s and (ii) $\Delta t = 0.002$ s [which are on either side of $\Delta t = 0.001$ s in part (b)]?

Solution

(a)   Now the original analog filter, being the inverse BLT of $H(z)$, must be in the s'-plane. Thus $\Omega'_c = 2\pi \cdot 250 = 1571$ rad/s and $\Omega'_r = 2\pi \cdot 350 = 2199$ rad/s. Using (8.69a) divided by 2π [$f_c = (1/\pi)\tan^{-1}(\Omega_c \Delta t/2)$], the digital cutoff and rejection frequencies are

$$f_c = \left(\frac{1}{\pi}\right)\tan^{-1}\left\{\frac{2\pi \cdot 250 \cdot (0.001)}{2}\right\} = 0.2119 \text{ sample/cycle}$$

$$f_r = \left(\frac{1}{\pi}\right)\tan^{-1}\left\{2\frac{\pi \cdot 350 \cdot (0.001)}{2}\right\} = 0.2651 \text{ sample/cycle.}$$

That these digital critical frequencies are correct is pictorially verified in Figure 8.47 of the solution of Example 8.8 (e.g., we use the routine `ginput` with axes zooming to accurately read the graph). When used in a sampled-data system with no aliasing, the actual critical analog frequencies for the design filter are as follows.

The attained cutoff frequency is $F_c = f_c/\Delta t = 0.2119/0.001 = 211.92$ Hz, which differs substantially from $F_{c,d} = 250$ Hz, the desired analog filter cutoff frequency. The attained rejection frequency is $F_r = f_r/\Delta t = 265.08$ Hz, which again differs substantially from $F_{r,d} = 350$ Hz, the desired analog filter rejection frequency.

(b)   We use prewarping of the critical frequencies as in (8.69a). First, we set the desired $F_{c,d} = 250$ Hz, $F_{r,d} = 350$ Hz, giving $f_{c,d} = F_{c,d}\Delta t = 0.2500$ (compare with 0.2119 that we obtained without prewarping) and $f_{r,d} = F_{r,d}\Delta t = 0.3500$ (compare with 0.2651 without prewarping). Thus with prewarping in the s'-plane where we will redesign the analog filter and then BLT to obtain the desired digital filter, $\Omega'_c = (2/\Delta t)\tan(\pi f_{c,d}) = (2/0.001)\tan(\pi \cdot 0.25) = 2000.0$ rad/s. Thus $F'_c = \Omega'_c/(2\pi) = (1/[\pi \Delta t])\tan(\pi f_{c,d}) = 318.31$ Hz, which differs from the desired $F_{c,d} = 250$ Hz. Also, $\Omega'_r = (2/\Delta t)\tan(\pi f_{r,d}) = 3925.2$ rad/s, or $F'_r = \Omega'_r/(2\pi) = 624.72$ Hz, which again differs from the desired $F_{r,d} = 350$ Hz. As ϵ is based only on A_c and not on critical frequencies, it remains at 0.3493. However, N changes because the critical frequencies have changed. The new value of N is 11. It is so much lower than the non-warping value of 21 because under prewarping, $F'_r/F'_c = 1.96$, whereas $F_r/F_c = 1.40$: As previously noted, F_c and F_r spread apart relative to each other under BLT warping.

The new analog filter poles are as follows and are shown in Figure 8.49: $s'_k = -313.2 \pm j2178.3, -914.2 \pm j2001.8, -1441.1 \pm j1663.2, -1851.3 \pm j1189.8, -2111.5 \pm j620.0$, and -2200.7 rad/s, where again we have selected

Left-half-plane Butterworth pole plot: $N = 11$, $\epsilon = 0.3493$, $F_c = 318.3$ Hz, $F_r = 624.7$ Hz

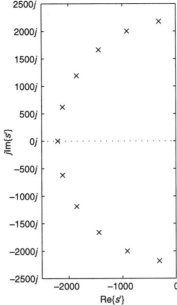

Figure 8.49

for $H_c(s')$ only the left-half-plane poles. Using the BLT, the digital filter poles are obtained as follows: $z_k = 0.8671 \angle \pm 95.53°$, $0.6441 \angle \pm 96.01°$, $0.4591 \angle \pm 97.22°$, $0.2975 \angle \pm 100.04°$, $0.1515 \angle \pm 108.77°$, and -0.0478. This time, we obtain in (8.78a) the constant $G = 2.6953 \cdot 10^{-3}$. There are $N = 11$ zeros at $z = -1$ [and no other zeros in $H(z)$]. This design data completely specifies $H(z)$, as given in (8.77e).

The digital frequency response is as shown in Figure 8.50, and the pole–zero plot of $H(z)$ is in Figure 8.51 (again, the only zeros are at $z = -1$). We see that the above digital filter specifications have been met or exceeded: ≤ -50 dB at $f_{r,d} = 0.35$ (as seen in the upper log-magnitude plot) and -1 dB at $f_{c,d} = 0.25$ (as seen in the lower log-magnitude plot). With $\Delta t = 0.001$ s, these critical frequencies translate to the actual analog frequencies $F_{c,d} = f_{c,d}/\Delta t = 250$ Hz and $F_{r,d} = 350$ Hz. We have thus succeeded in using the BLT to produce a digital filter that in a sampled-data system will effectively have the desired analog cutoff and rejection frequencies.

(c) If we use $\Delta t = 0.0005$ s, the procedure will still work, but we will require a higher-order filter ($N = 18$) under prewarping. N is larger when Δt is reduced because as $\Delta t \to 0$, there is no prewarping distortion, so N would be the value it would take without prewarping ($N = 21$ in this example for $\Delta t \to 0$). For $\Delta t = 0.0005$ s, the desired digital cutoff frequency is $f_{c,d} = 250 \cdot 0.0005 = 0.125$, and with $\Delta t_{\text{blt}} = \Delta t$, $F_c' = \{1/(\pi\Delta t)\}\tan(\pi f_{c,d}) = \{1/(0.0005\pi)\}\tan(\pi \cdot 0.125) = 263.7$ Hz. Similarly, $f_{r,d} = 350 \cdot 0.0005 = 0.175$, so $F_r' = 390.1$ Hz. We see that F_c' and F_r' are, respectively, much closer to the original 250 Hz and 350 Hz, respectively, than was the case for $\Delta t = 0.001$ s.

If, on the other hand, we attempt to use $\Delta t = 0.002$ s, we find that $f_{c,d} = F_c/F_s = F_c\Delta t = 250 \cdot (0.002) = 0.5$ and thus $F_c' = (2/\Delta t_{\text{blt}})\tan(\pi f_{c,d}) = \infty$, and $F_r' = -219.1$ Hz—obviously garbage. The reason is that the sampling frequency is only $F_s = 1/\Delta t = 500$ Hz $= 2 \cdot 250$ Hz $= 2 \cdot F_c$ and $F_s < 2F_r = 700$ Hz —clearly there is major aliasing. We know that under the bilinear

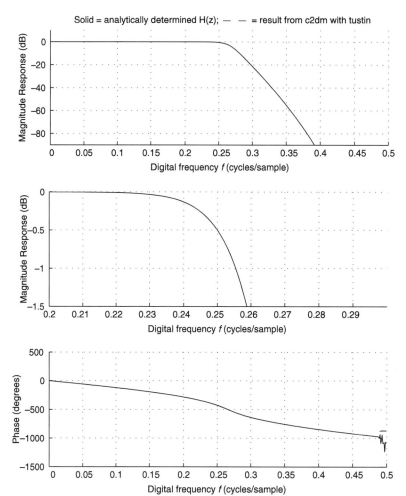

Figure 8.50

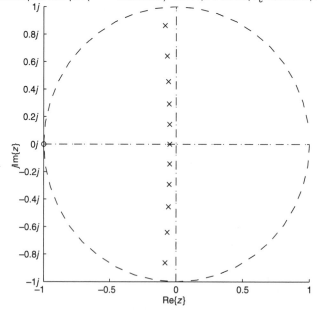

Figure 8.51

transform, $F' = \infty$ translates to half the sampling frequency $\frac{1}{2}F_s$ or equivalently, $f = \frac{1}{2}$. Although there is no aliasing problem if we *begin* with an analog filter and use the BLT to discretize it, there is still aliasing of cutoff frequencies under prewarping. The problem resides in specifying an original analog cutoff frequency that is greater than or equal to half the actual sampling frequency—no digital filter can possibly perform such filtering. As we are asking the impossible, the procedure produces garbage.

EXAMPLE 8.10

In this example, we qualitatively investigate digitally Butterworth filtering a sampled analog signal, using Matlab's Simulink. Simulink simulates "continuous-time" signals that can be "sampled" and then digitally filtered. The output of the digital filter is automatically ZOH "reconstructed" (i.e., turned into a "continuous-time" staircase signal) and available for plotting. We can also use a ZOH block to obtain a staircase Δt-sampled version of "continuous-time" signal, if desired. (The quotes above refer to the fact that Simulink "continuous-time" signals are still discrete; they are just evaluated on a dense time grid.)

Make a block diagram in Simulink for the analog system

$$H_c(s) = 260.13 \, \frac{(s^2 + 2.32s + 257.35)(s + 3.33)(s + 0.84)}{(s + 4)(s^2 + 0.2s + 36.01)(s^2 + 4s + 1300)} \, .$$

Let the input to the system be a boxcar beginning at 0.1 s and ending at 20 s, and carry the simulation out to 70 s. The constant 260.13 in $H_c(s)$ was chosen as the product of the poles divided by the product of the zeros, so that the dc gain is zero dB; and consequently, the "final value" of the step of the boxcar will be 1, not some other value.

Filter the sampled analog signal with a digital filter, within the Simulink block diagram; let the ZOH have $\Delta t = 0.05$ s.[15] Design the digital filter in Matlab to be a 10th-order Butterworth (i) bandstop filter blocking the lower-frequency pole in $H_c(s)$ and passing the higher-frequency pole (ii) lowpass filter blocking both oscillating poles as much as possible but without losing the boxcar shape.[16]

Plot on the same axes the digital spectra for the discrete-time signals both before and after digital filtering. Below this plot, also plot the magnitude frequency response of the digital filter, thereby verifying where the output spectrum is modified. Account quantitatively for the rapid oscillations in the spectra, and find their period. Also plot the continuous-time output of $H_c(s)$ and the ZOH-reconstructed filtered discrete-time signal. Interpret the time plots, in view of the filtering performed.

Solution

To make the boxcar, we add two step functions, the latter having value -1 and occurring at $t = 20$ s; the Simulink block diagram is as shown in Figure 8.52. (The various names `tsim, xorig, xsim, xsimz, xsimza` are just Matlab array variables

---

[15] The results are the same if ZOH in Figure 8.52 is removed, because the digital filter has a built-in Δt-ZOH. However, it will be convenient to have a staircase version of the output of the analog filter, so we retain the ZOH.

[16] There are, of course, much more accurate inverse filtering techniques that would better restore the original pulse; our goals here are to investigate the effects of frequency-selective filtering of a sampled signal and to gain practice with a modern simulator such as Simulink.

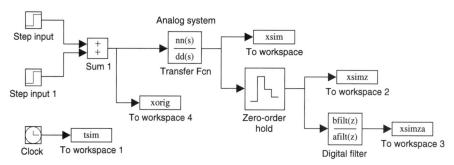

Figure 8.52

made available by Simulink to the Matlab command line.) The poles of the analog system $H_c(s)$ are at $s = -4$ Np/s, -0.1 Np/s $\pm j6$ rad/s, and -2 Np/s $\pm j36$ rad/s.

The digital filter is obtained as follows. First, we determine the resonant frequencies of the analog filter by either examining the spectrum of the continuous-time filter or, if $H_c(s)$ is rational (as in this example), by factoring $H_c(s)$ to determine its poles. We then translate these analog frequencies to digital ones via the nonaliasing sampling relation $f = \Omega \Delta t/(2\pi)$, yielding for this example $f_{\text{high}} = 36 \cdot 0.05/(2\pi) = 0.287$ cycle/sample and $f_{\text{low}} = 6 \cdot 0.05/(2\pi) = 0.048$ cycle/sample. After accounting for the width of the spectral peaks, we select the breakpoint frequencies as seen in the following `butter` Matlab commands and explained below (in this example, we let Matlab do the work, which is BLT with prewarping):

```
[bfilt, afilt] = butter(10,[0.02 * 2 0.17 * 2],'stop') for the bandstop filter,
[bfilt, afilt] = butter(10,0.02 * 2) for the lowpass filter.
```

It should be noted that Matlab digital filter routines have a strange notation for frequency in the digital domain. They use the "normalized frequency" discussed fully in Example 8.3. Thus their digital cutoff frequency Wn is neither in rad/sample (the proper units of ω) nor in cycles/sample (the proper units of f). Instead, it assigns to Wn the value 1 for 0.5 cycle/sample; thus, Wn $= \omega_c/\pi = 2f_c$. Thus with normalized digital frequencies in Matlab from 0 to 1, we must multiply our f_c values by 2, as shown in the Matlab commands above. We choose the bandstop and lowpass cutoff frequency $f = 0.02$ to be well below $f_{\text{low}} = 0.05$ and not too far below 0.05 so that the low-frequency content of the step is retained. We choose the $f = 0.17$ upper bandstop cutoff as being well above f_{low} but also well below f_{high} so as to pass the high-frequency oscillating mode.

In Figure 8.53, we show the analog-filter-distorted signal before sample-holding. Notice the effects of the two poles of $H_c(s)$: The lower-frequency pole has oscillations that take a very long time to attenuate, due to σ being only -0.1 Np/s, whereas the higher-frequency pole with $\sigma = -2$ Np/s dies out almost immediately when excited on the step-pulse discontinuities.

The required spectrum plots for the bandstop filter are in Figure 8.54. The bottom plot is the digital filter frequency response, clearly showing the stop-band and passbands. The top plot shows the full spectrum height, whereas the second plot limits the vertical axis to 40, to show more clearly the effect on the higher-frequency complex poles of $H_c(s)$. The solid curve is the unfiltered spectrum. The filtered spectrum is shown with a dashed curve, and to distinguish it from the unfiltered spectrum, X's were additionally periodically plotted for it (otherwise, e.g., when the output spectrum is nearly zero, it would be hard to see on the plot).

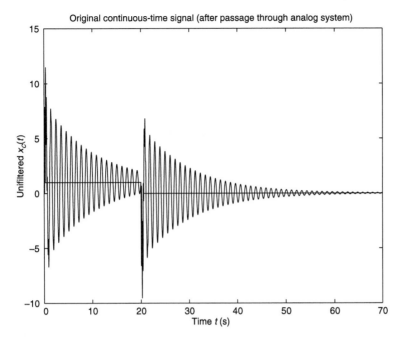

Figure 8.53

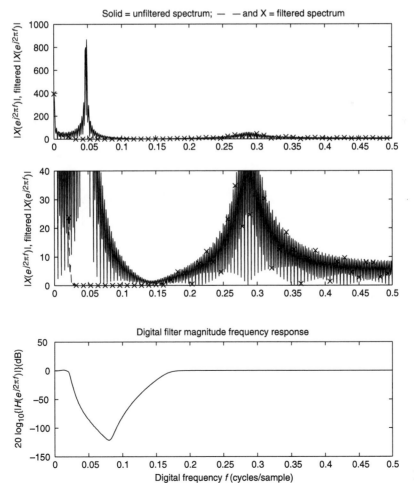

Figure 8.54

The spectra have rapid oscillations arising from the sinc spectrum of the boxcar. The "period"—that is, the digital frequency interval covering one complete oscillation in the spectrum— of those oscillations is found as follows. Set $2\pi = \Omega_{per} \cdot [\frac{1}{2} \cdot$ duration of boxcar], or $\Omega_{per} \approx 4\pi/20$, or $f_{per} = 2\Delta t/$duration $\approx 2 \cdot (0.05)/20 = 0.005$ cycle/sample. This result for the oscillation period has been verified by zooming in on a few oscillations using `axis` and is also seen when the original boxcar spectrum is plotted.

It can also be seen in Figure 8.54, as follows. We count, for example twenty verticle lines (half of which do not extend down very far due to plot resolution) from $f = 0.25$ to 0.3 cycle/sample. Because it is the magnitude that is being plotted, it takes two downward lines to traverse one period. Thus one "period" is equal to $\Delta f/$[number of oscillations] $= 0.05/10 = 0.005$ cycle/sample.

The bandstop filter filters out much of the lower-frequency oscillating pole of $H_c(s)$, but does not at all filter out the higher-frequency pole of $H_c(s)$. Consequently, the square discontinuities excite the high-frequency pole and there remain the huge, high-frequency, highly damped oscillations in the temporal output, shown in Figure 8.55.

Consider next the lowpass filter. Now we totally eliminate the higher-frequency oscillation, as is seen in the spectral plots in Figure 8.56. As noted, we do not completely eliminate the low-frequency oscillation because we do not want to completely distort the boxcar response, which has significant energy in the lower-pole band. In particular, we do not want the response to take too long to rise, so we do not want to cut down too far toward zero frequency. Thus we still have large overshoot in our LPF output (see Figure 8.57), but it is tolerable and is a great improvement over the original in Figure 8.53—and has none of the high-frequency distortion at the on/off transitions as did the output of the bandstop filter in Figure 8.55. We could reduce the digital lowpass filter cutoff and the overshoot in Figure 8.57 would be reduced, but it would continue to oscillate during the entire high-time of the pulse (i.e., it would not reach the final value of 1 before the input pulse goes back to 0).

Finally, it is instructive to examine the step response of the same digital filters with $H_c(s)$ removed (no analog filter distorting the input step). We find similar behavior but with reduced overshoot due to the absence of the lower-frequency pole of $H_c(s)$.

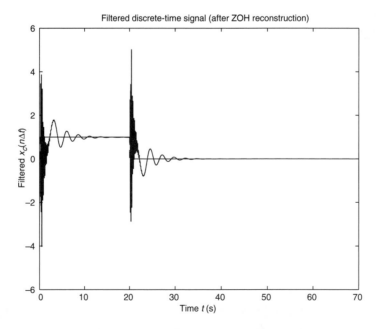

Figure 8.55

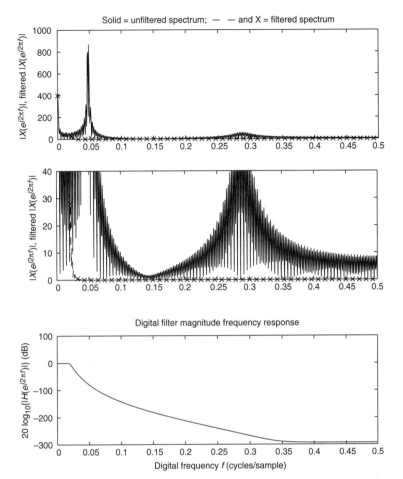

Figure 8.56

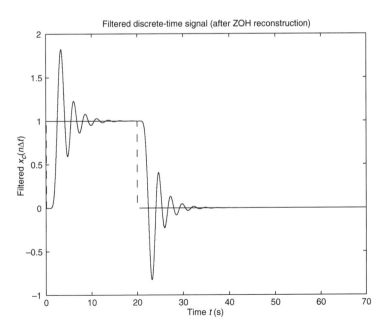

Figure 8.57

8.6 SPECTRAL TRANSFORMATIONS OF DIGITAL FILTERS

As previously mentioned, by simple transformations we may transform a good digital lowpass filter $H_{old}(z)$ into a good digital highpass filter $H_{new}(z)$ or other filter types, such as bandpass, bandstop, and lowpass filters having different cutoff frequencies. The design formulas in this section originate from A. G. Constantinides.[17] As noted in Section 8.4.1, an alternative approach to modifying the characteristics of a filter is to convert to state–space form and perform various array scalings, as Matlab does. Here, we will instead use the Constantinides approach.

The fundamental solution idea in this approach is to replace z^{-1} wherever it appears in the original filter transfer function $H_{old}(z)$ by an allpass function $G(z^{-1})$. An allpass system or filter is one whose magnitude is unity everywhere on the unit circle ($|z| = 1$); thus, all the $H_{old}(e^{j\omega})$ values will also appear in

$$H_{new}(z) = H_{old}(1/G(z^{-1}))\Big|_{|z|=1}. \tag{8.79}$$

The effect of $G(z^{-1})$ is to stretch, squash, and/or replicate the DTFT of the given digital filter $H_{old}(z)$ in a desirable, controllable way. The general form of $G(z^{-1})$ is the Kth-order allpass system function

$$G(z^{-1}) = \prod_{k=1}^{K} \frac{z^{-1} - \alpha_k}{1 - \alpha_k z^{-1}}. \tag{8.80}$$

In hardware, we may implement this transformation by replacing each delay element (shift register) by a structure implementing $G(z^{-1})$. Structures for digital filters are discussed in Section 9.6. In software or in hardware for a totally new structure, we mathematically substitute $G(z^{-1})$ for z^{-1} in $H_{old}(z)$ and "clear fractions" to form the modified rational function of z^{-1}, $H_{new}(z) = H_{old}(1/G(z^{-1}))$. The simplest case is modification of the cutoff frequency of a given digital lowpass filter, so we begin with it.

8.6.1 Changing the Cutoff of a Lowpass Filter

Suppose that we have a good digital lowpass filter $H_{old}(z)$ having cutoff frequency ω_c and we wish to obtain a lowpass filter $H_{new}(z)$ of similar quality with an adjustable cutoff frequency ω_c', controllable by adjusting a single parameter α. All we want to do is stretch the passband and squash the stopband if $\omega_c' > \omega_c$, or do the reverse if $\omega_c' < \omega_c$. $G(z^{-1})$ with $K = 1$ suffices for this purpose, where we desire a one-to-one correspondence between the DTFTs of the old and new digital filters. We want $\omega = 0$ to transform to $\omega' = 0$ and $\omega = \pi$ to transform to $\omega' = \pi$. From (8.79) and (8.80) with $K = 1$ we have

$$H_{new}(e^{j\omega'}) = H_{old}\left(\frac{1}{\frac{z^{-1} - \alpha}{1 - \alpha z^{-1}}}\Bigg|_{z=e^{j\omega'}}\right) \tag{8.81a}$$

$$= H_{old}(e^{j\omega}), \tag{8.81b}$$

where we can refer to the argument of H_{old} in (8.81a) as $e^{j\omega}$ in (8.81b) because for $z = e^{j\omega'}$, the magnitude of the argument of H_{old} is unity. The magnitude of the argu-

---

[17] A. G. Constantinides, "Spectral Transformations for Digital Filters," *Proc. IEE*, 117, 8 (August 1970): 1585–1590.

ment of H_{old} in (8.81a) is unity because $G(z^{-1})$ is allpass. Thus we can simply call that unity-magnitude argument in (8.81a) $e^{j\omega}$. By equating the arguments of H_{old} in (8.81a) and (8.81b) we have

$$e^{j\omega} = \frac{1 - \alpha e^{-j\omega'}}{e^{-j\omega'} - \alpha} \tag{8.82a}$$

or, rearranging,

$$e^{j(\omega - \omega')} - 1 = \alpha(e^{j\omega} - e^{-j\omega'}) \tag{8.82b}$$

and thus

$$\alpha = \frac{e^{j(\omega - \omega')} - 1}{e^{j\omega} - e^{-j\omega'}}. \tag{8.83}$$

Note that (8.82a) can also be solved for $e^{j\omega'}$: $e^{-j\omega'}[e^{j\omega} + \alpha] = 1 + \alpha e^{j\omega}$, or $e^{-j\omega'} = (1 + \alpha e^{j\omega})/(e^{j\omega} + \alpha)$. At $\omega = 0$, we have $(1 + \alpha)/(1 + \alpha) = 1$, giving $\omega' = 0$ as required. Also, for $\omega = \pi$ we have $(1 - \alpha)/(-1 + \alpha) = -1$, giving $\omega' = \pi$ as required. Beyond these values which hold for any α, we can match only one DTFT value using α, which we naturally choose to be the cutoff frequency. We thus obtain

$$\alpha = \frac{e^{j(\omega_c - \omega'_c)} - 1}{e^{j\omega_c} - e^{-j\omega'_c}} \tag{8.84a}$$

$$= \frac{\sin\{(\omega_c - \omega'_c)/2\}}{\sin\{(\omega_c + \omega'_c)/2\}\}} \tag{8.84b}$$

$$= \frac{\sin\{\pi(f_c - f'_c)\}}{\sin\{\pi(f_c + f'_c)\}}, \tag{8.84c}$$

where (8.84b) is obtained by factoring out $\exp\{j(\omega_c - \omega'_c)/2\}$ from the numerator and denominator of (8.84a). Note that for $\omega'_c > \omega_c$, we have $\alpha < 0$, and for $\omega'_c < \omega_c$, we have $\alpha > 0$.

It should also be noted that due to the nonlinear rearrangement of frequencies, there is no way to stipulate that the transition width be a specified value. Once α is selected, the transition width is fixed by (8.82), when the edge frequencies of the "old" filter transition band are substituted and the transformed frequencies calculated. However, in practice a narrow-transition-width filter will usually produce a good transformed filter.

EXAMPLE 8.11

Obtain a similar-quality digital IIR lowpass filter from the Chebyshev II digital IIR lowpass filter from Example 8.7 by using a spectral transformation on the result, but having a cutoff frequency f_c of (**a**) 0.1 and (**b**) 0.42 cycle/sample.

Solution

The result of Example 8.7 in (8.74d) and (8.75) was the digital lowpass filter ($f_c = 0.3$)

$$H_{old}(z) = \frac{0.1852z^5 + 0.8354z^4 + 1.5893z^3 + 1.5893z^2 + 0.8354z + 0.1852}{z^5 + 1.5435z^4 + 1.5565z^3 + 0.8273z^2 + 0.2584z + 0.03414}.$$

(a)   The required value of α is, from (8.84c)

$\alpha = \sin\{\pi(0.3 - 0.1)\}/\sin\{\pi(0.3 + 0.1)\} = 0.618$.

Thus

$$H_{new}(z) = \frac{0.185^2 z_1^5 + 0.8354 z_1^4 + 1.5893 z_1^3 + 1.5893 z_1^2 + 0.8354 z_1 + 0.1852}{z_1^5 + 1.5435 z_1^4 + 1.5565 z_1^3 + 0.8273 z_1^2 + 0.2584 z_1 + 0.03414},$$

where

$$z_1 = 1/G(z^{-1}) = \frac{1 - 0.618 z^{-1}}{z^{-1} - 0.618}$$

$$= \frac{z - 0.618}{-0.618 z + 1} = \frac{f(z)}{g(z)}.$$

With z_1 substituted into $H_{new}(z)$, and after multiplying the numerator and denominator by $\{g(z)\}^5$, we obtain

$$H_{new}(z) = \frac{0.1852 f^5 + 0.8354 f^4 g + 1.5893 f^3 g^2 + 1.5893 f^2 g^3 + 0.8354 fg^4 + 0.1852 g^5}{f^5 + 1.5435 f^4 g + 1.5565 f^3 g^2 + 0.8273 f^2 g^3 + 0.2584 fg^4 + 0.03414 g^5}$$
$$= \frac{0.00595 z^5 + 0.00531 z^4 + 0.00996 z^3 + 0.00996 z^2 + 0.00531 z + 0.00595}{0.4799 z^5 - 1.2677 z^4 + 1.5048 z^3 - 0.9425 z^2 + 0.3099 z - 0.04199}.$$

The frequency response of $H_{new}(z)$ is as shown in Figure 8.58, with that of $H_{old}(z)$ shown dashed. We see that the desired shift of f_c from 0.3 to 0.1 has been achieved as desired.

(b)   We merely recompute $\alpha = -0.4778$ and perform the same algebra (which for this example was automated by using the author's Matlab code). The results are shown in Figure 8.59, again against $H_{old}(z)$ shown dashed. The filter has been successfully transformed from $f_c = 0.3$ to $f_c = 0.42$ cycle/sample.

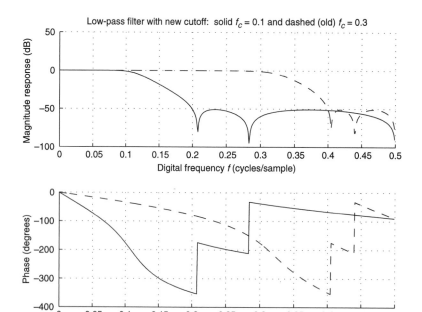

Figure 8.58

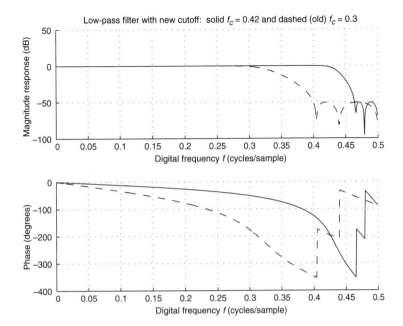

Figure 8.59

8.6.2   Converting a Lowpass Filter into a Highpass Filter

Now we want $\omega = 0$ to map to $\omega' = \pi$ and $\omega = \pi$ to map to $\omega' = 0$; also, we still may want to change ω_c to ω'_c. This is easily accomplished by replacing z^{-1} by $-z^{-1}$ in the $G(z^{-1})$ for the lowpass–lowpass transformation. The reason is that on the unit circle, $-z = -e^{j\omega} = e^{j(\omega + \pi)}$, which merely shifts the lowpass frequency response from being centered on $\omega = 0$ to being centered on π. As before, α in $G(z^{-1})$ controls the cutoff modification. Thus the new $G(z^{-1})$ is

$$G(z^{-1}) = -\frac{\alpha + z^{-1}}{1 + \alpha z^{-1}}. \tag{8.85}$$

It is left to the problems to prove that now

$$\alpha = -\frac{\cos\{\pi(f_c - f'_c)\}}{\cos\{\pi(f_c + f'_c)\}}. \tag{8.86}$$

EXAMPLE 8.12

Obtain a similar-quality digital IIR highpass filter having a cutoff frequency of 0.1 from the Chebyshev II digital IIR lowpass filter from Example 8.7 by using a spectral transformation.

Solution

Using (8.86) we obtain $\alpha = -2.618$. The results are shown in Figure 8.60 (again with the original filter frequency response shown dashed) and demonstrate the effectiveness of this technique. This frequency transformation technique is very convenient because by varying a single parameter α we may easily adjust the cutoff for a possibly very high-order filter.

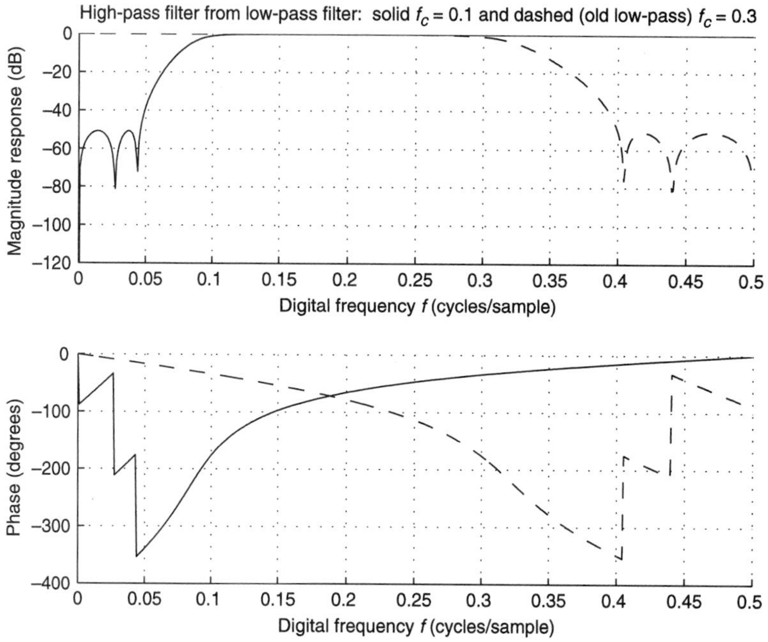

Figure 8.60

8.6.3 Converting a Lowpass Filter into a Bandpass Filter

The same ideas are used for converting a lowpass filter into a bandpass or bandstop filter; only the transformation formulas differ. For the bandpass filter, the formulas are

$$
G(z^{-1}) = -\frac{z^{-2} - \dfrac{2\alpha\beta}{\beta+1} z^{-1} + \dfrac{\beta-1}{\beta+1}}{\dfrac{\beta-1}{\beta+1} z^{-2} - \dfrac{2\alpha\beta}{\beta+1} z^{-1} + 1}, \tag{8.87}
$$

where

$$
\alpha = \frac{\cos\{\pi(f_2 + f_1)\}}{\cos\{\pi(f_2 - f_1)\}} \tag{8.88}
$$

and

$$
\beta = \cot\{\pi(f_2 - f_1)\}\tan(\pi f_c), \tag{8.89}
$$

where f_1 and f_2 are, respectively, the lower and upper cutoff frequencies of the desired passband.

We can intuitively see that for these filters, we need a second-order allpass transformation to be able to replicate the passband twice on the unit circle. That is, as z traverses the unit circle once for the new filter, it traverses it twice for the old filter, due to the z^{-2} terms in (8.87). Thus in general the required order of $G(z^{-1})$ is equal to the number of distinct passbands around the unit circle. This use of higher order of $G(z^{-1})$ to replicate the passbands combined with the α, β spectrum rotator/stretcher/squasher parameter values centers and scales the passbands as desired.

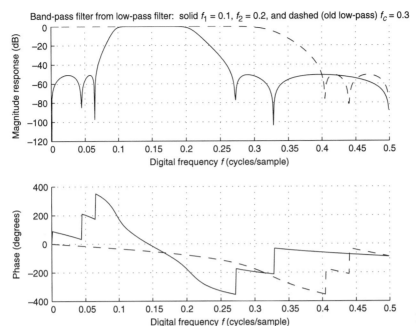

Band-pass filter from low-pass filter: solid $f_1 = 0.1$, $f_2 = 0.2$, and dashed (old low-pass) $f_c = 0.3$

Figure 8.61

EXAMPLE 8.13

Obtain a similar-quality digital IIR bandpass filter having passband from $f = 0.1$ to $f = 0.2$ from the Chebyshev II digital IIR lowpass filter from Example 8.7 by using a spectral transformation.

Solution

Using (8.88) we obtain $\alpha = 0.618$, and from (8.89) we have $\beta = 4.2361$. The results are shown in Figure 8.61 (again with the original filter frequency response shown dashed) and once again demonstrate the effectiveness of this technique.

PROBLEMS

IIR Filter Design: Analog Filters and Bilinear Transform

8.1. Design a minimal-order Butterworth filter having ≤ 1-dB attenuation at 60 Hz and ≥ 40-dB attenuation at 1 kHz. Plot the pole diagram, and write a fully numerically determined expression for $H_c(s)$, with only real-valued coefficients. Verify that specifications have been met by plotting the frequency response in Matlab.

8.2. A Chebyshev filter has $H_c(s) = \dfrac{2.64 \cdot 10^7}{(s + 186.3)(s^2 + 186.3s + 1.413 \cdot 10^5)}$. Plot only the log-

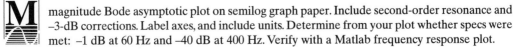

magnitude Bode asymptotic plot on semilog graph paper. Include second-order resonance and -3-dB corrections. Label axes, and include units. Determine from your plot whether specs were met: -1 dB at 60 Hz and -40 dB at 400 Hz. Verify with a Matlab frequency response plot.

8.3. Use Matlab to design a Chebyshev type I analog filter having cutoff frequency $\Omega_c = 2$ rad/s, rejection frequency $\Omega_r = 2.5$ rad/s, maximum attenuation of 0.5 dB in passband, and minimum attenuation of 30 dB in the stopband. Plot the frequency response of the result to verify that specifications have been met—plot the specifications as asterisks on top of the plot, and zoom in where necessary. Give the required filter order.

8.4. Repeat Problem 8.3 for a type II Chebyshev analog filter.

8.5. Design the transfer function $H_c(s)$ for a Chebyshev I analog lowpass filter having no more than 0.7 dB of ripple in the passband (extending up to 6 kHz) and at least 40 dB of attenuation beyond 8 kHz. Compare the order with that required for a Butterworth filter satisfying the same specifications. Express $H_c(s)$ so that the maximum magnitude of the frequency response is unity. Verify graphically that the specifications have been met. Compare your design results graphically using your $H_c(s)$ against the formula $|H_c(j\Omega)| = 1/\{1 + \epsilon^2 C_N^2(\Omega/\Omega_c)\}^{1/2}$; the results should match exactly (in the plots use a few points with asterisks for this formula). Produce one plot showing the overall frequency response and two other plots focusing on, respectively, Ω_c and Ω_r.

8.6. Design a third-order Butterworth filter with -2 dB at 1 kHz, using the passive networks in Figure P8.6. Figures P8.6a and b are, respectively, first- and second-order networks, and their combination is shown in Figure P8.6c. Let $L = 0.1$ H and $C_1 = 0.15$ µF. Verify that specifications are met with both Matlab and PSpice frequency responses (they should agree). Hints: Attempting the matching of first- and second-order transfer functions as in done in Example 8.5 will not work here. Why? What additional device would it take to make it work that way? Moreover, if you write the transfer function in first–second order, you will not get the required form (optional: show this). But if you use second–first order, you will succeed.

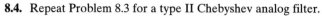

First-order section:

(a) $v_{s,c}(t)$ R C $v_{a,c}(t)$

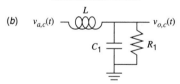

Second-order section:

(b) $v_{a,c}(t)$ L C_1 R_1 $v_{o,c}(t)$

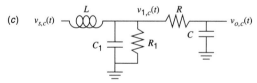

Entire filter:

(c) $v_{s,c}(t)$ L $v_{1,c}(t)$ R C_1 R_1 C $v_{o,c}(t)$

Figure P8.6

8.7. (a) If the input to a system is 1 V, what is the output if the gain is

    (i) -1 dB, (ii) -3 dB, (iii) -10 dB, (iv) -20 dB, (v) -40 dB, (vi) $-\ell \cdot 20$ dB?

(b) Design a Chebyshev I filter having $A_c = 1$ dB at 400 Hz and $A_r = 40$ dB at 450 Hz. Determine $|H_c(j\Omega)|$ in (8.25b) (include numerical values of all the polynomial coefficients) and $H_c(s)$ from its poles [using (8.36) – (8.38)]. Compare N for the Chebyshev I filter with N for the corresponding Butterworth filter. Plot $|H_c(j\Omega)|$ in Matlab using both $H_c(s)$ [e.g., bode] and the direct formula in (8.25b). Do they agree?

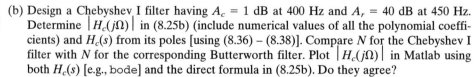

8.8. Make plots comparing Butterworth, Chebyshev I, Chebyshev II, and elliptic analog filters as in Figure 8.22, but for orders 4, 7, and 10.

8.9. (a) Using Matlab, plot all the poles of an analog Butterworth filter for which $\epsilon = 1$, $F_c = 10$ Hz, and (i) $N = 20$ and (ii) $N = 21$. Verify for these cases that real poles occur only for N odd and that the Butterworth poles are never imaginary. Use `axis('image')` to observe the circle on which the poles lie. In each case count the number of poles. (Matlab's `pzmap` is in the Controls toolbox.)

(b) Design an analog Butterworth filter whose passband cutoff is defined at the –1-dB attenuation level (rather than the usual –3-dB level) and occurs at $F_c = 30$ Hz, and that has minimum attenuation of 40 dB at $F_r = 50$ Hz. Plot all the poles, and write the explicit expression for $H_c(s)$, specifying all numerical values of the poles. Plot the magnitude and phase of the frequency response using $H_c(j\Omega)$ as obtained from the poles, and on the same axes, the magnitude as calculated from (8.9d). Show explicitly that the decibel attenuations at F_c and F_r are obtained (use `axis` for two separate plots to show this most clearly).

(c) Using the bilinear transform and $\Delta t = 0.005$ s, obtain the corresponding filter $H(z)$ from $H_c(s)$ in part (b). Use the shortcut formulas in the text, and compare $H(z)$ and its frequency-response plots with those obtained using the Matlab routine `c2dm`. Show the pole–zero plot of $H(z)$.

8.10. (a) Suppose that in Problem 8.9, part (c), the digital filter is used in a sampled-data system. In that case, neglecting aliasing effects the digital frequencies f correspond to actual analog frequencies $F = f/\Delta t$. Find the actual analog cutoff and rejection frequencies (in hertz) that you have designed using the bilinear transform of the analog Butterworth filter. As a check, in the process compare the digital critical frequencies with those observed in the final frequency response plot of Problem 8.9.

(b) You should find a small discrepancy with the analog filter critical frequencies. Redesign the filter using prewarping, so that the actual analog frequencies in the sampled-data system are $F_c = 30$ Hz and $F_r = 50$ Hz. Again use $\Delta t = 0.005$ s.

8.11. Repeat Problem 8.9 for the corresponding Chebyshev type I filters. In part (b), what modification to scaling must be used to obtain perfect agreement of the frequency response magnitude plots computed using the pole-constructed $H_c(s)$ versus using (8.25b)?

8.12. Repeat Problem 8.9 for the corresponding Chebyshev type II filters. In part (a), use $F_r = 1.2$ Hz; in all pole plots, also plot the zeros (remember to use column vectors for `pzmap`).

8.13. Suppose that we wish to use the bilinear transform to do bandpass filtering. Discuss precisely how the bilinear transform is used and what typical steps are involved in the design procedure. Do you obtain an FIR or an IIR filter?

8.14. (a) Show that in the bilinear transformation, the left half of the s'-plane maps to the interior of the unit circle in the z-plane and the right half of the s'-plane maps to the exterior of the unit circle in the z-plane.

(b) Prove that in Figure 8.41 the entire quarter-circle in the s'-plane occurs around the single point b in the z-plane, as claimed in the text. Hint: Analyze $s_1' = \text{BLT}(z_1)$ as z_1 travels along a circle of radius slightly less than 1, as $\angle z_1$ approaches π.

8.15. (a) Consider approximation of true differentiation d/dt by using the bilinear transformation. Give the input–output difference equation, with $y(n)$ alone on the left-hand side.

(b) Describe the operation that this BLT-derived filter performs *precisely* from a time-domain perspective (do not just say "it approximates differentiation;" say in what way it does). In particular, express the approximation for any $n > 0$ in terms of a rectangular-rule integration of an estimate of the second derivative discussed in the text. Be sure to explain the factor of 2 in the BLT definition.

(c) Find the true $h_c(t)$ for first-differentiation. Comment briefly on the applicability of impulse invariance on this filter design problem. If you can, give $h(n)$ and $H(z)$; if you cannot, state why not. Also comment on what happens to the infinite-Ω value in $H_c(j\Omega)$ under the BLT approximation.

(d) Return to this part after reading Chapter 9. Find the causal full-band differentiator $h(n)$ using the window method (N even, rectangular window), beginning with an ideal magnitude response of $|\omega|/\Delta t$ for $\omega \in [-\pi, \pi]$ (periodically extended, otherwise) and an appropriate phase function.

8.16. Using Matlab, on the same axes plot the frequency responses of $H_{\mathrm{BLT}}(z)$ and $H_{\mathrm{FD}}(z)$ where $H_{\mathrm{BLT}}(z)$ is the bilinear transform approximation of the first derivative, derived in Problem 8.15, and $H_{\mathrm{FD}}(z)$ is the first difference (FD) approximation of the first derivative, also arising in the solution of Problem 8.15 ($h_{\mathrm{fd}}(n) = [\delta(n) - \delta(n-1)]/\Delta t$). Superimpose the true first differentiation frequency response, bandlimited to half the sampling frequency; that is, $H_d(e^{j\omega}) = j\Omega|_{\Omega = \omega/\Delta t}$ for $|\omega| < \pi$ (and periodically extended). Let $\Delta t = 1$ s. Specific instruc tions for plots: Use nondecibel magnitude, and plot only the magnitude, not the phase; this will require making an altered form of `freqz`. Plot $H_{\mathrm{BLT}}(e^{j\omega})$ only out to $f = 0.4$ (because $H_{\mathrm{BLT}} \to \infty$ at $f = \frac{1}{2}$), but plot the others out to $f = \frac{1}{2}$. Note that you cannot use `freqz` for the true differentiation plot; use the `plot` command (with `hold on`). For what range of ω are the two approximations of differentiation accurate? Which of the two approximations (BLT, FD) do you prefer? Write expressions for the magnitude frequency responses of each approximation for general Δt values in the simplest forms possible, and comment on whether your graphical results are predictable from your expressions.

Plot stem plots of $h_{\mathrm{fd}}(n)$ and $h_{\mathrm{blt}}(n)$ (which you will have to first derive). As before, use general Δt in derivations and $\Delta t = 1$ s in Matlab plots. Compare the stem plots with $h_c(t) = d/dt\{\delta_c(t)\}$, and comment on stability.

8.17. What are the zeros $z_{z\ell}$ and poles $z_{p\ell}$ of $H(z)$ obtained by bilinear transformation from $H_c(s)$? What can you say in general about $h(n)$ versus $h_c(t)$, where $h(n) = \mathrm{ZT}^{-1}\{H(z)\}$ where $H(z)$ is obtained from $H_c(s)$ by the BLT?

8.18. Derive the first five terms of the Taylor (Maclaurin) series of the bilinear transform, cited in (8.70c).

8.19. Show that the poles of a Chebyshev filter fall on an ellipse with minor axis $\Omega_c\sinh(y_k/N)$ and major axis $\Omega_c\cosh(y_k/N)$.

8.20. Show, as cited in the text (footnote 6), that the real parts of the Chebyshev I poles are equal to $\mathrm{Re}\{s_k\} = \pm\Omega_c\sin\{(2k+1)\pi/[2N]\}\sinh\{\sinh^{-1}(1/\epsilon)/N\}$.

8.21. Beginning with (8.26c), show that $C_N(\Omega)$ indeed monotonically increases with Ω on the order of $2^{N-1}\Omega^N$ for $\Omega > 1$. To do this, use (and prove) the fact that $\cosh^{-1}(x) = \ln\{x + (x^2 - 1)^{1/2}\}$ to obtain a formula for $C_N(\Omega)$ that demonstrates that $C_N(\Omega)$ goes as $2^{N-1}\Omega^N$. For $N = 5$ and $1 < \Omega < 10^{10}$, plot $\log_{10}(C_N(\Omega))$ versus Ω on a log scale using the basic cosh formula, using the formula you derive, and also plot $2^{N-1}\Omega^N$. You should find very close agreement.

8.22. In the text, we showed using Taylor series that the BLT is approximately "like sampling" a pole at s_m—that is, that the transformation of a pole at s_m is similar under the BLT to what it is under true sampling. Now show that the s-plane and s'-plane (BLT) expansions of a value of z are also approximately the same, by expanding the two transforms ($z = e^{s\Delta t}$ solved for s and the BLT solved for s') about $z = 1$. For what digital frequencies is the approximation good?

8.23. Make a chart showing x and $\log_{10}(x)$ for integer values of x from 1 to 9. Show how to use this table to roughly plot $\log_{10}(\Omega)$ values on linear graph paper, given $\Omega = x \cdot 10^L$. Then refine to determine where to locate, for example, $x = 2.5$. Show an example use of this technique to locate the value $\Omega = 25$ rad/s on a log scale on linear graph paper having only uniformly spaced graticule lines at 0.1, 1, 10, 100, and so on. Determine a general formula, and invert it. The inverted form allows you, for example, to read off values in between graticule lines on a Matlab-produced Bode plot (which sometimes are sorely lacking). Demonstrate the technique with a selected analog filter frequency response. Produce a handy table for these purposes.

IIR Filter Design: Impulse Invariance (and Comparison with BLT)

8.24. Obtain the impulse-invariant digital filter from the analog Butterworth filter designed in

Example 8.5. First, analyze for general Ω_0 and Δt values. Check that your result gives the same
results as Matlab's impinvar command when you substitute numerical values of Ω_0 and Δt.
Then choose $\Omega_0 = 3935$ rad/s as in Example 8.5, choose Δt to map 500 Hz (-1 dB) to $f = 0.3$
cycle/sample, and evaluate and explain your results. Repeat for mapping 500 Hz to $f = 0.1$
cycle/sample, and compare.

8.25. (a) Using the bilinear transform, find $H(z)$ corresponding to the third-order analog

Butterworth filter developed in Example 8.5. Derive $H(z)$ for general Ω_0 and Δt. Then
make the $F_c = 500$-Hz frequency (-1 dB) map to $f_c = 0.3$ cycle/sample. Plot your results
in Matlab. Does the -1-dB point occur exactly at $f = 0.3$? If not, explain and modify to
achieve this. Repeat for mapping the -1-dB frequency to $f = 0.1$, and compare.

 (b) In sampled-data applications, we wish to design a digital filter that will perform effective
analog filtering of a sampled signal. Using your work in part (a), design a digital filter to
mimic an analog filter that passes through -2 dB at 1 kHz. Let the sampling frequency be
3 kHz. Verify your work with Matlab plots.

8.26. (a) A causal discrete-time filter with system function $H(z)$ is designed by transforming a causal
continuous-time filter with system function $H_c(s) = (s + a)/[s^2 + (b + c)s + bc]$.
Determine $H(z)$ under the assumptions of (i) impulse invariance and (ii) bilinear transformation relations between $H(z)$ and $H_c(s)$. Include determination of the poles and zeros of
$H(z)$ under the two design assumptions of impulse invariance and bilinear transformation.

 (b) Now let $a = 0.1$ Np/s, $b = 0.2$ Np/s, $c = 0.3$ Np/s, and (i) $\Delta t = 0.2$ s, (ii) $\Delta t = 2$ s, and (iii)
$\Delta t = 20$ s. Using the above values, find the numerical values of the poles and zeros of $H(z)$
under the two design assumptions. When, if ever, are they close?

8.27. Attempt to determine limiting values (for small Δt and for large Δt) of the zeros and poles of
the impulse-invariant and bilinear transform techniques in terms of the zeros and poles of the
original analog transfer function. Restrict your attention to $H_c(s)$ having simple poles and
zeros. Using these results, explain and interpret the results of part(b) of Problem 8.26.

8.28. (a) By your own calculations, find the impulse-invariant digital filter obtained from the ana-

log filter

$$H_c(s) = \frac{62(s + 0.3)}{(s + 2)(s + 3)(s + 3.1)}, \text{ with } \Delta t = 0.01 \text{ s.}$$

Plot the frequency response, and convert the Ω "corner" frequencies into corresponding
f frequencies. Attempt to at least generally locate these, by calling axis to zoom in on
the relevant frequency range. You will have trouble locating "corners;" explain why (in
particular, are there approximate straight-line asymptotes in the frequency response
plots?).

 (b) Repeat part (a) for the bilinear transform. Then compare the plots for parts (a) and (b)
for $f \in [0, \frac{1}{2}]$ and for the frequency-zoomed-in plots.

8.29. Suppose that a continuous-time lowpass filter has the system function

$H_c(s) = \dfrac{8(s + 2)}{(s + 4)^2(s + 1)}$. Find the discrete-time filter that is produced by impulse invariance.
Plot the analog filter response, and $h(n)$ for $\Delta t = 0.01$ s (first 20 points). Comment on what
might be practical values of Δt, and investigate results for $\Delta t = 0.001$ s and $\Delta t = 0.5$ s. Compare
frequency response results with Matlab's impinvar using plots of the frequency response.

8.30. Investigate the routine mpoles. Do so by examining the full commentary lines in the code

for mpoles (in the Signal toolbox or in Matlab\polyfun) with the example; the full version is
not printed out with help due to a blank line in the commentary. Try the example shown, and
verify that it makes sense. Then try mpoles with a given analog transfer function having multiple poles at the origin. In versions preceding Student Version 5, mpoles could not detect
these multiple poles, irrespective of the value of the tolerance parameter. It tested for
when the magnitude of the difference between two poles was less than the user-specified tolerance of the pole magnitude. What was the error, and how would you (and how did Matlab)
easily fix the problem (read the code)? When might this situation arise?

8.31. In the step invariance design method, we set $s(n) = s_c(n\Delta t)$, where $s_c(t) = h_c(t) *_c u_c(t)$, where $u_c(t)$ is the unit step function. To find $H(z)$ from $S(z)$, it is noted that $h(n) = s(n) - s(n - 1)$. Using *step* invariance, find the causal digital filter $H(z)$ based on the causal continuous-time-filter $H_c(s) = 1/\{s(s + 1)\}$. Express $H(z)$ as a factored rational function of z. Interpret the meaning of $H(z)$ in a practical physical situation.

Material on WP 8.3 will be helpful for Problem 8.32.

8.32. Equation (W8.3.4) gives one way of expressing the impulse-invariant transform of multiple-order poles. Show that an alternative expression is $\text{IMP}\{1/(s - s_m)^\ell\} = \{\partial/\partial s\}^{\ell-1}\{\Delta t/(1 - e^{s\Delta t}z^{-1})\}\big|_{s=s_m}/(\ell - 1)!$ Which is preferable in practice, and why?

Deconvolution

8.33. In (2.43), a discrete-time approximation of the second derivative is given by the central difference formula. Find the unit sample response of a causal discrete-time system that might be cascaded with the "second difference" system to recover the input. That is, find $h_i(n)$ so that $h_i(n) * y(n) = x(n)$, where $y(n) = h_{c.d.}(n) * x(n)$ where c.d. denotes central difference. What is a serious practical problem with this inverse filter? Hint: Use the z-transform. To check your answer, use direct convolution. Using a graph or explicit sum, show how individual terms sum to $\delta(n)$ in $h_i(n) * h_{c.d.}(n) = \delta(n)$.

Allpass and Minimum-Phase Systems/Group Delay

8.34. (a) Define a kth-order continuous-time "allpass" system $H_{\text{ap},k,c}(s)$ as one whose k poles and zeros are at negative-complex-conjugate locations s_ℓ (pole) and $-s_\ell^*$ (zero) with $\text{Re}\{s_\ell\} < 0$ (if s_ℓ is complex, then there is also a pole at s_ℓ^* and thus a zero at $-s_\ell$; for real-valued poles $s_m < 0$, the associated zeros are simply at $-s_m$). Recall that the poles and zeros must occur in complex-conjugate pairs if s_ℓ is complex, but for our purposes we will leave the quadratic factors in factored form. Thus we initially have

$$H_{\text{ap},k1,c}(s) = \frac{\prod_{\ell=1}^{L}(s + s_\ell^*)(s + s_\ell)\prod_{m=1}^{M}(s + s_m)}{\prod_{\ell=1}^{L}(s - s_\ell)(s - s_\ell^*)\prod_{m=1}^{M}(s - s_m)},$$

where $k = 2L + M$ is the order of the allpass system and the s_m in the first-order factors are real-valued (negative). However, the standard or canonical form for the allpass system is $(-1)^k H_{\text{ap},k1,c}(s)$, which of course does not affect the magnitude characteristic (but simplifies the proof of the allpass characteristic):

$$H_{\text{ap},k,c}(s) = \frac{\prod_{\ell=1}^{L}([-s_\ell^*] - s)([-s_\ell] - s)\prod_{m=1}^{M}([-s_m] - s)}{\prod_{\ell=1}^{L}(s - s_\ell)(s - s_\ell^*)\prod_{m=1}^{M}(s - s_m)}.$$

Show that an allpass system function is one whose magnitude is unity for all Ω; hence the name "allpass."

(b) Can an allpass filter have any poles or zeros at $s = 0$?

8.35. (a) Using the results of Problem 8.34, define the kth-order digital allpass filter, and prove that it has the allpass property.

(b) Show that if the denominator of the allpass filter is $\sum_{\ell=0}^{k} a_\ell z^{-\ell}$, then the numerator is $\sum_{\ell=0}^{k} a_{k-\ell}^* z^{-\ell}$. That is, the coefficients of the numerator of an allpass filter are in reverse order with respect to those of the denominator.

8.36. (a) Show that the group delay $\tau_G(\Omega) = -d\{\angle H_c(j\Omega)\}/d\Omega$ of a continuous-time system $H_c(s)$ has the following physical meaning: The group delay evaluated at a given frequency $\Omega = \Omega_1$ tells us how much a *narrow$-$band* signal (or portion of a signal) $x_c(t) = s_c(t)\cos(\Omega_1 t)$

centered on $\Omega = \Omega_1$ is temporally delayed by the system $H_c(s)$, where the signal $s_c(t)$ has a small bandwidth centered about $\Omega = 0$; that is, $s_c(t)$ is a low-frequency signal relative to Ω_1, the "carrier." Thus the signal component of the output $y_c(t)$ is approximately $s_c(t - \tau_G(\Omega_1))$.

(b) In the context of the discussion in part (a), show that the phase delay $\tau_P(\Omega) = \phi(\Omega)/\Omega$ evaluated at $\Omega = \Omega_1$ shows how much the carrier in the output $y_c(t)$ of the system is delayed with respect to the phase of the input carrier [and so it is called the "phase" delay; that is, the phase of the carrier sinusoid is delayed by $\tau_P(\Omega_1)$].

8.37. (a) Show that the group delay $\tau_G(\omega) = -d\{\angle H(e^{j\omega})\}/d\omega$ of a discrete-time system $H(z)$ has the following physical meaning: The group delay evaluated at a given frequency $\omega = \omega_1$ tells us how much a *narrow−band* signal (or portion of a signal) $x(n) = s(n)\cos(\omega_1 n)$ centered on $\omega = \omega_1$ is temporally delayed by the system $H(z)$, where the signal $s(n)$ has a small bandwidth centered about $\omega = 0$; that is, $s(n)$ is a low-frequency signal relative to ω_1, the "carrier." Thus the signal component of the output $y(n)$ is approximately $s(n - \tau_G(\omega_1))$. As most signal/carrier compositions are in practice done in continuous time, although the interpretation is correct, it is more conceptual than practical. A more practical use of group delay comes in the definition and discussion of minimum-phase systems and forming the minimum-phase/allpass decomposition. However, the analogy with continuous time is the origin of the concept.

(b) In the context of the discussion in part (a), show that the phase delay $\tau_P(\omega) = \phi(\omega)/\omega$ evaluated at $\omega = \omega_1$ shows how much the carrier in the output $y(n)$ of the system is delayed with respect to the phase of the input carrier [and so it is called the "phase" delay; that is, the phase of the carrier sinusoid is delayed by $\tau_P(\omega_1)$].

8.38. A further property of $H_{ap,k,c}(s)$ is that its phase on the imaginary axis $s = j\Omega$ is always non-positive, and minus the derivative of the phase of $H_{ap,k,c}(j\Omega)$ with respect to Ω, called the group delay $\tau_G = -d\{\angle H_{ap,k,c}(j\Omega)\}/d\Omega$, is always nonnegative (note the minus sign in τ_G!). Using the canonical continuous-time allpass transfer function given in Problem 8.34, argue that τ_G is nonnegative first for a first-order allpass section and then for a complex-conjugate-pole-pair second-order allpass section. With these results, extension to higher-order systems is obvious. Also show that the phase of $H_{ap,k,c}(j\Omega)$ is nonpositive for all Ω.

8.39. In this problem, we investigate minimum-phase systems. Minimum-phase systems are important in the theory of transfer functions. The term is actually short for "minimum phase-lag system," which in fact means maximum (absolute) phase! The terminology is, however, fixed. A major reason for studying minimum-phase systems is that, because of their very common appearance in practice, much of the digital signal processing and control theory relating in particular to frequency response assumes that the system under consideration is minimum-phase. For example, Bode asymptote rules, polar plot plotting rules, and stability margins in controls and electronic design have been conventionally defined for minimum-phase systems. These rules and definitions can be modified for nonminimum-phase systems, but care must be used and then the rules cannot be applied blindly or even at all. Minimum-phase systems also have importance as being "minimum-delay" systems in that the delay until their impulse response goes away is minimum for all systems having the same magnitude frequency response. We define a minimum-phase continuous-time system $H_{min,c}(s)$ (perhaps the more obvious term would be "non-right-half-plane system") as one whose zeros and poles are both all in the left half-plane or on the $j\Omega$-axis.

(a) Show that any stable continuous-time linear system $H_c(s)$ (e.g., perhaps having zeros in the right half-plane but all poles in the left half-plane—it is stable) can be decomposed into the cascade of a minimum-phase system $H_{min,c}(s)$ and an allpass system $H_{ap,c}(s)$: $H_c(s) = H_{min,c}(s) \cdot H_{ap,c}(s)$. In terms of poles and zeros, explicitly indicate the makeup of $H_{min,c}(s)$ and $H_{ap,c}(s)$.

(b) If the numerator of a general $H_c(s)$ is of order M and the denominator is of order N, how many different transfer functions of stable, causal systems have the same magnitude frequency response?

(c) Repeat part (a) for any stable discrete-time system $H(z)$.

8.40. Given $H(z)$ below, determine its minimum phase–allpass decomposition:

$$H(z) = \frac{(1 + 2.2z^{-1})(1 - 0.9z^{-1})(1 - 7.0z^{-1})}{(1 + 0.2z^{-1})(1 - 0.7z^{-1})(1 + 0.3z^{-1})} = H_{min}(z) \cdot H_{ap}(z).$$

8.41. Suppose that $H_{min}(z) = a + b \cdot z^{-1}$ and $H_{max}(z) = b + a \cdot z^{-1}$, where a and b are real and $|a| > |b|$ so that $H_{min}(z)$ is minimum phase and $H_{max}(z)$ is maximum phase.

(a) Demonstrate for this simple example that $H_{min}(z)$ is minimum phase and that $H_{max}(z)$ is maximum phase. That is, show that $H_{min}(z)$ has no poles or zeros outside the unit circle, whereas $H_{max}(z)$ has no poles or zeros inside the unit circle.

(b) $H_{min}(z)$ and $H_{max}(z)$ as given above are called *corresponding* minimum and maximum phase systems. In what sense do they correspond? Prove that they do in this sense.

(c) Find the phase angles $\theta_{min}(\omega)$ and $\theta_{max}(\omega)$, where in general, $H(e^{j\omega}) = |H(e^{j\omega})| \cdot e^{j\theta(\omega)}$. Show that $\theta_{min}(\omega) + \theta_{max}(\omega) = -\omega$ for all $\omega \in [-\pi, \pi)$. Hint: The identity $\tan(A) + \tan(B) = \tan(A + B) \cdot [1 - \tan(A) \cdot \tan(B)]$ may be of use.

(d) Without reference to the allpass decomposition, show explicitly that $\theta_{min}(\omega) > \theta_{max}(\omega)$ for all $\omega \in [0, \pi]$. (Recall that the "min" and "max" refer to "minimum lag" and "maximum lag," so these results are consistent.) Extend the result to all minimum- and maximum-phase systems.

8.42. Illustrate the results of Problem 8.41 with values for a and b of your choice, but satisfying $|a| > |b|$. In particular, show that $H_{min}(z)$ and $H_{max}(z)$ have the same magnitude frequency response, show that $\theta_{min}(\omega) + \theta_{max}(\omega) = -\omega$ for all $\omega \in [0, \pi]$, and show that $\theta_{min}(\omega) > \theta_{max}(\omega)$ for all $\omega \in [0, \pi]$.

Spectral Transformations

8.43. Prove (8.86), the expression for α in the spectral transformation of a lowpass digital filter to a highpass digital filter, by starting with (8.85).

8.44. Suppose that $H_c(s)$ is a continuous-time lowpass filter and that from it we design a digital filter obtained from the equation $H(z) = H_c((z + 1)/(z - 1))$. Then the passband of the digital filter is centered at one of (a) $\omega = 0$, (b) $\omega = \pi$, or (c) $\omega = a$ frequency other than 0 or π. Explain your choice.

8.45. (a) Plot the frequency response of the filter $H_c(s)$ in Problem 8.2. If possible, let the horizontal axis be F (hertz). Again, check against the specifications.

(b) Determine the impulse-invariant filter obtained using Matlab's `impinvar` for sampling frequency Fs = 5 kHz. Plot the frequency response of that $H(z)$ versus f (cycles/sample). Be sure to multiply by Δt if necessary (in the Student Edition 5 and before, Matlab's `impinvar` neglects this scaling) for 0 dB in the passband.

(c) Use (8.87) through (8.89) to determine the digital bandpass filter $H_{BP}(z)$ having lower and upper cutoff frequencies corresponding to analog frequencies 1 kHz and 2 kHz. Plot the frequency response of $H_{BP}(z)$. Are the digital passband cutoff frequencies where you predict them to be?

Chapter 9

Digital Filter Design: FIR Filters and Filter Structures

9.1 INTRODUCTION

In Chapter 8, we studied the design of digital filters whose unit sample response is theoretically infinite-length. This characteristic alone is not very important because in practice we can implement the filter with a small amount of hardware or software. However, their infinite duration places them in the category of filters that may be designed using analog filter design techniques.

In this chapter, we focus mainly on FIR filters—filters whose unit sample response is finite-length. In Section 9.2, we compare IIR versus FIR filters to put them in perspective and facilitate the decision of which to use in practice. We then examine in Sections 9.3 through 9.5 three popular ways of designing FIR filters. The chapter concludes with a discussion in Section 9.6 of structures for implementing digital filters in hardware and the effects of coefficient quantization.

9.2 FIR VERSUS IIR FILTERS

We have seen that useful digital IIR filters can be derived from analog counterparts. Design parameter specifications for the digital filter can be transformed into analog filter specifications. Analog design is carried out with these transformed parameters, and the resulting analog filter is transformed back to a digital filter satisfying the desired specifications. What about FIR filters?

Recall that FIR filter unit sample responses have finite length, whereas IIR filters have infinite length. Can the impulse-invariant or bilinear transform techniques be applied to obtain FIR digital filters? The answer is no, because there is no practical "FIR" counterpart in the analog domain. A general digital LSI system has the z-transfer function $H(z) = B(z)/A(z)$. For an FIR system, $A(z) = 1$ so $H(z) = B(z)$. [$A(z) = 1$ for causal filters if we assume that $B(z)$ is a polynomial in z^{-1}, not z; otherwise, $A(z)$ will have the form $A(z) = z^L$.] Because $B(z)$ is a finite-order polynomial in z^{-1}, $h(n) = b(n)$, a finite-length sequence. The corresponding analog filter would be $H_c(s) = B_c(s)$. With $B_c(s)$ a finite-order polynomial in s, say

$$B_c(s) = \sum_{\ell=0}^{M} b_\ell s^\ell, \tag{9.1}$$

we would obtain

$$h_c(t) = b_c(t) = \sum_{\ell=0}^{M} b_\ell \frac{d^\ell \delta_c(t)}{dt^\ell} \, . \tag{9.2}$$

Systems having weighted sums of high-order derivatives of Dirac delta functions for impulse responses are not commonly observed. Equivalently, systems whose outputs are weighted sums of high-order derivatives of the input are again not usually encountered. Continuous-time LSI systems therefore have transfer functions that are rational functions, and the inverse Laplace transforms of rational functions are infinite-duration. Moreover, if we construct well-behaved finite-duration continuous-time impulse responses, their Laplace transforms will involve exponential functions of s (infinite-order polynomials).

As an example, let $h_c(t) = \sum_{m=0}^{N-1} h(m) p_{\Delta t}(t - m\Delta t)$, so that $h_c(t) = h(m)$ for $m\Delta t \le t \le (m+1)\Delta t$ and 0 otherwise—that is, a zero-order-hold reconstruction of an FIR filter sequence $h(n)$. Then $H_c(s) = \sum_{m=0}^{N-1} h(m) e^{-sm\Delta t} \cdot \frac{(1 - e^{-s\Delta t})}{s}$, and we see that $H_c(s)$ is not a rational function of s.[1] For this reason, direct digital-domain design techniques must be used to design digital FIR filters.

A few of the relative comparisons of IIR and FIR filters are now reviewed. For a given quality of approximation of a given ideal filter, the digital IIR filter has lower numbers of coefficients and computations than those of the corresponding FIR filter. When hardware minimization and delay minimization are important, the IIR filter may be preferred.

The reason that the IIR is more "efficient" than the FIR is easy to see when we consider the pole–zero distributions of the two filter types. A causal FIR filter has the z-transfer function

$$H(z) = \sum_{\ell=0}^{M} h(\ell) z^{-\ell} \tag{9.3a}$$

$$= z^{-M} \sum_{\ell=0}^{M} h(\ell) z^{M-\ell} \tag{9.3b}$$

$$= z^{-M} B_p(z), \tag{9.3c}$$

where $B_p(z)$ is a polynomial in positive powers of z. All the poles of $H(z)$ are at the origin [the z^{-M} in (9.3c)], whereas for an IIR filter, the poles and zeros are distributed more flexibly. It is true that for a given total number of filter coefficients, the FIR filter can control more zeros than can the IIR. However, it is the presence of both zeros and poles that can lead to sharp transitions from passband to stopband.

In our examples of IIR digital filter design in Section 8.5 (see Figures 8.28, 8.34, 8.44, 8.48, and 8.51), the zeros tended to be in the "stop sector" of the z-plane and the poles in the "pass sector." The stop sector is the wedge whose vertex is the ori-

---

[1] More generally, if $h_{c,\text{FIR}}(t)$ is a continuous-time filter nonzero for $t \in [0, T]$, it can be expressed as a truncated version of an infinite-duration $h_{c,\text{IIR}}(t)$ as $h_{c,\text{FIR}}(t) = h_{c,\text{IIR}}(t) \cdot p_T(t)$, where as in Chapter 6, $p_T(t) = 1$ for $t \in [0, T]$ and 0 otherwise. Then from (6.31a) [in current notation, $P_T(s) = (1 - e^{-sT})/s$] and (5A.1.10a),

$$H_{c,\text{FIR}}(s) = \frac{1}{2\pi j} \int_C H_{c,\text{IIR}}(s') \frac{1 - \exp\{-(s - s')T\}}{s - s'} \, ds' = \sum_{\substack{m \text{ for which } s_m \\ \text{is enclosed by } C}} \text{Res}\{H_c(s') \cdot \frac{1 - \exp\{-(s-s')T\}}{s - s'}, s_m\}$$

where C is closed at $|s'| = \infty$ in the left half-plane. We see that $H_{c,\text{FIR}}(s)$ will be a sum of terms $H_c(s_m)(1 - \exp\{-(s - s_m)T\})/(s - s_m)$, which because of the exponential is not a rational function of s.

gin and that includes the stopband on the unit circle (denoted the unit stop arc). The zeros can be either inside or outside the unit stop arc, but tend to be near the stop arc to "stop" (annihilate) the undesired band noise. Similarly, the "pass sector" is the rest of the z-plane; the poles tend to be in the pass sector. For the filter to be stable, all the poles must be within the unit pass arc (magnitude < 1).

It is this flexibility of placing poles in the pass sector and zeros in the stop sector that allows a short transition band for an IIR filter of given low order. We cannot "beef up" the FIR response for a range of frequencies by placing a pole in the pass sector somewhat near the unit circle; all the poles are fixed at the origin. It should be noted that although the Chebyshev I digital filter has both poles and zeros, the original analog filter is all-pole (opposite or dual of FIR). Consequently, all the zeros are at $z = -1$ for the BLT design, and they are also fixed (to other locations) for the impulse-invariant design. Thus, the frequency response is not as narrow as it could be with, for example, the elliptic filter—for which the poles and zeros are both independently and advantageously positioned.

The above concern that all the poles of the IIR digital filter be within the unit circle is moot for an FIR filter, because as proved in (9.3), all the poles are at the origin. The guaranteed stability for the FIR filter may be considered an advantage of FIR over IIR filters. This is particularly true when IIR poles are *near* the unit circle but within. Quantization of coefficients may move the poles outside the unit circle, with the unintended result that the IIR digital filter is unstable even though the analog filter from which it was derived is stable (see Section 9.6). Certainly in a design situation, stability must be checked.

The guaranteed FIR stability is also extremely important in adaptive filter design. The mathematics quickly becomes complicated when optimization is performed, and the last thing we need is yet another constraint (stability) to worry about. Thus, FIR filters are usually used in adaptive filtering.

Another advantage of FIR filters over IIR filters is that they can easily be designed to have exactly linear phase. That is, the phase function of the filter frequency response changes linearly with ω. Recalling that the DTFT of $x(n - n_0)$ is $e^{-j\omega n_0}X(e^{j\omega})$, we see that linear phase is nothing more than a time delay. Assuming we can tolerate a fixed delay in the output, there is otherwise zero phase distortion. If in addition the passband is flat, there is no distortion at all produced by the filter, other than delay.

As noted and observed in Chapter 8, the IIR filter has nonlinear phase. In some of our examples, we saw that the nonlinearity was negligible in the passband. In such cases, there is no material advantage of FIR filters on this count. In other cases, the nonlinearity might be eliminated by cascading an allpass phase-equalizing digital filter with the original IIR filter. Recall that the magnitude of the allpass filter is unity for all ω, and now its phase is designed to cancel the nonlinear behavior of the IIR filter. However, this correction is complicated, and with an FIR filter exactly linear phase can be guaranteed without any such patching. We will look at linear phase in more detail in Section 9.5.1.

Yet another advantage of FIR filters over IIR filters is the flexibility in desired frequency response. IIR filters, based on analog filters, are for the most part limited to piecewise-constant magnitude functions. These piecewise-constant functions include the lowpass, highpass, bandpass, bandstop, notch, and related filters. FIR filter frequency response magnitudes can easily be designed to approximate virtually any specified function of frequency, with a sufficient number of filter coefficients. (There are, however, digital IIR optimization techniques that are exceptions to this rule.)

Often, the comparison between FIR and IIR or even between the various algorithms within each category may not be completely clear-cut. In the end, available hardware, software, and "brainware" (know-how) may override some of the considerations above in the decision concerning whether to use an IIR or an FIR filter.

9.3 THE WINDOW METHOD

There are two straightforward FIR design techniques: the window method and the frequency-sampling method (Section 9.4). These two techniques are duals. The window method samples the unit sample response of the desired filter, denoted $h_d(n)$, and the frequency-sampling method samples the desired frequency response itself, $H_d(e^{j\omega})$. The window method uses a gradual time-window to minimize the effects of sharp frequency-domain transitions, whereas the frequency-sampling approach uses *transition samples* for a similar purpose. We now consider the window method, which is the time-domain approach.

As just noted, the basis of the window method is to sample $h_d(n)$. Typically, $h_d(n)$ has infinite length because $H_d(e^{j\omega})$ has sharp corners (passband-stopband transitions). Consequently, it must be truncated if as, by assumption, $h(n)$ is to be an FIR filter. Mathematically, we have

$$h(n) = \begin{cases} h_d(n), & n \in [0, N-1] \\ 0, & \text{otherwise} \end{cases} \tag{9.4a}$$

$$= w_R(n)h_d(n), \tag{9.4b}$$

where the rectangular window is again

$$w_R(n) = \begin{cases} 1, & n \in [0, N-1] \\ 0, & \text{otherwise,} \end{cases} \tag{9.5}$$

and $h_d(n) = \text{DTFT}^{-1}\{H_d(e^{j\omega})\}$, where again $H_d(e^{j\omega})$ is the specified desired filter frequency response.

Let us pause a moment to discuss FIR filter order. With our N-length filter $h(n)$, $H(z)$ is an $(N-1)$st-order polynomial in z^{-1}, so the filter order is $N-1$. In our discussion of difference equations and their system functions, $H(z) = B(z)$ would be an Mth-order polynomial in z^{-1}, and $h(n)$ would have length $M+1$ [as in (9.3)]. However, because filter length plays such a prominent role in FIR analysis and design (for example, via windows and the DFT_N), we will usually speak of a length-N, order $N-1$ FIR filter $h(n)$. (Matlab defines an FIR filter of order N and length $N+1$.) With this in mind, we now consider the quality of our N-length truncation $h(n)$ of the infinite-length ideal filter $h_d(n)$.

The rectangular window has the property that it gives the minimum mean-squared error approximation of $H(e^{j\omega})$. The mean-squared error (MSE) is, by Parseval's theorem [put E in for X in (5.54)]

$$\text{MSE} = \frac{1}{2\pi} \int_{-\pi}^{\pi} |E(e^{j\omega})|^2 \, d\omega = \sum_{n=-\infty}^{\infty} |h_d(n) - h(n)|^2 = \|e(n)\|^2. \tag{9.6}$$

By its definition, $h(n) = 0$ for n outside the range $[0, N - 1]$; for n outside $[0, N - 1]$, the error terms in (9.6) are unavoidably $|h_d(n)|^2$; that is, we cannot reduce *these* $|e(n)|$ values at all. Thus, we can minimize the MSE by making those $e(n)$ terms of the sum in (9.6) for $n \in [0, N - 1]$ equal to zero: $h(n) = h_d(n)$ for $n \in [0, N - 1]$. This $h(n)$ is the rectangular window truncation given in (9.4).

Based on the above optimality property, it would appear that we can do no better than using the rectangular window method for approximating $H_d(e^{j\omega})$. For minimization of the MSE, this is true. There is, however, a problem in this optimization that bothers many users: There is no guaranteed limitation on the DTFT error magnitude *locally*, even though overall the accumulated MSE is minimum. In practice, there are oscillations both above and below unity in the passband. Consequently, for FIR filters Figure 8.4 is typically modified to show the maximum value $1 + \delta_1$ instead of 1 (and again a minimum value $1 - \delta_1$) in the passband. The problem of ripple near discontinuities is quantified for piecewise constant filters, which have discontinuities in $H_d(e^{j\omega})$, by examination of Gibbs's phenomenon.

9.3.1   Gibbs's Phenomenon

About the year 1900, J. W. Gibbs investigated and published the fact that truncation of a Fourier series representation of a discontinuous function introduces an overshoot error that cannot be reduced by increasing the length of that truncated series. That is, although the Fourier series exactly represents even a discontinuous function when all terms are kept, the error is finite and significant when truncated in the case of discontinuous functions, no matter how many finite-numbered terms we keep.

We will look at this phenomenon as applied to the important case of a lowpass filter with digital cutoff frequency ω_A [or $f_A = \omega_A/(2\pi)$]. A pictorial example will be given in Example 9.1 (in which Figures 9.5 and 9.6 will illustrate the invariance of Gibbs's phenomenon with increasing N). For simplicity, our analysis here will be for a noncausal, time-symmetrical implementation of the truncated filter. In Example 9.1, we easily obtain a causal filter by adding a linear phase onto the lowpass filter.

Our desired lowpass filter $H_d(e^{j\omega})$ is unity in the passband $\omega \in (-\omega_A, \omega_A)$ and zero elsewhere on $(-\pi, \pi)$. Thus, the DTFT^{-1} gives

$$h_d(n) = \frac{1}{2\pi} \int_{-\omega_A}^{\omega_A} 1 \cdot e^{j\omega n} d\omega = \frac{sin(\omega_A n)}{\pi n} = \left(\frac{\omega_A}{\pi}\right) sinc(\omega_A n), \tag{9.7}$$

where for $n = 0$ the above reduces to ω_A/π. We let $h(n) = h_d(n)$ for $n \in [-M, M]$; thus the actual filter order is $N = 2M$. Noting that $e^{j\theta} + e^{-j\theta} = 2\cos(\theta)$ and that $\omega_A \, sinc(\omega_A[-n]) = sin\{\omega_A(-n)\}/(-n) = sin(\omega_A n)/n$,

$$H(e^{j\omega}) = \frac{\omega_A}{\pi} + \frac{2}{\pi} \sum_{n=1}^{M} \frac{sin(\omega_A n) \cos(\omega n)}{n}. \tag{9.8}$$

The overshoot (greatest error) in the passband concerns us most, so let us restrict our attention to $\omega \in [-\omega_A, \omega_A]$. On this range, $H_d(e^{j\omega}) = 1$; thus, the error [called the remainder $R_M(\omega)$] in the truncation is

$$R_M(\omega) = H(e^{j\omega}) - H_d(e^{j\omega}) = \frac{\omega_A}{\pi} + \frac{2}{\pi} \sum_{n=1}^{M} \frac{sin(\omega_A n) \cos(\omega n)}{n} - 1. \tag{9.9}$$

The greatest error occurs for $R'_M(\omega)\big|_{\omega=\omega_x} = 0$, where ω_x is the root nearest to and less than ω_A. We find

$$\tag{9.10}$$

$$R'_M(\omega) = -\frac{2}{\pi} \sum_{n=1}^{M} sin(\omega_A n) \sin(\omega n) = \frac{1}{\pi} \sum_{n=1}^{M} \cos\{(\omega + \omega_A)n\} - \cos\{(\omega - \omega_A)n\}.$$

By writing $\cos(\alpha n) = \frac{1}{2}(e^{j\alpha n} + e^{-j\alpha n})$ and by using the geometric sum and trigonometric identities, we have (see the problems for algebraic details)

$$\sum_{n=1}^{M} \cos(\alpha n) = \frac{\sin\{\alpha(M + \frac{1}{2})\}}{2\sin(\frac{1}{2}\alpha)} - \frac{1}{2}. \tag{9.11}$$

Thus (9.10) becomes

$$R'_M(\omega) = \frac{1}{2\pi}\left\{ \frac{\sin\{(\omega + \omega_A)(M + \frac{1}{2})\}}{\sin\{\frac{1}{2}(\omega + \omega_A)\}} - \frac{\sin\{(\omega - \omega_A)(M + \frac{1}{2})\}}{\sin\{\frac{1}{2}(\omega - \omega_A)\}} \right\}. \tag{9.12}$$

By setting $R'_M(\omega) = 0$ and using trigonometric identities (again, see the problems), we can obtain the condition for the worst error, satisfied for $\omega = \omega_x$:

$$\frac{\sin\{\omega_x(M + 1)\}}{\sin(\omega_x M)} = K, \tag{9.13a}$$

where K, a constant with respect to ω, is

$$K = \frac{\sin\{\omega_A(M + 1)\}}{\sin(\omega_A M)}, \tag{9.13b}$$

and where we need the solution for $\omega = \omega_x$ [or $f_x = \omega_x/(2\pi)$] nearest *but not equal to ω_A* —which will give a maximum of $R_M(\omega)$.

Unfortunately, (9.13) is a transcendental equation that does not have a closed-form solution for ω_x. However, we can obtain ω_x numerically using a routine such as fmin in Matlab. Figure 9.1a shows $f_x - f_A$ versus f_A for the entire range of possible lowpass filters ($f_A \in [0, \frac{1}{2}]$), for $M = 25$. The difference $f_x - f_A$ is plotted because f_x is very close to f_A, particularly for large M. Figure 9.1b shows the same plot for $M = 50$. We see that as the filter order $2M$ increases, f_x becomes closer to f_A and stays within a tighter range as the filter cutoff frequency f_A is varied.

Figure 9.2 shows $H(e^{j\omega_x})$, the maximum value of the truncated filter, versus f_A. Again, Figure 9.2a is for $M = 25$, and Figure 9.2b is for $M = 50$. We conclude that the worst overshoot occurs for cutoff f_A near 0 and 0.5 and that for most f_A the overshoot is on the order of 9%. The greatest overshoot varies less with f_A for large M, but with greater rapidity of oscillation with respect to f_A.

The important thing to see is that the "9%" overshoot does not decrease as the filter order $N = 2M$ is increased; this is Gibbs's phenomenon. Gibbs's phenomenon is explicitly shown in Figure 9.3, which shows, for $f_A = 0.25$, the maximum value $H(e^{j\omega_x})$ for $1 \leq M \leq 60$. The flattening-out at 8.95% around $M = 40$ continues indefinitely.

Intuitively, the overshoot does not decrease because as the rectangular window transform $\sin(\frac{1}{2}N\omega)/\sin(\frac{1}{2}\omega)$ slides by the discontinuity, a peak ripple value equal to the area of the transform sidelobe always occurs. The transform sidelobe area does not decrease with N. Although $\sin(\frac{1}{2}N\omega)/\sin(\frac{1}{2}\omega)$ is a nasty function to try to analyze analytically, we can use Matlab's quad8 to perform the integration from the first zero (at $\omega = 2\pi/N$) to its second zero (at $4\pi/N$). The results are that for N greater than about 10 or 12, the integral is always roughly -0.88 or -0.89. This numerical study is perhaps simplistic, but is fundamental to the actual filter approximation overshoot problem.

The lack of reduction of overshoot with N can be proved for the simpler case of a sawtooth filter,[2] but can be observed only numerically for the variable-cutoff lowpass filter. Moreover, in Figure 9.3 we see that the overshoot can be *worse than 9%* for sufficiently low filter order, even with f_A at the central value of 0.25. To prove it

---

[2] A. Kufner and J. Kadlec, *Fourier Series* (London: Iliffe Books, 1971).

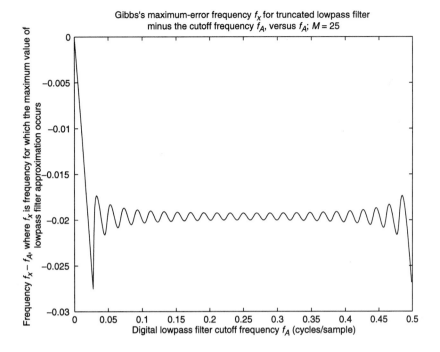

Figure 9.1*a*

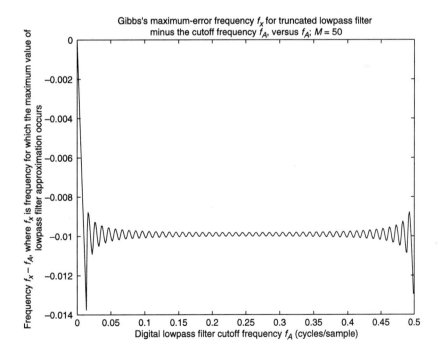

Figure 9.1*b*

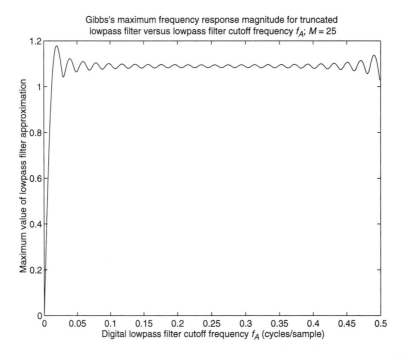

Figure 9.2a

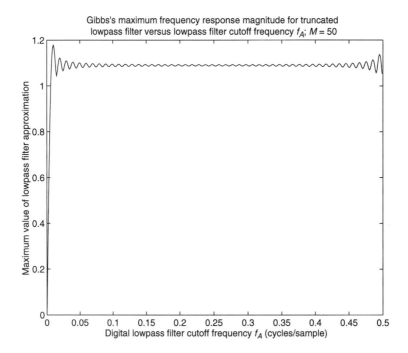

Figure 9.2b

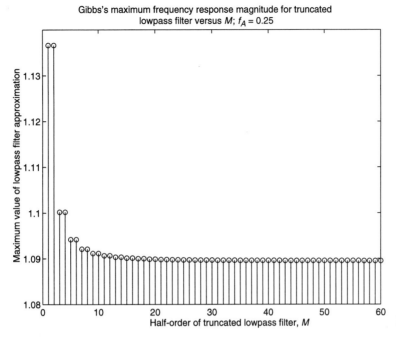

Figure 9.3

for the simpler sawtooth filter response, we proceed as we did above, solving for ω_x, which in this case can be determined analytically. Then we substitute ω_x into $H(e^{j\omega})$ and evaluate the resulting expression for $M \rightarrow \infty$, which is nearly the value of overshoot for any moderate value of M.

In addition, as M increases, the peak error frequency f_x approaches the cutoff frequency f_A from below. This phenomenon is illustrated in Figure 9.4, which shows f_x versus M for $f_A = 0.25$. We see that as M gets large, f_x approaches $f_A = 0.25$. We want f_x to be near f_A so that this worst error is confined away from the main passband area, where our desired signals are.

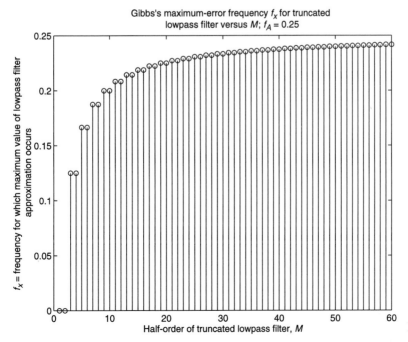

Figure 9.4

9.3.2 Windows and Design

To mitigate Gibbs's phenomenon, we can replace $w_R(n)$ in (9.4b) with alternative window functions $w(n)$. These window functions $w(n)$ have the effect of reducing the discontinuity overshoot oscillations (Gibbs's phenomenon), but also of widening the transition region relative to that occurring with $w_R(n)$ and increasing the overall MSE relative to using $w_R(n)$. In situations where a smooth passband may be of overriding importance, the alternative windows may be helpful. In other situations where the signal is of small magnitude below the appreciable onset of Gibbs's oscillations and the noise band is near the cutoff, the narrow transition band of the rectangular window may be desirable. Thus, the selection of filter design procedures as well as of filter specifications depend on the application.

Commonly used alternative windows are triangular (called the Bartlett window), raised cosine (Hann or "Hanning" and Hamming windows), modified raised cosine (Blackman window), and Bessel function (Kaiser window). They have varying degrees of sidelobe attenuation versus passband-to-stopband transition width. In the Kaiser window only is an adjustable β parameter that allows, for a single value of filter length N, variation of this sidelobe attenuation/transition width trade-off. The formulas are as follows, where for n outside $[0, N-1]$ all windows are zero. With

$$n_N \equiv \frac{n}{N-1} \text{ and } N_{12} \equiv \frac{N-1}{2}, \tag{9.14}$$

$$w_{\text{Bartlett}}(n) = \begin{cases} 2n_N, & n \in [0, N_{12}] \\ 2(1 - n_N), & n \in [N_{12} + 1, N-1] \end{cases} \tag{9.15}$$

$$w_{\text{Hann}}(n) = \tfrac{1}{2}\{1 - \cos(2\pi n_N)\} \tag{9.16}$$

$$w_{\text{Hamming}}(n) = 0.54 - 0.46 \cos(2\pi n_N) \tag{9.17}$$

$$w_{\text{Blackman}}(n) = 0.42 - \tfrac{1}{2}\cos(2\pi n_N) + 0.08 \cos(4\pi n_N) \tag{9.18}$$

$$w_{\text{Kaiser}}(n) = \frac{I_0(2\beta\sqrt{n_N - n_N^2})}{I_0(\beta)}, \tag{9.19}$$

where in (9.19), I_0 is the modified Bessel function of the first kind, order zero (i.e., the zeroth-order Bessel function J_0 with its argument scaled by j):

$$I_0(x) = J_0(jx) \tag{9.20a}$$

$$= \sum_{k=1}^{\infty} \left\{\frac{x^k}{k!2^k}\right\}^2, \tag{9.20b}$$

where the sum can be accurately truncated to about $k_{\text{max}} = 15$.

Now note that for FIR filters, $|H|$ ranges from $1 + \delta_1$ to $1 - \delta_1$, rather than from 0 to $1 - \delta_1$ as did our analog filters in Chapter 8. This is because $h(n)$ is a truncated Fourier series of the ideal flat response, and truncated Fourier series oscillate about the value they approximate. Contrast this behavior with that of the Butterworth and Chebyshev II filters, which have no passband ripple and just monotonically decline from 1 to $1 - \delta_1$ in the passband. Furthermore, we proved in Section 8.4.2.1 (page 550ff) that the Chebyshev I filter magnitude frequency response oscillates between 1 and $1 - \delta_1$ in the passband.

The Kaiser window design equations below were derived for stopband ripple $\delta_2 = \delta_1 = \delta = 10^{-A/20}$ [i.e., we define $A \equiv -20 \log_{10}(\delta)$] so that on the passband $1 - \delta < |H_d(e^{j\omega})| < 1 + \delta$ and on the stopband $|H_d(e^{j\omega})| < \delta$. Note that A is the minimum stopband attenuation in decibels. Because we usually want very large attenuation in the stopband, we expect that the passband ripple will appear negligible for the Kaiser design. For example, if we require $A = A_r = 40$ dB, then the passband ripple is only $\delta = 10^{-40/20} = 0.01$; thus the minimum passband value $1 - \delta = 0.99$ in decibels is only about $-A_c = 20 \log_{10}(1 - \delta) = -0.087$ dB rather than more common values such as -0.5 dB or -1 dB. Notice that the peak (not peak-to-peak) passband ripple amplitude in decibels is A_c, which is not equal to the minimum sidelobe (stopband) attenuation in decibels $\{A_r = A\}$. Rather, $A_c = -20 \log_{10}\{1 - 10^{-A/20}\}$ or, equivalently, $A = -20 \log_{10}\{1 - 10^{-A_c/20}\}$ [valid only for $\delta_1 = \delta_2$].

For $\delta_1 = \delta_2 = \delta$ and $A = -20 \log_{10}(\delta)$, Kaiser derived the formulas for minimum filter length N and required β value to meet specifications:

Kaiser window design parameter formulas
A = minimum stopband attenuation (dB)

$$N = \frac{A - 7.95}{2.285(\omega_r - \omega_c)} + 1 \tag{9.21}$$

$$\beta = \begin{cases} 0.1102(A - 8.7), & 50 \text{ dB} \leq A \\ 0.5842(A - 21)^{0.4} + 0.07886(A - 21), & 21 \text{ dB} \leq A \leq 50 \text{ dB} \\ 0 \text{ (rectangular window)}, & 0 \leq A \leq 21 \text{ dB}. \end{cases} \tag{9.22}$$

The reason $\beta = 0$ corresponds to the rectangular window is that in (9.19), $w_{\text{Kaiser},\beta=0}(n) = I_0(0)/I_0(0) = J_0(0)/J_0(0) = 1/1 = 1$ for $n \in [0, N - 1]$.

Plots of $w(n)$ and their DTFTs $W(e^{j\omega})$ for the windows in (9.15) through (9.19) are investigated in the problems. We will often see in the problems how advantageous the use of windows can be in filter design to "get what we want."

As defined in (9.15) through (9.19), the windows assume that a linear phase factor $\exp\{-jN_{12}\omega\}$ has been added to $H_d(e^{j\omega})$ so that the length–N $h(n)$ is centered on $n = N_{12}$ rather than on zero. This is done for the following reason.

A real-valued, even $H_d(e^{j\omega})$ has a real, even DTFT^{-1}, denoted $h_d(n)$. To avoid in (9.4b) $[h(n) = w_R(n) \cdot h_d(n)]$ the introduction of a huge erroneous discontinuity between the values $h_d(-1) w_R(-1) = h_d(-1) \cdot 0 = 0$ and $h_d(0) \cdot w_R(0) = h_d(0) \cdot 1 = h_d(0)$, and to avoid wasting half our FIR filter coefficients $h(n)$ on the nearly-zero values $h_d(\text{large } n)$, we shift the filter to the right by $N_{12} = (N - 1)/2$ for N odd (N even will be considered momentarily). Thus, $h_{d,\text{orig}}(-N_{12})$ becomes $h_d(0), \ldots,$ and $h_{d,\text{orig}}(N_{12})$ becomes $h_d(N - 1)$. However, this shift of N_{12} is equivalent to redefining $H_d(e^{j\omega})$ as $H_d(e^{j\omega}) = e^{-jN_{12}\omega}H_{d,\text{orig}}(e^{j\omega})$ and evaluating the DTFT^{-1} of the new $H_d(e^{j\omega})$ for $n \in [0, N - 1]$ to yield $h(n)$.

If we try to shift $h_{d,\text{orig}}(n)$ by an integer for N even, we find we are forced to make the left and right halves of $h(n)$ asymmetrical. Thus, instead, we use the same exponential formula as for N odd: $H_d(e^{j\omega}) = e^{-j(N-1)\omega/2}H_{d,\text{orig}}(e^{j\omega})$ and take the truncated DTFT^{-1} as $h(n)$. Because of the half-integral-shift involved, $h_d(n)$ will now be symmetrical about its center, which no longer falls on one of the integer index values n.

Let us now consider an example of FIR window digital filter design, which will clearly point out the transition width/sidelobe attenuation tradeoff.

EXAMPLE 9.1

Design (**a**) a rectangular window and (**b**) a Kaiser window FIR digital lowpass filter, with the same specifications as in the IIR examples: maximum attenuation in the passband $= -1$ dB, minimum attenuation in the stopband $= -40$ dB, $f_c = 0.3$, and $f_r = 0.4$ cycle/sample. These are the same specifications given in Examples 8.6 and 8.7.

Solution

(**a**) We will use the Matlab command `fir1` to design our filter, in which we set the window parameter to `boxcar(N)` and filter order (which, again, is one less than the filter length) to $N - 1$. However, the values have also been checked, for the rectangular-window unit sample response, against a straightforward numerical implementation of the DTFT definition

$$H(e^{j\omega}) = \sum_{n=0}^{N-1} h(n)e^{-j\omega n}, \text{ where}$$

$$h(n) = \frac{1}{2\pi}\int_{-\omega_A}^{\omega_A} 1 \cdot e^{j\omega(n-[N-1]/2)}d\omega$$

$$= \begin{cases} \dfrac{\sin(\omega_A\{n - [N-1]/2\})}{\pi\{n - [N-1]/2\}}, & \begin{array}{l} n \in [0, N-1] \text{ for } N \text{ even} \\ n \in [0, N-1] \text{ and } n \neq (N-1)/2 \text{ for } N \text{ odd} \end{array} \\ \omega_A/\pi, & n = (N-1)/2 \text{ for } N \text{ odd} \\ 0, & \text{otherwise.} \end{cases}$$

Note the linear phase $-\omega(N-1)/2$ added on to $H(e^{j\omega})$, designed to delay the filter enough $[(N-1)/2$ samples] so that $h(n)$ is casual.

With the FIR rectangular window design, our only completely controllable parameter is N, the filter length. Given that the higher-order IIR filter had $N = 8$ (impulse-invariance Chebyshev, in Example 8.6), let us start with $N = 16$ because we know FIR filters tend to be longer than IIR equivalents. We will find even $N = 16$ to be grossly low. A process of iteration is required to find the parameters that result in near-perfect matching of the stated specifications.

Although we have control over $f_A \equiv$ the brick-wall cutoff frequency of the ideal filter (i.e., the f_A in our Gibbs's phenomenon discussion), the obtained FIR filter magnitude response at $f = f_A$ is not determinable beforehand. We have defined the "cutoff" frequency of the designed approximating FIR filter $H(e^{j2\pi f})$, f_c, to be the minimum value of frequency f at which $H(e^{j2\pi f})$ is more than 1 dB down from unity. We will find that we need to make $f_A > f_c$ to satisfy this constraint. N will primarily control the transition width.

For example, suppose that for $N = 16$ we let $f_A = 0.33$. We find numerically that the minimum decibel gain value for $|f| < f_c = 0.300$ is -1.15 dB, which essentially meets the passband specification. However, we find that the minimum attenuation in the stopband ($|f| > f_r = 0.4$) is only -20.4 dB, way short of the specification; see Figure 9.5, middle plot. We conclude that we must substantially increase N. It takes going all the way to $\{N = 124$, for which we select $f_A = 0.304\}$ to meet both passband and stopband specifications using the rectangular window (-0.54 dB at $f_c = 0.3$, -39.9 dB at $f_r = 0.4$). Often, when we finally satisfy both specifications simultaneously, one of them will be "overdesigned"—will exceed the specification (here, $0.54 < 1$). If both are deemed close enough to expectations, the design is concluded.

The frequency response of the resulting filter, with both nondecibel and decibel magnitude plots, is shown in Figure 9.6. Having the nondecibel and decibel plots side by side facilitates getting a feel for "how far down" a given number of decibels is. For example, look for where the decibel gain is -20 dB or -30 dB, and you will see in the nondecibel plot above, at the same frequency, how small that is in the nondecibel stopband.

The first thing we notice is that by increasing N from 16 in Figure 9.5 to 124 in Figure 9.6, the passband ripple has not at all been reduced. This ripple is Gibbs's phenomenon; we showed in our discussion that the ripple does not decline with increasing N. Similarly, the first stopband sidelobe remains about -21 dB even as N is increased; it just moves closer to f_A.

Note: As investigated in the problems and presented in many textbooks on DSP, the rectangular window has -13 dB in the stopband. However, one cannot just transfer -13 dB of the window alone to a lowpass filter design based on the rectangular window; for example, in Figures 9.5 and 9.6, the first sidelobe has -21 dB. The convolution in (6.79c) [for this application, $H(e^{j\omega}) = H_d(e^{j\omega}) *_{c\omega} W_R(e^{j\omega})/(2\pi)$] is performed, which will alter the sidelobe level somewhat. [If ω_c is brought toward zero, we find as expected that the -21 dB sidelobe gradually changes to the window-only value -13 dB because $H_d(e^{j\omega})$ approaches an impulse function.] Thus, the comparison of window sidelobes in tables is only a relative guideline and does not translate directly into number of decibels down in a given filter design.

Next, looking at the nondecibel plot we would say that the practical transition width of the filter is way, way less than $f_r - f_c = 0.4 - 0.3 = 0.1$; it is more like 0.01. This is because most of the stopband decline short of the full 40 dB down occurs near f_c. Finally, we see in Figures 9.5 and 9.6 that the (unwrapped) phase (bottom plot) is perfectly linear in the passband, as we will discuss in more detail in Section 9.5.1.

(b)   We now replace the window parameter in `fir1` to `kaiser(N, β)`, where we use (9.21) and (9.22) to choose N and β. When we set $f_c = 0.3$, we find the resulting frequency response passband–stopband transition is way too far toward $f = 0$. This is because as the window slides by in (6.79b), the passband–stopband transition will be approximately centered on f_c rather than begin on it. Thus, selection of $f_c' = \frac{1}{2}(f_A + f_r) = \frac{1}{2}(0.3 + 0.4) = 0.35$ is appropriate.

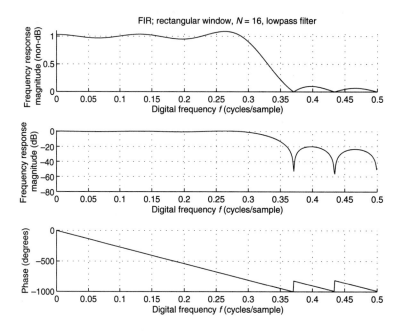

Figure 9.5

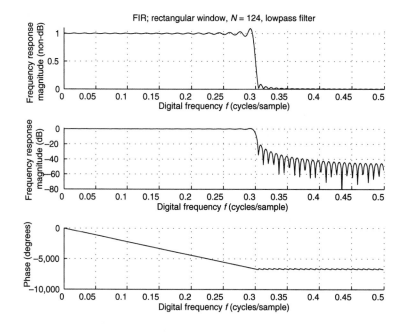

Figure 9.6

Also, our specification -1 dB in the passband would dictate $A = -20 \log_{10}\{1 - 10^{(-1/20)}\} = 19.3$ dB, which does not match our requirement $A = 40$ dB in the stopband. That is, contrary to the assumption behind formulas (9.21) and (9.22), we are asking for different passband and stopband ripples. Some juggling may thus be required to match specifications exactly. However, rather easily with $f_c' = 0.35$, $f_r' = 0.45 (= f_c' + 0.1)$, and $A = 40$ dB we succeed; see Figure 9.7, where from (9.21) and (9.22) we have $N = 23$, $\beta = 3.395$.

The first item of interest in Figure 9.7 is that we have essentially eliminated Gibbs's phenomenon by using the Kaiser window. Notice how

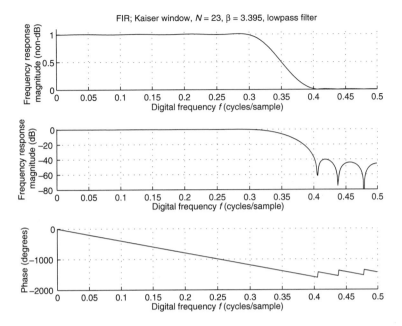

FIR; Kaiser window, $N = 23$, $\beta = 3.395$, lowpass filter

Figure 9.7

smooth the passband magnitude in Figure 9.7 is, relative to that in Figure 9.6. It might appear that the Kaiser window method has also given us a huge benefit—allowing $N = 23$ instead of $N = 120$ for the rectangular window approach. However, this statement is true only for precisely meeting the specifications. Comparison of the transition bands in Figures 9.5 through 9.7 shows that the $N = 23$ Kaiser frequency response resembles in shape the $N = 16$ rectangular window response (Figure 9.5) much more than it does the $N = 124$ rectangular window response (Figure 9.6).

As noted in part (a), the $N = 124$ nondecibel response looks like a filter with transition band 0.01 rather than 0.1. Contrarily, the transition width of the $N = 23$ nondecibel Kaiser response in Figure 9.7 does appear to be about 0.1. So does the nondecibel version of Figure 8.45—the Chebyshev II bilinear transform digital filter having only 11 nonunity coefficients; thus a fair IIR/FIR comparison is perhaps 11 versus 23, not 11 versus 124.

The price paid for low N of the Kaiser lowpass filter is that it takes a relatively wide frequency band to get down to near the stopband attenuation (gradual, soft "Kaiser roll"-off). For example, in Figure 9.6 it takes 12 sidelobes to get down to 40 dB, but not "far" at all (only first sidelobe) to drop below 20 dB or lower. For the Kaiser design in Figure 9.7, it takes about 80% of the way from $f_c = 0.3$ to $f_r = 0.4$ to reach -20 dB. With the rectangular window, we are "stuck with" an initially fast decline, followed by a very slow decline with f. The user tolerances will have to dictate which is preferable in a given application.

With Example 9.1, we see the severe extent to which filter order must sometimes be increased if an FIR filter is to be used to meet the same specifications as those of an equivalent IIR filter. Often, FIR filters are applied to approximate functions of frequency different from piecewise-constant ones, the latter of which are more effectively handled by IIR filters. An exception to this rule is where strict linear phase is desired, in which case an FIR filter is generally preferable.

9.4   FREQUENCY-SAMPLING METHOD

As noted in Section 9.3, the dual of time-sampling of $h_d(n)$ is frequency-sampling of $H_d(e^{j\omega})$. We have already performed an example of this design method in Example 7.9, where our goal was the annihilation of undesired spectrum replications arising from the zero-interlacing required for interpolating a sequence. Thus, our treatment here will be brief.

Frequency sampling is probably the most simple and obvious approach to filter design—just use exact values of the desired frequency response as the filter coefficients $H(k)$. If the unit sample response of the designed filter is desired, we merely compute $h(n) = \text{DFT}_N^{-1}\{H(k)\} = \text{DFT}_N^{-1}\{H_d(e^{j2\pi k/N})\}$ [where $H_d(e^{j\omega})$ includes the phase factor $e^{-j(N-1)\omega/2}$]. As noted in Example 7.9, when discontinuities in $H_d(e^{j\omega})$ exist, it is better to use a few transition samples at the discontinuity than to allow the full discontinuity of $H_d(e^{j\omega})$ to be reflected in the samples $H(k)$. Substantial intersample ripple in $H(e^{j\omega})$ will result if no transition samples are used. As is often the case in DSP and engineering in general, improving one aspect (here, the transition from 1 to 0 in a single sample) causes another aspect to deteriorate.

Rabiner, Gold, and McGonegal give tables of optimal transition samples for lowpass filters.[3] There are separate tables to use for one, two, or three transition samples and for filter lengths N that are powers of two from 16 to 256. A different optimal sample value is given for each bandwidth (i.e., each number of samples in the passband). The single-transition samples are mostly around 0.38, the double-transition samples are mostly around the values 0.1 (first transition sample) and 0.58 (second sample),[4] and the triple-transition samples are around 0.023, 0.25, and 0.72. These rules of thumb, assuming unity in the passband, are sufficient for most cases.

To illustrate the idea of transition samples for frequency-sampling filter design, consider the following numerical example.

EXAMPLE 9.2

Repeat the FIR filter design of Example 9.1 using the frequency-sampling method. Compare results without and with transition samples.

Solution

First, examine the case of no transition samples. Taking $N = 124$ from the rectangular window in Example 9.1 (Figure 9.6), we produce the $H(k)$ whose first half is shown in Figure 9.8. Notice that each passband magnitude sample is exactly 1 (0 dB) and each stopband sample is 0 ($-\infty$ dB). The phase is exactly linear in the passband. Filling in the frequency response between these perfect values, we obtain Figure 9.9.

Notice the considerable ripple in between samples, which was totally hidden from us in the "perfect" Figure 9.8. In fact, we find in Figure 9.9 that the sidelobes do not fall as low as -40 dB by $f_r = 0.4$, even though on the samples, the decibel values are all $-\infty$. Otherwise, Figure 9.9 fairly closely resembles Figure 9.6, except that the "Gibbs" ripple in frequency-sampling (Figure 9.9) seems to be a bit worse than for the rectangular window Gibbs ripple in Figure 9.6. Why do we care about in

---

[3]L. R. Rabiner, B. Gold, and C. A. McGonegal, "An Approach to the Approximation Problem for Nonrecursive Digital Filters," *IEEE Trans. on Audio and Electroacoustics*, AU-18, 2 (June 1970): 83–106.

[4] We used these values in Example 7.9, even though our filter order was $N = L \cdot 128 = 512$, whereas $N_{\text{max}} = 256$ in Rabiner, Gold, and McGonegal. The differences among transition values for different values of N are small.

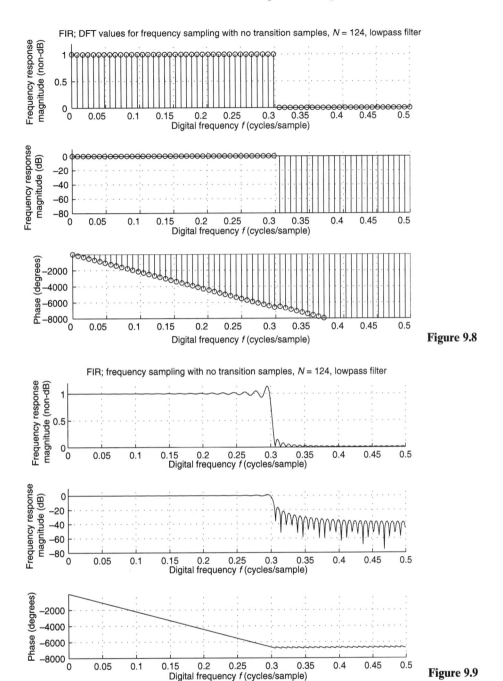

Figure 9.8

Figure 9.9

between the perfectly matching samples? We care because a general signal to be filtered will have energy there, so those frequency response ripples will distort that signal.

We repeat the procedure using two transition samples: 0.58 followed by 0.1 (using the Rabiner, Gold, and McGonegal rough average) for the two samples following $f = 0.3$. We find that we need only $N = 24$ to meet the specifications. The first half of the DFT sequence $H(k)$ (i.e., $k = [0, N/2] = [0, 12]$) is shown in Figure 9.10, and the filled-in frequency response is in Figure 9.11. The second half of $H(k)$ ($[N/2 + 1, N - 1]$) is obtained from the first half using $H(N - k) = H^*(k)$. As in the

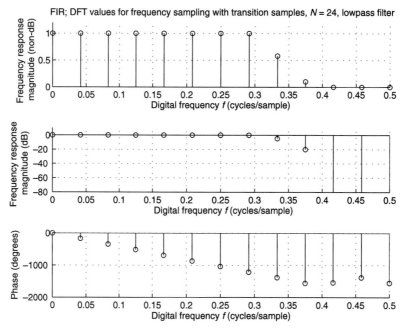

Figure 9.10

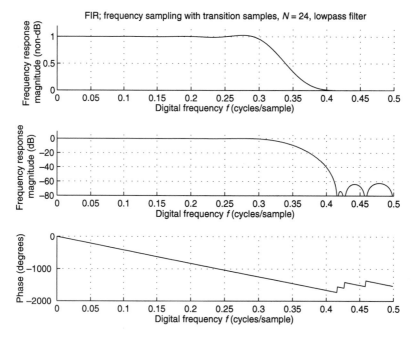

Figure 9.11

use of the Kaiser time-window, in Figure 9.11 we have essentially eliminated the passband ripple relative to that in Figure 9.9. Moreover, the frequency response passes through $f = 0.4$, $20\log_{10}\{H(e^{j2\pi \cdot 0.4})\} = -40$ dB, whereas the remaining side-lobes are below -60 dB. Insertion of the two transition samples appears to have had much the same effect as using the Kaiser window for about the same value of N (compare with Figure 9.7), with less effort from us.

9.5   CHEBYSHEV APPROXIMATION (McCLELLAN–PARKS) METHOD

In the desire that FIR filters be made as short as possible while achieving approximations of $H_d(e^{j\omega})$ that are as accurate as possible, researchers turned again to approximation theory, as they did before for analog filter design. The goal is to do better than the window or frequency-sampling methods. The most popular optimal design algorithm is the Chebyshev approximation or McClellan–Parks/Remez method, which produces an equiripple design (similar to but in application more general than the IIR Chebyshev and elliptic analog filters). Recall from Section 8.4.2.1 (page 550ff) that Chebyshev approximation is the minimax or equiripple design. For analog filters, the equiripple requirement resulted in Chebyshev approximating polynomials or Chebyshev rational functions. In the FIR case, Chebyshev polynomials are involved only at one place in the derivation, and the approximating functions are instead $\cos(\omega\ell)$.

Chebyshev approximation is implemented by what is called the Remez exchange algorithm, and to use it in Matlab requires only a single-line call to the routine `remez`. Because of its importance and popularity, it is worth trying to understand as much about it as possible. Moreover, understanding one involved algorithm will make the study of others easier. In the following, we will be assuming that $H_d(e^{j\omega})$ is either purely real or purely imaginary (the latter is the case, for example, for a first-order differentiator).

We all know that two points specify a line $y = c_0 + c_1 x$, three points specify a parabola $y = c_0 + c_1 x + c_2 x^2$, and so on. Specifically, we can determine the coefficients c_n of an Nth-order polynomial

$$y_N(x) = \sum_{n=0}^{N} c_n x^n \tag{9.23}$$

if given $N + 1$ $\{x, y\}$ pairs $\{x_\ell, y_\ell \equiv y_N(x_\ell)\}$:

$$\mathbf{y} = \begin{bmatrix} y_0 \\ y_1 \\ \cdot \\ \cdot \\ \cdot \\ y_N \end{bmatrix} = \begin{bmatrix} 1 & x_0 & x_0^2 & \cdots & x_0^N \\ 1 & x_1 & x_1^2 & \cdots & x_1^N \\ \cdot & \cdot & \cdot & & \cdot \\ \cdot & \cdot & \cdot & & \cdot \\ \cdot & \cdot & \cdot & & \cdot \\ 1 & x_N & x_N^2 & \cdots & x_N^N \end{bmatrix} \begin{bmatrix} c_0 \\ c_1 \\ \cdot \\ \cdot \\ \cdot \\ c_N \end{bmatrix} = \mathbf{Xc}. \tag{9.24}$$

The solution is $\mathbf{c} = \mathbf{X}^{-1}\mathbf{y}$. The form of the matrix $\mathbf{X}$ in (9.24) is known as Vandermonde. The determinant of a Vandermonde matrix is the product of the differences between the values whose powers are represented:[5]

$$|\mathbf{X}| = \prod_{0 \le j < i \le N} (x_i - x_j), \tag{9.25}$$

which is nonzero for distinct x_ℓ. Thus for $N + 1$ distinct points $\mathbf{X}$ is invertible, so indeed we can find the coefficients of the unique Nth-order polynomial passing through the given data points, by computing $\mathbf{c} = \mathbf{X}^{-1}\mathbf{y}$.

In FIR filter design, our function $H(e^{j\omega})$ approximating $H_d(e^{j\omega})$ is not an Nth-order polynomial but rather related to a function of the type

$$P_N(\omega) = \sum_{n=0}^{N} p(n) \cos(\omega n). \tag{9.26}$$

---

[5] For a proof outline, see S. Barnet and C. Storey, *Matrix Methods in Stability Theory* (New York: Barnes and Noble, 1970), p. 8, Exercise 1-7-1.

Just how $H(e^{j\omega})$ is related to $P_N(\omega)$ is the end result of the following discussions on linear phase FIR filters and coefficients conversion. Assuming (9.26) to be the form of the approximating $H(e^{j\omega})$, the x^n in (9.23) are replaced by $\cos(\omega n)$, the c_n are replaced by $p(n)$, and $y_N(x)$ is replaced by $H(e^{j\omega})$. Although not immediately apparent, a Vandermonde determinant is involved, allowing application of (9.25).

9.5.1 Linear-Phase FIR Filters

Linear phase or *affine phase* means that we can write $H(e^{j\omega})$ as $H(e^{j\omega}) = G(\omega)e^{j(A+B\omega)}$, where $G(\omega)$ is a real function of ω. We assume that $h(n)$ is real, so we know that $|H(e^{-j\omega})| = |H(e^{j\omega})|$ and thus $G(-\omega) = \pm G(\omega)$; that is, $G(\omega)$ is either an even or an odd function of ω if $H(e^{j\omega})$ has linear phase. We develop the linear phase concept here, even though we can already design linear-phase filters by either the window or frequency-sampling techniques. We do so because the time-domain conditions for linear phase we shall develop are used directly in deriving the McClellan–Parks design algorithm.

Again, because $h(n)$ is real, we may write

$$H(e^{j\omega}) = H^*(e^{-j\omega}), \tag{9.27a}$$

or

$$G(\omega)e^{j(A+B\omega)} = G(-\omega)e^{-j(A+B[-\omega])}. \tag{9.27b}$$

Suppose that $G(\omega)$ is even; then $G(\omega)e^{jB\omega}$ and $G(-\omega)e^{jB\omega}$ cancel in (9.27b), yielding $e^{jA} = e^{-jA}$, whose solutions are either $A = 0$ or $A = \pi$. Thus for $G(\omega)$ even, $H(e^{j\omega}) = \pm G(\omega)e^{jB\omega}$ (+ for $A = 0$, − for $A = \pi$), or equivalently,

$$G(\omega) = \pm e^{-jB\omega} H(e^{j\omega}) \tag{9.28a}$$

$$= \pm \sum_{n=0}^{N-1} h(n)e^{-j\omega(B+n)}. \tag{9.28b}$$

Noting that $G(\omega)$ is even and thus $G(\omega) = G(-\omega)$, (9.28b) can be equated to its version with ω replaced by $-\omega$:

$$G(\omega) = G(-\omega) = \pm \sum_{m=0}^{N-1} h(m)\, e^{j\omega(B+m)}, \tag{9.28c}$$

where to avoid confusion we have renamed the index n to become m.

We get our necessary constraints for this linear phase condition by equating (9.28b) with (9.28c), the latter rewritten in terms of the new index $n \equiv N - 1 - m$. With this index substitution, (9.28c) becomes

$$G(-\omega) = \pm \sum_{n=0}^{N-1} h(N - 1 - n)e^{j\omega(B+N-1-n)}. \tag{9.28d}$$

For (9.28b) to equal (9.28d) for all ω, we must have (for either $A = 0$ or π)

$$h(n) = h(N - 1 - n) \tag{9.29a}$$

and $-B = B + N - 1$, giving

$$B = -\frac{N-1}{2} = -N_{12}. \tag{9.29b}$$

Thus $H(e^{j\omega}) = \pm G(\omega)e^{-j(N-1)\omega/2} = \pm G(\omega)e^{-jN_{12}\omega}$ where as in (9.14), $N_{12} = \dfrac{N-1}{2}$.

Because $h(n)$ and $h(N-1-n)$ both have the same polarity in (9.29a), this situation is called positive symmetry. We designate positive symmetry as cases 1 for N odd and 2 for N even.

Notice that $H(e^{j\omega})$ here appears *not* to be a 2π-periodic function of ω when N is even. This is because N_{12} is not an integer for N even and because we have held A fixed at either 0 or π: $H(e^{j(\omega+2\pi)}) = \pm G(\omega + 2\pi) \, e^{-j\pi} \cdot e^{-jN_{12}\omega} = \mp G(\omega + 2\pi)e^{-jN_{12}\omega} = -H(e^{j\omega})$, assuming that $G(\omega)$ is 2π-periodic in ω. If this assumption were true, we would restore the periodicity by using $A = \pi$ for every other 2π interval. For example, for the usual case $\omega \in [-\pi, \pi]$ or $\omega \in [0, 2\pi]$ we may use $A = 0$ and not worry about the adjacent intervals where $A = \pi$. However, for case 2 (N even), we find on WP 9.1 [see (W9.1.5): $G(\omega) = \sum\limits_{n=1}^{N/2} b(n) \cos\{\omega(n - \tfrac{1}{2})\}$ where $b(n)$ is determined from $h(n)$ as on WP 9.1.] that $G(\omega + 2\pi) = -G(\omega)$, so G alone takes care of the problem and we can and will therefore use $A = 0$ for all ω.

Now consider the case $G(\omega) = -G(-\omega)$; that is, $G(\omega)$ is odd. Now cancellation of common terms in (9.27b) gives $-e^{jA} = e^{-jA}$ or $e^{j\pi} = e^{j2A}$, whose solutions are $A = \pi/2$ and $A = -\pi/2$. For the same reasons and conditions for which we selected $A = 0$ before, we select $A = +\pi/2$ here. Aside from this constant phase $\pi/2$ for all ω in $H(e^{j\omega}) = G(\omega)e^{j(A+B\omega)}$, we still have linear phase $B\omega$ (the total phase $A + B\omega$ is properly called affine). Thus for $G(\omega)$ odd, $H(e^{j\omega}) = jG(\omega)e^{jB\omega}$, or, equivalently,

$$jG(\omega) = e^{-jB\omega} H(e^{j\omega}) \tag{9.30a}$$

$$= \sum_{n=0}^{N-1} h(n)e^{-j\omega(B+n)}. \tag{9.30b}$$

Noting that $G(\omega)$ is odd, (9.30b) can be equated to minus its version with ω replaced by $-\omega$:

$$jG(\omega) = -jG(-\omega) = -\sum_{m=0}^{N-1} h(m) \, e^{j\omega(B+m)}, \tag{9.30c}$$

where we have renamed the index to m.

We get our necessary constraints for this linear phase condition by equating (9.30b) with (9.30c), the latter rewritten using $n \equiv N - 1 - m$. With this index substitution, (9.30c) becomes

$$jG(\omega) = \sum_{n=0}^{N-1} [-h(N-1-n)]e^{j\omega(B+N-1-n)}. \tag{9.30d}$$

For (9.30b) to equal (9.30d) for all ω, we must have

$$h(n) = -h(N-1-n), \tag{9.31a}$$

giving for N odd the additional constraint

$$h(N_{12}) = -h(N_{12}) \rightarrow h(N_{12}) = 0, \tag{9.31b}$$

and as before, for all N, $-B = B + N - 1$, giving

$$B = -N_{12}. \tag{9.31c}$$

Thus $H(e^{j\omega}) = jG(\omega)e^{-jN_{12}\omega}$ for all N.

Because $h(n)$ and $h(N - 1 - n)$ have opposite polarity in (9.31a), this situation is called negative symmetry. These are designated cases 3 for N odd and 4 for N even. Notice that for all four cases of linear phase filters,

$$H(e^{j\omega}) = G(\omega)e^{j(A-N_{12}\omega)}. \tag{9.32}$$

Specific forms of $G(\omega)$ for each case are given on WP 9.1 (that for case 2 was cited above).

To help visualize the symmetry starting and ending points, refer to Figure 9.12. For perspective, Figure 9.12a shows symmetry point pairs not for positive or negative symmetry, but rather for conjugate symmetry $h(n) = h^*(N - n)$, which is the condition for $H(e^{j\omega})$ to be real. Of course, the condition $h(n) = h^*(N - n)$ is not the situation currently of interest, because in our discussions $H(e^{j\omega})$ will have a linear phase factor and thus it will be complex. We examine the conjugate-symmetrical sequences to contrast them with positive- and negative-symmetrical sequences, which we will examine later, when we consider Figure 9.12b. Thus instead, the pairs indicated in Figure 9.12 are the conjugate-symmetry pairs that apply for making conjugate-symmetric (Hermitian) N-length sequences $h(n)$ that have real DFT$_N$s. Notice that for N odd, the index of the lower member of the pairs has values on the range $n \in [1, N_{12}]$; $n = 0$ is an unpaired value. For N even, both $n = 0$ and $n = N/2$ are unpaired, and the index of the lower pair member ranges over $n \in [1, N/2 - 1]$. An example sequence is $h(n) = [10 \ -7 \ 6 \ -5 \ 4 \ -5 \ 6 \ -7]$. This sort of sequence, though conjugate-symmetric and thus having a real-valued DFT$_N$ will now be shown to have nonlinear phase.

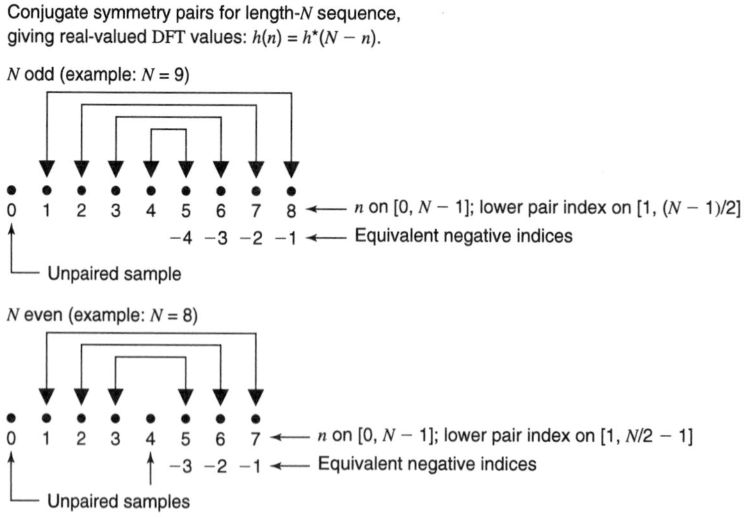

Conjugate symmetry pairs for length-N sequence, giving real-valued DFT values: $h(n) = h^*(N - n)$.

N odd (example: $N = 9$)

n on $[0, N - 1]$; lower pair index on $[1, (N - 1)/2]$

$-4 \ -3 \ -2 \ -1$ ◀—— Equivalent negative indices

Unpaired sample

N even (example: $N = 8$)

n on $[0, N - 1]$; lower pair index on $[1, N/2 - 1]$

$-3 \ -2 \ -1$ ◀—— Equivalent negative indices

Unpaired samples

Figure 9.12a

Positive/negative symmetry pairs for length-N sequence,
giving linear-phase DTFTs: $h(n) = \pm h(N - 1 - n)$

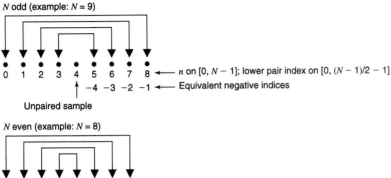

Figure 9.12b

From Figure 9.13a, we might conclude that the phase of $H(e^{j\omega})$ for this example $h(n)$ is zero; indeed, the phase of each of its DFT$_N$ coefficients $H(k)$ is zero. However, if we call Matlab's `freqz`, we do not obtain zero phase for the DTFT, $H(e^{j\omega})$ (except at the $N/2 + 1$ nonnegative frequency points shown in Figure 9.13a, $\omega_k = 2\pi k/N$). We obtain a continuous nonlinear, generally nonzero phase; see Figure 9.13b. Why? Recall that the zero-phase requirement for the DTFT is $h(n) = h^*(-n)$; that requirement becomes $h(n) = h^*(N - n)$ only when the sequence has finite length N and is assumed to be periodically extended for the DFT$_N$. The routine `freqz` zero-pads $h(n)$ out to $L = 512$ points. From the DFT$_L$ viewpoint, the

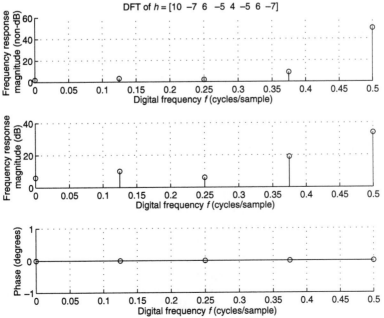

Figure 9.13a

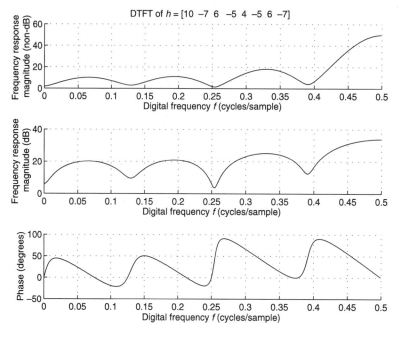

Figure 9.13*b*

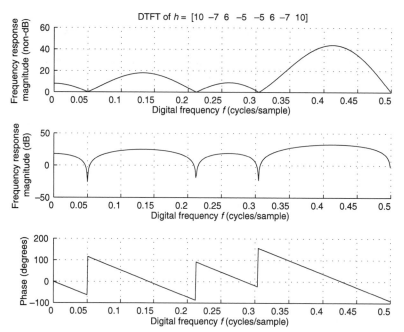

Figure 9.13*c*

zero-padded sequence does not satisfy $h(n) = h^*(L - n)$, and so the phase of the DFT_L is nonzero. From the DTFT viewpoint, $h(n) \neq h^*(-n)$, so again the phase is nonzero. Furthermore, the phase is nonlinear because $h(n)$ does not satisfy $h(n) = \pm h(N - 1 - n)$.

For our purposes now in studying linear-phase sequences, Figure 9.12*b* is relevant: $h(n) = \pm h(N - 1 - n)$. For real-valued $h(n)$ (the usual case) and for the case $h(n) = +h(N - 1 - n)$, the only difference from a (zero or π)-phase filter [even $h(n)$] to a linear-phase filter is the "-1" in $h(N - 1 - n)$, so in a plot the linear-phase sequence $h(n)$ has an appearance similar to that of an even sequence. Notice in Figure 9.12b that for N odd, the lower member of index pairs has values $n = [0, N_{12} - 1]$; $n = N_{12}$ is an unpaired value. For N even, all points are paired, and the lower member pair index values are $n = [0, N/2 - 1]$. An example sequence quite similar to the conjugate-symmetric one discussed above and having the same length is $h(n) = [10 \ -7 \ 6 \ -5 \ -5 \ 6 \ -7 \ 10]$. This sequence has positive symmetry and thus linear phase.

Moreover, if we call Matlab's `freqz`, we still get perfectly linear phase even though it zero-pads the sequence; see Figure 9.13c. [The phase discontinuities are all $180°$ and merely represent $G(\omega)$ changing sign.] The reason is that unlike the conjugate-symmetric condition, the linear-phase condition is not based on an assumption of N-periodicity of $h(n)$ (negative n being equivalent to $N - n$); it is based only on symmetries of the N nonzero terms forming the DTFT [refer to the development (9.28) and (9.29) to verify this]. Understanding all behaviors observed of this sort is a good check on one's overall understanding of the DTFT versus the DFT.

9.5.2   Examples of McClellan–Parks/Remez Filter Design

With the understanding of linear-phase filters in Section 9.5.1, the details of the McClellan–Parks algorithm may be studied. Determination of the cosine coefficients $p(n)$ in (9.26) is a fundamental part of the algorithm. Another major task is conversion of these coefficients $p(n)$ to FIR filter coefficients $h(n)$ for the various cases of N and symmetry. On WP 9.1, we give complete details on all these matters as well as on the overall strategy of the algorithm, and provide a numerical example with graphical illustrations of how the error as a function of frequency f changes from iteration to iteration (this example shows the iterations behind the filter designed in Example 9.3 below). It is well worth reading these, for this algorithm is very important in DSP. The algorithm is also fascinating for its combination of numerical and analytical techniques. The study on WP 9.1 also prepares one for modifying the Matlab code `remez` if necessary. Here, for brevity we will go directly to three typical application examples.

EXAMPLE 9.3

Plot the FIR Chebyshev approximation frequency response for the lowpass filter $H_d(e^{j\omega})$ used for Examples 9.1 and 9.2 ($f_c = 0.3, f_r = 0.4$).

Solution

See WP 9.1 for complete details of McClellan –Parks algorithm.

In this very simple example, we can just call `remez` with the following input parameters: `nfilt = 22`, `ff = 2.*[0  0.3  0.4  0.5]`, and `aa = [1  1  0  0]`. Remember, Matlab has that normalized definition of digital frequency which necessitates the "$2*$" factor in the frequency bands parameter `ff`. The parameter `aa` gives the value of the desired response at each end point of the frequency bands defined in `ff`. After that subroutine call, we merely plot the frequency response, which is shown in Figure 9.14a. The results are comparable to Kaiser windowing, but perhaps even better: The passband is quite smooth, and the transition is just as we specified.

Without the `wtx` input parameter, the default is uniform weighting of every band. This is the case for Figure 9.14a. To illustrate the idea of weighting the passband and stopband differently, compare the results for `wtx = [0.01  1]` (Figure 9.14b) and

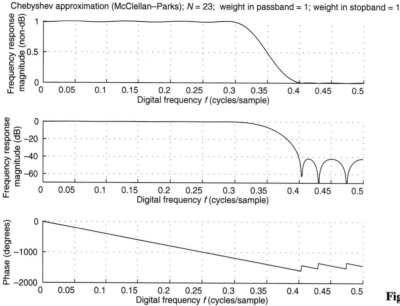

Figure 9.14a

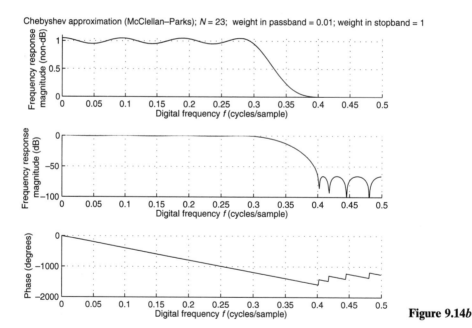

Figure 9.14b

wtx = [1 0.01] (Figure 9.14c). We clearly see that the band with the higher weight is given priority for accuracy. A low weight relative to other bands means that we tolerate more error for that band. Thus, Figure 9.14b has very poor passband ripple because we put all the emphasis (weight) on stopband rejection—notice that the sidelobes are down by over 70 dB, compared with only −40 dB for the uniform weighting in Figure 9.14a. Finally, in Figure 9.14c we see a beautifully flat passband. However, there are only about −22 dB of stopband attenuation; the sidelobes can be seen even in the nondecibel magnitude plot.

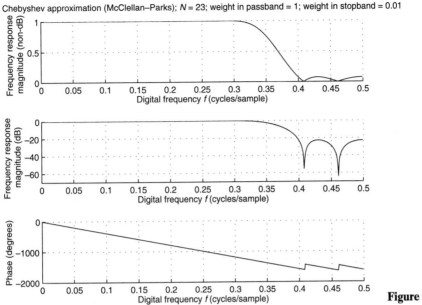

Chebyshev approximation (McClellan–Parks); $N = 23$; weight in passband = 1; weight in stopband = 0.01

Figure 9.14c

EXAMPLE 9.4

In Section 1.4, we considered the removal of broken-record thumping by manual processing. We found that there was a resonance centered around 8 kHz due to record surface noise (not due to the crack in the record). In this example, we investigate a simple approach to reducing that noise: Partially filter out those frequencies. Use the McClellan–Parks `remez` algorithm to do this filtering.

Solution

This solution will not be ideal, because when we filter out that noise we will also be filtering out the signal on that band. Thus, although the surface noise will be reduced and the overall signal-to-noise ratio increased, the music on that band will be attenuated. Instead of making a real stopband, we therefore choose instead to merely reduce the level moderately on the noise band.

Notice in Figure 1.8 that the resonance is triangular-shaped. Although the `remez` algorithm can be reprogrammed to inversely imitate this shape, a simpler approach is just to have a piecewise constant $H_d(e^{j\omega})$ that deepens in the center of its stopband. We select the bands as indicated in Table 9.1 and use a filter order of 500 and uniform error weighting. (A much lower-order filter would also work.) The frequency response of the designed filter is as shown in Figure 9.15.

It should be noted that a certain amount of experimentation is required for a filter as complicated as this one. The band edges and values in bands must be juggled or the algorithm will either not converge or will converge to an entirely erroneous solution. Note that transition bands that are too wide can be just as problematic as transition bands that are too narrow. Always check the frequency response of the result before proceeding with processing.

Table 9.1   Filter Frequency-Band Magnitude Values for Example 9.4

Range of F (Hz)	Range of f (cycles/sample)	Value in band
0–4,430	0–0.1005	1
4,810–5,800	0.1091–0.1315	0.32
6,200–9,000	0.1406–0.2041	0.06
9,400–12,200	0.2132–0.2766	0.23
12,600–14,000	0.2857–0.3175	0.45
14,400–22,050	0.3265–0.5000	1

The result of filtering is as shown in Figure 9.16. The filter was purposely designed so that the result generally follows the rest of the decline in the adjacent frequency bands. Notice at the top of the figure is shown the frequency response of the filter, which is 0 dB up to the stopband, where it appears as the dashed curve. The stopband is seen to appear where it should be to reduce the noise, and indeed the noise resonance in the signal spectrum is gone. Notice how in Figure 9.15 the stopband appears "too wide" and not centered at a high enough frequency, but is "just right" in Figure 9.16; this apparent difference is due to the logarithmic frequency scale in Figure 9.16.

To complete the experiment, we inverse-transform the data and listen to a playback of it. The result, as predicted above, is that the annoying high-frequency hissing is mostly gone, but also gone (or reduced) is the portion of the music in that band. Because the filter is linear phase, there was no other audible distortion caused by this operation. Other more sophisticated algorithms may be applied for the purpose of noise removal, but this example demonstrated both the power and limitations of frequency-selective filtering.

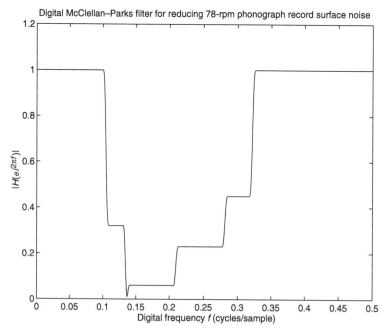

Digital McClellan–Parks filter for reducing 78-rpm phonograph record surface noise

Figure 9.15

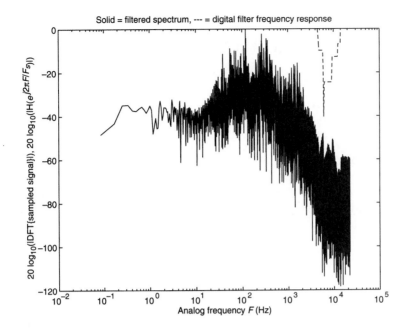

Solid = filtered spectrum, --- = digital filter frequency response

Figure 9.16

EXAMPLE 9.5

In the author's studies of wave propagation for ultrasonic medical imaging,[6] numerical estimations of the second derivative of the acoustic field with respect to time are required. Compare several FIR filters performing this estimation.

Solution

As noted in Section 2.4.1, the simplest and traditional method of computing an estimate of the second derivative is to use the central difference approximation from the theory of numerical analysis [we here cite the bicausal version of (2.43)]:

$$\frac{\partial^2}{\partial t^2} f_c(t)\Big|_{t=n\,\Delta t} \approx \frac{f_c([n+1]\,\Delta t) - 2f_c(n\,\Delta t) + f_c([n-1]\,\Delta t)}{\Delta t^2}. \tag{9.33}$$

This approximation is valid to order Δt^2. It is simply the backward difference of the backward difference, where each backward difference approximates a first-order time derivative. We note that the central difference approximation (9.33) is actually a bicausal FIR filter of length 3:

$$h_{c.d.}(n) = \frac{1}{\Delta t^2}\{\delta(n+1) - 2\delta(n) + \delta(n-1)\}, \tag{9.34}$$

which has DTFT

$$H_{c.d.}(e^{j\omega}) = \frac{1}{\Delta t^2}\{e^{j\omega} - 2 + e^{-j\omega}\} \tag{9.35a}$$

$$= \frac{2}{\Delta t^2}\{\cos(\omega) - 1\}. \tag{9.35b}$$

---

[6] T. J. Cavicchi, "Matrix solution of Transient High-Order Ultrasonic Scattering," *J. Acoust. Soc. Am.*, 90, 2 (August 1991): 1085–1092; T. J. Cavicchi, "Transient High-Order Ultrasonic Scattering in Soft Tissue," *J. Acoust. Soc. Am.*, 88, 2 (August 1990): 1132–1141.

However, the DTFT of the true analog second-differentiator is

$$H_c(j\Omega) = (j\Omega)^2 = -\Omega^2, \tag{9.36}$$

which has impulse response

$$h_c(t) = \frac{\partial^2}{\partial t^2} \delta_c(t). \tag{9.37}$$

How good an approximation of (9.36) is the simple central difference in (9.35b)? Although the high-pass nature of $H_c(j\Omega)$ certainly does not allow impulse invariance, we may still meaningfully compare the behavior of $H_{c.d.}(e^{j\Omega \Delta t})$ with that of $H_c(j\Omega)$ for $|\omega = \Omega \Delta t| < \pi$ [see discussion on the relation $\Omega = \omega/\Delta t$ in the paragraphs following (6.18b)]. Notice that for $|\omega| << \pi$, the Taylor expansion of (9.35b) evaluated at $\omega = \Omega \cdot \Delta t$ becomes

$$H_{c.d.}(e^{j\Omega \Delta t}) = \frac{2}{\Delta t^2} \{1 - \tfrac{1}{2}(\Omega \Delta t)^2 + \ldots - 1\} \tag{9.38}$$

$$= H_c(j\Omega) + \text{terms of order } (\Omega \Delta t)^\ell, \; \ell \geq 4.$$

(For simplicity in all figures for this example, we will take $\Delta t = 1$ s.) Thus, from (9.38) we expect that for low digital frequencies, central difference is good, whereas for higher digital frequencies, it is inaccurate. This is borne out in Figure 9.17, which shows the central difference DTFT (solid curve) and the ideal DTFT (dashed). We might even call the central difference a sort of "lowpass version" of the true differentiation. Such insights are more easily reasoned in the frequency response than in the time-samples domain.

Because the ever-increasing values of $|H_c(j\Omega)|$ with frequency could result in high-frequency noise overwhelming the signal, and because in practice we can oversample the signal content so that it is confined to lower frequencies, it is sensible to

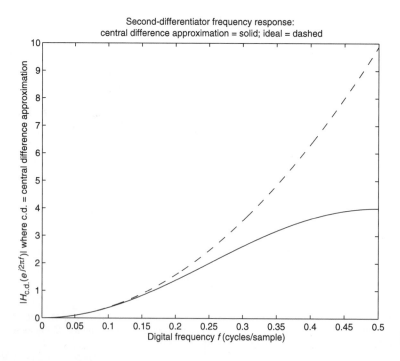

Second-differentiator frequency response:
central difference approximation = solid; ideal = dashed

Digital frequency f (cycles/sample)

Figure 9.17

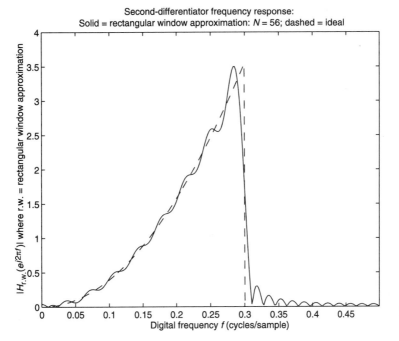

Second-differentiator frequency response:
Solid = rectangular window approximation: N = 56; dashed = ideal

Figure 9.18

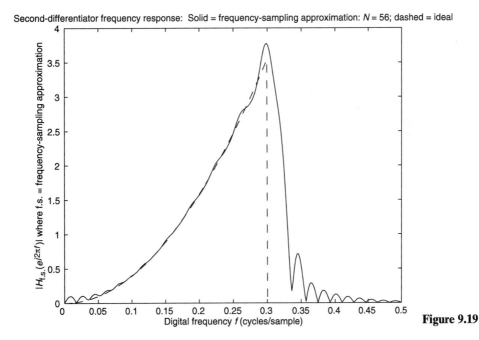

Second-differentiator frequency response: Solid = frequency-sampling approximation: N = 56; dashed = ideal

Figure 9.19

attempt only a lowpass version of the differentiator—for example, $f_c = 0.3$—as will be seen in the dashed curve depicting the ideal bandlimited second-differentiator in each of the following plots.

Using the techniques previously outlined, and in each case showing the ideal magnitude response as the dashed curve, the various approximations are rectangular window (Figure 9.18), frequency samples (Figure 9.19), and Chebyshev approximation (McClellan–Parks/Remez) (Figure 9.20). Below are included specific

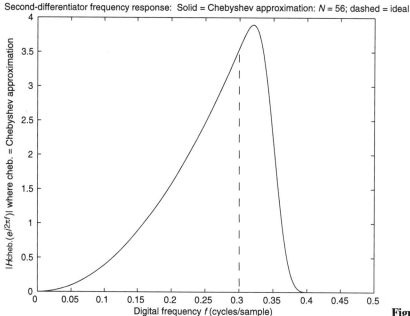

Second-differentiator frequency response: Solid = Chebyshev approximation: $N = 56$; dashed = ideal

Figure 9.20

comments about each design. Each of these approximations is for $N = 56$, so natural-ly all will outperform the $N = 3$ central difference approximation.[7] We see that the win-dow and frequency-sample results are comparable, but the Chebyshev (McClellan–Parks) results are superior. Examination of the log-magnitude plot would further show that the Chebyshev approximation has significantly more stopband attenua-tion than either the window or frequency-sampling methods, for the same filter order N.

It is left to the problems to show that the unit sample response for the low-band second derivative for use in generating the rectangular window approximation (Figure 9.18), before shifting by $N_{12} = (N - 1)/2$ and where $\omega_c = 2\pi f_c$, is

$$
h(n) = \begin{cases} \dfrac{-1}{\pi n}\left[\left\{\omega_c^2 - \dfrac{2}{n^2}\right\} \sin(\omega_c n) + \dfrac{2\omega_c}{n} \cos(\omega_c n)\right], & n \neq 0 \\ \dfrac{-\omega_c^3}{3\pi}, & n = 0. \end{cases}
\tag{9.39}
$$

Note that for our case of N even ($N = 56$), the "$n = 0$" case does not occur when the right-hand side of (9.39) is shifted by $(N - 1)/2$.

The frequency-sampling approach (Figure 9.19) uses $H_{\text{f.s.}}(k) = -(2\pi k/N)^2$ for k such that $2\pi k/N < \omega_c$ and zero otherwise, except for two transition samples where again 0.58 and 0.1 times the ideal $(-\omega^2)$ response are used. As we know, in practice, appropriate linear phase is also added to obtain a causal filter. Although the $N = 56$ values at $2\pi k/N$ are exact, in between the interpolation function $\sin(N\omega/2)/[N \sin(\omega/2)]$ does not yield $-\omega^2$ but instead the ripples in Figure 9.19. The DTFT was obtained by zero-padding the DFT_N^{-1} of the N exact samples and then taking the DFT of the zero-padded sequence. See (7.4) and (7.5) for details on the

---

[7]An interesting alternative approach would to be to try increasingly higher-order approximations derived from the "numerical analysis" method used to derive the central difference and compare these with the "DSP" methods described here. For example, examine approximations of differentiation based on Stirling's interpolation formula or Taylor series, as described in numerical analysis textbooks.

equivalent interpolation performed and the discussion following (7.13b) for why the interpolations are not exact [$h_d(n)$ is not finite-length].

In the Chebyshev approximation/McClellan–Parks approach, we must modify the Matlab code for remez to approximate the function $-\omega^2$ that is not built into the remez routine. In the old Fortran routine, the change was easily done by modifying a subroutine. In the Matlab routine, we must look for where the desired function is defined. We insert a few lines redefining des(sel) on the first (pass) band to be $-\omega^2$. We could here insert any desired real-valued response to be approximated.

As a final note, in the "off-line" calculations performed in the ultrasonic imaging algorithm, it may be permissible to omit the $N_{12} = (N - 1)/2$ shift and have a "π-phase" filter that is noncausal. This is because the entire time sequence is available, so there is no constraining issue of causality.

9.6   STRUCTURES FOR IMPLEMENTING DIGITAL FILTERS: COEFFICIENT QUANTIZATION

In the previous sections of this chapter and in Chapter 8, we have introduced the principles of digital filter design, both for FIR and IIR filters. In the remainder of this chapter, we consider a few implementation issues. For example, we address questions such as: What types of digital circuits/computational algorithms may most effectively be used to implement a digital filter, once designed? and What happens when we use finite-precision hardware to realize the filter?

As is true of filter design theory, a complete study of all the issues involved in digital filter structures would go well beyond the scope of an introduction to filter design. There is a huge variety of alternative digital filter structures; we will consider only a few that are straightforward to derive and that have greatest importance in practice. Moreover, the complete study of quantization effects would also fill volumes. Here we will restrict our attention to filter coefficient quantization. (In Section 10.8, we address the effects of signal/calculation quantization; we defer it until then so that the necessary statistical background can be reviewed first.) As technology continues to improve and quantization effects are reduced (e.g., the number of bits available in inexpensive chips increases), most users have less need for highly detailed and sophisticated quantitative analysis of quantization. When problems arise, one helpful trick is to simply increase the number of bits (if possible); if doing so produces no improvement, then the error is probably not due to quantization. Matlab is a very accurate computational system, so is ideal for such checking. In fact, in Chapter 10, we even use Matlab's Simulink to evaluate actual nonlinear rounding. One can also supply quantized values for the Matlab coefficient variables and thus simultaneously simulate coefficient and calculation rounding effects.

As previously noted, digital filters are fairly simple computational networks, involving only (a) delay elements, each of which implements "z^{-1}"; (b) constant coefficient signal multipliers or amplifiers, which implement, for example, the a_ℓ or b_ℓ in a transfer function; and (c) summers/subtractors that implement the sum/difference of two signals. The reason for the variety of filter structures is that different structures require different amounts of hardware or have different degrees of parallelization potential, require different amounts computation/storage, and have different susceptibilities to quantization effects that are unavoidable due to finite wordlengths in all computer circuits.

In this section, we limit our attention to the following basic structures: direct form, cascade, parallel, and lattice.

9.6.1   Direct Form Structures and Coefficient Quantization

Direct form structures not only follow from the transfer function most directly (hence their name), but also serve as building blocks for other structures. One of the direct forms is obtainable by writing $H(z) = B(z)/A(z) = Y(z)/X(z)$ as

$$H(z) = H_1(z) \cdot H_2(z), \tag{9.40}$$

where

$$H_1(z) = B(z) \tag{9.41a}$$

and

$$H_2(z) = \frac{1}{A(z)}. \tag{9.41b}$$

Defining $W(z) = H_2(z) \cdot X(z)$, we have

$$A(z) \cdot W(z) = X(z), \tag{9.42a}$$

or, upon inverse z-transformation and without loss of generality setting $a_0 = 1$,

$$w(n) = x(n) - \sum_{\ell=1}^{N} a_\ell w(n - \ell). \tag{9.42b}$$

It follows from the above definitions that

$$Y(z) = H_1(z) \cdot W(z) \tag{9.43a}$$

$$= B(z) \cdot W(z), \tag{9.43b}$$

or, upon inverse z-transformation,

$$y(n) = \sum_{\ell=0}^{M} b_\ell w(n - \ell). \tag{9.43c}$$

The key feature is to notice that both (9.42b) and (9.43c) involve weighted sums of $w(n - \ell)$. Thus, the common usage of $w(n - \ell)$ allows savings in delay elements. The straightforward implementation of these two difference equations, and thus the filter, is shown in Figure 9.21. It is called direct form II. Recall that the transposed direct form II was derived in Section 2.4.6.2 (page 56ff). A transposed structure is obtained by reversing all branch arrows and making all old input–output nodes the new output– input nodes.

The dual of direct form II, named direct form I, is even easier to derive using the same approach as above but instead writing $H(z) = H_2(z) \cdot H_1(z)$ and defining $V(z) = H_1(z) \cdot X(z) = B(z) \cdot X(z)$ and then $Y(z) = H_2(z) \cdot V(z)$. It is pursued in the problems. Whereas direct form II has N delay elements, direct form I has $N + M$ delay elements, so it is more costly to implement than direct form II.

The direct form structure is generally not used for implementing IIR filters, but is commonly used for implementing low-order FIR filters (for which all $a_\ell = 0$ in Figure 9.21). The reasons are as follows. When the filter coefficients a_ℓ and/or b_ℓ are

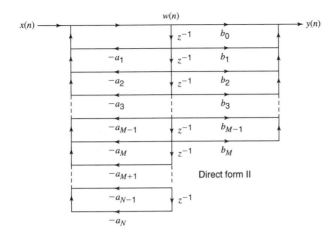

Figure 9.21

quantized, they of course each have a small shift in value. This shift would be tolerable if the coefficients a_ℓ and/or b_ℓ were the filter parameters whose values are most crucial. Instead, we generally care about the pole and zero locations, for it is these that directly affect the pass/stop frequency characteristics of the filter and directly determine filter stability. Consequently, it is the roots of $A(z)$ and $B(z)$ that are most crucial, not the a_ℓ and b_ℓ.

The roots z_m of $A(z)$ are functions of the a_ℓ via the operation of factorization or root-finding of the polynomial $A(z)$. It has been found that as the order N of $A(z)$ increases, the sensitivity of z_m to $\{a_\ell\}$ increases. That is, under increased N, small changes in the $\{a_\ell\}$ result in very large changes in z_m. The changes (errors) in a_ℓ as a result of quantization are fixed by the number representation (e.g., number of bits used to represent a_ℓ in a coefficients register). The effects of movement of the zeros under quantization in an FIR filter end up being less severe, and there is no instability concern. Thus, because of its convenience, the direct form—just a tapped and weighted delay line for FIR—is frequently used for moderate-order FIR filters.

To better understand coefficient quantization, let us consider more specifically what might happen in a fixed-point hardware realization of a digital filter. For a floating-point processor, usually there are so many bits in the representation (16 or 32) that coefficient rounding tends not to be a significant issue. Thus, we focus on the fixed-point representation. Fixed-point processors with low numbers of bits are used for minimizing production costs when large numbers of the given product are being fabricated or for minimizing size and/or maximizing processing speed.

The coefficients of the fixed-point filter will each be stored in binary form as b bits (zeros and ones) in the memory locations; usually, two's complement is used. Two's complement is used to allow the subtraction operation to be implemented by an adder.[8] Two's complement for positive numbers is the same as for the intuitive sign-magnitude representation, and it does include a sign bit to the left of the MSB which is set to 0. We obtain two's complement for negative numbers x by the following conversion operation, which we will designate symbolically as 2's$\{\cdot\}$: In $|x|$, change all 0s to 1s, all 1s to 0s, and add one to the LSB of the result, which is equivalent to computing the fundamental definition of two's complement, 2's$\{x\}$ = $2^b - |x|$. The sign bit of the result will of course be 1 for negative numbers.

The arithmetic results are as follows, for two's complement representations. Suppose that we wish to compute $x + y$. Let $X = \begin{cases} x, & x \geq 0 \\ 2\text{'s}\{|x|\}, & x < 0 \end{cases}$, and similarly

[8]An explanation of the rationale for the "complements" idea is in N. R. Scott, *Computer Number Systems and Arithmetic* (Englewood Cliffs, NJ: Prentice Hall, 1985).

define Y. We obtain $x + y$ by performing the binary addition $Z = X + Y$. To obtain $x - y$, we perform $Z = X + 2$'s$\{Y\}$. In the addition, we add the sign bits exactly the same way the other bits are added, and we always ignore any carry from addition of the sign bits. If the sign bit of Z is 1, we know we have a negative number for our result. The magnitude of that negative number is 2's$\{Z\}$ and thus is easily retrievable from Z.

The same rules are used for adding mantissas in floating-point arithmetic as are used for fixed-point addition and subtraction. If a floating-point number is negative, we use the two's complement of the mantissa and add the mantissas.

EXAMPLE 9.6

Perform the following operations using only addition and using three magnitude bits (plus a sign bit): **(a)** $3 - 5$, **(b)** $3 - (-2)$, **(c)** $-2 + (-3)$, **(d)** $-4 - 2$.

Solution

(a) We seek $3 - 5 = -2$. To perform this operation in base two, use $X = 0011$ for 3_{10} and $Y = 2$'s$\{|-5|\} = 2$'s$\{0101\} = 1011$ for -5_{10}. Adding 0011 and 1011 gives $Z = 1110$, which is negative; the magnitude is 2's$\{Z\} = 2$'s$\{1110\} = 0010 = 2_{10}$, and we thus obtain -2_{10}.

(b) We seek $3 - (-2) = 5$. We (again) use $X = 0011$ for 3_{10} and now use $Y = 2$'s$\{|-2|\} = 2$'s$\{0010\} = 1110$ for -2_{10} [incidentally, our result for Z in part (a)]. For subtraction, we now take 2's$\{Y\} = 2$'s$\{1110\} = 0010 (= +2_{10})$ and add to $X = 0011$ to obtain $Z = 0101 = 5_{10}$. The sign bit is 0, so we are done. (Of course, we could have just added 2 in the first place, but we are showing what is done in a computer.)

(c) We seek $-2 + (-3) = -5$. Again, we use $X = 1110$ for -2_{10}, and for -3_{10} we use $Y = 2$'s$\{|-3|\} = 2$'s$\{0011\} = 1101$. Adding X and Y gives $Z = 1011$, which is a negative number. The magnitude is 2's$\{Z\} = 0101 = 5_{10}$. Thus, we obtain -5_{10}.

(d) We seek $-4 - 2 = -6$. Now, $X = 2$'s$\{|-4|\} = 1100$ and $Y = 0010$. To subtract, we need 2's$\{Y\} = 1110$. Thus, we compute $Z = X + 2$'s$\{Y\} = 1010$, which is negative. The magnitude is 2's$\{1010\} = 0110 = 6_{10}$, so our result is -6_{10}.

For multiplication, suppose that the two numbers are positive. Then the ℓth partial product is set equal to the multiplicand if the ℓth LSB of the multiplier is 1; otherwise, the ℓth partial product is set to 0. To the ℓth partial product we add the $(\ell + 1)$st partial product shifted left one bit. The final product is the sum of all the partial products so added. If exactly one of the operands is negative (and in two's complement form), its two's complement is taken, the product of the resulting two positive numbers is taken, and the sign bit of the result is set to 1 if sign-magnitude form is desired. Or for further processing, the two's complement of the result is taken. If both operands are negative (and in two's complement form), we take the two's complement of both, multiply these, and leave the sign bit 0.

Note that to represent the product of two b-bit numbers, we need $2b$ bits. However, if we scale so that all numbers are less than 1, we may merely truncate the b LSBs. Or, we may use the floating-point representation, which requires somewhat different procedures (e.g., adding exponents and multiplying mantissas).

EXAMPLE 9.7

Multiply **(a)** 6×5 and **(b)** $6 \times (-5)$ in binary.

Solution

   (a)   110

            101

            110

           000

          110

          $11110 = 2^4 + 2^3 + 2^2 + 2^1 = 30_{10}.$

   (b)   We will be given -5 in two's complement form, 1011. Taking 2's{1011} gives 0101, and the process is exactly as in part (a). We merely take the two's complement of the result for further binary processing or for sign-magnitude presentation just set the sign bit to 1.

For simplicity, in discretizing our filter coefficients we assume here that N_L bits are designated as being to the left of the binary point. Thus, $2^{N_L} - 2^{-(b-N_L)}$ is the largest number that may be represented without overflow. If overflow occurs, we must either rescale all the coefficients and start over or (equivalently) move the binary point one place to the right. Either way, we still end up keeping only b significant bits.

There are several ways to discard the rest of the possibly infinite number of bits that exactly represent the original number. One approach is to round to the nearest b-bit binary number. In the following, "rounding up" means add 1 to the LSB, and "rounding down" means truncate the extra bits and leave the LSB unchanged.

In the rounding procedure, we first compute a $(b + 2)$-bit estimate and then examine the final two bits. If the final two bits are 00 or 01, we round down because they are less than halfway to 100. For example, if the $b = 4$-bit number with $N_L = 1$ (+2 extra bits for rounding determination) is 0.11101, we round down to 0.111. If the final two bits are 10 (in this example, 0.11110), the $(b + 2)$-bit estimate is exactly halfway from one b-bit estimate to the next higher b-bit estimate (in this example, halfway from 0.111 to 1.000). To average out the error of choosing one way versus the other, we select either the rounded-up or the rounded-down number, whichever has LSB $= 0$. Thus 0.11110 is rounded up to 1.000 (but 0.11010 would be rounded down to 0.110). (This latter idea, which we will be using here, is from Scott in footnote 8. But when the $b + 2$-bit estimate is obtained by truncation, those truncated bits tell us to always round up for "10.") Finally, if the two LSBs are 11, we round up; for example, 0.11111 is rounded up to 1.000. For negative numbers, this rounding procedure can be done with the magnitude and then converted back to two's complement.

We now see these ideas in action, in the quantization of a direct form structure for the poles of two example filters.

EXAMPLE 9.8

Draw pole plots for the original filter and for the b-bit quantized filter as represented in a direct form structure, for the following two examples: **(a)** the Chebyshev II bilinear transform IIR filter of Example 8.7, namely

$$H(z) = \frac{0.1852z^5 + 0.8354z^4 + 1.5893z^3 + 1.5893z^2 + 0.8354z + 0.1852}{z^5 + 1.5435z^4 + 1.5565z^3 + 0.8273z^2 + 0.2584z + 0.03414},$$

and **(b)** a filter whose poles have been designed to be at the following locations: $0.3 \angle \pm 70°, 0.6 \angle 0°, 0.7 \angle \pm 120°, -0.85, 0.9 \angle \pm 45°, 0.91 \angle \pm 50°$, and $0.95 \angle \pm 10°$. In both cases, plot for $b = 16$ bits and for $b = 8$ bits.

(c) Draw the direct form II structure for the filter in part (a).

Solution

(a)   The original unquantized coefficients for the denominator polynomial $A(z)$ are (to higher accuracy)

$a = [1 \quad 1.5435490 \quad 1.5565343 \quad 0.82727746 \quad 0.25841604 \quad 0.034141179]$;

that is, $a_0 = 1, \ldots, a_5 = 0.034141179$. Remember that a_ℓ is the coefficient of $z^{-\ell}$ in $A(z)$ expressed as a polynomial in z^{-1} (not z). Multiply $H(z)$ as given in the problem statement by z^{-5}/z^{-5} to obtain this form. We need one bit to the left of the binary point to handle all the numbers. Although in practice we would usually first scale all the coefficients to be less than 1, for illustrative convenience we will retain the original a values in this example. Conversion to $b = 16$ bits is as shown in Table 9.2.

The resulting pole plot is as shown in Figure 9.22a. Each original "exact" pole before quantization is shown as "x" and each quantized pole is shown as "∗". For $b = 16$, the poles before and after quantization are indistinguishable; thus the quantized filter will perform essentially identically to the original filter. Additionally, when we set $b = 8$ we obtain the pole plot in Figure 9.22b; again, there is only a small movement of the poles even with an only 8-bit representation.

(b)   For this second case, expanding out the pole factors yields a as shown in the first column of Table 9.3. We require $N_L = 3$ bits to the left of the binary point, and we have a somewhat high-order IIR filter. (Again, in actual implementation we could pull out factors from the numerator and from denominator of H so that all a and b coefficients are less than 1, but for simplicity here we will keep the original values.) With $b = 16$, there is essentially no degradation (see Figure 9.23a).

However, this time for $b = 8$ bits, we obtain the rounding situation presented in Table 9.3. As far as the coefficients go, there does not appear to be any more serious of a problem than for part (a) with 8 bits. However, when we examine the pole plot (Figure 9.23b), we find that not only is there very significant pole-estimation error, but some of the poles of the rounded system are unstable (outside the unit circle)! We can rectify this situation without

Table 9.2   Rounding of Filter Coefficients for Example 9.8, Part (a)

Exact base 10	Sign bit	Base 2 (16 + 2 = 18 bits)	Sign bit	Rounded base 2 (16 bits)	Base-10 equivalent of rounded number
	(Magnitude only; negative numbers converted to 2's complement afterwards)				
1	= 0	1.00000000000000000	0	1.000000000000000	1
1.54354900	≈ 0	1.10001011001001100	0	1.100010110010011	1.54354858
1.55653430	≈ 0	1.10001110011110010	0	1.100011100111100	1.55651855
0.82727746	≈ 0	0.11010011110010000	0	0.110100111100100	0.82727051
0.25841604	≈ 0	0.01000010001001111	0	0.010000100010100	0.25842285
0.03414118	≈ 0	0.00001000101111010	0	0.000010001011110	0.03411865

Table 9.3   Rounding of Filter Coefficients for Example 9.8, Part (b)

Exact base 10	Sign bit	Base 2 (8 + 2 = 10 bits)	Sign bit	Rounded base 2 (8 bits)	Base-10 equivalent of rounded number
		(Magnitude only; negative numbers converted to 2's complement afterwards)			
1	= 0	001.0000000	0	001.00000	1
−3.56901247	≈ 1	011.1001000	1	011.10010	−3.56250000
5.43734448	≈ 0	101.0110111	0	101.01110	5.43750000
−4.04874897	≈ 1	100.0000110	1	100.00010	−4.06250000
0.77021347	≈ 0	000.1100010	0	000.11000	0.75000000
1.11151884	≈ 0	001.0001110	0	001.00100	1.12500000
−1.26213641	≈ 1	001.0100001	1	001.01000	−1.25000000
1.23451273	≈ 0	001.0011110	0	001.01000	1.25000000
−1.27262705	≈ 1	001.0100010	1	001.01000	−1.25000000
0.91021659	≈ 0	000.1110100	0	000.11101	0.90625000
−0.36888799	≈ 1	000.0101111	0	000.01100	−0.37500000
0.08712507	≈ 0	000.0001011	0	000.00011	0.09375000
−0.01361519	≈ 1	000.0000001	1	000.00000	0

using any more bits for representing the a_ℓ by using a different structure involving only first-order and/or quadratic sections. Two structures for doing this are the cascade and parallel forms, which we consider next.

(c)   The direct form is particularly straightforward; just substitute the values of a_ℓ and b_ℓ for this example into Figure 9.21. The resulting structure is shown in Figure 9.24.

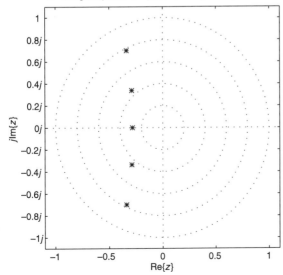

Direct form II poles—coefficient quantization for Chebyshev II lowpass filter; b = 16 bits
x = designed poles, ✳ = poles resulting from quantization

Figure 9.22a

Direct form II poles—coefficient quantization for Chebyshev II lowpass filter; *b* = 8 bits
x = designed poles, * = poles resulting from quantization

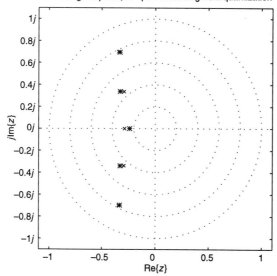

Figure 9.22*b*

Direct form II poles—coefficient quantization for high-order filter; *b* = 16 bits
x = designed poles, * = poles resulting from quantization

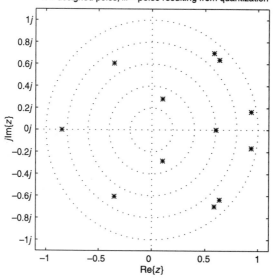

Figure 9.23*a*

(*a*)

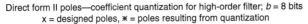

Direct form II poles—coefficient quantization for high-order filter; $b = 8$ bits
x = designed poles, ∗ = poles resulting from quantization

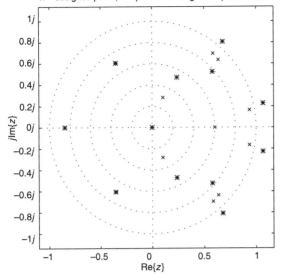

Figure 9.23*b*

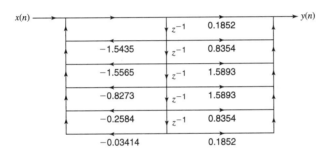

Figure 9.24

The direct form II for an FIR filter is obtained by just setting all the a_ℓ for $\ell > 0$ in Figure 9.21 to zero. The b_ℓ are just $h(\ell)$, and all coefficients are quantized according to the same rule. Therefore, if $h(n)$ has linear phase $[h(n) = \pm h(N - 1 - n)]$, then under quantization the linear phase is preserved.

9.6.2   Cascade Structures

We found in Example 9.8 that attempting to implement a high-order filter using the direct form can be catastrophic—the filter with rounded coefficients can become unstable as well as highly inaccurate. As noted, this is because high-order polynomial roots are very sensitive to coefficient quantization. An obvious way around this problem is to factor the transfer function before quantization and implement/round only first- or second-order sections, each of which is much less sensitive to coefficient quantization effects for its pole(s).

For the cascade form, we simply factor the given filter transfer function $H(z)$ into first-order sections $H_{1,\ell}(z)$ for real poles and/or zeros and/or into second-order sections $H_{2,m}(z)$ for complex pole and/or zero pairs, rather than use the direct form:

$$H(z) = A \prod_{\ell=1}^{N_1} H_{1,\ell}(z) \cdot \prod_{m=1}^{N_2} H_{2,m}(z), \tag{9.44}$$

where A = constant, N_1 = number of simple real poles, and N_2 = number of complex pole pairs. The first- and second-order sections are

$$H_{1,\ell}(z) = \frac{1 - z_\ell z^{-1}}{1 - p_\ell z^{-1}} \tag{9.45a}$$

$$= \frac{1 + \beta_\ell z^{-1}}{1 + \alpha_\ell z^{-1}}, \tag{9.45b}$$

and

$$H_{2,m}(z) = \left(\frac{1 - z_m z^{-1}}{1 - p_m z^{-1}} \right) \cdot \left(\frac{1 - z_m{}^* z^{-1}}{1 - p_m{}^* z^{-1}} \right) \tag{9.46a}$$

$$= \frac{1 + b_{1m} z^{-1} + b_{2m} z^{-1}}{1 + a_{1m} z^{-1} + a_{2m} z^{-1}}. \tag{9.46b}$$

For extra unpaired real or complex poles or zeros, we simply use 1 for the corresponding numerator or denominator for that section. Note that

$$\beta_\ell = -z_\ell, \qquad \alpha_\ell = -p_\ell, \tag{9.47a}$$

$$b_{1m} = -2\mathrm{Re}\{z_m\}, \qquad b_{2m} = |z_m|^2, \tag{9.47b}$$

$$a_{1m} = -2\mathrm{Re}\{p_m\}, \qquad a_{2m} = |p_m|^2. \tag{9.47c}$$

We use the symbols α_ℓ and β_ℓ for the first-order sections to avoid confusion with the difference-equation a_ℓ and b_ℓ. Notice that $|\beta_\ell|$, $|\alpha_\ell|$, $|b_{2m}|$, and $|a_{2m}|$ are all ≤ 1 for minimum-phase filters (i.e., for filters having all poles and zeros on or within the unit circle). For fixed point, we thus automatically have all significant bits to the right of the binary point. Unfortunately, however, there is no such guarantee on $|a_{1m}|$ and $|b_{1m}|$; we will find in Example 9.9b that $|b_{1m}|$ are greater than 1 because the absolute values of the real parts of those zeros exceed 0.5. In practice, the coefficients may all be scaled down to force the scaled coefficients to all be less than 1.

The graphical appearance of the cascade form is as shown in Figure 9.25. The choice of pairings of poles and zeros is to an extent arbitrary, although empirical results show that complex zeros should be paired with the nearest complex poles.

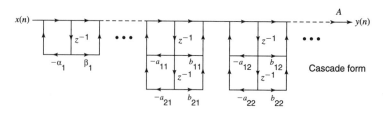

Cascade form

Figure 9.25

EXAMPLE 9.9

(a) Repeat Example 9.8b, with b = 8 bits, now using the cascade structure with first- and second-order sections.

(b) Draw the cascade structure for the IIR filter of Example 9.8a, namely

$$H(z) = \frac{0.1852z^5 + 0.8354z^4 + 1.5893z^3 + 1.5893z^2 + 0.8354z + 0.1852}{z^5 + 1.5435z^4 + 1.5565z^3 + 0.8273z^2 + 0.2584z + 0.03414}.$$

Solution

(a) Again, we focus on the poles; the zeros will have similar shifts. The coefficient percent errors are naturally about the same—on the order of 2^{-b}. However, the pole estimates are far more accurate because of the reduced sensitivity of first- and second-order sections to coefficient quantization, as evidenced by Figure 9.26 in which the "exact" and estimated (from the rounded coefficients) pole locations are visually indistinguishable. The problem of the system becoming unstable in Example 9.8b has been eliminated without using any more bits.

(b) For cascade implementation, we factor $H(z)$ as follows. First, factor out b_0 = 0.1852, which will be A in (9.44). We will have one first-order and two second-order sections. We pair up the poles $\{p_a, p_1, p_2\}$ = $\{-0.2843,$ 0.7788 $\angle\pm115.828°$, 0.445 $\angle\pm130.728°\}$ and zeros $\{z_a, z_1, z_2\}$ = $\{-1,$ 1 $\angle\pm145.69°$, 1 $\angle\pm158.34°\}$ as follows (pairing for this example is arbitrary due to the pole and zero locations):

$$H(z) = 0.1852 \cdot H_a(z) \cdot H_1(z) \cdot H_2(z),$$

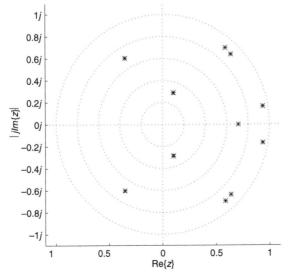

Cascade poles—coefficient quantization for high-order filter; b = 8 bits

x = designed poles, $*$ = poles resulting from quantization

Figure 9.26

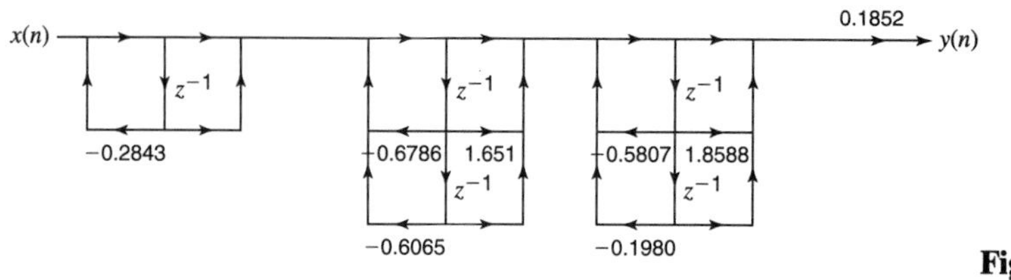

Figure 9.27

where

$$H_a(z) = \frac{1 + z^{-1}}{1 + 0.28427z^{-1}},$$

$$H_1(z) = \frac{1 - 2\text{Re}\{z_1\}z^{-1} + |z_1|^2 z^{-2}}{1 - 2\text{Re}\{p_1\}z^{-1} + |p_1|^2 z^{-2}}$$

$$= \frac{1 + 1.65135z^{-1} + z^{-2}}{1 + 0.67857z^{-1} + 0.60648z^{-2}},$$

and

$$H_2(z) = \frac{1 - 2\text{Re}\{z_2\}z^{-1} + |z_2|^2 z^{-2}}{1 - 2\text{Re}\{p_2\}z^{-1} + |p_2|^2 z^{-2}}$$

$$= \frac{1 + 1.85922z^{-1} + z^{-2}}{1 + 0.5807z^{-1} + 0.19803z^{-2}}.$$

The signal flowgraph, with the second-order sections implemented using direct form II, is as shown in Figure 9.27.

One last remark on the quadratic-section poles. From (9.47c), we see that because a_{1m} and a_{2m} are uniformly quantized, so will be twice the real part and the magnitude squared of the estimated p_m. The appearance of this type of quantization is shown in Figure 9.28a for $b = 4$ and Figure 9.28b for $b = 6$ (sign bits not included in b). The dashed lines and circles are, respectively, the loci of possible $\text{Re}\{p_m\}$ and $|p_m|$; the intersections are the only possible quantized poles available for that value of b. Only the upper half-plane is depicted (but is extended slightly below the real axis to be certain to include real poles). Note that the unit circle is excluded because such systems would not be stable.

Note the unfortunate characteristic that the poles are least dense around $z = 1$, which is where all the poles will be when processing oversampled signals. Notice that even for $b = 6$ bits, there are no accessible pole locations in the vicinity of $z = 1$. We thus conclude that although for aliasing Δt should be as small as possible, Δt too small (which drives filter poles to near $z = 1$) will result in large quantization error under filter coefficient rounding. Distinct but nearby poles may be quantized all to the same location, possibly quite far from the desired locations. They may also be canceled by nearby zeros quantized to the same location. In general, a filter with poles (and zeros) widely spaced will suffer least under coefficient quantization.

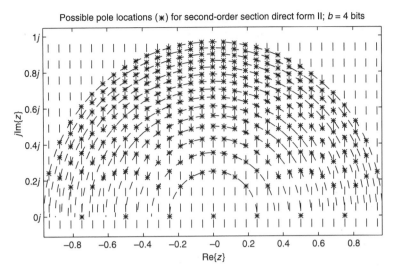

Figure 9.28a

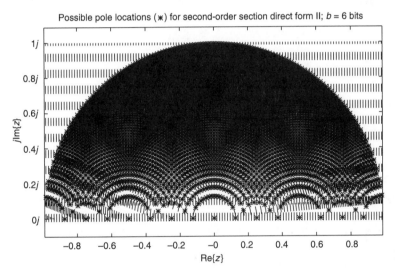

Figure 9.28b

9.6.3   Parallel Structures

Another way to implement an IIR filter using first- and second-order sections is by partial fraction expansion of $H(z)$. As we show explicitly below, each complex-pole term is added to its corresponding conjugate-pole term so that the coefficients of the resulting second-order term are all real. An advantage of the parallel structure is that the filter speed can be faster due to the parallel processing in each partial fraction term. As for the cascade structure, the complex pole terms are easily implemented using a direct form II structure with minimal coefficient quantization error.

For simplicity, assume that all poles are distinct. Let p_ℓ be the N_1 real poles and $\{p_m, p_m^*\}$ be the N_2 complex pole pairs, as in Section 9.6.2. Analogously, the partial fraction expansion coefficients are A_ℓ for p_ℓ and A_m for p_m. Then $H(z)$ can be written

$$H(z) = \sum_{\ell=1}^{N_1} \frac{A_\ell z}{z - p_\ell} + \sum_{m=1}^{N_2} \frac{A_m z}{z - p_m} + \frac{A_m^* z}{z - p_m^*}. \tag{9.48a}$$

Making a common denominator in the second sum gives

$$H(z) = \sum_{\ell=1}^{N_1} \frac{A_\ell z}{z - p_\ell} + z \sum_{m=1}^{N_2} \frac{z(A_m + A_m^*) - A_m p_m^* - A_m^* p_m}{(z - p_m)(z - p_m^*)} \tag{9.48b}$$

$$= \sum_{\ell=1}^{N_1} \frac{A_\ell z}{z - p_\ell} + 2z \sum_{m=1}^{N_2} \frac{\text{Re}\{A_m\}z - \text{Re}\{A_m p_m^*\}}{z^2 - 2\text{Re}\{p_m\}z + |p_m|^2}, \tag{9.48c}$$

and expressing $H(z)$ in terms of z^{-1} gives

$$H(z) = \sum_{\ell=1}^{N_1} \frac{A_\ell}{1 - p_\ell z^{-1}} + 2 \sum_{m=1}^{N_2} \frac{\text{Re}\{A_m\} - \text{Re}\{A_m p_m^*\}z^{-1}}{1 - 2\text{Re}\{p_m\}z^{-1} + |p_m|^2 z^{-2}}, \tag{9.48d}$$

where under our assumption of distinct poles,

$$A_\ell = (z - p_\ell)\frac{H(z)}{z}\Big|_{z=p_\ell} \tag{9.49}$$

for both real and complex poles. The appearance of the parallel implementation is seen in the signal flowgraph in Figure 9.29, in which we make the following identifications:

$$\alpha_\ell = -p_\ell, \quad \beta_\ell = A_\ell, \quad b_{0m} = 2\text{Re}\{A_m\}, \quad b_{1m} = -2\text{Re}\{A_m p_m^*\},$$
$$a_{1m} = -2\text{Re}\{p_m\}, \quad a_{2m} = |p_m|^2.$$

(The α_ℓ and β_ℓ are not to be confused with those defined for the cascade structure.) Note that the parallel realization is irrelevant for FIR filters, which do not have a nontrivial partial fraction expansion. Contrarily, the cascade form is also useful for

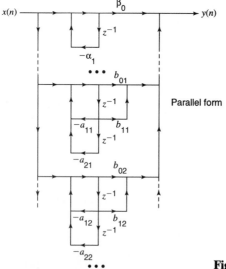

Parallel form

Figure 9.29

high-order FIR filters. As with direct form II, the cascade form for FIR filters can be used in a way that preserves linear phase.

EXAMPLE 9.10

Draw the parallel structure for the IIR filter of Example 9.8b, namely

$$H(z) = \frac{0.1852z^5 + 0.8354z^4 + 1.5893z^3 + 1.5893z^2 + 0.8354z + 0.1852}{z^5 + 1.5435z^4 + 1.5565z^3 + 0.8273z^2 + 0.2584z + 0.03414}.$$

Solution

First, we note that $H(z)$ as shown has numerator order M equal to denominator order N [which is common in practice, when discretizing analog filters as done for this $H(z)$]. We factor $H(z)$ and perform the partial fraction expansion. Because $N = M (= 5)$, we must use the special approach explored in Problem 65, part (b) in Chapter 3. The required expressions for the PFE coefficients as derived in that problem are

$$A_\ell = \left. \frac{\sum\limits_{k=0}^{M} b_k z^{-k}}{\prod\limits_{m=1, m \neq \ell}^{N} \left(1 - \frac{p_m}{z}\right)} \right|_{z = p_\ell}$$

for $\ell \in [1, N = 5]$, and

$$A_0 = \frac{b_N}{\prod\limits_{m=1}^{N} (-p_m)}$$

so that

$$H(z) = A_0 + \sum_{\ell=1}^{N} \frac{A_\ell z}{z - p_\ell}.$$

The numerical values of the coefficients are $A_0 = 5.42453$, $A_1 = -2.79914$, $A_2 = -0.007232 + j0.34531$, $A_3 = A_2^*$, $A_4 = -1.21287 + j0.69877$, $A_5 = A_4^*$ and $p_1 = -0.28427$, $p_2 = -0.33929 + j0.70097$, $p_3 = p_2^*$, $p_4 = -0.29035 + j0.33723$, $p_5 = p_4^*$. The complex pole pairs can be further simplified, using (9.48d), giving

$$H(z) = 5.42453 + \frac{-2.79914}{1 + 0.28427z^{-1}} + \frac{-0.014464 - 0.48902z^{-1}}{1 + 0.67857z^{-1} + 0.60648z^{-2}}$$

$$+ \frac{-2.42573 - 1.17561z^{-1}}{1 + 0.58070z^{-1} + 0.19803z^{-2}},$$

and the resulting parallel implementation is as shown in Figure 9.30. Note the extra branch for the constant term 5.42453, not included in the canonical Figure 9.29.

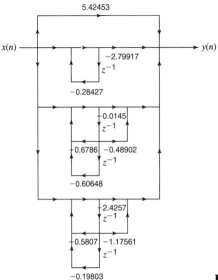

Figure 9.30

9.6.4 Lattice Structures

In digital communications, speech processing, adaptive digital filtering, and other applications, a filter structure called the lattice filter has been developed that has many advantageous properties over those we have considered above. For example, its modularity is such that in prediction filtering applications, changing the filter order may require changing only the coefficients of the last "stage" instead of all the coefficients. That is, a family or sequence of filters may be derived such that the optimal filter for the next higher order is the same as that for the current order, but with a new stage tacked on. This modularity also offers hardware advantages, because each stage has the same form, and each stage can be applied sequentially.

The coefficients of the lattice filter are called reflection coefficients because the lattice structures are in some ways reminiscent of models of multimedia transmission lines or other one-dimensional media in which waves propagate forward and backward. At interfaces, the reflection coefficients determine the proportions of signal that are transmitted and reflected. An important property of IIR lattice filters is that as long as all the reflection coefficients K_ℓ satisfy $|K_\ell| < 1$, the filter is stable. In addition, the lattice filters tend to be robust to quantization error, relative to other structures.

A more advanced treatment of our introductory study of prediction in Section 10.9 shows that certain signals in a lattice filter structure have physical meaning as *prediction errors*, and they will therefore here be labeled as such. The lattice structure is thus closely associated with the theory of linear prediction of stochastic signals. It is also used in system identification algorithms such as that used by Siglab (recall Example 4.11), which again are based on statistical least-squared-error considerations. The study of lattice filters thus serves as a bridge between the familiar (digital filter structures) and the new (statistical signal processing).

9.6.4.1   *FIR Lattice Filters*

The ℓth stage of an FIR lattice filter is shown in Figure 9.31. From the diagram, we read off

$$e_\ell^f(n) = e_{\ell-1}^f(n) + K_\ell\, e_{\ell-1}^b(n-1) \overset{\text{ZT}}{\longleftrightarrow} E_\ell^f(z) = E_{\ell-1}^f(z) + K_\ell z^{-1} E_{\ell-1}^b(z) \tag{9.50a}$$

and

$$e_\ell^b(n) = e_{\ell-1}^b(n-1) + K_\ell e_{\ell-1}^f(n) \overset{\text{ZT}}{\longleftrightarrow} E_\ell^b(z) = z^{-1} E_{\ell-1}^b(z) + K_\ell E_{\ell-1}^f(z). \tag{9.50b}$$

We set $e_0^f(n) = e_0^b(n) = x(n)$, or $E_0^f(z) = E_0^b(z) = X(z)$. We also set $y(n) = e_M^f(n)$ for the Mth-order FIR filter. Thus an ℓth-order FIR lattice filter has the appearance of Figure 9.32; the reason for the name "lattice" should there be evident. From (9.50) we obtain for $\ell = 1$

$$E_1^f(z) = X(z) + K_1 z^{-1} X(z) \tag{9.51a}$$
$$= (1 + K_1 z^{-1})X(z)$$
$$= B_1(z)X(z)$$

$$E_1^b(z) = z^{-1} X(z) + K_1 X(z) \tag{9.51b}$$
$$= (z^{-1} + K_1)X(z)$$
$$= \beta_1(z)X(z),$$

where

$$B_1(z) = 1 + K_1 z^{-1} \tag{9.52a}$$
$$\beta_1(z) = K_1 + z^{-1}, \tag{9.52b}$$

and for $\ell = 2$

$$E_2^f(z) = E_1^f(z) + K_2 z^{-1} E_1^b(z) \tag{9.53a}$$
$$= \{1 + K_1 z^{-1} + K_2 z^{-1}(z^{-1} + K_1)\}X(z)$$
$$= B_2(z)X(z)$$

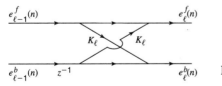

Figure 9.31

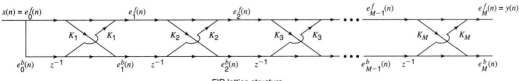

FIR lattice structure

Figure 9.32

$$E_2^b(z) = z^{-1}E_1^b(z) + K_2E_1^f(z) \tag{9.53b}$$
$$= \{z^{-1}(z^{-1} + K_1) + K_2(1 + K_1z^{-1})\}X(z)$$
$$= \beta_2(z)X(z),$$

where

$$B_2(z) = 1 + K_1(1 + K_2)z^{-1} + K_2z^{-2} \tag{9.54a}$$

$$\beta_2(z) = K_2 + K_1(1 + K_2)z^{-1} + z^{-2}. \tag{9.54b}$$

Because each recursion in (9.50) involves a linear combination of the next-lower-order $E_\ell^f(z)$ and $E_\ell^b(z)$ and because $E_0^f(z) = E_0^b(z) = X(z)$, it follows in general that $E_\ell^f(z) = B_\ell(z)X(z)$ and $E_\ell^b(z) = \beta_\ell(z)X(z)$, which is consistent with (9.53) and (9.54). Thus, with these substitutions for $E_\ell^f(z)$ and $E_\ell^b(z)$, the z-transform expressions in (9.50) can be rewritten, upon division by $X(z)$, as

$$B_\ell(z) = B_{\ell-1}(z) + K_\ell z^{-1}\beta_{\ell-1}(z) \tag{9.55a}$$

$$\beta_\ell(z) = z^{-1}\beta_{\ell-1}(z) + K_\ell B_{\ell-1}(z). \tag{9.55b}$$

Notice that $H(z) = \dfrac{Y(z)}{X(z)} = \dfrac{E_M^f(z)}{E_0^f(z)} = \dfrac{B_M(z)X(z)}{X(z)} = B_M(z).$

We see the pattern that

$$\beta_\ell(z) = z^{-\ell}B_\ell(z^{-1}), \tag{9.56a}$$

which implies that the coefficients of $\beta_\ell(z)$ are the same as those of $B(z)$ except that they are in the reverse order. That is, if $b_\ell(n) = \text{ZT}^{-1}\{B_\ell(z)\}$ and if $d_\ell(n) = \text{ZT}^{-1}\{\beta_\ell(z)\}$, then by the time-reversal and delay properties proved in Chapter 5,

$$d_\ell(n) = b_\ell(\ell - n). \tag{9.56b}$$

The proof of (9.56a) can be rigorously completed by using (9.51), (9.52), and (9.55), assuming that (9.56) holds true for "stage" ℓ, and then showing that (9.56) is preserved for stage $\ell + 1$.

Note that for all ℓ, the coefficient of z^0 in B_ℓ is always unity $[b_\ell(0) = 1]$, an important fact in the prediction interpretation of the lattice filter. Thus, additionally, $d_\ell(\ell) = 1$. Also, for the Mth-order FIR filter, by definition $b_M(n) = h(n)$; thus $B_M(z) = H(z)$. The question arises, however, how the $\{K_\ell\}$ are related to $b_M(n)$. That is, given $b_M(n) = h(n)$, what are $\{K_\ell\}$ of the corresponding lattice filter? This question is easily answered by using $b_\ell(0) = 1$, (9.55), and (9.56). The inverse z-transform of (9.55a) is

$$b_\ell(n) = b_{\ell-1}(n) + K_\ell d_{\ell-1}(n - 1) \tag{9.57a}$$

which when (9.56b) is used to obtain $d_{\ell-1}(n - 1) = b_{\ell-1}(\ell - 1 - [n - 1]) = b_{\ell-1}(\ell - n)$ becomes

$$b_\ell(n) = b_{\ell-1}(n) + K_\ell b_{\ell-1}(\ell - n), \quad n \in [1, \ell - 1], \tag{9.57b}$$

where (9.57b) is derived for use later; we will presently be using (9.57a).

The inverse z-transform of (9.55b) is [recalling $d_\ell(n) = \text{ZT}^{-1}\{\beta_\ell(z)\}$]

$$d_\ell(n) = d_{\ell-1}(n - 1) + K_\ell b_{\ell-1}(n), \tag{9.58a}$$

or, equivalently,

$$d_{\ell-1}(n - 1) = d_\ell(n) - K_\ell b_{\ell-1}(n). \tag{9.58b}$$

Substitution of (9.58b) into (9.57a) gives

$$b_\ell(n) = b_{\ell-1}(n)\{1 - K_\ell^2\} + K_\ell d_\ell(n), \tag{9.59a}$$

or, again using (9.56b),

$$b_{\ell-1}(n) = \frac{b_\ell(n) - K_\ell b_\ell(\ell - n)}{1 - K_\ell^2}, \quad n \in [1, \ell - 1]. \tag{9.59b}$$

We now have a simple procedure to obtain the reflection coefficients $\{K_\ell\}$ from $b_M(n)$ [where to match notation in the literature we here let the highest power of z^{-1} in $H(z)$ be M]:

1. $K_M = b_M(M) = h(M)$. This equality holds because in (9.55a) with $\ell = M$, the highest power of z^{-1} in $B_M(z)$ is $K_M d_{M-1}(M - 1) = K_M \cdot 1 = K_M$ and also because $H(z) = B_M(z)$, as shown above. Examples of this identity are seen in (9.52a) and (9.54a).
2. Use (9.59b) to obtain the $\ell = (M - 1)$st lattice coefficients $\{b_{M-1}(n)\}$ from $\{b_M(n)\}$. In particular, from these we thus obtain $K_{M-1} = b_{M-1}(M - 1)$.
3. Repeat step 2, stepping down ℓ one at a time, to obtain the remaining K_ℓ.

EXAMPLE 9.11

Implement the following FIR filter using a lattice structure:

$$H(z) = 1 + 0.8z^{-1} + 0.2z^{-2} - 0.1z^{-3} + 0.05z^{-4}.$$

Solution

$M = 4$: $\{b_4\} = \{1, 0.8, 0.2, -0.1, 0.05\}$; $K_4 = b_4(4) = 0.05$. Using (9.59b), we obtain $\{b_3\} = \{1, 0.807, 0.1905, -0.1404\}$; $K_3 = b_3(3) = -0.1404$; $\{b_2\} = \{1, 0.8505, 0.3098\}$; $K_2 = b_2(2) = 0.3098$; $\{b_1\} = \{1, 0.6493\}$; $K_1 = b_1(1) = 0.6493$. Thus the lattice filter for this $H(z)$ is as shown in Figure 9.33.

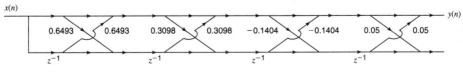

Figure 9.33

Now suppose that we have by some means obtained a set of reflection coefficients $\{K_\ell\}$, and we wish to obtain the usual (direct form) $b(n)$ coefficients. This task is also easily done, this time using $b_\ell(0) = 1$ for all ℓ, $b_\ell(\ell) = K_\ell$, and (9.57b):

---

1. $b_1(0) = 1$, $b_1(1) = K_1$.
2. $b_2(0) = 1$. Then use (9.57b) with $\ell = 2$ to obtain $b_2(n)$, $n \in [1, \ell - 1] = [1]$ and $b_2(2) = K_2$.
3. Repeat step 2, stepping up ℓ one at a time, to obatin $\{b_\ell(n)\}$ for all ℓ up to M. The last set, $\ell = M$, is the desired direct form filter coefficients, $h(n) = b_M(n)$.

If we use Steps 1 through 3 in Example 9.11 for $\{K_\ell\} = \{0.6493, 0.3098, -0.1404, 0.05\}$, we indeed reverse the process and obtain $\{b_4\} = \{1, 0.8, 0.2, -0.1, 0.05\} = \{h(n)\}$.

9.6.4.2 IIR Lattice Filters

For simplicity, we first consider the all-pole filter, $H(z) = 1/A(z)$; then later we will generalize to $H(z) = B(z)/A(z)$. For the all-pole filter, $Y(z) = X(z)/A(z)$, or, equivalently, $X(z) = A(z)Y(z)$. If we have a lattice structure for the FIR filter defined by $Y(z) = B(z)X(z)$, it should be a simple matter to convert it to an all-pole structure using the same rules for determining reflection coefficients. The only essential difference is that we are changing the roles of input and output.

Because in an IIR filter there must be both a forward path from $x(n)$ to $y(n)$ and also a feedback path from $y(n)$ back toward $x(n)$, a reasonable choice for the all-pole lattice structure is as shown in Figure 9.34. However, the reason for the minus sign in the upward-pointing K_ℓ must be explained. Because the path direction has been reversed in the upper path (after the reversal of the entire filter in Figure 9.32), the output of the ℓth stage will be (9.50a) in reverse:

$$e^f_{\ell-1}(n) = e^f_\ell(n) - K_\ell e^b_{\ell-1}(n-1). \tag{9.60}$$

Thus to have the K_ℓ be defined by the same set of equations as for the FIR case, we must have the K_ℓ negated that contributes to $e^f_{\ell-1}(n)$ in Figure 9.34. This allows us to use the same algorithms for converting between direct-form coefficients and K_ℓ as presented above.

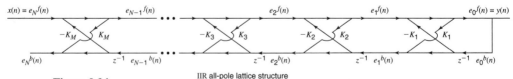

Figure 9.34

IIR all-pole lattice structure

EXAMPLE 9.12

If $H(z) = 1/(1 - 3.4z^{-1} + 4.46z^{-2} - 2.674z^{-3} + 0.6205z^{-4})$, draw the lattice implementation of this IIR all-pole filter.[9] The poles, by factorization of $H(z)$, are at $0.8544 \angle \pm 20.56°$ and $0.922 \angle \pm 12.53°$; thus $H(z)$ is a stable filter.

---

[9] Note that in this discussion, the term *all-pole* ignores the zeros at $z = 0$, just as the *all-zero* term ignores all the poles at $z = 0$.

Solution

Using the algorithm for obtaining K_ℓ from b_ℓ (which we might rename a_ℓ for the all-pole application), we obtain $\{K_\ell\} = \{-0.9754 \quad 0.98 \quad -0.9176 \quad 0.6205\}$. The lattice realization is as shown in Figure 9.35.

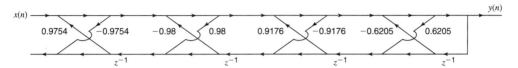

Figure 9.35

An important feature in Example 9.12 is that $|K_\ell| < 1$ for this stable filter. Although not proved here,[10] the result holds in general: The IIR system is stable if and only if $|K_\ell| < 1$ for all ℓ.

For the pole–zero IIR filter $H(z) = B(z)/A(z)$, we can use the all-pole lattice filter to realize $1/A(z)$ and take a weighted sum of taps $e_\ell^b(n)$ to implement $B(z)$. Let $W(z) = X(z)/A(z)$, the output of the all-pole lattice. Also, rename $B_\ell(z)$ in the lattice algorithm to be $A_\ell(z)$ because the all-pole filter is implementing $A(z)$, not $B(z)$. Similarly, rename $\beta_\ell(z)$ to be $\alpha_\ell(z)$. In the all-pole lattice, the FIR relation $E_\ell^b(z) = \beta_\ell(z)X(z)$ becomes

$$E_\ell^b(z) = \alpha_\ell(z)W(z), \tag{9.61}$$

where again from (9.56a), adapted to the all-pole structure,

$$\alpha_\ell(z) = z^{-\ell}A_\ell(z^{-1}). \tag{9.62}$$

If we form for $y(n)$ a linear combination of the $e_\ell^b(n)$, we can obtain the required additional degrees of freedom necessary to implement the zeros of $H(z)$; see Figure 9.36, which is called a lattice/ladder structure. Specifically, for a filter of denominator order N and numerator order M,

$$y(n) = \sum_{\ell=0}^{M} \gamma_\ell e_\ell^b(n), \text{ [ladder portion of Figure 9.36 is based on this equation]} \tag{9.63a}$$

or, equivalently,

$$Y(z) = \sum_{\ell=0}^{M} \gamma_\ell E_\ell^b(z) \tag{9.63b}$$

$$= \sum_{\ell=0}^{M} \gamma_\ell \alpha_\ell(z) \cdot W(z) \tag{9.63c}$$

$$= C_M(z) \cdot \frac{X(z)}{A(z)}, \tag{9.63d}$$

---

[10] See J. D. Markel and A. H. Gray, "On Autocorrelation Equations Applied to Speech Analysis," *IEEE Trans. on Audio and Electroacoustics*, AU-21, 2 (April 1973): 69–79.

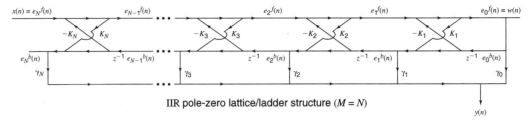

IIR pole-zero lattice/ladder structure ($M = N$)

Figure 9.36

where

$$C_m(z) = \sum_{\ell=0}^{m} \gamma_\ell \alpha_\ell(z) \qquad (9.64)$$

and thus $B(z) = C_M(z)$.

Our final task is to determine the γ_ℓ. Remember that $\alpha_\ell(n) = a_\ell(\ell - n)$ [and $a_N(n) = a(n) = \mathrm{ZT}^{-1}\{A(z)\}$ so $a_\ell(0) = 1$ in order that we match the canonical assumption $a(0) = 1$], so $\alpha_\ell(\ell) = 1$ for all ℓ. Consequently, if we write

$$C_m(z) = \sum_{\ell=0}^{m-1} \gamma_\ell \alpha_\ell(z) + \gamma_m \alpha_m(z) \qquad (9.65a)$$

or

$$C_m(z) = C_{m-1}(z) + \gamma_m \alpha_m(z), \qquad (9.65b)$$

we know from matching coefficients of z^{-m} that $\gamma_m \cdot 1 = c_m(m)$. Our procedure is therefore as follows:

1. $\gamma_M = b(M)$ where $B(z) = \sum_{n=0}^{M} b(n) z^{-n}$; also, $C_M(z) = B(z)$.
2. From (9.65b), use $C_{m-1}(z) = C_m(z) - \gamma_m \alpha_m(z)$, where $\alpha_m(n) = \mathrm{ZT}^{-1}\{\alpha_m(z)\} = a_m(m - n)$, where $\alpha_m(n)$ is obtained using the all-pole procedure.
3. From $C_{m-1}(z)$, pick off $\gamma_{m-1} = c_{m-1}(m - 1)$. Repeat for all m from $M - 1$ down to $m = 1$. The result is the desired $\{\gamma_m\}$ coefficients in Figure 9.36.

EXAMPLE 9.13

Determine the lattice structure for the Chebyshev II lowpass filter designed in Example 8.7, namely

$$H_{\mathrm{old}}(z) = \frac{0.1852 z^5 + 0.8354 z^4 + 1.5893 z^3 + 1.5893 z^2 + 0.8354 z + 0.1852}{z^5 + 1.5435 z^4 + 1.5565 z^3 + 0.8273 z^2 + 0.2584 z + 0.03414} = \frac{B(z)}{A(z)}.$$

Solution

Using the lattice reflection coefficients recursion algorithm derived above, we obtain $\{K_\ell\} = \{0.6146, 0.7528, 0.4790, 0.2059, 0.0341\}$, and from the algorithm above

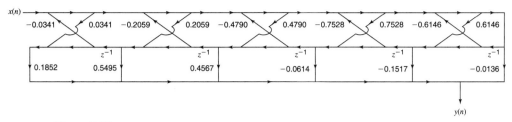

Figure 9.37

for γ_ℓ, we obtain $\{\gamma_\ell\} = \{-0.0136, -0.1517, -0.0614, 0.4567, 0.5495, 0.1852\}$. The order of appearance of these numerical coefficients is from right to left and, following Figure 9.36, is as shown in Figure 9.37.

PROBLEMS

FIR Linear Phase Filters: Basics and Window Method

9.1 If the length of a causal rectangularly windowed FIR filter is N, why is the required additional phase shift on $H_d(e^{j\omega})$ set to $-\omega(N-1)/2$?

9.2 (a) For window lengths $N = 32$, plot both windows $w(n)$ versus n and their frequency responses $W(e^{j2\pi f})$ versus f for the following windows: rectangular (i.e., `boxcar`), Bartlett, Hanning, Hamming, and Blackman; use the Matlab routines with these names. Compare the width of the main lobe and the minimum attenuation (in decibels) of the first sidelobe. Express the main lobe width as a multiple α of $1/N$, as it roughly follows α/N for all N; note that you will be measuring only the half-width of the main lobe.

(b) Let $\omega_c = 0$ in the Kaiser formula (9.21), and choose ω_r to be the mean of the main lobe half-widths for the rectangular and Hamming windows. Specify attenuation = 35 dB, and plot the frequency response. You would expect to obtain a compromise between the two in your frequency response but you do not. For an explanation, see the caution in the text on the interpretation of the first sidelobe attenuation.

9.3 Design a bandpass filter with $F_L = 100$ Hz, $F_H = 300$ Hz, and a sampling frequency of $F_s = 800$ Hz. Try rectangular and Hamming windows for $N = 10$, $N = 128$, and $N = 1024$ (this last perhaps being impractical, but interesting).

9.4 (a) Analytically design a rectangular-window-based $(N - 1)$st-order (length N) causal lowpass filter $h_N(n)$ with cutoff $f_c = 0.4$.

(b) Use the Matlab command `fir1` with filter length $N = 40$ (order = 39) to design a lowpass filter. The `fir1` command is supposed to replicate the rectangular window design when given a boxcar window input. However, you will find a slight difference in the DTFT of the result of `fir1` and the DTFT of the result of the window-designed $h_N(n)$, even when all the input parameters for `fir1` are correctly chosen. Go into the source code, and determine the reason for the difference.

(c) In this part, we investigate use of `fir2` to design a lowpass filter to learn more about exactly what `fir2` does (the main purpose of `fir2` is the design of arbitrary-shaped-magnitude filters). First, go into the `fir2.m` file, and under the heading `if narg = 4` change `npt = 512` to `npt = nn/2`. This change will simplify explanations of the effects we seek to examine by giving the same number of nonnegative frequency samples (npt) as in the DFT^{-1} used to produce the designed filter $h_{fir2}(n)$. Now try the command `fir2` (again with $N = 40$) with inputs $[0\ \ 2f_c\ \ 2f_c\ \ 1]$ for `ff` and $[1\ \ 1\ \ 0\ \ 0]$ for `aa`, again with a boxcar window. You should find disagreement between DTFT$\{h_{fir2}\}$ and DTFT$\{h_N(n)\}$. Why? Use a result from the text to independently duplicate the Matlab result. Hints: (i) Recall Section 7.3 on inverse DTFT approximation. Again, you will have to go into the code to see what `fir2` does. (ii) Discontinuities in spectra expanded in Fourier series converge to a sample midway in between the two discontinuous spectrum values when approximated by an infinite-order Fourier series.

9.5 Consider the use of a rectangular window in the approximation of an ideal bandstop filter:

$$H_d(e^{j\omega}) = \begin{cases} 0, & \pi/2 \le |\omega| \le 3\pi/4 \\ 1, & \text{otherwise} \end{cases}, \quad |\omega| \le \pi.$$

(a) (i) State and use two properties discussed in Chapter 5 concerning the given $H_d(e^{j\omega})$ to simplify the *general* DTFT formula for $h_d(n)$ for all n in terms of $H_d(e^{j\omega})$. Hint: Your answer may be the definition of a "new" (inverse) transform we have so far not discussed, which is just a special case of the DTFT.

(ii) A further simplification is possible if a new function $H'_d(e^{j\omega})$ is defined in terms of $H_d(e^{j\omega})$ that ends up requiring only one interval of integration (the bandstop filter requires two intervals of nonzero integration). What is the required relation between $H'_d(e^{j\omega})$ and $H_d(e^{j\omega})$? What is the corresponding relation between $h'_d(n)$ and $h_d(n)$? Using this relation and the results from (i), find $h_d(n)$ for *all* n.

(b) If $|h_d(n)| \to 0$ for $|n| \to \infty$, we can use rectangular truncation to obtain a finite-length sequence:

$$h_M(n) = \begin{cases} h_d(n), & |n| \le M/2 \\ 0, & \text{otherwise,} \end{cases}$$

where M is an even integer.

(i) To obtain a causal unit sample response $h_{M,c}(n)$, what must we do [working only with $h_d(n)$, not $H_d(e^{j\omega})$]? What are the lengths of $h_M(n)$ and $h_{M,c}(n)$? Are their lengths odd or even? Write a simple expression for $h_{M,c}(n)$ in terms of $h_M(n)$.

(ii) What are the first and last indices and values of the FIR filter sequence $h_{M,c}(n)$?

9.6 Determine the unit sample response of the N-length rectangular window design of a differentiator, with a cutoff frequency of ω_c.

9.7 (a) In Problem 8.16, we explored bilinear and forward-difference approximations of differentiation. Augment the Matlab magnitude plot in Problem 8.16 by adding the DTFT magnitude of the window-method differentiator $h_{win}(n)$ derived in Problem 9.6 but with $\omega_c = \pi$ (full band); let $N = 40$ (rectangular window). Comment on the results. Should you expect much better results than for the BLT or FD methods? Why?

(b) Suppose that you try to reproduce your results using the command `fir2`, which designs an FIR digital filter with user-specified magnitude. Do you find agreement with your $h_{win}(n)$ in part (a)? If not, why not? If not, can you alter `fir2` to produce the same result as in part (a)? Hint: `fir2` does not take a phase input.

9.8 Using Matlab and the results of Problem 9.6, plot the frequency response of a digital differentiator, where $f_c = 0.4$ by a direct evaluation of the frequency response at, say, 300 points to make a smooth curve. On the same axis, plot $|j\omega|$ and compare. Then plot the DFT magnitude for $N = 64$ and $N = 128$ (using asterisks for these points). Comment on your results (for example, Gibbs's phenomenon, where the ripple is worst, where on the ripple the DFT values fall, and so on).

9.9 A doctor wishes to remove the annoying 60-Hz line interference (60 Hz) from an electroencephalogram (EEG—brain waves) signal (0–100 Hz) as well as all signals above 150 Hz, using a notch–lowpass digital filter combination (the two digital filters are in cascade). Even though this processing will remove a portion of the ECG (that around 60 Hz), it will be well worth it, because the ECG is "drowned out" by the 60-Hz noise. Select a reasonable sampling interval. At a block diagram level, design the overall system (whose final result is stored in a computer memory). Then do the following detailed designs.

(a) Design a 32-length (31st-order) FIR lowpass filter first analytically (carry out the integral yourself), using a rectangular window and appropriate phase. Plot the unit sample response in Matlab. Then compare frequency responses in Matlab (`freqz`) using a Hamming window versus using a rectangular window. (You may use the `hamming` command.) You will not find much difference with the decibel plot (why?), so plot the non-decibel DTFT magnitude and show the difference between usage of Hamming and rectangular windows.

(b) Design a 60-Hz notch filter. This filter is to have zeros at 60 Hz (and the corresponding conjugate location) and poles "right behind" the zeros (radius $r < 1$). The closer the poles are to the zeros, the flatter the rest of the response will be (experiment with r —what are the tradeoffs?). Draw a pole–zero diagram. This example is one in which pole–zero placement is exactly what is desired.

(c) In both parts (a) and (b), ensure that the passband gain is unity. Now, cascade the two digital filters (with Hamming) and arrive at an overall $H(z)$. Plot its frequency response and show both stopbands (notch and high-frequency). You may be a bit disappointed in the result. However, try adding on an identical notch stage, plot its frequency response, and you will get a good improvement (quantitatively answer why).

9.10 Design a length-N highpass causal filter using a rectangular window with cutoff $\omega_c = 2\pi f_c$; obtain the analytical expression for $h_N(n)$. With $f_c = 0.2$, plot the frequency response for $N = 128$ and for $N = 127$. Do you prefer N even or odd? Why?

9.11 Invert the second-order low-band derivative frequency response to obtain the ideal infinite-length second-order differentiator unit sample response (9.39).

9.12 (a) Consider the rectangular window design for a digital lowpass filter of cutoff frequency ω_c. For simplicity, we deal with the unshifted $h_N(n)$ and its DTFT $H_N(e^{j\omega})$ (i.e., the bicausal filter approximation) with N odd. We wish to quantify the accuracy of our approximation for a given value of N. Write down $h_N(n)$, and determine the filter frequency response error $E(\omega)$: $E(\omega) = H_d(e^{j\omega}) - H_N(e^{j\omega})$ in terms of an infinite sum of real-valued functions.

(b) A standard performance criterion is the integral squared error ϵ:

$$\epsilon = \int_{-\pi}^{\pi} |E(\omega)|^2 d\omega.$$

Using the result of part (a), express ϵ as an infinite sum of terms, each of which is a function of n and a *single* trigonometric function of n. You may need the identity

$$\int_{-\pi}^{\pi} \cos(\omega n)\cos(\omega m) \, d\omega = \pi\delta(n - m) \text{ for } n > 0. \text{ Does } \epsilon \to 0 \text{ as } N \to \infty?$$

(c) Demonstrate the validity of (P9.12.1) and (P9.12.2):

$$\beta = \int_{-\pi}^{\pi} |H_d(e^{j\omega})|^2 d\omega = 2\omega_c \tag{P9.12.1}$$

$$= \frac{2\omega_c^2}{\pi} + \frac{4}{\pi} \cdot \sum_{n=1}^{\infty} \frac{\sin^2(\omega_c n)}{n^2}. \tag{P9.12.2}$$

Thereby obtain an identity for the value of the infinite sum in (P9.12.2). Hint for (P9.12.2): Use the method you used in part (b), and the fact that $\sin(\omega_c n)/(\pi n)\big|_{n\to 0} = \omega_c/\pi$.

(d) Using the results developed above, find $\rho(N) = \epsilon/\beta$. What are two advantages of this expression as compared with the other expression above (for ϵ)?

9.13 (a) Prove (9.11), which is useful in the discussion of Gibbs's phenomenon.

(b) Prove that (9.13) follows, for the value ω_x for which in (9.12) we have $R'_M(\omega_x) = 0$, a fact also useful in discussing Gibbs's phenomenon.

FIR Filter Design: Frequency Sampling and McClellan–Parks

9.14 Suppose that we set $H(k)$ to samples of a desired $H_d(e^{j\omega})$, where the latter is multiplied by the linear phase factor $e^{-j\omega(N-1)/2}$ for causal $h(n)$. Show that $H(e^{j\omega})$, the frequency response for $H(k)$, has linear phase for all ω and thus by this simple approach we obtain a linear-phase filter.

9.15 (a) Design a 32-length (31st-order) frequency-sampling differentiator filter, which cuts off at $f = 0.2$ cycle/sample. Use a single transition sample, with value 0.4 times the discontinuity from differentiating band (high end) and zero. Plot $|H(k)|$ for all k versus corresponding f_k. Obtain $h(n)$, and plot it after Hamming windowing. Then plot the magnitude nondecibel frequency response from $h(n)$ for $f \in [0, \frac{1}{2}]$, and superimpose $|H(k)|$ to see the agreement at f_k and ripple in between—do this for both the rectangularly windowed $h(n)$ and the Hamming-windowed $h(n)$ and compare the results. Then make a discrete-time triangle sequence

$$x(n) = \begin{cases} 2n/N_t, & n \in [0, N_t/2] \\ 2(1 - n/N_t), & n \in [N_t/2 + 1, N_t-1], \end{cases}$$

where $N_t = 256$ and filter it with your filter by using the fft (correctly!), using a power of two for the DFT size. Plot the output (use plot rather than stem if the DFT size is too large); have you obtained approximate differentiation? Quantitatively state what the ideal result would be and why you think the Hamming window is very helpful in this case. Explain any discrepancies from the ideal. Comment on why this filter might be overkill for this particular application $x(n)$, and suggest a very simple filter that in this case would actually work better.

(b) Now add to $x(n)$ a sampled cosine of amplitude $\frac{1}{2}$ and for which $f = 0.4$ cycle/sample. Plot $x(n)$, its DTFT magnitude, and the new output $y(n)$. To verify that the differentiator still works on the triangle, try using axis to zoom in on the desired vertical range. *Now* is your filter overkill?

Problems 9.16 through 9.19 Directly Involve Material on WP 9.1.

9.16 Derive (W9.1.5) and (W9.1.6) for Case 2, $h(n) = +h(N - 1 - n)$, N even, using (W9.1.1) through (W9.1.4) as a model.

9.17 Derive (W9.1.9a), from it (W9.1.9b), and derive (W9.1.10) and (W9.1.11) for Case 3, $h(n) = -h(N - 1 - n)$, N odd, using the derivations on WP 9.1 as a model.

9.18 Derive (W9.1.12a), from it (W9.1.12b), and derive (W9.1.13) and (W9.1.14) for Case 4, $h(n) = -h(N - 1 - n)$, N even, using the derivations on WP 9.1 as a model.

9.19 Show that (W9.1.29) simplifies to (W9.1.30). Hints: Consider Δ_{M+1} first, then later generalize by a mere change of notation. Let $x_\ell = \cos(\omega_\ell)$, so that $\omega_\ell = \cos^{-1}(x_\ell)$. Use the recursion relation for Chebyshev polynomials $\cos(m\omega_\ell) = \cos\{m \cos^{-1}(x_\ell)\} = C_m(x_\ell)$ and properties of determinants to express Δ_ℓ in Vandermonde form. Once in Vandermonde form, show and use $2 \cdot 4 \cdot 8 \cdot \ldots \cdot 2^{M-1} = 2^{(M-1)M/2}$ and the Vandermonde determinant property (9.25) to obtain (W9.1.30).

Structures for Digital Filters (Including Lattice):

9.20 For the signal flow diagram shown in Figure P9.20, find $H(z)$ and from it the causal $h(n)$. You may define one additional constant parameter for convenience. Also, you may define up to two intermediate variables. Simplify all expressions as much as possible.

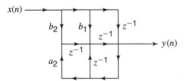

Figure P9.20

9.21 Suppose that we are given the signal flow diagram in Figure P9.21 and are asked to determine the direct form II structure that will implement that same system.

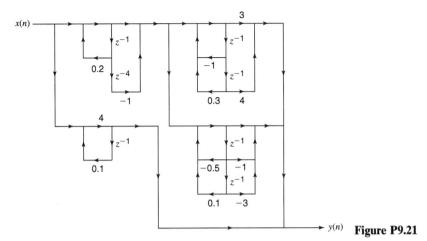

Figure P9.21

(a) Write $H(z)$ as a straightforward reading of the flow diagram would dictate.

(b) For $y(n) = -\sum_{m=1}^{N} a_m y(n-m) + \sum_{m=0}^{M} b_m x(n-m)$, what are N and M for this system? Justify your answer with reasoning/calculations.

(c) *Without calculating the coefficients*, and using your results from part (a), state explicitly and completely what calculations, operations, identifications, and so forth must be done in order to determine the direct form II parameters of this system. Draw the direct form II diagram, showing only branches for which there may be nonzero coefficients [your answer to part (b) could help here].

(d) Find the direct form II parameters and draw the final diagram.

9.22 Given the flow diagram in Figure P9.22 in which $c \neq b$:

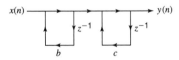

Figure P9.22

(a) Give the difference equation.

(b) Give a closed-form expression for the frequency response in terms of b, c, and ω (complex exponentials are allowed).

(c) For what b and c (possibly complex) is the net system minimum phase?

(d) Give the direct form II and parallel realizations.

9.23 Derive direct form I, using the derivation of direct form II in the text as a model. Draw the complete structure, and compare the number of delay elements required for direct form I with that for direct form II.

9.24 Suppose that we were to perform no algebraic simplifications on a spectral-transformed digital filter in which z^{-1} has been replaced by $G(z^{-1})$ in (8.80), with $K = 1$. If the input to the delay element is $x_D(n)$ and the output is $y_D(n)$, write the difference equation relating $x_D(n)$ and $y_D(n)$. Then indicate using both direct form I and direct form II the structure with which each delay element in the original structure would be replaced.

9.25 Given the spectrum $X'(e^{j\omega})$ of $x'(n)$, show the spectrum of the output $x(n)$ of the system shown in Figure P9.25. Show all your reasoning and also show the spectra of the intermediate sequences [i.e., the spectra of $x_d(n)$, $v(n)$, $w(n)$, and $y(n)$]. Describe the overall result of this processing.

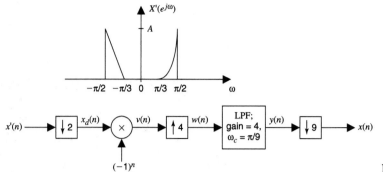

Figure P9.25

9.26 For each of the lattice filters in Figure P9.26, find the corresponding system function $H(z)$; express as a rational function of z^{-1}.

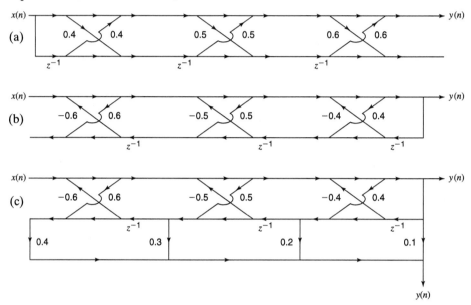

Figure P9.26

9.27 For the network shown in Figure P9.27, find the overall system function $H(z)$; simplify as much as possible. What kind of filter is this (FIR, IIR all-pole, or IIR pole–zero) and why, and what are its poles and/or zeros?

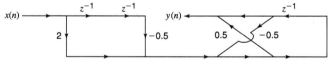

Figure P9.27

9.28 Part of the structure in Problem 9.27 leads to an interesting result. Notice that it is a first-order version of Figure 9.34 (drawn upside-down and backwards), but with the output selected to be $e_N^b(n)$ rather than $e_0^f(n)$. With $y(n) = e_N^b(n)$ in Figure 9.34, what type of transfer function do we obtain? Answer for general-order case.

9.29 (a) For the filter in Figure P9.29, find $p_2(n)$ and $q_2(n)$ in terms of the input signal $x(n)$ and K_1 and K_2.

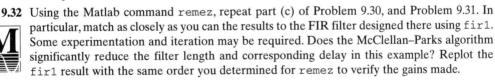

$q_1(n)$ $q_2(n)$ **Figure P9.29**

(b) Suppose that $p_2(n)$ is considered the output $y(n)$. Draw the flow diagram for the direct form filter that also produces $y(n)$ as its output.

(c) Repeat part (b) for the case in which $y(n)$ is instead $q_2(n)$.

(d) Find the magnitude of the frequency response of the filter whose output is $y(n)$ as defined in part (b) for $K_1 = 1, K_2 = 2$. Express your result in terms of only real-valued quantities.

Practical/Computer Applications

9.30 In Example 8.1, we briefly considered the removal of 60-Hz noise from a signal, by subtraction of a sample of the noise. Suppose that instead we eliminate as much of the 60-Hz signal as possible by using a notch filter. Suppose that the sampling frequency is $F_s = 400$ Hz. Design such a filter in Matlab by the following approaches.

(a) Pole–zero placement. Remember that we want the passband to be as smooth as possible. Let the digital filter have two poles and two zeros. Comment on the effects of placement of the pole.

(b) Use the `butter` command. Experiment with different orders and stopbands. Plot the frequency response of your final result.

(c) Use the `fir1` command. How high must the filter order be to achieve results similar to those in (b)? Plot this last result on the same axes as the plot in (b). State exactly what the `fir1` result represents.

9.31 Use your filters designed in Problem 9.30 [parts (b) and (c) only] to actually filter out a continuous-time noise signal from, for example, a sample of speech or other audio signal. Questions to answer first are: What is your sampling frequency? Therefore, what analog frequency will be filtered out by the digital filters designed in Problem 9.30? Use Simulink. Corrupt your speech signal with a "loud" noise signal (sinusoid) of the frequency you calculated. Filter it out, and compare the original signal, the corrupted signal, and the filtered signal in both time and frequency domains. In your magnitude spectrum plot, also superimpose the (appropriately scaled) filter response, and use `axis` to focus on the region of interest. Carefully note and explain the delay, if any, caused by the digital filter. To do this, plot and explain the group delay of the filter under consideration (using the command `grpdelay`), and examine the filter response phase plot. Of course, plot stem plots showing in detail the delay observed. What do you conclude about the use of group delay for IIR versus FIR linear-phase filters?

9.32 Using the Matlab command `remez`, repeat part (c) of Problem 9.30, and Problem 9.31. In particular, match as closely as you can the results to the FIR filter designed there using `fir1`. Some experimentation and iteration may be required. Does the McClellan–Parks algorithm significantly reduce the filter length and corresponding delay in this example? Replot the `fir1` result with the same order you determined for `remez` to verify the gains made.

Chapter 10

Stochastic Signal Processing: Concepts and Applications

10.1 INTRODUCTION

Up to this point in our discussion of digital signal processing, we have focused on two major studies: development of the fundamental digital–signal processing theory and use of those principles for basic practical processing of data. To complete an essential exposure to modern DSP requires an additional study: the processing, modeling, and spectral estimation of stochastic discrete–time signals. (The terms *stochastic* and *random* are synonyms.)

Why are so–called stochastic methods important? The reason is that we live in a world of aberrational processes. For example, it is clear that an electronic sine–wave oscillator does not produce a perfect sine wave; rather, it produces a slightly distorted sine wave with "random" noise added on. By its very nature, such noise cannot necessarily be pinned down to a localized spectral region for frequency–selective–filter removal, nor is it localized in time. It does not follow easily identifiable patterns either. With stochastic methods, however, it may be possible to reduce this noise, by using the statistical dissociation of the noise from the known or modeled signal data.

The same idea is even more powerful when we consider the processing of a signal–plus–random–noise for which the signal is unknown. Without stochastic methods, the special random nature of noise cannot be exploited to enhance noise reduction. Also, because all components of the data in the passband are considered "signal," simple frequency-selective approaches are bound to be inaccurate. Just as frequency–selective filtering takes advantage of knowledge about the frequency content of noise to remove it, stochastic filters can take advantage of this special random nature of noise to remove it. Keep in mind that to various extents, random noise corrupts all measurements and so is always present in raw data.

To begin our study of stochastic processing, we must first step back and consider the meaning of terms such as *deterministic, random,* and *correlation.* In the first part of this chapter, we review in detail the mathematical methods of expressing stochastic concepts. We also introduce some basic parameter estimation ideas and the concept of entropy. A text on probability theory and stochastic processes may prove to be a helpful supplement to this material.

The difference between *deterministic* and *stochastic* is often a philosophical one. When a coin is flipped, is it possible to calculate on which side it will fall, given all the forces and torques the fingers exhibit on the coin in the process of flipping? Is coin flipping "random" only because the person flipping does not have either accurate control over his or her fingers from one flip to the next nor sufficient ability to measure the forces and torques and to calculate the trajectory? Is it possible to calculate the exact price of a stock if the large number of personal decisions and influences involved were possible to document? Is the stock price "random" only in that it is impossible for anyone to possess all the information necessary for making an exact calculation? How about the water level as a function of time at a given location in the ocean? Think of the complexity of forces in this case!

Before losing too much sleep over these worrisome questions, it is better to consider the alternatives. The deterministic approach does not account for the reality of inaccurate measurements or the unavailability of other information necessary for predicting, say, tomorrow's closing stock price or the weather. The stochastic approach recognizes that in practice fully determined calculations are impossible or at best impractical. In this approach, we are content to obtain approximate predictions or estimates of expected values (or trends) rather than absolutely precise values for a particular instance under consideration. The advantages are that the required calculations are now possible and that within the context of proper qualifications, they give reliable and useful results.

The reader may already have had a course on probability and random processes. For this reason as well as brevity, we relegate our review of probability and single and joint random variables to WP 10.1. That review begins at the very concept of probability and builds with both intuition and quantitative analysis the major quantities in random variable studies. We review there the meaning of probability, samples spaces, probability mass and density functions (PDF/PMF), cumulative distribution functions, expected value/mean, variance/standard deviation, characteristic functions, indicator functions, functions of random variables, Markov's and Chebyshev's inequalities, joint PDFs, correlation, and covariance.

The reviews in both this chapter and WP 10.1 should extend and deepen understanding of the rationales behind all the stochastic theory. They may also serve as a handy and concise reference of the results most commonly used as well as basic concepts and important derivations fundamental to stochastic signal processing. It is absolutely essential that the stochastic concepts be clearly understood if their application in DSP—and even the decision to use them—is to be correct. A thorough understanding of this chapter will make a more detailed graduate–level study of the subject more comprehensible.

To conclude the chapter and the book, we put the stochastic theory to work in discrete–time signal–processing systems. Although entire books are devoted to these applications, the space and scope of this book allow only a small sampling. We look at the effects of rounding in specific IIR digital filter structures, least–squares filter theory, deconvolution, n_0–step prediction (methods include least–squares prediction, trend regression and sinusoidal approximation; data sequences include ARMA sequences, the Dow Jones Industrial Average, and a small variety of other economic time series), data whitening, and adaptive filtering.

10.2   RANDOM VECTORS AND RANDOM PROCESSES

With the material on WP 10.1, we now have the essential tools with which to analyze individual random variables as well as pairs of random variables. In digital signal processing, we are often interested in entire sequences of random variables all

Sample space; i = sample index, x = variable value

$x = 5.0$	$x = 4.1$	$x = 5.0$	$x = -1.0$
•	•	•	•
$i = 1$	$i = 2$	$i = 3$	$i = 4$

$x = -1.0$	$x = 5.0$	$x = 3.7$	$x = -1.0$
•	•	•	•
$i = 5$	$i = 6$	$i = 7$	$i = 8$

$x = -1.0$	$x = 2.0$	$x = 4.1$	$x = 5.0$
•	•	•	•
$i = 9$	$i = 10$	$i = 11$	$i = 12$

Figure 10.1

generated by the same physical process. These sequences may be either finite–length or infinite–length, depending on the situation. In the former case they are arranged in vectors **X** and are called random vectors. In the latter case, they are written as sequences $X(n)$ and are called random processes.

Recall that the sample space of a univariate random variable X is the real line or complex plane, whereas that of a real bivariate random variable pair (X, Y) is the (x, y)–plane. Analogously, the N–length random vector **X** has an N–dimensional sample space and $X(n)$ has an infinite–dimensional space. Expectation, probability density functions, cumulative distribution functions, and characteristic functions can all be defined for **X** and $X(n)$ in ways analogous to those for the bivariate random variable pair.

To help clarify the meaning of a random process, we recall the meaning of sample space for a univariate random variable [for details, see the first section of WP 10.1]. In Figure 10.1, we show a simple example sample space with number of elements $N_i = 12$. With a finite number of possible values, X in this case should be recognized as a discrete random variable. (In practice, N_i = the number of realizations from which we may sample is always infinite, but the same ideas apply.)

Notice that the value of X, denoted x, is actually a function of the sample index i: $x = x(i)$; see Figure 10.2, which merely displays the information in Figure 10.1 in a linear graphical form. Thus a value of the random variable X is associated with each point i in the sample space. With each value of i equally likely to be selected, the PMF of X in Figure 10.1 has especially high peaks at 5.0 and –1.0 due to their occurring for more values of i; this PMF is shown in Figure 10.3.

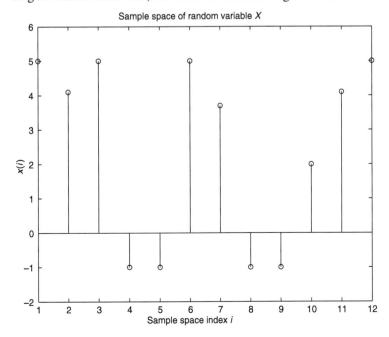

Figure 10.2

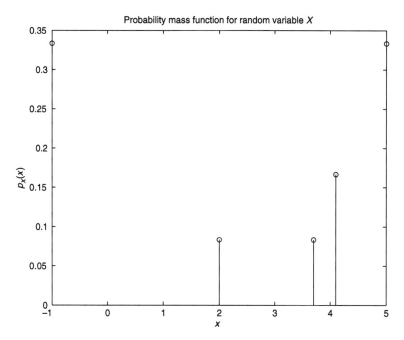

Figure 10.3

Similarly, random vectors **X** or random processes $X(n)$ can also be considered functions of the stochastic (sample space) index i: $\mathbf{X}(i)$ and $X(n, i)$, where we now let i be any integer value. The picture analogous to Figure 10.2, but which instead applies to a random vector, appears in Figure 10.4a and for a random process in Figure 10.4b (the data are speech samples). Now random selection of the value of index i produces a particular realization i of $\mathbf{X}(i)$ [denoted **x** or $\mathbf{x}(i)$] or of $X(n, i)$ [denoted $x(n)$ or $x(n, i)$]. For brevity, the index i is in practice usually suppressed. It should be noted that the most common use for random vectors is for the case in which the elements X_n, $n \in [1, N]$ of the random vector **X** comprise a length–N sample of a random process $X(n)$ (as is true for Figure 10.4a).

From Figure 10.4a and b, it is clear that for a given randomly selected value of i (e.g., $i = 3$ in Figure 10.4a and b), a sample vector **x** is selected (for random vectors) or an entire sample "time" waveform $x(n)$ is selected (for random processes). In Figure 10.4a and b, these are respectively labeled **x**(3) and $x(n, 3)$. This selection process is our way of modeling the random nature of experimentally obtained realizations of physical processes such as speech signals, echocardiograms, weather data, and so forth. It is nature, not we, that selects the value of i.

Alternatively, if n is selected (i.e., the nth element of the random vector or the time index n for the random process), we obtain a random variable as a function of the sample–space realization index i. In Figure 10.4, these are indicated with the horizontal dotted boxes and respectively labeled $x_6(i)$ ($n = 6$) in Figure 10.4a and $x(40, i)$ ($n = 40$) in Figure 10.4b. These random variables collectively have joint distributions analogous to $f_{XY}(x, y)$ for two random variables, as discussed on WP 10.1, but of dimension N for **X** and ∞ for $X(n)$.

The concept of correlation is a essential for all statistical studies. The degree of correlation between two things is, informally, how closely their behaviors are tied together. For example, is the direction of the deviation of one variable from its mean usually the same as that of the other, on the statistical average? If so, they have a large positive correlation. If the deviations of the two variables from their means typically have opposite signs, they have a large negative correlation. If they have no connection at all, they are called uncorrelated.

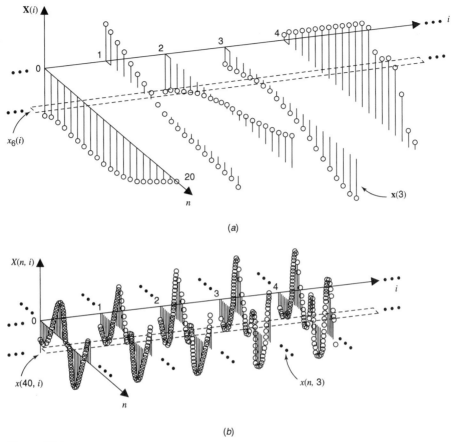

(a)

(b)

Figure 10.4

Covariances and correlations can also be calculated between any pair of random variables within the random vector or random process sequence, or between elements of different vectors or processes. For random vectors, these covariances make up an $N \times N$ matrix called the (auto– or cross–) covariance matrix. The cross–covariance matrix $\mathbf{C}_{XY}$ has nmth element $\text{cov}\{X_n, Y_m\} = \text{E}\{(X_n - \mu_{X,n})(Y_m - \mu_{Y,m})^*\}$ where $\text{E}\{\cdot\}$ is the expectation operator (see below). The matrix itself can be written as $\mathbf{C}_{XY} = \text{E}\{(\mathbf{X} - \boldsymbol{\mu}_X)(\mathbf{Y} - \boldsymbol{\mu}_Y)^{*T}\}$. Similarly, the cross–correlation matrix $\mathbf{R}_{XY}$ has nmth element $\text{E}\{X_n Y_m^*\}$. The autocovariance ($\mathbf{C}_{XX}$) and autocorrelation ($\mathbf{R}_{XX}$) matrices are obtained from above by replacing Y by X.

For random processes, the cross–covariance function [of $X(n)$ and $Y(n)$] is analogously defined as $c_{XY}(n, m) = \text{cov}\{X(n), Y(m)\} = \text{E}\{[X(n) - \mu_X(n)][Y(m) - \mu_Y(m)]^*\} = \text{E}\{[X(m + k) - \mu_X(m + k)][Y(m) - \mu_Y(m)]^*\}$, where $k \equiv n - m$ so $n = m + k$; the cross–correlation function is $r_{XY}(n, m) = \text{E}\{X(n)Y^*(m)\} = \text{E}\{X(m + k)Y^*(m)\}$; the autocovariance function is $c_{XX}(n, m) = \text{E}\{[X(n) - \mu_X(n)][X(m) - \mu_X(m)]^*\} = \text{E}\{[X(m + k) - \mu_X(m + k)][X(m) - \mu_X(m)]^*\}$; and the autocorrelation function is $r_{XX}(n, m) = \text{E}\{X(n)X^*(m)\} = \text{E}\{X(m + k)X^*(m)\}$.

As a practical example of a random vector, we could consider a random vector consisting of the hourly temperature over a single day. The length of this vector would be 24, and the index n represents the hour of the day. Each day could be represented by the index i, indexing from the beginning until the end of time. If we randomly select the date i for which the hourly temperatures are desired, then the analogy is perfect.

An example of a random process could be equispaced samples of an individual's speech over the telephone line. Different phone calls from the same person would bring forth different realizations of this same random process; each call (indexed by i) has such a long sequence of samples that it is more practical to think of it and treat it as a random process than as a random vector (even though the phone call is finite–duration).

Because of our lack of control over which value of i nature selects as well as our ignorance of the entire set of waveforms or vectors associated with each value of i, it is impractical to attempt to characterize the random process or vector by drawing complete graphs such as in Figure 10.4. Instead, we can finitely parameterize the vector or process by so–called moments, or averages of the stochastic entity over all values of i.

The average of, say, $X(n, i)$ over i is the expected value of the random variable $X(n)$, denoted $\mu_X(n) = E\{X(n)\}$. Recall that this stochastic average over i is merely a reordering of the terms to be summed in (W10.1.12) $[\mu_X = \int_{-\infty}^{\infty} xf_X(x)\,dx]$—which is ordered according to increasing x—to be written in the form of (W10.1.13) $[\mu_X = (1/N_i)\sum_{i=1}^{N_i} x_i]$—which is ordered according to increasing i. The former expression for μ_X is a probabilistically (and nonuniformly) weighted average of $X(n)$ over the possible range of values $X(n)$ may take on, whereas the latter is a uniformly weighted average of $X(n, i)$ values averaged over i.

In computing such averages by which to characterize the random process/random vector, two simplifying assumptions are frequently made: The random process/vector is (wide–sense) stationary (WSS) and it is ergodic. A process is WSS if

---

1. $\mu_X(n) = \mu_X(m) = \mu_X$ (10.1a)

Conditions for wide–sense stationarity

2. $r_{XX}(n, m) = E\{X(m + k)X^*(m)\} = r_{XX}(k)$, where $k = n - m$ (10.1b)

---

for all n, m. The first condition is that the mean is invariant with respect to "time" [i.e., to the value of index n in $X(n)$]. In other words, in averaging $X(n, i)$ over all realizations i, the result is the same for all values of n. The second condition analogously is that the autocorrelation function is independent of the absolute values of the indexes n and m in $r_{XX}(n, m) = E\{X(n)X^*(m)\}$, and depends only on the difference k between them.

A large proportion of stochastic signal processing is successfully based on the assumption of normal (Gaussian)/WSS random processes, which are completely characterized by first and second moments μ_X and $r_{XX}(k)$. Thus the condition of wide–sense stationarity not overly restrictive in practice. Also, non–WSS process are often locally WSS. Furthermore, the understanding of WSS processing is prerequisite to non–WSS processing, the latter of which is gaining in popularity. Finally, the overwhelming mathematical convenience of this pair of assumptions has caused many people to adopt it.

An example of a WSS sequence might be a sampled heartbeat pulse signal (electrocardiogram) of a healthy person at rest. Its characteristics are unlikely to drastically change from time to time; it is a rather monotonous sequence. Contrarily, speech signals are notoriously non–WSS, as was already demonstrated in Figures 7.4 and 7.5 in Example 7.2. Over a time duration of less than about 10 ms, drastic spectral (and temporal) changes occur, which violate the WSS assumptions.

Another convenience in stochastic signal processing is the assumption of ergodicity. How are we supposed to carry out these sums over i, when we have no control over the values of i that nature selects? Moreover, more often than not, we are given a single realization (only one value of i) with which to work. Under such conditions, numerical calculation of μ_X and $r_{XX}(k)$ is impossible, even assuming the process is WSS. The only recourse is to make the ergodic assumption: the assumption that averaging over n produces the same results as averaging over i. That is, we assume that temporal averages are equal to stochastic averages.

The ergodic assumption, which is for calculation of *averages*, is frequently more reasonable than it might first appear. A single natural phenomenon (e.g., an individual's voice) may be responsible for all possible realizations of the process $X(n)$ (that is, for all possible waveforms) as well as all values of $X(n)$ for a given realization. Thus the stochastic averages over all waveforms (realizations) may be similar to averages taken over a single realization. Consequently, in performing an average over i of $x(n_1, i)$ with n_1 fixed, we might well obtain a similar result if instead we perform an average over n of $x(n, i_1)$, where i_1 is the (only) realization available. The latter is certainly easier to do!

An example of an ergodic sequence might again be the electrocardiogram signal. The different realizations could be different days. The phases between days will be randomly distributed, so that averaging over i at fixed n will produce the same ballpark average as time–averaging over a single day.

Consider, however, the random vector discussed previously: the 24–hour hourly temperature vector. The "time" variation of this vector ranges from the cold of night to the heat of midday. That variation is quite unrelated to the variation of the temperature at 2 P.M. over each day for a month. In this case, the assumption of ergodicity would be inappropriate. The key is to compare the cause of variation with respect to "n" with that for variation with respect to "i." If they are similar, ergodicity may hold; if they are drastically different, it does not.

Some data illustrating these points is shown in Figures 10.5 through 10.7. Figures 10.5a and b show hourly temperatures (°F) in Cleveland, Ohio, from 1 October 1991 through 16 November 1991. One week's worth of data is plotted per graph (one day per curve), to avoid clutter. Figures 10.5a and b show, respectively, the first and last

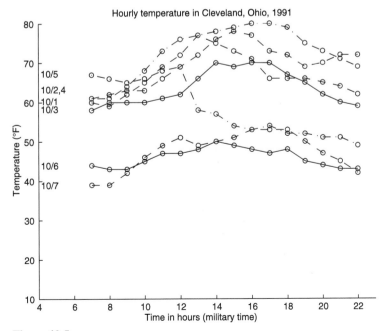

Figure 10.5a

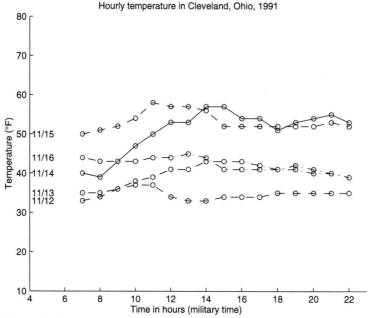

Figure 10.5*b*

weeks; for brevity, the five intermediate–week plots (also used in all averaging calculations that follow) are on WP 10.2. Only hours 7 A.M.. through 10 P.M. were recorded. This time of year is especially interesting because the temperatures are fairly rapidly changing from day to day. We may consider the data set for each as a single realization of a random process.

Figure 10.6 shows the average temperature for each day from 1 October 1991 through 16 November 1991 (including the days illustrated in Figure 10.5); we might call these "temporal averages." As we expect, when November comes there is a fall in daily average temperature (but notice the "heat wave" in mid–October). Figure 10.7 shows the average hourly temperature over all the "realizations" (days). For

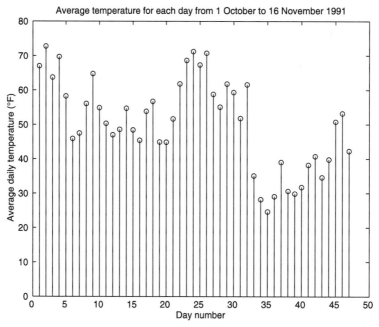

Figure 10.6

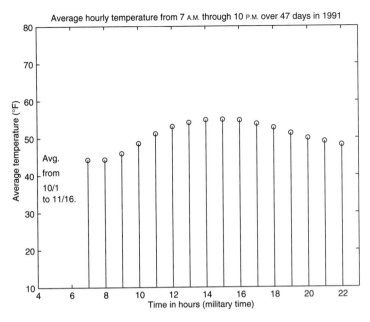

Figure 10.7

example, at 12 P.M. the average temperature over the 47 days was 52°F. Averaging the data these two ways clearly produces unrelated results, and we would not call the weather process ergodic.

Pictorial illustrations of stationarity and ergodicity for the mean appear in Figure 10.8a for speech data. In this figure, we make the additional practical assumption that all temporal averages will be over only finite intervals. This assumption gives meaning to the concept of *temporal stationarity* as it is commonly defined—performing a temporal average centered at one value of time (n) will yield nearly the same results

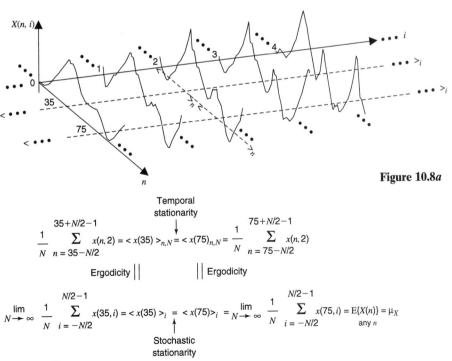

Figure 10.8a

Temporal
stationarity

$$\frac{1}{N} \sum_{n=35-N/2}^{35+N/2-1} x(n,2) = <x(35)>_{n,N} = <x(75)>_{n,N} = \frac{1}{N} \sum_{n=75-N/2}^{75+N/2-1} x(n,2)$$

Ergodicity $\|$ $\|$ Ergodicity

$$\lim_{N\to\infty} \frac{1}{N} \sum_{i=-N/2}^{N/2-1} x(35,i) = <x(35)>_i = <x(75)>_i = \lim_{N\to\infty} \frac{1}{N} \sum_{i=-N/2}^{N/2-1} x(75,i) = E\{X(n)\} = \mu_X$$
any n

Stochastic
stationarity

The symbol $\|$ means equality of quantities above and below it.
Words next to equality symbols are assumptions under which the equalities hold.

Figure 10.8b

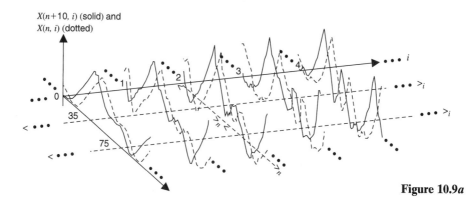

X(n+10, i) (solid) and
X(n, i) (dotted)

Figure 10.9a

Temporal
stationarity
↓

$$\frac{1}{N}\sum_{n=35-N/2}^{35+N/2-1} x(n+10,2)\,x^*(n,2) = <x(35+10)\,x^*(35)>_{n,N} = <x(75+10)\,x^*(75)>_{n,N} = \frac{1}{N}\sum_{n=75-N/2}^{75+N/2-1} x(n+10,2)\,x^*(n,2)$$

Ergodicity ‖ ‖ Ergodicity

$$\lim_{N\to\infty}\frac{1}{N}\sum_{i=-N/2}^{N/2-1} x(35+10,i)\,x^*(35,i) = <x(35+10)\,x^*(35)>_i = <x(75+10)\,x^*(75)>_i = \lim_{N\to\infty}\frac{1}{N}\sum_{i=-N/2}^{N/2-1} x(75+10,i)\,x^*(75,i)$$

Stochastic
stationarity
↑

$$= E\{X(n+10)X^*(n)\} = r_{XX}(10)$$
for any n.

The symbol ‖ means equality of quantities above and below it.
Words next to equality symbols are assumptions under which the equalities hold.

Figure 10.9b

as a temporal average of the same realization centered at another time (m). Temporal averages are denoted $\langle\ \rangle_n$ and stochastic averages are denoted $\langle\ \rangle_i$. For clarity, the discrete–time (n) values (for fixed i) have been connected with curves. The conditions for stationarity and ergodicity are given in Figure 10.8b.

Ideas similar to those in Figure 10.8 apply for correlation/covariance; see Figure 10.9, made from the same speech database as was Figure 10.8. Calculations in Figure 10.9a are of $r_{XX}(k)$ for $k = 10$; in Figure 10.9a we see why k is called the *lag*: For a given realization i, $x(n, i)$ lags behind $x(n + 10, i)$ by $k = 10$. The conditions for stationarity and ergodicity are given in Figure 10.9b.

Physically, autocorrelation quantifies how regular a random process $X(n)$ is. The more correlated that distant values are to each other [the more "regular" $X(n)$ is], the "longer" the significantly nonzero portion of $r_{XX}(k)$ will be. Two extremes are white noise and the sinusoid. Each value of a WSS white noise sequence $U(n)$ is completely independent of all others. Therefore, $r_{UU}(k) = E\{U(n + k)U^*(n)\} =$
$$\begin{cases}E\{U(n+k)\}E^*\{U(n)\} \\ E\{U(n)U^*(n)\}\end{cases} = \begin{cases}|\mu_U|^2, & k\neq 0 \\ E\{|U(n)|^2\}, & k=0.\end{cases}$$ If $\mu_U = 0$, which is often applicable in signal modeling without loss of generality, then we have $r_{UU}(k) = \delta(k)\cdot E\{|U(n)|^2\} = \delta(k)\cdot r_{UU}(0)$, where $r_{UU}(0) = E\{|U(n)|^2\}$. Contrarily, a sinusoid is perfectly regular, and a period infinitely far into the future is completely correlated with the present period. It will be proved in the problems that if $X(n) = A\cos(\omega_0 n + \theta)$, where θ is uniformly distributed on $(-\pi, \pi)$, then $r_{XX}(k) = \frac{1}{2}A^2\cos(\omega_0 k)$, which does not diminish in magnitude even as $|k| \to \infty$.[1]

As noted, figures 10.8 and 10.9 were created using actual speech data ($F = 11.025$ kHz), from different sections of the same recording. In practice, there may be a huge

---

[1]It is also shown in the problems that for the deterministic sequence $x(n) = A\cos(\omega_0 n + \theta)$ (i.e., θ is deterministic), then $r_{xx; n}(k) = \frac{1}{2}A^2\cos(\omega_0 k)$, where $r_{xx; n}(k)$ is as given in (7.53), and where the subscript notation is defined following (10.76b) below.

variation between realizations—for example, different people speaking over a communications link. Even over a single realization, voice characteristics are uniform (voice is WSS) over only periods of 10 ms or less. This stationarity over limited periods is called "local wide–sense stationarity," where temporal, not stochastic invariance is implied.

In Figure 10.9a, for example, time averages taken over finite segments centered on time indexes $n = n_1$ and $n = n_2$ separated less than 10 ms/Δt = 0.01/(1/11025) = 110 samples will not depend on the value of the segment center (n_1 or n_2). Such stationarity might seem reasonable when we can visually picture such time–averaging as in Figure 10.9a, if the averaging window is, for example, greater than 1000 (not the small intervals shown for convenience in Figure 10.9a). It is harder to estimate WSS duration when the averages are taken over all realizations, with only one term for each realization contributing to the average: Even adjacent realizations may have little in common. This aspect may or may not limit the duration of validity of stochastic wide–sense stationarity. Ergodicity is seemingly even less tenable: The seemingly unrelated temporal and stochastic averages are assumed to be equal.

EXAMPLE 10.1

Compare temporal and stochastic approximations of $r_{XX}(k)$ using speech data. Is the assumption of ergodicity valid?

Solution

Several sentences of speech, all spoken by the author in one recording, were broken into N_i = 512 sections each of duration N =512 samples. Thus each section might be considered a separate speech realization. The data might be termed *favorable* because these realizations are actually contiguous speech and so might be considered unrealistically related to the other realizations. Nevertheless, all data in each "realization" is distinct, coming from different portions of the spoken nonrepetitive paragraph (from this book).

In Figure 10.10, we show in the solid curve the estimate of autocorrelation function $r_{XX}(k)$ using a simple time average $\dfrac{1}{N}\displaystyle\sum_{n=0}^{N-1-k} x(n+k,i)x(n,i)$ with i fixed at the value 175 [this formula, (10.10) for real data and $k \geq 0$, is fully discussed there].

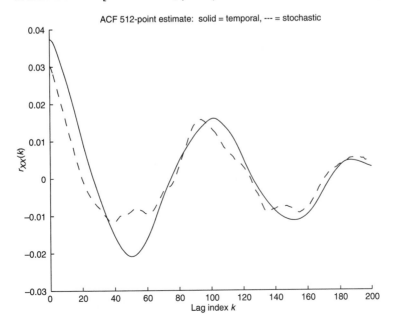

Figure 10.10

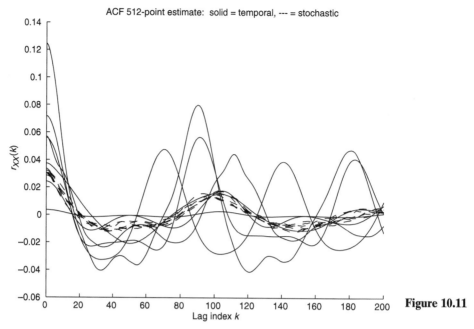

Figure 10.11

The dashed curve is a similarly computed average but taken over i, not n: $\frac{1}{N}\sum_{i=0}^{N_i} x(n+k,i)x(n,i)$ with n fixed at the value 100. The amazing agreement would seem to justify the assumption of ergodicity.

However, in Figure 10.11, we overlay several computations for different values of the fixed i (solid) or n (dashed). We find that depending on the realization, the temporal autocorrelation average (solid) varies widely. Contrastingly, for a wide range of n values selected, the stochastic average over all 512 realizations is reliably almost the same; all the dashed curves in Figure 10.11 fall nearly on top of each other. This suggests that the stochastically averaged autocorrelation, were it available to be computed, is more reliable than the temporal autocorrelation, because it involves a cross section of data over a very broadly spanning set of data ("realizations"). Contrarily, the entire temporal autocorrelation estimate is formed from one 512–point contiguous set of data from one "realization," which will change radically from one realization to the next.

Unfortunately, we usually do not have the luxury of performing the "stochastic" average over realizations, because we often have only one time–realization with which to work. We must then perform the temporal autocorrelation estimate and hope that it is representative. That is, we hope it is the one shown in Figure 10.10; we *hope* ergodicity holds.

As we saw in Example 7.13, cross–correlation of similar but differing sequences can also be revealing. Consider the further example of ultrasonic blood flow measurements. Echo sequences $X(n)$ and $Y(n)$ are obtained from two sites a known distance apart. During the time that the blood has moved down the artery, its scatterers have slightly changed orientations relative to each other. This, as well as the corruption by a different noise pattern, causes the scattering pattern received at the transducer to change. However, most of the original scattering patterns are preserved in the new echo sequence. Thus the downstream echo sequence $X(n)$ is a delayed and distorted version of the upstream signal $Y(n)$. By numerically determining the value of k that maximizes $<X(n+k)Y(n)>_n$, the time it takes the blood to move the known distance, and thus the blood flow rate, can be inferred.

Let us now show a few important properties of the autocorrelation sequence, $r_{XX}(k)$. First, under the WSS assumption [for which $r_{XX}(k)$ is defined],

$$r_{XX}(-k) = \mathrm{E}\{X(n-k)X^*(n)\} \tag{10.2a}$$

$$= [\mathrm{E}\{X^*(n-k)X(n)\}]^* \tag{10.2b}$$

$$= [\mathrm{E}\{X(m+k)X^*(m)\}]^* \quad \text{(where } m = n - k\text{)} \tag{10.2c}$$

or

$$r_{XX}(-k) = r_{XX}{}^*(k), \tag{10.2d}$$

so $r_{XX}(k)$ is a conjugate–even (Hermitian) sequence and thus has a real–valued DTFT (important in Section 10.3).

Next we show that Schwarz's inequality (W10.1.27) $[\,|\mathrm{E}\{X^*Y\}|^2 \le \mathrm{E}\{|X|^2\}\mathrm{E}\{|Y|^2\}]$ can be used to prove that $r_{XX}(0) \ge |r_{XX}(k)|$ for all k. In (W10.1.27), let X be $X^*(n+k)$ and Y be $X^*(n)$:

$$|r_{XX}(k)|^2 = |\mathrm{E}\{X(n+k)X^*(n)\}|^2 \tag{10.3a}$$

$$\le \mathrm{E}\{|X(n+k)|^2\}\mathrm{E}\{|X(n)|^2\}. \tag{10.3b}$$

By using the assumption that $X(n)$ is WSS so that $\mathrm{E}\{|X(n+k)|^2\}$ can be replaced by $\mathrm{E}\{|X(n)|^2\}$, (10.3b) reduces to

$$|r_{XX}(k)|^2 \le \mathrm{E}^2\{|X(n)|^2\} \tag{10.3c}$$

$$= r_{XX}{}^2(0). \tag{10.3d}$$

Because $r_{XX}(-k) = r_{XX}{}^*(k)$, from $k = 0$ we know that $r_{XX}(0)$ is real–valued. Therefore, we have the result

$$r_{XX}(0) \ge |r_{XX}(k)| \text{ for all } k. \tag{10.3e}$$

We will show in Section 10.3 that $r_{XX}(0)$ is the expected time–average power in the random process $X(n)$ [the time–average aspect is not obvious from $r_{XX}(0) = \mathrm{E}\{|X(n)|^2\}$]. If $\mu_X = 0$, we have $\sigma_X{}^2 = \mathrm{E}\{X(n)X^*(n)\} = \mathrm{E}\{|X(n)|^2\} = r_{XX}(0)$; that is, the variance of $X(n)$ is equal to the expected time–average power in $X(n)$ and also the autocorrelation function (ACF) evaluated at $k = 0$.

Being the stochastic inner product of two random process sequences (via $\mathrm{E}\{\cdot\}$), the ACF is nothing other than the stochastic inner product of the "vectors" whose respective ith "components" are $X(n+k, i)$ and $X^*(n, i)$. This stochastic inner product, $r_{XX}(k)$, is equal to the product of the magnitudes of these (infinite–dimensional) "vectors" $|X(n+k)| = \{\mathrm{E}\{|X(n+k)|^2\}^{1/2}$ and $|X(n)| = \{\mathrm{E}\{|X(n)|^2\}^{1/2}$ times the cosine of the angle between these "vectors." Because the magnitude of the cosine is always less than 1, $|r_{XX}(k)| \le r_{XX}(0)$ $[= \mathrm{E}\{|X(n)|^2\}]$, which is an intuitive proof of (10.3e). Thus, when looking at plots of ACFs, the maximum will always be at $k = 0$.

Now consider a second, related property of $r_{XX}(k)$ that typically holds for practical random processes. Except for sequences such as sinusoids, it is frequently the case that for extremely large k, $X(n) - \mu_X$ and $X(n+k) - \mu_X$ are nearly independent. In such cases, $r_{XX}(k) = \mathrm{E}\{X(n+k)X^*(n)\} \to \mathrm{E}\{X(n+k)\}\mathrm{E}\{X^*(n)\} = |\mu_X|^2$ as $k \to \infty$; if $\mu_X = 0$ (again, often possible to assume without loss of generality), then $r_{XX}(k) \to 0$ as $k \to \infty$, a fact we will use to derive the power spectral density function.

It is fitting to close this section with Table 10.1, which shows the definitions of various correlations.

Table 10.1   Correlation

Random Quantity	Self–Correlation	Cross–Correlation
Random Variables	Second moment: $E\{|X|^2\}$	Mixed moment: $E\{XY^*\}$
	Variance: $\sigma_X^2 = E\{|X - \mu_X|^2\}$	Correlation coefficient: $$\rho_{XY} = \frac{E\{(X - \mu_X)(Y - \mu_Y)^*\}}{\sigma_X \sigma_Y}$$
Notes:	X deterministic if $\sigma_X = 0$.	$|\rho_{XY}| \le 1$; Iff X and Y are uncorrelated, then $\rho_{XY} = 0$.

Random Vectors

Autocorrelation matrix:

$$\mathbf{R}_{XX} = \begin{bmatrix} E\{|X_1|^2\} & \cdots & E\{X_1 X_N^*\} \\ \vdots & & \vdots \\ E\{X_N X_1^*\} & \cdots & E\{|X_N|^2\} \end{bmatrix}$$
$$= \mathbf{R}_{XX}^{T*}$$

Cross–correlation matrix:

$$\mathbf{R}_{XY} = \begin{bmatrix} E\{X_1 Y_1^*\} & \cdots & E\{X_1 Y_N^*\} \\ \vdots & & \vdots \\ E\{X_N Y_1^*\} & \cdots & E\{X_N Y_N^*\} \end{bmatrix}$$
$$= \mathbf{R}_{YX}^{T*}$$

Autocovariance matrix ($\mu_{X,\ell} = E\{X_\ell\}$):

$$\mathbf{C}_{XX} = \begin{bmatrix} E\{|X_1 - \mu_{X,1}|^2\} & \cdots & E\{(X_1 - \mu_{X,1})(X_N - \mu_{X,N})^*\} \\ \vdots & & \vdots \\ E\{(X_N - \mu_{X,N})(X_1 - \mu_{X,1})^*\} & \cdots & E\{|X_N - \mu_{X,N}|^2\} \end{bmatrix}$$
$$= \mathbf{C}_{XX}^{T*}$$

Cross–covariance matrix:

$$\mathbf{C}_{XY} = \begin{bmatrix} E\{(X_1 - \mu_{X,1})(Y_1 - \mu_{Y,1})^*\} & \cdots & E\{(X_1 - \mu_{X,1})(Y_N - \mu_{Y,N})^*\} \\ \vdots & & \vdots \\ E\{(X_N - \mu_{X,N})(Y_1 - \mu_{Y,1})^*\} & \cdots & E\{(X_N - \mu_{X,N})(Y_N - \mu_{Y,N})^*\} \end{bmatrix}$$
$$= \mathbf{C}_{YX}^{T*}$$

Random Quantity	Self–Correlation	Cross–Correlation
Notes:	If $\mathbf{X}$ is white, then $\mathbf{C}_{XX} = \text{diag}\{\sigma_{X,1}^2 \ldots \sigma_{X,N}^2\}$.	Iff X and Y are uncorrelated, then $\mathbf{C}_{XY} = [0]$.

WSS Random Process

Random Quantity	Self–Correlation	Cross–Correlation
	Autocorrelation function: $r_{XX}(k) = E\{X(n + k)X^*(n)\}$ $= r_{XX}^*(-k)$	Cross–correlation function: $r_{XY}(k) = E\{X(n + k)Y^*(n)\}$ $= r_{YX}^*(-k)$
	Autocovariance function: $c_{XX}(k) = E\{(X(n + k) - \mu_X)(X(n) - \mu_X)^*\}$ $= c_{XX}^*(-k)$ $= r_{XX}(k) - |\mu_X|^2$	Cross–covariance function: $c_{XY}(k) = E\{(X(n + k) - \mu_X)(Y(n) - \mu_Y)^*\}$ $= c_{YX}^*(-k)$ $= r_{XY}(k) - \mu_X \mu_Y^*$
Notes:	If $X(n)$ is white, then $r_{XX}(k) = \sigma_X^2 \delta(k)$.	$X(n)$ and $Y(n)$ are uncorrelated iff $c_{XY}(k) = 0$ for all k.

10.3   POWER SPECTRAL DENSITY AND THE PERIODOGRAM

Throughout our study of digital signal processing, the frequency domain has played an important role in the design and analysis of signals and systems. The fact that our current studies involve signals that are random and may have unpredictable "noise" components does not diminish the importance of spectral analysis.

However, in the previous section, it was found that stationarity is an essential assumption to make in elementary stochastic analysis. Without it, for example, neither the single–parameter (k)–dependent $r_{XX}(k)$ nor $c_{XY}(k)$ are defined, and a different μ_X would have to be defined separately for every instant of time. Wide–sense stationarity means that, at least to second order, a random process signal looks much the same whether we look at it today, yesterday, tomorrow, or 10 years from now. Mathematically, this stationarity continues on to infinity.

A stationary sequence $X(n)$ presents a problem for frequency domain analysis: $X(n)$ generally does not possess a Fourier transform, because it is infinite–energy! Typically, $X(n)$ is in the same category as a sinusoid: It possesses finite power, but not finite energy. Recall that the Fourier transform of a sinusoid does not exist either, in the usual sense—it involves Dirac delta functions. Random processes are even more unmanageable, because their infinite energy is spread out all over the spectrum. Thus it is better to examine how the finite power of random processes is distributed over the spectrum in analogy to how the finite energy of finite–energy signals is spread over the spectrum.

The magnitude squared of the Fourier transform is an energy spectral density. Recall from Section 5.9 that if $x(n)$ has DTFT $X(e^{j\omega})$, then $|X(e^{j\omega})|^2/\pi$ is the energy spectral density of $x(n)$ at frequency ω, and $|X(e^{j\omega})|^2 \, d\omega/\pi$ is the energy of $x(n)$ over the frequency range $(\omega - d\omega/2, \omega + d\omega/2)$ as $d\omega \to 0$. Similarly, we will define the power spectral density (PSD) $P_{XX}(\omega)$ so that $P_{XX}(\omega) \, d\omega/\pi$ is the power of $X(n)$ over the same frequency range.

Let $x_i(n)$ now denote realization i of random process $X(n)$. Then we know that its energy $\sum_{n=-\infty}^{\infty} |x_i(n)|^2 = \infty$, but assuming that $X(n)$ is finite–power, we have the time–average power $<p(n)> = \dfrac{1}{N} \sum_{n=-N/2+1}^{N/2} |x_i(n)|^2 < \infty$, even as $N \to \infty$. Here $<p(n)>$ is analogous to the continuous–time case where $<p_c(t)> = \dfrac{1}{T} \int_{-T/2}^{T/2} p_c(t)dt = \dfrac{1}{T} \int_{-T/2}^{T/2} [dw_c(t)/dt]dt = \dfrac{1}{T} [w_c(T/2) - w_c(-T/2)] = \dfrac{1}{T} \int_{-T/2}^{T/2} |x_c(t)|^2 dt$ is the time–average power in $x_c(t)$ over $t \in [-T/2, T/2]$ and $w_c(t)$ is the energy in $x_c(t)$ accumulated from $-\infty$ to t. For convenience, let $x_{N,i}(n) = x_i(n)$ for $n \in [0, N-1]$ (rather than $-N/2 + 1$ to $N/2$) and zero otherwise, for the single realization i of $X(n)$, $x_i(n)$. Then $X_{N,i}(e^{j\omega})$ exists, and from Section 5.9 we know that $|X_{N,i}(e^{j\omega})|^2 \, d\omega/\pi$ represents the energy of $x_{N,i}(n)$ over the frequency range $(\omega - d\omega/2, \omega + d\omega/2)$; thus from above, $|X_{N,i}(e^{j\omega})|^2 \, d\omega/[N\pi]$ is the time–averaged power on that range.

We can extend this principle to the infinite–energy realization $x_i(n)$ of the random process $X(n)$ by simply letting $N \to \infty$:

$$P_{XX,i}(\omega) \, d\omega/\pi = \lim_{N\to\infty} |X_{N,i}(e^{j\omega})|^2 d\omega/[\pi N] \tag{10.4}$$

is the time–averaged power on $(\omega - d\omega/2, \omega + d\omega/2)$ of the *single realization $x_i(n)$* $= x(n, i)$.

However, every realization $x_i(n)$ of $X(n)$ will be different. For the purpose of characterizing the random process $X(n)$ itself, the statistical average should be taken over all possible realizations. Not doing so would be the same error as taking a single realization x_i of a random variable X to be the expected value of X, when in fact the expected value is the average of x_i over the sample–space index i. Equivalently, the expected value is a probabilistically weighted average of X over the continuous range of possible of values of X—that is, the range of x_i. The PSD is still formed by taking the temporal average power, but the temporal average power in turn is averaged over all realizations and becomes an *expected* (stochastically averaged) PSD rather than a single–realization–based PSD.

Thus to obtain the PSD from (10.4), we first divide by $d\omega/\pi$ to obtain a power *density* at ω (in power per cycle/sample) and then take the expected value of the result:

$$P_{XX}(\omega) = \lim_{N\to\infty} \frac{\mathrm{E}\{|X_N(e^{j\omega})|^2\}}{N}. \tag{10.5}$$

Notice that now we have replaced $X_{N,i}(e^{j\omega}) = \mathrm{DTFT}\{x_i(n)\}$ by $X_N(e^{j\omega})$, the DTFT of a truncated version $X_N(n)$ of the abstract random process $X(n)$. Because $X_{N,i}(e^{j\omega})$ is just a deterministic function of ω based on data from a single realization, its expected value is itself, so taking the expected value would not perform an average. Hence we must replace $X_{N,i}(e^{j\omega})$ by $X_N(e^{j\omega})$.

Now, an amazing relationship called the Wiener–Kinchin theorem shall be proved: $P_{XX}(\omega) = \mathrm{DTFT}\{r_{XX}(k)\}$. This identity will take a bit of work to show. First, let us examine $|X_N(e^{j\omega})|^2/N$:

$$\frac{|X_N(e^{j\omega})|^2}{N} = \frac{1}{N}\{\sum_{n=-\infty}^{\infty} X_N(n)\,e^{-j\omega n}\}^*\{\sum_{m=-\infty}^{\infty} X_N(m)e^{-j\omega m}\}, \tag{10.6}$$

where because $X_N(n)$ is zero outside $[0, N-1]$ no error is introduced by running the sums from $-\infty$ to ∞, and it is convenient to do so. Proceeding, we obtain by rearranging the sums

$$\frac{|X_N(e^{j\omega})|^2}{N} = \frac{1}{N}\sum_{n=-\infty}^{\infty}\sum_{m=-\infty}^{\infty} X_N^*(n)X_N(m)e^{-j\omega(m-n)}. \tag{10.7a}$$

Defining $k = m - n$, we have

$$\frac{|X_N(e^{j\omega})|^2}{N} = \frac{1}{N}\sum_{k=-\infty}^{\infty}\sum_{n=-\infty}^{\infty} X_N^*(n)X_N(n+k)\,e^{-j\omega k}. \tag{10.7b}$$

We now write

$$\frac{|X_N(e^{j\omega})|^2}{N} = \sum_{k=-\infty}^{\infty} r_{N,n}(k)e^{-j\omega k} = \mathrm{DTFT}\{r_{N,n}(k)\}, \tag{10.7c}$$

where we introduce the temporal (n –averaged) ACF $r_{N,n}(k)$ defined by[2]

$$r_{N,n}(k) = \frac{1}{N}\sum_{n=-\infty}^{\infty} X_N(n+k)X_N^*(n) \tag{10.8a}$$

$$= \frac{1}{N}\sum_{n=0}^{N-1} X^*(n)X_N(n+k), \tag{10.8b}$$

where in (10.8b) we recall that $X_N(n) = X(n)$ for $n \in [0, N-1]$ and zero otherwise, so that by reducing the limits on n to $[0, N-1]$, the subscript N can be removed from the first factor in (10.8b), $X_N^*(n)$.

Now consider the second factor in (10.8b), $X_N(n + k)$. We must have $0 \le n + k \le N - 1$ for $X_N(n + k)$ to be possibly nonzero, and we also know that $0 \le n \le N - 1$. Together, these imply that $|k| \le N - 1$ for $r_{N,n}(k)$ to be possibly nonzero. However, even with $|k| \le N - 1$, there will still be terms for which $n + k$ is not on

---

[2] The subscript N,n on $r_{N,n}(k)$ means that it is based on the finite–length $X_N(n)$ and that the average is temporal n rather than stochastic i, even though $X_N(n)$ is itself a truncated abstract random process, not a realization. Note that (10.7c) is a deterministic version of the Wiener–Kinchin theorem. In practical calculations, we can replace $X(n)$ in all expressions for $r_{N,n}(k)$ by a measured realization of $X(n)$, for example, $x(n)$.

$[0, N - 1]$ unless we further reduce the range on n as follows. For $k \geq 0$, we simply reduce the upper limit on n from $N - 1$ to $N - 1 - k$. For $k < 0$, we increase the lower limit on n from 0 to $-k$. By eliminating these terms, we can remove the subscript N from $X_N(n + k)$ in (10.8b). These index limit changes also allow us to compute ACF estimates under the ergodic assumption from a single length–N array without having the array index ever go out of bounds. Thus

$$r_{N,n}(k) = \frac{1}{N} \begin{cases} \displaystyle\sum_{n=0}^{N-1-k} X^*(n)X(n + k), & N - 1 \geq k \geq 0 \\ \displaystyle\sum_{n=-k}^{N-1} X^*(n)X(n + k), & -(N - 1) \leq k < 0, \end{cases} \tag{10.9a}$$

where for brevity we will usually omit repeating "and is zero for $|k| \geq N$." When we define $m = n + k$ in the lower sum, (10.9a) becomes

$$r_{N,n}(k) = \frac{1}{N} \begin{cases} \displaystyle\sum_{n=0}^{N-1-k} X^*(n)X(n + k), & N - 1 \geq k \geq 0 \\ \displaystyle\sum_{m=0}^{N-1-(-k)} X^*(m - k)X(m), & -(N - 1) \leq k < 0. \end{cases} \tag{10.9b}$$

Noting that for $k < 0$, $(-k) = |k|$, we see that both expressions have the same limits: $[0, N - 1 - |k|]$. The only difference is that the result for $k < 0$ is the complex conjugate of that for $k \geq 0$. Finally, we obtain

$$r_{N,n}(k) = \begin{cases} \dfrac{1}{N} \displaystyle\sum_{n=0}^{N-1-|k|} X(n + k)\, X^*(n), & N - 1 \geq k \geq 0 \tag{10.10a} \\[2mm] r_{N,n}^*(-k), & -(N - 1) \leq k < 0. \tag{10.10b} \end{cases}$$

The next step towards proving the Wiener–Kinchin theorem is to take the expected value of (10.7c), with (10.10) eventually substituted for $r_{N,n}(k)$:

$$E\{|X_N(e^{j\omega})|^2/N\} = E\left\{ \sum_{k=-\infty}^{\infty} r_{N,n}(k)e^{-j\omega k} \right\} \tag{10.11a}$$

$$= \sum_{k=-\infty}^{\infty} E\{r_{N,n}(k)\}e^{-j\omega k}, \tag{10.11b}$$

where (10.11b) holds because $E\{\cdot\}$ is a linear integration/sum operation, so that $E\{\sum\} = \sum(E\{\cdot\})$ [see (W10.1.33d): $E\{X + Y\} = E\{X\} + E\{Y\}$], and because the only stochastic quantities reside in $r_{N,n}(k)$. The expected value of (10.10) is

$$E\{r_{N,n}(k)\} = \begin{cases} \dfrac{1}{N} \displaystyle\sum_{n=0}^{N-1-|k|} E\{X(n + k)X^*(n)\}, & N - 1 \geq k \geq 0 \\[2mm] [E\{r_{N,n}(-k)\}]^*, & -(N - 1) \leq k < 0. \end{cases} \tag{10.12a}$$

Note that $E\{X(n + k)X^*(n)\} = r_{XX}(k)$ is independent of n (WSS) so that it can be pulled out of the sum. Fundamentally, stationarity here means that $E\{X(n + k)X^*(n)\}$ is equal to its time average; it is valid only because $X(n)$ rather than a realization of $X(n)$ is used in (10.12a). Thus, (10.12a) becomes

$$E\{r_{N,n}(k)\} = \begin{cases} \dfrac{1}{N} r_{XX}(k) \displaystyle\sum_{n=0}^{N-1-|k|} 1, & N - 1 \geq k \geq 0 \\[2mm] \left[\dfrac{1}{N} r_{XX}(-k) \displaystyle\sum_{n=0}^{N-1-|k|} 1 \right]^*, & -(N - 1) \leq k < 0. \end{cases} \tag{10.12b}$$

Evaluating the sum of ones gives

$$E\{r_{N,n}(k)\} = \left(1 - \frac{|k|}{N}\right) \cdot \begin{cases} r_{XX}(k), & N-1 \geq k \geq 0 \\ r_{XX}^*(-k), & -(N-1) \leq k < 0. \end{cases} \tag{10.12c}$$

Now, because both $r_{N,n}(k)$ and $r_{XX}(k)$ *already* satisfy conjugate symmetry,

$$E\{r_{N,n}(k)\} = \left(1 - \frac{|k|}{N}\right) r_{XX}(k), \quad -(N-1) \leq k \leq N-1, \tag{10.12d}$$

and is equal to zero for $|k| \geq N$. We have from (10.5)

$$P_{XX}(\omega) = \lim_{N \to \infty} \frac{E\{|X_N(e^{j\omega})|^2\}}{N}, \tag{10.13a}$$

which with (10.11b) and (10.12d) becomes

$$P_{XX}(\omega) = \lim_{N \to \infty} \sum_{k=-(N-1)}^{N-1} \left(1 - \frac{|k|}{N}\right) r_{XX}(k)\, e^{-j\omega k}. \tag{10.13b}$$

Finally, assuming that $\lim_{k \to \infty} r_{XX}(k) \to 0$ as is true for most finite–power signals (except sinusoids, where distributional interpretations of steps in the derivation will confirm the same following result), the $-|k|/N$ term can be removed for $N \to \infty$:

$$\boxed{P_{XX}(\omega) = \sum_{k=-\infty}^{\infty} r_{XX}(k)e^{-j\omega k} = \text{DTFT}\{r_{XX}(k)\}, \quad \begin{array}{l} \text{Wiener–Kinchin theorem:} \\ \text{PSD = DTFT\{ACF\}.} \end{array}} \tag{10.13c}$$

which is what we sought to prove. By the uniqueness of the DTFT, we also have

$$r_{XX}(k) = \frac{1}{2\pi} \int_{-\pi}^{\pi} P_{XX}(\omega)\, e^{j\omega k}\, d\omega. \tag{10.14a}$$

Physical arguments were made in deriving $P_{XX}(\omega)$ to indicate that it (divided by π) is the spectral density of expected time–average power in the random process $X(n)$ at frequency ω. Therefore, the expected total time–average power in $X(n)$ is the integral of $P_{XX}(\omega)$ over all frequencies $[\pi, \pi]$. By (10.14a) evaluated at $k = 0$, this integral is none other than $r_{XX}(0)$, as hinted after (10.3e). [$P_{XX}(\omega)/\pi$ includes positive- and negative-frequency components, so for integration from $-\pi$ to π in (10.14a) we must divide $P_{XX}(\omega)$ by 2π.] By setting $k = 0$ in (10.14a), we have

$$r_{XX}(0) = \frac{1}{2\pi} \int_{-\pi}^{\pi} P_{XX}(\omega)\, d\omega = \text{expected time–average power in } X(n). \tag{10.14b}$$

This double averaging means the stochastic average of all the temporal average powers $<p(n, i)>$ of each of the realizations $x(n, i)$.

The Wiener–Kinchin theorem (10.13c), which says that $P_{XX}(\omega)$ is the DTFT of $r_{XX}(k)$, should seem reasonable intuitively. After all, $r_{XX}(k)$ is a second–order moment of $X(n)$ [$r_{XX}(k) = E\{X(n + k)X^*(n)\}$], and its DTFT, the power spectral density $P_{XX}(\omega)$, is also second–order in $X(n)$ [$P_{XX}(\omega) = \lim_{N \to \infty} E\{|X_N(e^{j\omega})|^2\}/N$]. If $X(n)$ has periodicities, these will also be evident in $r_{XX}(k)$ with the same periods, because when $X(n)$ is shifted relative to itself by any integral number of those periods, there is a lining up that boosts $r_{XX}(k) = E\{X(n + k)X^*(n)\}$. Thus the spectral information of $X(n)$ is also present in $r_{XX}(k)$, but $r_{XX}(k)$ is finite–energy and has a DTFT, through which we can study the spectral properties of $X(n)$.

EXAMPLE 10.2

Find the PSD of real–valued WSS so–called white noise $U(n)$, for which each sample is stochastically independent of all the others. First, let the mean of the noise be μ_U, and then set $\mu_U = 0$.

Solution

We extend here our previous results on white noise [see the discussion three paragraphs before Example 10.1, namely $r_{UU}(k)$ for real $U(n)$ is

$$r_{UU}(k) = \begin{cases} \mu_U^2, & k \neq 0 \\ E\{U^2(n)\}, & k = 0. \end{cases}]$$

First, recall that the variance of $U(n)$ is $\sigma_U^2 = E\{(U(n) - \mu_U)^2\} = E\{U^2(n)\} - 2\mu_U^2 + \mu_U^2$, or $E\{U^2(n)\} = \sigma_U^2 + \mu_U^2$. Thus we have

$$r_{UU}(k) = \begin{cases} \mu_U^2, & k \neq 0 \\ \sigma_U^2 + \mu_U^2, & k = 0 \end{cases} = \mu_U^2 + \sigma_U^2\delta(k).$$

Consequently, the PSD is $P_{UU}(\omega) = \text{DTFT}\{r_{UU}(k)\} = \mu_U^2\delta(\omega) + \sigma_U^2$. For zero–mean white noise this reduces to $P_{UU}(\omega) = \sigma_U^2$, for all ω.

Perfect flatness of the PSD over all ω means that all frequencies are equally present. As this is the case with white light, where the frequencies represent different colors, such noise is given the name "white" noise. It is called noise because it is characteristic of measurement noise, not of signals. Notice that a nonzero mean merely adds an impulse at zero frequency to $P_{UU}(\omega)$. This dc term is just what we would predict for a constant μ_U added to a zero–mean signal.

EXAMPLE 10.3

Find the PSD of $X(n)$ if $X(n)$ is a real–valued zero–mean Markov process; that is, $X(n) = X(n - 1) + U(n)$, where $U(n)$ is a zero–mean white noise process having variance σ_U^2.

Solution

To first find $r_{XX}(k)$, multiply both sides of the recursion relation by $X(n + k)$ and take expectations:

$$E\{X(n)X(n + k)\} = E\{X(n - 1)\, X(n + k)\} + E\{U(n)X(n + k)\}.$$

Noting that $E\{X(n - 1)\, X(n + k)\} = E\{X(n)X(n + k + 1)\}$ because $X(n)$ is WSS,

$$r_{XX}(k) = r_{XX}(k + 1) + r_{XU}(k).$$

We can eliminate $r_{XU}(k)$ by multiplying the original recursion relation by $U(n + k)$ and again taking expectations:

$$E\{X(n)U(n + k)\} = E\{X(n - 1)U(n + k)\} + E\{U(n)U(n + k)\},$$

or

$$r_{UX}(k) = r_{UX}(k + 1) + \sigma_U^2\delta(k).$$

All the auto– and cross–correlation sequences are finite–energy, and so z–transforms can easily be applied. The z–transform of the above equation is $R_{UX}(z)[1 - z] = \sigma_U^2$, or, because $r_{UX}(k) = r_{XU}(-k)$ for real–valued sequences so that $R_{UX}(z) = R_{XU}(1/z)$, it follows from $R_{UX}(z) = \sigma_U^2/(1 - z)$ that $R_{XU}(z) = \sigma_U^2/(1 - z^{-1})$.

The ACF equation becomes, in the z–domain,

$$R_{XX}(z)(1 - z) = R_{XU}(z),$$

or

$$R_{XX}(z) = \frac{\sigma_U^2}{(1 - z)(1 - z^{-1})}.$$

We now note the following important fact: If $P_{XX}(\omega)$ is the DTFT of $r_{XX}(k)$, and if $R_{XX}(z)$ is the z–transform of $r_{XX}(k)$, it must be true that $P_{XX}(\omega) = R_{XX}(e^{j\omega})$—that is, $R_{XX}(z)$ evaluated on the unit circle. We therefore finally obtain, recalling that $e^{j\omega} = (e^{-j\omega})^*$,

$$P_{XX}(\omega) = \frac{\sigma_U^2}{|1 - e^{-j\omega}|^2} = \frac{\sigma_U^2}{2[1 - \cos(\omega)]}.$$

This Markov $X(n)$ is a special case of an autoregressive process, to which we will devote more attention in Section 10.9.

One additional intuitive fact may now be formally established: that $P_{XX}(\omega)$ is real–valued and positive. Recall from WP 5.1 that we studied convolution operators as particularly relevant for digital signal processing. A new revelation can be obtained if we note that an IIR unit sample response in a convolution operator for a shift–invariant system $h(n, m) = h(n - m)$ is identical to the operation of a linear, infinite–dimensional matrix operation, where the nmth matrix element is $h(n - m)$. Analogously, the ACF for WSS processes $r_{XX}(n, m) = E\{X(n)X^*(m)\} = r_{XX}(k) = E\{X(m + k)X^*(m)\}$ which depends on only $k = n - m$ can be thought of as elements of an infinite–dimensional *circulant* matrix R_{XX}, whose nmth element is $r_{XX}(n - m)$.

Because of this shift–invariance due to stationarity, we can let $r_{XX}(n - m)$ play the role of a unit sample response $h(n - m)$. Even though convolution is physically not involved, we can make inferences from $r_{XX}(k)$ on the eigenvalues of "convolution," which under this role playing is the Fourier transform of $r_{XX}(k)$, namely $P_{XX}(\omega)$. We already know that $r_{XX}(k)$ is a Hermitian sequence; that is, that because $r_{XX}(-k) = r_{XX}^*(k)$, the "convolution" matrix operator has real–valued eigenvalues—$P_{XX}(\omega)$ is real–valued.

Furthermore, if we can show that $r_{XX}(k)$ in the matrix operator is positive semi-definite, then we immediately know that all the corresponding eigenvalues, $P_{XX}(\omega)$, are nonnegative. Making direct use of (W5.1.36) $[Q(\mathbf{x}) = \sum_{n=1}^{N} x_n^* \sum_{m=1}^{N} A_{nm} x_m]$ and setting $\mathbf{x}$ to be a general sequence $g(n)$ and A_{nm} to $r_{XX}(n - m)$, we have (in which we can let N go to ∞ if desired)

$$Q(\mathbf{g}) = \sum_{n=1}^{N} g^*(n) \sum_{m=1}^{N} r_{XX}(n - m)g(m), \tag{10.15a}$$

which upon substitution of the definition of $r_{XX}(k)$ gives

$$Q(\mathbf{g}) = \sum_{n=1}^{N} g^*(n) \sum_{m=1}^{N} E\{X(n)X^*(m)\}g(m), \tag{10.15b}$$

and the linearity of the expectation operator allows us to rearrange to

$$Q(\mathbf{g}) = E\left\{ \sum_{n=1}^{N} g^*(n)\, X(n) \sum_{m=1}^{N} g(m) X^*(m) \right\}. \tag{10.15c}$$

Finally, we see in (10.15c) a sum multiplied by its conjugate, which is the magnitude squared of the sum:

$$Q(\mathbf{g}) = E\left\{ \left| \sum_{n=1}^{N} g^*(n) X(n) \right|^2 \right\} \geq 0 \quad \text{for any sequence } g(n). \tag{10.15d}$$

Therefore, $r_{XX}(k)$ is a positive semidefinite sequence. We thus obtain the desired and intuitive result that $P_{XX}(\omega)$ is not only real–valued but also nonnegative [see WP 5.1 for how (10.15d) proves this].

In practice, we are given a single–realization data sequence $x(n)$, for which $P_{XX}(\omega)$ is impossible to obtain exactly. However, samples of the squared magnitude of the DTFT of a finite section of $x(n)$ are easily computed and provide a reasonable estimate [recall (10.5): $P_{XX}(\omega) = \lim_{N \to \infty} E\{|X_N(e^{j\omega})|^2\}/N]$. This estimator of $P_{XX}(\omega)$ is called the periodogram:

$$\hat{P}_{\text{per}}(\omega) = \frac{1}{N} \left| \sum_{n=0}^{N-1} x(n)\, e^{-j\omega n} \right|^2. \tag{10.16}$$

Note that $\hat{P}_{\text{per}}(2\pi k/N) = (1/N) \cdot |X(k)|^2$ where $X(k) = \text{DFT}_N\{x(n)\}$. For different random processes $X(n)$, the accuracy of $\hat{P}_{\text{per}}(\omega)$ will vary. It is shown in the problems that the periodogram is a truncated version of the Wiener–Kinchin theorem, but with $r_{XX}(k)$ replaced by the sample ACF $r_{N,n}(k)$. In practice, it is helpful to average several periodograms of separate data blocks. This is because a single periodogram tends to fluctuate wildly (and its variance does not decrease with N), so the averaging provides a smoothed version more representative of the underlying process due to averaging out the short–time transient trends. For brevity, we leave to the problems several studies of the properties of the periodogram for specific example random processes.

10.4   GAUSSIAN (NORMAL) RANDOM VARIABLES, VECTORS, AND PROCESSES

The Gaussian or normal probabilistic distribution has been named in honor of Carl Friedrich Gauss. It is called "normal" because in life, it well describes what normally happens, especially when sums of large numbers of random variables are involved or when a single random variable can be so decomposed. It is fortunate that the Gaussian distribution has such wide applicability, especially for people in the business of playing statistical number games. Like wide-sense stationarity, once normality can be assumed, the mathematics simplifies and reduces to a well–known and well–tabulated formalism. Later, we will make the Gaussian assumption whenever convenient, especially whenever we desire to work with the PDF, such as in maximum likelihood estimation. We devote a large section to Gaussian entities because they are the foundation of basic stochastic signal processing.

The discoverer of the Gaussian (normal) distribution was Abraham de Moivre, in 1733. However, it was first practically applied by Gauss to arithmetic mean errors in 1809. The slope of the Gaussian is the proverbial bell curve; an example plot is presented later, in Figure 10.12 of Example 10.4. A Gaussian or normal (real–valued) random variable X is one whose PDF is of the following form:

$$f_X(x) = \frac{e^{-(x-\mu_X)^2/(2\sigma_X^2)}}{\sqrt{2\pi}\,\sigma_X}. \tag{10.17}$$

We will now systematically study the rationales for this form. First, it may be asked, why the independent–of–x constant $1/\{\sqrt{2\pi}\,\sigma_x\}$? This constant is necessary so that $f_X(x)$ is a valid PDF; that is, so that $\int_{-\infty}^{\infty} f_X(x)dx = 1$. Let us prove this. The integral of the Gaussian PDF is

$$\frac{1}{\sqrt{2\pi}\,\sigma_X}\int_{-\infty}^{\infty} e^{-(x-\mu_X)^2/(2\sigma_X^2)}dx = \frac{1}{\sqrt{2\pi}}\int_{-\infty}^{\infty} e^{-v^2/2}dv, \tag{10.18}$$

where on the right–hand side the substitution $v = (x - \mu_x)/\sigma_X$ was made. Defining I to be the right–hand side, except for the constant $1/\sqrt{2\pi}$,

$$I = \int_{-\infty}^{\infty} e^{-v^2/2}dv = 2\int_{0}^{\infty} e^{-v^2/2}dv. \tag{10.19}$$

We now introduce what has been called "the dirtiest trick in all of mathematics." $(I/2)^2$ can be written in terms of a two–dimensional integral of $\exp\{(-x^2 - y^2)/2\}$ over the first quadrant and subsequently converted to two–dimensional polar coordinates:

$$\frac{I^2}{4} = \int_{0}^{\infty} e^{-x^2/2}dx \cdot \int_{0}^{\infty} e^{-y^2/2}dy, \tag{10.20a}$$

which upon rearrangement gives

$$\frac{I^2}{4} = \int_{0}^{\infty}\int_{0}^{\infty} e^{-(x^2+y^2)/2}dx\,dy \tag{10.20b}$$

and conversion into polar form,

$$\frac{I^2}{4} = \int_{0}^{\pi/2} d\theta \int_{0}^{\infty} e^{-r^2/2}r\,dr \tag{10.20c}$$

which because the r factor is proportional to the derivative of the exponent allows us to use elementary integration to obtain

$$\frac{I^2}{4} = -e^{-r^2/2}\Big|_{0}^{\infty}\,\frac{\pi}{2} = \pi/2. \tag{10.20d}$$

Therefore, $I = 2 \cdot (\pi/2)^{1/2} = \sqrt{2\pi}$, which when divided by $\sqrt{2\pi}$ in the integral of the PDF gives unity, as required.

The next question that may be asked is whether the parameter μ_X is in fact the mean of X, if X has the PDF defined in (10.17). We find out by taking the expected value of X and then adding and subtracting $\mu_X \cdot 1 = \mu_X \cdot \int_{-\infty}^{\infty} f_X(x)dx$:

$$E\{X\} = \frac{1}{\sqrt{2\pi}\,\sigma_X}\int_{-\infty}^{\infty} xe^{-(x-\mu_X)^2/[2\sigma_X^2]}dx \tag{10.21a}$$

$$= \mu_X \cdot 1 + \frac{1}{\sqrt{2\pi}\,\sigma_X}\int_{-\infty}^{\infty} (x-\mu_X)\,e^{-(x-\mu_X)^2/[2\sigma_X^2]}dx. \tag{10.21b}$$

Now we define $u = (x - \mu_X)/\sigma_X$, for which $dx = du \cdot \sigma_X$ and $x - \mu_X = u \cdot \sigma_X$. These substitutions give

$$E\{X\} = \mu_X + \frac{\sigma_X}{\sqrt{2\pi}} \int_{-\infty}^{\infty} u \, e^{-u^2/2} du. \tag{10.21c}$$

The integrand of (10.21c) is seen to be an odd function of u and thus the integral is zero; therefore, indeed $E\{X\} = \mu_X$.

What about the σ_X^2 parameter? Is it really the variance of X, if X has the PDF defined in (10.17)? Again, using the definition of the variance of X, we have

$$\text{var}\{X\} = E\{(X - E\{X\})^2\} \tag{10.22a}$$

$$= \frac{1}{\sqrt{2\pi}\,\sigma_X} \int_{-\infty}^{\infty} (x - \mu_X)^2 \, e^{-(x - \mu_X)^2/[2\sigma_X^2]} dx. \tag{10.22b}$$

We again define $u = (x - \mu_X)/\sigma_X$, in which now we use $(x - \mu_X)^2 = u^2 \sigma_X^2$ and $dx = \sigma_X \, du$. Thus

$$\text{var}\{X\} = \frac{\sigma_X^2}{\sqrt{2\pi}} \int_{-\infty}^{\infty} u^2 e^{-u^2/2} du. \tag{10.22c}$$

The integral in (10.22c) can be evaluated by parts: $U = u, dU = du, dv = ue^{-u^2/2}du$, $v = -e^{-u^2/2}$ so that

$$\text{var}\{X\} = \frac{\sigma_X^2}{\sqrt{2\pi}} \left(-ue^{-u^2/2} \Big|_{-\infty}^{\infty} + \int_{-\infty}^{\infty} e^{-u^2/2} du \right). \tag{10.23}$$

The first (boundary) term in (10.23) is zero at both limits due to $e^{-u^2/2}$, and the results of (10.20) indicate that the integral is equal to $\sqrt{2\pi}$, which indeed gives $\text{var}\{X\} = \sigma_X^2$. We thus say that a random variable is normal with mean μ_X and variance σ_X^2, and write this as $X \sim \eta(\mu_X, \sigma_X^2)$.

One might even ask the purpose of the $\frac{1}{2}$ in the exponent. Its purpose is to guarantee that $\text{var}\{X\} = \sigma_X^2$ without any other constants involved. For example, if we defined $f_{X1}(x) = \frac{1}{\pi^{1/2}\sigma_X} e^{-(x - \mu_X)^2/\sigma_X^2}$ (show for yourself that the new constant out front is needed for normalization of the PDF), then indeed the mean would still be μ_X but the variance would be $\sigma_X^2/2$. The proofs of these results are left as exercises.

The far more important and compelling question is, Why the negative, squared exponential? First of all, the normal PDF is fairly simple. There are only two parameters necessary to completely specify the PDF: μ_X and σ_X^2. Consequently, if μ_X and σ_X^2 are known, then any higher moment or other probabilistic calculation can in principle be calculated, assuming that one has the required tenacity. This, however, may also be said of other simple models. There are also additional computational advantages of the negative, squared exponential.

However, the overriding reason for the preeminence of the normal distribution lies in the universality of its validity in describing real–world events and processes. This dominance is made explicit in the family of *central limit theorems*, where *central* refers to their central importance. The central limit theorem was first proposed by Laplace in 1810.

To study the limit theorems in their full generality, particularly their proofs, would take us far afield in a book on digital signal processing. However, because the

Gaussian assumption is so commonly used in stochastic signal processing, it is important to know the reasons for this ubiquity. Here we will state a general form of central limit theorem and prove it for a simple, common case. A general form of the central limit theorem is:

> If $X_\ell, 1 \le \ell \le N$ are independent random variables with means μ_ℓ and standard deviations σ_ℓ, not necessarily having identical PDFs (i.e., not necessarily *identically distributed*), then the scaled sum of the random variable
>
> $$S_N = \frac{\sum\limits_{\ell=1}^{N} (X_\ell - \mu_\ell)}{\left\{ \sum\limits_{\ell=1}^{N} \sigma_\ell^2 \right\}^{1/2}}$$
>
> approaches a zero–mean, unit–variance normal random variable as $N \to \infty$.[3]

In the special but important case of identically distributed random variables X_ℓ, the proof is straightforward using characteristic functions. Recall from WP 10.1 that the characteristic function $\phi_X(u)$ of random variable X is just the Fourier transform of $f_X(x)$, except that by convention the exponent sign is reversed. Specifically, recall (W10.1.17) $[\phi_X(u) = E\{e^{juX}\} = \int_{-\infty}^{\infty} f_X(x)\, e^{jux} dx]$. If X is a normal random variable with mean μ and variance σ^2, denoted $\eta(\mu, \sigma^2)$, then we can easily find its characteristic function (no ℓ subscripts on μ and σ^2 are needed because all the X_ℓ are identically distributed):

$$\phi_X(u) = \frac{1}{\sigma\sqrt{2\pi}} \int_{-\infty}^{\infty} e^{\frac{-(x-\mu)^2}{2\sigma^2}} e^{jux} dx. \tag{10.24}$$

By expanding out and then completing the square, the exponent in (10.24) may be written as

$$-\frac{(x-\mu)^2}{2\sigma^2} + jux = -\frac{1}{2}\left\{ \left(\frac{x}{\sigma}\right)^2 - \frac{2x\mu}{\sigma^2} + \left(\frac{\mu}{\sigma}\right)^2 - j2ux \right\} \tag{10.25a}$$

$$= -\frac{1}{2}\left\{ \left(\frac{x}{\sigma}\right)^2 - 2\left(\frac{x}{\sigma}\right)\left(\frac{\mu}{\sigma} + ju\sigma\right) + \left(\frac{\mu}{\sigma}\right)^2 \right\} \tag{10.25b}$$

$$= -\frac{1}{2}\left\{ \frac{x}{\sigma} - \frac{\mu}{\sigma} - ju\sigma \right\}^2 + j\mu u - \frac{u^2\sigma^2}{2}. \tag{10.25c}$$

Resubstitution of (10.25c) into (10.24), and noting that only the squared term involves x so the rest can be pulled outside the integral, gives

$$\phi_X(u) = e^{-(u\sigma)^2/2 + ju\mu} \cdot \frac{1}{\sigma\sqrt{2\pi}} \left\{ \int_{-\infty}^{\infty} e^{-\{x/\sigma - \mu/\sigma - ju\sigma\}^2/2} dx \right\}. \tag{10.26a}$$

The integral in $\{\cdot\}$ in (10.26a) [that is, not including the constant $1/(\sigma\sqrt{2\pi})$ outside the $\{\cdot\}$] is just a shifted form of (10.19)[4] in which $x/\sigma = v$ and so $dx = \sigma dv$, and thus the integral in $\{\cdot\}$ in (10.26a) has the value $\sigma\{2\pi\}^{1/2}$. Therefore,

> $$\phi_X(u) = e^{-(u\sigma)^2/2 + ju\mu} \qquad \text{for } X \sim \eta(\mu, \sigma^2). \tag{10.26b}$$

---

[3] For a proof of this general case, see W. Feller, *An Introduction to Probability Theory and Its Applications* (New York: Wiley, 1966), p. 256.

[4] Because $\exp\{\cdot\}$ is an analytic function everywhere, even though the shift is complex it will not affect the integral result, by the Cauchy integral theorem, (3.9).

For zero–mean, unit–variance normal random variables, (10.26b) reduces to

$$\phi_X(u) = e^{-u^2/2} \qquad \text{for } X \sim \eta(0, 1). \tag{10.26c}$$

Now we use (10.26) to prove the central limit theorem for independent, identically distributed random variables. We merely need to show that the characteristic function of S_N approaches that in (10.26c) as $N \to \infty$. We know that for independent random variables, the joint PDF is the product of the marginal PDFs, which in turn implies that the expected value of $X_\ell X_m$ is equal to $\text{E}\{X_\ell\}\text{E}\{X_m\}$. The same holds for the expected values of any functions of the X_ℓ, including e^{juX_ℓ} for the characteristic function. Thus, for example, $\phi_{X_\ell+X_m}(u) = \text{E}\{e^{ju(X_\ell+X_m)}\} = \text{E}\{e^{juX_\ell}e^{juX_m}\}$, which because X_ℓ and X_m are independent is $\phi_{X_\ell+X_m}(u) = \text{E}\{e^{juX_\ell}\}\text{E}\{e^{juX_m}\} = \phi_{X_\ell}(u) \cdot \phi_{X_m}(u)$. We now repeat these steps for the more general sum S_N of identically distributed random variables:

$$S_N = \frac{\sum_{\ell=1}^{N} X_\ell - N\mu}{\sigma\sqrt{N}},$$

and the characteristic function of S_N becomes

$$\phi_{S_N}(u) = \text{E}\{e^{juS_N}\} \tag{10.27a}$$

$$= \text{E}\left\{e^{j\frac{u}{\sigma\sqrt{N}}\sum_{\ell=1}^{N}(X_\ell-\mu)}\right\}, \tag{10.27b}$$

which, because $e^{a+b} = e^a \cdot e^b$ as described above for the two–variable case, becomes

$$\phi_{S_N}(u) = \text{E}\left\{\prod_{\ell=1}^{N} e^{j\frac{u}{\sigma\sqrt{N}}(X_\ell-\mu)}\right\}. \tag{10.27c}$$

Furthermore, because the X_ℓ are independent, (10.27c) becomes

$$\phi_{S_N}(u) = \prod_{\ell=1}^{N} \text{E}\left\{e^{j\frac{u}{\sigma\sqrt{N}}(X_\ell-\mu)}\right\}. \tag{10.28a}$$

When we recall that $\phi_A(u) = \text{E}\{e^{juA}\}$, where here we let $A = X_\ell - \mu$, (10.28a) can be written

$$\phi_{S_N}(u) = \prod_{\ell=1}^{N} \phi_{X_\ell-\mu}\left(\frac{u}{\sigma\sqrt{N}}\right). \tag{10.28b}$$

Because all the X_ℓ are identically distributed and thus have identical characteristic functions, (10.28b) becomes

$$\phi_{S_N}(u) = \phi_{X-\mu}^N\left(\frac{u}{\sigma\sqrt{N}}\right). \tag{10.28c}$$

In the special case that the individual random variables X_ℓ are Gaussian, substitution of (10.26b) into (10.28c) yields (upon recognition that the mean of $X - \mu$ is zero, so that μ in (10.26b) is zero and noting that N disappears when all powers are taken)

$$\phi_{S_N}(u) = e^{-u^2/2}, \tag{10.29}$$

and thus S_N is truly $\eta(0, 1)$ for any value of N, which is stronger than what was to be proved.

In the more general case where X_ℓ are not necessarily Gaussian but are still independent and identically distributed, with means μ and variances σ^2, we proceed as follows. Consider first one of the random variables, X. By the definition of the characteristic function, its value evaluated at $u/(\sigma\sqrt{N})$ for the random variable $X - \mu$ is

$$\phi_{X-\mu}\left(\frac{u}{\sigma\sqrt{N}}\right) = \mathrm{E}\left\{e^{j\frac{u}{\sigma\sqrt{N}}(X-\mu)}\right\}. \tag{10.30a}$$

Note that by the Maclaurin expansion and $\mathrm{E}\{a + b\} = \mathrm{E}\{a\} + \mathrm{E}\{b\}$, (10.30a) becomes

$$\phi_{X-\mu}\left(\frac{u}{\sigma\sqrt{N}}\right) = 1 + j\frac{u}{\sigma\sqrt{N}}\,\mathrm{E}\{X - \mu\} - \frac{u^2}{2\sigma^2 N}\,\mathrm{E}\{(X-\mu)^2\} + \ldots \tag{10.30b}$$

$$= 1 - \frac{u^2}{2\sigma^2 N}\sigma^2 + \ldots \tag{10.30c}$$

$$\approx 1 - \frac{u^2}{2N}, \tag{10.30d}$$

with increasingly perfect equality in (10.30d) as $N \to \infty$. Next, we raise the right–hand side of (10.30d) to the Nth power to obtain [by (10.28c)] $\phi_{s_N}(u) = \phi_{X-\mu}^N (u/[\sigma\sqrt{N}]$. By Newton's binomial theorem,

$$(1 + a)^N = \sum_{k=0}^{N} \binom{N}{k} a^k, \tag{10.31}$$

where in our application $a = -u^2/(2N)$ and Euler's notation is

$$\binom{N}{k} = \frac{N(N-1)(N-2)\ldots(N-k+1)}{k!} \tag{10.32}$$

and thus

$$\left(1 - \frac{u^2}{2N}\right)^N = \sum_{k=0}^{N} \binom{N}{k}\frac{\left(-\frac{1}{2}u^2\right)^k}{N^k}. \tag{10.33a}$$

As $N \to \infty$, the $N - \ell$ in (10.32) are all essentially just N for $k \ll N$, making the right-hand side of (10.32) very large, and for k near ∞ the term is relatively very small because of the $k!$ in the denominator. Thus the numerator of (10.32) is approximately N^k for the dominating terms, which when (10.32) is substituted into (10.33a) cancels the N^k in the denominator of (10.33a). Thus (10.33a) becomes

$$\left(1 - \frac{u^2}{2N}\right)^N \xrightarrow[N\to\infty]{} \sum_{k=0}^{\infty}\frac{\left(-\frac{1}{2}u^2\right)^k}{k!} = e^{-u^2/2} = \lim_{N\to\infty}\phi_{s_N}(u), \tag{10.33b}$$

and the central limit theorem is proved for the sum of a large number of non–Gaussian, independent, identically distributed random variables X_ℓ.

In passing, we recall that multiplication in the Fourier domain (multiplication of characteristic functions) is equivalent to convolution of the PDFs. Thus, if $W = X_1 + X_2$ where X_1 and X_2 are independent random variables so that $\phi_w(u) = \phi_{X_1+X_2}(u) = \phi_{X_1}(u) \cdot \phi_{X_2}(u)$, then $f_W(w) = (f_{X_1} * f_{X_2})(w)$. Moreover, the "central limit theorem" can be applied to *any* functions $f(x)$ having finite moments $\int_{-\infty}^{\infty} x^\ell f(x)dx$ and zero average value, not just PDFs, to conclude that $f(x)$ convolved with itself many, many times is approximately equal to a function of the Gaussian form $e^{-(w/\sigma)^2/2}$, where $\sigma^2 = \int_{-\infty}^{\infty} x^2 f(x)dx$.[5] In short, the math does not know or care

---

[5] P. A. Jansson, *Deconvolution with Applications in Spectroscopy* (New York: Academic Press, 1984).

whether or not we are applying it to probability; this kind of generalization is an important concept.

A couple of other specific central limit theorems may also be of common interest. If a "trial" is defined whose probability of "success" is p, let N_S = number of successes in N independent (Bernoulli binary) trials. Then, the probability that N_S is between two selected numbers N_1 and N_2 (e.g., the probability that in $N = 100$ trials you will have between $N_1 = 15$ and $N_2 = 20$ successes) is

$$\text{prob}\{N_1 \leq N_S \leq N_2\} = \sum_{N_S=N_1}^{N_2} \binom{N}{N_S} p^{N_S}(1-p)^{N-N_S}, \tag{10.34a}$$

which as the number of trials N goes to ∞ becomes[6]

$$\text{prob}\{N_1 \leq N_S \leq N_2\} \xrightarrow[N\to\infty]{} \frac{1}{\sqrt{2\pi}} \int_A^B e^{-y^2/2}dy, \tag{10.34b}$$

where the integrand is equal to the Gaussian PDF $\eta(0,1)$, $A = (N_1 - Np + \frac{1}{2})/\{Np(1-p)\}^{1/2}$ and $B = (N_2 - Np + \frac{1}{2})/\{Np(1-p)\}^{1/2}$. The integral is equal to the difference between two Gaussian cumulative distribution functions (CDFs), and is tabulated for easy lookup.

Similarly, the probability that the number of successes N_S in N trials is equal to k (e.g., the probability that in $N = 100$ trials there are $k = 27$ successes) is

$$\text{prob}\{N_S = k\} = \binom{N}{k} p^k(1-p)^{N-k}, \tag{10.35a}$$

which as $N \to \infty$ becomes[7]

$$\text{prob}\{N_S = k\} \xrightarrow[N\to\infty]{} \frac{e^{-(k-Np)^2/[2Np(1-p)]}}{\{2\pi\, Np(1-p)\}^{1/2}}, \tag{10.35b}$$

which is the PDF of a Gaussian random variable $\eta(Np, Np(1-p))$, valid for about $Np(1-p) > 10$. Note, however, that if $p \to 0$, we have[8]

$$\text{prob}\{N_S = k\} = \text{binomial PMF} = \binom{N}{k} p^k(1-p)^{N-k} \tag{10.35c}$$

$$\xrightarrow[\substack{N\to\infty \\ p\to 0 \\ Np = \lambda \text{ fixed}}]{} \frac{e^{-\lambda}\lambda^k}{k!} = \text{Poisson PMF}, \tag{10.35d}$$

which is approximately valid for $Np = \lambda < 0.1$. If λ is large as $N \to \infty$, the Poisson distribution is approximately Gaussian.

10.4.1   Gaussian (Normal) Random Vectors

So far, all our discussion on Gaussian random variables has concerned only the univariate distribution $\eta(\mu, \sigma^2)$. In signal–processing practice, jointly normal random vectors and random processes are more commonly encountered. To derive the general normal random vector, we begin by considering the simple, special case of N independent $\eta(0, 1)$ random variables Y_ℓ. Their joint PDF is just the product of N

---

[6]E. Parzen, *Modern Probability Theory and its Applications* (New York: Wiley, 1960).

[7]P. Meyer, *Introductory Probability and Statistical Applications* (Reading, MA: Addison–Wesley, 1970), pp. 247–249.

[8]S. Ross, *Introduction to Probability Models* (New York: Academic Press, 1972), p. 25.

univariate $\eta(0, 1)$ PDFs because of their independence and thus the exponent becomes a sum:

$$f_{\mathbf{Y}}(\mathbf{y}) = \frac{1}{(2\pi)^{N/2}} e^{-\frac{1}{2} \sum_{\ell=1}^{N} y_\ell^2}, \tag{10.36a}$$

which when we recall that the exponent of (10.36a) includes the inner product of a column vector $\mathbf{y}$ with itself may be written

$$f_{\mathbf{Y}}(\mathbf{y}) = \frac{e^{-\mathbf{y}^T\mathbf{y}/2}}{(2\pi)^{N/2}}, \tag{10.36b}$$

and, again invoking the independence of the Y_ℓ, the joint characteristic function is

$$\phi_{\mathbf{Y}}(\mathbf{u}) = \prod_{\ell=1}^{N} \phi_{Y_\ell}(u_\ell) \tag{10.37a}$$

$$= e^{-\frac{1}{2} \sum_{\ell=1}^{N} u_\ell^2} \tag{10.37b}$$

$$= e^{-\mathbf{u}^T\mathbf{u}/2}. \tag{10.37c}$$

Certainly for this case we have $E\{\mathbf{Y}\} = \mathbf{0}$, $C_{YY_{\ell m}} = \delta(\ell - m)$; that is, $\mathbf{C_{YY}} = E\{\mathbf{YY}^T\} = \mathbf{I}_N$. (To eliminate conjugation symbols, we assume real-valued random variables.)

The easiest way to derive the PDF of the general multivariate Gaussian $\mathbf{X}$ is to view it as a linear transformation of $\mathbf{Y}$, which can always be done.[9] Thus, we write

$$\mathbf{X} = \mathbf{AY} + \mathbf{b}, \tag{10.38}$$

where $\mathbf{A}$ is an $N \times N$ matrix. The mean of $\mathbf{X}$ is

$$E\{\mathbf{X}\} = \mathbf{A}E\{\mathbf{Y}\} + \mathbf{b} = \mathbf{b}. \tag{10.39}$$

For the conventional notation, we thus see we can replace the constant vector $\mathbf{b}$ by $\boldsymbol{\mu}$. Thus, $\mathbf{X} = \mathbf{AY} + \boldsymbol{\mu}$. The covariance matrix of $\mathbf{X}$ is [noting that for any arrays $\mathbf{U}$ and $\mathbf{V}$, $(\mathbf{UV})^T = \mathbf{V}^T\mathbf{U}^T$]

$$\mathbf{C}_{XX} = E\{(\mathbf{X} - \boldsymbol{\mu})(\mathbf{X} - \boldsymbol{\mu})^T\} \tag{10.40a}$$

$$= E\{\mathbf{AYY}^T\mathbf{A}^T\} \tag{10.40b}$$

$$= \mathbf{AA}^T. \tag{10.40c}$$

To obtain the PDF of $\mathbf{X}$, we need a theorem relating PDFs of sets of random variables related in a known manner. This theorem is obtained merely by performing a change of variables on the integral of the joint density function. First consider two random variables:

$$\int \int_{\substack{\text{region} \\ \text{of } Y_1, Y_2 \text{ plane}}} f_{Y_1 Y_2}(y_1, y_2) dy_1 dy_2 = \int \int_{\substack{\text{(transformed)} \\ \text{region in } X_1, X_2 \text{ plane}}} f_{X_1 X_2}(x_1, x_2) dx_1 dx_2. \tag{10.41}$$

If we can perform a coordinate transformation on the left–hand side of (10.41) from variables y_1, y_2 to variables x_1, x_2, then everything we obtain except for the $dx_1 dx_2$ in that result will, by the right–hand side of (10.41), be $f_{X_1 X_2}(x_1, x_2)$. Expanding out $dy_1 dy_2$, we have

$$dy_1 dy_2 = \left\{ \frac{\partial y_1}{\partial x_1} dx_1 + \frac{\partial y_1}{\partial x_2} dx_2 \right\} \left\{ \frac{\partial y_2}{\partial x_1} dx_1 + \frac{\partial y_2}{\partial x_2} dx_2 \right\} \tag{10.42a}$$

---

[9]B. W. Lindgren, *Statistical Theory* (New York: Macmillan, 1962), pp. 129–131.

$$= \frac{\partial y_1}{\partial x_1}dx_1 \frac{\partial y_2}{\partial x_1}dx_1 + \frac{\partial y_1}{\partial x_1}dx_1 \frac{\partial y_2}{\partial x_2}dx_2 + \frac{\partial y_1}{\partial x_2}dx_2 \frac{\partial y_2}{\partial x_1}dx_1 + \frac{\partial y_1}{\partial x_2}dx_2 \frac{\partial y_2}{\partial x_2}dx_2. \qquad (10.42\text{b})$$

The first and last terms in (10.42b) yield zero, essentially because they form a one–dimensional area, which has zero area. With increments $\{dy_1, dy_2\}$ as governed by the original integral, the third term is equal to $-\dfrac{\partial y_1}{\partial x_2}\dfrac{\partial y_2}{\partial x_1}\, dx_1 dx_2.$[10] Thus

$$dy_1 dy_2 = \begin{vmatrix} \partial y_1/\partial x_1 & \partial y_1/\partial x_2 \\ \partial y_2/\partial x_1 & \partial y_2/\partial x_2 \end{vmatrix} dx_1 dx_2. \qquad (10.42\text{c})$$

Using these facts, as well as the reasoning for the appearance of the absolute value in (W10.1.28) $\left[f_Y(y) = f_X(x)\, |\, dx/dy\, |\, \big|_{x=g^{-1}(y)} \right]$ for the one–dimensional case, (10.42c) becomes

$$dy_1 dy_2 = |\mathbf{J}_{yx}|\, dx_1 dx_2, \qquad (10.43)$$

where the absolute value of the determinant of the Jacobian matrix $\mathbf{J}_{yx}$ is

$$|\mathbf{J}_{yx}| = \text{absolute value of the determinant of matrix } \mathbf{J}_{yx}$$
$$\text{having } \ell m\text{th element } \partial y_\ell /\partial x_m. \qquad (10.44)$$

The result for two dimensions extends to any number of dimensions. The same procedure as above may be used. Any cross products analogous to those in (10.42b) are zero if any dx_ℓ are repeated. Let L be the number of instances in which dx_ℓ comes after dx_m for $\ell < m$. Then the partial derivative term written in numerical order $dx_1 \ldots dx_N$ will be multiplied by $(-1)^{\,L}$. The sum of these signed permutations of $\partial y_\ell/\partial x_m$ is precisely the basic definition of the determinant $\mathbf{J}_{yx}$.[11] Consequently,

$$dy_1 dy_2 \ldots dy_N = |\mathbf{J}_{yx}|\, dx_1 dx_2 \ldots dx_N. \qquad (10.45)$$

Therefore, on the left–hand side of the N–dimensional version of (10.41), we may substitute (10.45) to obtain

$$f_\mathbf{Y}(\mathbf{y}) |\mathbf{J}_{yx}| = f_\mathbf{X}(\mathbf{x}) \qquad (10.46\text{a})$$

or, equivalently, noting that by role reversal we have from (10.44) that $|\mathbf{J}_{xy}| = 1/|\mathbf{J}_{yx}|$, and by defining $h(\cdot)$ to be the function $\mathbf{X} = h(\mathbf{Y})$,

$$f_\mathbf{X}(\mathbf{x}) = \frac{f_\mathbf{Y}(h^{-1}(\mathbf{x}))}{|\mathbf{J}_{xy}|}. \qquad (10.46\text{b})$$

The above derivation holds for any functional relation h between $\mathbf{Y}$ and $\mathbf{X}$. For our special case of the linear transformation $\mathbf{X} = \mathbf{AY} + \boldsymbol{\mu} = h(\mathbf{Y})$, we may solve for $\mathbf{Y}$ to obtain $\mathbf{Y} = h^{-1}(\mathbf{X}) = \mathbf{A}^{-1}(\mathbf{X} - \boldsymbol{\mu})$. Also, $|\mathbf{J}_{xy}|$ is the absolute value of the determinant of the matrix whose ℓmth element is $\partial x_\ell /\partial y_m = a_{\ell m} = \ell m$th element of $\mathbf{A}$. Thus $|\mathbf{J}_{xy}| = \text{abs. val.}\{|\mathbf{A}|\}$. Making these substitutions, where $f_\mathbf{Y}(\mathbf{y}) = \exp\{-\tfrac{1}{2}\mathbf{y}^\mathrm{T}\mathbf{y}\}/\{2\pi\}^{N/2}$ as in (10.36b),

---

[10]M. E. Munroe, *Modern Multidimensional Calculus* (Reading, MA: Addison–Wesley, 1963).

[11]E. T. Browne, *Introduction to the Theory of Determinants and Matrices* (Richmond, VA: University of North Carolina Press, 1958).

$$f_{\mathbf{X}}(\mathbf{x}) = \frac{e^{-\{[\mathbf{A}^{-1}(\mathbf{x} - \boldsymbol{\mu})]^{\mathrm{T}}\mathbf{A}^{-1}(\mathbf{x} - \boldsymbol{\mu})\}/2}}{(2\pi)^{N/2}|\mathbf{A}|}. \tag{10.47}$$

Now again recall that $(\mathbf{UV})^{\mathrm{T}} = \mathbf{V}^{\mathrm{T}}\mathbf{U}^{\mathrm{T}}$, and also that for any invertible matrices $\mathbf{K}$ and $\mathbf{L}$ we have $\mathbf{L}^{-1}\mathbf{K}^{-1} = (\mathbf{KL})^{-1}$ and $(\mathbf{L}^{-1})^{\mathrm{T}} = (\mathbf{L}^{\mathrm{T}})^{-1}$, so the exponent in (10.47) may be written as

$$-\tfrac{1}{2}\{[\mathbf{A}^{-1}(\mathbf{x} - \boldsymbol{\mu})]^{\mathrm{T}}\mathbf{A}^{-1}(\mathbf{x} - \boldsymbol{\mu})\} = -\tfrac{1}{2}\{(\mathbf{x} - \boldsymbol{\mu})^{\mathrm{T}}(\mathbf{A}^{-1})^{\mathrm{T}}\mathbf{A}^{-1}(\mathbf{x} - \boldsymbol{\mu})\} \tag{10.48a}$$

$$= -\tfrac{1}{2}\{(\mathbf{x} - \boldsymbol{\mu})^{\mathrm{T}}(\mathbf{AA}^{\mathrm{T}})^{-1}(\mathbf{x} - \boldsymbol{\mu})\}, \tag{10.48b}$$

which by (10.40c) $[\mathbf{C}_{XX} = \mathbf{AA}^{\mathrm{T}}]$ reduces to

$$-\tfrac{1}{2}\{[\mathbf{A}^{-1}(\mathbf{x} - \boldsymbol{\mu})]^{\mathrm{T}}\mathbf{A}^{-1}(\mathbf{x} - \boldsymbol{\mu})\} = -\tfrac{1}{2}\{(\mathbf{x} - \boldsymbol{\mu})^{\mathrm{T}}\mathbf{C}_{XX}^{-1}(\mathbf{x} - \boldsymbol{\mu})\}. \tag{10.48c}$$

Next, note that because $|\mathbf{A}^{\mathrm{T}}| = |\mathbf{A}|$, we have

$$|\mathbf{A}| = \{|\mathbf{A}|^2\}^{1/2} = \{|\mathbf{A}||\mathbf{A}^{\mathrm{T}}|\}^{1/2}, \tag{10.49a}$$

so that, using $|\mathbf{U}||\mathbf{V}| = |\mathbf{UV}|$,

$$|\mathbf{A}| = \{|\mathbf{AA}^{\mathrm{T}}|\}^{1/2} \tag{10.49b}$$

$$= \{|\mathbf{C}_{XX}|\}^{1/2}. \tag{10.49c}$$

Thus we may finally write (10.47) as

$$f_{\mathbf{X}}(\mathbf{x}) = \frac{e^{-\{(\mathbf{x} - \boldsymbol{\mu})^{\mathrm{T}}\mathbf{C}_{XX}^{-1}(\mathbf{x} - \boldsymbol{\mu})\}/2}}{(2\pi)^{N/2}\{|\mathbf{C}_{XX}|\}^{1/2}}. \tag{10.50}$$

The PDF in (10.50) is valid for any Gaussian ("normal") random vector. It is left as an exercise to prove that the corresponding characteristic function is

$$\phi_{\mathbf{X}}(\mathbf{u}) = e^{j\mathbf{u}^{\mathrm{T}}\mathbf{u} - \mathbf{u}^{\mathrm{T}}\mathbf{C}_{XX}\mathbf{u}/2}. \tag{10.51}$$

Also, by some laborious calculations[12], we may prove that if $\mathbf{X}$ is defined by the PDF in (10.50), then $E\{\mathbf{X}\} = \boldsymbol{\mu}$ and $E\{(\mathbf{X} - E\{\mathbf{X}\})(\mathbf{X} - E\{\mathbf{X}\})^{\mathrm{T}}\} = \mathbf{C}_{XX}$. However, we have very directly derived $f_{\mathbf{X}}(\mathbf{x})$ by simply viewing each of its components as a linear combination of independent $\eta(0, 1)$ random variables.

Another important property of Gaussian random vectors is that their marginals are also Gaussian. For example, the ith marginal has the PDF $f_{\mathrm{marg}} = \int_{-\infty}^{\infty} f_{\mathbf{X}}(\mathbf{x})dx_i$, and is the joint PDF of the random variables $m = [1, \ldots, i - 1, i + 1, \ldots, N]$. Noting in the definition just given of f_{marg} that nothing is changed by adding the factor $\exp\{j0 \cdot x_i\}$, we see that the characteristic function of the marginal PDF is just that of the entire N–variable PDF evaluated at $u_i = 0$. This setting of u_i to zero will not change the form of (10.51) for the marginal except to make the final array dimensions be effectively $N - 1$ rather than N. Thus, the marginal will also be Gaussian.

Finally, an important property satisfied by Gaussian random vectors that does not hold in general for non–Gaussian random variables is that if all the X_i are uncorrelated, then the X_i are independent. With all X_i uncorrelated, then by definition $\mathbf{C}_{XX} = \mathbf{I}_N$, the $N \times N$ identity matrix. Substitution of $\mathbf{C}_{XX} = \mathbf{I}_N$ into (10.50)

---

[12]See S. Wilks, *Mathematical Statistics* (New York: Wiley, 1962), pp. 163–166.

proves that in this case $f_{\mathbf{X}}(\mathbf{x})$ is just the product of the individual univariate PDFs $f_{X_\ell}(x_\ell)$ and thus they are independent. Although independence always implies uncorrelatedness, the reverse implication does not hold, except if the random variables are Gaussian.

Other properties, such as the fact that a linear transformation of a general Gaussian random vector is another Gaussian random vector, are explored in the problems.

10.4.2 Gaussian (Normal) Random Processes

A random process is Gaussian if every finite set of $\{X(n)\}$ is a Gaussian random vector. The typical random process originates from or may be decomposed into a multitude of independently occurring phenomena. Examples are Brownian/thermal motion, electromagnetic noise, and so on. We also assume that none of the contributing phenomena dominates over the random process $X(n)$; otherwise, $X(n)$ would be predominantly characterized by that one contributing phenomenon, which in general may not be Gaussian. (In the general central limit theorem, the requirement that all X_ℓ have the same mean and variance helps guarantee the validity of this assumption.) From the central limit theorem, we know that under these assumptions, the random variables $\{X(n)\}$ will be Gaussian, and if jointly Gaussian will together constitute a Gaussian random vector $\mathbf{X}$ or process $X(n)$.

In Section 10.7, we will study the relations between outputs and inputs of LSI systems having random signal inputs. Given that a linear transformation of a Gaussian random vector is also Gaussian, it is a matter of taking limits of sequences to prove that if an input signal is a Gaussian random process, then so will be the output of the LSI system.

The Gaussian random process is known as a second–order process because its PDF and therefore all its statistical properties are completely determined by μ and $r_{XX}(k)$ [or equivalently by μ and $c_{XX}(k)$], which are, respectively, the first and second moments and are the sole parameters of the process. While the Gaussian random process still plays a great role in stochastic signal processing, non–Gaussian processes and higher–order moments called cumulants are of increasing importance to the researcher. Like all science, as the field progresses, more accurate and sophisticated approaches become possible. We have limited this presentation to a basic understanding of second–order processes.

10.5 INTRODUCTION TO ESTIMATION THEORY

Because random processes $X(n)$ are fundamentally nonrepeatable, the only way to attempt to characterize them is by means of unchanging parameters or functions associated with $X(n)$. An important example of this characterization is estimation of the power spectrum (recall the periodogram in Section 10.3). Other examples we have already considered are estimation of the mean and ACF, given data $x(n)$ produced by the random process $X(n)$. At a more advanced level, we may propose a model for the process generating our data and use specialized algorithms to estimate the parameters of our model.

In each of these cases, we want to know how well we are doing in our parameter estimation. The very terms *random* and *stochastic* are rather disconcerting, making us feel more unsure about accuracy and repeatability because of the nature of the signal. We need some measures and rules to give us confidence that the estimates we calculate are meaningful. Suppose that we wish to estimate a real parameter θ using the estimate $\hat{\theta}$. In our limited second–order studies, two measures are most important. These are

Bias:

$$B_\theta(\hat{\theta}) = E\{\hat{\theta}\} - \theta. \tag{10.52}$$

Variance:

$$\sigma_{\hat{\theta}}^2 = E\{(\hat{\theta} - E\{\hat{\theta}\})^2\}. \tag{10.53}$$

An estimate $\hat{\theta}$ is called unbiased if $B_\theta(\hat{\theta}) = 0$ so that $E\{\hat{\theta}\} = \theta$. We see that bias is associated with accuracy, whereas variance indicates the variability of $\hat{\theta}$. We want both bias and variance to be as small as possible, so that in every estimation, $\theta \approx \hat{\theta}$. The mean–squared error combines bias and variance:

$$MSE_\theta(\hat{\theta}) = E\{(\theta - \hat{\theta})^2\}. \qquad \text{Mean–squared error.} \tag{10.54a}$$

To express the MSE in terms of $B_\theta(\hat{\theta})$ and $\sigma_{\hat{\theta}}^2$, we add and subtract $E\{\hat{\theta}\}$:

$$MSE_\theta(\hat{\theta}) = E\{(\hat{\theta} - E\{\hat{\theta}\} + E\{\hat{\theta}\} - \theta)^2\}. \tag{10.54b}$$

Multiplying out the quadratic, we obtain

$$\tag{10.54c}$$
$$MSE_\theta(\hat{\theta}) = E\{(\hat{\theta} - E\{\hat{\theta}\})^2\} + E\{(E\{\hat{\theta}\} - \theta)^2\} + 2E\{(\hat{\theta} - E\{\hat{\theta}\})(E\{\hat{\theta}\} - \theta)\}.$$

Using (10.52) and (10.53) and noting that $E\{\hat{\theta}\} - \theta$ is a deterministic constant which can thus be pulled outside the expectation operator gives

$$MSE_\theta(\hat{\theta}) = \sigma_{\hat{\theta}}^2 + E\{B_\theta^2(\hat{\theta})\} + 2E\{(\hat{\theta} - E\{\hat{\theta}\})\}(E\{\hat{\theta}\} - \theta). \tag{10.54d}$$

Noting that $B_\theta(\hat{\theta})$ is a deterministic constant and that $E\{\hat{\theta} - E\{\hat{\theta}\}\} = 0$, we finally obtain

$$MSE_\theta(\hat{\theta}) = \sigma_{\hat{\theta}}^2 + B_\theta^2(\hat{\theta}). \tag{10.54e}$$

(It also follows that the RMS error is $\{\sigma_{\hat{\theta}}^2 + B_\theta^2(\hat{\theta})\}^{1/2}$.)

We conclude that if $MSE_\theta(\hat{\theta})$ is small, then both the bias magnitude and variance must be small, which means that the estimate is very good. In such cases, $\hat{\theta}$ for different–data–realization estimations are all very tightly clustered around θ. To illustrate these comments, let us review Figure 10.11 in Example 10.1. Recall that in Figure 10.11 we estimated the ACF by finite time averages and by finite (truncated) stochastic averages. In Figure 10.11, the temporal ACF estimate obviously has a large variance, whereas the stochastic estimate has a small variance (for each lag k). Assuming that the "true" ACF is covered by the dashed curves, we could also conclude that the stochastic estimate has a small bias; from the small number of plotted curves, it would be hard to conclude about the bias of the temporal ACF.

In the lingo of mathematical statistics, an estimate that can be made arbitrarily accurate by using more and more data to form the estimate is called a consistent estimate. (Perhaps this name is used because the estimate consistently gives good results when based on large amounts of data.) Mathematically, an estimate based on N data points is called consistent if

$$\lim_{N\to\infty} \text{prob}\{|\hat{\theta} - \theta| > \epsilon\} = 0 \qquad \text{for any } \epsilon > 0. \tag{10.55}$$

A sufficient condition for $\hat{\theta}$ to be consistent is that $\lim_{N\to\infty} \text{MSE}_\theta(\hat{\theta}) = 0$.[13]

10.5.1   Cramer–Rao Bound

By scaling or shifting, we can often produce an unbiased estimator from a biased one. If $\hat{\theta}$ is unbiased, then $\sigma_{\hat{\theta}}^2$ is a direct measure of the quality of $\hat{\theta}$, for it indicates on the average how tightly packed around θ the estimates $\hat{\theta}$ are for different data sets. In what follows, we assume that $\hat{\theta}$ is formed from N data points—an N–vector **x**. The Cramer–Rao bound provides a limit on how well we can do by indicating a minimum value that $\sigma_{\hat{\theta}}^2$ can have for a given parameter θ and PDF $f_{\mathbf{X}}(\mathbf{x})$, which applies for both biased and unbiased estimates $\hat{\theta}$.

Although the Cramer–Rao bound does not tell us *how* to choose the minimum–variance estimate, nor even whether it exists, it does tell us when to look no further. An estimate having the Cramer–Rao variance is called efficient and is generally as good as one can do, provided the bias is zero or small. The variance will be affected by any shifting or scaling to subsequently reduce or eliminate bias.

The Cramer–Rao bound is based on the Schwarz inequality and manipulative tricks to extract the variance. To explicitly show the dependence on θ, we write the PDF as $f_{\mathbf{X}}(\mathbf{x}; \theta)$. For brevity, the proof is relegated to the problems; the Cramer–Rao bound for the variance $\sigma_{\hat{\theta}}^2$ of an estimator $\hat{\theta}$ of parameter θ is

$$\sigma_{\hat{\theta}}^2 \geq \frac{-\left(1 + \frac{\partial}{\partial\theta}B_\theta(\hat{\theta})\right)^2}{E\left\{\frac{\partial^2 L}{\partial\theta^2}\right\}} \qquad \text{Cramer–Rao bound (general).} \tag{10.56a}$$

$$= \frac{-1}{E\left\{\frac{\partial^2 L}{\partial\theta^2}\right\}} \qquad \text{Cramer–Rao bound for unbiased estimates.} \tag{10.56b}$$

where

$$L = \ln\{f_{\mathbf{X}}(\mathbf{x}); \theta)\} \tag{10.57}$$

is called the log–likelihood function and has great importance in parameter estimation and mathematical statistics. The importance of the log–likelihood L arises because if the X_ℓ are independent, then $f_{\mathbf{X}}(\mathbf{x})$ involves a product and thus L involves a sum. A sum is usually more convenient than a product when, for example, differentiation is required for maximization purposes.

Note that the Cramer–Rao bound is the same for all estimators having the same bias. In the case of nonzero bias (10.56a), the Cramer–Rao bound gives the lowest possible variance for that θ, for any $\hat{\theta}$ having the given bias. An interesting question is, Under what condition is the bound in (10.56) on variance achieved? In the proof of (10.56), we find from Schwarz's inequality that equality holds in (10.56) when $A = \hat{\theta} - E\{\hat{\theta}\}$ and $B = \partial L/\partial\theta$ are "collinear" or proportional, with respect to the integration variables **x** in the stochastic expectation averages. Because the proportionality

---

[13]For a proof of this based on Chebyshev's inequality, see M. B. Priestly, *Spectral Analysis and Time Series* (New York: Academic Press, 1981), p. 301.

may (or may not) depend on the parameter θ being estimated, for generality we write this constant as $K(\theta)$, but $K(\theta)$ is a constant with respect to $\mathbf{x}$. Thus, for equality in (10.56) and thus an efficient estimator—that is, for $\hat{\theta} = \hat{\theta}_{\text{minvar}} = \theta_{\text{eff}}$,

$$\frac{\partial L}{\partial \theta} = K(\theta)(\hat{\theta}_{\text{eff}} - \text{E}\{\hat{\theta}_{\text{eff}}\}) \qquad \text{Condition for Cramer–Rao bound} \qquad (10.58a)$$
attainment.

$$= K(\theta)(\hat{\theta}_{\text{eff}} - \theta). \qquad \text{Condition for Cramer–Rao bound} \qquad (10.58b)$$
attainment; unbiased $\hat{\theta}$.

It is shown in the problems that for an unbiased, efficient estimator,

$$\sigma_{\hat{\theta},\min}^2 = 1/K(\theta). \qquad \text{Unbiased, efficient estimator.} \qquad (10.59)$$

10.5.2 Maximum Likelihood Estimation

So far in this discussion of estimation, we have said much about how to characterize an estimate we already have, but have not discussed how to find the estimate $\hat{\theta}$ itself. An intuitive approach commonly used is maximum likelihood estimation (MLE). It was developed by R. A. Fisher in 1912, although apparently Gauss had already thought of it more than a century earlier.[14] For generality, in our initial discussion we will consider not just one variable θ as in our discussion of Cramer–Rao, but instead a vector of parameters $\boldsymbol{\theta}$.

Suppose that we observe N values of the random process under consideration, designated $\mathbf{x} = x(n)$, $n \in [0, N - 1]$. Because $\boldsymbol{\theta}$ characterizes the random process $X(n)$, it will be a parameter vector of the joint PDF: $f_{\mathbf{X}}(\mathbf{x}) = f_{\mathbf{X}}(\mathbf{x}; \boldsymbol{\theta})$. We wish to estimate $\boldsymbol{\theta}$ by $\hat{\boldsymbol{\theta}}$, where $\hat{\boldsymbol{\theta}}$ here represents an estimate *vector*. The MLE principle is to choose $\hat{\boldsymbol{\theta}}$ to maximize $f_{\mathbf{X}}(\mathbf{x}; \boldsymbol{\theta})$, where the "$\mathbf{x}$" in $f_{\mathbf{X}}(\mathbf{x}; \boldsymbol{\theta})$ is the actual N–point data vector from which we form $\hat{\boldsymbol{\theta}}$. We assume that the functional form of the dependence of the PDF on $\boldsymbol{\theta}$ is either known or postulated.

To find $\hat{\boldsymbol{\theta}}$, we simply (multidimensionally) differentiate the PDF with respect to $\boldsymbol{\theta}$ and set the result to zero. The solution is $\hat{\boldsymbol{\theta}}_{\text{MLE}}$. This maximization procedure is both straightforward and intuitively appealing. Substitution of different $\boldsymbol{\theta}$ values into the (N–dimensional) PDF changes the appearance of the PDF. The MLE method selects the PDF model parameters for which the probability that $\mathbf{X}$ took on the value of $\mathbf{x}$ actually observed is maximum. Unfortunately, there is no guarantee that the MLE estimate is unbiased.

However, the MLE estimate has several desirable properties. Consider now the simpler case of estimating only one parameter θ. Because the PDF is nonnegative

---

[14]See R. Deutsch, *Estimation Theory* (Englewood Cliffs, NJ: Prentice Hall, 1965), p. 135. A look in the *Companion Encyclopedia of the History and Philosophy of the Mathematical Sciences* (New York: Routledge, 1994) edited by I. Grattan–Guinness shows the breadth of Gauss's discoveries: the FFT (1805), least squares and the Gaussian probability distribution (1795), quadratic forms (1801), Gaussian elimination (1810), Gauss–Seidel solution (1823), variance and sample variance, and unbiased–minimum variance solution is best (1823), basis for Kalman filter, electromagnetics (e.g., Gauss's law), asteroid orbits, rules for Taylor series–based linearization, approximation methods, inverse of symmetric matrix is symmetric (60 years before the formalization of matrix algebra), first usable proofs of the fundamental theorem of algebra, noneuclidean geometry, the Cauchy integral theorem (1811–14 years before Cauchy), coining of the terms *complex* and *imaginary* number and *norm*, geometrical view of complex numbers, he was the "father of differential geometry," simple form of the divergence theorem (1813), energy approach to explaining capillary action (1830), map–making projection technique still used today (1822), work on elliptic integrals and elliptic functions, work on number theory and approximation theory, solution of $x^n = 1$ for n an integer, background for determinants (first used term in 1801), stochastic versus deterministic errors, number theory (1801), maximum likelihood estimation, and so forth.

and the natural log is a monotonically increasing function of its argument, we know that maximizing $f_{\mathbf{X}}(\mathbf{x}; \theta)$ is equivalent to maximizing $L = \ln\{f_{\mathbf{X}}(\mathbf{x}; \theta)\}$—that is, to setting $\frac{\partial}{\partial\theta}\ln\{f_{\mathbf{X}}(\mathbf{x}; \theta)\}\big|_{\theta=\hat{\theta}_{\text{MLE}}} = 0$. When, for example, Gaussian random variables are involved, maximizing L is certainly the most convenient way of finding $\hat{\theta}_{\text{MLE}}$ because for Gaussian random variables, the natural log of the PDF is just the exponent of the PDF.

One should be cautioned not to associate the log derivative being zero in MLE with the left–hand side of (10.58a) being zero to make conclusions about MLE efficiency as done in some books [e.g., Deutsch (see footnote 14)]. The reason is that in MLE, the derivative of the log–likelihood is set to zero with θ set to $\hat{\theta}_{\text{MLE}}$, not to the true value of θ. Contrarily, the derivative of the log–likelihood in (10.58a) is evaluated at the true value of θ. Thus, conclusions about the right–hand side of (10.58) cannot be inferred for $\hat{\theta} = \hat{\theta}_{\text{MLE}}$; that is, we cannot conclude that $\theta_{\text{MLE}} = \hat{\theta}_{\text{eff}}$.

However,[15] the MLE estimate is asymptotically efficient as the number of data values N used goes to ∞, and the MLE estimate is also asymptotically Gaussian; that is, it asymptotically has the distribution $\eta(\theta, \sigma^2_{C-R})$.

EXAMPLE 10.4

Suppose that $\mathbf{X}$ is truly a Gaussian random vector whose elements are independent Gaussian random variables identically distributed, each with mean μ and variance σ^2. Given an N–vector of data $\mathbf{x}$ (that is, a realization of $\mathbf{X}$), estimate μ using the MLE method. Also find the bias and variance of this estimator; is it efficient? Illustrate the results with plots made using actual Gaussian data.

Solution

First, let us see what the Cramer–Rao inequality has to say on this example. The PDF of $\mathbf{X}$ is known to have the form

$$f_{\mathbf{X}}(\mathbf{x}; [\mu \ \sigma^2]) = \frac{1}{\{\sigma^2(2\pi)\}^{N/2}} \, e^{-\frac{1}{2}\sum_{\ell=1}^{N}\left\{\frac{x_\ell - \mu}{\sigma}\right\}^2}$$

(where we use the convenient $[\mu \ \sigma^2]$ notation even though in this example we are estimating only μ; we will estimate σ^2 in Example 10.5) and the log–likelihood function is

$$L = \ln\{f_{\mathbf{X}}(\mathbf{x}; [\mu \ \sigma^2])\} = -\left(\frac{N}{2}\right)(\ln\{\sigma^2\} + \ln\{2\pi\}) - \frac{1}{2}\sum_{\ell=1}^{N}\left\{\frac{x_\ell - \mu}{\sigma}\right\}^2.$$

The first derivative of L with respect to μ is

$$\frac{\partial L}{\partial\mu} = \sum_{\ell=1}^{N}\left\{\frac{x_\ell - \mu}{\sigma^2}\right\},$$

and the second derivative is

$$\frac{\partial^2 L}{\partial\mu^2} = -\frac{N}{\sigma^2}.$$

---

[15]Kendall, M. G., A. Stuart, and J. K. Ord, *Kendall's Advanced Theory of Statistics*, Vol. 2 (New York: Oxford University Press, 1991).

Thus, the Cramer–Rao inequality (10.56a) becomes

$$\sigma_{\hat{\mu}}^2 \geq \frac{\left\{1 + \frac{\partial B_\mu(\hat{\mu})}{\partial \mu}\right\}^2 \sigma^2}{N}.$$

The maximum likelihood estimate of μ is obtained by setting the partial derivative of the log–likelihood with respect to μ equal to zero:

$$\left.\frac{\partial L}{\partial \mu}\right|_{\mu = \hat{\mu}_{\text{MLE}}} = \sum_{\ell=1}^{N} \left.\left\{\frac{x_\ell - \mu}{\sigma^2}\right\}\right|_{\mu = \hat{\mu}_{\text{MLE}}} = 0,$$

which gives

$$\hat{\mu}_{\text{MLE}} = \frac{1}{N} \sum_{\ell=1}^{N} x_\ell.$$

This estimate of μ is called the sample mean.[16] The expected value of the sample mean, expressed in terms of the random variables X_ℓ, is

$$E\{\hat{\mu}_{\text{MLE}}\} = \frac{1}{N} \sum_{\ell=1}^{N} E\{X_\ell\} = \mu.$$

Thus the sample mean, which is the MLE estimator, is an unbiased estimator of the true mean: $B_\mu(\hat{\mu}_{\text{MLE}}) = 0$. Consequently, the Cramer–Rao bound from the above calculation reduces to the lower bound: $\sigma_{\hat{\mu}_{\text{MLE}}}^2 \geq \sigma^2/N$.

The variance of $\hat{\mu}_{\text{MLE}}$ is found by the definition of variance, where we recall that $\hat{\mu}_{\text{MLE}}$ is unbiased and that the X_ℓ are independent:

$$\sigma_{\hat{\mu}_{\text{MLE}}}^2 = E\{(\hat{\mu}_{\text{MLE}} - \mu)^2\}$$

$$= \frac{1}{N^2} E\left\{\sum_{\ell=1}^{N} (X_\ell - \mu) \sum_{m=1}^{N} (X_m - \mu)\right\}$$

$$= \frac{1}{N^2} \sum_{\ell=1}^{N} \sum_{m=1}^{N} E\{(X_\ell - \mu)(X_m - \mu)\}$$

$$= \sigma^2/N.$$

The variance of $\hat{\mu}_{\text{MLE}}$ is thus equal to the lower bound of the Cramer–Rao inequality as determined above. Thus, for this case, the MLE estimator is efficient for any N; this will not in general be true for other MLE estimation problems. Moreover, because the sum of two or more independent Gaussian random variables is itself Gaussian (a fact most easily proved using characteristic functions), then from our above results, $\hat{\mu}_{\text{MLE}} = \eta(\mu, \sigma^2/N)$.

We now illustrate the above results for some actual Gaussian data. The data was generated by calling `randn` in Matlab and then scaling and shifting to yield $\eta(\mu = 4, \sigma^2 = 2)$. Figure 10.12 shows the $\eta(\mu = 4, \sigma^2 = 2)$ PDF. Dashed lines show the mean and standard deviation to either side of the mean (mean = 4, standard deviation = variance$^{1/2}$ = $\sqrt{2}$). We could verify that about 68% of the values tend to be within the outer dashed lines by finding the area under the curve on this range. This area is tabulated in Gaussian CDF tables. We will examine values of N ranging from 10 to 990. The data may be considered as either N realizations or trials of a single random variable X or as a single realization of N independent random variables

---

[16]If σ^2 is also being estimated, in $\partial L/\partial \mu = 0$, σ^2 should also be set equal to its MLE estimate, but by multiplying both sides of $\partial L/\partial \mu$ by σ^2 we see the equations decouple in this case.

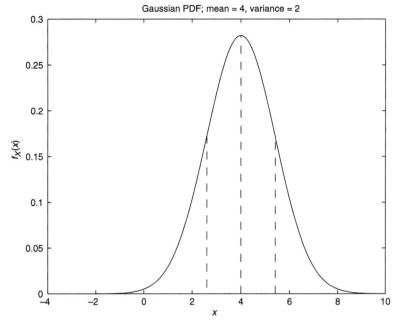

Figure 10.12

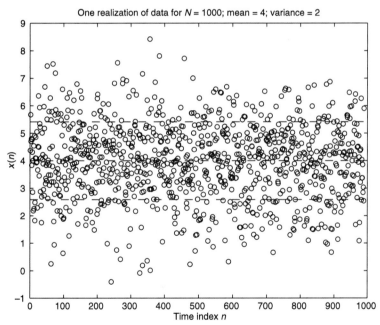

Figure 10.13

of a random vector **X**. In the latter case, which we will assume here, the autocovariance function of the associated random process is proportional to $\delta(k)$ and thus the PSD is constant—the sequence is white Gaussian noise.

For $N = 1000$, a single realization of these N random variables $x(n)$ ($n \in [0, N-1]$) is shown in Figure 10.13. Again, the dashed lines indicate the mean and a standard deviation on either side of it. We see that nearly all sample values are likely to be within a few standard deviations of the mean. Figure 10.14 shows the stochastic average over 512 realizations of the sample mean for $N = 10$ to 990 in steps of 20. We see that even for small N = number of values used to form the sample mean

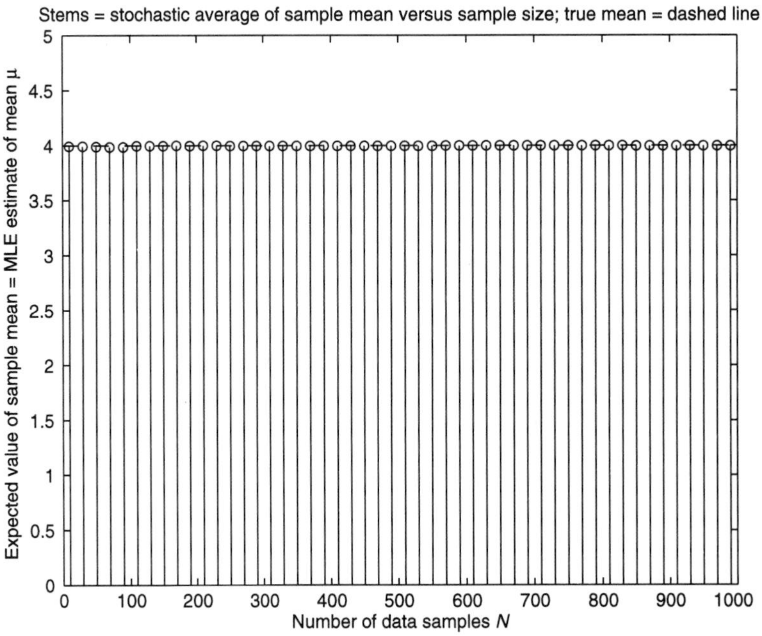

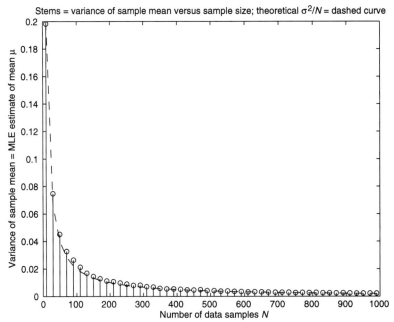

Figure 10.14

Figure 10.15

for each realization, the agreement with $\mu = 4$ is very good—reflecting that the sample mean is unbiased. In Figure 10.15, we show the variance of the sample mean versus N. The dashed curve is the theoretical Cramer–Rao bound σ^2/N, which we showed is attained by the sample mean. Attainment of the Cramer–Rao bound by the sample mean is borne out in Figure 10.15, where the data–calculated stochastically averaged variance of the sample mean (over 512 realizations—the stems) agrees well with σ^2/N.

EXAMPLE 10.5

For the Gaussian random vector in Example 10.4, find the maximum–likelihood estimator of the variance σ^2. Find the bias. Illustrate the results with plots made using actual Gaussian data.

Solution

Consideration of the MLE of the variance of N independent identically distributed Gaussian random variables is more demanding but equally revealing. Again, let us calculate the Cramer–Rao bound. For reference, we repeat here the log–likelihood:

$$L = \ln\{f_{\mathbf{X}}(\mathbf{x}; [\mu \ \sigma^2])\} = -\left(\frac{N}{2}\right)(\ln\{\sigma^2\} + \ln\{2\pi\}) - \frac{1}{2}\sum_{\ell=1}^{N}\left\{\frac{x_\ell - \mu}{\sigma}\right\}^2.$$

The first derivative of L with respect to σ^2 in terms of the x_ℓ is

$$\frac{\partial L}{\partial \sigma^2} = -\frac{N}{2\sigma^2} + \frac{1}{2}\sum_{\ell=1}^{N}\frac{(x_\ell - \mu)^2}{(\sigma^2)^2},$$

and the second derivative is

$$\frac{\partial^2 L}{(\partial \sigma^2)^2} = \frac{N}{2(\sigma^2)^2} - \sum_{\ell=1}^{N}\frac{(x_\ell - \mu)^2}{(\sigma^2)^3}.$$

Thus,

$$\mathrm{E}\left\{\frac{\partial^2 L}{(\partial \sigma^2)^2}\right\} = \frac{N}{2(\sigma^2)^2} - \sum_{\ell=1}^{N}\mathrm{E}\left\{\frac{(x_\ell - \mu)^2}{(\sigma^2)^3}\right\}$$

$$= \frac{N}{2(\sigma^2)^2} - \frac{N}{(\sigma^2)^2} = -\frac{N}{2\sigma^4}.$$

Thus, the Cramer–Rao inequality (10.56a) becomes

$$\sigma^2_{\hat{\sigma}^2} \geq \frac{2\left\{1 + \dfrac{\partial B_{\sigma^2}(\hat{\sigma}^2)}{\partial \sigma^2}\right\}^2 \sigma^4}{N}.$$

The maximum likelihood estimate of σ^2 is obtained by setting the partial derivative of the log–likelihood with respect to σ^2 equal to zero, where we assume that our only knowledge of the value of μ is through μ_{MLE}:

$$\left.\frac{\partial L}{\partial \sigma^2}\right|_{[\mu,\ \sigma^2] = [\hat{\mu}_{\mathrm{MLE}}, \hat{\sigma}^2_{\mathrm{MLE}}]} = -\frac{N}{2\sigma^2} + \frac{1}{2}\sum_{\ell=1}^{N}\left\{\frac{(x_\ell - \mu)^2}{(\sigma^2)^2}\right\}\Bigg|_{[\mu,\ \sigma^2] = [\hat{\mu}_{\mathrm{MLE}}, \hat{\sigma}^2_{\mathrm{MLE}}]} = 0,$$

which gives

$$\boxed{\hat{\sigma}^2_{\mathrm{MLE}} = \frac{1}{N}\sum_{\ell=1}^{N}(x_\ell - \hat{\mu}_{\mathrm{MLE}})^2,}$$

which is called the sample variance. Its expected value is

$$\mathrm{E}\{\hat{\sigma}^2_{\mathrm{MLE}}\} = \frac{1}{N}\sum_{\ell=1}^{N}\mathrm{E}\{(X_\ell - \frac{1}{N}\sum_{m=1}^{N}X_m)^2\}.$$

We play a trick now within the ℓ sum, by subtracting μ and then adding μ/N back N times in the m sum:

$$\mathrm{E}\{\hat{\sigma}^2_{\mathrm{MLE}}\} = \frac{1}{N}\sum_{\ell=1}^{N}\mathrm{E}\{[X_\ell - \mu - \frac{1}{N}\sum_{m=1}^{N}(X_m - \mu)]^2\}.$$

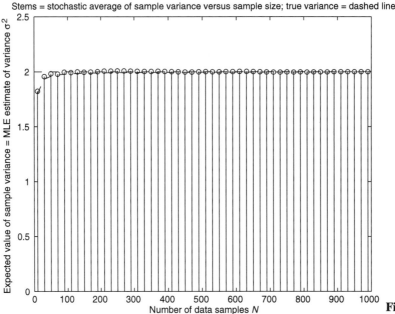

Stems = stochastic average of sample variance versus sample size; true variance = dashed line

Figure 10.16

We now expand out the square (in the following discussion, we omit the $1/N$ outside the ℓ sum). For the first term, we have the sum over ℓ of $\mathrm{E}\{(X_\ell - \mu)^2\}$, which because the variables are all identically distributed gives $N\sigma^2$. For the cross terms, we have the sum over ℓ of $-2\mathrm{E}\{(X_\ell - \mu)\dfrac{1}{N}\sum\limits_{m=1}^{N}(X_m - \mu)\}$. Because the X_ℓ are independent, the ℓth cross–product term reduces to $-2\sigma^2 \sum\limits_{m=1}^{N} \delta(m - \ell)/N$, which when summed over ℓ gives $-2\sigma^2$. Finally, the last term squared is just $N\sigma^2/N = \sigma^2$; we thus obtain

$$\mathrm{E}\{\hat{\sigma}^2_{\mathrm{MLE}}\} = \frac{1}{N}\{N\sigma^2 - 2\sigma^2 + \sigma^2\}$$

$$= \left(1 - \frac{1}{N}\right)\sigma^2.$$

Thus the sample variance, which is the MLE estimator, is a biased estimator of the true variance. However, it is obviously asymptotically unbiased, as $N \to \infty$. As the variance of $\hat{\sigma}^2_{\mathrm{MLE}}$ involves fourth–order moments, we will omit its calculation. However, the Cramer–Rao bound is easily calculated, now that our estimator $\hat{\sigma}^2_{\mathrm{MLE}}$ and thus its bias are known. The bias is $B_{\sigma^2}(\hat{\sigma}^2) = (1 - 1/N)\sigma^2 - \sigma^2 = -\sigma^2/N$ and thus $\partial B/\partial \sigma^2 = -1/N$. Thus, the Cramer–Rao inequality becomes

$$\sigma^2_{\hat{\sigma}^2} \geq \frac{2\left\{1 - \dfrac{1}{N}\right\}^2 \sigma^4}{N} = \frac{2(N-1)^2}{N^3}\sigma^4.$$

For estimation of $\sigma^2 = 2$ in our example of $\eta(\mu = 4, \sigma^2 = 2)$ data, Figure 10.16 shows the stochastic average of the sample variance versus N. The bias $-\sigma^2/N$ is essentially negligible for N greater than about 100 (the sample variance is asymptotically unbiased), but for $N < 100$, the bias is noticeable. Moreover, the stochastic average based on the actual data agrees very well with the theoretical stochastic average $(1 - 1/N)\sigma^2$. The former is shown in stems in Figure 10.16, and the latter is

Stems = variance of sample variance versus sample size; theoretical $2(N-1)^2 \sigma^4/N^3$ = dashed curve

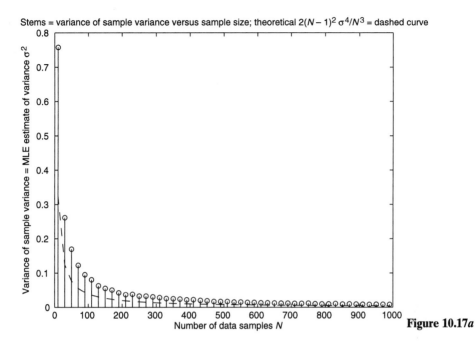

Figure 10.17a

Stems = variance of sample variance versus sample size; theoretical $2(N-1)^2 \sigma^4/N^3$ = dashed curve

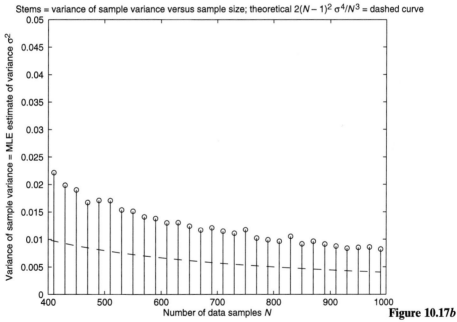

Figure 10.17b

shown in the dashed curve passing through (or nearby) the stems. The true variance, $\sigma = 2$, is also shown with the horizontal dashed line.

Finally, Figure 10.17a shows the stochastic variance of the actual data–computed variance estimate (stems) and the Cramer–Rao bound (dashed curve), both versus N. Notice that unlike the sample mean in Figure 10.15, the sample variance does not attain the Cramer–Rao variance bound; the MLE variance exceeds the Cramer–Rao bound. However, again the variance is asymptotically negligibly small. In Figure 10.17b, we focus on $N \in [410, 990]$ to show more clearly the behavior for fairly large N. Figure 10.17b shows that the Cramer–Rao bound is proportionately no more attainable on this range than for smaller N.

Table 10.2 Events, Probabilities, and Information

Event	Probability p_ℓ	$\log_2 (1/p_\ell)$ [$= \log_{10}(1/p_\ell)/\log_{10}(2)$]
A	$\frac{1}{8}$	3 bits
B	$\frac{1}{16}$	4 bits
C	$\frac{1}{2}$	1 bit
D	$\frac{5}{16}$	1.68 bits

In summary, the maximum likelihood method is a very commonly used and effective technique of statistical parameter estimation. As we have seen, it also produces intuitively meaningful estimates, such as sample mean and sample variance. It also includes many least–squares estimations.

10.6 MAXIMUM ENTROPY

In this section, we investigate the basic principle of entropy. Maximization of entropy is an important technique in the estimation of the digital PSD. Entropy is also important in maximizing the efficiency of digital communication channels.

Consider two equally probable events a_1 and a_2. We need only one binary digit (bit) to encode the information concerning which event occurred. For example, we could let 0 signify that a_1 had occurred and 1 signify that a_2 had occurred. The probability of either event occurring is $\frac{1}{2}$. Similarly, if there are four equally probable events a_1, a_2, a_3, and a_4, we could indicate that a_1 had occurred by 00, a_2 by 01, a_3 by 10, and a_4 by 11. That is, we need two bits to represent all these possibilities. For this case, the probability of each event is $\frac{1}{4}$. For eight equally probable events we would need three bits; the probability of each of these events occurring is $\frac{1}{8}$.

From the above discussion, we see that the rule giving the number of bits required for the transmission of the message specifying the trial outcome is $\log_2(1/p)$, where $p = 1/N$ is the probability of occurrence of one of the N equally probable events. We may say that information can be measured by the length of the message required for its conveyance.

When the N events are not equally probable, the average information required to specify which event occurred is defined and called the entropy H. The information (i.e., minimum number of bits) required to specify that event ℓ occurred out of a collection of N events is $\log_2 (1/p_\ell) = -\log_2 (p_\ell)$. We weight the information of event ℓ, $-\log_2 (p_\ell)$, by fractionally how often we expect to be conveying that event on the average—that is, by the probability of occurrence of the ℓth event, p_ℓ. The sum over all events is the total average information, or entropy:

$$H = -\sum_{\ell=1}^{N} p_\ell \log_2 (p_\ell). \tag{10.60}$$

Now notice that $-\log_2(p) = \log_2(1/p)$ is small if $0 < p \le 1$ is large (near 1). Thus, the most likely events require the fewest numbers of bits. Suppose that we have fixed "begin" and "end" markers on our message stream. Consider the four events described in Table 10.2. Indeed, we see that event C is most probable and requires the fewest bits (1 bit). The message "C occurred" is three bits shorter than the message "B occurred." Noting that Table 10.2 is complete (the p_ℓ sum to unity), using (10.60) we calculate $H = 1.65$ bits. The number of bits $n = \log_2(1/p_\ell)$ to be sent for event ℓ is in this case a discrete random variable. Examination of (10.60) shows that H is just the expected value of this discrete random variable, the number of bits n to be sent:

$$H = E\{\log_2(1/p)\} \tag{10.61a}$$

$$= E\{n\}. \tag{10.61b}$$

The base of the logarithm used in the definition of entropy H as a measure of information content depends on the encoding scheme used for transmitting or identifying the message. Thus, it is quite arbitrary—for example, we can use the natural log (base e) if we measure information in "e-ary" units as opposed to binary units (bits).

Let us further examine the meaning of entropy. Consider the extreme example in which there are 10 events, but $p_{\ell \neq 3} = 0$ and $p_3 = 1$. Then $H = 0 + 0 + 1 \cdot \log_2(1) + 0 + \ldots + 0 = 0$. That is, event 3 is certain, so there is no information in the message "event 3 occurred"—we already know that. Such a case is deterministic—no uncertainty, perfectly ordered. In all other nontrivial cases, $H > 0$ and is a measure of the "disorder" or "randomness" of the system. We may also view entropy H as a measure of our state of ignorance or uncertainty about the actual structure of a system. When we choose a probability assignment that describes the available information but is maximally noncommittal to the unavailable information, we regard such an assignment as *maximum entropy* because it maximizes H.[17] With maximum entropy, we impose no more constraints than the minimum required for consistency with the knowledge and data we do have. Alternatively, Norbert Wiener said[17] that a transmitted message can be regarded as a random variable drawn from a universe whose distribution function reflects our *a priori* knowledge of the situation.

If the random variable takes on a continuous range of values, then the sum in (10.60) must be replaced by an integral and the p_ℓ by a probability density function. We can think of N samples of a continuous–time random process as a collection of N random variables having a joint PDF $f_{\mathbf{X}}(\mathbf{x})$. Thus, in that case, the entropy is (expressed conventionally in e–ary units—which we will henceforth use exclusively):

$$H = -\int f_{\mathbf{X}}(\mathbf{x}) \ln\{f_{\mathbf{X}}(\mathbf{x})\}d\mathbf{x} \tag{10.62a}$$

$$= -E\{L\}, \tag{10.62b}$$

where L is the log–likelihood function, introduced in our discussion of the Cramer–Rao bound in Section 10.5.1. Thus, the log–likelihood function is a measure of "information" associated with the point $\mathbf{x}$ in the N–dimensional sample space, and its expected value is (minus) the entropy, $-H$.

Finally, the entropy of a random process may also be defined. B. McMillan makes use of an "alphabet" and a "message," which is an N–length sequence of "letters."[18] When N is finite, the entropy H (information content) of the message is finite, but for $N \to \infty$, H becomes infinite. In this case—the situation for random processes $X(n)$ that go on "forever"—*the entropy rate* or information content per letter, h, is defined as

$$h = \lim_{N \to \infty} \frac{H}{N}. \tag{10.63}$$

---

[17]T. J. Ulrych and T. N. Bishop, "Maximum Entropy Specral Analysis and Autoregressive Decomposition," *Review of Geophysics and Space Physics*, 13, 1 (1975): 183–200.

[18]B. McMillan, *Annals of Mathematical Statistics*, 24 (1953): 196–205.

The "noncommittal with respect to unknown data" characteristic of high entropy has been shown to be useful for reliable parameter estimation. A full discourse on this "philosophy" is beyond the scope of this book, but may be gained by reading the interesting papers of Edwin Jaynes, a pioneer in the field of information theory. Essentially, anything other than maximally noncommittal imposes a structure on the data that based on our known data we have no right to impose. Consequently, the non–maximum–entropy estimate is far less likely than the maximum entropy solution, because the latter includes all the possibilities, whereas the former is more restrictive (and probably erroneously so). Let us instead here confine our attention to answering the question, What distribution is maximum entropy?

We might guess from our above examples and reasoning that the maximum entropy distribution for a PMF is one that is uniform for all permitted values. Recall our example where we knew that $p_3 = 1$ and $p_{\ell \neq 3} = 0$—the exact opposite of a uniform distribution—for which we have the minimum possible entropy, $H = 0$. We may show that the uniform PMF is maximum entropy as follows. Now let p_ℓ quantify the probability that $X = x_\ell$, where x_ℓ is one of the N possible values for X. It is postulated that if $H(X)$ for random variable X can be written in the form

$$H(X) = -\sum_{\ell=1}^{N} p'_\ell \ln\{p_\ell\}, \tag{10.64}$$

where p'_ℓ is associated with the PMF of any random variable Y other than X (whose PMF is specified by p_ℓ), then $H(X) \geq H(Y)$. Assuming that we can "massage" $H(X)$ into the form (10.64), then

$$H(X) - H(Y) = -\sum_{\ell=1}^{N} p'_\ell (\ln\{p_\ell\} - \ln\{p'_\ell\}) \tag{10.65a}$$

$$= -\sum_{\ell=1}^{N} p'_\ell \ln\{p_\ell/p'_\ell\}. \tag{10.65b}$$

Using Taylor's expansion with the Lagrangian form of remainder (in turn based on integrations of the mean value theorem), it is proved in the problems that

$$\ln\{x\} = x - 1 - \frac{(x-1)^2}{2[1 + \alpha(x-1)]^2}, \quad \alpha \in [0, 1]. \tag{10.66}$$

Because the remainder (third) term in (10.66) is always negative, we have $\ln\{x\} \leq x - 1$, or $-\ln\{x\} \geq 1 - x$. Using this fact in (10.65b), as well as noting that all p'_ℓ are nonnegative and that $\sum_{\ell=1}^{N} p_\ell = \sum_{\ell=1}^{N} p'_\ell = 1$, gives

$$H(X) - H(Y) \geq \sum_{\ell=1}^{N} p'_\ell \left(1 - \frac{p_\ell}{p'_\ell}\right) \tag{10.67a}$$

$$= \sum_{\ell=1}^{N} p'_\ell - \sum_{\ell=1}^{N} p_\ell \tag{10.67b}$$

$$= 1 - 1 = 0, \tag{10.67c}$$

or

$$H(X) \geq H(Y) \quad (X \text{ has the maximum–entropy PMF}). \tag{10.67d}$$

Let us try out our hypothesis that the uniform distribution is maximum entropy. If $p_\ell = 1/N$ for all ℓ where N is the number of possible values of X (and of Y), we again use $\sum_{\ell=1}^{N} p_\ell = \sum_{\ell=1}^{N} p'_\ell = 1$ to show

$$H(X) = -\sum_{\ell=1}^{N} p_{\ell} \ln\{p_{\ell}\} \tag{10.68a}$$

$$= -\ln\left\{\frac{1}{N}\right\} \sum_{\ell=1}^{N} p_{\ell} \tag{10.68b}$$

$$= -\ln\left\{\frac{1}{N}\right\} \sum_{\ell=1}^{N} p_{\ell}' \tag{10.68c}$$

$$= -\sum_{\ell=1}^{N} p_{\ell}' \ln\{p_{\ell}\}, \tag{10.68d}$$

which is the form in (10.64); thus X is maximum–entropy–distributed.

It can be shown using exactly the same reasoning that the uniform PDF is the maximum–entropy PDF of all continuous random variable PDFs defined on a range of x designated B. Similarly to (10.64), the PDF of a continuous–value random variable X is maximum entropy over a class C of PDFs and thus $H(X) \geq H(Y)$ where Y is any other random variable, if $H(X)$ can be written in the form

$$H(X) = -\int_{B} f_Y(x) \ln\{f_x(x)\} \, dx, \tag{10.69}$$

where $f_Y(y)$ is the PDF of Y.

EXAMPLE 10.6

Show that the Gaussian PDF of variance σ^2 is the maximum–entropy PDF over all PDFs whose variance is σ^2, on the range $B = (-\infty, \infty)$.

Solution

First of all, the property $f_{X-a}(x) = f_X(x - a)$ can be used to show that $H(X - a) = H(X)$, and, in particular, $H(X - \mu_X) = H(X)$. Therefore, without loss of generality, let X be $\eta(0, \sigma^2)$. Thus,

$$H(X) = -\int_{-\infty}^{\infty} \frac{e^{-x^2/(2\sigma^2)}}{\sigma(2\pi)^{1/2}} \left\{ -\frac{1}{2}\ln\{2\pi\sigma^2\} - \frac{1}{2}\left(\frac{x}{\sigma}\right)^2 \right\} dx. \tag{10.70a}$$

The first term in (10.70a) is $\frac{1}{2}\ln\{2\pi\sigma^2\}$, because the rest of that term other than that constant is just the integral of the PDF over all x (namely unity). The second term is, aside from the constant $\frac{1}{2}(1/\sigma^2)$, just the variance of x (namely σ^2). These same results, namely $H(X) = \frac{1}{2}(\ln\{2\pi\sigma^2\} + 1)$, would still be obtained if the Gaussian PDF in (10.70a) were replaced by any other PDF for which the variance is σ^2. Therefore, we can replace the PDF in (10.70a) for both terms by $f_Y(x)$ of any other random variable Y having variance σ^2.

$$H(X) = -\int_{-\infty}^{\infty} f_Y(x)\left\{ -\frac{1}{2}\ln\{2\pi\sigma^2\} - \frac{1}{2}\left(\frac{x}{\sigma}\right)^2 \right\} dx \tag{10.70b}$$

$$= -\int_{-\infty}^{\infty} f_Y(x) \ln\{f_X(x)\} \, dx, \tag{10.70c}$$

which is the form (10.69). Thus we have verified that the Gaussian PDF is maximum entropy of all PDFs having the same variance. It is left to the problems to show that of all random N–vectors having the same autocorrelation matrix R_{XX}, the one that is Gaussianly distributed has maximum entropy. The reason the Gaussian PDF is so special in this context is that its log–likelihood function is a quadratic—the terms in H of which are associated with zeroth, first, and second moments of the random variable. More advanced analysis shows that of all zero–mean random processes having the same first N autocorrelation values $r_{XX}(k)$, the one with maximum entropy is the purely autoregressive process having those lags.

In this review of random variables and random processes, we have covered most of the major ideas important in their application to statistical digital signal processing. Although the theory of random variables is quite old, the concept of random signal processing emerged in the World War II era. With the increases in the computing power of small computers, techniques are now feasible that were originally considered esoteric or impractical. In the remainder of this chapter, we consider some of the most basic techniques of the processing of random signals. The problems tackled will involve quantization error in digital filters, deconvolution, prediction, whitening, and adaptive filtering/interference cancellation. Many of the problems examined will be seen to be related. We will finally use the theory in some very practical examples. We begin our study of applications of stochastic theory to signal processing by considering the effects of passing random signals through linear shift–invariant systems.

10.7   LSI SYSTEMS WITH STOCHASTIC INPUTS

We have seen that for characterizing LSI systems, two fundamental inputs are $\delta(n)$ and $e^{j\omega n}$. The respective system outputs are $h(n)$ and $H(e^{j\omega}) \cdot e^{j\omega n}$, namely the unit sample ("impulse") response and the input complex sinusoid scaled by the frequency response evaluated at the input frequency ω. We can always view an energy signal as being the output of a particular LSI system having $\delta(n)$ as its input.

Perhaps the most basic stochastic relation between input $X(n)$ and output $Y(n) = X(n) * h(n)$ is that between the means μ_X and μ_Y. Taking the expectation of convolution, we directly have

$$\mu_Y = E\{ \sum_{m=-\infty}^{\infty} h(m)X(n-m)\} = \sum_{m=-\infty}^{\infty} h(m)E\{X(n-m)\}. \tag{10.71}$$

Because $X(n)$ is assumed to be WSS, $E\{X(n-m)\} = E\{X(n)\} = \mu_X$. Thus

$$\mu_Y = \mu_X \sum_{m=-\infty}^{\infty} h(m). \tag{10.72}$$

In particular, if $\mu_X = 0$, then $\mu_Y = 0$ for stable systems.

In Section 10.3, we noted that a stochastic signal does not have a Fourier transform. Instead, we may characterize its spectral content by its (nonnegative, real) PSD $P_{XX}(\omega)$, which converges for all WSS stochastic signals. Under the assumption of wide–sense stationarity and Gaussian origin[19], we may statistically characterize a

---

[19]In this book, we confine our attention to first– and second–order moments. Although for a given $r_{XX}(k)$ and μ_X we cannot distinguish whether the process $X(n)$ is Gaussian [for other random processes with the same $r_{XX}(k)$ and μ_X may have higher–order moments different from those for the Gaussian process $\eta(\mu_X, r_{XX}(k))$], in this case, we will assume that $X(n)$ is Gaussian and thus completely characterized by μ_X and $r_{XX}(k)$. We will also often assume in this chapter that all time sequences, and thus their statistics, are real–valued (although for generality we begin by assuming complex–valued sequences).

random signal $X(n)$ completely by its mean μ_X and its ACF $r_{XX}(k)$. Moreover, if we first subtract off the mean μ_X from $X(n)$ [e.g., define $X'(n) = X(n) - \mu_X$, and then redefine $X(n)$ to be $X'(n)$], then all that will be left of the statistical characterization of $X(n)$ will be $r_{XX}(k)$. After processing, we can always add μ_X back on; thus without loss of generality we usually assume $\mu_X = 0$ unless we wish to show explicitly how the nonzero mean is handled.

In Section 10.3 we proved the Wiener–Kinchin theorem: $P_{XX}(\omega) =$ DTFT$\{r_{XX}(k)\}$. Thus, if we desire information on how the spectral characterization of a WSS stochastic signal $X(n)$ changes as it is filtered by an LSI system, we must be able to relate the ACF $r_{YY}(k)$ of the output $Y(n)$ to $r_{XX}(k)$. Then $P_{YY}(\omega) =$ DTFT$\{r_{YY}(k)\}$ is the PSD of the output. We thus suppose that the realization $x(n)$ of random process $X(n)$ is filtered by a filter having unit sample response $h(n)$, yielding an output $y(n)$—a realization of a random process $Y(n)$. The ACF $r_{YY}(k)$ is determined by definition:

$$r_{YY}(k) = E\{Y(n + k)Y^*(n)\}, \tag{10.73a}$$

which when the convolution expression for $Y(n + k)$ in terms of $\{X(n)\}$ is substituted gives

$$r_{YY}(k) = \sum_{m=-\infty}^{\infty} h(m)E\{X(n + k - m)Y^*(n)\}. \tag{10.73b}$$

By the definition of cross–correlation (see Table 10.1), (10.73b) may be rewritten as

$$r_{YY}(k) = \sum_{m=-\infty}^{\infty} h(m)r_{XY}(k - m). \tag{10.73c}$$

Recognizing the convolution in (10.73c), we have

$$r_{YY}(k) = h(k) * r_{XY}(k). \tag{10.73d}$$

We thus have the intermediate result $r_{YY} = h * r_{XY}$, and thus by the convolution theorem the cross–spectral density $P_{XY}(\omega) =$ DTFT$\{r_{XY}(k)\}$ and $P_{YY}(\omega)$ are related by $P_{YY}(\omega) = H(e^{j\omega}) P_{XY}(\omega)$. However, we seek r_{YY} in terms of r_{XX}, not r_{XY}. Let us then substitute the convolutional expression for $Y(n)$ into the definition of $r_{XY}(k)$:

$$r_{XY}(k) = E\{X(n + k)Y^*(n)\} \tag{10.74a}$$

$$= E\{X(n + k) \sum_{\ell=-\infty}^{\infty} h^*(\ell)X^*(n - \ell)\}, \tag{10.74b}$$

which upon rearrangement becomes

$$r_{XY}(k) = \sum_{\ell=-\infty}^{\infty} h^*(\ell)E\{X(n + k)X^*(n - \ell)\} \tag{10.74c}$$

$$= \sum_{\ell=-\infty}^{\infty} h^*(\ell)r_{XX}(k + \ell). \tag{10.74d}$$

Upon defining $m = -\ell$, (10.74d) becomes

$$r_{XY}(k) = \sum_{m=-\infty}^{\infty} h^*(-m)r_{XX}(k - m) \tag{10.74e}$$

$$= h^*(-k) * r_{XX}(k). \tag{10.74f}$$

Putting (10.73d) and (10.74f) together and using the associative property of convolution gives

$$r_{YY}(k) = \{h(k) * h^*(-k)\} * r_{XX}(k). \tag{10.75}$$

Noting that

$$\{h(k) * h^*(-k)\}(m) = \sum_{k=-\infty}^{\infty} h(k)h^*(-[m-k]), \tag{10.76a}$$

which under the substitution $\ell = k - m$ becomes

$$\{h(k) * h^*(-k)\}(m) = \sum_{\ell=-\infty}^{\infty} h(\ell + m)h^*(\ell), \tag{10.76b}$$

we recognize the right–hand side of (10.76b) as the temporal ACF that we studied in Section 7.8. We will henceforth denote it by $r_{hh;n}(m)$ to distinguish it from a stochastically averaged ACF.[20] Thus we have

$$r_{YY}(k) = \{r_{hh;n} * r_{XX}\}(k). \tag{10.77}$$

The relation corresponding to (10.75) in the frequency domain is obtained by recalling that $\mathrm{DTF}\{h^*(-n)\} = H^*(e^{j\omega})$. Using this DTFT pair and applying the convolution theorem to (10.75) give

$$P_{YY}(\omega) = H(e^{j\omega})H^*(e^{j\omega})P_{XX}(\omega) \tag{10.78a}$$

$$= |H(e^{j\omega})|^2 P_{XX}(\omega). \tag{10.78b}$$

More generally, recalling Example 10.3, we may define $R_{YY}(z) = \mathrm{ZT}\{r_{YY}(k)\}$ where $P_{YY}(\omega) = R_{YY}(e^{j\omega})$. Because $\mathrm{ZT}\{h^*(-n)\} = H^*(1/z^*)$, we have

$$R_{YY}(z) = H(z)H^*\left(\frac{1}{z^*}\right)R_{XX}(z). \tag{10.79}$$

Equation (10.79), with $R_{XX}(z)$ set to unity, is useful for determining the ACF of certain WSS signals via the inverse z–transform from the difference equation model coefficients that the signal satisfies, as we shall demonstrate in Example 10.7.

An important special case of (10.77) and (10.78) is when $X(n)$ is zero–mean white noise of variance σ^2: $r_{XX}(k) = \sigma^2\delta(k)$ and $P_{XX}(\omega) = \sigma^2$ for all ω. In this case, (10.77) becomes

$$r_{YY}(k) = \sigma^2 r_{hh;n}(k) \tag{10.80}$$

and thus for $\sigma^2 = 1$, the stochastic ACF of $Y(n)$ is equal to the temporal (deterministic) ACF of $h(n)$. Also when $X(n)$ is zero-mean white noise, (10.78) becomes $P_{YY}(\omega) = \sigma^2 |H(e^{j\omega})|^2$.

Because in this special case $Y(n)$ is the output of an LSI system having white noise input and thus it satisfies an ARMA difference equation, it is called an

---

[20]We distinguish $r_{hh;n}(k)$ from $r_{N,n}(k)$ in (10.8a) [or equivalently, (10.10)] because the latter is formed from only a finite portion of the original sequence and is divided by N. Also, $r_{hh;n}(k)$ conveniently allows specification of the signal h in its symbol. We may use $r_{N,n}(k)$ to implement $r_{hh;n}(k)$ in practice.

ARMA sequence. Even when $Y(n)$ does not exactly satisfy this model, it often approximately does. In fact, E. A. Robinson[21] showed that any $Y(n)$ for which the spectral distribution $\Lambda(\omega)$ [which is defined by $\partial\Lambda(\omega)/\partial\omega = P_{YY}(\omega)$] is absolutely continuous can be decomposed into the convolution of the deterministic $h(n)$ with the random white noise $U(n)$, where $h(n)$ is causal and minimum–phase. Recall that a minimum–phase system is one for which no poles or zeros are in the right–half plane. Of course, the true $h(n)$ may be infinite–length, and the corresponding ARMA difference equation may have arbitrarily high order.

The ARMA *parametric* model of a stochastic sequence is the most general and versatile model possible by a linear transformation of a white noise sequence. The above definition of an ARMA sequence is analogous to defining an LSI sequence as any (deterministic) sequence that can be produced by an LSI system having a unit sample as its input. Remember that the ACF of white noise of variance σ^2 is $r_{UU}(k) = \sigma^2\delta(k)$, whereas the ACF of a general ARMA sequence $X(n)$ is $r_{XX}(k) = r_{hh;n}(k)$, the temporal ACF of the filter impulse $h(n)$ response producing $X(n)$ at its output from white noise input $U(n)$ having unit variance $\sigma^2 = 1$.

Other interesting relations between correlation quantities and filter quantities can be proved. One such relation is that the cross–correlation $r_{XU}(k)$ between a signal $X(n)$ and the unit–variance white noise sequence $U(n)$ that produced it via the LSI system $h(n)$ $(X = U * h)$ is just $r_{XU}(k) = h(k)$ (see the problems).

EXAMPLE 10.7

Find and plot the PSD and ACF of the following ARMA sequence: $X(n)$ satisfies the difference equation $X(n) = 0.5X(n-1) - 0.2X(n-2) + 0.2X(n-3) + U(n) - 0.3U(n-1)$, where $U(n)$ is a zero–mean unit–variance white noise sequence.

Solution

By direct substitution of the coefficients, we have for the filter that produces $X(n)$ when white noise is applied

$$H(z) = \frac{1 - 0.3z^{-1}}{1 - 0.5z^{-1} + 0.2z^{-2} - 0.2z^{-3}}$$

$$= \frac{z^2(z - 0.3)}{(z - 0.658)(z - z_1)(z - z_1^*)},$$

where $z_1 = -0.0790 + j0.5456 = 0.5513 \angle 98.24°$. From the first expression for $H(z)$, we immediately obtain

$$H^*(1/z^*) = \frac{1 - 0.3z}{1 - 0.5z + 0.2z^2 - 0.2z^3}.$$

The PSD $P_{XX}(e^{j\omega}) = |H(e^{j\omega})|^2$ is as shown in Figure 10.18, where for convenience we use digital frequency f rather than ω on the abscissa.

---

[21]E. A. Robinson, "Predictive Decomposition of Time Series with Application to Seismic Exploration," *Geophysics*, 32, 3 (June 1967): 418–484.

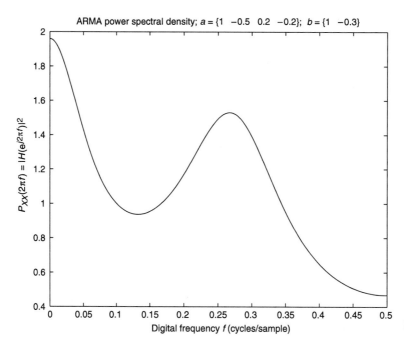

Figure 10.18

The ACF $r_{XX}(k)$ is most easily obtained by finding the inverse z–transform of $R_{XX}(z)$ found using (10.79), which in this case reads $R_{XX}(z) = H(z)H^*(1/z^*)$. We use the method of residues:

$$r_{XX}(k) = \text{ZT}^{-1}\{R_{XX}(z)\}$$

$$= \text{ZT}^{-1}\left\{H(z)\, H^*\!\left(\frac{1}{z^*}\right)\right\}$$

$$= \frac{1}{2\pi j}\oint_C \frac{z^2(z-0.3)}{(z-0.658)(z-z_1)(z-z_1^*)}\cdot\frac{(1-0.3z)}{(1-0.5z+0.2z^2-0.2z^3)}\, z^{k-1}dz.$$

To evaluate the integral, note that the poles of $H(z)$ are at 0.658, z_1, and z_1^*, all of which are within the unit circle. The filter $h(n)$ is stable and causal, so the region of convergence of $H(z)$ is everywhere outside the origin–centered circle of radius 0.658 (the radius of the pole farthest from the origin). The poles of $H^*(1/z^*) = \text{ZT}\{h^*(-n)\}$ are at the conjugate–reciprocal locations of the poles of $H(z)$. Because $h^*(-n)$ is an anticausal sequence, the region of convergence of $H^*(1/z^*)$ is everywhere within the origin–centered circle of radius $1/0.658 = 1.5198$ [the radius of the pole of $H^*(1/z^*)$ nearest the origin].

The inverse z–transform integral is defined on the ring $0.658 < |z| < 1/.0658 = 1.5198$; we thus conveniently choose the unit circle for C. There are three poles of the integrand enclosed by C, namely the poles of $H(z)$. The last two inverse z-transform terms are complex conjugates of each other and thus sum to twice the real part of either term. Thus, for $k \geq 0$,

$$r_{XX}(k) = \frac{(0.658-0.3)(1-0.3\cdot 0.658)}{|0.658-z_1|^2\,(1-0.5\cdot 0.658+0.2(0.658)^2-0.2(0.658)^3)}(0.658)^{k+1}$$

$$+\ 2\text{Re}\left\{\frac{(z_1-0.3)(1-0.3z_1)}{(z_1-0.658)\, j2\text{Im}\{z_1\}(1-0.5z_1+0.2z_1^2-0.2z_1^3)}\, z_1^{k+1}\right\},$$

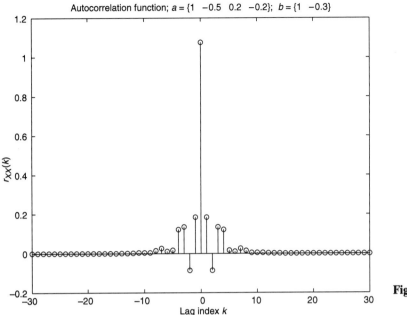

Figure 10.19

which when the complex arithmetic is computed gives

$$r_{XX}(k) = 0.3209 \cdot (0.658)^k + 2\mathrm{Re}\{(0.3789 \ \angle{-4.5233°})(0.5513 \ \angle 98.24°)^k\}$$

$$= 0.3209 \cdot (0.658)^k + 0.7578 \cdot (0.5513)^k \cos\{(\pi/180°) \cdot (98.24° \cdot k - 4.5233°)\}$$

for $k \geq 0$. For $k < 0$, we use $r_{XX}(-k) = r_{XX}^*(k)$. This ACF, for the first 30 positive and negative lags, is shown in Figure 10.19. We see that although technically $r_{XX}(k)$ is infinite–length, its values become insignificant for $|k|$ greater than about 12. It is easily verified in Matlab that the DTFT of $r_{XX}(k)$ as just expressed does indeed reproduce $P_{XX}(\omega)$ plotted in Figure 10.18.

10.8   ANALYSIS OF ROUNDOFF IN DIGITAL FILTERS

We can make immediate practical use of stochastic signal theory in the quantification of the effects of arithmetic roundoff in digital filters. In Section 9.6, we analyzed the effects of quantization of filter coefficients. Here we study the effects of round-off of products in b–bit implementations of digital filters.

As noted in Section 9.6.1, depending on the complexity of the processor, either fixed– or floating–point processors may be used. Except for the situation of over-flow handling, addition in fixed point can be done exactly; in floating point, there may be small rounding error in the process of lining up the binary points. However, the main rounding problem occurs in multiplication. That is because the result, which has $2b$ significant bits, has to be truncated to one having only b significant bits. This loss of information occurs in floating–point as well as in fixed–point processors, but for simplicity we will focus on fixed–point implementations.

The operations of rounding and truncation are nonlinear—the quantization of $2x$ is not in general equal to 2 times the quantization of x. Consequently, in assessing how much effect these operations have on the output of a digital filter, we

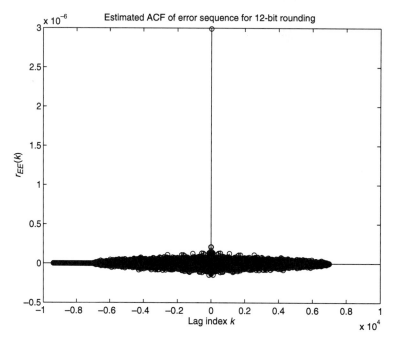

Figure 10.20

cannot directly use linear analysis. However, we can replace quantization error by equivalent noise sources inserted into the linear model, wherever the quantization occurs. Then we can analyze the propagation of these noise sources from their origin to the output. The total of all the individual errors will then be an estimate of the overall quantization error from all the filter computations.

The fundamental idea is based on the appearance of the rounding error $e(n) = x(n) - \text{round}\{x(n)\}$ back in Figure 1.3b: It is approximately white noise [each random variable $E(n)$ is independent of all other $E(m)$]. Referring to Figure 1.3a, we see that the fraction of the way from one quantization level to the next at which our sample at time index n lands is nearly unpredictable and independent of the corresponding fraction for $n + 1$ or for $n - 1$.

Let N_L be the number of bits that the MSB is to the left of the binary point; for conventional fixed–point systems, $N_L = -1$ (the MSB is just on the right of the binary point; N_L can be positive or negative, but never zero). The quantization interval of such a b–bit representation is $\Delta = 2^{-(b-N_L-1)}/\alpha$, where α is the factor by which $x(n)$ is scaled to avoid overflow. In practice, where Δ is much smaller than that shown in Figure 1.3a, the assumption that $E(n)$ is white noise is more valid than Figure 1.3b implies. Thus, the quantization error random signal $E(n)$, whose values range from $-\frac{1}{2}\Delta$ to $\frac{1}{2}\Delta$, is approximately a white noise sequence. It is also equally likely that $E(n)$ could take on any value between $-\frac{1}{2}\Delta$ and $\frac{1}{2}\Delta$; hence $E(n)$ is a uniformly distributed random variable.

In Example W10.1.2, it is shown that the variance of a random variable X uniformly distributed between $x = A$ and $x = B$ is equal to $(B - A)^2/12$. We here set $A = -\frac{1}{2}\Delta$ and $B = \frac{1}{2}\Delta$, we thus conclude that $E(n)$ has variance $\sigma^2 = \Delta^2/12$ and zero mean. Recall that zero–mean white noise has ACF $\sigma^2\delta(k)$. Using $r_{ee;n}(k)$ calculated by the FFT method in Section 7.8 for $r_{EE}(k)$ (with division of the results by N) and $\{b = 12, N_L = 3, \alpha = 1\}$, we obtain Figure 10.20; clearly, it is reasonably approximated as $\sigma^2\delta(k)$—that is, white noise. In fact, for this case $\Delta = 2^{-(12-3-1)} = 0.0078$, so that ideally $\sigma^2 = (0.0078)^2/12 \approx 5 \cdot 10^{-6}$, which is reasonably close to $r_{ee;n}(0) \approx 3 \cdot 10^{-6}$ in Figure 10.20. The fact that the temporal ACF estimate $r_{ee;n}(0)$ is a bit too small might be explained by "leakage" to $k \neq 0$ lags in Figure 10.20, which ideally are zero.

In the example just discussed, we merely rounded the signal $x(n)$, which is actual speech samples. Suppose now instead that we quantize every multiplication occurring within the digital filter. Now, each rounding error signal such as $E(n)$ above is further processed (added, multiplied, rounded) by later stages in the filter. In the following example, we estimate the total error due to all the instances of product rounding within the digital filter.

EXAMPLE 10.8

For the Chebyshev II digital filter studied in the examples of Section 9.6, namely

$$H(z) = \frac{0.1852z^5 + 0.8354z^4 + 1.5893z^3 + 1.5893z^2 + 0.8354z + 0.1852}{z^5 + 1.5435z^4 + 1.5565z^3 + 0.8273z^2 + 0.2584z + 0.03414},$$

investigate the rounding effect for $b = 12$ bits and the direct form II structure in Figure 9.24. Compare results based on theoretical considerations with those based on a nonlinear simulation of rounding performed in Simulink. Also compare with the results for cascade and parallel structures.

Solution

In Figure 10.21, we add in individual noise sequences that are equivalent to the quantizations of each corresponding multiplication. (For simplicity, with the exception of variance labeling, here we will use e to refer to either a particular realization or the abstract random process behind it). Noting that $e_{a1}(n)$ through $e_{a5}(n)$ are all summed, we might as well replace them with a single $e_a(n) = \sum_{m=1}^{5} e_{am}(n)$ added to $x(n)$. Because $e_{am}(n)$ are all independent, their variances simply add. Thus, assuming that the variance of $e_{am}(n)$ is $\sigma^2 = \Delta^2/12$, then the variance of $e_a(n)$ is $\sigma_{E_a}^2 = N\sigma^2 = 5\sigma^2$. Note that $e_a(n)$ added to $x(n)$ without the $e_{am}(n)$ shown in Figure 10.21, is equivalent to the original Figure 10.21 with the $e_{am}(n)$. We see that $e_a(n)$ travels through the entire now nonquantizing and linear digital filter, exactly as $x(n)$ does. Consequently, the filtered version of $e_a(n)$—that is, the noise output contribution due to $e_a(n)$, which we denote $e_a'(n)$—is zero–mean colored (no longer white) noise with variance $\sigma_{E_a'}^2 = r_{E_a'E_a'}(0) = \sigma_{E_a}^2 r_{hh;n}(0)$ [see (10.80)].

Similarly, if we define $e_b(n) = \sum_{\ell=0}^{M} e_{b\ell}(n)$, then $e_b(n)$ has variance $\sigma_{E_b}^2 = (M + 1)\sigma^2 = 5\sigma^2$. Now, $e_b(n)$ properly enters the final summation producing $y(n)$ and the $e_{b\ell}(n)$ are removed; $e_b(n)$ undergoes no filtering and so its contribution remains white in the output [in the notation above, $e_b'(n) = e_b(n)$]. We conclude that the total error in the output has variance equal to $\sigma_{dir\ II}^2 = N\sigma^2 r_{hh;n}(0) + (M + 1)\sigma^2$ for the direct form II structure. It is left to the problems to show that for the cascade structure for this filter, $\sigma_{casc}^2 = \sigma^2\{r_{hh;n}(0) + A^2[2r_{h_{12}h_{12};n}(0) + 3r_{h_2h_2;n}(n) + 1] + 1\}$, where

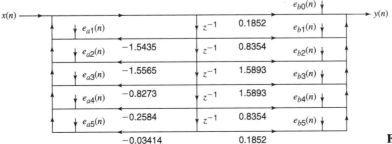

Figure 10.21

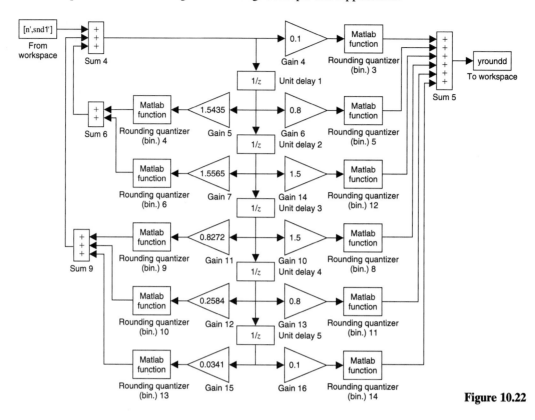

Figure 10.22

$A = 0.185217$, $H_{12}(z)$ is the cascade of the two second order cascade modules and $H_2(z)$ is the second second–order module. For the parallel form, it is shown in the problems that $\sigma_{par}^2 = \sigma^2\{B^2 + 2r_{h_{1p}h_{1p};n}(0) + 2r_{h_{2p}h_{2p};n}(0) + 6\}$, where $B = -2.79963$, and $H_{1p}(z)$ and $H_{2p}(z)$ are, respectively, the transfer functions of the first and second parallel second–order structures.

We carry out the numerical evaluation of the various ACFs evaluated at $k = 0$ needed for the variances above in Matlab, using the numerical quadrature routine

quad 8 applied to (7.53) $[\text{DTFT}\{r_{hh;n}(k)\} = R_{hh;n}(e^{j\omega}) = |H(e^{j\omega})|^2]$: $r_{hh;n}(0) = \frac{1}{2\pi}\int_{-\pi}^{\pi} |H(e^{j\omega})|^2 d\omega$, where $H(z)$ is as given in the problem statement. The final numerical result for the direct form II structure is $\sigma_{dirII}^2 = 4.661 \cdot 10^{-5}$. [We used $N_L = 3$, and $x(n)$ was prescaled by $\alpha = 0.5$ to avoid overflow.]

To check our work, in Figure 10.22 we implement the nonlinear N_b–bit rounding operation for the direct form II structure using fcn in Simulink and a homemade rounding routine. Again our input $x(n)$ is speech data. The output error is obtained by running the simulation with versus without quantization performed. As shown, when quantization is performed, it is done after every multiplication. When we simulate and compute the sample variance of the actual error, we obtain $\sigma_{dirII}^2 = 5.0633 \cdot 10^{-5}$, which is quite close to the theoretical value, $4.661 \cdot 10^{-5}$.

For the cascade structure for this filter, we obtain the following results: $\sigma_{casc}^2 = 1.413 \cdot 10^{-5}$ for the theoretical value and $\sigma_{casc}^2 = 1.850 \cdot 10^{-5}$ for the sample variance. For the parallel structure for this filter, we obtain the following results: $\sigma_{par}^2 = 1.373 \cdot 10^{-4}$ for the theoretical value and $\sigma_{par}^2 = 1.973 \cdot 10^{-4}$ for the sample variance. Moreover, the ratio of variances using 5000 points was $\sigma_{par}^2/\sigma_{casc}^2 = 10.666$ using Simulink, whereas using the formulas we predict 9.718, which is quite close. We also obtain $\sigma_{par}^2/\sigma_{dirII}^2 = 3.8976$ using Simulink, whereas using the formulas we predict

2.95, which again is at least reasonably close. Evidently, the parallel structure *in this particular example* has worse rounding error than the cascade or direct form II, and cascade has the least. However, no general conclusions can be gleaned from this analysis; it must be repeated for each filter we wish to implement.

In conclusion, we find that the theoretical estimator based on stochastic considerations is useful if simulation with binary rounding quantization is not available. It is nice to see the stochastic estimates actually verified in real data obtained using true rounding.

However, if Simulink is available, it would appear that it is less work to simply calculate the variances from actual data for a specific filter than derive the stochastic formulas for generalized–coefficients structures. Remember that the latter involves determination of the transfer function of *each* error signal from its point of entry to the output. Then either numerical quadrature or laborious analytical inverse z–transformation as in Example 10.7 is further required to obtain the particular $r_{hh;n}(0)$ that gives the variance contribution due to that error signal.

10.9 LEAST–SQUARES OPTIMAL FILTERING

We have already encountered optimal filtering when we studied the McClellan–Parks/Remez algorithm for frequency–selective filter design in Section 9.5. We now return to optimal filtering from a time–domain perspective. Instead of approximating an ideal frequency response, we approximate either a simple unit sample response [e.g., $\delta(k)$] or a signal. In the latter case, we have for the first time filters that are dependent on the signals they filter. Each of our designs in this section is based on minimization of a squared error. The result of the design is thus called a least–squares filter. We begin our discussion by considering the minimization of mean–squared error in a fairly general setting. We then specialize those results to particular problems, in which the signals may be either deterministic or stochastic.

For all our discussions, we will assume that all signals are real–valued, for simplicity, and that all stochastic signals are WSS. We will also assume the practical case that the filter $h_{LS}(n)$ to be designed is FIR, with length N_f. The general configuration we examine here is shown in Figure 10.23. The input signal (or other sequence) $s(n)$ is to be filtered by the least–squares filter $h_{LS}(n)$ to produce $\hat{d}(n)$, an estimate of the desired output signal $d(n)$. The estimation error is $e(n) = d(n) - \hat{d}(n)$. The coefficients of $h_{LS}(n)$ are selected to minimize the mean–squared error J:

$$J = \text{avg}\{\,|e(n)|^2\} \tag{10.81a}$$

$$= \begin{cases} E\{E^2(n)\}, & S(n) \text{ and } E(n) \text{ are random processes} \\ \displaystyle\sum_{n=-\infty}^{\infty} e^2(n), & s(n) \text{ and } e(n) \text{ are deterministic sequences.} \end{cases} \tag{10.81b}$$

Least-squares filter: minimize $J = \text{avg}\{|e(n)|^2\}$ **Figure 10.23**

The symbol "avg" in (10.81a) and used below thus refers to the two possibilities of averaging (stochastic or temporal) indicated in (10.81b). When the signals are random, the argument of the expected value operator is always the (capitalized) random process producing the actual sequences. For brevity, when writing the "avg" operator we often use just lowercase letters to mean whichever signal type applies. When we substitute $\hat{d}(n) = (s * h_{LS})(n)$ into $e(n)$, and note that it is combinations of $d(\cdot)$ and/or $s(\cdot)$ that end up being averaged in computing $\mathrm{avg}(\,|e(n)|^2)$, we obtain for (10.81)

$$J = \mathrm{avg}\left\{\,|d(n) - \sum_{m=0}^{N_f-1} h_{LS}(m)s(n-m)|^2\right\} \tag{10.81c}$$

$$= \mathrm{avg}\left\{d^2(n) - 2d(n)\sum_{m=0}^{N_f-1} h_{LS}(m)s(n-m)\right. \tag{10.81d}$$

$$\left. + \sum_{m=0}^{N_f-1} h_{LS}(m)s(n-m)\sum_{\ell=0}^{N_f-1} h_{LS}(\ell)s(n-\ell)\right\}$$

$$= \mathrm{avg}\{d^2(n)\} - 2\sum_{m=0}^{N_f-1} h_{LS}(m)\,\mathrm{avg}\{d(n)s(n-m)\} \tag{10.81e}$$

$$+ \sum_{m=0}^{N_f-1} h_{LS}(m)\sum_{\ell=0}^{N_f-1} h_{LS}(\ell)\,\mathrm{avg}\{s(n-m)s(n-\ell)\}.$$

It proves very helpful to express J in vector–matrix form. To do this, we recognize the first sum in (10.81e) as an inner (dot) product of two vectors. We also recognize the double sum in (10.81e) as the product of a vector times a matrix times the same vector; this double sum was defined on WP 5.1 as a quadratic form $[(W5.1.36)\colon \mathbf{x}^{*T}\mathbf{A}\mathbf{x} = \sum_{n=1}^{N} x_n^* \sum_{m=1}^{N} A_{nm}x_m]$. With these recognitions, we have

$$J = \mathrm{avg}\{d^2(n)\} - 2\,\mathrm{avg}\{[d(n)s(n)\;\; d(n)s(n-1)\;\; \ldots\;\; d(n)s(n-(N_f-1))]\}\begin{bmatrix} h_{LS}(0) \\ h_{LS}(1) \\ \cdot \\ \cdot \\ \cdot \\ h_{LS}(N_f-1) \end{bmatrix} \tag{10.81f}$$

$$+ [h_{LS}(0)\;\; h_{LS}(1)\;\; \ldots\;\; h_{LS}(N_f-1)]$$

$$\cdot \mathrm{avg}\left\{\begin{bmatrix} s^2(n) & s(n)s(n-1) & \cdots & s(n)s(n-(N_f-1)) \\ s(n-1)s(n) & s^2(n-1) & \cdots & s(n-1)s(n-(N_f-1)) \\ \cdot & \cdot & \cdot & \\ \cdot & \cdot & & \cdot \\ s(n-(N_f-1))s(n) & & \cdots & s^2(n-(N_f-1)) \end{bmatrix}\right\}\begin{bmatrix} h_{LS}(0) \\ h_{LS}(1) \\ \cdot \\ \cdot \\ h_{LS}(N_f-1) \end{bmatrix}$$

$$= \alpha - 2\boldsymbol{\beta}^T\mathbf{h}_{LS} + \mathbf{h}_{LS}^T\mathbf{R}\mathbf{h}_{LS}, \tag{10.81g}$$

where

$$\alpha = \mathrm{avg}\{d^2(n)\} \tag{10.82}$$

$$\boldsymbol{\beta} = \mathrm{avg}\left\{\begin{bmatrix} d(n)s(n) \\ d(n)s(n-1) \\ \cdot \\ \cdot \\ \cdot \\ d(n)s(n-(N_f-1)) \end{bmatrix}\right\} \tag{10.83}$$

$$\mathbf{R} = \text{avg}\left\{\begin{bmatrix} s^2(n) & s(n)s(n-1) & \cdots & s(n)s(n-(N_f-1)) \\ s(n-1)s(n) & s^2(n-1) & \cdots & s(n-1)s(n-(N_f-1)) \\ \vdots & & & \vdots \\ \vdots & & & \\ s(n-(N_f-1))s(n) & \cdots & & s^2(n-(N_f-1)) \end{bmatrix}\right\} \quad (10.84)$$

$$\mathbf{h}_{\text{LS}} = \begin{bmatrix} h_{\text{LS}}(0) \\ h_{\text{LS}}(1) \\ \vdots \\ \vdots \\ h_{\text{LS}}(N_f-1) \end{bmatrix}. \quad (10.85)$$

Notice, under the wide–sense–stationarity assumption (and always holding for deterministic signals, because the temporal average is assumed to run over all time), that kth element of the vector $\boldsymbol{\beta}$ is

$$\beta_k = \text{avg}\{d(n)s(n-k)\} \quad (10.86a)$$

$$= \begin{cases} r_{DS}(k), & S(n) \text{ stochastic} \\ r_{ds;\,n}(k), & s(n) \text{ deterministic} \end{cases} \quad (10.86b)$$

with $0 \le k \le N_f - 1$, and the ℓmth element of the matrix $\mathbf{R}$ is

$$r_{\ell m} = \text{avg}\{s(n-\ell)s(n-m)\} \quad (10.87a)$$

$$= \begin{cases} r_{SS}(m-\ell), & S(n) \text{ stochastic} \\ r_{ss;\,n}(m-\ell), & s(n) \text{ deterministic,} \end{cases} \quad (10.87b)$$

with ℓ and m ranging from 0 to N_f-1.

To find the $\{h_{\text{LS}}(n)\}$ that minimize J, it is most straightforward to differentiate (10.81e) with respect to the kth filter coefficient $h_{\text{LS}}(k)$:

$$\frac{\partial J}{\partial h_{\text{LS}}(k)} = -2\,\text{avg}\{d(n)s(n-k)\} + 2\sum_{\substack{m=0 \\ m \ne k}}^{N_f-1} h_{\text{LS}}(m)\text{avg}\{s(n-m)s(n-k)\}$$
$$+ 2h_{\text{LS}}(k)\text{avg}\{s^2(n-k)\} \quad (10.88a)$$

$$= -2\,\text{avg}\{d(n)s(n-k)\} + 2\sum_{m=0}^{N_f-1} h_{\text{LS}}(m)\text{avg}\{s(n-m)s(n-k)\}. \quad (10.88b)$$

We now recognize that the first term in (10.88b) is $-2\beta_k$ and the second term is $2(\mathbf{Rh}_{\text{LS}})_k$, which holds because $\mathbf{R}$ is a symmetric matrix. Thus

$$\frac{\partial J}{\partial h_{\text{LS}}(k)} = 2[(\mathbf{Rh}_{\text{LS}})_k - \beta_k], \quad (10.88c)$$

or, putting this set of equations for all k into one vector–matrix equation,

$$\frac{\partial J}{\partial \mathbf{h}_{\text{LS}}} = 2(\mathbf{Rh}_{\text{LS}} - \boldsymbol{\beta}). \quad (10.89)$$

Setting all these derivatives to zero gives

$$\mathbf{Rh}_{\text{LS}} = \boldsymbol{\beta}. \quad (10.90a)$$

Equivalently, from (10.88b) set to zero, we can represent the matrix solution in (10.90a) as a set of linear equations:

$$\sum_{m=0}^{N_f-1} h_{\text{LS}}(m) \, \text{avg}\{s(n-m)s(n-k)\} = \text{avg}\{d(n)s(n-k)\}, \quad 0 \le k \le N_f - 1 \quad (10.90b)$$

or

$$\sum_{m=0}^{N_f-1} h_{\text{LS}}(m) \, r_{ss}(k-m) = r_{ds}(k), \quad 0 \le k \le N_f-1, \quad (10.90c)$$

where in (10.90c) the ACFs are either stochastic or temporal, depending on the application.

We showed in Section 10.3 [see (10.15)] that $r_{XX}(k)$ is a positive semidefinite sequence. Also, letting A_{ij} in (W5.1.36) (cited above) be $R_{ij} = r_{XX}(i-j)$, we conclude that $\mathbf{R}$ is a positive semidefinite matrix. In practice, as long as $x(n)$ does not consist of $N_f - 1$ or fewer sinusoids,[22] $\mathbf{R}$ is generally positive definite and thus invertible. Note from Table W5.1 that $\mathbf{R}$ being positive definite implies that the eigenvalues of the finite–dimensional matrix $\mathbf{R}$ are all positive. Recall from Section 10.3 that the eigenvalues of the infinite–dimensional version of $\mathbf{R}$ are $P_{XX}(\omega)$.

Because $\mathbf{R}^{-1}$ exists, we may solve (10.90a) for $\mathbf{h}_{\text{LS}}$:

$$\mathbf{h}_{\text{LS}} = \mathbf{R}^{-1}\boldsymbol{\beta}. \quad (10.91)$$

We see the advantage of the matrix form, which allows us to very concisely write the solution in symbolic form. Furthermore, the matrix form allows us to easily evaluate the minimum value of J obtained by setting $\mathbf{h}_{\text{LS}}$ to the value $\mathbf{R}^{-1}\boldsymbol{\beta}$. We can manipulate J in (10.81g) into an illuminating form by completing the square:

$$J = \alpha - 2\boldsymbol{\beta}^{\text{T}}\mathbf{h}_{\text{LS}} + \mathbf{h}_{\text{LS}}{}^{\text{T}}\mathbf{R}\mathbf{h}_{\text{LS}} \quad (10.92a)$$

$$= \alpha - \boldsymbol{\beta}^{\text{T}}\mathbf{R}^{-1}\boldsymbol{\beta} + \boldsymbol{\beta}^{\text{T}}\mathbf{R}^{-1}\boldsymbol{\beta} - 2\boldsymbol{\beta}^{\text{T}}\mathbf{h}_{\text{LS}} + \mathbf{h}_{\text{LS}}{}^{\text{T}}\mathbf{R}\mathbf{h}_{\text{LS}}. \quad (10.92b)$$

Now, using the fact that $(\mathbf{R}^{-1})^{\text{T}} = \mathbf{R}^{-1}$, valid because $\mathbf{R}^{\text{T}} = \mathbf{R}$, we obtain

$$J = \alpha - \boldsymbol{\beta}^{\text{T}}\mathbf{R}^{-1}\boldsymbol{\beta} + (\mathbf{h}_{\text{LS}} - \mathbf{R}^{-1}\boldsymbol{\beta})^{\text{T}}\mathbf{R}\,(\mathbf{h}_{\text{LS}} - \mathbf{R}^{-1}\boldsymbol{\beta}). \quad (10.92c)$$

Because $\mathbf{R}$ is positive (semi)definite, the quadratic form in (10.92c) is always nonnegative. We can guarantee that it is zero, and thus minimize J, by requiring $\mathbf{h}_{\text{LS}} = \mathbf{R}^{-1}\boldsymbol{\beta}$ as in (10.91). We then obtain the minimum value of mean–squared error, obtained when $\mathbf{h}_{\text{LS}}$ is the least–squares filter in (10.91):

$$J_{\min} = \alpha - \boldsymbol{\beta}^{\text{T}}\mathbf{R}^{-1}\boldsymbol{\beta} \quad (10.93a)$$

$$= \alpha - \boldsymbol{\beta}^{\text{T}}\mathbf{h}_{\text{LS}}. \quad (10.93b)$$

One other fact is helpful to notice. We may write (10.88b) as

$$-\frac{1}{2}\left(\frac{\partial J}{\partial h_{\text{LS}}(k)}\right) = \text{avg}\{d(n)s(n-k)\} - \sum_{m=0}^{N_f-1} h_{\text{LS}}(m)\text{avg}\{s(n-m)s(n-k)\} \quad (10.94a)$$

$$= \text{avg}\{s(n-k)[d(n) - \sum_{m=0}^{N_f-1} h_{\text{LS}}(m)s(n-m)]\} \quad (10.94b)$$

$$= \text{avg}\{s(n-k)e(n)\} = 0, \quad 0 \le k \le N_f - 1. \quad (10.94c)$$

---

[22]S. M. Kay, *Modern Spectral Estimation*, (Englewood Cliffs, NJ: Prentice Hall, 1988), p. 116.

When we recall that in the stochastic case (10.94c) translates to

$$E\{S(n)E(n)\} = 0, \tag{10.95a}$$

$$E\{S(n-1)E(n)\} = 0, \tag{10.95b}$$

.

.

.

$$E\{S(n-[N_f-1])E(n)\} = 0, \tag{10.95c}$$

we see that the error at time index n is "orthogonal" or "normal" to past and present data [i.e., to past and present values of the input $S(n)$ to the LS filter]. Use of the synonyms *orthogonal* and *normal* is justified when we recall that, for example, (10.95a) can be written

$$E\{S(n)E(n)\} = \sum_i s(n, i)e(n, i) \tag{10.96a}$$

$$= \mathbf{s}_n \cdot \mathbf{e}_n = 0, \tag{10.96b}$$

where orthogonality is clearly identified between the two (infinite–length) data vectors, whose elements are values of $s(n)$ and $e(n)$ at a fixed value of n for all realizations i.

Another important interpretation of (10.95) and (10.96) is that the error in estimation is limited to the component of $d(n)$ that cannot be produced by linear combinations of the data $s(n)$. In other words, the filter output $\hat{d}(n)$ is the projection of $d(n)$ onto the signal space formed by the data $s(n)$. When we discuss least–squares prediction, we will express this view of least–squares estimation as follows: The filter produces that component of the future value(s) of the signal that is predictable based on the known past and present values of the input sequence.

In the deterministic case, (10.94c) can be written

$$\sum_{n=-\infty}^{\infty} s(n)e(n) = 0 \tag{10.97a}$$

$$\sum_{n=-\infty}^{\infty} s(n-1)e(n) = 0 \tag{10.97b}$$

.

.

.

$$\sum_{n=-\infty}^{\infty} s(n-[N_f-1])e(n) = 0. \tag{10.97c}$$

In this case, the error sequence $\{e(n)\}\big|_{n=-\infty}^{\infty}$ or the associated infinite–length vector $\mathbf{e}$ is orthogonal (or normal) to delayed versions of the data sequence $\{s(n-1)\}$, $\{s(n-2)\}, \ldots, \{s(n-[N_f-1])\}$. For these reasons, (10.90a) and (10.90c) are called the normal equations.

10.9.1   Deconvolution

Many special cases of the normal equations can be identified that solve specialized problems. One of these basic problems is deconvolution. In deconvolution, we would desire $h_{\text{deconv}}(n) = \text{ZT}^{-1}\{1/H(z)\}$, where $h(n) = \text{ZT}^{-1}\{H(z)\}$ is the distorting filter we wish to equalize so that $(h_{\text{deconv}} * h)(n) = \delta(n)$. However, $\text{ZT}^{-1}\{1/H(z)\}$ may be unstable if $H(z)$ has any zeros outside the unit circle, or it may be impractically long (of length $>> N$ where N is the length of the data record). In cases such as this, as in other cases where "exact" solutions of the relevant equations are ill–conditioned, we instead take the least–squares approach described above. In this case, $d(n) = \delta(n)$ and $s(n) = h(n)$. Here we see that the input "signal" $s(n)$ is actually a unit sample response $[h(n)]$. The mathematics does not care, so we go ahead and make the required identifications:

$$d(n) = \delta(n), \qquad \text{Deconvolution with zero delay.} \tag{10.98}$$

$$s(n) = h(n), \tag{10.99}$$

$$\beta_k = r_{ds;n}(k) \tag{10.100a}$$

$$= \sum_{n=-\infty}^{\infty} \delta(n)h(n-k) \tag{10.100b}$$

$$= h(0)\delta(k), \tag{10.100c}$$

where (10.100c) is due to the assumption that $h(n)$ is a causal sequence, and finally

$$r_{\ell m} = r_{ss;n}(m - \ell) \tag{10.101a}$$

$$= \sum_{n=-\infty}^{\infty} h(n-\ell)h(n-m) \tag{10.101b}$$

$$= r_{hh;n}(m - \ell), \tag{10.101c}$$

$$\alpha = \sum_{n=-\infty}^{\infty} \delta^2(n) = 1. \tag{10.102}$$

To help visualize the normal equations, we write them in matrix/vector form with individual elements shown:

$$\begin{bmatrix} r_{hh;n}(0) & r_{hh;n}(1) & \cdots & r_{hh;n}(N_f-1) \\ r_{hh;n}(1) & r_{hh;n}(0) & \cdots & r_{hh;n}(N_f-2) \\ \cdot & \cdot & & \cdot \\ \cdot & \cdot & & \cdot \\ \cdot & \cdot & & \cdot \\ r_{hh;n}(N_f-1) & r_{hh;n}(N_f-2) & \cdots & r_{hh;n}(0) \end{bmatrix} \begin{bmatrix} h_{LS}(0) \\ h_{LS}(1) \\ \cdot \\ \cdot \\ \cdot \\ h_{LS}(N_f-1) \end{bmatrix} = \begin{bmatrix} h(0) \\ 0 \\ \cdot \\ \cdot \\ \cdot \\ 0 \end{bmatrix} \tag{10.103}$$

and

$$J_{\min} = 1 - [h(0) \quad 0 \quad \cdots \quad 0] \, \mathbf{R}^{-1} \begin{bmatrix} h(0) \\ 0 \\ \cdot \\ \cdot \\ \cdot \\ 0 \end{bmatrix} \tag{10.104}$$

$$= 1 - h^2(0)R_{11}^{-1}.$$

An interesting practical application of this sort of deconvolution would be the restoration of old phonograph recordings. In this application, $s(n) = h(n)$ is the unit sample response of the ancient phonograph recording instrument. Let $\hat{x}(n) = x(n) * h(n)$, where $\hat{x}(n)$ is the tinny sound of a recording produced using $h(n)$ and

$x(n)$ is the desired original music. For example, $h(n)$ may be a soft narrow–band filter, relative to the complete aural spectrum, attenuating both low and high frequencies, and in some cases including annoying resonances. The distorted sound $\hat{x}(n)$ would be sampled and then filtered by $h_{LS}(n)$. The resulting sequence $y(n) \approx x(n)$ (we hope!) would be reconstructed and amplified for listening.

We now, however, consider a more mundane but simpler example of finding a least–squares inverse filter $h_{LS}(n)$.

EXAMPLE 10.9

A distorting filter has unit sample response

$$h(n) = \sum_{\ell=1}^{3} A_\ell p_\ell{}^n \cos(\omega_\ell n + \theta_\ell),$$

where

$$A_1 = 1, \qquad p_1 = 0.7, \qquad \omega_1 = 20°/\text{sample}, \qquad \theta_1 = 70°$$

$$A_2 = 0.5, \qquad p_2 = 0.95, \qquad \omega_2 = 40°/\text{sample}, \qquad \theta_2 = 150°$$

$$A_3 = 2, \qquad p_3 = 0.97, \qquad \omega_3 = 70°/\text{sample}, \qquad \theta_3 = -30°.$$

This unit sample response is shown in Figure 10.24. (Note that the ℓth upper–half–plane pole of the filter is $z_\ell = p_\ell \angle \omega_\ell$.) Find the least–squares filter that deconvolves $h(n)$, so that if $x(n) \rightarrow h(n) \rightarrow h_{LS}(n) \rightarrow y(n)$, the output $y(n)$ is approximately equal to $x(n)$.

Solution

We use (10.98) through (10.104). The ACF $r_{hh;n}(k)$ in (10.101c) (temporal average in any data–driven computational setting) is obtained using the FFT as discussed in Section 7.8 in which we here set $x = h$. The result of following the procedure described there is the ACF $r_{hh;n}(n)$ shown in Figure 10.25. From this sequence, the ACF values called for in (10.103) are extracted, and (10.103) is solved for $\mathbf{h}_{LS}$. The

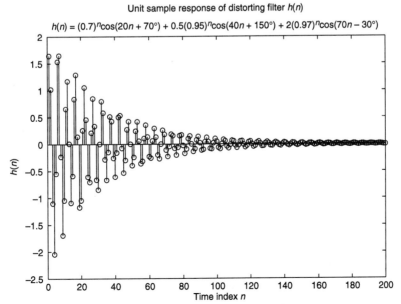

Unit sample response of distorting filter $h(n)$

$h(n) = (0.7)^n \cos(20n + 70°) + 0.5(0.95)^n \cos(40n + 150°) + 2(0.97)^n \cos(70n - 30°)$

Figure 10.24

Temporal ACF of distorting filter $h(n)$

$h(n) = (0.7)^n\cos(20n + 70°) + 0.5(0.95)^n\cos(40n + 150°) + 2(0.97)^n\cos(70n - 30°)$

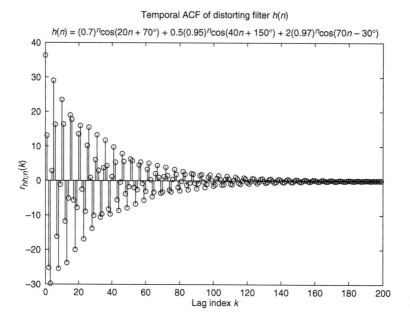

Figure 10.25

Least-squares deconvolution filter unit sample response $h_{LS}(n)$ of length $N_f = 150$; distorting filter is $h(n)$

$h(n) = (0.7)^n\cos(20n + 70°) + 0.5(0.95)^n\cos(40n + 150°) + 2(0.97)^n\cos(70n - 30°)$

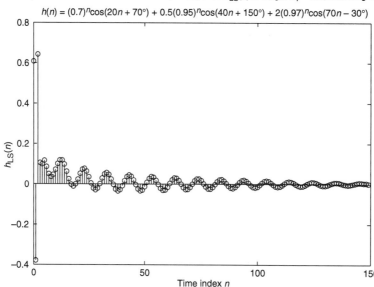

Figure 10.26

unit sample response of this optimal filter $h_{LS}(n)$ is shown in Figure 10.26, where we chose a filter length of $N_f = 150$.

We verify that indeed deconvolution is achieved by convolving h with h_{LS}; the result is in Figure 10.27. We clearly see that the result is unity at $n = 0$ and is very small elsewhere. The ripple is due to the edge effects resulting from finite N_f. Notice, for example, in Figure 10.26 that $h_{LS}(n)$ has not fully "died away" by its end. If we decrease N_f to 50, the result is as shown in Figure 10.28; the increased end effect ripple is to be expected.

The results of the quality of deconvolution, most easily observed by examining $h * h_{LS}$, depend on the coefficients $h(n)$ of the distorting filter. For example, if we change only the single parameter θ_1 from 70° to –70°, the results are significantly worse, as shown in Figure 10.29 for $N_f = 50$. In cases of high ripple, researchers have

Convolution of $h(n)$ with approximate inverse filter $h_{LS}(n)$ of length $N_f = 150$ designed for delay $n_0 = 0$

$$h(n) = (0.7)^n \cos(20n + 70°) + 0.5(0.95)^n \cos(40n + 150°) + 2(0.97)^n \cos(70n - 30°)$$

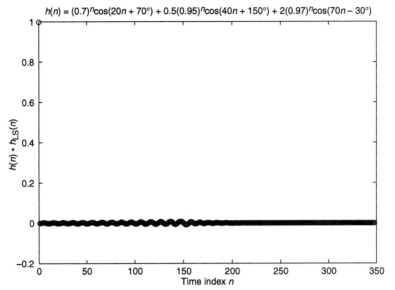

Figure 10.27

Convolution of $h(n)$ with approximate inverse filter $h_{LS}(n)$ of length $N_f = 50$ designed for delay $n_0 = 0$

$$h(n) = (0.7)^n \cos(20n + 70°) + 0.5(0.95)^n \cos(40n + 150°) + 2(0.97)^n \cos(70n - 30°)$$

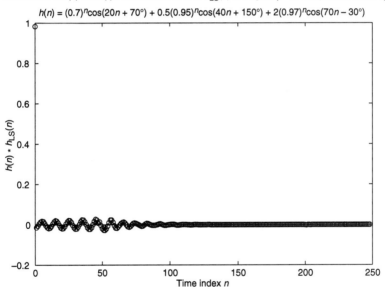

Figure 10.28

found that aiming for $\delta(n - n_0)$ rather than $\delta(n)$ improves the results. Such a modification merely produces a delay in the output of the deconvolution filter, which is usually more tolerable than the ripple distortion. With delay n_0, (10.100) becomes

$$\beta_k = r_{ds;n}(k) \tag{10.105a}$$

$$= \sum_{n=-\infty}^{\infty} \delta(n - n_0)h(n - k) \tag{10.105b}$$

$$= \begin{cases} h(n_0 - k), & 0 \le k \le n_0 \\ 0, & \text{otherwise.} \end{cases} \qquad \text{Deconvolution with delay } n_0. \tag{10.105c}$$

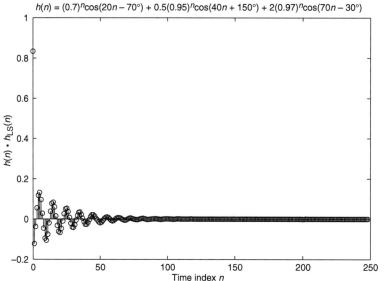

Convolution of $h(n)$ with approximate inverse filter $h_{LS}(n)$ of length $N_f = 50$ designed for delay $n_0 = 0$

$h(n) = (0.7)^n \cos(20n - 70°) + 0.5(0.95)^n \cos(40n + 150°) + 2(0.97)^n \cos(70n - 30°)$

Figure 10.29

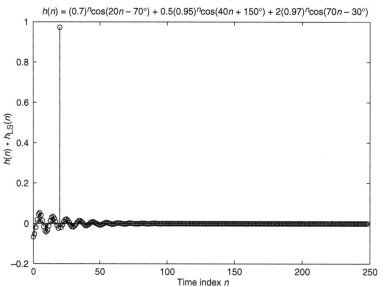

Convolution of $h(n)$ with approximate inverse filter $h_{LS}(n)$ of length $N_f = 50$ designed for delay $n_0 = 20$

$h(n) = (0.7)^n \cos(20n - 70°) + 0.5(0.95)^n \cos(40n + 150°) + 2(0.97)^n \cos(70n - 30°)$

Figure 10.30

If we let $n_0 = 20$, for example, the result for the $h(n)$ used in Figure 10.29 is as shown in Figure 10.30 ($N_f = 50$). Although there is still significant ripple, it is somewhat reduced compared with Figure 10.29. Also, the spike is nearer unity in Figure 10.30. Similar results apply for $N_f = 150$.

10.9.2   Prediction (n_0–Step)

The history and application of prediction is rich and long. This is not surprising, for the ability to accurately predict the future has always been a sought–after but elusive goal. In many of today's applications, however, prediction is not an end in itself, but rather a means to evaluate other quantities of interest. For example, in speech processing, linear prediction methods are used for breaking down speech into fundamental components that can be transmitted more efficiently than the sampled waveform itself. Quantitative "prediction" is not relevant for the aural reproduction; only the resulting model parameter values are relevant. In geophysical prospecting, the theory of prediction may be used to perform a specialized form of deconvolution that applies to the imaging of the subsurface structure of the earth. Nevertheless, prediction or "forecasting" is a highly developed subject under active use and research in such fields as stock market analysis and weather forecasting.

The difficulty or feasibility of accurately predicting future values of a signal depends fundamentally on the sort of signal we attempt to project. A trivial case is the sinusoid. If we know just three parameters of a sinusoid (amplitude, frequency, and phase), we may "predict" its value at an arbitrarily distant point in the future with zero error. The other extreme is white noise: We have no hope at all of predicting white noise with any more accuracy than blind guessing within a few standard deviations, even just for the very next sample. These two cases illustrate the essential idea in prediction: The more we know about and accurately model the structure of the signal and the factors influencing the signal, the better chance we have of making good predictions.

For example, if we were privy to all the factors affecting the stock price of a given company, we might be able know whether to buy or to sell on which particular days. In short, we could become rich! In practice, the amateur player of the market does not know these things, and even the pros are not in a much better situation. Often, the only factor we might have access to that affects the future stock prices is the set of past and present stock price values. The challenge, then, is to take a small amount of data and provide the most dependable prediction information possible or feasible. Naturally, the farther into the future we attempt to predict, the less reliable are the estimates because all the intermediate factors are necessarily ignored. Such factors may include fundamental changes in the dynamic structure of the process generating the sequence itself (e.g., sell–offs, change in management, restructuring, and so on).

Because of the general interest in the accuracy of predictions, we now consider some numerical examples both from synthesized data and from real–world data. The study of prediction accuracy also gives an intuitive confirmation of the validity of imposing certain models on the data. In digital signal processing, there are many other uses for signal modeling (diagnostics, spectral analysis, digital data communication, and so on). In our series of examples of prediction, we first we consider synthesized data.

10.9.3   Prediction of ARMA Sequences (Parametric Models)

Consider the linear prediction of sequences generated by ARMA processes. We will perform linear predictions on three ARMA sequences in Example 10.10. In preparation for that study, we now summarize the basic equations of n_0–step linear prediction and two important specializations of ARMA processes (the AR and MA processes). In our least–squares filtering approach, under the assumption that $x(n)$ is a realization of the WSS process $X(n)$, we have for the n_0–step prediction

$$d(n) = x(n + n_0) \tag{10.106}$$

$$s(n) = x(n), \tag{10.107}$$

$$\beta_k = r_{DS}(k) \tag{10.108a}$$

$$\quad = E\{X(n + n_0)\, X(n - k)\} \tag{10.108b}$$

$$\quad = r_{XX}(k + n_0) \tag{10.108c}$$

$$r_{\ell m} = r_{XX}(m - \ell) \tag{10.109}$$

$$\alpha = E\{X^2(n + n_0)\} = r_{XX}(0). \tag{10.110}$$

Thus, the normal equations take the form

$$
\begin{bmatrix}
r_{XX}(0) & r_{XX}(1) & \cdots & r_{XX}(N_f - 1) \\
r_{XX}(1) & r_{XX}(0) & \cdots & r_{XX}(N_f - 2) \\
\cdot & \cdot & & \cdot \\
\cdot & \cdot & & \cdot \\
\cdot & \cdot & & \cdot \\
r_{XX}(N_f - 1) & r_{XX}(N_f - 2) & \cdots & r_{XX}(0)
\end{bmatrix}
\begin{bmatrix}
h_{\mathrm{LS};n_0}(0) \\
h_{\mathrm{LS};n_0}(1) \\
\cdot \\
\cdot \\
\cdot \\
h_{\mathrm{LS};n_0}(N_f - 1)
\end{bmatrix}
=
\begin{bmatrix}
r_{XX}(n_0) \\
r_{XX}(n_0 + 1) \\
\cdot \\
\cdot \\
\cdot \\
r_{XX}(n_0 + N_f - 1)
\end{bmatrix}
\tag{10.111}
$$

and [see (10.39b)]

$$
J_{\min} = r_{XX}(0) - [r_{XX}(n_0) \ldots r_{XX}(n_0 + N_f - 1)] \mathbf{h}_{\mathrm{LS};n_0}. \tag{10.112}
$$

The above relations hold for prediction of any WSS process. We can use the above equations in practice by replacing the stochastic averages by temporal averages.

Recall that if $X(n)$ is an ARMA process, its realizations $x(n)$ may be generated by an ARMA difference equation, excited by realizations of a white noise sequence $u(n)$:

$$
\sum_{\ell=0}^{p} a_\ell x(n - \ell) = \sum_{\ell=0}^{q} b_\ell u(n - \ell). \tag{10.113}
$$

In Example 10.10, we will investigate the following distinct cases:

1. $AR(p)$, called an autoregressive process of order p
2. $MA(q)$, called a moving average process of order q
3. $ARMA(p, q)$, called an autoregressive moving average process of order $\{p, q\}$.

Let us carefully inspect the convolution equation for generation of the predictions using the n_0–step LS prediction filter $h_{\mathrm{LS};n_0}(n)$,

$$
\hat{x}(n + n_0) = \sum_{\ell=0}^{N_f - 1} h_{\mathrm{LS};n_0}(\ell) x(n - \ell) \tag{10.114}
$$

for the special case $n_0 = 1$, and with n on both sides of (10.114) replaced by $n - 1$:

$$
\hat{x}(n) = \sum_{\ell=0}^{N_f - 1} h_{\mathrm{LS};1}(\ell) x(n - 1 - \ell). \tag{10.115a}
$$

Defining $k = \ell + 1$, (10.115a) becomes

$$
\hat{x}(n) = \sum_{k=1}^{N_f} h_{\mathrm{LS};1}(k - 1) x(n - k). \tag{10.115b}
$$

We now compare (10.115b) with the difference equation for an autoregressive process [$q = 0$ in (10.113)],

$$
x(n) = \sum_{k=1}^{p} (-a_k) x(n - k) + u(n), \tag{10.116}
$$

where without loss of generality we have assumed $a_0 = 1$ and $b_0 = 1$. We can always obtain $a_0 = b_0 = 1$ by first dividing (10.113) by a_0 if it is nonunity and then absorbing the new b_0' into the variance of $u(n)$.

Comparing (10.116) with (10.115b), we see that linear prediction implicitly incorporates the assumption of an $AR(N_f)$ model for $x(n)$, but with zero excitation noise $u(n)$. The general transfer function of a system producing $x(n)$ with white noise input is $H(z) = B(z)/A(z)$, where $B(z) = \sum_{\ell=0}^{q} b_\ell z^{-\ell}$ and $A(z) = \sum_{\ell=0}^{p} a_\ell z^{-\ell}$.

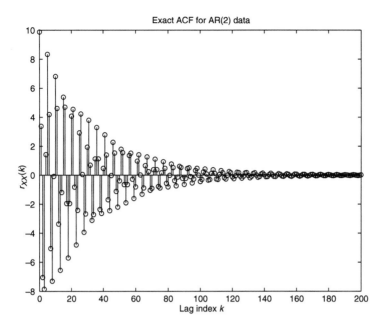

Figure 10.31

Thus, an AR model is an all–pole model for $X(n)$ (aside from zeros at $z = 0$). Similarly, a MA model is an all–zero model, and an ARMA model is a pole–zero model for $X(n)$. From these considerations, we would expect our least–squares linear predictions to be most accurate when $X(n)$ producing $x(n)$ is an AR(p) process and $N_f \geq p$.

EXAMPLE 10.10

Let us test out the above hypothesis with some data. Form predictions out to a time $n_{0\max} = 40$ samples into the "future" using computer–generated AR, MA, and ARMA data. For the ACF, in each case, calculate the exact ACF analytically by computing the inverse z–transform of $R_{XX}(z) = H(z)H^*(1/z^*)$, or, alternatively, use the temporal ACF of the actual data.

Solution

As noted in Section 10.5.2, white Gaussian noise is easily obtained in Matlab using the command `randn`. To obtain AR, MA, or ARMA data, we can simply use (10.113) solved for $x(n)$, with the generated white noise as $u(n)$.[23] Alternatively, in this example, the Fortran routine `gendata` in Kay's text was translated into Matlab code and used.

For the AR system, we select for an example: AR(p = 2): a = [1 −0.6635 0.9409], which has poles at $0.97\angle\pm70°$, and for $u(n)$ let $\sigma^2 = 1$. The corresponding exact ACF is (see the problems)

$$r_{XX}(k) = 9.8721 \cdot (0.97)^k \cdot \cos\{(70k - 0.6448) \cdot \pi/180\}, \ k \geq 0,$$

which is plotted in Figure 10.31. As is clear from both the expression and Figure 10.31, AR systems have infinite–length, decaying, and typically oscillating ACFs.

---

[23]For this method, we must keep only the values following several "start–up" samples to avoid including any erroneous transients arising from not having the previous samples of $u(n)$ and $x(n)$ required in (10.113).

Predictions using LS filter with deterministic data and exact r_{XX}

O = true $x(n)$ and * = predicted $x(n)$; $X(n)$ is AR(2)

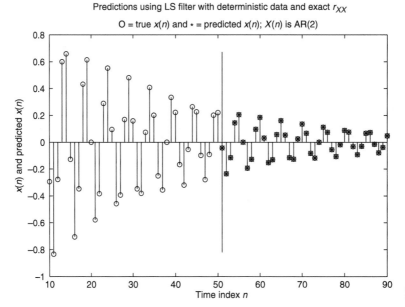

Figure 10.32

One-step predictions using LS filter with deterministic data and exact r_{XX}

O = true $x(n)$, * = predicted $x(n)$ using exact $x(n)$, solid = prediction using previous predictions; $X(n)$ is AR(2)

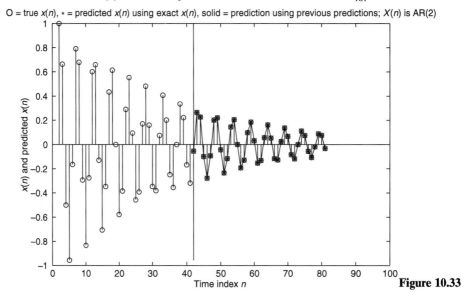

Figure 10.33

Because linear prediction follows the AR model without excitation, one–step prediction of an AR system having only initial condition "excitation" should be perfect, if $h_{LS;1}(n) = -a(n + 1)$ for $1 \leq n + 1 \leq p$ and 0 otherwise. Moreover, because the one–step prediction is nearly perfect, so should be successive one–step predictions based on previous predictions. Essentially, we are just using the correct difference equation to generate successive values. Somewhat more surprising is that n_0–step predictions are also nearly perfect for $n_0 > 1$. We set the filter order to $N_f = 5$, which amply exceeds $p = 2$. We do in fact numerically obtain $h_{LS;1}(n) = -a(n + 1)$ for $1 \leq n \leq 2$ and exactly zero otherwise when we use $\mathbf{h}_{LS;1} = \mathbf{R}^{-1}\boldsymbol{\beta}$ with $\mathbf{R}$ and $\boldsymbol{\beta}$ as in (10.111) and (10.108) with $n_0 = 1$ and the exact $r_{XX}(k)$ values as in the equation for $r_{XX}(k)$ given above. The results are also nearly exact when the estimates $r_{xx;n}(k)$ of $r_{XX}(k)$ are used instead.

These results are illustrated in Figures 10.32 and 10.33. Figure 10.32 shows the true $x(n)$ for the AR system excited by a single unit sample ("impulse response") in O stems and n_0–step predictions using the least–squares filter in * stems, where each * stem is

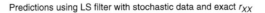

Predictions using LS filter with stochastic data and exact r_{XX}

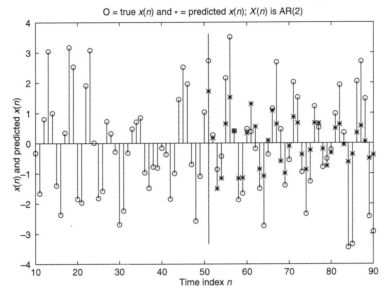

O = true $x(n)$ and ∗ = predicted $x(n)$; $X(n)$ is AR(2)

Figure 10.34

One-step predictions using LS filter with stochastic data and exact r_{XX}

O = true $x(n)$, ∗ = predicted $x(n)$ using exact $x(n)$, solid = prediction using previous predictions; $X(n)$ is AR(2)

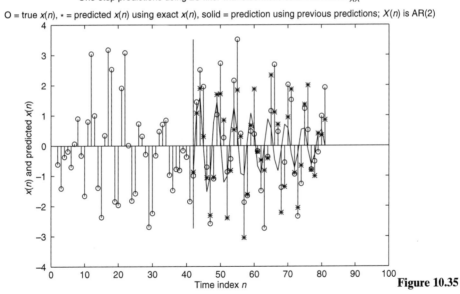

Figure 10.35

the prediction for a different value of n_0. We "pretend" that we do not know the exact values in the prediction interval when calculating the predictions, but in the simulation situation we know them exactly and show them for comparison as the O stems.

In Figure 10.33, again O stems are exact values, whereas ∗ stems show successive one–step predictions using exact $x(n)$ ("as they come in"). The solid line joins successive one–step predictions using previous one–step predictions; that is, for the points joined by the solid line, there is no new true data coming in after the first prediction. The vertical line in the center of the figures passes through the first predicted sample. Similarly high accuracy is seen when the exact $r_{XX}(k)$ values are replaced by $r_{xx;n}(k)$ computed from the data according to the DFT method given in Section 7.8.

On the other hand, if the data $x(n)$ is a realization of an AR(p) stochastic process [$u(n) \neq 0$], the accuracy of predictions deteriorates. Plots equivalent to those in Figures 10.32 and 10.33 are shown in, respectively, Figures 10.34 and 10.35. Results using the temporal $r_{xx;n}(k)$ are similar and are omitted for brevity.

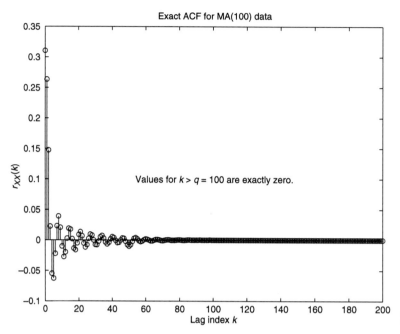

Figure 10.36

The essential point to be noticed in these figures is that although the overall pattern of predictions does correspond to that of the true $x(n)$, the accuracy of individual predictions is quite unreliable. This is exactly as would be expected: $u(n)$, being white noise, continually introduces an unpredictable component into $x(n)$. Without knowledge of $u(n)$, this component is impossible to predict. The overall pattern of periodicities, however, is predictable, and it is that component that is represented in the least–squares predictions.

Another ramification of the n_0–prediction model not being continually excited is that the predictions decay to zero for large n_0. Contrarily, $x(n)$ continues unabated in magnitude indefinitely. This is another reason why only short–term predictions using ARMA models are ever reliable.

Consider next the MA process. For this example we select for the MA coefficients $b(n)$, somewhat arbitrarily and, for convenience, a designed FIR digital filter unit sample response: MA($q = 100$): $b(n)$ is chosen to be the $h(n)$ of a McClellan–Parks/Remez lowpass filter of length $N_f = 101$ (shifted for central symmetry), with cutoff $f_c = 0.15$ (thus, $0.15 \cdot 2 = 0.3$ for routine `remez`) and rejection frequency $f_r = 0.165$.

The corresponding exact ACF of the MA signal $X(n)$ is (see the problems)

$$r_{XX}(k) = \sum_{\ell=0}^{q-k} b_\ell b_{\ell+k}, \qquad k \in [0, q]; r_{XX}(k) \text{ is zero otherwise,} \qquad (10.117)$$

which holds for any MA(q) process; here $q = 100$. This ACF has finite length equal to $q + 1$, as can be seen in (10.117) and as is illustrated in Figure 10.36. The distinction between finite–length $r_{XX}(k)$ for MA processes and infinite–length $r_{XX}(k)$ for AR processes is commonly used by forecasters to decide which type of model is more appropriate. If, for example, the ACF goes to nearly zero after about seven samples, MA(6) might be assumed. An MA(q) model can be represented exactly by an AR(∞) model [and an AR(p) model can be represented exactly by an MA(∞) model]; see Kay's book. Knowing this, we increase N_f to 150 because the prediction filter imposes an AR(N_f) model on the data. However, even letting N_f be $1.5q$ is not enough, as we shall see.

Predictions using LS filter with deterministic data and exact r_{XX}

O = true $x(n)$ and * = predicted $x(n)$; $X(n)$ is MA(100)

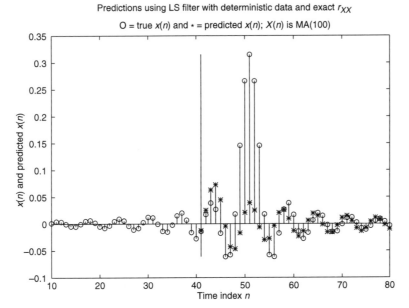

Figure 10.37

One-step predictions using LS filter with deterministic data and exact r_{XX}

O = true $x(n)$, * = predicted $x(n)$ using exact $x(n)$, solid = prediction using previous predictions; $X(n)$ is MA(100)

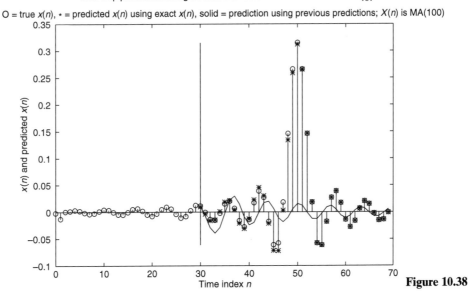

Figure 10.38

As just noted, the predictions are nowhere nearly as accurate as they were for the AR process. Figures 10.37 and 10.38 are comparable with, respectively, Figures 10.32 and 10.33 for the AR case; that is, the only excitation for $X(n)$ is a single unit sample at $n = 0$. The results for stochastic $[u(n) \neq 0]$ data are shown in Figures 10.39 and 10.40. For variety, we use the temporal $r_{xx;\,n}(k)$ rather than the exact $r_{XX}(k)$ in Figures 10.39 and 10.40; results are similar in either case. If all we are attempting is successive one–step predictions, always using true past $x(n)$ values for forming the predictions, naturally the results are reasonably good (Figures 10.38 and 10.40, * stems). If, however, we seek long–term forecasts, we are bound to be disappointed. Remember, we are still using an AR–type model inherent in our linear prediction convolution, which decays to zero for long–range predictions.

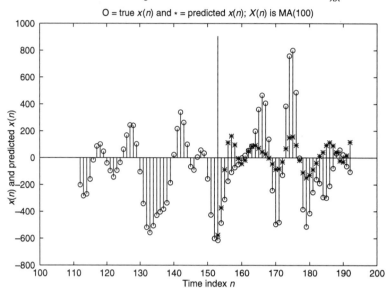

Figure 10.39

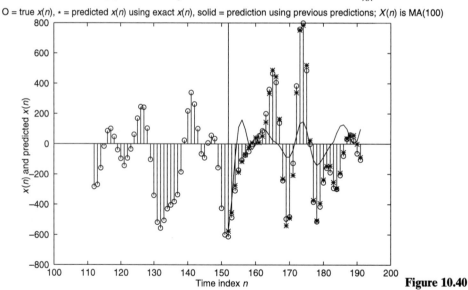

Figure 10.40

If by viewing the temporal ACF of a sequence we decide it is well–modeled as an MA(q) process, we may use specialized nonlinear algorithms to solve for the b_n, rather than just assume an AR(N_f) structure as we have done here. The description of such algorithms is, however, beyond the scope of this introduction.

Finally, we try predicting ARMA data: ARMA($p = 3$, $q = 1$): $a = [1 \; -2.5 \; 2.2 \; -0.67]$, $b = [1 \; -0.3]$, in which the poles are at $0.9247\angle\pm21.86°$ and 0.7836. The corresponding exact ACF is (see the problems)

$$r_{XX}(k) = 33.6139 \cdot (0.7836)^k + 86.6156 \cdot (0.9247)^k \cdot \cos\{(21.858k - 25.5727)\pi/180\},$$

which is plotted in Figure 10.41. As the ARMA model is just differently weighted pole terms, we expect the result that the ARMA $r_{XX}(k)$ is similar in appearance to the AR $r_{XX}(k)$. That is, $r_{XX}(k)$ oscillates and gradually decays to zero. In this exam-

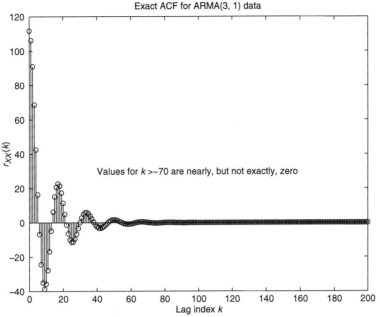

Figure 10.41

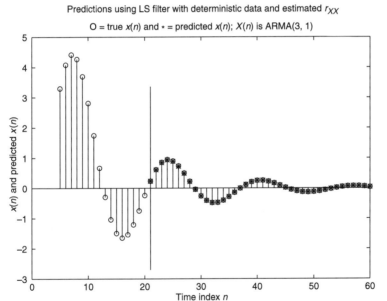

Figure 10.42

ple, if the data is excited only by a nonzero initial condition, the linear predictions using either exact $r_{XX}(k)$ or temporal $r_{xx;\,n}(k)$ give excellent results; those using $r_{xx;n}(k)$ are shown in Figures 10.42 and 10.43 for $N_f = 5$. However, when $x(n)$ is stochastic (Figures 10.44 and 10.45), again the predictions are disappointing. As for the MA process, there are (nonlinear) algorithms in existence that may be used for the determination of the $\{a_\ell,\ b_\ell\}$ parameters of an ARMA process, which should improve the predictions up to about the accuracy of our predictions of the AR random process. We again omit any detailed discussion of these nonlinear algorithms.

When our predictions fail, we may attempt to use other models to predict $x(n)$, but in this set of examples we know that $u(n)$ introduces an aspect of these sequences that is fundamentally impossible to predict. In other examples, other models may actually be quite helpful to try.

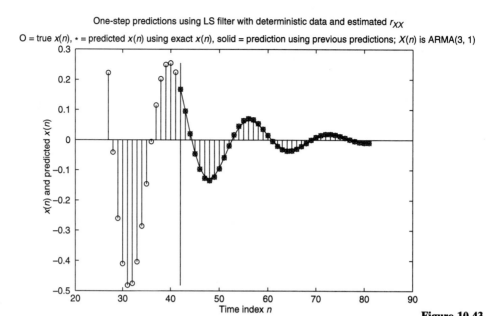

Figure 10.43

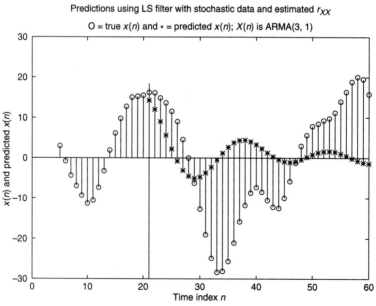

Figure 10.44

One-step predictions using LS filter with stochastic data and estimated r_{XX}

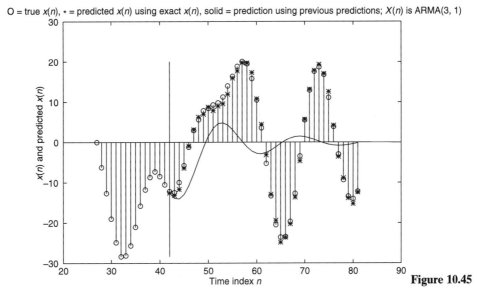

O = true $x(n)$, * = predicted $x(n)$ using exact $x(n)$, solid = prediction using previous predictions; $X(n)$ is ARMA(3, 1)

Figure 10.45

EXAMPLE 10.11

We return to the data presented in Example 7.6, the Dow Jones Industrial Averages. In Figure 7.18a, we showed the Dow Jones averages from January 1920 through June 1947. In this example, we reconsider the comment made there about the difficulty of predicting the Dow Jones averages. In Example 7.6, we concluded that they are hard to predict because of the "whiteness" of the spectrum of samples of the averages plotted in Figure 7.18b. Here we will actually attempt predictions, again beginning our predictions well within the known range of past Dow closings, so that we already know the exact values we are attempting to "predict," but basing our predictions on only values preceding the prediction index range.

Solution

We have 8192 samples from the one available realization of the process generating the Dow Jones averages. Arbitrarily, we select index $n_{\text{beg}} = 2440$ (in early 1928) to begin our first predictions, which we base on the previous 100 data points (closings). Using the linear least squares prediction method presented above with $N_f = 20$, we obtain the results in Figure 10.46. For clarity, the true sequence is shown by the solid curve. The prediction obtained (O stems) is virtually useless; whatever the value of n_{beg}, the prediction is mostly constant and equal to the most recent correct data value. The prediction may be close for the first few samples if the market is not rapidly changing then, but simply "remaining constant" is hardly a valuable prediction tool.

The reason for this failure can be seen in Figure 10.47, which shows the temporal ACF of the Dow Jones sequence (with points connected by a solid curve). From

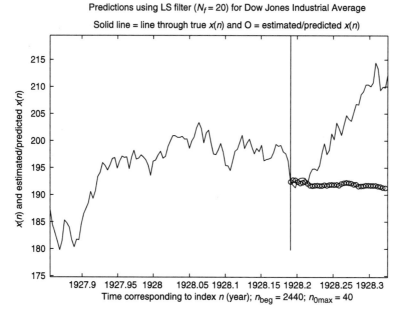

Predictions using LS filter ($N_f = 20$) for Dow Jones Industrial Average

Solid line = line through true $x(n)$ and O = estimated/predicted $x(n)$

Time corresponding to index n (year); $n_{beg} = 2440$; $n_{0max} = 40$

Figure 10.46

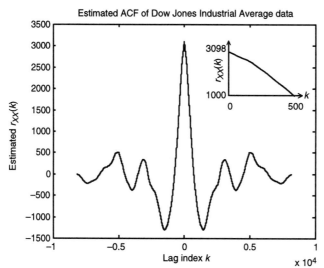

Estimated ACF of Dow Jones Industrial Average data

Lag index k

$\times 10^4$ **Figure 10.47**

its maximum at $k = 0$, $r_{xx;n}(k)$ decreases only by 64% by lag 500 (see inset, which shows only the first 500 lags), without any oscillation. This is commonly interpreted as signifying that the assumption of wide–sense stationarity is violated for this process, because the locally computed ACF averages and the means vary gradually and monotonically from one long averaging data segment to the next over the entire 27.5 years in our 8192 samples.

Alternatively, in looking at Figure 7.18a we see that the major "oscillations" are separated by years or even decades. Consequently, if we hold $x(n)$ as in Figure 7.18a and shift its replica by it, we do not expect the ACF to alternate in sign any more frequently than once per decade or so—a fact borne out in viewing the entire roughly

16,400–point temporal ACF sequence in Figure 10.47. Specifically, the first peak beyond $k = 0$ is found at about $k = 3108$ days, or about 8.5 years.

The implication of the nearly constant ACF over the lag range included in **R** and **β** for $N_f = 20$ (days) is that **R** in (10.111) and **β** in (10.108) have nearly uniformly valued elements. In **R**, the values $r_{XX}(m - \ell)$ gradually and monotonically decrease with $|m - n|$ and in **β**, the elements $\beta_m = r_{XX}(m + 1)$ gradually and monotonically decrease with m. The numerical consequence is that $\mathbf{h}_{LS}$ is nearly, though not exactly, equal to $[1\ 0 \ldots 0]^T$. From (10.114) with $n_0 = 1$, the one-step prediction is thus $x(n + 1) \approx x(n)$. Because for $n_0 > 1$ the situation is essentially the same as just described, again $\mathbf{h}_{LS} \approx [1\ 0 \ldots 0]^T$; thus, the predictions remain nearly equal to $x(n)$, the last "known" value.

The exceedingly slow and large–amplitude oscillations in $x(n)$ can be eliminated by differencing $x(n)$ before prediction. After prediction, the result is inverse–differenced ("integrated") to obtain a prediction of the given sequence. The use of ARMA models with differencing/"integrating" is called ARIMA forecasting, the "I" standing for "integration." We now implement this modified prediction technique for our stock market example.

The first difference of $x(n)$ is $x_d(n) = x(n) - x(n - 1)$, where we throw out the first value $x_d(0)$ when $x(-1)$ is unavailable. In a sequence of length thousands of samples, this omission is not problematic. To invert the differencing given $x_d(n)$, we provide a starting value $x(1)$ and recursively build up $x(n) = x(n - 1) + x_d(n)$. For our predictions, we replace $x(1)$ by $x(n)$, the last "known" value of the sequence being predicted. Thus, the algorithm is to first–difference $x(n)$, then predict $x_d(n)$, then inverse–difference to obtain the predictions of $x(n + n_0)$.

Unfortunately, as we see in Figure 10.48, the results of this procedure are no better than before. We again look to the ACF for an explanation. This time, aside from a few low–amplitude oscillations for small k, the ACF is roughly a Kronecker delta function at $k = 0$, plus some noise for $k > 0$; see Figure 10.49. Recalling that the ACF of a white noise sequence is $A\delta(k)$, Figure 10.49 confirms our comment in Example 7.6 that except for the large low-frequency peak, the spectrum of $x(n)$ was white. In other words, recalling that differencing is a form of highpass filter, the higher–frequency variations in $x(n)$ appear to be unpredictable. The reason for the unpredictability is undoubtedly the huge number of external factors influencing the stock market other than its previous values. Because these factors are unaccounted for in simple models of $x(n)$, they are impossible to predict.

Another view of the failure of differencing is that **β**, lacking the large $r_{XX}(0)$ as a component, has only small–magnitude components, while **R** is nearly diagonal. Thus, $\mathbf{h}_{LS} \approx r_{XX}(0)\boldsymbol{\beta}$ is small and thus so is $h_{LS} * x$. Therefore, the predicted differenced value provides little added contribution to the starting value $x(n)$, giving the rather horizontal appearance of the predictions in Figure 10.48. That is, the "derivative" of $\hat{x}(n)$ is nearly "zero," giving the nearly constant value when "integrated."

One more technique sometimes used to improve prediction is taking the log of the sequence, predicting the log sequence, and then exponentiating the result. The intention is to reduce the dynamic range of the sequence values to be predicted. However, this technique did not improve results, either with or without differencing.

Is there any hope for quantitative and useful short–term prediction of the Dow Jones averages? Our next discussion will allow us to conclude a guarded and qualified "yes," at least for the major lower–frequency components.

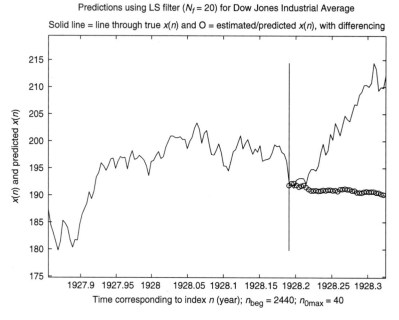

Figure 10.48

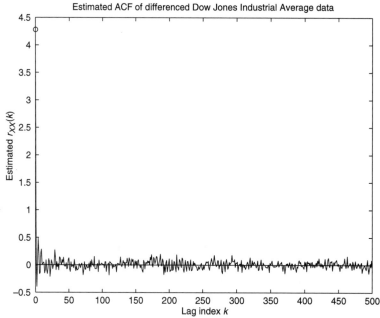

Figure 10.49

10.9.4   Trend Regression and Prony Sinusoidal Estimation

We will now show that use of Prony sinusoidal estimation and linear trend regression allows for some stock market predictions that more often than not are at least fairly reasonable. Prony sinusoidal estimation is a specialization of the Prony exponential estimation algorithm. Linear trend regression improves our accuracy by directly modeling linear trends in $x(n)$. First, we consider trend regression.

To be general, suppose first that we wish to approximate a function $y(n)$ by $\hat{y}(n)$, composed of a linear combination of M functions $f_m(n)$:

$$\hat{y}(n) = a_0 f_0(n) + a_1 f_1(n) + \ldots + a_{M-1} f_{M-1}(n). \tag{10.118}$$

Here we are assuming that the y values are taken for equally spaced "time" values (n) or x values [if $y = f(x)$]. However, all results are easily generalized for nonuniformly spaced data points. The error $e(n)$ in this approximation is

$$e(n) = y(n) - \hat{y}(n) \tag{10.119a}$$

$$= y(n) - a_0 f_0(n) - a_1 f_1(n) - \ldots - a_{M-1} f_{M-1}(n). \tag{10.119b}$$

We wish to minimize

$$J = \sum_{n=0}^{N-1} e^2(n), \tag{10.120a}$$

where N is the number of data points over which the approximation is made. Substituting (10.119b) into (10.120a) gives

$$J = \sum_{n=0}^{N-1} \{y(n) - a_0 f_0(n) - a_1 f_1(n) - \ldots - a_{M-1} f_{M-1}(n)\}^2 \tag{10.120b}$$

$$= \sum_{n=0}^{N-1} \{y^2(n) - 2y(n)[a_0 f_0(n) + a_1 f_1(n) + \ldots + a_{M-1} f_{M-1}(n)] \tag{10.120c}$$

$$+ [a_0 f_0(n) + a_1 f_1(n) + \ldots + a_{M-1} f_{M-1}(n)]^2\}.$$

We then obtain M equations in the M unknowns $\{a_m\}$ by setting to zero each of the partial derivatives of $\frac{1}{2}J$ with respect to the $\{a_m\}$:

$$\tag{10.121}$$

$$\frac{1}{2}\frac{\partial J}{\partial a_m} = \sum_{n=0}^{N-1} -y(n)f_m(m) + f_m(n)[a_0 f_0(n) + a_1 f_1(n) + \ldots + a_{M-1} f_{M-1}(n)] = 0,$$

which gives the matrix equation

$$\tag{10.122}$$

$$\begin{bmatrix} \sum\limits_{n=0}^{N-1} f_0^2(n) & \sum\limits_{n=0}^{N-1} f_0(n)f_1(n) & \cdots & \sum\limits_{n=0}^{N-1} f_0(n)f_{M-1}(n) \\ \sum\limits_{n=0}^{N-1} f_1(n)f_0(n) & \sum\limits_{n=0}^{N-1} f_1^2(n) & \cdots & \sum\limits_{n=0}^{N-1} f_1(n)f_{M-1}(n) \\ \vdots & \vdots & & \vdots \\ \sum\limits_{n=0}^{N-1} f_{M-1}(n)f_0(n) & \sum\limits_{n=0}^{N-1} f_{M-1}(n)f_1(n) & \cdots & \sum\limits_{n=0}^{N-1} f_{M-1}{}^2(n) \end{bmatrix} \begin{bmatrix} a_0 \\ a_1 \\ \vdots \\ a_{M-1} \end{bmatrix} = \begin{bmatrix} \sum\limits_{n=0}^{N-1} y(n)f_0(n) \\ \sum\limits_{n=0}^{N-1} y(n)f_1(n) \\ \vdots \\ \sum\limits_{n=0}^{N-1} y(n)f_{M-1}(n) \end{bmatrix}$$

Notice the similarities among and differences between (10.120) through (10.122) and (10.81a), (10.81b), (10.81c), (10.88b), and (10.90a) with (10.83) and (10.84). Also, (10.122) is very general. For example, for data $y(x_n)$ instead of $y(n)$, where the x_n are not equally spaced, just replace all (n) arguments in (10.122) by (x_n).

In linear trend regression, we have $M = 1$, $f_0(n) = 1$, and $f_1(n) = n$. In this case, (10.122) reduces to

$$\begin{bmatrix} \sum\limits_{n=0}^{N-1} 1 & \sum\limits_{n=0}^{N-1} n \\ \sum\limits_{n=0}^{N-1} n & \sum\limits_{n=0}^{N-1} n^2 \end{bmatrix} \begin{bmatrix} a_0 \\ a_1 \end{bmatrix} = \begin{bmatrix} \sum\limits_{n=0}^{N-1} y(n) \\ \sum\limits_{n=0}^{N-1} ny(n) \end{bmatrix}, \tag{10.123a}$$

or, using the results for sums over powers of n,

$$\begin{bmatrix} N & \frac{1}{2}N(N-1) \\ \frac{1}{2}N(N-1) & \frac{N(N-1)(2N-1)}{6} \end{bmatrix} \begin{bmatrix} a_0 \\ a_1 \end{bmatrix} = \begin{bmatrix} \sum_{n=0}^{N-1} y(n) \\ \sum_{n=0}^{N-1} ny(n) \end{bmatrix}. \qquad (10.123b)$$

In the following technique of prediction, in which $y(n)$ is our data sequence $x(n)$, first (10.123b) is solved for $\{a_0, a_1\}$. Then the trend defined by $\hat{y}(n)$ in (10.118), specializing in this case to $\hat{y}(n) = a_0 + a_1 n$, is subtracted from $y(n)$ to yield $e(n)$. The sequence $e(n)$ has no linear trend, because the linear trend has been subtracted out of $y(n)$. Next, a sinusoidal approximation $\hat{e}(n)$ of $e(n)$ is made, using the Prony technique[24] to be discussed next. Finally, we have $y(n) = \hat{y}(n) + e(n) \approx \hat{y}(n) + \hat{e}(n)$ in the form of a model equation, which may be evaluated for any n, including $n \geq N$ for prediction. Thus, to complete this algorithm, we need to discuss the sinusoidal estimation of $e(n)$.

It may be immediately asked why we need any additional sinusoidal estimation procedures, given that we already have the DFT. However, the DFT would not be useful in the prediction example, because we know it will only make the generally incorrect prediction that $x(n)$ repeats itself exactly with period N = length of the given $x(n)$ sequence = DFT length. Instead, with Prony sinusoidal estimation we perform a least–squares best fit of N_s sinusoids to the data, where $N_s \ll N$; it gives a "sinusoidal trend" or "major sinusoidal components" model of $x(n)$. We already know that there is no point in micromodeling $x(n)$, given the unpredictability of many of its more subtle variations. Our goal is to predict as much as the given data suggest and hope that that will at least point us in the right direction more often than not.

First, consider the approximation of $e(n)$ by N_e complex exponentials $z_\ell^n = e^{w_\ell n}$, where $w_\ell = \ln\{z_\ell\}$:

$$\hat{e}(n) = \sum_{\ell=1}^{N_e} c_\ell z_\ell^n. \qquad (10.124)$$

We may easily show that $\hat{e}(n)$ in (10.124) is a general solution of the difference equation (in which without loss of generality we set $a_0 = 1$):

$$\sum_{k=0}^{N_e} a_k \hat{e}(n-k) = 0, \qquad (10.125)$$

provided that $A(z_\ell) = 0$ [where $A(z) = \sum_{k=0}^{N_e} a_k z^{-k}$], by merely substituting the right-hand side of (10.124) for $\hat{e}(n)$:

$$\sum_{k=0}^{N_e} a_k \sum_{\ell=1}^{N_e} c_\ell z_\ell^{n-k} = 0, \qquad (10.126a)$$

or, rearranging the sums and multiplying both sides by z_ℓ^{-n},

$$\sum_{\ell=1}^{N_e} c_\ell \sum_{k=0}^{N_e} a_k z_\ell^{-k} = 0, \qquad (10.126b)$$

or

$$\sum_{\ell=1}^{N_e} c_\ell A(z_\ell) = 0, \qquad (10.126c)$$

which holds for all the z_ℓ in (10.124) if, as stipulated, $A(z_\ell) = 0$ for each z_ℓ.

---

[24]In this section we clarify, put in our notation, and apply the exponential and sinusoidal Prony methods given by F. B. Hildebrand in *Introduction to Numerical Analysis*, 2nd Ed. (New York: McGraw-Hill, 1974), pp. 457–465.

This result was also shown in Section 2.4.6.5 (page 64ff). There, in (2.89) the solution of the zero–input form of (2.58) [same as (10.125)] was found to be (in present notation)

$$y_{ZIR}(n) = \sum_{\ell=1}^{N_e} A_\ell z_\ell^n \text{ for } n \geq 0, \text{ where from (2.90e) we know that } \sum_{k=0}^{N_e} a_k z_\ell^{-k} = 0,$$

or $A(z_\ell) = 0$. That is, the modes of the system are the poles of its transfer function, as we concluded in Section 2.4.6.5.

We now solve (10.125) for $\{a_k\}$ by substituting actual data values $e(n - k)$ for $\hat{e}(n - k)$ in (10.125). Let us have n range from N_e to $N - 1$ so that in (10.125) no $e(n < 0)$ are called. We obtain the set of equations

$$e(N_e) + e(N_e - 1)a_1 + \ldots + e(0)a_{N_e} = 0 \tag{10.127}$$

$$e(N_e + 1) + e(N_e)a_1 + \ldots + e(1)a_{N_e} = 0$$

$$e(N - 1) + e(N - 2)a_1 + \ldots + e(N - N_e - 1)a_{N_e} = 0.$$

Relations (10.127) form a linear set of equations for the $\{a_\ell\}$. The equations are inconsistent because the actual data values $e(n)$ are not exactly of the form in (10.124) and so do not exactly satisfy (10.125) and (10.127). When overdetermined ($N - N_e > N_e$, the usual case for reliable answers with inconsistent equations), it may be solved using least-squares equation solution methods. Once the $\{a_\ell\}$ known, we solve $A(z_\ell) = 0$ for $\{z_\ell\}$ using a polynomial root–finder. Finally, with $\{z_\ell\}$ known, (10.124) with at least N_e distinct values of n and $\hat{e}(n)$ replaced by the known values of $e(n)$ becomes a linear set of equations for the $\{c_\ell\}$, which again can be solved by least squares.

The above exponential Prony modeling technique is in the end not really different from linear one–step prediction (AR) modeling—both are decaying sinusoidal models. Consequently, where AR modeling fails, we expect Prony exponential modeling to also fail. In particular, the Prony decaying sinusoids "prediction" over the long term goes to zero, as does the AR modeling technique that we already observed in Example 10.10. If, however, we let the N_e exponentials be $N_s = N_e/2$ non-decaying sinusoids (N_e even), we obtain a prediction that does not die away far from the end of the given data interval. (The linear trend also does not die out with n.) We hope that, at least reasonably near the beginning of the prediction interval, the approximation will be good.

We specialize the Prony exponential algorithm to sinusoids by enforcing the properties of sinusoids on the Prony equations. The complex exponentials z_ℓ of the N_e total number of exponentials always occur in complex–conjugate pairs. We let all the N_e exponentials have imaginary exponents. Thus, if $z_\ell = e^{j\omega\ell}$ is one of the exponentials, then so also is $z_{\ell 1} = e^{-j\omega\ell}$. Thus, for N_s sinusoids, we have $N_e = 2N_s$ exponentials in our model. From $A(z_\ell) = 0$, we have

$$\sum_{k=0}^{N_e} a_k e^{-j\omega\ell k} = 0, \tag{10.128}$$

and also for $z_{\ell 1}$,

$$\sum_{k=0}^{N_e} a_k e^{j\omega\ell k} = 0. \tag{10.129}$$

Multiplying (10.129) by $e^{-j\omega\ell N_e}$ gives

$$\sum_{k=0}^{N_e} a_k e^{-j\omega\ell(N_e - k)} = 0, \tag{10.130}$$

or, defining $m = N_e - k$,

$$\sum_{m=0}^{N_e} a_{N_e - m} e^{-j\omega\ell m} = 0. \tag{10.131}$$

If both (10.128) and (10.131) are to be true for all $\{\omega_\ell\}$, it follows that $a_{N_s-\ell} = a_\ell$ or, equivalently, $a_{2N_s-\ell} = a_\ell$. Note that in particular (for $\ell = 2N_s$), $a_{2N_s} = a_0 = 1$, and also that a_{N_s} has no "$2N_s-\ell$ partner." Consequently, for N_s sinusoids, (10.127) specializes to

(10.132)

$$\{e(2N_s) + e(0)\} \cdot 1 + \{e(2N_s - 1) + e(1)\} \cdot a_1 + \ldots$$
$$+ \{e(N_s + 1) + e(N_s - 1)\} \cdot a_{N_s-1} + e(N_s) \cdot a_{N_s} = 0$$

$$\{e(2N_s + 1) + e(1)\} \cdot 1 + \{e(2N_s) + e(2)\} \cdot a_1 + \ldots$$
$$+ \{e(N_s + 2) + e(N_s)\} \cdot a_{N_s-1} + e(N_s + 1) \cdot a_{N_s} = 0$$

$$\{e(N - 1) + e(N - 2N_s - 1)\} \cdot 1 + \{e(N - 2) + e(N - 2N_s)\} \cdot a_1 + \ldots$$
$$+ \{e(N - N_s) + e(N - N_s - 2)\} \cdot a_{N_s-1} + e(N - N_s - 1) \cdot a_{N_s} = 0.$$

Thus, (10.132) is a set of linear equations for determination of the $\{a_\ell\}$. Again, we substitute actual data values in for the $e(\ell)$.

Next we specialize (10.128) for the case of N_s real sinusoids, to determine the frequencies $\{\omega_\ell\}$ of those sinusoids. With $a_{2N_s-\ell} = a_\ell$ and in particular $a_{2N_s} = a_0 = 1$ and noting that a_{N_s} is "unpaired," (10.128) becomes

$$\sum_{m=0}^{N_s-1} a_m \{e^{-jm\omega_\ell} + e^{-j(2N_s-m)\omega_\ell}\} + a_{N_s} e^{-jN_s\omega_\ell} = 0, \tag{10.133}$$

or, multiplying by $e^{jN_s\omega_\ell}$,

$$\sum_{m=0}^{N_s-1} a_m \{e^{j(N_s-m)\omega_\ell} + e^{-j(N_s-m)\omega_\ell}\} + a_{N_s} = 0 \tag{10.134a}$$

or

$$a_{N_s} + 2 \sum_{m=0}^{N_s-1} a_m \cos\{(N_s - m)\omega_\ell\} = 0, \tag{10.134b}$$

where again $a_0 = 1$.

Although (10.133) may be solved with a polynomial root–finder (with $e^{j\omega_\ell}$ replaced by the unknown z_ℓ and $e^{-j\omega_\ell}$ replaced by $1/z_\ell$), it has twice the order of the polynomial we have to solve if we instead express (10.134b) in terms of Chebyshev polynomials. For the fairly large N_s we will be using in our example (on the order of 30 or so), use of the Chebyshev polynomials makes the difference between the Matlab root–finder succeeding and failing (Matlab will fail for order over about 40). Recalling from (8.20) that $C_m(x_\ell) = \cos(m\omega_\ell)$, where $x_\ell = \cos(\omega_\ell)$, we may write (10.134b) as

$$a_{N_s} + 2 \sum_{m=0}^{N_s-1} a_m C_{N_s-m}\{x_\ell\} = 0. \tag{10.134c}$$

Further recall from (8.23b) that $C_{m+1}(x_\ell) = 2x_\ell C_m(x_\ell) - C_{m-1}(x_\ell)$, and that with $C_0(x_\ell) = 1$ and $C_1(x_\ell) = x_\ell$, we may express (10.134c) as a polynomial in x_ℓ of order N_s. A simple routine implementing the recursion was written in Matlab code and can be used to generate the Chebyshev polynomial coefficients of any desired order. These Chebyshev polynomials are then substituted into (10.134c) and net coefficients of x_ℓ^m are obtained. A root–finder then gives x_ℓ, from which we obtain $\omega_\ell = \cos^{-1}\{x_\ell\}$. Finally, we express the estimate of $e(n)$, $\hat{e}(n)$, in terms of sines and cosines:

$$\hat{e}(n) = \sum_{\ell=1}^{N_s} A_\ell \cos(\omega_\ell n) + B_\ell \sin(\omega_\ell n). \tag{10.135}$$

Now that the ω_ℓ are known, (10.135) is a linear least–squares problem for the coefficients $\{A_\ell, B_\ell\}$, where in (10.122) the $y(n)$ are measured $e(n)$ values, and half the $f_\ell(n)$ are $\cos(\omega_\ell n)$, while the other half are $\sin(\omega_\ell n)$.

EXAMPLE 10.12

Repeat the Dow Jones Industrial Average predictions for the case of linear trend regression followed by Prony sinusoidal estimation.

Solution

As before, a 100–point section of the Dow Jones averages ending one sample before a chosen index n_{beg} (at which predictions begin) is selected as the basis from which to make predictions. First, a linear trend regression is performed, and that linear trend is subtracted. The residual $e(n)$ is then modeled using the Prony technique by a weighted sum of $N_s = 19$ sinusoids having different magnitudes, phases, and frequencies.[25]

The sinusoidal model equation is used to extend $e(n)$ for $1 \leq n_0 \leq n_{0\text{max}} = 40$ points beyond the 100–point data set. To this, the extension of the linear trend over the same 40 points is added. One additional correction is added. To make the prediction begin at the right level, the difference between the last "known" value $x(n)$ and the first predicted value $\hat{x}(n + 1)$ is added to all predicted values. Thus, with the bias approximately eliminated, the final prediction is complete.

For plotting, careful conversion of sample index to time in years is done. The entire 8192 samples stretch over approximately 27.5 years, beginning in January 1920. To account for the holidays, Δt is $27.5/8192 = 0.00336$ rather than $1/365 = 0.00274$, which would incorrectly assume a Dow Jones average on all days of the year, including holidays.

Results were examined visually for approximately 200 separate trials, with the first predicted sample n_{beg} of each trial 40 ahead of that for the previous trial. Qualitatively, we may define "success" as visually good accuracy for at least the first 20 predicted values, after which the agreement may or may not continue over the remaining 20 predictions. "Questionable success" is where the general trend (up, down, oscillatory) is reproduced in the predictions, but the accuracy is low. "Failure" is, for example, when the predictions head downward while the true values head upward.

Under these categories, 53% were "successes," 16% were "questionable successes," and the remaining 31% were "failures." Although these results may not be viewed as highly dependable, the Dow Jones averages are notoriously difficult to forecast because of such a large number of unpredictable internal and external influencing factors over a wide range of commodities.

For brevity, only three of the more interesting results are shown in Figure 10.50. Two of the three plots here are successful, illustrating how the Prony/linear trend regression prediction technique can often do better than eyeballing would do—knowing only data to the left of the vertical line. That, after all, is an important comparison—if eyeballing works better than the computer technique, the computer technique is worthless. It is granted that this technique could not predict the crash of 1929 (Figure 10.50b), but it is not alone in having this limitation. Not only are the predictions shown, but also shown for general interest are the sinusoidal/linear trend fit estimations over the period of "known" data preceding the prediction interval. In general, these back–estimations are quite good, including cases where the predictions are failures.

---

[25] The value $N_s = 19$ is somewhat arbitrary, but was empirically found to work fairly well. In practice, N_s can be selected by "predicting" known past values using still–earlier data and examining the mean–squared error in the "predictions."

Predictions using Prony sinusoidal estimation for Dow Jones Industrial Average; $N_s = 19$

Solid line = line through true $x(n)$ and O = estimated/predicted $x(n)$

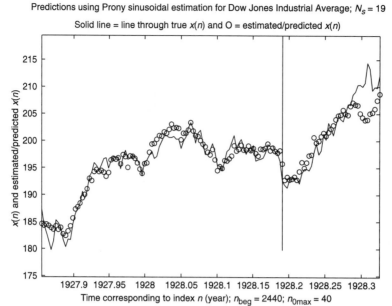

Time corresponding to index n (year); $n_{beg} = 2440$; $n_{0max} = 40$

Figure 10.50a

Predictions using Prony sinusoidal estimation for Dow Jones Industrial Average; $N_s = 19$

Solid line = line through true $x(n)$ and O = estimated/predicted $x(n)$

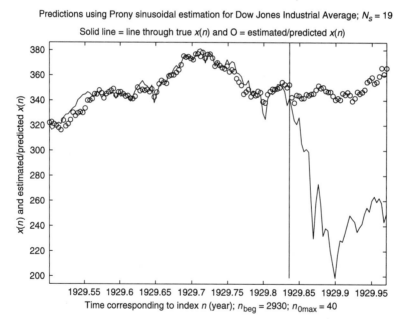

Time corresponding to index n (year); $n_{beg} = 2930$; $n_{0max} = 40$

Figure 10.50b

As noted above, there are many cases in this application in which mere eyeballing would not predict as well as the Prony/regression technique. For example, for $n_{beg} = 2440$ (Figure 10.50a; the value of n_{beg} for each plot is found in the time–axis label), the strong rise seems unpredictable from the past data given the leveling off just preceding $n = n_{beg}$, yet the Prony method produces it well. Note that this was the same time-window for which prediction failed in Example 10.11. Another example of a steeper, farther drop than one might have predicted from the previous data is seen for $n_{beg} = 7120$ (Figure 10.50c); it even succeeds in predicting the rebound toward the end of the forecasting period. Several additional predictions appear on WP 10.3; they also show how this technique can predict bumpy/oscillatory patterns as well as trends. Overall, the results of this approach are fairly encouraging and

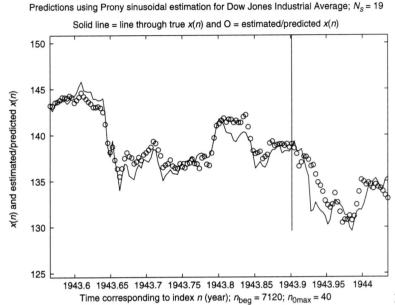

Predictions using Prony sinusoidal estimation for Dow Jones Industrial Average; $N_s = 19$

Solid line = line through true $x(n)$ and O = estimated/predicted $x(n)$

Time corresponding to index n (year); $n_{beg} = 7120$; $n_{0max} = 40$

Figure 10.50c

perhaps could be refined for further improvement, though it cannot predict the consequences of external events and factors that are not represented in the data on which the predictions are based.

EXAMPLE 10.13

Using the same technique as in Example 10.12, predict several other economic time series that have a more specialized trade involved than does the Dow Jones data.

Solution

Several sequences whose values are printed in numerical form are found in *Historical Monthly Energy Review, 1973–1988.*[26] For clarity in these examples, the estimated/predicted sequence is joined by dashed lines, whereas the true sequence is joined by solid lines. The length of each sequence is 192, including the portion "predicted" from the previous known values. The period studied is from 1973 to 1988, and the number of predicted values is $n_{0max} = 52$.

The time interval between samples is one month. In all cases, we will obtain fairly high quality predictions over most or all of this interval that, impressively, is more than four years into the future. Perhaps the predictions can be successful because each series represents a single business commodity or price and thus has a simpler dynamic than the Dow Jones averages. The method used is the same as in Example 10.12 except that no artificial level–shifting was used to correct the first predicted value and level thereafter.

---

[26]Published by the Energy Information Administration (Washington, DC): U.S. Department of Energy, 1991.

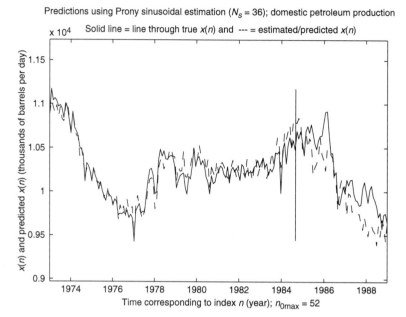

Predictions using Prony sinusoidal estimation ($N_s = 36$); domestic petroleum production

Solid line = line through true $x(n)$ and --- = estimated/predicted $x(n)$

Figure 10.51

Figure 10.51 shows estimated/predicted domestic oil production in thousands of barrels/day ($N_s = 36$). The downward trend (which eyeballing would probably miss) as well as some of the oscillations are predicted well, and after 52 months (end of plot) the prediction is nearly equal to the true value. Petroleum imports (thousands of barrels/day) are shown in Figure 10.52 ($N_s = 32$). Although hidden from eye-balling, the information on when the next rise would occur is apparently embedded in the past data, for the Prony technique correctly predicts both the timing and the magnitude of the resurgence in imports.

For general interest, the normalized–to–variances cross–covariance of the exact sequences in Figures 10.51 and 10.52 is shown in Figure 10.53 (solid curve). We also show a lowpass version (dashed curve). There is a large negative peak near lag $k = 0$, which merely indicates that whenever one sequence is high, the other tends to be low. That is, decreases in petroleum imports are made up for by increased domestic production. It is interesting to consider which sequence is really "ahead" of the other—that is, which is the sequence "causing" changes in the other. The negative peak and overall symmetrical point in Figure 10.53 is for k roughly –4 (months) (shown with dashed vertical line in Figure 10.53). Recall from (7.51) that

$$r_{xh}(k) = \sum_{n=-\infty}^{\infty} x(n + k)h^*(n)$$ and from (7.52), $R_{xh}(e^{j\omega}) = X(e^{j\omega})H^*(e^{j\omega})$, which holds

for finite–length segments of the random process realizations. The sequence whose DFT was conjugated in computing the cross-covariance in Figure 10.53 was the oil imports. Thus, the time–shifted sequence in $r_{xh}(k)$ [$x(n)$] was the domestic produc-tion sequence. (Because of the nonzero means, here we plot cross–covariances, not cross–correlations.)

Suppose we assume that the oil import sequence leads (changes ahead of) the domestic sequence. Then, if overlaid, the oil import sequence would appear as roughly an inverted version of the domestic sequence, shifted "to the left" of the domestic sequence. If true, the time–shifted sequence in r_{xh} above (domestic pro-duction) would have to be shifted left for a maximum cross–covariance, which would require a positive shift ($k > 0$). Because in fact the negative peak in the cross–covariance is for $k < 0$, we conclude from this data that the domestic sequence

Predictions using Prony sinusoidal estimation ($N_S = 32$); petroleum imports

Solid line = line through true $x(n)$ and --- = estimated/predicted $x(n)$

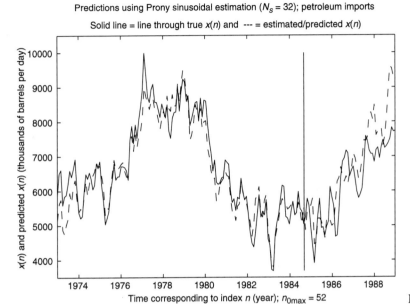

Time corresponding to index n (year); $n_{0max} = 52$

Figure 10.52

Normalized cross-covariance of domestic oil production versus oil imports

Solid = original cross-covariance; --- = lowpass filtered version; vertical = negative peak

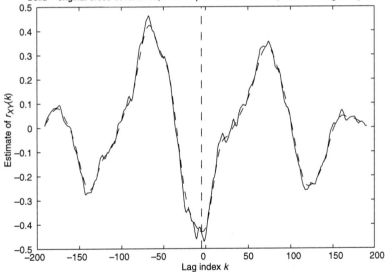

Lag index k

Figure 10.53

leads and that our imports change on a delayed basis as a consequence of our domestic production. Such shifts are difficult to assess from eyeballing Figures 10.51 and 10.52 because the trend is not always followed and the shift is small; it has to be estimated taking all the data into account.

Several additional related practical economic time series predictions over the same time period are found and explained on WP 10.4. These include domestic oil prices, at–port import oil prices, electric generation by coal, and total electric generation. We also examine the cross–covariances of (a) at–port import and domestic oil prices, (b) the amount of oil imports and their prices, and (c) coal–produced electricity and total electricity production. The predictions are quite accurate, especially given the long periods (four years) into the future; also, six–month (seasonal) energy usage cycles are evident in those plots.

10.9.5 Data Whitening

In matched filtering for communication systems and other applications such as speech processing, it may be desired to whiten the data. As we already know, whitening data removes the predictable part of the sequence. For this reason, the white noise sequence generating the general ARMA time series is known as the innovations sequence. The approximation of the innovations sequence can be used to reconstruct the original ARMA sequence if the ARMA model coefficients are known.

Consider linear filtering of the data by $h_{LS}(n)$, with $h_{LS}(n)$ designed for one–step prediction. If we let $h_{LS}(n)$ operate on $\{x(n-1), \ldots, x(n-N_f)\}$, then the output will be an estimate of $x(n)$. If we subtract this estimate (predictable component) from the true (known) value $x(n)$, we obtain a sequence that is approximately white. The closer $X(n)$ is to an $AR(N_f)$ process, the more white the sequence will be, because the least–squares filter produces an $AR(N_f)$ model for the data.

EXAMPLE 10.14

Whiten the ARMA data used in Example 10.10.

Solution

Using a least–squares $N_f = 40$ filter according to the instructions above, we obtain the following results. Figure 10.54 shows a 128–point sample of a realization of the ARMA sequence $X(n)$, namely $x(n)$. The first 40 lags of the temporal ACF of $x(n)$ are shown in Figure 10.55 as O stems, whereas the exact ACF is shown in * stems (the exact values were also shown in Figure 10.41). Figure 10.56 shows the resulting whitened sequence, and we verify in Figure 10.57 that the sequence is indeed white, as its ACF is a Kronecker delta function at $k = 0$. Notice that around $k = N_f$ there are some small erroneous nonzero values due to end effects.

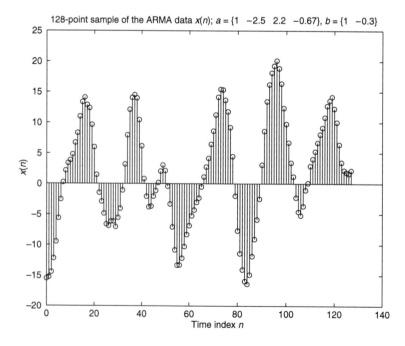

128-point sample of the ARMA data $x(n)$; $a = \{1 \quad -2.5 \quad 2.2 \quad -0.67\}$, $b = \{1 \quad -0.3\}$

Figure 10.54

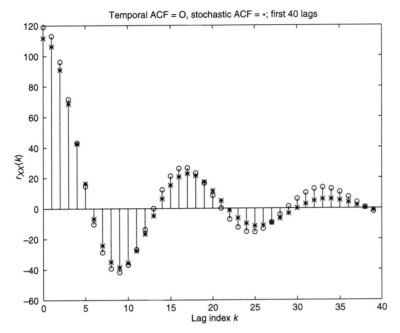

Figure 10.55

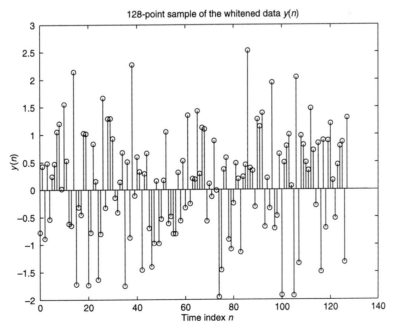

Figure 10.56

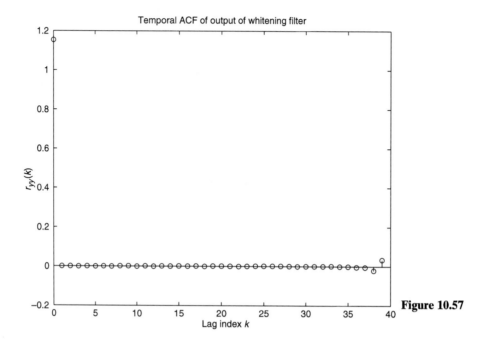

Figure 10.57

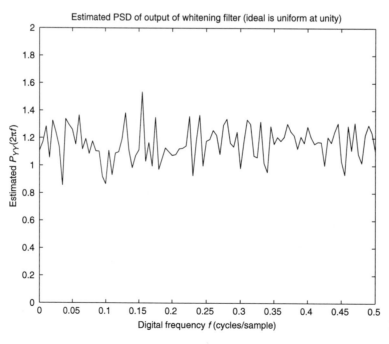

Figure 10.58

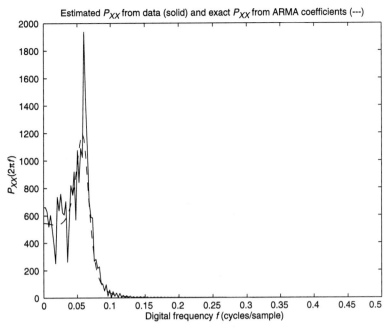

Figure 10.59

The DTFT of the ACF is the PSD function, as proved in Section 10.3. An estimate of the PSD can be obtained by using the temporal ACF. Thus in Figure 10.58, we show the DTFT of the temporal ACF of the whitened data, which is nearly white (ideally, it would be constant for all f). Finally, for reference, the estimated PSD from the DTFT of the temporal ACF of the input ARMA sequence is shown in Figure 10.59 (solid), along with the DTFT of the exact ACF of $X(n)$ (dashed). Again we find overall agreement, but due to the lack of stochastic averaging over many realizations, there are deviations.

Table 10.3 summarizes our least–squares procedures based on the standard normal equations.

Table 10.3   Some Least–Squares Problems and Solutions

Desired system $h_d(n)$	Desired output signal $d(n)$	Input signal $s(n)$ (measured)	Name of LS problem $\mathbf{R}h_{LS} = \boldsymbol{\beta}$	Quantity to be minimized
$ZT^{-1}\{1/H(z)\}$	$\delta(n)$	$h(n)$	Deconvolution	$\lvert\delta(n) - h_{LS} * h\rvert^2$
$\delta(n + n_0)$	$x(n + n_0)$	$s(n) = x(n)$	n_0-step prediction	$\lvert x(n + n_0) - h_{LS} * x\rvert^2$
$FT^{-1}\{1/[P_{XX}(\omega)]^{1/2}\}$	$u(n)$ (white noise)	$s(n) = x(n)$	Whitening	$\lvert x(n) - h_{LS} * x(n - 1)\rvert^2$ output $= x(n) - h_{LS} * x(n - 1)$

10.10   INTRODUCTION TO ADAPTIVE FILTERS

When the properties of a signal or system are unknown or vary with time, or the desired output of the filter varies with time, adaptive filters are required. An adaptive filter is a (digital) filter whose coefficients change with time. A typical example is telephone echo cancellation, in which the required parameters for estimating and canceling the return echo vary with each call—the system must adapt to the different phone lines and paths connecting different users. A wide range of algorithms for updating the coefficients with time has been developed to suit particular application problems and computational requirements. One of the simplest is the Widrow noisy gradient or least mean squares algorithm, which we briefly examine here.

The least mean squares (LMS) algorithm is a practical simplified implementation of the "steepest descent" algorithm. The essential idea of the steepest descent method is that we should update the adaptive digital filter weight vector $\mathbf{h}$ in a manner such that the mean–squared error descends, as n increases, down the steepest of all possible paths generated by all possible sequences of filters $\mathbf{h}_n$.

Let the desired filter output be $d(n)$. Then if the data vector at time index n is $\mathbf{s}_n = [s(n) \quad s(n-1) \ldots s(n - [N_f - 1])]$ and the filter output is $\hat{d}(n) = \mathbf{h}_n^{\mathrm{T}} \mathbf{s}_n$, then the error is $e(n) = d(n) - \hat{d}(n) = d(n) - \mathbf{h}_n^{\mathrm{T}} \mathbf{s}_n$, so that

$$e^2(n) = d^2(n) - 2d(n)\mathbf{s}_n^{\mathrm{T}}\mathbf{h}_n + (\mathbf{h}_n^{\mathrm{T}}\mathbf{s}_n)(\mathbf{s}_n^{\mathrm{T}}\mathbf{h}_n). \tag{10.136}$$

The true steepest descent algorithm would travel down the gradient of $\mathrm{E}\{E^2(n)\}$. In practice, only one realization $e(n)$ for $E(n)$ is available, so we instead minimize $e^2(n)$ itself. The least–squares solution would require matrix inversion of the normal equations, with the $\mathbf{R}$ matrix and $\boldsymbol{\beta}$ vector recalculated at every time step, if $\mathbf{R}$ and $\boldsymbol{\beta}$ are changing with time. It would go all the way down to the true minimum in one "step" if the stochastic gradient were available and if $\mathbf{R}$ and $\boldsymbol{\beta}$ were not changing with time. Instead, for time–adaptive filtering an iterative, approximate solution of the normal equations is proposed. This approach, without matrix calculations or inversions, makes the filter coefficients computation algorithm very rapid. We let

$$\mathbf{h}_{n+1} = \mathbf{h}_n + \mu \mathbf{g}, \tag{10.137}$$

where $\mathbf{g}$ is the search direction and μ is an empirically chosen constant. The larger μ is, the more rapid the changes in $\mathbf{h}_n$, but also the greater the possibility of overshoot, instability, and divergence of $\mathbf{h}_n$. Thus, in practice, μ is empirically chosen by trial and error.

As implied in its name, in the steepest descent algorithm, the search direction $\mathbf{g}$ is ideally chosen to be minus the gradient of $\mathrm{E}\{E^2(n)\}$ with respect to $\mathbf{h}_n$. Instead, because $\mathrm{E}\{E^2(n)\}$ is unavailable, we use the gradient of $e^2(n)$ itself. To find it, we differentiate $-e^2(n)$ with respect to the filter coefficients vector $\mathbf{h}_n$:

$$\mathbf{g} = -\frac{\partial e^2(n)}{\partial \mathbf{h}_n} = -2e(n) \cdot \frac{\partial e(n)}{\partial \mathbf{h}_n} \tag{10.138}$$

$$= 2e(n)\mathbf{s}_n.$$

Thus, (10.137) becomes

$$\mathbf{h}_{n+1} = \mathbf{h}_n + 2\mu e(n)\mathbf{s}_n. \tag{10.139}$$

Once an algorithm such as (10.139) is selected, the major challenges in adaptive filtering are the selection of $d(n)$—equivalently, the selection of what combination of signals to call the error $e(n)$—the selection of the value of μ, and sometimes the initial guess $\mathbf{h}_0$. In the examples below, we simply set $\mathbf{h}_0 = \mathbf{0}$.

EXAMPLE 10.15

As an introduction to adaptive filtering, suppose that we wish to alter the shape of an input signal $s(n)$. Let $s(n)$ be a triangle wave, and let $d(n)$ be at first a sinusoid and then change to a square wave.

Solution

With $N_f = 30$ and $\mu = 0.035$, the results are as shown in Figure 10.60. The desired sequence values $d(n)$ are joined by solid lines, the input sequence values $s(n)$ are joined by –.–.–. (a triangle wave throughout), and the filter output values $\hat{d}(n)$ are joined by – – – –. It takes only a modest number of iterations (increments in n) for $\hat{d}(n)$ to follow $d(n)$ even when $d(n)$ changes its form so radically. This is the adaptive filtering idea. The sequences of three of the weights $h_n(1), h_n(8)$, and $h_n(15)$ are shown in Figure 10.61. We see that to continually convert the triangle into the sinusoid or into the square wave, there must be periodic patterns in the coefficients themselves, and that once the "learning period" is past, the regularity of the patterns increases. With μ set to a smaller value, we can see $\hat{d}(n)$ begin appearing as a triangle and gradually follow the sinusoid, but the accuracy is reduced; with μ set too large, there is too much overshoot in the square–wave estimation.

Finally, the error $e(n)$ is plotted in Figure 10.62. The sinusoid, being smooth, is more easily followed using an adaptive filter than is the square wave. Notice how the error begins very large [in fact is equal to $d(n)$ until $n = N_f = 30$ because until then the filter output is zero] and quickly settles down to the lowest amplitude achievable by this algorithm. The sudden changes of the square wave cannot be followed instantaneously and thus result in spikes in the error sequence.

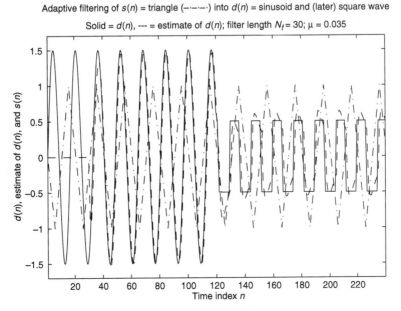

Adaptive filtering of $s(n)$ = triangle (–·–·–·) into $d(n)$ = sinusoid and (later) square wave

Solid = $d(n)$, --- = estimate of $d(n)$; filter length N_f= 30; μ = 0.035

Figure 10.60

Superimposed time-dependent weight vector elements; filter length $N_f = 30$, $\mu = 0.035$

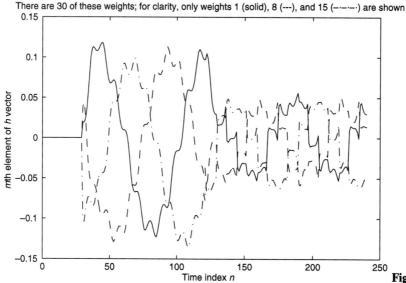

There are 30 of these weights; for clarity, only weights 1 (solid), 8 (---), and 15 (---·---·) are shown

Figure 10.61

Error in adaptive filtering; filter length $N_f = 30$, $\mu = 0.035$

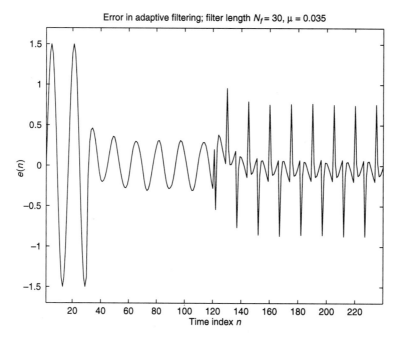

Figure 10.62

10.10.1 Adaptive Interference Cancellation

Often it is desired to reduce random noise in a system. If the nature of the interference changes with time, a time–varying adaptive filter is required. Suppose that the input signal $p(n)$ consists of a regular sequence $x(n)$ such as a quasi–periodic signal, corrupted by random noise $v(n)$; $p(n) = x(n) + v(n)$. As noted above, an adaptive filter acts as an adaptive estimator. The error signal will be the inestimable component in $d(n)$.

Let $d(n) = p(n)$ and $s(n) = p(n - n_0)$, where here n_0 is a delay parameter. In other words, let the "input signal" for the adaptive filter be a delayed version of the "desired signal," the latter being the input data sequence $p(n)$.[27] Then in this case, the adaptive filter attempts to approximate $p(n)$ using past values $\{p(n - n_0), \ldots, p(n - [n_0 + N_f - 1])\}$. Assume that n_0 is sufficiently large that the delayed noise $v(n - n_0)$ in $s(n) = p(n - n_0)$ is uncorrelated with both the noise $v(n)$ in $d(n) = p(n)$ and with $x(n)$, whereas the signal $x(n)$ *is* correlated with its past values. Then as always the filter output will be the "predictable part"—namely, the original $x(n)$. The least–squares projection, after all uncorrelations are recognized, is seen to be that of $x(n)$ onto $\{x(n - n_0), \ldots, x(n - [n_0 + N_f - 1])\}$. If $x(n)$ is regular (highly correlated for long lags) and n_0 is reasonably small (but still large enough to satisfy the requirement above), this approximation can be quite good.

EXAMPLE 10.16

A triangle wave $x(n)$ is corrupted by noise $v(n)$ that changes from white noise over to AR noise, where $\mathbf{a} = [1 \quad -0.6156 \quad 0.81]$. Use an adaptive filter to try to extract the triangle from the noise.

Solution

We use an adaptive filter of order $N_f = 24$, and let $\mu = 0.0008$. The delay is selected to be $n_0 = 24$. If the noise were only white noise and not changing over to AR noise, $n_0 = 1$ or 2 would be sufficient because white noise is perfectly uncorrelated with its past values. However, in this example, the AR noise has significant correlation with its past samples. One might try to use the temporal ACF of the data to try to estimate n_0; however, this task is complicated by the presence of the regular signal $x(n)$. An empirical compromise must be found in which $x(n)$ is correlated with $x(n - n_0)$, but $v(n)$ is approximately uncorrelated with $v(n - n_0)$.

The input signal $p(n)$ is as shown in Figure 10.63. The changeover from white to AR noise at $n = 476$ is evident, and the noise variance 0.5 is seen to be large enough

Input signal $p(n)$ = triangle wave + white Gaussian noise (0 mean, 0.5 variance) for $1 < n < 476$

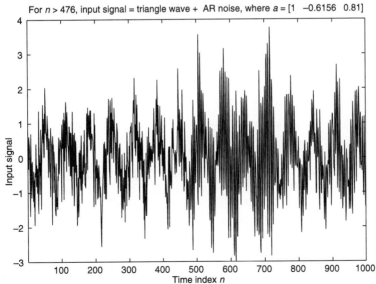

For $n > 476$, input signal = triangle wave + AR noise, where $a = [1 \quad -0.6156 \quad 0.81]$

Figure 10.63

---

[27]Notice that in this case, the signal that we have called the "desired" signal, $p(n)$, is not the signal we actually seek $[x(n)]$, but rather the available signal that the adaptive filter estimates..

Solid = estimate of triangle wave extracted from noise; --- = exact triangle wave; filter length N_f = 24, μ = 0.0008, n_0 = 24

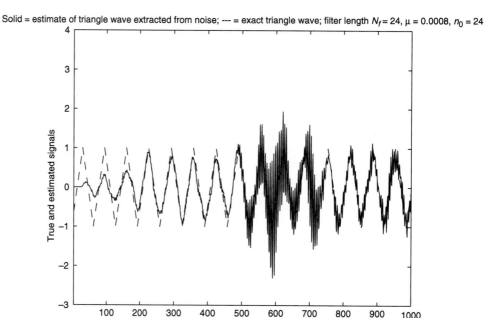

Figure 10.64

Superimposed time-dependent weight vector elements; filter length N_f = 24, μ = 0.0008, n_0 = 24

There are 24 of these weights; for clarity, only weights 1 (solid), 8 (---), and 16 (--·--·) are shown

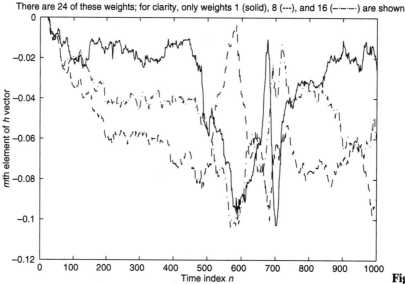

Figure 10.65

to nearly "swamp" the triangle wave buried in it. The results of adaptive filter are presented in Figure 10.64. The original triangle wave $x(n)$ that was corrupted by noise (and is unavailable to the filter) is shown dashed, whereas the adaptive filter output is shown with the solid curve.

Finally, in Figure 10.65 three of the filter coefficients $h_n(1), h_n(8)$, and $h_n(16)$ are shown as functions of n. The filter stabilizes for n between about 200 and 400; at $n = 476$ the input noise changes and the coefficients are seen to change radically so as to adapt to the new situation.

Overall, the adaptation shown in Figure 10.64 is seen to be quite effective, in comparison with the corrupted input signal in Figure 10.63.

PROBLEMS

Problems 10.1 through 10.3 involve materials on WP 10.1.
Random Variables

10.1. Show example PDFs having (a) large variance σ_X^2 and small mean μ_X, (b) large σ_X^2 and large μ_X, and (c) large μ_X and small σ_X^2.

10.2. Prove the Cauchy–Schwarz inequality for the stochastic case. Using (W6.2.8) as a model, obtain the result in (W10.1.27).

10.3. Show that (W10.1.28) is true. Hint: Let $g^{-1}(Y) = X$, and assume that on the interval $y \in [y_0, y_0 + \Delta]$ that $g(x)$ is a monotonic function of x (which is not a limiting assumption for infinitesimal Δ). If $g(x)$ is not monotonic, we can break up the interval into subintervals in which it is.

Wide–Sense Stationarity/Ergodicity, Correlation

10.4. The temporal autocorrelation function $r_{xx;n}(k)$, which is defined only for a stable signal $x(n)$, is given by (5.43):

$$r_{xx;n}(k) = \sum_{n=-\infty}^{\infty} x(n+k)x^*(n).$$

(a) Determine the (bilateral) z–transform $R_{xx;n}(z)$ of $r_{xx;n}(k)$ in terms of $X(z)$.

(b) If $x(n)$ is real, a slightly different (and simpler) form for $R_{xx;n}(z)$ results. What is $R_{xx;n}(z)$ in this case?

(c) For complex $x(n)$, what is $R_{xx;n}(e^{j\omega})$ in terms of $X(e^{j\omega})$? Is the result different for real $x(n)$? If so, specify the new form.

For the rest of this problem [parts (d) through (f)], assume that $x(n)$ is real.

(d) What is the region of convergence for $R_{xx;n}(z)$? Be quantitative. For example, the end result could be the allowed range for $|z|$. Here do not assume that $x(n)$ is causal.

(e) Suppose that $x(n) = a^n \cdot u(n), a > 0$ and real. Show the pole–zero plot for $R_{xx;n}(z)$, including the region of convergence. Also, find $r_{xx,n}(k)$ by evaluating the inverse z–transform of $R_{xx;n}(z)$.

(f) Specify another sequence $x_1(n)$ having the same autocorrelation sequence $r_{xx;n}(k)$ found in part (e). If you could not find $r_{xx;n}(k)$ in part (e), still try to relate $x(n)$ to $x_1(n)$.

10.5. Recall that a sequence may be considered wide–sense stationary if the mean $\mu_X(n) = E\{X(n)\}$ is independent of the time index and if the autocorrelation depends only on the time index difference m for $r_{XX}(n+m, n) = E\{X(n+m), X(n)\}$ for $X(n)$ real. Determine whether the following deterministic sampled sinusoid may be considered wide–sense stationary: $X(n) = A\sin(2\pi Fn\Delta t + \theta)$, where the sampling interval Δt, amplitude A, frequency F, and phase θ are all fixed values. Repeat for the case in which θ is uniformly distributed on $[0, 2\pi)$.

10.6. (a) Show that for the deterministic sinusoid $x(n) = A\cos(\omega_0 n + \theta)$, then $r_{xx;n}(k) = \frac{1}{2}A^2\cos(\omega_0 k)$, which does not diminish in magnitude even as $|k| \to \infty$.

(b) Show, as cited in the text, that the sinusoid $X(n) = A\cos(\omega_0 n + \theta)$, where θ is a random variable uniformly distributed over $(-\pi, \pi)$, has autocorrelation function $r_{XX}(k) = \frac{1}{2}A^2\cos(\omega_0 k)$.

10.7. In your own words, briefly define ergodicity, both in qualitative and quantitative terms (e.g., consider $E\{X\}$).

10.8. Someone proposes to use the statistical "autoconvolution" function $v_{XX}(k) = E\{X(n)X(k-n)\}$ in place of the usual statistical autocorrelation function. Why does the autoconvolution function have no place in basic statistical processing?

10.9. Does ergodicity imply wide-sense stationarity? Explain fully your reasoning; consider ergodicity in the mean and, briefly, autocorrelation ergodicity.

Power Spectral Density

10.10. Suppose that we are told that the power spectral density of a data sequence $x(n)$ is

$$P_{XX}(\omega) = \frac{A^2(1 - ae^{-j\omega})(1 - ae^{j\omega})}{(1 - be^{-j\omega})(1 - be^{j\omega})}, \qquad |a| < 1, |b| < 1.$$

Find the minimum–phase filter $H(z)$ and its unit sample response $h(n)$ that when driven by zero–mean, unit–variance white noise $u(n)$ has an output $x(n)$ whose power spectral density function is as given above. Also find the autocorrelation function $r_{XX}(k)$.

Normal Random Entities

10.11. Define $f_{X1}(x) = Ae^{-(x-\mu_X)^2/\sigma_X^2}$.

(a) Find the required constant A for normalization of the PDF, where μ_X and σ_X are assumed known and fixed.

(b) Show that the mean is μ_X but the variance is $\sigma_X^2/2$. Thus, we now see the need for the $\frac{1}{2}$ in the exponential of the normal PDF—so that the variance parameter is indeed equal to the variance.

10.12. Show that the linear transformation $\mathbf{W} = \mathbf{GX} + \mathbf{g}$ of a Gaussian random vector $\mathbf{X}$ is another Gaussian random vector $\mathbf{W}$. In the process, determine the mean vector and covariance matrix of $\mathbf{W}$. Hint: Use the characteristic function.

10.13. The PDF in (10.50) is valid for any Gaussian (normal) random vector. Prove that the corresponding characteristic function is as given in (10.51).

Estimation

10.14. Consider the estimator of the autocorrelation function $r_{XX}(k)$,

$$\hat{r}_{XX}(k) = \frac{1}{N} \sum_{n=0}^{N-1} X(n)X(n + k).$$

(a) Find the mean and variance of $\hat{r}_{XX}(k)$ for general $X(n)$ and for $X(n)$ a zero–mean, WSS, and normal random process [assume that $r_{XX}(k \to \infty) \to 0$].

Hint: A normal process satisfies the following identity for its fourth moment:

$$E\{X(k)X(\ell)X(m)X(n)\} = E\{X(k)X(\ell)\}E\{X(m)X(n)\} + E\{X(k)X(m)\}E\{X(\ell)X(n)\} + E\{X(k)X(n)\}E\{X(\ell)X(m)\}.$$

Try to eliminate any double sums from your answer; in so doing, you will prove an interesting and useful identity.

(b) Is $\hat{r}_{XX}(k)$ unbiased? Is it asymptotically consistent? Explain in as much detail as possible what would need to be done to determine whether $\hat{r}_{XX}(k)$ is an asymptotically efficient estimator of $r_{XX}(k)$ and where the difficulty lies.

10.15. N random variables are independent and identically distributed according to the exponential distribution:

$$f_X(x \mid \theta) = \frac{1}{\theta}e^{-x/\theta}, \qquad \text{for } x \geq 0 \text{ and } 0 \text{ otherwise, and } \theta > 0.$$

(a) Find $E\{X\}$.

(b) What is the joint PDF $f_{\mathbf{X}}(\mathbf{x} \mid \theta)$?

(c) Find the maximum likelihood estimate of θ, $\hat{\theta}_{\text{MLE}}$.

(d) Is $\hat{\theta}_{\text{MLE}}$ unbiased? Prove your claim.

(e) Find the Cramer–Rao bound for $\hat{\theta}$.

(f) Is $\hat{\theta}_{\text{MLE}}$ efficient? Prove your assertion.

10.16. Suppose that we have a real–valued deterministic sequence $h(\gamma, n)$, where γ is a parameter we wish to estimate from N noisy real–valued measurements $y(n) = h(\gamma, n) + u(n)$, $n \in [0, N-1]$, where $U(n)$ is a non–wide–sense–stationary zero–mean white normal sequence.

(a) If $E\{U^2(n)\} = \sigma^2(n)$, find $E\{U(n)U(m)\}$ and give the elements of $\mathbf{C}_{UU}$, the autocovariance matrix of U.

(b) Is $Y(n)$ wide–sense stationary? Is it normal? Find the expected value of the measurement $Y(n)$, find $E\{[Y(n) - E\{Y(n)\}][Y(m) - E\{Y(m)\}]\}$, and find the probability density function of $\mathbf{Y}$, where $\mathbf{y}$ is the N–vector of measurements. Simplify as much as possible; no matrices are allowed in the final form of your PDF.

(c) Find the coefficients of the cubic equation which when solved gives the maximum likelihood estimate of γ, if $h(\gamma, n) = n\gamma(\gamma + n)$ [assume that the $\sigma^2(n)$ are known]. Simplify as much as possible.

10.17. (a) Prove the Cramer–Rao bound (10.56), using the following directions.

Let $g(\theta) = \int f_{\mathbf{X}}(\mathbf{x}; \theta)\, d\mathbf{x} = 1$, where we know $g(\theta) = 1$ because $f_{\mathbf{X}}$ is a valid PDF. Express $\partial g/\partial\theta$ in terms of L and $f_{\mathbf{X}}$. Substitute $g(\theta) = 1$ into your expression for $\partial g/\partial\theta$ to obtain $E\{[..]\} = 0$ (where you must determine $[..]$). Then write $E\{-E\{\hat\theta\}[..]\} = 0$, which is valid because $\hat\theta$ is independent of the expectation integration variables $\mathbf{X}$. Then by the same arguments write $\partial E\{\hat\theta\}/\partial\theta = E\{\hat\theta[..]\}$. Adding to $E\{-E\{\hat\theta\}[..]\} = 0$, obtain $\partial E\{\hat\theta\}/\partial\theta = E\{(\hat\theta - E\{\hat\theta\})[..]\}$. Apply Schwarz's inequality to obtain an initial form of the Cramer–Rao bound. Take $\partial/\partial\theta$ of your $E\{[..]\} = 0$, and use one of your previous results to show that $E\{[..]^2\} = -E\{\partial/\partial\theta[..]\}$. Then differentiate with respect to θ the relation $E\{\hat\theta\} = \theta + B_\theta(\hat\theta)$, and put this and all your results together to obtain the Cramer–Rao bound, (10.56).

(b) Concerning the condition for attainment of the Cramer–Rao bound in (10.58): In the text it is stated that Schwarz's inequality dictates that equality holds in Cramer–Rao bound when $A = \hat\theta - E\{\hat\theta\}$ and $B = \partial L/\partial\theta$ are "collinear" or proportional, with respect to the integration variables $\mathbf{x}$ in the stochastic expectation averages. Show where in your proof in part (a) that A and B may be identified as shown, as the entities that must be proportional for Cramer–Rao attainment.

(c) Show that for Cramer–Rao bound attainment, $\sigma_{\hat\theta}^2 = 1/K(\theta)$.

Characteristic Function

10.18. Consider again the exponentially distributed random variable X of Problem 10.15 and the estimate $\hat\theta_{\text{MLE}}$ of the parameter θ.

(a) Find the characteristic function of X.

(b) Find the PDF of $Y = X/N$, first for a general random variable X and then in particular for the exponential random variable.

(c) Find the characteristic function of $Y = X/N$, first for a general random variable X and then in particular for the exponential random variable.

(d) Use the results in part (c) to find the characteristic function of $\hat\theta_{\text{MLE}}$.

(e) Use the moment–generating property of the characteristic function to find the mean and variance of $\hat\theta_{\text{MLE}}$, and compare with the values obtained in Problem 10.15.

Maximum Entropy

10.19. Prove that of all zero–mean random N–vectors having the same autocorrelation matrix $\mathbf{R}_{XX}$, the one that is Gaussianly distributed has maximum entropy.

10.20. Prove (10.66), used in proving that the PDF that is uniform over all permitted values is the one having maximum entropy. Integrate the mean–value theorem.

Linear Shift–Invariant Systems with Stochastic Inputs/Roundoff Effects

10.21. Prove that $r_{XU}(k) = h(k)$ (see the text just preceding Example 10.7).

10.22. In this problem, we determine how to obtain the unit sample response of an LSI digital filter $h(n)$ by passing a real allpass sequence $h_{\text{ap}}(n)$ through it. With the system output $y_1(n) = h(n) * h_{\text{ap}}(n)$, determine the allpass property in the z–domain [what is the value of $H_{\text{ap}}(z)H_{\text{ap}}(1/z)$?] to determine $h(n)$ in terms of $y_1(n)$ and $h_{\text{ap}}(n)$. Relate to a deterministic cross–correlation. Then produce as a special case of your result the well–known method for determining $h(n)$ when $h_{\text{ap}}(n)$ is replaced by a Kronecker delta function (which is also allpass).

10.23. Derive the output noise variances for the cascade structure, for $H(z)$ in Example 10.8. Verify with Simulink (at least approximately) the actual roundoff error figures quoted in the text (in Example 10.8) for the cascade structure.

10.24. Repeat Problem 10.23 for the parallel structure for the filter in Example 10.8.

AR, MA, and ARMA Processes

10.25. Find the autocorrelation function of an AR(1) process, satisfying the difference equation $x(n) + a_1 x(n - 1) = u(n)$ where $u(n)$ is zero–mean Gaussian white noise with variance σ^2.

10.26. Show that the autocorrelation function of a real AR(2) process, satisfying the difference equation $x(n) + a_1 x(n - 1) + a_2 x(n - 2) = u(n)$, where $u(n)$ is zero–mean Gaussian white noise with variance σ^2, is as given below. Assume that the process modes are oscillatory; designate the modes by $z_0 = r\angle\pm\theta$, where $r < 1$; show that

$$r_{XX}(k) = \frac{2\sigma^2 r^k \sin\{k\theta + \tan^{-1}(w)\}}{\sin(2\theta)(1 - r^2)^2\{1 + w^2\}^{1/2}} \quad \text{where } w = \frac{1 + r^2}{1 - r^2}\tan(\theta) \text{ and } k \geq 0.$$

[For $k < 0$, use $r_{XX}(-k) = r_{XX}(k)$, where $x(n)$ is real–valued.]

10.27. By using the results of Problem 10.26, show that $r_{XX}(k)$ for the AR(2) sequence in Example 10.10 is as given in the solution of Example 10.10.

10.28. Find an exact, closed–form expression for the autocorrelation function of a purely AR(p) process in terms of the poles z_ℓ and the noise power σ^2. Provide the special case $p = 1$ as an example.

10.29. A real WSS signal $X(n)$ is modeled as the output of a system $H(z)$ having the pole–zero diagram for the upper half–plane $\{\text{Im}(z) \geq 0\}$ shown in Figure P10.29 and white noise input. We are told that the PSD of $X(n)$ has value 2 at $f = 0$.

(a) What is the order of the zero of $H(z)$ at $z = 0$ if the numerator order does not exceed the denominator order? Find $H(z)$ explicitly.

(b) Determine $P_{XX}(\omega)$. Along the way, develop some useful general results for handling pole terms in the calculation of the PSD; your resulting expression should involve only real–valued functions of ω.

(c) What difference equation does $X(n)$ satisfy? Identify quantitatively any new quantities you may introduce. Provide numerical values of all model parameters involved.

(d) Classify $X(n)$ by giving the model type and order.

(e) What is the z–transform of the impulse response that when given white noise would produce $X(n)$? Express only in terms of numbers and powers of z^{-1}.

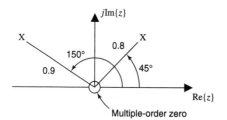

Figure P10.29

(f) Find the z–transform of the autocorrelation sequence, and then invert it to find the autocorrelation sequence. Your final expression should involve only real quantities and should be as simple as possible. Plot the first 50 lags.

(g) Plot the PSD determined in part (b) against the (truncated) DTFT of your $r_{XX}(k)$ found in part (f). They should agree for all ω.

10.30. Suppose that $X(n) = \exp\{-j(\pi n/4 + \phi)\}$, where ϕ is uniformly distributed on $[0, 2\pi]$.

(a) Write a difference equation that $X(n)$ satisfies, and determine $r_{XX}(k)$.

(b) Write the difference equation that $r_{XX}(k)$ would satisfy were $X(n)$ an AR(1) sequence. Generalize to AR(p). Attempt to model $X(n)$ in part (a) as an AR(1) sequence. What are a_1 and σ [use your results from part (a)]? Interpret the results in terms of prediction.

(c) Using Matlab, plot the power spectral density for $a_1 = -e^{-j\pi/4}$ and $\sigma = 1$, for $|f| \le \frac{1}{2}$. Include your initial result and that obtained using axis ([−0.5 0.5 0 10]). Comment on the values at $f = 0$ and $f = \frac{1}{2}$, comparing them (and all f) with what you expect for the PSD of a pure sinusoid. In particular, speculate on instead using the value of σ^2 obtained in part (b). Draw the pole diagram of $1/A(z)$. Interpret both results to help explain how the transition from AR to pure sinusoid can occur.

10.31. Prove that the general ACF expression (10.117) is valid for any MA signal. This result was used in Example 10.10.

10.32. Prove for the ARMA(3, 1) sequence in Example 10.10 that the ACF is as given there. You may use Example 10.7 as a model calculation.

10.33. A manufacturer models its production as

$$d(n) = \alpha_1 - \beta_1 p(n-1) + u(n) \qquad d(n) = \text{demand}, p(n) = \text{price}$$
$$s(n) = \alpha_2 - \beta_2 p(n) + w(n) \qquad s(n) = \text{supply},$$

where n is the month number, $\alpha_1 =$ base demand, $\alpha_2 =$ base supply, β_1 and β_2 are respectively the dependencies of demand and supply on price, and $u(n)$ and $w(n)$ are random disturbances independent of the past (white Gaussian zero–mean noise). For simplicity, suppose that the supply is always perfectly matched to the demand: $d(n) = s(n) = q(n)$. The intuitive equations above may thus be rewritten as $q(n) = \alpha_1 - \beta_1 p(n-1) + u(n) = \alpha_2 - \beta_2 p(n) + w(n)$ so that $p(n) = bp(n-1) + M + z(n)$, where $b = \beta_1/\beta_2$, $M = (\alpha_2 - \alpha_1)/\beta_2$, and $z(n) = \{w(n) - u(n)\}/\beta_2$. The ultimate goal is to forecast the price $p(n)$ and supply/demand $q(n)$ from values already observed. But first, to obtain $E\{p(n)\}$ we would like to express $p(n)$ as the output of a MA filter with $z(n)$ as the input.

(a) What is the required order of this MA filter?

(b) Let $M = 0$ ($\alpha_1 = \alpha_2$). Express $p(n)$ as the output of a MA filter with $z(n)$ as the input. What is the requirement on β_1 and β_2 for $p(n)$ to be a stable sequence?

(c) Repeat part (b) for $\alpha_1 \ne \alpha_2$. Simplify as much as possible.

(d) In part (c), what is the expected value of $p(n)$ in terms of the originally–specified parameters?

(e) For $\alpha_1 \ne \alpha_2$, what are the standard one–step predictors for $p(n)$ and $q(n)$, assuming that the values of all parameters are known?

Least–Squares Filtering and Related Topics

10.34. Suppose that

$$s(n) = \left\{\frac{a^n}{1 - b/a} + \frac{b^n}{1 - a/b}\right\} u(n), \text{ where } u(n) \text{ is the unit step sequence, } |a| < 1, |b| < 1,$$

a and b real.

(a) Find $r_{\ell m}$ and β_k for the deconvolution problem, in terms of a, b, and $k = m - \ell$. Minimize $J = ||\delta(n) - (s * h)(n)||^2$. Using Matlab, plot $r_{ss;n}(k)$ for $|k| \le 29$, $a = 0.5, b = 0.75$ using your formula and by using a truncated sum; plot both calculations of $r_{ss;n}(k)$ on the same axes.

(b) Using either a Matlab routine you write or a hand calculation, find $h_{LS}(n)$ for (i) $N = 2$ and (ii) $N = 3$ with $a = 0.5, b = 0.75$. To do this by Matlab, write a routine that, given a sequence $r(k)$ for $k \ge 0$, produces the **R** matrix.

(c) Suppose that someone suggests that rather than setting up the least squares problem, all we need to do is set $H(z) = 1/S(z)$. Is this approach possible, if $h(n)$ is restricted to being FIR? If so, find that $h(n)$ by calculating $H(z) = 1/S(z)$, and comment on why it will not always work. Based on your previous results in this problem, which (if either) is better, the $N = 2$ or the $N = 3$ solution, and how could you demonstrate that? What would you expect the $N = 4$ least–squares solution to be? Perform the calculation that verifies your claim.

10.35. Consider the least–squares prediction of $x(n)$ based on (i) $x(n - 1)$ only and on (ii) $\{x(n - 1)$ and $x(n - 2)\}$; that is, let (i) $\hat{x}(n) = -\alpha_{11}x(n - 1)$ and (ii) $\hat{x}(n) = -\alpha_{21}x(n - 1) - \alpha_{22}x(n - 2)$. Assume that perfect ACF estimates of $x(n)$ are available.

(a) Determine the optimal least–squares filter coefficients for each case. Find the (minimum) prediction error power $J_{\min}$ in each case.

(b) Specialize your results in part (a) to the case in which $x(n)$ is an AR(1) process; use the expression for the autocorrelation function of an AR(1) process found in Problem 10.25. What conclusions can you draw?

10.36. To see how your conclusions of Problem 10.35 are changed for higher order, consider an AR(2) signal. We determined its autocorrelation function in Problem 10.26. Suppose that the modes are at $z_0 = 0.9 \cdot e^{j\pi/4}$ and $z_0{}^*$, and that the driving white noise sequence again has variance σ^2. Then $r_{XX}(k)$ takes the form $r_{XX}(k) = 5.783865\sigma^2 (0.9)^{|k|}\cos(45° \cdot |k| - 6°)$, and the true AR parameters are [by simple expansion of $A(z)$] $\alpha_{21} = -1.2728$ and $\alpha_{22} = 0.81$. Based on the above equation for $r_{XX}(k)$, specialize part (a) of Problem 10.35 for this particular AR(2) process. What results for AR coefficients and for minimum MSE (prediction error power) do you expect for cases (i) and (ii); do they agree with your numerical results here?

10.37. (a) In the so–called method of moments (not to be confused with the electromagnetic computation technique with the same name), unknown parameters are estimated by replacing any required moments with their sample estimates (computed from actual data). Thus, for example, we replace the autocorrelation function by the unbiased estimate

$$r_{N, n/\text{unb}}(k) = \frac{1}{N - k} \sum_{n=0}^{N-1-k} x(n)x(n + k).$$

First, show that this estimator of the autocorrelation function is unbiased. Then find the method of moments estimate of a_1 and σ^2 for a real AR(1) process, based on $\{x(n), n \in [0, N - 1]\}$ and the unbiased ACF estimator. Hint: Use relations between the true ACF values and the model parameters for an AR(1) process.

(b) In Matlab, generate AR(1) data for $a_1 = 0.8, \sigma^2 = 1$. Use the method of moments estimates from part (a) to estimate a_1 and σ^2 from the actual data. Repeat for $a_1 = -0.5$, $\sigma^2 = 4$. Let $N = 1024$.

10.38. (a) Determine the normal equations for minimization of $J = \| \mathbf{d} - \mathbf{s} * \mathbf{h} \|^2$, where $H(z)$ is an FIR filter of length N and

$$d(n) = \begin{cases} 1, & n \in [0, N - 1] \\ 0, & \text{otherwise} \end{cases} \quad \text{and} \quad s(n) = \begin{cases} \alpha^n, & n \ge 0 \\ 0, & \text{otherwise} \end{cases}, \quad \alpha \in (0, 1).$$

Your answer should be only in terms of α and N; eliminate all fractions from both sides of the normal equations. Finally, find the minimum value of J in terms of α, N, and the elements $h_{LS}(n)$ of $\mathbf{h}_{LS}$, the optimal filter.

(b) Check these results numerically in Matlab by letting $\alpha = 0.1$ and then $\alpha = 0.8$ for $N = 100$; for which value of α is the result better, and why? Repeat for $N = 20$ (which N is better?). Plot $h_{LS}(n) * s(n)$, and comment on closeness to $d(n)$. Also compute $J_{\min} = \| d(n) - h_{LS}(n) * s(n) \|^2$, and compare against the formula you obtained in part (a) (they should agree exactly).

10.39. We wish to fit some measured data $\{x(n)\}, n \in [0, N - 1]$ by the sum of two sinusoids: $x(n) \approx A_0\cos(2\pi f_0 n) + A_1\cos(2\pi f_1 n)$, where $f_0 \ne f_1$. The sinusoidal amplitudes A_0 and A_1 are unknown, whereas f_0 and f_1 are assumed known. The measured data clearly constitutes an overdetermined set of equations for A_0 and A_1. Find the least-squares solution for A_0 and A_1. In the process, you should determine, for example, the values of the following sums:

$$\sum_{n=0}^{N-1} \cos(n\theta)\cos(n\phi) \text{ which for } \phi = \theta \text{ becomes } \sum_{n=0}^{N-1} \cos^2(n\theta), \text{ simplifying fully.}$$

(a) Assume that f_0 and f_1 are not necessarily integers divided by N.

(b) Assume that f_0 and f_1 are integers divided by N. What familiar simple operation can be used to find A_0 and A_1 in this case?

(c) Suppose that $x(n) = \cos(2\pi f_0 n) + 3\cos(2\pi f_1 n) + x_t(n)$, where $x_t(n)$ is a unit–amplitude triangle waveform of period 40. Let $N = 200$. Find the estimates of A_0 and A_1 using Matlab and your results in part (a) for $f_0 = 0.0729, f_1 = 0.302$ (note that neither of these is equal to m/N for m an integer). Plot the error in estimation versus n and comment. Repeat for $f_0 = 0.07, f_1 = 0.3$; use the results of part (b) to compute for this case, as $f_0 = 14/N$ and $f_1 = 60/N$. Use automatic determination of whether f_0 is m/N by using the rem command, and determine the value of m by another call to rem. Then add phases to the cosines making up $x(n)$, and note the collapse of accuracy due to the poor model selection of zero–phase cosines.

10.40. Suppose that a prediction–error filter (or *any* FIR filter) has coefficients $h(\ell)$, $\ell \in [0, N]$, where we use N rather than $N - 1$ for the upper index to simplify notation here; thus h has order N and length $N + 1$. The temporal autocorrelation of $\{h(\ell)\}$ [see (10.10)] is

$$r_{N,n}(k) = \frac{1}{(N+1)} \sum_{n=0}^{N-k} h(n+k)h^*(n), k \in [0, N] \text{ and } r_{N,n}(k) = r_{N,n}^*(-k) \text{ for } k \in [-N, -1].$$

Show that $\displaystyle\sum_{k=-N}^{N} r_{N,n}(k) = \frac{1}{(N+1)} \sum_{m=0}^{N} h^*(m) \sum_{\ell=0}^{N} h(\ell).$

Hint: Compare coefficients of $h(\ell)$ in both expressions, breaking up sums as convenient.

10.41. In reference to Problem 10.40, show that the mean–squared prediction error $J = E\{|E^2(n)|\}$

$= E\{|x(n) - \hat{x}(n)|^2\}$ is equal to $(N+1)\displaystyle\sum_{k=-N}^{N} r_{XX}(k)r_{N,n}(k)$. Again, for simplicity assume that h has length $N + 1$.[28]

10.42. It is shown[29] that the optimal prediction–error filter $\{h(n)\}$ is minimum–phase. Let us show this, filling in a few missing steps (as were also filled in, in Problems 10.40 and 10.41). We will also add some clarifications to the paper. In this problem, as in prediction–error filtering, we assume that $h(0) = 1$.

(a) Following the notation of Orfanidis, we write $H(z) = \displaystyle\sum_{n=0}^{N} h(n)z^{-n} = \prod_{\ell=1}^{N}(1 - z_\ell z^{-1})$ for a length–$(N + 1)$ FIR filter sequence [the right–hand side would have $h(0)$ out front if $h(0) \neq 1$]. Show that reversing any one of the zero factors in $H(z)$—that is, replacing $(1 - z_\ell z^{-1})$ by $(-z_\ell^* + z^{-1})$ for some ℓ—results in $r_{N,n}(k)$ not being changed. How many length–$(N + 1)$ sequences have the same temporal autocorrelation function? Comment on the special case for $h(n)$ real [Orfanidis uses complex notation in his paper, even though at the outset he assumes that $x(n)$ is real–valued].

(b) Let one of the sequences having the same autocorrelation as $\{h(n)\}$ be designated $h'(n)$ $= h'(0) \cdot \{c(n)\}$, where $c(0) = 1$. What is the autocorrelation $r_{N,n}(k)$ for $c(n)$ in terms of $r_{N,n}(k)$ for $h(n)$?

(c) Find $J(c)$ in terms of $J(h_{LS})$, where $h = h_{LS}$ is the optimal prediction–error filter and $c(n)$ $= h'(n)/h'(0)$ and where $h'(n)$ was obtained from h_{LS} as in part (a).

(d) Given that $J(h_{LS}) < J(c)$ by the assumption that h_{LS} is the least–squares optimal prediction error filter, it follows from part (c) that $|h'(0)| < 1$. Given how $h'(n)$ was obtained in part (a), what is the value of $h'(0)$? What implication does $|h'(0)| < 1$ therefore have on the zeros of $H_{LS}(z)$? Why might the minimum–phase property be useful in this case?

---

[28]Refer to S. J. Orfanidis, "A Proof of the Minimal–Phase Property of the Prediction–Error Filter," *Proc. IEEE*, 71, 7 (July 1983): 905, eq. (4) (which, however, is not proved there).

[29]Ibid.

10.43. Program in Matlab the prediction algorithm of Dow Jones Industrial Average data presented in the text. For data, look up the 100 most recent closings; make predictions up to 40 days in the future, as done in the text. Then, 40 days later, look up those 40 days you predicted and see how good the predictions were. This problem is obviously a longer–term project than others. Of course, if time runs short, just use data beginning 140 days ago, and use the first 100 data points to predict the most recent 40 closings.

Periodogram

10.44. (a) Find the periodogram estimate $\hat{P}_{per}(2\pi f)$ of the complex sinusoid $x(n) = Ae^{j(2\pi f_0 n + \phi)}$ given $x(n)$ for $n \in [0, N-1]$. The expression should be as simple as possible—only real terms involving A, f_0, and N, and no sums. Write in terms of the digital frequency $f = \omega/(2\pi)$.

(b) Evaluate $\hat{P}_{per}(2\pi f_0)$.

10.45. 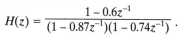 A random sequence $X(n)$ is filtered by a digital filter $h(n)$. The signal $X(n)$ is ARMA$(3, 2)$ with parameters $a_1 = 1.9, a_2 = 1.79, a_3 = 0.801, b_1 = 0.4, b_2 = -0.32$, and $\sigma^2 = 1$. The filter has transfer function

$$H(z) = \frac{1 - 0.6z^{-1}}{(1 - 0.87z^{-1})(1 - 0.74z^{-1})} \cdot$$

Find the PSD of the output $Y(n) = X(n) * h(n)$ in closed form, with its parameters fully numerically determined. Classify $Y(n)$ [e.g., specify p and q in ARMA(p, q)]. Verify your result by generating actual ARMA$(3, 2)$ data for $X(n)$ with the parameters above and filtering it with $H(z)$. Then generate stochastic data using the correct model for $Y(n)$. Plot the periodograms of the calculated and modeled outputs on the same axes, and compare with the analytically determined PSD calculated directly from $R_{YY}(z)$ (also plot on the same axes).

10.46. (a) If two real random processes $X_1(n)$ and $X_2(n)$ are uncorrelated, show that if $Y(n) = X_1(n) + X_2(n)$, then $P_{YY}(\omega) = P_{X_1 X_1}(\omega) + P_{X_2 X_2}(\omega)$.

(b) Suppose that $X(n)$ is a real–valued MA(1) process, with model parameters b_1 and σ^2. If $Y(n) = X(n) + U(n)$, where $U(n)$ is zero–mean white noise with variance σ_U^2 and $U(n)$ is uncorrelated with $X(n)$, find the appropriate model type and the values of the model parameters of $Y(n)$. Express your results in terms of b_1, σ^2, and σ_U^2, simplifying your results as much as possible.

(c) Using Matlab, generate pure MA data $x(n)$ for $b_1 = 0.8, \sigma^2 = 1$, and white noise $u(n)$ with $\sigma_U^2 = 0.5$ (zero mean). Calculate $y(n) = x(n) + u(n)$ for this data, and plot the periodogram (use $N = 1024$). Then generate data according to the model you obtained in part (b), and plot its periodogram on the same axes. Finally, plot the exact PSD of the model you obtained in part (b) [directly use $R_{YY}(z)$]. It should be a sort of average of all the wild swings of the periodogram. The two periodograms should be very close.

10.47. Show that we may write the periodogram formula from the intuitive truncated version of the PSD, but with $r_{XX}(k)$ replaced by the estimate $r_{N,n}(k)$ given in (10.10):

$$\hat{P}_{per}(\omega) = \sum_{k=-(N-1)}^{N-1} r_{N,n}(k)e^{-j\omega k}.$$

For simplicity, let $X(n)$ be real–valued. Thus, the periodogram is a truncated version of the Wiener–Kinchin formula, but with $r_{XX}(k)$ replaced by the biased estimator $r_{N,n}(k)$.

10.48. Suppose that we have perfect ACF estimates for a real random process $X(n)$: $\hat{r}(k) = r_{XX}(k)$, $k \in [0, N-1]$, and of course $r_{XX}(-k) = r_{XX}(k)$. We use them in place of the biased ACF estimate in the periodogram (see Problem 10.47). Find this improved periodogram for $X(n)$ a real AR(1) process, for which $-1 < a_1 < 0$. The exact $r_{XX}(k)$ for AR(1) was found in Problem 10.25.

(a) Express $\hat{P}_{per}(\omega)$ in simplest possible terms—only real terms involving a_1 and N, and no sums. Final form: $\hat{P}_{per}(\omega) = P_{AR(1)}(\omega) + P_\Delta(\omega)$. Does $\hat{P}_{per}(\omega) \to P_{AR(1)}(\omega)$ as $N \to \infty$?

(b) Find the error in $\hat{P}_{per}(\omega)$ at the peak frequency of $P_{AR(1)}(\omega)$ [i.e., find P_Δ(peak freq)] for $a_1 > 0$. Again, simplify as much as possible.

10.49. (a) Suppose that we have been given $N \geq 3$ exact values of the ACF of a real ARMA(1, 1) process $\{r(0), r(1), \ldots r(N - 1)\}$. A good side exercise is to show that the exact ACF values are

$$r_{XX}(k) = \frac{\sigma^2}{1 - a^2} \begin{cases} 1 + b^2 - 2ab, & k = 0 \\ (1 - ab)(1 - b/a)(-a)^{|k|}, & k \neq 0 \end{cases}$$

(where for brevity we drop all the 1 subscripts, as there is only one a and only one b). We know that this process is actually ARMA(1, 1) and thus $r_{XX}(k)$ has infinite duration, but someone suggests for convenience that we calculate the periodogram based on only these N ACF values. Determine the result, simplifying as much as possible. Final form: $\hat{P}_{per}(\omega) = P_{ARMA(1,1)}(\omega) + P_\Delta(\omega)$. [Thus you will also need to find $P_{ARMA(1,1)}(\omega)$ in terms of a and b.] Use any intermediate results from Problem 10.48 you may find helpful. Does the estimate become exact as $N \to \infty$?

(b) Show how you could use just three of these exact ACF values to obtain $P_{ARMA(1,1)}(\omega)$ exactly.

10.50. (a) Find the expected value of the periodogram of a real AR(1) process in terms of the exact $P_{AR}(\omega) = \sigma^2/|1 + ae^{-j\omega}|^2$ and the parameters $a_1 = a$, σ^2, N, and ω. Try to express your result in the form

$$E\{\hat{P}_{per}(\omega)\} = P_{XX,AR(1)}(\omega)\{1 - \sum_{k=k_1}^{k_2} \gamma_k(-a)^k \cos(\omega k)\}$$

(giving values for the finite–valued constants k_1, k_2, γ_k). What would you predict about the bias as $|a| \to 1$?

Hint: Begin with the following result from Problem 10.47:

$$\hat{P}_{per}(\omega) = \sum_{k=-(N-1)}^{N-1} r_{N,n}(k)e^{-j\omega k}; \text{ show that } r_{N,n}(k) \text{ is a biased estimator of } r_{XX}(k).$$

(b) Plot on the same axes: the true PSD for the AR(1) process, the expected value of the periodogram given AR(1) data [from your result in part (a)], and the absolute value of the bias, all as functions of digital frequency f. Plot for $a_1 = -0.7$, for $N = 1000$ and for $N = 20$. Repeat for $a_1 = 0.9$. Comment on the PSD shapes for $a_1 > 0$ and $a_1 < 0$, and on for what a_1 and for what N the bias is worst.

(c) Plot the expected value of the periodogram of an AR(1) process [as found in part (b)] on the same axes as the periodogram of an actual realization of AR(1) data [not the expected value found in part (b), which is based on model parameters, not actual data]. Contrast the two. Use $a_1 = -0.7$, $N = 100$.

Bibliography

PRIMARY JOURNALS

IEEE Transactions on Signal Processing
IEEE Signal Processing Magazine
IEE Proceedings-Vision, Image, and Signal Processing
ICASSP (International Conference on Acoustics, Speech, and Signal Processing)

BOOKS

Analog Devices. *Analog-Digital Conversion Handbook*. Prentice Hall, Englewood Cliffs, NJ, 1986.

Berberian, S. K. *Linear Algebra*. Oxford University Press, Oxford, 1992.

Berger, M. A. *An Introduction to Probability and Stochastic Processes*. Springer-Verlag, New York, NY, 1993.

Bode, H. W. *Network Analysis and Feedback Amplifier Design*. D. Van Nostrand, Princeton, NJ, 1945.

Burrus, C. S., R. A. Gopinath, and H. Guo. *Introduction to Wavelets and Wavelet Transforms: A Primer*, Prentice Hall, Upper Saddle River, NJ, 1998.

Churchill, R. V. *Operational Mathematics*. McGraw-Hill, New York, NY, 1958.

Churchill, R. V., and J. W. Brown. *Fourier Series and Boundary Value Problems*. McGraw-Hill, New York, NY, 1987.

Couch, L. W. *Modern Communication Systems*. Prentice Hall, Englewood Cliffs, NJ, 1995.

Evans, G. *Practical Numerical Analysis*. Wiley, New York, NY, 1995.

Feller, W. *An Introduction to Probability Theory and Its Applications*. Wiley, New York, NY, 1950.

Folland, G. B. *Fourier Analysis and Its Applications*. Wadsworth & Brooks, Pacific Grove, CA, 1992.

Frerking, M. E. *Digital Signal Processing in Communication Systems*. Van Nostrand Reinhold, New York, NY, 1994.

Gill, P. E., W. Murray, and M. H. Wright. *Numerical Linear Algebra and Optimization*, Vol. 1, Addison-Wesley, Reading, MA, 1991.

Gradshteyn, I. S., and I. M. Ryzhik. *Table of Integrals, Series, and Products*. Academic Press, New York, NY, 1980.

Granlund, G. H. (Ed.). *Signal Processing for Computer Vision*. Kluwer, Boston, 1994.

Hanselman, D., and B. Littlefield, *The Student Edition of Matlab, Version 5*. Prentice Hall, Upper Saddle River, NJ, 1998.

Held, G., and T. R. Marshall. *Data and Image Compression Tools and Techniques.* Wiley, New York, NY, 1996.

Helstrom, C. W. *Probability and Stochastic Processes for Engineers.* Macmillan, New York, NY, 1984.

Hildebrand, F. B. *Introduction to Numerical Analysis.* McGraw-Hill, New York, NY, 1974.

Jahne, B. *Digital Image Processing: Concepts, Algorithms, and Scientific Applications.* Springer-Verlag, New York, NY, 1997.

Jakowatz, C. V. *Spotlight-Mode Synthetic Aperture Radar: A Signal Processing Approach.* Kluwer, Boston, 1996.

Klaassen, K. B. *Electronic Measurement and Instrumentation.* Cambridge University Press, New York, NY, 1996.

Kleijn, W. B. (Ed.), and K. K. Paliwa. *Speech Coding and Synthesis.* Elsevier-Science, New York, NY, 1995.

Levinson, N., and R. M. Redheffer. *Complex Variables.* Holden-Day, San Francisco, 1970.

Lim, J. S. *Two-Dimensional Signal and Image Processing.* Prentice Hall, Englewood Cliffs, NJ, 1990.

Liu, K. R., and K. Yao. *High Performance VLSI Signal Processing.* IEEE Standards Office, Piscataway, NJ, 1997.

Marsden, J. E. *Basic Complex Analysis.* W. H. Freeman, San Francisco, 1973.

Marston, R. M. *Modern CMOS Circuits Manual.* Newnes/Butterworth, Oxford, 1996.

Mathews, J., and R. L. Walker. *Mathematical Methods of Physics.* W. A. Benjamin, New York, NY, 1964.

Mix, D. F. *Random Signal Processing.* Prentice Hall, Englewood Cliffs, NJ, 1995.

Montgomery, D. C., and G. C. Runger. *Applied Statistics and Probability for Engineers.* Wiley, New York, NY, 1994.

Mortensen, R. E. *Random Signals and Systems.* Wiley, New York, NY, 1987.

Nixon, F. E. *Handbook of Laplace Transformation.* Prentice Hall, Englewood Cliffs, NJ, 1965.

Oppenheim, A. V., and R. W. Schafer. *Discrete-Time Signal Processing.* Prentice Hall, Englewood Cliffs, NJ, 1989.

Owens, F. J. *Signal Processing of Speech.* McGraw-Hill, New York, NY, 1993.

Pallas-Areny, R., and J. G. Webster. *Sensors and Signal Conditioning.* Wiley, New York, NY, 1991.

Papoulis, A. *Probability, Random Variables, and Stochastic Processes.* McGraw-Hill, New York, NY, 1984.

Pirsch, P. *Architecture for Digital Signal Processing.* Wiley, New York, NY, 1998.

Poularikas, A. D. *The Transforms and Applications Handbook.* IEEE Press/CRC Press, Boca Raton, FL, 1996.

Proakis, J. G., and D. G. Manolakis. *Digital Signal Processing: Principles, Algorithms, and Applications.* 3rd ed. Prentice Hall, Upper Saddle River, NJ, 1996.

Proakis, J. G., and M. Salehi. *Communication Systems Engineering.* Prentice Hall, Englewood Cliffs, NJ, 1994.

Rabiner, L. R., and R. W. Schafer. *Digital Processing of Speech Signals.* Prentice Hall, Englewood Cliffs, NJ, 1978.

Russ, John C. *The Image Processing Handbook.* CRC Press, Boca Raton, FL, 1995.

Rutkowski, G. B. *Operational Amplifiers.* Wiley, New York, NY, 1993.

Sandige, R. S. *Modern Digital Design.* McGraw-Hill, New York, NY, 1990.

Shankar, R. *Principles of Quantum Mechanics.* Plenum Press, New York, NY, 1994.

Sorensen, H. V., and J. Chen. *A Digital Signal Processing Laboratory Using the TMS320C30.* Prentice Hall, Upper Saddle River, NJ, 1997.

Suter, B. W. *Mulitrate and Wavelet Signal Processing.* Academic Press, New York, NY, 1997.

Thomson, W. T. *Laplace Transformation.* Prentice Hall, Englewood Cliffs, NJ, 1960.

Van Der Pol, B., and H. Bremmer. *Operational Calculus Based on the Two-Sided Laplace Integral.* Cambridge University Press, Cambridge, 1959.

Van Valkenburg, M. E. *Analog Filter Design.* Holt, Rinehart, and Winston, New York, NY, 1982.

Watkinson, J. *The Art of Digital Audio.* Focal Press, London, 1988.

Weaver, H. J. *Applications of Discrete and Continuous Fourier Analysis.* Wiley, New York, NY, 1983.

Yaroslavsky, L. P., and M. Eden. *Fundamentals of Digital Optics: Digital Signal Processing in Optics and Holography.* Birkhauser-Boston, Cambridge, MA, 1996.

Zwillinger, D., Ed. *Standard Mathematical Tables and Formulae.* CRC Press, Boca Raton, FL, 1996.

Index